Estuarine Ecology

Estuarine Ecology

Third Edition

Edited by

BYRON C. CRUMP
College of Earth, Ocean, and
Atmospheric Sciences
Oregon State University
Corvallis, OR 97330, USA

JEREMY M. TESTA
Chesapeake Biological Laboratory
University of Maryland Center for
Environmental Science
Solomons, MD 20688, USA

KENNETH H. DUNTON
Marine Science Institute
The University of Texas at Austin
Port Aransas, TX 78373, USA

This third edition first published 2023

Edition History: John Wiley & Sons Inc (1e, 1989; 2e, 2012)

Registered Office
John Wiley & Sons, Inc., 111 River Street, Hoboken, NJ 07030, USA

Editorial Office
Boschstr. 12, 69469 Weinheim, Germany

For details of our global editorial offices, customer services, and more information about Wiley products visit us at **www.wiley.com**.

Wiley also publishes its books in a variety of electronic formats and by print-on-demand. Some content that appears in standard print versions of this book may not be available in other formats.

Library of Congress Cataloging-in-Publication Data
Names: Crump, Byron C., editor. | Testa, Jeremy M., editor. | Dunton, Kenneth H., editor. | Day, John W., 1945- author of introduction.
Title: Estuarine ecology / edited by Byron C. Crump (College of Earth, Ocean, and Atmospheric Sciences, Oregon State University), Jeremy M. Testa (Chesapeake Biological Laboratory, University of Maryland Center for Environmental Science), Kenneth H. Dunton (Marine Science Institute, the University of Texas at Austin).
Description: Third edition. | Hoboken, NJ : Wiley, 2023. | Includes bibliographical references and index.
Identifiers: LCCN 2022017156 (print) | LCCN 2022017157 (ebook) | ISBN 9781119534655 (cloth) | ISBN 9781119534624 (pdf) | ISBN 9781119534563 (epub)
Subjects: LCSH: Estuarine ecology.
Classification: LCC QH541.5.E8 E849 2023 (print) | LCC QH541.5.E8 (ebook) | DDC 577.7/86–dc23/eng/20220706
LC record available at https://lccn.loc.gov/2022017156
LC ebook record available at https://lccn.loc.gov/2022017157

Cover Design: Wiley
Cover Image: © Brent Durand/Getty Images (front cover); courtesy of Kenneth H. Dunton & Byron C. Crump (back cover)

Set in 9.5/12.5pt SourceSansPro by Straive, Pondicherry, India
Printed and bound by CPI Group (UK) Ltd, Croydon, CR0 4YY

C9781119534655_260922

We dedicate this book to the memory of our dear colleagues and friends Alejandro Yáñez-Arancibia and W. Michael Kemp. Alex and Michael were editors and authors of the first and second edition of this book, and their spirit and energy live on in the third edition. Many of the authors of this book were influenced beyond measure by Alex and Michael in many ways, through collaboration, inspiration, and mentorship. We hope this book continues to serve as a living reflection of their innumerable contributions to the field of estuarine science.

Contents

Preface

This book is a textbook for a course in estuarine ecology designed to introduce students to the interdisciplinary study of estuarine ecosystems. The textbook is not only designed for upper-level undergraduate students and introductory-level graduate students, but it is also a useful reference for estuarine scientists across the fields of physical, biological, chemical, and ecological sciences. This book is the third edition of *Estuarine Ecology*; the first edition was published in 1989 and was the first book focused on estuarine ecology; the second edition was published in 2012 and expanded in length and scope to account for the explosion of literature and knowledge related to estuarine ecology; this third edition not only builds upon the expanded scope of the second edition but also represents a transformation to a textbook geared toward a wider range of students (upper-level undergraduates) and a more complete set of teaching tools (study questions, exercises).

We continued the spirit of the second edition by engaging with a broad range of experts to sufficiently communicate the latest knowledge in a wide array of topics, and we also engaged a new cohort of early career scientists who carry on the tradition and knowledge of their graduate advisers and senior colleagues. The book begins with an introduction to estuaries and their ecology, with a new emphasis on how estuaries are in many ways a network of linked environments over salinity gradients, land–water interfaces, water columns, and underlying sediments, and from enclosed waters out to the sea. This introduction is followed by chapters that emphasize the hydrodynamic and chemical features of estuaries. These introductory perspectives are followed by a collection of chapters on estuarine primary producer communities, including phytoplankton, benthic algae, seagrasses, coastal marshes, and mangrove ecosystems; these chapters give students and faculty alike a detailed view of the key ecological features of these ecosystems that serve as the foundation of estuarine productivity. Next, several chapters detail the key consumer communities in estuaries, ranging from microscopic communities through zooplankton and on to benthos, nekton, and wildlife. The third edition includes a combination of the two previous microbial chapters, an expanded zooplankton chapter and a reorganization of the book into a sequence of consumers beginning with microbes and ending at megafauna. The final section of the book includes chapters that either provide synthetic treatments of key ecological topics or address more holistic aspects of estuarine ecosystem ecology. These include chapters on food webs and ecosystem metabolism, followed by tools-focused chapters of ecosystem modeling and fisheries sciences, and culminating in a broad summary of how estuaries respond to global changes.

The third edition of *Estuarine Ecology* continues the challenge of providing a comprehensive treatment of a diverse array of topics relevant for the field of estuarine science. In the preface to the second edition, those authors aimed to consider the book "a work in progress" that provides a framework that can be routinely updated to keep pace with both the fast pace of global change and the scientific literature that describes it. We took on the task of this first rapid iteration and also took the opportunity to make the book more widely adaptable for undergraduate and graduate coursework. At this time of publication, we can already see the light of the fourth edition emerge.

Byron C. Crump
Jeremy M. Testa
Kenneth H. Dunton

Acknowledgments

This book is a product of years of sustained efforts by the chapter authors, who we thank for their voluntary contributions to the community. We fondly recall our initial meeting with the authors at the Providence 2017 meeting of the Coastal and Estuarine Research Federation to launch the third edition. It is no small feat that all chapter authors contributed substantially to the process of writing the book, working with us to edit their previous chapters toward a broader audience, generating review questions for their chapters, and offering their time to review other chapters to make sure they are as widely understood as possible. We would also like to recognize the previous editors and authors of this book, John Day, Jr., Charles Hall, W. Michael Kemp, and Alejandro Yáñez-Arancibia, whose intellectual contributions and spirit live on in the third edition. We also recognize the Coastal and Estuarine Research Federation, who has strongly supported the publication of this book.

List of Contributors

Anderson, Iris, Virginia Institute of Marine Sciences, College of William and Mary, Gloucester Point, VA, USA

Aoki, Lillian, Department of Ecology and Evolutionary Biology, Cornell University, Ithaca, NY, USA

Benfield, Mark C., Louisiana State University, College of the Coast and Environment, Baton Rouge, LA, USA

Bianchi, Thomas S., Department of Geological Sciences, University of Florida, Gainesville, FL, USA

Blum, Linda K., Laboratory of Microbial Ecology, Department of Environmental Sciences, University of Virginia, Charlottesville, VA, USA

Borum, Jens, Department of Biology, University of Copenhagen, Copenhagen, Denmark

Cable, Jaye E., Department of Earth, Marine, and Environmental Sciences, University of North Carolina, Chapel Hill, NC, USA

Cowan, James H., Department of Oceanography and Coastal Sciences, Louisiana State University, Baton Rouge, LA, USA

Crump, Byron C., College of Earth, Ocean, and Atmospheric Sciences, Oregon State University, Corvallis, OR, USA

Day, John W., Department of Oceanography and Coastal Sciences, College of the Coast and Environment, Louisiana State University, Baton Rouge, LA, USA

Dunton, Kenneth H., University of Texas at Austin, Marine Science Institute, Port Aransas, TX, USA

Elphick, Chris S., Department of Ecology and Evolutionary Biology, University of Connecticut, Storrs, CT, USA

Fleeger, John W., Department of Biological Sciences, Louisiana State University, Baton Rouge, LA, USA

Fong, Peggy, Department of Ecology and Evolutionary Biology, University of California Los Angeles, Los Angeles, CA, USA

Greenberg, Russel, Smithsonian Migratory Bird Center, National Zoological Park, Washington, DC, USA

Gregory Shriver, W., Department of Entomology and Wildlife Ecology, University of Delaware, Newark, DE, USA

Gruber, Renee K., Australian Institute of Marine Science, Townsville, QLD, Australia

Gurbisz, Cassie, St. Mary's College of Maryland, St. Marys City, MD, USA

Hagy, James D., Center for Environmental Measurement and Modeling, Atlantic Coastal Environmental Sciences Division, Narragansett, RI, USA

Hopkinson, Charles S., Department of Marine Sciences, University of Georgia and Georgia Sea Grant, Athens, Georgia, USA

Ibáñez, Carles, Department of Climate Change, EURECAT, Technological Centre of Catalonia, Amposta, Catalonia, Spain

Justić, Dubravko, Department of Oceanography & Coastal Sciences, School of the Coast & Environment, Louisiana State University, Baton Rouge, LA, USA

Kjerfve, Björn, School of the Earth, Ocean and Environment, University of South Carolina, Columbia, SC, USA

Krauss, Ken W., Wetland and Aquatic Research Center, U.S. Geological Survey, Lafayette, LA, USA

Lovejoy, Connie, Department of Biology, Laval University, Québec, Canada

Martínez-Eixarch, Maite, Marine and Continental Waters, IRTA Institute of Agrifood Research and Technology, Sant Carles de la Ràpita, Spain

McGlathery, Karen J., Department of Environmental Sciences, University of Virginia, Charlottesville, VA, USA

Mendelssohn, Irving A., Department of Oceanography and Coastal Sciences, College of the Coast and Environment, Louisiana State University, Baton Rouge, LA, USA

Michael Kemp, W., Horn Point Laboratory, University of Maryland, Center for Environmental Science, Cambridge, MD, USA

Mills, Aaron L., Laboratory of Microbial Ecology, Department of Environmental Sciences, University of Virginia, Charlottesville, VA, USA

Morris, James T., University of South Carolina, Belle W. Baruch Institute for Marine and Coastal Sciences, Columbia, SC USA

Morrison, Elise S., Department of Environmental Engineering Sciences, University of Florida, Gainesville, FL, USA

Nesslage, Geneviève, Chesapeake Biological Laboratory, University of Maryland Center for Environmental Science, Solomons, MD, USA

Nolte, Stefanie, School of Environmental Sciences, University of East Anglia, Norwich, UK; Centre for Environment, Fisheries and Aquaculture Science, Lowestoft, UK

Paerl, Hans W., University of North Carolina at Chapel Hill, Institute of Marine Sciences, Morehead City, NC, USA

Paerl, Ryan W., North Carolina State University, Department of Marine, Earth, and Atmospheric Sciences, Raleigh, NC, USA

Pauly, Daniel, Institute for the Oceans and Fisheries & Department of Zoology, The University of British Columbia, Vancouver, BC, Canada

Pierson, James J., University of Maryland Center for Environmental Science, Horn Point Laboratory, Cambridge, MD, USA

Reyes, Enrique, Department of Biology, East Carolina University, Greenville, NC, USA

Rose, Kenneth A., University of Maryland Center for Environmental Science, Horn Point Laboratory, Cambridge, MD, USA

Rovai, Andre, Department of Oceanography and Coastal Sciences, Louisiana State University, Baton Rouge, LA, USA

Rybczyk, John M., Department of Environmental Science, Western Washington University, Bellingham, WA, USA

Snedden, Gregg A., Wetland and Aquatic Research Center, U.S. Geological Survey, Baton Rouge, LA, USA

Stæhr, Peter A., Department of Ecoscience, Marine Diversity and Experimental Ecology, Roskilde, Denmark

Sundbäck, Kristina, Department of Marine Ecology, University of Gothenburg, Göteborg, Sweden

Testa, Jeremy M., Chesapeake Biological Laboratory, University of Maryland Center for Environmental Science, Solomons, Maryland, USA

Twilley, Robert R., Department of Oceanography and Coastal Sciences, Louisiana State University, Baton Rouge, LA, USA

Weston, Nathaniel, Department of Geography and the Environment, Villanova University, Villanova, Pennsylvania, USA

Wilson, James G., Zoology Department, Trinity College, Dublin, Ireland

Woodland, Ryan J., Center for Environmental Science, Chesapeake Biological Laboratory, University of Maryland, Solomons, MD, USA

Yáñez-Arancibia, Alejandro, Red Ambiente y Sustentabilidad, Instituto de Ecologia, A.C. (CPI-CONACYT), Xalapa, Veracruz, México

About the Companion Website

This book is accompanied by a companion website.

www.wiley.com/go/crump/estuarine3

This website include:

- Figures PPTs
- MCQs
- Short Questions and Answers

CHAPTER 1

Introduction to Estuarine Ecology

Kenneth H. Dunton[1], Byron C. Crump[2], Jeremy M. Testa[3], and John W. Day[4]
[1]University of Texas at Austin, Marine Science Institute, Port Aransas, TX, USA
[2]College of Earth, Ocean, and Atmospheric Sciences, Oregon State University, Corvallis, OR, USA
[3]Chesapeake Biological Laboratory, University of Maryland Center for Environmental Science, Solomons, MD, USA
[4]Department of Oceanography and Coastal Sciences, Louisiana State University, Baton Rouge, LA, USA

Whooping Cranes (Grus americana) along the marsh edge within the Aransas National Wildlife Refuge on the coast of Texas, USA. These endangered native North American birds over-winter in South Texas and migrate annually to their summer breeding and nesting grounds at Wood Buffalo National Park in northern Canada. Photo credit: K.H. Dunton.

1.1 Background, Theory, and Issues

We begin this description of estuaries and their functions by defining estuaries very broadly as that portion of the earth's coastal zone where there is interaction of ocean water, fresh water, land, and atmosphere. Large estuarine zones are most common in low-relief coastal regions such as the expansive coastal plains of Europe and the east coast of North America. On glaciated coastlines at higher latitudes and on uplifted coastlines such as the Pacific coasts of Asia and the Americas, we refer to these estuarine systems as fjords. We begin our assessment as widely as possible to include all portions of the earth that interact at the edge of the sea and have produced a wide diversity of estuarine types, from coastal plain salt marshes to fjords (Figure 1.1).

From the vantage point of an orbiting satellite, several of the most basic attributes of estuaries are observable. Plumes of sediment-laden water float seaward on the ocean surface from the largest rivers, such as the Amazon, the Ganges, and the Mississippi. Color differences among various water masses, representing waters of different histories and different biotic richness, are often apparent. Coastal waters in areas with significant riverine input and broad shelf areas generally appear more greenish-brown than the deep blue waters adjacent to many other coastlines (Figure 1.1a). There are also atmospheric features of importance to estuaries obvious from space. Clouds commonly form directly over the edges of continents as one manifestation of the atmospheric "thermal engine" that maintains the freshwater cycle on which estuaries depend (Figure 1.1d). At the altitude of a satellite, the dense human populations that proliferate in coastal zones are outlined at night by their lights.

The most recent geological epoch, the Holocene, which started approximately 11,650 years before present, could be called the age of the estuary, for estuaries are abundant today even though they may be ephemeral on geologic timescales. It is interesting to note that all of the estuaries discussed in this book did not exist 10,000–15,000 years ago and that they will cease to exist in the near geological future. Many present-day estuaries are less than about 5000 years old, representing the

Estuarine Ecology, Third Edition. Edited by Byron C. Crump, Jeremy M. Testa, and Kenneth H. Dunton.

Companion website: www.wiley.com/go/crump/estuarine3

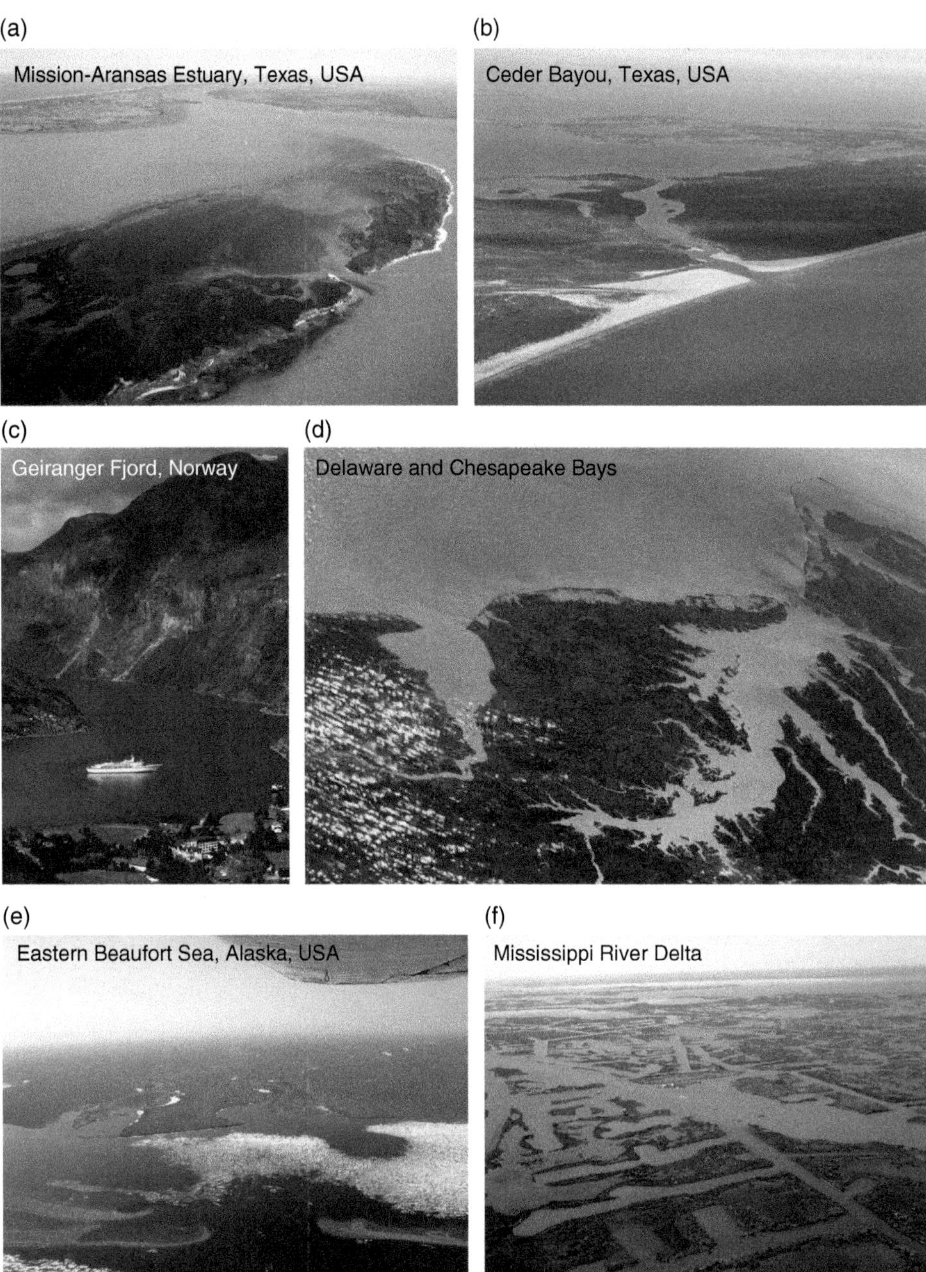

FIGURE 1.1 Examples of common estuary types across the globe: (a) the Mission-Aransas coastal plain salt marsh estuary includes seagrass, marsh and mangrove wetlands behind San Jose Island, a barrier island in the Gulf of Mexico, (b) the classic bar-built estuary of Cedar Bayou (Texas coast) connects Matagorda Bay with Gulf of Mexico waters, (c) the Geiranger Fjord in western Norway, (d) an astronaut view of drowned river valleys Delaware Bay (left) and Chesapeake Bay (center), as well as coastal lagoon Pamlico Sound (top right). *Source*: Earth Science and Remote Sensing Unit, NASA Johnson Space Center (e) barrier island lagoons along the eastern Alaskan Beaufort Sea coast at breakup in June, (f) a deltaic estuary as exemplified by the vast expanses of wetlands of the Mississippi River Delta. *Source*: All photos except (d) by K.H. Dunton.

time since sea level reached near its present level following the last ice age. Since that time, they have progressively filled with sediments and that process will continue. Consequently, our present-day estuaries will either fill with sediment or will change dramatically as sea level continues to rise.

Many estuaries are drowned river valleys (Figure 1.1d). Their formation began as rivers carved their way to the ocean when sea levels were considerably lower. As sea levels rose, the valleys flooded. At high latitudes, river valleys were further eroded by glaciers, resulting in profoundly deep fjords (Figure 1.1c) that became linked to the ocean through glacial melt, and the formation of a shallow entrance sill.

Human populations flourished during this same time period, in no small measure owing to exploitation of the rich estuarine resources of the coastal margin. Most "cradles of civilization" arose in deltaic and lower floodplain areas where natural biota was abundant and where flooding cycles produced the rich bottomland soils and readily available fresh water supplies on which agriculture flourished (Kennett and Kennett 2006; Day et al. 2007). Early centers of civilization that developed in estuarine or deltaic environments include the Tabascan lowlands of Mexico; the valley of the Nile, the Tigris-Euphrates, Huang He (Yellow), and Indus River deltas, and along the Andean coast of western South America where upwelling systems bordered estuarine systems.

As human populations expanded, so did human pressure on natural environments. Today, we have entered the Anthropocene, a new epoch, defined as the period in which humans have significantly altered Earth's environments and climate (Syvitski et al. 2020). These changes include excessive nutrient loading of our oceans (eutrophication), global climate change, rapid sea level rise, habitat loss, animal extinctions, and changes in the chemical composition of atmosphere, oceans, and soil.

Let us now continue our aerial survey of estuaries, but this time at a much lower altitude, about 1000 m, in a light airplane following the course of a coastal plain river in the temperate zone from its headwaters to the ocean. The headwater river is narrow with rapids and falls, but changes near the coast to a larger meandering form with broad marshy areas where the actual edge of the river is not always clearly evident. The color of the water changes from clear blue to yellowish-brown as the river picks up silt. As the river water nears the coast, tidal currents become apparent and, moving seaward, the influence of tidal currents becomes greater and greater.

Along the banks of the estuary, fresh and brackish water marsh plants grow at the edges of embayments. These marshes are often flanked by rows of houses with backyards that border the bank of the estuary, often with narrow piers that extend from the bank to provide access to deeper water. Among these marshes, a variety of wading birds may be observed stalking their prey at the water's edge. Where the water is shallow and relatively clear, dark-colored patches indicate the presence of submersed seagrasses.

As we travel seaward, tides become more important, and the intertidal zone becomes more extensive. Larger piers and bulkheads interrupt the banks of the estuary, and brown mud flats come into view, as well as greenish-gray oyster reefs fringing the banks or dotting the mud flats. Various birds such as oystercatchers feed on the reefs, along with an occasional raccoon. The mud flats are peppered with mud snails, and just beneath the surface are teeming communities of small worms and crustaceans. Common shore birds, such as oystercatchers and skimmers, are feeding at the water's edge. Skimmers fly along in quiet areas, each plowing a furrow in the water with their lower bill as they fish for silversides and other small fish. The darker colored path of a deep shipping channel maintained by dredging is evident toward the middle of the estuary and contrasts with the lighter colored shallows.

The mouth of the estuary takes the form of a broad sound that opens up behind a barrier island (Figure 1.1b). The sound is shallow, and we can see porpoises herding schools of juvenile menhaden, followed by gulls trying to get in on the action. Crab pot buoys and fishing boats are much in evidence. On either side of the barrier island are narrow passes with visible eddies and strange wave patterns, indicating rapid and complex currents. In high-latitude estuaries, river flow begins with an enormous flush at ice-breakup, dispersing ice that formed over the previous 9 months in the estuarine lagoons protected by barrier islands (Figure 1.1e)

Along the ocean beach, several shrimp boats raise long spiraling muddy plumes of sediment as they drag their trawls along the bottom. A kilometer or so offshore of the tidal passes the water changes color from dark brownish green to a lighter, less turbid green. Further offshore it is a darker and bluer color.

On the landward side of most such barrier islands, there are flat intertidal and shallow subtidal areas colonized by salt marsh plants that are bisected with meandering tidal creeks that have developed over centuries (Figure 1.1a). In low-latitude estuaries, these same areas are colonized with mangrove trees. Moving inland from the marsh, the highest part of the island includes larger trees. The seaward side of the island may include a series of dunes, the farthest from the ocean covered with vegetation, the nearer dunes less and less vegetated. The beach has much less vegetation because wave energy from storms makes it difficult for plants to survive. In parts of the beach–barrier system, vacation houses have replaced dunes and straight navigation channels have replaced twisting tidal channels.

Behind barrier islands, river deltas often form that support enormous wetland areas of salt marsh grasses or mangrove trees. Sometimes these deltaic estuaries have been hydrologically altered by dredging to allow access for commercial activities (Figure 1.1f). In addition to their ecological importance, marsh and mangrove vegetation often form critical natural barriers to storm surges.

In summary, estuaries are complex, dynamic, and biotically rich environments dominated by physical forces and impacted by human activity. Their study requires a consideration and knowledge of geology, hydrology, chemistry, physics, and biology. Ideally, we can integrate knowledge gained through these specific disciplines using what we call systems science, a fundamentally interdisciplinary venture. This book is an introduction to the specifics of estuarine science and its integration into a coherent view of estuaries as ecosystems. We will show how estuaries are different from one another and how they are similar, and why we need to preserve them while enhancing their value to society.

We will begin by describing a very generalized estuary, to provide the reader with an introduction to the geology, physics, chemistry, and biology of estuaries. This is done with a certain danger because, as the rest of the book will show, estuaries are characterized as much by differences as by similarities. Nevertheless, in this chapter, we attempt to describe a generalized estuary. But before we proceed further, we will define an estuary.

1.2 Definitions, Terms, and Objectives

1.2.1 Definitions of Estuary and Ecology, and Difficulties in Applying These Definitions to Estuaries

The term estuary comes from the Latin *aestus* meaning heat, boiling, or tide. Specifically, the adjective *aestuarium* means tidal. Thus, the *Oxford Dictionary* defines estuary as "the tidal mouth of a great river, where the tide meets the current." *Webster's Dictionary* is more specific: "(a) a passage, as the mouth of a river or lake where the tide meets the river current; more commonly, an arm of the sea at the lower end of a river; a firth. (b) in physical geography, a drowned river mouth, caused by the sinking of land near the coast."

Perhaps the most widely quoted definition of an estuary in the scientific literature is given by Pritchard (1967): "An estuary is a semi-enclosed coastal body of water which has a free connection with the open sea and within which sea water is measurably diluted with fresh water derived from land drainage." Certainly, one important characteristic attribute of most coastal areas is the action of the tide. Pritchard's definition makes no specific mention of tide, although it is implied by reference to the mixing of seawater and fresh water. There are, however, many nontidal or minimally tidal seas, such as the Mediterranean Sea and the Black Sea, where fresh water and salt water mix.

There are also estuaries in semiarid regions that may not receive any fresh water for long periods; and sometimes, as on the Pacific coast of California and Mexico, Western Australia, and several parts of Africa, estuaries may become blocked by alongshore sand drift, so that they are ephemerally isolated from the sea for months to perhaps years. In other regions the tidal limit, sometimes with a tidal bore, may reach 100 km or more above the limits of saltwater intrusion. So Pritchard's definition of estuary excludes some coastal areas where estuarine ecology is studied today.

In an attempt to address the limitations of Pritchard's definition, Fairbridge (1980) gave a more comprehensive definition of an estuary: An estuary is an inlet of the sea reaching into a river valley as far as the upper limit of tidal rise, usually being divisible into three sectors; (a) a marine or lower estuary, in free connection with the open sea, (b) a middle estuary subject to strong salt and freshwater mixing, and (c) an upper or fluvial estuary, characterized by fresh water but subject to daily tidal action. The limits between these sectors are variable, and subject to constant changes in the river discharge.

Fairbridge's definition excludes some coastal geomorphic features such as lagoons, deltas, sounds, and nontidal estuaries. The distinctions among these different terms are treated in detail in Chapter 2, but characteristic estuarine ecosystems have developed in all these coastal systems. Therefore, when the terms estuary and estuarine are used in this book, unless specifically stated otherwise, they are meant in a general ecological sense rather than any specific narrower geological sense.

All of the definitions of estuaries given above reflect, for the most part, physical and geological characteristics of estuaries. But why is this so? The people who first defined and classified estuaries were geologists and physical oceanographers, because in many respects the most salient features of estuaries are physical and geomorphic. And the ecosystems that exist in estuaries are often physically dominated. We can illustrate this point by comparing an estuarine ecosystem with a tropical forest ecosystem. A visitor to a rain forest is immediately struck by the richness of the vegetation. If the visitor stays in the forest for some time, he or she will notice that it rains a lot and the temperature is warm. If the visitor is a careful observer, he or she will perhaps learn about the soils of the forest. But the most striking characteristic is the vegetation. Rain forests are biologically dominated systems and have been described primarily by their biological characteristics.

In contrast, a visitor to an estuary cannot escape noticing the impact of abiotic characteristics. These include the rise and fall of the tide, complex water movements, high turbidity levels, and different salt concentrations. The nature of landforms such as beaches, barrier islands, mud flats, and deltas and the geometry of the basin are also very noticeable. There are, of course, outstanding biotic characteristics of estuaries such as salt marshes, mangrove wetlands, seagrass beds, and oyster reefs. But, in general, one must look carefully to obtain even an idea of the biological structure of estuaries.

The visibility of the abiotic attributes of estuaries reflects the fact that estuaries are, to a large degree, physically dominated ecosystems. They can be hydrologically very complex, since estuaries include not only riverine inflow, but groundwater inputs, direct precipitation, and influx of human waste loads through septic seepage and wastewater treatment facilities. The water residence time (or replacement time) of water within an estuary becomes a complex product of tidal exchanges and density currents, which are basic elements of estuarine circulation (Figure 1.2).

These physical attributes greatly influence and structure the ecological attributes of estuaries. Thus, in order to begin to understand how estuarine ecosystems function, an estuarine ecologist must have a good understanding of the geology, physical oceanography, and chemistry of estuaries. The integration of the physical attributes of estuaries at the land–ocean interface, combined with exchanges, linkages, inputs, and the biogeochemical processes of transformations, result in highly dynamic systems (Figure 1.3). The goal of Chapters 2 and 3 are to provide the physical and chemical bases for understanding how biotic processes in estuaries are driven by the ever-changing physics and chemistry of these systems.

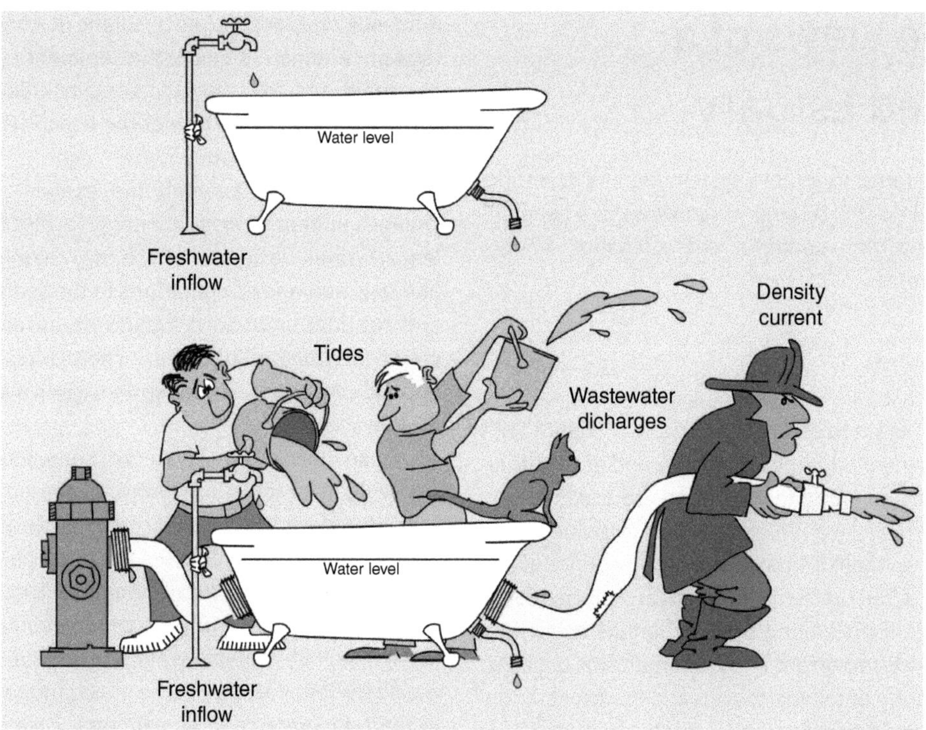

FIGURE 1.2 A simplified "bathtub" model of estuarine exchanges. Above, the conventional model, from which the residence time (a measure of dilution) is computed as $T = V/Q$, where V is estuary volume and Q freshwater inflow. Below, a more complete depiction of mechanisms of dilution in an estuary, including freshwater inflow, tides, the salinity-driven density current, and wastewater discharges, illustrating the complexity of estuarine circulation (drawing: G. Ward).

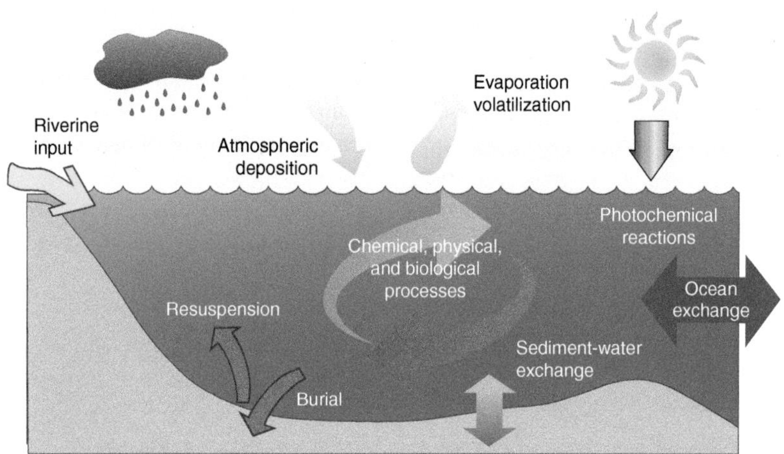

FIGURE 1.3 A diagrammatic view of the major physical, biological, and biogeochemical processes that control the physiochemical properties of estuaries and ultimately their productivity and roles in the global carbon cycle.

Before we go further it is necessary to define ecology, since this book is about estuarine ecology. Usually, ecology is defined as the study of the relation of organisms or groups of organisms to each other and to their environment. Margalef (1968) gives a definition of ecology that is, perhaps, more appropriate to the way we will approach estuarine ecology. He stated that "ecology is the study of systems at the level in which individuals or whole organisms can be considered as elements of interaction, either among themselves, or with a loosely organized environmental matrix. Systems at this level are called ecosystems, and ecology is the biology of ecosystems." Thus, in this book, we will consider the environmental matrix of estuaries, the interactions among specific organisms and the environment, and the structure and functioning of whole estuarine ecosystems.

1.3 Three Views of a Generalized Estuary

We now discuss estuaries in a more systematic and scientific manner, and we will do this through three views of a generalized estuary, emphasizing common characteristics among these divergent systems.

1.3.1 Top view

The first view of a generalized estuary is from above (Figure 1.4). At one end there is a river entering a large bay and at the other end a barrier island separating the bay from the ocean. Wave energy along the barrier island is high and there is a wide sandy beach. On the beaches there is little immediate evidence of life other than a few birds, but there are abundant and diverse communities of tiny organisms among the sand grains below the surface. Wave energy is important to these organisms because the waves pump water containing oxygen and food constantly through the sand while carrying away waste products. The beach is an area of very high physical energy and the sediments are completely oxidized. This is an example of the importance of physical processes in determining the biotic characteristics of an estuarine environment.

In the tidal pass through the barrier island, the water is still clear and the salinity high. Important exchanges, driven by tidal action, direct water movement in and out of the estuary. Normally, one to two tidal cycles occur over the course of a lunar day (24.8 hours), introducing saltwater into the estuary. Wave activity is somewhat reduced, but currents are still strong. At the entrance to the pass the sediments are still sandy and completely oxidized near the surface, but they are beyond the influence of strong currents, so anoxic conditions occur a few tens of centimeters below the sediment surface. The biota is abundant and diverse in the pass, including epifauna, in contrast to the barren surface of the beach. The pass is often the deepest part of the estuary.

As we move through the estuary, there are distinct changes in depth, physical energy levels (currents and turbulence), water clarity, salinity, biota, chemical concentrations, and oxic and anoxic conditions in the sediments. In intertidal and subtidal areas with significant currents, there are often worm flats or mollusk beds. These filter-feeding organisms depend on currents to transport oxygen and food and to carry away wastes.

In shallow subtidal waters of somewhat reduced currents where light reaches the bottom, marine meadows of submersed aquatic vegetation or seagrass beds often occur. Water clarity is high and sediments are finer, partially as a result of the grass's ability to trap sediments. The anoxic zone of the sediments extends to within a centimeter or less of the surface. In high-salinity tropical waters these submersed grass beds are often dominated by turtle grass, *Thalassia testudinum* and in the temperate zone by eel grass, *Zostera marina*. In lower salinity waters (less than 5) genera such as *Ruppia, Potamogeton,* and *Valisneria* are common (Chapter 5).

In this northern hemisphere example, fresh water entering the bay moves toward the barrier island but then is directed northward toward the tidal pass. Coriolis forces acting on river water entering the estuary and incoming seawater from the ocean produce flows in a counterclockwise pattern of circulation in the bay. This circulation is often interrupted by prevailing winds and periods of low river inputs, but such counterclockwise flow (clockwise in the southern hemisphere) is an important feature of estuarine circulation. Bordering the estuary in areas of mild to sluggish currents are intertidal

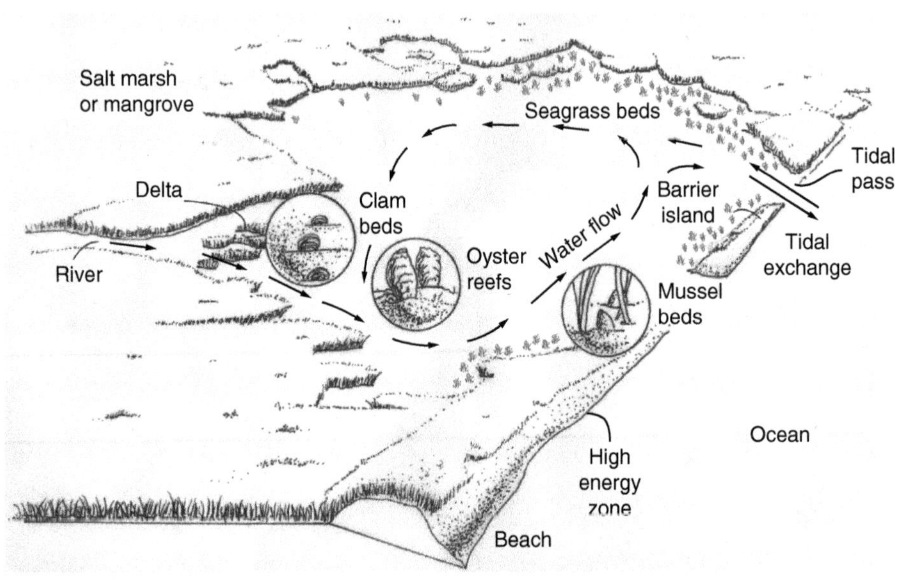

FIGURE 1.4 Idealized oblique aerial view of a typical estuarine system showing some of the major subsystems and circulation patterns that result from river inputs and exchange with the coastal ocean.

wetlands where salt marshes occur in the temperate zone and mangrove wetlands occur in the tropics. These are areas of turbid waters and very fine sediments that are highly anoxic. The vegetation has very high growth rates and the animals tend to be deposit feeders.

1.3.2 Cross-Section View

A cross-section view of an estuary (Figure 1.5) illustrates several vertical attributes of estuarine ecosystems. Vertical gradients are important in determining the nature of these ecosystems. Perhaps the most obvious is the intertidal zone that is alternately flooded and exposed. In the intertidal zone there may be salt marsh or mangrove wetlands, algal beds, sand or mud flats, or reefs of oysters, mussels, or clams. Most organisms that live in the intertidal zone have developed special adaptations, which are discussed in numerous places in this book.

Another important vertical gradient is the light environment and the natural attenuation of light with depth, which ultimately defines the lighted *euphotic zone* versus the unlighted *aphotic zone*. Obviously, photosynthesis occurs only in the euphotic zone, and, where sufficient light reaches the bottom, benthic microalgae (Chapter 8), macroalgae (seaweeds), and rooted plants such as seagrasses (Chapter 5), can survive and grow. Water clarity tends to be much lower in estuarine waters compared to ocean waters because light is absorbed by abundant particulate and dissolved organic matter, suspended sediments, and chlorophyll. Benthic microalgae and plankton can photosynthesize at light levels of about 1% of surface irradiance, which allows them to live at greater depths than shallow water seaweeds and seagrasses that need more light (>15% surface irradiance).

A third important vertical gradient in estuaries is the change in chemical conditions from oxidizing (where oxygen is present) to reducing (where oxygen is absent). An oxidizing environment is also called *oxic* or *aerobic* and a reducing one *anoxic* or *anaerobic*. Estuarine water is normally oxidizing, but estuarine sediments are usually reducing a short distance below the sediment surface. The amount of oxygen in sediments is related both to the rate at which oxygen moves into sediments and the rate at which it is consumed by the metabolic activity of microbes. In areas with high physical energies, waves or strong currents stir up sediments, add oxygen, and wash away finer, organic-rich particles, leaving behind coarser, oxidizing sediments (Figure 1.5). The opposite condition exists in areas of fine sediments where currents are too weak to sort the sediments and the organic-rich particles accumulate. These organic-rich particles enhance microbial activity that consumes oxygen and creates reducing conditions where anaerobic bacteria produce toxic sulfides and methane. In some cases (such as highly anoxic sediments in marshes), reducing conditions extend up to the sediment surface. Since most estuaries are underlain with fine sediments, this zone of reduced sediments can be widespread. However, organisms that inhabit sediments can manipulate these chemical gradients through their movement and metabolism (see Chapters 5 and 11).

Benthic organisms that create oxidizing conditions in these reducing sediments include worms and fiddler crabs, which construct burrows that allow the exchange of water and oxygen to occur with the overlying water column (see Figure 11.15, Chapter 11). In addition, many vascular plants that grow in reduced conditions, including emergent marsh grasses and submersed seagrasses, provide photosynthetically produced oxygen that is transported from leaf tissues to roots and rhizomes to create an oxidized rhizosphere (Figure 1.6; Chapters 5–7). These oxidized sediments support aerobic microbial communities that facilitate nutrient regeneration. The ability of vascular plants to maintain oxidizing conditions in sediments is largely linked to light availability. In seagrass beds, low light conditions can prevent plants from supporting the respiratory demands of their belowground tissues, leading to increased

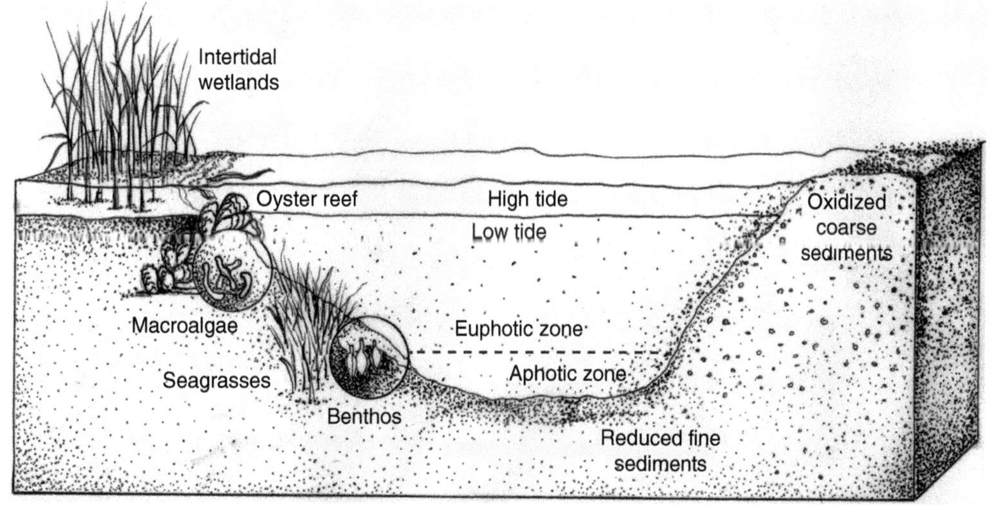

FIGURE 1.5 Idealized cross-section through a typical estuary showing vertical distribution of several important elements; note that organism sizes and vertical scale are exaggerated.

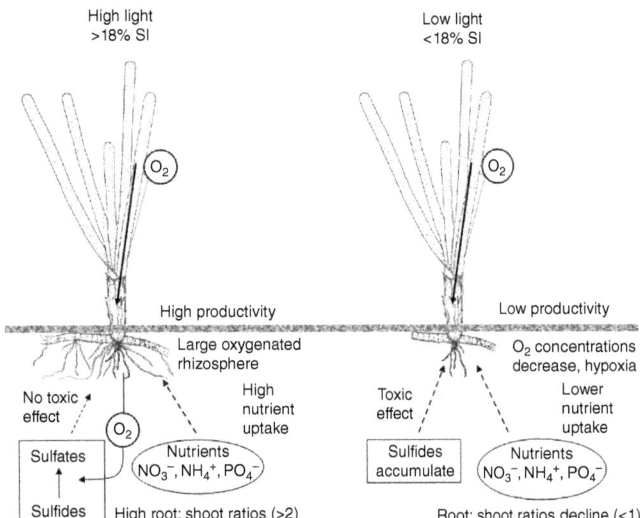

FIGURE 1.6 Seagrasses ventilate sediments by facilitating the movement of photosynthetically produced oxygen from leaves to belowground tissues to support the respiratory demands of roots and rhizomes. Under low light conditions, less oxygen is produced for belowground tissue respiration, which results in the accumulation of toxic sulfides and ammonium. *Source*: From Mateo et al. (2006) with permission.

sulfides which lead to root loss (Lee and Dunton 2000), and decline of the meadow (Dunton 1994; Morris et al. 2022).

1.3.3 Longitudinal Section

A longitudinal cross-section of an estuary (Figure 1.7) demonstrates some of the attributes that result from the mixing of fresh water and seawater. Salinity gradually increases seaward along the length of estuaries, and the isohalines (lines of equal salinity) show that salinity generally increases with depth (unless turbulence is sufficient to mix the water column). This salinity distribution results from the density difference between salt and fresh water. Freshwater from the river tends to flow on top of the salt water because it is less dense. As the freshwater flows to the sea, salt water from below mixes with it, making it saltier. The resulting water column average salinity therefore increases along the estuary, from 0 to 5 in the oligohaline zone near the river mouth to 30–35 in the polyhaline zone near the ocean. The resulting pressure gradient forces an inward flow in the lower layer compensated by an outward flow in the upper layer. This circulation pattern, and the resulting salinity distribution, is a general feature of many estuaries that have significant river input and will be discussed in more detail in Chapter 2.

In many estuaries, we observe a turbidity maximum at the upper reaches of the estuary (salinity between 1 and 5) at the convergence of an upper estuarine layer of lower salinity moving seaward with a lower layer of more salty water moving landward. The physical and chemical changes that occur as river water and sea water mix is addressed in Chapter 3.

1.4 Estuarine Food Webs and Energy Flow

Thus far our observations about estuaries have focused on their physical and chemical structure and, as such, have been rather static. We now discuss more dynamic aspects of estuaries by considering a typical estuarine food web (Figure 1.8). This allows for a more detailed consideration of some of the organisms that live in estuaries. For illustrative purposes, we will compare estuarine and marine systems.

We begin by listing several terms with definitions derived from E.P. Odum (1971). The transfer of food energy from the

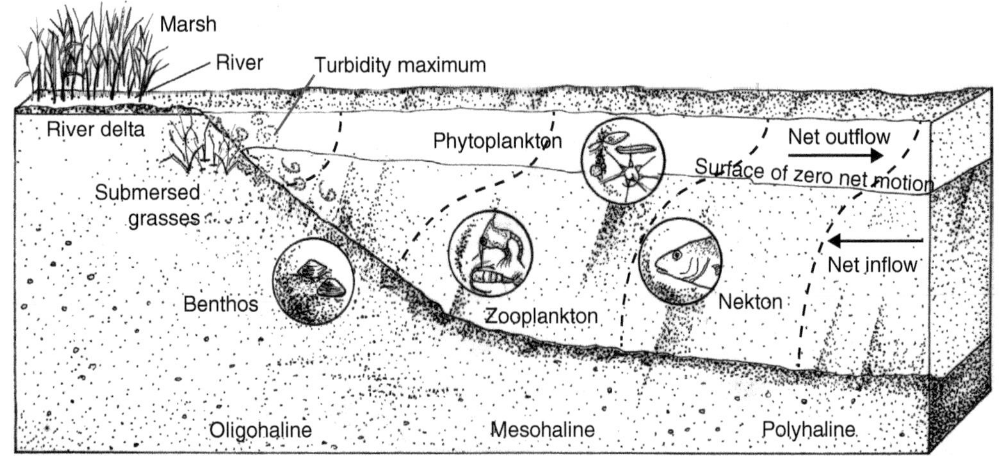

FIGURE 1.7 Idealized longitudinal-vertical section through a typical partially mixed estuary from river (left) to the sea (right). Note isohalines (lines of equal salinity) that curve landward in the lower saltier layer and seaward in the upper less salty layer. The tidal-mean currents are similarly layered, flowing seaward in the upper layer and landward in the lower layer, as indicated by the arrows. This layered counter-flowing circulation is exhibited even if the estuary is well mixed (i.e., vertical isohalines).

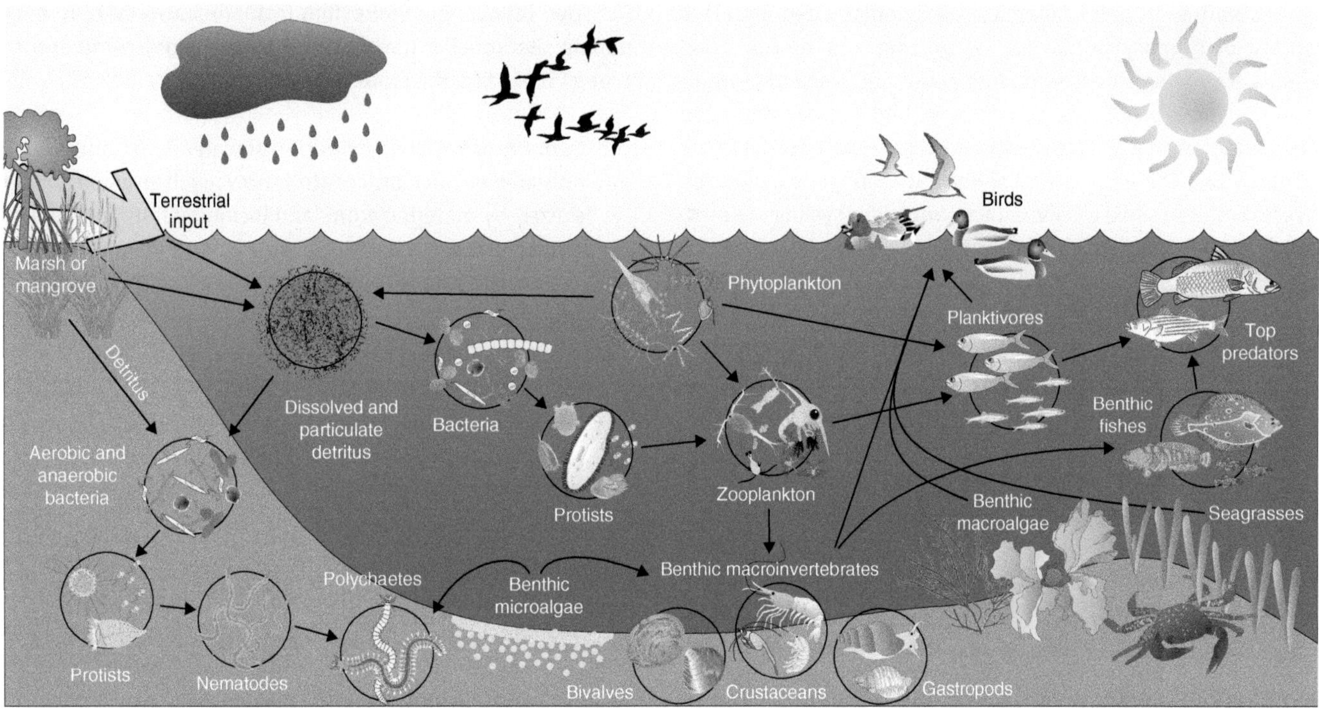

FIGURE 1.8 Food web diagram for a typical estuarine ecosystem showing important links among some of the major trophic groupings in both the water column and sediments. Black lines and arrows indicate flow of food from source to consumer. Note the high diversity of primary carbon sources that are largely responsible for the high faunal abundance, productivity, and enormous fisheries value of estuarine systems.

source in plants (or more appropriately, primary producers) through a series of organisms eating one another is referred to as a *food chain,* or, more properly, a *food web,* which describes many interacting food chains. The word *trophic* is used interchangably with food, and trophic dynamics refers to the pattern of food production and consumption as it occurs and changes over time.

The trophic dynamics of estuaries tends to be complex. Figure 1.8 illustrates a number of important characteristics of estuarine trophic dynamics. First, there are almost always several different types of primary producers in estuaries, including phytoplankton, salt marsh plants, mangroves, submersed aquatic vegetation, and benthic macroalgae and microalgae. By contrast, the open ocean has only phytoplankton (albeit with some exceptions such as *Sargassum*). There are other important distinctions between oceans and estuaries. For example, in oceans, practically all phytoplankton are consumed alive. A food web that begins with consumption of live plants is called a *grazing food web*. In estuaries, many important plants are not grazed while alive, but instead die and decompose. This decomposing material is called *organic detritus* and the food web it supports is called a *detrital food web*. Organic detritus is an important carbon source in estuaries. Considerable research over the past four decades has addressed detrital dynamics, a topic considered in a number of chapters, especially in Chapters 9 (microbial ecology) and 15 (estuarine food webs).

Nevertheless, the food web most recognized and understood by students is the grazing based food web, which is based on phytoplankton as the ultimate carbon source. These small primary producers are eaten by zooplankton which are then consumed by small planktivorous fish. In Figure 1.8, these fish are menhaden and anchovy, but in other estuaries these fish may be herring or sardines. Zooplankton are also grazed by larvae of larger fish. At the top of the food web are larger carnivores such as striped bass and barramundi. Phytoplankton and the organisms that consume them are also part of the microbial food web (see Figure 1.8) that are discussed later in this chapter and in Chapters 9 (estuarine microbial ecology) and 10 (estuarine zooplankton).

The diversity of feeding strategies used by animals that live in and on estuarine sediments is enormous. Epibenthic animals such as oysters, clams, and mussels are suspension feeders; that is, they remain in one place and concentrate food that flows past them in the water currents. There are other benthic organisms that live in areas of weak currents. They move over and through the sediments and take food from the sediment itself. These are called deposit feeders and include worms, amphipods, and a host of other small organisms (see Chapter 11). There are also a large number of non-bottom-dwelling organisms that feed on the bottom. These include a variety of invertebrates, fish, and birds. In fact, many fish species found in estuaries have adaptations for bottom feeding (see Chapter 12).

All of this flow of food energy from primary producers and detritus, and through planktonic and benthic food webs converges upon a group of top carnivores that are generalist feeders on a wide variety of organisms. These top carnivores include many species of fish including sea trout, striped bass, and flounder, birds such as ducks and sea gulls, and mammals such as seals and dolphins (see Chapters 12 and 13).

Estuarine trophic dynamics are characterized by (1) a variety of primary producers, (2) grazing and detrital food chains, (3) a high degree of interaction between the water column and bottom, (4) a complex, highly interconnected food web, and (5) a large number of generalist feeders (see Chapter 15). The importance of detrital food webs cannot be overestimated, since the dynamics of nutrient cycling are intricately tied to detritus decomposition. Sunlight is critical for the diversity of estuarine primary producers. Periods of high turbidity, which reduce the amount of light reaching the bottom, can have substantial consequences for marine plants (see Figure 1.6), resulting decomposition pathways, and the amount of oxygen available for animals, both on the bottom and in the water column.

Throughout this book, the concepts we have introduced here will be developed in much more detail to provide a fuller general understanding of estuaries. At this point, however, the reader should have a good foundation to begin a more in-depth study of estuaries. In the next section, we discuss a classification of estuarine ecosystems.

1.5 The Ever-Changing Dynamic Properties of an Estuary

We defined estuary and we defined ecology, but because these definitions are so broad, it is difficult to answer the question, "what geomorphic characteristics define an estuary?" We can easily describe, however, the abiotic factors that are most important in determining the specific nature of estuarine ecosystems as well as, the salient ecological characteristics of these ecosystems.

The biota that develops in many systems is influenced by the range of physical features discussed earlier (Figure 1.2) including tidal range, exchange with the coastal ocean, local and regional climate, and freshwater inputs. However, in many estuaries, temporal variability in these conditions can become more important than the conditions themselves. Biotic assemblages are more determined by the degree of change over various time intervals (e.g., diel, tidal, seasonal), than by the absolute level of such factors. But the effects of these temporal fluctuations are dampened where various plant assemblages offer extensive habitat protection. These include foundation species representing the extensive marsh, mangrove, and seagrass habitats that are found in most estuarine systems (and kelp beds in fjords). Foundation species *create habitat* and are defined as those species which are disproportionately important to the continued maintenance of the existing community structure. In contrast, sandy beach ecosystems are strongly impacted by physical events (e.g., waves, tidal range, storm frequency, and sediment transport) that often result in frequent changes in the character of the biotic system.

The following abiotic features are important in determining the specific nature of estuarine ecosystems and the type of biota that develops include:

1. *Quantity of freshwater input.* Freshwater flux controls estuarine salinity, but it also controls how much an estuary is subsidized by materials from land including nutrients, organic matter, and sediment. Freshwater flow influences the flushing rate of estuaries and retention of materials in estuaries.
2. *Water circulation patterns.* Estuarine circulation is determined by riverine and tidal currents, winds, exchange with the coastal ocean, and geomorphology. Tides are particularly important because they exert a profound influence on estuarine circulation and biological processes. Circulation determines how strongly the water is layered, or *stratified.*
3. *Bathymetry.* The depth of estuaries controls the degree of benthic–pelagic coupling. Where estuaries are shallow there is a stronger interaction between surface waters and the bottom. This allows for strong benthic–pelagic coupling which influences important ecological processes such as nutrient availability and carbon sequestration. For example, when benthic–pelagic coupling is strong, nutrients released by decomposition of organic matter in sediments can become available for phytoplankton in surface waters and organic detritus produced in surface waters can rapidly settle to the bottom.
4. *Physiography.* The shape and orientation of the estuary, especially with respect to prevailing winds, determines the degree to which habitats are protected and buffered from meteorological and oceanic forces. Prevailing wind and wave direction, along with tidal amplitude, influences the stability of physical conditions for sustained development of foundation species (e.g., salt marshes, oyster reefs). The rate at which sediment and detritus are retained in estuaries or washed into the ocean is also determined by the physical characteristics of the inlets that allow exchange.
5. *Rate of geomorphological change.* The physiography of estuaries change over time and those changes are much more rapid than for most terrestrial systems. Sand banks and mud flats form, degrade, and migrate within estuaries. Wetlands form at the mouths of rivers as sediments are deposited, and then degrade as relative sea level increases. Also, some biogenic processes, such as reef formation by oysters and corals, contribute to the changing geomorphology of estuaries.

Most estuarine ecosystems are open systems dominated by physical processes, resulting in large exchanges of materials with neighboring terrestrial and oceanic ecosystems including water, salt, nutrients, sediments, and organisms. The exchange of organisms over millions of years has resulted in a rich biological diversity with biota derived from marine, freshwater, and terrestrial sources. Over hundreds to thousands of years, deltas grow and erode and barrier islands shift. On shorter timescales, salinity changes with tide and river flow, and water levels fluctuate so that intertidal regions are subjected to wetting and drying and extremes of temperature. But estuarine organisms have developed physiologies and behaviors that allow them to thrive in this dynamic environment, and some organisms are able to directly modify the physical environment to suit their needs. Many organisms use the intense and variable physical energies in estuaries as subsidies, as, for example, in the case of an oyster reef "using" the flow of tides to exploit phytoplankton produced elsewhere. Nevertheless, despite these adaptations, the ever-changing physical conditions in estuaries can impose considerable stress because large changes in salinity, temperature, oxygen, and light availability can be deleterious, even lethal, for estuarine organisms.

1.6 High Productivity: An Estuarine Focal Point

High productivity makes estuaries great resources for spawning fishes, migrating animals, commercial fishing, and aquaculture, and it has made estuaries centers of human civilization. But in many ways, this high productivity also makes estuaries particularly vulnerable to the negative impacts of eutrophication and climate change (see Chapter 18). One of the things that makes the study of estuarine ecology so exciting is the lively discussion about what makes estuaries so highly productive. In this section, we outline some of these ideas, so that we have them in mind as we address the information in each chapter. The following discussion summarizes some current thinking about estuarine production. Perhaps future generations of estuarine scientists will view the idea very differently as the Anthropocene progresses.

1.6.1 Reasons for High Estuarine Primary Productivity

In a well-known paper titled "Mechanisms maintaining high productivity of Georgia estuaries" (Schelske and Odum 1962), the authors stated that Georgia estuaries were among the most productive natural ecosystems in the world. They listed several reasons for this high productivity: (1) transport of materials on to and out of marshes via tidal action, (2) abundant supplies of nutrients and the associated retention and regeneration of nutrients by benthic microorganisms and filter feeders, and (3) the existence of three distinct sources of primary production (marsh grass, benthic algae, and phytoplankton), which ensures a steady output of organic carbon year-round in response to light and nutrient availability. Remarkably, each of these reasons involves some kind of interaction between pelagic and benthic environments.

Now let us examine these points in more detail. Tidal action is considered to be an important factor contributing to high productivity. E.P. Odum once defined estuaries as "tidally subsidized fluctuating water level ecosystems." Others have expanded on this to include other physical energies, including wind, waves, and riverine currents. These factors produce complex water movements in estuaries. In much of this book we consider the nature of these physical factors and how they affect biogeochemical and ecological processes.

The next factor is the rapid regeneration and conservation of nutrients. Since most estuaries are relatively shallow and well-mixed, there are persistent interactions between water and sediments, unlike the deep ocean. This means that food in the water is available to organisms on the bottom and that nutrients released by benthic organisms are mixed into the water (Figure 1.9). Because there is a positive relationship between the supply of regenerated nutrients from decomposition of pelagic organic matter in sediments, and the production of new organic matter by primary producers using those nutrients in the water, many estuarine ecologists argue that the combination of benthic nutrient regeneration and a shallow, well-mixed water column are the most important factors producing high estuarine productivity. This benthic–pelagic coupling is an important process unique to shallow water systems that accelerates the recycling and reuse of nutrients associated with detritus produced in surface waters and with detritus delivered from land by rivers (Dollar et al. 1991) (Figure 1.9). There are abundant supplies of nutrients in the Georgia estuaries studied by Schelske and Odum, and nutrient concentrations in estuaries are almost always higher than in adjacent oceans and freshwater systems. This likely arises from the retention of nutrients from rivers and oceans via rapid benthic–pelagic exchange and uptake by highly productive autotrophic communities.

Finally, there are indeed several distinct groups of primary producers in estuaries, many of which have high rates of primary production. A more general statement is that there is a diversity of sources of organic matter. In Georgia there are three important producer groups: salt marsh grass, phytoplankton, and benthic algae. In other estuaries, we find seagrasses, mangrove wetlands, and macroalgae. Epiphytic algae grow on most surfaces in the euphotic zone. A variety of these plants grow on the bottom in shallow waters (e.g., marsh grasses, seagrasses, and benthic macroalgae) and also contribute to pelagic–benthic coupling by exuding dissolved organic matter that can account for a large proportion of bacterial carbon metabolism in the water (Zieglar and Benner 1999). According to Schelske and Odum, the existence of these primary producer types provides enough diversity to allow significant year-round production. This has been observed in a number of estuaries at lower latitudes, but it is not true in higher latitude estuaries, which are often very productive but only during specific seasons. However, high productivity by primary producers in estuaries often creates a surplus of organic matter, and many scientists argue that this organic matter is an important food source because it is available to consumers year-round as organic detritus.

In summary, high primary production measurements in Georgia and elsewhere led Schelske and Odum to conclude that "estuaries are among the most productive natural ecosystems in the world." Chapters 4 through 8 address four major

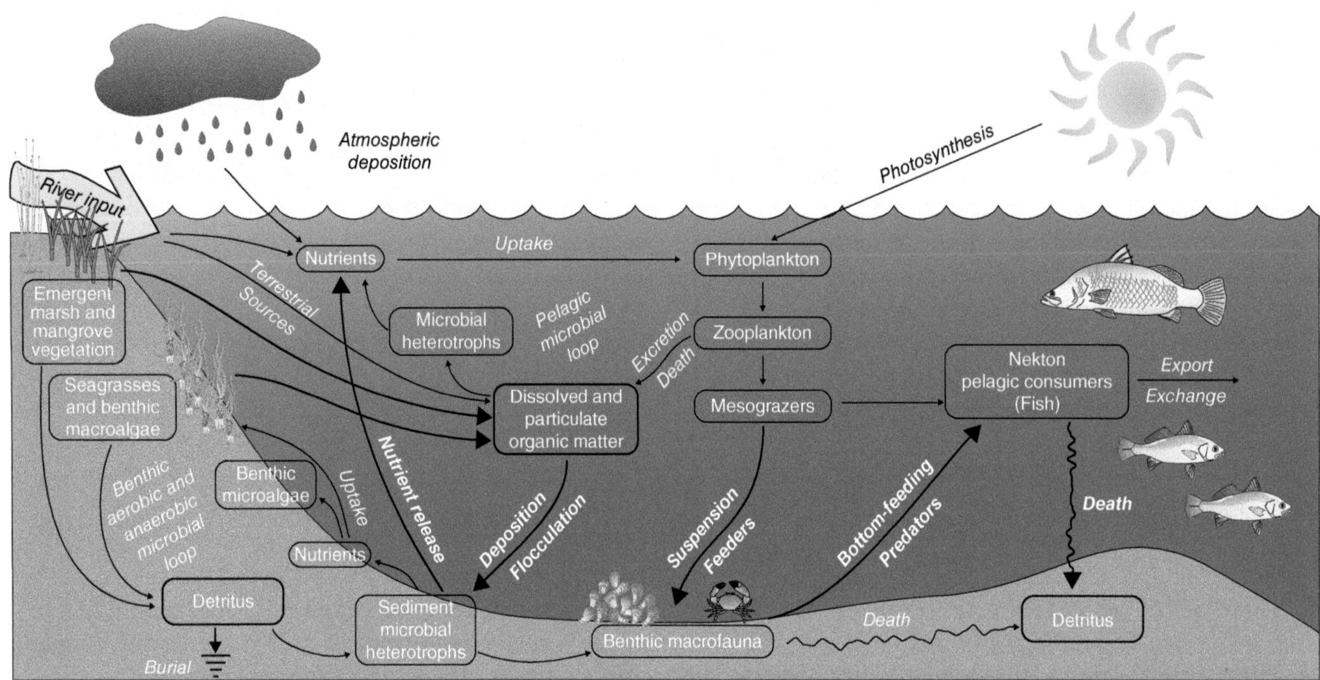

FIGURE 1.9 An energy-flow diagram illustrating benthic–pelagic coupling in estuarine ecosystems and the active presence of both grazing- and detrital-based food webs. Black thick arrows denote major connections between the shallow benthos and water column. Microbial loops, both in the water column and sediments, are largely responsible for the regeneration of nutrients for primary producers that include vascular plants, macroalgae, and microalgae.

sources of fixed carbon in estuaries: phytoplankton, seagrasses, marshes, and mangroves. These four chapters address the factors that regulate primary productivity and demonstrate the enormous contribution made by these communities. Their combined inputs show that estuaries possess extremely high rates of productivity relative to other ecosystems. Chapters 9 through 13 (estuarine consumers) and 15 (estuarine food webs), describe the trophic dynamics associated with high estuarine productivity in detail. Chapter 14 describes the role of high estuarine productivity in the global carbon budget. As we indicated in the previous paragraphs, the factors listed by Schelske and Odum are still discussed 40 years after their proposal. These issues have been an incentive for considerable research since then and are still a source of inspiration and controversy today.

1.6.2 Other Important Hypotheses about Estuarine Ecology

A number of other important hypotheses about estuarine function have been proposed that help to unify the study of estuaries. They are interrelated, but we will separate them for the sake of clarity.

Intertidal wetlands are important to estuarine productivity because they (a) produce large quantities of organic detritus that is an important source of energy in estuaries and (b) serve as an important nursery to the young of many marine and estuarine species. Many important chemical reactions take place in wetlands and these biogeochemical processes are critical to the cycling of carbon and nitrogen. Salt marshes and mangrove wetlands are covered in Chapters 6 and 7, respectively, and their roles in ecosystem metabolism are discussed in Chapter 14.

Organic detritus is exported from wetlands and serves as an important food source for a wide variety of estuarine consumers. The importance of organic detritus is covered in many chapters in this book, especially Chapters 9 (microbial ecology) and 15 (estuarine food webs).

Estuaries, especially those with extensive wetlands, support rich fisheries. There is considerable evidence showing both functional and empirical relationships between wetlands and fisheries. For example, the abundance of wetlands in estuarine regions is strongly correlated with regional fish catch, even though very few fish are absolutely estuarine dependent. This topic is discussed in Chapters 12 (Nekton) and 17 (Fisheries and Aquaculture).

1.7 Human Impacts and Management of Estuarine Ecosystems

Humans have lived in and around estuaries for tens of thousands of years. Early peoples harvested the rich primary and secondary productivity of estuaries. Archaeological evidence of their presence includes village sites and middens or large accumulations of shells of harvested mollusks. About 5000 years ago, the first human civilizations developed adjacent to estuaries and lower river valleys. It is thought that the rich resources of coastal margins provided the energy subsidies necessary for village-based agricultural societies to transform into the complex social organization of civilization.

Throughout the Holocene, humans congregated near the coast and many of the world's current large cities such as New York, London, Amsterdam, Venice, Alexandria, Kolkata, Singapore, and Shanghai developed near estuaries and deltas. These areas were important sources of food, and rivers provided important routes for navigation. Lower river valleys also supported rich agricultural areas.

Humans have modified estuarine areas since the beginning of civilization. Notable preindustrial impacts include the draining of much of the Rhine River delta and the elimination of most river input to Venice lagoon. But massive changes in estuaries mostly occurred in the twentieth century when human populations grew dramatically in the coastal zone, leading scientists to proclaim the start of the Anthropocene geological period. Human activity has physically changed coastal systems by draining and filling areas and by dredging channels for navigation, drainage, and access to minerals such as oil. Industrial, agricultural, and urban growth have introduced many toxic materials such as heavy metals and pesticides that poisoned estuarine organisms, and nutrients and organic matter that led to eutrophication. Overharvest of commercially important organisms, and introduction of new species, either accidentally or purposely, has changed the composition of the biota.

As human impacts grew, so did the study of these impacts and efforts to reduce or mitigate them. Early efforts on estuarine management often dealt with solutions to specific problems such as reduction of the use of certain pesticides, advanced treatment of sewage, and restoration of wetlands. More recent management has focused on more comprehensive approaches such as ecosystem-based management, integrated coastal management, and the incorporation of social and economic science in the study of estuarine ecosystems. These topics are covered in more detail in subsequent chapters in this book.

1.8 The Potential Impacts of Future Trends on Estuarine Ecosystems

The twenty-first century will see the intensification of several major global trends that will strongly impact estuaries, many of which are linked to global climate change. For example, sea level rise will measurably effect coastal wetlands and humans living near the coast, and it is thought that many, if not most, wetlands in estuaries may eventually disappear (Blum and Roberts 2009; Syvitski et al. 2009; Vorosmarty et al. 2009). In many coastal areas, sea levels are rising more rapidly than sediment deposition can compensate. On the Coast of the Gulf of Mexico, relative sea level rise (up to 7.9 mm $year^{-1}$) is more than twice the global average. As sea level rises, ghost forests (Figure 1.10) are becoming clear visual reminders that salt water incursion is rapidly transforming our coastal landscape while threatening urban civilizations across the globe. In areas of rapid sea level rise, marshes may drown from the rising sea levels faster than new marshes can form, eventually leading to open water, which translates to losses in habitat and deep carbon sequestration.

There are also several other aspects of climate change that will impact estuaries. Changes in precipitation will affect the quantity and seasonality of freshwater inflow to estuaries affecting such processes as circulation, plant growth, and fisheries. Tropical storms (hurricanes and typhoons) may become more intense and will frequently lead to changes in estuaries

FIGURE 1.10 A Ghost Forest in Goose Creek State Park in North Carolina near Pamlico Sound. The forested wetland is dying due to pulses of saltwater during storms. The dead trees are water tupelo (*Nyssa aquatica*). Ghost forests are occurring rapidly across the southeastern coastal plains of the United States and represent a loss of vegetated habitats that sequester atmospheric carbon dioxide. Photo credit: M. Martinez.

and human activity adjacent to them. Rising temperatures will impact rates of material cycling and oxygen consumption, and also affect the distributions of many estuarine species. For example, during this century, the Gulf of Mexico is expected to become completely tropical as mangroves replace salt marshes. Also, ocean acidification will affect the ability of some mollusks and corals to grow and survive. A detailed description of forecasted impacts of climate change is presented in Chapter 18.

1.9 How We Will Proceed Through the Book

This book is designed to systematically carry readers through the science of estuarine ecology. Chapters 2 and 3 introduce estuarine physical oceanography, geomorphology, and chemistry. Knowledge of these subjects is essential for a thorough understanding of the ecology of estuaries, because estuarine organisms respond to such factors as water movement, sediment distribution, and chemical gradients. These two chapters are designed to both introduce the subjects and place the information in the context of ecological processes.

The next sections of the book cover various aspects of the ecology of organisms that live in estuaries. In Chapters 4 through 8, we cover, respectively, phytoplankton, seagrasses, coastal marshes, mangrove wetlands and macroalgae of estuaries. In each chapter, we discuss the composition and distribution of the vegetation, spatial and seasonal patterns of productivity, and the factors regulating productivity. Chapter 9 introduces the reader to the important topic of estuarine microbial ecology and the critical role of microorganisms in the biogeochemical cycling of carbon and nitrogen.

The sources, transport, and use of organic matter are topics that have generated considerable research and discussion over the last two to three decades and we try to capture some of this excitement. We cover zooplankton, benthos, nekton, and wildlife in Chapters 10 through 13. In these chapters, we discuss topics such as composition and distribution of biological communities, rates of secondary production, food habits, and factors regulating these communities. The next section of the book introduces topics in ecosystem metabolism (Chapter 14) and estuarine food webs (Chapter 15). The use of modeling to aid in understanding these systems is covered in Chapter 16. Finally, we review the interactions of people with estuaries with a focus on fisheries in Chapter 17 and on how global climate change is altering coastal systems and will impact human populations that reside on coastlines (Chapter 18).

Further Reading

Alongi, D.M. (1998). *Coastal Ecosystem Processes*. Boca Raton, FL (USA), 419: CRC Press.

Day, J.W., Hall, C.A.S., Kemp, W.M., and Yanez-Aracibia, A. (ed.) (1989). *Estuarine Ecology*, 1e. New York, 558: Wiley.

Dyer, K.R. (1997). *Estuaries: A Physical Introduction*. Chichester (UK): Wiley.

Hobbie, J.E. (ed.) (2000). *Estuarine Science: A Synthetic Approach to Research and Practice*. Washington (DC): Island Press.

Kennish, M.J. (ed.) (2005). *Estuarine Research, Monitoring, and Resource Protection*. Boca Raton, FL (USA), 297: CRC Press.

Lauff, G.H. (ed.) (1967). *Estuaries*. Washington, DC, 757: AAAS Publication, No. 83.

McLusky, D.S. and Elliott, M. (2004). *The Estuarine Ecosystem: Ecology, Threats, and Management*. Oxford: Oxford Press.

Odum, H.T. (1983). *Systems Ecology*. New York, 664: Wiley-Interscience.

Parsons, T.R., Takahashi, M., and Hargrave, B. (1984). *Biological Oceanographic Processes*, 3e. Oxford (UK), 330: Pergamon Press.

Valiela, I. (1995). *Marine Ecological Processes*, 2e. Berlin, 686: Springer.

References

Blum, M. and Roberts, H. (2009). Drowning of the Mississippi Delta due to insufficient sediment supply and global sea-level rise. *Nature Geoscience* 2: 488–491.

Day, J., Gunn, J., Folan, W. et al. (2007). Emergence of complex societies after sea level stabilized. *EOS* 88: 170–171.

Dollar, S.J., Smith, S.V., Vink, S.M. et al. (1991). Annual cycle of benthic nutrient fluxes in Tomales Bay, California, and contribution of the benthos to total ecosystem metabolism. *Marine Ecology Progress Series* 79: 115–125.

Dunton, K.H. (1994). Seasonal growth and biomass of the subtropical seagrass *Halodule wrightii* in relation to continuous measurements of underwater irradiance. *Marine Biology* 120: 479–489.

Fairbridge, R.W. (1980). The estuary: its definition and geodynamic cycle. In: *Chemistry and Biochemistry of Estuaries* (ed. E. Olausson and I. Cato), 1–35. New York: Wiley.

Kennett, D.J. and Kennett, J.P. (2006). Early state formation in southern mesopotamia: sea levels, shorelines, and climate change. *Journal of Island and Coastal Archaeology* 1: 67–99.

Lee, K.-S. and Dunton, K.H. (2000). Diurnal changes in pore water sulfide concentrations in the seagrass *Thalassia testudinum* beds: the effects of seagrasses on sulfide dynamics. *Journal of Experimental Marine Biology and Ecology* 255: 201–214.

Margalef, R. (1968). *Perspectives in Ecological Theory*. Chicago: University of Chicago Press.

Morris, J.M., Hall, L.M., Jacoby, C.A. et al. (2022). *Seagrass in a Changing Estuary, the Indian River Lagoon*. Florida, United States: Frontiers in Marine Science https://doi.org/10.3389/fmars.2021.789818.

Mateo, M.A., Cebrián, J., Dunton, K., and Mutchler, T. (2006). Carbon flux in seagrass ecosystems. In: *Seagrasses: Biology, Ecology and Conservation* (ed. A.W.D. Larkum, R.J. Orth and C.M. Duarte), 159–192. Berlin: Springer.

Odum, E.P. (1971). *Fundamentals of Ecology*. Philadelphia, 574: W.B. Saunders.

Pritchard, D.W. (1967). What is an estuary: physical viewpoint. In: *Estuaries*, vol. 83 (ed. G.H. Lauff), 3–5. Washington, D.C.: American Association for the Advancement of Science Publication.

Schelske, C.L. and Odum, E.P. (1962). Mechanisms maintaining high productivity in Georgia estuaries. *Proceedings – Gulf and Carribbean Fisheries Institute* 14: 75–80.

Syvitski, J., Kettner, A., Overeem, I. et al. (2009). Sinking deltas due to human activities. *Nature Geoscience* 2: 681–686.

Syvitski, J., Walters, C., Day, J. et al. (2020). An extraordinary outburst of human energy consumption and resultant geological impacts beginning around 1950 CE initiated the proposed Anthropocene Epoch. *Nature Communications* https://doi.org/10.1038/s43247-020-00029-y.

Vorosmarty, C., Syvitski, J., Day, J. et al. (2009). Battling to save the world's river deltas. *Bulletin of the Atomic Scientists* 65: 31–43.

Zieglar, S. and Benner, R. (1999). Dissolved organic carbon cycling in a subtropical seagrass-dominated lagoon. *Marine Ecology Progress Series* 180: 149–160.

CHAPTER **2**

Estuarine Geomorphology, Circulation, and Mixing

The Rio de la Plata estuary empties into the Atlantic Ocean between Argentina (on the right) and Paraguay (on the left). Approximately 2 billion cubic feet of silt is carried into the estuary each year. Photo credit: NASA.

Gregg A. Snedden[1], Jaye E. Cable[2], and Björn Kjerfve[3]
[1] Wetland and Aquatic Research Center, U.S. Geological Survey, Baton Rouge, LA, USA
[2] Department of Earth, Marine, and Environmental Sciences, University of North Carolina, Chapel Hill, NC, USA
[3] School of the Earth, Ocean and Environment, University of South Carolina, Columbia, SC, USA

2.1 Introduction

To understand the processes affecting the distribution and cycles of particulates, pollutants, nutrients, and organisms in estuaries, it is insufficient to focus solely on the biological and chemical aspects of the processes. Water sources and movements (e.g., evaporation, precipitation, riverine discharge, submarine ground water discharge, wetland hydrology, and tidal exchange) as well as other hydrodynamic aspects of coastal systems, including circulation patterns, stratification, mixing, and flushing, must also be considered. When hydrodynamic changes occur quickly relative to biological, geological, and chemical transformations, they become the dominant controlling factors of many ecological processes in estuaries (Officer 1980), and it is now widely recognized that a thorough understanding of the marine estuarine ecology requires comprehensive knowledge and integration of physical processes affecting these systems. Using the terminology of a shallow-water oceanographer, this chapter aims to organize, classify, and describe some of these important physical characteristics and processes.

2.2 Glaciation Cycles

Present-day estuaries are geologically ephemeral coastal features. They formed during the last interglacial stage as sea level rose 120 m to the present level, which was reached approximately 5000 years ago (Milliman and Emery 1968; Figure 2.1). Such glaciation and deglaciation events have occurred regularly during the past few million years, causing shifts in the position of coastlines worldwide. The locations of estuaries have shifted accordingly.

Modern day estuaries are common coastal features, constituting as much as 80–90% of the coasts along the east and Gulf coasts of North America, but as little as 10–20% along the

Estuarine Ecology, Third Edition. Edited by Byron C. Crump, Jeremy M. Testa, and Kenneth H. Dunton.

Companion website: www.wiley.com/go/crump/estuarine3

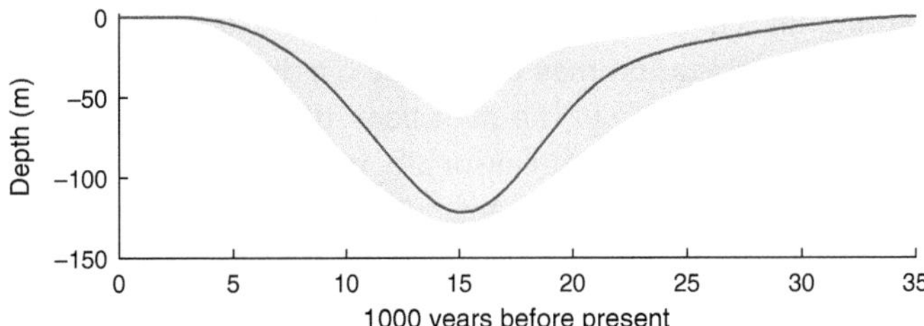

FIGURE 2.1 Variation of mean sea level for the past 35,000 years based on data from the Atlantic continental shelf of the United States. The solid line is the mean sea level and the shaded region is the uncertainty envelope based on all observed values. *Source*: Modified after Milliman and Emery (1968).

United States' Pacific coast (Emery 1967). Typically, estuaries are more abundant on coasts with broad flat continental margins (such as the east coasts of North and Central America) than on coasts with narrow, steep continental margins (such as west coasts of N. and S. America; Schubel and Hirschberg 1978).

During glaciated periods, a considerable fraction of the world's oceans were frozen into continental glaciers, and sea level was much lower than now. Coastlines were then located on what is now the continental slope, and estuaries were both smaller and rarer. During interglacial periods, the glaciers melted, global sea level increased, and estuaries became large and abundant (Schubel and Hirschberg 1978).

The present situation of high sea level and extensive estuaries has existed for only 10–20% of the time during the past million years, because, once they form, estuaries quickly fill with sediments and eventually disappear. The sediment sources are river-borne terrestrial materials from eroding continents and net up-estuary movement of sand-sized materials from the continental shelf (Meade 1969). From a geological viewpoint, the time scale of this infill is extremely short. Emery and Uchupi (1972) estimated that if sea level remained constant and all sediments were deposited into today's U.S. estuaries, they would be filled within 9500 years, even if the load of the Mississippi River, which is half the nation's total, was not counted. But all estuaries obviously do not infill at this rapid rate, which is why estuaries of various stages of geological development exist around the shorelines of the world (Schubel and Hirschberg 1978). One major reason is that sea level has been rising for the past 10,000 years. Another is that coastal erosional forces remove sediments from estuaries.

2.3 Definition

Cameron and Pritchard (1963) put forth perhaps the most widely accepted definition of an estuary. They defined an estuary as "a semi-enclosed coastal body of water which has a free connection with the open sea within which sea water is measurably diluted with fresh water derived from land drainage." While helpful, this definition emerged from early studies of estuaries situated at the lower reaches of coastal rivers, and we have since come to understand that fresh water derived from land drainage is not necessarily required for a coastal embayment to manifest the fundamental physical processes typically associated with estuaries. Many estuaries are periodically isolated from the open ocean by bathymetric features such as sand bars during seasonal periods of low water, and thus only exhibit a connection to the open ocean intermittently. Estuaries in arid regions often lack a significant source of freshwater inputs, and evaporation inside these estuaries exceeds the freshwater supply brought about by runoff or precipitation, leaving the estuaries more saline than the open ocean, and setting up longitudinal density gradients that drive circulation. Thus, we will use a definition that applies to the broad spectrum of estuarine types, including lagoons, river mouths, and deltas in temperate, tropical, and arid settings alike: *An estuary is a semi-enclosed coastal body of water which is at least intermittently connected with the open sea, and within which water salinity differs measurably from that in the open ocean.*

2.4 Classification of Estuaries

2.4.1 Geomorphic Classification

2.4.1.1 Coastal Plain Estuaries Coastal plain estuaries have been studied most extensively, because they are the most common type in regions where studies began. They formed during the last eustatic sea level rise (i.e., the increase in sea level resulting from a change in the volume of the ocean), when river valleys became increasingly more flooded by the melting glaciers. Thus, they exhibit the geomorphic characteristics of river channels and flood plains and are sometimes called "drowned river valley estuaries." The typical cross section of a classical coastal plain estuary consists of a V-shaped channel, seldom deeper than 20 m, bordered by broad shallow flats (Figures 2.2 and 2.3). Coastal plain estuaries vary greatly in size, with Chesapeake Bay, some 300 km long and on average 22 km wide, being the largest U. S. coastal plain estuary. Other good examples of large coastal plain estuaries are the Hudson R. estuary in the United States, the Thames R. estuary in Great Britain, the Pearl R. estuary in Hong Kong, and the Murray R. estuary in Australia. Others have been described by Officer (1976) and Dyer (1973). Coastal plain estuaries typically contain large expanses of salt marsh, such as those found along much of the east coast of North America, and the west and north coasts of Europe.

2.4.1.2 Deltaic Estuaries In contrast to the drowned river valley estuaries described above, deltaic systems provide home for expansive fresh, brackish, and salt marshes and mangroves. These systems are situated at

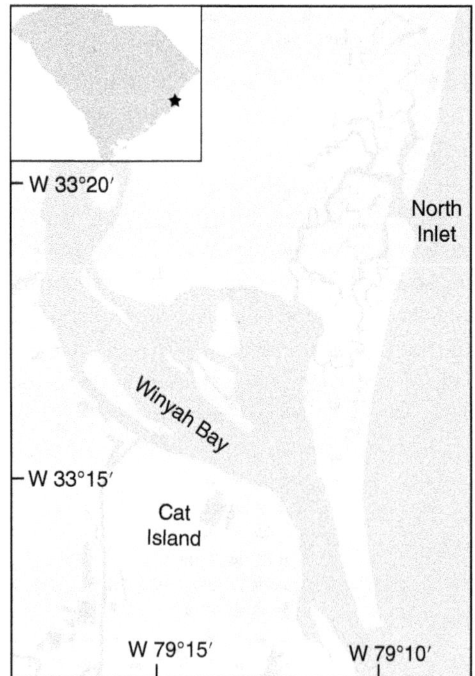

FIGURE 2.2 Examples of coastal plain estuaries: Winyah Bay and North Inlet, South Carolina. Winyah Bay is a classical coastal plain estuary, whereas the North Inlet system is a coastal plain salt marsh estuary.

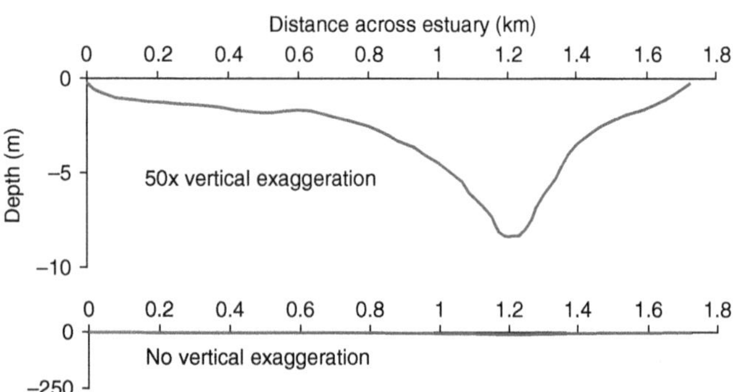

FIGURE 2.3 Cross-sectional profile of an idealized typical coastal plain estuary, showing the same cross section in 50 times vertical exaggeration (top) and without any vertical exaggeration (bottom). Coastal plain estuaries have very wide and shallow cross sections. When these sections are presented with vertical exaggeration (which is most often the case), this gives a misleading impression of greater depth in estuaries than is realistic.

the seaward terminus of major river systems, such as the Ganges–Brahmaputra–Meghna (India), Mekong (southeast Asia), Nile (Egypt), and Mississippi (USA) Rivers. As the main river approaches the coastline, water velocities slow and water is moved out across the deltaic system. Distributary channels can facilitate this movement into estuarine wetland environments, or the river may overflow its banks and flow as overland flow across the landscape. These distributary channels radiate from the main river source in a dendritic network and form large and generally shallow (less than 5 m) estuaries to support primary productivity and large nursery grounds for marine and freshwater fisheries. Spring discharge following winter snowmelt or tropical wet season precipitation may also cause overbank flooding and crevasses (breaches in the river bank through which floodwaters can escape) that seasonally recharge these systems with nutrients and sediments. Rivers that sustain these deltaic estuaries are typically greater than 15 m in depth with a wide range of average annual discharge (e.g., 3000 $m^3 s^{-1}$, Nile R.; 200,000 $m^3 s^{-1}$, Amazon R.; Vorosmarty et al. 1998). Deltas are often classified as wave-dominated, river-dominated, tide-dominated, or some combination thereof based on their gross morphology and the primary forces that drive their morphology (Galloway 1975; Figure 2.4). Deltaic estuarine areas can range in size from <100 km^2 in the salt marshes of the Santee River, South Carolina, USA (Prevost 2004) to >10,000 km^2 in the mangrove wetlands of the Ganges–Brahmaputra Delta (Islam and Gnauck 2008), and represent some of the most productive, yet fragile, estuaries worldwide. High population growth, coastal urbanization, and upstream hydrologic manipulation of rivers (e.g., dams) have led to diminished sediment supplies which, when combined with subsidence rates sometimes exceeding 25 mm $year^{-1}$ and accelerating sea level rise rates, have led to deterioration of deltas in the last 100 years (e.g., Giosan et al. 2014; Figure 2.5).

2.4.1.3 Bar-Built Estuaries

Bar-built estuaries form when sandbars or barrier islands are delivered and deposited by littoral drift in such a way as to semi-enclose existing coastal embayments. Typically, a series of these bars or islands delineate the seaward boundary of bar-built estuaries, usually separated by tidal passes on the order of 100 m wide through which estuary-ocean exchange occurs (Figure 2.6). The water bodies between coastlines and bars or islands are often called lagoons, although some of these water bodies feature extensive salt marsh and mangrove wetlands. The islands attenuate the energy of ocean waves before they can reach estuaries and coastlines, and are occasionally bisected or completely eroded away during energetic events such as hurricanes. Whereas coastal plain estuaries are most often oriented perpendicular to the coastline (Fairbridge 1980), bar-built estuaries are typically oriented parallel to the coastline. Bar-built estuaries are common to regions of the east and Gulf coasts of North America such as Texas, New Jersey, North Carolina, and Florida; the Pacific coast of Mexico; and the southern coast of Brazil. The physical characteristics and dynamics of a few were described by Smith (1987, 1993), Möller et al. (2001), Lankford (1976), Castanares and Phleger (1969), and Collier and Hedgepeth (1950).

2.4.1.4 Fjords

Fjords are found in locations where glaciation extended below current sea level. When continental glaciers advance toward the sea during the last glaciation cycle, they carved steep, narrow valleys. During this advance, the leading ice edge scoured out many river valleys in latitudes above 45° (Fairbridge 1980; Dyer 1979). Terminal moraines

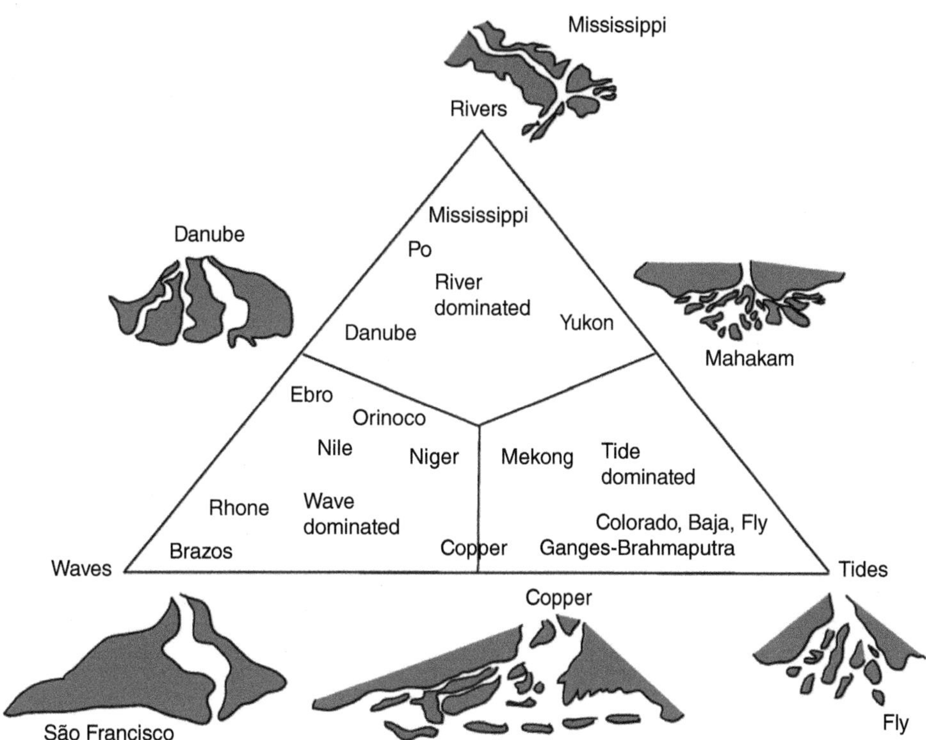

FIGURE 2.4 Classification of deltas into river-dominated, wave-dominated, and tide-dominated types. Under this approach, the relative importance of sediment supply, wave energy, and tidal energy are the primary geomorphic factors that determine the delta's spatial and stratigraphic configurations. *Source*: Syvitski and Saito (2007)/With permission of Elsevier.

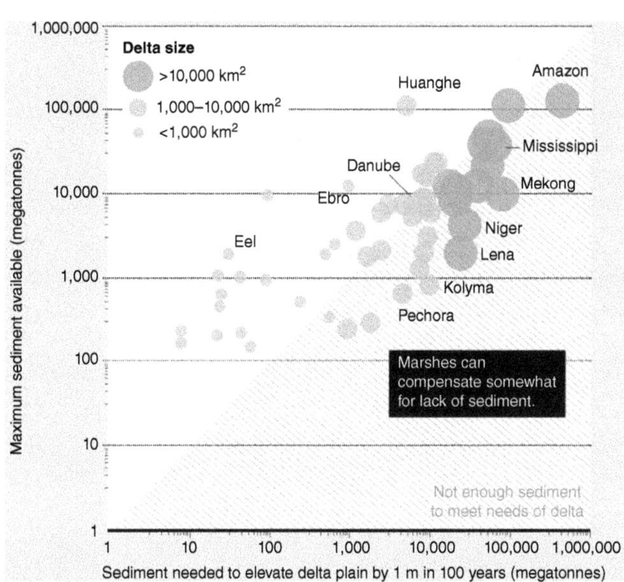

FIGURE 2.5 Sediment needed to elevate various delta plains by 1 m over the next 100 years vs. their respective sediment supplies. *Source*: Taken from Giosan et al. (2014)/With permission of Springer Nature. Sediment supply for deltas in the hashed region is insufficient to keep up with 1 meter of sea level rise over the next 100 years.

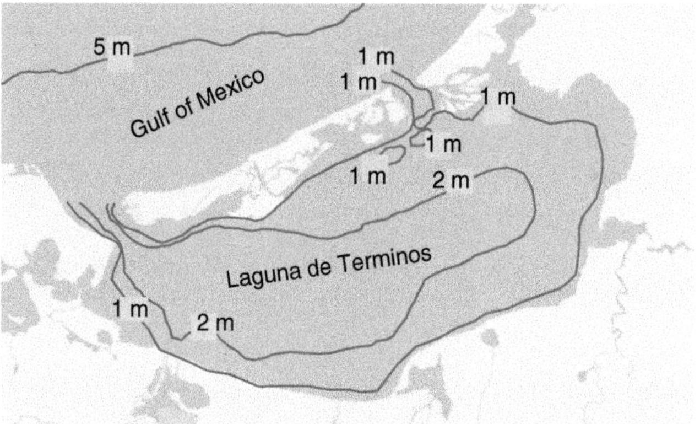

FIGURE 2.6 Example of a lagoon: Laguna de Terminos, Campeche, Mexico. Isobaths (lines denoting constant depth) in meters show that most of the lagoon exhibits depths <2 m. *Source*: Modified after Gierloff-Emden (1977).

(the deposits representing the furthest extent of the glacier), consisting primarily of boulders and gravel, were deposited where the ice edges reached their most seaward extent prior to retreat. When the glaciers retreated, the moraines remained, forming sills that restrict water exchange (Figure 2.7). Fjord estuaries are typically shallowest near their mouths (10–100 m deep) where these sills are situated, whereas the inland regions can be quite deep, exceeding 300 m in some cases. Water exchange with the open ocean is generally restricted to the shallow upper layer above the sill, and this restricted exchange causes the waters in the deeper, interior regions below the sill to remain largely stagnant with long residence times.

One consequence of the long residence times of isolated deep-water regions in fjords is that they can become oxygen-depleted and nutrient enriched, particularly in watersheds under heavy agricultural influence. These conditions

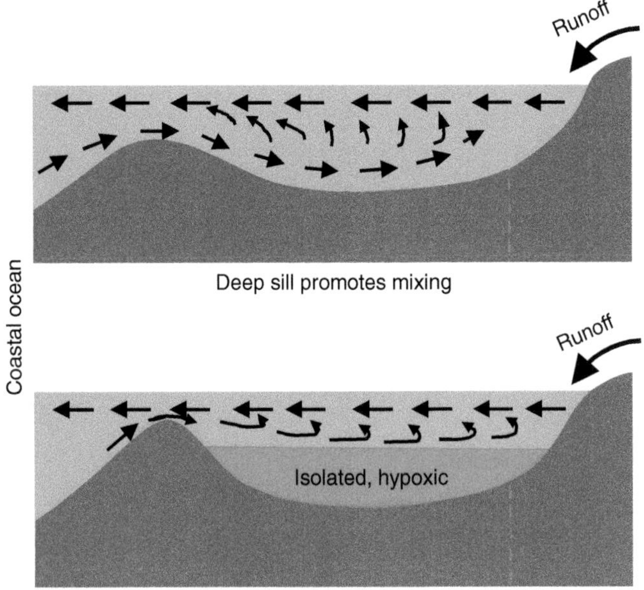

FIGURE 2.7 Comparison of circulation in fjords with deep (top) and shallow (sill) formations. Estuary-ocean exchange in fjord systems with shallow sills is inhibited relative to those with deeper sills. Reduced estuary-ocean exchange can isolate deeper water regions from the exchange flows, promoting hypoxic conditions.

can permeate into the underlying sediment layer (Aure and Stigebrandt 1989; Gillibrand et al. 1996) if water at the sediment-water interface becomes anoxic. The lack of routine flushing of deep water in fjords can facilitate the buildup of high levels of particulate organic matter (POM), because export of decaying phytoplankton and other sources of POM from the fjord is largely inhibited. Microbial decomposition of POM consumes oxygen in the water column and sediment, after which continued decomposition under anaerobic, low-sulfate conditions can lead to the production of methane (CH_4) through a process called methanogenesis. Ultimately this methane may be released to the atmosphere, where it is 25 times more effective as a greenhouse gas than carbon dioxide on a per mass basis.

Fjords and glacial inlets are common in both hemispheres where there has been glacial activity. They are particularly spectacular on coasts that serve as leading edges of tectonic plate margins, called subducted coasts. Good examples are the fjord inlets of southern Chile (Pickard 1971), Alaska, British Columbia (Pickard 1956; Figure 2.8), and New Zealand. Fjords also occur on the present or formerly glaciated coasts of Denmark, Norway, Svalbard, Greenland, and Graham Land in Antarctica (Fairbridge 1980). A useful description of the oceanographic features of Norwegian fjords can be found in Saelen (1967).

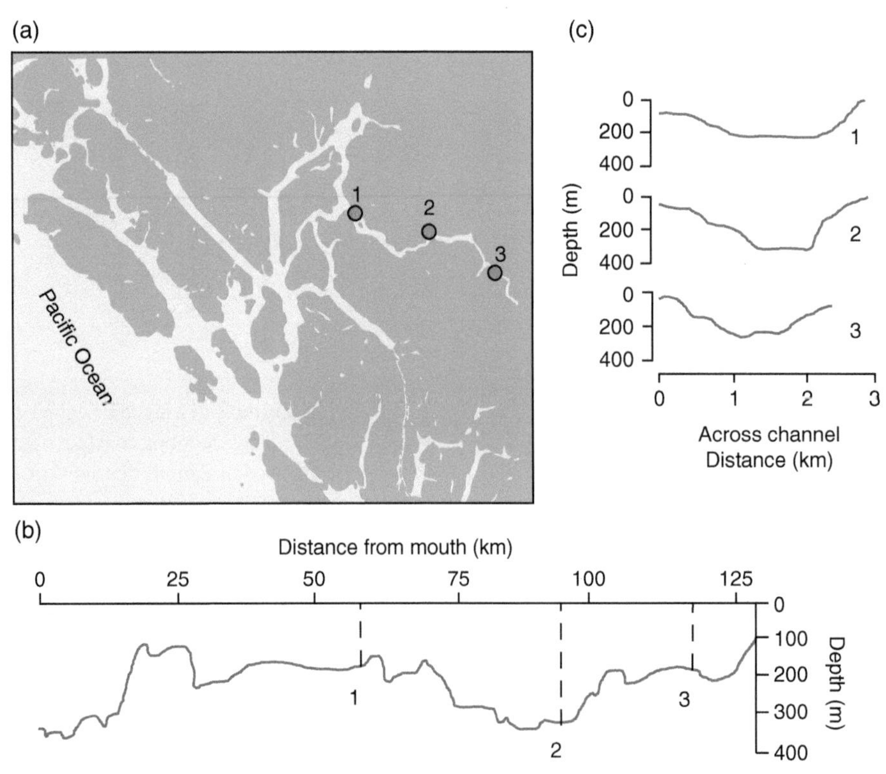

FIGURE 2.8 (a) Example of a fjord: The Gardner system in British Columbia, Canada. (b) Bathymetric profile along the length axis of the system (bottom) shows the presence of the sill near the mouth of the estuary around 25 km inland from the mouth of the system. *Source:* Modified after Pickard (1956). (c) Bathymetric cross-sectional profiles for locations 1, 2, and 3 are shown in panel (a).

2.4.1.5 Tectonic Estuaries

Tectonic estuaries are caused by faulting, graben formation, landslide, or volcanic eruption. The best and most extensively studied estuary in this group is San Francisco Bay (see Officer 1976; Conomos 1979), although many estuaries are tectonically influenced such as Puget Sound in the United States (Barnhardt and Sherrod 2006) and Manukau Harbor in New Zealand (Marra et al. 2006). Tectonically caused estuaries exhibit much variability and different ones may behave dynamically similar to coastal plain estuaries, fjords, or bar-built estuaries, depending on the local constraints.

2.4.1.6 Subterranean Estuaries

Analogous to salt wedge surface water estuaries, subterranean estuaries are coastal aquifers where groundwater derived from land drainage measurably dilutes sea water that has invaded aquifers through a free connection to the sea (Moore 1999; Figure 2.9). The groundwater plume is less dense and rises as it encounters infiltrated seawater in coastal sediments. This subterranean estuarine system exists wherever aquifers hydraulically intersect coastal water bodies (Cooper 1959; Johannes 1980; Moore 1999) and is characterized by longer residence times, greater water–particle interactions, lower dissolved oxygen concentrations, and strong biogeochemical gradients. Consequently, chemical reactions (e.g., diagenesis) and water circulation patterns within this porous and/or permeable media will be more complex than in surface estuaries. Submarine groundwater discharge (SGD) may include both meteoric groundwater from continental aquifers and pore water exchange from marine physical and biological processes (Burnett et al. 2003; Martin et al. 2007). The magnitude of meteoric sources to coastal waters vary greatly, depending upon local climate, geology, and consumptive groundwater use by coastal communities, while the magnitude of marine sources to SGD varies depending upon local wind stress and tides, resident benthic burrowing organisms, bottom topography, and sediment type (Cable and Martin 2008).

Subterranean estuaries can exist adjacent to surface water, and consequently, contribute to their hydrologic and biogeochemical budgets. Subterranean estuaries have been identified and studied in many places including Waquoit Bay (Michael et al. 2005) and Indian River Lagoon (Martin et al. 2007) in the United States, Flamengo Bay in Brazil (Burnett et al. 2006; Cable and Martin 2008; Povinec et al. 2008), Cockburn Sound in Perth, Australia (Burnett et al. 2006), and Osaka Bay in Japan (Taniguchi and Iwakawa 2004). Subterranean estuaries are also prominent in Arctic regions, where freshwater inputs have been increasing in recent decades due to increased permafrost melt (Connolly et al. 2020). Most of these subterranean estuaries are located within 10–20 m of the shoreline of their neighboring surface water estuary. In some cases, groundwater fluxes of constituents such as salt and nutrients in subterranean estuaries may exceed surface-water flux (Moore and Wilson 2005). Charette et al. (2005) and Charette and Sholkovitz (2006) demonstrated that the fresh-saline seepage front caused by the subterranean estuary is also responsible for an "iron curtain" of metal diagenesis (Chambers and Odum 1990), while other studies have shown the significance of this discharge for nutrient fluxes to coastal water bodies and the ocean (e.g., Krest et al. 2000; Santos et al. 2008). Groundwater inputs and their dissolved constituents are also considered responsible for local biological zonation in coastal water bodies by either preferentially excluding or encouraging flora and fauna growth (Kohout and Kolipinski 1967; Miller and Ullman 2004).

2.4.1.7 Geomorphic Variation Along the Fluvial–Marine Continuum

The type and rate of geologic development of estuaries depend not only on glaciation cycles and local variations in sediment supply, but also on a combination of other factors including climatic variability,

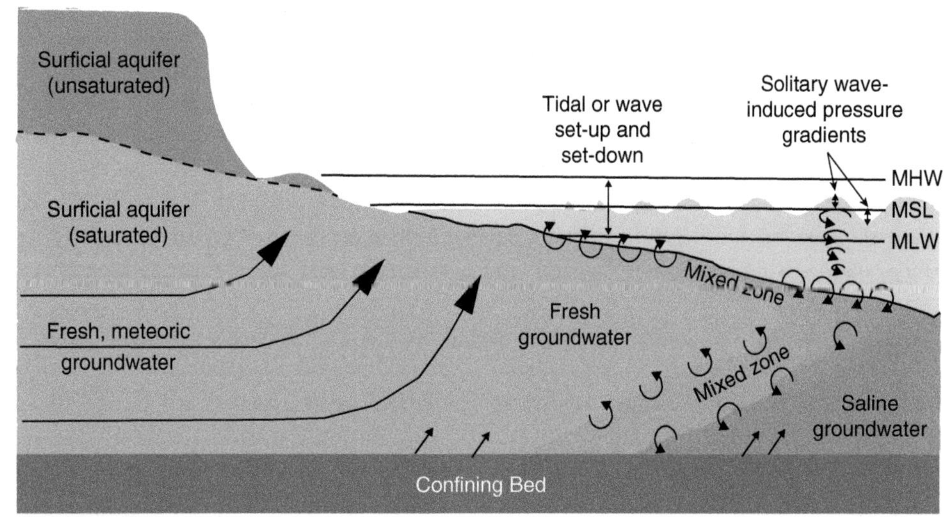

FIGURE 2.9 Example of a generalized subterranean estuary, where fresh meteoric groundwater mixes with infiltrated seawater. MSL is mean sea level, MHW is mean high water, and MLW is mean low water. *Source*: Modified from Smith et al. (2008).

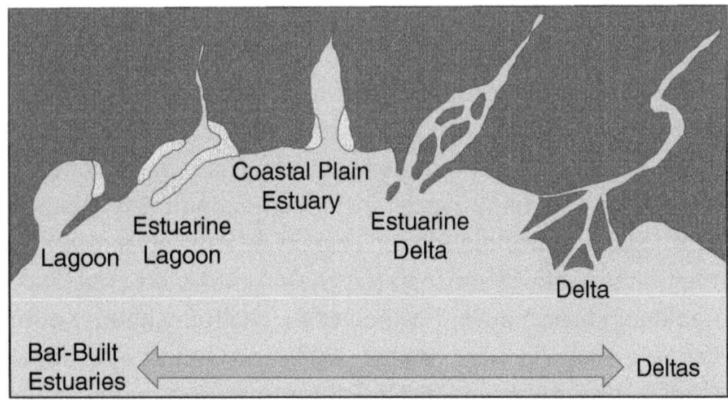

FIGURE 2.10 Schematic representation of the continuum of inlet types from lagoons to deltas. *Source*: Modified from Davies (1973).

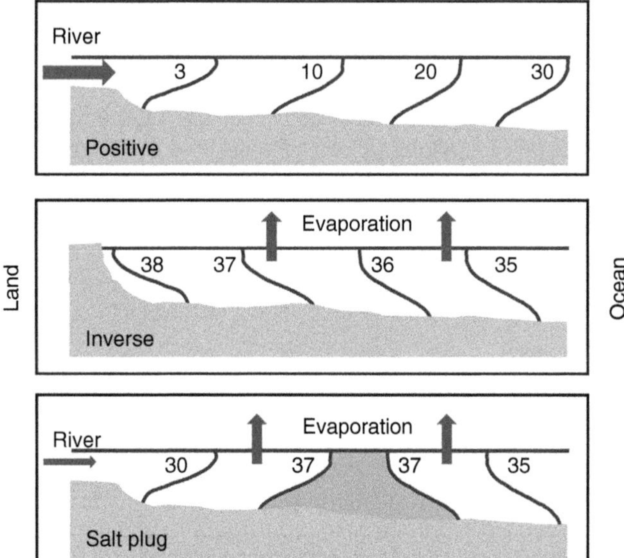

FIGURE 2.11 Classification of estuaries based on water balance showing salinity isohalines. "Typical" estuaries are positive (top); inverse estuaries can arise in arid systems that lack significant runoff, which can allow for development of hypersaline conditions which can reverse the salinity gradient and associated gravitational circulation (middle); salt plug (or neutral) estuaries arise when evaporation is sufficient to reduce the salinity gradient in the seaward reaches of an estuary, but salinity in the landward reaches is reduced by modest river input (bottom). Under these conditions, the region of maximum salinity (the salt plug, indicated by blue shading) is in the middle reaches.

regional and local geology, and the relative influence of fluvial and marine processes.

According to the scheme of Davies (1973), there is a continuum of estuarine types (Figure 2.10). At one end of the spectrum exist lagoons produced by marine (wave) action, found typically behind a barrier, and characterized by sand-sized sediments. Good examples of this type of estuary are the lagoons of the southern Texas and Mexican Gulf coasts (Lankford 1976). At the opposite end of the spectrum lie deltas. They are produced by river processes rather than by marine activities. They typically protrude into a receiving basin and are characterized by fine-grained silty sediments from terrestrial runoff. In between lagoons and deltas lie estuarine lagoons, estuaries, and estuarine deltas, representing a mixture and gradation of the two extreme coastal environments.

2.4.2 Water Balance Classification and Gravitational Circulation

Estuaries satisfying the classical definition put forth by Cameron and Pritchard (1963) (see Section 2.3), where seawater is diluted by fresh water derived from land drainage are called *positive* estuaries. They are called so because they have a positive ratio of the volume of freshwater inputs over a tidal cycle through river discharge and precipitation, *R*, to the volume of water brought into estuaries on the flood tide, *V* (flow ratio; (***R***/***V***) > 0; Figure 2.11, top). Positive estuaries exhibit a longitudinal density (and salinity) gradient whereby waters near the estuary mouth are denser and more saline than those found near the estuary head. This density gradient results in increased ebb (seaward flowing) tide velocities near the surface and increased flood (landward flowing) tide velocities near the bed that, when averaged over a tidal cycle, result in residual currents that transport saline waters landward in deeper regions of the water column underlying a seaward transport layer of fresher waters near the surface. This two-layer flow pattern (Figure 2.12a) is called *gravitational* (or *estuarine*) circulation because it is driven by gravity acting on density differences within estuaries. Gravitational circulation was first described by Prichard (1956; but see also Prichard and Kent 1956; Dyer 1973) and was represented in an elegant mathematical solution by Rattray and Hansen (1962) and Hansen and Rattray (1965).

In reality, most estuarine basins are not shaped like the estuary in Figure 2.12a with its flat, horizontal channel bottom and vertical walls at the channel edge. Rather, most estuaries possess a channel with sloped bathymetry between the thalweg (the deepest part of the channel) and the lateral banks of the basin. Under these circumstances, friction between the water and the estuary bottom causes gravitational circulation to vary across the channel cross section and tends to confine the inflows of dense saline water to the deeper regions of the central channel and surface outflows of buoyant fresh water to the shallow regions near the banks (Wong 1994; Figure 2.12b). Thus, deeper estuaries tend to have more developed gravitational circulation than shallow ones, and channel dredging activities can enhance saltwater intrusion and increase circulation. Through the Coriolis effect, the earth's rotation can further modify gravitational circulation whereby the surface net outflow is stronger along the left side of an estuary in the northern hemisphere (or the right side in the southern hemisphere), looking up-estuary (Figure 2.12c). Likewise, the net bottom inflow is more pronounced along the opposite side. The degree to which this occurs depends on how important the Coriolis term is relative to frictional effects in the overall dynamics of the flow and thus this phenomenon is more prominent in higher latitude

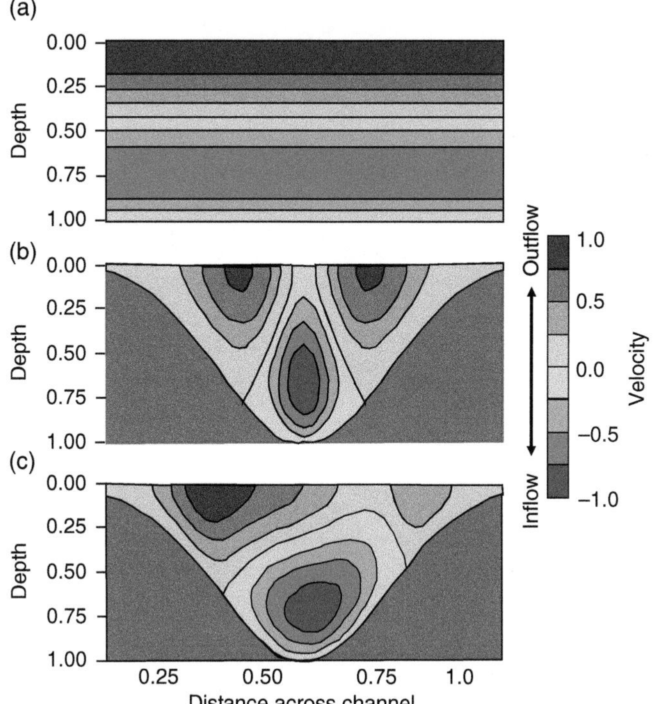

FIGURE 2.12 (a) Idealized gravitational circulation in an estuary with rectangular cross-section. *Source*: Modified from Wong (1994). (b) Idealized gravitational circulation taking into account interactions between bathymetry and friction when rotational effects are negligible. *Source*: Modified from Wong (1994). (c) Idealized gravitational circulation taking into account, effects of rotation and interactions between bathymetry and friction. *Source:* Modified from Valle-Levinson (2008). All channel cross sections are viewed looking up-estuary in the northern hemisphere. Values for depth, distance, and velocity are normalized to their maximum values.

estuaries that are deep, wide and stratified (Kasai et al. 2000; Valle-Levinson 2008).

Inverse (or *negative*) *estuaries* occur in arid regions under little or no river influence whereby losses of fresh water through evaporation can markedly exceed inputs from rivers and precipitation and, thus, (***R***/***V***) **<<0**. As a result, salinities inside inverse estuaries exceed those in the coastal ocean, and the longitudinal density gradient and the ensuing two-layer circulation is the reverse of that observed in positive estuaries, such that there is landward flow of less saline oceanic water overlying a seaward flow of hypersaline water near the bottom (Figure 2.11, middle). Due to the lack of freshwater inputs associated with inverse estuaries, they often exhibit prolonged residence and flushing times and diminished water quality (Luoma et al. 1985). Examples of inverse estuaries include the Red Sea, San Diego Bay in California, Laguna Madre in Texas, the lagoons in Baja California (Mexico), and Shark Bay and Spencer Gulf in Australia.

Estuaries that are intermediate to positive and inverse estuaries are those where evaporation exceeds freshwater inputs, but only slightly ((***R***/***V***) **<0**). These estuaries are called *salt plug* (or *neutral*) *estuaries*. Salt plug estuaries feature a density (and salinity) maximum that extends from the bottom to the surface waters somewhere along their longitudinal axis (Figure 2.11, bottom). Essentially, they exhibit salinity distributions similar to those of inverse estuaries down-estuary of the density maximum where salinities increase landward from the mouth up to a point where small quantities of river inflows begin to reduce them. The location of this density maximum (the "salt plug") depends on the relative importance of river inflows to evaporation; increased river inflows will move the plug seaward whereas increased evaporation will move it landward. Seaward of the salt plug, estuaries act as an inverse estuaries with inflow near the surface and outflow at the bottom, while landward of the salt plug they act as positive estuaries with seaward flow at the surface and landward flow near the bottom. Thus, there is very little flow across the plug—it is a region of intense downwelling where the flow simply reverses near the bottom—and as such there is little or no advective transport of constituents across the plug. As a result, constituents brought in by river inflow accumulate in the upper reaches of salt plug estuaries until the magnitude of the inflow is sufficient to push the plug out to sea.

2.4.3 Classification by Salinity Stratification

It can be helpful to further categorize positive estuaries (i.e., those where (***R***/***V***) **>0**) on the basis of their vertical salinity structure according to the value of (***R***/***V***) (see Section 2.4.2). On this basis, estuaries can be classified as *salt wedge*, *highly stratified*, *partially mixed* (sometimes called *weakly stratified*), or *vertically mixed*. Many estuaries transition between two or more of these classes through time based on seasonal changes in river discharge, variations in tidal forcing throughout the fortnightly tidal cycle, or even within tidal cycles.

Salt wedge estuaries occur when the volume of freshwater inflow *R* is large compared to the volume of the tidal prism *V* ((***R***/***V***) **>1**), allowing the fresh water to float as a buoyant layer over the underlying saline layer. The relatively weak tidal currents in salt wedge estuaries are insufficient to promote mixing between the two layers, which are separated by a sharp halocline (a region in the water column in which salinity changes rapidly with depth). The result is a tongue (or wedge) of dense oceanic water that underlies a turbulent layer of swiftly outflowing fresh water (Figure 2.13a). Currents in the layer below the halocline are typically much less energetic owing to the relatively small tidal prism. Salt wedge estuaries are commonly found at the mouths of major rivers that discharge into with micro- or mesotidal coastal regions. Examples include the Mississippi (USA), Ebro (Spain), and Rio de la Plata (Argentina) rivers.

Moderate increases in tidal volume *V* or decreases in river inflows *R* will both have the effect of reducing (***R***/***V***). When **0.1 <** (***R***/***V***) **< 1**, estuaries may be classified as *highly stratified*. Under these conditions, the velocity shear (difference in flow velocity between two layers of fluid) that occurs between the

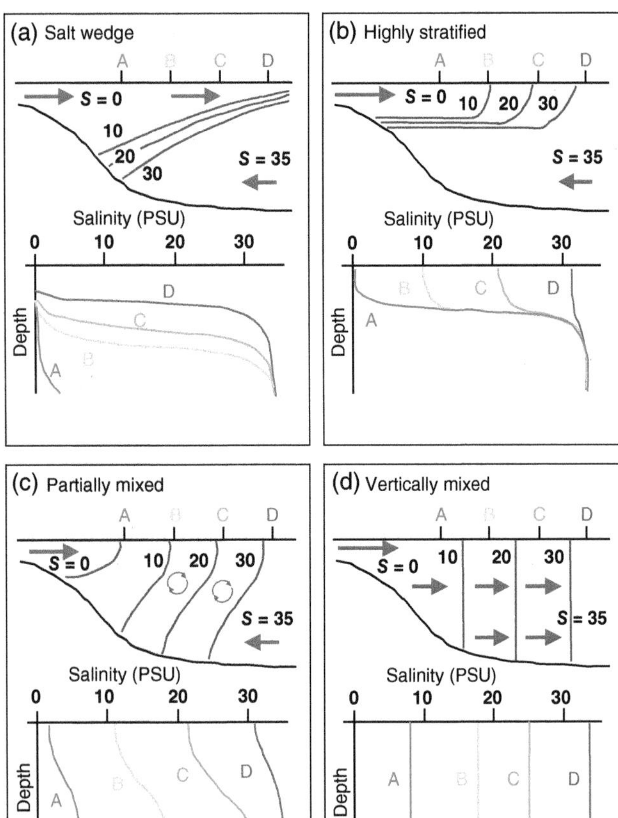

FIGURE 2.13 Salinity isohalines (top) and salinity vs. depth for locations A, B, C, and D along an estuary's length axis (bottom) for salt wedge (a), highly stratified (b), partially mixed (c) and vertically mixed (d) estuaries.

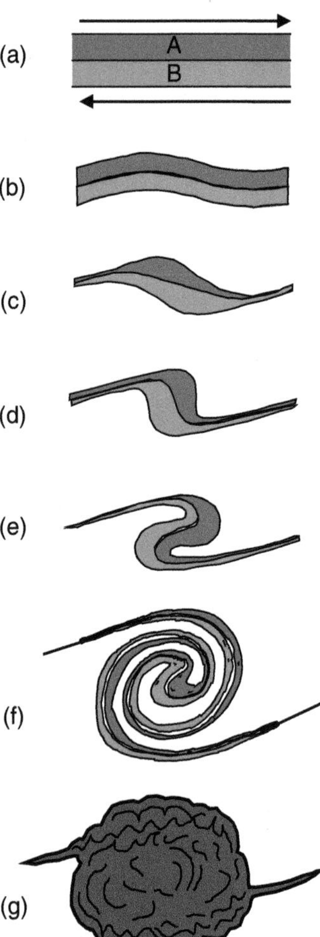

FIGURE 2.14 Development of internal waves in vertically sheared flows that ultimately break and facilitate entrainment of salt into the upper layer. The red/blue layers are the dense/buoyant and saline/fresh layers below/above the halocline. *Source*: Modified from Thorpe (1969). Velocity shear occurs at the surface/bottom layer boundary (a) which creates internal waves (b). The internal waves then grow (c,d) until they break (e,f), which results in infusion of bottom layer (saline) water into the surface layer (g).

seaward flowing surface layer of fresh water and the stronger currents of the bottom layer during the flood tide (Figure 2.14a) create instabilities in the form of internal waves (Figure 2.14b). The internal waves grow (Figure 2.14c,d), and when they break (Figure 2.14e,f), salt water from below the halocline is infused into the upper layer (Figure 2.14g). This vertical salt transport process, called *entrainment*, is unidirectional—it is not balanced by a downward transport of fresh water into the lower layer. Salinity above the halocline gradually increases toward the ocean, while that in the lower layer typically remains relatively constant with values close to those at the oceanic boundary. The entrainment process occurs throughout the length axis of highly stratified estuaries, and as a result, more and more salt water is injected to the surface layer, resulting in progressively increased volume transport in the seaward flowing layer above the halocline (Figure 2.13b). Thus, to conserve mass, a corresponding volume must be supplied to the landward flowing lower layer to account for the entrainment loss into the upper layer (Figure 2.15). For example, suppose the river discharge into an estuary is R m^3 s^{-1}, and the longitudinal density gradient forces $9R$ m^3 s^{-1} of saline water to flow landward into the estuary below the halocline. If all ($9R$) of the landward flowing bottom layer is eventually entrained into

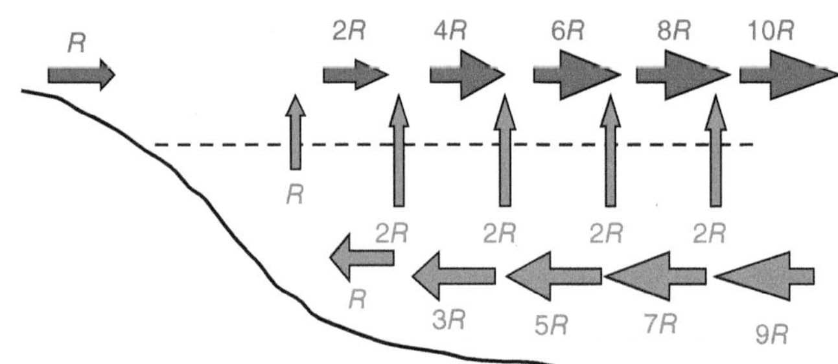

FIGURE 2.15 Illustration of how entrainment affects mass transport in a stratified estuary. River runoff of volume R enters an estuary at the landward end. As saline water (red) flows inland below the halocline, a portion of it is entrained into the seaward flowing surface layer (and thus removed from the bottom layer, represented by the upward arrows). Entrainment of saline water from the lower layer into the upper layer progressively increases the volume transport of the upper layer, such that (in this example) the seaward flowing volume at the estuary mouth is 10× that of the inflowing fresh water at the estuary head.

the seaward flowing surface layer on its landward journey, the near-surface layer discharge at the estuary mouth must be $10R$ m^3 s^{-1} ($1R$ from river input plus $9R$ from entrainment) for the estuary's volume to remain constant (i.e., if the surface layer discharge was less than that, the estuary's volume would be constantly increasing).

Further reductions in river inflows R or increases in tidal volume V create conditions where $\mathbf{0.005 < (R/V) < 0.1}$, the range under which estuaries are classified as *partially mixed* (or *weakly stratified*). Under these circumstances, tidal currents become swift enough that friction at the bed creates turbulent eddies with sufficient energy to cross the halocline. This turbulence adds to the turbulence generated by velocity shear across the halocline, and together cause mixing to occur. Unlike entrainment, turbulent mixing is bidirectional, and water and salt are exchanged upward and downward across the halocline. This mixing modifies the halocline in such a way that, although still present, it is less pronounced than in strongly stratified estuaries (Figure 2.13c). Similar to strongly stratified estuaries, the salinity in the upper layer increases toward the ocean due to the upward transport of salt from the lower layer. However, unlike strongly stratified estuaries, the salinity in the lower layer decreases in a landward direction, due to the downward transport of fresh water from above the halocline. Thus, a primary distinction between strongly stratified and partially mixed estuaries is that turbulence in strongly stratified estuaries is not strong enough to cause bi-directional mixing across the halocline, and thus, there is only upward salt transport through entrainment. In contrast, turbulence in partially mixed estuaries becomes important such that upward transport of the salt is coincident with downward transport of fresh water (indicated by the mixing loops in Figure 2.13c). Chesapeake Bay, the James River Estuary, and Delaware Bay are examples of partially mixed estuaries.

Further decreases in river inputs R or increases in tidal volume V allow for $\mathbf{0 < (R/V) < 0.005}$. Under these conditions, turbulent mixing becomes so effective that the vertical salinity structure becomes nearly homogenized and estuaries are categorized as *vertically mixed*. In these conditions salinity increases seaward along the estuary's length axis, but little or no variation occurs with depth (Figure 2.13d). The residual flow (i.e., flow averaged over a tidal cycle) is weak compared to tidal flow and is often directed seaward at all depths in the water column. It is not unusual for landward salt transport in vertically mixed estuaries to occur nearly entirely through tidal diffusion (see Section 2.5). Many estuaries that are classified as partially mixed during times of high river inflows become vertically mixed during parts of the year when river discharge is low, or during drought conditions. Very shallow estuaries such Barataria Bay, Louisiana, are often vertically mixed, owing in part to the fact that bottom-generated turbulence can influence the entire shallow water column before dissipation renders it ineffective at mixing across the halocline (although note that wind is also important in vertically mixing shallow estuaries).

2.4.4 Classification Based on Dynamics

Hansen and Rattray (1966) proposed a dynamical classification of estuaries based on two nondimensional hydrodynamic parameters, the circulation parameter and the stratification parameter. The circulation parameter (u_s/U_f) is a ratio of the tidally averaged (averaged over a complete tidal cycle) near surface velocity, u_s, to the tidally averaged, depth-averaged velocity U_f. This provides an index of vertical velocity shear for which high values typically indicate strong gravitational circulation. The stratification parameter ($\delta S/S_o$) is the ratio of the tidally-averaged salinity difference between the bottom and surface layers, δS, to the tidally averaged, depth-averaged salinity S_o. This provides an index of vertical salinity stratification for which high values indicate strong stratification.

The fraction of total upstream salt transport occurring through diffusion (υ) can be superimposed on the stratification–circulation diagram to better understand the mechanisms driving the salt transport in estuaries (Figure 2.16). Under conditions where $\upsilon = 1$, the upstream salt transport is driven solely by diffusion and gravitational circulation does not occur. As υ approaches zero, salt transport via gravitational circulation becomes increasingly important and the diffusive component vanishes. For intermediate values ($0.1 < \upsilon < 0.9$), which are typical for most estuaries, advective and diffusive components are both important to salt transport.

Using stratification–circulation diagrams, Hansen and Rattray (1966) identified several types of estuaries (Figure 2.16). Type 1 estuaries exhibit tidally averaged seaward flow at all depths and can be further subdivided into Type 1a estuaries which exhibit vertically mixed salinity profiles, and Type 1b which show appreciable stratification, but exhibit no tidally averaged landward flow in the bottom layer. In general, landward salt transport in type 1 estuaries is dominated by diffusive processes ($\upsilon \approx 1$). Type 2 estuaries show vertically sheared current profiles, with tidally averaged inflow at depth beneath a seaward-flowing surface layer. Advection and diffusion are both important to the salt balance in Type 2 estuaries, and, similar to Type 1, they can be further divided into those with salinity stratification (Type 2b) and those that are weakly stratified (Type 2a). Type 3 estuaries have very high circulation parameters; that is, a surface layer of swift outflow overlying an otherwise relatively stagnant water column. Such conditions are typically found in fjords. A distinguishing feature of Type 3 estuaries is that advection via gravitational circulation generally accounts for 99% of the total up-estuary salt flux, and strongly stratified estuaries often fall under this category. Finally, Type 4 estuaries occur where vertical flow structure is relatively homogenous, but salinity stratification is great. These conditions are typical to salt wedge estuaries where, because of the presence of a strong halocline, very little mixing occurs between the surface and bottom layers.

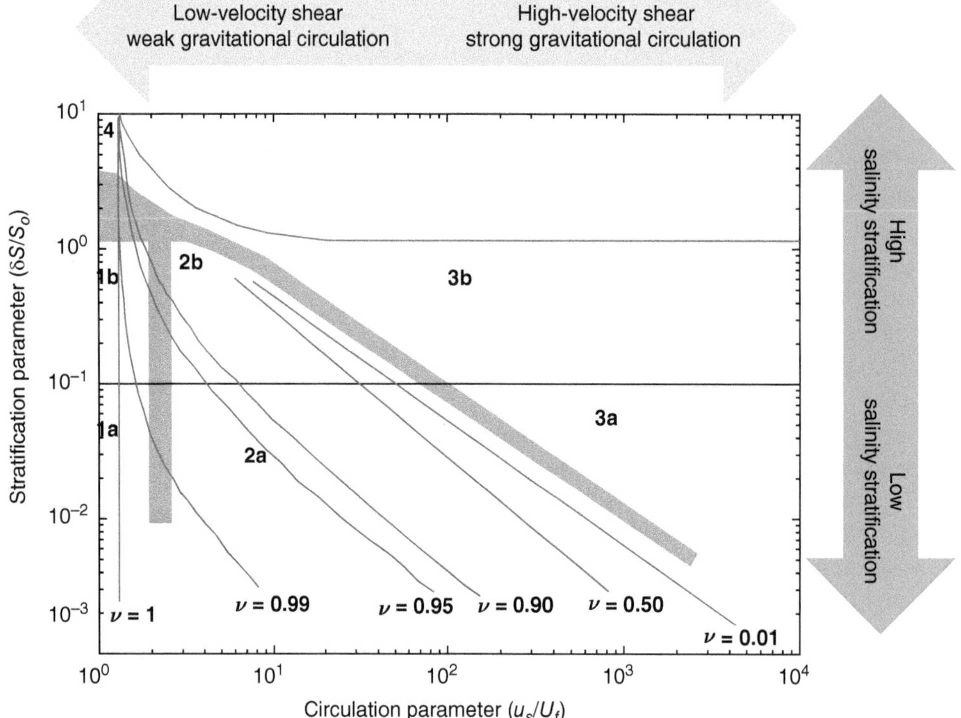

FIGURE 2.16 Estuarine classification diagram according to nondimensional stratification and circulation parameters. Contours of the fraction ν of upstream salt transport that occurs via diffusion (as opposed to gravitational circulation) are shown by the lines. *Source*: Modified from Hansen and Rattray (1965).

It is important to recognize that different regions of the same estuary often fall on different regions of the circulation–stratification diagram. Similarly, a given location may undergo transition from one type of estuary to another in response to fortnightly changes in tidal amplitude (Miranda et al. 1998; Miranda et al. 2005) or seasonal variation in fresh water inflows (Kumari and Rao 2009). As such, Geyer and MacCready (2014) proposed an alternative approach to classify estuaries based on nondimensionalized versions of the same principal forcing variables—freshwater inflow (*R*) and tidal velocity (*V*). Scaling *R* to the maximum speed at which fresh water can move through an estuary of given size and shape provides the nondimensional *freshwater Froude number* (Fr_f), while scaling *V* to the time required for complete vertical mixing of the water column provides the nondimensional *mixing parameter* (*M*), which can also be conceptualized as a number that quantifies the effectiveness of tidal mixing. Estuaries can then be mapped onto the two-parameter Fr_f–M space (Figure 2.17), with partially mixed estuaries falling in the middle of the diagram, salt wedge estuaries near the top, Fjords near the lower left, and well-mixed estuaries in the lower right. The diagonal line through the parameter space separates permanently stratified estuaries (upper/left) from those that remain mixed to some degree (lower/right).

2.4.5 Characterizing Conditions with Reynolds and Richardson Numbers

In addition to the nondimensional parameters R/V, u_s/U_f, $\delta S/S_0$ and υ discussed above, which form the basis of classifying estuaries based on water balance, stratification, and hydrodynamics, a few additional nondimensional parameters are useful in characterizing flow in estuaries. The Reynolds number *Re* is a ratio of the of the inertial force (i.e., the force arising from particles of the fluid resisting changes in momentum) to the viscous force (i.e., the force arising from friction between layers of the fluid) acting on a fluid, which can be calculated as

$$Re = \frac{uD}{\upsilon}$$

where *u* is water velocity, *D* is depth, and υ is viscosity. When $Re \lesssim 2000$, the viscous force acting on the fluid particles tend to restrict their flow along parallel paths, with each layer moving relative to the other layers in a way that facilitates little or no mixing. This kind of flow is said to be *laminar*, as the flow can be thought of as consisting of thin layers, or *laminae*. When $Re \gtrsim 10^5$, the viscous forces are no longer sufficient to promote

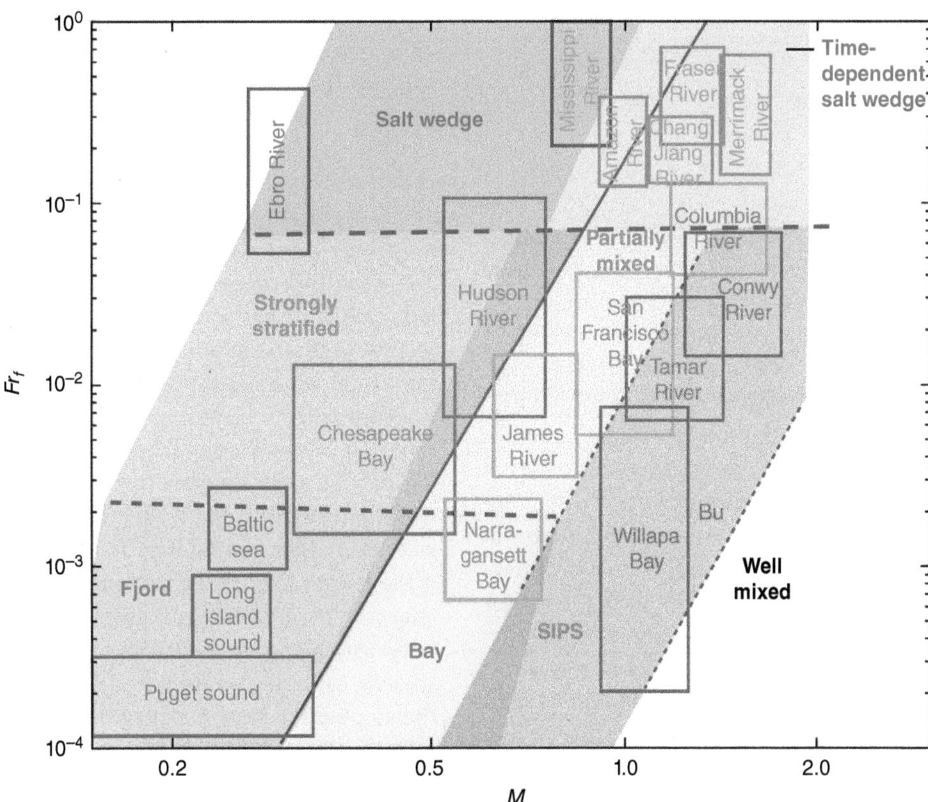

FIGURE 2.17 Classification of estuaries based on the freshwater Froude number (Fr_f) and mixing number (M). The red diagonal line separates persistently stratified estuaries (upper/left) from those that are mixed to some degree (lower/right). The various estuaries are represented by rectangles rather than spaces to indicate the variation that can result from changing tidal and river flow conditions. *Source*: Geyer and MacCready (2014)/With permission of Annual Reviews Inc.

laminar flow and as a result the velocity of the fluid at a given point continuously undergoes changes in both magnitude and direction. Flow under these conditions is said to be *turbulent*. In estuaries, a primary source of turbulence is friction at the bed. For $2000 \lesssim Re \lesssim 10^5$, the flow is said to be transitional, and the point at which it becomes turbulent is largely determined by the roughness of the bed. In larger estuaries with a large wind fetch (the length of water surface over which wind from a given direction can blow), wind-generated waves can be another important source of turbulence. Turbulence promotes a "tumbling" motion of the water column and tends to destratify the water column.

Counteracting the mixing effects of turbulence is stratification since strong density gradients tend to impede the vertical movement of water. This competition between stratification and mixing is a fundamental principle of estuarine dynamics, and the relative importance of the two processes can be assessed with the gradient Richardson number Ri, calculated as

$$Ri = \frac{-(g/\rho)(\partial\rho/\partial z)}{(\partial u/\partial z)^2}$$

where g is gravitational acceleration, ρ is fluid density, u is velocity, and z is depth. The numerator assesses the density stratification whereas the denominator is a measure of the vertical velocity shear. When $Ri > 0$, stratification at depth z exists and is stable; when $Ri = 0$ the fluid is unstratified at depth z; when $Ri < 0$ the water column is unstable at depth z and vertical mixing will occur to stabilize it. Often the gradient Richardson number is simplified to a *layer* Richardson number Ri_L, which treats the water column as two discrete layers, one above and one below the halocline, such that

$$Ri_L = \frac{(\Delta\rho/p)gD}{u^2}$$

where D is the depth of the upper layer, u is the velocity difference between the two layers, and $\Delta\rho$ is the density difference between the two layers. $Ri_L > 20$ indicates strong water column stability with negligible mixing, while $Ri_L < 2$ indicates isotropic turbulence and fully developed mixing. When Ri_L carries values between 2 and 20, the water column is weakly stable and moderate mixing occurs.

2.5 Tidal Circulation

In the absence of river discharge, density gradients, and wind stress, the transport of water in and out of an estuary is driven by tides in response to gravitational interactions between the sun, earth, and moon. Though the tide range varies from one tidal cycle to the next as the earth's oceans respond to ever-varying sun–moon–earth configurations, it generally exhibits a fortnightly cycle where it exhibits maximum and minimum ranges approximately every other week. This cycle is called the spring-neap cycle in semi-diurnal tidal regimes, and it repeats itself every 14.8 days in response to the relative positioning of the sun and the moon (Figure 2.18a). In diurnal tidal regimes, a 13.6-day, fortnightly, cycle called the tropic–equatorial cycle exists that is physically distinct from the 14.8-day spring-neap cycle. The tropic–equatorial cycle is caused by fortnightly changes in the maximum declination of the moon's orbit relative to the earth's equator (Figure 2.18b). Because strong currents can, through friction, generate turbulence at the bed that can mix water at the halocline and inhibit stratification, the fortnightly cycle in tidal current amplitude can exert strong control over gravitational circulation and its associated salt transport. As a result, during neap/equatorial tides the gravitational circulation is enhanced; during spring/tropic tides it is inhibited, and upstream salt transport is dominated by diffusive processes (see below).

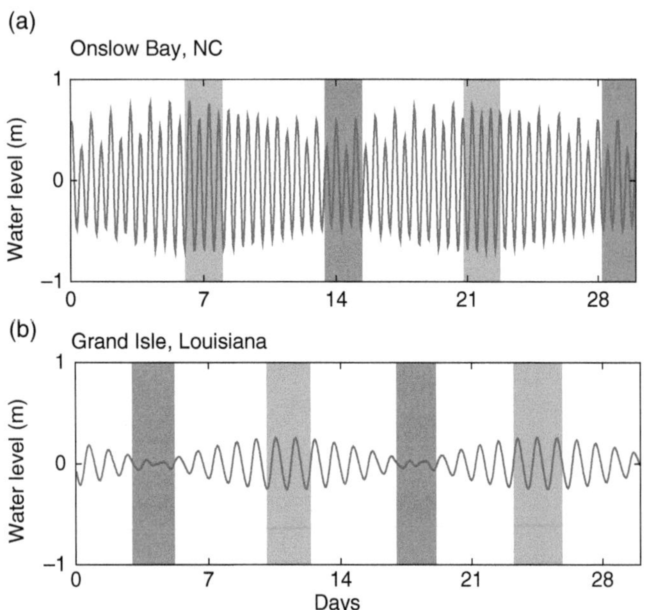

FIGURE 2.18 (a) Semi-diurnal water-level signal at Onslow Bay, NC, showing spring-neap variability in tide range. (b) Diurnal tidal water-level signal at Grand Isle, LA, showing tropic-equatorial variability in tide range. Red shading denotes neap and equatorial tides in Onslow Bay and Grand Isle, respectively; green shading denotes spring and tropic tides in Onslow Bay and Grand Isle, respectively.

The rise and fall of water in coastal oceans propagates into estuaries as a long wave. In an unstratified estuary, the speed, or celerity c, of this wave is reduced by friction at the bottom of the estuary and thus, tidal propagation is slower in shallow water according to $\mathbf{c} = \sqrt{\mathbf{gh}}$, where g is gravitational circulation and h is depth. Because tides take time to progress up an estuarine basin, high tide at the mouth of an estuary may occur much earlier than high tide near the estuary head. The time difference between high tide at the mouth and high tide at some point up-estuary is the *tidal phase lag*. In relatively deep estuaries where bottom friction is unimportant, tidal waves reach the estuary head unattenuated, and then are reflected and propagated back down-estuary. If this reflected wave returns to the estuary mouth at the exact time the next tidal wave enters from the coastal ocean, a *standing wave* will occur. Under these conditions high and low tide will occur when the tidal current is slack (Figure 2.19a), and if the length axis of such an estuary is short relative to the length of the tidal wave, high (or low) tides will occur simultaneously along the estuary's length. On the other hand, in shallow estuaries where bottom friction becomes important, the amplitude of the incoming tide diminishes along the estuary's length, and the amplitude of the reflected wave is either severely attenuated or eliminated altogether, resulting in a *progressive wave* whereby the time of high (or low) tide gets progressively later with increasing distance up the estuary. In these cases, the tidal current velocity and water level are in phase (Figure 2.19b). Friction dissipates tides to some degree in most estuaries, and the resulting wave is somewhere in between a perfectly progressive and standing wave.

Although tidal currents are oscillatory, they can lead to net (averaged over a tidal cycle) salt transport in estuaries

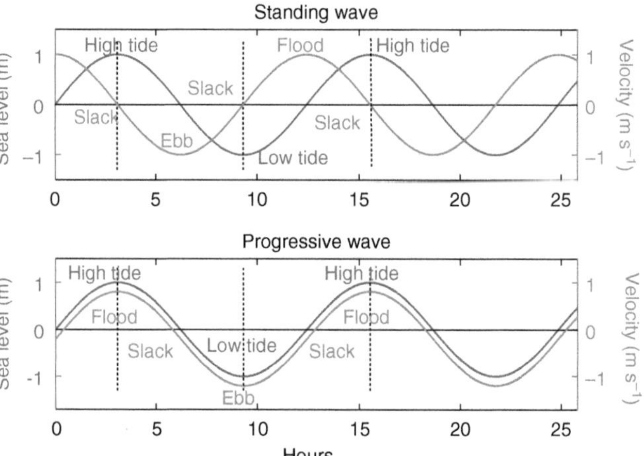

FIGURE 2.19 Purely standing (top) and progressive (bottom) tidal waves. For standing waves, the sea level signal (blue) is 90° out of phase with the velocity signal (orange), such that high and low tides occur at slack tidal currents (the instant in time at which the direction of the tidal current reverses). For progressive tidal waves, the sea level and velocity signals are in phase, such that the tide will begin to drop even before the tidal current reverses. Most real tidal waves are in between these two extremes.

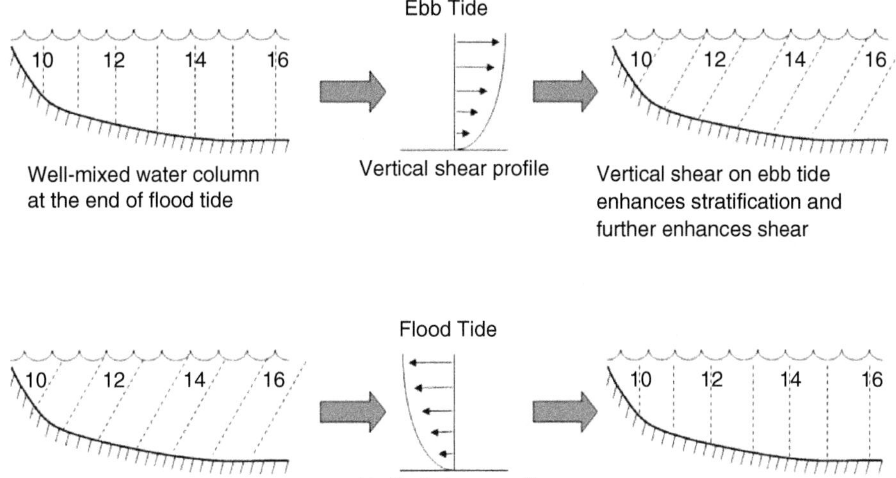

FIGURE 2.20 Changes in stratification brought about by *strain-induced periodic stratification* (SIPS, or tidal straining). Prior to onset of the ebb tide, stratification is often negligible (upper left). Due to vertical shear in the tidal current, near-surface waters are advected further seaward than near-bottom waters (upper right). This circumstance has the overall effect of increasing stratification. The water column then destratifies by the end of the flood tide (lower right).

through multiple mechanisms. One mechanism, called *tidal straining*, occurs when there is a vertical shear in the magnitude of the tidal current, where bottom friction causes tidal current velocities near the bed to be slower than at the surface (Simpson et al. 1990). This tidal velocity shear causes fresher water to be advected out over more saline water during ebb, enhancing stratification through a process called *strain-induced periodic stratification*, or *SIPS* (Figure 2.20). Increasing stratification has the overall effect of stabilizing the water column and restricting turbulent mixing to the dense underlayer, which further enhances the shear. On the flood tide the opposite happens: stratification is diminished, and turbulent mixing occurs throughout the water column, which tends to mix the momentum of the two layers, causing a reduction in the vertical current shear (Monismith et al. 1996). Thus, these circumstances can cause the water column to undergo intratidal cycles of stratification and mixing, which can enhance gravitational circulation when the column is stratified during the ebb tide or suppress it when the column is mixed during the flood tide.

Current velocity and salinity vary approximately sinusoidally throughout a given tidal cycle and thus, the net transport of water over the course of a tidal cycle is typically close to zero. However, even with a net water flux of zero, strong correlations between velocity and salinity over a tidal cycle can lead to transport of salt in the form of *tidal diffusion* (sometimes called tidal oscillatory flux, or tidal pumping). This can occur when high salinity seawater enters an estuary on the flood tide and is subsequently diluted with (relatively) low salinity water inside the estuary. On the ensuing ebb tide, the water discharged from the estuary to the sea has a lower salt concentration than the water imported during the flood tide, with the balance being retained within the estuary until it is advected out of the estuary by the mean flow. During the spring/tropic phase of the fortnightly tidal cycle when gravitational circulation tends to be diminished, tidal diffusion can become the dominant mechanism for salt transport in estuaries.

2.6 Wind-Driven Circulation

In addition to circulation driven by tides and density gradients, currents in estuaries can also be driven by wind. Atmospherically forced currents in estuaries are important, because they often are the most significant factors governing the long-term transport and distribution of constituents like salt and nutrients. These currents can also cause changes in the volume of water in estuaries by up to an order of magnitude greater than changes driven by tides (Wiseman 1986). This is especially true in microtidal systems (those with very little tidal variation).

Numerous studies have shown how estuaries respond to a combination of remote and local wind effects (Garvine 1985, Wong 1991; Wong and Moses-Hall 1998; Janzen and Wong 2002; Snedden et al. 2007a). The remote effect occurs when winds blowing parallel to the regional coastline immediately outside an estuary cause marked changes in shelf water levels. The water level response on the shelf occurs through the Coriolis effect, where water acceleration in response to wind stress over the surface is directed 90° to the right of the wind direction in the northern hemisphere, a phenomenon called *Ekman transport*. Thus, in the northern hemisphere, winds blowing to the

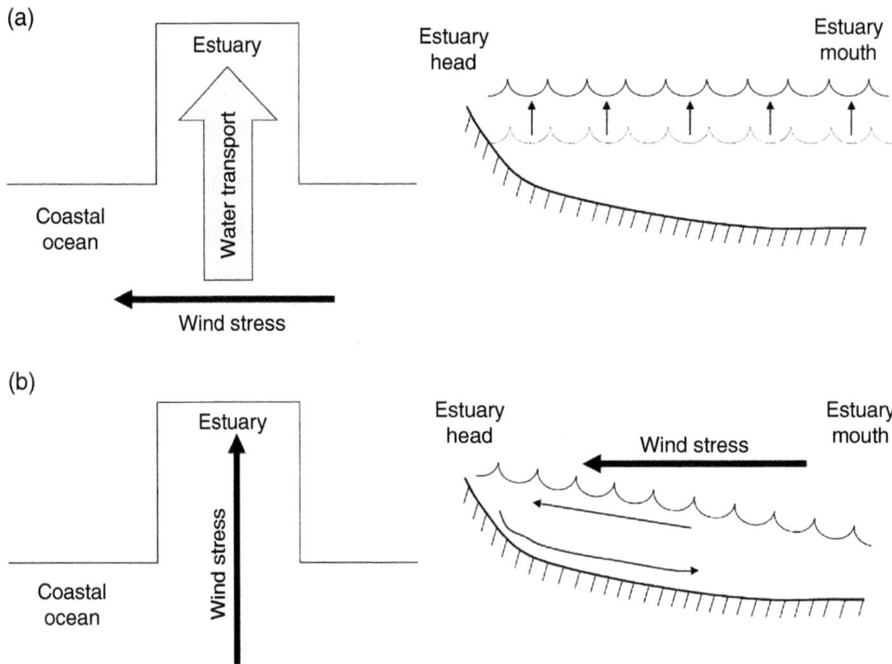

FIGURE 2.21 (a) Remote effect resulting from Ekman transport of shelf waters into an estuary in the northern hemisphere driven by westward winds. Water levels increase throughout the estuary. (b) Local effect resulting from winds blowing landward over an estuary surface. Water levels increase in the landward reaches of the estuary, and currents flow upwind near the surface, and downwind near the bottom.

south tend to transport water to the west and increase shelf sea levels along a western boundary of ocean basins (e.g., the Atlantic coast of North America). The opposite happens in the southern hemisphere, where Ekman transport is to the *left* of the wind direction. The increased shelf sea levels produce a pressure gradient between the shelf and adjoining estuaries in the region, ultimately forcing shelf waters into the estuaries (Figure 2.21a). The opposite process occurs when winds over the shelf reverse; shelf waters are transported offshore and coastal sea levels are thus reduced, releasing water from adjoining estuaries onto the shelf.

The local effect of wind on estuaries occurs when winds blowing directly over the surface of an estuary transfers momentum in the atmosphere directly to the water and induces a surface current that elevates sea level in the downwind region of the estuary. The resulting sea surface slope produces an opposing current in the bottom portion of the water column that is directed upwind (Figure 2.21b). When wind blows up an estuary and generates a sea surface slope along the estuary, sea level responses to this mechanism increase linearly with distance from the estuary mouth. Thus, upper reaches of longer estuaries, such as the north end of Chesapeake Bay, exhibit greater sensitivity to the local effect of wind (Garvine 1985).

Regardless of the mechanism for wind-forced estuarine motions, their ecological significance can be profound. Exchange flows between estuaries and coastal shelves can provide mechanisms for ingress and egress of fish and invertebrates that use estuaries for nursery habitat as juveniles (Pietrafesa and Janowitz 1988; Tilburg et al. 2005). These exchanges also provide a mechanism for the export of constituents including carbon (Das et al. 2010). Currents induced by wind forcing also induce changes in estuarine sea levels, which can inundate or drain vast tracts of coastal wetlands (Snedden et al. 2007a) and deliver mineral sediments to these landscapes (Snedden et al. 2007b). Wind-forced sea level fluctuations can also impact plant species composition and productivity through influences on wetland hydroperiod (Baldwin et al. 2001) or soil porewater salinities (Sharpe and Baldwin 2009).

2.7 Concluding Remarks

Estuaries are highly dynamic systems and their physical, chemical, and biological characteristics vary over a broad range of temporal and spatial scales. Physical processes in estuaries can be best understood by examining the interplay between hydrodynamics. These physical processes can, in turn, have profound impacts to ecological structure and function. For example, the long water residence times and limited mixing in deep regions of fjords can promote hypoxic conditions which limit the diversity of taxa to those capable of thriving under such conditions. Often, ecological processes are linked to physical processes through feedback loops. For example, an estuary's physical characteristics influence species distribution, richness, and production, which often control water quality and chemical characteristics that then, in turn, can impact geomorphological characteristics through modification by ecological processes such as growth of marshes, mangroves, seagrass beds, and shellfish reefs.

Physical processes in estuaries are highly sensitive to variations in freshwater inflows and tidal velocity, as either can alter the flow (*R*/*V*) ratio (see Sections 2.4.2 and 2.4.3), and these variations can vary in frequency and magnitude. For example, enhanced up-estuary intrusion of a salt wedge can develop temporarily due to brief, punctuated events such as hurricane landfalls or persist permanently as a result of sea level rise. Likewise, ephemeral (e.g., 1–2 years) changes in freshwater inflows may result from El Nino or La Nina events, whereas persistent changes in freshwater inflows may result from climate change impacts or human alterations to drainage basins such as dams, channelization, and flood protection levees. Though short in duration, ephemeral events may have long lasting consequences, such as when a flood or storm event opens new connections to the coastal ocean or closes existing ones and results in changes to an estuary's physical processes that have cascading impacts on chemistry and biota.

The impacts of climate change, through its impacts to freshwater inflows and conditions such as sea level at the marine boundary, to estuarine physical processes is becoming increasingly understood (Bakun 1990; Kennedy 1990; Michener et al. 1997; O'Gorman and Schneider 2009). Sea level rise will likely increase saltwater intrusion lengths in estuaries (Savenije 1993; Hilton et al. 2008), potentially increasing porewater salinities in coastal wetlands (Wang et al. 2020) and altering wetland vegetation community zonation (Snedden and Steyer 2013; Snedden 2019). Increased saltwater intrusion may also impact groundwater resources via salinization of coastal aquifers, placing municipal water supplies in jeopardy (Barlow and Reichard 2010). In addition to salinity effects, sea level rise may also increase tidal current velocities (Pethick 1993), which could lead to increased erosion and elevated turbidity in estuarine waters (Uncles et al. 2002) and negatively impact seagrass distribution and productivity (Longstaff and Dennison 1999), and ultimately diminish the abundance of essential fish habitat (Raposa and Oviatt 2000; Lazzari and Stone 2006).

Climate change is expected to alter patterns of freshwater inflow to estuaries around the globe. As air warms, its water-holding capacity increases, but this increased atmospheric moisture content will not fall as precipitation evenly across the planet. Forecasts suggest that increased precipitation is very likely in high-latitude regions, while decreases are likely in most subtropical land regions (IPCC 2007; Figure 2.22). Decreased freshwater supply can change the effective size of an estuary by increasing intrusion of oceanic water, and also increasing water residence times. Changing water residence times can have cascading impacts across trophic levels. For example, phytoplankton populations are typically only capable of doubling once or twice each day. In systems with short (<1 day) water residence times, phytoplankton can be flushed from estuaries as fast as they can grow, which mitigates the ecosystem's susceptibility to eutrophication. Increasing water residence times will have the overall effect of making estuaries more susceptible to blooms and eutrophication, as has been observed in the Hudson River estuary (Howarth et al. 2000).

Freshwater inflows also play a crucial role in determining the extent of vertical stratification in estuaries. Stratification generally increases with increasing flow ratio (***R***/***V***), and thus freshwater inflows. Estuarine stratification conditions have

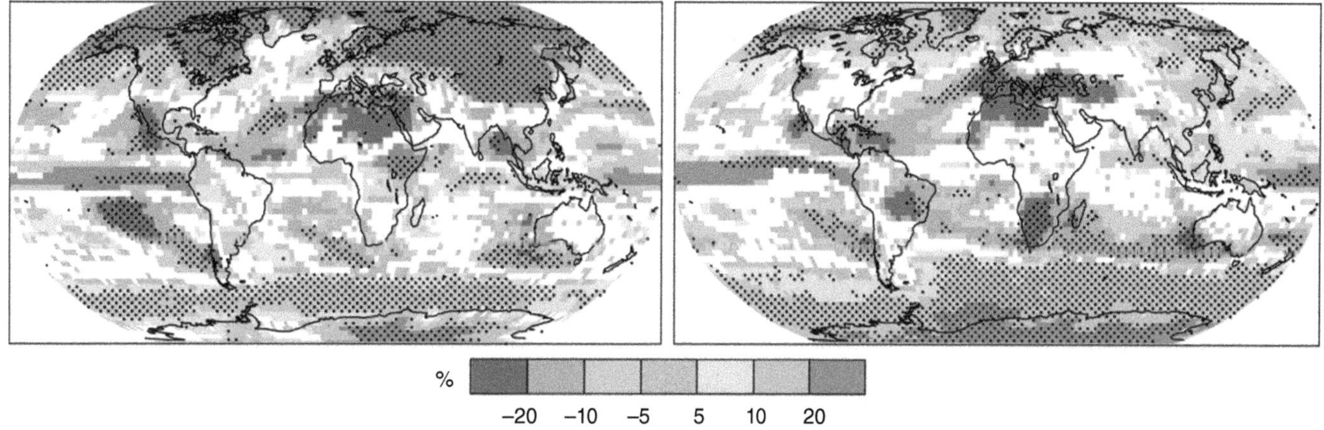

FIGURE 2.22 Relative changes in precipitation (in percent) for the period 2090–2099, relative to 1980–1999. Values are multi-model averages for December to February (left) and June to August (right). White areas are where less than 66% of the models agree in the sign of the change and stippled areas are where more than 90% of the models agree in the sign of the change. *Source*: IPCC AR4 Synthesis Report/The Nobel Foundation.

cascading impacts throughout the food web. For example, in the San Francisco Bay estuary, periods of elevated phytoplankton biomass tend to coincide with stratification episodes (Cloern 1984). This correspondence arises because sediment particles sink out of the surface layer and their replacement from particle resuspension at the bed is inhibited by the presence of the halocline. Together, these circumstances act to simultaneously increase phytoplankton production by increasing light availability in the surface layer (Haas et al. 1981) and decrease grazing pressure from benthic feeding while phytoplankton are confined to the surface layer by stratification (Cloern 1982).

Estuaries are dynamic environments that can vary tremendously through time and space. Most estuarine processes are interconnected, whereby physical processes are connected to chemical processes (e.g., water quality) which are, in turn, connected to ecological processes. Ultimately, these processes are all governed by the climatic and geomorphologic characteristics of estuaries. Natural variability in climate over a broad range of timescales (e.g., days, months, seasons, years, decades) adds to the variability of estuarine ecosystems and facilitates a corresponding natural range of estuarine processes that set the stage for the unique ecology of every estuary.

Review Questions

Multiple Choice

1. An estuary formed when a river valley is flooded by rising sea levels associated melting glaciers and the end of the last glaciation period is called a
   a. Deltaic estuary
   b. tectonic estuary
   c. coastal plain estuary
   d. lagoon

2. An estuary that forms when sandbars or barrier islands are delivered and deposited by littoral drift in such a way as to semi-enclose an existing coastal embayment is called a
   a. deltaic estuary
   b. fjord
   c. bar-built estuary
   d. inverse estuary

3. Dissolved oxygen depletion common to deeper regions of fjords is caused by
   a. extremely high tide ranges
   b. vertical mixing of the water column
   c. long water residence times
   d. excessive shading from mountains that typically flank the fjord

4. The kind of estuary with the largest river inflow relative to its tidal current velocity is a
   a. salt wedge estuary
   b. vertically mixed estuary
   c. inverse estuary
   d. salt plug estuary

5. The relative importance of stratification and mixing can be represented by a nondimensional number called the
   a. Reynolds number
   b. circulation parameter
   c. Coriolis number
   d. Richardson number

6. In the northern hemisphere, wind blowing to the south will result in a net transport of water to the
   a. north
   b. east
   c. south
   d. west.

7. In situations where the length axis of a deep estuary is short relative to the length of the tidal wave, the tide acts as a/an
   a. internal wave
   b. standing wave
   c. progressive wave

8. The velocity shear between the seaward flowing surface layer and the landward flowing bottom layer can result in the generation of
   a. laminar flow
   b. internal waves
   c. Ekman transport
   d. tidal straining

9. In vertically mixed estuaries, landward transport of salt occurs nearly entirely through
   a. gravitational circulation
   b. tidal diffusion
   c. entrainment
   d. breakdown of the halocline

10. Inverse estuaries result when
    a. gravitational circulation exceeds tidal diffusion
    b. evaporation exceeds freshwater inflows
    c. the Reynolds number exceeds the Richardson number
    d. the freshwater Froude number exceeds the mixing parameter

Short Answer

1. Why do more estuaries exist along the U.S. Atlantic and Gulf of Mexico coasts than on the U.S. Pacific coast?
2. What is the current definition of an estuary put forth in this chapter, and why was it changed from the original definition put forth by Cameron and Pritchard in 1963?
3. What geomorphic feature is responsible for the general circulation pattern in fjords? What are some of the ecological consequences of this feature and how do they arise?
4. For which class of estuary is R/V less than zero? What kinds of circumstances would cause R/V to be negative? What is the unusual characteristic of the circulation for estuaries in this class?
5. Sketch the 0, 10, 20 and 30 psu salinity isohalines for salt wedge, highly stratified, partially mixed, and well-mixed estuaries.
6. Describe the differences in tide propagation in deep vs. shallow estuaries. How is the amplitude and speed of an incoming tidal wave different in deep vs. shallow estuaries? What are the key differences between progressive and standing tidal waves, and which type is more likely to occur in deep estuaries?
7. How does the fortnightly tidal cycle impact the primary mechanism of upstream salt transport in estuaries?
8. Compare and contrast the remote and local mechanisms of wind-forced water transport in estuaries.
9. Describe or draw a simple diagram of a situation where opposing water level responses could be induced by the two mechanisms in response to the same wind direction.
10. Estuaries often serve as ports for ocean going vessels, and as such many estuaries contain dredged channels to facilitate navigation of these vessels. How would channel dredging impact estuaries?
11. What factors can determine water residence time in estuaries? How can residence time impact water quality?

References

Aure, J. and Stigebrandt, A. (1989). On the influence of topographic factors upon the oxygen consumption rate in sill basin fjords. *Estuarine, Coastal and shelf Science* 28: 59–69.

Bakun, A. (1990). Global climate change and intensification of coastal ocean upwelling. *Science* 247: 198–201.

Baldwin, A.H., Egnotovich, M.S., and Clarke, E. (2001). Hydrologic change and vegetation of tidal freshwater marshes: field, greenhouse, and seed bank experiments. *Wetlands* 21: 519–531.

Barlow, P.M. and Reichard, E.G. (2010). Saltwater intrusion in coastal regions of North America. *Hydrogeology Journal* 18: 247–260.

Barnhardt, W.A. and Sherrod, B.L. (2006). Evolution of a Holocene delta driven by episodic sediment delivery and coseismic deformation, Puget Sound, Washington, USA. *Sedimentology* 53: 1211–1228. doi: 10.1111/j.1365-3091.2006.00809.x.

Burnett, W., Bokuniewicz, H., Huettel, M. et al. (2003). Groundwater and pore water inputs to the coastal zone. *Biogeochemistry* 66: 3–33.

Burnett, W.C., Aggarwal, P.K., Aureli, A. et al. (2006). Quantifying submarine groundwater discharge in the coastal zone via multiple methods. *The Science of the Total Environment* 367: 498–543.

Cable, J.E. and Martin, J.B. (2008). In situ evaluation of nearshore marine and fresh pore water transport in Flamengo Bay, Brazil, special issue. *Estuarine Coastal and Shelf Science* 76: 473–483.

Cameron, W.M. and Pritchard, D.W. (1963). Estuaries. In: *The Sea*, vol. 2 (ed. M.N. Hill), 306–324. New York. Wiley.

Castanares, A.A. and Phleger, F.B. (eds.) (1969). Coastal Lagoons, A Symposium. Universidad Nacional Autonoma de Mexico. Ciudad Universitaria, Mexico 20, D.F., 686 pp.

Chambers, R.M. and Odum, W.E. (1990). Porewater oxidation, dissolved phosphate and the iron curtain. *Biogeochemistry* 10: 37–52.

Charette, M.A. and Sholkovitz, E.R. (2006). Trace element cycling in a subterranean estuary: part 2. Geochemistry of the pore water. *Geochimica et Cosmochimica Acta 70*: 811–826.

Charette, M.A., Sholkovitz, E.R., and Hansel, C.M. (2005). Trace element cycling in a subterranean estuary: part 1. Geochemistry of the permeable sediments. *Geochimica et Cosmochimica Acta* 69: 2095–2109.

Cloern, J.E. (1982). Does the benthos control biomass in South San Francisco Bay (USA)? *Marine Ecology Progress Series* 9: 191–202.

Cloern, J.E. (1984). Temporal dynamics and ecological significance of salinity stratification in an estuary (South San Francisco Bay, USA). *Oceanologica Acta* 7: 137–141.

Collier, A. and Hedgepeth, J.W. (1950). An introduction to the hydrography of tidal waters of Texas. *Publications of the Institute of Marine Science* 1: 125–194. Texas A&M University, Galveston Campus, Galveston, TX.

Connolly, C.T., Cardenas, M.B., Burkart, G.A. et al. (2020). Groundwater as a major source of organic matter to Arctic coastal waters. *Nature Communications* 11: 1479.

Conomos, T.J. (ed.) (1979). *San Francisco Bay: The Urbanized Estuary*. San Francisco: Pacific Division of the American Association for the Advancement of Science, 483 pp.

Cooper, J.H.H. (1959). A hypothesis concerning the dynamic balance of fresh water and salt water in a coastal aquifer. *Journal of Geophysical Research 64*: 461–467.

Das, A., Justic, D., and Swenson, E. (2010). Modeling estuarine-shelf exchanges in a deltaic estuary: implications for coastal carbon budgets and hypoxia. *Ecological Modelling* 221: 978–985.

Davies, J.L. (1973). *Geographical Variation in Coastal Development*. New York: Hafner.

Dyer, K.R. (1973). *Estuaries: A Physical Introduction*. New York: Wiley.

Dyer, K.R. (1979). Estuaries and estuarine sedimentation. In: *Hydrography and Sedimentation in Estuaries* (ed. K.R. Dyer), 118. London: Cambridge University Press.

Emery, K.O. (1967). Estuaries and lagoons in relation to continental shelves. In: *Estuaries*, vol. 83 (ed. G.H. Lauff), 9–11. Washington, D.C.: American Association for the Advancement of Science.

Emery, K.O. and Uchupi, E. (1972). *Western North Atlantic Ocean: Topography, Rocks, Structure, Water, Life and Sediments*, vol. 17. Tulsa, Oklahoma: American Association of Petroleum Geologists, Memoir.

Fairbridge, R.W. (1980). The estuary: its definition and geodynamic cycle. In: *Chemistry and Biochemistry of Estuaries* (ed. E. Olausson and I. Cato), 1–35. New York: Wiley.

Galloway, W.E. (1975). Process framework for describing the morphologic and stratigraphic evolution of delta depositional systems. In: *Deltas, Models for Exploration* (ed. M.L. Broussard), 87–98. Houston, TX: Houston Geological Society.

Garvine, R.W. (1985). A simple model of estuarine subtidal fluctuations forced by local and remote wind stress. *Journal of Geophysical Research-Oceans* 90: 1945–1948.

Geyer, W.R. and MacCready, P. (2014). The estuarine circulation. *Annual Review of Fluid Mechanics* 46: 175–197.

Gierloff-Emdem, H.G., 1977. Laguna de Terminos and Campeche Bay, Gulf of Mexido. In: Gierloff-Emden, H.G., Ed., Orbital Remote Sensing Coast and Offshore Environment, Walter de Gruyter, Berlin. pp 77–89. doi: https://doi.org/10.1515/9783111456706.

Gillibrand, P.A., Turrell, W.R., Moore, D.C., and Adams, R.D. (1996). Bottom water stagnation in a Scottish sea loch. *Estuarine, Coastal and Shelf Science* 43: 217–235.

Giosan, L., Syvitski, J.P.M., Constantinescu, S., and Day, J.W. (2014). Climate change: protect the world's deltas. *Nature* 516: 31–33.

Haas, L.W., Hastings, S.J., and Webb, K.L. (1981). Phytoplankton response to a stratification-mixing cycle in the York River estuary during late summer. In: *Nutrients and Estuaries* (ed. B.J. Neilson and L.E. Cronin), 619–636. Clifton, NJ: Humana Press.

Hansen, D.V. and Rattray, M. Jr. (1965). Gravitational circulation in straits and estuaries. *Journal of Marine Research* 23: 104–122.

Hansen, D.V. and Rattray, M. Jr. (1966). New dimensions in estuary classification. *Limnology and Oceanography* 11: 319–326.

Hilton, T.W., Najjar, R.G., Zhong, L., and Li, M. (2008). Is there a signal of sea-level rise in Chesapeake Bay salinity? *Journal of Geophysical Research* 113: C9002.

Howarth, R.W., Swaney, D., Butler, T.J., and Marino, R. (2000). Climatic control on eutrophication of the Hudson River estuary. *Ecosystems* 3: 210–215.

IPCC (2007). Climate change 2007: impacts, adaptation and vulnerability. In: *Contribution of Working Group II to the Fourth Assessment Report of the Intergovernmental Panel on Climate Change* (ed. M.L. Parry, O.F. Canziani, J.P. Palutikof, et al.). Cambridge, UK, 976: Cambridge University Press.

Islam, S.N. and Gnauck, A. (2008). Mangrove wetland ecosystems in Ganges-Brahmaputra delta in Bangladesh. *Frontiers of Earth Science in China* 2 (4): 439–448.

Janzen, C.D. and Wong, K.C. (2002). Wind-forced dynamics at the estuary-shelf interface of a large coastal plain estuary. *Journal of Geophysical Research-Oceans* 107 (C10): 3138. doi: 10.1029/2001JC000959.

Johannes, R. (1980). The ecological significance of the submarine discharge of groundwater. *Marine Ecology Progress Series* 3: 365–373.

Kasai, A., Hill, A.E., Fujiwara, T., and Simpson, J.H. (2000). Effect of the Earth's rotation on the circulation in regions of freshwater influence. *Journal of Geophysical Research-Oceans* 105: 16961–16969.

Kennedy, V.S. (1990). Anticipated effects of climate change on estuarine and coastal fisheries. *Fisheries* 15: 16–24.

Kohout, F. and Kolipinski, M. (1967). Biological zonation related to ground water discharge along the shore of Biscayne Bay, Miami, Florida. *Conference on Estuaries*, March 31–April 3, Jekyll Island, Georgia, ed. G. Lauff, 488–499. AAAS No. 83. Washington, D.C.: American Association for the Advancement of Science.

Krest, J., Moore, W., Gardner, L.R., and Morris, J. (2000). Marsh nutrient export supplied by ground water discharge: evidence from radium measurements. *Global Biogeochemical Cycles* 14: 167–176.

Kumari, V.R. and Rao, I.M. (2009). Estuarine characteristics of the lower Krishna river. *Indian Journal of Geo-Marine Sciences* 39: 215–223.

Lankford, R.R. (1976). Coastal lagoons of Mexico; their origin and classification. In: *Estuarine Processes*, Vol. II (ed. M.L. Wiley), 182–215. New York: Academic.

Lazzari, M.A. and Stone, B.Z. (2006). Use of submerged aquatic vegetation as habitat by young-of-the-year epibenthic fishes in shallow Maine nearshore waters. *Estuarine, Coastal and Shelf Science* 69: 591–606.

Longstaff, B.J. and Dennison, W.C. (1999). Seagrass survival during pulsed turbidity events: the effects of light deprivation on the seagrasses *Halodule pinifolia* and *Halophila ovalis*. *Aquatic Botany* 65: 105–121.

Luoma, S.N., Cain, D., and Johansson, C. (1985). Temporal fluctuations of silver, copper and zinc in the bivalve mocomabalthica at five stations in south San Francisco Bay. *Hydrobiologia* 129: 109–120.

Marra, M.J., Alloway, B.V., and Newnham, R.M. (2006). Paleoenvironmental reconstruction of a well-preserved Stage 7 forest sequence catastrophically buried by basaltic eruptive deposits, northern New Zealand. *Quaternary Science Reviews* 25: 2143–2161. doi: 10.1016/j.quascirev.2006.01.031.

Martin, J., Cable, J., Smith, C. et al. (2007). Magnitudes of submarine groundwater discharge from marine and terrestrial sources: Indian River Lagoon, Florida. *Water Resources Research* 43: W05440. doi:10.1029/2006WR005266.

Meade, R.H. (1969). Landward transport of bottom sediments in estuaries of the Atlantic coastal plain. *Journal of Sedimentary Petrology* 39: 222–234.

Michael, H.A., Mulligan, A., and Harvey, C. (2005). Seasonal oscillations in water exchange between aquifers and the coastal ocean. *Nature 436*: 1145–1148.

Michener, W.K., Blood, E.R., Bildstein, K.L. et al. (1997). Climate change, hurricanes and tropical storms, and rising sea level in coastal wetlands. *Ecological Applications* 7: 770–801.

Miller, D. and Ullman, W. (2004). Ecological consequences of groundwater discharge to Delaware Bay, United States. *Ground Water* 42 (7) Oceans Issue: 959–970.

Milliman, J.D. and Emery, K.O. (1968). Sea levels during the past 35,000 years. *Science* 162: 1121–1123.

Miranda, L.B., Castro, B.M., and Kjerfve, B. (1998). Circulation and mixing due to tidal forcing in the Bertioga Channel, Sau Paulo, Brazil. *Estuaries* 21: 204–214.

Miranda, L.B., Bergamo, A.L., and Castro, B.M. (2005). Interactions of river discharge and tidal modulation in a tropical estuary, NE Brazil. *Ocean Dynamics* 55: 430–440.

Möller, O., Castaing, P., Salomon, J.-C., and Lazure, P. (2001). The influence of local and non-local forcing effects on the subtidal circulation of Patos Lagoon. *Estuaries* 24 (2): 297–311.

Monismith, S.G., Burau, J., and Stacey, M. (1996). Stratification dynamics and gravitational circulation in northern San Francisco Bay. In: *San Francisco Bay: The Ecosystem* (ed. T. Hollibaugh), 123–153. Washington, DC: American Association for the Advancement of Science.

Moore, W.S. (1999). The subterranean estuary: a reaction zone of groundwater and seawater. *Marine Chemistry* 65: 111–125.

Moore, W.S. and Wilson, A.M. (2005). Advective flow through the upper continental shelf driven by storms, buoyancy, and submarine groundwater discharge. *Earth and Planetary Science Letters* 235: 564–576.

O'Gorman, P.A. and Schneider, T. (2009). The physical basis for increases in precipitation extremes in simulations of 21st-century climate change. *Proceedings of the National Academy of Sciences* 106: 14773–14777.

Officer, C.B. (1976). *Physical Oceanography of Estuaries and Associated Coastal Waters*. New York: Wiley.

Officer, C.B. (1980). Box models revisited. In: *Estuarine and Wetland Processes with Emphasis on Modeling* (ed. P. Hamilton and K.B. McDonald). New York: Plenum.

Pethick, J. (1993). Shoreline adjustments and coastal management: physical and biological processes under accelerated sea-level rise. *The Geographical Journal* 159: 162–168.

Pickard, G.L. (1956). Physical features of British Columbia inlets. *Transactions of the Royal Society Canada L, (III)* 50: 47–58.

Pickard, G.L. (1971). Some physical oceanographic features of inlets of Chile. 1. *Journal of the Fisheries Research Board of Canada* 28 (605-616): 1077–1106.

Pietrafesa, L.J. and Janowitz, G.S. (1988). Physical oceanographic processes affecting larval transport around and through the North Carolina inlets. In: *Larval Fish and Shellfish Transport Through Inlets* (ed. M.P. Weinstein), 34–50. Bethesda, MD: American Fisheries Society.

Povinec, P.P., Bokuniewicz, H., Burnett, W.C. et al. (2008). Isotope tracing of submarine groundwater discharge offshore Ubatuba, Brazil: results of the IAEA-UNESCO SGD project. *Journal of Environmental Radioactivity* 99 (10): doi: 10.1016/j.jenvrad.2008.06.010.

Prevost, M. (2004). Sewee to Santee, The Nature Conservancy in South Carolina, nature.org/southcarolina.

Pritchard, D.W. (1956). The dynamic structure of a coastal plain estuary. *Journal of Marine Research* 15: 33–42.

Pritchard, D.W. and Kent, R.E. (1956). A method for determining mean longitudinal velocities in a coastal plain estuary. *Journal of Marine Research* 15: 81–91.

Raposa, K.B. and Oviatt, C.A. (2000). The influence of contiguous shoreline type, distance from shore, and vegetation biomass on nekton community structure in eelgrass beds. *Estuaries* 23: 46–55.

Rattray, M. Jr. and Hansen, D.V. (1962). A simularity solution for circulation in an estuary. *Journal of Marine Research* 20: 121–133.

Saelen, O.H. (1967). Some features of the hydrography of Norwegian fjords. In: *Estuaries* (ed. G.H. Lauff), 63–70. Washington, DC: American Association for the Advancement of Science.

Santos, I., Burnett, W.C., Chanton, J.P. et al. (2008). Nutrient biogeochemistry in a Gulf of Mexico subterranean estuary and groundwater-derived fluxes to the coastal ocean. *Limnology and Oceanography* 53 (2): 705–718.

Savenije, H. (1993). Predictive model for sal intrusion in estuaries. *Journal of Hydrology* 148: 203–218.

Schubel, J.R. and Hirschberg, D.J. (1978). Estuarine graveyards, climatic change and the importance of the estuarine environment. In: *Estuarine Interactions* (ed. M.L. Wiley), 285–303. New York: Academic.

Sharpe, P.J. and Baldwin, A.H. (2009). Patterns of wetland plant species richness across estuarine gradients of Chesapeake Bay. *Wetlands* 29: 225–235.

Simpson, J.H., Brown, J., Matthews, J., and Allen, G. (1990). Tidal straining, density currents, and stirring in the control of estuarine stratification. *Estuaries* 13: 125–132.

Smith, N.P. (1987). An introduction to the tides of Florida's Indian River Lagoon, I. water levels. *Florida Scientist* 50: 49–61.

Smith, N.P. (1993). Tidal and nontidal flushing of Florida's Indian River Lagoon. *Estuaries* 16: 739–746.

Smith, C.G., Cable, J.E., and Martin, J.B. (2008). Episodic high-intensity mixing events in a subterranean estuary: impacts of tropical systems. *Limnology and Oceanography* 53: 666–674.

Snedden, G.A. (2019). Patterning emergent marsh vegetation assemblages in coastal Louisiana, USA, with unsupervised artificial neural networks. *Applied Vegetation Science* 22: 213–229.

Snedden, G.A. and Steyer, G.D. (2013). Predictive occurrence models for coastal wetland plant communities: delineating hydrologic response surfaces with multinomial logistic regression. *Estuarine, Coastal and Shelf Science* 118: 11–23.

Snedden, G.A., Cable, J.E., and Wiseman, W.J. (2007a). Subtidal sea level variability in a shallow Mississippi River deltaic estuary, Louisiana. *Estuaries and Coasts* 30: 802–812.

Snedden, G.A., Cable, J.E., Swarzenski, C., and Swenson, E.M. (2007b). Sediment discharge into a subsiding Louisiana deltaic estuary through a Mississippi River diversion. *Estuarine, Coastal and Shelf Science* 71: 181–193.

Syvitski, J.P.M. and Saito, Y. (2007). Morphodynamics of deltas under the influence of humans. *Global and Planetary Change* 57: 251–282.

Taniguchi, M. and Iwakawa, H. (2004). Submarine groundwater discharge in Osaka Bay, Japan. *Limnology* 5: 25–32.

Thorpe, S.A. (1969). Experiments on the stability of stratified shear flows. *Radio Science* 4: 1327–1331.

Tilburg, C.E., Reager, J.T., and Whitney, M.M. (2005). The physics of blue crab larval recruitment in Delaware Bay: a model study. *Journal of Marine Research* 63: 471–495.

Uncles, R.J., Stephens, J.A., and Smith, R.E. (2002). The dependence of estuarine turbidity on tidal intrusion length, tidal range and residence time. *Continental Shelf Research* 22: 1835–1856.

Valle-Levinson, A. (2008). Density-driven exchange flow in terms of the Kelvin and Ekman numbers. *Journal of Geophysical Research* 113: C04001. doi: 10.1029/2007JC004144.

Vorosmarty, C.J., Fekete, B.M., and Tucker, B.A. (1998). Global River Discharge, 1807–1991, V[ersion]. 1.1 (RivDIS). Data set. Available on-line [http://www.daac.ornl.gov] from Oak Ridge National Laboratory Distributed Active Archive Center, Oak Ridge, Tennessee, USA. doi: https://doi.org/10.3334/ORNLDAAC/199.

Wang, H., Krauss, K.W., Noe, G.B. et al. (2020). Modeling soil porewater salinity responses to drought in tidal freshwater forested wetlands. *Journal of Geophysical Research—Biogeosciences* 125: e2018JG004996.

Wiseman, W.J. (1986). Estuarine-shelf interactions. In: *Baroclinic Processes on Continental Shelves* (ed. C.N.K. Moeers), 109–115. Washington, DC: American Geophysical Union.

Wong, K.C. (1991). Sea-level variability in Long Island Sound. *Estuaries* 13: 362–372.

Wong, K.C. (1994). On the nature of transverse variability in a coastal plain estuary. *Journal of Geophysical Research-Oceans* 99: 14209–14222.

Wong, K.C. and Moses-Hall, J.E. (1998). On the relative importance of remote and local wind effects on the subtidal variability in a coastal plain estuary. *Journal of Geophysical Research-Oceans* 103: 18393–18404.

CHAPTER **3**

Estuarine Chemistry

Thomas S. Bianchi[1] and Elise S. Morrison[2]
[1] Department of Geological Sciences, University of Florida, Gainesville, FL, USA
[2] Department of Environmental Engineering Sciences, University of Florida, Gainesville, FL, USA

Collecting water samples in Fjordland, New Zealand using a CTD/Rosette water sampler. A conductivity, temperature, depth (CTD) rosette is a water sampler that contains Niskin sampling bottles and instruments to measure vertical profiles of physical, chemical, and biological parameters. Photo credit: Thomas S. Bianchi

3.1 Basics in Biogeochemical Cycles and Chemical Principles

3.1.1 Linking Estuarine Chemistry with Estuarine Ecology

Estuaries are dynamic systems with distinctive chemical characteristics due to interactions between ocean, land, and freshwater systems. Although chemistry may seem very distant from ecology, processes in estuarine chemistry can directly and indirectly influence many aspects of estuarine ecology. Chemical processes can alter environmental conditions that impact the ecology of estuarine organisms. For instance, in the case of an estuarine turbidity maximum (ETM), the distinct chemical characteristics that arise when salt and freshwater mix can result in conditions that impact primary producers and food webs from the bottom up (discussed more in this and subsequent chapters). Chemistry also lies at the heart of processes that impact pertinent issues such as greenhouse gas emissions and acidification of estuarine waters. This chapter introduces fundamental concepts in chemistry as they relate to estuarine processes and ecology.

3.1.2 Global Biogeochemistry

Some of the first applications of the integrative field of "Biogeochemistry" are derived from organic geochemical studies where organisms and their molecular biochemistry were used as an initial framework for interpreting sources of sedimentary organic matter (SOM; Abelson and Hoering 1960; Eglinton and Calvin 1967). Biogeochemical cycles involve the interaction of biological, chemical, and geological processes that determine sources, sinks, and fluxes of elements through different reservoirs within ecosystems (Bianchi 2007). For example, a *reservoir* is the amount of material (M), as defined by its chemical, physical, and/or biological properties.

Estuarine Ecology, Third Edition. Edited by Byron C. Crump, Jeremy M. Testa, and Kenneth H. Dunton.

Companion website: www.wiley.com/go/crump/estuarine3

The units used to quantify material in a reservoir, or in the box or compartment of a box-model, are typically mass or moles. *Flux* (F) is defined as the amount of material that is transported from one reservoir to another over a particular time period (mass/time or mass/area/time). A *source* (S_i) is defined as the flux of material *into* a reservoir from a particular location, while a *sink* (S_o) is the flux of material *out* of the reservoir (and is often proportional to the size of the reservoir) to a particular location. The *turnover time* is the amount of time required to remove all the materials in a reservoir, or the average time that elements spend in a reservoir. Finally, a *budget* is essentially "checks and balances" of all the sources and sinks as they relate to the material turnover in reservoirs. For example, if the sources and sinks are the same and do not change over time, the reservoir is considered to be in a *steady state*. The term *cycle* refers to situations when there are two or more connecting reservoirs, whereby materials are cycled (i.e., produced, consumed, and transformed) through the system—generally with a predictable pattern of cyclic flow.

The spatial and temporal scales of biogeochemical cycles vary considerably depending upon the reservoirs considered. In the case of estuaries, most biogeochemical cycles are based on regional rather than global scales. However, with an increasing awareness of the importance of atmospheric fluxes of biogases (e.g., CO_2, CH_4, and N_2O) in estuaries and their impact on global budgets (Seitzinger 2000; Frankignoulle and Middelburg 2002), some budgets will involve both regional and global scales.

3.1.3 Thermodynamics and Kinetics

Before discussing the chemical dynamics of estuarine systems, it is important to briefly review some of the basic principles of *thermodynamic* or *equilibrium models* and *kinetics* that are relevant to upcoming discussions in estuarine/aquatic chemistry. Similarly, the fundamental properties of freshwater and seawater are discussed because of the importance of salinity gradients and their effects on estuarine chemistry.

Due to the complexity of natural systems, equilibrium models are often used to convey something about how chemical constituents (gases, dissolved species, and solids) behave (i.e., partition among different forms) under well-constrained conditions (no change over time, with fixed temperature and pressure, and homogeneous distribution of constituents). Equilibrium models can quantify the concentration of constituents under a given set of conditions, but they do not predict the rate at which the system reached an equilibrium state; the kinetics. The laws of thermodynamics are the foundation for chemical systems at equilibrium. The basic objectives in using equilibrium models in estuarine/aquatic chemistry is to calculate equilibrium compositions of particular compounds in natural waters, to determine the amount of energy needed to make certain reactions occur, and to ascertain how far a system may be from equilibrium. For more details on this see Stumm and Morgan (1996).

Chemical kinetic models provide information on reaction rates that cannot otherwise be obtained in chemical thermodynamics (Bianchi 2007). However, in many situations the information needed for such models, such as kinetic rate constants, are not available. The basic premise in kinetics is to relate the rate of a process to the concentration of reactants. For example, we can examine the formation and dissociation of species AB as it relates to reactants A and B in the following reactions:

$$A+B \xrightarrow{k_a} AB \quad \text{and} \quad AB \xrightarrow{k_b} A+B \tag{3.1}$$

where, k_a is the rate constant of formation and k_b is the rate constant of dissociation.

The units of the rate constants (e.g., seconds, days) will depend on the units of concentrations as well as the exponents. Temperature is another important factor that is critical in affecting rate constants. It is well established that temperature increases chemical reaction rates and biological processes, which is particularly important in estuarine biogeochemical cycles mediated by microbial reactions (discussed further in this chapter and Chapter 9). The *Arrhenius equation* summarizes the relationship between these factors:

$$k = Ae^{-E_a/RT} \tag{3.2}$$

where k is the rate constant, A is the frequency factor (number of significant collisions producing a reaction), E_a is the *activation energy* (amount of energy required to start a reaction, in Joules), and R is the *universal gas constant*=0.082057 ($cm^3\,atm\,mol^{-1}\,K^{-1}$) and T = *absolute temperature* (°K).

The rates of chemical processes usually increase in the range of 1.5–3.0, and biological processes by a factor of 2.0, for every 10 °C increase in temperature, respectively (Brezonik 1994). The most common way, although not necessarily the best way, of dealing with the temperature dependence of biological processes began with a study of fermentation rates (Berthollet 1803), where it was suggested that since rates increase with temperature, k at T+1 is greater than k at T−1 by a fixed proportion. It has become common to use a ratio over a 10-degree difference, or what is called the Q_{10} factor defined as follows:

$$Q_{10} = k(T+10)/k(T) \tag{3.3}$$

Because k does not go up linearly with temperature, Q_{10} will be dependent on the temperatures of comparison. The larger the Q_{10} the greater the effect temperature has on the reaction. A value of Q_{10} = 1 implies temperature has no effect on the reaction. Consequently, many of the global biogeochemical cycles mediated by biological processes are highly dependent on temperature, perhaps most notably trace gas (e.g., CO_2 and CH_4) cycles—as they relate to global warming. As climate warms, we expect a significant release of methane to the atmosphere from methane hydrates (clathrates) stored in ocean sediments (Shakhova et al. 2017) and from permafrost soils (Zimov et al. 2006). Similarly, CH_4 and CO_2 fluxes to the atmosphere are expected from most ecosystems via microbial decay processes—which are governed by the Arrhenius Rate function. The chemical kinetics of reactions in such processes in natural waters, soils, and sediments, might include catalytic reactions, kinetic isotope effects, and enzyme-catalyzed reactions (Butcher and Anthony 2000).

3.1.4 Redox Chemistry

The significant oxygen gradients commonly found in estuarine waters and surface sediments are largely controlled by tidal and wind mixing, and organic matter loading (Officer et al. 1984; Borsuk et al. 2001; Bianchi 2007). Consequently, redox and acid–base reactions are also important in determining the state of an ion in estuarine waters. Redox reactions involve an exchange of electrons, and half-reactions are typically used to describe reduction/oxidation (*redox*) reactions, as follows:

$$\mathrm{OX_1} + ne^- = \mathrm{RED_1}\ \text{half-reaction 1} \tag{3.4}$$

$$\mathrm{RED_2} = \mathrm{OX_2} + ne^-\ \text{half-reaction 2} \tag{3.5}$$

where OX is the oxidized form, RED is the reduced form, and n is the number of electrons (e^-) transferred. When combined, the half-reactions and the final redox equation are as follows:

$$\mathrm{OX_1} + \mathrm{RED_2} = \mathrm{RED_1} + \mathrm{OX_2} \tag{3.6}$$

This equation represents the exchange of electrons between OX_1 and RED_2 to produce RED_1 and OX_2. In this, OX_1 is considered to be the *oxidant* because it oxidizes RED_2 to OX_2, and RED_2 is the *reductant* because it reduces OX_1 to RED_1. A typical redox half-reaction that occurs in estuarine waters is the reduction of Fe^{3+} to Fe^{2+}:

$$\mathrm{Fe^{3+}} + e^- = \mathrm{Fe^{2+}} \tag{3.7}$$

Redox in sediments is typically defined by E_h (redox potential) where positive values indicate oxidizing conditions (Stumm and Morgan 1996). In redox chemistry, the half-reaction concept is used to understand E_h, which is the electron activity in volts relative to a standard hydrogen electrode (Chester 2003). Thus, the activity of electrons in solution is expressed by electron activity ($p\varepsilon$), a dimensionless term, or E_h (volts) by the following equation

$$p\varepsilon = \mathrm{F}/2.3\,\mathrm{RT}\left(E_h\right) \tag{3.8}$$

where F is the Faraday's constant, R is the gas constant, and T is the absolute temperature. E_h conditions are primarily controlled by the production and decomposition of organic matter; positive values indicate oxidizing conditions and negative values indicate reducing conditions (Figure 3.1; see Bianchi 2007). The depth of the *redox potential discontinuity* (RPD) layer in sediments is associated with distinct sediment coloration, indicative of differences between oxic and suboxic conditions (Fenchel and Riedl 1970; Santschi et al. 1990). The oxidized region at the surface of estuarine sediments (one to a few cm of sediment) is typically orange-brown in color (*Fe* and *Mn oxides*) below which is a reduced region that is grey-black in color (*mono-polysulfides*). The depth of the RPD is largely controlled by the amount of organic matter loading, physical mixing, and bioturbation. The RPD is shallow when organic loading is high and/or physical mixing and bioturbation are low. The location of the RPD depth is very similar when measured with electrodes in sediment cores or with sediment profile images (SPI; Rosenberg et al. 2001).

The decomposition of organic matter in marine/estuarine sediments proceeds through a sequence of *terminal electron acceptors* (e.g., O_2, NO_3^-, MnO_2, FeOOH, SO_4^- and CO_2; Froelich et al. 1979; Canfield 1993) as shown in Figure 3.1 (Deming and Baross 1993). The sequence of reactions is principally determined by the free energy yields (ΔG^0) per mole of organic C, which is the amount of energy released or required by a reaction. If ΔG^0 is less than 0, energy will be released by the reaction, whereas if ΔG^0 is greater than 0, energy is required. While this sequence provides a basic framework from which to work, the versatility of many bacteria may allow for several of these reactions to occur in the same zone (Brandes and Devol 1995).

3.2 Mixing and Particle Effects on the Chemistry of Estuarine Waters

3.2.1 Reactivity of Dissolved Constituents

The mixing of river water and seawater can be quite varied in different estuarine systems resulting in a water column that can be highly mixed or strongly stratified (Bianchi 2007). These intense gradients in mixing and ionic strength can significantly affect concentrations of both dissolved and particulate constituents in the water column through processes such as *sorption-desorption* (physical and chemical processes where a substance becomes attached or detached from another, respectively) *and flocculation* (fine particles adhere to each other in clumps, commonly in the context of colloid chemistry), as well as biological processes. The reactivity of a particular estuarine constituent within an estuary has been traditionally interpreted by plotting its concentration across a salinity gradient in a salinity mixing diagram, where salinity is assumed to reflect only the mixing of fresh and marine water. As shown by Wen et al. (1999), the simplest distribution pattern, in a one-dimensional, two end-member (river and ocean),

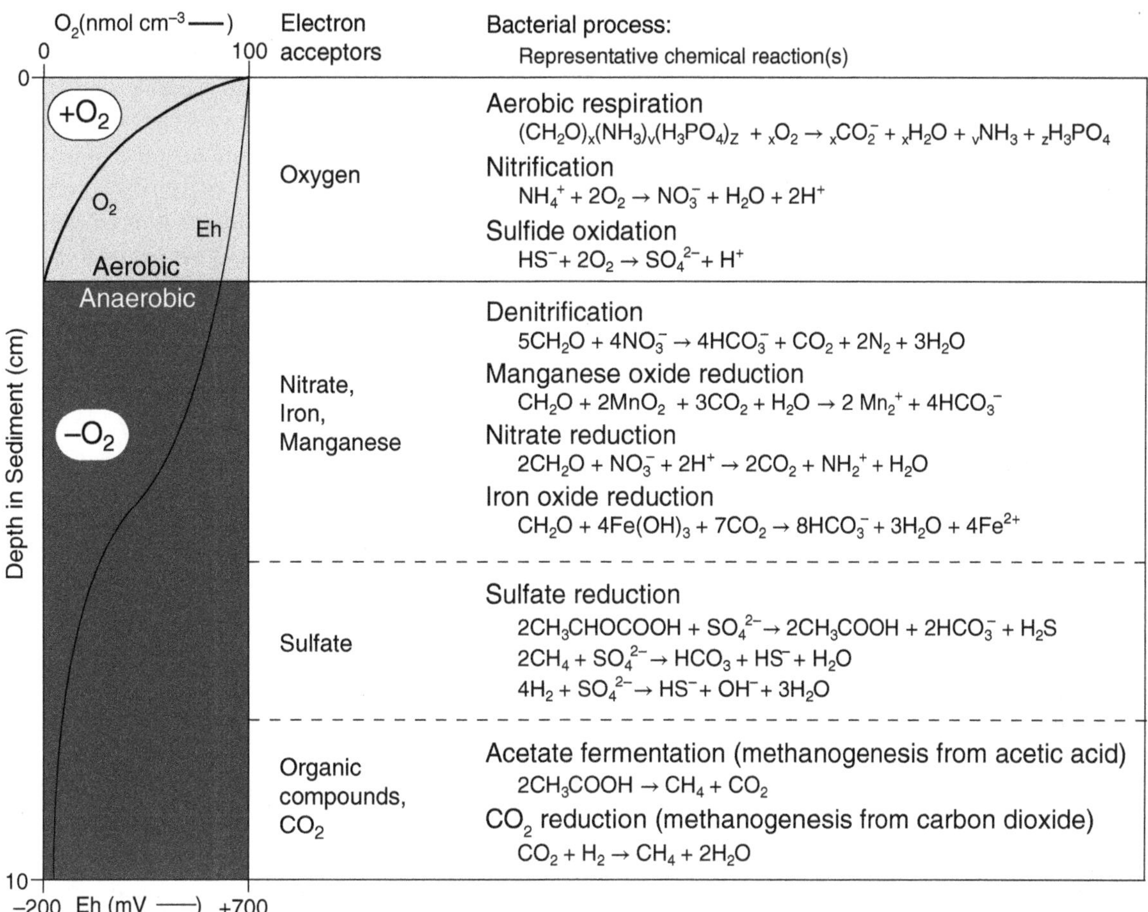

FIGURE 3.1 Bacterial decomposition of organic matter in marine/estuarine sediments through a sequence of terminal electron acceptors (e.g., O_2, NO_3^-, MnO_2, FeOOH, SO_4^- and CO_2) and changing redox. *Source*: Modified from Deming and Baross (1993).

steady-state system, would be for a *conservative (i.e., unreactive) constituent* to change linearly with salinity (Figure 3.2; see Bianchi 2007). For a *nonconservative constituent* (net loss or gain in concentration across a salinity gradient), extrapolation from high salinities can yield an "effective" river concentration (C*). This "effective" concentration can be used to infer reactivity of a constituent and can be used to determine total flux of the constituent to the ocean. For example, when $C^* = C_0$, the constituent is behaving conservatively, when $C^* < C_0$, there is removal of the constituent (nonconservative behavior) within the estuary, and when $C^* > C_0$, the constituent is being added (nonconservative behavior) within the estuary. River flux (F_{riv}) to an estuary and ultimately to the ocean is commonly estimated using this simplified mixing model. The flux of material into an estuary is $F_{riv} = RC_0$, where R is the river water flux. Similarly, flux from an estuary to the ocean (F_{ocean}) can be estimated by $F_{ocean} = RC^*$. Finally, the overall net internal flux (F_{int}) from input or removal can be estimated by $F_{int} = R(C^* - C_0)$.

Despite widespread application of the standard mixing model in estuarine systems, there are numerous problems when invoking these simple steady-state mixing assumptions. Early work showed that nonreactive constituents can display nonconservative mixing when estuarine mixing occurs on a

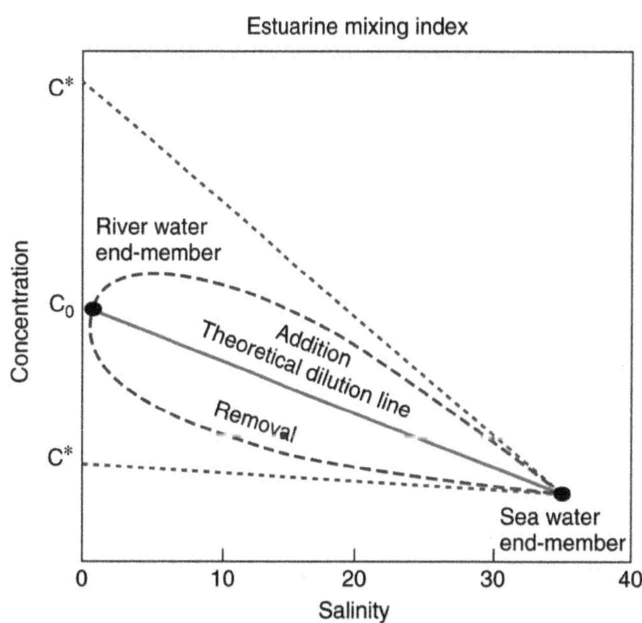

FIGURE 3.2 Illustration of the simplest distribution pattern in a one-dimensional, two end-member, steady-state system for a conservative constituent that changes linearly with salinity; a salinity mixing diagram. *Source*: From Wen et al. (1999)/John Wiley & Sons.

different timescale than changes in end-member concentrations in the adjacent river and ocean (Officer and Lynch 1981).

The relative importance of fluxes of dissolved constituents from groundwater inputs to estuarine and shelf waters is yet another transport mechanism that adds complexity to the estuarine mixing index. Groundwater inputs of nutrients to estuaries can be quite considerable (Kelly and Moran 2002) and can enhance the nonconservative behavior of constituents.

3.2.2 Sources and Mixing of Dissolved Salts in Estuaries

Prior to discussing the factors that control concentrations of the major dissolved components in rivers, estuaries, and oceans, it is important to discuss the operationally defined size-spectrum for different phases (dissolved, colloidal, and particulate) of an element. The conventional definition for dissolved materials is the fraction of total material that passes through a membrane filter with a nominal pore size of 0.45 μm (Figure 3.3; Wen et al. 1999), while colloids are an operationally defined fraction in the size range of 0.001–1 μm (Vold and Vold 1983). Although ultrafilters are available for isolation of colloidal materials, this fraction is traditionally included within the dissolved fraction. However, *colloids* are not considered to be truly dissolved even though some of them pass through a 0.45 μm filter.

All natural waters in the world have a certain amount of salts dissolved in them. From a chemical perspective, estuaries are places where seawater is measurably diluted by freshwater inputs from the surrounding drainage basin. The mixing of river water and seawater in estuarine basins is highly variable and typically characterized by sharp concentration gradients. In simple terms, estuaries contain a broad spectrum of mixing regimes between two dominant end-members; rivers and oceans. Rivers have highly variable amounts of salts in them, typically ranging within a few hundreds of milligrams per liter, while the oceans have more stable concentrations in the range of grams per liter.

The salts in rivers are primarily derived from the weathering of rocks in the drainage basins of rivers and estuaries, in addition to other human activities (e.g., agriculture; Burton and Liss 1976; Berner and Berner 1996; Bianchi 2007). When examining the relationship between drainage basin area and total sediment discharge in major rivers of the world it becomes clear that factors other than basin area are important. In addition to basin area, other factors typically include the following: relief (elevation) of the basin, amount of water discharge, influence of lakes/dams (e.g., storage) along river, geology of the basin, and climate (Milliman 1980; Milliman and Syvitski 1992). Consequently, the composition of suspended and dissolved materials in rivers is largely a function of the soil composition of the drainage basin. However, significant differences exist between the chemical composition of suspended

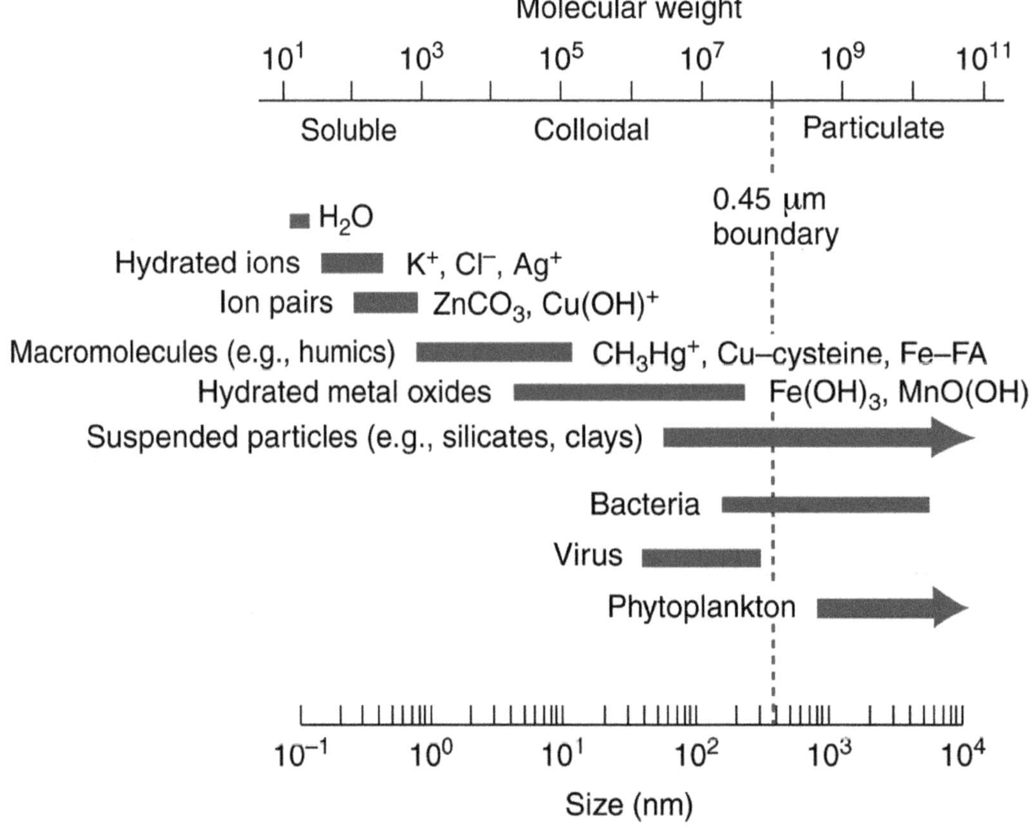

FIGURE 3.3 Conventional definition for dissolved materials shown as the fraction of total material that passes through a membrane filter with a nominal pore size of 0.45 μm. *Source*: From Wen et al. (1999)/John Wiley & Sons.

materials in rivers and the parent rock material. This is due to differences in the solubility of different elements in parent rock materials. For example, elements such as Fe and Al are less soluble than Na and Cl, which makes them less abundant in dissolved materials and more abundant in the suspended load of rivers (Berner and Berner 1996). This enrichment of Fe and Al in the suspended load is further supported by the element weight ratio, which if greater than 1 indicates elemental enrichment. This is because chemical differences in the weathering and solubility of crustal elements, combined with reprecipitation of the more insoluble elements, result in river particulates being dominated by less soluble elements (e.g., Fe and Al) relative to parent rock material. Thus, the element weight ratio is simply used as an indicator of this selective processes in weathering relative to parent crustal rocks, as it relates to transport of particulates to aquatic systems. In some cases, rivers may receive the majority of salt inputs through precipitation and evaporation processes. This relationship was established when Gibbs (1970) plotted the concentration of total dissolved solids (TDS) versus the compositional indices of $Na^+/(Na^+ + Ca^{2+})$ and $Cl^-/(Cl^- + HCO_3^-)$ for rivers with different levels of precipitation and evaporation.

A historical account of measurements of major dissolved components of seawater indicates that the most abundant elements (in order of decreasing abundance), are Cl^-, Na^+, Mg^{2+}, SO_4^{2-}, Ca^{2+}, and K^+ (Table 3.1; Millero 1996). In contrast to rivers, the major constituents of seawater are found in relatively constant proportions in the oceans, indicating that the residence times of these elements are long (thousands to millions of years); highly indicative of nonreactive behavior (Millero 1996). This relative constancy of major (and many minor) elements in seawater is referred to as the *Rule of Constant Proportions* or *Marcet's Principle*. More specifically, these elements are considered to be *conservative elements*, whereby changes in the concentrations of these elements reflect the addition or loss of water through physical processes. While these elements may be involved in other chemical or biological reactions, concentration changes from these processes are too small to change the constancy of the elemental ratios (Wangersky and Wangersky 1980; Libes 1992). The remaining elements in seawater are termed *nonconservative* because they do not remain in constant proportion due to biological (e.g., uptake via photosynthesis) or chemical (e.g., hydrothermal vent inputs) processes. In estuaries and other oceanic environments (e.g., anoxic basins, hydrothermal vents, and evaporated basins), the major components of seawater can be altered quite dramatically due to numerous processes (e.g., precipitation, evaporation, freezing, dissolution, and oxidation).

TABLE 3.1 **Relative composition of major components of seawater (pH_{sws} = 8.1, *S* = 35, and 25 °C).**

Solute	g_i/Cl(‰)
Na^+	0.55661
Mg^{2+}	0.06626
Ca^{2+}	0.02127
K^+	0.02060
Sr^{2+}	0.00041
Cl^-	0.99891
SO_4^{2-}	0.14000
HCO_3^-	0.00552
Br^-	0.00347
CO_3^{2-}	0.00083
$B(OH)_4^-$	0.000415
F^-	0.000067
$B(OH)_3$	0.001002
Σ =	1.81540

Source: Modified from Millero (1996).

3.2.3 Measurement of Salinity

Salinity was first rigorously defined by Knudsen (1902) as "the weight in grams of the dissolved inorganic matter in one kilogram of seawater after all bromide and iodide have been replaced by the equivalent amount of chloride and all carbonate converted to oxide." The relative constancy of the major ions in seawater is constant enough that determination of one major component could be used to determine the other components in a sample.

Due to the accuracy and reproducibility of the measurement, chloride was chosen as the ion of choice. Libes (1992) defines *chlorinity* as "the mass in grams of halides (expressed as chloride ions) that can be precipitated from 1000 g of seawater by Ag^+." This is referred to as the Mohr titration, where silver nitrate is used to titrate seawater with potassium chromate as an indicator. Chlorinity can also be estimated using density and conductivity measurements (Cox et al. 1967). Thus, salinity is now commonly measured using an inductive *salinometer*, where the conductivity of water is measured. In essence the electrical current is controlled by the movement and abundance of ions—the more dissolved salts, the greater the conductivity. Since much of the earlier work was presented in terms of salinity and chlorinity, the two units are related, by definition, according to the following equation:

$$S(‰) = 1.80655Cl(‰) \tag{3.9}$$

3.2.4 Dissolved Gases and Atmosphere–Water Exchange

Dissolved gases are critically important in many of the biogeochemical cycles of estuaries and coastal waters (Bianchi 2007). However, only recently have large-scale collaborative efforts addressed the importance of air–water gas exchange in estuaries. For example, the Biogas Transfer in Estuaries (BIOGEST) project spanned from 1996 to 1999, and focused on determining the distribution of biogases (CO_2, CH_4, CO, non-methane hydrocarbons, N_2O, DMS, COS, volatile halogenated organic compounds, and some biogenic volatile metals) in European estuaries and their impact on global budgets (Frankignoulle and Middelburg 2002). The role of estuaries and other coastal ocean environments as global sources and/or sinks of key greenhouse gases, such as CO_2, has also been a subject of intense interest in recent years (Raymond et al. 1997, 2000; Cai 2003, 2011; Wang and Cai 2004; Cai 2011). Similarly, O_2 transfer across the air–water interface is critical for the survival of most aquatic organisms. Unfortunately, many estuaries around the world are currently undergoing eutrophication, which commonly results in low O_2 concentrations in deeper waters (or *hypoxic* conditions, where $O_2 < 2\,mg\,L^{-1}$), due to microbial respiration of organic matter produced by phytoplankton responding to excessive nutrient loading in these systems (Rabalais and Turner 2001; Rabalais and Nixon 2002).

To understand how gases are transferred across the air-water boundary, we will first examine the dominant atmospheric gases and physical parameters that control their transport and solubility in natural waters. The atmosphere is also composed of *aerosols*, which are defined as condensed phases of solid or liquid particles, suspended in state, that do not undergo gravitational separation over a period of observation (Charlson 2000). The chemical composition and speciation of atmospheric aerosols is important for understanding their behavior after deposition and is strongly linked with the dominant sources of aerosols (e.g., windblown dust, sea-salt, and combustion). Aerosol deposition to estuaries and coastal waters, via precipitation (rain and snow) and/or dry particle deposition, has received considerable attention in recent years. For example, *dry and wet deposition* of nutrients (Paerl et al. 2002; Pollman et al. 2002) and metal contaminants (Siefert et al. 1998; Guentzel et al. 2001) have proven to be significant in biogeochemical budgets in wetlands and estuaries.

The direction of gas exchange across the air-water interface will change accordingly as atmospheric and aqueous concentrations of respective gases change over time. When rates of exchange for a particular gas are equal across the air–water interface the gas is considered to be in *equilibrium;* this is when the concentration of the gas in both aqueous (P_A) and gas phases (P_i) are equal. The equilibrium concentration of a gas in the aqueous phase is directly proportional to the pressure of that gas. Referred to as *Henry's Law of Equilibrium Distribution,* this is defined by the following equation:

$$P_A = K_H P_i \tag{3.10}$$

where K_H is the Henry's Law constant, which is a unique value for different compounds, and P_A is the concentration of gas in the aqueous phase (expressed as $mol\,kg^{-1}$).

The solubility of gases is influenced by their molecular weight. In general, heavier molecules are more soluble, but there are cases where certain molecules interact more strongly with water. In addition to molecular weight, there can be changes observed in the solubility of gases over different seasons, particularly with N_2 and O_2. These *normal atmospheric equilibrium concentrations* (NAEC) are based on expected equilibrium conditions, between water and atmosphere, at a particular pressure, temperature, salinity, and humidity. However, there are many physical and biological factors that cause many gases to behave in a non-conservative manner resulting in deviations from predicted NAEC. For example, phytoplankton can rapidly alter O_2 concentrations via photosynthetic production of O_2. Similarly, bacteria can increase CO_2 concentrations through decomposition and can influence N_2 concentrations through denitrification and N_2 fixation processes, which will be discussed further in Chapter 9. Gas concentrations can also be altered by nonbiological processes, such as the production of radon (Rn) through radioactive decay.

In situations where equilibrium conditions are not applicable, rates of gas exchange across the atmosphere–water boundary can be calculated using kinetic models (Broecker and Peng 1974; Kester 1975). The most common kinetic model used is the *Stagnant Film Model* (Figure 3.4). This model essentially has three regions of importance: (1) a well-mixed *turbulent atmospheric zone*; (2) a well-mixed *turbulent bulk liquid* zone; and (3) a *laminar zone* separating the two turbulent regions.

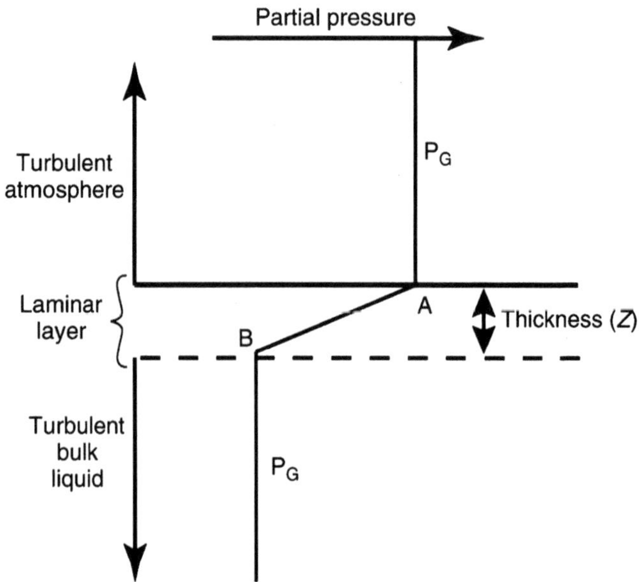

FIGURE 3.4 The most common kinetic model used to estimate rates of gas exchange across the atmosphere–water boundary is the Stagnant Film Model. This model essentially has three regions of importance: a well-mixed turbulent atmospheric zone (P_G); a well-mixed turbulent bulk liquid zone (P_G); and a laminar zone (A–B) separating the two turbulent regions. The thin film is considered permanent with a thickness defined as Z. *Source*: Based on Broecker and Peng (1974).

In this model, the thin-film laminar layer is considered permanent with a thickness defined as *z*. The average thickness of this film for the ocean has been estimated to be 17 μm (Murray 2000), using isotope measurements (Broecker and Peng 1974; Peng et al. 1979). Simply put, the turbulent atmospheric zone and the turbulent bulk liquid zone have multidirectional flow that is not governed by molecular diffusion, unlike the laminar zone—which is assumed to be driven by molecular diffusion in this model. This model can be used to determine rates of gas flux between water and the atmosphere, which is of particular importance for understanding numerous processes, including greenhouse gas fluxes.

By combining Henry's Law (Eq. 3.10) with *Fick's First Law* described by the following equation, we obtain an equation that describes the rate of gas flux through the atmosphere–water interface:

$$dC_i/dt = -D_i\left[dC_i/dz\right] \tag{3.11}$$

where C_i is the concentration of species *i*, t is the time, D_i is the diffusion coefficient, and dC_i/dz is the concentration gradient between the top and bottom of the thin film as represented by a vertical depth of *z* (thickness of the thin-film). The following equation can be used to describe the rate of gas flux through the atmosphere–water interface during disequilibrium:

$$dC_i/dt = \left(AD_i/zK_H\right)\left[P_i(\text{gas}) - P_A(\text{soln})\right] \tag{3.12}$$

where A is the interfacial area and K_H is Henry's Law constant.

This model assumes that molecular diffusivity is directly proportional to the gas flux across the atmosphere–water interface, and that Henry' Law constant is inversely related to the atmosphere–water gas flux. Molecular diffusion coefficients typically range from 1×10^{-5} to 4×10^{-5} $cm^2\,s^{-1}$ and typically increase with temperature and decreasing molecular weight. Other factors such as thickness of the thin-layer and wind also have important effects on gas flux. For example, wind creates shear that results in a decrease in the thickness of the thin layer. The sea *surface microlayer* consists of films 50–100 μm in thickness (Libes 1992). This layer can change in composition depending on biological activity (such as phytoplankton blooms) and understanding the variability of gas exchange across these layers is critical for understanding greenhouse gas fluxes between water and the atmosphere. Wind forcing can also be a very important factor for influencing gas transfer and near-surface turbulence in estuaries. Previously, the flux, F, of a gas has been estimated as the product of the gas transfer velocity and the concentration difference between air and water, using the following equation from Zappa et al. (2007):

$$F = k\left(C_w - sC_a\right)$$

where s is the solubility coefficient, k is the gas transfer velocity, and C_w and C_a are the gas concentrations in the water and air, respectively.

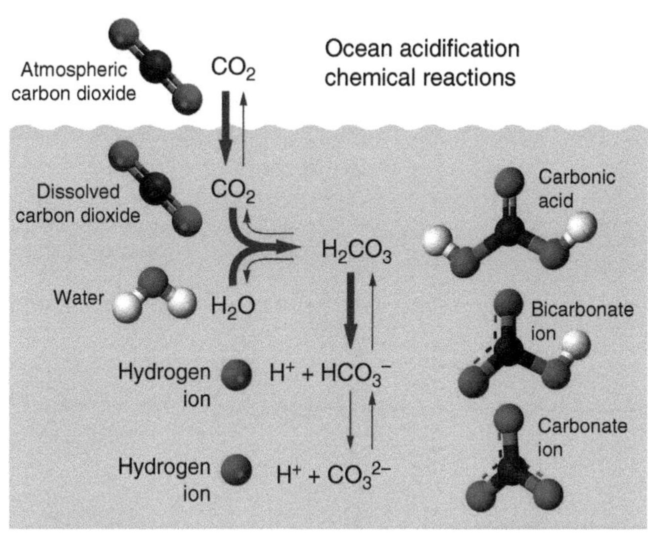

FIGURE 3.5 The chemical reactions involved in ocean and estuarine acidification. Figure modified from NOAA's Pacific Marine Environmental Laboratory's Carbon Program.

3.2.5 Dissolved CO_2 and Carbonate Chemistry

Dissolved gases can influence the chemistry of estuarine waters, as evidenced by the relationship between dissolved CO_2 and carbonate chemistry. This relationship drives the increasing acidification of estuarine and ocean waters as atmospheric CO_2 levels increase. Atmospheric CO_2 readily dissolves in water, and when it does, it forms carbonic acid (H_2CO_3). This carbonic acid then forms a bicarbonate ion (HCO_3^-) and a hydrogen ion (H^+), which drops the pH (increases the acidity) of the system (Figure 3.5). In the open ocean, this process is driven predominantly by air-water gas exchange, while in estuaries, biological production of CO_2 can also provide a considerable contribution to dissolved CO_2 (Feely et al. 2010) and estuarine acidification can be exacerbated by eutrophication (Cai et al. 2011). In the ocean, bicarbonate can further react to form carbonate (CO_3^{2-}), but this does not occur as frequently in estuaries. As water becomes more acidic and carbonate concentrations decline, organisms such as oysters, corals, and clams, have more difficulty producing calcium carbonate structures ($CaCO_3$), which may have severe consequences for coastal fisheries and economies (although different species have different responses in calcification rates, and this can be influenced by temperature as well).

3.2.6 Effects of Suspended Particulates and Chemical Interactions

Particulates in estuarine systems are composed of both *seston* (discrete biological particles) and inorganic lithogenic components. They are commonly defined operationally as the material that does not pass through a 0.45 μm filter. The highly dynamic character of estuarine systems (e.g., tides, wind, and resuspension) can result in considerable variability in particle

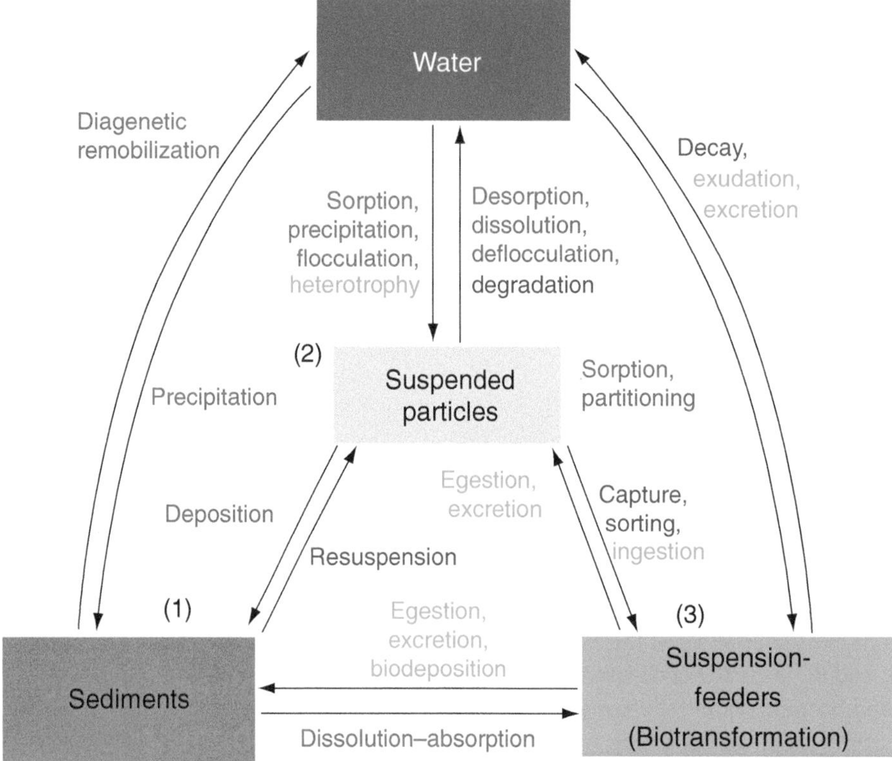

FIGURE 3.6 Processes critical in controlling the partitioning of chemical species in estuaries with particular emphasis on metals and hydrophobic organic micropollutants (HOMs). *Source*: Turner and Millward (2002)/with permission of Elsevier. Processes can be chemical/physical (red), biological (blue), or both (purple).

concentrations over weekly, daily, and diurnal time intervals (Fain et al. 2001). Moreover, the reactivity of these particles can change over short spatial intervals due to rapid changes in salinity, pH, and redox conditions (Herman and Heip 1999). Water column particulates in estuaries, primarily derived from rivers, adjacent wetland systems, and resuspension events, are important in controlling the fate and transport of chemicals in estuaries (Turner and Millward 2002). A review by Turner and Millward (2002), with particular emphasis on metals and hydrophobic organic micropollutants, showed that processes such as ion exchange, sorption–desorption, and precipitation–dissolution were critical in controlling the partitioning of chemical species in estuaries (Figure 3.6). Biological processing of particulates (e.g., respiration, consumption, and subsequent excretion) by both pelagic and benthic micro- and macroheterotrophs is also critical in estuaries. Chemical and biological processes can influence the transfer of chemical species between different compartments within an estuary (i.e., sediment, water, suspended particles, and suspension-feeders). By following the arrows in Figure 3.6, you can trace the movement of a constituent from one compartment to another. For example, a constituent that may originate from estuarine sediment (1) could be resuspended and transported into the suspended particulate pool (2), which could then be ingested by suspension feeders, such as oysters (3).

Lithogenic particles are derived from weathering of crustal materials and are mostly comprised of primary minerals like quartz and feldspar, secondary silicate minerals such as clays, and hydrogenous components (Fe and Mn oxides, sulfides, and humic aggregates) formed *in situ* by chemical processes (Turner and Millward 2002). Mineral surfaces on these particles are important in binding organic molecules, gels, and microaggregates (Mayer 1994a, b; Aufdenkampe et al. 2001). The fate of organic molecules in estuarine systems largely depends on whether or not they are sorbed to mineral surfaces (Keil et al. 1994a, b; Baldock and Skjemstad 2000). For example, selective partitioning of basic amino acids (positively charged) to clay particles (negatively charged) significantly affects the composition of dissolved amino acids in river waters of the Amazon Basin (Aufdenkampe et al. 2001), potentially influencing amino acid bioavailability. Similarly, concentrations of many trace metals (e.g., Cu, Cd, and Zn) in estuarine waters are also influenced by sorption–desorption interactions with suspended particles (Santschi et al. 1997).

Biogenic particulates derived from fecal pellets (e.g., zooplankton) and planktonic and terrestrial detrital materials are also important in controlling chemical interactions in estuaries (Bianchi 2007). Other suspended particulates composed of complex aggregates of biogenic and lithogenic materials have similar effects. Many of these biogenic particles are organic carbon-rich and are degraded and converted to DOM (dissolved organic matter), which can then be sorbed to lithogenic particles and provide an organic coating. These organic coatings are important in controlling the surface chemistry of particulates in aquatic environments (Loder and Liss 1985; Wang and Lee 1990). For example, adsorbed DOM can reduce oxidation and hydrolysis, and increases reduction of contaminants on mineral surfaces in natural waters (Polubesova and Chefetz 2014). Organic detrital particulates also affect the adsorption of amines (Wang and Lee 1990) and ammonium (Mackin and Aller 1984) in sediments, which is important for sediment nitrogen dynamics.

3.2.7 Estuarine Turbidity Maximum, Benthic Boundary Layer, and Fluid Muds

The Estuarine Turbidity Maximum (ETM) is defined as a region in estuaries where the suspended particulate matter (SPM) concentrations are considerably higher (10–100 times) than adjacent river or coastal end-members (Schubel 1968; Dyer 1986). Some of the most extensive early work on the ETM, conducted in Chesapeake Bay, suggested that the primary mechanisms of particle trapping in ETMs were due to simple convergence of flows from opposing directions at the limit of salt intrusion, in addition to slow particle settling (Schubel 1968; Schubel and Biggs 1969; Schubel and Kana 1972; Nichols 1974). A conceptual diagram illustrates that while the ETM roughly tracks the limit of salt, it is often de-coupled from the salt front (defined here as the isohaline with a salinity of 1) due to a lag between ETM sediment resuspension/transport and rapid meteorologically driven movement of the salt front (Figure 3.7; Sanford et al. 2001). Studies focused on sediment transport dynamics in the ETM have typically used acoustic Doppler profiles (ADP) of velocity, acoustic backscatter (ABS), and optical backscatter sensors (OBS; Fain et al. 2001), in addition to remote sensing tools such as the Sea-viewing Wide Field-of-view Sensor (SeaWiFS; Uncles et al. 2001). Due to rapid and high sedimentation rates in the ETM, the accumulation of particles in the *benthic boundary layer* (BBL) can result in the formation of *mobile* and *fluid muds*. The BBL is defined by Boudreau and Jorgensen (2001) as "those portions of sediment and water columns that are affected directly in the distribution of their properties and processes by the presence of the sediment-water interface."

(a) Full ebb tide

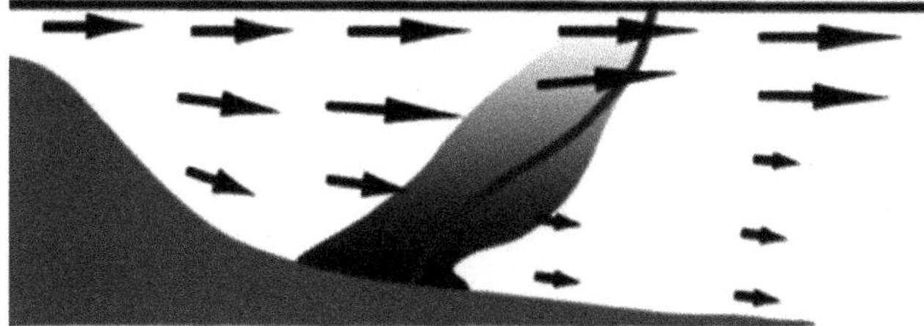

(b) Slack before flood

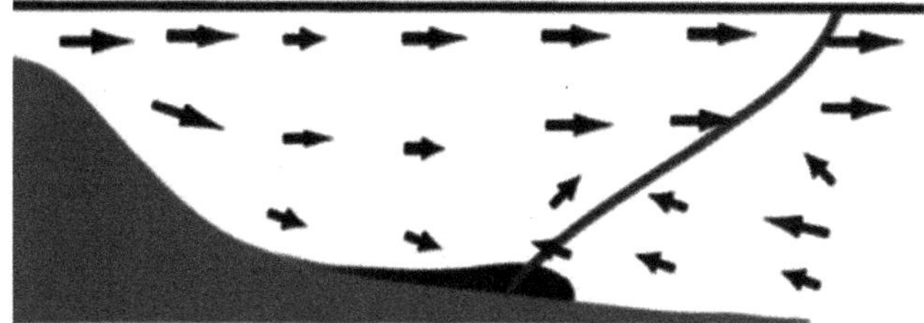

(c) Full flood tide

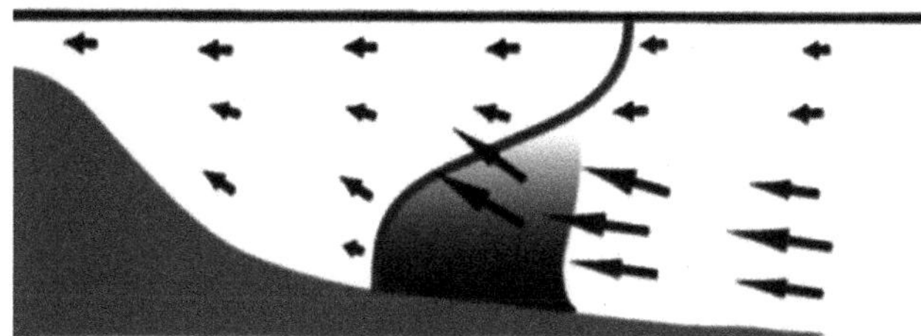

(d) Slack before ebb

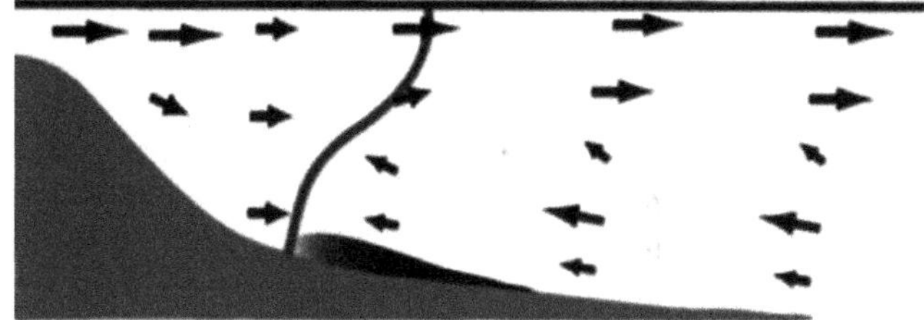

(e) Flood tide + seaward wind

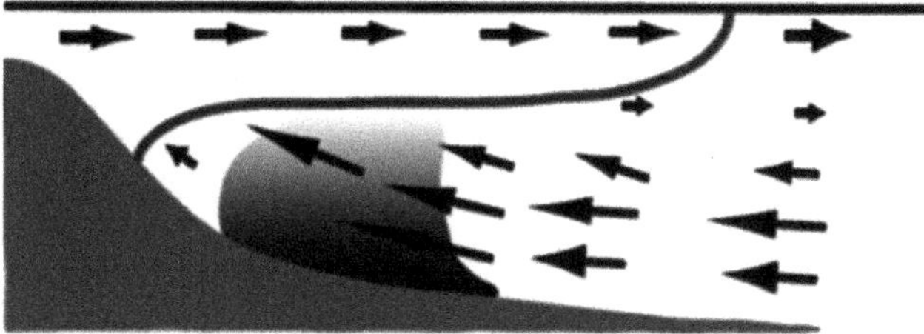

FIGURE 3.7 A conceptual diagram showing that while the ETM roughly tracks the limit of salt, it was often de-coupled from the salt front (defined here at the isohaline with a salinity of 1), due to a lag of ETM sediment resuspension/transport from rapid, meteorologically driven movement of the salt front. This process is shown during changes in the tidal cycle, from (a) full ebb tide, (b) slack before flood tide, (c) full flood tide, (d) slack before ebb, and (e) flood tide, with a seaward wind. *Source*: From Sanford et al. (2001)/Springer Nature.

3.3 Biogeochemistry of Organic Matter

3.3.1 Particulate and Dissolved Organic Matter in Estuaries

The photosynthetic fixation of inorganic carbon and nutrients into plant biomass is the primary source of organic matter within estuaries (Bauer and Bianchi 2011; Bianchi and Bauer 2011). Therefore, it is critical to have some basic understanding about the primary producer community in estuaries (e.g., phytoplankton, benthic macroalgae, microphytobenthos, seagrasses, and wetland plants), in addition to the habitats and conditional constraints needed for growth (see Chapters 4–8).

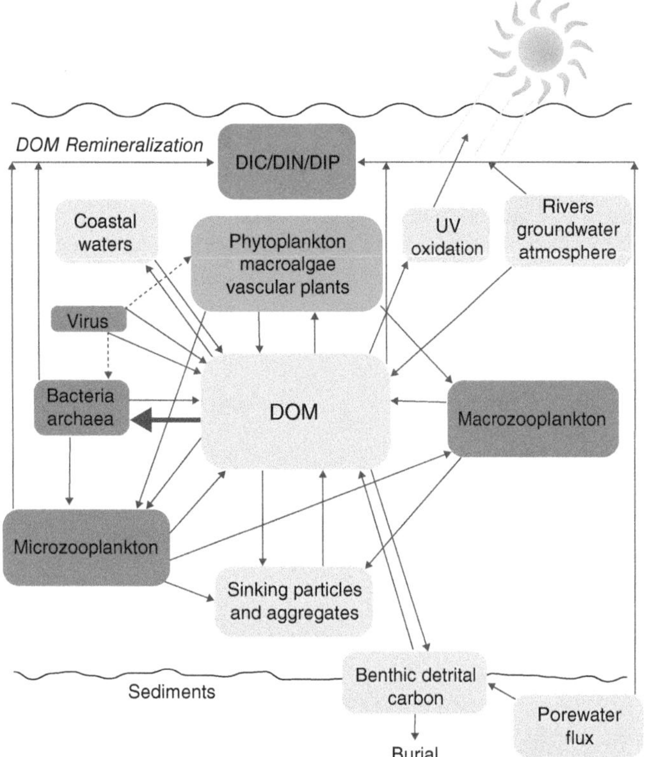

FIGURE 3.8 Major sources of dissolved organic matter (DOM) to estuaries, primarily composed of riverine inputs, autochthonous production from algal and vascular plant sources, benthic fluxes, groundwater inputs, and exchange with adjacent coastal systems. *Source*: Modified from Hansell and Carlson (2002).

Inputs of DOM, particulate organic matter (POM), and nutrients to estuaries can occur from both coastal ocean and riverine end-members, which support both autotrophic and heterotrophic production in different estuarine regions.

Estuarine DOM is comprised of a diverse array of allochthonous (externally produced) and autochthonous (internally produced) sources (see reviews by Cauwet 2002; Findlay and Sinasbaugh 2003; Bauer and Bianchi 2011; Ward et al. 2017). These major sources are primarily comprised of riverine inputs, autochthonous production from algal and vascular plant sources, benthic fluxes, groundwater inputs, and exchange with adjacent coastal systems (Figure 3.8). In coastal and open ocean waters, DOM is generally positively correlated with phytoplankton biomass (Hansell and Carlson 2002; Santschi et al. 1995). It has been estimated that phytoplankton release approximately 12% of total primary production as DOM over a range of freshwater to marine environments (Baines and Pace 1991). However, water column DOM in many estuarine systems does not show positive correlations with phytoplankton biomass because of allochthonous inputs of DOM from soils and vascular plants that are hydrologically coupled to these wetland systems and rich in DOM (Argyrou et al. 1997; Harvey and Mannino 2001; Jaffé et al. 2004).

Past work has shown that DOM from river/estuarine systems is primarily derived from terrestrial vegetation and soils (Malcolm 1990; Opsahl and Benner 1997). In fact, some of the highest DOM estuaries in the U.S. border the Gulf of Mexico (Guo et al. 1999; Engelhaupt and Bianchi 2001), a region that also has some of the highest rates of fresh litter decomposition in soils (Meentemeyer 1978).

Riverine/estuarine DOM is generally considered to be recalcitrant and transported conservatively to the ocean (Moore et al. 1979; van Heemst et al. 2000). However, spatial variability in the abundance of phytoplankton exudates (Aminot et al. 1990; Fukushima et al. 2001), uptake of DOM by bacteria (Zweifel 1999; Pakulski et al. 2000), chemical removal processes (e.g., *flocculation, deflocculation, adsorption, aggregation,* and *precipitation*; Sholkovitz 1976, Sholkovitz et al. 1978; Lisitzin 1995), inputs from porewaters during resuspension events (Burdige and Homestead 1994; Middelburg et al. 1997), atmospheric inputs (Velinsky et al. 1986), and photodegradation (Dalzell et al. 2009) can all contribute to nonconservative behavior in estuaries.

As described in Section 3.2, DOM exists in different size fractions in aquatic systems (Sharp 1973) and a large portion of estuarine DOM is composed of colloidal organic matter (Filella and Buffle 1993; Guo and Santschi 1997; Cauwet 2002; Bianchi 2007). The characterization of different size classes of DOM is established by physical separation through filters/membranes of differing pore sizes; thus, colloids are an operationally defined fraction of the total DOM in the size range of 0.001–1 μm (Vold and Vold 1983). The colloidal fraction of DOM is important biogeochemically due to its high surface area, and can influence biogeochemical cycles in aquatic systems (Engelhaupt and Bianchi 2001). For example, trace contaminants and metals can remain bound to the colloidal fraction of DOM (sometimes referred to as high molecular weight DOM, HMW DOM; Santschi et al. 1997), which may decrease their bioavailability.

Another term commonly used in association with DOM is humic substances. *Humic substances* are defined as complex assemblages of molecules that have a yellow-to-brown color and are derived from plants and soils (Hatcher et al. 2001). Humic substances represent a large fraction of what is termed *chromophoric dissolved organic matter* (CDOM) in aquatic systems around the world (Blough and Green 1995). Aquatic humic substances can further be categorized as *fulvic acids, humic acids,* and *humin* based on the solubility in acid and base solutions (Schnitzer and Khan 1972; McKnight and Aiken 1998). More specifically, humic acids typically have a molecular weight of greater than 100,000 daltons (Da) and are soluble above a pH of 2; fulvic acids are smaller molecules (approximately 500 Da) and are soluble at any pH; and humin is not soluble across a full pH range (McKnight and Aiken 1998).

Sediments may also represent an important source of DOM to the water column of shallow estuarine systems (Burdige 2006; Bianchi 2007). The accumulation of DOM in porewaters results from the diagenesis of sedimentary organic carbon into a complex mixture of macromolecules (e.g., humic substances) and smaller monomeric forms (e.g., amino acids; Orem et al. 1986; Burdige 2002, 2006). The flux of porewater DOM across the sediment–water interface represents an important source to the total DOM pool in estuaries (Alperin et al. 1992; Argyrou et al. 1997; Middelburg et al. 1997; Burdige 2001, 2002). It has been suggested that benthic fluxes are a significant source to coastal and estuarine systems (0.9×10^{14} g C year^{-1}) and may represent about the same magnitude as riverine inputs of DOC to the ocean ($2–2.3 \times 10^{14}$ C year^{-1}; Burdige et al. 1992; Burdige and Homstead 1994). Porewater DOM that gets entrained in the *benthic nepheloid layer*, from resuspension events of estuarine and shelf sediments, may also be important in affecting the composition and age of DOM in shelf and deeper slope waters (Guo and Santschi 2000; Mitra et al. 2000). Similarly, porewater DOM contributes to the protein or amino acid-like fluorescence signature of the water column and maximum concentrations of these signals have been observed at the sediment–water interface (Coble 1996).

3.3.2 Decomposition of Organic Detritus

Organic detritus has long been recognized for its importance as a food resource and its influence on overall biogeochemical cycling in coastal systems (Tenore et al. 1982; Rice 1982; Mann and Lazier 1991). Major contributors of organic detritus in estuarine systems are decaying plant materials and animal fecal pellets. The biosynthetic pathways of the compounds found in organic detritus can be divided into primary (involved directly with growth and metabolism) and secondary (end-products from primary metabolism not involved in key metabolic functions) metabolism (Figure 3.9). Vascular plant detritus is

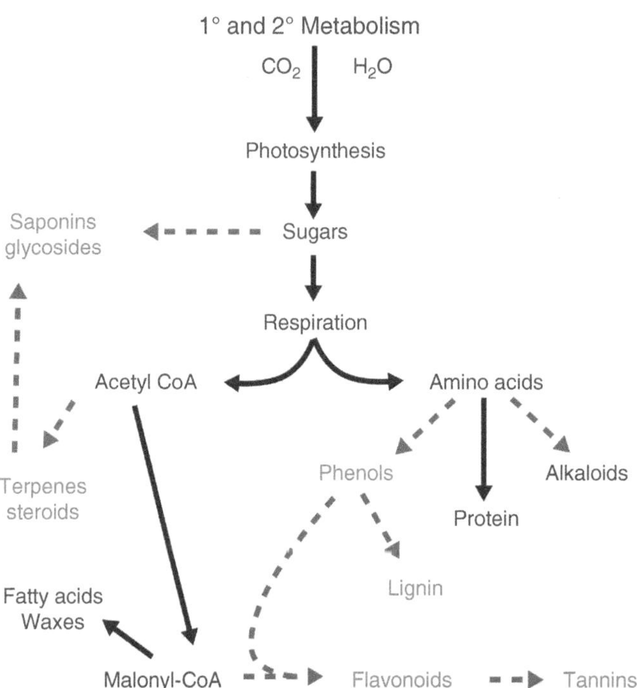

FIGURE 3.9 Pathways of biochemical processes, the sum of which is called metabolism. It is distinguished between primary and secondary metabolism. Primary metabolism (solid blue arrows) contains all pathways necessary to keep a cell alive while in secondary metabolism (dashed red arrows), compounds are produced and broken down that are essential for the entire organism.

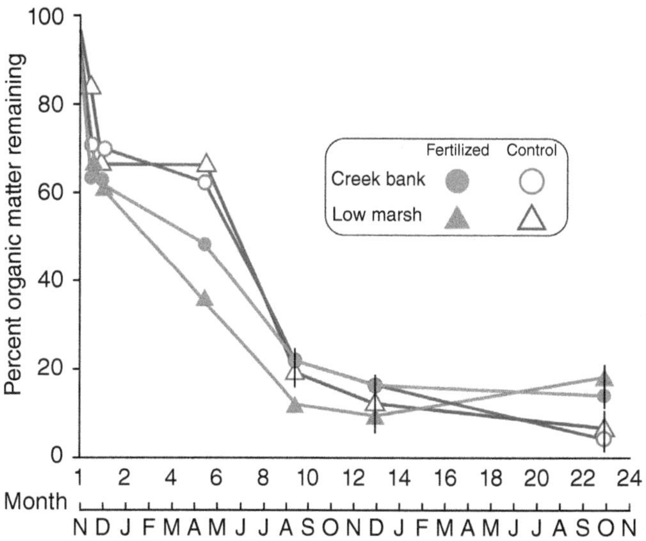

FIGURE 3.10 Percent organic matter remaining during a 23-month decay experiment with fertilized (solid icons) and control (open icons) *S. alterniflora* detritus from a creek bank (circles) and low marsh (triangles). *Source*: Modified from Wilson et al. (1986).

particularly important in many estuarine systems (Chapters 5–8); this refractory material typically requires the activity of microbial communities to convert lignocellulosic polymers into more labile available food resources for higher consumers (Moran and Hodson 1989a, b). The decay of aquatic organic detritus is generally divided into (1) *leaching*, (2) *decomposition*, and (3) *refractory* phases (Odum et al. 1973; Rice and Tenore 1981; Webster and Benfield 1986).

In the *leaching phase*, soluble compounds are rapidly lost from fresh detritus over a scale of minutes to weeks (Figure 3.10; Wilson et al. 1986). In the case of the marsh plant *S. alterniflora*, as much as 20–60% of the original material can be lost during this phase (Wilson et al. 1986). These soluble DOM compounds released from detrital particles are rapidly used by bacteria in the surrounding water column (Aneiso et al. 2003). Much of this leached material is likely comprised of short-chain carbohydrates, proteins and fatty acids (Dunstan et al. 1992; Harvey et al. 1995). There has also been considerable debate about the rate of decay of these labile leached substrates in oxic versus anoxic conditions, with some studies showing faster decay under oxic conditions (Bianchi et al. 1991; Sun et al. 1993; Harvey et al. 1995) and others showing no effects of redox (Henrichs and Reeburgh 1987; Andersen 1996).

The *decomposition phase* involves the heterotrophic breakdown of detritus by microbes and metazoans (e.g., *detritivores* and *deposit-feeders*). Nitrogenous compounds of micro- and macroalgae are more likely to contribute to the nutrition of detritivores than those of vascular plants (Findlay and Tenore 1982). As microorganisms colonize detritus during this phase, there is a relative increase in the nitrogen content of the "aging" detritus (Darnell 1967; Tenore et al. 1982; Rice and Hanson 1984). Earlier studies proposed that most of the nitrogen could be equated with "protein enrichment" from microbes (Newell 1965; Odum et al. 1973). However, further work showed that the presence of nonprotein nitrogen in plant detritus (Suberkropp et al. 1976) and enrichment from complexation (physical and chemical) in *humic geopolymers* (Hobbie and Lee 1980; Rice 1982; Rice and Hanson 1984) suggested that protein-N was not the only source. Adsorption of NH_4^+ by detritus may have also added to the total N (Mackin and Aller 1984). The attached bacteria are considered to be more active than the surrounding free bacteria (Griffith et al. 1994) and are critical in the breakdown of refractory compounds.

Vascular plants contain more ligneous and phenolic compounds, which can also reduce resource availability to consumers (Valiela et al. 1979; Rice and Tenore 1981; Bianchi 2007). *Secondary plant compounds* such as phenols, alkaloids, tannins, organic acids, saponins, terpenes, steroids, essential oils, and glycosides can deter herbivory and detritivory (Rietsma et al. 1982). For example, tannins are phenolic compounds known to inhibit microbial activity through the precipitation of enzyme proteins (Janzen 1974). Similarly, C_4 plants (e.g., S. *alterniflora, S. patens,* and *D. spicata*) produce cinnamic acids (e.g., ferulic and *p*-coumaric acids), which inhibit herbivory more effectively than comparable C_3 marsh plants (*Juncus* spp.; Haines and Montague 1979; Valiela 1995). The presence of these secondary compounds generally results in slower decay rates for vascular versus nonvascular plants (Valiela et al. 1985). The final *refractory phase* is characterized by detritus composed of lignin and cellulose, which decay very slowly (Maccubbin and Hodson 1980; Wilson et al. 1986). The source of detritus will clearly affect the time period of refractory decay.

Berner (1980) first introduced the following concept of a "one-G" model for determining first-order decay constants (*k*) of organic matter decomposition:

$$G_t = G_0 e^{-kt} \tag{3.13}$$

where G_t is the mass of detritus at time *t* and G_0 is the initial mass of detritus. Although this model appeared to work well in calculating decay constants for predominantly labile sources of organic matter such as macro- and microalgae, there were problems with more refractory sources of detritus such as *Spartina* (Rice and Hanson 1984). To better describe the decay dynamics of refractory detritus, which generally contain both labile and refractory biochemical components, the following "two-G" model was developed by Rice and Hanson (1984):

$$G_t = G^* + G_{10} e^{-k_1 t} \tag{3.14}$$

where G^* is the constant mass of refractory material, defined by $G^* = G - G_1$ (*G* is the total mass of the detritus and G_1 is the labile material), G_{10} is the initial mass of labile material, *and* k_1 is the decay constant of labile material.

In a decomposition experiment comparing decay dynamics of macroalgal (e.g., *Gracilaria foliifera*) and vascular plant (e.g., *Spartina alterniflora*) detritus, it was found that the two-G model proved more precise in calculating decay constants over shorter time intervals (days) compared to the one-G model, which works relatively well over weeks to months (Rice and Hanson 1984). Similarly, Westrich and Berner (1984) performed laboratory experiments to show that decomposition of phytodetritus in sediments can be separated into two decomposable fractions—with significant differences in reactivity, and a highly refractory (nonmetabolizable) fraction. This work established, for the first time, that organic matter decomposition in sediments can be described by a multi-G model.

3.3.3 Early Diagenesis

Most modern estuarine systems have been filling with sediments since their formation about 5000–6000 y B.P. (Bianchi 2007); this invokes long-term accumulation and storage of organic matter in these systems. The fate of SOM depends on the amount of early diagenesis that occurs in the upper sediments, which is largely controlled by the "quality" of organic detrital inputs and redox conditions of the sedimentary environment. Leeder (1982) defined *diagenesis* as "the many chemical and physical processes which act upon sediment grains in the subsurface"; which is further distinguished from *halmyrolysis*, defined as "a more restricted aspect of chemical changes operative at the sediment-water interface." Berner (1980) defined early diagenesis "as the changes occurring during burial to a few hundred meters where elevated temperatures are not encountered and uplift above sea level does not occur." This is the stage where key microbial and chemical transformations occur at low temperatures in recently deposited sediments, which has major effects on biogeochemical cycling in nearshore and estuarine environments.

Early diagenesis is typically described as a *steady-state phenomenon*; however, unless very long-term geological timescales are considered, steady-state conditions are generally not common in shallow turbid environments such as estuaries. There are many factors that contribute to these non-steady-state conditions, such as variations in sedimentation rate, inputs of organic matter, the chemistry of bottom waters and sediments, bioturbation rates, and resuspension (Lasagna and Holland 1976). Consequently, numerous attempts have been made to examine *non-steady-state diagenesis* over shorter time periods in estuarine systems (Mortimer et al. 1998; Deflandre et al. 2002).

As a result of particle settlement to the sediment–water interface, there is a mass accumulation of sediments, which results in compaction of sediments and the physical upward transport or *advection* of solutes in porewaters to the overlying water. Similarly, solutes in porewaters can also move by *diffusion* as a result of concentration gradients. *Biological pumping* or irrigation can also be an important transport mechanism, which is when benthic organisms pump fluids for respiration, feeding, and metabolite flushing, resulting in particulate and DOM fluxes (Aller 2001). Thus, porewaters can be transported by advection from burial, molecular diffusion, and biological pumping or irrigation (Aller 2001; Jørgensen and

Boudreau 2001). Diffusion in aqueous environments occurs according to *Fick's laws of diffusion* (Berner 1980).

Fick's First law, used for steady- and non-steady-state conditions, is as follows:

$$J_i = -D_i\left(\delta C_i / \delta x\right) \quad (3.15)$$

where J_i is the diffusional flux of component i in mass per unit area per time; D_i is the diffusion coefficient of component i in area per unit time; C_i is the concentration of component i in mass per unit area; x is the direction of maximum concentration gradient, negative sign indicates flux is in opposite direction of gradient. *Fick's Second law*, used for steady- and non-steady-state conditions, is as follows:

$$\delta c_i / \delta t = c_i\left(\delta^2 c_i\right) / \delta x \quad (3.16)$$

To apply Fick's laws to solute fluxes in sediments, these equations have to be adjusted to account for the negative interference effects that sediment particles have on the diffusion of solutes in pore waters (Lerman 1979; Berner 1980).

After making these adjustments for diffusion in sediments, the mass balance and vertical concentration patterns of nonconservative solutes in saturated sediments can be described by the following one-dimensional *advective-diffusive* general diagenetic equation (GDE; Berner 1980; Aller 2001; Jørgensen and Boudreau 2001):

$$\delta\varphi C / \delta t = \varphi D_s\left(\delta^2 c / \delta x^2\right) - \delta\omega_p C / \delta x - \Sigma R_i \quad (3.17)$$

where C is the concentration of solute; t is the time; D_s is the whole sediment diffusion coefficient of solute C; x is the sediment depth relative to surface ($x = 0$); ω_p is the pore water advection velocity relative to the sediment–water interface; ΣR_i is the sum of all reactions affecting solute C. These types of diagenetic models have been commonly used to describe the distribution of redox-sensitive metals (e.g., Mn and Fe) because of their close association with the mineralization of organic matter in sediments (Burdige 1993; Jørgensen and Boudreau 2001). In particular, a number of steady-state models have been used to describe the diagenesis of Mn and Fe (more details on this later in chapter) in sediments (Aller 1990; Boudreau 1996; Overnell 2002). In a study of sediments in the Loch Etive estuary (Scotland), diagenetic processes based on reactive Mn and Fe oxides as electron acceptors were compared using the diagenetic models of Van Cappellen and Wang (1996) and Slomp et al. (1997). Such comparative application of multiple diagenetic models is strongly encouraged (Overnell 2002). When examining biological mixing as a one-dimensional diffusive process, bioturbation and biodiffusion coefficients are used in the GDE similar to standard Fickian diffusivity (Wheatcroft et al. 1991). The justification for using a biodiffusion coefficient (D_B) is because it is assumed that all biological mixing activities are integrated over time—thereby making it a diffusion-like process.

3.3.4 Characterization of Organic Matter Using Biomarker Techniques

In general, estuarine organic matter is derived from a multitude of natural and anthropogenic allochthonous and autochthonous sources that originate across a freshwater to seawater continuum (Ward et al. 2017). Knowledge of the sources, reactivity, and fate of organic matter are critical in understanding the role of estuarine and coastal systems in global biogeochemical cycles (Hedges and Keil 1995; Bianchi and Canuel 2001; Bianchi and Canuel 2011). Due to a wide diversity of organic matter sources and the dynamic mixing that occurs in estuarine systems, it remains a significant challenge to determine the relative importance of these inputs to biogeochemical cycling in the water column and sediments. In recent years, there have been significant improvements in our ability to distinguish between organic matter sources in estuaries using tools such as elemental, isotopic (bulk and compound/class-specific), and chemical biomarker methods.

Bulk Organic Matter Techniques—The abundance and ratios of important elements in biological cycles (e.g., C, H, N, O, S, and P) provide the basic foundation of information on organic matter cycling. For example, concentrations of total organic carbon (TOC) provide the most important indicator of organic matter since approximately 50% of most organic matter is comprised of C. TOC in estuaries is derived from a broad spectrum of sources with very different structural properties and decay rates (Bianchi 2007; Bianchi and Bauer 2011; Bianchi and Canuel 2011). While this approach can provide essential information on organic matter dynamics, it lacks any specificity to source or age of the material.

When bulk C information is combined with additional elemental information, as in the case of the C to N (C:N) ratio, basic source information can be inferred about algal and terrestrial source materials (see review in Meyers 1997). The broad range of C:N ratios across divergent sources of organic matter in the biosphere demonstrate how such a ratio can provide an initial proxy for determining source information (Table 3.2). The basic reason for such differences in C:N ratios between vascular plants (>17) and microalgae (5–7) is that vascular plants are carbohydrate-rich (e.g., cellulose) and protein-poor, while microalgae are protein-rich and carbohydrate-poor. The most abundant carbohydrates supporting this high C content in vascular plants are structural polysaccharides such as cellulose, hemicellulose, and pectin (Aspinall 1970). However, C:N ratios can be very misleading in determining organic matter sources in the absence of additional source proxies. In some cases, selective utilization of N, due to N limitation in a system, can result in artificially high C:N ratios and misidentification of source materials. Finally, artifacts from the standard procedure of removing carbonate carbon when measuring TOC can also alter C:N ratios (Meyers 2003), but this artifact is relatively minor in sediments with more than 1% organic matter, such as estuarine sediments.

TABLE 3.2 Approximate carbon to nitrogen ratios in some terrestrial and marine producers.[a]

		C/N
Terrestrial	Leaves	35–100
	Wood	1000
Marine vascular plants	*Zostera marina*	17–70
	Spartina alterniflora	24–45
	Spartina patens	37–41
Marine macroalgae	Browns (*Fucus, Laminaria*)	30 (16–68)
	Greens	10–60
	Reds	20
Microalgae and microbes	Diatoms	6.5
	Greens	6
	Blue-greens	6.3
	Peridineans	11
	Bacteria	5.7
	Fungi	10

[a] Data compiled in Fenchel and Jorgensen (1977), Alexander (1977), Fenchel and Blackburn (1979), and data of I. Valiela and J.M. Teal.

Source: Modified from Valiela (1995).

TABLE 3.3 A summary of $\delta^{13}C$ and $\delta^{15}N$ values for various estuarine end members.

End member	$\delta^{13}C$	$\delta^{15}N$	Source
Terrigenous (vascular plant)	−26 to −30	−2 to +2	Fry and Sherr 1984, Deegan and Garritt 1997
Terrigenous soils (surface)/forest litter	−23 to −27	2.6–6.4	Cloern et al. 2002, Richter et al. 1999
Freshwater phytoplankton	−24 to −30	5–8	Sigleo and Macko 1985
Marine/estuarine phytoplankton	−18 to −24	6–9	Fry and Sherr 1984, Currin et al. 1995
C-4 salt marsh plants	−12 to −14	3–7	Fry and Sherr 1984, Currin et al. 1995
Benthic microalgae	−12 to −18	0–5	Currin et al. 1995
C-3 freshwater/brackish marsh plants	−23 to −26	3.5–5.5	Fry and Sherr 1984

3.3.4.1 Isotopic Mixing Models End-member mixing models are often used to evaluate sources of dissolved inorganic nutrients (C, N, S; Day et al. 1989; Fry 2002) and organic matter (POM and DOM) in estuaries (Raymond and Bauer 2001a, b; Gordon and Goñi 2003; McCallister et al. 2004; Bauer and Bianchi 2011; Bianchi and Bauer 2011). However, due to significant overlap in stable isotopic signatures it has proven difficult to discern multiple sources of dissolved and particulate constituents when using single and dual bulk isotopes in complex systems (Cloern et al. 2002). To address this problem, approaches were developed that couple end-member mixing models of multiple-isotopic tracers with chemical biomarker measurements (e.g., lignin-phenols and lipids) to determine carbon sources in coastal systems (Goñi et al. 1998; Raymond and Bauer 2001a, b; McCallister et al. 2004). The application of stable isotopes as tracers of organic matter sources in aquatic systems has been quite extensive (Lajtha and Michener 1994; Michener and Schell 1994), and the isotopic values for various organic matter sources are given in Table 3.3. Many of the early investigations were based on a two end-member mixing model (binary) in which the more depleted terrestrial $\delta^{13}C$ end-member could be used along with the more enriched marine phytoplankton end-member to establish their relative abundance throughout an estuary. For example, the following binary equation could be used to determine percent terrestrial organic matter in an estuary:

$$\%OC_{Terr} = \left(\delta^{13}C_{sample} - \delta^{13}C_{marine}\right) / \left(\delta^{13}C_{riverine} - \delta^{13}C_{marine}\right) \quad (3.18)$$

where $\delta^{13}C_{sample}$ is the isotopic composition of a sample, $\delta^{13}C_{marine}$ is a published isotopic value of marine phytoplankton (Table 3.3), and $\delta^{13}C_{riverine}$ is a published isotopic value of riverine POM (Table 3.3).

Studies in the Atchafalaya and Mississippi River estuaries compared the effectiveness of binary and three-end-member models for determining the relative abundance of marine versus terrestrial sources (Gordon and Goñi 2003; Bianchi et al. 2011) and concluded that it was important to use both biomarkers and stable isotopes for three organic carbon source end-members (soils and/or marsh, riverine, and marine) to separate terrestrial end-members into two different sources: vascular plants and soils.

Molecular Biomarkers—Due to the complexity of organic matter sources in estuaries and the aforementioned problems associated with bulk measurements, the application of chemical biomarkers has become widespread in estuarine research (Bianchi and Canuel 2001; Bianchi 2007; Bianchi and Canuel 2011, and references therein). The term "*biomarker molecule*" was defined by Meyers (2003) as "compounds that characterize certain biotic sources and that retain their source information after burial in sediments, even after some alteration." This molecular information is more specific and sensitive than bulk elemental and isotopic techniques in characterizing sources of organic matter, and further allows for identification of multiple sources (Meyers 1997, 2003).

3.4 Macronutrient Cycling

3.4.1 Sources of Nitrogen in Estuaries

Elemental nitrogen (N_2) makes up 80% of the atmosphere (by volume) and represents the dominant form of atmospheric nitrogen gas. Despite its high atmospheric abundance, N_2 is generally nonreactive, due to strong triple bonding between the N atoms, making much of the global N pool unavailable to organisms. In fact, only 2% of this N pool is believed to be available to organisms at any given time (Galloway 1998). Consequently, N_2 must be "fixed" into ionic forms such as NH_4^+ before it can be used by plants. Since N is essential for the synthesis of amino acids and proteins and because it is often in low concentrations, N is usually considered to be limiting to organismal growth in many ecosystems. Nitrogen has five valence electrons and can occur in a broad range of oxidation states that range from +V to −III, with NO_3^- and NH_4^+ being the most oxidized and reduced forms, respectively. Some of the most common N compounds that exist in nature, along with their boiling points, ΔH^0 and ΔG^0, are shown in Table 3.4 (Jaffe 2000); these thermodynamic data can be used to calculate equilibrium concentrations.

Nitrogen cycling in estuaries is affected by inputs of N from surface and groundwaters, atmospheric wet and dry fallout, as well as N recycling in both the water column and sediments (Figure 3.11; Paerl et al. 2002). Dominant inputs of N to estuaries are linked with freshwater inputs from rivers (Nixon et al. 1995, 1996; Boynton and Kemp 2000; Seitzinger et al. 2002; Bouwman et al. 2005). Many of these N inputs have increased in rivers and estuaries around the world as a direct result of expanded human impact on the landscape (Peierls et al. 1991; Howarth et al. 1996; Bouwman et al. 2005), and inputs of N to the Atlantic and Gulf coast U.S. estuaries are two to twenty times higher than in the preindustrialized periods (Howarth et al. 1996; Goolsby 2000).

TABLE 3.4 **Chemical data on important nitrogen compounds in the environment.**

Oxidation state	Compound	b.p. (°C)	$\Delta H^0(f)$	$\Delta G^0(f)$ (kJ mol^{-1}, 298 K)
+5	N_2O_5(g)	11	115	+5
+5	HNO_3(g)	83	−135	−75
+5	$Ca(NO_3)_2$(s)		−900	−720
+5	HNO_3(aq)		−200	−108
+4	NO_2(g)	21	33	51
+4	N_2O_4		9	98
+3	HNO_2(g)		−80	−46
+3	HNO_2(aq)		−120	−55
+2	NO(g)	−152	90	87
+1	N_2O(g)	−89	82	104
0	N_2(g)	−196	0	0
−3	NH_3(g)	−33	−46	−16.5
−3	NH_4(aq)		−72	−79
−3	NH_4Cl(s)		−201	−203
−3	CH_3NH_2(g)		−28	28
−3	H_2O(g)	100	−242	−229

Source: Modified from Jaffe (2000).

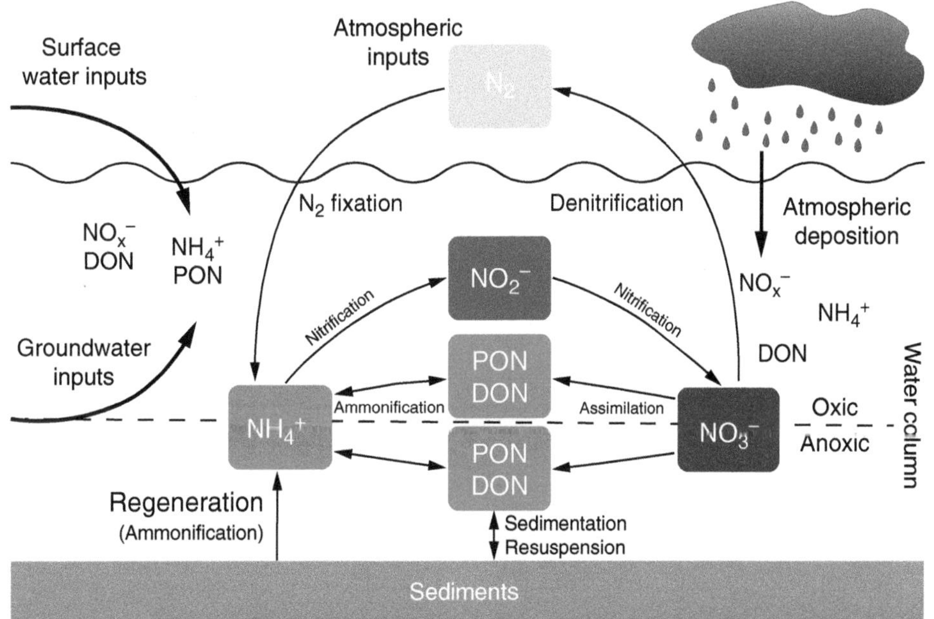

FIGURE 3.11 Schematic of nitrogen sources and cycling in estuaries. These sources include the atmosphere, and a diverse group of diffuse point and non-point sources (e.g., wastewater, industrial discharges, stormwater, and overflow discharges) across a broad spectrum of watersheds (e.g., urban, agricultural, upland, and lowland forests). *Source*: Modified from Paerl et al. (2002).

Nitrogen inputs to watersheds, which result in dissolved inorganic nitrogen (DIN) export to coastal systems, often lead to enhanced primary production, since many estuaries are N-limited (Nixon 1986, 1995; D'Elia et al. 1992; Howarth et al. 2000). This can result in the formation of harmful algal blooms (HABs; see Chapter 4) as well as *hypoxic* and *anoxic* water columns (Boynton et al. 1995; Paerl 1997; Richardson 1997). Submarine groundwater discharge (SGD) can also contribute N to estuaries. One such area is in northern Yucatan (Mexico), where concentrations of NO_3^- and NH_4^+ in SGD entering local lagoons range from 20 to 160 μM and 0.1 to 4 μM, respectively (Herrera-Silveira 1994). However, seasonal pulsing of SGD, driven by precipitation events, has not caused extensive eutrophication in these lagoons (Pennock et al. 1999). Waquoit Bay, Massachusetts, USA, is another well-studied system that receives groundwater N inputs that result in mixing zones with coupled N transformations, driven by both bacteria and archaea (Spiteri et al. 2008; Kroeger and Charette 2008; Rogers and Casciotti 2010).

Dissolved organic nitrogen (DON) can represent a significant fraction of the total dissolved nitrogen (TDN) in rivers and estuaries (Berman and Bronk 2003). Sources of DON to estuaries include actively growing phytoplankton communities, which release small molecules such as dissolved free amino acids (DFAA) that are highly bioavailable but represent only a small percent of the total DON (ca. < 10%; Diaz and Raimbault 2000; Bronk 2002). DON typically represents about 60–69% of the TDN in rivers and estuaries (Berman and Bronk 2003). The major components of DON include urea, dissolved combined amino acids (DCAA), DFAA, proteins, nucleic acids, amino sugars, and humic substances (Berman and Bronk 2003). However, less than 20% of DON is chemically characterized.

3.4.2 Transformations of Inorganic and Organic Nitrogen

The major processes involved in the biogeochemical cycling of N in estuaries and the coastal ocean are: (1) biological N_2 fixation (BNF); (2) ammonia assimilation; (3) nitrification; (4) assimilatory NO_3^- reduction; (5) ammonification or N remineralization; (6) anaerobic ammonium oxidation (i.e., anammox); (7) denitrification and dissimilatory NO_3^- reduction to NH_4^+; and (8) assimilation of DON (Figure 3.12; Libes 1992). Nitrogen transformations are driven by bacteria and archaea and biological nitrogen transformations are discussed in depth in Chapter 9. These transformations can result in either the addition (e.g., N_2-fixation) or loss of reactive nitrogen forms (e.g., denitrification, anammox) within estuaries.

Biological N_2-fixation (BNF) is the enzyme-catalyzed reduction of N_2 to NH_3, NH_4^+, or organic nitrogen compounds (Jaffe 2000). The high activation energy required to break N_2's triple bond makes BNF an energetically expensive process for organisms—thereby restricting BNF to select groups of organisms, called *diazotrophs*. BNF is an important process in aquatic systems (Howarth et al. 1988a, b) and is discussed further in Chapter 9.

Ammonia assimilation is where NH_4^+ is taken up and incorporated into organisms as organic nitrogen molecules (Jaffe 2000), as shown below for NH_4^+:

$$NH_4^+ \rightarrow organic \quad N \tag{3.19}$$

The ability to take up these reduced forms of N is a distinct energetic advantage for organisms because it is a direct means of getting a reduced form of N, unlike the uptake of NO_3^- which requires the extra step of NO_3^- reduction. However, some

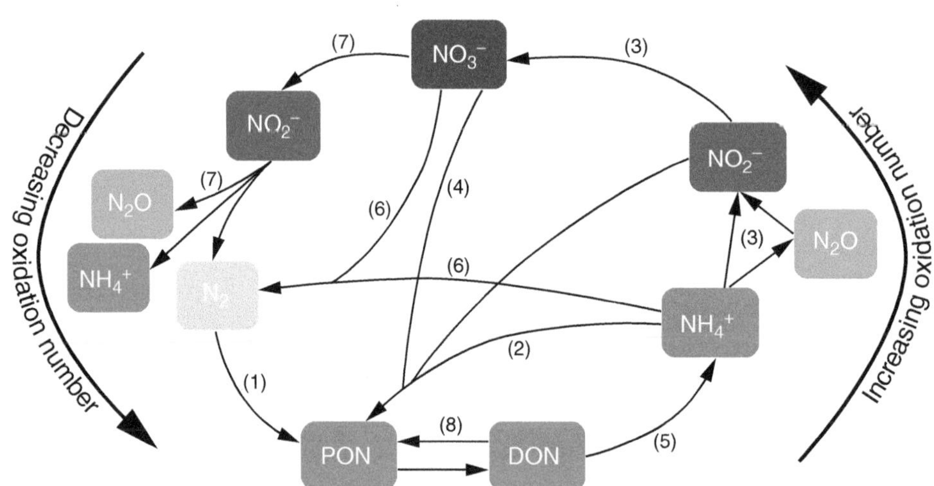

FIGURE 3.12 Major processes involved in the biogeochemical cycling of N in estuaries and the coastal ocean: (1) biological N_2-fixation; (2) ammonia assimilation; (3) nitrification; (4) assimilatory NO_3^- reduction; (5) ammonification or N remineralization; (6) anaerobic ammonium oxidation or anammox; (7) denitrification and dissimilatory NO_3^- reduction to NH_4^+; and (8) assimilation of dissolved organic nitrogen (DON). *Source:* Modified from Libes (1992).

phytoplankton prefer NO_3^- over NH_4^+ as a N source. In estuaries, NH_4^+ is generally the most dominant form of DIN in sediments, however, some pools of NH_4^+ are unavailable for uptake because they are bound within the internal matrix of minerals (Rosenfield 1979; Krom and Berner 1980).

Nitrification is where NH_4^+ is oxidized to NO_2^- or NO_3^- through the following two energy-producing reactions (Delwiche 1981):

$$NH_4^+ + 3/2(O_2) \rightarrow NO_2^- + H_2O + 2H^+, \Delta G^0 = -290 \text{ kJ mol}^{-1} \quad (3.20)$$

$$NO_2^- + 1/2(O_2) \rightarrow NO_3^-, \Delta G^0 = -89 \text{ kJ mol}^{-1} \quad (3.21)$$

Equation 3.20 involves the oxidation of NH_4^+ to NO_2^- and is primarily performed by bacteria of the genus *Nitrosomonas* (some species of the genus *Nitrocystis* sp. are also capable; Day et al. 1989), and Equation 3.21 is the continued oxidation of NO_2^- to NO_3^-, performed by bacteria of the genus *Nitrobacter spp.* More details about nitrifying organisms can be found in Chapter 9. Finally, nitrification requires O_2, thus the activity of these nitrifying bacteria in sediments is particularly sensitive to the availability of dissolved oxygen (Kemp et al. 1992, 1990).

Assimilatory NO_3^- *reduction to* NH_4^+ (ANRA) involves the simultaneous reduction of NO_3^- and uptake of N by the organism into biomass, as shown below:

$$NO_3^- + H^+ \rightarrow organic \quad N \quad (3.22)$$

This pathway is particularly dominant when reduced forms of N are low, such as oxic water columns in estuaries. This represents an important pathway for the uptake of N by estuarine organisms capable of uptake of both reduced and oxidized forms of N.

Ammonification is the process where NH_4^+ is produced during the breakdown of organic nitrogen by organisms (also called N remineralization). In the case of the most abundant nitrogen-containing organic compounds (proteins) peptide linkages are broken down followed by *deamination* of amino acids to produce NH_4^+. Ammonification is the conversion of DON to ammonium (DON → NH_4^+). This process occurs through the decay of plants and animals mediated by heterotrophic bacteria (Nowicki and Nixon 1985; Henriksen and Kemp 1988), or through excretion from animals (Bianchi and Rice 1988; Gardner et al. 1993).

Anaerobic ammonium oxidation (anammox), which involves the anaerobic NH_4^+ oxidation with NO_3^-, may be another mechanism whereby N is lost from estuarine/coastal systems. This process has been shown to represent from 20 to 67% of N_2 production in temperate continental shelf sediments (Thamdrup and Dalsgaard 2002); however, only limited knowledge currently exists on the pathways of anammox in natural systems. It does appear that the optimal conditions for anammox are different than for denitrification (Rysgaard and Glud 2004). Anammox is likely to occur in many estuarine systems because estuarine sediments are sites of intense organic matter remineralization, which can yield high concentrations of NO_3^- and NH_4^+ (Blackburn and Henriksen 1983; Nowicki and Nixon 1985) which are the required substrates for anammox. Other processes, such as Mn(II) oxidation by NO_3^- or Mn(IV) reduction by NH_4^+ (Luther et al. 1997), can also result in the evolution of N_2 in suboxic sediments.

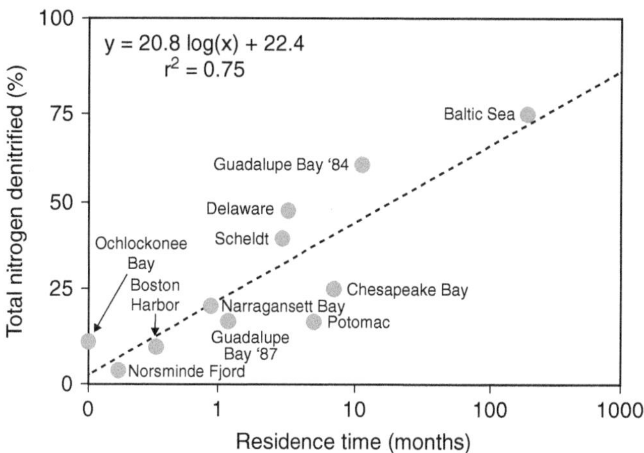

FIGURE 3.13 Fraction (%) of total N input from land and atmosphere that is denitrified in different estuaries as a function of residence time (months). *Source*: Modified from Nixon et al. (1996).

Denitrification, also known as *dissimilatory* NO_3^- or NO_2^- reduction, is the reduction of NO_3^- or NO_2^- to any gaseous form of N such as NO, N_2O, and N_2 by microorganisms, as in the case of N_2 production shown below:

$$CH_2O + NO_3^- + 2H^+ \rightarrow CO_2 + 1/2N_2 + 2N_2O \quad (3.23)$$

The denitrification pathway involves the production of intermediates such as NO_2^-, NO, and N_2O. Denitrification represents the major pathway from which N is lost from estuaries and has been shown to be proportional to the log mean water residence time ($r^2 = 0.75$), across a range of rivers and estuaries, which suggests that with increasing residence time, N will be more extensively recycled in the water column and sediments, thereby resulting in greater denitrification (Nixon et al. 1996; Figure 3.13). Although denitrification is largely driven by microorganisms, abiotic denitrification, or *chemodenitrification*, has been recognized as another mechanism of N_2O production (Samarkin et al. 2010; Zhu-Barker et al. 2015). Chemodenitrification involves the reduction of inorganic nitrogen species to gaseous nitrogen forms (NO or N_2O) and is coupled with the oxidation of iron (II) or humic acids.

In general, NO_3^- reduction produces N_2, which decreases the total availability of N in a system (Howarth et al. 1988a, b). However, if we take another pathway of *dissimilatory* NO_3^- *reduction to* NH_4^+ (DNRA), we find that N will be retained in a system, thereby increasing the total available N for organisms (Koike and Hattori 1978; Jørgensen 1989; Patrick et al. 1996). Although very little is known about the ecological consequences of DNRA (Sørensen 1987; Cornwell et al. 1999), rates of DNRA can be as high as those of denitrification in shallow estuaries and tidal flats (Bonin et al. 1998; Tobias et al. 2001). One study in Laguna Madre/Baffin Bay (USA) demonstrated that while sulfides can inhibit denitrification, they may stimulate DNRA by providing an electron donor, which likely results in the retention of available N in estuaries (An and Gardner 2002). Studies have reported the existence of chemolithotrophic bacteria that use sulfur compounds as an electron donor to convert NO_3^- to NH_4^+ (Schedel and Truper 1980).

Assimilation of DON involves the uptake and incorporation of organic forms of N, such as amino acids, into biomass. The DON pool represents an important component of the N cycle in estuarine systems (Sharp 1983; Jackson and Williams 1985), with a wide range of sources and sinks (Bronk and Ward 2000; Bronk 2002). Most research to date has focused on DIN, but we know that increased loading of DIN to estuaries is accompanied by increased DON (Correll and Ford 1982). In fact, rivers can have over 80% of their total N in the form of DON (Meybeck 1982; Seitzinger and Sanders 1997; Bronk 2002). The major "sinks" or pathways of DON consumption are mediated by bacteria (Anita et al. 1991; Bronk 2002), archaea (Ouverney and Fuhrman 2000), and, to a lesser degree, protists (Tranvik et al. 1993), although some phytoplankton can use DON compounds (see Chapter 4).

3.4.3 Sediment–Water Exchange of Dissolved Nitrogen

Estuarine and coastal sediments are important environments for the bacterial remineralization of nutrients (Rowe et al. 1975; Nixon 1981; Boynton et al. 1982; Blackburn and Henriksen 1983; Nowicki and Nixon 1985; Sundbäck et al. 1991; Warnken et al. 2000) that can subsequently support primary production via fluxes across the sediment–water interface (Cowan and Boynton 1996). The rates of remineralization and nutrient fluxes are typically highest with increasing water temperatures (Hargrave 1969; Kemp and Boynton 1984). Many other factors also contribute to variability of sediment–water exchange of nutrients such as the redox status of the sediments and overlying water column, sorption-desorption processes, microbial respiration, bioturbation, and macro-meiobenthic excretion (Kemp and Boynton 1981; Cowan and Boynton 1996).

The major pathways of the N cycle in sediments are strongly influenced by the redox conditions in bottom waters and sediments (Figure 3.14). Both diffusive and advective processes strongly control the distribution of O and N compounds, which ultimately affect the coupling between nitrification and denitrification (Jørgensen and Boudreau 2001). For example, anaerobic degradation of N-containing organic matter will contribute to the formation of NH_4^+ (ammonification; Figure 3.14 a, process i), which can then either efflux to the water column (Figure 3.14 a, process ii) or be oxidized to NO_2^- and NO_3^- (nitrification; Figure 3.14 a, process iii). Depending on the diffusive gradient, NO_3^- may efflux to the water column, or move downward where it can support denitrification just below the oxic–anoxic interface (Figure 3.14 a, process iv; Henriksen and Kemp 1988; Kristensen 1988; Rysgaard et al. 1994). Experimental loading of N in sediments showed that moderate N loading resulted in increased denitrification (due to associated enhancement of nitrification), but high N loading decreased rates of nitrification and denitrification (Sloth et al. 1995). The incorporation and/or loss of NO_3^- from sediments also depends on the presence and type of bioturbation (Aller 2001). Ammonium also accounts for most of the DIN secreted by invertebrates, as an end-product of protein catabolism (Le Bornge 1986).

Estuaries are active sites of nitrous oxide (N_2O) production (Bange et al. 1996; Seitzinger and Kroeze 1998; Usui et al. 2001; Bauza et al. 2002), which is one of the major greenhouse gases in the Earth's atmosphere (Wang et al. 1976; Khalil and Rasmussen 1992). In fact, the global warming potential of N_2O is 310 times higher than that of CO_2 (Houghton et al. 1995). Furthermore, N_2O also plays a role in stratospheric ozone depletion (Hahn and Crutzen 1982). Enhanced nutrient loading to estuaries stimulates microbial processes that produce N_2O (Seitzinger and Nixon 1985; Seitzinger 1988; Seitzinger and Kroeze 1998; Usui et al. 2001). Nitrification and denitrification

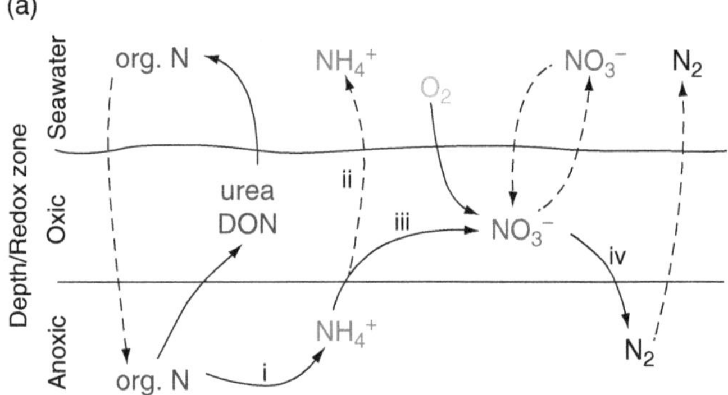

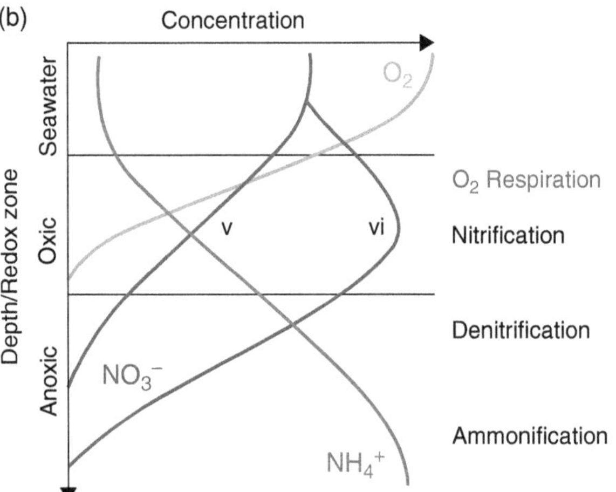

FIGURE 3.14 Major pathways of the N cycle in sediments (a), and as a function of redox conditions in bottom waters and sediments (b). Both diffusive and advective processes strongly control the distribution of O and N compounds which ultimately affect the coupling between nitrification and denitrification. *Source*: Modified from Jørgensen and Boudreau (2001). (i) mineralization, (ii) ammonium release, (iii) nitrification, and (iv) denitrification. Dashed lines indicate diffusive and advective processes. With low nitrification, sediments can act as a NO_3^- sink (*v*), and with high nitrification, sediments can act as a NO_3^- source (*vi*).

are the primary processes that produce N_2O in natural systems (Yoshinari et al. 1997), and it has been well documented that estuarine sediments have high spatial and temporal variability in N_2O production (Middelburg et al. 1995; Usui et al. 2001).

3.4.4 Sources of Phosphorus to Estuaries

Phosphorus (P) is one of the best-studied nutrients in aquatic ecosystems because of its role in limiting primary production on ecological and geological timescales (van Cappellen and Ingall 1996). Other key linkages to biological systems include the role of P as an essential constituent of genetic material (RNA and DNA), cellular membranes (phospholipids), and energy transforming molecules (e.g., ATP, etc.). Consequently, P has received considerable attention in recent decades, with particular emphasis on source and sink terms in budgets (Froelich et al. 1982; Meybeck 1982; Sutula et al. 2004). Excessive loading of N to estuaries that are N-limited can result in P-limitation of primary production. In such cases, N:P ratios are expected to exceed the Redfield value of 16:1, indicating P limitation (discussed further in Chapter 4, Section 4.4.2).

The cycling and availability of P in estuaries is largely dependent on P speciation. Consequently, total P (TP) has traditionally been divided into *total dissolved P* (TDP) and *total particulate P* (TPP) fractions (Juday et al. 1927), which can further be divided into *dissolved* and *particulate organic P* (DOP and POP) and *dissolved* and *particulate inorganic P* (DIP and PIP) pools (Figure 3.15). Another defined fraction within the TP pool is *reactive phosphorus (RP)*, which has been used to describe the *potentially bioavailable P* (BAP; Duce et al. 1991;

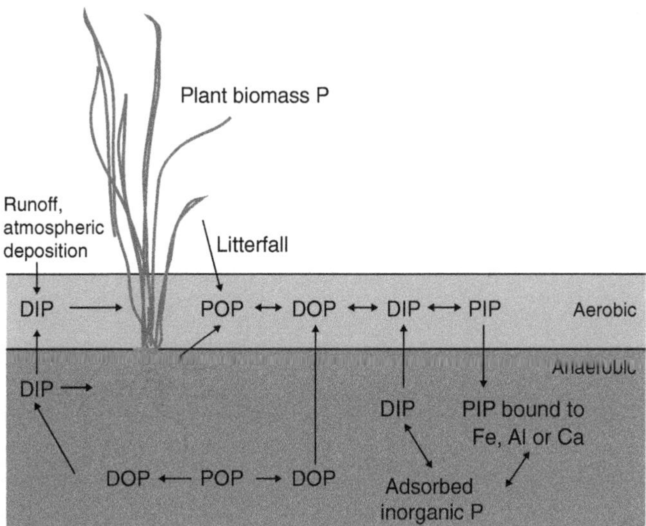

FIGURE 3.15 Phosphorus transformations in aerobic and anaerobic zones, showing the cycling of dissolved inorganic phosphorus (DIP), dissolved organic phosphorus (DOP), particulate inorganic phosphorus (PIP), and particulate organic phosphorus (POP). *Source*: Modified from DeLaune and Reddy (2008).

TABLE 3.5 Dissociation constants of phosphoric acid at 25 °C.

	Distilled water[a]	Seawater[b]
$H_3PO_4 \leftrightarrow H^+ + H_2PO_4^-$	2.2	1.6
$H_2PO_4^- \leftrightarrow H^+ + HPO_4^{2-}$	7.2	6.1
$HPO_4^{2-} \leftrightarrow H^+ + PO_4^{3-}$	12.3	8.6

[a] Stumm and Morgan (1996).
[b] Atlas (1975).

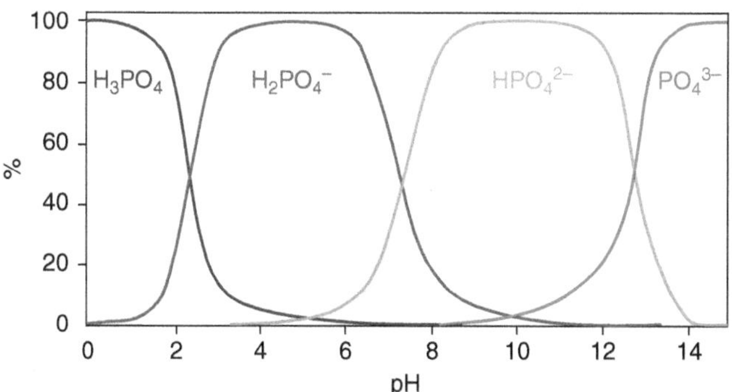

FIGURE 3.16 Relative distribution of phosphate species as a function of pH. $H_2PO_4^-$ and HPO_4^{2-} are the more common species in freshwater and seawater, respectively. *Source*: Modified from Reddy and Delaune (2008).

Delaney 1998). Much of the work to date has focused on the *soluble reactive P* (SRP), which is considered bioavailable and characterized as the P fraction which forms a phosphomolybdate complex under acidic conditions (Strickland and Parsons 1972). A significant fraction of the SRP is comprised of *orthophosphate* ($H_2PO_4^-$) and acid-labile organic compounds such as simple phosphate sugars (McKelvie et al. 1995). The DIP fraction is composed of phosphate (PO_43^-), phosphoric acid (HPO_4^{2-}), orthophosphate ($H_2PO_4^-$), and triprotic phosphoric acid (H_3PO_4); dissociation constants in freshwater and seawater are shown in Table 3.5 (Atlas 1975; Stumm and Morgan 1996).

The relative abundance of these species will vary with pH in aquatic systems making $H_2PO_4^-$ and HPO_4^{2-} the more common species in freshwater and seawater, respectively (Figure 3.16; Reddy and DeLaune 2008; Morel 1983). The difference between TDP and SRP provides an estimate of the DOP pool, which has more recently been referred to as *soluble nonreactive P* (SNP; Benitez-Nelson and Karl 2002). This can represent a much larger pool than SRP and can also be an important source of P to oceanic organisms (Benitez-Nelson and Karl 2002).

Using ^{31}P NMR spectroscopy, the dominant groups of P found in DOP in oceanic systems are phosphonates, phosphate monoesters, orthophosphate, phosphate diesters, pyrophosphates, tri- and tetra polyphosphates (Clark et al. 1998; Kolowith et al. 2001), and it is becoming increasingly evident that organic P plays an important role in aquatic systems (Turner et al. 2006; Baldwin 2013). It should be noted that phosphonates are a group of compounds that have a C-P bond, often associated with phosphoproteins (Quin 1967) and phospholipids (Hori et al. 1984).

Rivers are the major source of P to the ocean, via estuaries, where major chemical and biological transformations of P occur before it is delivered to the ocean (Froelich et al. 1982; Conley et al. 1995). The major source of P to rivers is from weathering of rock materials, and this is the major pathway from which P is lost from terrestrial systems (Jahnke 2000). Phosphorus is the tenth most abundant element on Earth with an average crustal abundance of 0.1% (Jahnke 2000). Apatite is the most abundant phosphate mineral in the Earth's crust, representing more than 95% of all the crustal P. Thus, the yield of P from weathering processes will vary depending upon the rock type. For example, P is generally low in granites (e.g., 0.13–0.27%), higher in shales (0.15–0.40%), and highest in basalts (0.40–0.80%; Kornitnig 1978). The fact that the uptake of P into organic matter per year is greater than the amount lost by land or supplied by rivers emphasizes the importance of P recycling in natural systems (Berner and Berner 1996), and the importance of organic P forms in aquatic systems.

Inputs of atmospheric sources of P are generally considered to be insignificant to coastal systems. In fact, they represent <10% of the riverine flux of reactive P (Duce et al. 1991; Delaney 1998). Only in highly oligotrophic (low nutrient) systems, such as oceanic gyres and the eastern Mediterranean (Krom et al. 1991, 1992), can such inputs have a significant impact on primary production. In fact, atmospheric deposition of DIP may account for as much as 38% of the new production during summer and spring in the Levantine Basin, eastern Mediterranean (Markaki et al. 2003). While gaseous forms of N and S are well studied in estuarine systems, the ecological importance of the gaseous form of P, phosphine, is not fully understood (Han et al. 2010; Li et al. 2010).

3.4.5 Phosphorus Fluxes Across the Sediment–Water Interface

The release of P from estuarine sediments is a common and important process that varies spatially and temporally. Early studies showed, using *in situ* benthic flux chambers, that P fluxes ranged from 30 to 230 mg P m^{-2} day^{-1} in estuaries such as Narragansett Bay (Elderfield et al. 1981a, b), the Potomac River estuary (Callender and Hammond 1982), San Francisco Bay (Hammond et al. 1985), and Guadalupe Bay (Montagna 1989). One key factor controlling the release of P from sediments is temperature. In many temperate systems, for example, much of the regeneration of P in sediments occurs via microbial processes, which are typically highest in summer months. A second key factor is salinity, which tends to be associated with reduced P release from sediments. Spatial variability in the extent of subtidal and intertidal areas may also have an impact on the spatio-temporal variability in P concentrations.

Numerous studies in freshwater (Roden and Edmunds 1997; Hupfer et al. 2004) and marine (Sundby et al. 1992; Gunnars and Blomqvist 1997; Sutula et al. 2004) systems have examined mechanisms controlling the release and efflux of P from sediments. Since total sediment P does not reflect the exchange capacity or bioavailability of P, sequential chemical extraction techniques have been useful in separating out different pools of P (Ruttenberg 1992; Jensen and Thamdrup 1993). Sedimentary pools of P have generally been divided into the following fractions: (1) organic P, (2) Fe-bound P, (3) *authigenic P minerals* (e.g., carbonate fluoroapatite (CFA), struvite, and vivianite), and (4) *detrital P minerals* (e.g., feldspar; see Ruttenberg and Berner 1993). More specifically, it is the organic P (Ingall and Jahnke 1997) and Fe-bound P (Krom and Berner 1981) fractions that are considered to be the most reactive, and most likely to be released from sediments during P regeneration. While it is generally accepted that much of the inorganic P is bound with Fe and Ca in sediments, the cycling of organic phosphorus is less clear. Some of the organic P components occur as phytic acid, phosphomonoesters, phosphodiesters, nucleic acids, and humic substances (Ogram et al. 1978; de Groot 1990). In the case of organic P release, PO_4^{3-} is remineralized during diagenesis of organic matter. In the Fe-bound fraction of P, release of P can occur when Fe-oxides that bind P are reduced as sediments become anaerobic (McManus et al. 1997). For example, experimental work has shown that this results in release of P from estuarine and freshwater sediments following the reduction of FeOOH (Gunnars and Blomqvist 1997). Dissolved inorganic phosphorus in bottom waters was shown to be negatively correlated with oxygen concentration, indicating the release of DIP from sediments in the Baltic Sea (Figure 3.17; Conley et al. 2002).

When compared to freshwater systems, interactions between Fe and S (Canfield 1989; Kostka and Luther 1995) in marine and estuarine systems can further complicate the mechanisms controlling P release. Returning to the experimental work by Gunnars and Blomqvist (1997), which indicated that the reduction of FeOOH was most important in controlling the release and efflux of PO_4^{3-} across the sediment–water interface, it was also found that the dissolved Fe:P ratio of the efflux was equal to 1 in freshwater sediments but less than 1 in marine

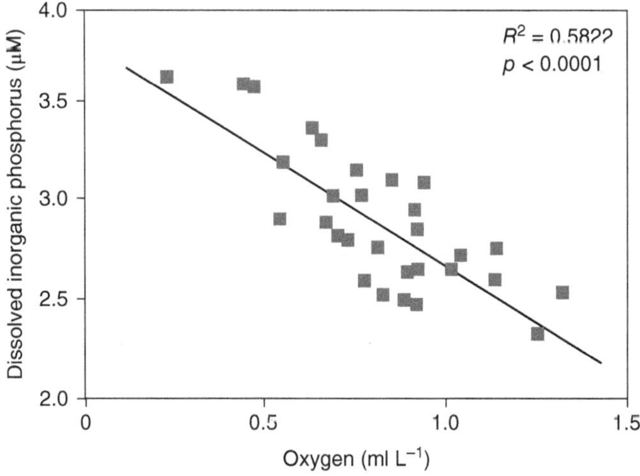

FIGURE 3.17 Negative correlation between dissolved inorganic phosphorus (µM) and oxygen (mL L^{-1}) concentrations in bottom waters of the Baltic Sea. *Source*: Modified from Conley et al. (2002).

sediments. It was further shown that carbon-normalized P remobilization from sediments was approximately five-fold higher in marine systems with higher SO_4^{2-} concentrations than in freshwater systems with low SO_4^{2-} concentrations (Caraco et al. 1990). These differences in ratios are likely explained by quick scavenging of Fe^{2+} by Fe-sulfides in marine and estuarine systems, which can result in the precipitation of FeS and FeS_2 (pyrite; Taillefert et al. 2000). The interactions between P, Fe, and S cycles shown here further highlight how biogeochemical cycles can be linked within estuaries.

3.4.6 Cycling of Inorganic and Organic Phosphorus

In both freshwater and estuarine systems, concentrations of DIP have been strongly linked with the suspended sediment load. In fact, a stable or "equilibrium" concentration range of DIP between 0.5 and 2 μM has been reported for a number of estuarine systems (Liss 1976; Froelich 1988; Ormaza-Gonzalez and Statham 1991). These stable DIP concentrations are believed to be controlled by a "buffering" of DIP through the adsorption and desorption onto metal oxide surfaces (Mortimer 1941; Carritt and Goodgal 1954; Stirling and Wormald 1977). This "P-buffering" is believed to balance the low availability of SRP in higher salinity waters, which occurs from phytoplankton uptake and anionic competition for surface adsorption sites (Froelich 1988; Fox 1989). For example, TPP concentration decreases with increasing salinity in the Delaware Bay (Lebo 1991) and Sheldt (The Netherlands) estuaries, suggesting the importance of desorption of DIP from aluminum and iron oxides with salinity. Finally, in the high salinity reaches of estuaries, calcite can serve as a carrier phase for adsorbed P (de Jonge and Villerius 1989). However, other work has shown that inorganic exchange processes were not able to "buffer" DIP concentrations across different regions of Chesapeake Bay (Conley et al. 1995). When examining the ratio of Fe:P in citrate–dithionate–bicarbonate (CDB-P) extracts in surface and bottom samples of suspended particles, there is a clear decrease in the ratio with increasing salinity (Conley et al. 1995). This decrease in CBD-Fe:P has also been observed in the St. Lawrence estuary (Canada) (Lucotte and d'Anglejan 1993).

Another geochemical/physical mechanism controlling P concentrations in estuarine waters may involve particle sorting and/or particle colloid interactions. For example, negative correlations were found between particulate P and SPM and the partition coefficient (K_d) for orthophosphate (K_d = [TPP $[mg\,kg^{-1}]$]/[orthophosphate $[mg\,L^{-1}]$]) and (SPM) in the Galveston estuary (USA) (Santschi 1995). Much of the total P in this system is composed of orthophosphate as indicated by the strong positive relationship between orthophosphate and TP. The negative correlation between K_d and SPM is commonly referred to as the "particle concentration effect," an effect well supported by radionuclide and trace metal work (Honeyman and Santschi 1988; Baskaran et al. 1992; Benoit et al. 1994). This effect occurs when a fraction of the P, or trace elements and radionuclides, is associated with the colloidal fraction, which is less than 0.45 μm and greater than 1 kD (Benoit et al. 1994). In fact, the colloidal P represented 30–80% of the filter-passing organic P concentration, an amount significant enough to account for such an effect, and further indicating the importance of colloids within estuarine biogeochemistry.

Although the characterization of DOP in rivers and estuaries has not been as thorough as the characterization of DIP, some work in the Mississippi River indicated that the composition of soluble nonreactive P primarily consisted of diester and monoester phosphate, phosphonates, orthophosphates, and/or tri-tetrapolyphosphates (Nanny and Minear 1997)—essentially the same as that found in the ocean (Kolowith et al. 2001). Phosphonates and refractory P esters are also abundant in marine sediments and waters and may represent a significant sink for organic P (Ingall et al. 1990). Conversely, other work has shown that phosphonates may actually be preferentially removed relative to other bioavailable P esters in anoxic waters (Benitez-Nelson et al. 2004). From the perspective of soil inputs at the river end-member of estuaries, 90% of the total organic P in some soils was represented in the form of the monoester phosphate fraction (Condron et al. 1985), with phosphonates also being present (Hawkes et al. 1984). These comparisons of possible oceanic and terrigeneous end-member inputs to estuaries provide at least some insight on the potential composition of DOP in estuaries.

3.4.7 Sources of Silica to Estuaries

Although silicon (Si) is the second most abundant element in the Earth's crust, it has relatively limited importance to biogeochemical cycles (Conley 2002; Ragueneau et al. 2005a, b). Much of the work to date has focused on the weathering of Si (Wollast and Mackenzie 1983), and the oceanic Si cycle (DeMaster 1981; Tréguer et al. 1995). Only recently has the cycling of Si in terrestrial ecosystems been shown to be important on a global scale (Figure 3.18; Conley 2002). The majority of inputs to the oceans occur via rivers (80%), with much of the losses controlled by sedimentation of biogenic silica or opal (Tréguer et al. 1995). The average global concentration of dissolved SiO_2 (DSi) in rivers is 150 μm L^{-1} (Conley 1997); the majority of this DSi is in the form of silicic acid (H_4SiO_4) in rivers, which typically have a pH in the range of 7.3–8.0.

The primary source of DSi (80%) to the global ocean, via estuaries, is riverine. However, anthropogenic alterations have begun to change to abundance and sources of riverine Si (Tréguer et al. 1995; Conley 2002). For example, decreases in the delivery of DSi loading of the Mississippi River to coastal waters were attributed to enhanced N loading to the river, which is believed to have increased diatom production and sedimentation in the watershed (Turner and Rabalais 1991; Turner et al. 2003). However, increases in the number of dams can also account for such decreases (Conley 1997). Decreases in DSi in the Danube River (Humborg et al. 1997) and the Swedish rivers (Humborg et al. 2000, 2002) have also been attributed to enhanced uptake of DSi from enhanced diatom production.

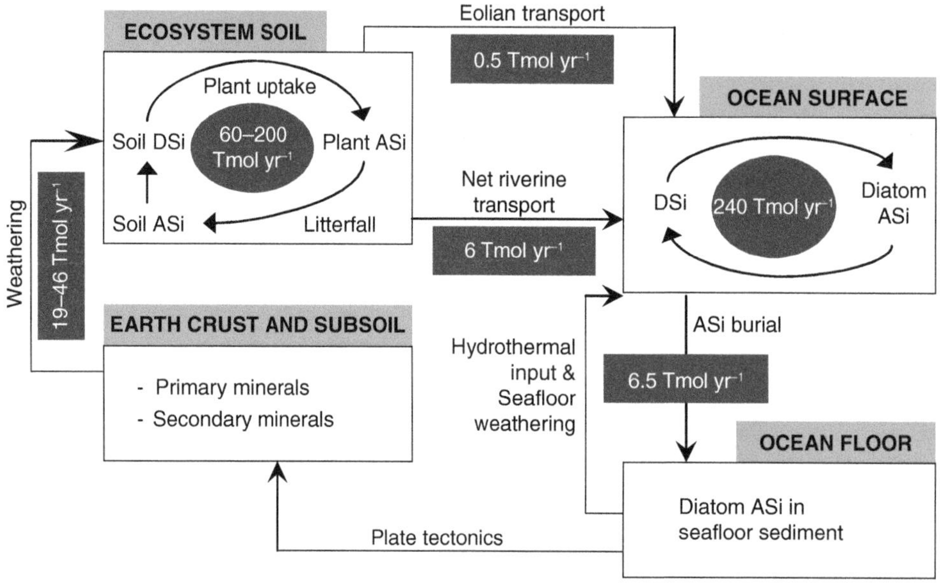

FIGURE 3.18 Global biogeochemical cycling of silicon. *Source*: Struyf et al. (2009)/Springer Nature.

Other less important sources of DSi to estuaries and the coastal ocean include submarine groundwater discharge (SGD), atmospheric inputs, and in the case of estuaries, import of DSi from coastal upwelling (Ragueneau et al. 2005a, b). For SGD, current estimates of the global flux range from 0.01 to 10% of surface runoff (Taniguchi et al. 2002). More work is needed on DSi in groundwaters before any conclusive statements can be made about its effects on estuarine biogeochemistry. As for atmospheric inputs of Si, fluxes to the global ocean have been estimated to be approximately 0.5 Tmol Si $year^{-1}$ (Duce et al. 1991), although these inputs only account for 0.2% of the annual oceanic gross biogenic Si production and 10% of the river inputs.

3.4.8 Cycling of Silica

The cycling of DSi in estuaries is a function of river inputs and biological uptake. In Chesapeake Bay, river input is the main source of DSi in winter and spring (Figure 3.19; Conley and Malone 1992), when the concentration of DSi decreases from the head to the mouth of the estuary. This DSi supports spring diatom blooms that reduce the concentration of DSi (Figure 3.19). Then in summer, regeneration of DSi via dissolution of diatom frustules in sediments becomes an important source of Si to the water column (Kamatani 1982). In fact, much of the diatom demand for DSi in summer can be supported by flux of this regenerated Si (D'Elia et al. 1983). From a spatial perspective, the highest DSi concentrations occurred in bottom waters in the mesohaline and lower bay regions—where regeneration rates were the highest. However, maximum uptake rates of DSi by diatoms also occurred in the mesohaline region in spring—resulting in significant DSi limitation (Figure 3.19; Conley and Malone 1992).

3.4.9 Sources of Sulfur to Estuaries

Sulfur (S) is an important redox element in estuaries because of its linkage with biogeochemical processes such as SO_4^{2-} reduction (Howarth and Teal 1979; Luther et al. 1986; Roden and Tuttle 1993a, b), pyrite (FeS_2) formation (Giblin 1988; Hsieh and Yang 1997), metal cycling (Tang et al. 2000), ecosystem energetics (King et al. 1982; Howes et al. 1984), and atmospheric S emissions (Dacey et al. 1987; Simo et al. 1997). The range of oxidation states for S intermediates formed in each of these processes is between +VI and −II, and the main sources of S to sediments are SO_4^{2-} and detrital sulfur inputs to the sediment–water interface.

On a global scale, most of the S is located in the lithosphere; however, there are important interactions between the hydrosphere, biosphere, and atmosphere where important transfers of S occur (Charlson 2000). For example, coal and biomass burning, along with volcano emissions inject SO_2 into the atmosphere, which can then be further oxidized in the atmosphere and removed as SO_4^{2-} in rainwater (Galloway 1985). An example of biogenic sulfur formation is the reduction of seawater SO_4^{2-} to sulfide by phytoplankton and eventual incorporation of the S into dimethylsulfoniopropionate (DMSP). DMSP, in turn, is converted to volatile dimethylsulfide (DMS; CH_3SCH_3,) which is emitted to the atmosphere. In seawater, SO_4^{2-} represents one of the major ions, with concentrations that range from 24 to 28 mM, which is considerably higher than the concentrations found in freshwaters (~0.1 mM). This marked difference makes seawater the major input to estuaries and sets up an important gradient in estuarine biogeochemical cycling.

Approximately 50% of the global flux of S to the atmosphere is derived from marine emissions of DMS. Although there have only been a few studies to date, it has been suggested that

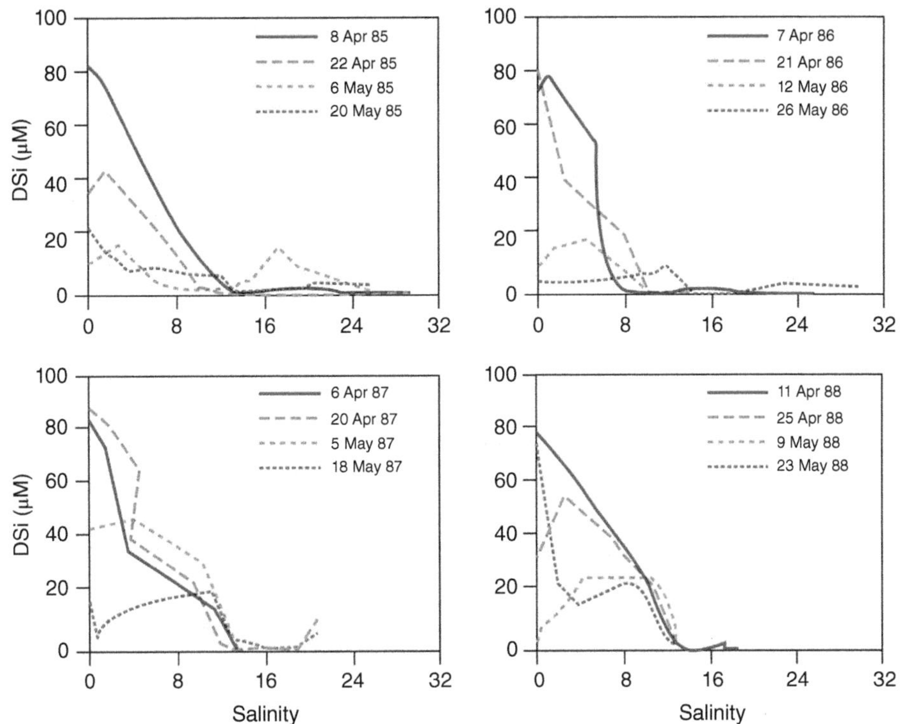

FIGURE 3.19 Temporal variability of concentrations of DSi (μM) in Chesapeake Bay across a salinity gradient. *Source*: Conley and Malone (1992)/Inter-Research Science Publisher.

coastal plumes (Turner et al. 1996; Simo et al. 1997) and estuaries (Iverson et al. 1989; Cerqueira and Pio 1999) may be important atmospheric sources of DMS. DMS, a compound produced by certain phytoplankton, has possible implications for climate control once released into the atmosphere (Charlson et al. 1987). DMS is formed by cleavage of dimethylsulfoniopropionate (DMSP; Kiene 1990). In fact, DMSP may provide as much as 100% of the sulfur, and 3.4 % of the carbon required for bacterial growth in oceanic waters (Kiene and Linn 2000). Other sulfur compounds such as carbonyl sulfide (OCS) and carbon disulfide (CS_2) are also possible sources of S in estuaries. Oxidation of DMS in the atmosphere leads to production of SO_4^{2-} aerosols, which can influence global climate patterns (Charlson et al. 1987; Andreae and Crutzen 1997). The key processes controlling DMS emissions from the euphotic zone in the ocean are bacterial metabolism, water column mixing, and photochemistry (Kieber et al. 1996; Kiene and Linn 2000). Bacterial mechanisms driving DMS emissions are discussed further in Chapter 9.

3.4.10 Transformations of Inorganic and Organic Sulfur

Just like N, S undergoes a series of redox-sensitive transformations within estuaries. These transformations include: (1) assimilatory sulfate reduction; (2) sulfur oxidation; and (3) dissimilatory sulfate reduction. These processes are microbially mediated and are discussed in further detail in Chapter 9.

Assimilatory sulfate reduction is the process through which microorganisms take up sulfate (SO_4^{2-}) and assimilate it into amino acids (see Chapter 9).

Anerobic metabolism represents a significant pathway for carbon and sulfur cycling in estuarine sediments (Jørgensen 1977, 1982; Crill and Martens 1987; Roden and Tuttle 1992). In particular, *dissimilatory sulfate (SO_4^{2-}) reduction* is the process through which microorganisms use SO_4^{2-} as a terminal electron acceptor to oxidize organic matter, leading to the formation of hydrogen sulfide (H_2S; Capone and Kiene 1988), which gives salt marsh soils their distinctive "rotten egg" smell. Sulfate reduction is particularly important in the S and C chemistry of highly productive shallow-water subtidal and salt marsh environments (Howarth and Teal 1979; King et al. 1985; Kostka and Luther 1994). Sulfate reduction is driven by sulfate reducing bacteria (SRB), which are described in more detail in Chapter 9 (Rooney-Varga et al. 1997; King et al. 2000). Some of these SRB are also important in methyl mercury (CH_3Hg) formation in contaminated sediments, whereby methylation occurs during the respiration processes of SRB (King et al. 2000; Gilmour et al. 2013). Organic matter decomposition via sulfate reduction can be represented by the following equation (Richards 1965; Lord and Church 1983):

$$\begin{aligned}&2\left[(CH_2O)_c(NH_3)_n(H_3PO_4)_p\right]+c\left(SO_4^{2-}\right)\\&\rightarrow 2c\left(HCO^{3-}\right)+2n\left(NH_3\right)+2p\left(H_3PO_4\right)+c\left(H_2S\right)\end{aligned} \tag{3.24}$$

where *c*, *n*, and *p* represent the C:N:P ratio of the decomposing organic matter.

Sulfide oxidation is another important microbially mediated process that results in the oxidation of inorganic sulfur compounds, such as H_2S. There are a suite of sulfur-oxidizing microorganisms within estuaries that are discussed in more detail in Chapter 9.

3.4.11 Sulfur at the Sediment–Water Interface

A significant fraction of the sulfides formed by sulfate reduction are reoxidized to SO_4^{2-} at the oxic–anoxic interface in sediments. Dissolved sulfides (DS = S^{2-} + HS^- + H_2S) formed in porewaters during sulfate reduction can diffuse into overlying bottom waters and contribute to O_2 depletion in estuaries (Tuttle et al. 1987). Sulfides can also be removed from porewaters via reactions with Fe oxyhydroxides to form pyrite (FeS_2) (Berner 1970, 1984). In addition to vertical molecular diffusion as a controlling factor in dissolved sulfur transport, gas bubble ebullition, derived from CH_4 production beneath the sulfate reduction zone, can be important in stripping dissolved sulfur from porewaters (Roden and Tuttle 1992). Although sulfate reduction has been viewed as the dominant pathway of organic matter oxidation in anaerobic salt marsh sediments (see reviews by Howarth 1993; Alongi 1998), other work has suggested that microbial Fe (III) reduction (FeR) is also key in controlling the oxidation of organic C in these sediments and is inherently linked with dissolved sulfide formed from sulfate reduction (Kostka and Luther 1995).

The flux of dissolved sulfur across the sediment–water interface can in some cases be strongly influenced by the presence of chemoautotrophic bacterial mats in estuaries. These sulfur oxidizing bacteria occur at the oxic–anoxic interface. The dominant colorless bacteria that live within the microzone of the O_2 and H_2S interface (ca. 1–2 mm) are *Beggiatoa* and *Thiovulum* spp. (Jørgensen and Revsbech 1983; Jørgensen and Des Marais 1986). This early work used microelectrodes to show that these bacteria live within the steep microgradient of O_2 and H_2S (usually between 0 and 0.5 mm thick; Figure 3.20; Jørgensen and Revsbech 1983). In the presence of O_2, these bacteria can oxidize H_2S to S^0, which can be further oxidized to SO_4^{2-} (Nelson and Castenholz 1981). These bacteria enzymatically oxidize H_2S in the presence of O_2, despite rapid abiotic oxidation of H_2S. The common pigmented sulfur bacteria are the green sulfur bacteria (GSB; e.g., *Prosthecochloris aestuarii*), which are obligate anerobes that use H_2S as the dominant electron acceptor (Massé et al. 2002) and are commonly found in occurrence with brown colored GSB (e.g., *Chlorobium vibriforme*) in brackish to hypersaline environments (Pfennig 1989).

Within sediments, the biogeochemical cycling of S and iron (Fe) are often coupled. In sediments, Fe sulfides are typically divided into the following two groups: *acid volatile sulfides* (AVS), which are evolved via acid distillation and generally include amorphous forms (e.g., mackinawite [FeS], greigite [Fe_3S_4], and pyrrhotite [FeS]; Morse and Cornwell 1987). In some cases, AVS may represent the dominant pool of sulfides in estuarine sediments (Oenema 1990). Pyrite (FeS_2) is the dominant Fe-sulfur mineral in most estuarine systems, particularly in salt marsh sediments (Hsieh and Yang 1997). In sediments

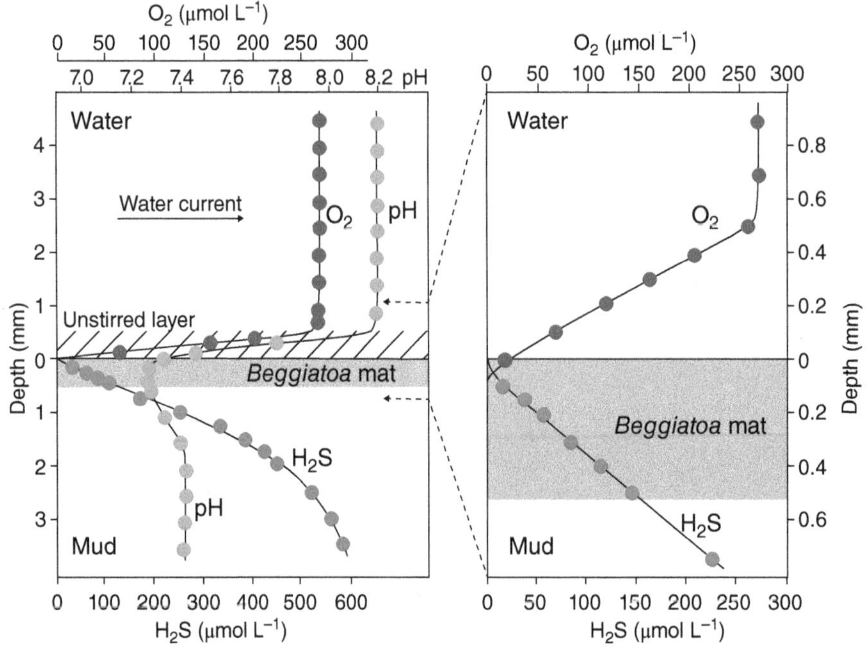

FIGURE 3.20 Depth profile of the microzone of O_2 and H_2S at the sediment-water interface (c. 1–2 mm) where *Beggiatoa* spp. and *Thiovulum* spp. thrive. *Source*: Modified from Jørgensen and Revsbech (1983).

with high concentrations of sulfide, reductive dissolution of Fe (III) oxyhydroxide phases results in the formation of FeS_{aq} and FeS_s as shown below:

$$Fe^{2+} + HS^- \rightarrow H^+ + FeS_{aq} \quad (3.25)$$

Pyritization can then occur under anoxic conditions relatively quickly by reaction of H_2S and S(0) (as S_8 or polysulfides $[S_x^{2-}]$) with FeS (Rickard and Luther 1997; Rozan et al. 2002) in the following equations:

$$FeS_{aq} + H_2S_{aq} \rightarrow FeS_2 + H_2 \quad (3.26)$$

$$FeS_{aq} + S_x^{2-} \text{ or } [S_8] \rightarrow FeS_2 + S^2_{(x-1)} \quad (3.27)$$

Pyrite is considered to be very stable under reducing conditions where it can be preserved over geological timescales, thereby retaining high amounts of energy (Howarth 1984), however it decomposes rapidly under oxidizing conditions. Although there can be considerable differences in the rates of pyritization of different precursor iron hydroxide minerals (Canfield et al. 1992; Raiswell and Canfield 1996), the actual effects of mineralogy appear to only be important in the initial iron sulfidization and not in pyritization rates (Canfield et al. 1992; Morse and Wang 1997).

The degree of pyritization (DP) is a parameter first used by Berner (1970) to distinguish environments where FeS_2 is Fe or carbon limited (Raiswell and Berner 1985). It is a measure of the extent to which the original reducible or reactive Fe has been converted to FeS_2. The original equation defining DP has been modified in recent years because of the operational definition of reactive non-sulfidic Fe. The following equation incorporates some of these modifications and was used by Rozan et al. (2002):

$$DP = [FeS_2] / ([FeS_2] + [AVS - Fe] + [Dithionite - Fe]) \quad (3.28)$$

where FeS_2 is the concentration of reduced S as pyrite; AVS is the acid volatile sulfur composed of both aqueous and solid FeS; and Dithionite—Fe is a measure of non-sulfidic Fe.

The emergence of dissolved sulfur from sediments into stratified bottom waters can contribute significantly to bottom water oxygen depletion in estuaries (Tuttle et al. 1987; Roden and Tuttle 1992). Hydrogen sulfide's toxicity stems from its ability to combine with the Fe-heme of blood cells, thereby replacing O_2 and inhibiting respiration (Smith et al. 1977). The persistence and concentrations of H_2S in the bottom waters of estuaries are controlled by many factors such as the rate of sulfate reduction, which is affected by organic matter/nutrient loading, temperature, and parameters controlling the flux of H_2S across the sediment–water interface and the transfer to surface waters (e.g., diffusive/advective processes and S-oxidizing bacteria).

3.4.12 Carbon Cycling in Estuaries

As discussed earlier, water-to-air fluxes of carbon dioxide are significant in estuaries (Frankignoulle and Borges 2001; Wang and Cai 2004; Bianchi 2007; Bauer and Bianchi 2011; Bianchi and Bauer 2011). In the Satilla and Altamaha estuaries in Georgia, USA, calculated values of the partial pressure of CO_2 (pCO_2), based on dissolved inorganic carbon (DIC) and pH data, showed the highest pCO_2 (1000 to > 6000 μatm) at the lowest salinities (<10) (Cai and Wang 1998). Corresponding CO_2 water-to-air fluxes in these low salinity waters were also high, ranging from 20 to >250 mol m^{-2} year^{-1}. The high pCO_2 and CO_2 water-to-air fluxes in these systems were attributed to inputs of CO_2 from organic carbon respiration in tidally flooded salt marshes and from groundwater (Cai and Wang 1998). Groundwater entering the South Atlantic Bight (southeastern coast of the USA) was shown to have high pCO_2 (0.05–0.12 atm; Cai et al. 2003), which is consistent with other work suggesting that high pCO_2 in rivers is, in part, due to groundwater inputs (Kempe et al. 1991; Mook and Tan 1991). However, export of DIC from marshes to coastal waters of the South Atlantic Bight was found to rival that of riverine export (Wang and Cai 2004). This work further proposed that the pathway of CO_2 being fixed by marsh grasses and then exported to coastal waters in the form of organic and inorganic carbon can be described as a "marsh CO_2 pump." Other studies have shown that marsh-influenced estuaries are important sources of DIC to adjacent coastal waters (Raymond et al. 2000; Neubauer and Anderson 2003) and other studies have documented the marsh pump in mangrove systems as well (Bouillon et al. 2003). Average pCO_2 for other estuarine systems in Europe and the United States indicate that estuaries are net sources of CO_2 to the atmosphere (Table 3.6). Highly dynamic regions in estuaries, such as the ETM, can be particularly important as net sources of CO_2 (Abril et al. 1999). While estuarine plumes can be net sinks of CO_2, particularly in outer-plume regions (Frankignoulle and Borges 2001), overall estuarine systems are generally net sources (Borges and Frangnoulle 2002).

Methane is an important atmospheric greenhouse gas that has a mean atmospheric concentration of 1.7 ppm. While this is considerably lower than that of CO_2 (350 ppm), it has a greater *radiative forcing* capability (Cicerone and Oremland 1988). Despite the smaller global surface area of estuaries relative to the global ocean, the contribution to total global CH_4 emissions from estuaries (ca. 7.4%) is within the range found for oceanic environments (1–10 %; Bange et al. 1994). It has been estimated that estuarine water-to-air fluxes contribute 1.1–3.0 Tg CH_4 year^{-1} to the global budget (Middelburg et al. 2002). Tidal creeks and marshes are the dominant source of CH_4 in estuaries (Middelburg et al. 2002). Groundwater also tends to be highly enriched in CH_4 (Bugna et al. 1996) and is likely responsible for the general increase in riverine CH_4 with increasing river size (Jones and Mulholland 1998). Methane concentrations in rivers are typically one to two orders of magnitude higher than open ocean waters (Scranton and McShane 1991; Jones and Amador 1993; Middelburg et al. 2002). There exists a wide range of spatial and

TABLE 3.6 **Average pCO_2 ranges for various United States and European estuaries. The average range was obtained by averaging the low and high concentrations for each transect. Estuaries are ranked by the high average range.**

Estuary	Number of transects	Average pCO_2 range (ppmv)
Altamaha (USA)[a]	1	380–7800
Scheldt (Belgium/The Netherlands)[b]	10	496–6653
Sada (Portugal)[b]	1	575–5700
Satilla (USA)[a]	2	420–5475
Thames (UK)[b]	2	485–4900
Ems (Germany/The Netherlands)[b]	1	560–3755
Gironde (France)[b]	5	499–3536
Douro (Portugal)[b]	1	1330–2200
York (USA)[c]	12	352–1896
Tamar (UK)[b]	2	390–1825
Hudson (NY, USA)[d]	6	517–1795
Rhine (The Netherlands)[b]	3	563–1763
Rappahannock (USA)[c]	9	474–1613
James (USA)[c]	10	284–1361
Elbe (Germany)[b]	1	580–1100
Columbia (USA)[e]	1	590–950
Potomac (USA)[c]	12	646–878
	Average	531–3129

[a] From Cai and Wang (1998) and Cai et al. (1999).
[b] From Frankignoulle et al. (1998).
[c] From Raymond et al. (2000).
[d] From Raymond et al. (1997).
[e] From Park et al. (1969).

Source: Modified from Raymond et al. (2000).

temporal variability in methane concentrations in rivers and estuaries (De Angelis and Scranton 1993; Bianchi et al. 1996). Methane oxidation can be an important methane sink in estuaries as well, and is highly dependent on temperature and salinity—with lower oxidation rates at higher salinities (De Angelis and Scranton 1993; Pulliam 1993). In fact, turnover of the dissolved CH_4 pool can occur as fast as 1.4 to 9 days in the upper Hudson River estuary (USA) (De Angelis and Scranton 1993) and < 2 h to 1 day in the Ogeechee River (USA) (Pulliam 1993). In conclusion, there is considerable temporal and spatial variability in the sources and sinks of methane, water-to-air fluxes, as well as mechanisms of transport (e.g., *ebullition, diffusion,* and *plant-mediated*) in estuarine systems (Bianchi 2007; Bauer and Bianchi 2011; Bianchi and Bauer 2011, and references therein).

3.4.13 Transformations and Cycling of Dissolved and Particulate Organic Carbon (DOC and POC)

Sources of particulate organic carbon (POC) and dissolved organic carbon (DOC) in estuaries consist of a diverse mixture of allochthonous and autochthonous sources. Hydraulic residence time, river discharge, tidal exchange, and frequency of resuspension events are important physical controlling variables that determine the fate and reactivity of organic carbon in estuaries. Once again, in this section we provide only a very brief overview of POC and DOC cycling dynamics in the waters and sediments of estuaries, relying more on the background of organic carbon sources, chemical biomarkers, and bulk isotopic measurements (Bianchi 2007; Bauer and Bianchi 2011; Bianchi and Bauer 2011, and references therein).

Concentrations of POC in many estuaries are strongly coupled with suspended sediment particles, which may depend on river discharge and sediment resuspension. POC concentration generally increases with the concentration of suspended sediments, but the percent of the mass of suspended sediment that is composed of organic matter (% POC) tends to decrease at higher suspended sediment concentrations. For example, in the Sabine-Neches estuary (USA), there is a significant increase % POC when total suspended particulates are lower than 20–30 mg L^{-1} (Figure 3.21; Bianchi et al. 1997a). This relationship has been found in large river systems; however, because of the higher suspended loads in many rivers, the increase in % POC of TOC, typically occurs at less than 50 mg L^{-1} (Meybeck 1982).

This general pattern is attributed to a dilution effect of sediment load on % POC at high river discharge, and to an decrease in phytoplankton production during high TSP and low light availability. Seasonal ranges of POC concentrations

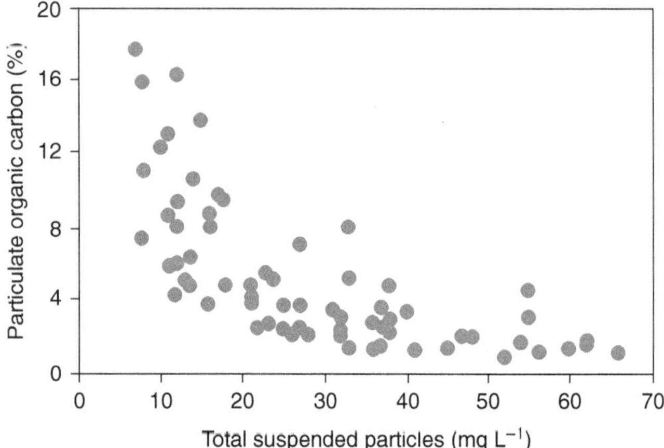

FIGURE 3.21 Percent particulate organic carbon (POC) vs. total suspended particle concentration (TSP) in the water column from three regions of the Sabine-Neches estuary, sampled from March 1992 to October 1993. *Source:* Bianchi et al. (1997)/Springer Nature.

in San Francisco Bay and Chesapeake Bay estuaries are typical of many estuaries and are not significantly different between systems (Canuel 2001). Seasonal and temporal differences in POC in both systems are generally controlled by river discharge and light availability. However, further analyses using fatty acid biomarkers revealed that phytoplankton represent a greater fraction of POC in Chesapeake Bay than in San Francisco Bay (Canuel 2001). Thus, while bulk POC provides a general index of the overall loading of allochthonous and autochthonous C in the system, POC alone can be very misleading in terms of overall C cycling dynamics.

Salinity gradients of DOC concentrations have commonly been used to examine conservative and nonconservative behavior of DOC in estuaries (see Guo et al. 1999, and references therein). The typical DOC mixing gradients for six different estuaries in the Gulf of Mexico region clearly indicated that DOC is not conservatively mixed across the salinity gradient, and is removed or transformed within the estuaries. (Figure 3.22; Benoit et al. 1994). High concentrations of DOC at the low salinity ends of these estuaries are likely due to riverine inputs of DOC. DOC concentrations drop rapidly between salinities of 0 and 5, suggesting that the lower salinity regions are important sinks for DOC. These regions feature ETMs where fractions of DOC can be removed due to coagulation, flocculation, and other processes.

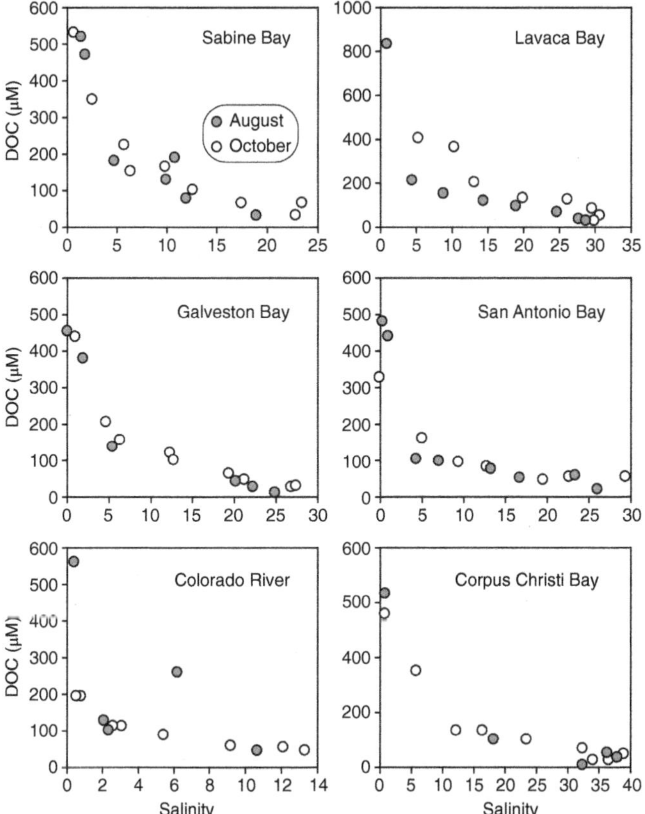

FIGURE 3.22 Salinity mixing diagrams showing concentrations of dissolved organic carbon (DOC) across salinity gradients in six Texas estuaries in August and October. *Source*: Modified from Benoit et al. (1994).

Other important sub-components of the total DOC pool are the high and low molecular weight dissolved organic carbon (HMW DOC and LMW DOC); the HMW DOC is sometimes referred to as colloidal organic carbon (COC). Sources of HMW DOC could be derived from old soil materials, fresh litter from terrestrial runoff, or perhaps more labile algal sources (both benthic and pelagic) in the estuary. Radiocarbon values of two size fractions of HMW DOC reveal that the smaller size fraction (> 1 kDa) was younger than the larger (>10 kDa), with both being higher in abundance in the lower salinity region of Galveston Bay estuary (USA) (Guo and Santschi 1997). These differences were attributed to inputs of older HMW DOC (more geopolymerized) from porewaters in resuspended sediments. The importance of older HMW DOC from porewaters to estuarine waters has also been shown to be important in coastal bottom waters along shelf regions in the Mid Atlantic Bight (Mitra et al. 2000).

This does not support the open ocean Size-Reactivity model proposed by Amon and Benner (1996), which, simply stated, predicts that the "older" fraction of DOC will reside in the lower molecular weight classes, typically at deeper water depths. However, the Size-Reactivity model is generally observed in oceanic waters because of the selective removal of labile "younger" C as POC produced in the euphotic zone sinks through the water column, accumulating the "older" smaller molecular weight DOC at depth (Amon and Benner 1996; Guo et al. 1996).

Bulk radiocarbon and ^{13}C signatures of both total POC and DOC pools have also proven useful in constraining what their sources might be in estuaries (McCallister et al. 2004, and references therein). For example, $\Delta^{14}C$-DOC is always more ^{14}Cenriched compared to $\Delta^{14}C$-DOC along riverine margins of U.S. East Coast estuaries (Raymond and Bauer 2001a, b). Since the ^{14}C-enriched DOC is also more depleted in ^{13}C, it appears that these differences are because of greater contribution of ^{14}C-enriched soil and litterfall organic matter that is leached from soils. Similarly, studies have shown, using a dual-isotopic tracer approach of isotopic signatures ($\delta^{13}C$ and $\Delta^{14}C$) for bacterial nucleic acids collected from different regions and potential C source materials, that there was good delineation of C sources from aquatic and terrestrial systems in the York River estuary (McCallister et al. 2004). In general, the results from these dual-isotopic tracer studies indicate that the broad classification of organic C, and the interchangeable use of the terms "old" and "refractory," are in many cases not valid. In fact, highly depleted ^{14}C (1000–5000 years old) in the Hudson River estuary appears to be an important labile source fueling heterotrophy (Cole and Caraco 2001). Thus, the storage of organic matter for centuries and millennia in soils and rocks can actually become available to aquatic microbes over periods of weeks to months (Petsch et al. 2001), completing a unique linkage between river metabolism and the history of organic matter preservation in the drainage basin (Cole and Caraco 2001), and prompting a re-evaluation of the age versus bioavailability paradigm (Guillemette et al. 2017).

In estuarine sediments, down-core concentrations of bulk C indices, such as total organic C (TOC), $\delta^{13}C$, and C:N ratios can be used as a general index of the loading and sources of POC to sediments. There are a number of problems with using the C:N ratio as the only diagnostic indicator of organic matter source, as mentioned previously; however, if coupled with stable isotope and chemical biomarker tools, sources are better constrained. If we simply use bulk C tools to examine down-core sediment profiles, we see that TOC and atomic C:N ratios in sediments from the York River estuary (USA) clearly reflect a dominance of phytoplankton inputs, with a typical %TOC in surface sediments of estuaries, decreasing with diagenetic "burn-off" at depth (Arzayus and Canuel 2004). The bulk $\delta^{13}C$ signal also reflects inputs of phytoplankton sources. Down-core profiles of porewater bulk DOC and C:N ratios (of DOM) should also reflect changes in rates of POM remineralization. This is illustrated in down-core profiles of DOC, C:N ratios, and dissolved inorganic carbon at three stations in Chesapeake Bay (Figure 3.23; Burdige and Zheng 1998).

Differences in these DOC profiles are controlled primarily by differences in the physical and biogeochemical processes in sediments in the upper, mid- and lower regions of the bay. More specifically, the higher porewater DOC concentrations at the mid-Bay station are due to higher remineralization and a greater storage of the remineralization "signature" (e.g., DOC and DIC), due to lower O_2 conditions and less physical mixing than at stations in the upper and lower regions of the bay. Changes in the C:N ratios of porewater DOM at these stations are likely due to differences in the selective utilization of N-rich DOM, and not due to differences in source inputs (e.g., terrestrial versus marine), as originally thought (Burdige 2001). Thus, redox (more consistently anoxic versus fluctuating redox) and the presence/absence of macrofauna were the most important controlling factors in determining the remineralization of POM to DOM and overall storage of DOM in Chesapeake Bay sediments.

3.5 Concluding Remarks

A major recent realization by earth, aquatic, and marine scientists is that virtually all systems have already been altered both directly and indirectly a by human activity. That is, our findings of the ecological and biogeochemical function of these systems are increasingly biased away from natural and toward altered states. Human activities can have a variety of effects on watersheds, rivers, and estuaries, sometimes in opposite senses. Deforestation, tillage, hydrological alteration, and irrigation-enhanced erosion have increased loads of sediments and particulate carbon to rivers and estuarine systems. Widespread construction of dams, however, has resulted in significant retention of carbon and sediments in reservoirs (see reviews, Bauer and Bianchi 2011; Bianchi and Bauer 2011). Furthermore, the reduction of river suspended loads by dams around the world has generally resulted in increasing light availability and phytoplankton biomass in rivers and estuaries, subsequently influencing their biogeochemistry. To better understand anthropogenic effects on estuaries, long-term monitoring studies of rivers and estuaries will continue to be valuable tools for assessing change due to both natural and human factors.

Estuaries are tremendously sensitive to long-term global climate change. Changes in river discharge in a warming climate are predicted to be particularly large in Arctic regions as snow pack and glacial water retention decreases and precipitation and runoff increase. Rising sea levels that outpace the growth of estuary mouth sills (or bars) could also subject previously protected estuaries to increased wave-energy regimes, increasing coastal wetland erosion, and altering the depositional and redox environment on estuary floors. Dramatic fluctuations in major biogenic nutrient (N, P, and Si) loading over the last 50–75 years have undoubtedly led to fluctuations in C and OM fluxes and cycling in rivers and

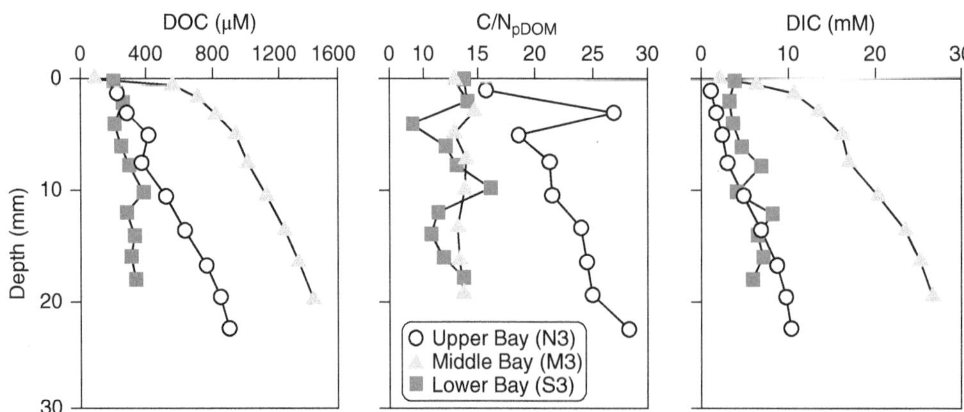

FIGURE 3.23 Down-core profiles of sediment porewater dissolved organic carbon (DOC), C/N ratios of porewater dissolved organic matter (pDOM), and dissolved inorganic carbon (DIC) at three stations in Chesapeake Bay. *Source*: Modified from Burdige and Zheng (1998).

estuaries that may be unprecedented over at least the Holocene, if not longer. Nonetheless, it is imperative that monitoring studies of rivers and estuaries continue—not only to document changing hydrological conditions, but also to assess changes in soil OM losses and concomitant changes in river and estuarine carbon and OM inputs. These studies are critical for providing baseline information for future studies of anthropogenic effects on river, estuarine, and coastal ocean systems and their biogeochemical cycles, metabolic balances, and ecosystem functions.

Review Questions

Multiple Choice

1. The *Arrhenius equation* describes:
   a. The relationship between the equilibrium concentration of a gas and its pressure
   b. The rate of gas flux between the atmosphere–water interface
   c. The relationship between chemical reaction rates and temperature
   d. The distribution of nonconservative constituents within sediments

2. If something passes through a 0.45 micron filter it is considered to be:
   a. a phytoplankton
   b. dissolved
   c. a particle
   d. nonconservative

3. Redox chemistry is important in estuaries because (select all that apply):
   a. Estuarine water and sediments have fluctuating oxygen gradients
   b. Estuarine bacteria can use alternate terminal electron acceptors when oxygen is not present
   c. Changes in redox potential can affect greenhouse gas (CO_2, CH_4, and N_2O) emissions from estuaries
   d. Electrons are not active in anoxic estuarine sediments

4. If you are studying a one-dimensional, two end-member, steady state system (such as that in Figure 3.2) and notice that the concentration of a constituent has a linear relationship with salinity, it is likely:
   a. a nonconservative constituent
   b. a conservative constituent
   c. not being added to the estuary from a groundwater source
   d. both a and c
   e. both b and c

5. The estuarine turbidity maximum is NOT:
   a. a site of high suspended particulate matter concentrations
   b. responsible for the formation of mobile and fluid muds in the benthic boundary layer
   c. characterized by low sedimentation rates
   d. decoupled from the estuaries' salt front due to a lag in resuspension

6. Organic matter in estuaries can come from (select all that apply):
   a. phytoplankton biomass transported from ocean water
   b. terrestrial plants
   c. resuspension of sediments into the water column
   d. denitrification
   e. *in situ* primary production

7. Which of the following is NOT a phase in the decomposition of organic detritus in estuaries?
   a. leaching phase, where soluble compounds are quickly lost
   b. decomposition phase, where heterotrophic organisms break down detritus
   c. decomposition phase, where autotrophic organisms break down detritus
   d. refractory phrase, where compounds like lignin decay slowly

8. Which of the following is NOT a microbial process that occurs in anaerobic environments?
   a. denitrification
   b. sulfate reduction
   c. nitrification
   d. methanogenesis
   e. iron reduction

9. Which TWO microorganisms commonly perform nitrification in estuaries?
   a. *Nitrosomonas*
   b. *Desulfovibrio*
   c. *Chlorobium*
   d. *Nitrobacter*
   e. *Nitrosovibrio*

10. What greenhouse gases can be emitted from marsh soils? (select all that apply)
    a. CO_2
    b. N_2O
    c. CH_4
    d. CO

Short Answer

1. Use the Arrhenius equation (Eq. 3.2) to figure out how the rate constant (k) of sulfate reduction changes seasonally. Assume that the frequency factor (A) is 36.1, the activation energy (Ea) is 80.3 KJ mol^{-1}, and the universal gas constant (R) is 0.082057 cm^3 atm mol^{-1} K^{-1} for both cases.
   a. What is the rate constant in July, when the temperature is 26 °C (78.8 K)?
   b. What is the rate constant in February, when the temperature is 2 °C (35.6 K)?
2. You are using a mixing model to study nitrate concentrations along an estuarine salinity gradient. You find that nitrate concentrations are higher than predicted by the theoretical dilution line. Is nitrate acting as a conservative or nonconservative constituent? Explain why it might be acting this way.
3. Describe the different factors that influence the concentration of dissolved salts in an estuary. If an estuary is in a more temperate climate (higher precipitation, lower evaporation) will it have a higher or lower concentration of dissolved salts relative to an estuary that has a more arid climate (lower precipitation, higher evaporation)? Explain why.
4. Give an example of a case where dissolved gas exchange can influence the organisms living in an estuary.
5. Explain how gas exchange between the atmosphere and estuarine water can change the pH of water.
6. Use the equation $F = k\,(C_w{-}sC_a)$ to determine the flux (F) of CO_2 from the water to atmosphere in an estuary. In the equation, F is the flux of CO_2 (mol cm^{-2} s^{-1}), *s* is the solubility coefficient of CO_2 at the *in situ* temperature and salinity, k is the gas transfer velocity, and C_w and C_a are the concentration of CO_2 in the water and air, respectively. $(C_w{-}sC_a)$ is the concentration gradient (mol cm^{-3}). Calculate the flux of CO_2 in an estuary if the concentration of CO_2 in water is 378.11 ppm and pCO_2 in air is 403.26 ppm. Assume a solubility coefficient of 0.7 mol cm^{-3}, and a gas transfer velocity of 0.42 cm s^{-1}.
7. Describe the importance of the estuarine turbidity maximum and what factors drive it.
8. Explain how processes in the nitrogen cycle are linked between aerobic and anaerobic zones in an estuary.
9. Describe the trends seen in DOC concentrations, C:N, and DIC in Figure 3.23. What do these profiles tell us about the source and fate of carbon in this system?
10. Discuss three ways that anthropogenic activities can influence estuarine chemistry.

References

Abelson, P.H. and Hoering, T.C. (1960). The biogeochemistry of stable isotopes of carbon. *Carnegie Inst. Wash.* 59: 158–165.

Abril, G., Etcheber, H., Le Hir, P. et al. (1999). Oxic-anoxic oscillations and organic carbon mineralization in an estuarine maximum turbidity zone (The Gironde, France). *Limnol. Oceanogr.* 44: 1304–1315.

Alexander, M. (1977). *Introduction to Soil Microbiology.* New York: John Wiley and Sons.

Aller, R.C. (1990). Bioturbation and manganese cycling in hemipelagic sediments. *Phil. Trans. Royal Soc. London* 331: 51–68.

Aller, R.C. (2001). Transport and reactions in the bioirrigated zone. In: *The Benthic Boundary Layer* (ed. B.P. Boudreau and B.B. Jorgensen), 269–301. New York: Oxford University Press.

Alongi, D.M. (1998). *Coastal Ecosystem Processes.* New York: CRC Press.

Alperin, M.J., Blair, N.E., Albert, D.B. et al. (1992). Factors that control isotopic composition of methane produced in an anaerobic marine sediment. *Global Biogeochem. Cycles* 6: 271–329.

Aminot, A., El-Sayed, M.A., and Kerouel, R. (1990). Fate of natural and anthropogenic dissolved organic carbon in the macro-tidal Elorn estuary. *Mar. Chem.* 29: 255–275.

Amon, R.M.W. and Benner, R. (1996). Photochemical and microbial consumption of dissolved organic carbon and dissolved oxygen in the Amazon River system. *Geochim. Cosmochim. Acta* 60: 1783–1792.

An, S. and Gardner, W.S. (2002). Dissimilatory nitrate reduction to ammonium (DNRA) as a nitrogen link, versus denitrification as a sink in a shallow estuary (Laguna Madre/Baffin Bay, Texas). *Mar. Ecol. Prog. Ser.* 237: 41–50.

Andersen, F.O. (1996). Fate of organic carbon added as diatom cells to oxic and anoxic marine sediment microcosms. *Mar. Ecol. Prog. Ser.* 134: 225–233.

Andreae, M.O. and Crutzen, P.J. (1997). Atmospheric aerosols: biogeochemical sources and role in atmosphere chemistry. *Science* 276: 1052–1058.

Aneiso, A.M., Abreu, P.C., and Biddanda, B.A. (2003). The role of free and attached microorganisms in the decomposition of estuarine macrophyte detritus. *Estuar. Coastal Shelf Sci.* 56: 197–201.

Anita, N.J., Harrison, P.J., and Oliveira, L. (1991). Phycological reviews: the role of dissolved organic nitrogen in phytoplankton nutrition, cell biology, and ecology. *Phycologia* 30: 1–89.

Argyrou, M.E., Bianchi, T.S., and Lambert, C.D. (1997). Transport and fate of dissolved organic carbon in the Lake Pontchartrain estuary, Louisiana, USA. *Biogeochemistry* 38: 207–226.

Arzayus, K.M. and Canuel, E.A. (2004). Organic matter degradation of the York River estuary: effects of biological vs. physical mixing. *Geochim. Cosmochim. Acta* 69: 455–463.

Aspinall, G.O. (1970). Pectins, plant gums and other plant polysaccharides. In: *The Carbohydrates; Chemistry and Biochemistry,* 2e (ed. W. Pigman and D. Horton), 515–536. New York: Academic Press.

Atlas, E.L. (1975). Phosphate equilibria in seawater and interstitial waters. Ph.D thesis. Oregon State University.

Aufdenkampe, A., Hedges, J.I., and Richey, J.E. (2001). Sorptive fractionation of dissolved organic nitrogen and amino acids onto fine sediments within the Amazon Basin. *Limnol. Oceanogr.* 46: 1921–1935.

Baines, S.B. and Pace, M.L. (1991). The production of dissolved organic matter by phytoplankton and its importance to bacteria: patterns across marine and freshwater systems. *Limnol. Oceanogr.* 36: 1078–1090.

Baldock, J.A. and Skjemstad, J.O. (2000). Role of soil matrix and minerals in protecting natural organic material against biological attack. *Org. Geochem.* 31: 697–710.

Baldwin, D.S. (2013). Organic phosphorus in the aquatic environment. *Environ. Chem.* 10: 439–454. http://dx.doi.org/10.1071/EN13151.

Bange, H.W., Bartell, U.H., Rapsomanikis, S., and Andreae, M.O. (1994). Methane in the Baltic and North Seas and a reassessment of the marine emissions of methane. *Global Biogeochem. Cycles* 8: 465–480.

Bange, H.W., Rapsomanikis, S., and Andreae, M.O. (1996). The Agean Sea as a source of atmospheric nitrous oxide and methane. *Mar. Chem.* 53: 41–49.

Baskaran, M., Santschi, P.H., Benoit, G., and Honeyman, B.D. (1992). Scavenging of Th isotopes by colloids in seawater of the Gulf of Mexico. *Geochim. Cosmochim. Acta* 56: 3375–3388.

Bauer, J.E. and Bianchi, T.S. (2011). Dissolved organic carbon cycling and transformation. In: *Treatise on Estuarine and Coastal Marine Science, Volume 5: Biogeochemistry* (ed. J. Middelburg and R. Laane), 7–67. Elsevier Press.

Bauza, J.F., Morell, J.M., and Corredor, J.E. (2002). Biogeochemistry of nitrous oxide production in red mangrove (*Rhizophora mangle*) forest sediments. *Estuar. Coastal Shelf Sci.* 55: 697–704.

Benitez-Nelson, C.R. and Karl, D.M. (2002). Phosphorus cycling in the North Pacific Subtropical Gyre using cosmogenic ^{32}P and ^{33}P. *Limnol. Oceanogr.* 47: 762–770.

Benitez-Nelson, C.R., O'Neill, L., Kolowith, L.C. et al. (2004). Phosphonates and particulate organic phosphorus cycling in an anoxic marine basin. *Limnol. Oceanogr.* 49: 1593–1604.

Benoit, G., Oktay-Marshall, S., Cantu, A. et al. (1994). Partitioning of Cu, Pb, Ag, Zn, Fe, Al, and Mn between filter-retained particles, colloids, and solution in six Texas estuaries. *Mar. Chem.* 45: 307–336.

Berman, T. and Bronk, D.A. (2003). Dissolved organic nitrogen: a dynamic participant in aquatic ecosystems. *Aquat. Microb. Ecol.* 31: 279–305.

Berner, R.A. (1970). Sedimentary pyrite formation. *Amer. J. Sci.* 268: 1–23.

Berner, R.A. (1980). *Early Diagenesis: A Theoretical Approach.* New Jersey: Princeton University Press.

Berner, R.A. (1984). Sedimentary pyrite formation. *Am. J. Sci.* 268: 1–23.

Berner, R.A. and Berner, R.A. (1996). *Global Environment: Water, Air, and Geochemical Cycles.* New York: Prentice Hall.

Berthollet, C.L. (1803). *Essai de statique chimique, 1.* Paris: Demonville-Et-Soeurs.

Bianchi, T.S. (2007). *Biogeochemistry of Estuaries.* Oxford: Oxford University Press.

Bianchi, T.S. and Bauer, J.E. (2011). Particulate organic carbon cycling and transformation. In: *Treatise on Estuarine and Coastal Marine Science, Volume 5: Biogeochemistry* (ed. J. Middelburg and R. Laane), 69–117. Elsevier Press.

Bianchi, T.S. and Canuel, E.A. (2001). Organic geochemical tracers in estuaries. *Org. Geochem.* 32: 451–452.

Bianchi, T.S. and Canuel, E.A. (2011). *Chemical Biomarkers in Aqautic Ecosystems.* Pronceton University Press. ISBN: 978-0-691-13414-7.

Bianchi, T.S. and Rice, D.L. (1988). Feeding ecology of *Leitoscoloplos fragilis* II. Effects of worm density on benthic diatom production. *Mar. Biol.* 99: 123–131.

Bianchi, T.S., Findlay, S., and Fontvieille, D. (1991). Experimental degradation of plant materials in Hudson River sediments. I. Heterotrophic transformations of plant pigments. *Biogeochemistry* 12: 171–187.

Bianchi, T.S., Freer, M.E., and Wetzel, R.G. (1996). Temporal and spatial variability and the role of dissolved organic carbon (DOC) in methane fluxes from the Sabine River floodplain (Southeast Texas, USA). *Arch. Hydrobiol.* 136: 261–287.

Bianchi, T.S., Baskaran, M., Delord, J., and Ravichandran, M. (1997). Carbon cycling in a shallow turbid estuary of southeast Texas: the use of plant pigments as biomarkers. *Estuaries* 20: 404–415.

Bianchi, T.S., Wysocki, L.A., Schneider, K.M. et al. (2011). Sources of terrestrial organic carbon in the Louisiana shelf (USA): evidence for the importance of coastal marsh inputs. *Aquat. Geochem.* 17: 431–456.

Blackburn, T.H. and Henriksen, K. (1983). Nitrogen cycling in different types of sediments from Danish waters. *Limnol. Oceanogr.* 28: 477–493.

Blough, N.V. and Green, S.A. (1995). Spectroscopic characterization and remote sensing of non-living organic matter. In: *The Role of Non-living Organic Matter in the Earth's Carbon Cycle* (ed. R.G. Zepp and C. Sonntag), 23–45. Chichester: Wiley.

Bonin, P., Omnes, P., and Chalamet, A. (1998). Simultaneous occurrence of denitrification and nitrate ammonification in sediments of the French Mediterranean coast. *Hydrobiologia* 389: 169–182.

Borges, A.V. and Frankignoulle, M. (2002). Distribution of surface carbon dioxide and air-sea exchange in the upwelling system off the Galician coast. *Global Biogeochem. Cycles* 16 (2): 1020. https://doi.org/10.1029/2000GB001385.

Borsuk, M., Stowe, C., Luettich, R.A. et al. (2001). Modeling oxygen dynamics in an intermittently stratified estuary: estimation of process rates using field data. *Estuar. Coastal Shelf Sci.* 52: 33–49.

Boudreau, B.P. (1996). The diffusive tortuosity of fine-grained unlithified sediments. *Geochim. Cosmochim. Acta* 60: 3139–3142.

Boudreau, B.P. and Jorgensen, B.B. (ed.) (2001). *The Benthic Boundary Layer: Transport Processes and Biogeochemistry.* Oxford: Oxford University Press.

Bouillon, S., Frankignoulle, M., Dehairs, F. et al. (2003). Inorganic and organic carbon biogeochemistry in the Gautami Godavari estuary (Andhra Pradesh, India) during pre-monsoon: the local impact of extensive mangrove forests. *Global. Biogeochem. Cycles* 17: 1114.

Bouwman, A.F., van Drecht, G., Knoop, J.M. et al. (2005). Exploring changes in river nitrogen export to the worlds oceans. *Global Biogeochem. Cycles* 19: doi: 1029/2004GB002314.

Boynton, W.R. and Kemp, W.M. (2000). Influence of river flow and nutrient loads on selected ecosystem processes—a synthesis of Chesapeake Bay data. In: *Estuarine Science: A Synthetic Approach to Research and Practice* (ed. J.E. Hobbie), 269–298. Washington, D.C.: Island Press.

Boynton, W.R., Kemp, W.M., and Keefe, C.W. (1982). A comparative analysis of nutrients and other factors influencing estuarine phytoplankton production. In: *Estuarine Comparisons* (ed. V.S. Kennedy), 69–90. New York: Academic Press.

Boynton, W.R., Garber, J.H., Summers, R., and Kemp, W.M. (1995). Inputs, transformations, and transport of nitrogen and phosphorus in Chesapeake Bay and selected tributaries. *Estuaries* 18: 285–314.

Brandes, J.A. and Devol, A.H. (1995). Isotopic fractionation of oxygen and nitrogen in coastal marine sediments. *Geochim. Cosmochim. Acta* 61: 1793–1801.

Brezonik, P.L. (1994). *Chemical Kinetics and Process Dynamics in Aquatic Systems*. London: Lewis Publishing.

Broecker, W.S. and Peng, T.H. (1974). Gas exchange rates between the air and the sea. *Tellus* 26: 21–35.

Bronk, D.A. (2002). Dynamics of organic nitrogen. In: *Biogeochemistry of Marine Dissolved Organic Matter* (ed. D.A. Hansell and C.A. Carlson), 153–231. San Diego: Academic Press.

Bronk, D.A. and Ward, B.B. (2000). Magnitude of DON release relative to gross nitrogen uptake in marine systems. *Limnol. Oceanogr.* 45: 1879–1883.

Bugna, G.C., Chanton, J.P., Cable, J.E. et al. (1996). The importance of groundwater discharge to the methane budgets of nearshore and continental shelf waters of the northwestern Gulf O Mexico. *Geochim. Cosmochim. Acta* 60: 4735–4746.

Burdige, D.J. (1993). The biogeochemistry of manganese and iron reduction in marine sediments. *Earth Sci. Rev.* 35: 249–284.

Burdige, D.J. (2001). Dissolved organic matter in Chesapeake Bay sediment pore waters. *Org. Geochem.* 32: 487–505.

Burdige, D.J. (2002). Sediment pore waters. In: *Biogeochemistry of Marine Dissolved Organic Matter* (ed. D.A. Hansell and C.A. Carlson), 612–653. New York: Academic Press.

Burdige, D.J. (2006). *Geochemistry of Marine Sediments*. New Jersey: Princeton University Press.

Burdige, D.J. and Homestead, J. (1994). Fluxes of dissolved organic carbon from Chesapeake Bay sediments Geochim. *Cosmochim. Acta* 58: 3407–3424.

Burdige, D.J. and Zheng, S. (1998). The biogeochemical cycling of dissolved organic nitrogen in estuarine sediments. *Limnol. Oceanogr.* 43: 1796–1813.

Burdige, D.J., Alperin, M.J., Homstead, J., and Martens, C.S. (1992). The role of benthic fluxes of dissolved organic carbon in oceanic and sedimentary carbon cycling. *Geophys. Res. Lett.* 19: 1851–1854.

Burton, J.D. and Liss, P.S. (1976). Basic properties and processes in estuarine chemistry. In: *Estuarine Chemistry* (ed. J.D. Burton and P.S. Liss), 1–36. New York: Academic Press.

Butcher, S.S. and Anthony, S.E. (2000). Equilibrium, rate, and natural systems. In: *Earth System Science, from Biogeochemical Cycles to Global Change* (ed. M.C. Jacobson, R.J. Charlson, H. Rodhe and G.H. Orians), 85–105. New York: Academic Press, International Geophysics Series.

Cai, W.J. (2003). Riverine inorganic carbon flux and rate of biological uptake in the Mississippi River plume. *Geophys. Res. Lett.* 30: 1032. https://doi.org/10.1029/2002GL016312.

Cai, W.-J. (2011). Estuarine and coastal ocean carbon aradox: CO_2 sinks or sites of terrestrial carbon incineration? *Annu. Rev. Mar. Sci.* 2011 (3): 123–145.

Cai, W.J. and Wang, Y. (1998). The chemistry, fluxes and sources of carbon dioxide in the estuarine waters of the Satilla and Altamaha Rivers. *Georgia. Limnol. Oceanogr.* 43: 657–668.

Cai, W.J., Pomeroy, L.R., Moran, M.A., and Wang, Y. (1999). Oxygen and carbon dioxide mass balance in the estuarine/intertidal marsh complex of five rivers in the southeastern U.S. *Limnol. Oceanogr.* 44: 639–649.

Cai, W.J., Wang, Y., Krest, J., and Moore, W.S. (2003). Te geochemistry of dissolved inorganic carbon in a surficial groundwater aquifer in North Inlet, South Carolina, and the carbon fluxes to the coastal ocean. *Geochem. Cosmochim. Acta* 67: 631–637.

Cai, W.J., Hu, X., Huang, W.J. et al. (2011). Acidification of subsurface coastal waters enhanced by eutrophication. *Nat. Geosci. 4* (11): 766.

Callender, E. and Hammond, D.E. (1982). Nutrient exchange across the sediment-water interface in the Potomac River estuary. *Estuar. Coastal Shelf Sci.* 15: 395–413.

Canfield, D.E. (1989). Reactive iron in marine sediments. *Geochim. Cosmochim. Acta* 53: 619–632.

Canfield, D.E. (1993). Organic matter oxidation in marine sediments. In: *NATO-ARW interactions of C, N, P and S biogeochemical Cycles and Global Change* (ed. R. Wollast, L. Chou and F. Mackenzie), 333–365. New York: Springer.

Canfield, D.E., Raiswell, R., and Bottrell, S. (1992). The reactivity of sedimentary iron minerals toward sulfide. *Am. J. Sci.* 292: 659–683.

Canuel, E.A. (2001). Relations between river flow, primary production and fatty acid composition of particulate organic matter in San Francisco and Chesapeake Bays: a multivariate approach. *Organic Geochem.* 32 (4): 563–583.

Capone, D.G. and Kiene, R.P. (1988). Comparison of microbial dynamics in marine and freshwater sediments: contrast in anaerobic carbon catabolism. *Limnol. Oceanogr.* 33: 725–749.

van Cappellen, P. and Ingall, E.D. (1996). Redox stabilization of the atmosphere and oceans by phosphorus-limited marine productivity. *Science* 271: 493–496.

van Cappellen, P. and Wang, Y. (1996). Cycling of iron and manganese in surface sediments: a general theory for the coupled transport and reaction of carbon, oxygen, nitrogen, sulfur, iron, and manganese. *Am. J. Sci.* 296: 197–243.

Caraco, N.F., Cole, J.J., and Likens, G.E. (1990). A comparison of phosphorus immobilization in sediments of freshwater and coastal marine systems. *Biogeochemistry* 9: 277–290.

Carritt, D.E. and Goodgal, S. (1954). Sorption reactions and some ecological implications. *Deep-Sea Res.* 1: 224–243.

Cauwet, G. (2002). DOM in the coastal zone. In: *Biogeochemistry of Marine Dissolved Organic Matter* (ed. D.A. Hansell and C.A. Carlson), 579–602. New York: Academic Press.

Cerqueira, M.A. and Pio, C.A. (1999). Production and release of dimethylsulphide from an estuary in Portugal. *Atmos. Environ.* 33: 3355–3366.

Charlson, R.J. (2000). The atmosphere. In: *Earth System Science, from Biogeochemical Cycles to Global Change* (ed. M.C. Jacobson, R.J. Charlson, H. Rodhe and G.H. Orians), 132–158. New York: International Geophysics Series, Academic Press.

Charlson, R.J., Lovelock, J.E., Andreae, M.O., and Warren, S.G. (1987). Oceanic phytoplankton atmospheric sulphur, cloud albedo and climate. *Nature* 326: 655–661.

Chester, R. (2003). *Marine Geochemistry*. United Kingdom: Blackwell.

Cicerone, R.J. and Oremland, R.S. (1988). Biogeochemical aspects of atmospheric methane. *Global Biogeochem. Cycles* 2: 299–327.

Clark, L.L., Ingall, E.D., and Benner, R. (1998). Marine phosphorus is selectively remineralized. *Nature* 393: 426.

Cloern, J.E., Canuel, E.A., and Harris, D. (2002). Stable carbon and nitrogen isotopic composition of aquatic and terrestrial plants in the San Francisco Bat estuarine system. *Limnol. Oceanogr.* 47: 713–729.

Coble, P.G. (1996). Characterization of marine and terrestrial DOM in seawater using excitation-emission matrix spectroscopy. *Mar. Chem.* 51: 325–346.

Cole, J.J. and Caraco, N.F. (2001). Carbon in catchments: connecting terrestrial carbon losses with aquatic metabolism. *Mar. Freshwat. Res.* 52: 101–110.

Condron, L.M., Goh, K.M., and Newman, R.H. (1985). Nature and distribution of soil phosphorus as revealed by a sequential extraction method followed by ^{31}P nuclear magnetic resonance analysis. *J. Soil Sci.* 36: 199–207.

Conley, D.J. (1997). Riverine contribution of biogenic silica to the oceanic silica budget. *Limnol Oceanogr* 42 (4): 774.

Conley, D.J. (2002). Terrestrial ecosystems and the global biogeochemical silica cycle. *Global Biogeochem. Cycle* 16: 68-1–68-7.

Conley, D.J. and Malone, T.C. (1992). Annual cycle of dissolved silicate in Chesapeake Bay: implications for the production and fate of phytoplankton biomass. *Mar. Ecol. Prog. Ser.* 81: 121–128.

Conley, D.J., Smith, W.M., and Boynton, W.R. (1995). Transformation of particle-bound phosphorus at the land sea interface. *Estuar. Coastal Shelf Sci.* 40: 161–176.

Conley, D.J., Humborg, C., Rahm, L. et al. (2002). Hypoxia in the Baltic Sea and basin-scale changes in phosphorus biogeochemistry. *Environmental Science and Technology* 36: 5315–5320. https://doi.org/10.1021/Es025763w.

Cornwell, J.C., Kemp, W.M., and Kana, T.M. (1999). Denitrification in coastal ecosystems: methods, environmental controls and ecosystem level controls, a review. *Aquat. Ecol.* 33: 41–54.

Correll, D.L. and Ford, D. (1982). Comparison of precipitation and land runoff as sources of estuarine nitrogen. *Estuar. Coastal Shelf Sci.* 15: 45–56.

Cowan, J.L. and Boynton, W.R. (1996). Sediment water oxygen and nutrient exchanges along the longitudinal axis of Chesapeake Bay: seasonal patterns, controlling factors and ecological significance. *Estuaries* 9: 562–580.

Cox, R.A., Culkin, E., and Riley, J.P. (1967). The electrical conductivity/chlorinity relationship in natural seawater. *Deep Sea. Res.* 14: 203–220.

Crill, P.M. and Martens, C.S. (1987). Biogeochemical cycling in an organic-rich coastal marine basin. 6. Temporal and spatial variation in sulfate reduction rates. *Geochim. Cosmochim. Acta* 51: 1175–1186.

Currin, C.A., Newell, S.Y., and Paerl, H.W. (1995). The role of standing dead *Spartina alterniflora* and benthic microalgae in salt marsh food webs: considerations based on multiple stable isotope analysis. *Mar. Ecol. Prog. Ser.* 121: 99–116.

D'Elia, C.F., Nelson, D.M., and Boynton, W.R. (1983). Chesapeake Bay nutrient and plankton dynamics III. The annual cycle of dissolved silicon. *Geochim. Cosmochim. Acta.* 47: 1945–1955.

D'Elia, C.F., Harding, L.W., Leffler, M., and Mackiernan, G.B. (1992). The role and control of nutrients in Chesapeake Bay. *Wat. Sci. Technol.* 26: 2635–2644.

Dacey, J.W.H., King, G.M., and Wakeham, S.G. (1987). Factors controlling emission of dimethylsulfide from salt marshes. *Nature* 330: 643–645.

Dalzell, B.J., Minor, E.C., and Mopper, K.M. (2009). Photodegradation of estuarine dissolved organic matter: a multi-method assessment of DOM transformation. *Org. Geochem.* 40: 243–257. https://doi.org/10.1016/j.orggeochem.2008.10.003.

Darnell, R.M. (1967). The organic detritus problem in estuaries: American Association for the Advancement of Science, Publication No. 83, 374–375.

Day, J., Hall, C.S., Kemp, W.M., and Yanez-Arancibia, A. (1989). *Estuarine Ecology.* New York: John-Wiley.

De Angelis, M.A. and Scranton, M.D. (1993). Fate of methane in the Hudson River and estuary. *Global Biogeochem. Cycles* 7: 509–523.

Deegan, C.E. and Garritt, R.H. (1997). Evidence for spatial variability in estuarine food webs. *Mar. Ecol. Prog. Ser.* 147: 31–47.

Deflandre, B., Mucci, A., Gagne, J. et al. (2002). Early diagenetic processes in coastal marine sediments disturbed by catastrophic sedimentation event. *Geochim. Cosmochim. Acta* 66: 2547–2558.

Delaney, M.L. (1998). Phosphorus accumulation in marine sediments and the oceanic phosphorus cycle. *Global Biogeochem. Cycles* 12: 563–572.

DeLaune, R.D. and Reddy, K.R. (2008). *Biogeochemistry of Wetlands: Science and Applications.* CRC press.

Delwiche, C.C. (1981). Atmospheric chemistry of nitrous oxide. In: *Denitrification, Nitrification, and Atmospheric Nitrous Oxide.* New York: John Wiley and Sons.

DeMaster, D.J. (1981). The supply and accumulation of silica in the marine environment. *Geochim. Cosmochim. Acta* 64: 2467–2477.

Deming, J.W. and Baross, J.A. (1993). The early diagenesis of organic matter: bacterial activity. In: *Organic Geochemistry* (ed. M.H. Engel and S.A. Macko), 119–114. New York: Plenum Press.

Diaz, F. and Raimbault, P. (2000). Nitrogen regeneration and dissolved organic nitrogen release during spring in a NW Mediterranean coastal zone (Gulf of Lions); implications for the estimation of new production. *Mar. Ecol. Prog. Ser.* 197: 51–65.

Duce, R.A., Liss, P.S., Merrill, J.T. et al. (1991). The atmospheric input of trace species to the world ocean. *Global Biogeochem. Cycle* 5: 193–259.

Dunstan, G.A., Volkman, J.K., Jefferey, S.W., and Barrett, S.M. (1992). Biogeochemical-composition of microalgae from the green algal classes Chlorophyceae and Prasinophyceae .2. Lipid classes and fatty acids. *J. Exp. Mar. Biol. Ecol.* 161: 115–134.

Dyer, K.R. (1986). *Coastal and Estuarine Sediment Dynamics.* Chichester: John Wiley and Sons.

Eglinton, G. and Calvin, M. (1967). Chemical fossils. *Sci. Am.* 216: 32–43.

Elderfield, H., Luedtke, N., McCaffrey, R.J., and Bender, M.L. (1981a). Benthic flux studies in Narragansett Bay. *Am. J. Sci.* 281: 768–787.

Elderfield, H., McCaffrey, R.J., Luedtke, N. et al. (1981b). Chemical diagenesis in Narragansett Bay sediments. *Am. J. Sci.* 281: 1021–1055.

Engelhaupt, E. and Bianchi, T.S. (2001). Sources and composition of high-molecular-weight dissolved organic carbon in a southern Louisiana tidal stream (Bayou Trepagnier). *Limnol. Oceanogr.* 46: 917–926.

Fain, A.M., Jay, D.A., Wilson, D. et al. (2001). Seasonal and tidal monthly patterns of particulate matter dynamics in the Columbia River estuary. *Estuaries* 24: 770–786.

Feely, R.A., Alin, S.R., Newton, J. et al. (2010). The combined effects of ocean acidification, mixing, and respiration on pH and carbonate saturation in an urbanized estuary. *Estuar. Coastal Shelf Sci. 88* (4): 442–449.

Fenchel, T.M. and Blackburn, T.H. (1979). *Bacteria and Mineral Cycling.* New York: Academic Press.

Fenchel, T.M. and Jorgensen, B.B. (1977). Detritus food chains of aquatic ecosystems: the role of bacteria. *Adv. Microb. Ecol.* 1: 1–57.

Fenchel, T.M. and Riedl, R.J. (1970). The sulfide system: a new biotic community underneath the oxidized layer of marine sand bottoms. *Mar. Biol.* 7: 255–268.

Filella, M. and Buffle, J. (1993). Factors affecting the stability of submicron colloids in natural waters. *Colloids Surfaces: Physico-Chem. Eng. Aspects* 73: 255–273.

Findlay, S.E.H. and Sinasbaugh, R.L. (ed.) (2003). *Aquatic Ecosystems-Interactivity of Dissolved Organic Matter.* New York: Academic Press.

Findlay, S.E.G. and Tenore, K.R. (1982). Effect of a free-living marine nematode (*Diplolaimella chitwoodi*) on detrital carbon mineralization. *Mar. Ecol. Prog. Ser.* 8: 161–166.

Fox, L.E. (1989). A model for inorganic control of phosphate concentrations in river waters. *Geochim. Cosmochim. Acta* 53: 417–428.

Frankignoulle, M. and Borges, I. (2001). European continental shelf as a significant sink for atmospheric CO_2. *Global Biogeochem. Cycles* 15: 569–576.

Frankignoulle, M. and Middelburg, J.J. (2002). Biogases in tidal European estuaries: the BIOGEST project. *Biogeochemistry* 59: 1–4.

Frankignoulle, M., Abril, G., Borges, A. et al. (1998). Carbon dioxide emission from European estuaries. *Science* 28.

Froelich, P.N. (1988). Kinetic control of dissolved phosphate in natural rivers and estuaries: a primer on the phosphate buffer mechanism. *Limnol. Oceanogr.* 33: 649–668.

Froelich, P.N., Klinkhammer, G.P., Bender, M.L. et al. (1979). Early oxidation of organic matter in pelagic sediments of the eastern equatorial Atlantic: suboxic diagenesis. *Geochim. Cosmochim. Acta* 43: 1075–1091.

Froelich, P.N., Bender, M.L., and Luedtke, N.A. (1982). The marine phosphorus cycle. *Am. J. Sci.* 282: 474–511.

Fry, B. (2002). Conservative mixing of stable isotopes across estuarine salinity gradients: a conceptual framework for monitoring watershed influences on down stream fisheries production. *Estuaries* 25: 264–271.

Fry, B. and Sherr, E.B. (1984). $\delta^{13}C$ measurements as indicators of carbon flow in marine and freshwater ecosystems. *Contrib. Mar. Sci.* 27: 13–47.

Fukushima, T., Ishibashi, T., and Imai, A. (2001). Chemical characterization of dissolved organic matter in Hiroshima Bay. *Japan. Estuar. Coastal Shelf Sci.* 53: 51–62.

Galloway, J.N. (1985). The deposition of sulfur and nitrogen from the remote atmosphere. In: *The Biogeochemical Cycling of Sulfur and Nitrogen in the Remote Atmosphere* (ed. J. Galloway, M. Charlson, O. Andreae and H. Rodhe). Dordrecht: Reidel.

Galloway, J.N. (1998). The global nitrogen cycle: changes and consequences. *in* Proceedings of the First International Nitrogen conference, p. 15–24, Elsevier Science, New York.

Gardner, W.S., Escobar-Broines, E., Cruz-Kaegi, E., and Rowe, G.T. (1993). Ammonium excretion by benthic invertebrates and sediment-water nitrogen flux in the Gulf of Mexico near the Mississippi River outflow. *Estuaries* 16: 799–808.

Gibbs, R.J. (1970). Mechanisms controlling world water chemistry. *Science* 170: 1088–1090.

Giblin, A.E. (1988). Pyrite formation in marshes during early diagenesis. *Geomicrobiol. J.* 6: 77–97.

Gilmour, C.C., Podar, M., Bullock, A.L. et al. (2013). Mercury methylation by novel microorganisms from new environments. *Environ. Sci. Technol.* 47: 11810–11820.

Goñi, M.A., Ruttenberg, K.C., and Eglinton, T.I. (1998). A reassessment of the sources and importance of land-derived organic matter in surface sediments from the Gulf of Mexico. *Geochim. Cosmochim. Acta* 62: 3055–3075.

Goolsby, D.A. (2000). Mississippi basin nitrogen flux believed to cause Gulf hypoxia. EOS Transactions 2000, 321.

Gordon, E.S. and Goñi, M.A. (2003). Sources and distribution of terrigenous organic matter delivered by the Atchafalaya River to sediments in the northern Gulf of Mexico. *Geochim. Cosmochim. Acta* 67: 2359–2375.

Griffith, P., Shiah, F.K., Gloersen, K. et al. (1994). Activity and distribution of attached bacteria in Chesapeake Bay. *Mar. Ecol. Prog. Ser.* 108: 1–10.

de Groot, C. (1990). Some remarks on the presence of organic phosphates in sediments. *Hydrobiologia* 207: 303–309.

Guentzel, J.L., Landing, W.M., Gill, G.A., and Pollman, C.D. (2001). Processes influencing rainfall deposition of mercury in Florida: the FAMS Project (1992–1996). *Env. Sci. Technol.* 35: 863–873.

Guillemette, F., Bianchi, T.S., and Spencer, R.G. (2017). Old before your time: ancient carbon incorporation in contemporary aquatic foodwebs. *Limnol. Oceanogr. 62* (4): 1682–1700.

Gunnars, A. and Blomqvist, S. (1997). Phosphate exchange across the sediment-water interface when shifting from anoxic to oxic conditions: an experimental comparison of freshwater and brackish-marine systems. *Biogeochemistry* 37: 203–226.

Guo, L. and Santschi, P.H. (1997). Composition and cycling of colloids in marine environments. *Rev. Geophys.* 35: 17–40.

Guo, L. and Santschi, P.H. (2000). Sedimentary sources of old high molecular weight dissolved organic carbon from the ocean margin benthic nepheloid layer. *Geochim. Cosmochim. Acta* 64: 651–660.

Guo, L., Santschi, P.H., Cifuentes, L.A. et al. (1996). Cycling of high molecular-weight dissolved organic mater in the Middle Atlantic Bight as revealed by carbon isotopic (^{13}C and ^{14}C) signatures. *Limnol. Oceanogr.* 41: 1242–1252.

Guo, L., Santschi, P.H., and Bianchi, T.S. (1999). Dissolved organic matter in estuaries of the Gulf of Mexico. In: *Biogeochemistry of Gulf of Mexico Estuaries* (ed. T.S. Bianchi, J. Pennock and R.R. Twilley), 269–299. New York: John Wiley and Sons.

Hahn, J. and Crutzen, P.J. (1982). The role of fixed nitrogen in atmosphere photochemistry. *Phil. Trans. R. Soc. Lond.* 296: 521–541.

Haines, E.B. and Montague, C.L. (1979). Food sources of estuarine invertebrates analyzed using $^{13}C/^{12}C$ ratios. *Ecology* 60: 48–56.

Hammond, D.E., Fuller, C., Harmon, D. et al. (1985). Benthic fluxes in San Francisco Bay. *Hydrobiologia* 129: 69–90.

Han, C., Gu, X., Geng, J. et al. (2010). Production and emission of phosphine gas from wetland ecosystems. *J. Environ. Sci. 22* (9): 1309–1311.

Hansell, D.A. and Carlson, C.A. (ed.) (2002). *Biogeochemistry of Marine Dissolved Organic Matter.* New York: Academic Press.

Hargrave, B.T. (1969). Similarity of oxygen uptake by benthic communities. *Limnol. Oceanogr.* 14: 801–805.

Harvey, R.H. and Mannino, A. (2001). The chemical composition and cycling of particulate and macromolecular dissolved organic matter in temperate estuaries as revealed by molecular organic tracers. *Org. Geochem.* 32: 527–542.

Harvey, R.H., Tuttle, J.H., and Bell, J.T. (1995). Kinetics of phytoplankton decay during simulated sedimentation: changes in biochemical composition and microbial activity under oxic and anoxic conditions. *Geochim. Cosmochim. Acta* 59: 3367–3377.

Hatcher, P.H., Dria, K.J., Kim, S., and Frazier, S.W. (2001). Modern analytical studies of humic substances. *Soil Sci.* 166: 770–794.

Hawkes, G.E., Powlson, D.S., Randall, E.W., and Tate, K.R. (1984). A ^{31}P nuclear magnetic resonance study of the phosphorus species in alkali extracts from long-term field experiments. *J. Soil Sci.* 35: 35–45.

Hedges, J.I. and Keil, R. (1995). Sedimentary organic matter preservation; an assessment and speculative synthesis. *Mar. Chem.* 49: 81–115.

van Heemst, J.D.H., del Rio, J.C., Hatcher, P.G., and de Leeuw, J.W. (2000). Characterization of estuarine and fluvial dissolved organic matter by themochemicalysis using tetramethylammonium hydroxide. *Acta Hydrochim. Hydrobiol.* 28: 69–76.

Henrichs, S.M. and Reeburgh, W.S. (1987). Anaerobic mineralization of marine sediment organic matter: rates and the role of anaerobic processes in the oceanic carbon economy. *Geomicrobiol. J.* 5: 191–237.

Henriksen, K. and Kemp, W.M. (1988). Nitrification in estuarine and coastal marine sediments. In: *Nitrogen Cycling in Coastal Marine Environments. SCOPE* (ed. T.H. Blackburn and J. Sørensen), 207–249. New York: John Wiley and Sons.

Herman, P.M. and Heip, C.H.P. (1999). Biogeochemistry of the MAximum TURbidity zone of Estuaries (MATURE): some conclusions. *J. Mar. Syst.* 22: 89–104.

Herrera-Silveira, J.A. (1994). Nutrients from underground water discharges in a coastal lagoon (Celestun, Yucatan, Mexico) Verh. *Int. Ver. Limnol.* 25: 1398–1401.

Hobbie, J.E. and Lee, C. (1980). Microbial production of extracellular material: importance in benthic ecology. In: *Marine Benthic Dynamics* (ed. K. Tenore and B. Coull), 341–346. Columbia: Belle W. Baruch Institute for Marine and Biology, University of South Carolina Press.

Honeyman, B.D. and Santschi, P.H. (1988). Critical review: metals in aquatic systems. Predicting their scavenging residence times from laboratory data remains a challenge. *Environ. Sci. Technol.* 22: 862–871.

Hori, T., Horiguchi, M., and Hayashi, A. (1984). *Biogeochemistry of Natural C-P Compounds.* Shiga: Maruzen.

Houghton, J.T., Meiro-Filho, L.G., Bruce, J. et al. (1995). *Climate Change 1994, Radiative Forcing of Climate Change and an Evaluation of the IPCC IS92 Emission Scenarios: Reports of Working Groups I and II of the International Panel on Climate Change.* New York: Cambridge University Press.

Howarth, R.W. (1984). The ecological significance of sulfur in the energy dynamics of salt marsh and coastal sediments. *Biogeochemistry* 1: 5–27.

Howarth, R.W. (1993). Microbial processes in salt-marsh sediments. In: *An Ecological Approach* (ed. T.E. Ford). Cambridge, MA: Blackwell Publishing.

Howarth, R.W. and Teal, J.M. (1979). Sulfate reduction in a New England salt marsh. *Limnol. Oceanogr.* 24: 999–1013.

Howarth, R.W., Marino, R., Lane, R., and Cole, J.J. (1988a). Nitrogen fixation in freshwater, estuarine, and marine ecosystems. 1. Rates and importance. *Limnol. Oceanogr.* 33: 669–687.

Howarth, R.W., Marino, R., and Cole, J.J. (1988b). Nitrogen fixation in freshwater, estuarine, and marine ecosystems. 2. Biogeochemical controls. *Limnol. Oceanogr.* 33: 688–701.

Howarth, R.W., Billen, G., Swaney, D. et al. (1996). Regional nitrogen budgets and riverine inputs of N & P for the drainages to the North Atlantic Ocean: natural and human influences. *Biogeochem.* 35: 75–139.

Howarth, R.W., Jaworski, N., Swaney, D. et al. (2000). Some approaches for assessing human influences on fluxes of nitrogen and organic carbon to estuaries. In: *Estuarine Science: A Synthetic Approach to Research and Practice* (ed. J.E. Hobbie), 17–42. Washington, D.C.: Island Press.

Howes, B.L., Dacey, J.W.H., and King, G.M. (1984). Carbon flow through oxygen and sulfate reduction pathways in salt marsh sediments. *Limnol. Oceanogr.* 29: 1037–1051.

Hsieh, Y. and Yang, C. (1997). Pyrite accumulation and sulfate depletion as affected by root distribution in a *Juncus* (needlerush) salt marsh. *Estuaries* 20: 640–645.

Humborg, C., Ittekot, V., Cociasu, A., and von Bodungen, B. (1997). Effect of Danube river on Black Sea biogeochemistry and ecosystem structure. *Nature* 386: 385–388.

Humborg, C., Conley, D.J., Rahm, L. et al. (2000). Silica retention in river basins: far-reaching effects on biogeochemistry and aquatic food webs in coastal marine environments. *AMBIO* 29: 45–50.

Humborg, C., Blomqvist, S., Avsan, E. et al. (2002). Hydrological alterations with river damming in northern Sweden: implications for weathering and river biogeochemistry. *Global Biogeochem. Cycle* 16: 1039.

Hupfer, M., Rube, B., and Schmieder, P. (2004). Origin and diagenesis of polyphosphate in lake sediments: a ^{31}P-NMR study. *Limnol. Oceanogr.* 49: 1–10.

Ingall, E.D. and Jahnke, R. (1997). Influence of water column anoxia on the elemental fractionation of carbon and phosphorus during sediment diagenesis. *Mar. Geol.* 139: 219–229.

Ingall, E.D., Schroeder, P.A., and Berner, R.A. (1990). The nature of organic phosphorus in marine sediments: new insights from ^{31}P NMR. *Geochim. Cosmochim. Acta* 54: 2617–2620.

Iverson, R.L., Nearhof, F.L., and Andreae, M.O. (1989). Production of dimethylsulfonium proprionate and dimethylsulfide by phytoplankton in estuarine and coastal waters. *Limnol. Oceanogr.* 34: 53–67.

Jackson, G.A. and Williams, P.W. (1985). Importance of dissolved organic nitrogen and phosphorus to biological nutrient cycling. *Deep-Sea Res.* 32: 223–235.

Jaffe, D.A. (2000). The nitrogen cycle. In: *Earth System Science—From Biogeochemical Cycles to Global Change* (ed. M.C. Jacobson, R.J. Charlson, H. Rodhe and G.H. Orians), 322–342. New York: Academic Press.

Jaffé, R., Boyer, J.N., Lu, X. et al. (2004). Source characterization of dissolved organic matter in a subtropical mangrove-dominated estuary by fluorescence analysis. *Mar. Chem.* 84: 195–210.

Jahnke, R.A. (2000). The phosphorus cycle. In: *Earth System Science—From Biogeochemical Cycles to Global Change* (ed. M.C. Jacobson, R.J. Charlson, H. Rodhe and G.H. Orians), 360–376. New York: Academic Press.

Janzen, D.H. (1974). Tropical blackwater rivers, animals, and mast fruiting by the Dipterocarpaceae. *Biotropica* 6 (2): 69–103.

Jensen, H.S. and Thamdrup, B. (1993). Iron-bound phosphorus in marine sediments as measured by bicarbonate-dithionite extraction. *Hydrobiologia* 252: 47–59.

Jones, R.D. and Amador, J.A. (1993). Methane and carbon monoxide production, oxidation and turnover times in the Caribbean Sea as influenced by the Orinoco river. *J. Geophys. Res.* 98: 2353–2359.

Jones, J.B. and Mulholland, P.J. (1998). Influence of drainage basin topography and elevation on carbon dioxide and methane supersaturation of stream water. *Biogeochemistry* 40: 57–72.

de Jonge, V.N. and Villerius, L.A. (1989). Possible role of carbonate dissolution in estuarine phosphate dynamics. *Limnol. Oceanogr.* 34: 332–340.

Jørgensen, B.B. (1977). The sulfur cycle of coastal marine sediment (Limfjorden, Denmark). *Limnol. Oceanogr.* 28: 814–822.

Jørgensen, B.B. (1982). Mineralization of organic matter in the sea—the role of sulfate reduction. *Nature* 296: 643–645.

Jørgensen, B.B. (1989). Biogeochemistry of chemoautotrophic bacteria. In: *Autotrophic Bacteria* (ed. H.G. Shlegel and B. Bowien), 117–146. Madison: Science and Technology Publications and Springer-Verlag.

Jørgensen, B.B. and Boudreau, B.P. (2001). Diagenesis and sediment-water exchange. In: *The Benthic Boundary Layer* (ed. B.P. Boudreau and B.B. Jorgensen), 211–244. New York: Oxford University Press.

Jørgensen, B.B. and Des Marais, D.J. (1986). Competition for sulfide among colorless and purple sulfur bacteria in cyanobacterial mats. *FEMS Microbiol. Ecol.* 38: 179–186.

Jørgensen, B.B. and Revsbech, N.P. (1983). Colorless sulfur bacteria, *Beggiatoa* spp. and *Thiovulum* spp., on O_2 and H_2S microgradients. *Appl. Environ. Microbiol.* 45: 1261–1270.

Juday, C., Birge, E.A., Kemmerer, G.I., and Robinson, R.J. (1927). Phosphorus content of lake waters in northwestern Wisconsin. *Trans. Wis. Acad. Arts. Lett.* 23: 233–248.

Kamatani, A. (1982). Dissolution rates of silica from diatoms decomposing at various temperatures. *Mar. Biol.* 68: 91–96.

Keil, R.G., Montlucon, D.B., Prahl, F.G., and Hedges, J.I. (1994a). Sorptive preservation of labile organic matter in marine sediments. *Nature* 370: 549–552.

Keil, R.G., Tsamakis, E., Fuh, C.B. et al. (1994b). Mineralogical and textural controls on the organic composition of coastal marine sediments: hydrodynamic separation using SPLITT-fractionation. *Geochim. Cosmochim. Acta* 58: 879–893.

Kelly, R.P. and Moran, S.B. (2002). Seasonal changes in groundwater input to a well-mixed estuary estimated using radium isotopes and implications for coastal nutrient budgets. *Limnol Oceanogr.* 47: 1796–1807.

Kemp, W.M. and Boynton, W.R. (1981). External and internal factors regulating metabolic rates of an estuarine benthic community. *Oecologia* 51: 19–27.

Kemp, W.M. and Boynton, W.R. (1984). Spatial and temporal coupling of nutrient inputs to estuarine primary production: the role of particulate transport and decomposition. *Bull. Mar. Sci.* 35: 522–535.

Kemp, W.M., Sampou, P., Caffrey, J. et al. (1990). Ammonium recycling versus denitrification in Chesapeake Bay sediments. *Limnol. Oceanogr.* 35: 1545–1563.

Kemp, W.M., Sampou, P., Garber, J. et al. (1992). Seasonal depletion of oxygen from bottom waters of Chesapeake Bay—roles of benthic and planktonic respiration and physical exchange processes. *Mar. Ecol. Prog. Ser.* 85: 137–152.

Kempe, S., Pettine, M., and Cauwet, G. (1991). Biogeochemistry of Europe rivers. In: *Biogeochemistry of Major World Rivers* (ed. E.T. Degens, S. Kempe and J.E. Richey), 169–211. New York: John Wiley.

Kester, D.R. (1975). Dissolved gases other than CO_2. In: *Chemical Oceanography*, 2e (ed. J.P. Riley and G. Skirrow), 497–556. New York: Academic Press.

Khalil, M.A. and Rasmussen, R.A. (1992). The global sources of nitrous oxide. *J. Geophys. Res.* 97: 14651–14660.

Kieber, R.J., Jiao, J., Kiene, R.P., and Bates, T.S. (1996). Impact of dimethylsulfide photochemistry on methyl sulfur cycling in the Equatorial Pacific Ocean. *J. Geophys. Res.* 101: 3715–3722.

Kiene, R.P. (1990). Dimethyl sulfide production from dimethylsulfonioproprionate in coastal seawater samples and bacterial cultures. *Appl. Environ. Microbiol.* 56: 3292–3297.

Kiene, R.P. and Linn, L. (2000). The fate dissolved dimethylsulfoniopropionate (DMSP) in seawater: tracer studies using ^{35}S-DMSP. *Geochim. Cosmochim. Acta* 64: 2797–2810.

King, G.M., Klug, M.J., Wiegert, R.G., and Chalmers, A.G. (1982). Relation of soil water movement and sulfide concentration to *Spartina alterniflora* production. *Science* 218: 61–63.

King, G.M., Howes, B.L., and Dacey, J.W.H. (1985). Short-term endproducts of sulfate reduction in a salt marsh: the significance of acid volatile sulfide, elemental sulfur, and pyrite. *Geochim. Cosmochim. Acta* 49: 1561–1566.

King, J., Kostka, J., Frischer, M., and Saunders, F. (2000). Sulfate-reducing bacteria methylate mercury at variable rates in pure cultures and in marine sediments. *Appl. Environ. Microbiol.* 66: 2430–2437.

Knudsen, M. (1902) Berichte uber die Konstantenbestimmungen zur Aufstellung der hydrographischen Tabellen. Kon Danske Videnskab. Selsk. Skrifter, 6 Raekke, Naturvidensk. Mathemat. Vol. XII, 1–151 pp.

Koike, I. and Hattori, A. (1978). Denitrification and ammonia formation in aerobic coastal sediments. *Appl. Environ Microbiol.* 35: 278–282.

Kolowith, L.C., Ingall, E.D., and Benner, R. (2001). Composition and cycling of marine phosphorus. *Limnol. Oceanogr.* 46: 309–320.

Kornitnig, S. (1978). Phosphorus. In: *Handbook of Geochemistry* Vol. 2 (ed. K.H. Wedephol), 15E1–15E9. New York: Springer-Verlag.

Kostka, J.E. and Luther, G.W. III (1994). Partitioning and speciation of solid phase iron in saltmarsh sediments. *Geochim. Cosmochim. Acta* 58: 1701–1710.

Kostka, J.E. and Luther, G.W. III (1995). Seasonal cycling of reactive Fe in salt-marsh sediments. *Biogeochemistry* 29: 159–181.

Kristensen, E. (1988). Benthic fauna and biogeochemical processes in marine sediments: microbial activities fluxes. In: *Nitrogen Cycling in Coastal Marine Environments*. SCOPE (ed. T.H. Blackburn and J. Sørensen), 275–299. Wiley.

Kroeger, K.D. and Charette, M.A. (2008). Nitrogen biogeochemistry of submarine groundwater discharge. *Limnol. Oceanogr. 53* (3): 1025–1039.

Krom, M.D. and Berner, R.A. (1980). Adsorption of phosphate in anoxic marine sediments. *Limnol. Oceanogr.* 25: 797–806.

Krom, M.D. and Berner, R.A. (1981). The diagenesis of phosphorus in a near shore marine sediment. *Geochim. Cosmochim. Acta* 45: 207–216.

Krom, M.D., Kress, N., Brenner, S., and Gordon, L.I. (1991). Phosphorus limitation of primary productivity in the eastern Mediterranean Sea. *Limnol. Oceanogr.* 36: 424–432.

Krom, M.D., Brenner, S., Kress, N. et al. (1992). Nutrient dynamics and new production in a warm-core eddy from the eastern Mediterranean Sea. *Deep-Sea Res.* 39: 467–480.

Lajtha, K. and Michener, R.H. (ed.) (1994). *Stable Isotopes in Ecology and Environmental Science*. Oxford: Blackwell Scientific.

Lasagna, A.C. and Holland, H.D. (1976). Mathematical aspects of non-steady state diagenesis. *Geochim. Cosmochim. Acta* 40: 257–266.

Le Bornge, R. (1986). The release of soluble end products of metabolism. In: *The Biological Chemistry of Marine Copepods* (ed. D.S. Corner and S.C.M. O'Hara), 109–164. Oxford: Oxford University Press.

Lebo, M.E. (1991). Particle-bound phosphorus along an urbanized coastal plain estuary. *Mar. Chem.* 34: 225–246.

Leeder, M. (1982). *Sedimentology: Process and Product*. London: George Allen and Unwin.

Lerman, A. (1979). *Geochemical processes: Water and Sediment Environments*. New York: Wiley Interscience.

Li, J.B., Zhang, G.L., Zhang, J. et al. (2010). Matrix bound phosphine in sediments of the Changjiang Estuary and its adjacent shelf areas. *Estuar. Coastal Shelf Sci. 90* (4): 206–211.

Libes, S.M. (1992). *An Introduction to Marine Biogeochemistry*. New York: Wiley.

Lisitzin, A.P. (1995). The marginal filter of the ocean. *Oceanol.* 34: 671–682.

Liss, P.S. (1976). Conservative and non-conservative behavior of dissolved constituents during estuarine mixing. In: *Estuarine Chemistry* (ed. J.D. Burton and P.S. Liss), 93–130. London: Academic Press.

Loder, T.C. and Liss, P.S. (1985). Control by organic coatings of the surface-charge of estuarine suspended particles. *Limnol. Oceanogr.* 30: 418–421.

Lord, C.J. III and Church, T.M. (1983). The geochemistry of salt marshes: sedimentary iron diffusion. Sulfate reduction, and pyritization. *Geochem. Cosmochim. Acta* 47: 1381–1391.

Lucotte, M. and d'Anglejan, B. (1993). Forms of phosphorus and phosphorus-iron relationships in the suspended matter of the St. Lawrence estuary. *Can. J. Fish. Aquat. Sci.* 20: 1880–1890.

Luther, G.W. III, Church, T.M., Scudlark, J.R., and Cosman, M. (1986). Inorganic and organic sulfur cycling in salt-marsh pore waters. *Science* 232: 746–779.

Luther, G.W. III, Sundby, B., Lewis, B.L. et al. (1997). Interactions of manganese with nitrogen cycle: alternative pathways to dinitrogen. *Geochim. Cosmochim. Acta* 61: 4043–4052.

Maccubbin, A.E. and Hodson, R.E. (1980). Mineralization of detrital lignocelluloses by salt marsh sediment microflora. *Appl. Environ. Microbiol.* 40: 735–740.

Mackin, J.E. and Aller, R.C. (1984). Ammonium adsorption in marine sediments. *Limnol. Oceanogr.* 29: 250–257.

Malcolm, R.I. (1990). The uniqueness of humic substances in each of soil, stream, and marine environments. *Anal. Chim. Acta* 232: 19–30.

Mann, K.H. and Lazier, J.R.N. (1991). *Dynamics of Marine Ecosystems—Biological-Physical Interactions in the Oceans.* Boston: Blackwell Scientific Publications.

Markaki, Z., Oikonomou, K., Kocak, M. et al. (2003). Atmospheric deposition of inorganic phosphorus in the Levantine Basin, eastern Mediterranean: spatial and temporal variability and its role in seawater productivity. *Limnol. Oceanogr.* 48: 1557–1568.

Massé, A., Pringault, O., and de Wit, R. (2002). Experimental study of interactions between purple and green sulfur bacteria in sandy sediments exposed to illumination deprived of near-infrared wavelengths. *Appl. Environ. Microbiol.* 68: 2972–2981.

Mayer, L.M. (1994a). Surface area control or organic carbon accumulation on continental shelf sediments. *Geochim. Cosmochim. Acta* 58: 1271–1284.

Mayer, L.M. (1994b). Relationships between mineral surfaces and organic carbon concentrations in soils and sediments. *Chem. Geol.* 114: 347–363.

McCallister, S.L., Bauer, J.E., Cherrier, J.E., and Ducklow, H.W. (2004). Assessing sources and ages of organic matter supporting river and estuarine bacterial production: a multiple-isotope ($\Delta^{14}C$, $\delta^{13}C$, and $\delta^{15}N$) approach. *Limnol. Oceanogr.* 49: 1687–1702.

McKelvie, I.D., Peat, D.M., and Worsfold, P.J. (1995). Techniques for the quantification and speciation of phosphorus in natural waters. *Anal. Proc. Icnl. Anal. Comm.* 32: 437–445.

McKnight, D.M. and Aiken, G.R. (1998). Sources and age of aquatic humus. In: *Aquatic Humic Substances: Ecology and Biogeochemistry* (ed. D.O. Hessen and L.J. Tranvik), 9–39. Berlin: Springer-Verlag.

McManus, J., Berelson, W.M., Coale, K.H. et al. (1997). Phosphorus regeneration in continental margin sediments. *Geochim. Cosmochim. Acta* 61: 2891–2902.

Meentemeyer, V. (1978). Climate regulation of decomposition rates of organic matter in terrestrial ecosystems. In: *Environmental Chemistry and Cycling Processes* (ed. D.C. Adriand and I.L. Brisbin), 779–789, Conf. 760429. National Technical Information Service.

Meybeck, M. (1982). Carbon, nitrogen, and phosphorus transport by world rivers. *Am. J. Sci.* 282: 401–450.

Meyers, P.A. (1997). Organic geochemical proxies of paleoceanographic, paleolimnologic, and paleoclimatic processes. *Org. Geochem.* 27: 213–250.

Meyers, P.A. (2003). Applications of organic geochemistry to paleolimnological reconstructions: a summary of examples from the Laurentian Great Lakes. *Org. Geochem.* 34: 261–290.

Michener, R.H. and Schell, D.M. (1994). Stable isotope ratios as tracers in marine aquatic food webs. In: *Stable Isotopes in Ecology and Environmental Science* (ed. K. Lajtha and R. Michener), 138–157. Oxford: Blackwell Scientific.

Middelburg, J.J., Klaver, G., Niewenhuize, J. et al. (1995). Nitrous oxide emissions from estuarine intertidal sediments. *Hydrobiologia* 311: 43–55.

Middelburg, J.J., Soetaert, K., and Herman, P.M.J. (1997). Empirical relationships for use in global diagenetic models. *Deep Sea Res.* 44: 327–344.

Middelburg, J.J., Nieuwenhuize, J., Iverson, N. et al. (2002). Methane distribution in European tidal estuaries. *Biogeochemistry* 59: 95–119.

Millero, F.J. (1996). *Chemical Oceanography*, 2e. Boca Raton, Florida: CRC Press.

Milliman, J.D. (1980). Sedimentation in the Fraser River and its estuary, southwestern British Columbia (Canada). *Estuar. Coastal Shelf Sci.* 10: 609–633.

Milliman, J.D. and Syvitski, J.P.M. (1992). Geomorphic tectonic control of sediment discharge to the ocean – the importance of small mountainous rivers. *J. Geol.* 100: 525–554.

Mitra, S., Bianchi, T.S., Guo, L., and Santschi, P.H. (2000). Terrestrially-derived dissolved organic matter in Chesapeake Bay and the Middle Atlantic Bight. *Geochim. Cosmochim. Acta* 64: 3547–3557.

Montagna, P.A. (1989). Nitrogen process studies (NIPS): the effects of freshwater inflow on benthos communities and dynamics. Final report to the Texas Water Development Board, Austin, TX, UT Marine Science Institute Technical Report No. TR/89-011.

Mook, J.G. and Tan, F.C. (1991). Stable carbon isotopes in rivers and estuaries. In: *Biogeochemistry of Major World Rivers* (ed. E.T. Degens, S. Kempe and J.E. Richey), 245–264. SCOPE.

Moore, R.M., Burton, J.D., Willimas, P.L., and Young, M.L. (1979). The behavior of dissolved organic material, iron, and manganese in estuarine mixing. *Geochim. Cosmochim. Acta* 43: 919–926.

Moran, M.A. and Hodson, R.E. (1989a). Formation and bacterial utilization of dissolved organic carbon derived from detrital lignocellulose. *Limnol. Oceanogr.* 34: 1034–1037.

Moran, M.A. and Hodson, R.E. (1989b). Bacterial secondary production on vascular plant detritus: relationships to detritus composition and degradation rate. *Appl. Environ. Microbiol.* 55: 2178–2189.

Morel, F.M. (1983). *Principles of Aqautic Chemistry.* New York: Wiley.

Morse, J.W. and Cornwell, J.C. (1987). Analysis and distribution of iron sulfide minerals in recent anoxic marine sediments. *Mar. Chem.* 22: 55–69.

Morse, J.W. and Wang, Q. (1997). Pyrite formation under conditions approximating those in anoxic sediments: II. influence of precursor iron minerals and organic matter. *Mar. Chem.* 57: 187–193.

Mortimer, C.H. (1941). The exchange of dissolved substances between mud and water in lakes. *J. Ecol.* 29: 280–320.

Mortimer, R.J., Krom, M.D., Watson, P.G. et al. (1998). Sediment-water exchange of nutrients in the intertidal zone of the Humber Estuary, U.K. *Mar. Pollut. Bull.* 37: 261–279.

Murray, J.W. (2000). The oceans. In: *Earth System Science, from Biogeochemical Cycles to Global Change* (ed. M.C. Jacobson, R.J. Charlson, H. Rodhe and G.H. Orians), 230–278. New York: Academic Press, International Geophysics Series.

Nanny, M.A. and Minear, R.A. (1997). Characterization of soluble unreactive phosphorus using ^{31}P nuclear magnetic resonance spectroscopy. *Mar. Geol.* 139: 77–94.

Nelson, D.C. and Castenholz, R.W. (1981). Organic nutrition of *Beggiatoa* sp. *J. Bacteriol.* 147: 236–247.

Neubauer, S.C. and Anderson, I.C. (2003). Transport of dissolved inorganic carbon from a tidal freshwater marsh to the York River estuary. *Limnol. Oceanogr.* 48: 299–307.

Newell, R. (1965). The role of detritus in the nutrition of two marine deposit feeders, the prosobranch *Hydrobia ulvae* and the bivalve *Macoma balthica*. *Proc. zool. Soc.* 144: 25–45.

Nichols, M.N. (1974). Development of the turbidity maximum in the Rappahannock estuary. *Summary. Mem. Inst. Geol. Bassin d'Aquitaine*. 7: 19–25.

Nixon, S.W. (1981). Remineralization and nutrient cycling in coastal marine ecosystems. In: *Estuaries and Nutrients* (ed. B.J. Neilson and L.E. Cronin), 111–138. New York: Humana.

Nixon, S.W. (1986). Nutrient dynamics and productivity of marine coastal waters. In: *Coastal Eutrophication* (ed. B. Clayton and M. Behbehani), 97–115. Oxford: The Alden Press.

Nixon, S.W. (1995). Coastal marine eutrophication: a definition, social causes, and future concerns. *Ophelia* 4: 199–219.

Nixon, S.W., Granger, S.L., and Nowicki, B.L. (1995). An assessment of the annual mass balance of carbon, nitrogen, and phosphorus in Narragansett Bay. *Biogeochemistry* 31: 15–61.

Nixon, S.W., Ammerman, J.W., Atkinson, L.P. et al. (1996). The fate of nitrogen and phosphorus at the land-sea margin of the north Atlantic Ocean. *Biogeochemistry* 35: 141–180.

Nowicki, B.L. and Nixon, S.W. (1985). Benthic nutrient remineralization in a coastal lagoon ecosystem. *Estuaries* 8: 182–190.

Odum, W.E., Zieman, J.C., and Heald, E.J. (1973). The importance of vascular plant detritus to estuaries. In: *Coastal Marsh and Estuary Symposium* (ed. R.H. Chabreck), 91–135. Baton Rouge, Louisiana: LSU.

Oenema, O. (1990). Sulfate reduction in fine-grained sediments in the Eastern Scheldt, southwest Netherlands. *Biogeochemistry* 9: 53–74.

Officer, C.B. and Lynch, D.R. (1981). Dynamics of mixing in estuaries. *Estuar. Coastal Shelf Sci.* 12: 525–534.

Officer, C.B., Biggs, R.B., Taft, J.L. et al. (1984). Chesapeake Bay anoxia: origin, development, and significance. *Science* 223: 22–26.

Ogram, A., Sayler, G.S., Gustin, D., and Lewis, R.J. (1978). DNA adsorption to soils and sediments. *Environ. Sci. Technol.* 22: 982–984.

Opsahl, S. and Benner, R. (1997). Distribution and cycling of terrigenous dissolved organic matter in the ocean. *Nature* 386: 480–482.

Orem, W.H., Hatcher, P.G., and Spiker, E.C. (1986). Dissolved organic matter in anoxic pore waters from Mangrove Lake, Bermuda. *Geochim. Cosmochim. Acta* 50: 609–618.

Ormaza-Gonzalez, F.I. and Statham, P.J. (1991). The occurrence and behavior of different forms of phosphorus in the waters of four English estuaries. In: *Estuaries and Coasts: Spatial and temporal intercomparisons* (ed. M. Elliott and J.P. Ducrotoy), 77–83. Denmark: Olsen and Olsen.

Ouverney, C.C. and Fuhrman, J.A. (2000). Marine planktonic Archaea take up amino acids. *Appl. Env. Microbiol.* 66: 4829–4833.

Overnell, J. (2002). Manganese and iron profiles during early diagenesis in Loch Etive, Scotland. Application of two diagenetic models. *Estuar. Coastal Shelf Sci.* 54: 33–44.

Paerl, H.W. (1997). Coastal eutrophication and harmful algal blooms: Importance of atmospheric deposition and groundwater as "new" nitrogen and other nutrient sources. *Limnol. Oceanogr.* 42: 1154–1112.

Paerl, H.W., Dennis, R.L., and Whitall, D.R. (2002). Atmospheric deposition of nitrogen: implications for nutrient over-enrichment of coastal waters. *Estuaries* 25: 677–693.

Pakulski, J.D., Benner, R., Whitledge, T. et al. (2000). Microbial metabolism and nutrient cycling in the Mississippi and Atchafalaya River plumes. *Estuar. Coastal Shelf Sci.* 50: 173–184.

Park, P.K., Gordon, L.I., Hager, S.W., and Cissel, M.C. (1969). Carbon dioxide partial pressure in the Columbia River. *Science* 166: 867–868.

Patrick, O., Slawayk, G., Garcia, N., and Bonin, P. (1996). Evidence of denitrification and nitrate ammonification in the river Rhone plume (northwest Mediterranean Sea). *Mar. Ecol. Prog. Ser.* 141: 275–281.

Peierls, B.L., Caraco, N.F., Pace, M.L., and Cole, J.J. (1991). Human influence on river nitrogen. *Nature* 350: 386–387.

Peng, T.H., Broecker, W.S., Mathieu, G.G., and Li, Y.H. (1979). Radon evasion rates in the Atlantic and Pacific oceans as determined during the GEOSECS program. *J. Geophys. Res.* 84: 2471–2486.

Pennock, J.R., Boyer, J.N., Herrera-Silveira, J.A. et al. (1999). Nutrient behavior and phytoplankton production in Gulf of Mexico estuaries. In: *Biogeochemistry of Gulf of Mexico Estuaries* (ed. T.S. Bianchi, J.R. Pennock and R.R. Twilley), 109–162. New York: Wiley.

Petsch, S., Eglinton, T.I., and Edwards, K.J. (2001). ^{14}C-dead living biomass: evidence for microbial assimilation of ancient organic carbon during shale weathering. *Science* 292: 1127–1131.

Pfennig, N. (1989). Ecology of phototrophic purple and green sulfur bacteria. In: *Autotrophic Bacteria* (ed. H.G. Schlegel and B. Bowien), 97–116. Berlin: Springer.

Pollman, C.D., Landing, W.M., Perry, J.J., and Fitzpatrick, T. (2002). Wet deposition of phosphorus in Florida. *Atmos. Environ.* 36: 2309–2318.

Polubesova, T. and Chefetz, B. (2014). DOM-Affected transformation of contaminants on mineral surfaces: a review. *Crit. Rev. Environ. Sci. Technol.* 44: 223–254.

Pulliam, W.M. (1993). Carbon dioxide and methane exports from a southeastern floodplain swamp. *Ecol. Monograph.* 63: 29–53.

Quin, L.D. (1967). The natural occurrence of compounds with the carbon-phosphorus bond. In: *Topics in Phosphorus Chemistry*, vol. 4 (ed. M. Grayson and E.J. Griffith), 23–48. New York: Wiley.

Rabalais, N.N. and Nixon, S.W. (2002). Preface: nutrient over-enrichment of the coastal zone. *Estuaries* 25: 639.

Rabalais, N.N. and Turner, R.E. (ed.) (2001). *Coastal Hypoxia: Consequences for Living Resources and Ecosystems*, Coastal and Estuarine Studies 58. Washington, D.C.: American Geophysical Union.

Ragueneau, O., Conley, D.J., Longphuirt, S. et al. (2005a). A review of the Si biogeochemical cycle in coastal waters, I: diatoms in coastal food webs and the coastal Si cycle. In: *Land-Ocean Nutrient Fluxes: Silica Cycle* (ed. V. Ittekot, C. Humborg and L. Garnier). SCOPE.

Ragueneau, O., Conley, D.J., Longphuirt, S. et al. (2005b). A review of the Si biogeochemical cycle in coastal waters II: anthropogenic perturbation of the Si cycle and responses of coastal ecosystems. In: *Land-Ocean Nutrient Fluxes: Silica Cycle* (ed. V. Ittekot, C. Humborg and L. Garnier). SCOPE.

Raiswell, R. and Berner, R.A. (1985). Pyrite formation in euxinic and semi-euxinic sediments. *Am. J. Sci.* 285: 710–724.

Raiswell, R. and Canfield, D.E. (1996). Rates of reaction between silicate iron and dissolved sulfide in Peru Margin sediments. *Geochim. Cosmochim. Acta* 60: 2777–2787.

Raymond, P.A. and Bauer, J.E. (2001a). Use of ^{14}C and ^{13}C natural abundances for evaluating riverine, estuarine and coastal DOC and

POC sources and cycling: a review and synthesis. *Org. Geochem.* 32: 469–485.

Raymond, P.A. and Bauer, J.E. (2001b). Riverine export of aged terrestrial organic matter to the North Atlantic Ocean. *Nature* 409: 497–500.

Raymond, P.A., Caraco, N.F., and Cole, J.J. (1997). Carbon dioxide concentration and atmospheric flux in the Hudson River. *Estuaries* 20: 381–390.

Raymond, P.A., Bauer, J.E., and Cole, J.J. (2000). Atmospheric CO_2 evasion, dissolved inorganic carbon production, and net heterotrophy in the York River estuary. *Limnol. Oceanogr.* 45: 1707–1717.

Rice, D.L. (1982). The detritus nitrogen problem. New observations and perspectives from organic geochemistry. *Mar. Ecol. Prog. Ser.* 9: 153–162.

Rice, D.L. and Hanson, R.B. (1984). A kinetic model for detritus nitrogen: role of the associated bacteria in nitrogen accumulation. *Bull. Mar. Sci.* 35: 326–340.

Rice, D.L. and Tenore, K.R. (1981). Dynamics of carbon and nitrogen during the decomposition of detritus derived from estuarine macrophytes. *Estuar. Coastal Shelf Sci.* 13: 681–690.

Richards, F.A. (1965). Anoxic basins and fjords. In: *Chemical Oceanography*, vol. 1 (ed. J.P. Riley and G. Skirrow), 611–645. New York: Academic Press.

Richardson, T.I. (1997). Harmful or exceptional phytoplankton blooms in the marine ecosystem. *Adv. Mar. Biol.* 31: 302–385.

Richter, D.D., Markewitz, D., Trumbore, S.E., and Wells, C.G. (1999). Rapid accumulation and turnover of soil carbon in a re-establishing forest. *Nature* 400: 56–58.

Rickard, D.T. and Luther, G.W. III (1997). Kinetics of pyrite formation by the H_2S oxidation of iron (II) monosulfide in aqueous solutions between 25 and 125°C; the mechanism. *Geochim. Cosmochim. Acta* 61: 135–147.

Rietsma, C.S., Valiela, I., and Sylvester-Serianni, A. (1982). Food preferences of dominant salt marsh herbivores and detritivores. *Mar. Ecol.* 3: 179–189.

Roden, E.E. and Edmonds, J.W. (1997). Phosphate mobilization in iron-rich anaerobic sediments: microbial Fe(III) oxide reduction versus iron-sulfide formation. *Arch. Hydrobiol.* 139: 347–378.

Roden, E.E. and Tuttle, J.H. (1992). Sulfide release from estuarine sediments underlying anoxic bottom water. *Limnol. Oceanogr.* 37: 725–738.

Roden, E.E. and Tuttle, J.H. (1993a). Inorganic sulfur cycling in mid- and lower Chesapeake Bay sediments. *Mar. Ecol. Prog. Ser.* 93: 101–118.

Roden, E.E. and Tuttle, J.H. (1993b). Inorganic sulfur turnover in oligohaline estuarine sediments. *Biogeochemistry* 22: 81–105.

Rogers, D.R. and Casciotti, K.L. (2010). Abundance and diversity of archaeal ammonia oxidizers in a coastal groundwater system. *Appl. Environ. Microbiol. 76* (24): 7938–7948.

Rooney-Varga, J.N., Devereux, R., Evans, R.S., and Hines, M.E. (1997). Seasonal changes in the relative abundance of uncultivated sulfate-reducing bacteria in salt marsh sediments and in the rhizosphere of *Spartina alterniflora. Appl. Environ. Microbiol.* 63: 3895–3901.

Rosenberg, R., Nilsson, H.C., and Diaz, R.J. (2001). response of benthic fauna and changing sediment redox profiles over a hypoxic gradient. *Estuar. Coastal Shelf Sci.* 53: 343–350.

Rosenfield, J.K. (1979). Amino acid diagenesis and adsorption in nearshore anoxic sediments. *Limnol. Oceanogr.* 24: 1014–1021.

Rowe, G.T., Clifford, C.H., Smith, K.L., and Hamilton, P.L. (1975). Benthic nutrient regeneration and its coupling to primary productivity in coastal waters. *Nature* 255: 215–217.

Rozan, T.F., Taillefert, M., Trouwborst, R.E. et al. (2002). Iron-sulfur-phosphorus cycling in the sediments of a shallow coastal bay: implications for sediment nutrient release and benthic macroalgal blooms. *Limnol. Oceanogr.* 47: 1346–1354.

Ruttenberg, K.C. (1992). Development of a sequential extraction method for different forms of phosphorus in marine sediments. *Limnol. Oceanogr.* 37: 1460–1482.

Ruttenberg, K.C. and Berner, R.A. (1993). Authigenic apatite formation and burial in sediments from non-upwelling continental margin environments. *Geochim. Cosmochim. Acta* 57: 991–1007.

Rysgaard, S. and Glud, R.N. (2004). Anaerobic N_2 production in Arctic sea ice. *Limnol. Oceanogr.* 49: 86–94.

Rysgaard, S., Risgaard-Petersen, N., Sloth, N.P. et al. (1994). Oxygen regulation of nitrification and denitrification in sediments. *Limnol. Oceanogr.* 39: 1634–1652.

Samarkin, V.A., Madigan, M.T., Bowles, M.W. et al. (2010). Abiotic nitrous oxide emission from thehypersaline Don Juan pond in Antarctica. *Nature Geoscience* 3: 341–344.

Sanford, L.P., Suttles, S.E., and Halka, J.P. (2001). Reconsidering the physics of the Chesapeake Bay estuarine turbidity maximum. *Estuaries* 24: 655–669.

Santschi, P.H. (1995). Seasonality in nutrient concentrations in Galveston Bay. *Mar. Environ. Res.* 40: 337–362.

Santschi, P.H., Hohener, P., Benoit, G., and Buchholtzen, M. (1990). Chemical processes at the sediment-water interface. *Mar. Chem.* 30: 269–315.

Santschi, P.H., Guo, L., Baskaran, M. et al. (1995). Isotopic evidence for the contemporary origin of high-molecular weight organic matter in oceanic environments. *Geochim. Cosmochim. Acta* 59: 625–631.

Santschi, P.H., Lenhart, J.J., and Honeyman, B. (1997). Heterogeneous processes affecting trace contaminant distribution in estuaries: the role of natural organ matter. *Mar. Chem.* 58: 99–125.

Schedel, M. and Truper, H. (1980). Anaerobic oxidation of thiosulfate and elemental sulfur in *Thiobacillus denitrificans. Arch. Microbiol.* 124: 205–210.

Schnitzer, M. and Khan, S.U. (1972). *Humic Substances in the Environment.* N.Y.: Marcel Dekker.

Schubel, J.R. (1968). Turbidity maximum of the northern Chesapeake Bay. *Science* 161: 1013–1015.

Schubel, J.R. and Biggs, R.B. (1969). Distribution of seston in upper Chesapeake Bay. *Ches. Sci.* 10: 18–23.

Schubel, J.R. and Kana, T.W. (1972). Agglomeration of fine-grained suspended sediment in northern Chesapeake Bay. *Power Technol.* 6: 9–16.

Scranton, M.I. and McShane, K. (1991). Methane fluxes in the southern North Sea: the role of European rivers. *Cont. Shelf Res.* 11: 37–52.

Seitzinger, S.P. (1988). Denitrification in freshwater and coastal marine ecosystems: ecological and geochemical significance. *Limnol. Oceanogr.* 33: 702–724.

Seitzinger, S.P. (2000). Scaling up: site-specific measurements to global estimates of denitrification. In: *Estuarine Science: A Synthetic Approach to Research and Practice* (ed. J.E. Hobbie), 211–240. Washington, D.C.: Island Press.

Seitzinger, S.P. and Kroeze, C. (1998). Global distribution of nitrous oxide production and N inputs in freshwater and coastal marine ecosystems. *Global Biogeochem. Cycles* 12: 93–113.

Seitzinger, S.P. and Nixon, S.W. (1985). Eutrophication and the rate of denitrification and N2O production in coastal marine sediments. *Limnol. Oceanogr.* 30: 1332–1339.

Seitzinger, S.P. and Sanders, R.W. (1997). Contribution of dissolved organic nitrogen from rivers to estuarine eutrophication. *Mar. Ecol. Prog. Ser.* 159: 1–12.

Seitzinger, S.P., Kroeze, C., Bouman, A.F. et al. (2002). Global patterns of dissolved inorganic and particulate nitrogen inputs to coastal systems: recent conditions and future projections. *Estuaries* 25: 640–655.

Shakhova, N., Semiletov, I., Gustafsson, O. et al. (2017). Current rates and mechanisms of subsea permafrost degradation in the East Siberian Arctic Shelf. *Nat. Commun.* 8: 15872. https://doi.org/10.1038/ncomms15872.

Sharp, J.H. (1973). Size classes of organic carbon in seawater. *Limnol. Oceanogr.* 18: 441–447.

Sharp, J.H. (1983). The distribution of inorganic nitrogen and dissolved and particulate organic nitrogen in the sea. In: *Nitrogen in the Marine Environment* (ed. E.J. Carpenter and D.G. Capone). New York: Academic Press.

Sholkovitz, E.R. (1976). Flocculation of dissolved organic and inorganic matter during the mixing of river water and seawater. *Geochim. Cosmochim. Acta* 40: 831–845.

Sholkovitz, E., Boyle, E., and Price, N. (1978). Removal of dissolved humic acids and iron during estuarine mixing, Earth Planet. 20 Sci. *Lett.* 40: 130–136. 10.1016/0012-821X(78)90082-1.

Siefert, R.L., Johansen, A.M., Hoffmann, M.R., and Pehkonen, S.O. (1998). Measurements of trace metal (Fe, Cu, Mn, Cr) oxidation states in fog and stratus clouds. *Air and Waste Manag.* 48 (2): 128–143. https://doi.org/10.1080/10473289.1998.10463659.

Sigleo, A.C. and Macko, S.A. (1985). Stable isotope and amino acid composition of estuarine dissolved colloidal material. In: *Marine and Estuarine Geochemistry* (ed. A.C. Sigleo and A. Hattori), 29–46. Boca Raton, Florida: Lewis Publishing.

Simo, R., Grimalt, J.O., and Albaiges, J. (1997). Dissolved dimethylsulfide, dimethylsulphonioproprionate and dimethylsulphoxide in western Mediterranean waters. *Deep-Sea Res.* II: 929–950.

Slomp, C.P., Malschaert, J.F.P., Lohse, L., and van Raaphorst, W. (1997). Iron and manganese cycling in different sedimentary environments on the Morth Sea continental margin. *Cont. Shelf Res.* 17: 1083–1117.

Sloth, N.P., Blackburn, H., Hansen, L.S. et al. (1995). Nitrogen cycling in sediments with different organic loading. *Mar. Ecol. Prog. Ser.* 116: 163–170.

Smith, L., Kruszynah, H., and Smith, R.P. (1977). The effect of metheglobin on the inhibition of cytochrome c oxidase by cyanide, sulfide or azide. *Biochem. Pharmacol.* 26: 2247–2250.

Sørensen, J. (1987). Nitrate reduction in marine sediment: pathways and interactions with iron and sulfur cycling. *Geomicrobiol. J.* 5: 401–421.

Spiteri, C., Slomp, C.P., Charette, M.A. et al. (2008). Flow and nutrient dynamics in a subterranean estuary (Waquoit Bay, MA, USA): field data and reactive transport modeling. *Geochimica et Cosmochimica Acta* 72 (14): 3398–3412.

Stirling, H.P. and Wormald, A.P. (1977). Phosphate/sediment interaction in toto and Long Harbors, Hong Kong, and its role in estuarine phosphorus availability. *Estuar. Coastal Shelf Sci.* 5: 631–642.

Strickland, J.D.H. and Parsons, T.R. (1972). *A Practical Handbook of Seawater Analysis.* Flsherles Research Board of Canada.

Struyf, E., Smis, A., Van Damme, S. et al. (2009). The global biogeochemical silicon cycle. *Silicon 1* (4): 207–213.

Stumm, W. and Morgan, J.J. (1996). *Aquatic Chemistry, Chemical Equilibria and Rates in Natural Waters*, 3e. New York: Wiley.

Suberkropp, K., Godshalk, G., and Klug, M.J. (1976). Changes in the chemical composition of leaves during processing in a woodland stream. *Ecology* 57: 720–727.

Sun, M.Y., Lee, C., and Aller, R.C. (1993). Laboratory studies of oxic and anoxic degradation of cholorophyll-a in Long-Island Sound sediments. *Geochim. Cosmochim. Acta* 57: 147–157.

Sundbäck, K., Enoksson, V., Granéli, W., and Pettersson, K. (1991). influence of sublittoral microphytobenthos on the oxygen and nutrient flux between sediment and water: a laboratory continuous-flow study. *Mar. Ecol. Prog. Ser.* 74: 263–279.

Sundby, B., Gobeil, C., Silverburg, N., and Mucci, A. (1992). The phosphorus cycle in coastal marine sediments. *Limnol. Oceanogr.* 37: 1129–1145.

Sutula, M., Bianchi, T.S., and McKee, B. (2004). Effect of seasonal sediment storage in the lower Mississippi River on the flux of reactive particulate phosphorus to the Gulf of Mexico. *Limnol. Oceanogr.* 49: 2223–2235.

Taillefert, M., Bono, A.B., and Luther, G.W. III (2000). Reactivity of freshly formed Fe(III) in synthetic solutions and porewaters: voltammetric evidence of an aging process. *Environ. Sci. Technol.* 34: 2169–2177.

Tang, D., Chin-Chang, H., Warnken, K.W., and Santschi, P.H. (2000). The distribution of biogenic thiols in surface waters of Galveston Bay. *Limnol. Oceanogr.* 45: 1289–1297.

Taniguchi, M., Burnett, W.C., Cable, J.E., and Turner, J.V. (2002). Investigations of submarine groundwater discharge. *Hydrological Processes* 16: 2115–2129.

Tenore, K.R., Cammen, L., Findlay, S.E.G., and Phillips, N. (1982). Perspectives of research on detritus: do factors controlling the availability of detritus to macroconsumers depend on its source? *J. Mar. Res.* 40: 473–480.

Thamdrup, B. and Dalsgaard, T. (2002). Production of N_2 through anaerobic ammonium oxidation coupled to nitrate reduction in marine sediments. *Appl. Environ. Microbiol.* 68: 1312–1318.

Tobias, C.R., Anderson, I.C., Canuel, A.C., and Macko, S.A. (2001). Nitrogen cycling through a fringing marsh-aquifer ecotone. *Mar. Ecol. Prog. Ser.* 210: 25–39.

Tranvik, L.J., Sherr, E.B., and Sherr, B.F. (1993). Uptake and utilization of colloidal DOM by heterotrophic flagellates in seawater. *Mar. Ecol. Prog. Ser.* 92: 301–305.

Tréguer, P., Nelson, D.M., van Bennekom, A.J. et al. (1995). The silica balance in the world ocean: a re-estimate. *Science* 268: 375–379.

Turner, A. and Millward, G.E. (2002). Suspended particles: their role in estuarine biogeochemical cycles. *Estuar. Coastal Shelf Sci.* 55: 857–883.

Turner, R.E. and Rabalais, N.N. (1991). Changes in Mississippi River water quality this century: implications for coastal food webs. *Bioscience* 41: 140–147.

Turner, S.M., Malin, G., Nightingale, P.D., and Lis, P.S. (1996). Photochemical production and air-sea exchange of OCS in the eastern Mediterranean Sea. *Mar. Chem.* 53: 25–39.

Turner, R.E., Rabalais, N.N., Justic, D., and Dortch, Q. (2003). Global patterns of dissolved N, P, and Si in large rivers. *Biogeochemistry* 64: 297–317.

Turner, B.L., Newman, S., and Newman, J.M. (2006). Organic phosphorus sequestration in subtropical treatment wetlands. *Environ. Sci. Technol.* 40: 727. https://doi.org/10.1021/ES0516256.

Tuttle, J.H., Jonas, R.B., and Malone, T.C. (1987). Origin, development, and significance of Chesapeake Bay anoxia. In: *Contaminant Problems and Management of Living Chesapeake Bay Resources* (ed. S.K. Majumdar, L.W. Hall and M.A. Hebert), 442–472. Philadelphia: Pennsylvania Academy of Natural Sciences.

Uncles, C.M., Lavender, S.J., and Stephens, J.A. (2001). Remotely sensed observations of the turbidity maximum in the high turbid Humber Estuary, U.K. *Estuaries* 24: 745–755.

Usui, T., Koike, I., and Ogura, N. (2001). N_2O production, nitrification and denitrification in an estuarine sediment. *Estuar. Coastal Shelf Sci.* 52: 769–781.

Valiela, I. (1995). *Marine Ecological Processes*, 2e. New York: Springer.

Valiela, I., Koumjian, L., Swain, T. et al. (1979). Cinnamic acid inhibition of detritus feeding. *Nature* 280: 55–57.

Valiela, I., Teal, J.M., Allan, S.D. et al. (1985). Decomposition in salt marsh ecosystems: the phases and major factors affecting disappearance of above-ground organic matter. *J. Exp. Mar. Biol. Ecol.* 89: 29–54.

Velinsky, D.J., Wade, T.L., and Wong, G.T.F. (1986). Atmospheric deposition of organic carbon to Chesapeake Bay. *Atmos. Environ.* 20: 941–947.

Vold, R.D. and Vold, M.J. (1983). *Colloid and Interface Chemistry.* Reading, Massachusetts: Addison-Wesley.

Wang, Z.A. and Cai, W. (2004). Carbon dioxide degassing and inorganic carbon export froma marsh-dominated estuary (the Duplin River): a marsh CO_2 pump. *Limnol. Oceanogr.* 49: 341–354.

Wang, X.C. and Lee, C. (1990). The distribution and adsorption behavior of aliphatic amines in marine and lacustrine sediments. *Geochim. Cosmochim. Acta* 54: 2759–2774.

Wang, W.C., Yung, Y.L., Lacis, A.A. et al. (1976). Greenhouse effects due to man-made perturbations of trace gases. *Science* 194: 685–690.

Wangersky, P.J. and Wangersky, C.P. (1980). The Manna effect—A model of phytoplankton patchiness in a regenerative system. *Int. Revue Der Gesam. Hydrobio.* 65: 681–690.

Ward, N.D., Bianchi, T.S., Medeiros, P.M. et al. (2017). Where carbon goes when water flows: carbon cycling across the aquatic continuum. *Front. Mar. Sci. 4*: 7.

Warnken, K.W., Gill, G.A., Santschi, P.H., and Griffin, L.L. (2000). Benthic exchange of nutrients in Galveston Bay, Texas. *Estuaries* 23: 647–661.

Webster, J.R. and Benfield, E.F. (1986). Vascular plant breakdown in freshwater ecosystems. *Ann. Rev. Ecol. Syst.* 17: 567–594.

Wen, L., Shiller, A., Santschi, P.H., and Gill, G. (1999). Trace element behavior in Gulf of Mexico estuaries. In: *Biogeochemistry of Gulf of Mexico Estuaries* (ed. T.S. Bianchi, J.R. Pennock and R.R. Twilley), 303–346. New York: Wiley.

Westrich, J.T. and Berner, R.A. (1984). The role of sedimentary organic matter in bacterial sulfate reduction: the G model tested. *Limnol. Oceanogr.* 29: 236–249.

Wheatcroft, R.A., Jumars, P.A., Smith, C.R., and Nowell, A.R.M. (1991). A mechanistic view of the particulate biodiffusion coefficient: step lengths, rest periods and transport direction. *J. Mar. Res.* 48: 177–207.

Wilson, J.O., Buchsbaum, R., Valiela, I., and Swain, T. (1986). Decomposition in salt marsh ecosystems: phenolic dynamics during decay of litter of *Spartina alterniflora. Mar. Ecol. Prog. Ser.* 29: 177–187.

Wollast, R. and Mackenzie, F.T. (1983). The global cycle of silica. In: *Silicon Geochemistry and Biogeochemistry* (ed. S.R. Aston), 39–76. San Diego: Academic Press.

Yoshinari, T., Altabet, M.A., Naqvi, S.W.A. et al. (1997). Nitrogen and oxygen iso- topic composition of N20 from suboxic waters of the eastern tropical North Pacific and the Arabian Sea – Measurement by continuous-flow isotope-ratio monitoring. *Mar. Chem.* 56: 253–264.

Zappa, C.J., McGillis, W.R., Raymond, P.A. et al. (2007). Environmental turbulent mixing controls on air-water gas exchange in marine and aquatic systems. *Geophys. Res. Lett.* 34: L10601. https://doi.org/10.1029/2006GL028790.

Zhu-Barker, X., Cavazos, A.R., Ostrom, N.E. et al. (2015). The importance of abiotic reactions for nitrous oxide production. *Biogeochemistry* 126: 251–267.

Zimov, S.A., Schuur, E.A.G., and Chapin, F.S. III (2006). Permafrost and the global carbon budget. *Science* 312: 1612–1613.

Zweifel, U.L. (1999). Factors controlling accumulation of labile dissolved organic carbon in the Gulf of Riga. *Estuaries* 48: 357–370.

CHAPTER 4

Estuarine Phytoplankton

Hans W. Paerl[1], Dubravko Justić[2], Connie Lovejoy[3], and Ryan W. Paerl[4]
[1]University of North Carolina at Chapel Hill, Institute of Marine Sciences, Morehead City, NC, USA
[2]Department of Oceanography and Coastal Sciences, Louisiana State University, Baton Rouge, LA, USA
[3]Department of Biology, Laval University, Québec, Canada
[4]Department of Marine, Earth, and Atmospheric Sciences, North Carolina State University, Raleigh, NC, USA

Phytoplankton bloom in the Chowan R. estuary at the head of Albemarle Sound, North Carolina, USA. Photo credit: Abe Loven.

4.1 Introduction

Phytoplankton means "drifting plant" in Greek. These planktonic microalgae are comprised of many taxonomic groups (e.g., chlorophytes, chrysophytes, cryptophytes, cyanobacteria, diatoms, and dinoflagellates; Figure 4.1) that conduct a large share of photosynthesis and primary production, and play a central role in carbon, nutrient (i.e., N and P), and oxygen cycling in estuarine and coastal waters (jointly termed coastal waters). In most coastal ecosystems, phytoplankton account for at least half of ecosystem primary production (Malone et al. 1999; Cloern 2001; Sigman and Hain 2012). Hence, phytoplankton are of fundamental importance in supporting estuarine food webs, and play a central role in determining water quality.

Phytoplankton have fast growth rates, on the order of one doubling per day, and members of some groups (dinoflagellates, cryptophytes, haptophytes, and cyanobacteria) can proliferate in explosive ways, forming dense "blooms" that can discolor affected waters and cause water quality problems. When blooms die or "crash" they often decompose rapidly and sink to the bottom where they fuel high rates of oxygen consumption, leading to oxygen-depleted bottom waters (i.e., hypoxia and anoxia; Diaz and Rosenberg 2008). Some bloom species also produce foul odors and tastes, which can be problematic from water supply, recreational, and aquaculture perspectives (Davidson et al. 2011). Lastly, some species produce secondary metabolites that can be toxic to higher fauna (Carmichael 2001; Trainer et al. 2012), including zooplankton grazers, fish, and a variety of mammals, including humans.

Most phytoplankton exist as microscopic solitary cells, although some form multicelled chains and aggregates that are visible to the naked eye. They exist in several size classes, including the picoplankton (<3 μm), nanoplankton (3–20 μm), and microplankton (>20 μm). The relative contributions of these size classes to total phytoplankton biomass varies according to nutrient status, physical (temperature, irradiance, and mixing), and hydrologic (freshwater discharge) conditions, as well as climatic regimes.

Generally, phytoplankton cells are denser than water, having silica, cellulose, and/or carbonate components, and thus tend to sink. Various mechanisms allow phytoplankton to remain in the illuminated upper water column. Some cells remain easily suspended because of their diminutive size (i.e., picophytoplankton). Others glide or actively swim using eukaryotic flagella, while some adjust their buoyancy by altering cellular density, or forming intracellular vesicles containing air (Walsby 1994). Several major phytoplankton groups

Estuarine Ecology, Third Edition. Edited by Byron C. Crump, Jeremy M. Testa, and Kenneth H. Dunton.

Companion website: www.wiley.com/go/crump/estuarine3

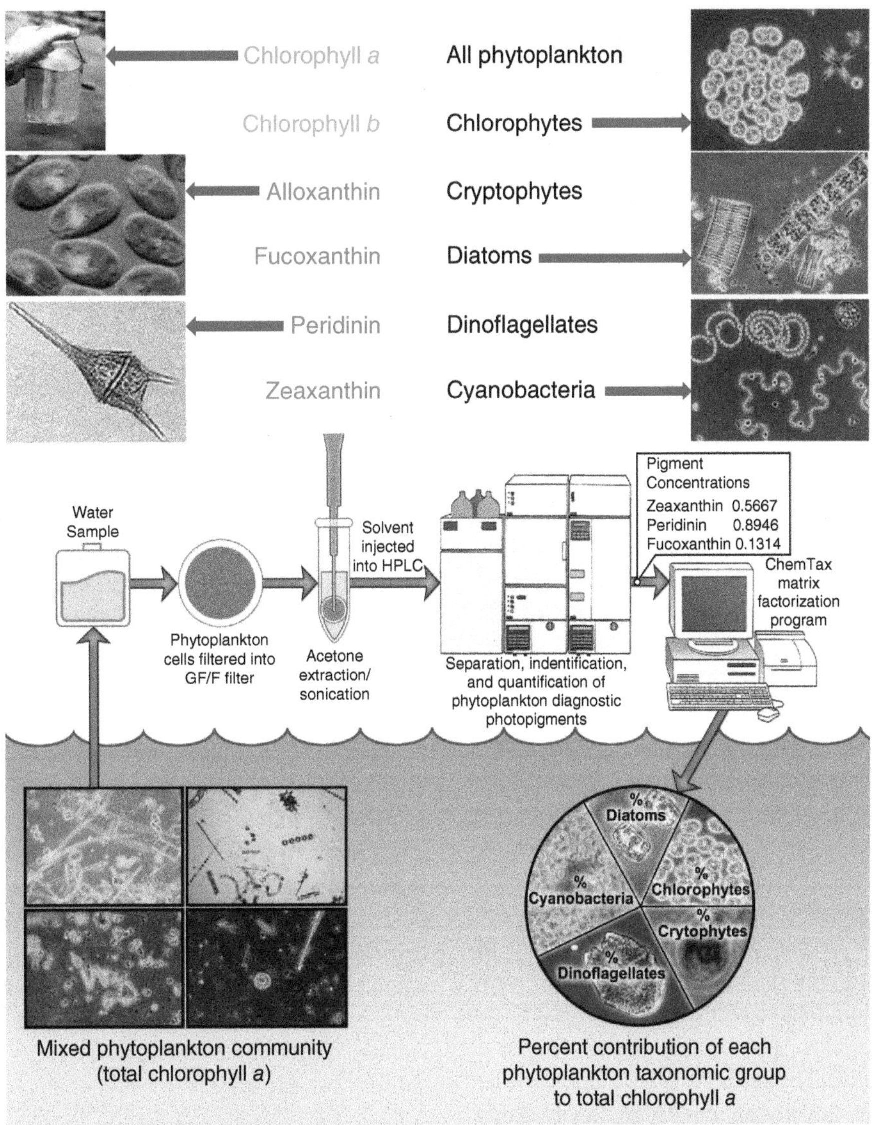

FIGURE 4.1 Upper figure. Examples of the major estuarine phytoplankton groups and their diagnostic photopigments. Lower figure. A schematic figure showing the procedure for preparing and quantifying algal diagnostic photopigments by high performance liquid chromatography. *Source*: From Paerl et al. (2003).

(e.g., diatoms, dinoflagellates, and cyanobacteria) can form cysts and inactive cells as "resting stages" and during unfavorable growth periods, sink to the sediments. These cells may be reactivated when favorable conditions reappear.

4.2 The Players: Phytoplankton Community Composition and Function

Estuarine and coastal phytoplankton communities are oxygenic photosynthetic plankton containing chlorophyll *a* that use light energy to convert ("fix") CO_2 into organic carbon and produce O_2. Phytoplankton can be grouped taxonomically and functionally. With the exception of cyanobacteria, phytoplankton are microbial eukaryotes, with membrane-bound organelles including the chloroplast, the main reactor for fixing inorganic carbon. The eukaryotic coastal phytoplankton are phylogenetically diverse and are categorized by their genetic differences. Their foundational classification is based roughly on pigmentation and morphology. For example, the chlorophytes (green algae) and the chrysophytes (yellow-green algae), which are mostly small flagellates, are distinguished based on their color. Groups identified by morphological characteristics include the Bacillariophyta (diatoms), the Dinophyta (dinoflagellates), the Cryptophyta (cryptomonads), and the Prymnesiophyta. Each group plays an important role in primary production, food web dynamics, and biogeochemical cycling. They also differentially (and at times uniquely) respond to nutrient enrichment, hydrologic forcing (freshwater discharge and salinity regimes), irradiance gradients, and grazing pressures.

4.2.1 Cyanobacteria

Cyanobacteria, like other bacteria, lack a well-defined nucleus and do not have membrane-bound organelles. Cyanobacteria can be spheroid single or aggregated cells, aggregated or solitary non-heterocystous filamentous groups, and heterocystous filamentous groups (Stanier and Cohen-Bazire 1977; Komarek and Anagnostidis 1986). Many cyanobacteria are diazotrophic, meaning that they are able to fix atmospheric nitrogen (N_2). In a number of estuarine species, specialized cells (heterocysts) house the oxygen-sensitive nitrogen fixing apparatus (Wolk 1982), an important physiological adaptation that enables these taxa to collectively photosynthesize and fix N_2, converting it to biologically available ammonia (NH_3) (Paerl and Zehr 2000). This biologically available N is useful during N-limited conditions that frequently characterize estuarine and coastal waters. Some spheroid and non-heterocystous filamentous cyanobacterial groups can also fix N_2, mostly during dark periods, temporarily segregating the O_2-sensitive N_2 fixation process from the O_2-producing process of photosynthesis, and in oxygen-depleted microenvironments such as aggregates, biofilms, mats, and as endosymbionts (Paerl and Kuparinen 2002; Zehr and Paerl 2008).

Interestingly, while N_2 fixing cyanobacteria are capable of meeting their N requirements under N-depleted conditions, these N_2 fixers rarely dominate the phytoplankton communities in N-depleted estuaries. Notable exceptions are brackish, lagoonal, and periodically stratified waters, where water column stability and low salinity appear to favor this group (Paerl 1990). In the brackish Baltic Sea, extensive blooms of heterocystous filamentous genera (*Aphanizomenon*, *Dolichospermum*, *Nodularia*) occur during vertically stratified summer conditions (Kononen et al. 1996). These blooms are significant sources of "new" N, supporting production and eutrophication of the Baltic Sea. Other examples include non- or micro-tidal lagoonal estuarine systems like Peel-Harvey Estuary in Australia (*Nodularia*), Lake Ponchartrain, LA, USA (*Dolichospermum*), and the sub-estuaries (Chowan, Neuse) of the lagoonal Pamlico Sound system, NC, USA (*Cylindrospermopsis*, *Dolichospermum*).

Salinity does not *a priori* appear to be a barrier to the expansion of diazotrophic cyanobacteria in estuaries (Moisander et al. 2002). Grazing pressure, likewise, does not explain the puzzling lack of diazotrophic cyanobacterial dominance. However, physical constraints, including excessive turbulence, persistent vertical mixing and high rates of flushing (i.e., short water residence time) that characterize many of the world's estuarine and coastal ecosystems, prevents dominance by these otherwise opportunistic diazotrophs (Moisander et al. 2002). This may help explain chronic N limitation in these systems where biological N_2 fixation fails to meet ecosystem N demands (Paerl 2017).

Blooms of the heterocystous N_2 fixing cyanobacteria can produce a variety of odor and taste compounds such as geosmin and 2-Methylisoborneol (MiB), rendering affected waters unsuitable for consumption, aquaculture, and recreation (Carmichael 2001; Stewart and Falconer 2008). Some cyanobacterial bloom species produce alkaloid, peptide, and other compounds that can be toxic to mammals upon ingestion or in contact with affected waters (Carmichael 2001; Stewart and Falconer 2008). The ecological relevance of these toxic compounds is debated, but potentially includes coping with reactive oxygen species during oxygen supersaturated bloom conditions and other cellular processes, including sequestration of iron, storage of nitrogen, and deterring grazers (e.g., Paerl and Otten 2013a, b; Paerl 2018).

The subtropical–tropical oceanic, non-heterocystous, filamentous N_2 fixing genus *Trichodesmium* can, at times, make excursions into coastal waters when it is carried in by major currents including the Gulf Stream in the Atlantic and Gulf of Mexico and the Kurushio current in the South China Sea. *Trichodesmium* blooms can provide a major source of "new" N supporting primary production in these waters (Capone et al. 1997). They also produce secondary metabolites that may deter grazing by dominant crustacean zooplankton (Hawser et al. 1992), thereby affecting food web dynamics.

Non-N_2 fixing cyanobacterial genera make up an important, and at times dominant, fraction of estuarine phytoplankton biomass. In particular, small (<3 μm) coccoid picocyanobacteria, belonging to the genus *Synechococcus*, can constitute more than 50% of estuarine phytoplankton during summer months (Gaulke et al. 2010). A second picocyanobacterial genus *Prochlorococcus* is generally more dominant in warmer oligotrophic waters and occurs in some tropical and subtropical estuaries (Mitbavkar et al. 2012; Zhang et al. 2013).

4.2.2 Green Algae

"Green algae" are green in part due to the presence of an additional form of the chlorophyll molecule, chlorophyll *b*. They are taxonomically diverse with several major evolutionary lineages (Tragin and Vaulot 2018). In estuaries and coastal regions, the green algal groups tend to be separated by salinity. Chlorophyceae (chlorophytes) are found at the node of an evolutionary branch leading to nonvascular plants and eventually land plants. They are mostly found in fresh water and at times dominant in the low salinity, upstream segments of estuaries (Tomas 1996). They exist as free-floating solitary or aggregated cells varying in size and shape; from small coccoid cells (2–5 μm diameter), in part composing the picophytoplankton, to larger ovoid and disk-shaped cells (the desmids) as small groups of stacked cells (e.g., *Scenedesmus*). The chlorophyta also contain flagellated genera (e.g., *Chlamydomonas* and *Dunaliella*) that can accumulate as bright-green or deep red blooms under high light due to the accumulation of carotenoids, which are photoprotective pigments. Chlorophytes have relatively fast growth rates, and many thrive in fast flowing, short residence time, and low-salinity waters. These waters tend to have elevated nutrient concentrations, favoring fast growing species. One group of chlorophytes, such as *Dunaliella*, are highly salt (NaCl) tolerant and are found in evaporative coastal systems. While the chlorophytes can form blooms, there are no known toxic species and blooms are not considered to be particularly "harmful". However, large accumulations of un-grazed cells can contribute to bottom water hypoxia by sinking out of the illuminated euphotic zone.

Other "green algae" are more common in marine waters, and there are several separate evolutionary branches that are frequently seen in coastal and estuarine systems. One of these branches, the Pyramonadales, can be bright green with 4–16 flagella. Although mostly marine, at least one genus, *Pyramimonas* has a wide salinity tolerance and is found from tide pools to the open ocean (Tragin and Vaulot 2018). Another group, the Mammiellales, is often reported in coastal waters and includes three picoplankton size genera (*Micromonas*, *Bathycoccus*, and *Ostreococcus*) that can "bloom" (~5×10^5 cells mL^{-1}; O'Kelly et al. 2003). These very small eukaryotes can exceed the abundances of similarly sized pico-cyanobacteria and exhibit different morphologies and behaviors. For example, *Micromonas* are motile, while *Bathycoccus* and *Ostreococcus* are nonmotile.

4.2.3 Cryptophytes

Cryptophytes all possess flagella, which help them maintain an optimal position in the upper water column relative to light and nutrient conditions (e.g., Tomas 1996; Reynolds 2006). The morphology of most cryptophytes is similar, with droplet shaped cells, two flagella, and a furrow-gullet complex referred to as an oral groove. The two unequal flagella serve as forward and reverse "thrusters," propelling them in a distinct and highly effective manner. Cryptophytes are also mixotrophic and able to capture small prey items, such as bacteria, as well as perform photosynthesis. This provides considerable metabolic flexibility, enabling the cells to function in photosynthetic autotrophic and heterotrophic modes. Some cryptophytes have "lost" their capacity to photosynthesize over time (e.g., *Goniomonas* spp.); hence, they derive all their nutrition heterotrophically. The photosynthetic cryptophytes contain chlorophylls *a* and c_2, xanthophylls, and blue or red phycobiliproteins. This imparts a brownish, reddish, or even bluish color to blooms. Their preferred habitat is fresh to brackish nutrient-enriched waters. As a result, blooms are often observed just upstream of strong salinity gradients in the oligohaline regions of estuaries. Cryptophyte blooms are not known to be toxic. The most "harmful" aspect of these blooms is that they can contribute significant amounts of organic matter to bottom waters when they die, potentially enhancing hypoxic conditions.

4.2.4 Chrysophytes

Chrysophytes contain chlorophyll *a*, chlorophyll *c*, and a yellowish-brown carotenoid pigment called fucoxanthin. Blooms typically appear light brown to golden in appearance. Most species are free-swimming (flagellar) and unicellular, but some form colonies with distinct forms such as *Dinobryon*. Chrysophytes have complex life cycles and some form distinct resting cysts, used in paleoclimate studies. The genus *Dinobryon*, with both marine and freshwater species, has individual cells that are surrounded by vase-shaped loricae, composed of chitin fibrils and other polysaccharides. The colonies grow as branched or unbranched chains. Another colonial form, *Synura*, has cells that are covered by silica scales. Some species are colorless, lack the capacity to photosynthesize, and primarily graze on bacteria. Most chrysophytes contain photosynthetic pigments and exhibit a great deal of metabolic flexibility. For example, they can be facultative heterotrophs in the absence of light, or when high concentrations of dissolved organic matter are present.

4.2.4.1 Pelagophytes

In recent decades, pelagophytes, which are distantly related to chrysophytes and also have chlorophylls *a* and *c*, were identified as agents of toxic algal blooms. In particular, *Aureococcus* sp. and *Aureoumbra* sp. have proliferated as "brown tides" in estuaries along the US Northeast and mid-Atlantic as well as some of the lagoonal estuaries in the Gulf of Mexico (Bricelj and Lonsdale 1997; Gobler et al. 2002), and Chinese coastal waters. These motile species reveal considerable genetic diversity and functional complexity. Under favorable conditions they can bloom at densities high enough to effectively "shade out" bottom-dwelling higher plant (seagrass) and benthic microalgal communities, leading to increases in hypoxic conditions, and adversely affecting bottom habitat for infauna, including commercially important shellfish species. The recent upsurge of pelagophyte blooms appears linked to biogeochemical changes brought about by droughts, excessive groundwater withdrawal, and nutrient enrichment in nearby coastal regions, which have locally increased salinity, nutrient, and organic matter concentrations. In addition, increased hypoxia and other adverse effects on the infauna in these habitats have reduced grazing on pelagophytes, further increasing its dominance (i.e., positive feedback; Gobler et al. 2002).

4.2.5 Diatoms

Diatoms in the phylum Bacillariophyta are among the most abundant, widespread, and productive phytoplankton in coastal waters, and serve a central role in planktonic food webs (Hasle et al. 1996). Most diatoms prefer waters with moderate to high nutrient concentrations. They are capable of very fast growth rates, on the order of two doublings per day or even faster, and because their growth optima occur at relatively low temperatures, they tend to bloom during late winter and spring, when relatively high nutrient loads often coincide with maximum freshwater runoff conditions or when upwelling might take place. Diatoms are an excellent source of food for grazers, including zooplankton, benthic filter feeders, larvae, and planktivorous fish (e.g., menhaden). Members of several diatom genera (*Rhizosolenia, Hemiaulus*) are capable of hosting endosymbiotic N_2-fixing cyanobacteria (Foster et al. 2011), and as a result these organisms do well under N-limiting conditions (Carpenter et al. 1999).

Morphologically, diatoms are diverse, as their siliceous cell walls or frustules take on many different shapes and sizes. Classes of diatoms are defined by the symmetry of the frustules, with bilaterally symmetric, longer than wide, sometimes boat-shaped "pennates" (Bacillariophyceae) and radially symmetric "centric" diatoms (Coscinodiscophyceae) that resemble round pill boxes. A third class, the Mediophyceae, previously considered centric (Round et al. 1990) was proposed in 2004 (Medlin and Kaczmarska 2004) and is now generally accepted by diatom taxonomists. The Mediophyceae, which include many common coastal marine genera including *Chaetoceros* and *Skeletonema*, are genetically distinct and less obviously symmetric due to bi- and multipolar symmetry. Centric diatoms and the Mediophyceae are generally planktonic, while pennate diatoms predominate benthic communities. However, in shallow coastal waters benthic pennate diatoms are often resuspended from the bottom and seen in plankton samples.

Diatoms occur as either solitary or joined (usually as chains) planktonic cells, or attached to a substratum by means of gelatinous extrusions and can form visible masses of cells within mucus-like material. Some species are capable of movement via "jet propulsion" accomplished through mucilaginous excretions, while other species are free floating and are dependent on currents for transport. Diatoms have complex life cycles that can involve benthic resting and sexual reproductive stages occurring after an extended pelagic growth phase. Diatom cells range from 2 μm to well over 100 μm in diameter. Because they rely on silicon for cell wall formation, diatoms may, at times, be controlled or "limited" by silicon supply (Flynn and Martin-Jézéquel 2000). Silicon limitation can be particularly evident when nitrogen and phosphorus supplies are elevated relative to silicon as a result of eutrophication (Humborg et al. 2007).

Diatoms are considered highly desirable phytoplankton in estuaries because they support planktonic and benthic food webs (Hasle et al. 1996; Reynolds 2006). However, one planktonic pennate genus, *Pseudo-nitzschia*, can exude a toxin that causes amnesia in vertebrates (D'agostino et al. 2017). This species appears to bloom in response to eutrophication in coastal waters, including the Pacific Northwest (Trainer et al. 2012), California (McCabe et al. 2016), and the Mississippi River plume in the Northern Gulf of Mexico (Parsons et al. 2002). The causes of *Pseudo-nitzschia* blooms are unclear but may be related to increases in nutrient loading and shifts in nutrient ratios (e.g., N:P; Parsons et al. 2002; McCabe et al. 2016).

4.2.6 Prymnesiophytes

Prymnesiophytes, or haptophytes, include about 500 species in 50 genera, with many additional fossil genera and species, most notably the coccolithophorids that are covered in calcium carbonate "coccoliths." These hard exteriors can lead to complex architectures and are the source of ancient deposits of calcareous earth, including the Cliffs of Dover in the United Kingdom. Prymnesiophytes are primarily unicellular and photosynthetic, and are consumed by zooplankton in coastal waters (Tomas 1996). Prymnesiophytes are often a golden-brown color because they contain the carotenoid accessory pigments diadinoxanthin and fucoxanthin in addition to chlorophylls *a* and *c*. Prymnesiophytes have a complex life cycle, altering between motile and nonmotile morphologies. The motile forms typically have two flagella and a haptonema, which is a filamentous appendage used for capturing bacterial and small phytoplankton prey. Apart from coccolithophorids covered in coccoliths, many prymnesiophytes are covered with organic scales and spines, and more rarely silica (Yoshida et al. 2006).

Organic-walled prymnesiophytes of the genera *Chrysochromulina* and *Phaeocystis* form blooms that can impact commercial fisheries. Large blooms of *Chrysochomulina*, like those reported in the Baltic Sea region, are problematic because of the mucilage surrounding the algal cells, which can clog fish gills and render them permeable to dissolved toxins (e.g., Richardson 1997). Blooms of *Phaeocystis* can form dense colonies and produce white foam that washes up on beaches. Both genera produce the gas dimethyl sulfide (DMS) which, at low concentrations, is associated with the "smell of the sea," but at high concentrations is noxious-smelling and can alter fish migration routes, adversely affecting the ecology, and sustainability of commercial and recreational fish species.

4.2.7 Dinoflagellates

Dinoflagellates are morphologically and functionally diverse. Their cell sizes range from less than 10 μm to over 1000 μm, and although often occurring as single cells, many species can also form long swimming chains (Hasle et al. 1996; Reynolds 2006). Dinoflagellate means "whirling flagella." Each dinoflagellate has two flagella, facilitating rapid forward and lateral movement. Dinoflagellates are surrounded by a complex covering called the amphiesma, which consists of outer and inner continuous membranes, between which lie a series of flattened vesicles. In armored forms, these vesicles contain cellulose plates called thecae. If this armor is lacking or shed under certain environmental conditions, the cells are "naked." Some dinoflagellate species are well known for their ability to produce light through bioluminescence, usually triggered by a physical disturbance, including breaking waves.

Around half of described dinoflagellates are photosynthetic, possessing chlorophyll *a*, other chlorophylls, and often the carotenoid peridinin. However, many dinoflagellate species have lost their peridinin-containing chloroplasts and acquired chloroplasts from other algae, including prasinophytes, haptophytes, cryptophytes (Keeling 2011), and even diatoms (Imanian et al. 2012). Dinoflagellates are important and at times dominant estuarine primary producers, sustaining the grazing component of the food web. Most photosynthetic dinoflagellates are facultative heterotrophs. These, along with dinoflagellates without chloroplasts, graze on bacteria and other phytoplankton. Several dinoflagellates host symbionts with diverse capabilities, including N_2-fixation, vitamin synthesis, or energy harvesting from light without photosynthesis (Farnelid et al. 2010; Wagner-Döbler et al. 2010). Other dinoflagellates are symbionts of animals, with one family, the Suessiaceae, that includes the *Symbiodinium* complex of species, which are responsible for much of the color of coral species (Cunning et al. 2018). In addition, taxonomically distinct dinoflagellates are parasites on zooplankton, fish, and other protists.

Even though their growth rates are generally slower than other phytoplankton, dinoflagellates can form large blooms in estuaries, and in some cases these blooms are harmful. Harmful dinoflagellate blooms have been linked to eutrophication. Examples include several red tide species (e.g., *Karenia brevis*, *Noctiluca* spp.) that are toxic to a wide variety of finfish, shellfish, and other fauna, including humans. These organisms are often of oceanic origin but can enter and proliferate in estuaries, especially during stratified summer months (Hasle et al. 1996). Red tide dinoflagellates produce several different types of neurotoxins that affect muscle function in susceptible organisms. Humans can also be affected by these toxins after eating fish or shellfish containing toxins produced by the dinoflagellates. The resulting diseases, including ciguatera (from eating affected fish) and paralytic shellfish poisoning (PSP; from eating affected shellfish, such as clams, mussels, and oysters) can be serious but are not usually fatal. Other toxic estuarine dinoflagellate genera include *Karlodinium*, *Alexandrium*, *Gyrodinium*, *Dinophysis*, and *Prorocentrum* species (Hasle et al. 1996; Verity 2010). These species produce toxins that have been implicated in shellfish poisoning and fish kills (Hall et al. 2008).

4.2.8 Assessing Phytoplankton Communities

Phytoplankton communities are dynamic multispecies assemblages that exhibit spatial patchiness on scales ranging from microns to meters, with populations fluctuating over scales ranging from minutes to days (Dustan and Pinckney 1989; Pinckney et al. 1999). Because these primary producers play a central role in the regulation of estuarine biogeochemical cycling, detailed characterizations of the community-level processes that structure phytoplankton communities are essential for understanding overall ecosystem dynamics. A critical prerequisite for characterizing these processes is the ability to reliably and accurately determine the taxonomic composition of natural phytoplankton assemblages.

High spatiotemporal variability of phytoplankton communities requires that analytical approaches for describing phytoplankton taxonomic diversity be applicable for processing large numbers of samples quickly with minimal cost. A reliable technique for enumerating individual species in mixed phytoplankton samples is microscopic counts, but these are tedious, require a high level of expertise, and are costly. In addition, species-level identification and enumeration may not be necessary for examining larger-scale phytoplankton impacts on biogeochemical cycling and food web dynamics. Often, examinations at higher taxonomic levels (i.e., class and group) are effective and make quantification easier.

Chemosystematics based on pigments can be used to classify broad phytoplankton groups (i.e., diatoms, chlorophytes, dinoflagellates, cyanobacteria, cryptomonads, etc., Figure 4.1; Roy et al. 2011). In particular, carotenoids and phycobilins, especially phycoerythrin and phycocyanin, serve as diagnostic biomarkers for determining the relative abundance of key phytoplankton groups, which can be optimized for a particular system. Photopigment extracts from natural microalgal samples can be separated and quantified by high performance liquid chromatography (HPLC; Wright et al. 1996; Paerl et al. 2003; Figure 4.1). Photopigments that have been used as markers are shown in Figure 4.1 and detailed in Paerl et al. (2003). When compared to total phytoplankton biomass (based on chlorophyll *a*) and cell counts, diagnostic photopigments can quantitatively determine the contributions of each phytoplankton group (Schlüter et al. 2006; Tamm et al. 2015).

Flow cytometry and cell imaging (e.g., FlowCam and Imaging Flow Cytobot) are now frequently employed to rapidly enumerate phytoplankton (Buskey and Hyatt 2006; Olson et al. 2017). The approach sends cells through a fluidic stream that passes by a sensor and/or laser that detects or visualizes cells. Flow cytometry commonly records cell size (light scatter), fluorescence properties, and counts of cells per sample volume. Cell imaging devices record cell counts and collect images of each cell that can then be assessed with image analysis software. The major advantage of these morphological approaches is high throughput analysis capacity. The disadvantage is that the systems (especially imaging devices) need to be trained on local phytoplankton, but machine learning algorithms are constantly being improved.

Molecular techniques can also be used to identify members of phytoplankton communities. These techniques use an informative molecule (nucleic acid) from phytoplankton cells, rather than morphotype (e.g., size, shape, and pigmentation; Figure 4.2). In particular, the DNA that codes for ribosomal RNA (rRNA) is a commonly targeted molecule used in diversity studies, since some regions of the rRNA genes are conserved to maintain function and other regions, termed hypervariable regions, are less conserved and drift over evolutionary time making these genes good taxonomic identifiers (Lane et al. 1985). More conserved regions of genes are good targets for primers in "universal" PCR (polymerase chain reaction) amplification of DNA from a broad diversity of phytoplankton species. Conversely, more unique regions of genes serve as good targets for amplifying DNA from specific populations, such as cyanobacteria or chlorophytes. The PCR products or "amplicons" are then DNA-sequenced and compared to a database to identify the organisms they represent.

In studies of mixed communities of phytoplankton, amplified DNA is usually sequenced directly using next generation sequencing (NGS). The millions of sequence "reads" produced

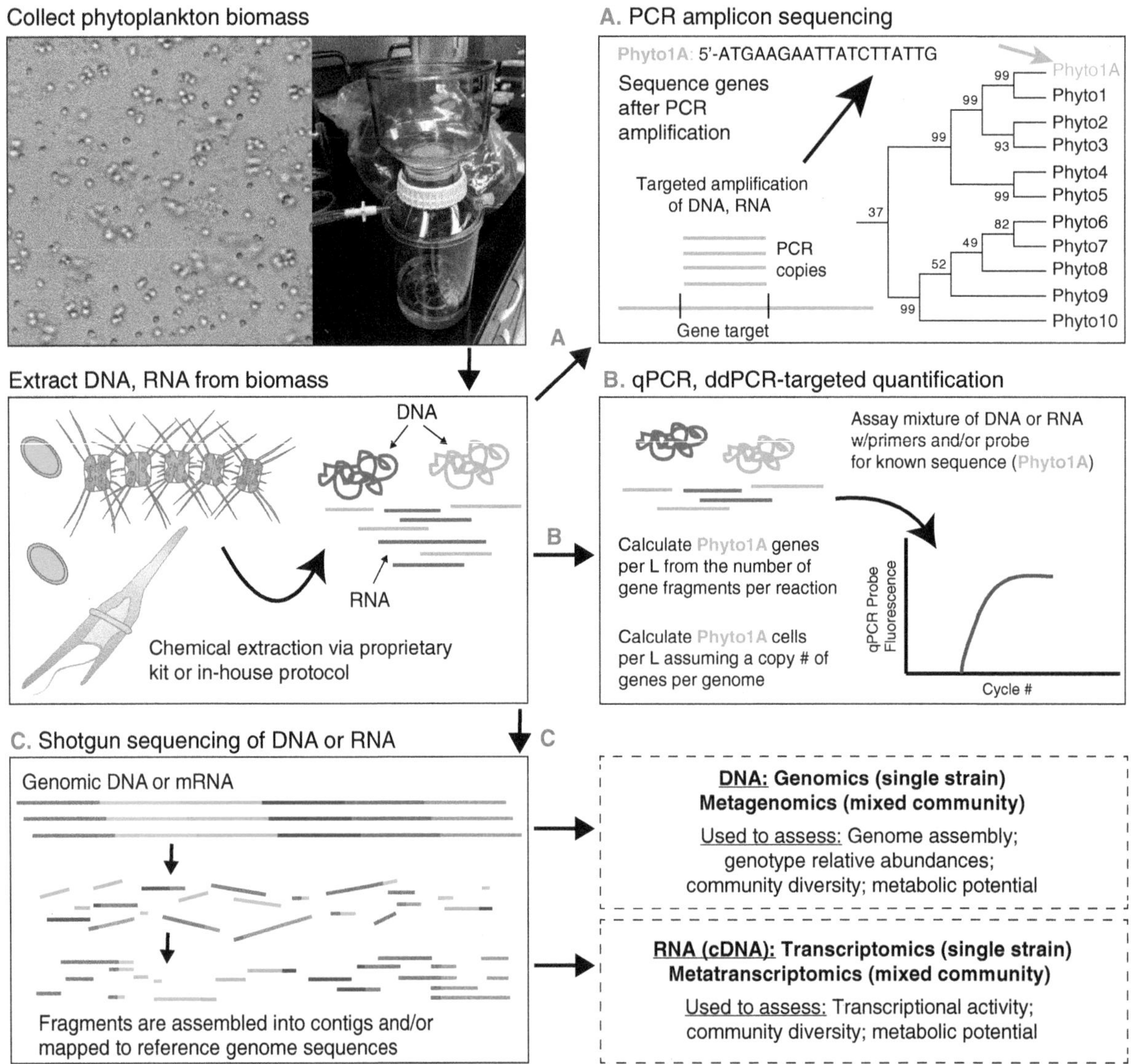

FIGURE 4.2 Multiple molecular techniques involving sequencing or PCR methods are often used to study estuarine phytoplankton. These techniques require DNA or RNA extraction from phytoplankton biomass prior to sequencing or amplification (RNA is also reverse transcribed to cDNA for downstream analysis). (A) PCR amplicon sequencing focuses on the study of a specific genomic region (e.g., a ribosomal RNA gene). (B) Quantitative PCR (qPCR) and digital droplet PCR (ddPCR) are used to quantify the abundance of target phytoplankton sequences (e.g., rRNA gene or protein encoding gene) from a mixture of DNA or cDNA. Copy numbers of target sequences per cell are either assumed or experimentally determined, then used to calculate the number of cells in an initial sampled volume (e.g., cells mL^{-1}). (C) Shotgun sequencing of DNA or cDNA uses little to no PCR amplification and is the foundation of "-omic" approaches to study diverse characteristics of a single strain (genomics, transcriptomics) or a community (metagenomics, metatranscriptomics; see dashed boxes). In order to be confident that changes in relative abundances of cDNA sequences (derived from rRNA or mRNA) reflect change in activity, the relationship between change in cDNA abundance and a direct measure of activity (e.g., growth rate, nutrient assimilation) should be assessed. *Source*: Phytoplankton illustration credit: Tracey Saxby (Integration and Application Network).

by NGS are clustered as individual operational taxonomic units (OTUs) based on a percent similarity (e.g., 97–99%) or as unique amplicon sequence variants (ASV; Callahan et al. 2017), or by mapping short reads onto robust phylogenetic trees. Most studies of phytoplankton communities use the small subunit rRNA gene, termed 16S rRNA in cyanobacteria and other bacteria, and 18S rRNA in eukaryotes (which is coded for in the nucleus). Popular use of this gene is facilitated by the vast number of publicly available rRNA gene sequences obtained since the original studies on the evolution of life (Pace et al. 1985). Analysis has been simplified following an effort to curate and collect the rRNA genes into publicly available databases such as SILVA (Quast et al. 2013), Greengenes (McDonald et al. 2012), and PR2 (Guillou et al. 2013). The sheer number of sequences available has also stimulated the bioinformatics community to make publicly available software packages and programs, to speed and simplify the analysis of small rRNA gene datasets (Figure 4.2).

Other genes can be used to further differentiate between species and to identify populations to a sub-genera level, such as large ribosomal subunit (LSU) genes or the intergenic spacer (ITS) region between the large and small ribosomal subunit genes (Urbach et al. 1998; Rocap et al. 2002). Protein encoding genes can be used to differentiate species as well, and are useful for focused assessments of population diversity based on a particular metabolic activity, such as nitrogen fixation based on the *nifH* gene for the iron-containing subunit of the nitrogenase enzyme or carbon fixation based on the *rbcL* gene for the large subunit of the RuBisCO enzyme (Zehr and McReynolds 1989; John et al. 2006).

Messenger (mRNA) and ribosomal RNA (rRNA) are also useful molecules for studying phytoplankton diversity and function. These molecules are usually converted to complementary DNA (cDNA) prior to DNA sequencing (Figure 4.2). Examining the sequence diversity of rRNA provides an indication of phytoplankton taxa that are presumably active in the environment because the rRNA is found in ribosomes that are needed for protein synthesis. Analogously, mRNA can be examined to identify populations that are transcriptionally active and likely to engage a particular metabolism, (e.g., carbon fixation, indicated by *rbcL* mRNA; Paul et al. 2000). Further, examining mRNA can provide information about the metabolic pathways and physiologies, some of which are not readily detected by common biochemical assays, used by phytoplankton to thrive in nature.

Another DNA analysis tool is quantitative PCR (qPCR); a method that relies on specific PCR primers (and at times a fluorescent DNA probe) which enables per milliliter estimation of a specific phytoplankton DNA sequence in a sample (DNA extracted from phytoplankton biomass). The abundance of the target sequence is calculated based on the volume of water filtered to obtain phytoplankton biomass, efficiency of DNA extraction from phytoplankton biomass, and the empirically determined or assumed number of DNA sequence (e.g., protein encoding gene) copies in the genome of the target phytoplankton cell. While this approach carries key calculation caveats, it notably enables simple, *en masse* quantification of specific phytoplankton (e.g., harmful taxa) from samples of mixed communities and quantification of populations unidentifiable by microscopy or other standard methods.

4.3 Spatial and Temporal Patterns of Phytoplankton Biomass and Productivity

Phytoplankton account for at least half of ecosystem primary production in coastal waters (Sigman and Hain 2012). Their rates of primary production are remarkably variable and range from near zero to several $\mathrm{gC\,m^{-2}\,day^{-1}}$ (Cloern et al. 2014). Because they have fast growth rates, phytoplankton can rapidly respond to diverse chemical (nutrients, toxicants), physical (light, temperature, turbulence), and biotic (grazing) impacts over a wide range of concentrations and intensities. Changes in phytoplankton communities often precede changes in ecosystem function, including shifts in material flux, oxygen, $p\mathrm{CO_2}$, and pH balance, food webs and fisheries, and potentially, permanent losses of higher plant and animal assemblages.

Coastal phytoplankton communities typically show strong seasonal and spatial distributions. This is illustrated for the Baltic Sea (Figure 4.3), as well as Pamlico Sound and Chesapeake Bay, the two largest estuarine systems in the United States (Figures 4.4 and 4.5) and for the Breton Sound estuary, Louisiana, part of the Mississippi River Delta system (Figure 4.6; Day et al. 2009). In all systems, the distribution of phytoplankton biomass, measured as chlorophyll *a*, is strongly influenced by the delivery of fresh water (i.e., discharge), which also delivers watershed-based nutrients to support new primary production and the development of phytoplankton biomass. In addition

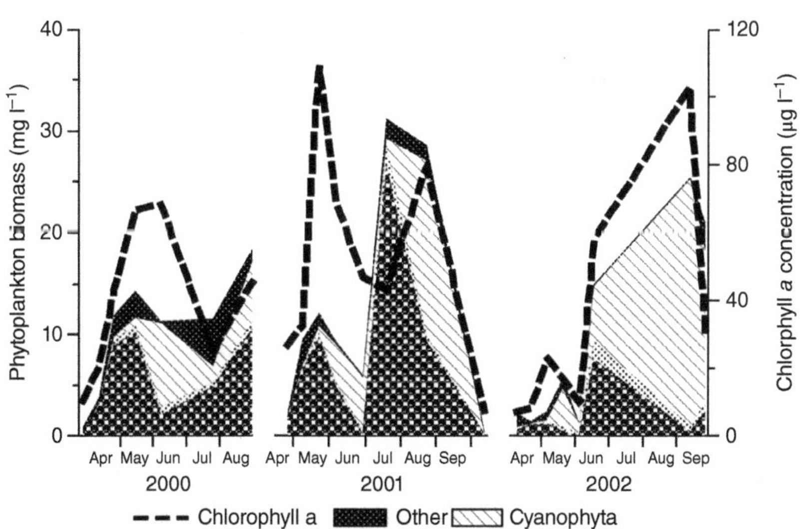

FIGURE 4.3 Seasonal changes in chlorophyll *a* concentrations and abundance of dominant phytoplankton taxa in the Curonian Lagoon, Baltic Sea. Diatoms (Bacillariophyceae) were the most abundant group during the spring while cyanobacteria (Cyanophyta) usually dominated during the summer. *Source*: Pilkaitytė and Razinkovas (2007)/Boreal Environment Research.

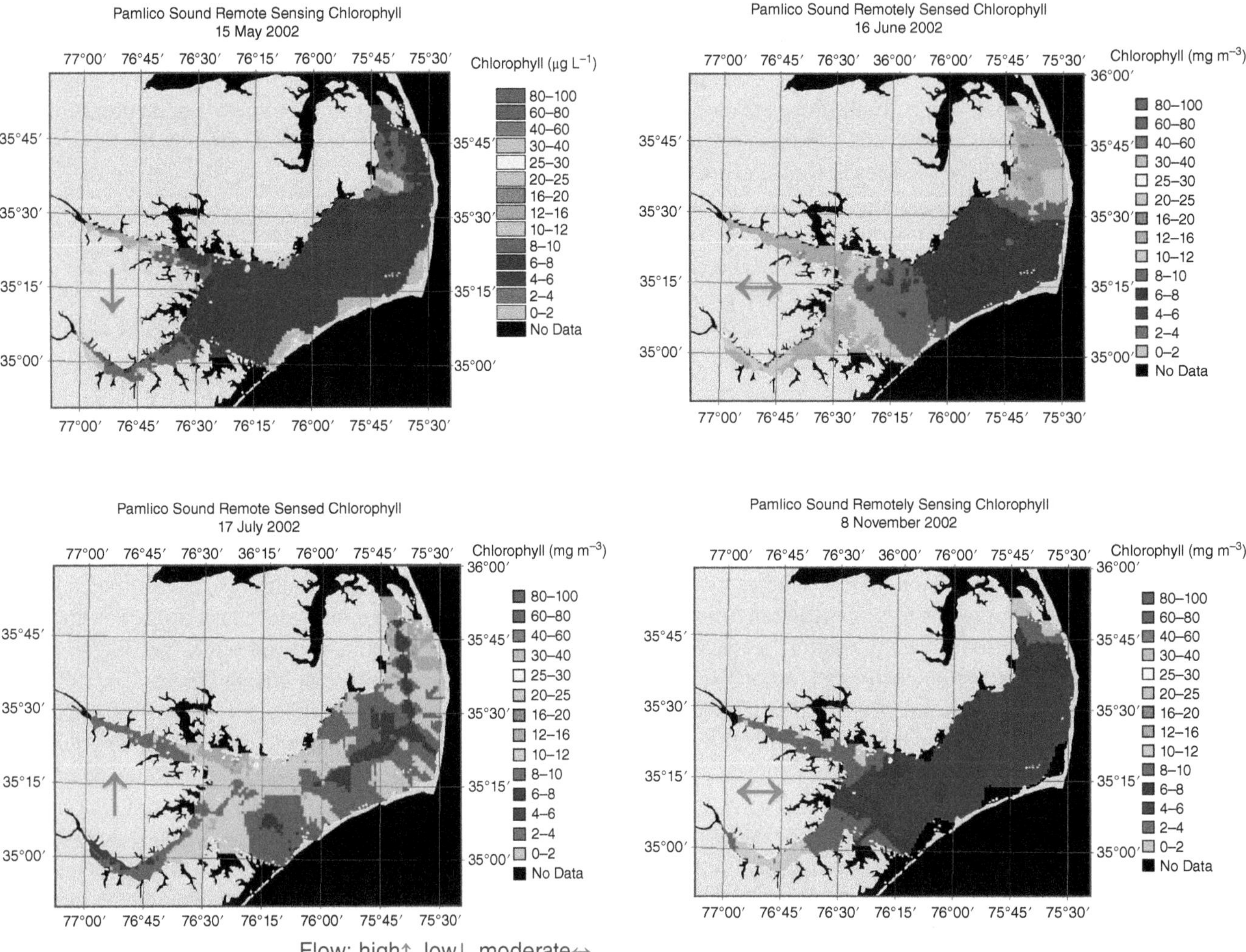

FIGURE 4.4 Spatial relationships of phytoplankton biomass, as chlorophyll *a*(Chl *a*), and freshwater discharge to the Pamlico Sound System, NC, USA. Surface water Chl *a* concentration was estimated using aircraft-based SeaWiFS remote sensing (Courtesy L. Harding, Univ. of Maryland), calibrated by field-based Chl *a* data. Under relatively low-flow (upper left), long-residence time conditions, phytoplankton biomass is concentrated in the upper reaches of the Neuse and Pamlico R. Estuaries. Under moderate flow (top right and bottom right), phytoplankton biomass maxima extend further downstream. Under high-flow (bottom left; short residence time) phytoplankton biomass maxima are shifted further downstream into Pamlico Sound. *Source*: Paerl et al. (2007)/John Wiley & Sons.

to being a source of new nutrients, freshwater discharge also influences flushing characteristics and resultant residence times in estuarine waters. Residence time is the amount of time it takes for fresh water, nutrients, or other allochthonous material (e.g., organic matter) delivered to an estuary to move through the system before exiting to the coastal ocean.

Residence times are strong determinants of where in an estuary the maximum amount of phytoplankton biomass can develop and build up in response to nutrient inputs. For example, during periods of high freshwater flow, nutrient delivery to an estuary will be high, potentially boosting phytoplankton growth. Simultaneously, high flow will create short residence time conditions, restricting phytoplankton growth and biomass buildup to the most downstream, widest, and longest residence time segments of an estuary (illustrated in Figures 4.4–4.6). In the Pamlico Sound system, high flow conditions typically lead to maximum phytoplankton biomass (as Chl *a*) in the downstream segments of the estuaries and open waters of the sound itself (Figure 4.4), while under low flow, long residence time conditions, phytoplankton biomass maxima are typically in the upper estuarine regions.

Temporal patterns in phytoplankton biomass and productivity are strongly controlled by temperature and the availability and quality of light (Cloern 1999; Reynolds 2006). Light is essential for photosynthetic activity, and temperature influences the activity rates of enzymes involved in carbon and nutrient assimilation, respiration, energy metabolism, and growth. As such, seasonality plays an important role in the control of photosynthetic production (primary production), biomass, and species composition (Pinckney et al. 1999). Typically, maximum rates of primary production and growth occur in spring when light and/or nutrient availability are both high, leading to "spring bloom" conditions. These blooms may persist during summer if nutrient availability remains high

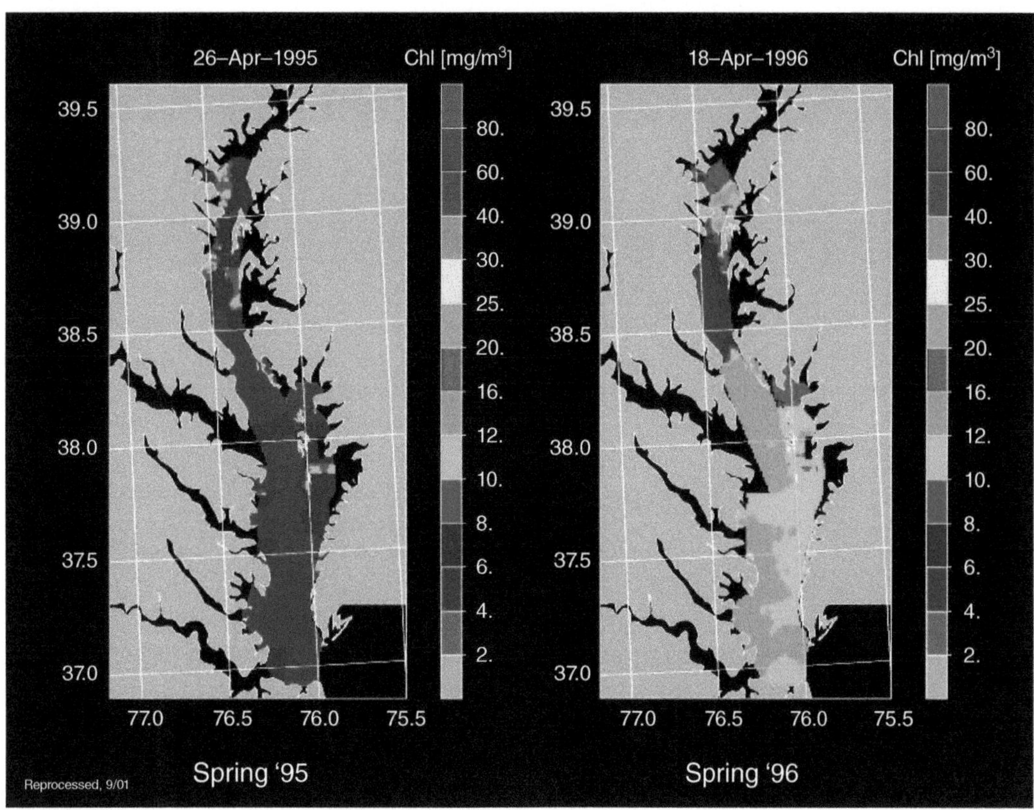

FIGURE 4.5 Contrasting spring phytoplankton biomass as chlorophyll a (Chl *a*) distributions in the Chesapeake Bay, in which a relatively dry year (1995) was compared with a wet year (1996). In 1996 (right panel), when freshwater discharge was high and residence time was short, fast-growing diatom blooms extended into the lower Bay. In 1995 (left panel), during lower flow (long residence time) conditions, blooms of slower growing phytoplankton (largely dinoflagellates) were confined to the riverine tributaries flowing into the Bay. Surface water Chl *a* concentration was estimated using aircraft-based SeaWiFS remote sensing. *Source*: L. W. Harding, personal communication.

through input and/or recycling. Summer also typically features maximum temperatures, favoring relatively high growth rates. Not all phytoplankton groups and individual species respond similarly to shifting light, temperature, and nutrient gradients; this is largely due to the individual energetic, nutrient, and other requirements of each species (e.g., water column stability, salinity, trace metals, pH, dissolved inorganic carbon, etc.).

Some dinoflagellate species show sudden, strong positive responses to increases in light levels and day length that occur in late winter. Many temperate estuaries experience increasing nutrient-enriched runoff in late winter at a time when the community of organisms that graze on phytoplankton is still at its wintertime low. These factors create ideal conditions for large blooms of dinoflagellates (e.g., *Heterocapsa* spp.) that can occur as early as January in temperate waters (Paerl et al. 1998; Litaker et al. 2002).

Other phytoplankton groups dominate during spring and summer. Typically, diatoms tend to bloom in early spring because they exhibit relatively fast growth rates, even at fairly low temperatures, and hence can take advantage of early spring freshwater runoff (Harding et al. 2016; Valdes-Weaver et al. 2006). However, if runoff is extremely high such as during spring floods, large segments of estuaries can turn fresh, creating conditions that favor chlorophytes and a variety of flagellates (e.g., cryptomonads and chrysophytes), which are better adapted to fluctuating freshwater flow, salinity, and nutrient supply conditions (Valdes-Weaver et al. 2006).

Motile taxa such as cryptomonads tend to bloom after diatoms, in part because they are capable of maintaining their position in the upper water column once spring diatom blooms have declined and sediment-laden runoff decreases water transparency (Hall and Paerl 2011). Cryptomonads are also quite efficient in sequestering nutrients once diatom blooms have reduced ambient inorganic nutrient levels, potentially turning to grazing of bacteria to meet their nutritional demands. During summer months when freshwater runoff is lowest, residence time is highest, and nutrients are depleted, temperature and vertical stratification are high. These conditions favor slower-growing, highly efficient nutrient utilizing taxa such as picocyanobacteria and small eukaryotic flagellates. Both picocyanobacteria and larger cyanobacteria exhibit strong summer maxima in biomass, likely because their maximum growth rates are optimized at relatively high temperatures (Gaulke et al. 2010). Picocyanobacteria also exhibit high surface to volume ratios (Chisholm 1992) and low nutrient requirements (Bertilsson et al. 2003) that make them superior competitors under low nutrient concentrations.

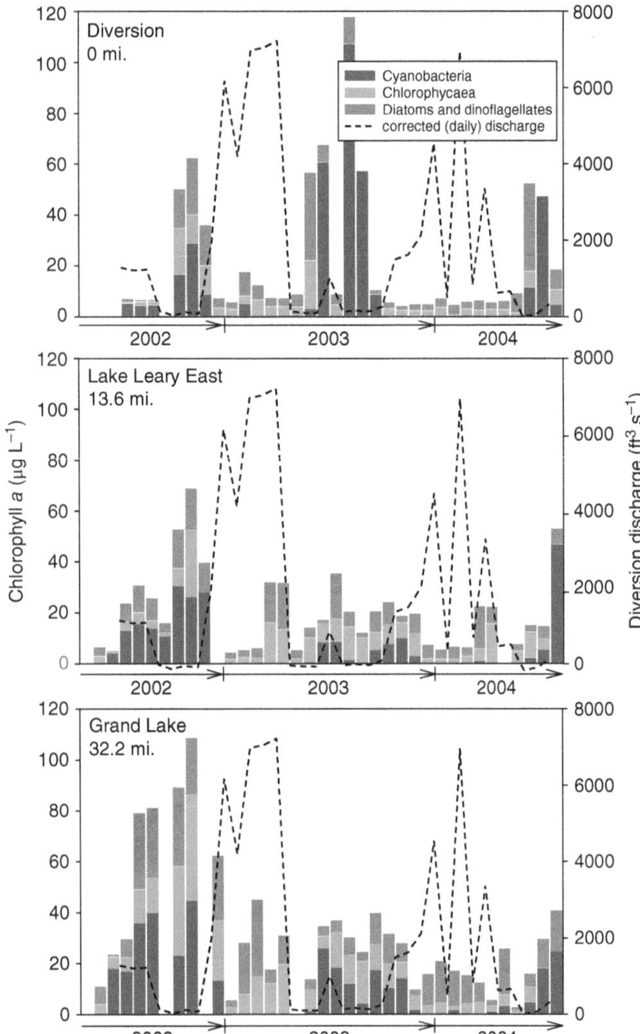

FIGURE 4.6 Spatial, seasonal, and interannual variability of the phytoplankton community structure at three stations in the Breton Sound Estuary, Louisiana, from February 2002 to September 2004. Chlorophyll *a* concentrations of the three major groups (cyanobacteria, chlorophyceae, and diatoms or dinoflagellates) are given as stacked bars for the monthly surveys along with diversion discharge (dashed line). *Source*: Day et al. (2009).

Lastly, the activities of microzooplankton, macrozooplankton, and benthic invertebrate grazers exert seasonally variable and taxa-specific "top down" controls on phytoplankton community biomass, composition, and succession (Sterner 1989; Cloern and Jassby 2012). In addition, biological interactions, such as microbial-phytoplankton consortia and symbioses affect phytoplankton community composition and activity. Grazing and predation interact seasonally and spatially (along the estuarine salinity gradient) with physical and chemical (nutrient) "bottom up" controls on phytoplankton community structure and function. These multiple, interacting controls promote high diversity of phytoplankton functional groups and species in hydrologically- and biogeochemically-variable coastal ecosystems.

4.4 Factors Controlling Phytoplankton Productivity and Community Composition

4.4.1 Light

Light is of fundamental importance in controlling the activity, biomass, distribution, and composition of phytoplankton. Phytoplankton use light in the 400–700 nm part of the visible light spectrum. This part of the spectrum is termed photosynthetically active radiation, or PAR. Both the instantaneous flux of light, or irradiance, and the total amount of light available during daylight tend to be excellent predictors of photosynthetic performance and primary production (Jassby and Platt 1976; Fee 1980). Light availability is controlled by the concentration of key light attenuating substances, including suspended sediments, chlorophyll and other algal photopigments, and colored dissolved organic matter (CDOM) content (Gallegos et al. 1990, 2011; Biber et al. 2005). Together with surface irradiance, these components determine the attenuation or extinction coefficient K_d. Vertical mixing rates and phytoplankton vertical migration capabilities also strongly affect photosynthetic rates (Mallin and Paerl 1992). Phytoplankton photosynthetic response to light can be described with photosynthesis-irradiance (P-I) curves.

Phytoplankton exhibit vertical photosynthetic rate patterns that closely follow the vertical distribution of ambient light. However, in clear waters and/or under conditions of high surface irradiance, phytoplankton incubated at the water surface frequently experience light inhibition. Because of this, maximum rates of photosynthesis often occur at subsurface depths experiencing from ~30 to 70% of surface irradiance. In highly turbid waters, however, near-surface light inhibition is less common and maximum photosynthetic rate occurs at the surface rather than subsurface.

Estuarine phytoplankton assemblages are adapted to the fluctuating light regimes experienced in well-mixed surface waters (Marra 1978; Mallin and Paerl 1992), with maximum rates often occurring under well-mixed conditions. Falkowski (1980) showed that under vertically mixed conditions, algal cells can rapidly adapt to changing light and maximize photosynthesis by varying their photosynthetic pigment composition and enzyme concentrations, with response times of one hour or less. Marra (1978) and Mallin and Paerl (1992) showed that algal cells that were experimentally rotated through a vertical series of depths often had significantly greater photosynthetic production, even though the total light available was the same in both cases. Platt and Gallegos (1980) developed a model simulating phytoplankton photoadaptation to demonstrate that a mixed-water column generated about 20% more photosynthesis than a static system. However, excessive vertical mixing, especially in deeper turbid waters, can be detrimental since it can force phytoplankton into completely dark waters, where photosynthesis ceases.

Under vertically stratified conditions, phytoplankton taxa that are able to adjust their position in the water column are frequently at an advantage. They can migrate to and maintain their position at light levels supporting optimal photosynthetic production and growth, while periodically migrating into deeper, nutrient-rich waters to access essential nutrients (Ralston et al. 2007; Hall et al. 2015). In particular, near-surface dwelling bloom taxa (some dinoflagellate, cryptophyte, and cyanobacterial species) use this strategy to maintain dominance when other factors conducive to bloom formation prevail (e.g., elevated temperatures, lack of grazing, optimal flow, and residence time conditions).

Light quality (i.e., light color determined by the wavelength of the light) also influences the composition of phytoplankton communities. For example, within the picocyanobacterial genus *Synechococcus*, phycocyanin-rich strains capture red light more efficiently and flourish in turbid waters, while phycoerythrin-rich strains better capture green and blue light that is more predominant in clearer waters (Olson et al. 1990; Stomp et al. 2004).

A useful index of light in estuaries is the ratio of photic zone depth to water column depth. When this index increases, a greater proportion of an estuarine water column has sufficient light for photosynthetic activity to occur. This index correlates with net phytoplankton production in weakly stratified San Francisco Bay and in more strongly stratified Chesapeake Bay and the Neuse River estuary (Cole and Cloern 1984; Mallin and Paerl 1992; Harding et al. 1987). These relationships demonstrate the important interactive roles that water clarity and vertical mixing play in regulating primary production and phytoplankton growth, even in very shallow but often turbid estuarine ecosystems (e.g., Cloern 1999, 2001; Gallegos et al. 2011).

4.4.2 Nutrients

Phytoplankton production requires a variety of inorganic and organic nutrients, including carbon (C), nitrogen (N), phosphorus (P), silicon (Si), trace metals, most importantly iron, and in some cases B-vitamins. The most important are the so-called "macronutrients" (carbon, nitrogen, phosphorus, and silicon), and among these nitrogen and phosphorus are most significant, largely because they are usually in shortest supply relative to demand. As a result, availability of these elements limits rates of primary production (Nixon 1992, 1995). Furthermore, because different algal species require these nutrients in different proportions due to different metabolic and enzymatic demands, the supply ratios of these and other elements (e.g., N:Si, N:Fe) as well as their chemical forms modulate the growth and competitive interactions among phytoplankton taxonomic groups (Dortch and Whitledge 1992; Justić et al. 1995; Altman and Paerl 2012; Glibert and Burford 2017).

Redfield (1934) found a remarkable consistency in the average of the relative proportions of major nutrients in deep ocean water and in marine phytoplankton biomass. These proportions of the elements C: N: Si: P, since named the "Redfield Ratios" of atomic weights, are 106: 16: 15: 1. In surface waters and estuaries, however, nutrient ratios are not constant over time and space. They can be altered by episodic nutrient inputs or depletions from a variety of human and natural sources, including rainfall and runoff, groundwater, and oceanic inputs. Nutrient delivery to estuaries generally occurs in a nonsteady, pulsed manner, with several nutrients simultaneously being close to limiting, creating "co-limited" conditions. Although in estuaries N has been identified as the most limiting nutrient (Nixon 1995; Paerl and Piehler 2008; Figures 4.7 and 4.8),

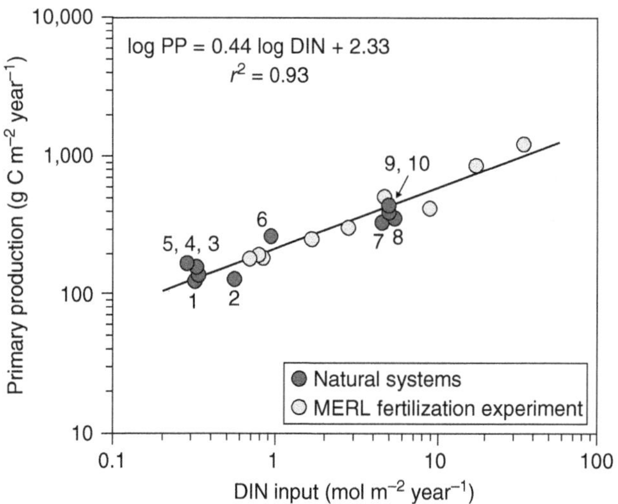

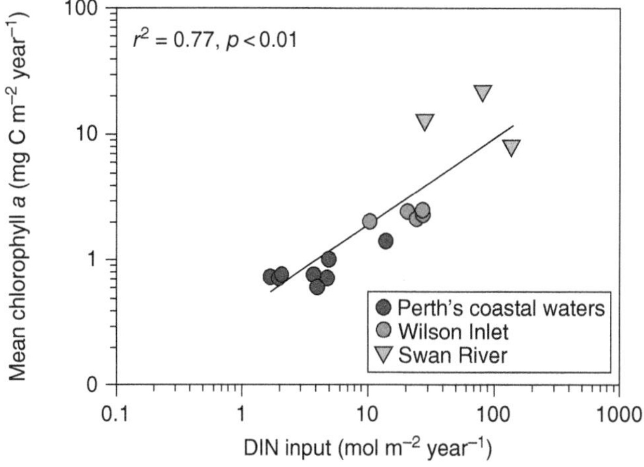

FIGURE 4.7 Upper: Relationships between dissolved inorganic N (DIN) input and primary production in a multiyear fertilization experiment conducted at the Marine Ecosystems Research Laboratory (MERL), and from several natural ecosystems; 1. Scotian Shelf; 2. Sargasso Sea; 3. North Sea; 4. Baltic Sea; 5. North Central Pacific; 6. Tomales Bay, California, USA; 7. Continental Shelf off New York, USA; 8. Outer continental shelf off southeastern USA; 9. Peru upwelling; 10. Georges Bank. Lower: Relationship between dissolved inorganic N input and phytoplankton biomass, as mean annual chlorophyll *a* content of several Western Australian estuarine systems. *Source*: Based on Twomey and Thompson (2001).

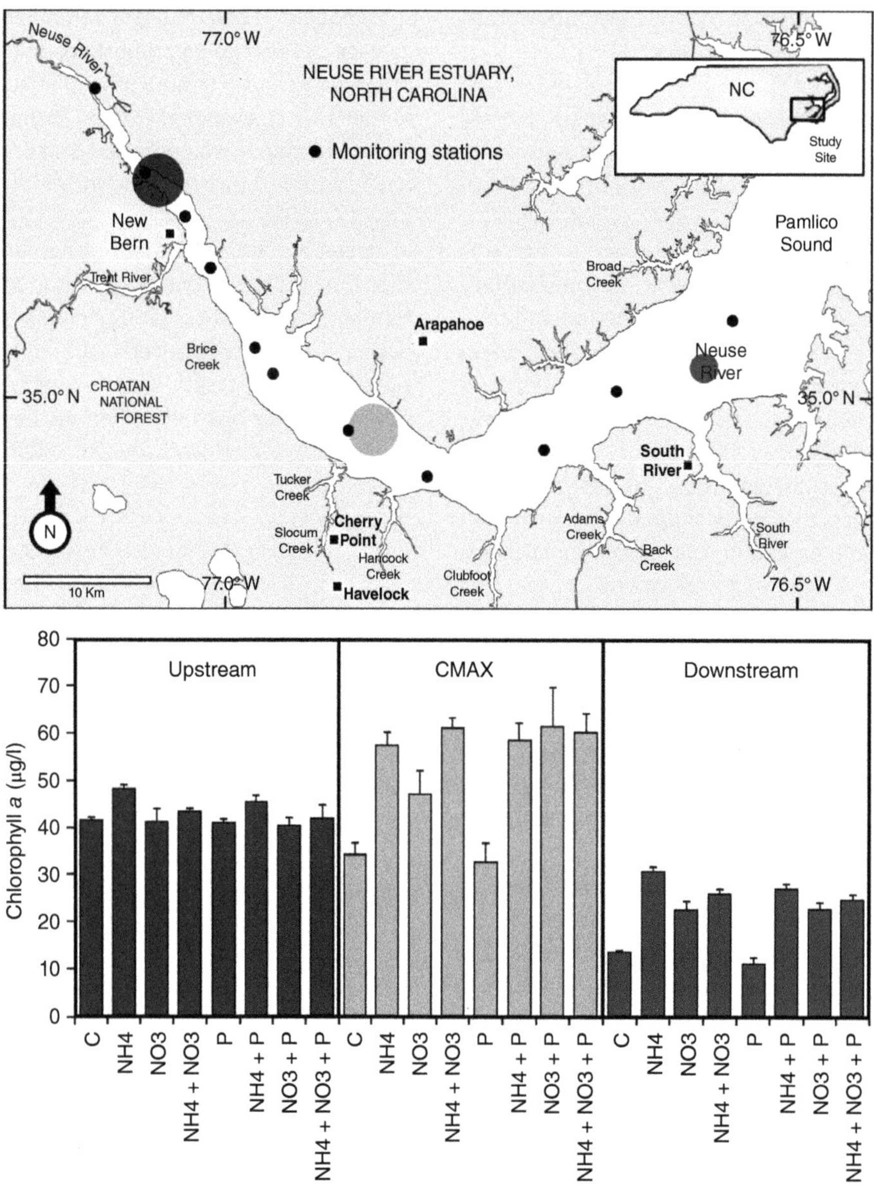

FIGURE 4.8 Results from nutrient addition bioassays conducted at 3 locations in the Neuse River Estuary, NC. Phytoplankton growth response was measured as accumulation of chlorophyll *a* after 3 days incubation under natural light and temperature conditions. "C" indicates controls, in which no nutrients were added. Nitrogen (N) was added as either ammonium (NH_4^+) or nitrate (NO_3^-) at a 10 µM final concentration. Phosphorus (P) was added as phosphate (PO_4^{3-}) at a 3 µM final concentration. These summertime bioassays indicate N limitation, which is most profound at more saline downstream locations. Note that per amount of N, ammonium was more stimulatory than nitrate.

N and P co-limitation is also commonly observed (Boynton and Kemp 1985; Malone et al. 1996), especially in the low salinity, upstream segments of estuaries (Figure 4.9; Fisher et al. 1988; Paerl and Piehler 2008), and in brackish water deltaic regions where large rivers (e.g., Mississippi) discharge to the coastal ocean (Sylvan et al. 2006). N and P co-limitation and exclusive P limitation are most evident during periods of elevated freshwater runoff, which tends to be N enriched (Fisher et al. 1988; Sylvan et al. 2006). Under these conditions, the molar "Redfield ratio" of N:P (16:1) can be greatly exceeded, sometimes reaching 200:1, leading to strong P limitation.

Certain phytoplankton groups have more specific nutrient requirements. The most notable are diatoms, which have siliceous cell walls or frustules, making this group reliant on adequate silicon (Si) supplies. Silicon is a product of weathering of upstream siliceous soils. If these soils are absent or sparse, Si supply may be limited. In addition, the supply ratio of Si to other potentially limiting nutrients (N, P) may dictate the relative availability of one or several nutrients in order to maintain balanced or "Redfield" growth (Redfield 1958). Hence, if human or climatic perturbations increase the flux of N and/or P to an estuary, the supply rate of Si may become limiting,

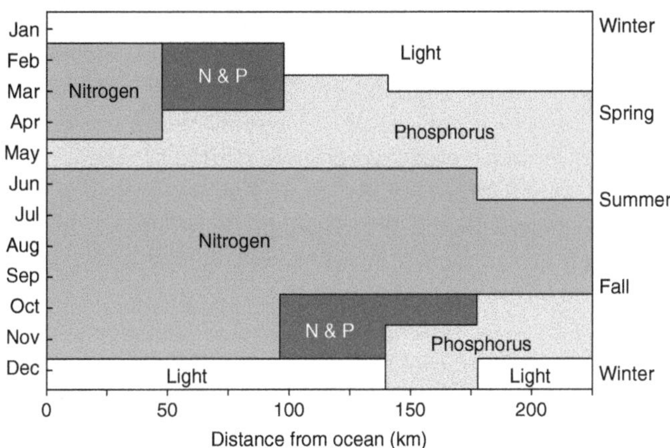

FIGURE 4.9 Seasonal and spatial patterns of N and P limitation determined from nutrient addition bioassays conducted on Chesapeake Bay from its relatively fresh headwaters to the saline entrance to the Atlantic Ocean. *Source*: Adapted from Fisher et al. 1988.

leading to changes in the phytoplankton community composition. This occurs in the Northern Gulf of Mexico region where Mississippi River discharge supplies excess N and P relative to Si (Justić et al. 1995; Turner and Rabalais 2013). Sedimentological evidence indicates that the abundance of the lightly silicified and hence relatively low Si requiring diatom *Pseudo-nitzschia* has increased since the 1950s, coincidentally with increasing riverine N flux and increasing N:Si ratios (Parsons et al. 2002; Trainer et al. 2012). Diverse (e.g., dinoflagellates, chlorophytes, and haptophytes) phytoplankton require supplied organic nutrients in the form of B-vitamins, particularly B_1 (thiamin), B_7 (biotin), and B_{12} (cobalamin); Provasoli and Carlucci 1974. Recent genomic analyses highlight that vitamin precursors—not just intact vitamins—are also important nutrients for cosmopolitan phytoplankton (McRose et al. 2014; Paerl et al. 2017).

4.4.3 Temperature

Temperature controls phytoplankton metabolism and growth by influencing the rate at which cellular enzymes function, but it also affects species composition because different phytoplankton have different optimal temperatures at which they grow and photosynthesize. For example, one study showed that temperature optima for five coastal phytoplankton species ranged from 20 to 25 °C (Goldman 1979). However, the shape of the curve between temperature and phytoplankton growth and photosynthesis is similar for most algal species (see Section 4.5.2 and Figure 4.13; Eppley 1972). These temperature response curves (e.g., growth vs. temperature) show similar increases in growth with temperature, and similar rapid declines in growth at temperatures in excess of their optima (Figure 4.13).

Some phytoplankton taxa are more strongly regulated by temperatures than others. For example, cyanobacterial growth typically exhibits temperature optima in the 25–30 °C range, which is substantially higher than other phytoplankton groups. As a result, estuarine and coastal cyanobacteria often show growth and biomass optima that closely follow temperature changes (Waterbury et al. 1986; Tai and Palenik 2009). The fact that cyanobacteria "like it hot" more so than other phytoplankton has been linked to an upsurge in cyanobacterial blooms and geographic expansion of such blooms (relative to other phytoplankton blooms) accompanying regional and global warming (Paerl and Huisman 2008).

4.4.4 "Top Down" Control: Herbivory

Zooplankton, benthic filter feeders, and some of their larvae, as well as larval, juvenile, and adult fish are the primary consumers of coastal phytoplankton. The zooplankton are commonly divided into several size classes, including microzooplankton (< 200 μm), mesozooplankton, (0.2–2 mm), macrozooplankton (2–20 mm), and megazooplankton (>20 mm; see Chapter 10). The relative contributions of these size classes to grazing of phytoplankton can vary substantially. The most abundant zooplankton in coastal ecosystems are microzooplankton including all heterotrophic protists, especially ciliates and dinoflagellates. The less numerous but larger bodied mesozooplankton includes calanoid copepods, cladocerans, and thaliacean tunicates.

Zooplankton grazing represents an important control of phytoplankton biomass and community composition. However, grazing varies with factors such as seasonality, vertical mixing, freshwater flushing, and residence time, underscoring the complex interactions between "bottom up" and "top down" controls of phytoplankton dynamics. Phytoplankton generally have faster growth rates than zooplankton, which can result in phytoplankton blooms that accumulate more quickly than the zooplankton that graze on them. As a result, some rapidly growing bloom taxa can proliferate in a seemingly unabated fashion. These blooms are primarily limited by the nutrient supply and physical factors, such as light availability and hydrodynamic conditions (flushing rate, vertical mixing rate). This is especially true for phytoplankton species that are capable of blooming during winter and early spring, when water temperatures are too low to support increases in macrozooplankton grazers, which are also constrained by life cycle stages. During these periods, phytoplankton biomass and composition are largely controlled by these physical–chemical factors. This type of control is termed "bottom up."

As water temperatures warm up in spring and summer, zooplankton growth and reproduction rates are enhanced, increasing overall zooplankton populations and biomass. At the same time, nutrient supplies to phytoplankton often decrease, largely because the main source of nutrients to estuaries, freshwater runoff, often decreases during these drier periods. All these combined effects cause herbivorous grazing to play a greater role as a control on phytoplankton biomass in late spring and summer. This type of control is termed "top down." Because zooplankton grazing is very dynamic and dependent on interacting physical, chemical, and biotic factors, its importance as a control on phytoplankton stocks is a topic that carries a great deal of uncertainty and has been hotly debated for over half a century.

High rates of grazer-induced mortality on phytoplankton occurs in many estuarine and coastal environments. For example, in the Northern Gulf of Mexico, the copepod community ingested 4–62 % of the daily phytoplankton production (Dagg 1995a). In a productive subtropical estuary (Fourleague Bay, Louisiana, USA), ingestion rates of phytoplankton by the microzooplankton community averaged 43–165% of the daily phytoplankton production (Dagg 1995b). In contrast, the grazing contribution from the mesozooplankton, comprised primarily of copepod *Acartia tonsa*, was negligible, presumably because of high advective losses (e.g., flushing) and predation on copepods by zooplanktivorus fish (Dagg 1995b).

While zooplankton herbivory may be a seasonally important control on coastal phytoplankton production under certain environmental conditions (Martin 1970), it is not likely to be a severe limitation overall. Often, grazing is insufficient to balance phytoplankton growth, which leads to the development of phytoplankton blooms. In Chesapeake and Narragansett Bays, high suspension feeding rates of ctenophores and medusae on zooplankton may serve to keep the zooplankton grazing in balance (Heinle 1974; Kremer 1979).

In some estuaries, suspension feeding benthic macrofauna can reduce phytoplankton abundance significantly (Cloern 1982; Officer et al. 1982). For several estuarine systems, it has been shown that a single dominant suspension-feeding bivalve population was capable of filtering the entire overlying water column in 1–4 days (Cohen et al. 1984; Nichols 1985; Doering et al. 1986). Such grazing rates can reduce phytoplankton standing stocks significantly. A well-documented example is the introduction and proliferation of zebra mussels (*Dreissena polymorpha*) throughout North America (MacIsaac 1996; Strayer 2008). For example, in the Hudson River Estuary, the invasion of zebra mussels in the 1990s lead to a 10-fold increase in grazing pressure and a corresponding 85% decline in phytoplankton biomass (Caraco et al. 1997).

4.5 Human and Climatic Impacts on Coastal Phytoplankton Dynamics

4.5.1 Effects of Nutrient Over-Enrichment on Estuarine Phytoplankton

Coastal primary production and phytoplankton biomass are strongly controlled by the availability and supply rates of nutrients, especially nitrogen (Paerl and Piehler 2008) and to a lesser extent phosphorus (Sylvan et al. 2006). There is a delicate balance between beneficial nutrient enrichment to sustain a productive and healthy food web, and over-fertilization which can greatly accelerate primary production, and promote excessive organic matter production—a process called eutrophication. If not effectively used by the food web, excess phytoplankton organic matter can accumulate, leading to water quality problems and habitat degradation (Figure 4.10).

Symptoms of eutrophication include phytoplankton blooms, loss of submerged aquatic vegetation, severe oxygen depletion (hypoxia), and fish kills (Paerl 1988, 2004; Bricker et al. 1999; Rabalais et al. 2014). Sudden changes in temperature, nutrient depletion, and light availability can cause blooms to die (Paerl 1988). When blooms die, they sink into deeper waters, where they decompose (Figure 4.10). This process consumes vast amounts of oxygen, and if the bottom waters are isolated by stratification from oxygen-rich surface waters, they eventually run out of oxygen, creating hypoxia. The lack of oxygen suffocates finfish and shellfish, and hypoxia is a major cause of habitat loss, as well as finfish and shellfish kills (Diaz and Rosenberg 2008).

Rapidly expanding human urban, agricultural, and industrial activities in coastal watersheds have greatly accelerated the production and delivery of nutrients to nutrient-sensitive estuarine and coastal waters. Anthropogenic and natural sources of N and P are delivered by (1) surface water discharge delivered via creeks and rivers, (2) subsurface discharge (groundwater), and (3) atmospheric deposition (rainfall or dryfall; mainly as N). The proportions of these nutrient sources vary geographically and demographically. In rural, agriculturally dominated regions (US Midwest and Southeast Coastal Plain; China's Yangtze River watershed; India's Ganges River watershed; Brazil's and Argentina's coastal watersheds and lagoonal systems), 50% to over 75% of N and P input originates from diffuse, non-point sources such as surface runoff, rainfall, and groundwater (Paerl 1997; Boesch et al. 2001; Moore 1999; U.S. EPA 2011). Point sources, including wastewater, industrial, and municipal discharges, account for the rest. In contrast, N and P loading in urban watersheds (e.g., USA's Narragansett Bay and Puget Sound, UK's Thames River, and France's Seine River watershed) are dominated (>50%) by point sources, while watersheds encompassing both urban centers and intensive agriculture (e.g., USA's Chesapeake Bay and San Francisco Bay, Baltic Sea coastal watersheds, and Rhine River watershed) exhibit a more even distribution of these source types (Castro et al. 2003; Paerl and Piehler 2008).

4.5.2 The Roles of Climatic Variability in Eutrophication Dynamics

Nutrient, sediment, and other contaminants in coastal waters are strongly influenced by climatic forcing features, especially freshwater discharge, which is the main delivery mechanism.

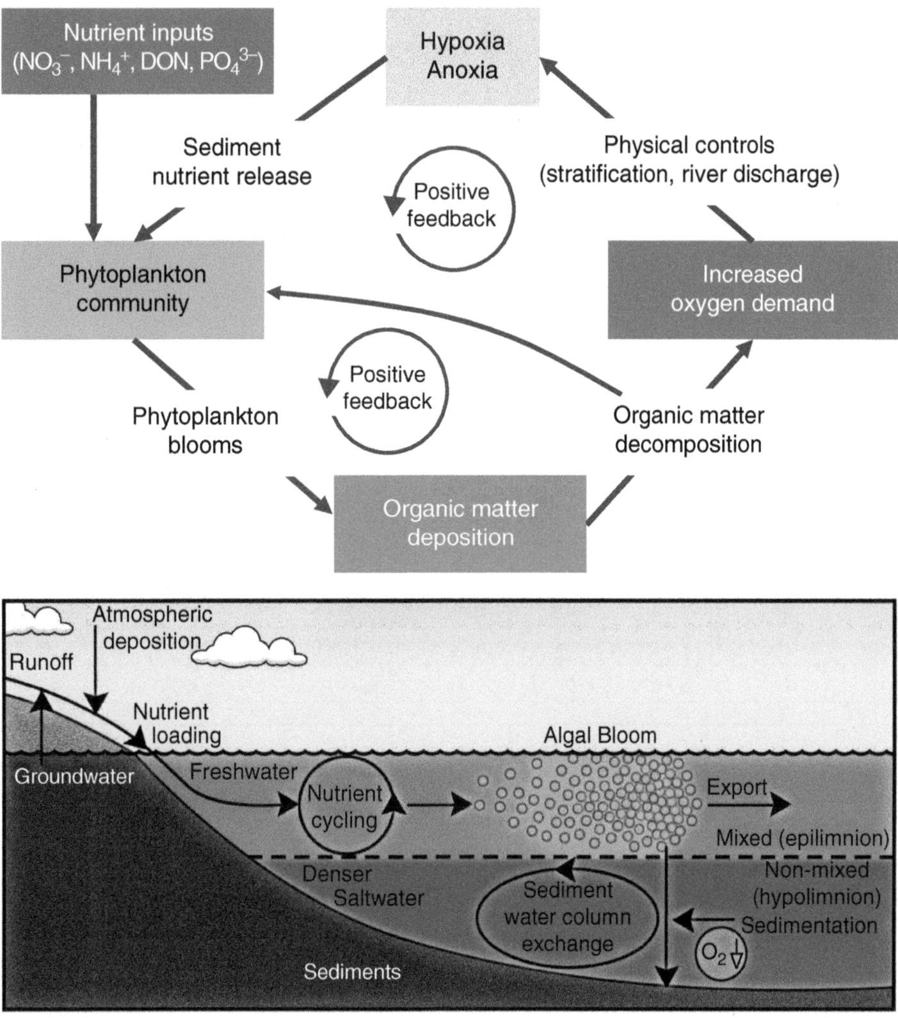

FIGURE 4.10 Upper: Conceptual linkages between external nutrient loading, internal nutrient cycling, nutrient-enhanced algal bloom formation, and hypoxia under salinity-stratified conditions. Lower: Illustration of the impact on hypoxia of phytoplankton that are not readily consumed in a stratified estuary, where sedimented organic matter, derived from algal blooms, enhances sediment-water nutrient and gas exchange and drives hypoxia. *Source*: Based on Paerl (2004).

Climatic shifts, including increasing tropical storm and hurricane frequencies (Webster et al. 2005; Bender et al. 2010) and changing drought conditions (Trenberth 2005) alter these inputs. It is therefore useful to develop ecological indicators that could help distinguish human impacts from natural perturbations (Paerl et al. 2007). This goal is compounded by the fact that human and natural perturbations may be similar, overlap, or act synergistically, potentially blurring this distinction.

Data from the mid-1990s to present show that coastal systems have experienced the combined stresses of anthropogenic nutrient enrichment, droughts (reduced flushing, minimal nutrient inputs), and floods (increased flushing, high nutrient inputs). These distinct perturbations have proven useful in examining impacts of anthropogenic and natural stressors on phytoplankton community structure. Seasonal and storm induced variations in river discharge, which affect residence times, strongly affect competition and relative dominance among different phytoplankton taxonomic groups as a function of their contrasting growth rates and doubling times. For example, the relative contribution of chlorophytes, cryptophytes, and diatoms to the total chlorophyll *a* pool was strongly controlled by variable freshwater discharge and flow rates, modulated by droughts and floods, in North Carolina's Neuse River Estuary and downstream Pamlico Sound (Figure 4.11; Valdes-Weaver et al. 2006; Paerl et al. 2009).

These effects are most likely due to differential nutrient uptake and growth rates among the different phytoplankton groups (Pinckney et al. 1999). Cyanobacteria, which have slow growth rates relative to other phytoplankton groups at ambient temperatures, were more abundant when residence times were longer during summer (Figure 4.11; Gaulke et al. 2010; Peierls et al. 2012; Paerl et al. 2013). Historic trends in dinoflagellate and chlorophyte abundance provide additional evidence that hydrologic changes have altered phytoplankton community structure in the Neuse River Estuary. Both decreases

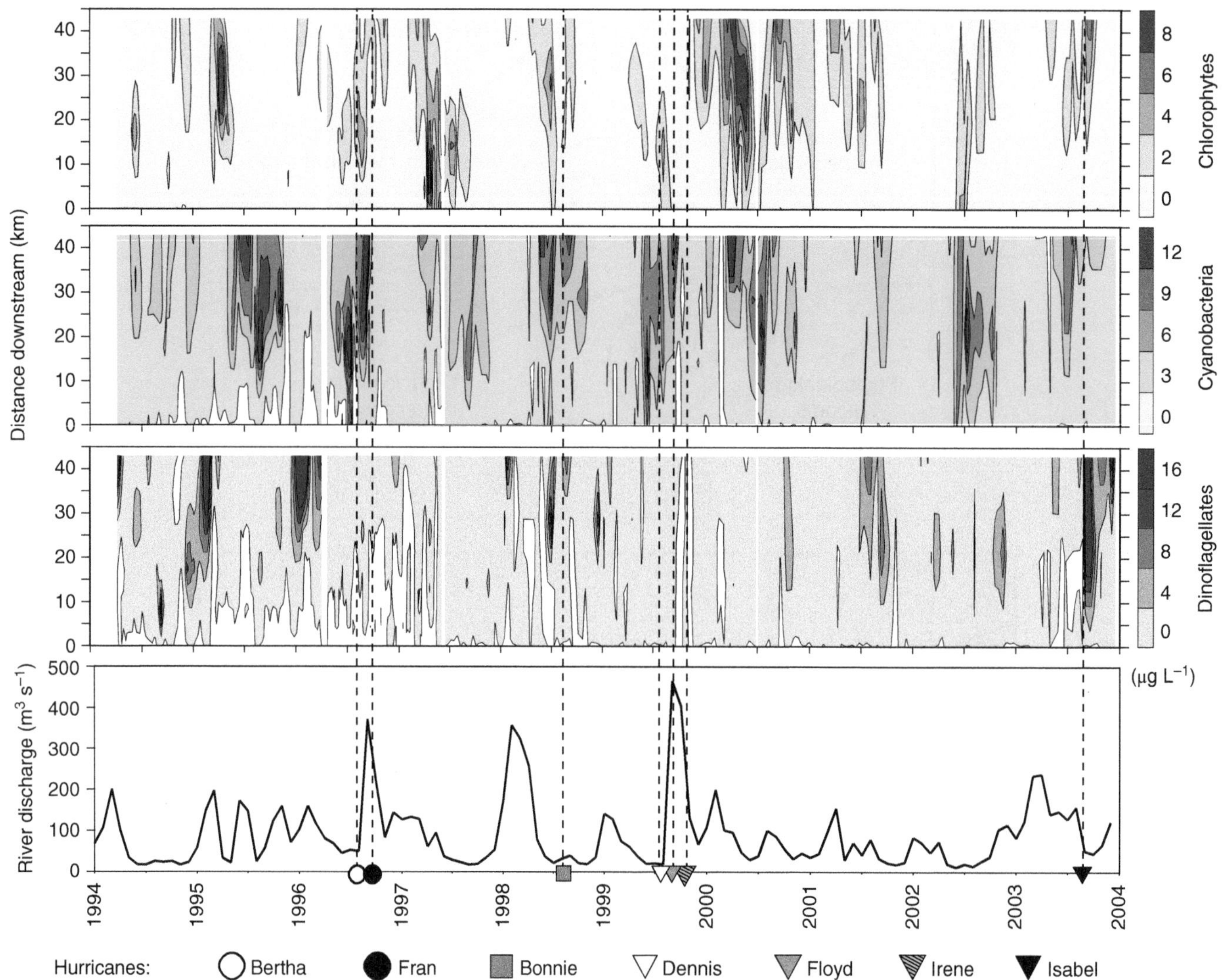

FIGURE 4.11 Phytoplankton taxonomic group biomass responses, based on measurements of pigments diagnostic of algal taxonomic groups (µg L^{-1}), to flow and nutrient enrichment in the Neuse River Estuary during 1994–2004 for chlorophytes (green algae), cyanobacteria (blue-green algae), and dinoflagellates. Chlorophytes are strongly stimulated by high discharge following major hurricanes (1996, 1999) and periods of high spring runoff (spring 1998). In contrast, the relative contribution of dinoflagellates to phytoplankton community biomass decreased during periods of high flow. Cyanobacterial biomass contributions also decreased during high flow but recovered noticeably during subsequent summer low flow periods.

in the occurrence of winter-spring dinoflagellate blooms and increases in the abundance of chlorophytes coincided with the increased frequency and magnitude of hurricanes since 1996 (Figure 4.11; Paerl et al. 2017; Valdes-Weaver et al. 2006). The relatively slow growth rates of dinoflagellates account for their reduced abundance during the ensuing high river discharge events. Overall, phytoplankton composition has been altered since 1994 following major hydrologic changes, specifically flooding from large hurricanes such as Fran and Floyd (Paerl et al. 2017). These phytoplankton community changes signal potential trophic and biogeochemical alterations.

There is a scientific consensus that the buildup of greenhouse gases in the atmosphere is warming the earth (IPCC 2014, and see Chapter 18). Increased global temperatures, combined with an enhanced hydrologic cycle, may influence estuarine and coastal eutrophication in three major ways (Figure 4.12). First, the magnitude and seasonal patterns of fresh water and nutrient inputs would be affected, which could impact nutrient-enhanced coastal productivity. Second, altered flushing and residence times would affect phytoplankton competitive interactions and hence dominance among major taxonomic groups. Third, increases in air and water temperatures will have a direct effect on phytoplankton physiology and growth. For example, cyanobacteria generally exhibit a higher growth maxima at higher temperatures than other taxonomic groups (e.g., diatoms, cryptophytes, and chlorophytes; Figure 4.13). Thus, in a warmer world, cyanobacterial growth and possibly bloom formation would be enhanced relative to other competitive taxonomic groups (Paerl et al. 2016a). Such taxonomic shifts would have ramifications for food web and nutrient cycling dynamics (Paerl and Huisman 2008, 2009).

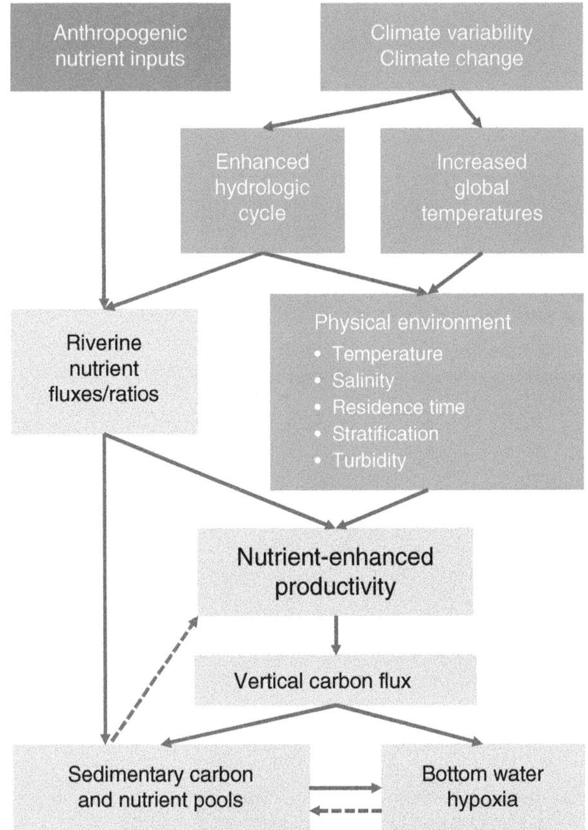

FIGURE 4.12 Coupling between climate variability, coastal eutrophication, and hypoxia. *Source*: Justić et al. (2005)/With Permission of Elsevier.

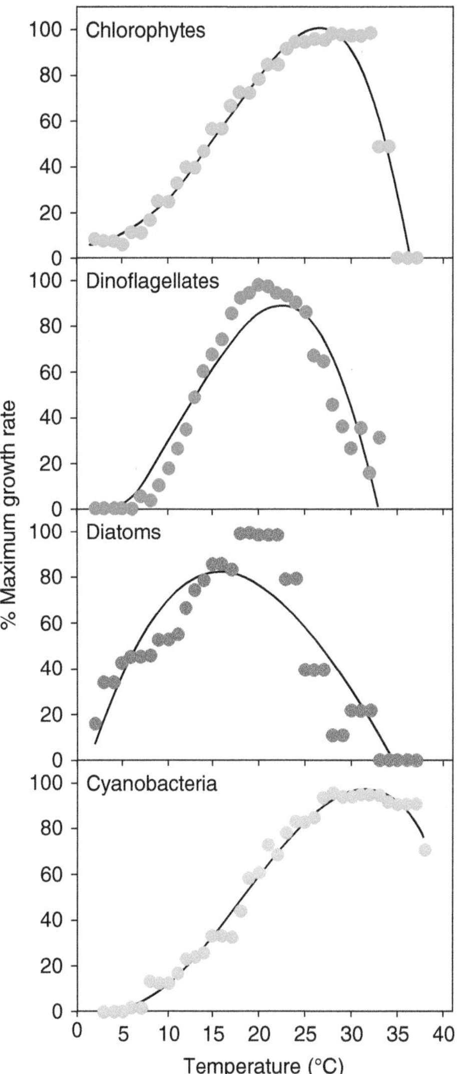

FIGURE 4.13 Effect of temperature on growth rates of major phytoplankton groups common to temperate freshwater and brackish environments. Data points are 5 °C running bin averages of percent maximum growth rates from 3 to 4 species within each class. The dashed line indicates optimal cyanobacterial growth temperature. Fitted lines are third-order polynomials and are included to emphasize the shape of the growth versus temperature relationship. *Source*: Paerl et al. (2016a)/With Permission of Elsevier.

4.6 Harmful Algal Blooms (HABs)

Phytoplankton blooms are a rapid and perceptible increase in phytoplankton biomass in an aquatic ecosystem (Figure 4.14), and when these blooms produce toxins, cause hypoxia, or greatly alter food webs they can be deemed "harmful algal blooms" (HABs) or harmful algal events (HAE) as when algae are at sub-lethal concentration. HABs constitute an environmental health hazard when they degrade water quality and habitat (Table 4.1). Species that form HABs include cyanobacteria, dinoflagellates, prymnesiophytes, and diatoms. During blooms of toxin-producing species, fish and shellfish may consume these phytoplankton, then accumulate and concentrate the biotoxins without apparent harm. This renders the fish and shellfish extremely toxic to whomever consumes them, including marine mammals, sea birds, and humans. In places where HAB monitoring and surveillance programs do not exist, these blooms may go unnoticed until they cause illnesses and/or death in humans who consume products from the sea or cause serious disruption to aquatic food webs via declines in a keystone species (Lefebvre et al. 1999).

4.7 Nutrient Management of Phytoplankton Production and Composition

When controlling and managing the effects of excessive nutrient loading on estuarine and coastal eutrophication and harmful algal bloom development, anthropogenic point and nonpoint sources are the most significant targets for nutrient reduction. The amounts, forms, and relative proportions of nutrients vary according to human activities, locations,

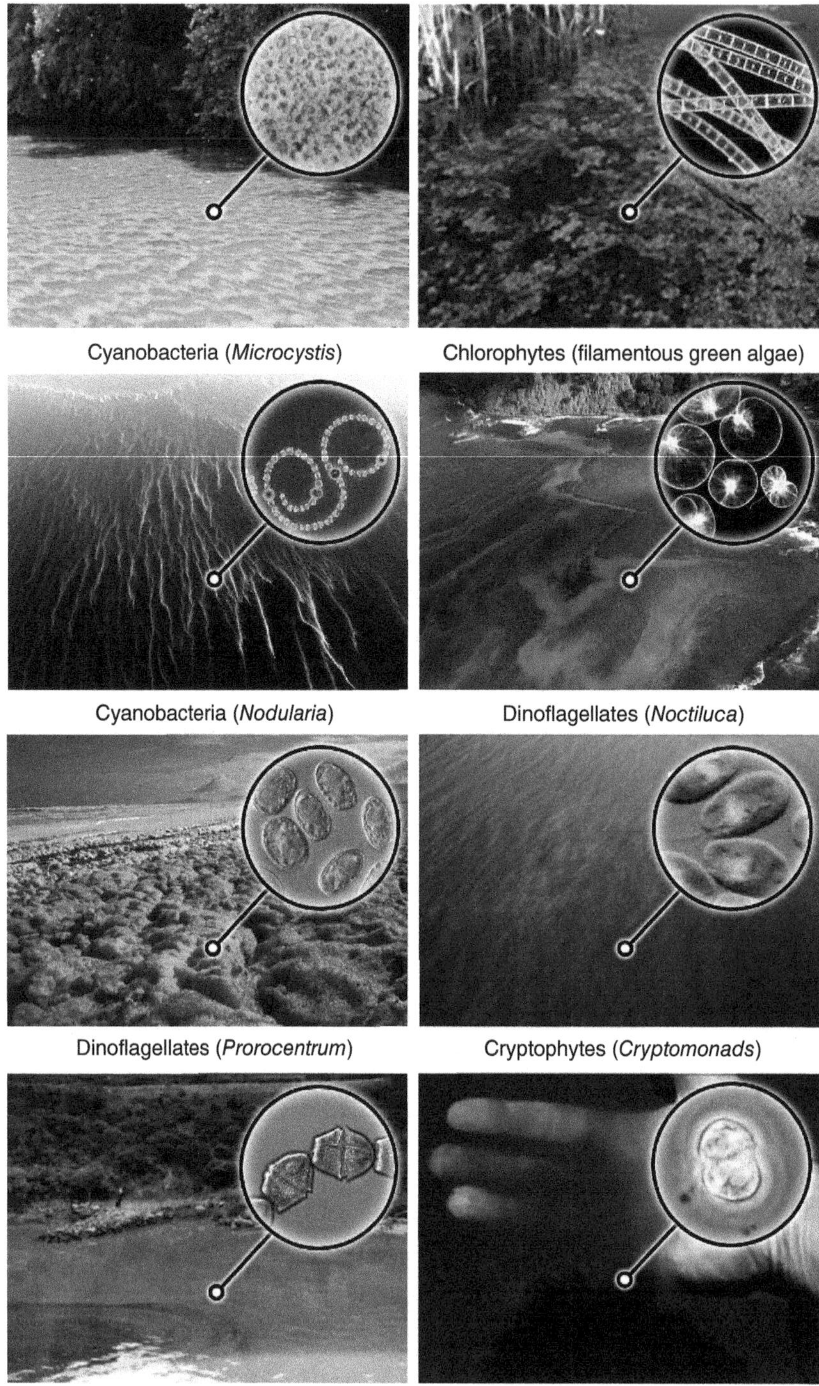

FIGURE 4.14 Algal blooms and micrographs of representative bloom-forming taxa in estuarine and coastal waters. Upper row (from left to right): Cyanobacterial (*Microcystis* sp.) bloom in the lagoonal Neuse River-Pamlico Sound, North Carolina. *Source*: Photo H. Paerl. Filamentous chlorophyte bloom in the Newport River, NC. *Source*: Photo H. Paerl. Second row: Cyanobacterial bloom in the Baltic Sea near the Finnish coast. *Source*: Courtesy Finnish Border Guard and Finnish Marine Research Institute. Dinoflagellate red tide, in coastal waters near Hong Kong. *Source*: Courtesy K. Yin. Third row: Dinoflagellate bloom, coastal North Sea. *Source*: Netherlands Institute of Ecology. Cryptophyte bloom, Neuse River Estuary/Pamlico sound, NC. *Source*: Photo, H. Paerl. Bottom row: Near-shore Dinoflagellate bloom, W. Florida. *Source*: Courtesy Florida Dept. of Environmental Protection. Toxic dinoflagellate (*Karlodinium*) bloom in Chesapeake Bay, MD/VA. *Source*: Courtesy, USEPA Chesapeake Bay Program.

TABLE 4.1 Economic and public health problems related to coastal harmful algae globally.

- Paralytic shellfish poisoning (PSP), which occurs in all coastal New England states as well as New York and along much of the west coast from Alaska to California as well as Singapore, China, France, Great Britain, South Africa, Australia and New Zealand. This problem has also extended to offshore areas in the northeast (causative species: the dinoflagellates *Alexandrium tamarense*, *A. fundyense*, and *A. catenella*).
- Neurotoxic shellfish poisoning (NSP) and fish mortalities in the Gulf of Mexico, and more recently extending along the Atlantic Coast from Florida to North Carolina, USA (causative species: the dinoflagellate *Karenia brevis*). It has also been observed in New Zealand.
- Mortalities of farmed salmonids in Washington and Oregon, USA (causative species: the diatoms *Chaetoceros convolutus* and *C. concavicornis* and the raphidophyte *Heterosigma akashiwo*).
- Recurrent brown tides causing mass mortalities of mussel populations in Rhode Island, USA, and massive recruitment failure of scallops, and reduction of eelgrass beds around Long Island, NY, USA (causative species: the chrysophyte, *Aureococcus anophagefferens*).
- Ciguatera fish poisoning (CFP), a malady associated with dinoflagellate toxins accumulated in tropical fish flesh, occurring in virtually all sub-tropical to tropical U.S. waters (Florida, Hawaii, Guam, U.S. Virgin Islands, Puerto Rico, Guam and other Pacific Territories; major causative species *Gambierdiscus toxicus, Prorocentrum* spp., *Ostreopsis* spp., *Coolia monotis*, *Thecadinum* sp., and *Amphidinium carterae*).
- Amnesic shellfish poisoning (ASP), which occurred first in southeastern Canada in 1987, is a problem for the U.S. Pacific coast states (causative species: the diatoms *Pseudonitzschia pungens* forma *multiseries* and *Pseudonitzschia australis*). This sometimes fatal illness is so named because one of its most severe symptoms is permanent loss of short-term memory. The ASP toxin, domoic acid, has been detected in shellfish from both the West and East Coasts of the U.S., and toxic *P. pungens* f. *multiseries* cells have been isolated from Gulf of Mexico waters, though a toxin has yet to be detected in the field. The threat to U.S. shellfish consumers from this alga covers a broad geographic area. The name "ASP" understates the severity of the problem, since domoic acid also accumulates in fish and in crab viscera along the west coast of the United States, where the impact of this toxin on non-molluscan fisheries may well exceed the loss to molluscan fisheries (e.g., razor clam).
- Diarrhetic shellfish poisoning (DSP) which some consider the most serious and globally widespread phytoplankton-related seafood illness. Major causative species are dinoflagellates *Dinophysis* and *Prorocentrum*. The first confirmed incidence of DSP in North America occurred in 1990 when these toxins were detected in shellfish from the southern coast of Nova Scotia, Canada following numerous human illnesses. Another DSP outbreak in Canada occurred in 1992. DSP-producing phytoplankton species occur throughout all temperate coastal waters of the U.S., and thus present a potential problem for the future, though no outbreaks of DSP have yet been confirmed.
- Cyanobacteria blooms are becoming more numerous, widespread and persistent in nutrient-enriched estuarine and coastal waters worldwide. Blooms have multiple negative impacts, such as toxicity, overgrowing and smothering seagrasses, coral reefs, and shellfish habitats, and food web shifts. Some species produce toxins and other bioactive metabolites. Of concern are blooms of toxic colonial non N_2 fixing genera, including *Microcystis* (oligohaline) and heterocystous N_2 fixing genera *Nodularia* (full salinity), *Cylindrospermopsis* and *Dolichospemum* (oligohaline). The filamentous, non-heterocystous N_2 fixing species *Lyngbya majuscula* produces both dermatotoxins and neurotoxins. Blooms of this subtropical/tropical species have fouled large segments of estuaries and bays in both Hawaii and Florida, USA. Some tropical harmful cyanobacterial species are moving into more temperate regions in part due to global warming, which has expanded their habitat (Paerl and Huisman 2008).

Source: Adapted from Anderson et al. (2000).

and distributions of population centers and routes of nutrient discharge (i.e., surface, subsurface, and atmospheric). Magnitudes and proportions of N and P input as well as forms on N (i.e., ammonium vs. nitrate vs. organic N) have been shown to control both phytoplankton community productivity and compositional responses in receiving waters (Altman and Paerl 2012; Paerl et al. 2016b; Glibert and Burford 2017; Paerl et al. 2018), and as such both aspects require careful assessment and management. In point source-dominated watersheds, the emphasis has been on improved wastewater treatment and removal of both N and P. In nonpoint source-dominated watersheds, surface runoff, especially those originating from agricultural operations and urban stormwater, are the main focus of N reduction strategies. These strategies include agricultural best management practices, including prudent and timely applications of fertilizers, soil conservation, establishment of riparian vegetative buffer zones, and use of wetlands to enhance "stripping" of runoff-based N (Lowrance et al. 1984; Mitsch et al. 2001).

Point sources of N and P are under intense local, state, and federal scrutiny. Government agency (U.S. EPA 1998, 1999, European Parliament: EEU-Water Framework 2003) and legislative action has led to strict N and P discharge limits from wastewater treatment plants, and those plants not able to meet these standards are under considerable and continuing pressure to upgrade. In addition, a phosphate detergent ban was enacted in the mid-1980s in North America, Europe, and parts of Australasia. This led to marked decreases in P loading in many watersheds. Current strategies aimed at reducing nonpoint N discharge, including riparian buffers, wetland construction, and soil conservation, also limit the movement of P to nutrient sensitive waters (Gilliam et al. 1997). Regionally and globally (China, India, Australasia, North and South America, Europe, and Africa), nonpoint strategies and targets for pursuing constraints on one nutrient have led to significant parallel reductions of the other.

In addition to surface runoff, atmospheric deposition and groundwater should also be recognized as significant sources of new N potentially stimulating primary production in estuarine and coastal waters (Paerl 1985, 1997). Local and regional studies have shown atmospheric deposition and groundwater inputs of N to be large, increasing, and of widespread

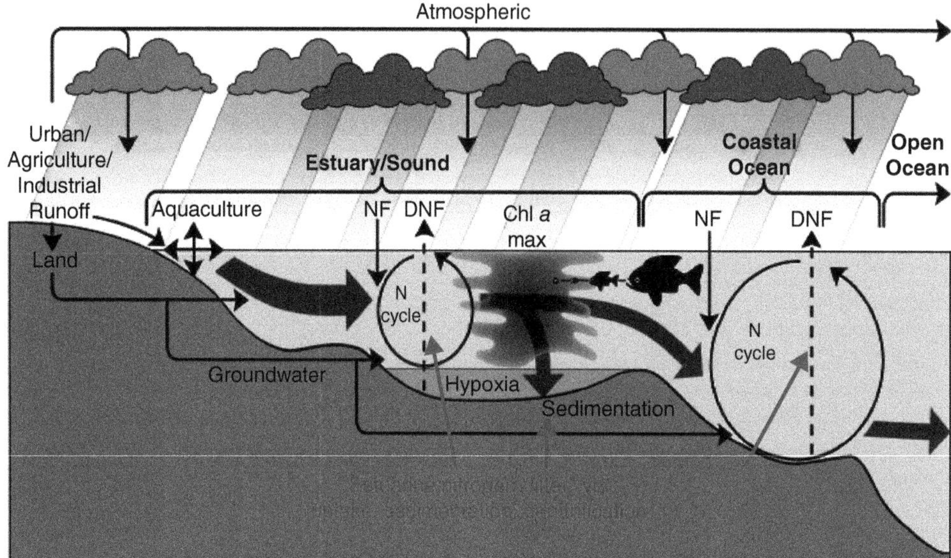

FIGURE 4.15 Illustration of the estuarine "filter" concept, where land-based nitrogen (N) is filtered by the estuary, while some portion of atmospheric N deposition bypasses the "filter," directly fertilizing coastal and oceanic waters. NF indicates nitrogen fixation, and DNF indicates denitrification.

importance in estuarine and coastal environments (Jaworski et al. 1997; Paerl et al. 2002; Moore 1999). For example, atmospheric deposition accounts for 10% to over 30% of externally supplied N sources along the US Eastern seaboard, the Baltic Sea region, and eastern Asian coastal regions (Paerl et al. 2002; Castro et al. 2003). Atmospheric deposition plays an increasingly important role in coastal waters, since surface runoff N is often effectively "filtered" in N limited estuarine systems as it transits to the ocean (Paerl and Whitall 1999). Atmospheric deposition (and some groundwater) directly enters coastal waters and bypasses this estuarine N filter (Paerl et al. 2002; Moore 1999; Figure 4.15). These N inputs may be key drivers of a reported increase of harmful algal blooms in coastal waters (Hallegraeff 1993; Glibert and Burford 2017; Paerl et al. 2018). This, combined with observations that N and P loading to these waters has increased steadily in comparison to silicon (Si), may also help explain observed shifts in phytoplankton communities away from Si-requiring diatoms to flagellates, dinoflagellates, and cyanobacteria (Dortch and Whitledge 1992).

Certain forms of N may be preferred by phytoplankton, including bloom-forming species. For example, in highly turbid estuarine waters where light availability may be limited, N sources that require the least amount of energy for uptake and assimilation may be preferred. The most readily-assimilated and least energy-requiring form of N is ammonium, which is quickly used over the oxidized forms nitrite and nitrate (Harrison and Turpin 1982). Experimental work has confirmed this for nutrient-enriched estuarine and coastal waters (Collos 1989; Stolte et al. 1994). In addition, some phytoplankton groups and species are more capable than others at utilizing organic forms of N and other nutrients (Antia et al. 1991; Altman and Paerl 2012; Glibert and Burford 2017). These findings stress the need for considering both the qualitative and quantitative influences of N (and other nutrient) loadings on phytoplankton biomass and compositional responses.

An additional challenge for managing estuarine primary production and phytoplankton community composition is the influence of climate change. Some symptoms of climate change, such as increased tropical storm activity, larger oscillations between extremely wet and drought conditions, and hence the more pulsed, episodic manner in which nutrients are supplied to coastal ecosystems, strongly influence phytoplankton compositional and biomass responses in estuaries (Adolf et al. 2006; Valdes-Weaver et al. 2006; Harding et al. 2016) and hence confound management strategies that are based on average or "normal" hydrologic and nutrient discharge patterns. Since entering the recent period of elevated tropical cyclone activity and more extreme hydrologic events (floods, droughts; Trenberth 2005), it seems that episodicity is now the norm rather than the exception. This calls for highly adaptive nutrient management strategies that take into consideration pulsed instead of more gradual patterns of nutrient and other pollutant input.

Much work lies ahead to further identify, characterize, and manage nutrient inputs controlling phytoplankton production and composition in estuarine and coastal waters. Research and monitoring are providing information essential for formulating long-term nutrient management strategies aimed at protecting and preserving the high standards of water quality and resource benefit that we expect. This means managing for enough, but not excessive phytoplankton production, and avoiding the promotion of harmful taxa. Fostering a process-based understanding of nutrient-water quality interactions and using this for effective management are key social, economic, and political responsibilities as we ensure long-term conservation and sustainability of coastal waters.

Acknowledgments

We appreciate the editorial and illustrative assistance of A. Joyner, L. Harding, B. Peierls, and K. Rossignol. Work discussed was supported by; the North Carolina Department of Environment and Natural Resources, the North Carolina Sea Grant Program, the University of North Carolina Water Resources Research Institute, the US Dept. of Agriculture-NRI-CRGP, the US EPA-STAR-EaGLe Program, the NOAA/CSCOR Northern Gulf of Mexico Hypoxia Assessment Program, and the National Science Foundation (Biological and Chemical Oceanography, Ecology of Infectious Diseases (EID), Environmental Engineering Technology (EET), and Dimensions in Biodiversity Programs), the NSF-NIH Oceans and Human Health Program, and NOAA Center for Sponsored Coastal Ocean Research. Remote sensing imagery was provided by NASA's SeaWiFS and MODIS Program.

Review Questions

1. Phytoplankton in coastal systems typically account for at least what fraction of total primary production?
   a. 1/4
   b. 1/3
   c. 1/2
   d. 2/3

2. Diagnostic photopigments are useful in phytoplankton taxonomy at the species level
   a. True
   b. False, they are useful at the taxonomic group level

3. Flow cytometry enables the assessment of estuarine phytoplankton based on what properties?
   a. Size
   b. Fluorescence
   c. Abundance
   d. All of the above

4. Which gene can be used to study eukaryotic phytoplankton diversity but not cyanobacteria diversity?
   a. *nifH*
   b. *rbcL*
   c. 18S rRNA gene
   d. 16S rRNA gene

5. Picocyanobacteria are favored under all of the following conditions, except
   a. Elevated temperatures
   b. Increased flow
   c. Strong water column stratification
   d. None of the above

6. Eutrophication is the ability of certain phytoplankton groups to exploit salinity regimes ranging from fresh water to the coastal ocean.
   a. True
   b. False

7. The supply of which of the following nutrient elements can control estuarine and coastal phytoplankton primary production? (Select all that apply)
   a. Calcium
   b. Iron
   c. Nitrogen
   d. Phosphorus
   e. Potassium
   f. Silicon
   g. Sodium

8. What is the "harm" in Harmful Algal bloom species? (Select all that apply)
   a. toxicity
   b. elevated predation
   c. oxygen depletion (hypoxia/anoxia)
   d. reduced detritus carbon
   e. food web disruption

9. How can changes in river flow control the location of maximum phytoplankton production in an estuary? (Select all that apply)
   a. By controlling the flux of river-borne nutrients
   b. By controlling the temperature of an estuary
   c. By influencing the residence time of water
   d. By controlling the salinity of an estuary
   e. By controlling the turbidity in an estuary.

10. How do phytoplankton communities adapt to changing light conditions? (Select all that apply)
    a. by varying their photosynthetic pigment composition
    b. by developing heterocysts
    c. by changing species composition
    d. by using adaptive nutrient management strategies
    e. by migrating vertically

11. Why do phytoplankton often bloom in spring in temperate estuaries? (Select all that apply)
    a. Nutrient inputs via river flow are often elevated in spring in temperate estuaries
    b. Turbidity in estuaries is lowest in spring
    c. Phosphorus is released from sediments when bottom waters become anoxic
    d. Zooplankton growth and grazing is limited in spring by colder waters
    e. Water temperature is highest in spring, enhancing phytoplankton growth

12. How are increased global temperatures, combined with an enhanced hydrologic cycle, forecast to impact phytoplankton community structure and function? (Select all that apply)
    a. Warming will favor cyanobacteria production.
    b. Changes in wind speeds will change the locations of blooms.
    c. Sea level rise will enhance productivity.
    d. Changes to fresh water and nutrient inputs will alter productivity.
    e. Changes to estuarine residence times will alter community composition.

Short Answer

1. N_2 fixing cyanobacteria are highly competitive under N-deplete conditions; however, rarely dominate estuarine phytoplankton communities. What conditions limit their dominance?
2. How does sequencing of mRNA (converted to cDNA) from phytoplankton offer insight on potential activity of specific phytoplankton?
3. Describe the primary sources of anthropogenic nutrient to estuaries in (a) rural, agriculturally dominated regions, and (b) urban watersheds.
4. Check all boxes that are correct.

	a. Chlorophytes	b. Cryptophytes	c. Chrysophytes	d. Cyanobacteria	e. Dinoflagellates	f. Diatoms	g. Pelagophytes	h. Prymnesiophytes
1. Eukaryotes								
2. Prokaryotes								
3. Euryhaline								
4. Motile								
5. Diazotrophic								
6. HAB forming								

References

Adolf, J.E., Yeager, C.L., Miller, W.D. et al. (2006). Environmental forcing of phytoplankton floral composition, biomass, and primary productivity in Chesapeake Bay, USA. *Estuarine Coastal and Shelf Science* 67: 108–122.

Altman, J.C. and Paerl, H.W. (2012). Composition of inorganic and organic nutrient sources influences phytoplankton community structure in the New River Estuary, North Carolina. *Aquatic Ecology* 42: 269–282.

Anderson, D.M., Kaoru, Y., and White, A.W. (2000). *Estimated Annual Economic Impacts from Harmful Algal Blooms (HABs) in the United States*, Woods Hole Oceanographic Institution, Technical report WHOI-2000-11, 97. Woods Hole Oceanographic Institution.

Antia, N., Harrison, O., and Oliveira, L. (1991). The role of dissolved organic nitrogen in phytoplankton nutrition, cell biology and ecology. *Phycologia* 30: 1–89.

Bender, M.A., Knutson, T.R., Tuleya, R.E. et al. (2010). Modeled impact of anthropogenic warming on the frequency of intense Atlantic hurricanes. *Science* 327: 454–458.

Bertilsson, S., Berglund, O., Karl, D., and Chisholm, S.W. (2003). Elemental composition of marine *Prochlorococcus* and *Synechococcus*: implications for the ecological stoichiometry of the sea. *Limnology and Oceanography* 48 (5): 1721–1731.

Biber, P.D., Paerl, H.W., Gallegos, C.L., and Kenworthy, W.J. (2005). Evaluating indicators of seagrass stress to light. In: *Proceedings of the Estuarine Indicators Workshop* (ed. S. Bartone), 193–209. Boca Raton, Florida: CRC Press, Orlando.

Boesch, D.F., Burreson, E., Dennison, W. et al. (2001). Factors in the decline of coastal ecosystems. *Science* 293: 629–638.

Boynton, W.R. and Kemp, W.R. (1985). Nutrient regeneration and oxygen consumption by sediments along an estuarine salinity gradient. *Marine Ecology Progress Series* 23: 45–55.

Bricelj, V.M. and Lonsdale, D.J. (1997). *Aureococcus anophagefferens*: Causes and consequences of brown tides in U.S. mid-Atlantic coastal waters. *Limnology and Oceanography* 42 (5): 1023–1038.

Bricker, S.B., Clement, C.G., Pirhalla, D.E. et al. (1999). *National estuarine eutrophication assessment: effects of nutrient enrichment in the Nation's estuaries*. Silver Spring, MD: NOAA, National Ocean Service, Special Projects Office, and the National Centers for Coastal Ocean Science.

Buskey, E.J. and Hyatt, C.J. (2006). Use of the FlowCAM for semi-automated recognition and enumeration of red tide cells (Karenia brevis) in natural plankton samples. *Harmful Algae* 5 (6): 685–692.

Callahan, B. J., McMurdie, P.J. and Holmes, S.P. (2017). Exact sequence variants should replace operational taxonomic units in marker-gene data analysis. *The ISME Journal* 11: 2639–2643. doi:https://doi.org/10.1038/ismej.2017.119

Capone, D.G., Zehr, J.P., Paerl, H.W. et al. (1997). *Trichodesmium*, a globally-significant marine cyanobacterium. *Science* 276: 1221–1229.

Caraco, N.F., Cole, J.J., Raymond, P.A. et al. (1997). Zebra mussel invasion in a large, turbid river: Phytoplankton response to increased grazing. *Ecology* 78 (2): 588–602.

Carmichael, W.W. (2001). Health effects of toxin producing cyanobacteria: the cyanoHABs. *Human Ecological Risk Assessment* 7: 1393–1407.

Caron, D.A., Alexander, H., Allen, A.E. et al. (2017). Probing the evolution, ecology and physiology of marine protists using transcriptomics. *Nature Reviews Microbiology* 15 (1): 6.

Carpenter, E.J., Montoya, J.P., Burns, J.A. et al. (1999). Extensive bloom of a N_2-fixing diatom/cyanobacterial association in the Tropical Atlantic Ocean. *Marine Ecology Progress Series* 185: 273–283.

Castro, M.S., Driscoll, C.T., Jordan, T.E. et al. (2003). Sources of nitrogen to estuaries in the United States. *Estuaries* 26 (3): 803–814.

Chisholm, S.W. (1992). Phytoplankton size. In: *Primary Productivity and Biogeochemical Cycles in the Sea* (ed. P.G. Falkowski, A.D. Woodhead and K. Vivirito), 213–237. Boston, MA: Springer.

Cloern, J.E. (1982). Does the benthos control phytoplankton biomass in south San Francisco Bay. *Marine Ecology Progress Series* 9: 191–202.

Cloern, J.E. (1999). The relative importance of light and nutrient limitation on phytoplankton growth: a simple index of coastal ecosystem sensitivity to nutrient enrichment. *Aquatic Ecology* 33: 3–16.

Cloern, J.E. (2001). Our evolving conceptual model of the coastal eutrophication problem. *Marine Ecology Progress Series* 210: 223–253.

Cloern, J.E. and Jassby, A.D. (2012). Drivers of change in estuarine-coastal ecosystems: discoveries from four decades of study in San Francisco Bay. *Reviews in Geophysics* 50, RG4001, doi:https://doi.org/10.1029/2012RG000397.

Cloern, J.E., Foster, S.Q., and Kleckner, A.E. (2014). Phytoplankton primary production in the world's estuarine-coastal ecosystems. *Biogeosciences* 11: 2477–2501.

Cohen, R.R.H., Dresler, P.Y., Phillips, E.J.P., and Cory, R.L. (1984). The effect of the Asiatic clam, *Corbiculafluminea*, on phytoplankton of the Potomic River, Maryland. *Limnology and Oceanography* 29 (1): 170–180.

Cole, B.E. and Cloern, J.E. (1984). Significance of biomass and light availability to phytoplankton productivity in San Francisco Bay. *Marine Ecology Progress Series* 15: 15–24.

Collos, Y. (1989). A linear model of external interactions during uptake of different forms of inorganic nitrogen by microalgae. *Journal of Plankton Research* 11: 521–533.

Cunning, R., Bay, R.A., Gillette, P. et al. (2018). Comparative analysis of the *Pocillopora damicornis* genome highlights role of immune systems in corals. *Nature Scientific Reports* 8: 16134.

Dagg, M.J. (1995a). Copepod grazing and the fate of phytoplankton in the northern Gulf of Mexico. *Continental Shelf Research* 15: 1303–1317.

Dagg, M.J. (1995b). Ingestion of phytoplankton by the micro- and mesozooplankton communities in a productive subtropical estuary. *Journal of Plankton Research* 17: 845–857.

D'Agostino, V.C., Degrati, M, Sastre, V., Santinelli, N., Krock, B., Krohn, T., Dans, S.L. and Hoffmeyer, M.S. (2017). Domoic acid in a marine pelagic food web: exposure of southern right whales *Eubalaena australis* to domoic acid on the Peninsula Valdes calving ground, Argentina. *Harmful Algae* 68:248–257 DOI: https://doi.org/10.1016/j.hal.2017.09.001

Davidson, K., Tett, P., and Gowen, R. (2011). Harmful algal blooms. In: *Marine Pollution and Human Health, Issues in Environmental Science and Technology* 33 (ed. R.E. Hester and R.M. Harrison), 95–167. Cambridge, UK: Publ. by Royal Society of Chemistry.

Day, J.W., Cable, J.E., Cowan, J.H. Jr. et al. (2009). The impacts of pulsed reintroduction of river water on a Mississippi Delta Coastal Basin. *Journal of Coastal Research* 54: 225–243.

Diaz, R.J. and Rosenberg, R. (2008). Spreading dead zones and consequences for marine ecosystems. *Science* 321: 926–929.

Doering, P.H., Oviatt, C.A., and Kelly, J.R. (1986). The effects of the filter-feeding clam *Mercenaria mercenaria* on carbon cycling in experimental marine mesocosms. *Journal of Marine Research* 44: 839–861.

Dortch, Q. and Whitledge, T.E. (1992). Does nitrogen or silicon limit phytoplankton production in the Mississippi River plume and nearby regions? *Continental Shelf Research* 12: 1293–1309.

Dustan, P. and Pinckney, J. (1989). Tidally induced estuarine phytoplankton patchiness. *Limnology and Oceanography* 34: 408–417.

Eppley, R.W. (1972). Temperature and phytoplankton growth in the sea. *Fisheries Bulletin* 70: 1063–1085.

European Parliament: EEU-Water Framework (2003). Common Implementation Strategy for the Water Framework Directive (2000/60/EC). Carrying forward the Common Implementation Strategy for the Water Framework Directive. Progress and Work Programme for 2003 and 2004. European Union Publications. Brussels, Belgium.

Falkowski, P.G. (1980). Light-shade adaptation in marine phytoplankton. In: *Primary Productivity in the Sea* (ed. P.G. Falkowski), 99–120. New York: Plenum.

Farnelid, H., Tarangkoon, W., Hansen, G. et al. (2010). Putative N_2-fixing heterotrophic bacteria associated with dinoflagellate–Cyanobacteria consortia in the low-nitrogen Indian Ocean. *Aquatic Microbial Ecology* 61: 105–117.

Fee, E.J. (1980). Important factors for estimating annual phytoplankton production in the Experimental Lakes Area. *Canadian Journal of Fisheries and Aquatic Science* 37: 513–522.

Fisher, T.R., Harding, L.W. Jr., Stanley, D.W., and Ward, L.G. (1988). Phytoplankton, nutrients, and turbidity in the Chesapeake, Delaware, and Hudson estuaries. *Estuarine Coastal and Shelf Science* 27: 61–93.

Flynn, K.J. and Martin-Jézéquel, V. (2000). Modelling Si-N limited diatom growth. *Journal of Plankton Research* 22: 447–472.

Foster, R.A., Kuypers, M.M., Vagner, T. et al. (2011). Nitrogen fixation and transfer in open ocean diatom–cyanobacterial symbioses. *The ISME Journal* 5 (9): 1484.

Gallegos, C.L., Correll, D.L., and Pierce, J.W. (1990). Modeling spectral diffuse attenuation, absorption, and scattering coefficients in a turbid estuary. *Limnology and Oceanography* 35: 1486–1502.

Gallegos, C. L., Werdell, P. J. and McClain, C. R. (2011). Long-term changes in light scattering in Chesapeake Bay inferred from Secchi depth, light attenuation, and remote sensing measurements. *Journal of Geophysical Research (Oceans)* 116, C00H08, doi: https://doi.org/10.1029/2011JC007160.

Gaulke, A.K., Wetz, M.S., and Paerl, H.W. (2010). Picophytoplankton: A major contributor to planktonic biomass and primary production in a eutrophic, river-dominated estuary. *Estuarine, Coastal and Shelf Science* 90: 45–54.

Gilliam, J.W., Osmond, D.L., and Evans, R.O. (1997). *Selected Agricultural Best Management Practices to Control Nitrogen in the Neuse River Basin*. NC Agricultural Research Service Technical Bull 311. NC State University, Raleigh, NC.

Glibert, P.M. and Burford, M.A. (2017). Globally changing nutrient loads and harmful algal blooms: Recent advances, new paradigms, and continuing challenges. *Oceanography* 30: 58–69.

Gobler, C.J., Renaghan, M., and Buck, N. (2002). Impacts of nutrients and grazing mortality on the abundance of *Aureococcus anophagefferens* during a New York Brown Tide bloom. *Limnology and Oceanography* 47: 129–141.

Goldman, J.E. (1979). Temperature effects on steady-state growth, phosphorus uptake, and the chemical composition of a marine phytoplankton. *Microbial Ecology* 5: 153–166.

Guillou, L., Bachar, D., Audic, M. et al. (2013). The Protist Ribosomal Reference database (PR2): a catalog of unicellular eukaryote Small Sub-Unit rRNA sequences with curated taxonomy. *Nucleic Acids Research* 41 (D1): D597–D604.

Hall, N.S. and Paerl, H.W. (2011). Vertical migration patterns of phytoflagellates in relation to light and nutrient availability in a shallow, microtidal estuary. *Marine Ecology Progress Series* 425: 1–19.

Hall, N.S., Litaker, R.W., Fensin, E. et al. (2008). Environmental factors contributing to the development and demise of a toxic dinoflagellate (*Karlodinium veneficum*) bloom in a shallow, eutrophic, lagoonal estuary. *Estuaries and Coasts* 31: 402–418.

Hall, N.S., Whipple, A.C., and Paerl, H.W. (2015). Vertical spatio-temporal patterns of phytoplankton due to migration behaviors in two shallow, microtidal estuaries: Influence on phytoplankton function and structure. *Estuarine, Coastal and Shelf Science* 162: 7–21.

Hallegraeff, G. (1993). A review of harmful algal blooms and their apparent global increase. *Phycologia* 32: 79–99.

Harding, L.W. Jr., Mallonee, M.M., Perry, E.S., Miller, W.D., Adolf, J.E., Gallegos, C.L. and Paerl, H.W. (2016). Variable climatic conditions dominate recent phytoplankton dynamics in Chesapeake Bay. *Nature Scientific Reports* 6:23773 DOI: https://doi.org/10.1038/srep23773

Harding, L.W., Meeson, B.W., and Fisher, T.R. (1987). *Photosynthesis patterns in Chesapeake Bay Phytoplankton: Short- and Long-Term Responses of P-I Curve Parameters to Light*. College Park, MD: University of Maryland Sea Grant Publication RS-87-05.

Harrison, P.J. and Turpin, D.H. (1982). The manipulation of physical, chemical, and biological factors to select species from natural phytoplankton communities. In: *Marine Mesocosms* (ed. G.D. Grice et al.), 275–289. New York: Springer.

Hasle, G.R., Syvertsen, E.E., Steidinger, K.A. et al. (1996). *Identifying Marine Diatoms and Dinoflagellates*. New York: Academic Press.

Hawser, S.P., O'Neil, J.M., Roman, M.R., and Codd, G.A. (1992). Toxicity of blooms of the cyanobacterium *Trichodesmium* to zooplankton. *Journal of Applied Phycology* 4: 79–86.

Heinle, D.R. (1974). An alternative grazing hypothesis for the Patuxent Estuary. *Chesapeake Science 15*: 146–150.

Humborg, C., Rahm, L., Conley, D.J. et al. (2007). Silicon in the Baltic Sea: Long-term Si decrease in the Baltic Sea- A conceivable ecological risk? *Journal of Marine Systems* 73: 221–222.

Imanian, B., Pombert, J.F., Dorrell, R.G. and Keeling, B.F. (2012). Endosymbiosis in two diatoms has generated little change in the mitochondrial genomes of their dinoflagellate hosts and diatom endosymbionts. *PLOS ONE* 7(8): e43763 DOI: https://doi.org/10.1371/journal.pone.0043763

Intergovernment Panel on Climate Change (IPCC) (2014). *Climate Change Synthesis Report. Contribution of Working Groups I, II and III to the Fifth Assessment Report of the Intergovernmental Panel on Climate Change* Core Writing Team (ed. R.K. Pachauri and L.A. Meyer). Geneva, Switzerland: IPCC, 151 pp.

Jassby, A.D. and Platt, T. (1976). Mathematical formulation of the relationship between photosynthesis and light for phytoplankton. *Limnology and Oceanography* 21: 540–547.

Jaworski, N., Howarth, R., and Hetling, L. (1997). Atmospheric deposition of nitrogen oxides onto the landscape contributes to coastal eutrophication in the Northeast United States. *Environmental Science and Technology* 31: 1995–2004.

John, D.E., Wawrik, B., Paul, J.H., and Tabita, F.R. (2006). Gene diversity and organization in *rbcL*-containing genome fragments from uncultivated *Synechococcus* in the Gulf of Mexico. *Marine Ecology Progress Series* 316: 23–33.

Justić, D., Rabalais, N.N., Turner, R.E., and Dortch, Q. (1995). Changes in nutrient structure of river-dominated coastal waters: Stoichiometeric nutrient balance and its consequences. *Estuarine, Coastal and Shelf Science* 40: 339–356.

Justić, D., Rabalais, N.N., and Turner, R.E. (2005). Coupling between climate variability and marine coastal eutrophication: Historical evidence and future outlook. *Journal of Sea Research* 54: 25–35.

Keeling, P.J. (2011). The endosymbiotic origin, diversification and fate of plastids.philosophical. *Transactions of the Royal Society B-Biological Sciences* 365 (1541): 729–748.

Komarek, J. and Anagnostidis, K. (1986). Modern approach to the classification of the Cyanophytes 2--Chroococcales. *Archives of Hydrobiology Supplement* 73: 157–226.

Kononen, K., Kuparinen, J., Mäkelä, J. et al. (1996). Initiation of cyanobacterial blooms in a frontal region at the entrance to the Gulf of Finland, Baltic Sea. *Limnology and Oceanography* 41: 98–112.

Kremer, P. (1979). Predation by the ctenophore *Mnemiopsis leidyi* in Narragansett Bay, R.I. *Estuaries* 2: 97–105.

Lane, D.J., Pace, B., Olsen, G.J. et al. (1985). Rapid determination of 16S ribosomal RNA sequences for phylogenetic analyses. *Proceedings of the National Academy of Sciences USA* 82 (20): 6955–6959.

Lefebvre, K.A., Powell, C.L., Busman, M. et al. (1999). Detection of domoic acid in northern anchovies and California sea lions associated with an unusual mortality event. *Natural Toxins 7* (3): 85–92.

Litaker, R.W., Tester, P.A., Duke, C.S. et al. (2002). Seasonal niche strategy of the bloom forming dinoflagellate *Heterocapsa triquetra*. *Marine Biology Progress Series* 232: 45–62.

Lowrance, R., Todd, R., Fair, J. Jr. et al. (1984). Riparian forests as nutrient filters in agricultural watersheds. *Bioscience* 34: 374.

MacIsaac, H.J. (1996). Potential abiotic and biotic impacts of zebra mussels on the inland waters of North America. *Integrative and Comparative Biology* 36(3): doi https://doi.org/10.1093/icb/36.3.287.

Mallin, M.A. and Paerl, H.W. (1992). Effects of variable irradiance on phytoplankton productivity in shallow estuaries. *Limnology Oceanography* 37: 54–62.

Malone, T.C., Conley, D.J., Fisher, T.R. et al. (1996). Scales of nutrient-limited phytoplankton productivity in Chesapeake Bay. *Estuaries* 19: 371–385.

Malone, T.C., Malej, A., Harding, L.W. Jr. et al. (ed.) (1999). *Coastal and Estuarine Studies*, Ecosystems at the Land-Sea Margin, Drainage Basin to Coastal Sea, 55. Washington, DC: American Geophysical Union.

Marra, J. (1978). Effect of short-term variations in light intensity on photosynthesis of a marine phytoplankter: a laboratory simulation study. *Marine Biology* 46: 191–202.

Martin, J.H. (1970). Phytoplankton-zooplankton relationships in Narragansett Bay. IV. The seasonal importance of grazing. *Limnology and Oceanography* 15: 413–418.

McCabe, R.M., Hickey, B.M., Kudela, R.M. et al. (2016). An unprecedented coastwide toxic algal bloom linked to anomalous ocean conditions. *Geophysical Research Letters* 43: 10366–10376.

McDonald, D., Price, M.N. Goodrich, J. et al. (2012). An improved Greengenes taxonomy with explicit ranks for ecological and evolutionary analyses of bacteria and archaea. *ISME Journal* 6: 610–618. doi:https://doi.org/10.1038/ismej.2011.139

McRose, D., Guo, J., Monier, A. et al. (2014). Alternatives to vitamin B 1 uptake revealed with discovery of riboswitches in multiple marine eukaryotic lineages. *The ISME Journal* 8 (12): 2517.

Medlin, L.K. and Kaczmarska, I. (2004). Evolution of the diatoms: V. Morphological and cytological support for the major clades and a taxonomic revision. *Phycologia* 43: 245–270.

Mitbavkar, S., Rajaneesh, K.M., Anil, A.C. and Sundar, D. (2012). Picophytoplankton community in a tropical estuary: Detection of Prochlorococcus-like populations. *Estuarine, Coastal and Shelf Science* 107:159–164, doi:https://doi.org/10.1016/j.ecss.2012.05.002.

Mitsch, W., Day, J., Gilliam, J. et al. (2001). Reducing nitrogen loading to the Gulf of Mexico from the Mississippi River basin: Strategies to counter a persistent problem. *BioScience* 51 (5): 373–388.

Moisander, P.H., McClinton, E. III, and Paerl, H.W. (2002). Salinity effects on growth, photosynthetic parameters, and nitrogenase activity in estuarine planktonic cyanobacteria. *Microbial Ecology* 43: 432–442.

Moore, W.S. (1999). The subterranean estuary: A reaction zone of groundwater and seawater. *Marine Chemistry* 65: 111–126.

Nichols, F.H. (1985). Increased benthic grazing: An alternative explanation for low phytoplankton biomass in northern San Francisco Bay during the 1976-1977 drought. *Estuarine, Coastal and Shelf Science* 21: 379–388.

Nixon, S.W. (1992). *Quantifying the relationship between nitrogen input and the productivity of marine ecosystems*, Advanced Marine Technology Conference, No. 5. Japan: Tokyo.

Nixon, S.W. (1995). Coastal marine eutrophication: a definition, social causes, and future concerns. *Ophelia* 41: 199–219.

Officer, C.B., Smayda, T.J., and Mann, R. (1982). Benthic filter feeding: a natural eutrophication control. *Marine Ecology Progress Series* 9: 203–210.

O'Kelly, C.J., Sieracki, M.E., Their, E.C. and Hobson, I.C. (2003). A Transient Bloom of *Ostreococcus* (Chlorophyta, Prasinophyceae) in West Neck Bay, Long Island, New York. *Journal of Phycology* 39(5): 850–854. doi:https://doi.org/10.1046/j.1529-8817.2003.02201.x

Olson, R.J., Chisholm, S.W., Zettler, E.R., and Armbrust, E.V. (1990). Pigments, size, and distribution of *Synechococcus* in the North Atlantic and Pacific Oceans. *Limnology and Oceanography* 35: 45–58.

Olson, R.J., Shalapyonok, A., Kalb, D.J. et al. (2017). Imaging FlowCytobot modified for high throughput by in-line acoustic focusing of sample particles. *Limnology and Oceanography Methods* 15 (10): 867–874.

Pace, N.R., Stahl, D.A., Lane, D.J., and Olsen, G.J. (1985). The analysis of natural microbial populations by ribosomal RNA sequences. *American Society of Microbiology News* 51: 4–12.

Paerl, H.W. (1985). Enhancement of marine primary production by nitrogen-enriched acid rain. *Nature* 316: 747–749.

Paerl, H.W. (1988). Nuisance phytoplankton blooms in coastal, estuarine, and inland waters. *Limnology and Oceanography* 33: 823–847.

Paerl, H.W. (1990). Physiological ecology and regulation of N_2 fixation in natural waters. *Advances in Microbial Ecology* 11: 305–344.

Paerl, H.W. (1997). Coastal eutrophication and harmful algal blooms: importance of atmospheric deposition and groundwater as "new" nitrogen and other nutrient sources. *Limnology and Oceanography* 42: 1154–1165.

Paerl, H.W. (2004). Estuarine eutrophication, hypoxia and anoxia dynamics: Causes, consequences and controls. Rupp, G.L. and M. D. White. *Proceedings of 7th International Symposium on Fish Physiology, Toxicology and Water Quality.* May, 2003, Tallinn, Estonia. U.S. EPA Office of Research and Development, Ecosystems Research Div. Athens, Georgia, USA. EPA/600/R-04/049, pp. 35–56.

Paerl, H.W. (2017). The cyanobacterial nitrogen fixation paradox in natural waters. *F1000 Faculty Review* 6:244 doi: https://doi.org/10.12688/f1000research.10603.1

Paerl, H.W. (2018). Mitigating toxic cyanobacterial blooms in aquatic ecosystems facing increasing anthropogenic and climatic pressures. *Toxins* 10(2) 76, doi:https://doi.org/10.3390/toxins10020076.

Paerl, H.W. and Huisman, J. (2008). Blooms like it hot. *Science* 320: 57–58.

Paerl, H.W. and Huisman, J. (2009). Climate change: a catalyst for global expansion of harmful cyanobacterial blooms. *Environmental Microbiology Reports* 1 (1): 27–37.

Paerl, H.W. and Kuparinen, J. (2002). Microbial aggregates and consortia. In: *Encyclopedia of Environmental Microbiology*, vol. 1 (ed. G. Bitton), 160–181. New York, NY: John Wiley and Sons, Inc.

Paerl, H.W. and Otten, T.G. (2013a). Harmful cyanobacterial blooms: causes, consequences and controls. *Microbial Ecology* 65: 995–1010.

Paerl, H.W. and Otten, T.G. (2013b). Blooms bite the hand that feeds them. *Science* 342: 433–434.

Paerl, H.W. and Piehler, M.F. (2008). Nitrogen and marine Eutrophication. In: *Nitrogen in the Marine Environment*, vol. 2 (ed. D.G. Capone, M. Mulholland and E. Carpenter), 529–567. Orlando: Academic Press.

Paerl, H.W. and Whitall, D.R. (1999). Anthropogenically-derived atmospheric nitrogen deposition, marine eutrophication and harmful algal bloom expansion: Is there a link? *Ambio* 28: 307–311.

Paerl, H.W. and Zehr, J.P. (2000). Nitrogen Fixation. In: *Microbial Ecology of the Oceans* (ed. D. Kirchman), 387–426. New York: Academic Press.

Paerl, H.W., Pinckney, J.L., Fear, J.M., and Peierls, B.L. (1998). Ecosystem responses to internal and watershed organic matter loading: Consequences for hypoxia in the eutrophying Neuse River Estuary, North Carolina, USA. *Marine Ecology Progress Series* 166: 17–25.

Paerl, H.W., Dennis, R.L., and Whitall, D.R. (2002). Atmospheric deposition of nitrogen: Implications for nutrient over-enrichment of coastal waters. *Estuaries* 25: 677–693.

Paerl, H.W., Valdes, L.M., Pinckney, J.L. et al. (2003). Phytoplankton photopigments as Indicators of Estuarine and Coastal Eutrophication. *BioScience* 53 (10): 953–964.

Paerl, H.W., Valdes, L.M., Joyner, A.R., and Winkelmann, V. (2007). Phytoplankton Indicators of Ecological Change in the nutrient and climatically-impacted neuse river-Pamlico Sound System, North Carolina. *Ecological Applications* 17 (5): 88–101

Paerl, H.W., Rossignol, K.L., Hall, N.S., Peierls, B.L. and Wetz, M.S. (2009). Phytoplankton community indicators of short- and long-term ecological change in the anthropogenically and climatically impacted Neuse River estuary, North Carolina, USA. *Estuaries and Coasts* DOI https://doi.org/10.1007/s12237-009-9137-0.

Paerl, H.W., Hall, N.S., Peierls, B.L., Rossignol, K.L. and Joyner, A.R. (2013). Hydrologic variability and its control of phytoplankton community structure and function in two shallow, coastal, lagoonal ecosystems: the Neuse and New River Estuaries, North Carolina, USA. *Estuaries and Coasts* 37(S1): 31–45. DOI https://doi.org/10.1007/s12237-013-9686-0

Paerl, H.W., Gardner, W.S., Havens, K.E. et al. (2016a). Mitigating cyanobacterial harmful algal blooms in aquatic ecosystems impacted by climate change and anthropogenic nutrients. *Harmful Algae* 54: 213–222.

Paerl, H.W., Scott, J.T., McCarthy, M.J. et al. (2016b). It takes two to tango: When and where dual nutrient (N & P) reductions are needed to protect lakes and downstream ecosystems. *Environmental Science & Technology* 50: 10805–10813.

Paerl, R.W., Bouget, F.Y., Lozano, J.C. et al. (2017). Use of plankton-derived vitamin B1 precursors, especially thiazole-related precursor, by key marine picoeukaryotic phytoplankton. *The ISME Journal* 11 (3): 753.

Paerl, H.W. Otten, T.G. and Kudela, R. (2018). Mitigating the expansion of harmful algal blooms across the freshwater-to-marine continuum. *Environmental Science & Technology* 52:5519–5529. DOI: https://doi.org/10.1021/acs.est.7b05950

Parsons, M., Dortch, Q., and Turner, R.E. (2002). Sedimentological evidence of an increase in *Pseudo-nitzschia* (Bacillariophyceae) abundance in response to coastal eutrophication. *Limnology and Oceanography* 47: 551–558.

Paul, J.H., Alfreider, A., and Wawrik, B. (2000). Micro- and macrodiversity in *rbcL* sequences in ambient phytoplankton populations from the southeastern Gulf of Mexico. *Marine Ecology Progress Series* 198: 9–18.

Peierls, B.L., Hall, N.S., and Paerl, H.W. (2012). Non-monotonic responses of phytoplankton biomass accumulation to hydrologic variability: a comparison of two coastal plain North Carolina estuaries. *Estuaries and Coasts* 35: 1376–1392.

Pilkaitytė, R. and Razinkovas, A. (2007). Seasonal changes in phytoplankton composition and nutrient limitation in a shallow Baltic lagoon. *Boreal Environmental Research* 12: 551–559.

Pinckney, J.L., Paerl, H.W., and Harrington, M.B. (1999). Responses of the phytoplankton community growth rate to nutrient pulses in variable estuarine environments. *Journal of Phycology* 35: 1455–1463.

Platt, T. and Gallegos, C.L. (1980). Modelling primary productivity. Primary productivity in the sea, Brookhaven Symp. Biol. 31, Plenum. pp. 339–362.

Provasoli, L. and Carlucci, A.F. (1974). Vitamins and growth regulators. In: *Algal Physiology and Biochemistry* (ed. W.D.P. Stewart), 741–787. Oxford: Blackwell Scientific Publications.

Quast, C., Pruess, E., Yilmaz, P. et al. (2013). The SILVA ribosomal RNA gene database project: improved data processing and web-based tools. *Nucleic Acids Research* 41 (D1): D590–D596.

Rabalais, N.N., Cai, W.J., Carstensen, J. et al. (2014). Eutrophication-driven deoxygenation in the coastal ocean. *Oceanography* 70: 123–133.

Ralston, D.K., Mc.Gillicuddy, D.J. Jr., and Townsend, D.K. (2007). Asynchronous vertical migration and bimodal distribution of motile phytoplankton. *Journal of Plankton Research* 29 (9): 803–821.

Redfield, A.C. (1934). On the proportions of organic derivations in sea water and their relation to the composition of plankton. In: *James Johnstone Memorial Volume* (ed. R.J. Daniel), 177–192. University Press of Liverpool.

Redfield, A.C. (1958). The biological control of chemical factors in the environment. *American Scientist* 46: 205–222.

Reynolds, C.S. (2006). *Ecology of Phytoplankton (Ecology, Biodiversity and Conservation)*. Cambridge, UK: Cambridge University Press.

Richardson, K. (1997). Harmful or exceptional phytoplankton blooms in the marine ecosystem. *Advances in Marine Biology* 31: 302–385.

Rocap, G., Distel, D.L., Waterbury, J.B., and Chisholm, S.W. (2002). Resolution of *Prochlorococcus* and *Synechococcus* ecotypes using 16S-23S rRNA internal transcribed spacer (ITS) region sequences. *Applied Environmental Microbiology* 68: 1180–1191.

Round, F.E., Crawford, R.M., and Mann, D.G. (1990). *The Diatoms Biology and Morphology of the Genera*, Cambridge University Press, Cambridge.

Roy, S., Llewellyn, C.A., Egeland, E.S., and Johnsen, G. (2011). *Phytoplankton pigments. Characterization, Chemotaxonomy and Applications in Oceanography*, 845. Cambridge University press.

Schlüter, L., Lauridsen, T.L., Krogh, G., and Jørgensen, T. (2006). Identification and quantification of phytoplankton groups in lakes using new pigment ratios – a comparison between pigment analysis by HPLC and microscopy. *Freshwater Biology* 51: 1474–1485.

Sigman, D.M. and Hain, M.P. (2012). The biological productivity of the ocean. *Nature Education* 3 (6): 1–16.

Stanier, R.Y. and Cohen-Bazire, G. (1977). Phototrophic prokaryotes: the cyanobacteria. *Annual Review of Microbiology* 31: 225–274.

Sterner, R.W. (1989). The role of grazers in phytoplankton succession. In: *Plankton Ecology: Succession in Plankton Communities* (ed. U. Somer), 107–170. Berlin: Springer-Verlag.

Stewart, I. and Falconer, I.R. (2008). Cyanobacteria and cyanobacterial toxins. In: *Oceans and Human Health: Risks and Remedies from the Seas* (ed. P.J. Walsh, S. Smith, L. Fleming, et al.), 271–296. San Diego: Academic Press.

Stolte, W., McCollin, T., Noodeloos, A., and Riegman, R. (1994). Effect of nitrogen source on the size distribution within marine phytoplankton populations. *Journal of Experimental Marine Biology and Ecology* 184: 83–97.

Stomp, M., Huisman, J., de Jongh, F. et al. (2004). Adaptive divergence in pigment composition promotes phytoplankton biodiversity. *Nature* 432: 104–107.

Strayer, D. (2008). Twenty years of zebra mussels: Lessons from the mollusk that made headlines. *Frontiers in Ecology and the Environment* 7 (3): 135–141.

Sylvan, J.B., Dortch, Q., Nelson, D.M. et al. (2006). Phosphorus limits phytoplankton growth on the Louisiana shelf during the period of hypoxia formation. *Environmental Science & Technology* 40: 7548–7553.

Tai, V. and Palenik, B. (2009). Temporal variation of *Synechococcus* clades at a coastal Pacific Ocean monitoring site. *The ISME Journal* 3 (8): 903.

Tamm, M., Freiberg, R., Tõnno, I., and Nõges, P. (2015). Pigment-based chemotaxonomy—a quick alternative to determine algal assemblages in large shallow eutrophic lake? *PLoS One* 10 (3): e0122526.

Tomas, C.R. (ed.) (1996). *Identifying Marine Phytoplankton*. New York: Academic Press.

Tragin, M. and Vaulot, D. (2018). Green microalgae in marine coastal waters: the Ocean Sampling Day (OSD) dataset. *Nature Scientific Reports* 8: 12.

Trainer, V.L., Bates, S.R., Lundholm, N. et al. (2012). *Pseudo-nitzschia* physiological ecology, phylogeny, toxicity, monitoring and impacts on ecosystem health. *Harmful Algae* 14: 271–300.

Trenberth, K.E. (2005). The impact of climate change and variability on heavy precipitation, floods, and droughts. In: *Encyclopedia of Hydrological Sciences* (ed. M.G. Anderson), 1–11. John Wiley and Sons, Ltd.

Turner, R.E. and Rabalais, N.N. (2013). Nitrogen and phosphorus phytoplankton growth limitation in the northern Gulf of Mexico. *Aquatic Microbial Ecology* 68: 159–169.

Twomey, L. and Thompson, P. (2001). Nutrient limitation of phytoplankton in Wlison Inlet, Western Australia: a seasonally opened bar built estuary. *Journal of Phycology* 37: 16–29.

U.S. EPA (1998). *Condition of the Mid-Atlantic Estuaries*. Washington, DC: Office of Research and Development.

U.S. EPA. (1999). *Total Maximum Daily Load (TMDL) Program*. Office of Water, Washington, DC (On line). http://www.epa.gov/OWOW/tmdl (20 December 1999).

U.S. EPA (2011). *Reactive nitrogen in the United States: An analysis of inputs, flows, consequences, and management options*; EPA-SAB-11-013. Washington, DC: Unites States of America Environmental Protection Agency.

Urbach, E., Scanlan, D.J., Distel, D.L. et al. (1998). Rapid diversification of marine picophytoplankton with dissimilar light-harvesting structures inferred from sequences of *Prochlorococcus* and *Synechococcus* (cyanobacteria). *Journal of Molecular Evolution* 46: 188–201.

Valdes-Weaver, L.M., Piehler, M.F., Pinckney, J.L. et al. (2006). Long-term temporal and spatial trends in phytoplankton biomass and class-level taxonomic composition in the hydrologically variable Neuse-Pamlico estuarine continuum, NC, USA. *Limnology and Oceanography* 51 (3): 1410–1420.

Verity, P.G. (2010). Expansion of potentially harmful algal taxa in a Georgia Estuary (USA). *Harmful Algae* 2:144–152 DOI: https://doi.org/10.1016/j.hal.2009.08.009.

Wagner-Döbler, I., Ballhausen, B., Berger, M. et al. (2010). The complete genome sequence of the algal symbiont *Dinoroseobacter shibae*: a hitchhiker's guide to life in the sea. *The ISME Journal* 4: 61–77. https://doi.org/10.1038/ismej.2009.94.

Walsby, A.E. (1994). Gas vescicles. *Microbiological Reviews* 58 (1): 94–144.

Waterbury, J.B., Watson, S.W., Valois, F.W., and Franks, D.G. (1986). Biological and ecological characterization of the marine unicellular cyanobacterium *Synechococcus*. *Canadian Bulletin of Fisheries and Aquatic Sciences* 214: 71–120.

Webster, P.J., Holland, G.J., Curry, J.A., and Chang, H.R. (2005). Changes in tropical cyclone number, duration, and intensity in a warming environment. *Science* 309: 1844–1846.

Wolk, C.P. (1982). Heterocysts. In: *The Biology of Cyanobacteria* (ed. N.G. Carr and B.A. Whitton), 359–387. Berkeley: University of California press.

Wright, S.W., Thomas, D.P., Marchant, H.J. et al. (1996). Analysis of phytoplankton of the Australian sector of the Southern Ocean: comparisons of microscopy and size frequency data with interpretations of pigment HPLC data using the "Chemtax" matrix factorization. *Marine Ecology Progress Series* 144: 285–298.

Yoshida, M., Noel, M.H., Nakayama, T. et al. (2006). A haptophyte bering siliceous scales: Ultrastructure and phylogenetic position of *Hyalolithus neolepis* gen. et sp nov (Pyrmnesiophyceaes, Haptophyta). *Protist* 157 (2): 213–234.

Zehr, J.P. and McReynolds, L.A. (1989). Use of degenerate oligonucleotides for amplification of the *nifH* gene from the marine cyanobacterium *Trichodesmium* spp. *Applied Environmenatal Microbiology* 55: 2522–2526.

Zehr, J.P. and Paerl, H.W. (2008). Molecular ecological aspects of nitrogen fixation in the marine environment. In: *Microbial Ecology of the Oceans* (ed. D. Kirchman), 481–525. New York: Academic Press.

Zhang, X., Shi, Z., Feng, Y. et al. (2013). Picophytoplankton abundance and distribution in three contrasting periods in the Pearl River Estuary, South China. *Marine and Freshwater Research* 64: 692–705.

CHAPTER 5

Estuarine Seagrasses

Vallisneria americana bed in Chesapeake Bay with in situ oxygen sensor. Photo credit: Cassie Gurbisz.

Renee K. Gruber[1], Cassie Gurbisz[2], Jens Borum[3], and W. Michael Kemp[4]
[1]Australian Institute of Marine Science, Townsville, QLD, Australia
[2]St. Mary's College of Maryland, St. Marys City, MD, USA
[3]Department of Biology, University of Copenhagen, Copenhagen, Denmark
[4]Horn Point Laboratory, University of Maryland, Center for Environmental Science, Cambridge, MD, USA

5.1 Introduction

Seagrasses and other submersed vascular plants are iconic and important primary producers in estuarine habitats. They play key roles in coastal ecosystems such as supporting fisheries, stabilizing the sediment, sequestering carbon, utilizing dissolved nutrients, serving as a food source for higher trophic orders, and providing cultural benefits to society (Barbier et al. 2011). Seagrasses are a diverse group of monocotyledonous flowering plants that can grow in marine and estuarine environments (Figure 5.1). All species have belowground rhizomes from which root and leaf bundles are formed. Many freshwater vascular plants have apical meristems that enable them to grow toward the light at the water's surface. In contrast, all seagrasses have basal leaf meristems that allow continuous replacement of old leaves with new clean leaves.

Estuarine seagrasses have terrestrial ancestors that invaded the marine environment at different times during their evolution (Les et al. 1997). Accordingly, seagrass species have different origins and are referred to as a polyphyletic group, where some species are more closely related to freshwater vascular plants than to other seagrasses. As a result of this evolutionary history, the morphology of seagrasses is quite diverse (Figure 5.1), ranging from small shoots (on the order of centimeters) with elliptic leaves to larger plants with broad, linear leaves (on the order of meters).

The evolution of terrestrial plants in the marine environment created a number of unique physiological properties of seagrasses (Olsen et al. 2016). Terrestrial plants are susceptible to desiccation and have evolved hairs or other leaf structures to reduce moisture loss, whereas submersed plants have lost most of these protections. The leaf cuticle (a waxy outer covering that helps prevent desiccation) is reduced in seagrasses compared to terrestrial plants. One advantage of this evolutionary change is that seagrasses can efficiently extract nutrients from the water column in addition to taking up belowground nutrients through their root systems (Hemminga and Duarte 2000). Submersed plants lack the ability to transpire from leaves, a process used by terrestrial plants to drive internal transport of solutes, and instead use the energy-consuming process of root pressure (Pedersen and Sand-Jensen 1993). Seagrasses have reduced structural tissues compared to terrestrial plants, and instead maintain their structure and buoyancy with air-filled lacunae in their leaves. The flexibility and strength of seagrass leaves is an important morphological feature necessary for withstanding the drag forces induced by waves and currents (Koch 2001).

Estuarine Ecology, Third Edition. Edited by Byron C. Crump, Jeremy M. Testa, and Kenneth H. Dunton.

Companion website: www.wiley.com/go/crump/estuarine3

FIGURE 5.1 Examples of submersed vascular plants that occur in estuarine systems: (a) *Myriophyllum spicatum* (*Source*: Photo by Ole Pedersen), (b) *Stuckenia pectinata* (*Source*: Photo by Ole Pedersen), and (c) *Vallisneria americana* (*Source*: Photo by C. Gurbisz) are found in the upper reaches of estuaries (salinities <15), while (d) *Thalassia testudinum* (*Source*: Photo by Ole Pedersen), (e) *Zostera marina* (*Source*: P.B. Christensen; NERI), and (f) *Halophila ovalis*. (*Source*: P. Lavery) are seagrasses that thrive in full-strength seawater.

5.2 Diversity and Global Distribution

Based on current seagrass taxonomy there are 65 species recognized globally, which belong to six botanical families: Cymodoceaceae, Hydrocharitaceae, Posidoniaceae, Potamogetonaceae, Ruppiaceae, and Zosteraceae (Larkum et al. 2018). Estuaries also contain a variety of freshwater plant species that can tolerate low salinity conditions. There are six main bioregions for seagrasses globally (Figure 5.2) that are delineated by latitude (temperate or tropical) and ocean basin including the Atlantic, Pacific, and Southern Oceans, and the Mediterranean Sea (Short et al. 2007). The most diverse bioregion is the Tropical Indo-Pacific, which encompasses the eastern coast of Africa and tropical portions of the Asia-Pacific and contains 24 seagrass species (Short et al. 2007). There are no submersed plants in Antarctica, but low diversity seagrass communities are found in the Arctic areas of North America and Europe. The geographical distribution of submersed vascular plants in estuaries varies substantially among species. Some seagrass species are confined to a relatively small area such as *Posidonia oceanica*, which only occurs in the Mediterranean bioregion. Other species such as *Halodule wrightii* are widely distributed and can occur in both temperate and tropical climates, while species such as *Zostera marina* occur in both Atlantic and Pacific Ocean basins (den Hartog 1970). The most widely distributed submersed plant is *Ruppia maritima*, which is a highly adaptable species found in all six global bioregions.

The abundance and species composition of submersed plant communities vary greatly along the estuarine salinity gradient, along with other environmental variables including water clarity and exposure to wave action. In the oligohaline (low salinity) upper reaches of estuaries, one may find species such as *Elodea canadensis*, *Myriophyllum spicatum*, and *Vallisneria americana* (salinity of 0–5), while *Zannichellia palustris*,

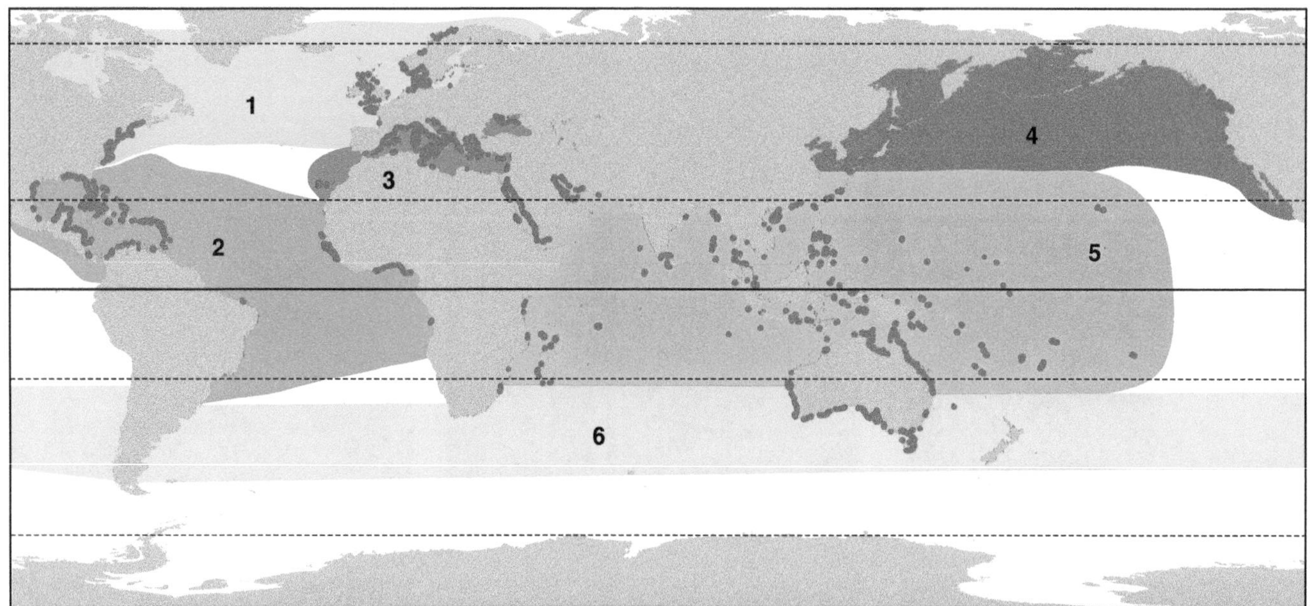

FIGURE 5.2 Map of global seagrass distribution (green coastal areas) and bioregions (1. Temperate North Atlantic, 2. Tropical Atlantic, 3. Mediterranean, 4. Temperate North Pacific, 5. Tropical Indo-Pacific, 6. Temperate Southern Oceans). *Source*: Short et al. (2007)/With permission of Elsevier.

Potamogeton perfoliatus, and *Stuckenia pectinata* can form dense stands in mesohaline areas (salinity up to 18). A few species such as *Ruppia maritima* are euryhaline, or able to tolerate a wide range of salinities, and occur throughout the salinity range from landward riverine sites to oceanic water at the mouth of an estuary. In contrast, most species are considered stenohaline, or having relatively narrow distributions along the estuarine salinity gradient.

The vertical distribution of submersed plants ranges from the intertidal region to depths of around 90 m (Duarte 1991). Seagrasses in the intertidal zone are regularly exposed to stresses such as desiccation, wave action, and high light levels and thus a limited number of species are able to thrive there including several in the *Halophila* and *Zostera* genera (den Hartog 1970). The depth range within which submersed plants can grow is controlled by a combination of factors including light availability, sediment type, and local hydrodynamics (Koch 2001). As an example, the depth distribution of *Zostera marina* in a large number of Danish coastal areas reflects a balance between exposure to physical stresses in shallow water and limited light availability in deep water (Figure 5.3), and these appear to be the most important factors controlling the upper and lower depth limits of these plants (Krause-Jensen et al. 2003).

In addition to environmental factors, the distribution of seagrasses at both global and local scales is also related to their reproductive strategies. Seagrasses and other submersed plants are capable of both sexual and asexual reproduction and are sometimes able to vary their reproductive strategy based on local stresses or other environmental conditions (Cabaço and Santos 2012). Sexual reproduction occurs through pollination of flowers, which produce fruits and seeds. Seagrass pollen is generally neutrally-buoyant and pollination occurs in the water column, although a small number of species including *Enhalus acoroides* (and a number of oligohaline species) have pollen that is dispersed on the water's surface (McMahon et al. 2014). Recent work has shown that marine animals such as invertebrates may assist in seagrass pollination in a manner similar to insect pollination in terrestrial plants (van Tussenbroek et al. 2016b). Seeds of some species can be transported hundreds of kilometers by currents or grazing animals, while other species retain most propagules as a "seed bank" within the meadow (McMahon et al. 2014). Seagrasses also reproduce asexually through clonal growth, whereby the rhizome extends laterally and produces new shoots. This process is slow (between centimeters and meters per year, depending on the species) relative to seed dispersal, but it allows individual clones to persist and expand over long time periods. Genetic studies have shown that clones in species such as *Posidonia oceanica* can be >100,000 years old and spread over 10 km (Arnaud-Haond et al. 2012). Many species of submersed plants are also able to reproduce asexually through fragmentation of aboveground biomass, which is dispersed by currents and can form new rhizomes and shoots (Riis et al. 2009).

5.3 Biomass and Productivity

Biomass refers to the weight of plant material growing in a given area and is a common metric used to quantify the productivity and health of a seagrass community. Submersed plant biomass varies over several orders of magnitude depending on species

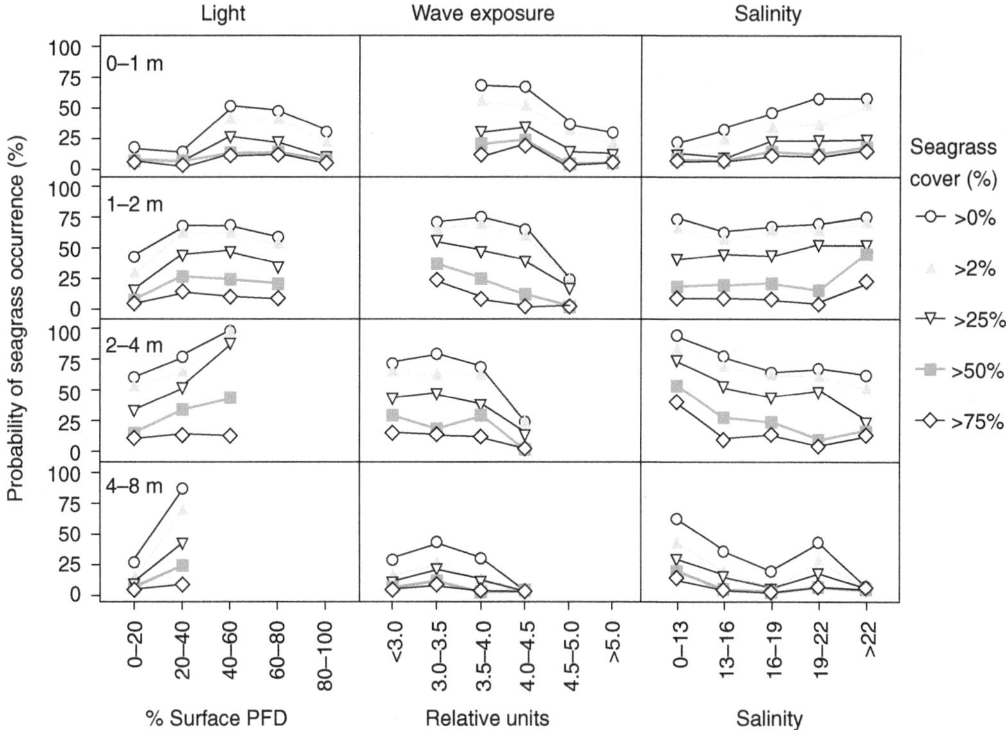

FIGURE 5.3 The probability of finding the seagrass *Zostera marina* at a cover (% of bottom area) of greater than 0%, greater than 2%, greater than 25%, greater than 50%, and greater than 75% at different depth intervals and as functions of light availability (% of surface photon flux density, PFD), wave exposure (in relative units as a function of fetch, wind direction, speed, and water depth), and salinity. *Source*: Krause-Jensen et al. (2003)/With permission of Springer Nature.

and local environmental factors (Figure 5.4). The aboveground biomass of species with small leaves, such as those in the *Halophila* genus (Figure 5.1f), is generally the lowest among seagrasses (<100 g dry weight m^{-2}). Such low-biomass species tend to be highly ephemeral, relying on prolific seed production and rapid growth rates to maintain populations (Marbà and Duarte 1998); as a result, they can rapidly colonize available habitat, including after disturbances (Fonseca et al. 2008). Aboveground biomass is high in seagrass species of the genera *Amphibolis, Phyllospadix*, and *Posidonia*, often exceeding 1000 g dry weight m^{-2} (Duarte and Chiscano 1999); these species tend to be long-lived and slow to colonize new habitat. The aboveground biomass of submersed plants growing in brackish and mesohaline areas is typically moderate but occasionally relatively high (~700 g DW m^{-2}) such as for canopy-forming species like *Stuckenia pectinata* (Figure 5.1b; Gruber and Kemp 2010) or *Hydrilla verticillata* (Shields et al. 2012). Root and rhizome material comprise the belowground biomass, which is generally high in long-lived species and low in ephemeral species. On average, seagrass species have similar amounts of above and belowground biomass or an above:below ratio of ~1 (Duarte and Chiscano 1999), although this varies with species, region, and environmental factors. Nutrient regime, sediment type, salinity, temperature, light availability, physical exposure, and other factors can alter seagrass morphology such that two communities growing under contrasting conditions can look completely different (Ferguson et al. 2016). This ability is known as phenotypic plasticity, which seagrasses use to reduce environmental stressors and increase production, as will be described later.

Submersed plant biomass is not constant over time, but rather displays seasonal and year-to-year variability like many terrestrial plant communities. In temperate systems such as the Dutch Wadden Sea, communities of *Zostera marina* display a dramatic seasonal cycle: their aboveground biomass disappears during winter and regrows from rhizomes or seeds in the spring. Submersed plants such as *Ruppia maritima* display a similar pattern in temperate zones, with shoot length and density becoming reduced during the winter months compared to the tall and dense canopies that form during summer; these patterns are primarily driven by environmental conditions including temperature, day length, and light availability. In subtropical and tropical waters, changes in temperature and light are less dramatic between seasons, and therefore seagrass biomass fluctuates far less seasonally than in higher-latitude climates (Hemminga and Duarte 2000). Year-to-year variability in biomass is generally related to drivers such as variability in river discharge, temperature, grazing pressure, human impacts, or episodic disturbances such as storms (Fonseca et al. 2008; Gurbisz et al. 2016).

Seagrass productivity refers to the rate at which new biomass is generated and is often estimated in experiments by measuring the uptake of dissolved carbon (C) or the release of oxygen (O_2) into the water column. Gross production (P_G, the total

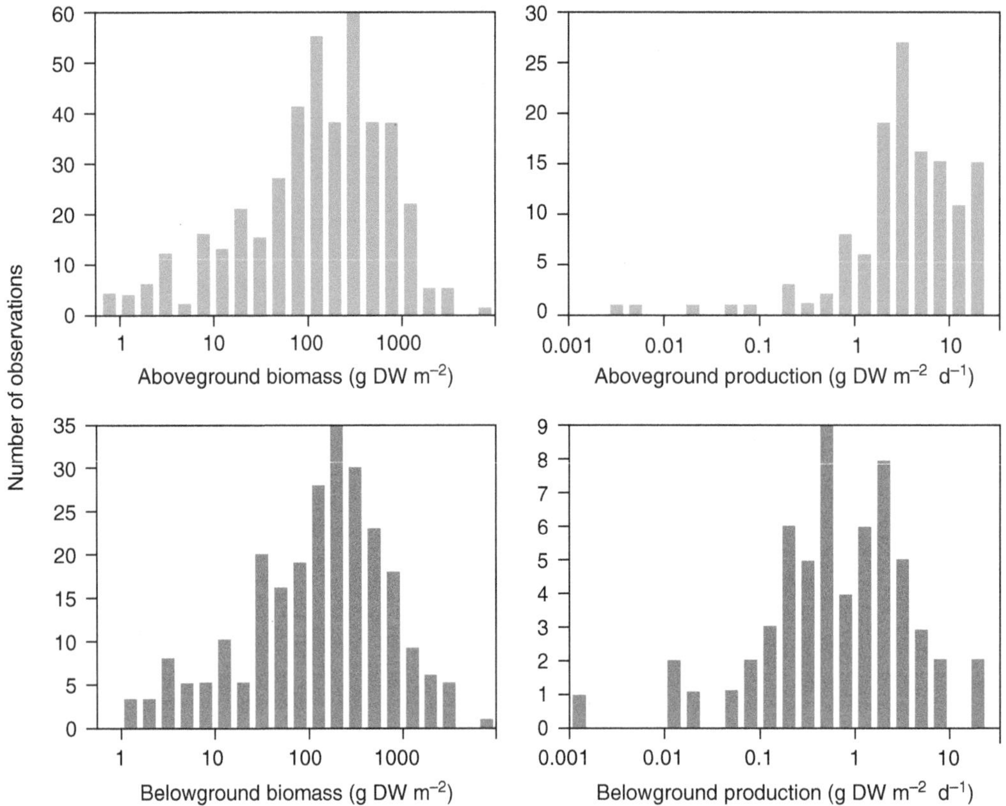

FIGURE 5.4 Frequency distribution of biomass and production for aboveground and belowground tissues compiled for a large number of seagrass species. *Source*: Duarte and Chiscano (1999)/With permission of Elsevier.

amount of carbon taken up by plants during photosynthesis) occurs simultaneously with respiration (*R*, the consumption of carbon to generate energy for metabolic processes). Therefore, net production (P_G–*R*, the amount of fixed carbon available to generate new biomass) is the process directly measured in experiments. Production has been the subject of considerable attention since early in the last century, when Ostenfeld (1908) and Petersen (1918) examined the depth distribution and biomass of *Z. marina* and estimated its importance as a source of organic matter to benthic invertebrates and fishes. It is often suggested that seagrass meadows are among the most productive aquatic ecosystems globally (Mann 1972; McRoy and McMillan 1977; Zieman and Wetzel 1980; Duarte et al. 2010). Estimates of above- and belowground biomass production from a large number of seagrass species suggest that seagrasses accumulate ~5 g dry weight m^{-2} day^{-1} (Figure 5.4) of new biomass on average (Duarte and Chiscano 1999). These growth rates refer to periods when seagrasses are growing, which varies with latitude; in temperate regions, submersed plants may lose aboveground biomass during colder months, while seagrasses in tropical regions tend to retain biomass throughout the year. Net community production tends to be greater in temperate regions and lower in tropical regions (Duarte and Chiscano 1999), which is due to greater rates of respiration in tropical regions as a function of higher average temperatures (Duarte et al. 2010). Submersed plant productivity is important not only for meadow growth, but also as a pathway to transform dissolved carbon and nutrients into living biomass that benefits other organisms in the ecosystem.

5.4 Factors Controlling Productivity and Community Composition

The biomass and production of all autotrophs are regulated by many different factors acting alone or in concert as multiple limitations. In early studies, environmental control of production was generally attributed to either light or nutrient limitation in accordance with Liebig's "Law of the Minimum," a theory developed for agricultural crops (von Liebig 1863). This theory assumes that at any time only one factor is the bottleneck that limits plant growth; whereas, multiple environmental factors can be interrelated in ecological systems. Recent decades have seen research into other controlling factors including sediment type, sediment chemistry, and physical disturbance by local hydrodynamics such as wave action (Koch 2001; Soissons et al. 2018). When regulating factors are examined, resource limitation of photosynthesis and growth is often the focus, but loss factors may be equally important in limiting plant production. For example, herbivore grazing can reduce photosynthetic potential, and under some conditions may be an important constraining factor (Heck and Valentine 2006). Bearing this in mind, we present the potentially limiting individual factors below, acknowledging that, although interdependent, a single factor is often most limiting to photosynthesis or growth at any given time.

5.4.1 Light

Irradiance within the wavelength range of 400–700 nm (photosynthetically active radiation, or PAR) is the ultimate limiting factor controlling the fixation of dissolved inorganic carbon through photosynthesis. The importance of light for seagrass distribution is immediately apparent when comparing the lower depth limits of seagrass growth with light attenuation, or the rate at which light dims as it passes through the water column (Figure 5.5; Duarte 1991; Duarte et al. 2007). The lower depth limit of seagrasses corresponds to irradiance levels ~10–25% of incoming surface irradiance; these light levels are often referred to as the minimum light requirements for seagrasses. In order to maximize light-harvesting capability in these low-light habitats, seagrasses vary their morphology including leaf length, width, and shoot density. Experimental work has also shown that leaf pigment concentrations (and thus leaf absorbance) change as a function of light level (Ralph et al. 2007). Many estuarine submersed plant species form tall canopies, which allow them to harvest sufficient light in conditions that would otherwise be too turbid (Fleming and Dibble 2015). In contrast, micro- and macroalgae need far less

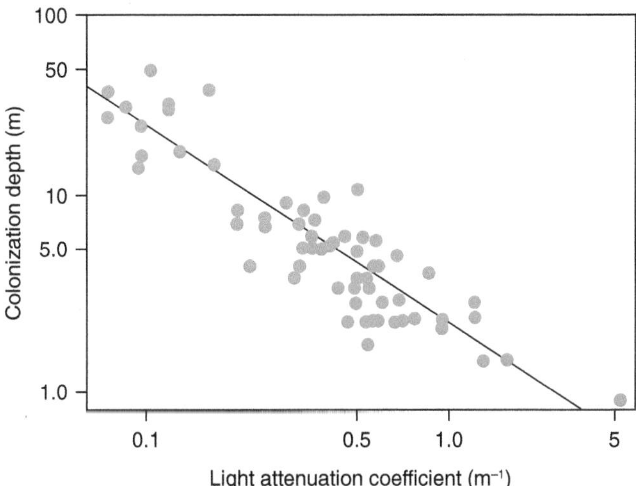

FIGURE 5.5 Colonization depth (lower depth limit) for seagrass communities distributed worldwide versus light attenuation coefficient. The relationship is described by the equation $\log(Z_c) = 0.26–1.07(\log(k))$, where Z_c is the colonization depth and k is the light attenuation coefficient. Large values of k mean that light is quickly attenuated (or dimmed) as it moves through the water column, whereas small values of k mean that light penetrates deeper in the water column. *Source*: Duarte (1991)/With permission of Elsevier.

light to grow than seagrasses and can survive with ~1% of the surface irradiance or even as little as 0.001% (Markager and Sand-Jensen 1996). The high light requirements of submersed plants might be explained by the energetic costs related to growth and maintenance of belowground tissues or the need for surplus oxygen transport to roots to modify sediment chemistry (Kemp et al. 2004). Unfortunately, these light requirements make seagrasses vulnerable to anthropogenic pressures that decrease water clarity including eutrophication, dredging, and shading from coastal structures (Ralph et al. 2007).

High irradiances can also control seagrass community composition and growth in shallow high-light areas such as the intertidal zone in tropical regions. High levels of ultraviolet (UV) radiation can damage the photosynthetic apparatus or prevent seeds from germinating (Ralph et al. 2007). Seagrasses compensate for this by growing smaller and thicker leaves and by producing UV-absorbing pigments, which cause leaves to turn reddish or purplish colors (Novak and Short 2010). In addition, some species such as *Zostera capricorni* are more tolerant of high UV conditions than other species (Dawson and Dennison 1996).

Photosynthesis versus irradiance (*P–I*) curves are commonly used in seagrass research to measure the production of plants in relation to a given light level. At the scale of an individual leaf, seagrasses follow the same hyperbolic response pattern as microalgae and terrestrial plants. At low irradiances, carbon fixation increases almost linearly with irradiance reflecting the light-harvesting capacity of the photosynthetic apparatus. At higher irradiances, the increase in the rate of carbon fixation begins to decline until the leaf becomes light saturated (i.e., no change in photosynthesis with increases in irradiance). Such *P–I* curves are derived under controlled conditions and may not necessarily reflect rates of *in situ* photosynthesis. In the field, light does not reach the leaf surface as a perpendicular beam, but rather passes down through the canopy while being partly attenuated and scattered and hits leaf surfaces from multiple angles often very different from 90°. Consequently, the *P–I* curves for well-developed canopies are very different from those of individual leaves, often with no clear saturation irradiances and a continuous increase in integrated photosynthesis even at irradiances exceeding maximum midday levels (Binzer and Sand-Jensen 2002). Accordingly, dense canopies of submersed plants are often light limited, and any increase in day length or average daily irradiance will increase photosynthesis and production.

5.4.2 Temperature

At a fundamental level, the survival of a plant depends on maintaining a positive carbon balance. This means that seagrasses must fix enough carbon through photosynthesis to meet the demands of respiration, physiological maintenance, growth, and reproduction. Photosynthesis and respiration are temperature-dependent processes, which have optimum temperatures that vary with species and latitude (Bulthuis 1987). Gross photosynthesis tends to increase exponentially with

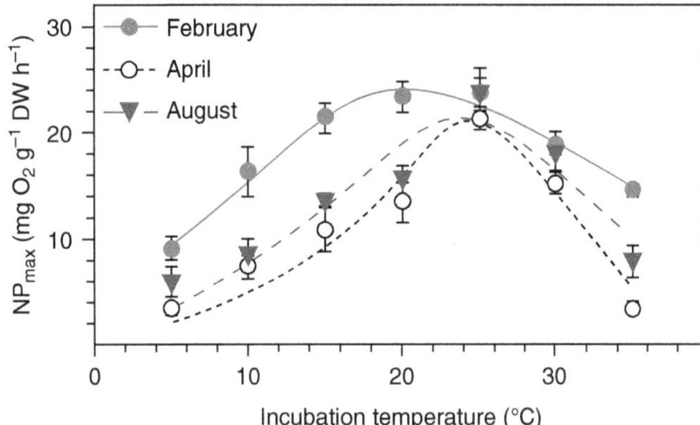

FIGURE 5.6 Temperature dependency of light saturated net photosynthesis (NP_{max}) for the seagrass *Zostera marina* collected in winter (February), spring (April), and late summer (August). Optimum temperature for light saturated net photosynthesis varied from 20 to 25°C depending on the season. *Source*: Staehr and Borum (2011)/With permission of Elsevier.

increasing temperature and gradually declines above an optimum temperature level (Staehr and Borum 2011), as enzymes begin to degrade and damage to photosynthetic apparatus can occur (York et al. 2013). Seagrass respiration is strongly controlled by temperature; respiration increases with temperature at a greater rate than photosynthesis, and maximum rates of respiration are reached at higher temperatures than those for photosynthesis (Berry and Raison 1981). Combining these effects of increasing temperature, net photosynthesis increases exponentially with moderate increases in temperature, then gradually declines above an optimum temperature level (Figure 5.6) and may become negative at very high temperatures (Greve et al. 2003). The consequences of exceeding this optimum temperature can be reductions in seagrass biomass or growth rate, or even complete mortality at high temperatures (Collier et al. 2011).

Experimental work done to determine the effects of temperature on seagrasses generally takes place in controlled environments such as tanks or chambers, where temperatures are held constant over periods of weeks to months. However, temperatures in a natural estuarine setting can be highly variable over short time scales (days to weeks), especially in shallow areas such as the intertidal zone. Seagrasses that are acclimated to routine short-term exposures to high temperatures (>35 °C) appear to be able to maintain moderate levels of productivity (Gruber et al. 2017). However, episodic events such as marine heat waves can cause temperatures of 10–15°C above the optimum for certain seagrass species, especially in shallow environments. Even short-term exposure to these extremes causes physiological damage to the tissues and mortality (Collier and Waycott 2014). Extreme events can cause widespread diebacks in species such as *Zostera marina*, which can give a competitive advantage to species that are more heat-tolerant or are able to rapidly colonize a new area (Moore et al. 2014). Heat waves may also trigger flowering events in species that do not typically flower prolifically (such as *Posidonia oceanica*), which is thought to be a stress response (Arias-Ortiz et al. 2018). The effect of marine heat waves on seagrass communities is of great concern because these events are predicted to increase in frequency and severity under climate change (Arias-Ortiz et al. 2018).

5.4.3 Salinity

Salinity is an important factor affecting submersed plants, as reflected by the changes in species composition of communities along the estuarine continuum. The salinity tolerance of seagrasses largely depends on their ability to accumulate sufficient solutes to maintain turgor pressure (Touchette 2007). To varying degrees, the freshwater species occurring in estuaries tolerate oligohaline (low salinity) to mesohaline (mid-salinity) conditions. The most widely distributed seagrass species globally, *Ruppia maritima*, has a euryhaline (able to tolerate salinities from 0 to full-strength seawater) distribution. Hypersalinity (i.e., salinity above that of seawater, >35) generally has negative effects on most seagrass species, severely disturbing their osmoregulatory capacity (Koch et al. 2007; Walker and McComb 1992). High salinities in Florida Bay, USA, resulting from high evaporation during summer and reduced freshwater discharge are thought to have contributed to widespread seagrass die-offs (Zieman et al. 1999). However, some species can tolerate salinities as high as 70 (twice the strength of seawater) when they are gradually acclimated to increasing salinities, which can occur during periods of low precipitation and high evaporation in shallow lagoons (Koch et al. 2007).

5.4.4 Nutrients

Seagrasses require nutrients to maintain their productivity and growth. Although they utilize 15 different elements, nitrogen (N) and phosphorus (P) are the most important of these nutrients. Seagrasses readily take up dissolved inorganic nutrients in the forms of ammonium (NH_4^+), phosphate (PO_4^{3-}), and nitrate (NO_3^-). Nutrient uptake occurs through both the leaves and belowground tissues, allowing seagrasses to access nutrients in the water column and the sediment porewater (Hemminga 1998). The balance between above and belowground uptake can vary depending on the species and overall nutrient enrichment in the ecosystem (Touchette and Burkholder 2000). Nutrient uptake occurs in a thin micro-layer at the tissue surface, referred to as the diffusive boundary layer. Water flow over leaves continually transports nutrients into this boundary layer, where they diffuse to the leaf surface. Rates of nutrient uptake by leaves are thus a function of nutrient concentration and flow speed (Thomas and Cornelisen 2003; Atkinson 2011).

Low concentrations of nitrogen and phosphorus in the environment can limit the growth of seagrasses. Decades of fertilization studies have shown that most seagrass communities are nitrogen-limited (Touchette and Burkholder 2000) although phosphorus-limitation can occur in certain conditions, especially for communities growing in carbonate sediment, which readily binds phosphate and prevents it from being used by seagrasses (Short et al. 1990). Seagrasses growing in carbonate sediments can overcome this limitation by exuding weak acids from their root systems, which helps to liberate phosphorus and allow it to be taken up (Jensen et al. 1998). The ratio of nutrient concentrations (i.e., N:P) present in the environment is referred to as the nutrient stoichiometry and can be an indicator of whether nitrogen or phosphorus is the more limiting factor for seagrass growth. For phytoplankton, the nutrient stoichiometry of the environment is often compared to the nutrient stoichiometry (the C:N:P ratio) of their biomass, referred to as the Redfield ratio (106 : 16 : 1; see Chapter 4), to determine which element is more limiting. Seagrasses have more structurally complex tissues than phytoplankton and as such have greater carbon and nitrogen demands. Their tissue C:N:P stoichiometry varies with species (Atkinson and Smith 1983) but the global median is ~474 : 24 : 1 (Duarte 1990). This tissue stoichiometry can be used to estimate nutrient limitation status (Demars and Edwards 2007), but care must be taken in the interpretation given the many other limiting factors that affect seagrasses.

Many estuaries globally are experiencing eutrophication as a result of increasing inputs of N and P from anthropogenic loading. Excess nutrient inputs can affect seagrasses both directly (through toxicity effects) and indirectly (through ecological shifts). There is evidence that ammonium can be toxic to many submersed plants (Britto and Kronzucker 2002) and some seagrass species when present at high concentrations in their tissues (Touchette and Burkholder 2000). Excess nutrient inputs enhance the growth of phytoplankton, algal epiphytes, and macroalgae, which shade seagrass meadows or reduce the amount of light reaching leaf surfaces (Burkholder et al. 2007). Due to the much lower light requirements of micro and macroalgae, these species are efficient competitors in low-light environments and can eventually displace seagrasses (McGlathery et al. 2007). Nutrient enrichment is also thought to increase the competitiveness of certain invasive seagrass and submersed plant species. Canopy-forming submersed plants such as *Hydrilla verticillata* and *Egeria densa* (both invasive to North America) form dense and tall stands under eutrophic conditions, which can shade and displace native species (Mony et al. 2007). The canopy growth strategy also increases the competitiveness of these species with micro and macroalgae. In nutrient-enriched systems, seagrass such as *Halophila stipulacea* (invasive in Tropical Atlantic and Mediterranean bioregions) can compete against native species by forming dense meadows that prevent natives from establishing (van Tussenbroek et al. 2016a).

5.4.5 Inorganic Carbon

The supply of inorganic carbon to fuel photosynthesis represents a major problem for submersed vegetation (Madsen and Sand-Jensen 1991; Beer and Koch 1996). Carbon dioxide (CO_2) is readily dissolved in water, but its diffusion through the boundary layer around leaf surfaces is very slow compared to that in air because the diffusion coefficients of gases in air are about 10,000 times faster than in water. Therefore, many submersed plants have developed the ability to use inorganic carbon in the form of bicarbonate (HCO_3^-), which is about 100 times more abundant than dissolved CO_2 in seawater (see Chapter 2).

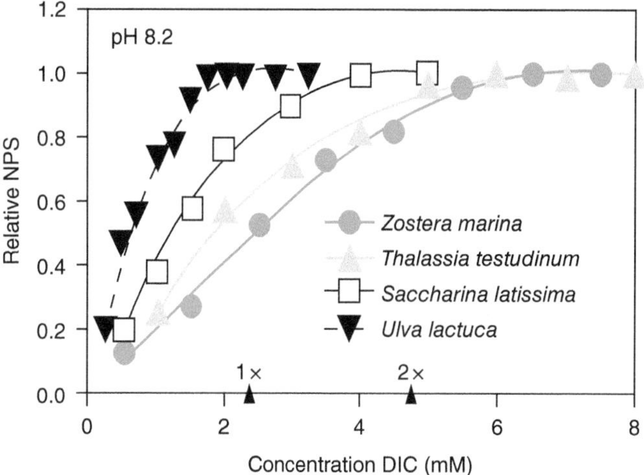

FIGURE 5.7 Net photosynthetic rates (NPS) of two macroalgae species, *Ulva lactuca* and *Saccharina latissima*, and two seagrass species, *Thalassia testudinum* and *Zostera marina*, at normal seawater pH but varying availability of inorganic carbon (DIC in mM). Rates of NPS are relative to maximum net photosynthesis. Normal (1×) and double (2×) seawater DIC levels are indicated on the *x*-axis. *Source*: Beer and Koch (1996)/With Permission of Inter-Research Science Publisher.

The role of inorganic carbon as a potentially limiting factor for photosynthesis of submersed vascular plants has become an issue of increasing interest due to the rising atmospheric CO_2 level and global climate change. Limited availability of inorganic carbon is potentially a bottleneck to the growth of submersed plants because of their terrestrial origin and their relatively short evolution period for adapting to all challenges of life in the aquatic environment (Madsen and Sand-Jensen 1991). Seagrasses readily take up CO_2 from the water, but (like most macroalgae and many submersed freshwater plants) most of them also have the capacity to extract the much higher concentrations of HCO_3^- (<20 μM CO_2 vs. 2.2 mM HCO_3^-; Madsen and Sand-Jensen 1991). The large total pool of DIC in seawater, however, does not seem to be sufficient to saturate photosynthesis of seagrasses when examined in single leaf experiments (Figure 5.7; Beer and Koch 1996). Accordingly, seagrass production may increase with rising atmospheric CO_2 levels (Koch et al. 2013). However, one should be cautious in extrapolating conclusions from single leaf measurements to fully developed plant beds in nature, where many factors vary simultaneously, and carbon chemistry may have high spatial and temporal variability related to release from sediments and other processes.

5.4.6 Oxygen and Sulfide Dynamics

Belowground tissues of submersed plants are often situated in anoxic sediments, meaning that oxygen concentrations are essentially zero; this is another major challenge for submersed

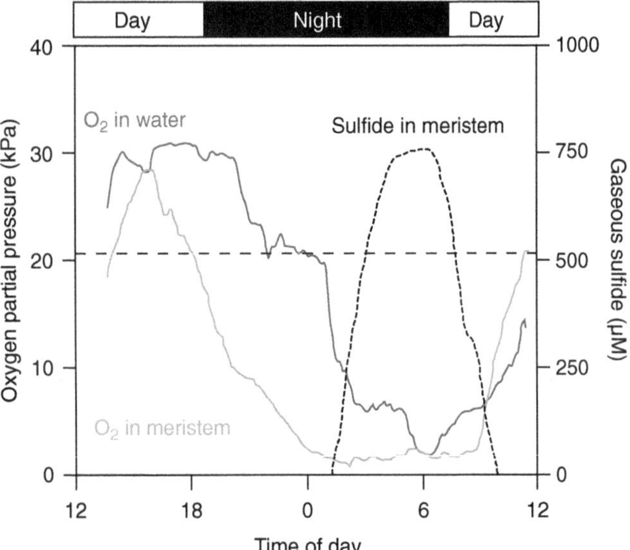

FIGURE 5.8 A major physiological challenge for submersed vascular plants is to supply rhizomes and roots with enough oxygen to support aerobic metabolism and prevent entry of anaerobic metabolites like sulfide from the sediment porewater. This figure shows how internal plant oxygen and gaseous sulfide, measured with microelectrodes, interact over a diel cycle in a seagrass meadow during a die-off period. The oxygen within the plant meristem changed from supersaturation (i.e., above the oxygen partial pressure of the atmosphere, indicated by the dashed line) during daytime to very low levels at night, when water column oxygen content was low. When internal oxygen is low, toxic sulfide can enter the plant from the sediment as seen shortly after midnight on this occasion. *Source:* Borum et al. (2005)/With Permission of John Wiley & Sons.

plants in the estuarine environment (Armstrong 1980) and seagrasses (Borum et al. 2006). Anoxia in sediments develops due to abiotic factors such as sediment grain size, permeability, and hydrodynamic regime (Santos et al. 2012) as well as inputs of organic matter that are respired by microbes, which is initially an oxygen-consuming process. Once the oxygen in sediment porewater has been depleted, anaerobic respiration processes such as sulfate reduction dominate and produce toxic hydrogen sulfide as a by-product (see Chapter 9). Porewater sulfide is a major challenge to submersed plants as it can enter the plant through roots and rhizomes when internal oxygen concentrations are low (Figure 5.8; Carlson et al. 1994; Raven and Scrimgeour 1997; Pedersen et al. 2004). The combined effect of plant anoxia and sulfide entry may be responsible for periodic massive plant mortality as has been suggested for *Thalassia testudinum* in Florida Bay (Zieman et al. 1989; Robblee et al. 1991; Borum et al. 2005) and for *Zostera marina* in Europe (Plus et al. 2003).

Seagrasses cope with the challenge of sulfide toxicity by supplying enough oxygen to belowground biomass to limit anaerobic respiration near plant tissues and to support microbial sulfide oxidation. It was once assumed that the oxygen supply to seagrass roots and rhizomes relied almost exclusively on the oxygen produced by leaf photosynthesis during the day, making roots and rhizomes susceptible to anoxia and anaerobic metabolism mainly at night (Smith et al. 1984). However, it has since been demonstrated that oxygen can passively diffuse from the water into the seagrass leaves and can then be transported to rhizomes and roots by diffusion through the air-filled lacunae connecting leaves with roots (Pedersen et al. 1998). This transport is rapid (within minutes) because gas-phase diffusion, as mentioned previously, is about 10,000 times faster than diffusion in water. Under "normal" conditions with high water-column oxygen, moderate plant respiration, and low oxygen consumption in the sediment, the oxygen transport is sufficient to support aerobic root and rhizome metabolism and even allow the release of oxygen to the surrounding sediment in the dark (Pedersen et al. 1998; Borum et al. 2006). This root oxygen release creates an oxygenated micro-layer around the roots, which prevents toxic sulfide from reaching the root surface and invading the plant tissue, because the sulfide is rapidly reoxidized by microbes within the micro-layer (Lee and Dunton 2000; Pedersen et al. 2004; Holmer et al. 2005). If root oxygen release becomes too low, sulfides and other anaerobic metabolites can enter the roots and poison plant tissues (Figure 5.8). This is most likely to occur when (i) water column oxygen is low at night, (ii) water column mixing is reduced, (iii) plant respiration is high, or (iv) sediment oxygen consumption by microbes is accelerated.

5.5 Ecosystem Benefits

Submersed vascular plants are conspicuous benthic primary producers that have important effects on many ecosystem processes. Their aboveground tissues, which vary in structural complexity with species and environmental conditions, influence water movement, leading to generally quiescent conditions within plant beds. They are also an important part of the food chain and serve as refuge from predation for fish and invertebrates and as substratum for epiphytic colonization. In addition, belowground root and rhizome networks influence sediment biogeochemistry. Submersed plants therefore substantially modify their local environment, producing many positive feedbacks that can increase the suitability of an area for further plant growth as well as negative feedbacks that can reduce plant production (Jones et al. 1997; Gruber and Kemp 2010).

5.5.1 Consequences of Physical Effects on Water Movement

Increased frictional drag associated with the physical structure of plant canopies can reduce water velocity (Gambi et al. 1990) and wave heights (Fonseca and Cahalan 1992) within a plant bed. These effects generally enhance the deposition of fine organic and inorganic particles as velocities drop below a critical threshold (Palmer et al. 2004). Quiescent conditions in submersed plant communities can result in substantially longer water residence times compared to unvegetated areas (Rybicki et al. 1997). This has important implications for particle accumulation and dissolved nutrient chemistry. The influence of plant beds on local hydrodynamics and sedimentology depends on key characteristics of the plant canopy including shoot density, shoot length, and aboveground biomass (Figure 5.9), which vary with species and over space and time (Hasegawa et al. 2008). Patch size also plays a key role in the magnitude of these interactions (Carr et al. 2016; Licci et al. 2019). Additionally, submersed plants are phenotypically plastic and can acclimate to a range of hydrodynamic conditions, with leaves, for example, minimizing drag by becoming shorter and thinner under higher flow conditions (Peralta et al. 2005).

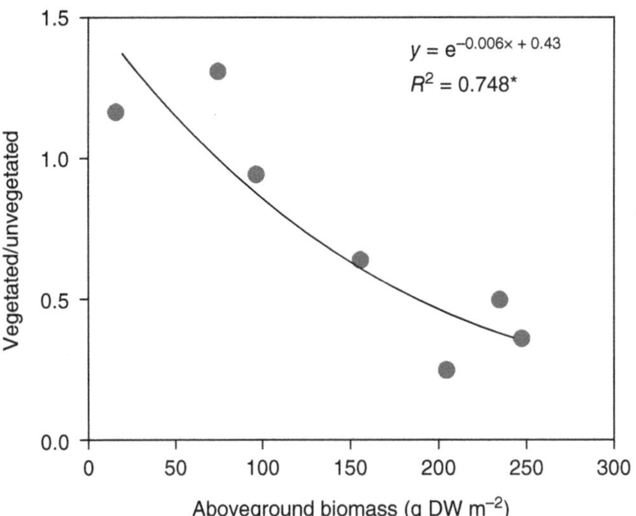

FIGURE 5.9 Ratio between current speeds measured inside (vegetated) and outside (unvegetated) a seagrass bed during periods with different aboveground biomass. The ratio between current velocities clearly declined with increasing biomass, reflecting the current-damping effect of the canopy. *Source*: Hasegawa et al. (2008)/With Permission of Springer Nature.

In addition to trapping particles, submersed plants tend to stabilize the sediment surface through sheltering by leaves and binding by roots and rhizomes (Liu and Nepf 2016). Resuspension of fine sediments is reduced within plant beds even during storm events (Gacia and Duarte 2001; Granata et al. 2001). As a result, water clarity in plant communities increases (Moore 2004), thereby increasing the available light for photosynthesis. This effect can be dramatic (e.g., ~halving in turbidity, ~doubling in dissolved oxygen) for dense beds of plants (Figure 5.10; Gruber et al. 2011). The extent to which this occurs depends on canopy architecture, the nature of the suspended material, and ambient hydrodynamics (Adams et al. 2016).

5.5.2 Biogeochemical Effects

At local scales, the damping of current and wave energy by submersed plant beds tends to result in accumulation of organic-rich material within vegetated sediments. The decomposition of this trapped organic material augments inorganic nutrient pools, especially in the sediment porewater (Kenworthy et al. 1982; Figure 5.11). Porewater concentrations may also remain elevated because there is reduced mixing of

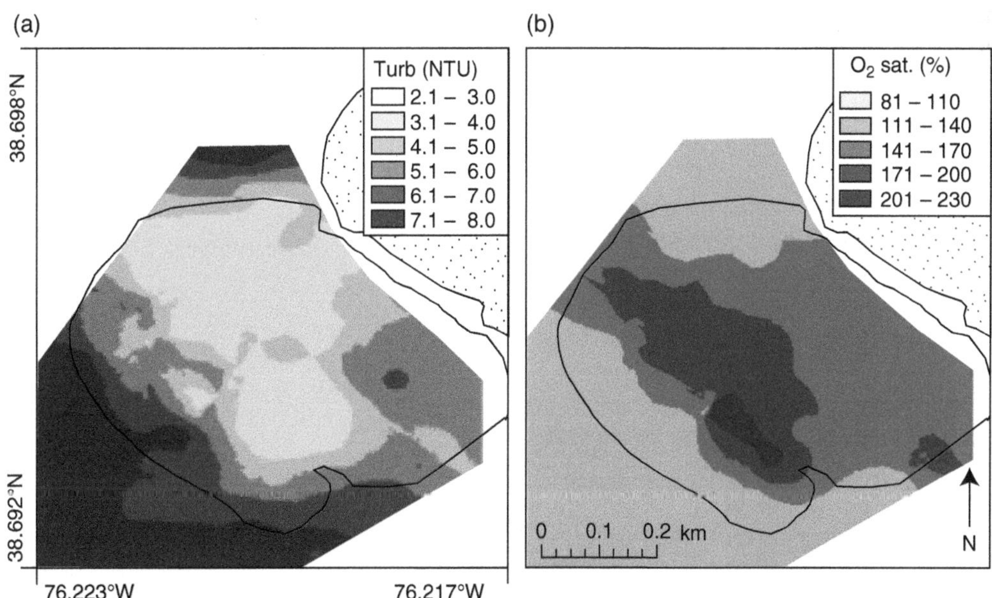

FIGURE 5.10 Maps of (a) turbidity and (b) oxygen saturation created from measurements across and outside a bed of the canopy-forming submersed macrophyte *Stuckenia pectinata* in Chesapeake Bay. The solid black line represents the perimeter of the bed, while the dotted area represents land. Turbidity was measured in NTUs (nephelometric turbidity units), a relative measure of water clarity. The maps show a reduction in turbidity and increase in oxygen saturation with distance into the meadow, which reflects the substantial impact submersed plant beds can have on water quality in shallow environments. *Source*: Gruber et al. (2011)/With Permission of Springer Nature.

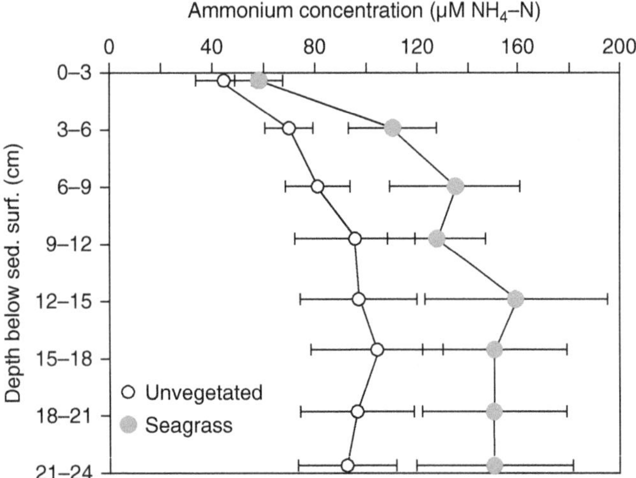

FIGURE 5.11 Sediment profiles of dissolved ammonium in vegetated sites consisting of mixtures of *Zostera marina* and *Halodule wrightii* (closed circles) and unvegetated sites (open circles). Error bars are the standard error. The higher concentrations of porewater ammonium in the seagrass bed reflect the retention and recycling of organic matter produced by the community and trapped from sources outside the plant bed. *Source*: Kenworthy et al. (1982)/With Permission of Springer Nature.

porewater with the overlying water column due to changes in sediment composition and reduced flow speeds within the plant bed (Koch 1999). These high nutrient concentrations tend to stimulate belowground plant uptake (Caffrey and Kemp 1990). Uptake of nutrients through seagrass leaves is also strongly influenced by local hydrodynamics, because uptake is a function of flow speed and concentration (Atkinson 2011). Reduced current speeds within seagrass beds can result in wider diffusive boundary layers around seagrass leaves, which would reduce nutrient uptake rates in theory (Thomas and Cornelisen 2003).

Interactions between plant roots and sediments also have substantial effects on other important sediment biogeochemical processes. As mentioned previously, submersed plants release oxygen in a micro-layer around their roots (Kemp and Murray 1986; Pedersen et al. 1998). This oxidized zone can stimulate coupled nitrification-denitrification (Caffrey and Kemp 1990), as ammonium in sediment porewater is reoxidized and nitrate then diffuses to the anoxic zone where denitrification occurs. Plant stimulation of nitrification and denitrification may be substantial for some submersed plants including freshwater isoetids (Christensen and Sørensen 1986; Ottosen et al. 1999). Denitrification "removes" nitrogen from the seagrass bed, but nitrogen is also being "added" in a simultaneous process called nitrogen fixation. Nitrogen fixation refers to the conversion of dissolved nitrogen gas (N_2) into ammonia by diazotropic microbes. This ammonia can then be readily used by bacteria and seagrasses. Rates of nitrogen fixation are enhanced in seagrass meadows, which is likely related to the increased levels of organic matter present (Cole and McGlathery 2012). Because both denitrification and nitrogen fixation are stimulated in seagrass beds, the net change in the bed nitrogen pool may be relatively small overall (Rysgaard et al. 1996; Risgaard-Petersen and Ottosen 2000).

5.5.3 Submersed Plants as Food and Habitat

Historically it was believed that relatively few organisms grazed directly on submersed vascular plants, and herbivory was of minor importance for seagrass biomass and production (den Hartog 1970). However, many recent studies show important direct and indirect effects of herbivory, especially in estuarine plant communities (Valentine and Heck 1999; Heck and Valentine 2006). Marine mammals such as dugongs and manatees feed on beds of seagrass, leaving long grazing tracks devoid of vegetation (Thayer et al. 1984), and sea turtles, fishes, and urchins can remove up to 50% of leaf production (Cebrian and Duarte 1998). In estuaries, waterfowl including geese, swans, and diving ducks feed extensively on submersed plant beds (Holm and Clausen 2006; Rybicki and Landwehr 2007). Consequently, reductions in size and density of submersed plant beds owing to human and natural disturbances can result in large declines in abundance of herbivorous ducks during fall migration periods (Figure 5.12; Kemp et al. 1984). Research

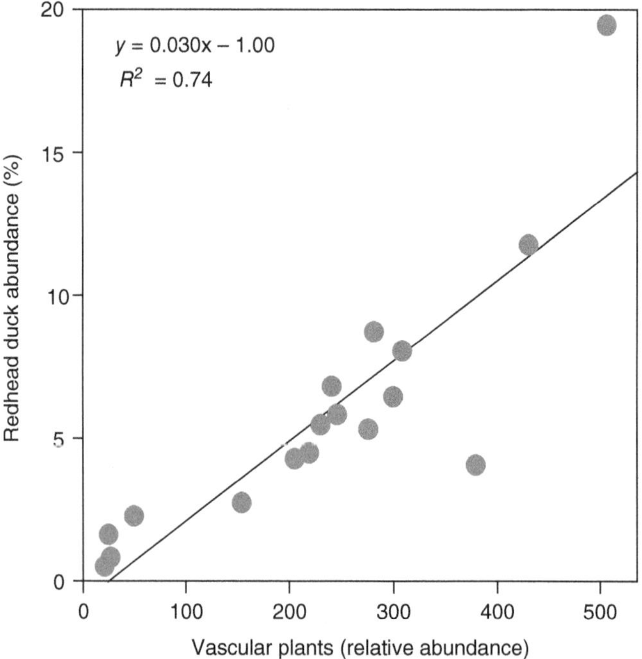

FIGURE 5.12 The abundance of the diving duck, Redhead (in % of the total count for North American breeding grounds), during winter in upper Chesapeake Bay versus the relative abundance of submersed vascular plants at upper Bay sites. The number of ducks correlated well with plant abundance during the period from 1961 to 1975. With gradually decreasing plant abundance up through the 1960 and 1970s, Redhead abundance declined in this area and the ducks moved to areas with seagrass further south in Chesapeake Bay. *Source*: Kemp et al. (1984)/With Permission of Elsevier.

using stable isotopes of carbon and nitrogen suggests that submersed plant primary production supports growth of animals at higher trophic levels; in one study, half the fauna in a tropical marine ecosystem appeared to depend directly or indirectly on these plants (Vonk et al. 2008).

Primary production by epiphytic algae living on submersed plant leaf surfaces can sometimes be as large as production by the host plants themselves (Borum et al. 1984). The growth form of most seagrass species is characterized by continuous production of new clean leaves that replace older leaves, which are often covered with heavy epiphyte growth. Thus, these species tend to maintain a balanced mixture of new, mature, and old leaves with an associated gradient of epiphyte accumulations. Epiphytic algae are a key food source for numerous gastropods, amphipods, isopods, and fishes, which leave distinct grazing tracks within the attached community and often show a preference for older leaves with dense coatings of epiphytes (Wressnig and Booth 2007).

Although not as well studied as many other aspects of submersed plant ecology, the production of dead (detrital) leaf material represents a significant source of nutrients for the wider estuarine community. Detritus is grazed directly by some detritivorous species, which ingest not only dead plant material but also its associated epiphytic community (François et al. 2018). Leaf material breaks down at widely varying rates, which can depend on the carbon content of the seagrass species and the surrounding environmental conditions. Seagrass detritus exposed to different conditions can degrade at rates ranging from months to years. Studies of the species *Amphibolis antarctica* and *Posidonia sinuosa* showed that detritus that remained within the seagrass bed could degrade in less than 80 days, while detritus deposited on the shoreline took >370 days to degrade (Oldham et al. 2010).

The abundance of organic matter and the protection offered by the submersed plant canopy make these plant beds attractive habitats for many animals. Beds support substantially higher abundance, diversity, and production of fishes and invertebrates compared to nearby unvegetated sites (Kemp et al. 1984; Lubbers et al. 1990; Lee et al. 2001). However, few faunal species are completely restricted to seagrass beds (e.g., the fan mussel, *Pinna nobilis*, which is restricted to Mediterranean *Posidonia* stands) with most species also occurring outside the beds. Some animal species use seagrass beds only during juvenile stages, but a number of species feed and remain within the beds throughout their life cycle (Gillanders 2007). For fish, submersed plant beds can serve as refuges for predator avoidance (Canion and Heck 2009) or can simply be the preferred habitat irrespective of predation pressure (Burfeind et al. 2009). Often, submersed plants and fauna have a mutualistic relationship. For example, increased abundance of ribbed mussels in seagrass beds stimulates plant growth through efficient nutrient recycling (Peterson and Heck 2001). For other bivalve species living in seagrass meadows, sulfur-oxidizing bacteria present in their gills can decrease toxic hydrogen sulfide levels locally (van der Heide et al. 2012; de Fouw et al. 2018). Plants have also been shown to benefit from epiphyte grazers that increase light availability to leaves (Hughes et al. 2004).

5.5.4 Ecosystem Services

Many of these processes provide society with "ecosystem services," or ecological functions from which humans derive a diverse array of benefits. As mentioned above, the physical structure of submersed plants dampens wave energy, which can reduce coastal erosion. For example, shoreline erosion rates increased in Danish coastal waters following the massive die-off of eelgrass (*Z. marina*) populations in the 1930s (Rasmussen 1973; Christiansen et al. 1981). Submersed plant beds provide food and habitat for commercially and recreationally important fish and invertebrates. Recent review work suggests that seagrass beds support ~20% of the world's largest 25 fisheries and that intertidal fishing activities for fauna such as bait, finfish, and prawns support many coastal communities globally (Unsworth et al. 2019). Submersed plant beds also act as reservoirs of nutrients, potentially suppressing excessive phytoplankton growth and improving water clarity. For example, estimates of submersed plant abundance at historical levels in the Chesapeake Bay, USA indicate that these communities could attenuate nearly half of the current nitrogen inputs to the estuary (Kemp et al. 2005). Similar estimates have been developed for the effect of plant beds on regional budgets for fluxes of suspended sediments, organic carbon, and total phosphorus (Kemp et al. 1984).

Interactions with the carbon cycle are particularly important as global CO_2 concentrations increase, and the effects of climate change intensify. For example, submersed plants are sinks for "blue carbon," or carbon stored in aquatic systems (Lavery et al. 2013). The capability of seagrasses to sequester carbon is the focus of a majority of current research, and findings have shown that rates vary by species and location. Some studies suggest that organic carbon storage rates per unit area of submersed plant meadows can be almost as high as those estimated for terrestrial forests (Fourqurean et al. 2012). Submersed plants can also locally modulate ocean acidification by assimilating CO_2 through photosynthesis (Hendriks et al. 2014). Ocean acidification, or decreased ocean pH, occurs as the ocean absorbs anthropogenic CO_2 emissions. This change in ocean chemistry is a problem for organisms that use carbonate to build their shells and skeletons, such as oysters and other shellfish, because decreasing pH and increasing CO_2 concentrations lead to decreased availability of carbonate for shell formation (Gazeau et al. 2007). However, CO_2 uptake via seagrass production may locally protect marine calcifiers from these effects (Koweek et al. 2018). Oyster farmers are investigating the potential to harness this ecosystem function by incorporating submersed plants into oyster aquaculture systems to buffer pH in regions where acidic seawater limits oyster growth (Clements and Chopin 2017).

Submersed plants provide a host of other ecosystem services, such as support for tourism and recreational fishing, bequest or nonuse value, and spiritual or religious value (Barbier et al. 2011). They have also been found to reduce bacterial pathogen concentrations in coastal waters, thereby protecting human and ecological health (Lamb et al. 2017). Quantification of the monetary value of ecosystem services provided by seagrass beds worldwide shows them to be one of the most valuable (per hectare) natural aquatic communities (Costanza et al. 1997). However, it should be stressed that the monetary value of the services provided by seagrasses such as shoreline protection is highly variable due to the variability in space and time of seagrass beds (Koch et al. 2009).

5.6 Human Impacts and Management

The growing human population and its increasing exploitation of space and natural resources in coastal regions worldwide result in continuous pressure on estuarine ecosystems. Because submersed plants are sensitive to a wide variety of coastal stresses, they serve as "sentinel species," or organisms that can provide early warning of broader ecological impairment (Orth et al. 2017). Pressures such as excessive nutrient enrichment (eutrophication), inputs of toxic contaminants, siltation, coastal construction, dredging, and fishery activities affect both submersed plants as well as broader ecosystem functions (Orth et al. 2006a). Quantifying changes in seagrass abundance can indicate general ecological impairment because their survival is a direct response to time-integrated water quality (Dennison et al. 1993). Furthermore, detailed analysis of stable isotope composition and elemental nitrogen, phosphorus, and carbon content in submersed plant tissue can indicate drivers of ecosystem degradation and sources of pollution. For example, values of the stable isotope, ^{15}N, in seagrass tissue relative to standard values can indicate human sewage or N fertilizer as chronic nitrogen sources, and ^{13}C values can be indicative of light limitation (Fourqurean et al. 2015).

Submersed plants also respond to several other environmental stresses. Climate change has negative effects on seagrass ecosystems due to rising temperatures, increasing runoff of nutrients with greater precipitation, higher frequency and strength of storm events, salinity stress during droughts, and rising sea level (Short and Neckles 1999; Najjar et al. 2010). Infection with pathogens, such as *Labyrinthula*, can also cause submersed plant die-offs (Sullivan et al. 2013), and blooms of macroalgae can out-compete submersed plants (Thomsen et al. 2012; Montefalcone et al. 2015). Deterioration of seagrass beds has been observed over wide geographical scales (Figure 5.13; Duarte 2002). In fact, it has been estimated that nearly 30% of worldwide seagrass area has been lost since the 1870s and that the rate of loss continues to increase (Waycott et al. 2009).

The loss of submersed plant beds has resulted in increased attention by estuarine resource managers who have implemented monitoring programs that track changes in submersed plant cover and density to detect and mitigate negative trends in these valuable living resources (Orth et al. 2002; Orth et al. 2010b). The process of mapping existing plant beds has evolved over the last century from labor-intensive sampling of plants along depth gradients (Ostenfeld 1908) to present surveys that apply an array of technologies. In-water acoustic instruments and video camera surveys detect the physical structure of submersed plants and can be used to estimate plant cover, canopy height, and biomass (Duarte 1987; Paul et al. 2011). In addition, high resolution aerial photography, unmanned aerial systems (UAS), and satellite imagery techniques facilitate seagrass mapping over large geographic areas (Orth et al. 2010a).

In light of widespread loss of submersed plants and the valuable ecosystem services they provide, reintroduction often becomes part of an estuary management plan (Fonseca et al. 1998). Developing effective restoration strategies requires an understanding of the life history of a plant species as well as the factors that led to its decline in a given system. When disease or physical disturbance from storms or boat damage is the culprit, reintroducing plant propagules may be sufficient. For some seagrasses, such as the Mediterranean *Posidonia oceanica*, growth occurs slowly through rhizome extension, and natural recovery may take decades (Hemminga and Duarte 2000). Therefore, seeding or planting can accelerate the recovery process, as long as environmental conditions are otherwise sufficient to support plant growth (Orth et al. 2010a). Selecting genetically diverse, locally-adapted donor plants is an important component of direct planting programs (Reynolds et al. 2012; Engelhardt et al. 2014). When chronic light limitation due to poor water quality is the cause of seagrass decline, restoration must focus on addressing water quality issues by reducing nutrient and sediment loads to the system. Considerable effort has gone into establishing habitat quality thresholds for submersed plants, which aim to define minimum light requirements or maximum nutrient levels needed to restore and protect submersed plant beds (Dennison et al. 1993; Kemp et al. 2004). These can be used in simulation models to predict plant response to alternative restoration and mitigation efforts (Madden and Kemp 1996; Cerco and Moore 2001; Latimer and Rego 2010).

As mentioned previously, a complex network of interactions exists between submersed plants, hydrodynamics, sediments, and associated organisms, which can result in positive feedback effects on plant growth. These feedbacks have implications for submersed plant restoration and recovery. Because positive feedbacks tend to improve local growing conditions and promote plant growth, submersed plant beds can remain relatively stable despite degrading environmental conditions.

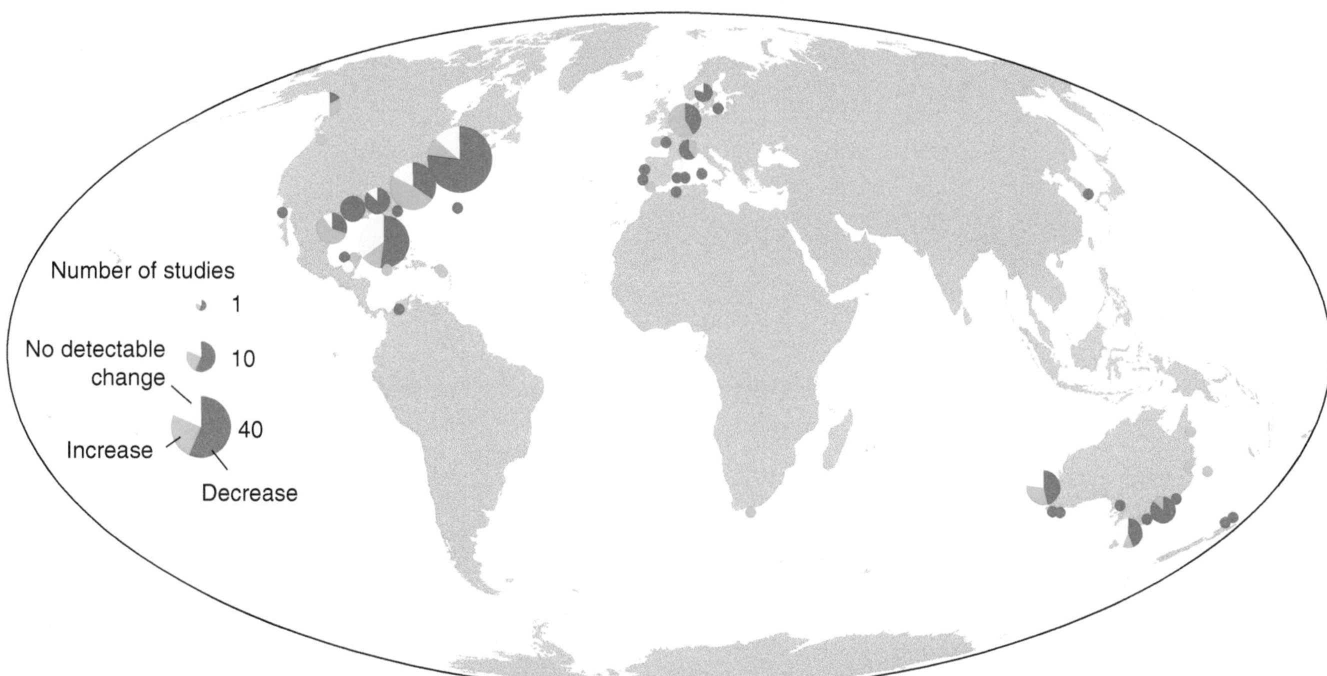

FIGURE 5.13 Global map indicating number of studies measuring change in seagrass areal extent since the earliest records available (1879 in one study). The change measured in these studies is indicated by pie chart colors, where seagrass extent is defined as increasing (final extent >110% of initial extent), decreasing (final extent <90% of initial extent), or not changing (±10% of initial extent). Monitoring of seagrasses is not evenly distributed globally, with the majority occurring in wealthy regions like North America, Europe, and Australia. Far less is known about changes in seagrass extent in the highly diverse and large meadows of the Tropical Indo Pacific bioregion. *Source*: Waycott et al. (2009)/With Permission of PNAS.

However, once a critical water quality threshold is reached, the system may collapse, and widespread loss can occur quite suddenly (McGlathery et al. 2007). Without submersed plants present to improve within-meadow environmental conditions, natural recovery is often difficult (Scheffer et al. 2001). This type of abrupt, persistent ecological change is known as a "regime shift" and is one of many examples of regime shifts that can occur in the estuarine environment (deYoung et al. 2008; Rocha et al. 2015).

Regime shifts can also occur in reverse. When recovering submersed plant beds reach a critical mass in response to reductions in nutrient or suspended sediment concentrations, feedbacks can accelerate growth and spatial expansion (Gurbisz et al. 2016). Restoration ecology via direct planting is beginning to incorporate some of these concepts in practical ways (Byers et al. 2006) to ensure that the self-catalyzing positive feedback effects of recovering beds are incorporated into the design of revegetation programs. Strategies include optimizing site choice (van Katwijk et al. 2009; Hengst et al. 2010), planting density (van der Heide et al. 2007), and transplant types and sizes (Bouma et al. 2009). For example, because the magnitude of feedback interactions is scale-dependent, large-scale plantings that include at least 1000–10,000 propagules tend to be the most successful (van Katwijk et al. 2016). Similarly, propagule planting sites can be colocated with existing seagrass beds or even adjacent to shellfish reefs to take advantage of the quiescent environment that these structures create (van Katwijk et al. 2009). As watershed-based restoration programs aimed at reducing nutrient inputs take effect and direct planting methodologies are refined, submersed plant abundance is now increasing in some locations. Widespread submersed plant recovery has been documented in Chesapeake Bay (Lefcheck et al. 2018), Tampa Bay (Greening et al. 2011), the Wadden Sea (Reise and Kohlus 2008), and coastal Virginia, USA (Orth et al. 2006b). These serve as promising examples that restoration efforts can be successful.

In order to conserve existing submersed plant populations and further promote recovery, seagrass researchers and managers are increasingly focusing on resilience, or the capacity of a system to withstand or recover quickly from disturbances (Walker et al. 2004). Resilience is particularly important as novel stressors associated with climate change, such as storms, droughts, heat waves, and other weather-driven perturbations become more frequent and intense. Although this area of research is still developing, an array of factors is thought to control submersed plant resilience, including genetic and species diversity, habitat fragmentation, reproductive capacity, presence of important symbionts, integrity of nearby sheltering habitats (e.g., coral reefs), and water quality (Unsworth et al. 2015). Environmental managers are now incorporating these factors into restoration plans and are developing resilience-based seagrass assessments to guide conservation area prioritization of these important and iconic species.

Review Questions

Multiple Choice

1. Seagrasses evolved from:
   a. Marine macroalgae
   b. Cyanobacteria
   c. Terrestrial plants
   d. Bryophytes

2. What is the most important environmental factor that determines how deep seagrasses can grow?
   a. Temperature
   b. Salinity
   c. Light availability
   d. Wave height

3. In which latitudes do seagrasses have the smallest change in biomass with season?
   a. Polar
   b. Temperate
   c. Tropical
   d. Seasonal biomass changes do not differ with latitude

4. Which aspects of plant morphology vary in order to increase the light-harvesting capability of seagrass?
   a. Shoot density
   b. Leaf length
   c. Leaf width
   d. All the above

5. Seagrass respiration is most strongly controlled by:
   a. Light
   b. Nitrogen availability
   c. Temperature
   d. Sediment grain size

6. Nutrient uptake in seagrasses is primarily a function of:
   a. Temperature and nutrient concentration
   b. Leaf length and light
   c. Flow speed and nutrient concentration
   d. Salinity and light

7. Which nutrient is most often limiting for seagrass communities globally?
   a. Phosphorus
   b. Sulfur
   c. Nitrogen
   d. Calcium

8. Submersed plant canopies reduce water flow within the meadow. Which of the following is a result of this interaction?
   a. Salinity increases within the meadow
   b. Suspended particle concentrations decrease within the meadow
   c. Grazing pressure on seagrass decreases
   d. All the above

9. Which of these is not an ecosystem service provided by seagrasses?
   a. Carbon sequestration
   b. Nursery for fish species
   c. Reduce coastal water temperatures
   d. Aesthetic and recreational values

10. Seagrass extent is monitored at many places worldwide. By what percentage is it estimated to have declined since the 1870s?
    a. 10%
    b. 25%
    c. 30%
    d. 50%

Short Answer

1. What are the mechanisms of asexual and sexual reproduction for seagrasses and what are some benefits of these strategies?
2. What is phenotypic plasticity and how does it benefit seagrasses?
3. Light is one of the most important factors regulating seagrass production. What are some strategies seagrasses and submersed plants use to capture light as efficiently as possible?
4. Hydrogen sulfide in sediments is a major challenge for seagrasses. What are some of the environmental conditions that would lead to high amounts of sulfide entering seagrass roots and rhizomes?
5. Seagrasses interact with and alter their surrounding physical and biogeochemical environment in several ways. Please discuss one such physical or biogeochemical interaction. Describe (a) the mechanisms driving the interaction, (b) the effects of the interaction on the surrounding environment, and (c) the ways in which these environmental changes help or hinder seagrass growth.
6. Seagrasses provide many ecosystem services. Define ecosystem service, and list at least five ecosystem services that seagrasses provide.
7. Environmental changes associated with climate change are predicted to be stressful for seagrass populations globally. Based on what you know about seagrass reproductive strategies, plasticity, and response to environmental stressors, what do you think are the key characteristics of seagrass species that may be most impacted by climate change?
8. List and describe one mechanism by which seagrasses can mitigate climate change and one mechanism by which they can locally mitigate an effect of climate change.
9. Imagine you are working as an estuarine natural resource manager and you are tasked with developing a seagrass restoration plan for your estuary. List and explain the factors you would consider in developing the plan. How would you monitor seagrass populations in the estuary to track restoration progress?

References

Adams, M.P., Hovey, R.K., Hipsey, M.R. et al. (2016). Feedback between sediment and light for seagrass: where is it important? *Limnol. Oceanog.* 61 (6): 1937–1955.

Arias-Ortiz, A., Serrano, O., Masqué, P. et al. (2018). A marine heatwave drives massive losses from the world's largest seagrass carbon stocks. *Nat. Clim. Change* 8: 338–344.

Armstrong, W. (1980). *Aeration in Higher Plants*, Advances in Botanical Research, 225–332. London, New York: Academic Press.

Arnaud-Haond, S., Duarte, C.M., Diaz-Almela, E. et al. (2012). Implications of extreme life span in clonal organisms: millenary clones in meadows of the threatened seagrass *Posidonia oceanica*. *PLoS One* 7: e30454.

Atkinson, M.J. (2011). Biogeochemistry of nutrients. In: *Coral Reefs: An Ecosystem in Transition* (ed. Z. Dubinsky and N. Stambler), 199–206. Netherlands: Springer.

Atkinson, M.J. and Smith, S.V. (1983). C:N:P ratios of benthic marine plants. *Limnol. Oceanogr.* 28: 568–574.

Barbier, E.B., Hacker, S.D., Kennedy, C. et al. (2011). The value of estuarine and coastal ecosystem services. *Ecol. Monogr.* 81: 169–193.

Beer, S. and Koch, E. (1996). Photosynthesis of marine macroalgae and seagrasses in globally changing CO_2 environments. *Mar. Ecol. Prog. Ser.* 141: 199–204.

Berry, J.A. and Raison, J.K. (1981). Responses of macrophytes to temperature. In: *Physiological Plant Ecology I* (ed. O.L. Lange, P.S. Nobel, C.B. Osmond and H. Ziegler), 277–338. Berlin Heidelberg: Springer.

Binzer, T. and Sand-Jensen, K. (2002). Production in aquatic macrophyte communities: a theoretical and empirical study of the influence of spatial light distribution. *Limnol. Oceanogr.* 47: 1742–1750.

Borum, J., Kaas, H., and Wium-Andersen, S. (1984). Biomass variation and autotrophic production of an epiphyte-macrophyte community in a coastal danish area: II. Epiphyte species composition, biomass and production. *Ophelia* 23: 165–179.

Borum, J., Pedersen, O., Greve, T.M. et al. (2005). The potential role of plant oxygen and sulphide dynamics in die-off events of the tropical seagrass, *Thalassia testudinum*. *J. Ecol.* 93: 148–158.

Borum, J., Sand-Jensen, K., Binzer, T. et al. (2006). Oxygen movement in seagrasses. In: *Seagrasses: Biology, Ecology and Conservation* (ed. A. Larkum, R.J. Orth and C.M. Duarte), 255–270. Dordrecht, the Netherlands: Springer.

Bouma, T.J., Friedrichs, M., Klaassen, P. et al. (2009). Effects of shoot stiffness, shoot size and current velocity on scouring sediment from around seedlings and propagules. *Mar. Ecol. Prog. Ser.* 388: 293–297.

Britto, D.T. and Kronzucker, H.J. (2002). NH_4^+ toxicity in higher plants: a critical review. *J. Plant Physiol.* 159: 567–584.

Bulthuis, D.A. (1987). Effects of temperature on photosynthesis and growth of seagrasses. *Aquat. Bot.* 27: 27–40.

Burfeind, D.D., Tibbetts, I.R., and Udy, J.W. (2009). Habitat preference of three common fishes for seagrass, *Caulerpa taxifolia*, and unvegetated substrate in Moreton Bay, Australia. *Environ. Biol. Fishes* 84: 317–322.

Burkholder, J.M., Tomasko, D.A., and Touchette, B.W. (2007). Seagrasses and eutrophication. *J. Exp. Mar. Biol. Ecol.* 350: 46–72.

Byers, J.E., Cuddington, K., Jones, C.G. et al. (2006). Using ecosystem engineers to restore ecological systems. *Trends Ecol. Evol.* 21: 493–500.

Cabaço, S. and Santos, R. (2012). Seagrass reproductive effort as an ecological indicator of disturbance. *Ecol. Indic.* 23: 116–122.

Caffrey, J.M. and Kemp, W.M. (1990). Nitrogen cycling in sediments with estuarine populations of *Potamogeton perfoliatus* and *Zostera marina*. *Mar. Ecol. Prog. Ser.* 66: 147–160.

Canion, C.R. and Heck, K.L. (2009). Effect of habitat complexity on predation success: re-evaluating the current paradigm in seagrass beds. *Mar. Ecol. Prog. Ser.* 393: 37–46.

Carlson, P.R., Yarbro, L.A., and Barber, T.R. (1994). Relationship of sediment sulfide to mortality of *Thalassia testudinum* in Florida Bay. *Bull. Mar. Sci.* 54: 733–746.

Carr, J.A., D'Odorico, P., McGlathery, K.J., and Wiberg, P.L. (2016). Spatially explicit feedbacks between seagrass meadow structure, sediment and light: habitat suitability for seagrass growth. *Adv. Water Resour.* 93: 315–325.

Cebrian, J. and Duarte, C.M. (1998). Patterns in leaf herbivory on seagrasses. *Aquat. Bot.* 60: 67–82.

Cerco, C.F. and Moore, K.A. (2001). System-wide submerged aquatic vegetation model for Chesapeake Bay. *Estuaries* 24: 522–534.

Christensen, P.B. and Sørensen, J. (1986). Temporal variation of denitrification activity in plant-covered, littoral sediment from Lake Hampen, Denmark. *Appl. Environ. Microbiol.* 51: 1174–1179.

Christiansen, C., Christoffersen, H., Dalsgaard, J., and Nørnberg, P. (1981). Coastal and near-shore changes correlated with dieback in eel-grass (*Zostera marina*, L.). *Sediment. Geol.* 28: 163–173.

Clements, J.C. and Chopin, T. (2017). Ocean acidification and marine aquaculture in North America: potential impacts and mitigation strategies. *Rev. Aquac* 9: 326–341.

Cole, L.W. and McGlathery, K.J. (2012). Nitrogen fixation in restored eelgrass meadows. *Mar. Ecol. Prog. Ser.* 448: 235–246.

Collier, C.J., Uthicke, S., and Waycott, M. (2011). Thermal tolerance of two seagrass species at contrasting light levels Implications for future distribution in the Great Barrier Reef. *Limnol. Oceanogr.* 56: 2200.

Collier, C.J. and Waycott, M. (2014). Temperature extremes reduce seagrass growth and induce mortality. *Mar. Pollut. Bull.* 83: 483–490.

Costanza, R., d'Arge, R., De Groot, R. et al. (1997). The value of the world's ecosystem services and natural capital. *Nature* 387: 253–260.

Dawson, S.P. and Dennison, W.C. (1996). Effects of ultraviolet and photosynthetically active radiation on five seagrass species. *Mar. Biol.* 125: 629–638.

de Fouw, J., van der Heide, T., van Belzen, J. et al. (2018). A facultative mutualistic feedback enhances the stability of tropical intertidal seagrass beds. *Sci. Rep.* 8: 12988.

Demars, B.O.L. and Edwards, A.C. (2007). Tissue nutrient concentrations in freshwater aquatic macrophytes: high inter-taxon differences and low phenotypic response to nutrient supply. *Freshw. Biol.* 52: 2073–2086.

den Hartog, C. (1970). *The Seagrasses of the World*. London: Northholland Publishing Company Amsterdam.

Dennison, W.C., Orth, R., Moore, K. et al. (1993). Assessing water quality with submersed aquatic vegetation: Habitat requirements as barometers of Chesapeake Bay health. *BioScience* 43: 86–94.

deYoung, B., Barange, M., Beaugrand, G. et al. (2008). Regime shifts in marine ecosystems: detection, prediction and management. *Trends Ecol. Evol.* 23: 402–409.

Duarte, C.M. (1987). Use of echosounder tracings to estimate the aboveground biomass of submerged plants in lakes. *Can. J. Fish. Aquat. Sci.* 44: 732–735.

Duarte, C.M. (1990). Seagrass nutrient content. *Mar. Ecol. Prog. Ser.* 67: 201–207.

Duarte, C.M. (1991). Seagrass depth limits. *Aquat. Bot.* 40: 363–377.

Duarte, C.M. (2002). The future of seagrass meadows. *Environ. Conserv.* 29: 192–206.

Duarte, C.M. and Chiscano, C.L. (1999). Seagrass biomass and production: a reassessment. *Aquat. Bot.* 65: 159–174.

Duarte, C.M., Marba, N., Gacia, E., and Fourqurean, J.W. (2010). Seagrass community metabolism: assessing the carbon sink capacity of seagrass meadows. *Glob. Biogeochem. Cycles* 24: GB4032.

Duarte, C.M., Marbà, N., Krause-Jensen, D., and Sánchez-Camacho, M. (2007). Testing the predictive power of seagrass depth limit models. *Estuaries and Coasts* 30: 652–656.

Engelhardt, K.A.M., Lloyd, M.W., and Neel, M.C. (2014). Effects of genetic diversity on conservation and restoration potential at individual, population, and regional scales. *Biol. Conserv.* 179: 6–16.

Ferguson, A.J.P., Gruber, R.K., Orr, M., and Scanes, P. (2016). Morphological plasticity in *Zostera muelleri* across light, sediment, and nutrient gradients in Australian temperate coastal lakes. *Mar. Ecol. Prog. Ser.* 556: 91–104.

Fleming, J.P. and Dibble, E.D. (2015). Ecological mechanisms of invasion success in aquatic macrophytes. *Hydrobiologia* 746: 23–37.

Fonseca, M., Kenworthy, W.J., and Thayer, G.W. (1998). *Guidelines for the Conservation and Restoration of Seagrasses in the United States and Adjacent Waters*, NOAA Coastal Ocean Program Decision Analysis Series 12, 222pp. Silver Spring, MD.

Fonseca, M.S. and Cahalan, J.A. (1992). A preliminary evaluation of wave attenuation by four species of seagrass. *Estuar. Coast. Shelf Sci.* 35: 565–576.

Fonseca, M.S., Kenworthy, W.J., Griffith, E. et al. (2008). Factors influencing landscape pattern of the seagrass *Halophila decipiens* in an oceanic setting. *Estuar. Coast. Shelf Sci.* 76: 163–174.

Fourqurean, J.W., Duarte, C.M., Kennedy, H. et al. (2012). Seagrass ecosystems as a globally significant carbon stock. *Nat. Geosci.* 5: 505.

Fourqurean, J.W., Manuel, S.A., Coates, K.A. et al. (2015). Water quality, isoscapes and stoichioscapes of seagrasses indicate general P limitation and unique N cycling in shallow water benthos of Bermuda. *Biogeosciences* 12: 6235–6249.

François, R., Thibaud, M., Marleen, D.T. et al. (2018). Seagrass organic matter transfer in *Posidonia oceanica* macrophytodetritus accumulations. *Estuar. Coast. Shelf Sci.* 212: 73–79.

Gacia, E. and Duarte, C.M. (2001). Sediment retention by a Mediterranean *Posidonia oceanica* meadow: the balance between deposition and resuspension. *Estuar. Coast. Shelf Sci.* 52: 505–514.

Gambi, M.C., Nowell, A.R.M., and Jumars, P.A. (1990). Flume observations on flow dynamics in *Zostera marina* (eelgrass) beds. *Mar. Ecol. Prog. Ser.* 61: 159–169.

Gazeau, F., Quiblier, C., Jansen, J.M. et al. (2007). Impact of elevated CO_2 on shellfish calcification. *Geophys. Res. Lett.* 34: https://doi.org/10.1029/2006GL028554

Gillanders, B.M. (2007). Seagrasses, Fish, and Fisheries. In: *Seagrasses: Biology, Ecology and Conservation* (ed. A. Larkum, R.J. Orth and C.M. Duarte), 503–505. Dordrecht: Springer.

Granata, T.C., Serra, T., Colomer, J. et al. (2001). Flow and particle distributions in a nearshore seagrass meadow before and after a storm. *Mar. Ecol. Prog. Ser.* 218: 95–106.

Greening, H.S., Cross, L.M., and Sherwood, E.T. (2011). A multiscale approach to seagrass recovery in Tampa Bay, Florida. *Ecol. Restor.* 29: 82–93.

Greve, T.M., Borum, J., and Pedersen, O. (2003). Meristematic oxygen variability in eelgrass (*Zostera marina*). *Limnol. Oceanogr.* 48: 210–216.

Gruber, R.K., Hinkle, D.C., and Kemp, W.M. (2011). Spatial patterns in water quality associated with submersed plant beds. *Estuaries and Coasts* 34: 961–972.

Gruber, R.K. and Kemp, W.M. (2010). Feedback effects in a coastal canopy-forming submersed plant bed. *Limnol. Oceanogr.* 55: 2285–2298.

Gruber, R.K., Lowe, R.J., and Falter, J.L. (2017). Metabolism of a tide-dominated reef platform subject to extreme diel temperature and oxygen variations. *Limnol. Oceanogr.* 62: 1701–1717.

Gurbisz, C., Kemp, W.M., Sanford, L.P., and Orth, R.J. (2016). Mechanisms of storm-related loss and resilience in a large submersed plant bed. *Estuaries Coasts* 39: 951–966.

Hasegawa, N., Hori, M., and Mukai, H. (2008). Seasonal changes in eelgrass functions: current velocity reduction, prevention of sediment resuspension, and control of sediment-water column nutrient flux in relation to eelgrass dynamics. *Hydrobiologia* 596: 387–399.

Heck, K.L. and Valentine, J.F. (2006). Plant-herbivore interactions in seagrass meadows. *J. Exp. Mar. Biol. Ecol.* 330: 420–436.

Hemminga, M.A. (1998). The root/rhizome system of seagrasses: an asset and a burden. *J. Sea Res.* 39: 183–196.

Hemminga, M.A. and Duarte, C.M. (2000). *Seagrass Ecology*. Cambridge University Press.

Hendriks, I.E., Olsen, Y.S., Ramajo, L. et al. (2014). Photosynthetic activity buffers ocean acidification in seagrass meadows. *Biogeosciences* 11: 333–346. https://doi.org/10.5194/bg-11-333-2014.

Hengst, A., Melton, J., and Murray, L. (2010). Estuarine restoration of submerged aquatic vegetation: the nursery bed effect. *Restor. Ecol.* 18: 605–614.

Holm, T.E. and Clausen, P. (2006). Effects of water level management on autumn staging waterbird and macrophyte diversity in three danish coastal lagoons. *Biodivers. Conserv.* 15: 4399–4423.

Holmer, M., Frederiksen, M.S., and Mollegaard, H. (2005). Sulfur accumulation in eelgrass (*Zostera marina*) and effect of sulfur on eelgrass growth. *Aquat. Bot.* 81: 367–379.

Hughes, A.R., Bando, K.J., Rodriguez, L.F., and Williams, S.L. (2004). Relative effects of grazers and nutrients on seagrasses: a meta-analysis approach. *Mar. Ecol. Prog. Ser.* 282: 87–99.

Jensen, H.S., McGlathery, K.J., Marino, R., and Howarth, R.W. (1998). Forms and availability of sediment phosphorus in carbonate sand of Bermuda seagrass beds. *Limnol. Oceanogr.* 43: 799–810.

Jones, C.G., Lawton, J.H., and Shachak, M. (1997). Positive and negative effects of organisms as physical ecosystem engineers. *Ecology* 78: 1946–1957.

Kemp, W.M., Batiuk, R.A., Bartleson, R.D. et al. (2004). Habitat requirements for submerged aquatic vegetation in Chesapeake Bay: water quality, light regime, and physical-chemical factors. *Estuaries* 27: 363–377.

Kemp, W.M., Boynton, W.R., Adolf, J.E. et al. (2005). Eutrophication of Chesapeake Bay: historical trends and ecological interactions. *Mar. Ecol. Prog. Ser.* 303: 1–29.

Kemp, W.M., Boynton, W.R., Twilley, R.R. et al. (1984). Influences of submersed vascular plants on ecological processes in upper Chesapeake Bay. In: *Estuaries as Filters* (ed. V.S. Kennedy), 367–394. Academic Press.

Kemp, W.M. and Murray, L. (1986). Oxygen release from roots of the submersed macrophyte *Potamogeton perfoliatus* L.: regulating factors and ecological implications. *Aquat. Bot.* 26: 271–283.

Kenworthy, W.J., Zieman, J.C., and Thayer, G.W. (1982). Evidence for the influence of seagrasses on the benthic nitrogen cycle in a coastal plain estuary near Beaufort, North Carolina (USA). *Oecologia* 54: 152–158.

Koch, E.W. (1999). Preliminary evidence on the interdependent effect of currents and porewater geochemistry on *Thalassia testudinum* Banks ex König seedlings. *Aquat. Bot.* 63: 95–102.

Koch, E.W. (2001). Beyond light: physical, geological, and geochemical parameters as possible submersed aquatic vegetation habitat requirements. *Estuaries Coasts* 24: 1–17.

Koch, E.W., Barbier, E.B., Kennedy, C.J. et al. (2009). Non-linearity in ecosystem services: temporal and spatial variability in coastal protection. *Front. Ecol.Environ.* 7: 29–37.

Koch, M., Bowes, G., Ross, C., and Zhang, X.-H. (2013). Climate change and ocean acidification effects on seagrasses and marine macroalgae. *Glob. Change Biol.* 19: 103–132.

Koch, M.S., Schopmeyer, S.A., Kyhn-Hansen, C. et al. (2007). Tropical seagrass species tolerance to hypersalinity stress. *Aquat. Bot.* 86: 14–24.

Koweek, D.A., Zimmerman, R.C., Hewett, K.M. et al. (2018). Expected limits on the ocean acidification buffering potential of a temperate seagrass meadow. *Ecol. Appl.* 28: 1694–1714.

Krause-Jensen, D., Pedersen, M.F., and Jensen, C. (2003). Regulation of eelgrass (*Zostera marina*) cover along depth gradients in Danish coastal waters. *Estuaries* 26: 866–877.

Lamb, J.B., Jeroen, A.J.M., Bourne, D., and Altier, C. (2017). Seagrass ecosystems reduce exposure to bacterial pathogens of humans, fishes, and invertebrates. *Science* 355: 731.

Larkum, A.W., Waycott, M., and Conran, J.G. (2018). Evolution and biogeography of seagrasses. In: *Seagrasses of Australia* (ed. A.W.D. Larkum, G.A. Kendrick and P.J. Ralph), 3–29. Springer.

Latimer, J.S. and Rego, S.A. (2010). Empirical relationship between eelgrass extent and predicted watershed-derived nitrogen loading for shallow New England estuaries. *Estuar. Coast. Shelf Sci.* 90: 231–240.

Lavery, P.S., Mateo, M.-Á., Serrano, O., and Rozaimi, M. (2013). Variability in the carbon storage of seagrass habitats and its implications for global estimates of blue carbon ecosystem service. *PLoS One* 8: e73748.

Lee, K.-S. and Dunton, K.H. (2000). Diurnal changes in pore water sulfide concentrations in the seagrass *Thalassia testudinum* beds: the effects of seagrasses on sulfide dynamics. *J. Exp. Mar. Biol. Ecol.* 255: 201–214.

Lee, S.Y., Fong, C.W., and Wu, R.S.S. (2001). The effects of seagrass (*Zostera japonica*) canopy structure on associated fauna: a study using artificial seagrass units and sampling of natural beds. *J. Exp. Mar. Biol. Ecol.* 259: 23–50.

Lefcheck, J.S., Orth, R.J., Dennison, W.C. et al. (2018). Long-term nutrient reductions lead to the unprecedented recovery of a temperate coastal region. *Proc. Natl. Acad. Sci. USA* 115: 3658.

Les, D.H., Cleland, M.A., and Waycott, M. (1997). Phylogenetic studies in alismatidae, II: evolution of marine angiosperms (Seagrasses) and hydrophily. *Syst. Bot.* 22: 443–463.

Licci, S., Nepf, H., Delolme, C. et al. (2019). The role of patch size in ecosystem engineering capacity: a case study of aquatic vegetation. *Aquat. Sci.* 81: 41.

Liu, C. and Nepf, H. (2016). Sediment deposition within and around a finite patch of model vegetation over a range of channel velocity. *Water Resour. Res.* 52: 600–612.

Lubbers, L., Boynton, W.R., and Kemp, W.M. (1990). Variations in structure of estuarine fish communities in relation to abundance of submersed vascular plants. *Mar. Ecol. Prog. Ser.* 65: 1–14.

Madden, C.J. and Kemp, W.M. (1996). Ecosystem model of an estuarine submersed plant community: Calibration and simulation of eutrophication responses. *Estuaries* 19: 457–474.

Madsen, T.V. and Sand-Jensen, K. (1991). Photosynthetic carbon assimilation in aquatic macrophytes. *Aquat. Bot.* 41: 5–40.

Mann, K.H. (1972). Macrophyte production and detritus food chains in coastal water. *Memorie dell'Istituto Italiano di Idrobiologia* 29: 353–383.

Marbà, N. and Duarte, C.M. (1998). Rhizome elongation and seagrass clonal growth. *Mar. Ecol. Prog. Ser.* 174: 269–280.

Markager, S. and Sand-Jensen, K. (1996). Implications of thallus thickness for growth-irradiance relationships of marine macroalgae. *Eur. J. Phycol.* 31: 79–87.

McGlathery, K.J., Sundback, K., and Anderson, I.C. (2007). Eutrophication in shallow coastal bays and lagoons: the role of plants in the coastal filter. *Mar. Ecol. Prog. Ser.* 348: 1–18.

McMahon, K., van Dijk, K., Ruiz-Montoya, L. et al. (2014). The movement ecology of seagrasses. *Proc. R. Soc. B: Biol. Sci.* 281: 20140878.

McRoy, C.P. and McMillan, C. (1977). Production ecology and physiology of seagrasses. In: *Seagrass Ecosystems* (ed. C.P. McRoy and C. Helfferich), 53–87. Marcel Dekker.

Montefalcone, M., Vassallo, P., Gatti, G. et al. (2015). The exergy of a phase shift: ecosystem functioning loss in seagrass meadows of the Mediterranean Sea. *Estuar. Coast. Shelf Sci.* 156: 186–194.

Mony, C., Koschnick, T.J., Haller, W.T., and Muller, S. (2007). Competition between two invasive Hydrocharitaceae (*Hydrilla verticillata* (L.f.) (Royle) and *Egeria densa* (Planch)) as influenced by sediment fertility and season. *Aquat. Bot.* 86: 236–242.

Moore, K.A. (2004). Influence of seagrasses on water quality in shallow regions of the lower Chesapeake Bay. *J. Coast. Res.* 45: 162–178.

Moore, K.A., Shields, E.C., and Parrish, D.B. (2014). Impacts of varying estuarine temperature and light conditions on *Zostera marina* (eelgrass) and its interactions with *Ruppia maritima* (widgeongrass). *Estuaries Coasts* 37: 20–30.

Najjar, R.G., Pyke, C.R., Adams, M.B. et al. (2010). Potential climate-change impacts on the Chesapeake Bay. *Estuar. Coast. Shelf Sci.* 86: 1–20.

Novak, A.B. and Short, F.T. (2010). Leaf reddening in seagrasses. *Bot. Mar.* 53: 93–97.

Oldham, C.E., Lavery, P.S., McMahon, K. et al. (2010). *Seagrass wrack dynamics in Geographe Bay*, 214. Western Australia: Department of Transport.

Olsen, J.L., Rouze, P., Verhelst, B. et al. (2016). The genome of the seagrass *Zostera marina* reveals angiosperm adaptation to the sea. *Nature* 530: 331.

Orth, R.J., Batiuk, R.A., Bergstrom, P.W., and Moore, K.A. (2002). A perspective on two decades of policies and regulations influencing the protection and restoration of submerged aquatic vegetation in Chesapeake Bay, USA. *Bull. Mar. Sci.* 71: 1391–1403.

Orth, R.J., Carruthers, T.J.B., Dennison, W.C. et al. (2006a). A global crisis for seagrass ecosystems. *BioScience* 56: 987–996.

Orth, R.J., Lefcheck, J.S., Moore, K.A. et al. (2017). Submersed aquatic vegetation in Chesapeake Bay: sentinel species in a changing world. *BioScience* 67: 698–712.

Orth, R.J., Luckenbach, M.L., Marion, S.R. et al. (2006b). Seagrass recovery in the Delmarva Coastal Bays, USA. *Aquat. Bot.* 84: 26–36.

Orth, R.J., Marion, S.R., Moore, K.A., and Wilcox, D.J. (2010a). Eelgrass (*Zostera marina* L.) in the Chesapeake Bay region of mid-Atlantic coast of the USA: Challenges in conservation and restoration. *Estuaries Coasts* 33: 139–150.

Orth, R.J., Williams, M.R., Marion, S.R. et al. (2010b). Long-term trends in submersed aquatic vegetation (SAV) in Chesapeake Bay, USA, related to water quality. *Estuaries Coasts* 33: 1144–1163.

Ostenfeld, C.H. (1908). On the ecology and distribution of the grass wrack (*Zostera marina* L.). In: *Danish Waters* (ed. C.H. Ostenfeld and C.G.J. Petersen), 1–62. Danish Biological Station.

Ottosen, L.D.M., Risgaard-Petersen, N., and Nielsen, L.P. (1999). Direct and indirect measurements of nitrification and denitrification in the rhizosphere of aquatic macrophytes. *Aquat. Microb. Ecol.* 19: 81–91.

Palmer, M.R., Nepf, H.M., and Pettersson, T.J.R. (2004). Observations of particle capture on a cylindrical collector: Implications for particle accumulation and removal in aquatic systems. *Limnol. Oceanogr.* 49: 76–85.

Paul, M., Lefebvre, A., Manca, E., and Amos, C.L. (2011). An acoustic method for the remote measurement of seagrass metrics. *Estuar. Coast. Shelf Sci.* 93: 68–79.

Pedersen, O., Binzer, T., and Borum, J. (2004). Sulphide intrusion in eelgrass (*Zostera marina* L.). *Plant Cell Environ.* 27: 595–602.

Pedersen, O., Borum, J., Duarte, C.M., and Fortes, M.D. (1998). Oxygen dynamics in the rhizosphere of *Cymodocea rotundata*. *Mar. Ecol. Prog. Ser.* 169: 283–288.

Pedersen, O. and Sand-Jensen, K. (1993). Water transport in submerged macrophytes. *Aquat. Bot.* 44: 385–406.

Peralta, G., Brun, F.G., Hernandez, I. et al. (2005). Morphometric variations as acclimation mechanisms in *Zostera noltii* beds. *Estuar. Coast. Shelf Sci.* 64: 347–356.

Petersen, C.G.J. (1918). *The Sea Bottom and its Production of Fish Food: A Survey of the Work Done in Connection with Valuation of Danish Waters from 1883-1917*, 1–82. Centraltrykkeriet Reports from the Danish Biological Station.

Peterson, B.J. and Heck, K.L. (2001). Positive interactions between suspension-feeding bivalves and seagrass - a facultative mutualism. *Mar. Ecol. Prog. Ser.* 213: 143–155.

Plus, M., Deslous-Paoli, J.M., and Dagault, F. (2003). Seagrass (*Zostera marina* L.) bed recolonisation after anoxia-induced full mortality. *Aquat. Bot.* 77: 121–134.

Ralph, P.J., Durako, M.J., Enriquez, S. et al. (2007). Impact of light limitation on seagrasses. *J. Exp. Mar. Biol. Ecol.* 350: 176–193.

Rasmussen, E. (1973). Systematics and ecology of the Isefjord marine fauna (Denmark) with a survey of the eelgrass (*Zostera*) vegetation and its communities. *Ophelia* 11: 1–495.

Raven, J.A. and Scrimgeour, C.M. (1997). The influence of anoxia on plants of saline habitats with special reference to the sulphur cycle. *Ann.Bot.* 79: 79–86.

Reise, K. and Kohlus, J. (2008). Seagrass recovery in the Northern Wadden Sea? *Helgol.Mar.Res.* 62: 77–84.

Reynolds, L.K., McGlathery, K.J., and Waycott, M. (2012). Genetic diversity enhances restoration success by augmenting ecosystem services. *PLoS One* 7: e38397.

Riis, T., Madsen, T.V., and Sennels, R.S.H. (2009). Regeneration, colonisation and growth rates of allofragments in four common stream plants. *Aquat. Bot.* 90: 209–212.

Risgaard-Petersen, N. and Ottosen, L.D.M. (2000). Nitrogen cycling in two temperate *Zostera marina* beds: seasonal variation. *Mar. Ecol. Prog. Ser.* 198: 93–107.

Robblee, M.B., Barber, T.R., Carlson, P.R. et al. (1991). Mass mortality of the tropical seagrass *Thalassia testudinum* in Florida Bay (USA). *Mar. Ecol. Prog. Ser.* 71: 297–299.

Rocha, J.C., Peterson, G.D., and Biggs, R. (2015). Regime shifts in the Anthropocene: drivers, risks, and resilience. *PLoS One* 10: e0134639.

Rybicki, N. and Landwehr, J.M. (2007). Long-term changes in abundance and diversity of macrophyte and waterfowl populations in an estuary with exotic macrophytes and improving water quality. *Limnol. Oceanogr.* 52: 1195–1207.

Rybicki, N.B., Jenter, H.L., Carter, V. et al. (1997). Observations of tidal flux between a submersed aquatic plant stand and the adjacent channel in the Potomac River near Washington, D.C. *Limnol. Oceanogr.* 42: 307–317.

Rysgaard, S., Risgaard-Petersen, N., and Sloth, N.P. (1996). Nitrification, denitrification, and nitrate ammonification in sediments of two coastal lagoons in Southern France. *Hydrobiologia* 329: 133–141.

Santos, I.R., Eyre, B.D., and Huettel, M. (2012). The driving forces of porewater and groundwater flow in permeable coastal sediments: a review. *Estuar. Coast. Shelf Sci.* 98: 1–15.

Scheffer, M., Carpenter, S., Foley, J.A. et al. (2001). Catastrophic shifts in ecosystems. *Nature* 413: 591–596.

Shields, E.C., Moore, K.A., and Parrish, D.B. (2012). Influences of salinity and light availability on abundance and distribution of tidal freshwater and oligohaline submersed aquatic vegetation. *Estuaries Coasts* 35: 515–526.

Short, F., Carruthers, T., Dennison, W., and Waycott, M. (2007). Global seagrass distribution and diversity: A bioregional model. *J. Exp. Mar. Biol. Ecol.* 350: 3–20.

Short, F.T., Dennison, W.C., and Capone, D.G. (1990). Phosphorus-limited growth of the tropical seagrass *Syringodium filiforme* in carbonate sediments. *Mar. Ecol. Prog. Ser.* 62: 169–174.

Short, F.T. and Neckles, H.A. (1999). The effects of global climate change on seagrasses. *Aquat. Bot.* 63: 169–196.

Smith, R.D., Dennison, W.C., and Alberte, R.S. (1984). Role of seagrass photosynthesis in root aerobic processes. *Plant Physiol.* 74: 1055–1058.

Soissons, L.M., Van Katwijk, M.M., Peralta, G. et al. (2018). Seasonal and latitudinal variation in seagrass mechanical traits across Europe: the influence of local nutrient status and morphometric plasticity. *Limnol. Oceanogr.* 63: 37–46.

Staehr, P.A. and Borum, J. (2011). Seasonal acclimation in metabolism reduces light requirements of eelgrass (*Zostera marina*). *J. Exp. Mar. Biol. Ecol.* 407: 139–146.

Sullivan, B.K., Sherman, T.D., Damare, V.S. et al. (2013). Potential roles of *Labyrinthula* spp. in global seagrass population declines. *Fungal Ecol.* 6: 328–338.

Thayer, G.W., Bjorndal, K.A., Ogden, J.C. et al. (1984). Role of larger herbivores in seagrass communities. *Estuaries* 7: 351–376.

Thomas, F.I.M. and Cornelisen, C.D. (2003). Ammonium uptake by seagrass communities: effects of oscillatory versus unidirectional flow. *Mar. Ecol. Prog. Ser.* 247: 51–57.

Thomsen, M.S., Wernberg, T., Engelen, A.H. et al. (2012). A meta-analysis of seaweed impacts on seagrasses: generalities and knowledge gaps. *PLoS One* 7: e28595.

Touchette, B.W. (2007). Seagrass-salinity interactions: physiological mechanisms used by submersed marine angiosperms for a life at sea. *J. Exp. Mar. Biol. Ecol.* 350: 194–215.

Touchette, B.W. and Burkholder, J.M. (2000). Review of nitrogen and phosphorus metabolism in seagrasses. *J. Exp. Mar. Biol. Ecol.* 250: 133–167.

Unsworth, R.K.F., Collier, C.J., Waycott, M. et al. (2015). A framework for the resilience of seagrass ecosystems. *Mar. Pollut. Bull.* 100: 34–46.

Unsworth, R.K.F., Nordlund, L.M., and Cullen-Unsworth, L.C. (2019). Seagrass meadows support global fisheries production. *Conserv. Lett.* 12: e12566.

Valentine, J.F. and Heck, K.L. (1999). Seagrass herbivory: evidence for the continued grazing of marine grasses. *Mar. Ecol. Prog. Ser.* 176: 291–302.

van der Heide, T., Govers, L.L., de Fouw, J. et al. (2012). A three-stage symbiosis forms the foundation of seagrass ecosystems. *Science* 336: 1432–1434.

van der Heide, T., van Nes, E.H., Geerling, G.W. et al. (2007). Positive feedbacks in seagrass ecosystems: implications for success in conservation and restoration. *Ecosystems* 10: 1311–1322.

van Katwijk, M.M., Bos, A.R., de Jonge, V.N. et al. (2009). Guidelines for seagrass restoration: Importance of habitat selection and donor population, spreading of risks, and ecosystem engineering effects. *Mar. Pollut. Bull.* 58: 179–188.

van Katwijk, M.M., Thorhaug, A., Marbà, N. et al. (2016). Global analysis of seagrass restoration: the importance of large-scale planting. *J. Appl Ecol.* 53: 567–578.

van Tussenbroek, B.I., van Katwijk, M.M., Bouma, T.J. et al. (2016a). Non-native seagrass *Halophila stipulacea* forms dense mats under eutrophic conditions in the Caribbean. *J. Sea Res.* 115: 1–5.

van Tussenbroek, B.I., Villamil, N., Márquez-Guzmán, J. et al. (2016b). Experimental evidence of pollination in marine flowers by invertebrate fauna. *Nat. Commun.* 7: 12980.

von Liebig, J.F. (1863). *The Natural Laws of Husbandry*. London: Walton & Maberly.

Vonk, J.A., Christianen, M.J.A., and Stapel, J. (2008). Redefining the trophic importance of seagrasses for fauna in tropical Indo-Pacific meadows. *Estuar. Coast. Shelf Sci.* 79: 653–660.

Walker, B., Holling, C.S., Carpenter, S.R., and Kinzig, A. (2004). Resilience, adaptability and transformability in social–ecological systems. *Ecol. Soc.* 9: 5.

Walker, D.I. and McComb, A.J. (1992). Seagrass degradation in Australian coastal waters. *Mar. Pollut. Bull.* 25: 191–195.

Waycott, M., Calladine, A., Duarte, C.M. et al. (2009). Accelerating loss of seagrasses across the globe threatens coastal ecosystems. *Proc. Natl. Acad. Sci. U. S. A.* 106: 12377–12381.

Wressnig, A. and Booth, D.J. (2007). Feeding preferences of two seagrass grazing monacanthid fishes. *J. Fish Biol.* 71: 272–278.

York, P.H., Gruber, R.K., Hill, R. et al. (2013). Physiological and morphological responses of the temperate seagrass *Zostera muelleri* to multiple stressors: investigating the interactive effects of light and temperature. *PLoS One* 8: e76377.

Zieman, J.C., Fourqurean, J.W., and Frankovich, T.A. (1999). Seagrass die-off in Florida Bay: long-term trends in abundance and growth of turtle grass, *Thalassia testudinum*. *Estuaries* 22: 460–470.

Zieman, J.C., Fourqurean, J.W., and Iverson, R.L. (1989). Distribution, abundance and productivity of seagrasses and macroalgae in Florida Bay. *Bull. Mar. Sci.* 44: 292–311.

Zieman, J.C. and Wetzel, R.G. (1980). Productivity in seagrasses: methods and rates. In: *Handbook of Seagrass Biology: An Ecosystem Perspective* (ed. R.C. Phillip and C.P. McRoy), 87–116. New York: *Garland Press*.

CHAPTER **6**

Coastal Marshes

Carles Ibáñez[1], Maite Martínez-Eixarch[2], Irving A. Mendelssohn[3], James T. Morris[4], Stefanie Nolte[5,6], and John W. Day[7]
[1]Department of Climate Change, Eurecat, Technological Centre of Catalonia, Amposta, Catalonia, Spain
[2]Marine and Continental Waters, IRTA, Institute of Agrifood Research and Technology, Sant Carles de la Ràpita, Spain
[3]Department of Oceanography and Coastal Sciences, College of the Coast and Environment, Louisiana State University, Baton Rouge, LA USA
[4]University of South Carolina, Belle W. Baruch Institute for Marine and Coastal Sciences, Columbia, SC USA
[5]School of Environmental Sciences, University of East Anglia, Norwich, UK
[6]Centre for Environment, Fisheries and Aquaculture Science, Lowestoft, UK
[7]Department of Oceanography and Coastal Sciences, College of the Coast and Environment, Louisiana State University, Baton Rouge, LA USA

Four images of salt marshes on the Ebro River delta, Spain.
Photo credit: Carles Ibáñez

6.1 Introduction

Coastal marshes with emergent vegetation are a common feature of most estuaries, deltas, and coastal plains all over the world. Salt marshes are the most abundant wetland type, but brackish and tidal freshwater marshes are also present in many estuarine systems (Day et al. 1989, 2014). Coastal marshes are beds of intertidal rooted vegetation that are alternately inundated and drained by the tides and non-tidal changes in sea level. The interaction between open water and marsh is mediated by a complex network of channels that facilitate the passage of water, suspended and dissolved material, and organisms. Grasses are normally the dominant primary producer, although sometimes low, shrubby vegetation replaces grasses. In addition, many types of algae grow on the plants and on the surface of the mud, as well as submerged macrophytes.

Marshes occur principally along temperate and boreal coasts (Figure 6.1), but often form in the tropics on salt flats not occupied by mangroves. The mangrove community occurs along tropical shores and is composed mainly of trees. The two communities are ecologically analogous because of their physical location, ecological processes, and trophic contribution to the overall estuarine ecosystem. Mangroves are the subject of Chapter 7, and here we focus on marshes.

Numerous studies show that coastal marshes are among the most productive plant communities in the world. They also carry out important functions and provide valuable ecosystem services, from which humankind benefits, such as climate change mitigation through carbon sequestration. In this chapter, we discuss the structure, composition, distribution, productivity, functions, values, assessment, and human impacts of coastal marshes and the factors that affect composition and productivity.

6.2 Diversity, Zonation, and Global Distribution

6.2.1 General Features and Typology of Coastal Marshes

Recent estimates of salt marsh distribution show a total area of circa five million hectares, but previous estimates range between 2.2 and 40 million hectares (Mcowen et al. 2017).

Estuarine Ecology, Third Edition. Edited by Byron C. Crump, Jeremy M. Testa, and Kenneth H. Dunton.

Companion website: www.wiley.com/go/crump/estuarine3

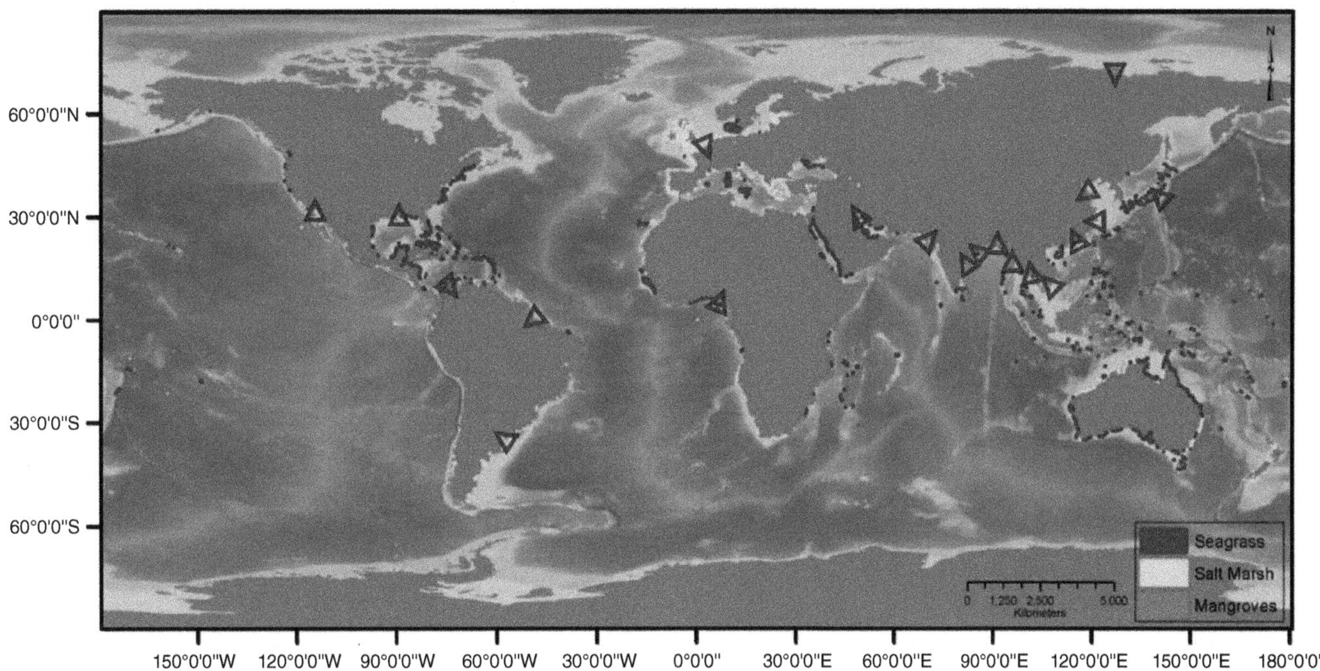

FIGURE 6.1 Worldwide distribution of coastal marshes, seagrasses, mangrove swamps, and major deltas (red triangles). *Source*: Moffett et al. (2015)/Public Domain CC BY 4.0.

In an attempt to explain the variability of salt marshes, Chapman (1960) divided marshes into nine broad groups based on community structure, floristic composition, and geographic distribution (Table 6.1). He also listed over six hundred species of plants that grow in salt marshes throughout the world. Species diversity is lowest in the Arctic and increases toward lower latitudes. A number of genera are broadly distributed and are major components of the flora in many areas (Figure 6.2). *Puccinellia* is common along northern coasts where ice and extremely low temperatures occur in winter. *Spartina, Juncus*, and *Distichlis* are common along non-rocky temperate coasts of Europe, Asia, and North and South America. *Limonium* is found along rugged coasts. *Salicornia* and related species have the broadest distribution; they are found in microtidal marshes and in the upper intertidal zone in practically all areas because of their ability to withstand desiccation and salinity stress.

The most extensive development of coastal marshes occurs in areas of low relief, abundant rainfall, muddy to sandy coasts, and moderate climate. In the United States, the broadest development occurs along the Northeast coast and the Gulf of Mexico. The largest single area of coastal marshes in the United States is in the delta of the Mississippi River in Louisiana, where about 37% of coastal herbaceous marshes ($\sim 2 \times 10^6$ ha) in the contiguous United States occur (NOAA 2010). Besides the Mississippi, extensive marshes are found in deltas worldwide, such as those of the Rhône (France), Nile (Egypt), Magdalena (Colombia), Niger (Nigeria), and Huang He Rivers (China). Coastal marshes are also abundant in estuaries and other coastal systems (i.e., coastal lagoons).

To understand the factors affecting the distribution and features of marsh types, it is important to contrast the differences between tidal (~1 m tidal range) and microtidal (typically <0.5 m tidal range) systems (Table 6.2). Tidal systems are widespread and abundant on most oceanic coasts, whereas microtidal systems are restricted to semi-enclosed seas, such as the Mediterranean, Black and Baltic Seas, and the Gulf of Mexico. A major portion of microtidal coastal marshes are located in deltaic areas. Tidal marshes often occur in coastal plain estuaries and along the shores of bays, where there is a mixture of tidal activity and riverine influence, leading to close coupling between wetlands and adjacent water bodies. In microtidal coasts, marshes are often not directly connected to estuaries but are frequently associated with coastal lagoons and bays. Tidal marshes occupy large areas often associated with coastal plain estuaries and are characterized by strong and regular tidal fluxes, which play a crucial role in ecosystem functioning. Microtidal estuaries are often salt-wedge estuaries that are part of a delta (Ibáñez et al. 1997a), and the role of the weak tide in the ecosystem is less pronounced. Sea level changes due to seasonal cycles and storms likely play a more important role than astronomical tides in microtidal marshes (Hiatt et al. 2019).

Microtidal deltas are often bounded by river levees and under natural conditions, are vegetated by riparian forests, which are flooded only during high discharge. In contrast, tidal marshes are located along the shores of estuarine systems, which often have a typical valley-shaped topography. In both cases, fresh, brackish, or saline marshes often occur, depending on factors such as elevation, inputs of upland runoff, riverine and tidal influence, and soil drainage. In most cases, there is a clear vegetation zonation mainly related to soil salinity and flooding regime, which is discussed in the following section.

TABLE 6.1 **Worldwide classification of salt marshes based on geographic distribution, floristic composition, and community structure.**

Group	Geographic distribution	Dominant plants	Important controlling parameters
Arctic	Greenland, Iceland, and *Arctic* coasts of North America, Scandinavia, and Russia, Iberian peninsula north to Norway, and southwestern Sweden	*Puccinellia phryganodes* *Carex* sp.	Ice, extreme low temperature
Northern Europe	West coast of Great Britain, Scandinavia Coasts of Baltic Sea Along English Channel and other muddy coasts	*Puccinelia maritime, Juncus gerardii, Salicornia* spp., *Spartina* spp., *Festuca rubra, Agrostis stolonifera, Carex palecea, Juncus bufonius, Desmoschoenus bottnica, Scirpus* sp., *Spartina townsendii*	Sufficient precipitation Sandy to muddy substrate Moderate climate Coast with a high proportion of sand Low to moderate salinity Muddy coasts
Mediterranean	Along Mediterranean coasts	Much low shrubby vegetation, *Arthrocnemum Limonium*. Also, *Juncus* and *Salicornia* spp.	Arid to semiarid, rocky to sandy coasts Generally high salinity Moderate temperatures
Eastern North America	Temperate Atlantic and Gulf of Mexico coasts of North America	*Spartina alterniflora, Spartina patens, Juncus roemerianus, Distichlis spicata, Salicornia* spp., *Puccinellia marituma* at higher latitudes	Generally muddy coasts and moderate climate Broad development of marshes
Western North America	Temperate Pacific coasts of North America	*Spartina gracilis, Spartina foliosa, Frankenia* spp., *Salicornia* spp.	Arid in south to high precipitation in north Rugged coasts Limited marsh development
Sino-Japanese	Temperate Pacific coasts of China, Japan, Russia, and Korea	*Triglochin martima, Limonium japonicum*; also, *Zoysia macrostachya, Salcornia* spp.	Rugged, uplifting coasts Limited marsh development Moderate precipitation
South America	South America coasts too cold for mangroves	Unique species of *Spartina, Limonium, Distichlis, Juncus, Heterostachys,* and *Allenrolfea*; also *Salicornia* spp.	Rugged coasts Geographic isolation
Australia	South Australia, New Zealand, Tasmania	*Hemichroa* spp., *Arthrocnemetum* spp., *Salicornia* spp.	Rainfall Geographic isolation
Tropical	Saline flats not occupied by mangroves	*Salicornia* spp., *Limonium* spp., *Spartina brasiliensis*	High salinity

Source: Based on Chapman (1960). Note scientific names are those used by Chapman.

6.2.2 Distribution and Zonation of Coastal Marshes

Distribution and zonation patterns occur at several different spatial scales. At the broadest level, there are latitudinal patterns where climate plays a major role in marsh distribution. At an intermediate regional scale, there are drainage basin patterns, where water salinity and coastal morphology are important in determining zonation. Finally, local patterns occur along and across estuaries because of elevation changes and variation in tidal water exchanges as one moves closer to, or further from, a tidal creek. In addition, some worldwide distribution patterns occur as a result of geographic isolation. For example, Australian and South American salt marshes host unique vegetation assemblages. Also, in all wetlands, there is local "patchiness" caused by adjustment of plants to various types of small-scale heterogeneity. Bertness and Pennings (2002) suggested that our current understanding of marsh zonation patterns is oversimplified and that the processes creating these patterns may vary in importance between marshes. We now consider distribution and zonation at these spatial scales (latitudinal, drainage basin, and local) in more detail.

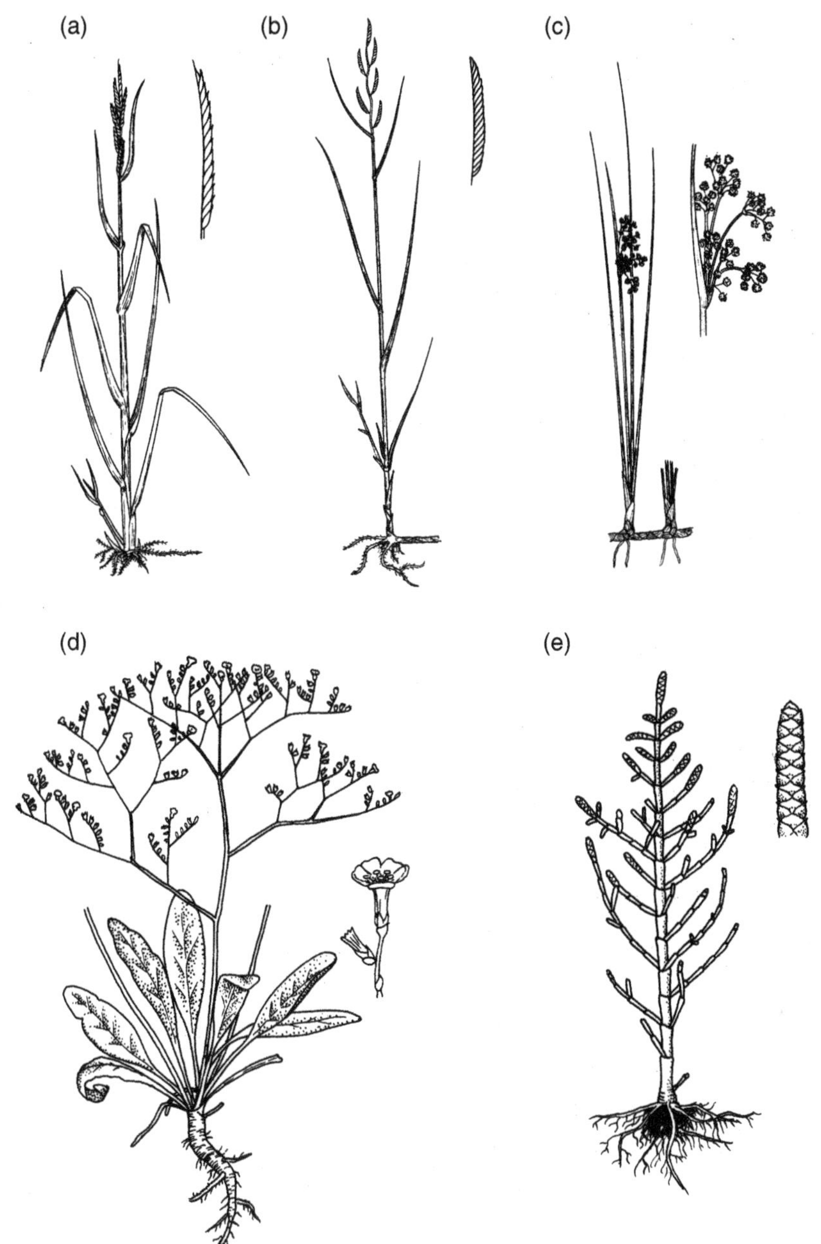

FIGURE 6.2 Several common species of salt marsh plants. (a) *Spartina alterniflora*, (b) *Spartina patens*, (c) *Juncus roemarinus*, (d) *Limonium carolinianum*, and (e) *Salicornia bigelovii*.

6.2.2.1 Latitudinal Patterns

On a broad latitudinal scale, distribution of coastal marshes in the northern hemisphere are affected primarily by climate (Day et al. 1989). Marshes are present along north temperate coasts from about 30° N up to 65° N latitude. Above this latitude, ice and extremes of temperature prevent marsh formation. Below about 30° N mangroves are present. Climate may also have important effects on marsh plant community organization. At lower latitudes, hotter climates lead to salt accumulation, elevated soil salinities, and marsh zonation patterns that are strongly driven by soil salinity (Bertness and Pennings 2002). In contrast, cool temperate marshes can have relatively low soil salinities, which limits the importance of salinity in maintaining marsh plant distributional patterns.

Rainfall and temperature also influence latitudinal patterns in marshes. Marshes that experience strong dry seasons may feature large barren flats within emergent wetlands, where plants do not have enough water or are too saline to maintain growth. Under conditions of extreme or prolonged dryness, there may be little or no vegetation in the intertidal zone. This is illustrated by patterns of wetland distribution around the Gulf of Mexico, which extends from the tropics in the south to the temperate zone in the north and includes a large variation in temperature and rainfall (Figure 6.3). These temperature and rainfall gradients affect the distribution of coastal vegetation. Mangrove distribution is mostly affected by temperature and in the United States they occur along the tropical and semitropical coasts south of Cedar Keys in Florida and generally south of

TABLE 6.2 Some differences between tidal temperate and microtidal temperate estuarine systems.

Tidal estuarine systems	Microtidal estuarine systems
Tidal range 1–10 m	Tidal range 0.1–1 m
Astronomical tides > meteorological tides	Astronomical tides < meteorological tides
Tide-dominated estuaries	River-dominated estuaries
Coastal plain estuaries	Deltas, coastal lagoons
Partially and well mixed estuaries	Salt-wedge estuaries
Extensive tidal marshes	Small and localized microtidal marshes
Presence of tidal freshwater marshes	Microtidal freshwater marshes rare
Marshes located along the shore of the estuary	Marshes located along coastal lagoons and outer coast
Marshes with regular flooding	Marshes with summer drought and salt stress; irregular flooding.
High biological productivity	Medium to low biological productivity
Maximum productivity usually from high to middle marsh	Maximum productivity usually from low to middle marsh
Marsh diversity gradient along the estuary	Marsh diversity gradient across the river delta
Plant zonation mainly due to tidal flooding	Plant zonation mainly due to maximum summer salinity
Higher topographic gradient, lower habitat patchiness, tidal creeks	Lower topographic gradient, higher habitat patchiness, no tidal creeks

Port Isabel, Texas. Except for isolated stands of black mangrove in the Mississippi deltaic plain, mangroves are absent from the northern Gulf Coast, which is dominated by salt marshes. This is changing, however, due to recently warmer temperatures because of climate change (Osland et al. 2014, see Chapter 18). For instance, Armitage et al. (2015) noted that marsh loss in Texas between 1990 and 2010 was 24%, compared to a 74% increase in mangroves, with 6% of that marsh loss related to mangrove expansion.

6.2.2.2 Drainage Basin Patterns

The composition and diversity of plants in coastal marshes often form distinct vegetative zones that are structured by salinity and tides (Zedler 1977; Bertness 1992; Pennings and Callaway 1992). In most salt marsh estuaries, if we travel landward from the coast there is a progressively diminishing tidal influence until, at some point, there is no longer a tide. Within this area affected by the tide, there is also a salinity gradient. Nearby the coast, marsh vegetation is dominated by salt marshes. The composition of this community changes with decreasing salinity until it is composed entirely of freshwater vegetation. In the Barataria Basin, Louisiana, there are several broad vegetation zones (Day et al. 1989). Nearby the coast there is a band of saline marsh, with zones of brackish marsh, freshwater marsh, and swamp forest progressively further inland. The plant diversity is lowest in the saline marshes and highest in the freshwater marshes; a pattern that holds for most estuaries. Similar patterns exist for the marshes of the Chesapeake Bay and other areas. However, a study in Chesapeake Bay (Sharpe and Baldwin 2009) found that species richness in the oligohaline regions of brackish marshes was as high or higher than that in tidal freshwater marshes, resulting in a distinctly nonlinear pattern of plant species richness along the relatively undisturbed Nanticoke River. In contrast, this peak in species richness was not detected in oligohaline regions of the more urbanized Patuxent River estuary, suggesting that the nonlinear pattern of plant

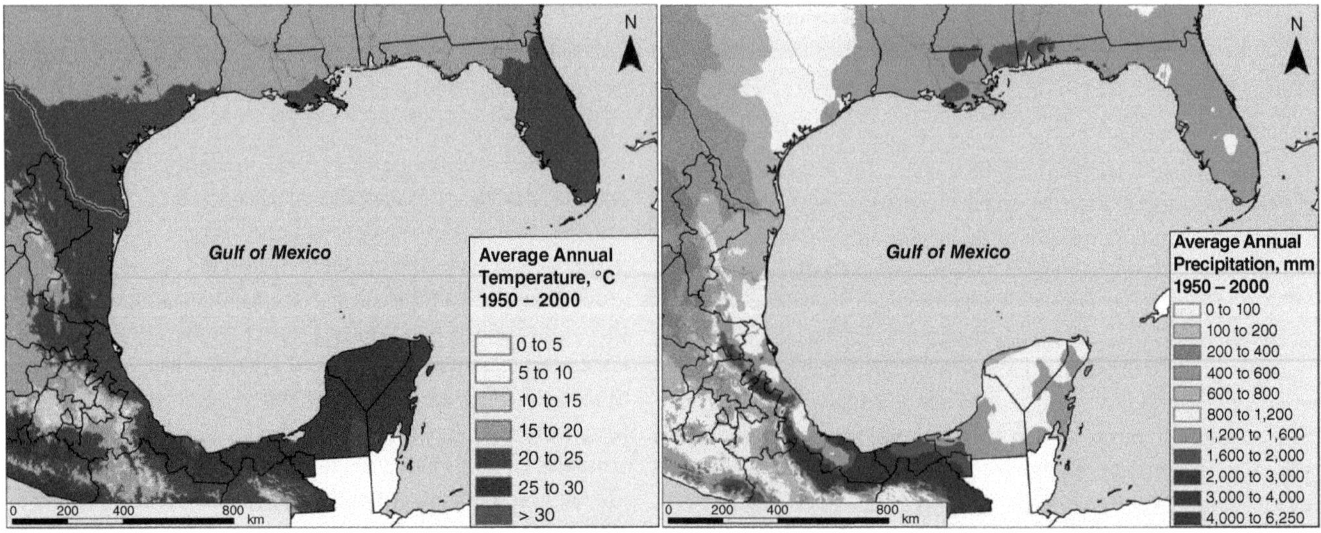

FIGURE 6.3 Average annual temperature and precipitation for terrestrial areas adjacent to the Gulf of Mexico. *Source:* Mendelssohn et al. (2017)/Springer Nature.

species richness observed along the Nanticoke River may be typical of relatively undisturbed estuaries.

6.2.2.3 Local Zonation Patterns

The striking patterns of marsh zonation (Figure 6.4) that occur along tidal creeks have attracted the attention of wetland scientists for many years (e.g., Mendelssohn and Morris 2000 and references therein). These zones are a result of differences in reproduction and growth in response to environmental gradients encountered from low to high water and plant competition. Some of the most important factors that cause this zonation are elevation, drainage, and soil type. Tidal exchange and soil type are important in determining the oxidation–reduction state of the soil (see Chapter 3) and the degree of hypersalinity in high marsh areas. In salt marshes, there is typically a distinct elevation gradient from the water's edge to the upland boundary and often a streamside levee is present along the tidal creek or bay shore. In high marsh areas, salt pans often form that prohibit establishment of vegetation because of anoxic soils and hypersaline conditions (Figure 6.4).

From the water's edge, a number of factors affect the distribution of vascular plant species along an elevational gradient, including salinity, frequency and duration of tidal inundation, and substrate composition (Odum 1988; Mendelssohn and Morris 2000; Figure 6.5). On tidal coasts, low marshes are flooded daily by seawater, producing a soil salinity regime independent of local climatic conditions. For this reason, the plants that grow in tidal marshes are similar across many climate regions including relatively dry Mediterranean-type climate coasts (e.g., California), and relatively wet coasts in the United States and Europe. In these tidal systems, low marsh communities are composed mainly of *Spartina* spp., and middle and upper marsh areas are dominated by halophytes, often succulent chenopods, adapted to high soil salinity. The main difference in tidal marsh vegetation patterns between climate regions is the widespread formation of salt flats where perennial vegetation is absent in systems that experience extended periods of hypersalinity.

On tidal coasts, the factors responsible for marsh plant zonation are essentially similar to those causing zonation in rocky intertidal habitats. In low marsh areas, the range of a species is limited by its tolerance to physical and chemical conditions including the frequency of submergence and the degree of hypoxia in marsh soils. In high marsh areas, however, species can be excluded by competition (Bertness and Pennings 2002). This gradient in factors creates a gradient in species diversity across tidal salt marshes. Few species are adapted to the physical and chemical conditions typical of low marsh areas, and thus species richness is low. Many more species are adapted to high marsh conditions, and thus the diversity is decided by competition. Mediterranean-type tidal salt marshes, however, do not exhibit a simple monotonic gradient in the severity of physical factors across marsh elevations; rather there is an interaction between flooding and salinity that creates a band of more optimal habitat in middle marsh areas, where both factors are moderate, a phenomenon not reported elsewhere (Pennings and Callaway 1992).

In temperate tidal salt marshes, vascular plants are typically found only in the upper two-thirds of the intertidal zone. The lower one-third consists of bare mud and, at times, a layer

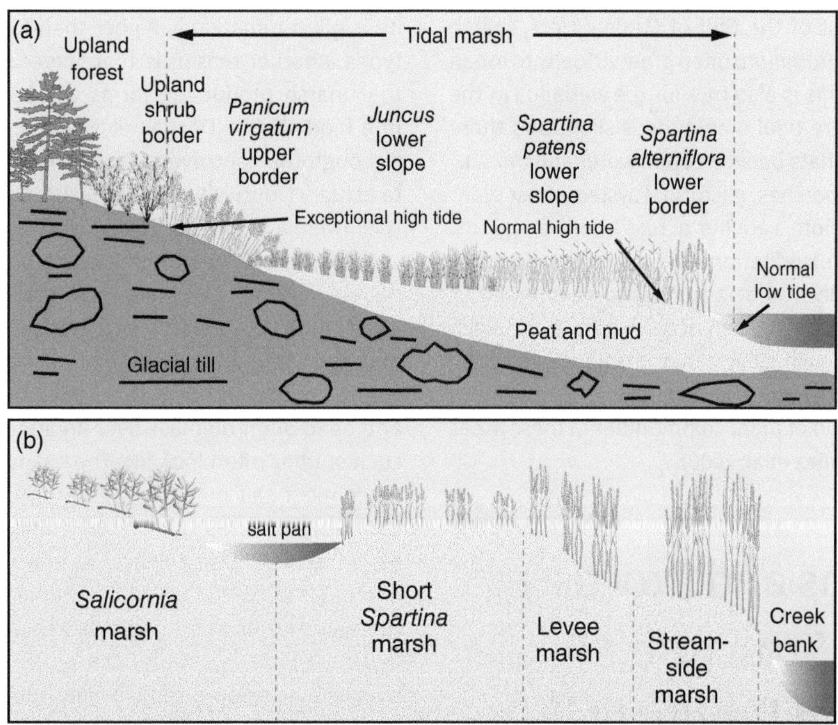

FIGURE 6.4 Zonation of vegetation in southern New England salt marshes: (a) typical zonation from a bay shore to an upland forest; (b) typical zonation from the tidal creek to the higher marsh. Note the salt pan on the boundary between the *Salicornia* and Short *Spartina* marsh assemblages. *Source*: Based on Bertness and Pennings (2002).

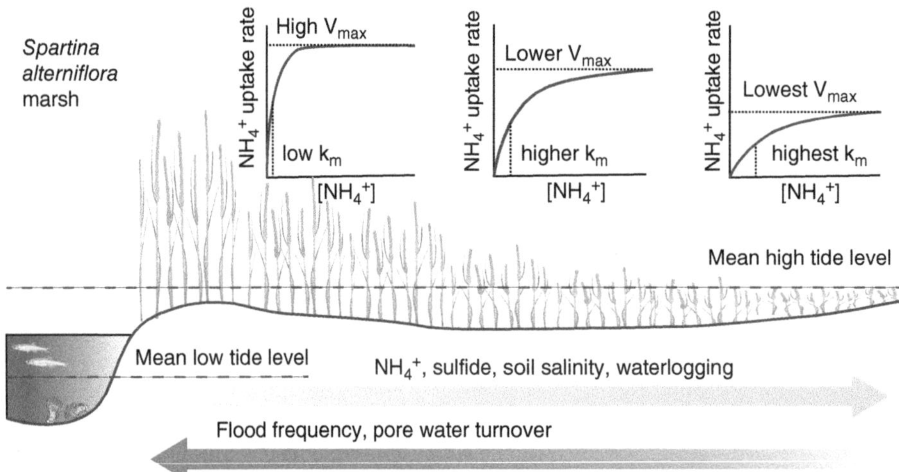

FIGURE 6.5 The tall to short *Spartina alterniflora* transition is a complex gradient characterized by increases in concentrations of interstitial ammonium, soluble sulfide, and, in the high marsh, salt. In contrast, inundation frequency and porewater turnover decrease along this gradient. In the tall (streamside) *Spartina* zone, relatively oxidized soil conditions and low sulfide concentrations promote aerobic root respiration. Under these conditions biomass production is greatest, and root ammonium uptake rate is high (high V_{max}) and efficient (low K_m). At intermediate sulfide levels, where anaerobic root metabolism is stimulated, ammonium uptake kinetics are only marginally affected. However, in the short (inland) *Spartina* zone, where sulfide concentrations are extreme, ammonium uptake rate and efficiency are severely reduced, resulting in low biomass production. See Mendelssohn and Morris (2000) for further description of mechanisms. *Source*: Based on Mendelssohn and Morris (2000).

of micro- and macroalgae. This lack of colonization is primarily a result of the high duration of flooding. Exceptions to this general pattern occur where the tidal amplitude is very slight, as along the northern coast of the Gulf of Mexico. Here, marsh plants such as *Spartina alterniflora* often grow virtually to mean low tide (Odum 1988). This is also true for the wetlands in the Mediterranean Sea, where tidal amplitude is slight and there are almost no tidal mud flats between open water and marsh.

In contrast to tidal marshes, microtidal systems host plant communities that are more variable across climate regions. In microtidal systems on Mediterranean-type coasts, *Spartina* spp. grasses are practically absent, and glasswort communities (dominated by *Sarcocornia* spp.) can develop all the way from the lower to the upper marsh. Several authors have shown the importance of summer drought and hypersaline conditions in explaining the distribution of plant communities in these areas (Callaway et al. 1990; Ibáñez et al. 2000).

6.3 Patterns and Processes Controlling Structure, Biomass, and Productivity

The factors affecting the productivity of salt marshes have been studied extensively. Most of the existing studies are for *S. alterniflora*, although there is considerable information available for other species (Morris et al. 1990; Ibáñez et al. 2000; Spalding and Hester 2007). One reason that the productivity of salt marshes has been examined so thoroughly is that productivity is often remarkably higher than for most other ecosystem types. Another reason is that there is considerable evidence that marsh production forms the basis of important estuarine food chains (Deegan et al. 2000; Winemiller et al. 2007), although the controversy about nutrient subsidy from marsh to estuary ("outwelling hypothesis"; Odum 1968) remains open (Weinstein and Kreeger 2000).

There have been numerous studies of net primary production (NPP) in salt marshes, but temporal and spatial variability limits our ability to generalize these rates across space and time (Odum 1988). Most of the existing productivity values are for net production calculated from changes in live and dead plant biomass over an annual cycle. The data from such studies often look like the examples given in Figure 6.6. Since most salt marshes are in temperate climates, the live biomass increases during the growing season (spring until fall), then flowers and dies. As the grass dies, live biomass decreases and dead organic material increases. In the spring, the quantity of dead grass decreases as it decomposes. In lower latitude marshes, such as in Louisiana and Georgia, USA, there is some growth year-round and live material is always present. In Louisiana, the live biomass of *S. patens* or *Juncus roemerianus* does not change in any regular manner. In addition to these seasonal differences in marsh grass biomass, all marshes display spatial variability in live and dead biomass between streamside and inland marsh locations,

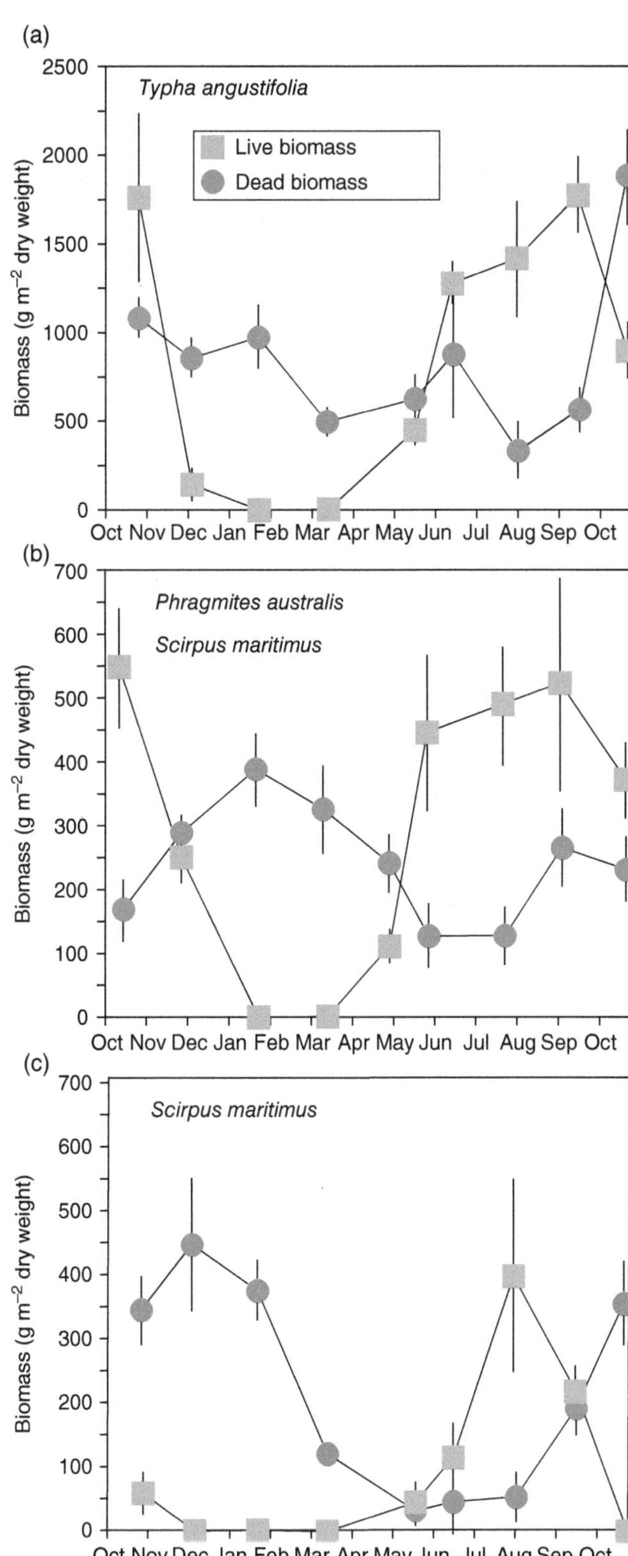

FIGURE 6.6 Seasonal patterns of aboveground live (teal squares) and dead biomass (red dots) from marshes dominated by (a) *Typha angustifolia*, (b) *Phragmites australis* and *Scirpus maritimus*, and (c) *Scirpus maritimus* in the Rhône Delta (France) dominated by different species (units in g m^{-2} dry weight). *Source*: Based on Ibáñez et al. (1999).

owing to factors such as variations in tidal flooding, nutrient chemistry, and oxygen content of the soil.

Marshes in North America exhibit distinct latitudinal patterns in aboveground primary productivity. Analyses from

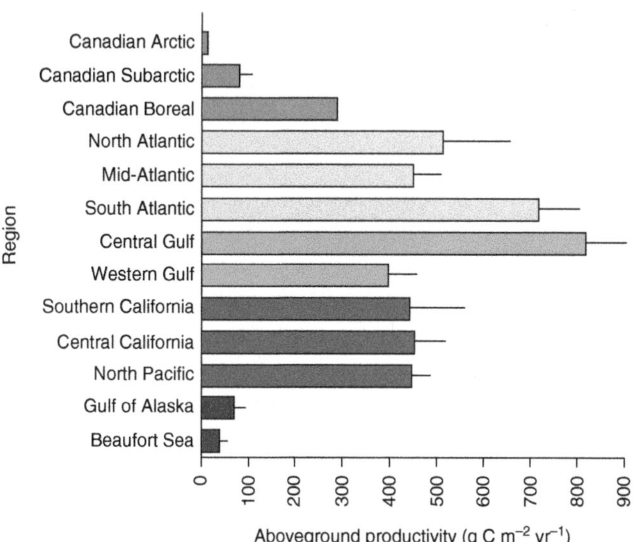

FIGURE 6.7 Salt marsh aboveground primary productivity by region in North America, north of Mexico. *Source*: Based on Mendelssohn and Morris (2000).

Mendelssohn and Morris (2000) show regional aboveground NPP values of less than 200 at high latitudes to greater than 800 g m^{-2} year^{-1} towards the tropics (Figure 6.7). NPP can also show large variation within a region based on marsh type and site-specific conditions. Estimates of aboveground NPP in *Spartina* marshes from the Atlantic Coast of North America range from 200 to 4000 g m^{-2} year^{-1}; similarly, belowground estimates range from 500 to 6200 g m^{-2} year^{-1} (Turner, 1976; Pierfelice et al., 2015). In the freshwater tidal wetlands of the middle Atlantic coast, NPP ranges from 1000 to 4000 g m^{-2} year^{-1}, with peak aboveground biomass ranging from 566 to 2312 g m^{-2}, and belowground biomass from 500 to over 7100 g m^{-2} (Whigham et al. 1978; Odum 1988; Spalding and Hester 2007). Marsh vegetation community composition also contributes to variation in NPP estimates. In coastal Louisiana, a microtidal area with high temperature and rainfall, estimates of NPP from seven individual marsh species ranged from 1355 to 6043 g m^{-2} year^{-1} (Hopkinson et al. 1978, 1980). By comparison, aboveground NPP in seagrass beds range from 239 to 2557 g m^{-2} year^{-1}, and peak biomass ranges from 50 to 854 g m^{-2} (Duarte 1989).

There have been fewer studies on NPP of coastal Mediterranean marshes (Ibáñez et al. 1999, 2000; Curcó et al. 2002; Scarton et al. 2002; Neves et al. 2007). Aboveground NPP in reed-type brackish marshes range from 452 g m^{-2} year^{-1} in a *Scirpus maritimus marsh to* 2989 g m^{-2} year^{-1} in a *Typha angustifolia* marsh (both in the Rhone delta), while shrubby saline marshes show lower values (with a minimum of 94 g m^{-2} year^{-1} in an *Arthocnemum macrostachyum* marsh). In this case, the variation was mainly due to grazing in the *S. maritimus* marsh (the *T. angustifolia* marsh was protected by an enclosure).

Belowground NPP of marsh plants has been reported less often than aboveground production primarily because it is more difficult to measure (Curcó et al. 2002; Darby and

Turner 2008). In contrast with aboveground vegetation, it is often very difficult to distinguish live from dead roots and rhizomes and to determine distinct seasonal patterns. Belowground NPP estimates for *S. alterniflora* range from 500 to 6200 g m^{-2} year^{-1} (Good et al. 1982; Darby and Turner 2008). The ratio of the belowground to aboveground biomass for *S. alterniflora* ranges from 0.3 to 48.9 with most values being higher than one, indicating that the belowground biomass is almost always considerably greater than aboveground biomass.

Another approach to measuring salt marsh production is to monitor the exchange of CO_2 between marsh plants and the atmosphere (Brix et al. 1996; Neubauer et al. 2000; Forbrich and Giblin 2015). There are two ways to measure gas exchange in marshes: one uses open or closed chambers that are deployed on marshes, and the other uses flux towers. Eddy covariance flux measurements allow for the examination of daily and even hourly dynamics of marsh grass production. Figure 6.8 shows CO_2 exchange between a marsh surface (including plants and soils) and the atmosphere over three days during the summer when productivity was high. There was a net uptake of atmospheric CO_2 during the day when the plants were photosynthesizing and a net release at night when there was only respiration. The height of the tide also affects the rate of CO_2 exchange with the atmosphere. When the tide is high, more of the plants and soils are submersed and CO_2 exchanges with the water rather than with the atmosphere. Therefore, flux measurements of photosynthetic uptake of CO_2 from the air during the day and respiratory release at night are both lower during high tide. This does not necessarily mean that photosynthesis or respiration is lower, because some of the CO_2 exchange is with the water, but it does mean that the data must be interpreted with care.

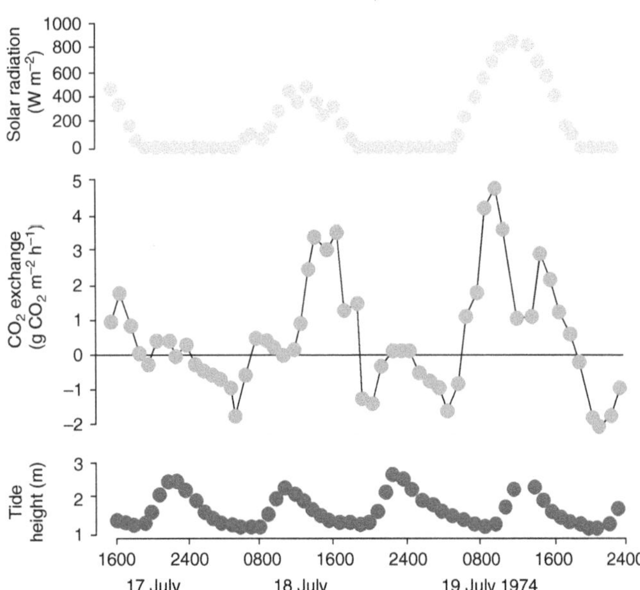

FIGURE 6.8 Solar radiation, atmospheric CO_2 exchange, and tide height for Flax Pond, New York, USA. *Source*: From Houghton and Woodwell (1980). Copyright by the Ecological Society of America, reprinted by permission.

6.3.1 The Effect of Measurement Method on Productivity Results

The most commonly used methods for measuring marsh productivity involve the harvest of live and dead plants, done most simply by the harvest of peak end-of-season live standing material (Morris 2007). Live material is continually dying, however, so this end-of-season technique is an underestimate. More sophisticated methods include determining the changes in both live and dead standing crop at regular time intervals and estimating the loss of both dead and live material between sampling dates (Wiegert and Evans 1964). Other techniques include different types of tagging to measure changes in height, diameter, and number of leaves (Hopkinson et al. 1978, 1980). For example, Hopkinson et al. (1980) found that different techniques for measuring the annual net production of salt marsh plants in Louisiana gave highly variable results (Figure 6.9). Kaswadji et al. (1990) also got variable results when they measured aboveground production of *S. alterniflora* in a Louisiana salt marsh using harvest methods and a nondestructive method based on measurement of stem density and longevity. The current consensus is that the most accurate estimates of production are measured with techniques that track changes in both live and dead standing material, and that nondestructive methods that account for stem turnover are preferred over harvest methods (Morris 2007).

As an alternative to harvest methods, fluxes of CO_2 and CH_4 have been used to estimate macrophyte productivity under ambient field conditions (Whiting and Chanton 1996; Forbrich and Giblin 2015) or under experimental conditions such as elevated atmospheric CO_2 concentrations (Langley et al. 2013). This carbon gas flux technique integrates processes that occur aboveground and belowground and therefore provides a more reliable estimate of the total production than harvest methods (Neubauer et al. 2000). If carbon fluxes from the sediments are measured *in situ*, the sediment microalgal production can also be calculated (Anderson et al. 1997). Morris and Jensen (1998) obtained an estimate of maximum gross belowground production of 1310 g m^{-2} year^{-1} from monthly measurements of canopy and soil CO_2 exchange in a Danish marsh.

6.3.2 Factors Affecting Marsh Productivity

A number of factors affect the productivity of coastal marshes. Flooding frequency and duration, as well as soil salinity, are perhaps the key factors, but temperature, rainfall, nutrient availability, oxygen levels, sediment type, and drainage are also important. These factors are interrelated and are in turn affected by plant growth (Mendelssohn and Morris 2000). Salt marshes can have significant interannual variation in productivity due to variation in the above factors (Teal and

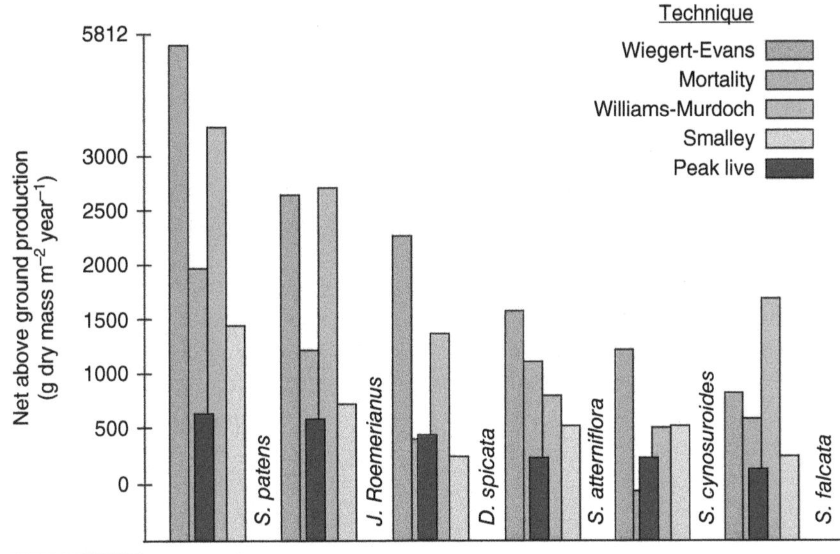

FIGURE 6.9 Estimates of annual aboveground production calculated from several widely used techniques. *Source*: Hopkinson et al. (1980). Copyright by the Ecological Society of America, reprinted by permission.

Howes 1996). In this section, we consider each of these factors separately, although obviously they act together.

6.3.2.1 Solar Radiation, Temperature, and Evapotranspiration

Solar radiation directly affects rates of photosynthesis, but solar radiation also affects plants indirectly because changes in radiation lead to changes in temperature, and temperature directly affects the rate of metabolic processes. Heat energy is used to evaporate water and vascular plants obtain essential nutrients as a result, because nutrients are moved from the soil to the leaves with this water. *Evapotranspiration*, as this process is called, is temperature dependent. Plants change their activity in response to changes in solar inputs, for example, as the day length increases or decreases.

Solar radiation, temperature, and evapotranspiration act together to produce differences in marsh productivity over a latitudinal gradient. Geographic variations in the growth of plants with similar physiology, litterfall, and net ecosystem production are closely correlated to factors that vary with latitude, especially temperature and sunlight (Bertness and Pennings 2002; Asaeda et al. 2005). Low winter temperatures and the presence of ice reduce the winter minimum biomass found in marshes. Along the Atlantic and Gulf coasts of North America, the minimum winter biomass decreases gradually from latitude 28 °N to 35°N. Above 38 °N, the presence of ice generally prevents any winter biomass (Turner 1976).

6.3.2.2 Tidal Range

Tides are one of the unique attributes of estuaries, and estuaries can be defined as fluctuating water level ecosystems, which are subsidized by tidal action. Differences in marsh production have been identified over broad areas with different tidal ranges as well as between streamside and inland marshes. Steever et al. (1976) reported a strong positive correlation between tidal range and the peak aboveground biomass of *S. alterniflora* along the Connecticut coast that was unrelated to the changes in climatic factors and soil conditions. In addition, they found that a "gated" marsh at Westport, Connecticut, had a lower peak standing crop than a nearby marsh where tidal exchange was not restricted. Steever et al. (1976) concluded that the "energy subsidy" provided by tidal action is a significant factor in the standing crop production of *S. alterniflora*. *S.alterniflora* production is not positively influenced by tidal action at very high tidal ranges, since the tide may become a stress (McKee and Patrick 1988). This is demonstrated in the Bay of Fundy where the tidal range exceeds 10 m and intertidal *S. alterniflora* is only 10–20 cm high. In addition, coastal Louisiana has a low tidal range (0.3 m) and very high productivity.

Tidal subsidies may also operate on a smaller scale. In almost all marshes, there is a striking difference in biomass between streamside (higher) and inland marshes (lower). Turner (1979) showed that the difference between peak streamside and inland biomass was greater with greater tidal range. Much of the difference between streamside and inland marshes is related to the movement of water in the two areas. At low tide, water drains almost completely from the streamside marsh surface. In inland marshes, there is often incomplete drainage, leading to ponding and high salt levels during periods with high evaporation (Odum 1980). Another important factor in the drainage of marsh soils is the percolation rate (the rate at which water flows through the sediment surface into the deeper sediments). In a Massachusetts salt marsh, Howes et al. (1981) found that the percolation of water into sediments was inversely related to grass height. They hypothesized that fine sediments probably filled the sediment pores, forming a relatively impermeable layer on the surface. In addition, they found that the subsurface sediments of the streamside region drained more rapidly at low tide than inland sediments. Thus, the combination of a relatively impermeable surface and well-drained sediments leads to high *Spartina* growth in the streamside zone. The situation in the inland marsh is just the opposite. Water flows freely through the surface, but the soils drain very poorly and growth of *Spartina* is less, which was shown experimentally by Mendelssohn and Seneca (1980). It is interesting that they also found tall *Spartina* in a narrow zone bordering the upland. Here, the water table is depressed because of the elevation of the site. They concluded that *Spartina* grows well where the root zone is not waterlogged. Many other factors that affect *Spartina* growth at the local level are also related to the differences in flooding and drainage. We examine these next.

6.3.2.3 Flooding Regime and Rainfall

Coastal marshes can be flooded either by marine water (meteorological and astronomical tides) or by freshwater (river floods, underground inputs, and rainfall). The timing and magnitude of seasonal oscillations in sea level seem to be the critical factors that influence salt marsh productivity (Morris et al. 1990). These authors found that the annual aboveground productivity of *S. alterniflora* in a South Carolina salt marsh varied by a factor of two and correlated positively with anomalies in mean sea level during the growing season.

In a 13-year study carried out in a salt marsh in the Netherlands (De Leeuw et al. 1990), the authors found that year-to-year variation in peak aboveground biomass of six annual angiosperm communities could be explained by the rainfall deficit during the growing season, while inundation frequency did not contribute to explain biomass variability. These authors suggested that rainfall may increase vegetation production by reducing soil salinity and increasing soil moisture content, and they concluded that this effect increases with marsh elevation, where soil salinity is determined by the mutually opposing effects of evapotranspiration and precipitation. At tidal elevations below mean high water, fluctuations in soil salinity are strongly related to the salinity of the inundation waters and not to the rainfall deficit. Productivity of coastal Mediterranean marshes also seems to be strongly influenced by rainfall, owing to its effect in lowering soil salinity (especially in poorly flooded marshes). Mediterranean-type salt marshes are less productive than the marshes in the Netherlands, due to low rainfall during summer leading to salt and water stress.

Temporary impounding can cause a decrease in production if rainfall is low and soil salinities become hypersaline, but it can also result in increased production if salinities return to normal because of high precipitation or upland runoff during the impounded period (Zedler et al. 1980). On the other hand, increased flooding may negatively affect production by decreasing sediment oxidation. Waterlogging is a key factor affecting redox potential, which in turn affects sulfide concentrations and the availability of nutrients in the soil to plants (see Chapter 2) (Pennings and Callaway 1992; Mendelssohn and Morris 2000).

6.3.2.4 Salinity

Adams (1963) reported that soil salinities of 70 and above prevented the establishment and survival of most salt marsh species. On the other hand, *S. alterniflora* grown in freshwater becomes chlorotic (i.e., lost green color in tissues). At high salinities (>25), osmotic stress (resulting in reduced water uptake) and cell membrane damage are likely to limit growth (Linthurst 1980; Stachelek and Dunton 2013). Halophytes seem to deal with osmotic stress by selectively concentrating preferred ions while making metabolic adaptations to the high concentrations of ions. In addition, some of these plants have the capacity for salt removal via salt glands and a mechanism in the roots for slowing the inward penetration of toxic ions. The result of this process is readily observed in the form of the salt deposits found on the leaves of *S. alterniflora* and in other salt-tolerant species (Ibáñez et al. 2000). Usually, the optimal salinity for maximum salt marsh growth is in the range of 7.2–14.4, with growth becoming significantly reduced as salinity increases or decreases. *Salicornia bigelovii* is a succulent annual species that occurs in coastal estuaries and is reported to have maximum growth at a salinity of 14. The deleterious effects of salinity are assumed to result from water stress, ion toxicities, ion imbalance, or a combination of these factors (Ayala and O'Leary 1995; Al Hassan et al. 2016).

In one greenhouse experiment, Rozema (1991) found that of 17 halophytic species, only those from the genera *Salicornia* and *Suaeda* showed an increase in the mean relative growth rate under saline conditions. Chenopodiaceae species such as *Atriplex nummularia, Suaeda maritima, Halimione portulacoides*, and *Salicornia dolichostachya* have been found to have maximum growth rates where the external salinity is 50–100 mM NaCl (3.6–7.2), where salinity is necessary to maintain the turgor pressure potential required for growth.

Zedler (1983) found that a short-term reduction in the salinity of normally hypersaline soils was followed by a 40% increase in the August biomass of *Spartina foliosa* at the Tijuana estuary (southern California) and that a longer period of brackish water influence on hypersaline soils of Los Peasquitos lagoon was followed by a 160% increase in the August biomass of *Sarcocornia pacifica*. The largest increase in salt marsh biomass occurred in a nontidal lagoon with a relatively small increase in stream discharge, while tidal marshes underwent lesser changes in biomass following major flooding events. This is because higher tidal range and rainfall implies more stable and lower salinity conditions preventing marsh vegetation growth from being limited by soil hypersalinity.

6.3.2.5 Nutrients and Other Factors Affecting Marsh Productivity at the Local Level

A number of factors affect the growth of salt marsh plants within local areas. These include nutrient levels (such as the concentration of N, P, and Fe), the pH and Eh of the soil, waterlogging and drainage of the soil, oxygen levels, sediment type, herbivory, and competition. These factors are interrelated and affect plant growth and in turn are affected by it (Ibáñez et al. 2000; Mendelssohn and Batzer 2006). Since nutrients are one of the most important factors limiting plant growth, we begin our discussion with this subject.

The most studied salt marsh plant, *S. alterniflora,* is generally limited by N availability. On a physiological level, N availability is dictated by the presence of salts, sulfides, and oxygen that modify the kinetics of N uptake (Morris 2007). Studies over a wide geographic range have shown that the addition of inorganic nitrogen (but seldom inorganic phosphorus) increases the growth of *S. alterniflora* and other species (Mendelssohn 1979; Morris 1991; Bertness et al. 2008).

However, other studies show that additional factors affect nitrogen uptake by plants. For example, in both Louisiana and Georgia marshes, nitrogen addition gave a greater response in inland marshes than it did in streamside marshes (Buresh et al. 1980). These and other studies in Louisiana, North Carolina, and Massachusetts have shown that ammonia levels were higher in inland marsh soils, although *Spartina* growth was less (Valiela and Teal 1979; Craft et al. 1991; Mendelssohn and Morris 2000). Thus, higher nitrogen does not necessarily mean greater plant growth (Turner et al. 2009). Some other factors can help explain these results.

The amount of oxygen present in marsh soils is an important factor affecting plant growth. Maximum growth of *S. alterniflora* occurs in more oxygenated soils, and the H_2S produced in anaerobic sediments inhibits respiration and nutrient uptake (Mendelssohn and Morris 2000). Marsh soils, however, are almost completely anaerobic. How then do *Spartina* and other marsh plants achieve such high growth? The answer to this question is a function of both plant metabolism and the drainage characteristics of marshes. The pH of the soil also affects plant growth. Linthurst (1980) reported that the growth of *Spartina* was optimal at pH 6 in comparison to a pH of either 4 or 8. Although *S. alterniflora* has a well-developed aerenchyma (airspace) system, it is not able to alleviate root oxygen deficiencies in the more biochemically reduced zones of the marsh (Mendelssohn et al. 1981). There is also evidence that H_2S toxicity may contribute to the development of growth forms of *S. alterniflora* by inhibiting ammonium uptake (Mendelssohn and McKee 1988; Bradley and Morris 1990). Furthermore, H_2S inhibits the anaerobic production of energy, thus providing a mechanism for the effect of H_2S on nitrogen uptake kinetics (Koch et al. 1990).

The nature of the soil in marshes affects the growth of marsh plants. Fine-grained clay and silty clay soils have higher nutrient levels than sandy soils and support greater growth of marsh grasses. Soil density is highest in streamside marsh soils because of the input of mineral sediments during high tide. DeLaune et al. (1979) found in Louisiana, USA, that the aboveground standing crop of *S. alterniflora* was correlated with soil bulk density. This correlation was apparently a result of the association of this property with the content of mineral matter in the soil. They also found that the input of nutrients with new sediments was the most important source of "new" nutrients for the salt marsh. Since the sedimentation rate is much higher on streamside marshes, this is a contributing factor causing higher productivity (due to better soil drainage) compared with marshes further inland. Craft (2007) found that tidal fresh and brackish marsh soils across all geographic regions of the contiguous United States had significantly lower bulk density and greater percent organic C, N, and P than salt marshes.

Competition is also a factor affecting marsh productivity. In southern California, USA, Pennings, and Callaway (1992) showed that the growth of *Sarcocornia pacifica* and *Arthrocnemum subterminalis* was negatively affected by flooding, salinity, and competition. However, the relative importance of these factors varied across marshes. Thus, the benefit of reduced flooding in the zone dominated by *A. subterminalis* outweighs the disadvantage of higher salinity, and both species grow better in the zone dominated by *A. subterminalis* than in the low marsh zone dominated by *S. pacifica*. In contrast, both species do poorer in the transition zone, even though flooding is greatly reduced, probably because of the high salinity of the transition zone soil. Similarly, *A. subterminalis* plants in the low marsh areas dominated by *S. pacifica* did not have high productivity, presumably because of harsh physical conditions.

Herbivory can also affect the growth of marsh grasses. The prevailing paradigm in marsh ecology for nearly five decades has been that bottom-up forces are the primary determinants of plant production (Mitsch and Gosselink 2001), but this has changed over the last two decades (see next section).

6.3.3 Top-down Control of Marsh Vegetation

Coastal marsh vegetation, as well as marsh ecosystem functions and services, are influenced by herbivory. In the 1960s, some studies in North American salt marshes mistakenly concluded that plant-herbivore interactions were of little consequence to community dynamics and productivity (Marples 1966). For five decades it has been the prevailing paradigm in marsh ecology that bottom-up forces are the primary determinants of plant production (Mitsch and Gosselink 2001). Grazer exclusion experiments conducted on the East Coast of the United States from Virginia to Louisiana (Silliman and Zieman 2001; Silliman and Bertness 2002; Silliman et al. 2005), challenged the bottom-up control marsh theory and suggested that powerful trophic interactions influence the high primary production observed in these communities. These authors showed that (i) the snail *Littoraria* strongly suppresses *Spartina* production (both short-form and tall-form) everywhere on marsh surfaces where it reached high densities and (ii) snail-grazing impacts were strongest in N-rich soils where tall-form *Spartina* dominates. A recent synthesis study using data from 443 experiments and observations found that salt-marsh herbivores, including insects, snails, crabs, waterfowl, and small mammals reduce plant survival, aboveground biomass, and height (Figure 6.10) (He and Silliman 2016). The response of other parameters such as plant density, belowground biomass, reproduction, and herbivory were, however, more variable, due to the differences in herbivore type. The authors point out that burrowing animals such as crab have a clear effect on

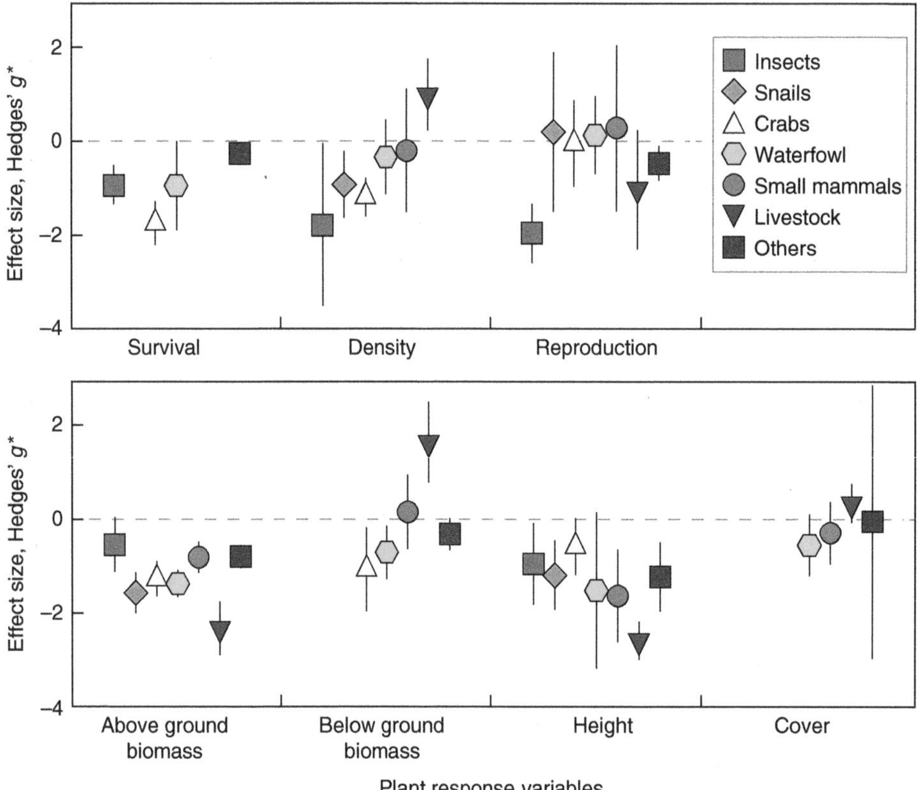

FIGURE 6.10 Effects of different types of herbivores on plant performances in salt marshes: (a) Demographic variables include survival, density and reproduction, and (b) growth variables included aboveground biomass, belowground biomass, height and cover. *Source*: He and Silliman (2016)/John Wiley & Sons.

belowground biomass (Coverdale et al. 2012), while insects that consume flowers consequently reduce reproduction (Bertness et al. 2008). The study by He and Silliman (He and Silliman 2016) also describes the effects of livestock grazing, which is an important anthropogenic influence on salt marshes in Latin America (Di Bella et al. 2013), Asia (Yang et al. 2017) and Europe (Bakker et al. 1993), where this land-use practice is a centuries old tradition. Big livestock such as cattle and horses consume large amounts of biomass per day (Duncan et al. 1990) and consequently have a stronger effect on aboveground biomass than smaller animals (He and Silliman 2016).

Herbivores, in turn, can experience top-down control by predators, which can have indirect positive effects on the plants (He and Silliman 2016). This type of indirect effect, known as a "trophic cascade" (Altieri et al. 2012), was illustrated by experiments in which exclusion of predators of the herbivorous snail *Littoraria* (e.g., Blue Crab) in a *Spartina* marsh in Georgia (USA) led to a strong increase in snail numbers (Silliman and Bertness 2002). A similar experiment in which predators for the herbivorous burrowing crab *Sesarma reticulatum* were excluded resulting in a >100% increase in herbivory and a >150% increase in bare soil regions (Figure 6.11) (Bertness et al. 2014). These studies concluded that predators indirectly facilitate high levels of primary production, and that loss of predators could be the cause of severe marsh die-offs such as those observed in New England salt marshes in the United States. These findings have important implications for the long-term conservation of salt marshes because intense fishing off the East Coast of the United States and elsewhere has led to depleted densities of predators including large, recreationally harvested predatory crabs and fish in estuarine communities (Silliman and Bertness 2002).

While predators can control the abundance of herbivores from the top-down, herbivores can in turn indirectly affect other groups by changing the abundance of resources or of abiotic conditions. This indirect effect, known as "exploitation competition," has been extensively studied in European salt marshes used for livestock grazing (van Klink et al. 2016). A strong reduction of biomass by intensive livestock grazing can reduce the availability of flowers and therefore reduce the abundance of pollinators (van Klink et al. 2016). Consistent with this finding, the abandonment of livestock grazing has been found to positively affect many invertebrate species (Rickert et al. 2012; van Klink et al. 2013). Cessation of grazing can also benefit small herbivores such as hare (Schrama et al. 2015) and geese (Mandema et al. 2014) by allowing a high-quality regrowth

(a)

(b)

FIGURE 6.11 Predator exclusion (a) and open control (b) plots after 3 months of predator exclusion. All plots had similar initial cordgrass cover. *Source*: Bertness et al. (2014). John Wiley & Sons.

of plants following intense plant biomass reduction. Livestock grazing also indirectly affects the microbial community in salt marshes by altering the source of organic material to soils (Mueller et al. 2017). In ungrazed marshes, a taller plant canopy changes patterns of sediment deposition and leads to a higher input of allochthonous organic material with a higher quality during flooding, thereby affecting the availability of nitrogen and the structure of the microbial community.

6.3.4 Factors Affecting Marsh Accretion and Habitat Change

Coastal marshes are known to have maintained an elevation in equilibrium with mean sea level for thousands of years (Redfield 1972). They do this by accumulating mineral sediment and organic matter (Mudd et al. 2009). Marshes in cold regions, common in New England, Scandinavia, and the coastal Arctic Ocean, are often peat dominated and apparently maintain elevation primarily by accumulating organic matter, while most of marshes worldwide primarily accrete mineral sediment.

The elevation of the surface of salt marshes is an important variable that controls productivity of salt marsh plant communities. In turn, the productivity or biomass density influences the rate of accretion of marsh surfaces (Morris et al. 2002; Kirwan et al. 2010). The elevation of marsh platforms relative to mean high water and mean sea level determines inundation frequency, duration and, consequently, wetland productivity (Figure 6.12). There is a species-specific optimum

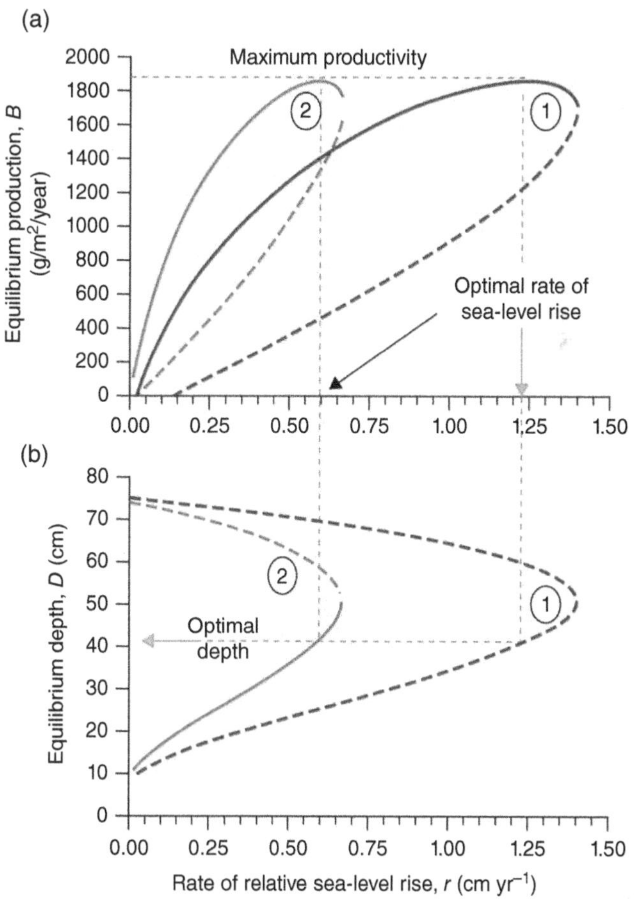

FIGURE 6.12 Equilibrium functions of (a) productivity and (b) depth below mean high tide (MHT) as functions of the rate of relative sea level rise, for estuaries characterized by (1) high and (2) low sediment loading. The optimal depth is the depth below MHT that results in maximum productivity. *Source*: Morris et al. (2002)/John Wiley & Sons.

elevation and hydroperiod for marsh vegetation, and, at elevations above that optimum (i.e., superoptimal), an increase in mean sea level relative to marsh surface height will stimulate growth (Morris and Haskin 1990). The sedimentation rate in marshes is affected positively by the biomass density of marsh vegetation due to the drag exerted by the plant canopy (Morris et al. 2002). Hence, in response to rising sea level, an increase in biomass density will increase sedimentation rate, thereby raising the elevation of a marsh (Kirwan and Megonigal 2013). Also, increased aboveground plant growth increases belowground productivity, which also contributes to elevation change (McKee et al. 2007). Thus, the equilibrium elevation is inversely related to the rate of sea level rise and subsidence. One implication of this feedback in this current era of rising sea level due to global warming is that a vegetated marsh is only stable when its elevation is superoptimal, because at suboptimal elevations an increase in sea level will depress growth and, thus, sedimentation.

Marsh species or communities typically segregate along gently sloping topographic gradients on marsh platforms. The competitive balance among species is determined largely by the relative elevation of a site (Bertness 1991; Pennings et al. 2005), and the outcome determines primary production as well as the sedimentation rate. There are intraspecific differences among marsh macrophytes in their effect on the sedimentation rate. For example, Leonard and Luther (1995) showed that *J. roemerianus* and *S. alterniflora* at equal densities have significantly different effects on the turbulence intensity of the floodwater over a marsh, which has a great effect on sedimentation and erosion. Thus, the net sedimentation rate is a function of species composition and, as noted above, productivity and biomass density. Therefore, the competitive interactions among plant species are mediated by the influence of the vegetation on sediment accretion and modifications to the relative elevation of marsh surfaces (Morris 2006).

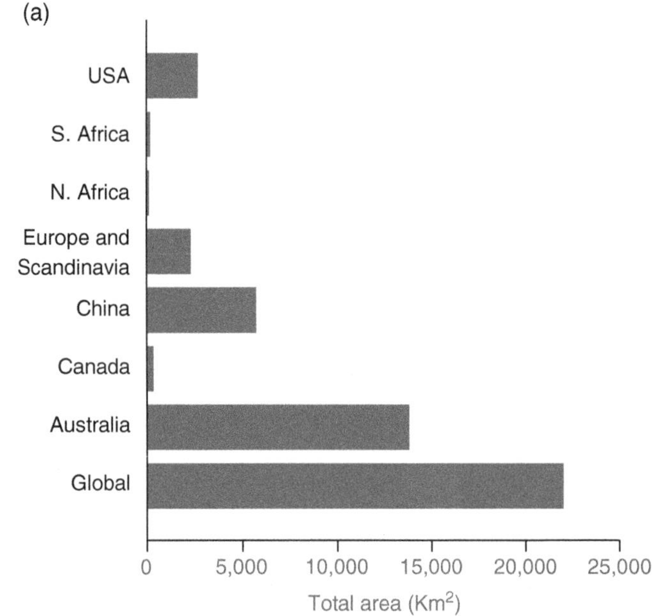

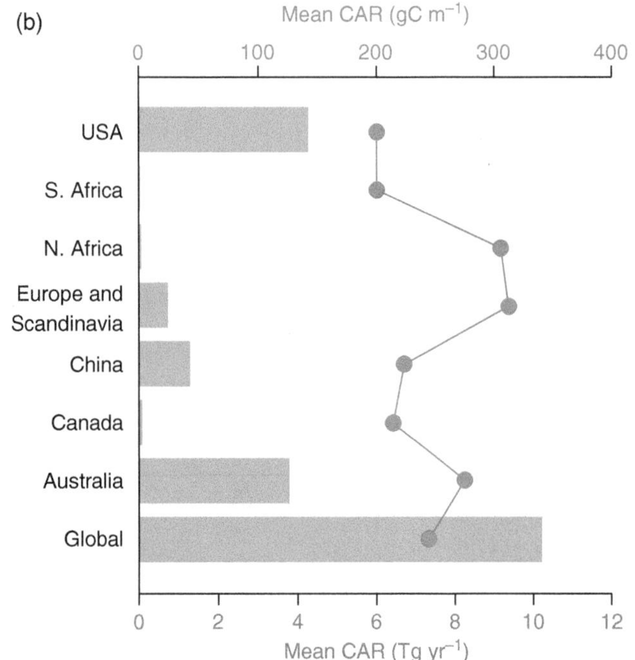

FIGURE 6.13 (a) Surface area of coastal marshes and (b) annual means of Carbon Accretion Rates (CAR), in g C m^{-2} year^{-1}, plus global mean values in Tg year^{-1} and per geographic area. CAR data obtained from Ouyang and Lee (2014) and surface area of coastal marshes in USA from Nahlik and Fennessy (2016).

6.4 Carbon Dynamics and Greenhouse Gas Emissions in Coastal Marshes

Coastal vegetated habitats, including mangrove forests, seagrass beds, and salt marshes, represent an important carbon sink globally. The carbon stored in marshes and other marine ecosystems is known as "blue carbon" (Hiraishi et al. 2014). The capacity of salt marshes for long-term storage of blue carbon is extraordinary (Figure 6.13), with carbon burial rates two orders of magnitude higher than terrestrial systems (McLeod et al. 2011).

Carbon sequestration is a result of the range of processes involved in the removal of carbon from the atmosphere and the subsequent storage of that carbon in the ecosystem. The imbalance between high primary production (Odum 1988) and low decomposition in coastal marshes provide them the capacity to act as carbon sinks (Mitsch et al. 2013). On a global scale, tidal range and latitude (Chmura et al. 2003; Ouyang and Lee 2014) are the major drivers of carbon sequestration, because these factors determine the biotic and biogeochemical factors that directly govern carbon and sediment accumulation rate, i.e., primary production and the dynamics of sediment and organic matter. A comprehensive review of primary production in coastal marshes is presented in previous sections of this chapter. This section presents a synthesis of the major drivers of decomposition processes and greenhouse gas emissions in coastal marshes.

6.4.1 Decomposition of Organic Matter in Coastal Marshes

Decomposition is a key process regulating carbon cycling (Aerts and Toet 1997) with effects on long-term carbon sequestration and soil accretion (Janousek et al. 2017). Recalcitrant fractions of the organic matter in marshes contribute to the build-up of large soil carbon stores while the relative proportion of the labile compounds modulates the stabilization of the organic matter in the long-term (Cotrufo et al. 2015; Haddix et al. 2016).

The degradability of soil organic matter (SOM) is determined by the substrate quality and environmental factors such as hydroperiod (Huertas et al. 2017), salinity (Stagg et al. 2017a) and nutrient concentration (Morris et al. 2013), that together determine the accessibility of this material to decomposers (Dungait et al. 2012). In coastal marshes, the combination of a high ratio of belowground to aboveground production (Chmura et al. 2003; Stagg et al. 2017b; Shiau et al. 2019), and the fact that root-derived carbon is more persistent in soils than leaf production (Hemminga et al. 1996; Bragazza et al. 2009) slows soil decomposition rates relative to other ecosystems.

The composition of plant communities also determines the degradability of plant litter (de Neiff et al. 2006; Laskowski and Berg 2006). Stagg et al. (2017b) reported a transition in carbon sequestration from a high oligohaline marsh to forested wetlands where decomposition rates decrease landwards, mostly due to lower salinity and slower decomposition of lignin in forests. Degradation studies showed that *Spartina* genera have the highest capacity of soil carbon accumulations and *Distichlis* the lowest (Ouyang and Lee 2014), and that *Spartina patens* decomposes more slowly than *Spartina alterniflora* (Frasco and Good 1982; Valiela et al. 1984). Furthermore, the effect of changes in plant community composition on carbon storage capacity has been addressed in studies on the effects of invasive species (Windham 2001; Liao et al. 2007) revealing a "home-field advantage" (Veen et al. 2015), whereby plant litter decomposes faster in the "home" site because of the presence of an adapted community of decomposers, and decomposes more slowly when introduced to a new environment.

Nutrient and carbon cycling are closely linked during decomposition, suggesting that nutrient supply is a crucial factor controlling decomposition rates. Microbial respiration and decomposition rates are strongly correlated with soil C, N, and P pools. Hill et al. (2018) related the slower decomposition rates in estuarine marshes in comparison to other types of aquatic wetlands to higher C/N ratios and recalcitrant carbon indices. The effects of nutrient enrichment on carbon budgets in wetlands is controversial as research indicate negative (Deegan et al. 2012; Pennings 2012), neutral (Graham and Mendelssohn 2014), or positive (Anisfeld and Hill 2012; Morris et al. 2013) feedbacks.

Salt concentrations in marshes also correlate with carbon sequestration. Sequestration rates are typically higher in salt marshes than in freshwater wetlands (especially peatlands), (Chmura et al. 2003; Bridgham et al. 2006; Cheng et al. 2020). The effect of salinity on soil carbon stock depends in part on the temporal scale (i.e., short vs. chronic exposure) and on the type of wetland (Howard and Mendelssohn 2000; Stagg et al. 2017a). Also, seawater-affected organic matter decomposition, a primary determinant of carbon sequestration, can vary with the quality and quantity of the organic matter, ambient nutrient and flooding regimes, species composition, and a suite of other factors (Herbert et al. 2015). Short-term carbon accumulation rates can decrease with increases in salinity (Baustian et al. 2017). Lower carbon accumulation may result from salinity constrains primary production, via nutrient limitation (Nixon et al. 1996), and salinity-associated toxicity (Lamers et al. 2013). In contrast, carbon sequestration may increase with elevated salinities as high sulfate concentrations in seawater favor anaerobic decomposition via sulfate reduction and thus limit CO_2 mineralization rates (Setia et al. 2010; Rath et al. 2016); also, elevated salinities can reduce litter quality and decomposition, potentially increasing carbon sequestration (Stagg et al. 2018).

Flooding regime largely influences carbon dynamics in wetland soils. Increased water saturation can shift decomposition from aerobic to anaerobic which slows rates of decay (Day and Megonigal 1993; Bridgham et al. 1998) via microbial decomposition (Keller et al. 2012). In the case of salt marshes, tidal flooding effects often interact with salinity (Howard and Mendelssohn 2000) and the individual effects of these two factors generally increase sequestration (Krauss and Whitbeck 2012).

6.4.2 Greenhouse Gas Emissions

Much of the carbon that is not sequestered in marshes is released as CO_2, but some is released as the potent greenhouse gas CH_4. In fact, wetlands represent 20–30% of current CH_4 global emissions (Metz et al. 2007). CH_4 is the final product of heterotrophic respiration by methanogens which occurs in highly reducing anoxic conditions that form in soils during temporal or permanent flooding periods (Ponnamperuma 1972). In salt marshes, methanogenesis accounts for less than 5% of total anaerobic respiration because these organisms are outcompeted by sulfate-reducing bacteria and iron-reducing bacteria (Tobias and Neubauer 2019). High concentrations of sulfate in seawater makes salinity the major controlling factor for CH_4 emissions, which decrease with increasing salinity.

Vegetation also influences the rate of methanogenesis in marshes. CH_4 emissions are positively correlated to plant productivity, which provides organic carbon for methanogen metabolism via plant litter and root exudates (Whalen 2005; Chaudhary et al. 2018). Plants also function as physical conduits through which CH_4 is released via aerenchyma tissues. On the other hand, plant root growth limits methanogenesis by enhancing rhizosphere-associated aerobic microenvironments in soils. These microenvironments can potentially inhibit CH_4 production (Neubauer et al. 2005) or induce aerobic methanotrophy by methane-oxidizing bacteria, which can convert up to 70% of total CH_4 production to CO_2 in some marsh soils (Megonigal et al. 2004).

6.5 Human Impacts, Management, and Assessment of Coastal Marshes

One of the major ways in which coastal marshes have been impacted by human activity is by physical alterations that have led to their direct and indirect destruction. Wetlands along the United States and the European coast in areas with high population densities have suffered the greatest proportional loss (Mitsch and Gosselink 2001), a reflection of the pressure of development in these areas. This is the case for most coastal areas globally where wetlands have been drained, filled, and "reclaimed." In almost all coastal cities, large areas of wetlands have been destroyed for development of one kind or another. This process has led to a significant loss of carbon stocks in coastal wetlands, known as blue carbon (Macreadie et al. 2013).

Deltaic wetlands have been particularly impacted because deltas have large areas of coastal wetlands, and deltas are used intensely for agricultural, industrial, urban, and navigation developments (Syvitski et al. 2009; Vörösmarty et al. 2009). The list of highly impacted deltas is long and includes the Rhine, Rhone, Ebro, Po, Nile, Mississippi, Colorado, San Joaquin-Sacramento, Tigris-Euphrates, Indus, Ganges, Mekong, and Yangtze. Only a few tropical (e.g., Orinoco) and arctic (e.g., Mackenzie and Lena) deltas have remained relatively unaffected.

The Mississippi and Ebro deltas serve as examples where human activities have resulted in dramatic regional changes in coastal systems. In the Ebro, the changes have proceeded deliberately for the most part to provide irrigation for agriculture and water supply to other parts of Spain. In the Mississippi Delta, many of the changes have followed because of a wide range of indirect and cumulative impacts of human activities. In both cases, large areas of coastal ecosystems have been altered and destroyed. In the Ebro River basin (85,530 km^2), which covers much of northeast Spain, nearly 190 dams were constructed, mainly in the twentieth century. These dams store approximately 60% of the annual runoff, which is used for hydroelectricity and irrigation. Two large dams (Mequinensa and Riba-roja) were constructed some 100 km upstream of the river mouth in the 1960s for hydropower. These large mainstem dams reduced sediment transport to the lower Ebro River and delta by up to 99% (Ibáñez et al. 1996). Because of this reduction in sediment, nearly half of the delta will be below sea level by 2100 due to sediment starvation combined with subsidence and sea level rise. In addition, the shoreline is retreating several meters per year near the river mouth threatening the Ebro Delta (Ibáñez et al. 1997b).

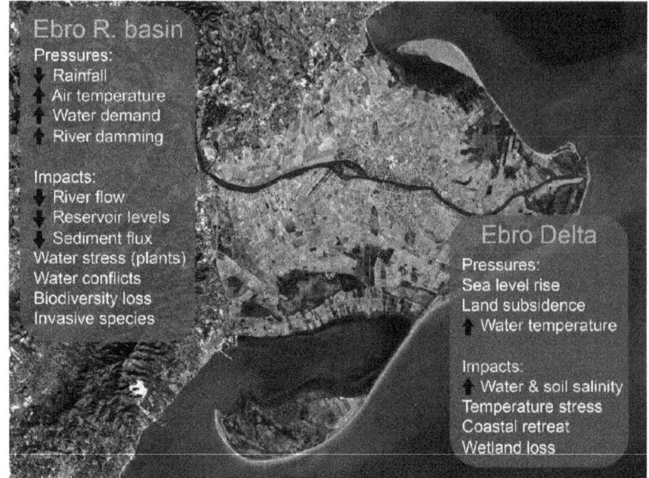

FIGURE 6.14 Overview of global change pressures and impacts, presently existing in the lower Ebro River basin and the Ebro Delta (Catalonia, Spain). *Source*: NASA Earth Observatory.

Rising sea level will flood Mediterranean coastal wetlands gradually, unless enough sediment supply exists to offset the rise (Day et al. 2011). Thus, some wetlands of the Ebro delta may disappear, and rice fields are at risk due to increasing flooding and salt intrusion (Figure 6.14). However, for deltas with high sediment loading, wetlands can survive sea level rise (Day et al. 1997). This stresses the importance of maintaining and restoring the sediment fluxes to the Ebro River delta and other deltas where sediment flux has been strongly reduced due to dam construction (Rovira and Ibáñez 2007).

Proposed solutions to mitigate the impacts of climate change and relative sea level rise in the Ebro delta range from hard engineering methods (e.g., levees, dikes, and jetties) to soft engineering defenses (e.g., artificial dunes and drainage systems) and ecological engineering approaches (e.g., restoration of fluvial sediment fluxes, controlled diversions, and wetland restoration; Ibáñez et al. 2014). These last approaches are the most sustainable in the long run, since the restoration of sediment delivery to the delta is the only solution to maintaining land elevation and wetland ecosystem integrity (Sánchez-Arcilla et al. 2008; Ibáñez et al. 2010). Rovira and Ibáñez (2007) showed that bypassing of sediments through the reservoirs of the lower Ebro River is feasible and represents a significant amount of sediments to offset the relative sea level rise in the delta plain and coastal retreat at the river mouth. Sustainable sediment management in reservoirs is becoming more common in the last decades as a consequence of problems

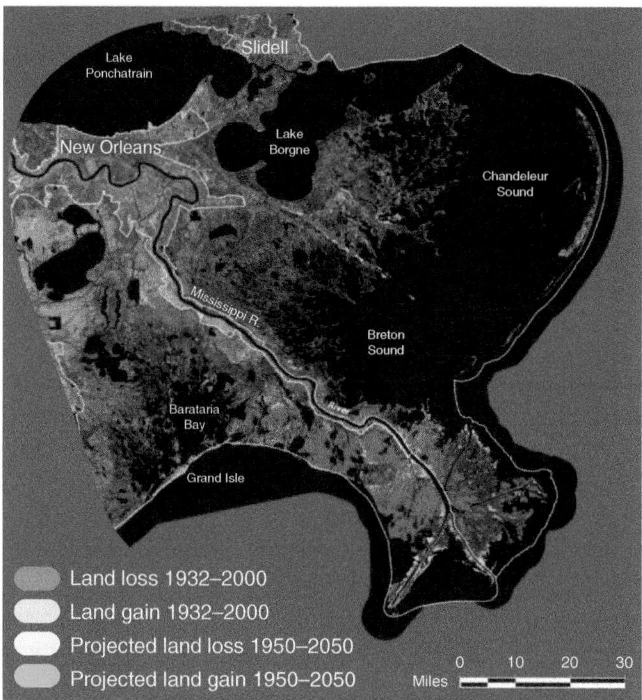

FIGURE 6.15 Historical and projected land change in coastal Louisiana, USA, in the Mississippi River delta. *Source*: USGS.

derived from silting and the environmental impacts caused downstream (Kondolf et al. 2014).

In the Mississippi Delta, dramatic changes have taken place, not because of deliberate planning, but due to the lack of it. Coastal wetland loss rates as high as 100 km^2 $year^{-1}$ occurred with a total loss of about 5000 km^2 or about 28% of the wetlands that existed at the beginning of the twentieth century (Figure 6.15, Day et al. 2014; Day and Erdman 2018; Jankowski et al. 2017). This was the result of a disastrous interaction of human actions and natural processes. The entire deltaic region is subsiding at a rate of about 1 cm $year^{-1}$ because of compaction, consolidation, and dewatering of sediments deposited by the river. The combination of sinking due to subsidence and rising sea level causes a high rate of relative sea level rise. To survive, wetlands must grow upward at the same rate as the water level rise. If not, they will be stressed by excessive flooding and other stresses and will ultimately die, as described earlier in this chapter and by Elsey-Quirk et al. (2019). Historically, seasonal flooding by the Mississippi River into coastal marshes without extensive alterations of hydrology allowed coastal marshes to survive.

A number of human impacts have altered the natural cycle in the delta. One of the most important is that the river has been diked essentially to its mouth. Thus, riverine input has ceased for most of the coastal zone. The wetlands themselves have also been altered dramatically. Large areas have been impounded (Day et al. 1990) and earlier impoundments were reclaimed for agriculture, but these projects mostly failed owing to subsidence caused by organic soil oxidation. Most can be identified today as large rectangular water bodies in the coastal marshes. A very damaging impact in the wetlands has been the construction of over 15,000 km of canals for drainage and navigation, mainly for petroleum exploration and production. These canals destroy wetlands directly, but cumulatively they have led to changes in the regional hydrology and salt water intrusion (Day et al. 2000, 2020; Shaffer et al. 2009). Finally, the spoil banks that border these canals are barriers to overland sheet flow of water and they alter wetland hydrology (Swenson and Turner 1987). Thus, the spoil banks impede water exchange, sediment deposition, and movement of nekton. Since sediments resuspended by winds are important for marsh surface accretion, canals with associated spoil banks are important factors contributing to land loss. Unless action is taken to reintroduce riverine sediments and restore hydrology, most of the wetlands of the Mississippi Delta will disappear within this century (Reyes et al. 2000; Day et al. 2007). Indeed, Blum and Roberts (2009) concluded that the combination of sea level rise and sediment starvation would result in the almost complete loss of coastal wetlands in the Mississippi delta by 2100.

Ecological processes in coastal wetlands are being altered in many ways by climate change, mostly due to warming and sea level rise (Saintilan et al. 2018; FitzGerald and Hughes 2019). Other human activities have accelerated elevation loss in large coastal areas that are going below sea level. The sinking of coastal wetlands can be caused by oxidation of organic soils after drainage or the extraction of belowground liquids or natural gas. Drainage of highly organic wetland soils in the metropolitan New Orleans area created parts of the city that are 4–5 m below sea level and 60–80% of the population living below sea level. Extraction of oil and gas in the Mississippi delta in some cases has doubled the rate of wetland subsidence, though it may have slowed down more recently (Morton and Bernier 2010; Day et al. 2020). Other areas have been similarly affected. For instance, most of the Sacramento-San Joaquin is below sea level due to oxidation of drained organic soils (Deverel et al. 2016), whereas in the Po delta (Italy), extraction of shallow deposits of natural gas has caused parts of the delta to subside 3–4 m below sea level (Corbau et al. 2019).

6.5.1 Upstream Alterations

Upstream changes in rivers can have pronounced effects on coastal wetlands (Sklar and Browder 1998). Two of the most important impacts are reduced freshwater discharge and reduced mineral sediment input. Construction of dams has resulted in the diversion of freshwater upstream for irrigation and for industrial and residential use. On the other hand, channelization of streams causes more rapid pulses of water to coastal systems. Construction of impoundments on the Mississippi River system has resulted in about a 50% decrease in suspended load (Blum and Roberts 2009). Freshwater diversions from the Colorado River in the western United States are causing hypersaline conditions and deterioration of the delta of the river at the northern extent of the Gulf of California. Because most of the water use occurs in the United States but the coastal problems are felt in Mexico, this issue has been an important point of contention and discussion between the two countries. Salinization is particularly a problem in arid and semiarid areas where freshwater input has been reduced, especially in arid deltas (Day et al. 2021). For example, large freshwater withdrawals from the Indus River have led to hypersaline soils and loss of tens of thousands of hectares of wetlands in the Indus delta (Khan and Akbar 2012). As indicated above, freshwater, and suspended sediments have been reduced by more than 40% and 99% in the Ebro River, respectively, resulting in parts of the delta being lost by coastal erosion or sinking below sea level (Rovira et al. 2015; Cozzi et al. 2019).

The construction of the Aswan High Dam is an excellent case study of the impact of dam construction on a coastal wetland ecosystem (Stanley 1988; Chen et al. 2021). Because of the loss of sediments, the shoreline of the Nile delta is now undergoing retreat. Stanley and Clemente (2017) reported that the delta is undergoing subsidence ranging from 3.7 to 8.4 mm $year^{-1}$ in the NW and NE, respectively. This rapid subsidence combined with reduction of sediment and freshwater input due to the Aswan Dam and rising sea level is likely to lead to flooding and salinization of a large part of the delta plain by the end of next century. As the Nile delta is the site of production for the majority of food in Egypt, this will have a strong societal impact in this century.

6.5.2 Functions and Values of Coastal Marshes

From an ecological standpoint, wetlands perform a wide variety of functions at a hierarchy of scales ranging from the specific (e.g., nitrogen retention) to the more encompassing (e.g., biogeochemical cycling) because of their physical, chemical, and biological attributes. At the highest level of this hierarchy is the maintenance of ecological integrity, the function that encompasses all ecosystem structure and processes. The link between function and condition lies in the assumption that ecological integrity is an integrating "super" function of wetlands. If the condition is excellent (i.e., equal to reference condition or some "ideal" condition), then the ecological integrity of the wetland is intact and the functions that are typical of that wetland type also occur at reference levels (Fennessy et al. 2004). A detailed list of functions, values, and services provided by coastal wetlands is shown in Table 6.3.

The assessment of the wetland condition, based on a set of indicators, is necessary to quantify the ecological status of wetland ecosystems and their deviation from historic conditions. Scientists and managers can use the results of such assessments to establish restoration and management measures to improve wetland conditions. Nowadays, there is a large number of assessment methods worldwide, especially in the United States and the European Union.

6.5.3 Indicators of Coastal Marsh Stability and Productivity

Many coastal marshes depend on sediments supplied by rivers to counteract the effects of land subsidence, sea level rise, and sediment compaction. In some areas, changes on the land have led to reduced riverine sediment supply to marshes, leading to a decrease in marsh surface height relative to the mean sea level. Where dams or levees have been constructed to prevent flooding, marshes have been cut off from their source of sediment, and the net effect is conversion of marsh habitat to open water (Day et al. 2007). Vertical elevation is a critical variable that determines the productivity and stability of salt marshes. The long-term existence of salt marshes depends on the success of the dominant plants, such as *Spartina* and *Juncus* spp., and their close relationship to sediment supply, sea level change, and tidal range. Researchers at the University of South Carolina and the Marine Biological Laboratory in Woods Hole have developed two coastal indicators that can be applied to assess the condition of coastal marshes (Morris et al. 2002; Morris 2007). One is the vertical elevation relative to the mean sea level (geomorphic) and the other is the level of stress of marsh vegetation (physiologic).

The vertical elevation relative to the mean sea level is an important geomorphic indicator of marsh productivity and stability and is determined using light detection and ranging (LIDAR) remote sensing (Rosso et al. 2006). This LIDAR elevation data is combined with high-resolution airborne data acquisition and registration (ADAR) digital camera images of a marsh landscape to construct a frequency distribution of marsh land cover with elevations relative to the elevation of the mean sea level. The frequency distribution is then compared to optimal distributions across the range of tolerance for the specific vegetation. The height of coastal marshes relative to sea level moves upward or downward toward equilibrium with the sea depending on factors such as the rate of sea level rise and sedimentation. When marsh elevation drops too far below the optimal level for vegetation either because of rapidly increasing sea level or rapidly decreasing marsh surface height caused by a reduced supply of mineral sediment and organic matter, then salt marsh vitality declines. Such a drop in the relative elevation of a marsh surface below an optimum suggests that coastal marshes are on a course leading to degradation.

The level of stress of marsh vegetation is another important indicator of marsh productivity and stability. Two complementary measurements, one ground-based and the other

TABLE 6.3 Ecosystem services provided by coastal wetlands, seagrasses, and coral reefs. The size of the dots indicates the importance of the service.

Services	Comments and Examples	Estuaries and Marshes	Mangroves	Lagoons, Including Salt Ponds	Inter tidal Flats, Beaches, and Dunes	Kelp	Rock and Shell Reefs	Seagrass Beds	Coral Reefs
Coastal Wetlands									
Provisioning									
Food	production of fish, algae, and invertebrates	●	●	•	●	•	●	•	●
Fresh water	storage and retention of water; provision of water for irrigation and for drinking	•		•					
Fiber, timber, fuel	production of timber, fuelwood, peat, fodder, aggregates	●	●	●					
Biochemical products	extraction of materials from biota	•	•			•			•
Genetic materials	medicine; genes for resistance to plant pathogens, ornamental species, and so on	•	•	•		●			•
Regulating									
Climate regulation	regulation of greenhouse gases, temperature, precipitation, and other climatic processes; chemical composition of the atmosphere	●	●	●	•		•	•	●
Biological regulation (C11.3)	resistance of species invasions; regulating interactions between different trophic levels; preserving functional diversity and interactions	●	●	●	•		•		•
Hydrological regimes	groundwater recharge/discharge; storage of water for agriculture or industry	•		•					
Pollution control and detoxification	retention, recovery, and removal of excess nutrients and pollutants	●	●	●		■	•	•	•
Erosion protection	retention of soils	●	●	•				•	•
Natural hazards	flood control; storm protection	●	●	•	•	•	●	●	●
Cultural									
Spiritual and inspirational	personal feelings and well-being	●	•	●	●	•	•	•	●
Recreational	opportunities for tourism and recreational activities	●	•	•	●	•			●
Aesthetic	appreciation of natural features	●	•	●	●				●
Educational	opportunities for formal and informal education and training	•	•	•	•		•		•
Supporting									
Biodiversity	habitats for resident or transient species	●	●	•	●	•	●	•	●
Soil formation	sediment retention and accumulation of organic matter	●	●	•	•				
Nutrient cycling	storage, recycling, processing, and acquisition of nutrients	●	●	●	•	•	•		●

Source: Millennium Ecosystem Assessment (2005)/Island Press.

remotely sensed, are usually applied to measure stress. The ground-based measurement is fluorescence emitted by leaves measured with a pulse-amplitude-modulated (PAM) fluorescence meter, which gives an estimate of the efficiency of energy utilization by leaves. A healthy leaf will have higher energy efficiency than a leaf that is stressed. The remotely sensed measurement detects different forms of xanthophyll pigments, which can be used as an indicator of stress. The stress of marsh vegetation, as measured by the spectral reflectance of plant pigments, is governed by nutrient and water availability, phytotoxins, salinity, and the relative sea level. Combining marsh elevation data with measurements of the level of stress of vegetation is an integrative indicator of marsh productivity, health, and stability.

Study Questions

Multiple Choice

1. Which type of primary producer is NOT found in coastal marshes?
   a. Grasses
   b. Shrubs
   c. Phytoplankton
   d. Algae
2. What is the greatest factor affecting zonation patterns in microtidal marsh systems?
   a. Tides
   b. Temperature
   c. Light availability
   d. Seasonal variation in sea level
3. What are the major drivers of global coastal marsh distribution? (Select all that apply)
   a. Precipitation
   b. Temperature
   c. Tides
   d. Elevation
   e. Soil type
4. What are the most important (but not the only) factors influencing productivity in marsh systems?
   a. Oxygen levels and drainage
   b. Rainfall and nutrient availability
   c. Temperature and nutrient availability
   d. Flooding frequency and soil salinity
5. What influences the ability of marshes to maintain equilibrium elevation? (Select all that apply)
   a. Sediment supply
   b. Organic matter supply
   c. Belowground biomass
   d. Groundwater supply
   e. Temperature
6. What is an example of top-down control affecting coastal marsh vegetation?
   a. The snail *Littoraria* strongly suppressing *Spartina* production via grazing
   b. Humans cutting the tops of *Juncus* to collect the biomass for animal feedstock
   c. Increased runoff from recent rains providing extra nutrients for *Salicornia* plants
   d. Greenhouse gases creating a hotter environment that is beneficial for mangrove growth, thereby replacing marsh plants
7. What type of correlation do methane emissions have with salinity and why?
   a. Positive, lower salinity kills methane-producing bacteria
   b. Negative, higher sulfate concentrations in seawater allow sulfate-reducing bacteria to outcompete
   c. Negative, higher salinity kills methane-producing bacteria
   d. Positive, lower sulfate concentrations in seawater allow sulfate-reducing bacteria to outcompete
8. Which are not examples of major human impacts on coastal marshes? (Select all that apply)
   a. Cattle grazing
   b. Draining
   c. Damming
   d. Impounding
   e. Burning
9. Which of the following statements regarding the ecosystem services offered by coastal marshes is true?
   a. Seagrass meadows offer more useable material that can be extracted from biota
   b. Coastal lagoons typically have higher biodiversity than marshes
   c. Kelp forests are better at regulating biogeochemical processes that affect the chemical composition of the atmosphere
   d. Marshes are more important for the storage and retention of water than coral reefs
10. Why might rainfall increase marsh vegetation production?
    a. It washes away surface soil
    b. It reduces soil salinity
    c. It leads to cooler soil temperatures
    d. It causes waterlogged soils

Short Answer

1. Describe the factors that affect the distribution of vascular plant species along an elevational gradient.
2. Define the "outwelling hypothesis."
3. Compare the common methods of measuring marsh productivity.
4. Briefly describe the factors that affect marsh productivity.
5. Describe why decomposition of organic matter is low in coastal marshes.
6. Compare and contrast physical, chemical, and biological characteristics of tidal and microtidal marsh systems.
7. Explain how you would combine indicators to give an overall view of marsh health/condition.
8. What steps might you take to restore marsh habitats worldwide? (Open-ended)
9. Provide a reasonable explanation for the regional pattern of aboveground productivity shown in Figure 6.7.
10. Show the patterns of salinity, tidal influence, and marsh plant diversity from coast to inland.

References

Adams, D.A. (1963). Factors influencing vascular plant zonation in North Carolina salt marshes. *Ecology* 44: 445–456.

Aerts, R. and Toet, S. (1997). Nutritional controls on carbon dioxide and methane emission from Carex-dominated peat soils. *Soil Biol. Biochem.* 29: 1683–1690.

Al Hassan, M., Chaura, J., López-Gresa, M.P. et al. (2016). Native-invasive plants vs. halophytes in Mediterranean salt marshes: stress tolerance mechanisms in two related species. *Front. Plant Sci.* 7: 473.

Altieri, A.H., Bertness, M.D., Coverdale, T.C. et al. (2012). A trophic cascade triggers collapse of a salt-marsh ecosystem with intensive recreational fishing. *Ecology* 93 (6): 1402–1410.

Anderson, I.C., Tobias, C.R., Neikirk, B.B., and Wetzel, R.L. (1997). Development of a process-based nitrogen mass balance model for a Virginia (USA) Spartina alterniflora salt marsh: implications for net DIN flux. *Mar. Ecol. Prog. Ser.* 159: 13–27.

Anisfeld, S.C. and Hill, T.D. (2012). Fertilization effects on elevation change and belowground carbon balance in a Long Island Sound tidal marsh. *Estuar. Coasts* 35: 201–211.

Armitage, A.R., Highfield, W.E., Brody, S.D., and Louchouarn, P. (2015). The contribution of mangrove expansion to salt marsh loss on the Texas Gulf Coast. *PLoS One* 10 (5): e0125404.

Asaeda, T., Hai, D.N., Manatunge, J. et al. (2005). Latitudinal characteristics of below- and above-ground biomass of Typha: a modelling approach. *Ann. Bot.* 96 (2): 299–312.

Ayala, F. and O'Leary, J.W. (1995). Growth and physiology of Salicornia bigelovii (Torr.) at suboptimal salinity. *Int. J. Plant Sci.* 156 (2): 197–205.

Bakker, J.P., Leeuw, J., Dijkema, K.S. et al. (1993). Salt marshes along the coast of the Netherlands. *Hydrobiologia* 265: 73–95.

Baustian, M. M., C. L. Stagg, C. L. Perry, L. C.Moss, T. J. B. Carruthers and M. Allison. 2017. Relationships Between Salinityand Short-Term Soil Carbon Accumulation Rates from Marsh Types Across aLandscape in the Mississippi River Delta. Wetlands, 37: 313–324.

Bertness, M.D. (1991). Zonation of Spartina patens and Spartina alterniflora in a New England salt marsh. *Ecology* 72: 138–148.

Bertness, M.D. (1992). The ecology of a New England salt marsh. *Am. Sci.* 80 (3): 260–268.

Bertness, M.D. and Pennings, S.C. (2002). Spatial variation in process and pattern in salt marsh plant communities in Eastern North America. In: *Concepts and Controversies in Tidal Marsh Ecology* (ed. M.P. Weinstein and D.A. Kreeger), 39–57. The Netherlands: Kluwer Academic Publishers.

Bertness, M.D., Crain, C., Holdredge, C., and Sala, N. (2008). Eutrophication and consumer control of New England salt marsh primary productivity. *Conserv. Biol.* 22 (1): 131–139.

Bertness, M.D., Brisson, C.P., Coverdale, T.C. et al. (2014). Experimental predator removal causes rapid salt marsh die-off. *Ecol. Lett.* https://doi.org/10.1111/ele.12287.

Blum, M.D. and Roberts, H.H. (2009). Drowning of the Mississippi Delta due to insufficient sediment supply and global sea-level rise. *Nat. Geosci.* 2: 488–491.

Bradley, P.M. and Morris, J.T. (1990). Influence of oxygen and sulfide concentration on nitrogen uptake kinetics in Spartina alterniflora. *Ecology* 71 (1): 282–287.

Bragazza, L., Buttler, A., Siegenthaler, A., and Mitchell, E.A. (2009). Plant litter decomposition and nutrient release in peatlands. *Geoph. Monog. Series.* 184: 99–110.

Bridgham, S.D., Updegraff, K., and Pastor, J. (1998). Carbon, nitrogen, and phosphorus mineralization in northern wetlands. *Ecology* 79: 1545–1561.

Bridgham, S.D., Megonigal, J.P., Keller, J.K. et al. (2006). The carbon balance of North American wetlands. *Wetlands* 26: 889–916.

Brix, H., Sorrell, B.K., and Schierup, H.-H. (1996). Gas fluxes achieved by in situ convective flow in Phragmites australis. *Aquat. Bot.* 54: 151–163.

Buresh, R.J., DeLaune, R.D., and Patrick, W.H. (1980). Nitrogen and phosphorus distribution and utilization by Spartina alterniflora in a Louisiana Gulf Coast marsh. *Estuaries* 3 (2): 111–121.

Callaway, R.M., Jones, S., Ferren, W.R., and Parikh, A. (1990). Ecology of a mediterranean-climate estuarine wetland at Carpinteria, California: plant distributions and soil salinity in the upper marsh. *Can. J. Bot.* 68: 1139–1146.

Chapman, V.J. (1960). *Salt Marshes and Salt Deserts of the World.* New York, USA: Interscience Press.

Chaudhary, D.R., Kim, J., and Kang, H. (2018). Influences of different halophyte vegetation on soil microbial community at temperate salt marsh. *Microb. Ecol.* 75: 729–738.

Chen, Z., Xu, H., and Wang, Y. (2021). Ecological degradation of the Yangtze and Nile delta-estuaries in response to dam construction with special reference to monsoonal and arid climate settings. *Water* 13 (9): 1145.

Cheng, C., M. Li, Z. Xue, Z. Zhang, X. Lyu, M. Jiang and H. Zhang. 2020. Impacts of Climate and Nutrients on Carbon Sequestration Rate byWetlands: A Meta-analysis. Chinese Geographical Science, 30: 483–492.

Chmura, G.L., Anisfeld, S.C., Cahoon, D.R., and Lynch, J.C. (2003). Global carbon sequestration in tidal, saline wetland soils. *Glob. Biogeochem. Cycles* 17: 22-1–22-12.

Corbau, C., Simeoni, U., Zoccarato, C. et al. (2019). Coupling land use evolution and subsidence in the Po Delta, Italy: Revising the past occurrence and prospecting the future management challenges. *Sci. Total Environ.* 654: 1196–1208.

Cotrufo, M.F., Soong, J.L., Horton, A.J. et al. (2015). Formation of soil organic matter via biochemical and physical pathways of litter mass loss. *Nat. Geosci.* 8: 776.

Coverdale, T.C., Altieri, A.H., and Bertness, M.D. (2012). Belowground herbivory increases vulnerability of New England salt marshes to die-off. *Ecology* 93: 2085–2094.

Cozzi, S., Ibáñez, C., Lazar, L. et al. (2019). Flow regime and nutrient-loading trends from the Largest South European watersheds: implications for the productivity of Mediterranean and Black Sea's Coastal Areas. *Water* 11 (1): 1.

Craft, C. (2007). Freshwater input structures soil properties, vertical accretion, and nutrient accumulation of Georgia and U.S. tidal marshes. *Limnol. Oceanogr.* 52 (3): 1220–1230.

Craft, C., Seneca, E.D., and Broome, S.W. (1991). Porewater chemistry of natural and created marsh soils. *J. Exp. Mar. Biol. Ecol.* 152: 187–200.

Curcó, A., Ibáñez, C., Day, J.W., and Prat, N. (2002). Net primary production and decomposition of salt marshes of the Ebre Delta (Catalonia, Spain). *Estuaries* 25 (3): 309–324.

Darby, F.A. and Turner, R.E. (2008). Below- and aboveground Spartina alterniflora production in a Louisiana salt marsh. *Estuar. Coasts* 31: 223–231.

Day, J. and Erdman, J. (ed.) (2018). *Mississippi Delta Restoration – Pathways to a Sustainable Future*, 261. New York: Springer.

Day, J.D. Jr., Boesch, D.F., Clairain, E.J. et al. (2007). Restoration of the Mississippi Delta: lessons learned from hurricanes Katrina and Rita. *Science* 315: 1679–1684.

Day, F.P. and Megonigal, J.P. (1993). The relationship between variable hydroperiod, production allocation, and belowground organic turnover in forested wetlands. *Wetlands* 13: 115–121.

Day, J.W., Hall, C.A.S., Kemp, W.M., and Yáñez-Arancibia, A. (1989). *Estuarine Ecology*. New York: Wiley.

Day, R., Holz, R., and Day, J.W. (1990). An inventory of wetland impoundments in the coastal zone of Louisiana, USA: historical trends. *Environ. Manag.* 14 (2): 229–240.

Day, J.W., Martin, J.F., Cardoch, L., and Templet, P.H. (1997). System functioning as a basis for sustainable management of deltaic ecosystems. *Coast. Manag.* 25: 115–153.

Day, J., Shaffer, G., Britsch, L. et al. (2000). Pattern and process of land loss in the Mississippi delta: a spatial and temporal analysis of wetland habitat change. *Estuaries* 23: 425–438.

Day, J.W., Ibáñez, C., Scarton, F. et al. (2011). Sustainability of Mediterranean deltaic and lagoon wetlands with sea-level rise: the importance of river input. *Estuar. Coasts* 34: 483–493.

Day, J., Kemp, P., Freeman, A., and Muth, D. (ed.) (2014). *Perspectives on the Restoration of the Mississippi Delta: The Once and Future Delta*, 194. New York: Springer.

Day, J.W., Clark, H.C., Chang, C. et al. (2020). Life cycle of oil and gas fields in the Mississippi River Delta: a review. *Water* 12 (5): 1492.

Day, J., Goodman, R., Chen, Z. et al. (2021). Deltas in arid environments. *Water* 13 (12): 1677.

De Leeuw, J., Olff, H., and Bakker, J.P. (1990). Year-to-year variation in peak above-ground biomass of six salt-marsh angiosperm communities as related to rainfall deficit and inundation frequency. *Aquat. Bot.* 36: 139–151.

De Neiff, A.P., Neiff, J.J., and Casco, S.L. (2006). Leaf litter decomposition in three wetland types of the Paraná River floodplain. *Wetlands* 26: 558–566.

Deegan, L.A., Hughes, J.E., and Rountree, R.A. (2000). Salt marsh ecosystem support of marine transient species. In: *Concepts and Controversies in Tidal Marsh Ecology* (ed. M.P. Weinstein and D.A. Kreeger), 333–365. The Netherlands: Kluwer Academic Publishers.

Deegan, L.A., Johnson, D.S., Warren, R.S. et al. (2012). Coastal eutrophication as a driver of salt marsh loss. *Nature* 490: 388.

DeLaune, R.D., Buresh, R.J., and Patrick, W.H. (1979). Relationship of soil properties to standing crop biomass of Spartina alterniflora in a Louisiana marsh. *Estuar. Coast. Mar. Sci.* 8: 477–487.

Deverel, S.J., Ingrum, T., and Leighton, D. (2016). Present-day oxidative subsidence of organic soils and mitigation in the Sacramento-San Joaquin Delta, California, USA. *Hydrogeol. J.* 24 (3): 569–586.

Di Bella, C.E., Jacobo, E., Golluscio, R.A., and Rodríguez, A.M. (2013). Effect of cattle grazing on soil salinity and vegetation composition along an elevation gradient in a temperate coastal salt marsh of Samborombón Bay (Argentina). *Wetl. Ecol. Manag.* 22: 1–13.

Duarte, C.M. (1989). Temporal biomass variability and production/biomass relationships of seagrass communities. *Mar. Ecol. Prog. Ser.*. Oldendorf 51 (3): 269–276.

Duncan, P., Foose, T.J., Gordon, I.J. et al. (1990). Comparative nutrient extraction from forages by grazing bovids and equids: a test of the nutritional model of equid bovid competition and coexistence. *Oecologia* 84: 411–418.

Dungait, J.A., Hopkins, D.W., Gregory, A.S., and Whitmore, A.P. (2012). Soil organic matter turnover is governed by accessibility not recalcitrance. *Glob. Chang. Biol.* 18: 1781–1796.

Elsey-Quirk, T., Graham, S., Mendelssohn, I. et al. (2019). Synthesis of wetland responses to freshwater, sediment, and nutrient inputs: Will Mississippi River sediment diversions enhance wetland sustainability? *Estuar. Coast. Shelf Sci.* https://doi.org/10.1016/j.ecss.2019.03.002.

Fennessy, M.B., Jacobs, A.D., and Kentula, M.E. (2004). *Review of Rapid Methods for Assessing Wetland Condition*. EPA/620/R-04/009. U.S. Environmental Protection Agency.

FitzGerald, D.M. and Hughes, Z. (2019). Marsh processes and their response to climate change and sea-level rise. *Annu. Rev. Earth Planet. Sci.* 47.

Forbrich, I. and Giblin, A.E. (2015). Marsh-atmosphere CO_2 exchange in a New England salt marsh. *J. Geophys. Res. Biogeosci.* 120 (9): 1825–1838.

Frasco, B.A. and Good, R.E. (1982). Decomposition dynamics of Spartina alterniflora and Spartina patens in a new jersey salt marsh. *Am. J. Bot.* 69: 402–406.

Good, R., Good, N., and Frasco, B. (1982). A review of primary production and decomposition dynamics of the belowground marsh component. In: *Estuarine Comparisons* (ed. V.S. Kennedy), 139–157. New York: Academic Press.

Graham, S.A. and Mendelssohn, I.A. (2014). Coastal wetland stability maintained through counterbalancing accretionary responses to chronic nutrient enrichment. *Ecology* 95: 3271–3283.

Haddix, M.L., Paul, E.A., and Cotrufo, M.F. (2016). Dual, differential isotope labeling shows the preferential movement of labile plant constituents into mineral-bonded soil organic matter. *Glob. Chang. Biol.* 22: 2301–2312.

He, Q. and Silliman, B.R. (2016). Consumer control as a common driver of coastal vegetation worldwide. *Ecol. Monogr.* https://doi.org/10.1002/ecm.1221.

Hemminga, M., Huiskes, A., Steegstra, M., and Van Soelen, J. (1996). Assessment of carbon allocation and biomass production in a natural stand of the salt marsh plant *Spartina anglica* using ^{13}C. *Mar. Ecol. Prog. Ser.* 130: 169–178.

Herbert, E. R.,P. Boon, A. J. Burgin, S. C. Neubauer, R. B. Franklin, M. Ardón, K. N.Hopfensperger, L. P. M. Lamers and P. Gell. 2015. A global perspective onwetland salinization: ecological consequences of a growing threat to freshwaterwetlands. Ecosphere, 6: 1–43.

Hiatt, M., Snedden, G., Day, J.W. et al. (2019). Drivers and impacts of water level fluctuations in the Mississippi River delta: implications for delta restoration. *Estuarine, Coastal and Shelf Science* 224: 117–137.

Hill, B.H., Elonen, C.M., Herlihy, A.T. et al. (2018). Microbial ecoenzyme stoichiometry, nutrient limitation, and organic matter decomposition in wetlands of the conterminous United States. *Wetl. Ecol. Manag.* 1–15.

Hiraishi, T., Krug, T., Tanabe, K. et al. (2014). *2013 supplement to the 2006 IPCC guidelines for national greenhouse gas inventories: Wetlands*. Switzerland: IPCC.

Hopkinson, C.S., Gosselink, J.G., and Parrondo, R.T. (1978). Above-ground production of seven marsh plant species in coastal Louisiana. *Ecology* 59 (4): 760–769.

Hopkinson, C., Gosselink, J.G., and Parrondo, R. (1980). Production of coastal Louisiana marsh plants calculated from phenometric techniques. *Ecology* 61 (5): 1091–1098.

Houghton, R.A. and Woodwell, G.M. (1980). The Flax Pond Ecosystem Study: Exchanges of CO_2 Between a Salt Marsh and the Atmosphere. *Ecology* 61 (6): 1434–1445.

Howard, R.J. and Mendelssohn, I.A. (2000). Structure and composition of oligohaline marsh plant communities exposed to salinity pulses. *Aquat. Bot.* 68: 143–164.

Howes, B.L., Howarth, R.W., Teal, J.M., and Valiela, I. (1981). Oxidation-reduction potentials in a salt marsh: spatial patterns and interactions with primary production. *Limnol. Oceanogr.* 26 (2): 350–360.

Huertas, I.E., Flecha, S., Figuerola, J. et al. (2017). Effect of hydroperiod on CO_2 fluxes at the air-water interface in the Mediterranean coastal wetlands of Donana. *J. Geophys. Res. Biogeosci.* 122: 1615–1631.

Ibáñez, C., Prat, N., and Canicio, A. (1996). Changes in the hydrology and sediment transport produced by large dams on the lower Ebro River and its estuary. *Regul. Rivers Res. Manag.* 12: 51–62.

Ibáñez, C., Pont, D., and Prat, N. (1997a). Characterization of the Ebre and Rhone Estuaries: a basis for defining and classifying salt wedge estuaries. *Limnol. Oceanogr.* 42 (1): 89–101.

Ibáñez, C., Canicio, A., Day, J.W., and Curcó, A. (1997b). Morphologic development, relative sea level rise and sustainable management of water and sediment in the Ebre Delta. *Spain. J. Coast. Conserv.* 3: 191–202.

Ibáñez, C., Day, J.W., and Pont, D. (1999). Primary Production and decomposition in wetlands of the Rhône Delta, France: interactive impacts of human modifications and relative sea level rise. *J. Coast. Res.* 15 (3): 717–731.

Ibáñez, C., Curcó, A., Day, J.W., and Prat, N. (2000). Structure and productivity of microtidal Mediterranean coastal marshes. In: *Concepts and Controversies in Tidal Marsh Ecology* (ed. M.P. Weinstein and D.A. Kreeger), 107–136. The Netherlands: Kluwer Academic Publishers.

Ibáñez, C., Sharpe, P.J., Day, J.W. et al. (2010). Vertical accretion and relative sea level rise in the Ebro Delta wetlands. *Wetlands* 30: 979–988.

Ibáñez, C., Day, J.W., and Reyes, E. (2014). The response of deltas to sea-level rise: natural mechanisms and management options to adapt to high-end scenarios. *Ecol. Eng.* 65: 122–130.

Jankowski, K.L., Törnqvist, T.E., and Fernandes, A.M. (2017). Vulnerability of Louisiana's coastal wetlands to present-day rates of relative sea-level rise. *Nat. Commun.* 8 (1): 1–7.

Janousek, C.N., Buffington, K.J., Guntenspergen, G.R. et al. (2017). Inundation, vegetation, and sediment effects on litter decomposition in Pacific Coast tidal marshes. *Ecosystems* 20: 1296–1310.

Kaswadji, R.F., Gosselink, J.G., and Turner, R.E. (1990). Estimation of primary production using five different methods in a Spartina alterniflora salt marsh. *Wetl. Ecol. Manag.* 1: 57–64.

Keller, J.K., Takagi, K.K., Brown, M.E. et al. (2012). *Soil Organic Carbon Storage in Restored Salt Marshes in Huntington Beach*. California: SPIE.

Khan, M.Z. and Akbar, G. (2012). In the Indus delta it is no more the mighty Indus. *River Conservation and Manag.* 69–78.

Kirwan, M.L. and Megonigal, J.P. (2013). Tidal wetland stability in the face of human impacts and sea-level rise. *Nature* 504: 53.

Kirwan, M.L., Guntenspergenm, G.R., D'Alpaos, A. et al. (2010). Limits on the adaptability of coastal marshes to rising sea level. *Geophys. Res. Lett.* 37: L23401. https://doi.org/10.1029/2010GL045489.

van Klink, R., Rickert, C., Vermeulen, R. et al. (2013). Grazed vegetation mosaics do not maximize arthropod diversity: Evidence from salt marshes. *Biol. Conserv.* 164: 150–157.

van Klink, R., Nolte, S., Mandema, F.S. et al. (2016). Effects of grazing management on biodiversity across trophic levels–The importance of livestock species and stocking density in salt marshes. *Agric. Ecosyst. Environ.* 235: 329–339.

Koch, M.S., Mendelssohn, I.A., and McKee, K.L. (1990). Mechanism for the hydrogen sulfide-induced growth limitation in wetland macrophytes. *Limnol. Oceanogr.* 35: 399–408.

Kondolf, G.M., Gao, Y., Annandale, G.W. et al. (2014). Sustainable sediment management in reservoirs and regulated rivers: Experiences from five continents. *Earth's Future* 2 (5): 256–280.

Krauss, K.W. and Whitbeck, J.L. (2012). Soil greenhouse gas fluxes during wetland forest retreat along the lower Savannah River, Georgia (USA). *Wetlands* 32: 73–81.

Lamers, L.P.M., Govers, L.L., Janssen, I.C.J.M. et al. (2013). Sulfide as a soil phytotoxin-a review. *Front. Plant Sci.* 4: 268–268.

Langley, A.J., Mozdzer, T.J., Shepard, K.A. et al. (2013). Tidal marsh plant responses to elevated CO_2, nitrogen fertilization, and sea level rise. *Glob. Chang. Biol.* 19 (5): 1495–1503.

Laskowski, R. and Berg, B. (2006). *Litter Decomposition: Guide to Carbon and Nutrient Turnover*. Amsterdam.

Leonard, L.A. and Luther, M.E. (1995). Flow dynamics in tidal marsh canopies. *Limnol. Oceanogr.* 40: 11474–11484.

Liao, C., Luo, Y., Jiang, L. et al. (2007). Invasion of Spartina alterniflora Enhanced Ecosystem Carbon and Nitrogen Stocks in the Yangtze Estuary, China. *Ecosystems* 10: 1351–1361.

Linthurst, R.A. (1980). An evaluation of aeration, nitrogen, pH and salinity as factors affecting Spartina alterniflora growth: a summary. In: *Estuarine Perspectives* (ed. V. Kennedy), 235–247. New York: Academic Press.

Macreadie, P.I., Hughes, A.R., and Kimbro, D.L. (2013). Loss of 'blue carbon' from coastal salt marshes following habitat disturbance. *PLoS One* 8 (7): e69244.

Mandema, F.S., Tinbergen, J.M., Stahl, J. et al. (2014). Habitat preference of geese is affected by livestock grazing - seasonal variation in an experimental field evaluation. *Wildl. Biol.* 20: 67–72.

Marples, T.G. (1966). A radionuclide study of arthropod food chains in a Spartina salt marsh ecosystem. *Ecology* 47: 270–277.

McKee, K.L. and Patrick, W.H. (1988). The relationship of smooth cordgrass (Spartina alterniflora) to tidal datums – a review. *Estuaries* 11 (3): 143–151.

McKee, K.L., Cahoon, D.R., and Feller, I.C. (2007). Caribbean mangroves adjust to rising sea level through biotic controls on change in soil elevation. *Glob. Ecol. Biogeogr.* 16: 545–556.

McLeod, E., Chmura, G.L., Bouillon, S. et al. (2011). A blueprint for blue carbon: toward an improved understanding of the role of vegetated coastal habitats in sequestering CO_2. *Front. Ecol. Environ.* 9: 552–560.

Mcowen, C.J., Weatherdon, L.V., Van Bochove, J.W. et al. (2017). A global map of saltmarshes. *Biodivers. Data J.* 5.

Megonigal, J.P., Hines, M., and Visscher, P. (2004). Anaerobic metabolism: linkages to trace gases and aerobic processes. *Biogeochemistry.*

Mendelssohn, I.A. (1979). The influence of nitrogen level, form, and application method on the growth response of Spartina alterniflora in North Carolina. *Estuaries* 2: 106–112.

Mendelssohn, I.A. and Batzer, D. (2006). Abiotic constraints for wetland plants and animals. In: *Ecology of Freshwater and Estuarine Wetlands* (ed. D.P. Batzer and R.R. Sharitz), 82–114. Berkley, CA: University of California Press.

Mendelssohn, I.A. and McKee, K.L. (1988). Spartina alterniflora dieback in Louisiana: time-course investigation of soil waterlogging effects. *J. Ecol.* 76: 509–521.

Mendelssohn, I.A. and Morris, J.T. (2000). Eco-physiological constraints on the primary productivity of Spartina alterniflora. In: *Concepts and Controversies of Tidal Marsh Ecolgy* (ed. M.P. Weinstein and D.A. Kreeger). Elsevier Press.

Mendelssohn, I.A. and Seneca, E.D. (1980). The influence of soil drainage on the growth of salt marsh cordgrass Spartina alterniflora in North Carolina. *Estuar. Coast. Mar. Sci.* 11: 27–40.

Mendelssohn, I.A., McKee, K.L., and Patrick, W.H. Jr. (1981). Oxygen deficiency in Spartina alterniflora roots: metabolic adaptation to anoxia. *Science* 214: 439–441.

Mendelssohn, I.A., Byrnes, M.R., Kneib, R.T., and Vittor, B.A. (2017). Coastal habitats of the Gulf of Mexico. In: *Habitats and Biota of the Gulf of Mexico: Before the Deepwater Horizon Oil Spill*, vol. 1 (ed. C.H. Ward) Chapter 6., 359–640. New York: Springer Open.

Metz, B., Davidson, O.R., Bosch, P.R. et al. (2007). *Climate Change 2007: Mitigation. Contribution of Working Group III to the Fourht Assessment Report of the Intergovernmental Panel on Climate Change*. Intergovernmental Panel on Climatic Change.

Millennium Ecosystem Assessment (Program) (2005). *Ecosystems and Human well-being: Wetlands and water synthesis: A Report of the Millennium Ecosystem Assessment*. World Resources Institute.

Mitsch, W. and Gosselink, J. (2001). *Wetlands*, 539. New York: Van Nostrand Reinhold.

Mitsch, W.J., Bernal, B., Nahlik, A.M. et al. (2013). Wetlands, carbon, and climate change. *Landsc. Ecol.* 28: 583–597.

Moffett, K.B., Nardin, W., Silvestri, S. et al. (2015). Multiple stable states and catastrophic shifts in coastal wetlands: progress, challenges, and opportunities in validating theory using remote sensing and other methods. *Remote Sens.* 7 (8): 10184–10226.

Morris, J.T. (1991). Effects of nitrogen loading on wetland ecosystems with particular reference to atmospheric deposition. *Annu. Rev. Ecol. Syst.* 22: 257–279.

Morris, J.T. (2006). Competition among marsh macrophytes by means of geomorphological displacement in the intertidal zone. *Estuar. Coast. Shelf Sci.* 69: 395–402.

Morris, J.T. (2007). Estimating net primary production of salt marsh macrophytes. In: *Principles and Standards for Measuring Primary Production* (ed. T.J. Fahey and A.K. Knapp), 106–119. Oxford University Press.

Morris, J.T. and Haskin, B. (1990). A 5-yr record of aerial primary production and stand characteristics of Spartina alterniflora. *Ecology* 71 (6): 2209–2217.

Morris, J.T. and Jensen, A. (1998). The carbon balance of grazed and non-grazed Spartina anglica saltmarshes at Skallingen, Denmark. *Journal of Ecology* 86 (2): 229–242.

Morris, J.M., Kjerfve, B., and Dean, J.M. (1990). Dependence of estuarine productivity on anomalies in mean sea level. *Limnol. Oceanogr.* 35 (4): 926–930.

Morris, J.T., Sundareshwar, P.V., Nietch, C.T. et al. (2002). Responses of coastal wetlands to rising sea level. *Ecology* 83: 2869–2877.

Morris, J.T., Shaffer, G.P., and Nyman, J.A. (2013). Brinson review: perspectives on the influence of nutrients on the sustainability of coastal Wetlands. *Wetlands* 33: 975–988.

Morton, R.A. and Bernier, J.C. (2010). Recent subsidence-rate reductions in the Mississippi Delta and their geological implications. *J. Coast. Res.* 555–561.

Mudd, S.M., Howell, S.M., and Morris, J.T. (2009). Impact of dynamic feedbacks between sedimentation, sea-level rise and biomass production on near-surface marsh stratigraphy and carbon accumulation. *Estuar. Coast. Shelf Sci.* 82: 377–389.

Mueller, P., Granse, D., Nolte, S. et al. (2017). Top-down control of carbon sequestration: Grazing affects microbial structure and function in salt marsh soils: Grazing. *Ecol. Appl.* 27: 1435–1450.

Nahlik, A.M. and Fennessy, M.S. (2016). Carbon storage in US wetlands. *Nat. Commun.* 7: 13835.

National Oceanic and Atmospheric Administration Office for Coastal Management(2010). *C-CAP Regional Land Cover. Coastal Change Analysis Program (C-CAP) Regional Land Cover.* Charleston, SC: NOAA Office for Coastal Management https://www.coast.noaa.gov/digitalcoast/.

Neubauer, S.C., Miller, W.D., and Anderson, I.C. (2000). Carbon cycling in a tidal freshwater marsh ecosystem: a carbon gas flux study. *Mar. Ecol. Prog. Ser.* 199: 13–30.

Neubauer, S.C., Givler, K., Valentine, S., and Megonigal, J.P. (2005). Seasonal patterns and plant-mediated controls of subsurface wetland biogeochemistry. *Ecology* 86: 3334–3344.

Neves, J.P., Ferreira, L.F., Simões, M.P., and Gazarini, L.C. (2007). Primary production and nutrient content in two salt marsh species, Atriplex portulacoides L. andLimoniastrum monopetalum L., in Southern Portugal. *Estuar. Coasts* 30 (3): 459–468.

Nixon, S., Ammerman, J., Atkinson, L. et al. (1996). The fate of nitrogen and phosphorus at the land-sea margin of the North Atlantic Ocean. *Biogeochemistry* 35: 141–180.

Odum, E.P. (1968). A research challenge: evaluating the productivity of coastal and estuarine water. Proceedings of the Second Sea Grant Congress, University of Rhode Island, USA, page 6364.

Odum, E.P. (1980). The status of three ecosystem-level hypothesis regarding salt marsh estuaries: tidal subsidy, outwelling and detritus-based food chains. In: *Estuarine Perspectives* (ed. V. Kennedy), 485–495. New York: Academic.

Odum, W.E. (1988). Comparative ecology of tidal freshwater and salt marshes. *Annu. Rev. Ecol. Syst.* 19: 147–176.

Osland, M.J., Enwright, N., and Stagg, C.L. (2014). Freshwater availability and coastal wetland foundation species: ecological transitions along a rainfall gradient. *Ecology* 95 (10): 2789–2802.

Ouyang, X. and Lee, S.Y. (2014). Updated estimates of carbon accumulation rates in coastal marsh sediments. *Biogeosciences* 11: 5057–5071.

Pennings, S.C. (2012). Ecology: The big picture of marsh loss. *Nature* 490: 352.

Pennings, S.C. and Callaway, R.M. (1992). Salt marsh plant zonation: the relative importance of competition and physical factors. *Ecology* 73 (2): 681–690.

Pennings, S.C., Grant, M.B., and Bertness, M.D. (2005). Plant zonation in low-latitude salt marshes: disentangling the roles of flooding, salinity and competition. *J. Ecol.* 93: 159–167.

Pierfelice, K.N., Lockaby, G.B., Krauss, K.W. et al. (2015). Salinity influences on aboveground and belowground net primary productivity in tidal wetlands. *J. Hydrol. Eng.* 22 (1): D5015002.

Ponnamperuma, F.N. (1972). The chemistry of submerged soils. *Adv. Agron.* 24: 29–96.

Rath, K.M., Maheshwari, A., Bengtson, P., and Rousk, J. (2016). Comparative Toxicities of Salts on Microbial Processes in Soil. *Appl. Environ. Microbiol.* 82: 2012–2020.

Redfield, A.C. (1972). Development of a New England salt marsh. *Ecol. Monogr.* 42 (2): 201–237.

Reyes, E., White, M.L., Martin, J.F. et al. (2000). Landscape modeling of coastal habitat change in the Mississippi delta. *Ecology* 81 (8): 2331–2349.

Rickert, C., Fichtner, A., van Klink, R., and Bakker, J.P. (2012). α- and β-Diversity in moth communities in salt marshes is driven by grazing management. *Biol. Conserv.* 146: 24–31.

Rosso, P., Ustin, S., and Hastings, A. (2006). Use of LIDAR to study changes associated with Spartina invasion in San Francisco Bay marshes. *Remote Sens. Environ.* 100: 295–306.

Rovira, A. and Ibàñez, C. (2007). Sediment management options for the lower Ebro River and its delta. *J. Soils Sediments* 7 (5): 285–295.

Rovira, A., Ibáñez, C., and Martín-Vide, J.P. (2015). Suspended sediment load at the lowermost Ebro River (Catalonia, Spain). *Quat. Int.* 388: 188–198.

Rozema, J., Dorel, F., Janissen, R. et al. (1991). Effect of elevated atmospheric CO2 on growth, photosynthesis and water relations of salt marsh grass species. *Aquat. Bot.* 39: 45–55.

Saintilan, N., Rogers, K., Kelleway, J.J. et al. (2018). Climate change impacts on the coastal Wetlands of Australia. *Wetlands*.

Sánchez-Arcilla, A., Jiménez, J.A., Valdemoro, H.I., and Gracia, V. (2008). Implications of climatic change on spanish mediterranean low-lying coasts: the Ebro delta case. *J. Coast. Res.* 24: 306–316.

Scarton, F.J., Day, J.W., and Rismondo, A. (2002). Primary production and decomposition of Sarcocornia fruticosa and Phragmites australis in the Po Delta, Italy. *Estuaries* 25: 325–336.

Schrama, M., Kuijper, D.P.J., Veeneklaas, R.M., and Bakker, J.P. (2015). Long-term decline in a salt marsh hare population largely driven by bottom-up factors. *Ecoscience* https://doi.org/10.1080/11956860.2015.1079409.

Setia, R., Marschner, P., Baldock, J., and Chittleborough, D. (2010). Is CO2 evolution in saline soils affected by an osmotic effect and calcium carbonate? *Biol. Fertil. Soils* 46: 781–792.

Shaffer, G., Day, J.W., Mack, S. et al. (2009). The MRGO navigation project: a massive human-induced environmental, economic, and storm disaster. *J. Coast. Res.* 54: 206–224.

Sharpe, P.J. and Baldwin, A.H. (2009). Patterns of wetland plant species richness across estuarine gradients of Chesapeake Bay. *Wetlands* 29 (1): 225–235.

Silliman, B.R. and Bertness, M.D. (2002). A trophic cascade regulates salt marsh primary production. *Proc. Natl. Acad. Sci.* 99 (16): 10500–10505.

Silliman, B.R. and Zieman, J.C. (2001). Top-down control of Spartina Alterniflora production by periwinkle grazing in a Virginia salt marsh. *Ecology* 82: 2830–2845.

Silliman, B.R., Van Koppel, J., De Bertness, M.D. et al. (2005). Drought, snails, and large-scale die-off of Southern U.S. salt marshes. *Science* 310: 1803–1806.

Sklar, F.H. and Browder, J.A. (1998). Coastal environmental impacts brought about by alterations to freshwater flow in the Gulf of Mexico. *Environ. Manag.* 22 (4): 547–562.

Spalding, E.A. and Hester, M.W. (2007). Interactive effects of hydrology and salinity on oligohaline plant species productivity: implications of relative sea-level rise. *Estuar. Coasts* 30 (2): 214–225.

Stachelek, J. and Dunton, K.H. (2013). Freshwater inflow requirements for the Nueces Delta, Texas: *Spartina alterniflora* as an indicator of ecosystem condition. *Tex. Water J.* 4 (2): 62–73.

Stagg, C. L., M. M. Baustian, C. L. Perry, T. J. B. Carruthers, C.T. Hall and A. Zanne. 2018. Direct and indirect controls on organic matterdecomposition in four coastal wetland communities along a landscape salinitygradient. Journal of Ecology, 106: 655–670.

Stagg, C.L., Schoolmaster, D.R., Krauss, K.W. et al. (2017a). Causal mechanisms of soil organic matter decomposition: deconstructing salinity and flooding impacts in coastal wetlands. *Ecology* 98: 2003–2018.

Stagg, C.L., Schoolmaster, D.R., Piazza, S.C. et al. (2017b). A landscape-scale assessment of above-and belowground primary production in coastal wetlands: implications for climate change-induced community shifts. *Estuar. Coasts* 40: 856–879.

Stanley, D.J. (1988). Subsidence in the northeastern Nile delta: rapid rates, possible causes, and consequences. *Science* 240: 497–500.

Stanley, J.D. and Clemente, P.L. (2017). Increased land subsidence and sea-level rise are submerging Egypt's Nile Delta coastal margin. *GSA Today* 27 (5): 4–11.

Steever, E.Z., Warren, R.S., and Niering, W.A. (1976). Tidal energy subsidy and standing crop production of Spartina alterniflora. *Estuar. Coast. Mar. Sci.* 4 (4): 473–478.

Swenson, E.M. and Turner, R.E. (1987). Spoil banks: Effects on a coastal marsh water level regime. *Estuar. Coast. Shelf Sci.* 24: 599–609.

Syvitski, J., Kettner, A., Overeem, I. et al. (2009). Sinking deltas due to human activities. *Nat. Geosci.* https://doi.org/10.1038/NGE0629.

Teal, J.M. and Howes, B.L. (1996). Interannual variability of a salt-marsh ecosystem. *Limnol. Oceanogr.* 41 (4): 802–809.

Tobias, C. and Neubauer, S.C. (2019). Salt marsh biogeochemistry—an overview. In: *Coastal Wetlands* (ed. G. Perillo, E. Wolanski, D. Cahoon and C. Hopkinson), 539–596. Elsevier.

Turner, R.E. (1976). Geographic variations in salt marsh macrophyte production: a review. *Contrib. Mar. Sci.* 20: 47–68.

Turner, R.E. (1979). A simple model of the seasonal growth of *Spartina alterniflora* and *Spartina patens*. *Contrib. Mar. Sci.* 22: 137–147.

Turner, R.E., Howes, B.L., Teal, J.M. et al. (2009). Salt marshes and eutrophication: an unsustainable outcome. *Limnol. Oceanogr.* 54 (5): 1634–1642.

Valiela, I. and Teal, J.M. (1979). The nitrogen budget of a salt marsh ecosystem. *Nature* 280: 652–656.

Valiela, I., Wilson, J., Buchsbaum, R. et al. (1984). Importance of chemical composition of salt marsh litter on decay rates and feeding by detritivores. *Bull. Mar. Sci.* 35: 261–269.

Veen, G., Freschet, G.T., Ordonez, A., and Wardle, D.A. (2015). Litter quality and environmental controls of home-field advantage effects on litter decomposition. *Oikos* 124: 187–195.

Vörösmarty, C., Syvitski, J., Day, J.W. et al. (2009). Battling to save the world's river deltas. *Bull. At. Sci.* 65: 31–43.

Weinstein, M.P. and Kreeger, D.A. (ed.) (2000). *Concepts and Controversies in Tidal Marsh Ecology*, 875. Boston: Kluwer Academic Publishers.

Whalen, S. (2005). Biogeochemistry of methane exchange between natural wetlands and the atmosphere. *Environ. Eng. Sci.* 22: 73–94.

Whigham, D.F., McCormick, J., Good, R.E., and Simpson, R.L. (1978). Biomass and primary production in freshwater tidal wetlands of the Middle Atlantic Coast. In: *Freshwater Wetlands: Ecological Processes and Management Potential* (ed. R.E. Good et al.), 3–20. New York, USA: Academic Press.

Whiting, G.J. and Chanton, J.P. (1996). Control of the diurnal pattern of methane emission from emergent aquatic macrophytes by gas transport mechanisms. *Aquat. Bot.* 54: 237–253.

Wiegert, R. and Evans, F. (1964). Primary production and the disappearence of dead vegetation on an old field in southeastern Michigan. *Ecology* 45: 49–63.

Windham, L. (2001). Comparison of biomass production and decomposition between Phragmites australis (common reed) and Spartina patens (salt hay grass) in brackish tidal marshes of New Jersey, USA. *Wetlands* 21: 179–188.

Winemiller, K.O., Akin, S., and Zeug, S.C. (2007). Production sources and food web structure of a temperate tidal estuary: integration of dietary and stable isotope data. *Mar. Ecol. Prog. Ser.* 343: 63–76.

Yang, Z., Nolte, S., and Wu, J. (2017). Tidal flooding diminishes the effects of livestock grazing on soil micro-food webs in a coastal saltmarsh. *Agric. Ecosyst. Environ.* 236: 177–186.

Zedler, J.B. (1977). Salt marsh community structure in the Tijuana Estuary, California. *Estuar. Coast. Mar. Sci.* 5: 39–54.

Zedler, J.B. (1983). Freshwater impacts in normally hypersaline marshes. *Estuaries* 6: 346–355.

Zedler, J.B., Winfield, T., and Williams, P. (1980). Salt marsh productivity with natural and altered tidal circulation. *Oecologia* 44: 236–240.

CHAPTER 7

Mangrove Wetlands

Robert R. Twilley[1], Andre Rovai[1], and Ken W. Krauss[2]
[1] Department of Oceanography and Coastal Sciences, Louisiana State University, Baton Rouge, LA, USA
[2] Wetland and Aquatic Research Center, U.S. Geological Survey, Lafayette, LA, USA

Red Mangrove (Rhizophora mangle) scrub forest on Taylor Slough in Everglades National Park, Florida, USA. Photo credit: Robert R. Twilley

7.1 Introduction

Mangroves are a unique group of wetlands dominated by trees that colonize nearly 150,000 km^2 of the intertidal zone in tropical, subtropical, and warm temperate climates (Figure 7.1a–f; Table 7.1) (Duke et al. 1998; Saintilan et al. 2014; Tomlinson 2016). Early definitions described mangroves as plant communities dominated by trees, shrubs, ground ferns, and palms below the high-tide mark, and the term tidal forest was a common, though not mutually exclusive synonym for mangrove forest. The current definition states that true mangrove trees must follow several unique criteria, including (1) complete fidelity to the saline intertidal zone (form a major role in structuring the vegetative community and do not occur in terrestrial environments), (2) morphological specialization representing adaptation to the intertidal environment including aerial roots and vivipary of embryo (see Section 7.4), and (3) physiological adaptations including salt exclusion, salt secretion, and salt accumulation (Tomlinson 2016). Trees, shrubs, and ferns with these characteristics are called "true" mangroves (Table 7.1).

The term *mangrove* may best define a specific type of tree, whereas *mangrove wetlands* refer to whole plant associations with other community assemblages in the intertidal zone (Duke et al. 1998). This is similar to the term "mangal" introduced by Macnae (1968) to refer to tropical, coastal swamp ecosystems. These are important distinctions since the biodiversity of mangrove wetlands may be considered species poor in some regions when exclusively considering tree species richness, but the biodiversity of mangrove wetlands may equal that of other tropical ecosystems when including all flora and fauna (Rützler and Feller 1996; Twilley et al. 1996). In addition, the habitats of mangrove-associated estuaries support diverse communities, consisting of a variety of primary producers and secondary consumers distributed in bays and lagoons. Thus, there is a hierarchical use of the term mangrove that can be used to describe individual trees, to coastal forested wetlands, and whole estuarine landscapes/seascapes. In this chapter, we will focus on the ecological characteristics of mangrove wetlands by describing individual, community, and ecosystem features of landscapes dominated by these unique wetland plants.

Estuarine Ecology, Third Edition. Edited by Byron C. Crump, Jeremy M. Testa, and Kenneth H. Dunton.

Companion website: www.wiley.com/go/crump/estuarine3

FIGURE 7.1 (a) Red mangrove trees (*Rhizophora mangle*) in the Everglades National Park, Florida (USA) (*Source*: Andrewtappert/ Wikimedia Commons/CC BY 3.0); (b) Mangrove zonation in savanna-dry winter climate along Pacific coast of Honduras in the Gulf of Fonseca dominated by taller red mangroves along the fringe (*R. mangle* and *R. racemosa* in background); note change in height of black mangroves in foreground due to salinity stress along interior (*Avicennia bicolor* and *Avicennia germinans*). (c) Basin mangroves of swamps and drainage basins with restricted hydrology in Puerto Rico which are often dominated by black mangrove trees (*A. germinans*); (d) Tidal creek mangroves in New Caledonia; (e) Fringe mangroves dominated by *R. mangle* along Gambia River, Africa; and (f) Basin mangroves on Kosrae island (Micronesia) dominated by *Bruguiera gymnorrhiza* and *Sonneratia alba* trees.(*Source*: (b), (c), (e), (f) by R. Twilley; (d) by K. Krauss)

7.2 Biogeography

7.2.1 Diversity

Mangroves are a group of halophytes with species from 28 genera in 20 different families and two plant divisions with approximately 70 species and/or putative hybrids recognized throughout the world (Duke et al. 1998). A total of at least 36 true species are recognized from the Indo-West-Pacific area, but fewer than 10 species are found in the new world (Macnae 1968) reflecting the old-world origin of mangroves (Table 7.1; Figure 7.2a). In addition, there is a strong effect of latitude on species richness in the eastern hemisphere with maximum richness along the equator (40 species) and reduction by half at 20 degrees north and south latitude (Ellison et al. 1999). The Gulf of Mexico and Caribbean have even lower species richness with only four mangroves commonly found in the region including *Rhizophora mangle*, *Avicennia germinans*, *Laguncularia racemosa*, and *Conocarpus erectus*. *Rhizophora* (Rhizophoracea) and *Sonneratia* (Sonneratiaceae) have nine species each (including three distinct hybrids), and *Avicennia* (Avicenniaceae) has eight species, collectively representing the most diverse genera of mangrove trees globally (Table 7.1). There are three other genera in the Rhizophoraceae family, all of which are located only in the old-world tropics (*Bruguiera* has the most at six species). In addition, rare mangrove species are almost exclusively located in the old-world tropics (nearly 19 species), reflecting speciation time and adaptation, with only one specially adapted species, *Pelliciera rhizophorae*, located in the new world tropics (Duke et al. 1998; Dangremond et al. 2015).

7.2.2 Global Patterns

Tree species richness of mangrove wetlands exhibit clear patterns between the eastern and western hemisphere, but the total mangrove wetland areas of these two regions is very similar (Table 7.1, Figure 7.2b; Ellison et al. 1999). Previous estimates of global mangrove wetland area ranged from 137,600 to 152,361 km^2 (Spalding 2010; Giri et al. 2011b), but recent updates using revised mangrove mapping techniques estimate that the area could be slightly less (Hamilton and Casey 2016). There is a general trend that about 75% of that total mangrove wetland area is found between 0 and 10 degrees north and south of the equator. Thus, both species richness and mangrove wetland distribution are highest near the equator. About 77,169 km^2 are in Asia, which along with Australia (10,287 km^2), make up over half of the global mangrove wetland area (Giri et al. 2011b). The most extensive region of mangrove wetlands in the Atlantic East-Pacific (AEP) region is along the coast of Brazil, followed by Mexico and Venezuela. Mangrove wetlands colonize large areas of depositional environments along large river delta estuaries with low coastal plain relief such as the Sundarbans mangrove forest in Bay of Bengal. Reduced disturbances from cyclones or strong winds promote the structural development of forests in regions such as Micronesia, where mangrove trees can survive for more than a century (see Section 7.3.1).

The distribution and abundance of mangrove wetlands is restricted to shores with warm temperate to tropical climates where the average air temperature of the coldest month is higher than 20°C and where the seasonal range does not exceed 10 °C (Chapman 1976). The world distribution of mangroves (Figure 7.2b), particularly at the northern and southern limits, appears to correlate reasonably well with the 16 °C isotherm for the air temperature of the coldest month (Chapman 1976). Northernmost populations occur at 32°N latitude (Bermuda, Japan) and southernmost populations occur at 37–38°S

TABLE 7.1 **Genera of mangroves and the distribution of species between the Atlantic Eastern Pacific (AEP) and the Indo-West Pacific (IWP) regions.**

Family	Genus	Mangrove species	List of species—AEP	List of species—IWP
Major components				
Avicenniaceae	*Avicennia*	8	*A. germinans, A. schaueriana, A. bicolor*	*A. marina, A. alba, A. officinalis, A. lanata, A. eucalyptifolia,*
Combretaceae	*Laguncularia*	1	*L. racemosa*	
Combretaceae	*Lumnitzera*	2		*L. littorea, L. racemosa*
Palmae	*Nypa*	1		*N. fruticans*
Rhizophoraceae	*Bruguiera*	6		*B. gymnorrhiza, B. cylindrica, B. parviflora, B. sexangula, B. hainesii, B. exaristata*
Rhizophoraceae	*Ceriops*	2		*C. tagal, C decandra*
Rhizophoraceae	*Kandelia*	1		*K. candel*
Rhizophoraceae	*Rhizophora*	8	*R. harrisonii, R. mangle, R. racemosa, R. x harrisonii,*	*R. mucronata, R. apiculata, R. stylosa, R. x lamarckii, R. samoensis, R. x selala*
Sonneratiaceae	*Sonneratia*	5		*S. alba, S. apetala, S. caseolaris, S. griffithii, S. ovata*
Subtotal	**9**	**34**	**7**	**28**
Minor components				
Bombacaceae	*Camptostemon*	2		*C. philippensis, C. schultzii*
Euphorbiaceae	*Exoecaria*	1		*E. agallocha*
Lythraceae	*Pemphis*	1		*P. acidula*
Meliaceae	*Xylocarpus*	2		*X. granatum, X. mekongensis*
Myrsinaceae	*Aegiceras*	2		*A. floridum, A. corniculatum*
Myrtaceae	*Osbornia*	1		*O. octodonta*
Pellicieraceae	*Pelliciera*	1	*P. rhizophorae*	
Plumbaginaceae	*Aegialitis*	2		*A. rotundifolia, A. annulata*
Pteridaceae	*Acrostichum*	3		
Rubiaceae	*Scyphiphora*	1		*S. hydrophyllacea*
Sterculiaceae	*Heritiera*	3		*H. littoralis, H. fomes, H. globatis*
Subtotal	**11**	**20**	**1**	**15**
Total	**20**	**54**	**8**	**43**

Source: Modified from Tomlinson (2016).

latitude (Australia and New Zealand) (Saintilan et al. 2014). While these general trends hold, the frequency, duration, and/or severity of freezing temperature is a prime factor governing the distribution and abundance of mangroves around the world (Stuart et al. 2007; Lovelock et al. 2016). Where freezing is routine, mangrove wetlands are replaced in the intertidal zone by salt marsh vegetation (Friess et al. 2012). The greater resprouting ability of *Avicennia* (Figure 7.1c) and *Laguncularia* mangroves results in greater recovery from freeze damage in some locations (Lugo and Patterson-Zucca 1977). Both of these

(a)

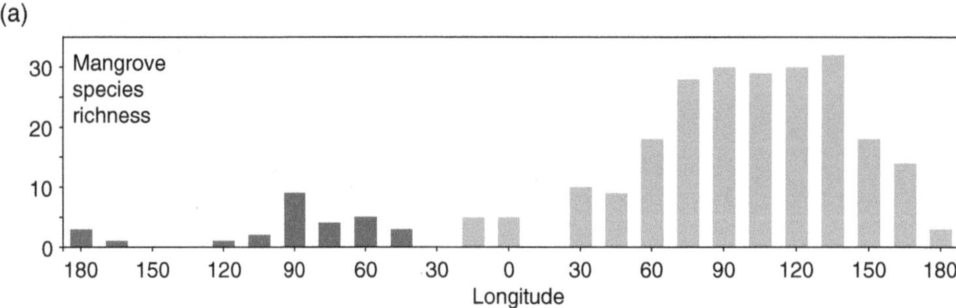

(b)

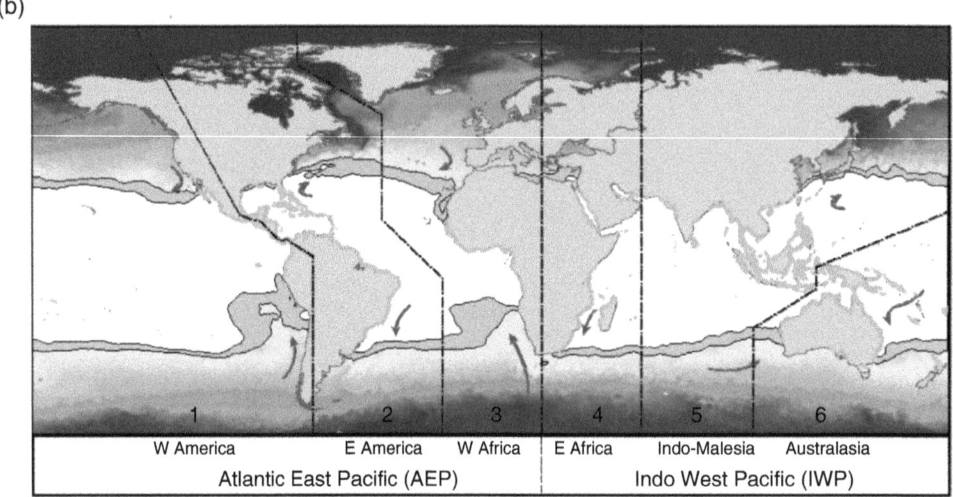

FIGURE 7.2 (a) Mangrove species richness by longitude (*Source*: Based on Ellison et al. 1999) depict the greater species richness in Indo-West Pacific (IWP) versus Atlantic East-Pacific (AEP) biogeographic regions. (b) Distribution of mangrove forests along coastlines globally (light green) along with the six biogeographic regions defined by Duke et al. (1998). Polygons (pink) over the oceans represent updated 20 °C isotherms for summer and winter along with the influential currents affecting those isotherms (arrows). *Source*: Based on Duke et al. (1998).

mangrove genera have extensive secondary meristematic tissues which allow rapid development of sprouts following defoliation of the canopy by frost (Snedaker et al. 1992). In warm temperate regions of the southeastern United States, mangroves are excluded where minimum temperatures fall below −8.9 °C (Osland et al. 2013) with significant stress leading to elimination of many species at −4 °C (Cavanaugh et al. 2014).

7.3 Ecogeomorphology and Ecosystem Processes

Mangrove wetlands are unique given that they are forested wetlands adapted to the intertidal zone of diverse prograding and transgressive coastal settings where rivers, tides, and waves determine many of the ecological processes shaping these ecosystems' structure and function (see Chapter 2). Coastal landforms where mangroves inhabit the interface of land and sea are diverse with geomorphology responding to sedimentary processes forming intertidal zones ranging from depositional environments shaped by rivers, waves, and tides, to carbonate environments, with sediments shaped through biological processes (e.g., coral reefs, Woodroffe 2002). The concept of ecogeomorphology is based upon the observation that complex interaction of river, tides, and waves are geomorphic processes that together with microtopography of the intertidal zone control patterns of environmental gradients that define habitat types and ecosystem properties. The combination of geophysical, geomorphic, and biologic processes defines coastal environmental settings leading to predictions of ecosystem attributes across diverse coastal landforms. Furthermore, the physical presence of mangrove wetlands in the intertidal zone creates "feedback effects" on geomorphic processes that directly facilitate development of the intertidal zone, clarifying our use of ecogeomorphology. An ecosystem model (Figure 7.3) describes how coastal processes of climate, tides, and river control conditions along with landform characteristics shape distinct coastal environmental settings, which are reflected in site-specific controls of hydrology, salt, sediment, and nutrient exchange between the estuary and mangrove platforms (*platform* describes that portion of intertidal zone colonized by mangrove wetland). The ecological processes (e.g., biomass and productivity, biogeochemistry, and food webs, Figure 7.3) associated with mangrove wetlands vary among these coastal environmental settings (Twilley 1995; Twilley et al. 2018).

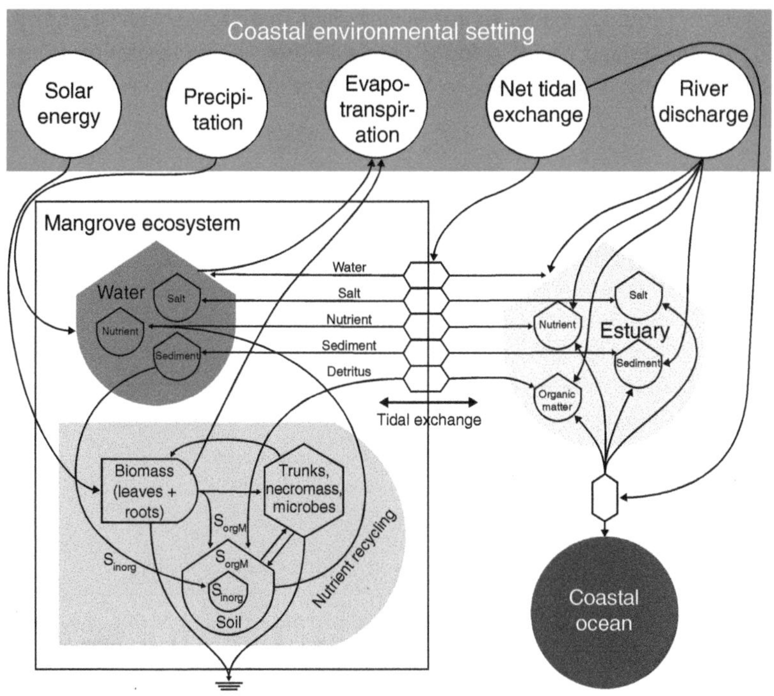

FIGURE 7.3 Energy flow model of mangrove wetlands capturing the influence of river and tides on ecological processes. *Source*: Modified from Cintrón et al. (1978). The model depicts biomass accumulations; the process of productivity, respiration, and cycling; input and output of salt, sediment, nutrients (net input), and organic matter (net output); and the factors (both internal and external) believed to control these rates. Soil includes inorganic (S_{inorg}) and organic (S_{orgM}) matter. Symbols are from Odum (1994), and emphasize stocks (tanks store raw materials, bullets store materials of primary producers, hexagons store materials of secondary consumers), flows (lines with arrows with direction of flow; fluxes are flows across boundaries), and sources (circles represent unlimited amounts).

7.3.1 Hierarchy of Ecosystem Patterns

Given the complexity of how coastal geomorphic and ecological processes interact in diverse coastal environmental settings (Figure 7.3), a hierarchical classification system was developed to describe patterns of mangrove ecosystem ecology based on biogeographic (global), geomorphological (regional), and ecological (local) factors that control the adaptation of mangrove wetlands to gradients in hydrology, regulators, and resources in mangrove soils (Figure 7.4; also see Section 7.4). Coastal environmental settings are defined by the source of sediment (see inputs to the soil compartment of mangrove wetlands in Figure 7.3), which is controlled by a combination of geomorphic processes and local runoff (Woodroffe 2002). At the regional scale, there are two major sedimentary settings: those formed by either terrigenous clastic sediments or by biogenic carbonate sediments. Terrigenous coastal settings are sediment-rich, depositional coastal environments where clastic sediments delivered by rivers are resuspended by tides and waves (sediment stock in soil water in Figure 7.3) to form coastal landforms including deltas (e.g., Mekong Delta, Vietnam), estuaries (e.g., Guayas River Estuary, Ecuador), and lagoons (e.g., Terminos Lagoon, Mexico). In contrast, carbonate coastal settings are formed by *in situ* processes of sediment deposition that result from a combination of biogenic carbonate formation and mangrove belowground productivity, that can generate large deposits of mangrove peat. Carbonate settings comprise karstic environments (such as southwestern Florida and many islands in the Caribbean) and Holocene reef tops, including mangroves bordering oceanic islands (such as Micronesia). As a result, these sedimentary settings can be classified into several distinct coastal environmental settings, including large river deltas, small deltas, tidal estuaries, and lagoons, as well as carbonate and arheic (isolated from water sources) coastal settings (Woodroffe 1992; Dürr et al. 2011).

Spatially explicit global typology of coastlines (Dürr et al. 2011) recognizes muddy coasts (including large rivers and small deltas), estuarine coasts (estuaries, bays, and lagoons), and carbonate coasts, aligning with the coastal environmental classifications by Thom (1982) and Woodroffe (1992). Based on this analysis, mangrove wetland area can be distributed among coastal environmental settings in both the Atlantic East Pacific (AEP) and Indo West Pacific (IWP) regions of the world (Figures 7.2b, 7.5). Most mangrove wetlands are in tidal systems of the IWP region, followed by the small deltas. Tidal systems and small deltas represent more than 80% of the total mangrove wetland area. Lagoons are evenly distributed between AEP and IWP, but AEP has a higher area of mangrove wetlands in carbonate settings. The global distribution of mangrove wetlands in distinct coastal morphologies is essential in determining global carbon budgets, and potentially other ecosystem processes (Rovai et al. 2018; Twilley et al. 2018; Jennerjahn 2020; Rovai et al. 2021).

In each of these types of coastal environmental settings, local variations in topography and hydrology result in the development of distinct ecological types of mangrove wetlands such as riverine, fringe, and basin forests (Figures 7.1, 7.4; Lugo and Snedaker 1974; Woodroffe 2002). Mangrove wetlands typically occur from slightly below mean sea level (MSL) to the upper intertidal zone; lower intertidal and sub-tidal zones are naturally limited in mangrove wetland distribution. In river-dominated environments, the mid-to-upper intertidal zone is classified as riverine mangrove wetlands, with zones of mangroves that are taller and more productive (Lugo and Snedaker 1974). With more tidal influence and less river input (Figure 7.3), fringe mangrove zones consist of sandier, less muddy sediment and more saline water inundating mangroves. Intertidal island environments dominated by tides are referred to as overwash mangroves (mangrove islands that are flooded by tides). All three of these ecological types may be considered fringe mangroves that occupy river-dominated (riverine) to tidal-dominated (overwash) environments (Figure 7.4, local controls).

The upper intertidal zone is less frequently inundated than fringe and riverine mangroves and have a different substrate and forest stature (Figure 7.4; Woodroffe 2002). These interior mangrove wetlands include the basin and scrub forests classified in Lugo and Snedaker (1974). Since river and tidal inundation decreases within interior mangroves, the balance of water (hydroperiod) is controlled more by regional climate than by the river and tidal properties of the coastal zone (Twilley and Chen 1998). In more arid coastal settings of AEP where evapotranspiration exceeds precipitation, interior mangrove zones are monospecific and dominated by more salt tolerant species, such as in the genus *Avicennia*, forming scrub mangrove forests at the most interior locations in proximity of salt pannes or salinas (< 3 m tall, Figure 7.4) (Castañeda-Moya et al. 2006). In contrast, interior mangrove zones in moist climate zones with significant upland runoff gradually mix with freshwater vegetation forming a continuous ecotone of vegetation with upland environments, such as on some high rainfall Pacific islands. There are coastal landscapes in arid regions where interior mangroves are not limited to scrub forests due to the presence of freshwater inputs to the coast from upland watersheds, such as the Ciénaga Grande de Santa Marta along the Caribbean Coast of Colombia. Freshwater subsidies to interior mangrove wetlands can also be provided by groundwater from upland watersheds, such as cenotes (natural sinkholes) in the Yucatán Peninsula of Mexico (Breithaupt et al. 2017; Beltram 2018), where mangroves can be highly productive and sequester considerable amounts of carbon in the soil (Adame et al. 2013; Adame et al. 2021).

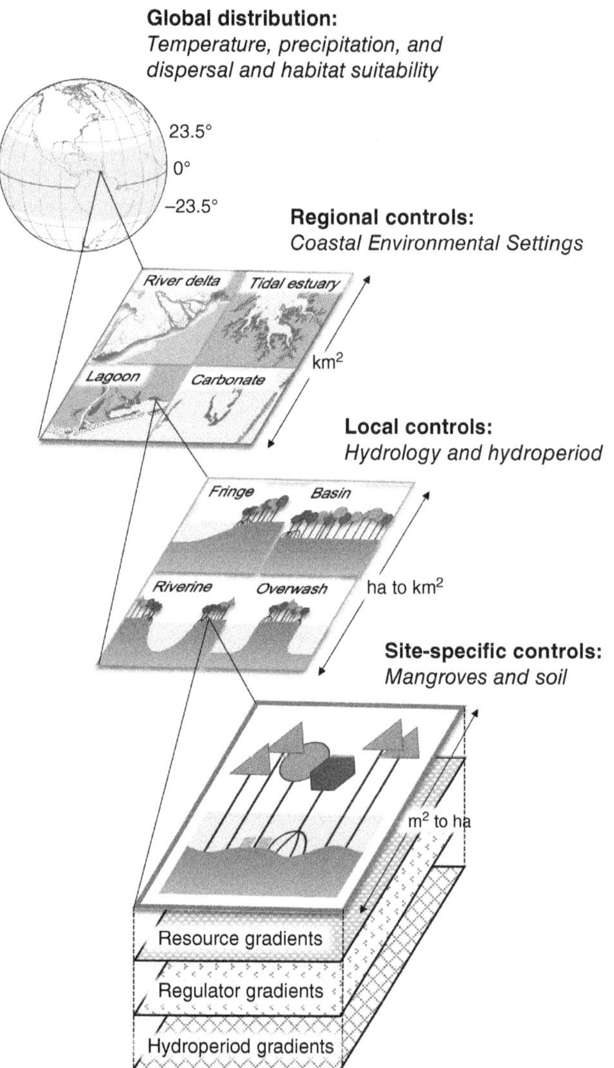

FIGURE 7.4 Hierarchical classification system to describe patterns of mangrove wetland structure and function based on global (biogeographic), morphological (regional), and local ecological factors (e.g., tidal amplitude, temperature, and precipitation) that control the hydrology and concentrations of resources and regulators in soil along gradients from fringe to more interior locations from shore. *Source*: Modified from Twilley et al. (2018) and Rovai et al. (2018).

7.3.2 Biomass and Productivity

Mangrove biomass is the sum of three major living compartments of a forest including the canopy, tree trunks, and roots (both aerial and in soil). The first two compartments constitute the aboveground biomass and roots in soil represent belowground biomass of mangrove wetlands. Dead standing tree trunks, coarse woody debris (e.g., fallen tree trunks and branches), and senescent roots also occupy mangrove soils and surfaces and are referred to as necromass. Mangrove biomass and productivity vary across coastal environmental settings and biogeographic provinces that provide a very interesting comparison on how mangrove wetlands adapt to diverse coastal environmental gradients. River discharge, a proxy for phosphorus loading to coastal oceans (Harrison et al. 2005), is an especially important determinant of biomass allocation between aboveground (trunk, branches, leaves, and aerial roots) and belowground (roots) compartments in mangrove ecosystems. In muddier river-dominated systems that feature low N:P ratios (nitrogen:phosphorus) due to high river-borne phosphorus supply, mangroves typically allocate more biomass aboveground than to roots (Twilley et al. 2017). In contrast, carbonate systems, where river discharge is negligible (or absent), phosphorus may be a limiting nutrient (i.e., high N:P ratios where N:P > 15), causing mangroves to increase biomass allocation to roots to enhance phosphorus foraging (McKee et al. 2007; Adame et al. 2013). In tide-dominated coastlines, where river inputs of sediment and nutrients are less, partitioning of aboveground and belowground biomass is more balanced.

Mangrove aboveground biomass can be estimated using allometric equations based on individual tree diameter and height (Komiyama et al. 2008) or stand-level structural attributes (basal area in $m^2 ha^{-1}$ and mean canopy height or tree density in stems ha^{-1}) (Rovai et al. 2016; Rovai et al. 2021). Interestingly, despite the higher number of species in the IWP region, these allometric scaling relationships are consistent between the AEP and IWP mangroves as well as among coastal environmental settings (Rovai et al. 2021). In addition, species richness and composition are not clearly related to aboveground biomass because trends in forest structure and allometric scaling relationships are consistent across IWP and AEP mangroves. However, there are significant differences in aboveground biomass across coastal morphologies from deltaic (mean of 134.2 Mg ha^{-1}) to arheic (30.0 Mg ha^{-1}) coastal environmental settings (Rovai et al. 2021). Deltas, small deltas, and tidal systems have significantly higher aboveground biomass than lagoonal, carbonate, and arheic coastal settings. High aboveground biomass values have been described for high oceanic islands, based largely on the observations from Micronesia (Cole et al. 1999), where values are nearly 200 Mg ha^{-1}. In fact, tall well-developed (e.g., basal area = 40 m^2 ha^{-1}; height = 20–25 m) mangrove wetlands can develop over thick peat deposits on carbonate settings, where there is very little hurricane disturbance. Conversely, several regions of the Caribbean contain the tallest mangrove forests in hurricane prone zones such as Shark River estuary (Everglades, Florida, USA). In this area, allochthonous mineral inputs from storm surges enhance soil nutrient concentrations and contribute to higher forest development (Castañeda-Moya et al. 2010). These carbonate regions maximize mangrove aboveground biomass through the positive feedback of hurricanes when a storm surge transports calcium bound phosphorus into mangrove wetlands. This process temporarily converts otherwise phosphorus-limited carbonate soils into phosphorus-enriched conditions allowing for greater forest structural development (Castañeda-Moya et al. 2010). Similarly, tall mangrove forests develop around cenotes which provide surface connection to subterranean nutrient-rich groundwater; this commonly occurs in the Yucatán peninsula (Adame et al. 2013).

Modeling approaches to estimate global patterns in aboveground biomass include variables such as temperature (mean temperatures of warmest and coldest seasons) and precipitation (precipitation of wettest and driest seasons) (Hutchison et al. 2014), with recent improvements adding coastal geomorphic processes such as tides and river discharge (Rovai et al. 2016). Based on global models to predict mangrove biomass, aboveground biomass ranges from 15 to 516 Mg ha^{-1}, with a mean (±SE) of 164 ± 0.36 Mg ha^{-1} (Rovai et al. 2021). By coupling these predicted biomass estimates with a high-resolution mangrove area map (Figure 7.5a–c), mangrove aboveground biomass globally is predicted at 1.7 Pg dry mass

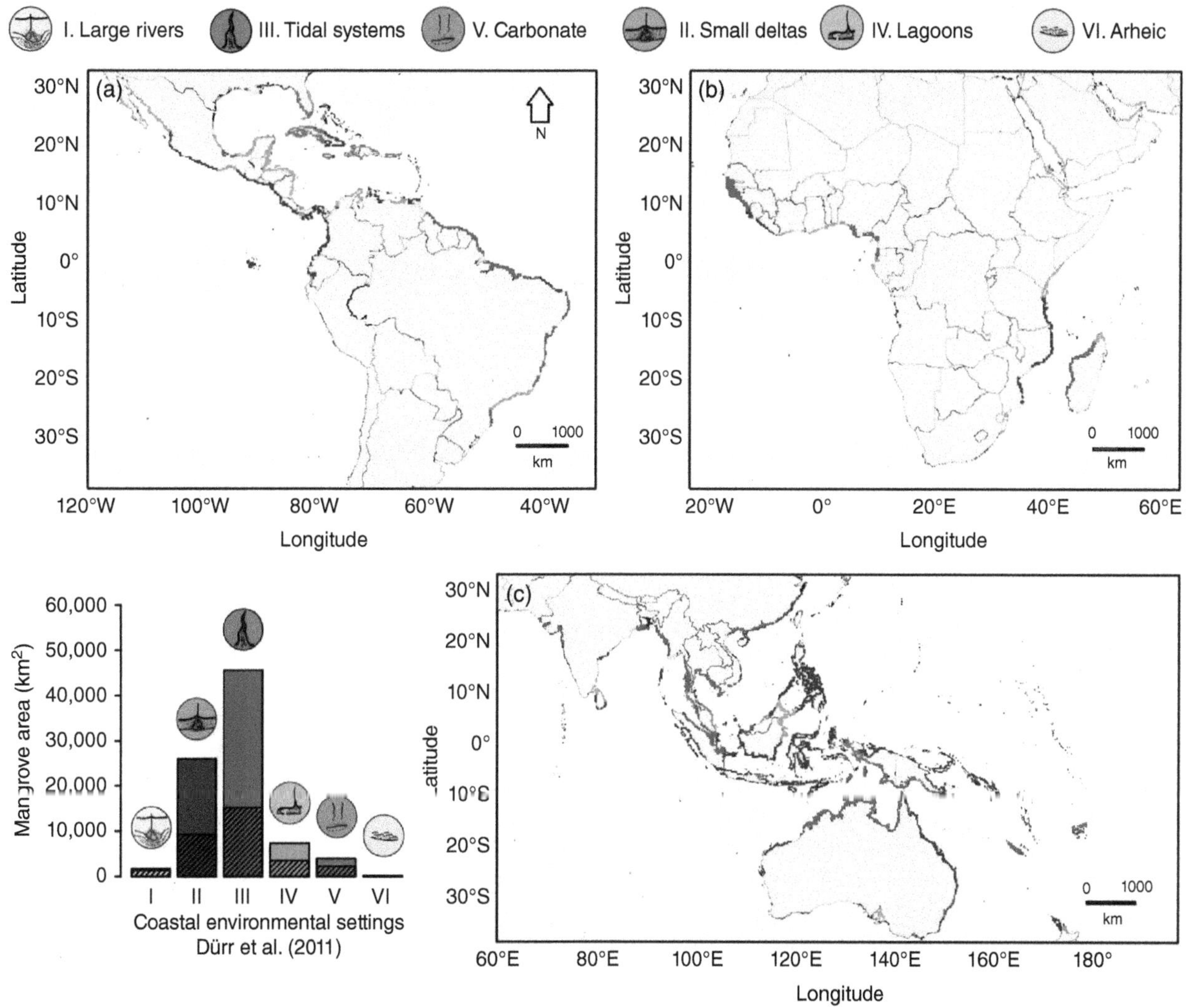

FIGURE 7.5 Global mangrove wetland area divided into distinct coastal environmental settings (a–c). The bar chart shows values for the Atlantic East Pacific (hashed) and Indo-West Pacific regions (solid), respectively (Twilley et al. 2018). Dürr et al. (2011) classification system: I: large rivers (deltas); II: small deltas; III: tidal systems; IV: lagoons; V: carbonate; VI: arheic. *Source*: Based on Twilley et al. (2018) and Dürr et al. (2011).

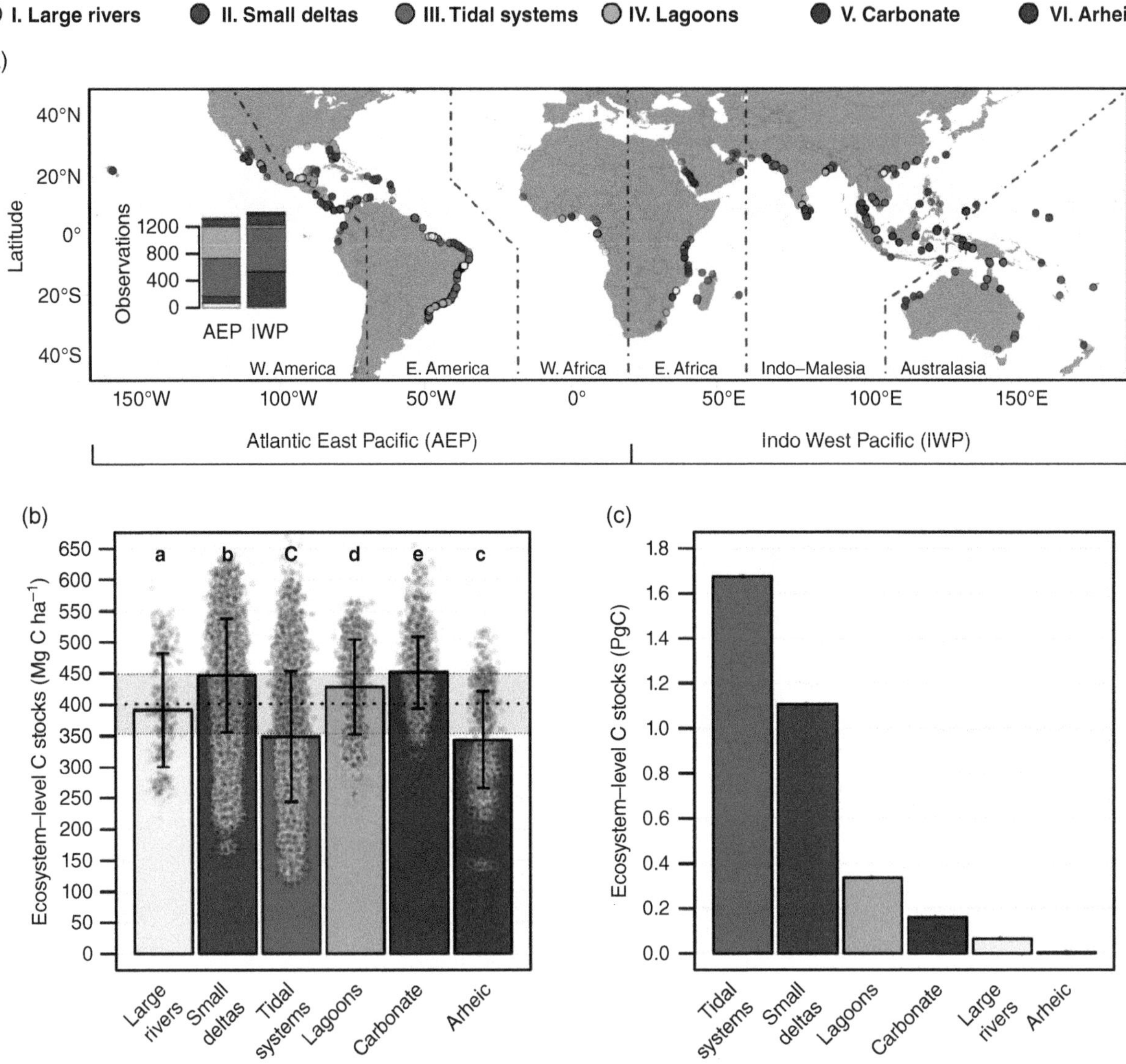

FIGURE 7.6 (a) Global distribution of mangrove aboveground biomass (Mg ha^{-1}) data across distinct coastal environmental settings (CES). (b) The mean global mangrove ecosystem-level carbon stocks (total biomass and soil combined) across distinct coastal environmental settings (CES); the mean of 401 ± 48 Mg C ha^{-1} as indicated by the red dotted line. Error bars show 1 standard deviation of the mean, and different letters show statistical difference ($P < 0.05$) among CES, respectively. (c) The total global mangrove ecosystem-level C stocks for each CES based on mean values of ecosystem-level carbon stocks in 'b", and area of each CES from Fig. 7.5. *Source*: Modified from Rovai et al. (2021); using classification of Dürr et al. (2011).

(Pg = Petagrams = 10^{15} g) and based on carbon content of 0.475 g C g^{-1} dry mass, the estimated carbon stored globally in mangrove aboveground biomass is ~0.8 Pg C (Figure 7.6).

Belowground biomass allocation in mangrove wetlands is considered significant to soil formation and surface elevation gain (Chen and Twilley 1999; McKee et al. 2007). Roots of mangrove wetlands are phenotypically plastic in their ability to exploit nutrients in response to environmental gradients (Feller et al. 2010; Reef et al. 2010). Mangroves can allocate on average 40–60% of their total biomass to belowground roots (Komiyama et al. 1987; Lugo 1990; Khan et al. 2009), but allocation can reach 90% in scrub mangroves in carbonate settings due to nutrient limitation and flooding stress (Castañeda-Moya et al. 2013). Nearly half of the belowground biomass estimates in the literature are <50 Mg ha^{-1}, and 27% of sites have belowground biomass >100 Mg ha^{-1}. For the majority of observations, there is no difference in belowground biomass between IWP and AEP except that the two greater estimates of belowground biomass occur in IWP (>200 Mg ha^{-1}). Currently, the lack of data prevents robust modeling of mangrove belowground biomass at the global scale. Current estimates of global mangrove belowground biomass use an aboveground to belowground biomass ratio of 0.5, and thus a global belowground biomass of 0.84 Pg dry mass and carbon stock of 0.4 Pg C (Rovai et al. 2021).

Estimating the productivity of mangrove wetlands is very complex given the extensive root biomass and tree architecture, along with the fate of litterfall from the forest canopy (see Clark et al. 2001 for forest biomass and productivity protocols). The canopy of mangrove wetlands captures sunlight and CO_2 that determine the gross primary productivity (GPP), from which the net primary productivity (NPP) is expressed in both litterfall (NPP_L), wood production (NPP_W), and root productivity (NPP_B) and can be calculated as follows:

$$NPP = NPP_L + NPP_W + NPP_B. \quad (7.1)$$

The aboveground net primary productivity of mangrove wetlands can be estimated by summing litterfall with wood production as follows:

$$NPP_A = NPP_L + NPP_W \quad (7.2)$$

There are very few sites where both the litterfall and wood productivity have been measured to estimate NPP_A (Twilley et al. 2017). When NPP_A estimates are distributed across the coastal environmental settings (Figure 7.7), values range from about 14–21 Mg ha^{-1} year^{-1} with a global mean of 16.1 Mg ha^{-1} year^{-1} (1610 gdm m^{-2} year^{-1}). Litterfall contributes from 45 to 68% of NPP_A with a global mean of 57%. Thus, if only litterfall is used to estimate NPP_A, it represents about half of the NPP that occurs aboveground in mangrove wetlands. Wood production varies from 4.9 to 11.4 Mg ha^{-1} year^{-1} across the four coastal settings demonstrating more variability than observed for litterfall. The global mean for litterfall is 9.2 Mg ha^{-1} year^{-1} compared to 6.9 Mg ha^{-1} year^{-1} for wood production. A global average of litterfall analyzed for the new world tropics was estimated at 10.2 Mg ha^{-1} year^{-1} (Ribeiro et al. 2019), very similar to the value used in this global summary of mangrove wetlands. A global estimate of belowground productivity (NPP_B in Eq. 1) is about 3 Mg ha^{-1} year^{-1} (Twilley et al. 2017) for a NPP of 19.1 Mg ha^{-1} year^{-1} (1910 gdm m^{-2} year^{-1}). These findings align with the average NPP_A and NPP_B summarized for salt marshes and seagrasses in previous chapters.

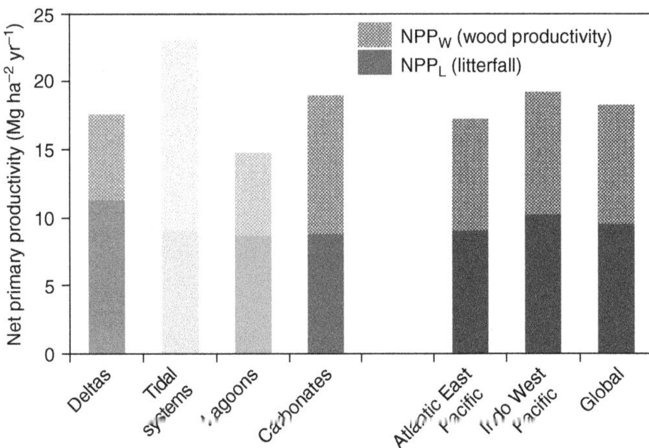

FIGURE 7.7 The net primary productivity of aboveground structures of mangrove wetlands (NPP_A) based on litter productivity (NPP_L, dark colored bars) and wood productivity (NPP_W, stippled bars) across four different types of coastal environmental settings along with a means for AEP region, IWP region, and global. These estimates of NPP_A are based on review by Twilley et al. (2017) using only those sites where both litter and wood productivity were measured. (Multiply values by 100 to convert to g dry mass m^{-2} year^{-1} from Mg ha^{-1} year^{-1} to compare with other primary producers in estuaries). *Source*: Twilley et al. (2017)/Springer Nature.

7.3.3 Litter Dynamics and Export

The dynamics of mangrove leaf litter and, thus, the fate of NPP_A is controlled by relative rates of decomposition and detritus export, which can determine the coupling of mangrove wetlands to the secondary productivity and biogeochemistry of estuarine waters (Twilley et al. 2019; Figure 7.3, exchange of organic matter between soil compartment and estuary). *Rhizophora* has slower decomposition rates than *Avicennia* and *Sonneratia*, which typically take as little as two to three months (residence time) to decompose (Twilley et al. 1997; Bosire et al. 2005; Mfilinge et al. 2005). The rates of leaf litter turnover (1/residence time) by inundation zone are usually of the order riverine>fringe>basin mangrove wetlands with greater litter export in sites with increasing tidal inundation (Twilley et al. 1997). However, several studies in higher energy coastal environments of Australia and Malaysia have emphasized the influence of crabs on the fate of mangrove leaf litter compared to mangrove hydrology (Robertson and Daniel 1989; Imgraben and Dittmann 2008). In coastal environments, where crabs consume 28–79% of the annual leaf fall, litter turnover rates are much higher than those without crabs, suggesting that biological factors may influence leaf litter turnover more than tidal exchange. Grapsid crabs can process significant amounts of mangrove leaf litter, especially in the Indo-West-Pacific region, and can serve as ecological engineers in mangrove ecosystems (Kristensen 2008; Lee 2008). This role of grapsid crabs as grazers change the physicochemical conditions of the soil by increasing drainage, affecting soil redox potentials, increasing soil surface area, and stimulating decomposition and nutrient cycling.

Several authors have suggested that organic matter exported from mangroves (Figure 7.3) is of global significance in the coastal zone (Dittmar and Lara 2001; Kristensen et al. 2008). Mangrove wetlands could be responsible for ~10% of the global export of terrigenous particulate and dissolved organic carbon (POC and DOC) to the coastal zone (Jennerjahn and Ittekkot 2002; Dittmar et al. 2006), and for ~10% of the global organic carbon burial along with seagrasses in the coastal ocean (Duarte et al. 2005). Organic carbon export from mangrove wetlands ranges from 0.5 to 138 g C m^{-2} year^{-1}, with an average rate of about 92 g C m^{-2} year^{-1} (Adame and Lovelock 2011). These estimates of organic carbon export from mangrove wetlands are similar to the average organic carbon export from salt marshes (Nixon 1980), about 100 g C m^{-2} year^{-1}, but much lower than previously published (Twilley 1988). Significant carbon export is particularly evident in river-dominated mangrove systems such as muddy coasts and deltas where organic material exchanged at the boundary of the forest is greater compared to other coastal settings, as observed for major river systems such as Fly River, Papua New Guinea, or systems with large tidal exchanges in Australia (Alongi 2009).

7.3.4 Mangrove Food Webs

The "outwelling hypothesis," or the idea that a significant export of organic matter from mangroves is an energy source to estuarine aquatic food webs, has been revised from the original paradigms described by Odum and Heald (1972) to reflect the diversity of food sources that support estuarine-dependent fisheries (Figure 7.8a). Early conceptual models based on gut content analyses of fish described how organic detritus from mangrove wetlands dominated the lower-level consumers in estuarine food webs (Odum and Heald 1975; Yáñez-Arancibia et al. 1993). In addition, fisheries statistics established a positive correlation between nearshore catch (shrimp or fish) and mangrove area in the vicinity of those harvests (Macnae 1974; Sasekumar et al. 1992; Primavera 1996). Such statistics support the claim that mangrove wetlands support of fisheries depends on the total mangrove wetland area compared to open water habitats (Twilley 1995); a similar finding to salt marshes (Nixon 1980). In regions with a low ratio of wetland to water area, mangrove carbon contribution can range from 2 to 52% of the total available carbon pool for secondary productivity (Twilley 1995; Wafar et al. 1997; Li and Lee 1998). Experimental tests of the outwelling hypothesis using isotope signatures in diets of consumers have commonly found sharp declines in the isotope signal of mangrove detritus with distance from the mouth of a mangrove-dominated creek, much lower signals in adjacent bays, and no signal > 2–4 km from mangrove sources (Rodelli et al. 1984; Bouillon et al. 2004).

Organic matter mass balance studies and isotope observations in mangrove dominated estuaries support a "dual-gradient hypothesis" (Figure 7.8b) of organic matter, as proposed by Odum (1984). The dual-gradient concept states that organic carbon distributions occur along two gradients including salinity along longitudinal axis of an estuary and stream order along marsh tidal creeks. The dual-gradient concept of detritus transport predicts that isotopic ratios in consumer tissues will change from predominantly terrestrial signals to predominantly phytoplankton (less negative) signals as salinity increases and stream order increases. Low stream orders represent habitat with high wetland:open water ratios, which are denser in the lower salinity (oligohaline and mesohaline) regions of estuaries than in the polyhaline regions near the mouth of an estuary (Odum 1984). In these small streams, where high exchange of organic matter and nutrients between open water bodies and surrounding intertidal wetlands such as mangroves is expected, consumers will digest larger proportion of wetland detritus for food. Such classification seems to be a very appropriate guide in the analysis and interpretation of mangrove detritus as an energy source, and this typology as a frame of reference is similar to the conceptual model that isotopic signature of mangrove carbon will vary among distinct ecogeomorphic settings (Twilley et al. 2019). Such trends support the claim that the significance of mangrove detritus to fisheries depends on the total mangrove wetland area compared to the area and/or volume of water habitats in the estuary (Abrantes et al. 2015). More expansive mangrove wetland areas along subtropical and tropical muddy coasts have a higher wetland to water area ratio, compared to lagoon and karstic coastal environmental settings. This pattern may explain variations in contributions of mangrove detritus to food webs.

7.3.5 Net Ecosystem Productivity and Nutrient Biogeochemistry

When respiration by microbes and animals that inhabit mangrove wetlands is subtracted from NPP, this represents net ecosystem productivity (NEP, see Chapter 14). However, NEP

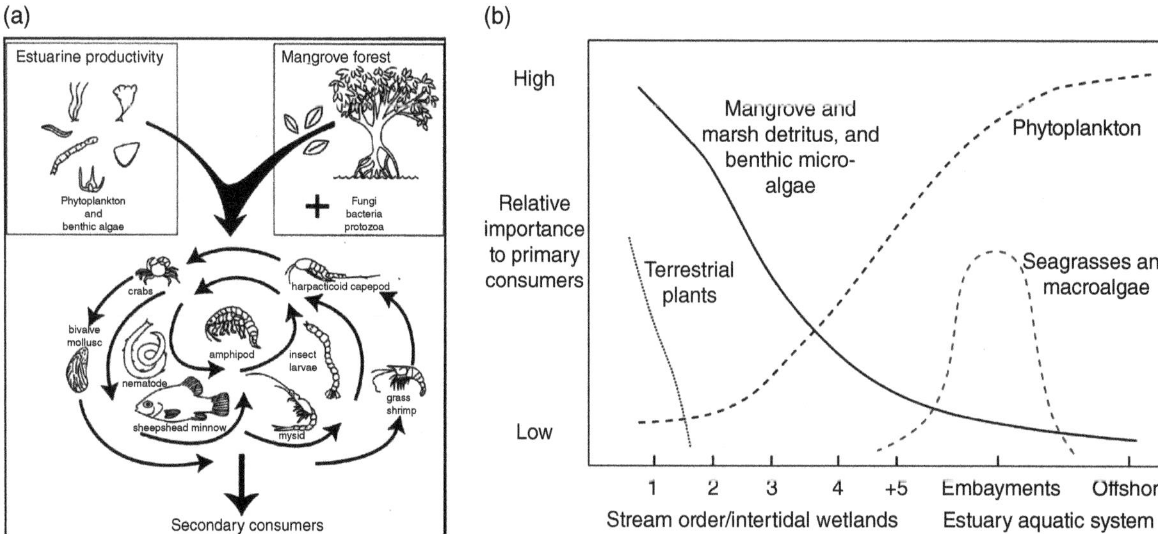

FIGURE 7.8 (a) Schematic of the detrital food chain of the North River Estuary in Everglades National Park. *Source*: Based on Odum and Heald (1975). (b) Conceptual diagram of the principal sources of organic carbon available to primary consumers along the marsh stream order gradient (relative importance values are hypothetical; *Source*: Odum 1984/Ingenta).

calculations of mangrove wetlands also must account for the significant amount of organic matter exported to the estuary via tidal exchange (see the flux with detritus in Figure 7.3 and discussion in Sections 7.3.3 and 7.3.4). Rather than rely on balancing the equation of NEP with complicated measures of tidal exchange at the mangrove-estuary boundary, measurements of the change in soil organic matter with time (the soil stock in Figure 7.3), or ΔS_{orgM}, integrate all the fluxes in and out of the forest floor of a mangrove wetland. The change in S_{orgM} accounts for inputs to the soil compartment from litter and root production that are lost to respiration and net tidal exchange. The other source of organic matter contributing to NEP is wood production, and therefore, the sum of ΔS_{orgM} and wood production (NPP_W) is the best estimate of mangrove NEP (Twilley et al. 2017; Twilley et al. 2019):

$$NEP = NPP_W + \Delta S_{orgM} \tag{7.3}$$

Accounting for NEP in mangrove wetlands provides insights to many ecosystem services of mangroves that would benefit humans in different ways, including detritus export that supports estuarine food webs, organic matter storage that contributes to carbon sequestration (offsetting of greenhouse gas emissions), soil elevation increases that adapt to relative sea level rise, and root interactions with wave energy leading to coastal protection (see Section 7.5).

Soil in mangrove wetlands is developed with a combination of inorganic and organic matter deposited during flooding. The flux of this allochthonous material is a function of net tidal exchange (Woodroffe et al. 2016). Analysis of 144 mangrove wetlands around the world found a global average of 2429 g m^{-2} year^{-1} of sedimentation with 87% consisting of inorganic sediments (Figure 7.9a). Inorganic sediment input is greater in those coastal environmental settings with significant river inputs (terrigenous coastal settings) at 2641 g m^{-2} year^{-1} compared to carbonate coastal settings with inorganic contributions of 331 g m^{-2} year^{-1}. Input of organic sediments across mangrove wetlands in different coastal environmental settings ranges much less, from 341 to 239 g m^{-2} year^{-1} in terrigenous vs carbonate coastal settings, respectively, with global average of 318 g m^{-2} year^{-1} (Figure 7.9a; Twilley et al. 2019). Of the coastal settings that are terrigenous, estuaries and deltas have much more inorganic sedimentation (1755–1078 g m^{-2} year^{-1}) than lagoons (872 g m^{-2} year^{-1}). The ratio of inorganic sediment relative to organic sediment deposition demonstrates how geomorphic processes control sedimentation patterns in mangrove wetlands, as has been observed in other comparisons across types of mangrove wetlands (Breithaupt et al. 2017).

Nutrient burial associated with soil development in mangrove wetlands (soil stock in Figure 7.3) also follows these trends in sedimentation rates, but there are different patterns for phosphorus (P) vs nitrogen (N). Phosphorus burial ranges from 0.11 to 0.78 g P m^{-2} year^{-1} and the higher rates occurred at terrigenous coastal settings with higher inorganic matter loading. These elevated P input rates are associated with

(a)

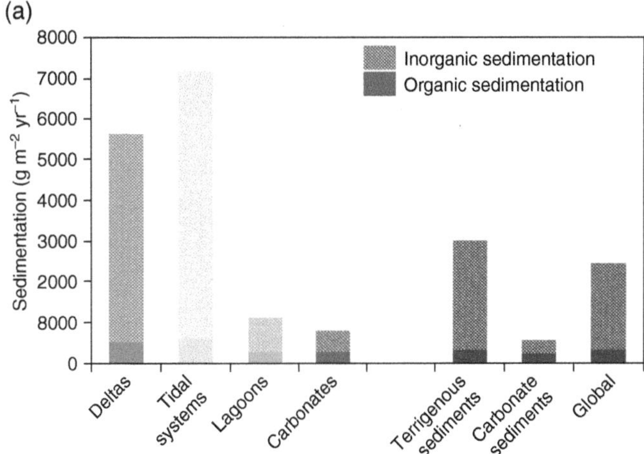

(b)

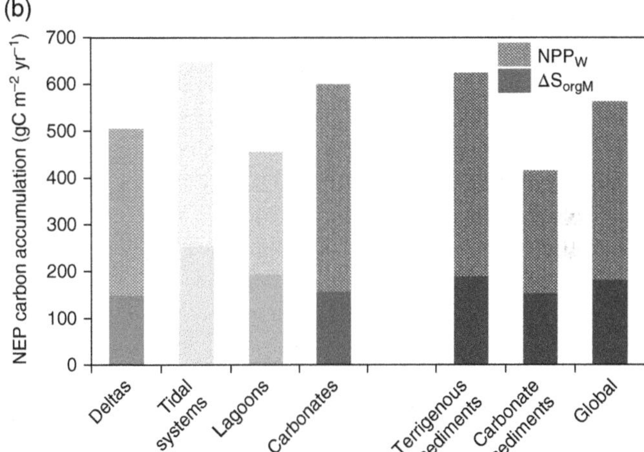

FIGURE 7.9 (a) Sedimentation rates in mangrove wetlands based on organic sedimentation (dark colored bars) and inorganic sedimentation (stippled colored bars) across four different types of coastal environmental settings along with a mean for sites with terrigenous sediments, sites with carbonate sediments, and global estimate. (b) Net ecosystem productivity (NEP) of mangrove wetlands converted to carbon fluxes based on organic matter sedimentation (ΔS_{orgM}, dark colored bars) and wood production (NPP_W, stippled colored bars) across four different types of coastal environmental settings along with a mean for sites with terrigenous sediments, sites with carbonate sediments, and global estimate.

higher levels of NPP_A of mangrove wetlands in deltas and estuaries compared to lagoons that have less phosphorus input and lower productivity. By contrast, the pattern of nitrogen accumulation across mangrove wetlands in coastal environmental settings appears to be more tightly coupled to biomass allocations and organic sedimentation. The accumulation of nitrogen among mangrove wetlands in terrigenous coastal settings is similar to carbonate settings, similar to what was observed for organic sedimentation. Riverine and basin forests are also similar at about 5.5 g N m^{-2} year^{-1}, which is higher than the range of denitrification and N fixation (see Section below), indicating the significant nitrogen burial in mangrove wetland soil.

Organic sedimentation in mangrove soils together with wood production represents the NEP of mangrove wetlands

(Figure 7.9b). Rates are converted to carbon equivalent rates for $NPP_W + \Delta S_{orgM}$ using 0.475 g C g^{-1} dry mass. NEP rates range from 416 to 647 g C m^{-2} year^{-1}, with a global average of 562 g C m^{-2} year^{-1}. Mangrove wetlands in carbonate coastal settings have average NEP of 416 g C m^{-2} year^{-1} compared to 625 g C m^{-2} year^{-1} for mangrove wetlands occurring in coastal settings with terrigenous sediment inputs. NPP_W contributes more to NEP based on a global average of 381 g C m^{-2} year^{-1}, with a NPP_W:ΔS_{orgM} ratio (carbon stored in wood production compared to carbon sequestration in sediments) of 2.4 for mangrove wetlands located in delta settings to 1.4 for mangrove wetlands in lagoons. The global average of carbon sequestration in mangrove wetlands soils is 181 g C m^{-2} year^{-1} in this review compared to a global estimate of 163 g C m^{-2} year^{-1} by Breithaupt et al. (2012). These estimates of carbon stored in NEP for mangrove wetlands in different coastal environmental settings is significant when evaluating the role of these coastal ecosystems in removing CO_2 from the atmosphere (see Section 7.5.5 on blue carbon ecosystems). In many cases, wood production (NPP_W) is not included in estimates of NPP of mangrove wetlands and measures are much less (124 global estimates) than global estimates for litterfall (NPP_L, 537 global estimates). More thorough global estimates of mangrove NPP_A are needed to understand the carbon sequestration potential of mangrove wetlands in all types of coastal environmental settings.

Measuring the nutrient exchange at the atmosphere and mangrove wetland boundary is complex as explained above for organic matter (see discussion of NEP in Section 7.3.2) largely due to the flux of nutrients at the mangrove-estuary (aquatic) boundary. Nutrient exchange is more complicated for nitrogen given the many processes in aerobic and anaerobic environments that control the forms of this nutrient in a mangrove wetland (Figure 7.10) (Reis et al. 2017). Nitrogen flux at the mangrove-estuary boundary includes both inflow and export with tides, where net exchange is nearly equal to 0 or up to −1 mg N m^{-2} day^{-1}. However, the range of fluxes for dissolved inorganic nitrogen species vary from mostly −890 to +1666 mg N m^{-2} day^{-1} (−325 to +608 g N m^{-2} year^{-1}), with evidence that there is a net flux of dissolved inorganic nitrogen into mangrove wetlands from estuaries. Other surveys of nitrogen tidal exchange in Mexico and Australia, which included two inorganic species (NH_4^+ and NO_3^-), particulate nitrogen, and dissolved organic N revealed that the largest nitrogen flux at both sites is particulate nitrogen export to the estuary (Boto and Wellington 1988; Rivera-Monroy et al. 1995a, Rivera-Monroy et al. 1995b). In summary, mangrove wetlands transform inorganic nutrients from tidal import into organic nutrients that are then exported to coastal waters. The most significant flux in the nitrogen budget is biological nitrogen fixation (BNF; Figure 7.10) at 168 mg N m^{-2} day^{-1} (61 g N m^{-2} year^{-1}). Several studies have shown that decaying

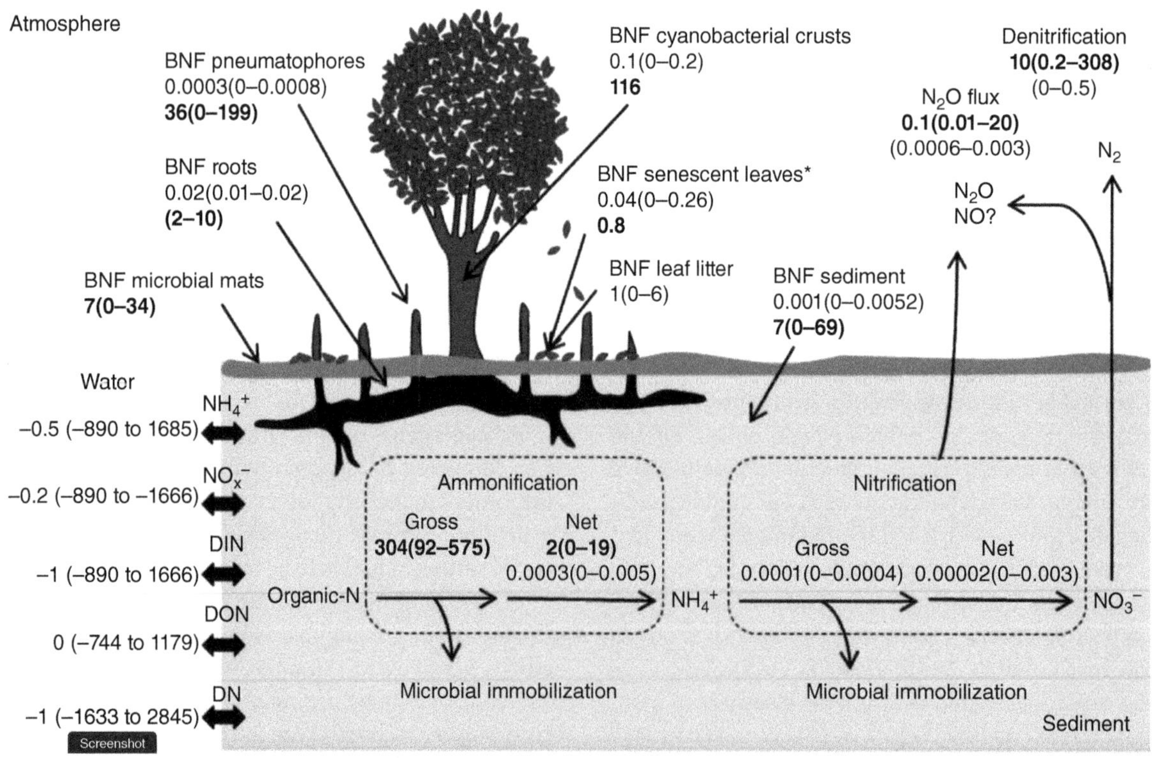

FIGURE 7.10 Nitrogen cycling rates in mangroves from Reis et al. 2017. Median rates (and ranges) in mg N g^{-1} day^{-1} (unbold values) and mg N m^{-2} day^{-1} (bold values) are presented, including biological N fixation (BNF), nitrous oxide (N_2O) flux, and fluxes of ammonium (NH_4^+), nitrite plus nitrate (NO_x^-), dissolved inorganic N (DIN), dissolved organic N (DON), and total dissolved N (DN) between mangrove soil and overlying water. Negative values indicate net flux into the soil. For BNF, mean rates are presented. * incubated on the forest floor. *Source*: Reis et al. (2017)/Springer Nature.

mangrove leaves are substrates for particularly high rates of nitrogen fixation and bacterial growth, and this incorporation of N into bacterial biomass may account for 15–100% of nitrogen immobilization during decomposition of mangrove leaf litter on the forest floor (Pelegri et al. 1997). This is consistent with isotope studies of $^{15}NO_3$ that found a significant amount of nitrogen immobilized rather than being denitrified (Rivera-Monroy and Twilley 1996). However, in one study of Shark River Estuary in south Florida, nitrogen fixation supplied only 7% (8.3 mg N m^{-2} day^{-1}) of the nitrogen required (53 mg N m^{-2} day^{-1}) for mangrove growth (Pelegri et al. 1997), and another study determined that an average nitrogen-fixation rate of 16.8 mg N m^{-2} day^{-1} would only support about 0.4% of the NPP of mangrove wetlands (Bouillon et al. 2008a). Other reviews have found that nitrogen fixation and denitrification occur at about equal rates in mangrove wetlands (Twilley et al. 2019). There is slight evidence that denitrification may dominate the net flux based on this review, but the summary in Figure 7.10 suggests that denitrification is only 10 mg N m^{-2} day^{-1} (3.7 g N m^{-2} year^{-1}). Furthermore, there is no evidence that either process may dominate nitrogen processing in a specific coastal environmental setting or ecological type of mangrove wetlands. However, there is evidence that sponges, tunicates, and a variety of other forms of epibionts on prop roots of mangrove wetlands along carbonate shorelines with little terrigenous sediment input are sites of nitrogen fixation that influence the nitrogen budget of mangrove canopies (Ellison et al. 1996).

7.4 Factors Controlling Productivity and Distribution

7.4.1 Hydroperiod and Waterlogged Soils

Hydrology, often defined operationally by a mangrove wetland's *hydroperiod,* provides the most critical environmental control of mangrove productivity and persistence. Hydroperiod is associated with frequency, depth, and duration of flooding, which can be described in specific inundation classes of mangroves (Friess 2017). These inundation classes explain that hydroperiod establishes patterns of zonation that control seedling establishment and growth. In Malaysia, five inundation classes described the distribution patterns of 17 mangrove species, ranging from inundation by all high tides to inundation by only spring tides. Nearly half of these mangrove species were included in a lower inundation class defined as two or fewer inundations per month associated with exceptionally high tides. While *R. mangle*, *A. germinans*, *L. racemosa*, and *C. erectus* have been suggested to transgress the inundation continuum from wettest to driest, respectively, all three species are present in sites flooded from 30 to 50% of a year (Krauss et al. 2006). However, populations of *Rhizophora* are found at higher inundation durations between 50 and 75%, with the higher inundation only present immediately adjacent to tidal creeks or bays (Krauss et al. 2006). Mangrove wetlands develop naturally in areas that are drained more often than flooded on an annual basis—many mangroves are flooded < 30% of the year (Lewis 2005). Mangrove wetlands cannot generally persist and function appropriately with permanent flooding, as is indicated by their normal intertidal positioning from just below MSL to upper intertidal areas.

Another notable environmental control on mangrove productivity is soil biogeochemistry. Intertidal environments often feature waterlogged soils with anaerobic zones, which contain elevated levels of sulfide and other reduced compounds. Mangroves produce aboveground and belowground roots that provide morphological adaptations due to the lack of oxygen in muddy soils (Figure 7.11). There are five types of aboveground root adaptations most of which have specialized cells, called lenticels, which allow the passage of air (including oxygen) to internal tissue and fine roots in otherwise low oxygen environments: (1) pneumatophores that are conical projections extending vertically upward from the cable root system extending above ground at a height proportional to average flooding depths (e.g., *Avicennia*, *Sonneratia*, see Figure 7.11), (2) knee-roots that grow aboveground and then back down into soil (e.g., *Bruguiera*), (3) stilt or prop roots that are branched roots from the main tree trunk extending downward to the soil surface forming fine roots once embedded in the soil (e.g., *Rhizophora*, see Figure 7.1a, d), (4) buttress or plank roots that are blade-like structures as modifications of lower tree trunk (e.g., *Xylocarpus*, *Heritiera, Pelliciera rhizophorae*), and (5) aerial roots, which are stilt roots growing from tree trunks and branches that extend downward but do not attach to the soil (e.g., *Rhizophora*) (Krauss et al. 2014). Physiological adaptations of belowground roots include ethylene accumulation, which promotes adventitious root production, and tannins that might reduce phytotoxic effects of reduced iron and sulfides in anoxic soils (Pezeshki and DeLaune 2012; Lamers et al. 2013). Also, metabolic adaptations allow mangroves to avoid accumulation of ethanol under anaerobic conditions, by producing less-toxic malic acid as the end-product of anaerobic respiration.

The level of anaerobic conditions in mangrove soil is linked to the duration and pulsing of flooding. The normally high concentration of organic matter in mangrove soils, due to both litter and root productivity, establishes high sediment oxygen demand (see Chapter 2). Oxygen diffusion into these soils is slow when they are flooded (0.227×10^{-6} cm^2 s^{-1} in water) but is greatly accelerated when they are exposed to air and soil pores are dry (0.205 cm^2 s^{-1} in air) (see Chapter 2). In addition, oxygen diffusion into soils is influenced by sediment texture (Hutchings and Saenger 1987), water table depth at lower tide, and soil bioturbation (Twilley and Chen 1998; Kristensen 2008; Williams et al. 2016) (see Chapter 11). As a result, the degree of

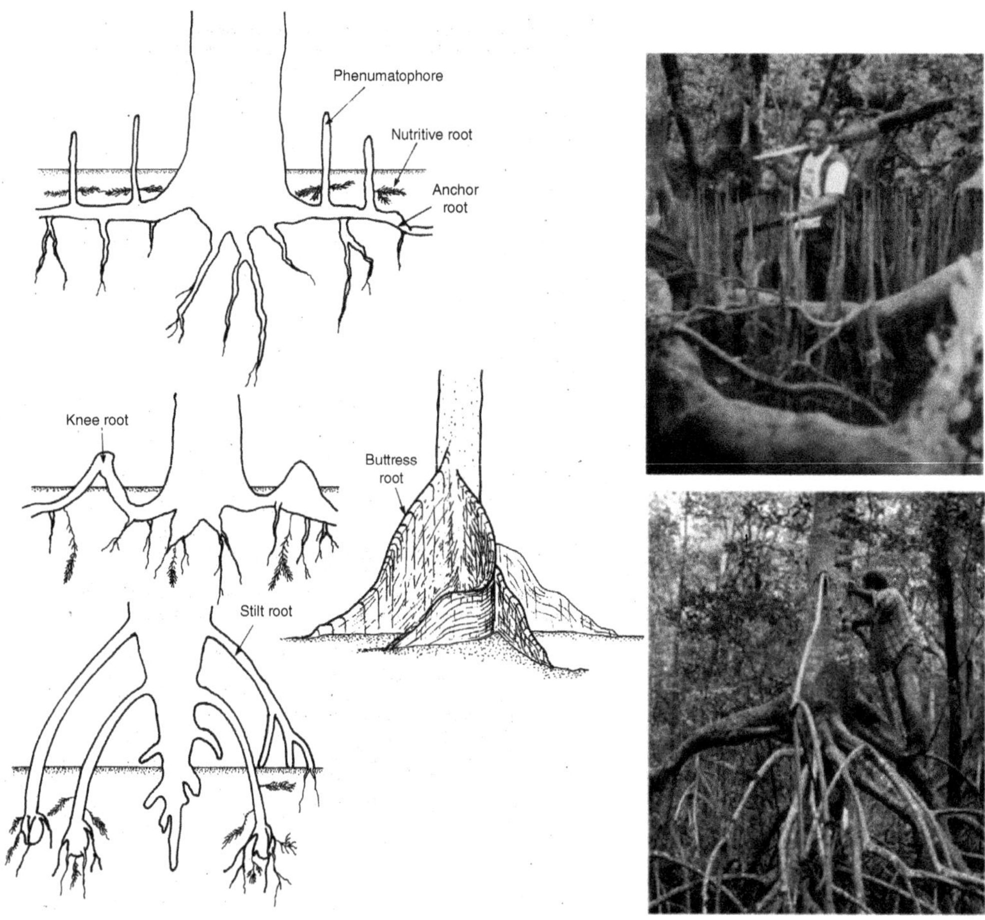

FIGURE 7.11 Adaptations of root structures to anaerobic environments. *Source*: Hutchings and Saenger 1987; Photos credit: R. Twilley.

anaerobiosis is often less in mangrove forests than in permanently flooded wetlands.

7.4.2 Resources and Regulators

Mangrove distribution in the intertidal zone is determined by a variety of stress conditions (Figure 7.12) that can be grouped along with hydroperiod as resources and regulators (Twilley and Rivera-Monroy 2005). Resources include attributes of the environment that control mangrove growth and are consumed, setting up competition with other organisms in mangrove wetlands, such as light, water, nutrients, and space (Huston 1994). Regulators also influence plant growth, such as salinity, sulfide, pH, and temperature. However, in contrast to resources, they are not consumed during the growth process and, thus, regulator condition is unaffected by plant responses (Tilman 1982). As described in Section 7.4.1, hydroperiod can control wetland plant growth (Gosselink and Turner 1978). It could be argued that hydroperiod is a regulator, an attribute that controls growth but is not consumed, but we specifically include this as a unique category since it can control the concentration and availability of resources and regulators in wetland environments. Mangroves have adaptations to promote specific tolerances within the intertidal zone in response to conditions of resources, regulators, and hydroperiod (Figure 7.12) (Lugo 1980).

Perhaps the most critical regulators controlling propagule establishment, seedling survival, and plant growth is salinity (Ball 1988b, Chen and Twilley 1998; Krauss et al. 2008). As a group, mangroves are facultative halophytes (Krauss and Ball 2013), meaning that they generally do not require salinity to survive, but rather, they proliferate with salinity and often experience growth or physiological enhancement at moderate salinities. While evidence suggests that most mangrove species do not have an obligate physiological salinity requirement (Wang et al. 2011), one recent study indicated that some mangrove species (e.g., *Avicennia marina*) may require salinity for proper hydraulic system functioning (Nguyen et al. 2015). Unlike most trees, mangroves exhibit substantial physiological tolerance to changes in salinity, which allows them to flourish under a range of saline conditions in coastal environments.

However, life in the intertidal marine environment requires special adaptation to water balance imposed by the presence of salt. To translocate water from the soil to the canopy tens of meters above the soil surface, the water potential in the plant must be more negative than that of the saline

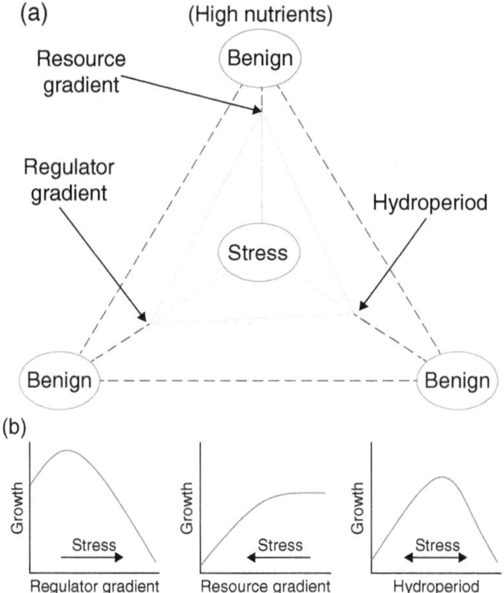

FIGURE 7.12 (a) Factorial interactions controlling the productivity of coastal wetlands including regulator gradients (temperature, salinity), resource gradients (nutrients, light, and carbon), and hydroperiod. (b) Relationships among these controls and mangrove growth rate. *Source*: Twilley and Rivera-Monroy (2005).

water around plant roots. For a tree in freshwater soils, water potentials in leaves are around −30 atm (range from −26.6 to −56.3 atm) and soil osmotic potential is zero, establishing a favorable gradient for water transport against gravity and to the canopy. But in a typical mangrove wetland, soil osmotic potential may average −30 atm, requiring mangrove osmotic potential to be more negative than −30 atm to transport water to the canopy. Mangroves use several physiological mechanisms to control water potential in their tissues, including salt exclusion, salt secretion, salt storage, succulence, reduced stomatal conductance, and ion compartmentalization, as well as osmo-compensation which may involve production of high molecular weight organic compounds that lower stem osmotic potentials (Lüttge 2007).

The salt content of soils in the interior regions of mangrove wetlands can even become hypersaline due to periodic tidal flooding followed by evapotranspiration. Mangrove species that dominate communities in these regions tend to have well developed salt glands that allow for secretion of salt from mangrove leaves. This adaptation varies among mangrove species and is more well developed in *Avicennia* and *Aegiceras* compared to *Rhizophora*, *Sonneratia*, and *Bruguiera* (Scholander et al. 1962). Mangrove species that grow in hypersaline soils also have adaptations associated with the exchange of water and CO_2 through stomata, which are specialized openings on leaf surface that control the delicate balance of water loss by transpiration versus CO_2 uptake for photosynthesis. Species more tolerant of higher salinity are typically more efficient in water use as described by the amount of carbon gained through stomata compared to the amount of water lost (Ball 1996; Lovelock et al. 2016). Given these constraints on water loss and carbon gain at the leaf surface for mangroves in saline soils, optimum photosynthesis has been documented at about 17% of full-strength seawater, or about 6 salinity, across a range of genera despite the capacity for mangrove survival at much higher salinities.

Another regulator in the intertidal zone is sulfide. Field experiments show that adult mangrove plants can grow in soils with high concentrations of sulfide. For example, soils near *R. mangle* trees have shown concentrations of 1.63–1.70 mmol L^{-1}, while soils near *A. germinans* have sulfide concentrations of 1.44–3.75 mmol L^{-1} (McKee et al. 1988; McKee 1993). Mangroves are also capable of releasing dissolved oxygen directly from their roots, which creates microzones, where hydrogen sulfide is oxidized to nontoxic compounds. Oxygen released by roots in sulfidic soils can also encourage the growth of sulfur oxidizing bacteria that oxidize sulfide to nontoxic chemical forms (see Chapter 5).

In the absence of significant control over plant growth by regulators in the intertidal zone, resources can set the rate of net primary productivity. Light is a key resource that defines plant growth in mature mangrove forests (McKee 1995a, Chen and Twilley 1998). A greater leaf area, for example, may benefit seedlings in closed-canopy mangrove wetlands where light near the forest floor can be reduced by 97% compared to ambient levels above the canopy (Sherman et al. 2001). Under such conditions, enhanced leaf area index (area of leaves per unit area of land) or greater seedling/sapling specific leaf areas (individual leaf area per unit biomass) allows competitive growth responses prior to and immediately after the stochastic occurrence of openings in the canopy ("light gaps") created by disturbances such as lightning, wind, or individual tree falls. The competitiveness of mangrove species in light gaps provides insight into the intolerance of many mangroves to shaded conditions. The light adaptations of mangrove species fall into three broad categories: (1) those that are somewhat shade tolerant as seedlings and adults, (2) those that are shade-intolerant, and (3) those with different shade tolerances as seedlings, saplings, and trees (Saenger 2013). There is no clear pattern in light adaptation among mangrove genera and some species have evidence of both shade tolerant and shade intolerant strategies in different environmental conditions or life stages (Krauss et al. 2008).

The formation of a light gap in a mangrove canopy enhances recruitment of mangrove saplings that occupy areas under a closed mangrove canopy (low light levels). This area now exposed to higher light levels in a mangrove wetland regenerates a new plant community depending on the soil conditions of resources, regulators, and hydroperiod in that space (Chen and Twilley 1998). Models and theories of plant growth indicate that, in response to resource variation, plants partition photosynthates to optimize resource capture and consequently maximize growth rate (Shipley and Meziane 2002). Implicit in these ideas is the prediction that plants will adjust their allocation in above- and belowground biomass to compensate for that resource which is in shortest supply (Chapin et al. 1990; Gleeson and Tilman 1992). For example, as more light becomes available after a gap in the forest is formed, the seedling produces more leaf area. These plant strategies in biomass allocation are also found in mangroves, but the effects of salinity complicate the relative compensation of light resource acquisition since salt uptake may regulate these processes. For example, the increased cost of carbon acquisition per unit of water and nutrient uptake in saline habitats suggest that mangroves exhibit an increased biomass allocation to shoots in low light conditions, and to roots with increasing salinity (Ball 1988a, b). Recent evidence highlights the interactive effects of light adaptation to salinity levels and conclude that many interactions with salinity may mask any specific definition of shade tolerance for mangroves (López-Hoffman et al. 2007; Krauss et al. 2008). Those mangrove species that demonstrate higher allocation to aboveground structures under high light conditions, such as *Laguncularia*, can be less competitive under higher soil salinity, but may thrive when soils have low salinity and high nutrients (Krauss et al. 2008).

In the absence of regulators and light controlling plant growth, soil nutrient availability is often implicated as an important factor determining variation in mangrove growth. Phosphorus availability is one of the major factors limiting annual growth of mangrove forests in many neotropical settings on carbonate platforms (Feller et al. 2003; McKee et al. 2007; Twilley and Rivera-Monroy 2009; Adame et al. 2013). A recent review showed that over one third (n = 57) of about 125 mangrove wetlands have soil with a total phosphorus density of <0.20 mg cm^{-3}. In fact, 90% of these mangrove sites have N:P > 15, indicating that phosphorus is potentially limiting in these types of coastal environmental settings. However, this survey also demonstrates that nitrogen is potentially limiting in locations where sediment deposition in estuarine and muddy coasts provide soil total phosphorus density sufficient to meet growth requirements. Comparison of soil nutrient densities in mangrove wetlands across coastal environmental settings reveal that N:P ratios are distinctly higher in sediment-poor systems such as lagoons and carbonate karst systems compared to lower N:P in muddy systems such as deltas and tidal systems. This trend in the relative content of nitrogen and phosphorus among mangrove sites in muddy, estuarine, and carbonate settings is probably one of the strongest linkages between geomorphological settings and ecology of mangrove wetlands (Twilley et al. 2018).

7.4.3 Vivipary

Vivipary is the condition whereby the embryo (the young plant within the seed) grows first to break through the seed coat then out of the fruit wall while still attached to the parent plant (Figure 7.13a). This condition is common among mangrove trees and can be found in the genera *Bruguiera*, *Ceriops*, *Kandelia*, and *Rhizophora*. Cryptovivipary (Greek *kryptos*, meaning "hidden") refers to the condition whereby the embryo grows to break through the seed coat but not the fruit wall before it splits open (Figure 7.13b). This condition is exhibited in *Aegiceras*, *Avicennia*, and *Nypa*. Vivipary, including cryptovivipary, is a rare phenomenon among higher plants, occurring mostly among plants of tropical shallow marine habitats, such as seagrasses (20 genera in 13 families, see Chapter 5) and mangroves (7 families). The storage material provides nutrition during early stages of seedling growth and development once propagules become stranded in the intertidal zone after dispersal. Accumulated stores of starch within the hypocotyls provide initial nutrition for rooting, erection of the propagule in the vertical position,

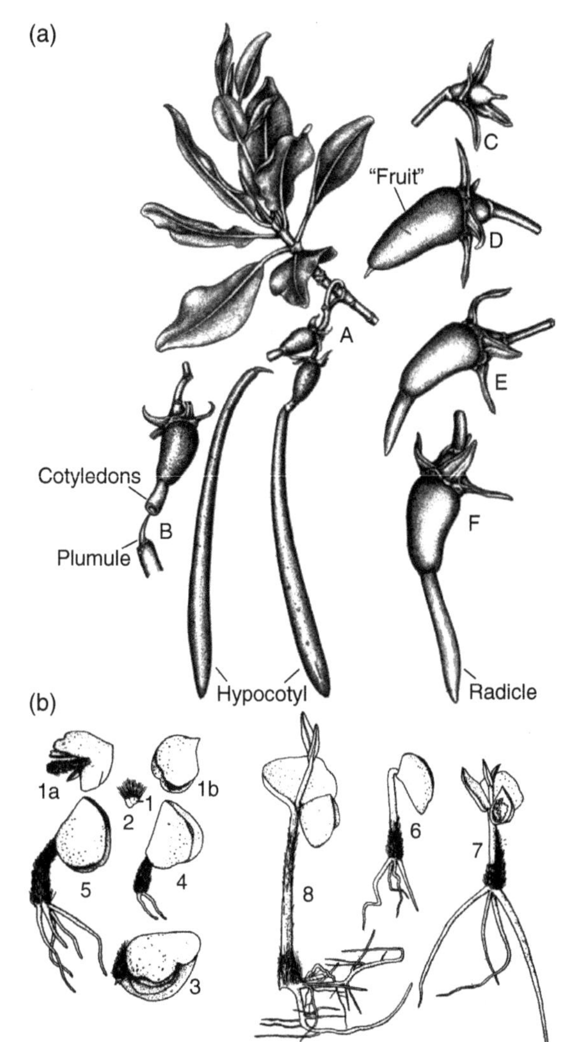

FIGURE 7.13 (a) Seedling development showing vivipary of *Rhizophora mangle* [A. Shoot with mature seedlings, one just detached (x1/3). B. Fruit (x1/2), with plumule of seedling detached from cotyledonary collar. C.–F. Fruit (x1) with stages in its development and protrusion of seedling hypocotyl]; and (b) cryptovivipary of *Avicennia germinans;* numbers refer to stages of development. *Source*: Panel A is from Tomlinson (2016)/Cambridge University Press and Panel B is from Chapman (1976)/Vaduz: J. Cramer.

and respiration and growth without being limited by photosynthesis; such benefits can last for months in some species (Smith and Snedaker 2000). Early growth in mangrove seedlings is linked to maternal reserves (Ellison and Farnsworth 1993; Ball 2002) and their strategies to capture and use soil resources and light (McKee 1995a, b). Therefore, seedling responses to environmental and biotic factors determine the success of its conversion to the adult phase (Krauss et al. 2008). The transport of propagules with tides and young propagule/seedling ecophysiology are important characteristics that favor mangroves living in intertidal zones.

7.5 Human Impacts, Conservation, and Carbon Sequestration

7.5.1 Human Exploitation

The ecological properties of mangrove wetlands described in the previous sections define the function of these ecosystems in productivity, biogeochemistry, and food web structuring of mangrove-dominated estuaries. Because of the value of these ecosystem processes in many coastal settings, there is increased human utilization of mangrove resources that vary throughout the tropics depending on economic and cultural constraints (Figure 7.14). The over-exploitation of mangrove resources by humans can negatively affect ecological processes of mangrove ecosystems (Twilley et al. 1998a) (Figure 7.14). These can include indirect impacts such as diversion of freshwater, which can cause changes in the productivity, litter export, and biogeochemistry of mangrove wetlands, among other processes. Direct impacts may include inputs of excessive nutrients beyond their accustomed levels or spills of toxic substances, such as crude oil. If these impacts limit the ecological processes and thus the function of mangrove wetlands, then the impact connects all the way back to the use and value of these natural resources. Thus, the properties of mangrove wetlands are influenced both by the environmental constraints (forcing functions in Figure 7.14) and social constraints (economics and culture in Figure 7.14) of a coastal region. Sustainable development of mangrove wetlands may be viewed as the proper balance between environmental and social constraints that allow for the continued use and value of these coastal resources by future generations.

Deforestation of mangrove wetlands is associated with many uses of coastal environments including urban, agriculture, and aquaculture reclamation (i.e., cultivation of land previously under water), as well as the use of forest timber for subsistence, furniture, energy, chip wood, and construction materials (Saenger et al. 1983). In the Gulf of Mexico and Caribbean, there are very few examples of mangrove use in

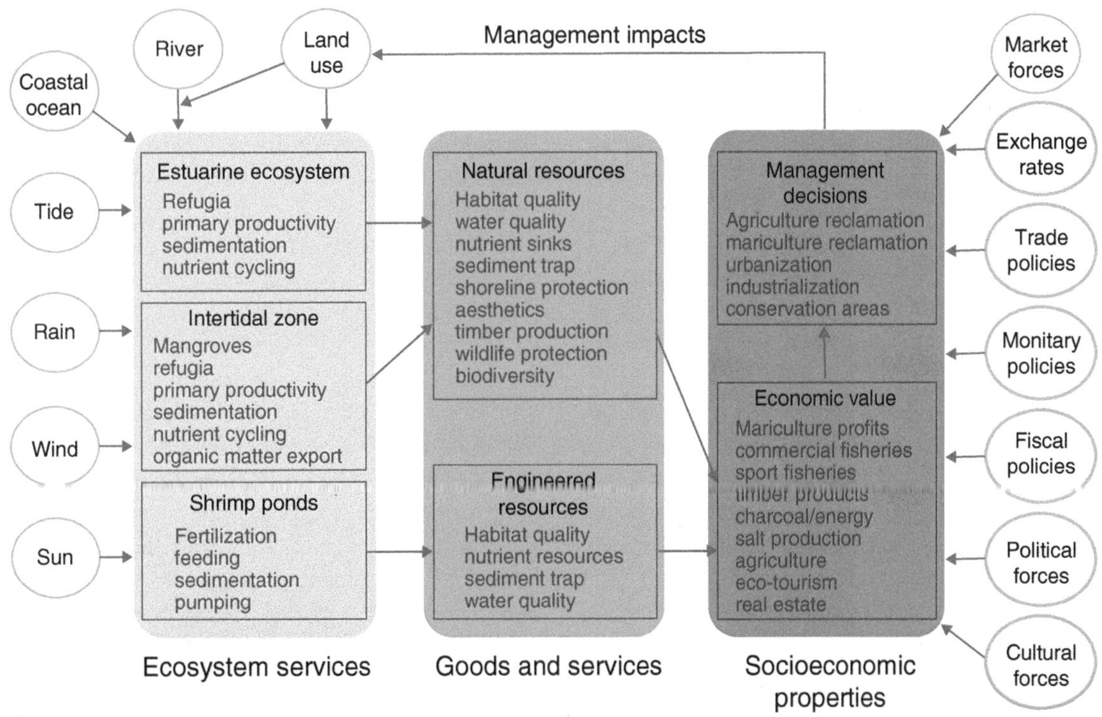

FIGURE 7.14 Linkages among environmental setting, ecosystem properties, ecological functions, and uses of mangrove ecosystems, including feedback effects of human utilization on ecosystem properties. *Source*: Modified from Twilley et al. (1998a).

silviculture (Palacios and Cantera 2017), but in Indonesia and other Southeast Asian countries, mangrove plantations have been in existence for several decades (Watson 1928; Richards and Friess 2016; Sillanpää et al. 2017). Agriculture and aquaculture enterprises have also led to massive mangrove deforestation in some regions. Agriculture impacts to mangrove wetlands have historically been aligned with rice culture and are most prominent in West Africa and southeast Asia, though more recently, oil palm expansion has also become a threat (Richards and Friess 2016). Many of the large agricultural uses are found in humid coastal areas or deltas where freshwater is abundant and intertidal lands are seasonally available for crop production. Mariculture use (or, propagating marine species such as shrimp, Warne 2011) of the tropical intertidal zones has become one of the fastest growing aquaculture industries, and can leave lots of destruction in its wake (Kauffman et al. 2017). One of the largest and most publicized is the expansion of shrimp farming in Ecuador, where in the late 1980s there were more shrimp produced in the ponds of this Pacific coastal zone than anywhere else in the world (Twilley et al. 2001; Kauffman et al. 2017). Ironically, the environmental impact of shrimp farming on mangrove wetlands and water quality of tropical estuaries may threaten the sustainability of this economic activity.

Human alterations of upland watersheds causing diversion of freshwater (dams and canals) can have severe impacts on mangrove wetlands, particularly in dry coastal environments, by altering their landscape distribution and reducing their productivity and habitat quality. Together, these disturbances can also alter the secondary productivity of tropical estuarine ecosystems due to degradation of habitat and environmental quality and alter the terrestrial and aquatic food webs that mangrove ecosystems support. The most dramatic impact of freshwater diversion in the New world tropics has been documented in Colombia, where the construction of highways and levees restricted the flow of the Magdalena River to the Ciénaga Grande de Santa Marta resulting in the loss by mortality of nearly 25,000 ha of mangrove wetlands over several decades (Botero 1990; Rivera-Monroy et al. 2011). The sensitivity of mangrove wetlands to slight changes in freshwater input in these dry coastal regions of the Caribbean basin has been well documented (Cintrón et al. 1978).

Other human impacts on mangrove wetlands can be more direct such as the introduction of contaminants that can disrupt the natural ecological processes of these forested wetlands (Figure 7.3). Deterioration of water quality includes inputs of excessive nutrients to coastal waters (eutrophication) and toxic materials (heavy metals, oil spills, pesticides) (Lewis et al. 2011). The increased attraction of coastal environments for humans has made coastal regions more vulnerable to these water quality problems. Real estate development in coastal regions not only causes direct impacts such as deforestation, but also increases the amount of nutrient discharge to coastal waters and leads to "coastal squeeze" (or, compressing areas available for mangrove colonization between sea and built structures). A recent study of mangrove wetlands in Puerto Rico demonstrated the negative and positive impacts of human activities on mangrove wetlands over a period of 200 years (Martinuzzi et al. 2009). During the period 1800–1940 the agricultural and urban expansion within the island caused a 45% decline in mangrove cover (from 12,000 to 6500 ha), mostly due to changes in hydrology, drainage, and excessive sedimentation. After late 1940 with the onset of the industrial economy, mangrove forests expanded by natural recovery due to reduced land use pressure. Nevertheless, between 1960 and 1970 mangrove land cover decreased again due to urban expansion. Since the establishment of legal protection for all mangrove wetlands in the early 1970s, reductions in mangrove wetland area in Puerto Rico have been prevented despite continuing urbanization.

Oil spills represent contaminants to mangrove wetlands that can alter their succession, productivity, and nutrient cycling, and thus reduce ecosystem services. These impacts have been well documented in ecological studies in Panama (Duke and Pinzón 1993; Garrity et al. 1994), the Gulf of Mexico (Getter et al. 1981; Giri et al. 2011a), and South Africa (Naidoo et al. 2010), among other places. An oil slick in a mangrove wetland can cause direct and acute tree mortality depending on the concentration of hydrocarbons and species of trees, as well as continued chronic stress levels as hydrocarbons (especially polycyclic aromatic hydrocarbons) persist in mangrove soils (Duke and Pinzón 1993). Efforts to minimize the damage of oil spills and enhance the recovery of mangrove wetlands are important to minimize the complex loss of ecological functions to the coastal zone.

There have been several reviews of mangrove restoration, and collectively they have alluded to the concept that since these forested wetlands are adapted to stressed environments, they are relatively amenable to restoration efforts if undertaken properly (Field 1999; Lewis 2005; Bosire et al. 2008; Feller et al. 2017; López-Portillo et al. 2017). Successful mangrove wetland restorations often include the establishment of the proper environmental settings (especially related to hydroperiod) that control the ecosystem structure and function (Lewis et al. 2016). The goal of ecological restoration is to return the degraded site back to either the natural condition (restoration) or to some other new condition (rehabilitation). The rates of change in the ecological characteristics of mangrove wetlands between natural, degraded, and some rehabilitated conditions are known as trajectories. Trajectories are important concepts in ecological restoration because they track both the structural and functional characteristics of sites relative to natural conditions, and forecast the time required to obtain a rehabilitated state or condition (Twilley et al. 1998b). Not all mangrove restoration efforts are effective, but because many are successful if planned properly, losses of mangroves from deforestation are being offset somewhat through restoration efforts (Feller et al. 2017).

7.5.2 Biodiversity

Mangrove ecosystems support a variety of marine and estuarine food webs involving an extraordinarily large number of animal species and a complex heterotrophic microbial food web (Yáñez-Arancibia et al. 1988; Demopoulos et al. 2007). In the new world tropics, extensive surveys of the composition and ecology of nekton in mangrove estuaries identified 26–114 species of fish (Robertson and Blaber 1992), and an expanded survey to 32 additional estuaries globally revealed between 22 and 197 fish species present in any given mangrove estuary (Blaber 2007). There are also a large number and variety of organisms that live in and/or feed directly on mangrove vegetation. These include sessile organisms such as oysters and tunicates, and arboreal birds and insects such as folivores and frugivores, and ground-level seed predators. Epibionts growing on the prop roots of mangroves form highly diverse communities including sponges and tunicates (Ellison and Farnsworth 1992; Rützler and Feller 1996), especially along mangrove shorelines with little terrigenous input. In addition, over 200 species of insects have been documented in mangrove wetlands in the Florida Keys (Simberloff and Wilson 1969) and other parts of the Caribbean (Rützler and Feller 1996).

One of the most published links between mangrove biodiversity and ecosystem function is the presence and role of crabs in mangrove wetlands, as described in Section 7.3.3 above (Jones 1984; Smith 1987). In general, the mangrove crab fauna is dominated by representatives of two families, the Ocypodidae and Grapsidae, and each family by one genus, *Uca* and *Sesarma*, respectively. Furthermore, the genus *Sesarma* accounts for over 60 species of crabs predominantly associated with mangrove wetlands (Jones, 1984), of which the tropical Americas have only three to five species compared to over 30 species in other tropical continents. Crabs can influence forest structure (Smith et al. 1989; Lindquist et al. 2009), litter dynamics and chemical changes (Robertson and Daniel 1989; Bakkar et al. 2017), nutrient cycling (Smith et al. 1991), organic material burial (Kristensen 2008), and soil salinity (Pestana et al. 2017). These ecological interactions suggest that crabs are a keystone guild in these wetland ecosystems. The selective consumption by crabs of mangrove seeds and propagules can influence rates of sapling recruitment by removing some species from zones of crab habitation.

7.5.3 Mangroves as Invasive Species

Conservation and management efforts must also consider situations where mangroves have recently invaded. While some debate surrounds how specific mangrove species may have arrived at a location (e.g., *Rhizophora samoensis* in Australasia and the Pacific; Duke 1992; Tomlinson 2016), several introductions were deliberate and well-documented. The mangrove palm, *Nypa fruticans*, was introduced to western Africa in 1906 and 1912 from populations maintained in Singapore (Wilcox 1985). *Sonneratia caseolaris* was introduced to mainland China from Hainan Island for afforestation purposes and is now propagated widely (Chen et al. 2009). *Laguncularia racemosa* was introduced to China from the new world tropics, and populations of *Bruguiera gymnorrhiza* and *Lumnitzera racemosa* have escaped confinement in Miami and are expanding their populations into native south Florida mangrove forests (Fourqurean et al. 2010).

The most well-known case of mangrove bioinvasion is currently occurring in Hawaii, USA, where no mangrove has ever colonized naturally (Allen 1998). While multiple mangrove species were introduced to Hawaii, two species have established: *Rhizophora mangle* from south Florida populations introduced starting in 1902 and *Bruguiera sexangula* from Philippine populations introduced in 1922. The *Bruguiera* taxonomy took time to sort, and fortunately *B. sexangula* emerged over what was originally assumed to be a more aggressive *Bruguiera* species (*B. gymnorrhiza*) (Allen et al. 2000). However, mangroves have since transformed the intertidal zone of Hawaii with incredible rates of productivity and unrestricted reproduction (Cox and Allen 1999) and are now present on six of the seven main Hawaiian Islands. More recently, an ornamental variety of *Conocarpus erectus* is also beginning to spread within higher intertidal elevations on the Hawaiian Islands. Mangrove expansion in Hawaii is causing shading and replacement of native (and introduced) salt marsh grasses, replacement of critical nesting habitat for endangered water birds, degradation of native archaeological structures, and clogging of water ways and critical drainage corridors. Mangroves are also altering food web structure and diversification of sediment bacterial communities in Hawaii (Demopoulos et al. 2007). However, positive influences of mangroves have also been suggested, including their potential role in food web support, facilitating sediment retention from upland erosion, and sequestering and storing C in both wood biomass and soils (Soper et al. 2019). On the Island of Oahu, 20 species of fish were found associated with mangroves at twelve sites, and of the 16 native fish species encountered, the abundance of two (and one nonnative) was correlated with lower salinity and higher mangrove cover (Goecke and Carstenn 2017). In Hawaii and elsewhere, negative impacts of mangrove introductions must be weighed against benefits; in China, mangrove introductions provide forestry value while in other places, eradication is considered as the nuisance strengthens. Eradication of established mangroves is expensive.

7.5.4 Sea Level Rise

An aspect of hydroperiod that nearly all tidal wetlands must accommodate is sea level rise. Mangroves adjust both to eustatic rates of sea level rise (from thermal expansion of water leading to increased ocean heights) and submergence (from shallow soil compaction and deep subsidence) associated with the local condition to persist as emergent vascular plants. However, mangroves are not passive members of an intertidal community. Like submerged seagrasses, their dense vegetative assemblage of aerial roots, stems and leaves serve to trap and retain sediments during tidal cycles, allowing them to continuously increase their soil elevations as result of sedimentation (see Section 7.3.5). Most prominently, this is accomplished by increasing belowground root volumes that enhance organic sedimentation and increase soil surface elevation (McKee et al. 2007), and by facilitating surface inorganic sedimentation with several different morphological root types (Krauss et al. 2003). Also, when flooding of mangrove wetlands increases and tidal flushing is incomplete, mangrove litter, and fine woody debris contribute more organic matter sedimentation on the surface with slower rates of organic matter decomposition because of lower oxygen diffusion into soils. These conditions also create an environment that supports algal and benthic mat development that help retain sediments. In fact, mangroves also help retain deposited sediments and organic matter through rapid colonization of new sediment by fine surface roots (Krauss et al. 2014). It is notable that sea level rise is the condition that facilitates the greatest carbon storage in tidal wetland soils (Rogers et al. 2019); much of this is a consequence of organic and inorganic matter accumulation as mangroves build vertical elevations to accommodate rising seas.

7.5.5 Mangroves as Blue Carbon Ecosystems

Mangrove wetlands, along with salt marshes and seagrasses can store significant amounts of carbon that represent a sink of excess CO_2 in the atmosphere (Mcleod et al. 2012). As such, they are considered "blue carbon" ecosystems, in contrast to "green carbon" ecosystems found on land in forests and grasslands around the world. As forested wetlands at the terrestrial–aquatic interface, mangrove wetlands have long been recognized for their potential role as a major global C sink, thereby helping to mitigate atmospheric CO_2 enrichment (Twilley et al. 1992). Mangrove wetlands are generally considered to be the most carbon-dense forests in the tropics (Donato et al. 2011). Moreover, only recently have specific models been developed to account for global variation in mangrove soil organic carbon stocks, where most of the carbon in mangrove systems is stored (Jardine and Siikamäki 2014; Atwood et al. 2017; Rovai et al. 2018). Although most estimates of global mangrove soil organic carbon are similar at 2.5 Pg C (Rovai et al., 2018; Atwood et al., 2017), the density of soil organic carbon varies markedly across different types of coastal environmental settings (Rovai et al. 2018).

Mangrove ecosystem-level C stocks can be calculated by combining carbon estimates from aboveground and belowground biomass along with soil carbon. Ecosystem-level C stocks vary narrowly across distinct coastal environmental settings with a global mean (±SD) of 401 ± 48 Mg C ha^{-1} (Figure 7.6b, Rovai et al. 2021). Ecosystem-level C stocks ranged from 343 ± 78 Mg C ha^{-1} in arheic systems to 451 ± 57 Mg C ha^{-1} in carbonate coastlines (Figure 7.6b). When these estimates are multiplied by the total area of each coastal environmental setting (Figure 7.5), nearly 85% (or 2.5 Pg C) of ecosystem-level C stocks are concentrated in tidal and river-dominated coastlines (Figure 7.6c) because of the larger area occupied by mangrove wetlands in these coastal environmental settings. As nations with coastlines focus on carbon inventories as part of international carbon accounting, these average ecosystem-level C stock estimates can be used to determine national carbon budgets. For equally diverse coastlines for which data are unavailable, including West Africa and the Indo-West Pacific regions. Such a framework provides a powerful approach to address global variability in mangrove global carbon stocks.

The global inventory of carbon in mangroves represents the stock of carbon at any one point in time, not the annual flux or sink of carbon from the atmosphere that is stored in mangrove wetlands around the world. Net ecosystem production (NEP), as described in Section 7.3.5, can also be used to estimate the net carbon stored annually in mangrove wetlands that represents a sink from the atmosphere. The average NEP of mangrove wetlands is about 562 g C m^{-2} year^{-1}, based on about 381 g C m^{-2} year^{-1} for wood production, and carbon sequestration in mangrove soils at about 181 g C m^{-2} year^{-1} (Figure 7.9b, see reviews in Twilley et al. 2017, 2019). This is very similar to the NEP estimates (corrected) by Bouillon et al. 2008b and Alongi 2009 (Table 7.2). This represents an annual removal of 84.3 Tg C year^{-1} from the atmosphere to mangrove wetlands globally (based on global mangrove area of 150,000 km^2). Bouillon et al. (2008b) estimates that more than 50% of the carbon fixed by mangrove vegetation, estimated at ~217 ± 72 Tg C year^{-1}, appears to be unaccounted for based on estimates of various carbon sinks. This missing carbon sink is conservatively estimated at ~112 ± 85 Tg C year^{-1}, or about 700 g C m^{-2} year^{-1}. However, estimates of NEP to balance carbon accounting in mangrove wetlands are difficult to obtain, especially when measuring the exchange of inorganic and organic forms of carbon with the atmosphere and across mangrove-estuarine boundaries (see discussion of NEP in 7.3.5). Based on updated estimates of carbon exchange between mangroves and estuarine waters, total organic export of about 250 g C m^{-2} year^{-1} (DOC + POC) is equal to export of dissolved inorganic carbon (DIC) at 250 g C m^{-2} year^{-1} for global estimates of 80 Tg C year^{-1}. It is unknown how much of that carbon (either organic or inorganic) fluxes back to the atmosphere from estuaries, but that flux is part of the NEP of estuaries that defines their role in carbon exchange with the atmosphere. Carbon accounting for mangrove wetlands as contributions by countries to global estimates is the mangrove NEP rate per unit area and the area of mangroves in that jurisdiction.

TABLE 7.2 Global mass balance estimates of carbon flow in mangrove ecosystems.

	Bouillon et al. (2008b) Tg C yr^{-1}	Alongi (2009) Tg C yr^{-1}	Bouillon et al. (2008b) g C m^{-2} yr^{-1}	Alongi (2009) g C m^{-2} yr^{-1}
Inputs				
GPP	NA	690 ± 264		4313
NPP	204 ± 68	290 ± 107	1275	1813
Wood	63 ± 40	63 ± 42	394	394
Litter	64 ± 20	64 ± 20	400	400
Roots	77 ± 56	163	481	1019
Outputs	82	508	512	3175
POC Export	20 ± 22	27 ± 25	125	169
DOC Export	23 ± 21	13 ± 12	144	81
Tree Respiration	NA	396 ± 151		2475
Soil + Water Respiration	39	72 ± 50	244	450
NEP (inputs–outputs)	122	182	763	1138
Soil Burial	17 ± 20	27 ± 20	106	169
*NEP (Soil Burial + Wood Production	80	90	500	563
Missing DIC (NEP—Soil Burial)	105	155	656	969
Missing DIC (NPP—(*NEP + Outputs))	42	92	263	575

The values in Tg C yr^{-1} are organized around estimates of net ecosystem production scaled to common total global area of 150,000 km^2. Estimates are also given on a per meter square basis of mangrove wetland area in intertidal zone (g C m^{-2} yr^{-1}). Values are ± 1 SD. POC = Particulate organic carbon, DOC = dissolved organic carbon, DIC = dissolved inorganic carbon.*Source*: Modified from Bouillon et al. (2008b) and Alongi (2009).

Study Questions

Multiple Choice

1. Mangrove refers to one of the following:
   a. A unique group of plants that colonize the intertidal and terrestrial environments of tropical and subtropical environments with unique root systems.
   b. A unique group of wetlands that are dominated mainly by trees that colonize the saline intertidal zone of tropical, subtropical, and warm temperate climates with morphological and physiological adaptations to these estuarine environments.
   c. A common group of wetland plants that dominate the fresh and saline intertidal environments of tropical to warm temperate climates with physiological adaptations to wetland environments.
   d. A group of wetlands trees that are adapted to saline environments restricted to tropical climates with unique physiological and morphological adaptations to estuarine environments.
2. Which of the following genera of mangroves tree are associated only with the old-world tropics (Indo-West Pacific biogeographic region):
   a. *Rhizophora, Laguncularia, Bruguiera*
   b. *Rhizophora, Avicennia, Bruguiera*
   c. *Ceriops, Sonneratia, Bruguiera*
   d. *Rhizophora, Laguncularia, Avicennia*
3. The distribution of mangrove wetlands vs salt marsh wetlands in the intertidal zone around the world is determined by which of the following conditions:
   a. Where the duration/severity of freezing temperature is frequent, mangrove wetlands are replaced by salt marsh vegetation, which correlates well near the 16°C isotherm for the air temperature of the coldest month.
   b. Where distinct wet and dry seasonal rainfall occurs in the tropics with temperatures above 20°C isotherm is the climate that favors mangrove wetlands over salt marsh vegetation.

c. Mangrove wetlands do not occur in the warm temperate to artic regions climates of the intertidal zone, which is dominated by salt marsh vegetation.
d. Mangrove wetlands are not restricted to climate zones but are replaced by salt marshes due to the lower adaptation of mangroves to the saline and redox conditions of estuarine environment.

4. The ecogeomorpholgy classification of wetlands is based upon the following conditions of the intertidal zone:
   a. The influence of tides on redistribution of sediments and formation of the intertidal topography that define environmental gradients such as salinity and nutrients, resulting in unique ecosystem properties.
   b. The influence of river and sedimentary processes that form depositional environments were mangrove colonize along salinity gradients, resulting in unique ecosystem properties.
   c. The microtopography of intertidal platforms that drive the ecology of mangrove wetlands include habitat formation and biogeochemical processes.
   d. Combination of geomorphic processes defined by river, tides, and waves that together with microtopography of the intertidal zone establishes environmental gradients that define coastal environmental settings that are also defined by feedback of mangrove wetlands on forming intertidal zone.

5. One of the patterns associated with mangrove biomass along different coastal environmental settings is the following:
   a. Soil conditions resulting from nutrient inputs from river discharge have greater aboveground relative to belowground biomass compared to carbonate settings with nutrient poor soils and more belowground than aboveground biomass.
   b. Soil conditions resulting from nutrient inputs from river discharge have lower aboveground relative to belowground biomass compared to carbonate settings with nutrient poor soils and less belowground than aboveground biomass
   c. Soil conditions resulting from dominate tides have greater aboveground relative to belowground biomass compared to riverine settings with nutrient poor soils and more belowground than aboveground biomass
   d. Soil conditions resulting in mesohaline gradients have greater belowground relative to aboveground biomass compared to dry climate settings with hypersaline soils and more aboveground than belowground biomass

6. Patterns of mangrove aboveground biomass across coastal environmental settings is very distinctly described as follows:
   a. Deltas, small deltas, and tidal systems have significantly higher aboveground biomass at 1150 $Mg\,ha^{-1}$ than lagoonal, carbonate, and arheic coastal settings at 250 $Mg\,ha^{-1}$.
   b. Deltas, small deltas, and tidal systems have significantly higher aboveground biomass at 130 $Mg\,ha^{-1}$ than lagoonal, carbonate, and arheic coastal settings at 30 $Mg\,ha^{-1}$.
   c. Deltas, small deltas, and tidal systems have significantly lower aboveground biomass at 30 $Mg\,ha^{-1}$ than lagoonal, carbonate, and arheic coastal settings at 130 $Mg\,ha^{-1}$.
   d. Deltas, small deltas, and tidal systems have nearly the same aboveground biomass at 130 $Mg\,ha^{-1}$ than lagoonal, carbonate, and arheic coastal settings.

7. Mangrove biomass and net primary productivity are higher in regions of hurricane disturbance in south Florida largely due to the following processes.
   a. Hurricanes knock down smaller trees allowing more light to penetrate the canopy and stimulating tree growth.
   b. Hurricane storm surge reduces salinity to oligohaline conditions and stimulates mangrove biomass production.
   c. Hurricane storm surge deposits sediments that are rich in phosphorus on mangrove soils that stimulates mangrove biomass production.
   d. None of the above

8. Hydroperiod is described as the most critical environmental control of mangrove net primary productivity and persistence. Why is this such an important factor in mangrove productivity and distribution?
   a. Hydroperiod is the volume of water that controls sediment and nutrient input and these regulators are critical to stimulating primary productivity.
   b. Hydroperiod is the frequency, depth and duration of flooding which controls the distribution of resources and regulators of mangroves soils that constrain net primary productivity.
   c. Hydroperiod is the duration of flooding and this restricts the amount of oxygen in soils and this redox condition is the most critical factor controlling productivity and distribution of mangrove wetlands.
   d. Hydroperiod is frequency and depth of inundation that controls salinity distribution, and this resource is the most critical factor controlling mangrove productivity and distribution.

9. The "outwelling hypothesis" and mangrove food webs across diverse coastal environmental settings of mangrove dominated estuaries can best be described by the following statement.
   a. Tidal exchange results in all mangrove leaf detritus exporting as particulate organic carbon that dominates all estuarine food webs in support of the outwelling hypothesis.
   b. Litter export is associated with rain events and export dominated by dissolved organic carbon that dominates estuarine food webs by stimulating microbial production and resulting in more nutritious detritus.
   c. Litter export by tides and rainfall is minimal in mangrove wetlands because all the leaf litter from the canopy decomposes in the forest floor or consumed by crabs contributing insignificant organic detritus to food webs.
   d. Litter export is controlled by frequency of tides and rainfall events that can be major source of organic detritus export to estuaries and depending on location of consumer relative to organic detritus export be significant to food webs.

10. Ecosystem-level carbon sequestration due to NEP of mangrove wetlands can be significant carbon sink due to the following processes.
    a. Litter fall from canopy is year-round and contributes to soil formation that represents major carbon sink in mangrove soils with similar rates of about 560 $gC\,m^{-2}\,year^{-1}$ in all coastal environmental settings.
    b. Carbon sequestration is about 560 $gC\,m^{-2}\,year^{-1}$ with annual carbon rates from wood production nearly double those of soil formation.
    c. Storage sequestration is about 60 $gC\,m^{-2}\,year^{-1}$ with annual rates from wood production nearly double those of soil formation.
    d. Carbon sequestration rate is about 560 $gC\,m^{-2}\,year^{-1}$ with rates from wood production nearly half those of soil formation.

Short Answer

1. Describe the distribution of mangrove tree diversity among the new world (Atlantic East Pacific) and old world (Indo West Pacific) biogeographic regions and also with latitude in each of these regions. Compare and contrast this diversity with Gulf of Mexico and Caribbean.
2. Describe the global contribution of annual blue carbon sinks in mangrove wetlands as result of NEP.
3. Describe the global contribution of mangrove wetlands to blue carbon storage due to ecosystem-level stocks.
4. Tree diversity of mangroves is very different in the new vs old world tropics. Does this also reflect the diversity of mangrove wetlands as an ecosystem? Explain biodiversity of mangrove wetlands around the world. What one group of organisms is known as a keystone guild in mangrove wetlands?
5. Describe how hydrology and waterlogged soils control the productivity and zonation of mangroves, along with distribution of resources and regulators. What are mangrove adaptations to these environmental gradients?

References

Abrantes, K.G., Johnston, R., Connolly, R.M., and Sheaves, M. (2015). Importance of mangrove carbon for aquatic food webs in wet-dry tropical estuaries. *Estuaries and Coasts* 38: 383–399.

Adame, M.F. and Lovelock, C.E. (2011). Carbon and nutrient exchange of mangrove forests with the coastal ocean. *Hydrobiologia* 663: 23–50.

Adame, M.F., Zaldívar-Jimenez, A., Teutli, C. et al. (2013). Drivers of mangrove litterfall within a Karstic Region affected by frequent hurricanes. *Biotropica* 45: 147–154.

Adame, M., Santini, N., Torres-Talamante, O., and Rogers, K. (2021). Mangrove sinkholes (cenotes) of the Yucatan Peninsula, a global hotspot of carbon sequestration. *Biology Letters* 17: 20210037.

Allen, J.A. (1998). Mangroves as alien species: the case of Hawaii. *Global Ecology & Biogeography Letters* 7: 61–71.

Allen, J.A., Krauss, K.W., Duke, N.C. et al. (2000). Bruguiera species in Hawai'i: Systematic considerations and ecological implications. *Pacific Science* 54 (Part 4): 331–344.

Alongi, D.M. (2009). *The Energetics of Mangrove Forests*. Springer Science & Business Media.

Atwood, T.B., Connolly, R.M., Almahasheer, H. et al. (2017). Global patterns in mangrove soil carbon stocks and losses. *Nature Climate Change* 7: 523.

Bakkar, T., Helfer, V., Himmelsbach, R., and Zimmer, M. (2017). Chemical changes in detrital matter upon digestive processes in a sesarmid crab feeding on mangrove leaf litter. *Hydrobiologia* 803: 307–315.

Ball, M.C. (1988a). Ecophysiology of mangroves. *Trees* 2: 129–142.

Ball, M.C. (1988b). Salinity tolerance in the mangrove *Aegiceras corniculatum* and *Avicennia marina*. I. Water use in relation to growth, carbon partitioning, and salt balance. *Australian Journal of Plant Physiology* 15: 447–464.

Ball, M.C. (1996). Comparative ecophysiology of mangrove forest and tropical lowland moist rainforest. In: *Tropical Forest Plant Ecophysiology* (ed. R.L. Chazdon, S.S. Mulkey and A.P. Smith), 461–496. Boston, MA: Springer.

Ball, M.C. (2002). Interactive effects of salinity and irradiance on growth: implications for mangrove forest structure along salinity gradients. *Trees* 16: 126–139.

Beltram, G. (2018). *Karst wetlands. The Wetland Book: II: Distribution, Description, and Conservation*, 313–329. Dordrecht: Springer.

Blaber, S.J. (2007). Mangroves and fishes: issues of diversity, dependence, and dogma. *Bulletin of Marine Science* 80: 457–472.

Bosire, J.O., Dahdouh-Guebas, F., Kairo, J.G. et al. (2005). Litter degradation and CN dynamics in reforested mangrove plantations at Gazi Bay, Kenya. *Biological Conservation* 126: 287–295.

Bosire, J.O., Dahdouh-Guebas, F., Walton, M. et al. (2008). Functionality of restored mangroves: a review. *Aquatic Botany* 89: 251–259.

Botero, L. (1990). Massive mangrove mortality on the Caribbean Coast of Colombia. *Vida Silvestre Neotropical* 2: 77–78.

Boto, K.G. and Wellington, J.T. (1988). Seasonal variations in concentrations and fluxes of dissolved organic and inorganic materials in a tropical, tidally-dominated, mangrove waterway. *Marine Ecology Progress Series* 50: 151–160.

Bouillon, S., Koedam, N., Baeyens, W. et al. (2004). Selectivity of subtidal benthis invertebrate communities for local microalgal production in an estuarine mangrove ecosystem during the post-monsoon period. *Journal of Sea Research* 51: 133–144.

Bouillon, S., Connolly, R.M., and Lee, S.Y. (2008a). Organic matter exchange and cycling in mangrove ecosystems: recent insights from stable isotope studies. *Journal of Sea Research* 59: 44–58.

Bouillon, S., Borges, A.V., Castaneda-Moya, E. et al. (2008b). Mangrove production and carbon sinks: a revision of global budget estimates. *Global Biogeochemical Cycles* 22: GB2013. https://doi.org/10.1029/2007GB003052.

Breithaupt, J.L., Smoak, J.M., Smith, T.J. et al. (2012). Organic carbon burial rates in mangrove sediments: strengthening the global budget. *Global Biogeochemical Cycles* 26: GB3011. https://doi.org/10.1029/2012GB004375.

Breithaupt, J.L., Smoak, J.M., Rivera-Monroy, V.H. et al. (2017). Partitioning the relative contributions of organic matter and mineral sediment to accretion rates in carbonate platform mangrove soils. *Marine Geology* 390: 170–180.

Castañeda-Moya, E., Rivera-Monroy, V.H., and Twilley, R.R. (2006). Mangrove zonation in the dry life zone of the Gulf of Fonseca, Honduras. *Estuaries and Coasts* 29: 751–764.

Castañeda-Moya, E., Twilley, R.R., Rivera-Monroy, V.H. et al. (2010). Sediment and nutrient deposition associated with Hurricane Wilma in mangroves of the Florida Coastal Everglades. *Estuaries and Coasts* 33: 45–58.

Castañeda-Moya, E., Twilley, R.R., and Rivera-Monroy, V.H. (2013). Allocation of biomass and net primary productivity of mangrove forests along environmental gradients in the Florida Coastal Everglades, USA. *Forest Ecology and Management* 307: 226–241.

Cavanaugh, K.C., Kellner, J.R., Forde, A.J. et al. (2014). Poleward expansion of mangroves is a threshold response to decreased frequency of extreme cold events. *Proceedings of the National Academy of Sciences* 111: 723–727.

Chapin, F.S. III, Schulze, E.D., and Mooney, H.A. (1990). The ecology and economics of storage in plants. *Annual Review of Ecology and Systematics* 21: 423–447.

Chapman, V.J. (1976). *Mangrove Vegetation*. Vaduz, Germany: J. Cramer.

Chen, R.G. and Twilley, R.R. (1998). A gap dynamic model of mangrove forest development along gradients of soil salinity and nutrient resources. *Journal of Ecology* 86: 37–51.

Chen, R.H. and Twilley, R.R. (1999). Patterns of mangrove forest structure and soil nutrient dynamics along the Shark River estuary, Florida. *Estuaries* 22: 955–970.

Chen, L., Zan, Q., Li, M. et al. (2009). Litter dynamics and forest structure of the introduced *Sonneratia caseolaris* mangrove forest in Shenzhen, China. *Estuarine, Coastal and Shelf Science* 85: 241–246.

Cintrón, G., Lugo, A.E., Pool, D.J., and Morris, G. (1978). Mangroves of arid environments in Puerto Rico and adjacent islands. *Biotropica* 10: 110–121.

Clark, D.A., Brown, S., Kicklighter, D.W. et al. (2001). Measuring net primary production in forests: concepts and field methods. *Ecological Applications* 11: 356–370.

Cole, T.G., Ewel, K.C., and Devoe, N.N. (1999). Structure of mangrove trees and forest in Micronesia. *Forest Ecology and Management* 117: 95–109.

Cox, E.F. and Allen, J.A. (1999). Stand structure and productivity of the introduced *Rhizophora mangle* in Hawaii. *Estuaries* 22: 276–284.

Dangremond, E.M., Feller, I.C., and Sousa, W.P. (2015). Environmental tolerances of rare and common mangroves along light and salinity gradients. *Oecologia* 179: 1187–1198.

Demopoulos, A.W., Fry, B., and Smith, C.R. (2007). Food web structure in exotic and native mangroves: a Hawaii–Puerto Rico comparison. *Oecologia* 153: 675–686.

Dittmar, T. and Lara, R.J. (2001). Do mangroves rather than rivers provide nutrients to coastal environments south of the Amazon River? Evidence from long-term flux measurements. *Marine Ecology Progress Series* 213: 67–77.

Dittmar, T., Hertkorn, N., Kattner, G., and Lara, R.J. (2006). Mangroves, a major source of dissolved organic carbon to the oceans. *Global Biogeochemical Cycles* 20.

Donato, D.C., Kauffman, J.B., Murdiyarso, D. et al. (2011). Mangroves among the most carbon-rich forests in the tropics. *Nature Geoscience* 4: 293–297.

Duarte, C.M., Middelburg, J.J., and Caraco, N. (2005). Major role of marine vegetation on the oceanic carbon cycle. *Biogeosciences* 2: 1–8.

Duke, N.C. (1992). Mangrove floristics and biogeography, Chapter 4. In: *Tropical Mangrove Ecosystems*, Coastal and Estuarine Studies Series, vol. 41 (ed. A.I. Robertson and D.M. Alongi), 63–100. Washington, DC: American Geophysical Union.

Duke, N. and Pinzón, Z. (1993). *Mangrove Forests*. New Orleans, LA: U.S. Dept of the Interior, Minerals Management Service, Gulf of Mexico OCS Regional Office.

Duke, N.C., Ball, M.C., and Ellison, J.C. (1998). Factors influencing biodiversity and distributional gradients in mangroves. *Global Ecology and Biogeography Letters* 7: 27–47.

Dürr, H.H., Laruelle, G.G., van Kempen, C.M. et al. (2011). Worldwide typology of nearshore coastal systems: defining the estuarine filter of river inputs to the oceans. *Estuaries and Coasts* 34: 441–458.

Ellison, A.M. and Farnsworth, E.J. (1992). The ecology of Belizean mangrove-root fouling communities: patterns of epibiont distribution and abundance, and effects on root growth. *Hydrobiologia* 20: 1–12.

Ellison, A.M. and Farnsworth, E.J. (1993). Seedling survivorship, growth, and response to disturbance in Belizean mangal. *American Journal of Botany* 80: 1137–1145.

Ellison, A.M., Farnsworth, E.J., and Twilley, R.R. (1996). Facultative mutualisms between red mangroves and root-fouling sponges in Belizian mangal. *Ecology* 77: 2431–2444.

Ellison, A.M., Farnsworth, E.J., and Merkt, R.E. (1999). Origins of mangrove ecosystems and the mangrove biodiversity anomaly. *Global Ecology and Biogeography* 8: 95–115.

Feller, I.C., Whigham, D.F., McKee, K.L., and Lovelock, C.E. (2003). Nitrogen limitation of growth and nutrient dynamics in a disturbed mangrove forest, Indian River Lagoon, Florida. *Oecologia* 134: 405–414.

Feller, I.C., Lovelock, C.E., Berger, U. et al. (2010). Biocomplexity in mangrove ecosystems. *Annual Review of Marine Science* 2: 395–417.

Feller, I.C., Friess, D.A., Krauss, K.W., and Lewis, R.R. (2017). The state of the world's mangroves in the 21st century under climate change. *Hydrobiologia* 803: 1–12.

Field, C. (1999). Rehabilitation of mangrove ecosystems: an overview. *Marine Pollution Bulletin* 37: 383–392.

Fourqurean, J.W., Smith, T.J., Possley, J. et al. (2010). Are mangroves in the tropical Atlantic ripe for invasion? Exotic mangrove trees in the forests of South Florida. *Biological Invasions* 12: 2509–2522.

Friess, D. (2017). JG Watson, inundation classes, and their influence on paradigms in mangrove forest ecology. *Wetlands* 37: 603–613.

Friess, D.A., Krauss, K.W., Horstman, E.M. et al. (2012). Are all intertidal wetlands naturally created equal? Bottlenecks, thresholds and knowledge gaps to mangrove and salt marsh ecosystems. *Biological Reviews* 87: 346–366.

Garrity, S.D., Levings, S.C., and Burns, K.A. (1994). The Galeta oil spill. I. Long-term effects on the physical structure of the mangrove fringe. *Estuarine, Coastal and Shelf Science* 38: 327–348.

Getter, C.D., Scott, G.I., and Michel, J. (1981). The effects of oil spills on mangrove forests: a comparison of five oil spill sites in the Gulf of Mexico and the Caribbean Sea. In: *International Oil Spill Conference*, vol. 1981, No. 1, 535–540. American Petroleum Institute.

Giri, C., Long, J., and Tieszen, L. (2011a). Mapping and monitoring Louisiana's mangroves in the aftermath of the 2010 gulf of Mexico Oil Spill. *Journal of Coastal Research* 27: 1059–1064.

Giri, C., Ochieng, E., Tieszen, L.L. et al. (2011b). Status and distribution of mangrove forests of the world using earth observation satellite data. *Global Ecology and Biogeography* 20: 154–159.

Gleeson, S.K. and Tilman, D. (1992). Plant allocation and the multiple limitation hypothesis. *American Naturalist* 139 (6): 1322–1343.

Goecke, S.D. and Carstenn, S.M. (2017). Fish communities and juvenile habitat associated with non-native Rhizophora mangle L. in Hawai'i. *Hydrobiologia* 803: 209–224.

Gosselink, J.G. and Turner, R.E. (1978). The role of hydrology in freshwater wetland ecosystems. In: *Freshwater Wetlands: Ecological Processes and Management Potential* (ed. R.E. Good, D.F. Whigham and R.L. Simpson), 63–78. New York: Academic Press.

Hamilton, S.E. and Casey, D. (2016). Creation of a high spatio-temporal resolution global database of continuous mangrove forest cover for the 21st century (CGMFC-21). *Global Ecology and Biogeography* 25: 729–738.

Harrison, J.A., Caraco, N., and Seitzinger, S.P. (2005). Global patterns and sources of dissolved organic matter export to the coastal zone: results from a spatially explicit, global model. *Global Biogeochemical Cycles* 19: GB4S04. https://doi.org/10.1029/2005GB002480.

Huston, M.A. (1994). *Biological Diversity: The Coexistence of Species*, 704pp. Cambridge University Press. ISBN: 9780521369305.

Hutchings, P. and Saenger, P. (1987). *Ecology of Mangroves*. Quensland University Press.

Hutchison, J., Manica, A., Swetnam, R.B. et al. (2014). Predicting global patterns in mangrove forest biomass. *Conservation Letters* 7: 233–240.

Imgraben, S. and Dittmann, S. (2008). Leaf litter dynamics and litter consumption in two temperate south Australian mangrove forests. *Journal of Sea Research* 59: 83–93.

Jardine, S.L. and Siikamäki, J.V. (2014). A global predictive model of carbon in mangrove soils. *Environmental Research Letters* 9: 104013.

Jennerjahn, T.C. (2020). Relevance and magnitude of 'Blue Carbon' storage in mangrove sediments: carbon accumulation rates vs. stocks, sources vs. sinks. *Estuarine, Coastal and Shelf Science* 247: 107027.

Jennerjahn, T.C. and Ittekkot, V. (2002). Relevance of mangroves for the production and deposition of organic matter along tropical continental margins. *Naturwissenschaften* 89: 23–30.

Jones, D.A. (1984). Crabs of the mangal ecosystem. In: *Hydrobiology of the Mangal* (ed. F.D. Por and I. Dor), 89–109. The Hague: Dr. W. Junk Publishers.

Kauffman, J.B., Arifanti, V.B., Trejo, H.H. et al. (2017). The jumbo carbon footprint of a shrimp: carbon losses from mangrove deforestation. *Frontiers in Ecology and the Environment* 15: 183–188.

Khan, M.N.I., Suwa, R., and Hagihara, A. (2009). Biomass and aboveground net primary production in a subtropical mangrove stand of *Kandelia obovata* (S., L.) Yong at Manko Wetland, Okinawa, Japan. *Wetlands Ecology and Management* 17: 585–599.

Komiyama, A., Ogino, K., Aksornkoae, S., and Sabhasri, S. (1987). Root biomass of a mangrove forest in southern Thailand,. 1. Estimation by the trench method and the zonal structure of root biomass. *Journal Tropical Ecology* 3: 97–108.

Komiyama, A., Ong, J.E., and Poungparn, S. (2008). Allometry, biomass, and productivity of mangrove forests: a review. *Aquatic Botany* 89: 128–137.

Krauss, K.W. and Ball, M.C. (2013). On the halophytic nature of mangroves. *Trees* 27: 7–11.

Krauss, K.W., Allen, J.A., and Cahoon, D.R. (2003). Differential rates of vertical accretion and elevation change among aerial root types in Micronesian mangrove forests. *Estuarine Coastal and Shelf Science* 56: 251–259.

Krauss, K.W., Doyle, T.W., Twilley, R.R. et al. (2006). Evaluating the relative contributions of hydroperiod and soil fertility on growth of south Florida mangroves. *Hydrobiologia* 569: 311–324.

Krauss, K.W., Lovelock, C.E., McKee, K.L. et al. (2008). Environmental drivers in mangrove establishment and early development: a review. *Aquatic Botany* 89: 105–127.

Krauss, K.W., McKee, K.L., Lovelock, C.E. et al. (2014). How mangrove forests adjust to rising sea level. *The New Phytologist* 202: 19–34.

Kristensen, E. (2008). Mangrove crabs as ecosystem engineers; with emphasis on sediment processes. *Journal of Sea Research* 59: 30–43.

Kristensen, E., Bouillon, S., Dittmar, T., and Marchand, C. (2008). Organic carbon dynamics in mangrove ecosystems: a review. *Aquatic Botany* 89: 201–219.

Lamers, L.P., Govers, L.L., Janssen, I.C. et al. (2013). Sulfide as a soil phytotoxin—a review. *Frontiers in Plant Science* 4: 268.

Lee, S.Y. (2008). Mangrove macrobenthos: assemblages, services, and linkages. *Journal of Sea Research* 59: 16–29.

Lewis, R.R. (2005). Ecological engineering for successful management and restoration of mangrove forests. *Ecological Engineering* 24: 403–418.

Lewis, M., Pryor, R., and Wilking, L. (2011). Fate and effects of anthropogenic chemicals in mangrove ecosystems: a review. *Environmental pollution* 159: 2328–2346.

Lewis, R.R., Milbrandt, E.C., Brown, B. et al. (2016). Stress in mangrove forests: early detection and preemptive rehabilitation are essential for future successful worldwide mangrove forest management. *Marine Pollution Bulletin* 109: 764–771.

Li, M. and Lee, S. (1998). The particulate organic matter dynamics of Deep Bay, eastern Pearl River estuary, China I. Implications for waterfowl conservation. *Marine Ecology Progress Series* 172: 73–87.

Lindquist, E.S., Krauss, K.W., Green, P.T. et al. (2009). Land crabs as key drivers in tropical coastal forest recruitment. *Biological Reviews* 84: 203–223.

López-Hoffman, L., Anten, N.P.R.R., Martínez-Ramos, M., and Ackerly, D.D. (2007). Salinity and light interactively affect neotropical mangrove seedlings at the leaf and whole plant levels. *Oecologia* 150: 545–556.

López-Portillo, J., Lewis, R.R., Saenger, P. et al. (2017). Mangrove forest restoration and rehabilitation. In: *Mangrove Ecosystems: A Global Biogeographic Perspective* (ed. V.H. Rivera-Monroy, S.Y. Lee, E. Kristensen and R.R. Twilley), 301–345. Springer.

Lovelock, C.E., Krauss, K.W., Osland, M.J. et al. (2016). The physiology of mangrove trees with changing climate. In: *Tropical Tree Physiology* (ed. G. Goldstein and L. Santiago), 149–179. Cham: Springer https://doi.org/10.1007/978-3-319-27422-5_7.

Lugo, A. (1980). Mangrove ecosystem: successional or steady state. *Biotropica* 12: 65–72.

Lugo, A.E. (1990). Fringe wetlands. In: *Forested Wetlands* (ed. A.E. Lugo, M. Brinson and S. Brown), 143–169. Amsterdam: Elsevier.

Lugo, A.E. and Patterson-Zucca, C. (1977). The impact of low temperature stress on mangrove structure and growth. *Tropical Ecology* 18: 149–161.

Lugo, A.E. and Snedaker, S.C. (1974). The ecology of mangroves. *Annual Review of Ecology and Systematics* 5: 39–64.

Lüttge, U. (2007). *Physiological ecology of tropical plants*. Springer Science & Business Media.

Macnae, W. (1968). A general account of the fauna and flora of mangrove swamps and forests in the Indo-West-Pacific region. *Advanced Marine Biology* 6: 73–270.

Macnae, W. 1974. Mangrove Forests and Fisheries. FAO/UNDP Indian OceanProgramme, IOFC/DEV /7434.

Martinuzzi, S., Gould, W.A., Lugo, A.E., and Medina, E. (2009). Conversion and recovery of Puerto Rican mangroves: 200 years of change. *Forest Ecology and Management* 257: 75–84.

McKee, K.L. (1993). Soil physiochemical patterns and mangrove species distribution—reciprocal effects? *Journal of Ecology* 81: 477–487.

McKee, K.L. (1995a). Interspecific variation in growth, biomass partitioning, and defensive characteristics of neotropical mangrove seedlings: response to light and nutrient availability. *American Journal of Botany* 82: 299–307.

McKee, K.L. (1995b). Seedling recruitment patterns in a Belizean mangrove forest: effects of establishment ability and physico-chemical factors. *Oecologia* 101: 448–460.

McKee, K.L., Mendelssohn, J.A., and Hester, M.W. (1988). Reexamination of pore water sulfide concentrations and redox potentials near the aerial roots of *Rhizpohora Mangle* and *Avicennia Germinans*. *American Journal of Botany* 75: 1352–1359.

McKee, K.L., Cahoon, D.R., and Feller, I.C. (2007). Caribbean mangroves adjust to rising sea level through biotic controls on change in soil elevation. *Global Ecology and Biogeography* 16: 545–556.

Mcleod, E., Chmura, G.L., Bouillon, S. et al. (2012). A blueprint for blue carbon: toward an improved understanding of the role of vegetated coastal habitats in sequestering CO_2. *Frontiers in Ecology and the Environment* 9: 552–560.

Mfilinge, P.L., Meziane, T., Bachok, Z., and Tsuchiya, M. (2005). Litter dynamics and particulate organic matter outwelling from a subtropical mangrove in Okinawa Island, South Japan. *Estuarine, Coastal and Shelf Science* 63: 301–313.

Naidoo, G., Naidoo, Y., and Achar, P. (2010). Responses of the mangroves Avicennia marina and Bruguiera gymnorrhiza to oil contamination. *Flora-Morphology, Distribution, Functional Ecology of Plants* 205: 357–362.

Nguyen, H.T., Stanton, D.E., Schmitz, N. et al. (2015). Growth responses of the mangrove Avicennia marina to salinity: development and function of shoot hydraulic systems require saline conditions. *Annals of Botany* 115: 397–407.

Nixon, S.W. (1980). *Between Coastal Marshes and Coastal Waters—A Review of Twenty Years of Speculation and Research on the Role of Salt Marshes in Estuarine Productivity and Water Chemistry*. Estuarine and Wetlands Processes, 437–525. New York: Plenum Publishing Corporation.

Odum, W.E. (1984). Dual-gradient concept of detritus transport and processing in estuaries. *Bulletin of Marine Science* 35: 510–521.

Odum, H.T. (1994). *Ecological and General Systems: An Introduction to Systems Ecology*, Revised Edition. Niwot, CO: University Press of Colorado. ISBN:0-87081-320-x (paper).

Odum, W.E. and Heald, E.J. (1972). Trophic analyses of an estuarine mangrove community. *Bulletin Marine Science* 22: 671–738.

Odum, W.E. and Heald, E.J. (1975). The detritus-based food web of an estuarine mangrove community. In: *Estuarine Research* (ed. L.E. Cronin), 265–286. New York: Academic Press.

Osland, M.J., Enwright, N., Day, R.H. et al. (2013). Winter climate change and coastal wetland foundation species: salt marshes vs. mangrove forests in the southeastern United States. *Global Change Biology* 19: 1482–1494.

Palacios, M.L. and Cantera, J.R. (2017). Mangrove timber use as an ecosystem service in the Colombian Pacific. *Hydrobiologia* 803: 345–358.

Pelegri, S.P., Rivera-Monroy, V.H., and Twilley, R.R. (1997). A comparison of nitrogen fixation (acetylene reduction) among three species of mangrove litter, sediments, and pneumatophores in south Florida, USA. *Hydrobiologia* 356: 73–79.

Pestana, D.F., Pülmanns, N., Nordhaus, I. et al. (2017). The influence of crab burrows on sediment salinity in a *Rhizophora*-dominated mangrove forest in North Brazil during the dry season. *Hydrobiologia* 803: 295–305.

Pezeshki, S. and DeLaune, R. (2012). Soil oxidation-reduction in wetlands and its impact on plant functioning. *Biology* 1: 196–221.

Primavera, J.H. (1996). Stable carbon and nitrogen isotope ratios of *Penaeid* juveniles and primary producers in a riverine mangrove in Guimaras, Philippines. *Bulletin of Marine Science* 58: 675–683.

Reef, R., Feller, I.C., and Lovelock, C.E. (2010). Nutrition of mangroves. *Tree Physiology* 30: 1148–1160.

Reis, C.R.G., Nardoto, G.B., and Oliveira, R.S. (2017). Global overview on nitrogen dynamics in mangroves and consequences of increasing nitrogen availability for these systems. *Plant and Soil* 410: 1–19.

de Ribeiro, R., A., Rovai, A.S., Twilley, R.R., and Castañeda-Moya, E. (2019). Spatial variability of mangrove primary productivity in the neotropics. *Ecosphere* 10: e02841.

Richards, D.R. and Friess, D.A. (2016). Rates and drivers of mangrove deforestation in Southeast Asia, 2000–2012. *Proceedings of the National Academy of Sciences* 113: 344–349.

Rivera-Monroy, V.H. and Twilley, R.R. (1996). The relative role of denitrification and immobilization on the fate of inorganic nitrogen in mangrove sediments of Terminos Lagoon, Mexico. *Limnology and Oceanography* 41: 284–296.

Rivera-Monroy, V.H., Day, J.W., Twilley, R.R. et al. (1995a). Flux of nitrogen and sediment in a fringe mangrove forest in Terminos Lagoon, Mexico. *Estuarine, Coastal and Shelf Science* 40: 139–160.

Rivera-Monroy, V.H., Twilley, R.R., Boustany, R.G. et al. (1995b). Direct denitrification in mangrove sediments in Terminos Lagoon, Mexico. *Marine Ecology Progress Series* 126: 97–109.

Rivera-Monroy, V.H., Twilley, R.R., Mancera-Pineda, J.E. et al. (2011). Salinity and chlorophyll a as performance measures to rehabilitate a mangrove-dominated deltaic coastal region: the Cienaga Grande de Santa Marta-Pajarales Lagoon Complex, Colombia. *Estuaries and Coasts* 34: 1–19.

Robertson, A.I. and Blaber, S.J.M. (1992). Plankton, epibenthos and fish communities. In: *Tropical Mangrove Ecosystems*, Coastal and Estuarine Series, 41 (ed. A.I. Robertson and D.M. Alongi), 173–224. American Geophysical Union ISBN 0-87590-255-3.

Robertson, A.I. and Daniel, P.A. (1989). The influence of crabs on litter processing in high intertidal mangrove forests in tropical Australia. *Oecologia* 78: 191–198.

Rodelli, M.R., Gearing, J.N., Gearing, P.J. et al. (1984). Stable isotope ratio as a tracer of mangrove carbon in Malaysian ecosystems. *Oecologia* 326–333.

Rogers, K., Kelleway, J.J., Saintilan, N. et al. (2019). Wetland carbon storage controlled by millennial-scale variation in relative sea-level rise. *Nature* 567: 91–95.

Rovai, A.S., Riul, P., Twilley, R.R. et al. (2016). Scaling mangrove aboveground biomass from site-level to continental-scale. *Global Ecology and Biogeography* 25: 286–298.

Rovai, A.S., Twilley, R.R., Castañeda-Moya, E. et al. (2018). Global controls on carbon storage in mangrove soils. *Nature Climate Change* 8: 534–538.

Rovai, A.S., Twilley, R.R., Castañeda-Moya, E. et al. (2021). Macroecological patterns of forest structure and allometric scaling in mangrove forests. *Global Ecology and Biogeography* 30: 1000–1013.

Rützler, K. and Feller, I.C. (1996). Caribbean mangrove swamps. *Scientific American* 274: 94–99.

Saenger, P. (2013). *Mangrove Ecology, Silviculture and Conservation*. Springer Science & Business Media.

Saenger, P., Hegerl, E.J., and Davie, J.D.S. (1983). Global status of mangrove ecosystems. *The Environmentalist* 3: 3–88.

Saintilan, N., Wilson, N.C., Rogers, K. et al. (2014). Mangrove expansion and salt marsh decline at mangrove poleward limits. *Global Change Biology* 20: 147–157.

Sasekumar, A., Chong, V., Leh, M., and D'cruz, R. (1992). Mangroves as a habitat for fish and prawns. *Hydrobiologia* 247: 195–207.

Scholander, P.F., Hammel, H.T., Hemmingsen, E., and Garey, W. (1962). Salt balance in mangroves. *Plant Physiology* 37: 722–729.

Sherman, R.E., Fahey, T.J., and Martinez, P. (2001). Hurricane impacts on a mangrove forest in the Dominican Republic: damage patterns and early recovery. *Biotropica* 33: 393–408.

Shipley, B. and Meziane, D. (2002). The balanced-growth hypothesis and the allometry of leaf and root biomass allocation. *Functional Ecology* 16: 326–331.

Sillanpää, M., Vantellingen, J., and Friess, D.A. (2017). Vegetation regeneration in a sustainably harvested mangrove forest in West Papua, Indonesia. *Forest Ecology and Management* 390: 137–146.

Simberloff, D.S. and Wilson, E.O. (1969). Experimental zoogeography of islands: the colonization of empty islands. *Ecology* 50: 278–289.

Smith, T.J. (1987). Seed predation in relation to tree dominance and distribution in mangrove forests. *Ecology* 68: 266–273.

Smith, S.M. and Snedaker, S.C. (2000). Hypocotyl function in seedling development of the red mangrove, *Rhizophora mangle L.*1. *Biotropica* 32: 677–685.

Smith, T.J., Chan, H.T., McIvor, C.C., and Robblee, M.B. (1989). Comparisons of seed predation in tropical, tidal forests from three continents. *Ecology* 70: 146–151.

Smith, T.J., Boto, K.G., Frusher, S.D., and Giddins, R.L. (1991). Keystone species and mangrove forest dynamics: the influence of burrowing by crabs on soil nutrient status and forest productivity. *Estuarine, Coastal and Shelf Science* 33: 419–432.

Snedaker, S.C., Brown, M.S., Lahmann, E.J., and Araujo, R.J. (1992). Recovery of a mixed-species mangrove forest in South Florida following canopy removal. *Journal of Coastal Research* 8: 919–925.

Soper, F.M., MacKenzie, R.A., Sharma, S. et al. (2019). Non-native mangroves support carbon storage, sediment carbon burial, and accretion of coastal ecosystems. *Global Change Biology* 25: 4315–4326.

Spalding, M. (2010). *World Atlas of Mangroves*. Routledge.

Stuart, S.A., Choat, B., Martin, K.C. et al. (2007). The role of freezing in setting the latitudinal limits of mangrove forests. *New Phytologist* 173: 576–583.

Thom, B.G. (1982). Mangrove ecology-a geomorphological perspective. In: *Mangrove Ecosystems in Australia* (ed. B.F. Clough), 3–17. Canberra: Australian National University Press.

Tilman, D. (1982). *Resource Competition and Community Structure*. Princeton University Press.

Tomlinson, P.B. (2016). *The Botany of Mangroves*. Cambridge University Press.

Twilley, R.R. (1988). Coupling of mangroves to the productivity of estuarine and coastal waters. In: *Coastal-Offshore Ecosystem Interactions* (ed. B.O. Janson), 155–180. Germany: Springer-Verlag.

Twilley, R.R. (1995). Properties of mangrove ecosystems related to the energy signature of coastal environments. In: *Maximum Power: The Ideas and Aplications of H. T. Odum* (ed. C.A.S. Hall), 43–62. Niwot, Colorado: University Press of Colorado.

Twilley, R.R. and Chen, R. (1998). A water budget and hydrology model of a basin mangrove forest in Rookery Bay, Florida. *Marine and Freshwater Research* 49: 309–323.

Twilley, R.R. and Rivera-Monroy, V.H. (2005). Developing performance measures of mangrove wetlands using simulation models of hydrology, nutrient biogeochemistry and community dynamics. *Journal of Coastal Research SPEC. ISS.* 40: 79–93.

Twilley, R.R. and Rivera-Monroy, V.H. (2009). *Ecogeomorphic models of nutrient biogeochemistry for mangrove wetlands.* In: *Coastal Wetlands: An Integrated Ecosystem Approach* (ed. G. M. E. Perillo, E. Wolanski, D. R. Cahoon, M. M. Brinson), 641–683. Amsterdam: Elsevier.

Twilley, R.R., Chen, R.H., and Hargis, T. (1992). Carbon sinks in mangroves and their implications to carbon budget of tropical coastal ecosystems. *Water, Air and Soil Pollution* 64: 265–288.

Twilley, R.R., Snedaker, S.C., Yanez-Arancibia, A., and Medina, E. (1996). Biodiversity and ecosystem processes in tropical estuaries: perspectives of mangrove ecosystems. *Scope-Scientific Committee on Problems of the Environment International Council of Scientific Unions* 55: 327–370.

Twilley, R.R., Pozo, M., Garcia, V.H. et al. (1997). Litter dynamics in riverine mangrove forests in the Guayas River estuary, Ecuador. *Oecologia* 111: 109–122.

Twilley, R.R., Gottfried, R.R., Rivera-Monroy, V.H. et al. (1998a). An approach and preliminary model of integrating ecological and economic constraints of environmental quality in the Guayas River Estuary, Ecuador. *Environmental Science and Policy* 1: 277–288.

Twilley, R.R., Rivera-Monroy, V.H., Chen, R.H., and Botero, L. (1998b). Adapting an ecological mangrove model to simulate trajectories in restoration ecology. *Marine Pollution Bulletin* 37: 404–419.

Twilley, R., Cárdenas, W., Rivera-Monroy, V. et al. (2001). *Ecology of the Gulf of Guayaquil and the Guayas River Estuary. Estuaries of Latin America*, 245–264. Berlin: Springer-Verlag.

Twilley, R.R., Castañeda-Moya, E., Rivera-Monroy, V.H., and Rovai, A. (2017). Productivity and carbon dynamics in mangrove wetlands. In: *Mangrove Ecosystems: A Global Biogeographic Perspective: Structure, Function, and Services* (ed. V.H. Rivera-Monroy, S.Y. Lee, E. Kristensen and R.R. Twilley), 113–162. Cham: Springer International Publishing.

Twilley, R.R., Rovai, A.S., and Riul, P. (2018). Coastal morphology explains global blue carbon distributions. *Frontiers in Ecology and the Environment* 16 (9): 503–508.

Twilley, R.R., Rivera-Monroy, V.H., Rovai, A.S. et al. (2019). *Mangrove Biogeochemistry at Local to Global Scales Using Ecogeomorphic Approaches*, In: Coastal Wetlands, second edition. (ed. G.M.E. Perillo, E. Wolanski, D.R. Cahoon, C. S. Hopkinson), 717–785. Amsterdam: Elsevier.

Wafar, S., Untawale, A., and Wafar, M. (1997). Litter fall and energy flux in a mangrove ecosystem. *Estuarine, Coastal and Shelf Science* 44: 111–124.

Wang, W., Yan, Z., You, S. et al. (2011). Mangroves: obligate or facultative halophytes? A review. *Trees* 25: 953–963.

Warne, K. (2011). *Let Them Eat Shrimp*. Washington, DC: Island Press.

Watson, J.G. (1928). Mangrove forests of the Malay Peninsula. *Malayan Science Bulletin* IV (6): 1–275.

Wilcox, B. (1985). Angiosperm flora of the mangrove ecosystem of the Niger Delta. In: *The Mangrove Ecosystems of the Niger Delta:*

Proceedings of a Workshop (ed. W. BHR and C.P. Powell), 34–44. Harcourt: University of Port.

Williams, T.M., Krauss, K.W., and Okruszko, T. (2016). Hydrology of flooded and wetland forests. In: *Forest Hydrology: Processes, Management and Assessment* (ed. T. Williams, D. Amatya, L. Bren and C. de Jong), 103–123. UK: CAB International.

Woodroffe, C. (1992). Mangrove sediments and geomorphology. In: *Coastal and Estuarine Studies* (ed. A.I. Robertson and D.M. Alongi), 7–41. Washington, DC: American Geophysical Union.

Woodroffe, C.D. (2002). *Coasts: Form, Process and Evolution.* Cambridge University Press.

Woodroffe, C.D., Rogers, K., McKee, K.L. et al. (2016). Mangrove sedimentation and response to relative sea-level rise. *Annual Review of Marine Science* 8: 243–266.

Yáñez-Arancibia, A., Lara-Dominguez, A.L., Rojas-Galaviz, J.L. et al. (1988). Seasonal biomass and diversity of estuarine fishes coupled with tropical habitat heterogeneity (southern Gulf of Mexico). *Journal of Fish Biology* 33: 191–200.

Yáñez-Arancibia, A., Lara-Domínguez, A.L., and Day, J.W. Jr. (1993). Interactions between mangrove and seagrass habitats mediated by estuarine nekton assemblages: coupling of primary and secondary production. *Hydrobiologia* 264: 1–12.

CHAPTER **8**

Estuarine Benthic Algae

Turbinaria ornata (top) and *Halimeda opuntia* (bottom) on a fringing reef in Opunohu Bay, Mo'orea, French Polynesia. Photo credit: Emily Ryznar

Karen J. McGlathery[1], Kristina Sundbäck[2], Peggy Fong[3], Lillian Aoki[4], and Iris Anderson[5]

[1] Department of Environmental Sciences, University of Virginia, Charlottesville, VA, USA
[2] Department of Marine Sciences, University of Gothenburg, Göteborg, Sweden
[3] Department of Ecology and Evolutionary Biology, University of California Los Angeles, Los Angeles, CA, USA
[4] Department of Ecology and Evolutionary Biology, Cornell University, Ithaca, NY, USA
[5] Virginia Institute of Marine Sciences, College of William and Mary, Gloucester Point, VA, USA

8.1 Introduction

Benthic algae play a key role in regulating carbon and nutrient turnover and in supporting food webs in shallow-water coastal environments. They are especially important in the wide diversity of estuarine habitats found worldwide. Benthic producers are generally divided into macroalgae and microalgae (also known as microphytobenthos). Macroalgae contribute between 4.8 and 5.9% of the total marine net primary productivity (calculated from Duarte and Cebrian 1996). Although lower than total oceanic (81.1%) and coastal (8.5%) phytoplankton, local macroalgal productivity (per m^2) is comparable to some of the most productive terrestrial ecosystems such as tropical forests. Benthic microalgae often form visible brown or green mats on the sediment surface. These mats are thin, as the microalgae are confined to the photic zone (1–3 mm) of the sediment, yet the density of algae and other microorganisms is usually high, often 100–1000 times higher than in the water column. In shallow estuaries, benthic microalgae normally account for 20–50% of the total primary production (Underwood and Kromkamp 1999). In this chapter, we discuss the benthic algal communities that inhabit the soft-bottom (mud/sandflat, seagrass bed, and marsh), hard-bottom (rocky intertidal, shallow subtidal), and coral reef habitats of estuarine ecosystems.

8.2 Taxonomy

All algae at some stages of their life cycles are unicellular (usually as reproductive stages such as spores or zygotes), and they are viewed as "primitive" photosynthetic organisms because of their relatively simple construction and their long evolutionary history. Prokaryotic "blue green algae," or cyanobacteria, are the oldest group with fossils dating back almost 3,000 million years. These first algal fossil remains include stromatolites, which are structures formed in shallow tropical waters on which cyanobacterial mats accrete layers by trapping, binding, and cementing sediment grains. Photosynthesis by early primary producers was responsible for much of the oxygen that eventually built up to the levels that occur today.

Estuarine Ecology, Third Edition. Edited by Byron C. Crump, Jeremy M. Testa, and Kenneth H. Dunton.

Companion website: www.wiley.com/go/crump/estuarine3

Evolution of eukaryotic algae occurred much later, about 700–800 million years ago, though this date is difficult to determine more exactly as most groups were composed of soft tissue that would not have been preserved reliably in the fossil record.

Marine macroalgae or "seaweeds" are a functional group comprised of members from two Kingdoms and at least four major Phyla (Divisions). There is a wide variation in algal classification schemes among systematists, but the traditional Divisions for macroalgae are the *Cyanobacteria* (prokaryotic blue-green algae, sometimes termed Cyanophyta), *Chlorophyta* (green algae), *Phaeophyta* (brown algae), and *Rhodophyta* (red algae; Littler and Littler 2000).

The marine green algae range from cold temperate to tropical waters. Green algae reach their highest diversity and abundance in tropical regions, with several families such as the *Caulerpaceae* and *Udoteaceae* that are very abundant in coral reef and seagrass habitats. Often overlooked, but very abundant, are filamentous green algae that bore into coral skeletons and proliferate widely, with high rates of productivity (Littler and Littler 1988). Opportunistic green algae form nuisance blooms in estuaries worldwide; in eutrophic systems, they form almost monospecific mats of extremely high biomass.

Brown algae are almost exclusively marine and are dominant in temperate waters where hard-bottomed habitats occur. Some genera of structurally robust forms such as *Laminaria* and *Sargassum* dominate in very high-energy zones. Kelps are foundation species that form extensive forests in coastal areas where nutrients are supplied by upwelling (Graham 2004; Teagle et al. 2017). Other groups of fast-growing and more opportunistic genera such as *Dictyota* may form seasonal blooms in tropical and sub-tropical regions.

Most seaweed species are in the *Rhodophyta*. At present, the approximately 4000 named species of red algae exceed the number of species in all other groups combined (Lee 1999). Although red algae are extremely speciose in tropical and subtropical regions, their biomass is low relative to temperate areas. The most common forms of red algae in the tropics include crustose members of the family *Corallinaceae* as well as a high diversity of small, less obvious filamentous species that comprise algal turfs. However, there are some genera of upright and branching calcifying forms such as *Galaxaura* and branching or flattened foliose red algae in the genera *Laurencia*, *Asparagopsis*, and *Halymenia* that can be quite conspicuous and abundant on reefs under certain conditions. The highest biomass of red algae is found in temperate and boreal regions. Some large fleshy members of the *Rhodophyta* with descriptive names such as "Turkish towels" blanket rocky intertidal and subtidal regions. Other genera, such as *Gracilaria*, form blooms in estuaries and lagoons.

In most cases the benthic microalgal community in estuarine habitats is a mixture of several taxonomical groups, although blooms tend to be dominated by one or a few species. The taxa that form typical visible microbial mats on the sediment surface in the photic zone are mostly diatoms (Phylum *Bacillariophyta*) and cyanobacteria (the prokaryotic phylum *Cyanophyta* or *Cyanobacteria*). The term microbial mat originally referred to consortia dominated by prokaryotic phototrophs (Stal and Caumette 1992), but is here used in a broader sense, including all types of mats consisting of microscopic phototrophic organisms.

Diatoms are by far the most common taxonomic group, giving the sediment a brown color because of the pigment fucoxanthin. Benthic diatoms are different from planktonic diatoms in that they mostly represent the pennate diatoms, with more or less bilateral symmetry (classes *Fragilariophyceae* and *Bacillariophyceae*; according to the systematics in Round et al. 1990). They are solitary, and when they have a raphe (a slit-like structure) on both valves of their silica frustule (or covering), they are motile. Those that have no raphe, or a raphe only on one valve, can form short colonies. Some centric diatoms (class *Coscinodiscophyceae*) can be common on sediments, such as *Paralia sulcata*. The size of benthic diatoms ranges from a few μm to 500 μm; the sigmoid cells belonging to the genera *Gyrosigma* and *Pleurosigma* are some of the largest. Because many benthic diatoms are small (< 10 μm), it is difficult to identify live cells to species level in a light microscope. When the organic cell contents are removed by oxidation to prepare "diatom slides," the taxonomically important ornamentation of the silica frustule can be viewed.

Cyanobacteria are widespread both on soft and hard substrates. They form mats along reef margins or on coral (Smith et al. 2008), may be epiphytic on other algae (Fong et al. 2006), rapidly colonize open space opportunistically after disturbances (Janousek et al. 2007), and may bloom in response to nutrient enrichment (Armitage and Fong 2004). Cyanobacterial mats are also common on salt marshes, and particularly in subtropical and tropical estuaries and extreme habitats, such as hypersaline lagoons. They are laminated systems that form interdependent layers of vertically stratified phototrophic, heterotrophic, and chemotrophic microorganisms. They function as laterally compressed ecosystems that support most of the major biochemical cycles within a vertical dimension of a few millimeters (Paerl and Pinckney 1996). Stromatolites are one type of cyanobacterial assemblage. Typical benthic genera are the filament-forming *Oscillatoria* and *Microcoleus* and the colony-forming *Merismopedia*. Many filamentous benthic cyanobacteria fix nitrogen gas (N_2; Paerl and Pinckney 1996). For more characteristics of cyanobacteria, see Chapter 4.

Flagellates such as dinoflagellates (*Dinophyta*), euglenophytes (*Euglenophyta*), chlorophytes (*Chlorophyta*), and cryptophytes (*Cryptophyta*) are also found in estuarine benthic microalgal mats. Dense populations of dinoflagellates of the genus *Amphidinium* can occasionally give the sediment surface a red brown color. For characteristics of the various phyla, see Chapter 4. As for phytoplankton, photopigments can be used to identify and quantify the presence of major taxonomical groups of benthic microalgae in surface sediments (see Chapter 4).

8.3 Functional Forms

Structurally, benthic algae include diverse forms that range from single cells to giant kelps over 45 m in length with complex internal structures analogous to vascular plants. Species diversity may be extremely high in benthic algal communities (Guiry and Guiry 2007; see **http://www.algaebase.org**) and can be simplified by classifying algae in functional form categories. Steneck and Dethier (1994) classified algae into the following groups based on productivity and susceptibility to grazing: microalgae, filamentous, crustose, foliose, corticated foliose, corticated macrophyte, leathery macrophyte, and articulated calcareous.

For macroalgae, a classification scheme based on a broader set of characteristics was put forward by Littler and Littler (1984): sheet-like, filamentous, coarsely branched, thick-leathery, jointed-calcareous, and crustose forms (Table 8.1; Figure 8.1). These groups of functional forms have characteristic rates of nutrient uptake and mass-specific productivity, turnover rates, and resistance to herbivory that allow them to perform similarly in response to environmental conditions, despite differences in taxonomy. A key characteristic that drives these differences in function is the ratio of surface area to volume (SA:V) of the thallus. Macroalgae are also grouped more generally into "ephemeral" or "perennial" forms. Ephemeral macroalgae typically have a high thallus SA:V and inherently rapid nutrient uptake and growth rates. Most of the bloom-forming macroalgae that occur in response to nutrient over-enrichment (eutrophication) are ephemeral (e.g., *Ulva* spp., *Chaetomorpha* spp., *Gracilaria* spp., and *Polysiphonia* spp.), and usually have simple sheet, filamentous, or coarsely branched thallus forms. They tend to live floating and unattached or loosely attached to hard surfaces (i.e., shells, worm tubes) on the sediment (Schories and Reise 1993; Thomsen and McGlathery 2005). Perennial species like crustose and calcareous macroalgae tend to live in low-nutrient or stressful habitats (e.g., low light, low temperature) where a slow growth rate or perennial life-form confers an advantage.

For benthic microalgae that live in sediments, the energy regime of the habitat determines the dominant life forms. In wave-exposed, well-sorted sands, life forms that are firmly attached to sand grains dominate and are called *epipsammic* (attached to sand grains). In high-energy sand, most epipsammic diatoms are found in the crevices of the sand grains, where they are protected from abrasion. Typical genera are *Achnanthes*, *Cocconeis* (with raphe on one valve), and small-sized *Navicula*-like genera and *Amphora* (raphe on two valves) (Figure 8.2a). These life forms attach by mucopolysaccharides extruded through their raphe. In less wave-exposed sands, solitary cells or colonies of small diatoms are found protruding from the sediment particles. Such genera include *Fragilaria* and *Opephora,* which lack raphes, but attach through a mucopolysaccharide pad extruded from apical pores. The more sheltered from wave exposure the sediment is, the more common are motile life forms (*epipelic;* originally meaning "living on mud"). These are the diatoms that form visible, cohesive microbial mats on the sediment surface, particularly on muddy tidal flats (Figure 8.2b). The cohesiveness is due to extracellular polymeric substances (EPS) that are extruded from the raphe

TABLE 8.1 **Functional-form groups of predominant macroalgae, their characteristics, and representative taxa.**

	Functional-form Group	External Morphology	Comparative Anatomy	Thallus Size/texture	Example Genera
1.	Sheet-like algae	Flattened or thin tubular (foliose)	One—several cell layers thick	Soft, flexible	*Ulva* *Halymenia*
2.	Filamentous algae	Delicately branched	Uniseriate, multiseriate, or lightly corticated	Soft, felxible	*Chaetomorpha* *Cladophora* *Gelidium* *Caulerpa*
3	Coarsely branched algae	Terete, upright, thicker branches	Corticated	Wiry to fleshy	*Acanthophora* *Laurencia*
4.	Thick leathery macrophytes	Thick blades and branches	Differentiated, heavily corticated, thick walled	Leathery-rubbery	*Sargassum* *Turbinaria*
5.	Jointed calcareous algae	Articulated, calcareous, upright	Calcified genicula, flexible	Stony	*Galaxaura* *Amphiroa*
6.	Crustose algae	Epilithic, prostrate, encrusting	Calcified, heterotrichous	Stony and tough	*Porolithon* *Hydrolithon*

Source: Adapted from Littler and Littler (1984).

FIGURE 8.1 Clockwise from top left: *Codium fraglie* (coarsely branched), *Macrocystis pyrifera* (thick leathery), *Galaxuara rugosa* (jointed calcareous), *Crustose coralline algae* (unknown species). *Source*: Emily Ryznar.

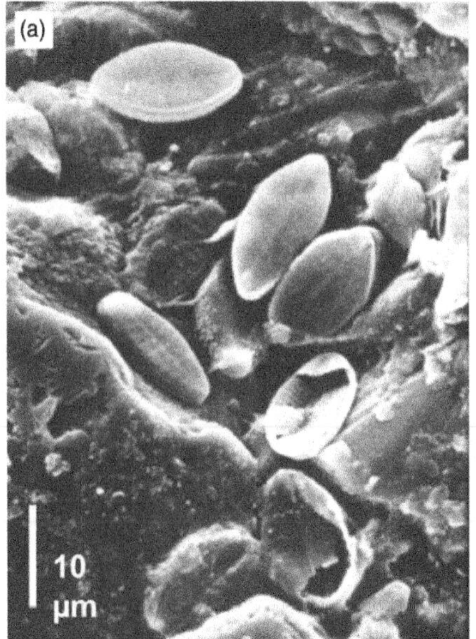

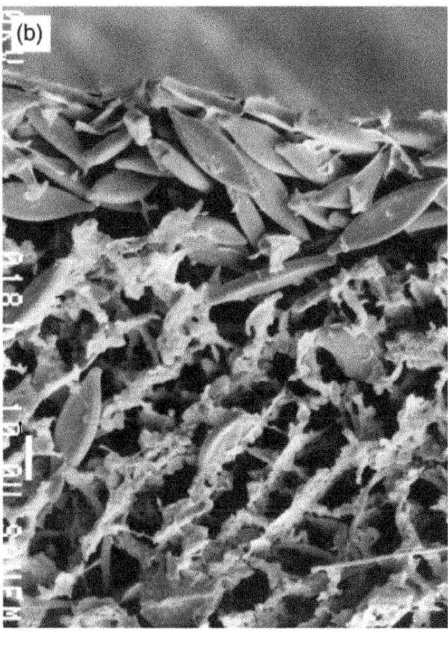

FIGURE 8.2 Benthic diatom assemblages viewed by scanning electron microscopy (SEM). (a) Epipsammic diatoms (*Achnanthes* and *Navicula*) in the cavity of a sand grain. *Source*: H. Håkansson and K. Sundbäck (author). (b) Vertical cut of a sediment surface at low tide photographed in a low-temperature SEM. Motile diatoms have moved to the sediment surface where they form a dense microalgal mat. Strings forming network-like pattern consist of extracellular polymeric substances (EPS) excreted by diatoms (the pattern is an artifact produced by the method). *Source*: A. Miles and D. Paterson, Sediment Ecology Research Group, University of St. Andrews, Scotland.

and from pores in the silica frustule. The production of EPS is related to the motility of raphe-bearing life forms (genera such as *Navicula*, *Amphora*, *Nitszchia*, *Gyrosigma*, etc.; for systematics of the diatom genera, see Round et al. 1990). In fine sediments, epipsammic life forms are found attached to flocs of organic matter consisting of detritus and fecal pellets.

8.4 Habitats

8.4.1 Soft-bottom: Mud/Sandflats, Seagrass Beds, and Marshes

Benthic algae are important primary producers in shallow soft-sediment systems worldwide where light penetrates to large areas of the seabed. In soft-sediment systems subject to low nitrogen loading rates, macroalgae occur in relatively low abundance attached to the seabed, are epiphytic on seagrass blades, or form drifting mats. Within tropical seagrass beds, calcareous and/or siphonaceous green macroalgae such as *Halimeda* and *Caulerpa* are commonly attached to the seabed; calcification and chemical defenses provide protection from most herbivores. Calcified macroalgae ultimately contribute significantly to the accumulation and stabilization of tropical sands (MacIntyre et al. 2004). In both tropical and temperate seagrass beds, macro- and microalgae attach epiphytically to seagrass blades. Although even low abundances of epiphytes can have negative effects on seagrasses due to shading and interference with gas and nutrient exchange, in seagrass systems with low nutrient loading and intact herbivore populations, epiphyte biomass accumulation is modest and contributions to food webs are significant (Williams and Ruckelshaus 1993). In low-nutrient soft-sediment systems, drift macroalgae are also present in low abundance, but are ecologically important because they can provide invertebrates and fishes with protection from predation and aid in dispersal (Salovius et al. 2005; Green and Fong 2016). However, when abundances increase, drifting mats can have negative effects on their seagrass hosts, reducing density through smothering and shading (Huntington and Boyer 2008).

A variety of macroalgae can occasionally be abundant in the lower salt marsh zone in estuaries. Many macroalgae in the understory of salt marsh vegetation are complexed with cyanobacteria and diatom mats. Productivity of these mixed algal communities can be very high, at times equaling or exceeding the productivity of the vascular plant canopy (Zedler 1982). When algae are abundant in the understory of a vascular plant community they may affect primary production rates, biogeochemical cycling, trophic interactions, and environmental conditions, such as evapotranspiration, infiltration, and sediment characteristics (Boyer and Fong 2005; Thomsen et al. 2009).

Like macroalgae, benthic microalgae are also important primary producers in illuminated shallow-water sediments. On sediments where there are no macroscopic primary producers, they are generally the dominant benthic primary producers, forming the base for benthic food-webs. Benthic microalgae are most well studied in tidal mud- and sandflats, but are also important in subtidal habitats, particularly in microtidal estuaries, where the water column often stays clear through most of the day because of the lack of strong tidally induced turbidity (see below). In large estuaries, such as Chesapeake Bay, a high percentage of the sea floor is within the photic zone, enabling primary production to occur over large areas.

8.4.2 Hard-bottom: Rocky Intertidal, Shallow Subtidal

Rocky intertidal and shallow subtidal zones worldwide are dominated by macroalgae, microalgae, and sessile invertebrates. While common along open coasts, some rocky areas exist in sheltered estuarine systems such as fjords and sounds. The rocky intertidal zone is characterized by environmental extremes (temperature, salinity, desiccation, nutrient supply), yet there are also strong biotic interactions that combine to produce striking patterns of zonation along elevational gradients. Both macro- and microalgae play important roles as *in situ* producers, forming the base of local food webs.

Macroalgae in rocky intertidal habitats are highly diverse and abundant, especially in temperate regions. Virtually every functional form of macroalgae can be found in hard-bottomed habitats, from delicate branching genera such as *Plocamium* to large and fleshy reds and browns such as *Mastocarpus* and *Egregia*. Many classical studies identified the importance of top-down forces in controlling the structure of rocky intertidal communities (Connell 1972; Paine 1974). More recent work focused on bottom-up processes and on the relationship between biodiversity and ecosystem functions such as productivity and nutrient retention (Worm and Lotze 2006; Bruno et al. 2008). Although rocky intertidal and shallow subtidal systems in tropical regions are far less studied, there is some evidence to suggest that macroalgal communities in some areas, such as the Pacific coast of Panama, are controlled by the same ecological processes as temperate systems (Lubchenco et al. 1984). In polar regions, the extreme and variable light climate and continuous near-freezing temperatures impose constraints on macroalgal production and depth distribution, as areas are typically ice-covered for all but 2 months of the year. Despite these harsh conditions, crustose coralline macroalgae can be found down to 50 m depth in Arctic estuaries where the light level is only 0.004% of surface irradiance (Rysgaard et al. 2001). Large brown algae in the genera *Laminaria* and *Fucus* often dominate the community in the 2–20 m depth region attached to rocks, stones, and even gravel in protected areas (Witman and Dayton 2001; Krause-Jensen et al. 2007). Physical disturbance by ice scouring often limits the distribution at shallow depths, and light-limitation and possibly also disturbance from walrus feeding activities sets the lower limit of distribution in Arctic waters (Borum et al. 2002; Krause-Jensen et al. 2007).

Although macroalgae are the most prominent vegetation of rocky shores, there is also another less evident, but ecologically important algal community, the microscopic *epilithic* (growing on rock) community. Rocky surfaces are covered by a biofilm comprised of microalgae, cyanobacteria, and newly germinated stages of macroalgae. This often-slippery biofilm also includes bacteria, protozoans, and meiofauna. In the upper intertidal zone, cyanobacteria give rock surfaces a dark, almost black color. Typical cyanobacterial genera are *Rivularia* and *Calothrix*. However, the black color can also be due to salt-tolerant lichens, such as *Verrucaria maura*. Epilithic diatoms often include life forms that are stalked and form colonies (for example the genera *Gomphonema*, *Fragilaria*). These biofilms are important food sources for limpets and periwinkles, and also influence the settlement of invertebrate larvae. In addition to top-down control by grazers, the algae in intertidal biofilms are also controlled by physical stress caused by high insolation (Thompson et al. 2004).

8.4.3 Coral Reefs

Coral reefs are productive, diverse, and economically important ecosystems that dominate hard-bottomed habitats in low-nutrient tropical and subtropical waters. They proliferate in open nearshore habitats and in sheltered lagoons along tropical and subtropical coasts. On pristine coral reefs, fleshy macroalgae are rarely spatially dominant (Littler and Littler 1984); rather, tropical reefs in low-nutrient waters are dominated by crustose coralline and turf-forming algae. These algae form the base of benthic food chains, contribute to biodiversity, and stabilize reef framework. Crustose coralline algae play an important role in reef accretion, cementation, and stabilization (Littler et al. 1995). Algal turfs, composed of filamentous algae and cropped bases of larger forms, are ubiquitous throughout tropical reefs and are characterized by high rates of primary productivity. An exception to dominance by corallines and turfs on pristine reefs can occur when mechanisms exist which limit the efficacy of herbivores. Physical or chemical defenses produced by macroalgae such as *Dictyota* and *Sargassum* and spatial refuges from herbivory, such as surrounding sand planes or the bases of branching corals, can support fleshy macroalgae (Hay 1984; Smith et al. 2006). Human impacts on reefs such as overfishing of herbivorous fishes (Hughes et al. 2003; Jackson et al. 2001; Munsterman et al. 2021) and increased nutrient supplies (Lapointe et al. 2005; Mörk et al. 2009) may also result in dominance by macroalgae.

8.5 Spatial Patterns of Biomass and Productivity

8.5.1 Broad Geographic Scale—Latitudinal Differences

Benthic macroalgae are found in subtidal and intertidal estuarine habitats from tropical to polar regions. Some macroalgal species have broad geographic distributions, indicating their ability to acclimate to climatic variations in temperature and irradiance. For example, the foliose *Laminaria saccharina* occurs in rocky subtidal habitats from Spain to North Greenland (Lüning 1990), and coralline algae inhabit waters throughout polar regions (Steneck 1986). Mesoscale differences in annual rates of productivity are related predictably to temperature and irradiance levels. Rates of productivity can be as high as 2500 g C m^{-2} year^{-1} (Valiela 1995), and highest annual rates are in tropical communities where growth and temperature conditions are favorable throughout the year. In regions where water column productivity is low, as in the tropics and some polar regions, benthic macroalgal production can exceed pelagic production on an aerial basis (Duarte and Cebrian 1996; Krause-Jensen et al. 2007; Gómez et al. 2009).

Benthic microalgae are found on every surface that is reached by light. While benthic microalgae exist in all climatic zones, their temporal and depth distribution varies with latitudinal light conditions and the transparency of the water column. The range of annual benthic microalgal primary production varies from 5 to over 3000 g C m^{-2} year^{-1}, but most values are within the range 20–500 g C m^{-2} year^{-1} (Cahoon 1999), and the highest values are from tropical regions. Areal values of both daily net primary production and chlorophyll *a* (a rough measure of microalgal biomass) are often similar in magnitude to those for phytoplankton in shallow, clear water of coastal regions. In shallow (1–3 m) estuaries, benthic microalgae can account for up to 70% of total primary production (Underwood and Kromkamp 1999; Baird et al. 2004). Even in temperate seagrass meadows, the contribution of benthic microalgae can be 20–25% of the total benthic primary production (Asmus and Asmus 1985). In tropical and subtropical seagrass-vegetated sediments, the biomass of benthic microalgae can be as high as in adjacent unvegetated sediments, even though the seagrass canopy reduces light availability at the sediment surface (Miyajima et al. 2001).

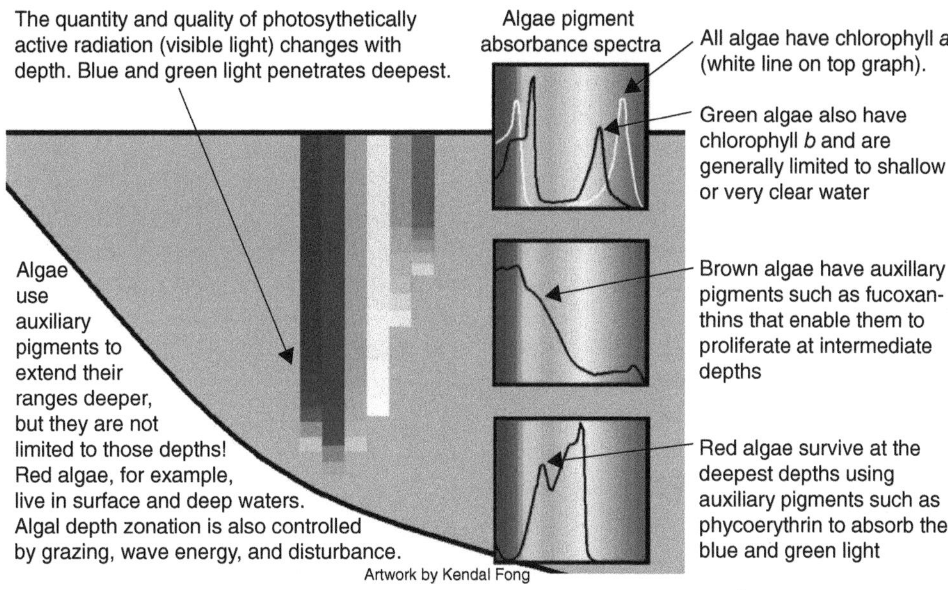

FIGURE 8.3 Different divisions of algae have adapted to the varied light regimes that occur along depth gradients in estuaries.

8.5.2 Depth Distribution

Benthic macroalgae contain accessory pigments that allow capture of different wavelengths of light and efficient use of varying light quantity and quality with depth (Figure 8.3). Red algae contain accessory pigments such as phycoerythrin that absorb green and blue-green wavelengths that penetrate deep in clear coastal waters. Crustose coralline red algae are the deepest-living marine macroalgae and have been found at depths of 268 m in the clear, tropical waters of the Bahamas where irradiances were <0.001% of surface irradiance (Lüning and Dring 1979). These macroalgae are characterized by slow growth rates. Light attenuation with depth and the changing quality of light limits the potential depth distribution of macroalgae, although disturbance and grazing losses also influence the actual depth distribution (Figure 8.3). The minimum light requirements for thick macroalgae (~0.12% of surface irradiance) and thin macroalgae (<0.005%) are substantially less than that for most seagrasses (~10–25%, see Chapter 5), and as a result their depth penetration is much greater than for the vascular plants (Figure 8.4).

The depth at which benthic microalgal primary production is important depends on the transparency of the water body. In clear waters, benthic microalgal net primary production can be substantial at depths > 20 m and can be sustained at a light level of 5–10 μmol photons m^{-2} s^{-1} (Cahoon 1999). At a site in NE Greenland, the maximum depth limit for benthic microalgae matched the 50 m depth limit for crustose coralline macroalgae, where light was 0.004% of surface irradiance during the open-water season (Rysgaard et al. 2001).

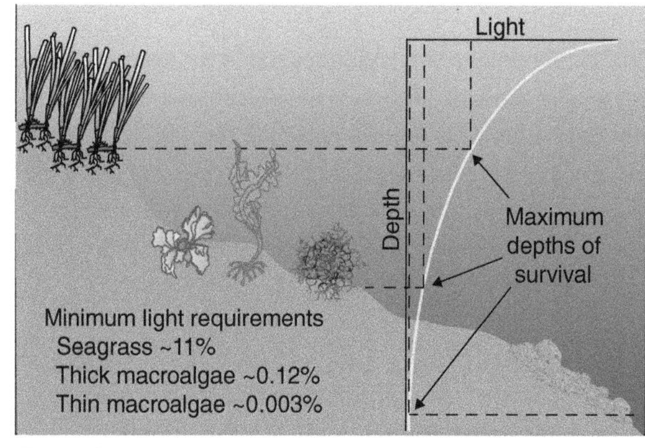

FIGURE 8.4 Comparison of minimum light requirements between macroalgae and seagrass. Depth limits are set by light attenuation in the water column.

8.5.3 Energy Regime

Flow velocities around macroalgal thalli can have positive effects on rates of nutrient uptake and photosynthesis by reducing the thickness of the boundary layer and increasing the exchange of gases and solutes across the thallus surface (Hurd 2000; Hepburn et al. 2007). In dense macroalgal communities, flow velocity is reduced due to the drag imposed by the thalli and this both increases particle deposition and decreases particle resuspension (Gaylord et al. 2007; Morrow and Carpenter 2008). Reduced flow rates can also feedback to influence macroalgal growth. For example, Stewart et al. (2009) showed that in dense beds of the giant kelp *Macrocystis pyrifera*, fronds on the seaward edge of the bed had faster elongation rates, larger blades, and greater carbon and nitrogen accumulation than interior fronds that were exposed to lower flows. Flow velocities also decrease with depth in subtidal areas, and maximum velocities typically occur at the subtidal-intertidal fringe (Denny et al. 1985). Biomechanical models for high-energy systems (up to 15 m s^{-1}) can be used to calculate velocities that cause macroalgal thalli to break and thus predict survival against hydrodynamic forces (Denny 1995). The two most important factors determining break forces for macroalgae are substrate type and thallus size (Malm et al. 2003). In low-energy systems, water speeds as low as 0.22 m s^{-1} can cause both breakage and "pruning" (thallus fragments left for regrowth) for attached algae with delicate thalli such as *Ulva* (Kennison and Fong 2013). Macroalgal form can vary in response to flow regimes, with exposed areas having algae with flatter, narrower fronds to reduce drag and protected areas having wider, undulate fronds to increase turbulent flow and decrease boundary layer effects. This can reflect both morphological plasticity and genetic differentiation (Hurd et al. 1997; Roberson and Coyer 2004).

The largest accumulations of macroalgae are found in poorly flushed systems, such as sheltered embayments and estuaries, with elevated nutrient concentrations where bloom-forming species occur (e.g., *Ulva* spp., *Chaetomorpha* spp., *Gracilaria* spp, and *Polysiphonia* spp.; Viaroli et al. 1996; Pihl et al. 1999). Advective transport of macroalgae, both between habitats within an estuary (i.e., subtidal to intertidal) and from the estuary to the coastal ocean can be important in terms of nutrient exchange. Macroalgal material moves either as bedload or as floating mats at the water surface, depending on their specific gravity (which is influenced by photosynthesis rates and invertebrates that colonize mat-forming macroalgae). Unattached, living macroalgae move at current velocities as low as 2 cm s^{-1} (Flindt et al. 2004), and these current velocities are common in estuaries where both winds and tides affect current speeds at the sediment surface (Lawson et al. 2007; Kennison and Fong 2013). Macroalgae also settle 1,000 to 5,000 times faster than phytoplankton, and if transported out of estuaries they usually settle on the ocean floor rather than being returned on the flood tide (Flindt et al. 2004). Macroalgae are therefore an important source of carbon that is buried in coastal and deep-sea sediments (Krause-Jensen and Duarte 2016). Few studies include mass transport of nutrients bound in plant material and as a result nutrient retention in estuaries can be overestimated.

Generally, benthic microalgal production is higher in fine sediments than in sandy sediments exposed to wave action (Underwood and Kromkamp 1999). The explanation for this finding is that fine sediments, with high organic matter concentrations and rapid bacterial mineralization, contain more nutrients in the pore-water than sandy sediments with less accumulation of organic matter. Moreover, the biomass of benthic microalgae (measured as chloropyll *a* content of the sediment) in the upper few mm of muddy sediments can be much higher than in sandy sediments. Occasionally, in sandy sediments with a thicker photic zone allowing photosynthesis at deeper depths, net primary production may be as high or higher than in muddy sediments. In addition, advective flushing can reduce nutrient limitation in inundated permeable sands (Billerbeck et al. 2007). In high-energy environments, such as tidal sand flats, a large part of the benthic microalgal community can be resuspended during the tidal cycle, contributing to pelagic production. Up to half of the food resource of filter feeders can consist of resuspended benthic microalgae (de Jonge and Beusekom 1992). The turnover of benthic microalgal biomass on sandy tidal flats is typically higher than on muddy flats (Middelburg et al. 2000).

8.5.4 Diel Cycle

The migratory behavior of motile diatoms in intertidal sediments is influenced by both tidal and diurnal cycles and in subtidal sediment mainly by the diurnal cycle. As shown in Figure 8.5 in the shallow subtidal Kattegat Sea, primary production responds directly to incident irradiance as benthic microalgal biomass changes during migration (Miles and Sundbäck 2000). Similarly, in the muddy intertidal Tagus Estuary, primary production initially increases with incident irradiation but declines several hours before the start of tidal inundation. Upward vertical migration during emersion during low tide takes advantage of the favorable light conditions; however, too much light can result in photoinhibition. Downward migration protects benthic microalgae from photoinhibition but also from resuspension by waves and currents. Benthic microalgae, thus, appear to be capable of rapidly shifting their location in the sediment light gradient by vertical migration to optimize photosynthetic performance and provide a more stable habitat avoiding disturbance by currents and predation (Cartaxana et al. 2016). Mechanisms responsible for the success of benthic microalgae in variable light environments are not well understood but include regulation of migratory behavior by endogenous internal clocks, shifts in composition of algal communities during the photoperiod (Consalvey et al. 2004; Underwood et al. 2005), and physiological photoprotection mechanisms such as that offered by the xanthophyll cycle, which allows for thermal dissipation of excess energy that protects the photosystem from damage during light stress events (Cartaxana et al. 2011).

8.5.5 Seasonal Cycle

Seasonality in benthic algal growth is highest in polar regions and decreases toward lower latitudes. For perennial macroalgae in polar regions, shade adaptation is one way of dealing with the dark winter months and the ice cover 10 months of the year. Laminarians that dominate these regions have inherently slow growth rates, storage of carbon reserves during the summer period of high productivity, a long-life span, and a high resistance to grazing (Lüning 1990). Macroalgae need to maintain low respiration to minimize carbon losses during the long winter months, high photosynthetic efficiency

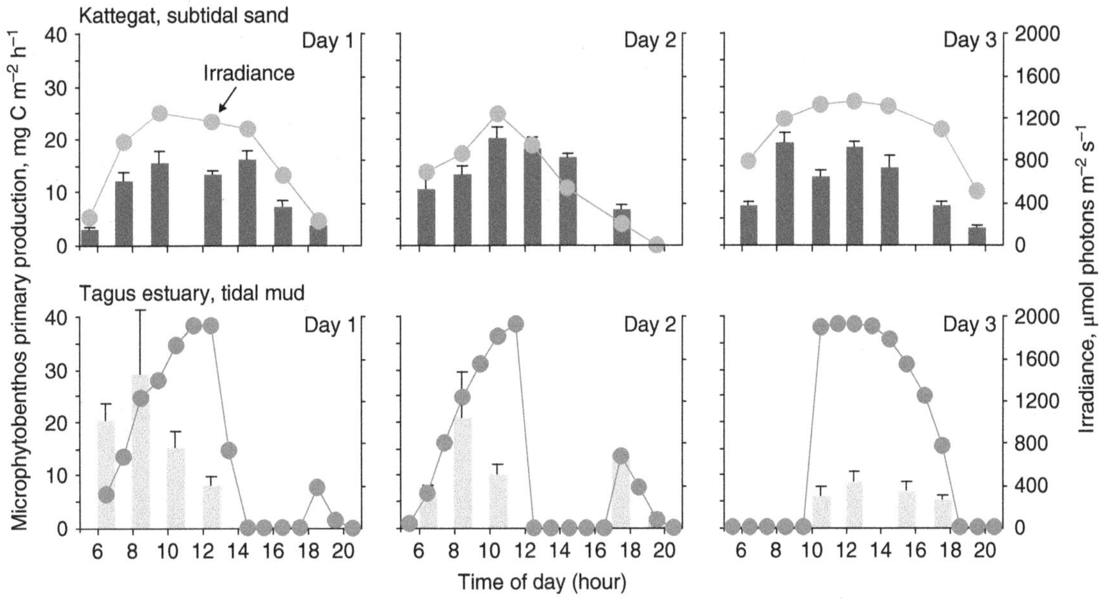

FIGURE 8.5 Diel pattern of benthic microalgal primary production (bars) on a shallow-water subtidal sandy site (Kattegat, microtidal west coast of Sweden) and an intertidal muddy site (Tagus estuary, Portugal). Measurements were made by the ^{14}C technique with subsequent 2-hour incubations during three full days. Filled circles show irradiance measured at the sediment surface. Because of high turbidity, no light penetrated to the sediment surface during high tide in the Tagus estuary. *Source*: Redrawn from Miles and Sundbäck (2000)/Inter-Research Science Publisher.

at low light when ice covers the water surface, and the ability to take advantage of continuous light in the summer. Laminarians can increase their blade surface area and produce new thin blades before the breakup of ice cover using stored reserves so that they are poised for rapid growth as soon as ice-free period begins (Chapman and Lindley 1980; Dunton 1985). A contrasting seasonal growth strategy adopted by some species, such as *Saccharina latissimi*, is to minimize growth under ice and to accelerate growth only when light is available (Wiencke et al. 2009). In temperate soft-bottom systems, different species have different thermal tolerances and optimum temperatures for photosynthesis, respiration, and growth so species composition can change with increasing temperature. In general, with rising temperature, respiration increases faster than photosynthesis leading to a decrease in the photosynthesis:respiration ratio (P/R) and the seasonal growth dynamics.

Ephemeral macroalgae go through boom-and-bust cycles in many regions, with population crashes occurring during the mid-growing season due to high temperatures and self-shading. In some areas such as southern California, floating blooms can occur during any season, with magnitude and frequency related to land use in watersheds and patterns of nutrient input (Kamer et al. 2001; Boyle et al. 2004). Intertidal macroalgae on rocky shores, especially in the high tidal zone, change biomass in response to monthly and seasonal changes in tidal amplitude, temperature, and light (Menge and Branch 2001). Seasonality is much less pronounced in tropical systems, although some exceptions occur. Some coral reef algae show marked seasonal variation in standing stocks of carbon (Lirman and Biber 2000). On Panamanian reefs, algal productivity and biomass accumulation is much higher during the upwelling season (Smith 2005), suggesting that algal communities are regulated by nutrients in the absence of upwelling.

Seasonal variability in thallus photosynthesis is higher for ephemeral macroalgae such as *Cladophora* and *Ulva* than for perennial genera that grow fast and store little nutrient reserves (Sand-Jensen et al. 2007). Respiration rates typically vary systematically over the year and increase as temperature and irradiance levels increase. For temperate species, this leads to higher minimum light requirements during summer than winter for photosynthesis to balance respiration. While there are significant changes in actual production as light and temperature vary with season, the maximum potential production (at saturating light) varies less (Middelboe et al. 2006). This is in part due to changes in species composition and abundance that favor species that have optimal performance at different times of the year.

Long-term studies of seasonality in benthic microalgae are rare. One of the longest seasonal studies (12 years, 14 stations) of benthic microalgal biomass and production is from a tidal flat on the coast of the Netherlands (Cadée and Hegeman 1977). This study shows that in temperate areas, biomass peaks during the warm season, although occasional dips in benthic chlorophyll *a* can occur during summer due to strong grazing pressure, particularly by mud snails (such as *Hydrobia ulvae*). This seasonal pattern agrees with patterns found in shallow subtidal and microtidal areas of temperate estuaries, although well-developed diatom mats have also been found on the sediment surface under sea ice during winter. Such under-ice proliferation of benthic diatoms can be explained by the presence of shade-tolerant species, good nutrient availability, and low grazing pressure. In tropical areas, seasonal variations in benthic microalgal abundance can be influenced by the monsoon, such that abundance is lowest during the monsoon, and highest during the post-monsoon period (Mitbavkar and Anil 2006). In polar regions, the length of the productive season for benthic primary production is only about 80–90 days. During this period daily benthic microalgal primary production (172–387 mg C m^{-2} day^{-1}) at depths ≤20 m has been found to exceed that of phytoplankton (Glud et al. 2002).

8.6 Methods for Determining Productivity

For macroalgae, various methods are used to estimate production, based either on short-term estimates of carbon assimilation or oxygen production, or on longer-term growth estimates.

1. *Growth measurements*. Frond-marking is a common technique to estimate growth rates of large, attached macroalgae such as kelps. The fronds, or blades, are marked with holes at the junction between the stipe and the blade where the growth zone is located. The holes are displaced upward as the blade grows and the distance between the hole and the stipe/blade junction represents new growth (Krause-Jensen et al. 2007). This is similar to the leaf-marking technique that is commonly used to measure seagrass growth. Rates of elongation are then converted to production as mass in grams of dry weight (gdw) or grams of carbon using conversion rates. For unattached mats or rafts of macroalgae with diffuse growth (growth throughout the thallus), biomass can be measured directly as wet weight. Algae are collected from a known area of seabed or volume of water, cleaned of mud and fauna, and, if desired and possible, separated into species. Techniques to assure a consistent wet weight are varied but include blotting the thalli dry or spinning in a salad spinner for a consistent time and rate. Change in average biomass over time estimates productivity.
2. *Photosynthesis measurements*. Measurements of net or gross production as gas (O_2 or CO_2) exchange provide shorter-term estimates of production, and these can be scaled spatially and temporally based on availability of incident irradiance. Changes in oxygen or dissolved inorganic carbon (DIC) can be measured on individual thalli in light or dark bottles or chambers to estimate net primary production and respiration. Photosynthesis-irradiance (P-I) curves can be constructed for each species to derive data on maximum photosynthetic rate (P_{max}), efficiency of light utilization (a), respiration (R), and the saturation and compensation irradiances (I_s and I_c). These parameters can be related to light availability at specific water depths using an equation (e.g., $P = P_{max}[1\text{-}\exp(aI/P_{max})]+R$; Platt et al. 1980) and scaled to areal rates of production.

Since macroalgae can grow in dense accumulations, models are often used to account for self-shading within macroalgal communities (McGlathery et al. 2001; Brush and Nixon 2003). These models scale production in macroalgal mats based on incident irradiance reaching the canopy, known patterns of light attenuation with depth in macroalgal mats, and P-I relationships for individual thalli. This approach is similar to terrestrial canopy models. Sand-Jensen et al. (2007) modified this approach further to consider variations in thallus light absorbance (ratio of absorbed radiant flux to incident radiant flux), canopy structure, and density. The model has been tested for single-species communities of the sheet-like *Ulva lactuca*, the leathery *Fucus serratus,* and for mixed-species communities. An interesting result of the model is the stabilizing effect of species richness, whereby multiple species complement each other in absorbing light.

The pulse-amplitude-modulation (PAM) chlorophyll *a* fluorescence technique is also used to determine relative measures of photosynthetic activity of algal species. These measurements typically provide a snapshot (seconds–minutes) of photosynthetic capacity to acclimate to light conditions. However, they do not easily translate into quantitative measures of photosynthesis in terms of oxygen production or carbon fixation. The PAM technique can be used *in situ* and measures the activity of photosystem II, giving an estimate of the electron transport rate, ETR. Although there is a good linearity between ETR, oxygen production, and ^{14}C incorporation, this relationship can become nonlinear at high irradiances. For a review on this technique, see Kromkamp and Forster (2006).

For benthic microalgae, the best methods for measuring primary production are those that do not disturb the natural physical, chemical, and biological micro-gradients in the sediment. There are four principal methods that are used for primary production measurements in sediments. All these techniques have their limitations, and primary production rates determined by the different methods can be slightly different because they measure different aspects of photosynthesis.

1. *Sediment-water O_2 or CO_2 exchange*. Oxygen flux is measured as changes in the concentration of oxygen in well-mixed overlying water in benthic chambers, sediment cores, or aquatic eddy covariance. The advantage of this method is that the measurements integrate over a sediment area and that measurements both in the light and dark give three ecologically relevant rates, net primary production (NPP), community respiration (CR), and gross primary production (GPP). Field measurements of benthic primary production on tidal flats during immersion is possible by measuring CO_2 fluxes in benthic chambers (Migné et al. 2004). Disadvantages are that cores and chambers do not replicate natural hydrodynamic and light conditions and may underestimate the flux. The aquatic eddy covariance method described below is the

best *in situ* method that avoids environmental disturbances (Berg et al. 2022).

2. *Uptake of radioactively labeled carbon (^{14}C-bicarbonate).* This is a suitable technique for tracking the fate of carbon in food webs. In sandy sediments, ^{14}C can be percolated a few mm into the photic zone of the sediment (Jönsson 1991). In muddy sediments, percolation does not usually work, and we rely on diffusion of the label into the sediments, which results in underestimation of the primary production. Slurry incubations disrupt the micro-gradients in the sediment but are suitable in experiments where maximum potential primary production is used as a variable. The stable isotope ^{13}C also can be used for tracking carbon through food webs (Middelburg et al. 2000).
3. *Oxygen microsensors (with a tip of only a few μm) and planar optodes.* These are used for high-resolution non-destructive measurements of oxygen microgradients in sediments, from which primary production can be calculated by modeling (Figure 8.6). The dark-light shift microelectrode technique can also be used to measure oxygen production (Revsbech et al. 1981). The classic oxygen microsensor is the oxygen electrode with a guard cathode (Revsbech 1989), but there are also oxygen optodes that are based on fiber optics (Glud 2006). The use of microsensors has contributed greatly to our understanding of the temporal and spatial variations among the processes operating in the top few millimeters of the sediment. Planar optodes make it possible to get a two-dimensional picture of the O_2 distribution and dynamics at a given area over many days (Figure 8.6; Glud 2006). Such measurements have provided new insights into oxygen dynamics around rhizospheres, faunal structures, structures in permeable sands, and within phototrophic communities.
4. *PAM chlorophyll a fluorescence.* The fluorescence technique can be used to estimate photosynthesis in sediment microalgal communities and to measure photosynthetic activity of individual cells using a modified fluorescence microscope (Underwood et al. 2005)

For all types of benthic producers, the *in situ* aquatic eddy covariance method based on the similar flux technique for terrestrial and intertidal ecosystems is increasingly being used and provides the most accurate benthic production estimates (Berg et al. 2022). This technique has been adapted to subtidal conditions and has the advantage of measuring community metabolism under natural hydrodynamic and light conditions, can capture short-term (minutes) variation in production/respiration, and integrates over a much larger area (>100 m²) than conventional techniques (<1 m²) (Berg et al. 2003, 2007). Recent studies have applied this aquatic eddy covariance technique to measure productivity in brown algae canopies in the Baltic Sea, in red coralline algae beds off the coast of Scotland (Attard et al. 2015, 2018), and in invasive intertidal macroalgal communities (Volaric et al. 2019).

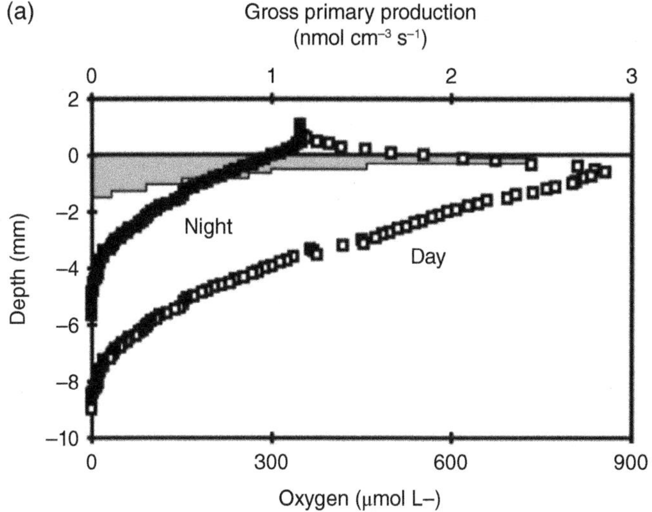

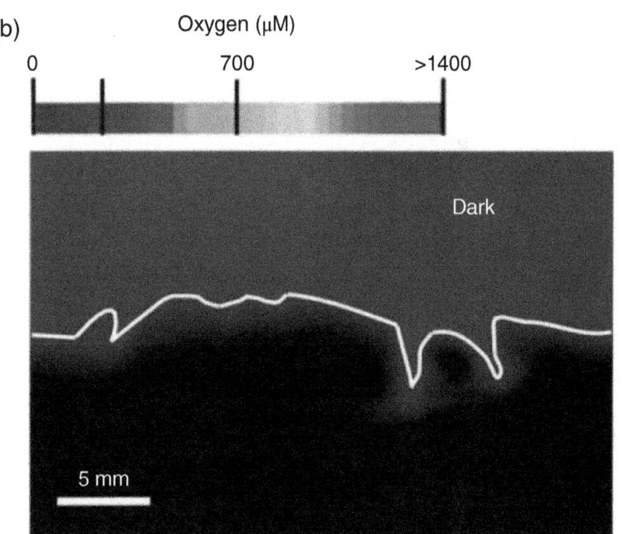

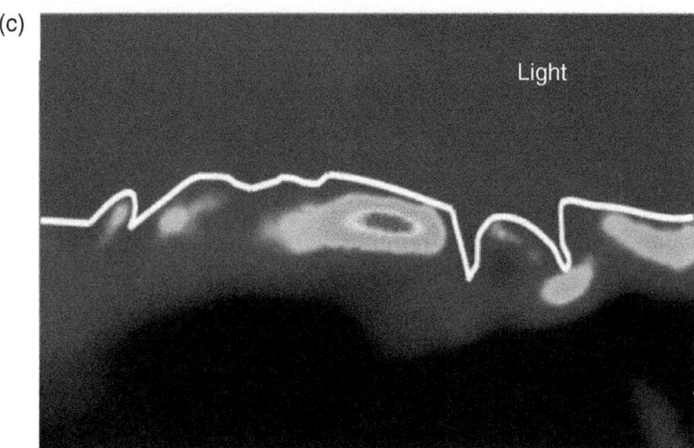

FIGURE 8.6 Influence of benthic microalgae on oxygen distribution in surface sediment. (a) Oxygen profiles in light (day) and dark (night) measured by oxygen microelectrodes. Shown also is a vertical profile of modeled rates of primary production (bars). (b, c) Oxygen distributions under light and dark conditions in bioturbated, sandy silt measured by planar oxygen optodes (see Section 8.7). The white line indicates the sediment surface, and the bright red spot indicates a high rate of photosynthetic oxygen production in light by an assemblage of benthic diatoms just below the sediment surface. *Source:* (a) Redrawn from Glud et al. (2009), (b) Based on Fenchel and Glud (2000).

8.7 Factors Regulating Productivity and Community Composition

The mechanisms that control net production of benthic algae in estuarine ecosystems are the same as for other primary producers: geographic limits for growth are set by temperature and light. Within the geographical limits, biomass accumulation is controlled by many interacting biotic and abiotic factors including light quantity and quality, nutrient availability, water motion, temperature, intra- and inter-specific competition, grazing, and physical disturbance. Here, we discuss light, nutrients, and grazing; additional factors are included in other sections.

8.7.1 Light

Like all primary producers, benthic macro- and microalgae use visible light in the 400–700 nm spectrum, which is termed "photosynthetically active radiation" (PAR). Accessory pigments in benthic algae allow certain groups to capture wavelengths that would otherwise be inaccessible by chlorophyll *a* alone (see Figure 8.3). Light is attenuated exponentially with depth in the water column following the Beer–Lambert law, $I_z = I_0 e^{-K_d z}$, where I_z is the irradiance at depth z, I_0 is the surface irradiance, and K_d is the attenuation coefficient for diffuse downwelling irradiance. This is described in detail in Chapter 4. The surface irradiance reaching the seabed at a given depth is influenced by properties of the water that affect the attenuation coefficient, including suspended sediment, organic (e.g., phytoplankton) and detrital particles, and colored dissolved organic matter (CDOM).

Light is attenuated rapidly within dense macroalgal communities. For the bloom-forming species *Cladophora prolifera* and *Chaetomorpha linum*, 90% of available irradiance hitting the surface of the algal mat can be absorbed in the top few cm (Krause-Jensen et al. 1996). The extent of this light attenuation will depend on the thallus form (canopy structure) and density of the algae, the absorbance of the algae, and species composition. Complementarity of light use between different algal species in mixed-species communities often means that total community production is higher for a given irradiance level than for single-species communities (Middelboe and Binzer 2004). Benthic macroalgae can acclimate rapidly (within minutes) to changing light conditions. This can be an adaptation to optimizing production in a short growing season (Borum et al. 2002) or to a variable light climate due to short-term changes in incident irradiance (e.g., cloud cover) or water column light attenuation (e.g., wind/wave induced sediment resuspension). A general adaptation to reduced light levels is an increase in pigment content and light utilization efficiency. This results in reduced compensation and saturation irradiances (I_c, I_s) for photosynthesis. These characteristics allow macroalgae to photosynthesize more effectively at low light levels. However, the production and maintenance of higher pigment content and enzyme activities increases respiratory costs in low-light plants. At high irradiance levels, pigment concentrations are typically lower, compensation and saturation irradiances are higher, and maximum photosynthesis is high. Photoinhibition may occur at high irradiances or under high UV stress, and the damage to the photosystem that causes this inhibition is not necessarily reversible (see Chapter 4).

Benthic microalgae also function in a wide range of light climates and can adapt to widely fluctuating light conditions. Diatoms can optimize their photosynthetic apparatus efficiently to current light conditions (within minutes; Glud et al. 2002). Moreover, as epipelic life forms are able to migrate vertically in the sediment, they can position themselves in favorable light conditions. In this way, physiological photoinhibition is avoided at high ambient light levels. Because of back-scattering effects, the light intensity at the sediment surface can be up to 200% higher than the ambient light above the sediment (Kühl et al. 1994). While the photic zone in the water column is generally measured in meters, the photic zone in shallow-water sediments is measured in micro- and millimeters. Fiber-optic microsensors enable high-resolution measurements of the light quantity and quality in microbial mats and surface sediments (Kühl et al. 1994). Light penetrates deepest in sandy sediments (~3 mm), while in fine sediments the photic zone is less than 1 mm (Cartaxana et al. 2016). The light climate in sediments also differs from that in the water column in that red light penetrates deepest in sediments. The vertical change in the light quantity and spectrum often results in stratified microbial mats consisting of layers of organisms with different optimal light requirements (e.g., a sequence of diatoms on the top, followed by cyanobacteria, and then by photosynthetic purple sulfur bacteria, which prefer anaerobic conditions). Live microalgae can be found far below the photic zone in the sediment. Epipsammic diatoms on sand grains can be mixed down to 10 cm depth in wave-exposed sediments. They can survive long periods (weeks, months) of darkness, and rapidly resume photosynthesis when transported up to the photic zone by sediment mixing.

8.7.2 Nutrients

8.7.2.1 Sources Benthic algae obtain nutrients from both underlying sediments and overlying water, and the sources of these nutrients can be both external (allochthonous) and recycled (autochthonous) sources. Sources of "new" or allochthonous nutrients include terrestrial inputs via rivers from coastal watersheds, nitrogen fixation, groundwater, atmospheric deposition, and upwelling (Nixon 1995; Smith et al. 1996; Whitall and Paerl 2001; Anderson et al. 2014). With the possible exception of upwelling and N-fixation, all these sources are rapidly increasing as a result of anthropogenic alterations of global nutrient cycles (see Howarth 2008 and Galloway et al. 2008). Autochthonous sources of nutrients include recycling from other biota and regeneration from the sediments during decomposition.

8.7.2.2 Uptake and Storage The relationship between nutrient concentration in water and the rate at which nutrients are taken up by benthic macro- and microalgae can be described by a hyperbolic function. The Michaelis–Menten equation for enzyme uptake kinetics is often used to describe this function. The Michaelis–Menten equation describes uptake as: $V = V_{max} * [S/(K+S)]$, where V is the uptake rate at a given substrate (nutrient) concentration, S is the substrate concentration, K is the half-saturation concentration—where uptake is equal to ½ of the maximum uptake rate, and V_{max} is the maximum uptake rate at high substrate concentration. From a biological perspective, there are two important components of this relationship: (1) nutrient uptake rates become saturated as nutrient concentration increases; and (2) the initial slope of the curve at low concentrations provides a useful index of an alga's affinity for nutrients at low levels. This latter characteristic may be important in determining the competitive abilities of algae when nutrients are in critically low supply. The V_{max} rate varies depending on the nutrient status of the plant. Algae that have lower nitrogen contents in more nitrogen limiting environments would be expected to have higher maximum nutrient uptake rates.

For macroalgae, uptake rates are higher for thin-structured sheet-like or filamentous thalli that have a high surface area to volume ratio than for more coarsely branched species (Figure 8.7; Wallentinus 1984). In addition, some nitrogen-starved ephemeral macroalgae have the capacity for very rapid ("surge") uptake during the first minutes of exposure to nutrients, which results in a three-phase pattern of nutrient uptake as a function of external supply (Pedersen 1994). The capacity for surge uptake is thought to be advantageous for ephemeral algae subject to short-term nutrient pulses that might result from animal excretion or temporarily high mineralization rates. For macroalgae, there are two primary ways to measure nutrient uptake: (1) the "multiple flask" method, where different substrate concentrations are added to individual flasks containing macroalgae, and the disappearance of substrate is measured over a short (15 min) incubation, and the data are pooled to obtain an uptake vs. concentration curve; and (2) the "perturbation" method, where thalli are exposed to a high concentration of substrate and a time series of substrate concentration is measured (Pedersen 1994).

Both macro-and micro-algae can store nutrients, and their capacity to do this varies depending on their inherent growth rates and the frequency and magnitude of nutrient pulses. The general pattern that emerges is that fast-growing ephemeral macroalgae such as *Ulva* and *Chaetomorpha* store little reserves compared to slow-growing perennial species such as *Fucus*, which has reserves that can support growth for longer periods of time. This influences the duration of nutrient limitation that often occurs in temperate regions in response to a seasonal (summer) depletion of nutrients. For example, Pedersen and Borum (1996) showed that different algal species, from phytoplankton to slow-growing perennial macroalgae, had similar storage reserves per unit biomass (mg N gdw^{-1}; Figure 8.8). Similar dynamics are evident in the storage reserves and storage capacity of phosphorous across fast- and slow-growing macroalgae (Pedersen et al. 2010).

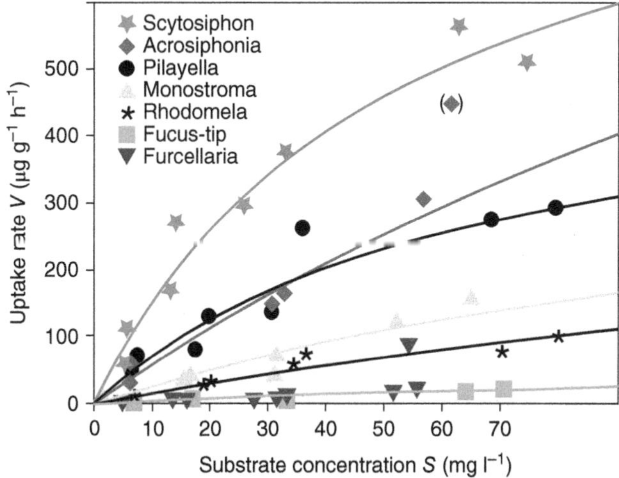

FIGURE 8.7 Nutrient uptake rates versus concentration for macroalgae with different ratios of surface area to volume. *Source*: From Wallentinus (1984)/Springer Nature.

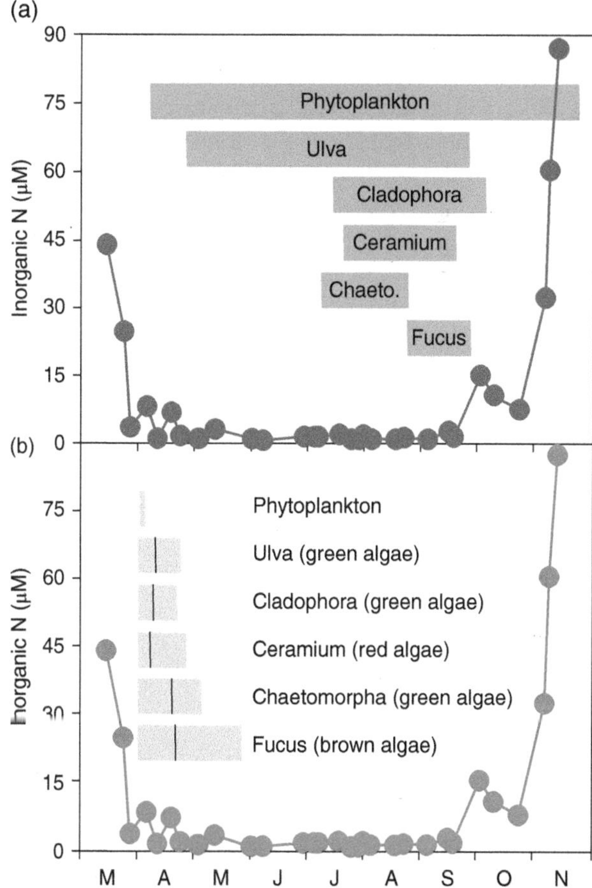

FIGURE 8.8 Periods of (a) nutrient-limited growth and (b) storage capacity (ability to support growth in the absence of an external nutrient supply) for phytoplankton and macroalgae in a Danish Fjord compared to ambient nutrient concentrations. *Source*: From Pedersen and Borum (1996)/Inter-Research Science Publisher.

On shorter time scales (hours–days), macroalgae have rapid but transient responses to variability in N availability, with uptake rates influenced by changing tissue N pools (McGlathery et al. 1996). Also, the coupling of C metabolism (photosynthesis and storage) and N metabolism over the short-term implies that algal growth rate acts as a feedback regulator to maintain balanced C-N metabolism, except under extreme conditions of high irradiance and low N supply. Such self-regulation may be especially beneficial to algae growing in estuarine environments that are characterized by high spatial and temporal variability in nutrient and light availability (McGlathery and Pedersen 1999).

8.7.2.3 Limitation of Growth

In most temperate and polar estuaries, nitrogen is considered to be the primary limiting nutrient for growth. For macroalgae in tropical/subtropical systems dominated by carbonate sediments, phosphorus often limits growth, largely due to the binding of phosphorus by the carbonate sands. There are exceptions to this generalization that are related to the ratio of N:P in external loading, variations in the relative rates of nitrogen fixation and denitrification, and the adsorption capacity of carbonate sediments in different systems. In addition, treatment of wastewater has more effectively reduced phosphorus pollution to coastal waters than nitrogen pollution, creating conditions in some areas with excess nitrogen where phosphorus may be a limiting nutrient (Kronvang et al. 2005). Large data sets on C:N:P ratios of marine plants indicate that macrophytes (macroalgae and seagrasses) have different characteristic atomic ratios of tissue nutrients than phytoplankton. Atkinson and Smith (1983) found a ratio of 550 : 30 : 1 for benthic marine macrophytes, and Duarte (1992) showed that the ratio for 46 macroalgal species was 800 : 49 : 1. These ratios compare to the "redfield ratio" of phytoplankton of 106 : 16 : 1 (see Chapter 4). As with phytoplankton, these ratios are often used to infer nutrient availability and possible nutrient (N vs. P) limitation.

Increases in nutrient supplies that relieve nutrient limitation in subtidal soft sediment habitats result in enhanced macroalgal production and potentially in the loss of seagrasses, which tend to dominate under low nutrient conditions (see Chapter 5 and Duarte 1995; Valiela et al. 1997; McGlathery et al. 2007). In salt marshes, nutrients sequestered by understory macroalgae are retained within the salt marsh community and become available to support vascular plant growth when macroalgae senesce (Boyer and Fong 2005). However, macroalgal mats can also smother salt marshes and damage new growth (Newton and Thornber 2013). Transitions from macroalgae to toxic cyanobacterial mats occur in estuaries on the west coast of the United States (Armitage and Fong 2004), perhaps due to higher tidal amplitudes, lower residence times, and broad areas of suitable intertidal mudflats. Nutrient-driven community shifts also occur in rocky intertidal areas (Worm et al. 2002; Worm and Lotze 2006). In these studies, experimental nutrient enrichment increased macroalgal abundance, and grazing by consumers moderated that response. However, macroalgae in areas with low ambient nutrients responded to enrichment with increased thallus complexity and diversity, while macroalgae in more enriched areas responded with decreases in diversity due to a shift in dominance to opportunists. These responses are context-dependent; another long-term nutrient enrichment study found a similar shift in community composition towards ephemeral algae under high-nutrient conditions but also found greater diversity, due in part to reduced dominance of brown algae under enriched conditions (Kraufvelin et al. 2010). Biotic processes such as grazing interact with bottom-up nutrient availability to structure macroalgae communities.

For coral reefs, there is considerable debate about whether algae are limited by nutrients. Effects of experimental nutrient additions have varied by orders of magnitude (Fong and Paul 2011; Koop et al. 2001). Interpretation of experimental results is limited, in part, by the difficulty of relating results of laboratory or microcosm studies to natural growth in the high-energy, high-flow environments with variable nutrient supply typical of coral reefs (Fong et al. 2006), and the related methodological challenge of effectively conducting *in situ* experiments in these same environments (reviewed in McCook 1999). In general, factors such as the type of substrate, amount of P-absorbing carbonate in sediments, and characteristics of terrestrial environments adjacent to coral reefs can influence the spatial and temporal variability in nutrient limitation (Fong and Paul 2011).

Benthic microalgae in fine sediments are not usually nutrient limited. Diatom-dominated benthic microalgae can experience sequential limitation, with productivity first limited by nitrogen and then by silicate (Porubsky et al. 2008). However, in sandy sediments exposed to wave action, efficient advective transport can increase nutrient availability, resulting in high microalgal productivity (Billerbeck et al. 2007). In nutrient-limited sediments, benthic microalgae are strong competitors for nitrate and can outcompete denitrifers (Joye and Anderson 2008).

8.7.3 Grazing

Preferential feeding by herbivores can influence the abundance and species composition of macroalgae. Many studies show an association between high nutrient content of primary producers and high consumption rates; however, other factors such as herbivore abundance, per capita grazing rates of the dominant herbivores, and feeding preferences also play important roles in determining patterns of herbivory (Cebrian 1999, 2002).

Herbivory has been shown to control algal biomass accumulation with community-level effects. For example, grazers such as amphipods and gastropods can control the abundance of algal epiphytes on seagrasses and hard substrates and can mediate the negative shading effects of epiphytes in response to nutrient loading (Williams and Ruckelshaus 1993; Hillebrand et al. 2000; Valentine and Duffy, 2006). Likewise, grazers can mediate the impact of macroalgal blooms in eutrophic environments (Worm et al. 2000). On coral reefs, at low to intermediate nutrient supplies, herbivores can control macroalgal abundance and maintain the competitive dominance of slower-growing community types (Hughes et al. 2003; Pandolfi et al. 2003). Rapid growth with nutrient enrichment allows algae to escape control by herbivores and accumulate biomass, which ultimately leads to a phase shift from a coral-dominated state to an algal-dominated state (Adams et al. 2021). Chemical and structural defenses also provide a refuge from control by herbivores, allowing algal biomass to accumulate even in heavily grazed areas. In soft-sediment communities, enhanced algal growth from high nutrient loading can saturate grazing potential and decrease per capita consumption rates. For example, grazer abundance can be impacted negatively by the change in physicochemical conditions (i.e., low oxygen, high sulfide and NH_4^+) that result from the decomposition of algal blooms (Hauxwell et al. 1998).

Benthic microalgal primary production can enter the food web through several pathways, including through grazers, detritus, and dissolved organic carbon (DOC). Field studies show that a large fraction of benthic microalgal biomass can pass through macrograzers (> 1 mm), particularly mud snails (*Hydrobia*) (Asmus and Asmus 1985; Andresen and Kristensen 2002). In addition, some benthic microalgal primary production can pass through the "small food chain" consisting of microfauna (unicellular fauna such as ciliates) and meiofauna (< 100 μm; Kuipers et al. 1981; Pinckney et al. 2003). Meiofauna can occasionally graze up to 100% or more of the benthic microalgal standing stock (Pinckney et al. 2003). Benthic microalgae have been shown to be the primary food source for a wide range of consumers in intertidal systems, including worms, molluscs, fish, crustaceans, and birds (Christianen et al. 2017), making benthic microalgal primary productivity an important driver of food web dynamics.

8.8 Energy Flow

The amount of energy sequestered in primary producers and the proportion flowing through algal-dominated benthic estuarine communities varies among soft sediment, coral reef, and rocky intertidal and subtidal ecosystems (Figure 8.9). In all systems, the amount of energy that is fixed into algal biomass and made available to higher trophic levels is dependent on resources, usually light and nutrients. Overall, external nutrient supplies strongly affect systems that are closely associated with terrestrial environments and that experience restricted water exchange. However, light availability can be as important as nutrient supply in some estuaries (Anderson et al. 2014).

8.8.1 *Recycling of Nutrients*

Across estuarine systems, recycling of nutrients from sediments and biota is most likely greatest in soft sediment and coral reefs and least in rocky intertidal and subtidal systems. In soft-sediment estuaries and lagoons, fluxes of nutrients regenerated from sediments are an important source of recycled nutrients (McGlathery et al 1997; Fong and Zedler 2000; Tyler et al. 2003). In some seasons, recycled nutrient fluxes meet up to 100% of macroalgal demand (Sundbäck et al. 2003). Interception of nutrients by macroalgae may reduce supplies to other producer groups and change the path and rate of carbon flow (Hardison et al. 2011). Macroalgal communities may also be sustained by nutrient release from senescent or self-shaded thalli, or by recycling of nutrients from nearby vascular plant communities such as seagrass beds (for a review see McGlathery et al. 2007). In coral reef ecosystems, recycling from sediments is only likely to be significant in enriched systems (Stimpson and Larned 2000). Other recycled sources in coral reefs include flocculent material settling on the surfaces of algal thalli (Schaffelke 1999) and animals that release nitrogenous waste products (Williams and Carpenter 1988). Release of waste products from closely associated animals is also a recycled source of nutrients in intertidal and subtidal systems (Hurd et al. 1994; Bracken 2004).

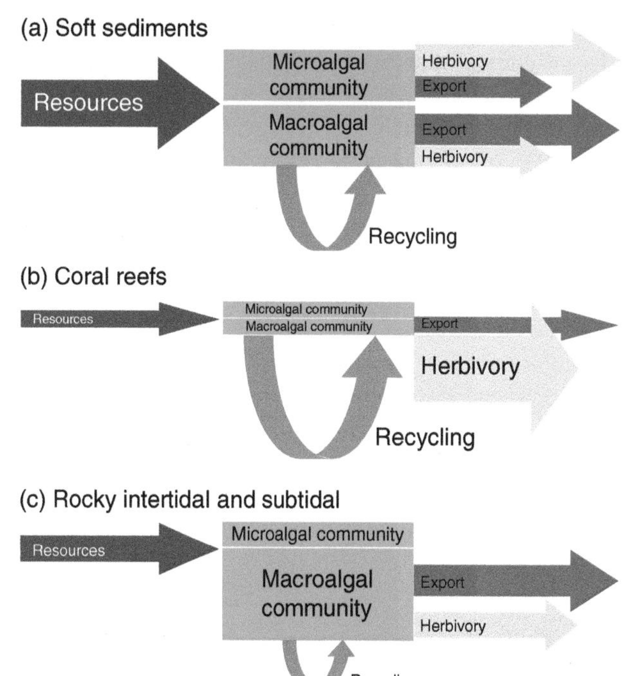

FIGURE 8.9 Energy flow diagrams for macroalgal communities in (a) soft sediment, (b) coral reef, and (c) rocky intertidal and subtidal habitats. The size of the arrow represents the magnitude of flow in the different pathways.

8.8.2 Carbon Storage

The capacity of ecosystems to store energy, represented by the amount of carbon in the biomass of producers and consumers, varies across estuarine ecosystems. An example is the tremendous difference in producer biomass between coral reef algae (0.03–0.6 kg wet weight m^{-2} based on a wet:dry weight ratio of 10 : 1, Foster 1987) and rocky subtidal reefs (3.5 kg wet weight m^{-2}, Manley and Dastoor 1987). Although both are highly productive in terms of gross primary productivity, they occupy different ends of a spectrum in terms of biomass accumulation and therefore carbon storage. In general, opportunistic species with simple thallus forms such as those that dominate coral reef turfs have low levels of biomass, therefore storing little carbon despite relatively rapid growth rates. Low biomass is due to short life spans, susceptibility to removal by physical disturbance, and high rates of grazing by herbivores. In contrast, more complex macroalgae that dominate rocky subtidal and intertidal areas often have high standing stocks of carbon. This is due to the longevity of individual thalli, investment in structure to withstand physical disturbances, and lower rates of carbon transfer to herbivores due to protection by chemical and physical defenses. One exception to this generalization is opportunistic macroalgae in shallow soft sediment estuaries that go through periods of boom and bust and cycle carbon quickly.

8.8.3 Herbivory

The amount of energy that is transferred from macroalgae through consumption can vary by orders of magnitude across estuary types. On a global scale, direct consumption by grazers was estimated as 33.6% of net carbon fixed by macroalgae, demonstrating their general importance as the base of food webs in coastal ecosystems (Duarte and Cebrian 1996). Herbivory on coral reefs is extremely high, with large herbivorous fishes and invertebrates acting as "lawn mowers" that keep algal standing stocks very low and carbon transfer very high (for a review see McCook 1999). However, due to overfishing of herbivorous fish, maximum herbivory rates may occur only at remote or well-protected reefs (Fong and Paul 2011; Jackson et al. 2001). In rocky subtidal habitats, grazing is important in maintaining diversity, but cannot always overcome the effects of increased resources (Worm and Lotze 2006). Within soft sediment systems, rates of primary consumption can vary greatly (Fong et al. 1997; Giannotti and McGlathery 2001), and consumption of benthic microalgae can be very important (Asmus and Asmus 1985; Armitage and Fong 2004).

Benthic microalgal production forms the basis for benthic food webs where macroscopic primary producers are lacking. The main pathway of this fixed carbon into the food web is generally considered to be through macrofaunal and meiofaunal grazers (Duarte and Cebrian 1996), although the detrital pathway is also important. The microbial film (epilithon) on rocky surfaces is an important component in the cycling of carbon, particularly on exposed and moderately exposed shores with large numbers of limpets and other grazers.

8.8.4 Detrital Pathway

Death and subsequent decomposition of macroalgal detritus results in release and recycling of stored carbon. Processing of carbon through detrital pathways comprises about a third of macroalgal net primary productivity globally (Duarte and Cebrian 1996). Macroalgal detritus can be a major source of energy for filter feeders, including hard clams raised in aquaculture farms (Hondula and Pace 2014). However, like grazing, the relative importance of recycling varies across habitats. For example, high-energy rocky areas recycle much less carbon from detritus within the system compared to lagoons, while the estimates for recycling within coral reef algae is very high (Duarte and Cebrian 1996). When algae decompose, they release organic matter to the water and, in soft sediments systems, to the sediments (for a review see McGlathery et al. 2007). Carbon and nitrogen derived from macroalgal biomass can be retained in the sediments and shuttled between benthic microalgae and the bacterial community (Hardison et al. 2010). The detrital pathway is also important for benthic production (Hardison et al. 2011). As much as 70% of net primary production by benthic microalgae on a tidal flat was found to enter the food web as detritus (POC) and was further transferred to bacteria and detritus-feeding fauna (Baird et al. 2004; Hardison et al. 2011).

8.8.5 Dissolved Organic Carbon

Substantial DOC also "leaks" from healthy macroalgal thalli (Tyler et al. 2003; Fong et al. 2003). Some organic compounds in the water can be taken up directly by consumers and may alter food webs toward heterotrophic bacterial pathways (Valiela et al. 1997). A large portion (> 50%) of the photosynthetic product of benthic microalgae can also be released as DOC, particularly under low-nutrient conditions. This DOC enters food webs rapidly through bacteria (Middelburg et al. 2000), forming the benthic equivalent of the pelagic microbial loop. About 40–75% of the carbon fixed by motile benthic diatoms can be released as colloidal organic matter or EPS (Goto et al. 1999; Smith and Underwood 2000) and rapidly transferred to sediment bacteria.

8.8.6 Export of Carbon

The rate of carbon exported from benthic estuarine communities varies tremendously across ecosystem types and is a function of standing stock, water motion, and algal morphology. Duarte and Cebrian (1996) calculated that a global average of 43.5% of macroalgal net primary production is exported; however, the range is perhaps the more important metric, varying from ~0–85% across macroalgal-dominated habitats. Rocky subtidal systems export a far larger amount of carbon than coral reefs despite vigorous wave action in both systems due to their vastly larger standing stock. In contrast, pristine soft sediment lagoonal systems export less carbon than rocky systems as there is both less physical disturbance and lower water exchange to detach and remove biomass. Thus, seagrasses in oligotrophic lagoons represent a large and relatively stable reservoir of carbon compared to more ephemeral macroalgae. In contrast, in eutrophic systems there is often accumulation of large floating algal rafts, resulting in faster turnover and more rapid export of carbon to the ocean (McGlathery et al. 2007). In sediments exposed to strong tidal or wave action, benthic microalgae is easily resuspended together with sediment particles and hence can be transported away. A large study in the Ems-Dollart Estuary (North Sea coast) showed that, on an annual basis, 14–25% of the microphytobenthic carbon is found in the water column as a result of resuspension (de Jonge and Beusekom 1995). Algal carbon that is exported from coastal systems may ultimately be buried in both coastal and deep sea (>1000 m depth) sediments. The presence of macroalgal carbon has been reported in temperate, tropical, and polar deep-sea sediments (Krause-Jensen and Duarte 2016). Macro- and microalgae also contribute to sediment carbon buried in shallow waters, such as in seagrass meadow sediments (Kennedy et al. 2010). Microalgae produced *in situ* within the meadow may be a major source of seagrass meadow sediment carbon (Oreska et al. 2017). Algal biomass is therefore an important contributor to "blue carbon," the carbon sequestered in coastal sediments (Krause-Jensen et al. 2018).

8.9 Feedbacks and Interactions

8.9.1 Feedbacks on Biogeochemical Cycling in Soft Sediment Estuaries

Both macro- and micro-algal mats have a large impact on biogeochemical cycling in shallow-water sediment habitats. This effect can be both direct, through nutrient assimilation, retention and release, and indirect through oxygen production and consumption, affecting mineralization and redox-sensitive processes in general. Despite large differences in biomass, the quantitative role of macroalgal and microalgal mats on biogeochemical processes (for example nitrogen assimilation), can be similar, making turnover time of algae-bound nutrients a key factor in nutrient retention in shallow lagoons (McGlathery et al. 2004). Nutrients that are assimilated by benthic algae are, for the most part, only temporarily retained within individual algal thalli on a time scale of days to months. Tissue turnover times vary for the different autotrophs, with seagrasses having longer retention time (weeks–months) than bloom-forming macroalgae (days–weeks) and microalgae (days). This suggests that nutrients will be recycled faster in systems dominated by microalgae and ephemeral macroalgae than in those dominated by perennial macrophytes (Duarte 1995).

Both types of benthic algal mats strongly influence the degree of benthic–pelagic coupling by reducing the flux of remineralized nutrients from the sediment pore-water (Fong and Zedler 2000; Anderson et al. 2003; Tyler et al. 2003; see also 8.9.5). Benthic algae can outcompete phytoplankton for nutrients if the major nutrient supply is internal flux from the sediments. Therefore, in shallow coastal systems, short-lived phytoplankton blooms often coincide with low benthic algal biomass (Sfriso et al. 1992; Valiela et al. 1992; McGlathery et al. 2001). The influence of benthic microalgae on sediment–water nutrient fluxes is often observed as lower fluxes—or no flux at all—out of the sediment in light when compared with in the dark (Sundbäck and McGlathery 2005; Anderson et al. 2014). When sediment nutrient sources are insufficient to meet the growth demand of benthic microalgae, there is a downward flux from the water to the sediment, such that the sediment functions as a temporary *nutrient sink* instead of a nutrient source. This applies particularly to autotrophic sediments in which oxygen production exceeds oxygen consumption (Engelsen et al. 2008). Also dissolved organic nutrients, such as dissolved nitrogen (DON) are influenced by both benthic micro- and macroalgae (Tyler et al. 2003; Veuger and Middelburg 2007).

Benthic algal mats influence the vertical profiles of oxygen and this affects biogeochemical cycling. The presence of dense macroalgal mats can move the location of the oxic–anoxic interface up from the sediments into the mat since only the upper few cm of the mat are in the photic zone where oxygen is produced by photosynthesis (Krause-Jensen et al. 1996; Astill and Lavery 2001). Below the photic zone, decomposing macroalgae release nutrients that can diffuse upwards to support production. Unattached macroalgal mats tend to be patchy and unstable, but oxygen and nutrient gradients develop quickly, in as little as 24 hours, suggesting that this filtering function occurs even in dynamic environments (Astill and Lavery 2001). Overall, sediment nutrient cycling is enhanced by the presence of macroalgae, presumably due to the input of organic matter and faster decomposition (Trimmer et al. 2000; Tyler et al. 2003). In the surface layer of the sediment, dynamic oxygen gradients created by benthic microalgal activity also control the rate and vertical position of the sequence of redox-sensitive processes in the sediment, such as nitrification, denitrification, sulfate, iron and manganese reduction, and methane production.

Benthic micro- and macroalgae have an important influence on rates of denitrification and nitrification. Rates tend to be low in sediments underlying macroalgal mats, likely due to algal competition with bacteria for NH_4^+ and NO_3^-. In addition, high free sulfide concentrations in organic-rich sediments underlying macroalgal accumulations may inhibit coupled nitrification–denitrification (Trimmer et al. 2000; Dalsgaard 2003). The combined use of oxygen and nitrogen microsensors has shown that benthic microalgae can both reduce and enhance denitrification. During photosynthesis, the oxygenated sediment layer gets deeper, and it takes longer for the NO_3^- to diffuse from the water column to the denitrification zone of the sediment, reducing the rate of denitrification and of dissimilatory nitrate reduction to ammonium, DNRA (Porubsky et al. 2008). Under low-nitrogen conditions, benthic microalgae will also compete with nitrifying bacteria for NH_4^+, reducing the availability of substrate (NO_3^-) for the denitrifiers. Such an effect by benthic microalgae can still occur down to a water depth of

15 m (Sundbäck et al. 2004). When ambient N concentrations are high, oxygen production will instead stimulate nitrification (an aerobic process), and thereby also stimulate coupled nitrification-denitrification (Risgaard-Petersen 2003). The alternative pathway of bacterial nitrogen removal, anaerobic ammonium oxidation (anammox) is negatively affected by the presence of active benthic microalgae (Risgaard-Petersen et al. 2004).

8.9.2 Feedbacks on Sediment Stabilization

Macroalgal mats may either stabilize or destabilize sediments, depending on algal abundance. Dense macroalgal mats stabilize sediments by decreasing shear flow at the sediment surface and thus sediment suspension (Venier et al. 2012). Thick macroalgal mats (equivalent to 3.5–6.2 kg wet wt. m^{-2}) displace velocities vertically and can deflect 90% of the flow over the mat, with only 10% of the flow traveling through the mat (Escartín and Aubrey 1995). However, macroalgae also exist at low densities and in patchy distributions that are often dependent on available substratum for attachment or on advection of drift algae (Thomsen et al. 2006) and may act to destabilize sediments. At low densities, flow causes macroalgae to move and scour the sediment increasing sediment suspension relative to bare sediments. This sediment destabilization is akin to the well-documented phenomenon of saltating or abrading particles increasing erosion in cohesive sediments (Houser and Nickling 2001; Thompson and Amos 2002, 2004). In cohesive beds, the critical stress required to initiate erosion is often greater than the stress required to maintain the sediment in suspension. Macroalgae that scrape the bed while moving across it can dislodge particles and increase sediment suspension/erosion.

The influence of benthic microbial mats on surface sediment stability is well studied (de Brouwer et al. 2006). The mechanism behind this stabilizing effect is the production and extrusion of EPS by diatoms through the raphe during their gliding movement. Its composition can be complex and varies between species, but mainly consists of carbohydrates, proteins, and sulfate groups. EPS binds sediment particles together so that the shear stress needed for erosion is increased, and therefore the sediment is less easily eroded and resuspended (Reidenbach and Timmerman 2019). Besides gluing sediment particles together, EPS also has physiological and ecological implications. On tidal flats there are large diel changes in environmental variables (light, temperature, salinity, water content, oxygen and erosive forces). By secreting EPS, the diatoms become embedded in a cohesive matrix that can create more stable conditions. This sediment stabilization is a seasonal phenomenon at least in temperate systems, and the deposited material may be resuspended at times of the year when the microalgae are less productive (Widdows et al. 2004).

8.9.3 Effects on Faunal Biomass, Diversity, and Abundance

Macroalgae can have positive and negative effects on associated organisms. On the positive side, they can provide a food source through direct assimilation or through detritus-based food chains, as well as a protective refuge from predators (Hondula and Pace 2014). The complex structure of some macroalgal species, such as *Gracilaria vermiculophylla*, could potentially create a predation refuge for commercially valuable blue crab and fish recruits as well as shrimps and amphipods in both the subtidal and lower intertidal zones of estuaries (Johnston and Lipcius 2012; Thomsen et al. 2009). Macroalgae also may provide an important link between estuarine habitats (subtidal to intertidal, sand flats to seagrass beds), as associated organisms are transferred with advecting macroalgae between habitats (Thomsen et al. 2009).

The negative effects of dense macroalgae on benthic fauna include harmful exudates that are toxic to some organisms (Peckol and Putnam 2017), low dissolved oxygen within and under dense macroalgal mats (Hull 1987), and high dissolved NH_4^+ and sulfide concentrations that also can be toxic (Hauxwell et al. 1998). These negative effects have been associated with reductions in abundance of various macrofauna, including bivalves, gastropods, amphipods, and fish, as well as increases in certain polychaetes, oligochaetes, and amphipods (Raffaelli 2000).

Both macro- and microalgal mats can cause negative upward cascades in estuarine food webs. For example, microalgal mats subjected to high nutrient loads shifted to dominance by cyanobacteria and purple sulfur bacteria, and therefore increased mortality of the dominant herbivore, the mudsnail *Cerithidia californica,* threefold over diatom-dominated mats subjected to lower nutrients (Armitage and Fong 2004). Changes in benthic fauna associated with macroalgal blooms may cause resident and migratory shorebirds to change foraging behavior. For example, sandpipers spend more time probing in their search for food when they are foraging on top of macroalgal mats, but more time repeatedly pecking when mats are absent (Green et al. 2015). As infauna are a major food source for birds and other secondary consumers, macroalgal impacts that reduce this link of the food chain may impair this vital ecosystem function.

8.9.4 Facilitation by Fauna

Macroalgae in soft-bottom environments can be facilitated by fauna that provide hard substrate for settlement and growth. For example, the tube-cap forming polychaete, *Diopatra cuprea*, facilitates macroalgal assemblages in shallow lagoons. These organisms create and maintain attachment sites by incorporating algal fragments into tube caps, thus increasing algal residence time on mudflats compared to unattached algae (Thomsen et al. 2005). This association increases population stability and resilience by providing a stable substrate that retains algae against tidal flushing and storm surges, and by providing new fragments to populate new areas or repopulate areas after storm disturbance. Oyster reefs, mussel, and clam beds likewise provide substrate for algal attachment (Murphy et al. 2015) and increase both biomass and diversity relative to nearby bare sediments (McCormick-Ray 2005). These biotic substrates also enhance algal growth by local fertilization effects. Grazers such as the California Horn Snail can facilitate the development of macroalgal mats by consuming microalgal competitors and releasing nutrients for uptake by macroalgae (Fong et al. 1997). Also, on coral reefs Carpenter and Williams (1993) showed that urchin grazing facilitated algal turf production by reducing self-shading within the dense algal turf community. Invertebrate grazers that remove microalgal films that sometimes form on macroalgal thalli, but do not damage the macroalgae, may facilitate macroalgal growth by enhancing nutrient and gas exchange.

8.9.5 Competition Between Benthic Algal Primary Producers

Competition between benthic and pelagic algae was mentioned briefly in Section 8.9.1. Here, we will discuss in more detail the interplay between mats of benthic microalgae and loose mats of opportunistic macroalgae in shallow water (Figure 8.10).

Benthic micro- and macroalgae interact, directly and indirectly, through their impact on light, oxygen, and nutrient conditions. The type and strength of the interactions change with the growth phase, physiological state, the spatial location of the mats in relation to each other, and the type of the water body as it relates to tidal amplitude. Shading by dense macroalgal mats generally decreases benthic microalgal production. However, benthic microalgae biomass, as measured by chlorophyll *a* in surface sediments, shows no consistent relationship with macroalgal biomass (Besterman and Pace 2018).

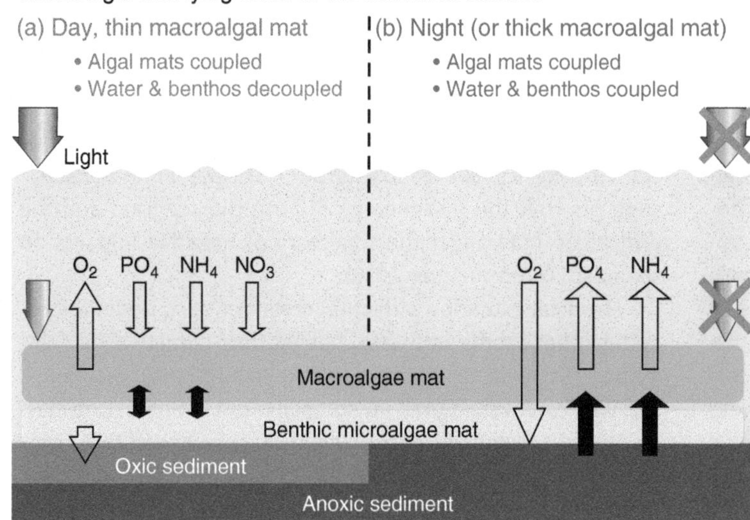

Macroalgal mat floating on water surface

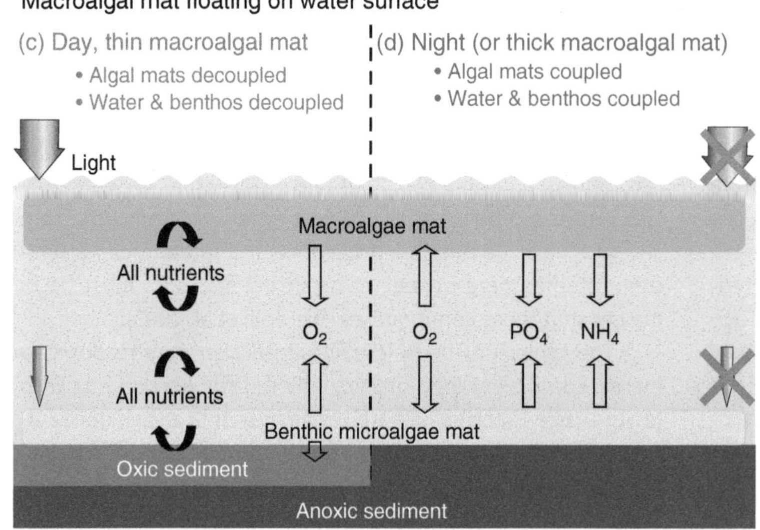

FIGURE 8.10 Conceptual models of interactions between mats of benthic microalgae and loose macroalgae. Upper panel (a and b): macroalgal mat lying close to the sediment surface, a situation common on tidal coasts during low tide. Lower panel (c and d): macroalgal mat floating at the water surface, a typical situation in microtidal waters. In (a), the two closely coupled mats intercept nutrient release from the sediment to the overlying water, whereas in (b) nutrients are released from the anoxic sediment and the coupled mats to the overlying water. When the two algal mats are spatially separated (c and d), there is no nutrient exchange between the two mats or between the sediment and water column. Instead, algal productivity is sustained by efficient recycling of nutrients within the mats themselves (c). This scenario applies particularly to autotrophic (often sandy) sediments in microtidal areas when nutrient levels in the overlying water column are low. At night, or when a thick floating macroalgal mat does not allow light to penetrate to the sediment surface (d), pore-water nutrients are released to the water column where they can be used by floating macroalgae. *Source*: Conceptual model drawings were inspired by the model in Astill and Lavery (2001) and redrawn from Sundbäck and McGlathery (2005).

Furthermore, shade-adapted benthic microalgal communities remain active—and even thrive—below macroalgal mats, provided that some light penetrates to the sediment surface (Sundbäck and McGlathery 2005). An example of such a shade-adapted species is the large sigmoid diatom *Gyrosigma balticum*. Benthic microalgae, particularly diatoms, are in fact often resistant to short-term (days, few weeks) hypoxia and anoxia and promote recovery of sediment systems by rapidly re-oxygenating the sediment surface after hypoxic and anoxic events (Larson and Sundbäck 2008).

When nutrient concentrations in the water column are low, primary producer groups compete for nutrients. In such a situation, the importance of sediments as a source of remineralized nutrients increases. At the same time, an active benthic microalgal mat will decrease the availability of sediment nutrients, limiting the growth of loose macroalgal mats, as well as phytoplankton. In Figure 8.10, two conceptual models are shown, one where the macroalgal mat (MA) exists close to the sediment surface (Figure 8.10a, b), and one where the macroalgal mat is floating at the water surface (Figure 8.10c, d). The coupling also depends on the thickness of the macroalgal mat and the time of the day (day/night, tidal cycle).

8.9.6 Interactions with Marine Diseases

Diseases are important drivers of population and community structure in marine ecosystems (Morton et al. 2020). Macroalgae can play important roles as reservoirs of marine pathogens. In coral reefs, algal-associated microbial activity can increase coral mortality. This effect may be due to the release of dissolved compounds by the algae that stimulate microbial activity, including by coral pathogens (Smith et al. 2006). However, additional stress or damage to the corals is likely necessary before algal-associated microbes cause damage (Sweet et al. 2013). A study in the Virginia coastal bays has also shown that the red algae *Gracilaria vermiculophylla* can serve as a reservoir of *Vibrio* bacteria, including the pathogenic *V. parahaemolyticus* and *V. vulnificus* (Gonzalez et al. 2014). In sufficient concentrations, these bacteria can cause symptoms in humans ranging from gastrointestinal infection to septicemia and even death. Understanding the interactions between macroalgae and marine pathogens therefore has implications for human health as well as ecological ramifications.

Marine disease can also affect macroalgae through a trophic cascade. In 2013, the northeast Pacific experienced a widespread outbreak of sea star wasting disease, leading to massive declines in sea star populations. The loss of this predator created a trophic cascade, with an increase in abundance of herbivorous sea urchins and a decrease in kelp density (Schultz et al. 2016). Declines in kelp, a foundation species that creates habitat, can have wide-ranging consequences for biodiversity, community structure, and ecosystem productivity (Teagle et al. 2017). In contrast, an earlier outbreak of urchin disease in the Caribbean devasted urchin populations, and the loss of urchin herbivory facilitated a transition from coral to algae dominated reefs (Hughes 1994). Outbreaks of marine diseases, which may have synergistic effects with stressors such as increasing temperatures and eutrophication, can therefore be important drivers of macroalgal populations.

8.10 Human Impacts

8.10.1 Eutrophication

As estuarine systems become increasingly eutrophied and macroalgal blooms occur, the decomposing macroalgal mats contribute significant amounts of organic matter to the water and sediments (Trimmer et al. 2000; Tyler et al. 2003). This decomposing organic matter turns over relatively fast and is a positive feedback mechanism that increases nutrient availability to sustain large algal standing stocks. Other consequences of algal bloom formation in eutrophic estuaries include decreases in fish/invertebrate abundance and diversity, anoxia, and loss of seagrasses, corals and perennial algae. Eutrophication can lead to phase shifts from seagrass communities to algal-dominated communities (Valentine and Duffy 2006), similar to what has been observed in lakes and several other ecosystems (Scheffer et al. 2001; Scheffer and van Nes 2004). It is essential to know whether these transitions are reversible phase shifts or if they represent an alternative state stabilized by negative feedbacks (Box 8.1). Recent meta-analysis suggests that the morphology of both the seagrass and the drift algae may affect the trajectory of negative impacts on seagrass (Thomsen et al. 2012).

Tropical estuaries with coral reefs may also undergo ecosystem phase shifts and state changes from coral to algal domination (Stimson 2018), although the link to eutrophication is less well established than for soft-sediment habitats. One classic study in Kaneohe Bay (Hawaii, USA) established that hard-bottomed communities shifted from coral to algae and back to coral again with changes in nutrient loading. In this system, sewage outfalls into the Bay increased nutrients and stimulated phytoplankton blooms. Lower light penetration stressed coral, shifting the competitive advantage to *Dictyospheria cavernosa*, the "green bubble algae" that was able to creep over the substrate and replace coral. After sewage was diverted to an offshore outfall, water clarity increased and coral gradually recolonized and replaced algae. However, while Kaneohe Bay is a clear example of a eutrophication-driven transition from coral to algal dominated tropical reef communities, several other studies suggest that co-occurring stressors must be in play to shift these communities (Pandolfi et al. 2003).

The growth of benthic microalgae also may respond to increased nutrient load, particularly in sandy sediments with lower concentrations of pore-water nutrients than finer sediments (Nilsson et al. 1991). It appears that benthic microalgal communities are highly resilient to eutrophication-related disturbances and may play an important role for the resilience of

Box 8.1 The Nature of Macroalgal Community Collapse: Phase Shifts or Alternative Stable States?

Macroalgae are taking over benthic estuarine communities worldwide. Catastrophic collapses from long-lived seagrass and coral-dominated ecosystems to opportunistic and shorter-lived macroalgae have focused research on the nature of these shifts. One important question is whether these are simple and reversible phase shifts, or if they represent alternative stable states with stabilizing mechanisms that inhibit recovery to the initial state (Scheffer et al. 2001; Beisner et al. 2003; Didham et al. 2005).

Shifts along an environmental gradients

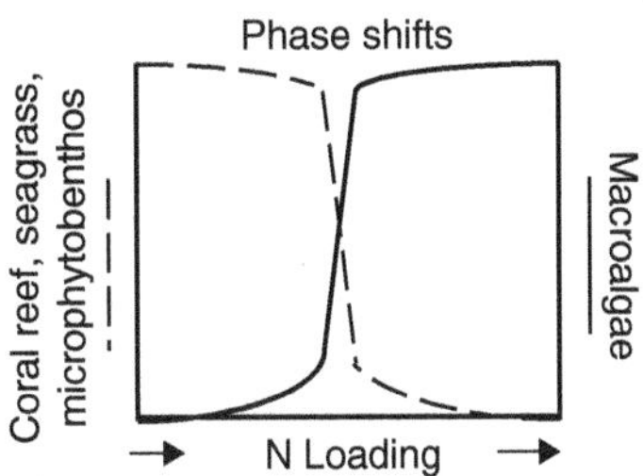

Phase shifts are rapid, catastrophic shifts from one dominant community to another over a very short range of environmental change. Incremental changes in environmental conditions cause little change until a critical threshold is reached.

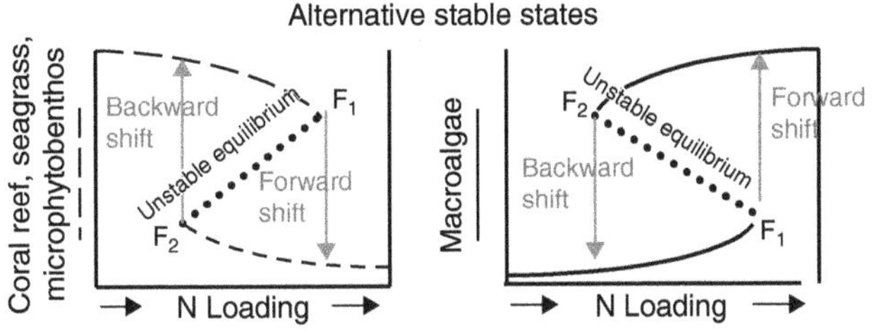

Communities that exist as alternative stable states also undergo rapid and catastrophic changes. However, forward transitions occur at different points than backward transitions resulting in a range of conditions (between F1 and F2) where either of two different states can occur. Thus, there is limited ability to predict the community state based on the condition of the environment.

Alternative stable states - shifts in response to disturbance

Why do we care?

Management strategies differ if communities exist as alternative stable states. Theory predicts there are only two ways to restore the initial, often more desirable state:

1) Reverse the environmental conditions (beyond F2. This can be difficult and expensive.
2) Cause a large disturbance that could push the community beyond the unstable equilibrium (dotted lines). This may have unknown consequences.

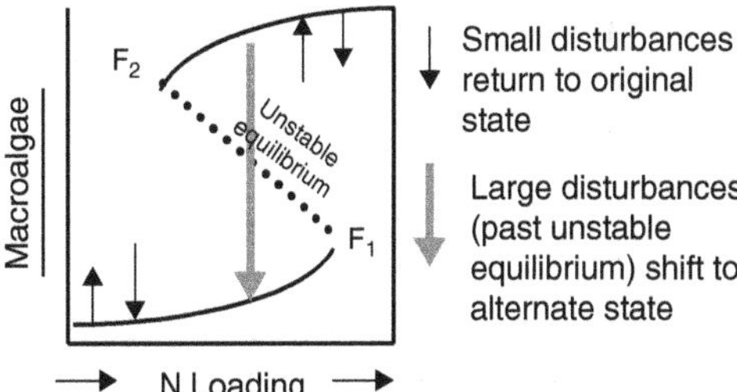

Distinguish phase shifts from alternate stable states

To distinguish phase shifts from alternative stable states scientists must search for positive feedback mechanisms that stabilize each state and are strong enough to buffer these states across a range of environmental conditions (between F1 and F2) and provide resilience to small disturbances.

the sediment community after, for example, hypoxic and anoxic events (McGlathery et al. 2007; Larson and Sundbäck 2008). The negative impact of eutrophication on the benthic microalgal community may be more gradual and slower than for benthic macroscopic primary producers, such as seagrasses. Thus, a partial beneficial "buffering" effect of benthic microalgae on shallow sediment systems may persist even in more heavily eutrophied systems. Benthic microalgal communities possess, due to high diversity and functional redundancy, a certain degree of plasticity, increasing the overall resilience of shallow-water sediment systems after pelagic bloom events. Benthic diatoms can survive periods of only a few % of incident light (or even darkness) and high sulfide levels, and can rapidly resume photosynthesis when exposed to light or after an anoxic event (Larson and Sundbäck 2008 and references therein). This scenario, with benthic microalgae surviving despite deteriorating conditions, may apply particularly to areas where macroalgal bloom events last only a few months (Pihl et al. 1999; McGlathery et al. 2001; Dalsgaard 2003), leaving the rest of the year open to benthic microalgal primary production. In warm, eutrophic microtidal systems with long-lasting macroalgal blooms, benthic microalgae can be outcompeted by shading. The benthic microalgae will collapse along with the macroalgae during dystrophic events (Viaroli et al. 1996).

8.10.2 Invasions

At least 346 non-native macroalgal species have been identified in marine waters worldwide, including 61 Chlorophyceae 77 Phaeophyceae, and 208 Rhodophyceae (Thomsen et al. 2016). The primary vectors for invasion are boat traffic and aquaculture, although the aquarium trade has also been an important invasion pathway (Padilla and Williams 2004). Most reports of invasion are from temperate regions; there is a general lack of information on introduced species from tropical waters, especially coral reef habitats (Coles and Eldredge 2002). The most invaded regions of the world are the Mediterranean and the NE Atlantic. Most successful invaders were foliose and filamentous forms, followed by leathery and siphonous forms (Williams and Smith 2007).

Several characteristics of successful invaders have been identified: rapid reproduction and the potential for successful evolution in new habitats, rapid colonization (including fragmentation as a source of new propagules), vegetative growth for population stability, rapid nutrient uptake and growth potentials ("weedy" species), anti-herbivore defenses, and a wide environmental tolerance (Nyberg and Wallentinus 2005). Both physical disturbance (by killing natives and opening space) and eutrophication (by relaxing resource competition) can make estuarine habitats more vulnerable to invasion. Some well-known invasions include *Caulerpa taxifolia, Gracilaria vermiculophylla, G. salicornia. Kappaphycus alvarezii, Hypnea musciformis, Sargassum muticum,* and *Codium fragile* (Figure 8.11). "Cryptic" invasions may also occur when introduced and native species share identical morphology and must be identified genetically (McIvor et al. 2001).

Dense accumulations of invasive macroalgae cause negative community-level effects, including shading of native algal species and seagrass, declines in fish abundance, diversity, and reproduction, and increased incidence of anoxia and hypoxia (Thomsen and McGlathery 2006; Grosholz and Ruiz 2009). However, the impacts of invasion are often context dependent and may affect different compartments of the community in contrasting ways. For example, biomass of the invasive green alga *C. taxifolia* was associated with increased abundance of epifauna but decreased abundance of infauna (Gribben et al. 2013). Invasive macroalgae can also increase habitat complexity and structure on unvegetated mudflats, as is the case for *G. vermiculophylla* in some North American and European mudflats and can have a positive effect on faunal abundances as long as algal populations stay below bloom proportions (Thomsen et al. 2009; Byers et al. 2012; Davoult et al. 2017). For example, juvenile blue crab survival was greater in invasive

FIGURE 8.11 Many species of invasive red algae proliferate on Hawaiian coral reefs including (a) *Acanthophora spicifera* (b) *Gracilaria salicornia*, and (c) *Kappaphycus alvarezii*. (d) In some areas, such as the coast of Maui, blooms of *Hypnea musciformis* become so large that they detach, form floating rafts, and deposit on the beach. *Source*: Jennifer Smith.

G. vermiculophylla habitat compared to native seagrass habitat or bare mud in Chesapeake Bay (Johnston and Lipcius 2012).

Changes in habitat complexity in turn can also affect foraging behavior by shorebirds (Haram et al. 2018). Invasive *Gracilaria* may also play a positive role in reducing the impacts of clam aquaculture on water quality by taking up from 20 to 77% of the nutrients released by the clams (Murphy et al. 2015). In areas where native foundation species such as seagrass have been lost, invasive macroalgae may act as a replacement foundation species, potentially supporting multiple ecosystem functions and services (Ramus et al. 2017). Macroalgae invasions are widespread and are likely to increase in coming years through continued human transport of invaders as well as impacts from anthropogenic stressors such as habitat disturbance. These invasions will impact the community structure and ecosystem functions of coastal ecosystems and may lead to homogenization of seaweed communities across regions (Mineur et al. 2015).

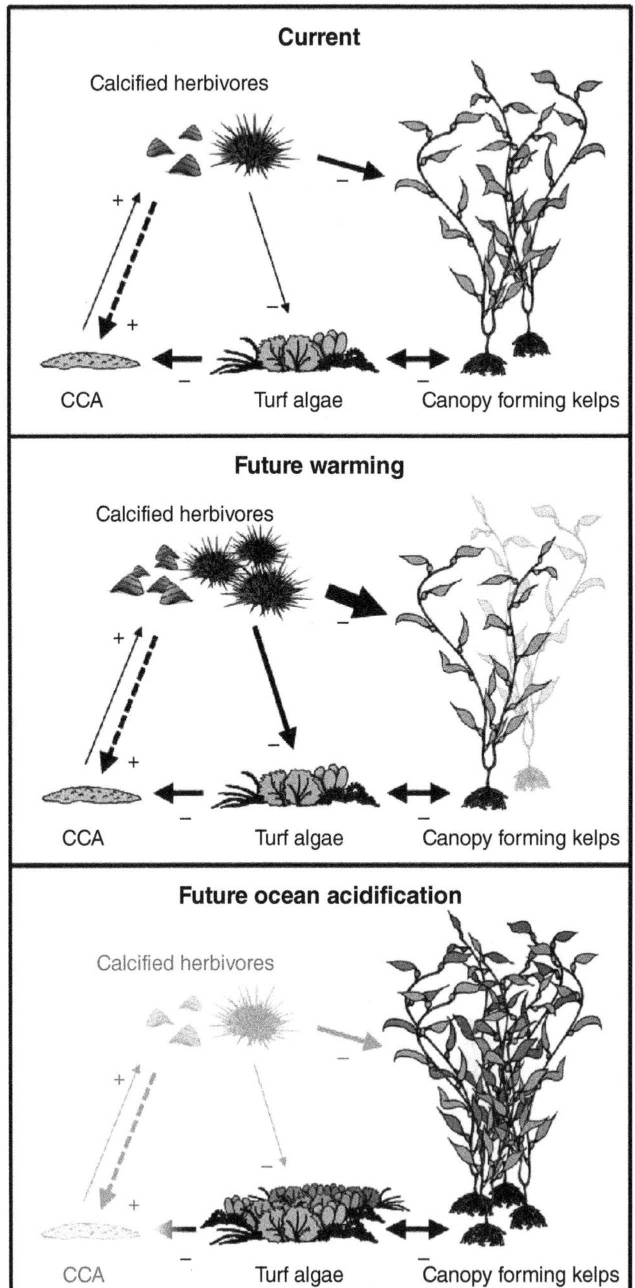

FIGURE 8.12 The elements of climate change, including increasing temperatures and ocean acidification, will have varying effects on different kinds of estuarine algae. Direct and indirect effects are shown with solid and dashed lines, respectively; faded symbols indicate functional groups with reduced ecological roles in future climate change scenarios. *Source*: Reproduced from Harley et al. (2012).

8.10.3 Climate Change

Climate change impacts on benthic algae include sea level rise, increased temperature, increased CO_2 in the air and water, ocean acidification, and changes in weather patterns (Parmesan and Yohe 2003; Doney et al. 2012). Rising temperatures have been predicted to cause poleward shifts in geographic ranges of species and ultimately alter the composition of marine communities (for a review see Hawkins et al. 2008), including benthic primary producers in estuaries. In Australia, these temperature-driven range shifts may push species beyond the southern coast of the continent, leading to the extinction of native seaweed species (Wernberg et al. 2011). In temperate latitudes, warming may result in deepening thermoclines and relaxation of cold upwelling, thus enabling warm water species to more easily jump gaps in distribution, especially along the western margins of major continents. These range shifts will influence herbivores as well as benthic algae; increased presence of tropical herbivores in temperate latitudes may lead to the "tropicalization" of temperate ecosystems, with cascading effects on biodiversity and ecosystem function (Verges et al. 2014). In upwelling zones where nutrient availability controls large-scale biogeography of kelp, suppression of upwelling due to increased temperatures may negatively affect kelp distribution, abundance, and size (Fong 2008). In tropical reef systems, continued temperature increases will likely result in further increases in dominance of benthic algae due to coral bleaching and mortality, especially for reefs experiencing additional stressors (Hoegh-Guldberg 1999; Graham et al. 2015). Rising temperatures will also cause physiological impacts on benthic algae, particularly for tropical species that are generally closer to their thermal limits under current conditions (Koch et al. 2013).

Changes in storm frequency may influence the distribution and abundance of benthic primary producers in estuaries by reducing or enhancing intertidal and shallow subtidal habitat via sedimentation, reducing water clarity and the depth of the photic zone, and increasing scouring (Zedler and West 2008; Anderson et al. 2014). For example, experimentally simulated scouring by sediments shifted the algal community from a diverse assemblage of macroalgae to more opportunistic forms (Vaselli et al. 2008) with high temporal variability (Bertocci et al. 2005) in rocky coastal habitats. Increased freshwater discharge during storms reduced uptake of nutrients by benthic microalgae, and thus the effectiveness of the estuarine benthic filter. Elevated CO_2 concentrations and ocean acidification are also likely to strongly impact estuarine benthic algae. Many macroalgae can use HCO_3^- as well as CO_2 for photosynthesis, and algal productivity may therefore increase under ocean acidification (Koch et al. 2013). However, experiments with elevated CO_2 and lowered pH have revealed strong negative effects on calcifying algae (Kroeker et al. 2010), including tropical crustose calcareous algae (Anthony et al. 2008; Jokiel et al. 2008).

Complex interactions between climate change factors will drive impacts on benthic algae (Figure 8.12). For example, while increased temperature and nutrients may increase algal recruitment and growth up to some threshold, they should also increase the number of grazers, driving intertidal communities toward opportunistic species and away from dominance by fucoids (Lotze and Worm 2002). O'Connor (2009) found that warming strengthens herbivore–algal interactions, shifting important trophic pathways. Recent multifactorial experiments show that certain stages in complex life cycles may be more sensitive to interactions among climate-related factors and may be important bottlenecks limiting the ability of algal dominants to survive climate change over the long term (Harley et al. 2012). For example, while adult kelp sporophytes in polar regions are relatively hardy to changes in temperature, UV light, salinity, and their interactions, germination of zoospores is much more sensitive (Fredersdorf et al. 2009). Aquaculture studies have identified an important interaction between rising CO_2 and nutrients, with positive CO_2 effects on growth of *Gracilaria* being accelerated with pulsed nutrient supplies (Friedlander and Levy 1995), suggesting that storms combined with rising CO_2 will facilitate algal blooms in estuaries. Other studies suggest that interactions among climate change factors may result in changes in habitats. For example, in sheltered embayments of the North Sea, rising sea level and eutrophication combined to cause shifts from *Zostera* to opportunistic green macroalgae (Reise et al. 2008). Overall, it appears that interacting factors associated with climate change will enhance algal blooms and may impact shift communities to greater algal dominance.

Study Questions

Multiple Choice

1. Which group of benthic macroalgae can grow in the deepest water?
   a. Rhodophyta
   b. Chlorophyta
   c. Phaeophyta
   d. Cyanophyta

2. Which taxa of microalga are the most dominant in shallow-water marine sediments?
   a. Dinoflagellates
   b. Cyanobacteria
   c. Diatoms
   d. Cryptophytes

3. Which aspects of macroalgal morphology affect light harvesting and photosynthetic rates? (Select all that apply)
   a. Chlorophyll
   b. Strength of attachment to surfaces
   c. Accessory pigments
   d. Calcification
   e. Surface area:volume ratio

4. Which types of macroalgae are most likely to exhibit surge uptake when exposed to nutrients?
   a. Fast-growing bloom-forming macroalgae
   b. Long-lived slow-growing macroalgae
   c. Macroalgae in deeper waters
   d. Macroalgae with a low surface area to volume ratio

5. In which of these ways do benthic microalgae influence marine sediments? (Select all that apply)
   a. Sediment stabilization
   b. Increased sediment nutrients
   c. Sediment oxygenation
   d. Limit of growth of heterotrophic bacteria in sediment
   e. Change sediment color to black

6. How are macroalgae in polar regions adapted to extreme seasonality in favorable growth conditions? (Select all that apply)
   a. Boom and bust life cycle
   b. Shade tolerance
   c. Carbon stores
   d. Rapid nutrient uptake
   e. Stronger holdfasts

7. Which of these are negative effects of dense macroalgae communities on benthic faunal communities?
   a. Increase sulfide concentrations
   b. Predation refuge
   c. Food source
   d. Increase biodiversity
   e. Low dissolved oxygen

8. In which macroalgal communities is the grazing food web more important than recycling via the detrital food web? (Select all that apply)
   a. Rocky intertidal
   b. Coral reef
   c. Soft sediment
   d. All of the above

9. Benthic microalgae can compete with phytoplankton for which resource(s)?
   a. Carbon Dioxide
   b. Nutrients
   c. Light
   d. Oxygen
   e. None of the above

Short Answer

1. How do high nutrient concentrations from eutrophication and reduced grazing from overfishing interact to affect macroalgae on coral reefs?
2. How does macroalgal morphology tend to vary due to differences in flow velocity in high-energy (e.g., rocky intertidal) and low-energy (e.g., protected embayments) systems?
3. What are the main biogeochemical processes that account for the difference between nitrogen limitation of marine algae in temperate waters and phosphorus limitation in tropical waters?
4. Macroalgae can contribute to "blue carbon" sequestration and storage in marine ecosystems. This is considered one type of "nature-based solution" to climate change by removing CO_2 from the atmosphere. Describe the ways in which marine algae contribute to long-term carbon storage in the ocean.
5. How do the climate change impacts of increased storm frequency and magnitude, increased CO_2 in the air and water, and ocean warming affect benthic macroalgal communities?
6. How do marine macroalgae affect the distribution and prevalence of marine diseases?
7. Eutrophication is a common phenomenon in coastal marine ecosystems, despite advances in reducing watershed nutrient inputs. What are the main characteristics of benthic algae that affect their response as systems becoming increasingly eutrophied?
8. Macroalgal invasions are widespread in coastal ecosystems. What makes a successful macroalgal invader? What are the most likely ecosystem impacts of marine invasions of macroalgae?

References

Adams, T.C., Burkepile, D.E., Holbrook, S.J. et al. (2021). Landscape-scale patterns of nutrient enrichment in a coral reef ecosystem: implications for coral to algae phase shifts. *Ecological Applications* 31 (1): e02227. https://doi.org/10.1002/eap.2227.

Anderson, I.C., McGlathery, K.J., and Tyler, A.C. (2003). Microbial mediation of 'reactive' nitrogen transformations in a temperate lagoon. *Marine Ecology Progress Series* 246: 73–84.

Anderson, I. C., M. J. Brush, M. F. Piehler, C. A. Currin, J. W. Stanhope, A. R. Smyth, J. D. Maxey, and M. L. Whitehead. (2014). Impacts of climate-related drivers on the Benthic nutrient filter in a shallow photic estuary. Estuaries and Coasts 37: 46–62. https://doi.org/10.1007/s12237-013-9665-5

Andresen, M. and Kristensen, E. (2002). The importance of bacteria and microalgae in the diet of the deposit-feeding polychaete *Arenicola marina*. *Ophelia* 56: 179–196.

Anthony, K.R.N., Kline, D.I., Diaz-Pulida, G., and Hoegh-Guldberg, O. (2008). Ocean acidification causes bleaching and productivity loss in coral reef builders. *Proceedings of the National Academy of Science* 105: 17442–17446.

Armitage, A.R. and Fong, P. (2004). Upward cascading effects of nutrients: shifts in a benthic microalgal community and a negative herbivore response. *Oecologia* 139: 560–567.

Asmus, H. and Asmus, R. (1985). The importance of grazing food chain for energy flow and production in three intertidal sand bottom communities of the northern Wadden Sea. *Helgoländer Meeresuntersuchungen* 39: 273–301.

Astill, H. and Lavery, P. (2001). The dynamics of unattached benthic macroalgal accumulations in the Swan-Canning Estuary. *Hydrological Processes* 15: 2387–2399.

Atkinson, M.J. and Smith, S.V. (1983). C:N:P ratios of marine benthic plants. *Limnology and Oceanography* 28: 568–574.

Attard, K.M., Stahl, H., Kamenos, N.A. et al. (2015). Benthic oxygen exchange in a live coralline algal bed and an adjacent sandy habitat: an eddy covariance study. *Marine Ecology Progress Series* 535: 99–115. https://doi.org/10.3354/meps11413.

Attard, K.M., Rodil, I.F., Berg, P. et al. (2018). Seasonal metabolism and carbon export potential of a key coastal habitat: the perennial canopy-forming macroalga Fucus vesiculosus. *Limnology and Oceanography* 9999: 1–16.

Baird, D., Asmus, H., and Asmus, R. (2004). Energy flow of a boreal intertidal ecosystem, the Sylt-Rømø Bight. *Marine Ecology Progress Series* 279: 45–61.

Beisner, B., Haydon, D., and Cuddington, K. (2003). Alternative stable states in ecology. *Frontiers in Ecology and the Environment* 1: 376–382. https://doi.org/10.1890/1540-9295(2003)001[0376:ASSIE]2.0.CO;2.

Berg, P.B., Roy, H., Janssen, F. et al. (2003). Oxygen uptake by aquatic sediments measured with a novel non-invasive eddy-correlation technique. *Marine Ecology Progress Series* 261: 75–83.

Berg, P., M. Huettel, R.N. Glud, C.E. Reimers, K.M. Attard (2022). Aquatic eddy covariance: The method and its contributions to defining oxygen and carbon fluxes in marine environments. Annual Review of Marine Science. http://www.annualreviews.org/eprint/Z2J6BCTFERENRSJR4CMZ/full/10.1146/annurev-marine-042121-012329

Berg, P.B., Roy, H., and Wiberg, P.W. (2007). Eddy correlation flux measurements: The sediment surface area that contributes to the flux. *Limnology and Oceanography* 52: 1672–1684.

Bertocci, I., Maggi, E., Vaselli, S., and Benedetti-Cecchi, L. (2005). Contrasting effects of mean intensity and temporal variation of disturbance on a rocky seashore. *Ecology* 86: 2061–2067.

Besterman, A.F. and Pace, M.L. (2018). Do Macroalgal Mats Impact Microphytobenthos on Mudflats? Evidence from a Meta-Analysis, Comparative Survey, and Large-Scale Manipulation. *Estuaries and Coasts* 41: 2304–2316. https://doi.org/10.1007/s12237-018-0418-3.

Billerbeck, M., Røy, H., Bosselmann, K., and Huettel, M. (2007). Benthic photosynthesis in submerged Wadden Sea intertidal flats. *Estuarine Coastal Shelf Science* 71: 704–716.

Borum, J., Pedersen, M.F., Krause-Jensen, D. et al. (2002). Biomass, photosynthesis and growth of *Laminaria saccharina* in a High-Arctic fjord, NE Grenland. *Marine Biology* 141: 11–19.

Boyer, K.E. and Fong, P. (2005). Macroalgal-mediated transfers of water column nitrogen to intertidal sediments and salt marsh plants. *Journal of Experimental Marine Biology and Ecology* 321: 59–69.

Boyle, K.A., Fong, P., and Kamer, K. (2004). Spatial and temporal patterns in sediment and water column nutrients in an eutrophic southern California estuary. *Estuaries* 27: 254–267.

Bracken, M.E. (2004). Invertebrate-mediated nutrient loading increases growth of an intertidal macroalga. *Journal of Phycology* 40: 1032–1041.

de Brouwer, J.F.C., Neu, T.R., and Stal, L.J. (2006). On the function of secretion of extracellular polymeric substances by benthic diatoms and their role intertidal mudflats. In: *Functioning of Microphytobenthos in Estuaries* (ed. J.C. Kromkamp, J.F.C. de Brouwer, G.F. Blanchard, et al.), 45–61. Amsterdam: Royal Netherlands Academy of Arts and Sciences.

Bruno, J.F., Boyer, K.E., Duffy, J.E., and Lee, S.C. (2008). Relative and interactive effects of plant and grazer richness in a benthic marine community. *Ecology* 89: 2518–2528.

Brush, M.J. and Nixon, S.W. (2003). Biomass layering and metabolism in mats of the macroalga *Ulva lactuca* L. *Estuaries* 26: 916–926.

Byers, J. E., P. E. Gribben, C. Yeager, and E. E. Sotka. (2012). Impacts of an abundant introduced ecosystem engineer within mudflats of the southeastern US coast. Biological Invasions 14: 2587–2600. https://doi.org/10.1007/s10530-012-0254-5

Cadée, G.C. and Hegeman, J. (1977). Distribution of primary production of the benthic microflora and accumulation of organic matter on a tidal flat area, Balgzand, Dutch Wadden Sea. *Netherlands Journal of Sea Research* 11: 24–41.

Cahoon, L.B. (1999). The role of benthic microalgae in neritic ecosystems. *Oceanography and Marine Biology: An Annual Review* 37: 47–86.

Carpenter, R.C. and Williams, S.L. (1993). Effects of algal turf canopy height and microscale substratum topography on profiles of flow speed in a coral forereef environment. *Limnology and Oceanography* 38: 687–694.

Cartaxana, P., Ruivo, M., Hubas, C. et al. (2011). Physiological versus behavioral photoprotection in intertidal epipelic benthic diatom communities. *Journal of Experimental Marine Biology and Ecology* 405: 120–127.

Cartaxana, P., Cruz, S., Gameiro, C., and Kuhl, M. (2016). Regulation of intertidal microphytobenthos photosynthesis over a diel emersion period is strongly affected by diatom migration patterns. *Frontiers in Microbiology* https://doi.org/10.3389/fmicb.2016.00872.

Cebrian, J. (1999). Patterns in the fate of production in plant communities. *American Naturalist* 154: 449–468.

Cebrian, J. (2002). Variability and control of carbon consumption, export, and accumulation in marine communities. *Limnology and Oceanography* 47: 11–22.

Chapman, A.R.O. and Lindley, J.E. (1980). Seasonal growth of *Laminaria solidongula* in the Canadian high Arctic in relation to irradiance and dissolved nutrient concentrations. *Marine Biology* 57: 1–5.

Christianen, M. J. A., J. J. Middelburg, S. J. Holthuijsen, J. Jouta, T. J. Compton, T. van der Heide, T. Piersma, et al. (2017). Benthic primary producers are key to sustain the Wadden Sea food web: stable carbon isotope analysis at landscape scale. *Ecology* 98: 1498–1512. https://doi.org/10.1002/ecy.1837

Coles, S.L. and Eldredge, L.G. (2002). Nonindigenous species introductions on coral reefs: a need for information. *Pac Sci* 56: 191–209.

Connell, J.H. (1972). Community interactions on marine rocky intertidal shores. *Annual Review of Ecology and Systematics* 3: 169–192.

Consalvey, M., Paterson, D.M., and Underwood, G.J.C. (2004). The ups and downs of life in a benthic biofilm: migration of benthic diatoms. *Diatom Research* 19: 181–202. https://doi.org/10.1080/0269249X.2004.9705870.

Dalsgaard, T. (2003). Benthic primary production and nutrient cycling in sediments with benthic microalgae and transient accumulation of macroalgae. *Limnology and Oceanography* 48: 2138–2150.

Davoult, D., Surget, G., Stiger-Pouvreau, V. et al. (2017). Multiple effects of a Gracilaria vermiculophylla invasion on estuarine mudflat functioning and diversity. *Marine Environmental Research* 131: 227–235. https://doi.org/10.1016/j.marenvres.2017.09.020.

Denny, M.W. (1995). Predicting physical disturbance: Mechanistic approaches to the study of survivorship on wave-swept shores. *Ecological Mongraphs* 55: 69–102.

Denny, M.W., Daniel, T.L., and Koehl, M.A.R. (1985). Mechanical limits to size in wave-swept organisms. *Ecological Mongraphs* 51: 69–102.

Didham, R.K., Watts, C.H., and Norton, D.A. (2005). Are systems with strong underlying abiotic regimes more likely to exhibit alternative stable states? *Oikos* 110: 409–416.

Doney, Scott C., Mary Ruckelshaus, J. Emmett Duffy, James P. Barry, Francis Chan, Chad A. English, Heather M. Galindo, et al. (2012). Climate change impacts on marine ecosystems. *Annual Review of Marine Science* 4: 11–37. https://doi.org/10.1146/annurev-marine-041911-111611

Duarte, C.M. (1992). Nutrient concentration of aquatic plants: Patterns across species. *Limnology and Oceanography* 37: 882–889.

Duarte, C.M. (1995). Submerged aquatic vegetation in relation to different nutrient regimes. *Ophelia* 41: 87–112.

Duarte, C.M. and Cebrian, J. (1996). The fate of marine autotrophic production. *Limnology and Oceanography* 41: 1758–1766.

Dunton, K.H. (1985). Growth of dark-exposed Laminaria saccharina (L.) Lamour. and Laminaria solidungula J. Ag. (Laminariales: Phaeophyta) in the Alaska Beaufort Sea. *Journal of Experimental Marine Biology and Ecology* 94: 181–189.

Engelsen, A., Hulth, S., Pihl, L., and Sundbäck, K. (2008). Benthic trophic status and nutrient fluxes in shallow-water sediments. *Estuarine Coastal Shelf Science* 78: 783–795.

Escartín, J., and D. G. Aubrey. (1995). Flow structure and dispersion within algal mats. Estuarine, Coastal and Shelf Science 40: 451–472. https://doi.org/10.1006/ecss.1995.0031

Fenchel, T. and Glud, R.N. (2000). Benthic Primary production and O_2-CO_2 dynamics in a shallow water sediment: spatial and temporal heterogeneity. *Ophelia* 53: 159–171.

Flindt, M.R., Neto, J., Amos, C.L. et al. (2004). Plant bound nutrient transport: mass transport in estuaries and lagoons. In: *Estuarine Nutrient Cycling: The Influence of Primary Producers* (ed. S.L. Nielsen, G.T. Banta and M.F. Pedersen), 93–128. Dordrecht: Kluwer Academic Publishers.

Fong, P. (2008). Macroalgal-dominated ecosystems. In: *Nitrogen in the Marine Environment*, 917–948. Elsevier.

Fong, P. and Paul, V.J. (2011). Coral Reef Algae. In: *Coral Reefs: An Ecosystem in Transition* (ed. Z. Dubinsky and N. Stambler), 241–272. Dordrecht: Springer Netherlands https://doi.org/10.1007/978-94-007-0114-4_17.

Fong, P. and Zedler, J.B. (2000). Sources, sinks, and fluxes of nutrients (N + P) in a small highly-modified estuary in southern California. *Urban Ecosystems* 4: 125–144.

Fong, P., Desmond, J.S., and Zedler, J.B. (1997). The effect of a horn snail on *Ulva expansa* (Chlorophyta): consumer or facilitator of growth? *Journal of Ecology* 33: 353–359.

Fong, P., Fong, J., and Fong, C. (2003). Growth, nutrient storage, and release ofDONby Enteromorpha intestinalis in response to pulses of nitrogen and phosphorus. *Aquatic Botany* 78: 83–95.

Fong, P., Smith, T., and Wartian, M. (2006). Protection by epiphytic cyanobacteria maintains shifts to macroalgal-dominated communities after the 1997–98 ENSO disturbance on coral reefs with intact herbivore populations. *Ecology* 87: 1162–1168.

Foster, S.A. (1987). The relative impacts of grazing by Caribbean coral reef fishes and *Diadema:* Effects of habitat and surge. *Journal of Experimental Marine Biology Ecology* 105: 1–20.

Fredersdorf, J., Müller, R., Becker, S. et al. (2009). Interactive effects of radiation, temperature, and salinity on different life history stages of the Artic kelp Alaria esculenta (Phaeophyceae). *Oecologia* 160: 483–492.

Friedlander, M. and Levy, I. (1995). Cultivation of *Gracilaria* in outdoor tanks and ponds. *Journal of Applied Phycology* 7: 315–324.

Galloway, J.N., Townsend, A.R., Erisman, J.W. et al. (2008). Transformation of the nitrogen cycle: recent trends, questions, and potential solutions. *Science* 320: 889–892.

Gaylord, B., Rosman, J.H., Reed, D.C. et al. (2007). Spatial patterns of flow and their modification within and around a giant kelp forest. *Limnology and Oceanography* 52: 1838–1852.

Giannotti, A.L. and McGlathery, K.J. (2001). Consumption of *Ulva lactuca* (Chlorophyta) by the omnivorous mud snail *Ilyanassa obsoleta. Journal of Phycology* 37: 1–7.

Glud, R.N. (2006). Microscale techniques to measure photosynthesis: a mini review. In: *Functioning of Microphytobenthos in Estuaries* (ed. J.C. Kromkamp, J.F.C. de Brouwer, G.F. Blanchard, et al.), 31–41. Amsterdam: Royal Netherlands Academy of Arts and Sciences.

Glud, R.N., Kühl, M., Wenzhöfer, F., and Rysgaard, S. (2002). Benthic diatoms of a high Arctic fjord (Young Sound, NE Greenland): importance of ecosystem primary production. *Marine Ecology Progress Series* 238: 15–29.

Glud, R.N., Woelfel, J., Karsten, U., and Kuhl, M. (2009). S. Rysgaard (submitted) Benthic microalgal production in the Arctic: status of the current database. *Botanica Marina.* 52: 559–571.

Gómez, I., A. Wulff, M. Y. Roleda, P. Huovinen, U. Karsten, M. L. Quartino, K. Dunton, and C. Wiencke. (2009). Light and temperature demands of marine benthic microalgae and seaweeds in polar regions. Botanica Marina 52. https://doi.org/10.1515/BOT.2009.073

Gonzalez, D., R. Gonzalez, B. Froelich, J. Oliver, R. Noble, and K. McGlathery. (2014). Non-native macroalga may increase concentrations of Vibrio bacteria on intertidal mudflats. Marine Ecology Progress Series 505: 29–36. https://doi.org/10.3354/meps10771

Goto, N.T., Kawamura, T., Mitamura, O., and Terai, H. (1999). Importance of extracellular organic carbon production in the total primary production by tidal-flat diatoms in comparison to phytoplankton. *Marine Ecology Progress Series* 190: 289–295.

Graham, M. H. (2001). Effects of local deforestation on the diversity and structure of Southern California giant kelp forest food webs. Ecosystems 7: 341–357. https://doi.org/10.1007/s10021-003-0245-6

Graham, N.A.J., Simon Jennings, M., MacNeil, A. et al. (2015). Predicting climate-driven regime shifts versus rebound potential in coral reefs. *Nature* 518: 94–97. https://doi.org/10.1038/nature14140.

Green, L. and Fong, P. (2016). The good, the bad and the *Ulva* : the density dependent role of macroalgal subsidies in influencing diversity and trophic structure of an estuarine community. *Oikos* 125: 988–1000. https://doi.org/10.1111/oik.02860.

Green, L., Blumstein, D.T., and Fong, P. (2015). Macroalgal mats in a eutrophic estuary obscure visual foraging cues and increase variability in prey availability for some shorebirds. *Estuaries and Coasts* 38: 917–926. https://doi.org/10.1007/s12237-014-9862-x.

Gribben, P.E., Byers, J.E., Wright, J.T., and Glasby, T.M. (2013). Positive versus negative effects of an invasive ecosystem engineer on different components of a marine ecosystem. *Oikos* 122: 816–824. https://doi.org/10.1111/j.1600-0706.2012.20868.x.

Grosholz, E.D. and Ruiz, G.M. (2009). Multitrophic effects of invasions in marine and estuarine systems. In: *Biological Invasions in Marine Ecosystems: Ecological, Management, and Geographic Perspectives* (ed. G. Rilov and J.A. Crooks), 305–324. Springer.

Guiry, M.D. and Guiry, G.M. (2007). *AlgaeBase Version 4.2. World-wide Electronic Publication*. Galway: National University of Ireland http://www.algaebase.org.

Haram, L.E., Kinney, K.A., Sotka, E.E., and Byers, J.E. (2018). Mixed effects of an introduced ecosystem engineer on the foraging behavior and habitat selection of predators. *Ecology.* https://doi.org/10.1002/ecy.2495.

Hardison, A., Canuel, E., Anderson, I., and Veuger, B. (2010). Fate of macroalgae in benthic systems: carbon and nitrogen cycling within the microbial community. *Marine Ecology Progress Series* 414: 41–55. https://doi.org/10.3354/meps08720.

Hardison, A.K., Anderson, I.C., Canuel, E.A. et al. (2011). Carbon and nitrogen dynamics in shallow photic systems: interactions between macroalgae, microalgae, and bacteria. *Limnology and Oceanography* 56: 1489–1503. https://doi.org/10.4319/lo.2011.56.4.1489.

Harley, C.D.G., Anderson, K.M., Demes, K.W. et al. (2012). Effects of climate change on global seaweed communities. *Journal of Phycology* 48: 1064–1078. https://doi.org/10.1111/j.1529-8817.2012.01224.x.

Hauxwell, J., McClelland, J., Behr, P.J., and Valiela, I. (1998). Relative importance of grazing and nutrient controls of macroalgal biomass in three temperate shallow estuaries. *Estuaries* 21: 347–360.

Hawkins, S.J., Moore, P.J., Burrows, M.T. et al. (2008). Complex interactions in a rapidly changing world: responses of rocky shore communities to recent climate change. *Climate Research* 37: 123–133.

Hay, M.E. (1984). Pattern of fish and urchin grazing on Caribbean coral reefs: are previous results typical? *Ecology* 65: 446–454.

Hepburn, C.D., Holborow, J.D., Wing, S.R. et al. (2007). Exposure to waves enhances growth rate and nitrogen status of the giant kelp *Macrocystis pyrifera. Marine Ecology Progress Series* 339: 99–108.

Hillebrand, H.B., Worm, B., and Lotze, H.K. (2000). Marine microbenthic community structure regulated by nitrogen loading and grazing pressure. *Marine Ecology Progress Series* 204: 27–38.

Hoegh-Guldberg, O. (1999). Climate change, coral bleaching, and the future of the world's coral reefs. *Marine and Freshwater Research* 50: 839–866.

Hondula, K., and M. Pace. (2014). Macroalgal support of cultured hard clams in a low nitrogen coastal lagoon. Marine Ecology Progress Series 498: 187–201. https://doi.org/10.3354/meps10644

Houser, C.A. and Nickling, W.G. (2001). The factors influencing the abrasion efficiency of saltating grains on a clay-crusted playa. *Earth Surface Processes and Landforms* 26: 491–505.

Howarth, R. W. (2008). Coastal nitrogen pollution: a review of sources and trends globally and regionally. Harmful Algae 8: 14–20. https://doi.org/10.1016/j.hal.2008.08.015

Hughes, T. P. (1994). Catastrophes, phase shifts, and large-scale degradation of a Caribbean Coral Reef. Science 265: 1547–1551. https://doi.org/10.1126/science.265.5178.1547

Hughes, T.P., Baird, A.H., Bellwood, D.R. et al. (2003). Climate change, human impacts, and the resilience of coral reefs. *Science* 301: 929–933.

Hull, S.C. (1987). Macroalgal mats and species abundance: a field experiment. *Estuar coast Shelf Sci* 25: 519–532.

Huntington, B.E. and Boyer, K.E. (2008). Effects of red macroalgal (*Grailariopsis* sp.) abundance on eelgrass *Zostera marina* in Tomales Bay, California, USA. *Marine Ecology Progress Series* 367: 133–142.

Hurd, C.L. (2000). Water motion, marine macroalgal physiology, and production. *Journal of Ecology* 36: 453–472.

Hurd, C.L., Durante, K.M., Chia, F.-.S., and Harrison, P.J. (1994). Effect of bryozoan colonization on inorganic nitrogen acquisition by the kelps *Agarum fimbiratum* and *Macrocyctis integrifolia*. *Marine Biology* 121: 167–173.

Hurd, C.L., Stevens, C.L., Laval, B.E. et al. (1997). Visualization of seawater flow around morphologically distinct forms of the giant kelp *Macrocystis integrifolia* from wave-sheltered and exposed sites. *Limnology and Oceanography* 42: 156–163.

Jackson, J.B.C., Kirby, M.X., Berger, W.H. et al. (2001). Historical overfishing and the recent collapse of coastal ecosystems. *Science* 293: 629–638.

Janousek, C. N., C. A. Currin, and L. A. Levin. (2007). Succession of microphytobenthos in a restored coastal wetland. Estuaries and Coasts 30: 265–276. https://doi.org/10.1007/BF02700169

Johnston, C., and R. Lipcius. (2012). Exotic macroalga Gracilaria vermiculophylla provides superior nursery habitat for native blue crab in Chesapeake Bay. Marine Ecology Progress Series 467: 137–146. https://doi.org/10.3354/meps09935

Jokiel, P.L., Rodgers, K.S., Kuffner, I.B. et al. (2008). Ocean acidification and calcifying reef organisms: a mesocosm investigation. *Coral Reefs* 27: 473–483.

de Jonge, V.N. and Beusekom, J.E.E. (1992). Contribution of resuspended microphytobenthos to total phytoplankton in the Ems estuary and its possible role for grazers. *Netherlands Journal of Sea Research* 30: 91–105.

de Jonge, V.N. and Beusekom, J.E.E. (1995). Wind- and tide-induced resuspension of sediment and microphytobenthos from tidal flats in the Ems Estuary. *Limnology and Oceanography* 40: 766–778.

Jönsson, B. (1991). A ^{14}C-incubation technique for measuring microphytobenthic primary productivity in intact sediment cores. *Limnology and Oceanography* 36: 1485–1492.

Joye, S.B. and Anderson, I.C. (2008). Nitrogen cycling in coastal sediments. In: *Nitrogen in the Marine Environment*, 867–916. Elsevier.

Kamer, K., Karleen, A., and Boyle, P.F. (2001). Macroalgal bloom dynamics in a highly eutrophic southern California estuary. *Estuaries* 24: 623–635.

Kennedy, H., J. Beggins, C. M. Duarte, J. W. Fourqurean, M. Holmer, N. Marbà, and J. J. Middelburg. (2010). Seagrass sediments as a global carbon sink: isotopic constraints. Global Biogeochemical Cycles 24: n/a-n/a. https://doi.org/10.1029/2010GB003848

Kennison, R., and P. Fong. (2013). High amplitude tides that result in floating mats decouple algal distribution from patterns of recruitment and nutrient sources. Marine Ecology Progress Series 494: 73–86. https://doi.org/10.3354/meps10504

Koch, M., Bowes, G., Ross, C., and Zhang, X.-H. (2013). Climate change and ocean acidification effects on seagrasses and marine macroalgae. *Global Change Biology* 19: 103–132. https://doi.org/10.1111/j.1365-2486.2012.02791.x.

Koop, K., Booth, D., Broadbent, A. et al. (2001). ENCORE: The effect of nutrient enrichment on coral reefs. Synthesis of results and conclusions. *Marine Pollution Bulletin* 41: 91–120.

Kraufvelin, P., Lindholm, A., Pedersen, M.F. et al. (2010). Biomass, diversity and production of rocky shore macroalgae at two nutrient enrichment and wave action levels. *Marine Biology* 157: 29–47. https://doi.org/10.1007/s00227-009-1293-z.

Krause-Jensen, D. and Duarte, C.M. (2016). Substantial role of macroalgae in marine carbon sequestration. *Nature Geoscience* 9: 737–742. https://doi.org/10.1038/ngeo2790.

Krause-Jensen, D., McGlathery, K., Rysgaard, S., and Christensen, P.B. (1996). Production within dense mats of the filamentous macroalga *Chaetomorpha linum* in relation to light and nutrient availability. *Marine Ecology Progress Series* 134: 207–216.

Krause-Jensen, D., Kuhl, M., Christensen, P.B., and Borum, J. (2007). Benthic primary production in Young Sound, Northeast Greenland. In: *Carbon cycling in Arctic marine ecosystems: Case study Young Sound*, Meddr. Gronland, Bioscience, vol. 58 (ed. S. Rysgaard and R.N. Glud), 160–173. Danish Polar Center.

Krause-Jensen, D., Lavery, P., Serrano, O. et al. (2018). Sequestration of macroalgal carbon: the elephant in the Blue Carbon room. *Biology Letters* 14: 20180236. https://doi.org/10.1098/rsbl.2018.0236.

Kroeker, K.J., Kordas, R.L., Crim, R.N., and Singh, G.G. (2010). Meta-analysis reveals negative yet variable effects of ocean acidification on marine organisms: biological responses to ocean acidification. *Ecology Letters* 13: 1419–1434. https://doi.org/10.1111/j.1461--0248.2010.01518.x.

Kromkamp, J.C. and Forster, R.M. (2006). Developments in microphytobenthos primary productivity studies. In: *Functioning of microphytobenthos in estuaries* (ed. J.C. Kromkamp, J.F.C. de Brouwer, G.F. Blanchard, et al.), 9–30. Amsterdam: Royal Netherlands Academy of Arts and Sciences.

Kronvang, B., E. Jeppesen, D. J. Conley, M. Søndergaard, S. E. Larsen, N. B. Ovesen, and J. Carstensen. (2005). Nutrient pressures and ecological responses to nutrient loading reductions in Danish streams, lakes and coastal waters. Journal of Hydrology 304: 274–288. https://doi.org/10.1016/j.jhydrol.2004.07.035

Kühl, M., Lassen, C., and Jørgensen, B.B. (1994). Light penetration and light intensity in sandy marine sediments measured with irradiance and scalar irradiance fiber-optic microprobes. *Marine Ecology Progress Series* 105: 139–148.

Kuipers, B.R., de Wilde, P.A.W.J., and Creutzberg, F. (1981). Energy flow in a tidal flat ecosystem. *Marine Ecology Progress Series* 5: 215–221.

Lapointe, B.E., Barile, P.J., Littler, M.M., and Littler, D.S. (2005). Macroalgal blooms on southeast Florida coral reefs II. Cross-shelf discrimination of nitrogen sources indicates widespread assimilation of sewage nitrogen. *Harmful Algae* 4: 1106–1122.

Larson, F. and Sundbäck, K. (2008). Role of microphytobenthos in recovery of functions in a shallow-water sediment system after hypoxic events. *Marine Ecology Progress Series* 357: 1–16.

Lawson, S.E., Wiberg, P.L., McGlathery, K.J., and Fugate, D.C. (2007). Wind-driven sediment suspension controls light availability in a shallow coastal lagoon. *Estuaries Coasts* 30: 102–111.

Lee, R.E. (1999). *Phycology*. Cambridge: Cambridge University Press.

Lirman, D. and Biber, P. (2000). Seasonal dynamics of algal communities in the northern Florida reef tract. *Botanica Marina* 43: 305–314.

Littler, M. M. and D. S. Littler. (1984). A relative-dominance model for biotic reefs. Proceedings of the Joint Meeting of the Atlantic Reef Committee Society of Reef Studies, Miami, Florida.

Littler, M.M. and Littler, D.S. (1988). Structure and role of algae in tropical reef communities. In: *Algae and Human Affairs* (ed. C.A. Lembi and J.R. Waaland), 29–56. Cambridge University Press.

Littler, D.S. and Littler, M.M. (2000). *Caribbean Reef Plants*. Washington, D.C.: Offshore Graphics.

Littler, M.M., Littler, D.S., and Taylor, P.R. (1995). Selective herbivore increases biomass of its prey: a chiton-coralline reefbuilding association. *Ecology* 76 (5): 1661–1681.

Lotze, H.K. and Worm, B. (2002). Complex interactions of climatic and ecological controls on macroalgal recruitment. *Limnology and Oceanography* 47: 1734–1741.

Lubchenco, J., Menge, B.A., Garrity, S.D. et al. (1984). Structure, persistence, and the role of consumers in a tropical rocky intertidal community (Taboguilla Island, Bay of Panama). *Journal of Experimental Marine Biology and Ecology* 78: 23–73.

Lüning, K. (1990). *Seaweeds – their Environment, Biogeography, and Ecophysiology*. New York: John Wiley & Sons, Inc.

Lüning, K. and Dring, M.J. (1979). Continuous underwater light measurement near Helgoland (North Sea) and its significance for characteristic light limits in the sublittoral region. *Helgoland Marine Research* 32: 403–424.

MacIntyre, I.G., Toscano, M.A., and Bond, G.B. (2004). Modern sedimentary environments, Twin Cays, Belize. *Central America. Atoll Research Bulletin.* 509: 1–12.

Malm, T., Kautsky, L., and Claesson, T. (2003). The density and survival of Fucus vesiculosus L. (Fucales, Phaeophyta) on different bedrock types on a Baltic Sea moraine coast. *Botanica Marina* 46: 256–262.

Manley, S.L. and Dastoor, M.N. (1987). Methyl halide (CH_3X) production from the giant kelp, Macrocystis, and estimates of global CH_3X production by kelp. *Limnology and Oceanograhy* 32: 709–715.

McCook, L.J. (1999). Macroalgae, nutrients and phase shifts on coral reefs: scientific issues and management consequences for the Great Barrier Reef. *Coral Reefs* 18: 357–367.

McCormick-Ray, J. (2005). Historical oyster reef connections to Chesapeake Bay – a framework for consideration. *Estuarine Coastal and Shelf Science* 64: 119–134.

McGlathery, K.J. and Pedersen, M.F. (1999). The effect of growth irradiance on the coupling of carbon and nitrogen metabolism in *Chaetomorpha linum* (Chlorophyta). *Journal of Phycology* 35: 721–731.

McGlathery, K.J., Pedersen, M.F., and Borum, J. (1996). Changes in intracellular nitrogen pools and feedback controls on nitrogen uptake in *Chaetomorpha linum* (Chlorophyta). *Journal of Phycology* 32: 393–401.

McGlathery, K.J., Anderson, I.C., and Tyler, A.C. (2001). Magnitude and variability of benthic and pelagic metabolism in a temperate coastal lagoon. *Marine Ecology Progress Series* 216: 1–15.

McGlathery, K.J., Sundbäck, K., and Anderson, I.C. (2004). The importance of primary producers for benthic nitrogen and phosphorus cycling. In: *The influence of Primary Producers on Estuarine Nutrient Cycling* (ed. S.L. Nielsen, G.T. Banta and M.F. Pedersen). Kluwer Academic Publishers.

McGlathery, K.J., Sundbäck, K., and Anderson, I.C. (2007). Eutrophication patterns in shallow coastal bays and lagoons. *Marine Ecology Progress Series* 348: 1–18.

McIvor, L., Maggs, C.A., Provan, J., and Stanhope, M.J. (2001). rbcL sequences reveal multiple cryptic introductions of the Japanese red alga Polysiphonia harveyi. *Molecular Ecology* 10: 911–919. https://doi.org/10.1046/j.1365-294X.2001.01240.x.

Menge, B.A. and Branch, G.M. (2001). Rocky intertidal communities. In: *Marine Community Ecology* (ed. M.D. Bertness, S.D. Gaines and M.E. Hay), 221–254. Massachusetts: Sinauer Associates, Inc.

Middelboe, A.L. and Binzer, T. (2004). The importance of canopy structure on photosynthesis in single- and multi-species assemblages of marine macroalgae. *Oikos* 107: 442–432.

Middelboe, A.L., Sand-Jensen, K., and Binzer, T. (2006). Highly predictable photosynthetic production in natural macroalgal communities from incoming and absorbed light. *Oecologia* 150: 464–476.

Middelburg, J.J., Barranguet, C., Boschker, H.T.S. et al. (2000). The fate of intertidal microphytobenthos carbon: an in situ 13C-labelling study. *Limnology and Oceanography* 45: 1224–1234.

Migné, A., Spilmont, N., and Davoul, D. (2004). In situ measurements of benthic primary production during emersion: seasonal variations and annual production in the Bay of Somme (eastern English Channel, France). *Continental Shelf Research* 24: 1437–1449.

Miles, A. and Sundbäck, K. (2000). Diel variation of microphytobenthic productivity in areas with different tidal amplitude. *Marine Ecology Progress Series* 205: 11–22.

Mineur, F., Arenas, F., Assis, J. et al. (2015). European seaweeds under pressure: Consequences for communities and ecosystem functioning. *Journal of Sea Research* 98: 91–108. https://doi.org/10.1016/j.seares.2014.11.004.

Mitbavkar, S. and Anil, A.C. (2006). Diatoms of the microphytobenthic community in a tropical intertidal sand flat influenced by monsoons: spatial and temporal variations. *Marine Biology* 148: 693–709.

Miyajima, T., Suzumura, M., Umezawa, Y., and Koike, I. (2001). Microbiological nitrogen transformation in carbonate sediments of a coral-reef lagoon and associated seagrass beds. *Marine Ecology Progress Series* 217: 273–286.

Mörk, E., Sjöö, G.L., Kautsky, N., and McClanahan, T.R. (2009). Top–down and bottom–up regulation of macroalgal community structure on a Kenyan reef. *Estuarine, Coastal and Shelf Science* 84: 331–336. https://doi.org/10.1016/j.ecss.2009.03.033.

Morrow, K.M. and Carpenter, R.C. (2008). Macroalgal morphology mediates particle capture by the corallimorpharian *Corynactis californica. Marine Biology* 155: 273–280.

Morton, J.P., Silliman, B.R., and Lafferty, K.D. (2020). Disease can shape marine ecosystems. In: *Marine Disease Ecology* (ed. D.C. Behringer, K.D. Lafferty and B.R. Silliman), 61–70. Oxford University Press.

Munsterman, K. S., J. E. Allgeier, J. R. Peters, D. E. Burkepile. (2021). A view from both ends: Shifts in herbivore assemblages impact top-down and bottom-up processes in coral reefs. Ecosystems 1-14, https://doi.org/10.1007/s10021-021-00612-0

Murphy, A., J. Anderson, and M. Luckenbach. (2015). Enhanced nutrient regeneration at commercial hard clam (Mercenaria mercenaria) beds and the role of macroalgae. Marine Ecology Progress Series 530: 135–151. https://doi.org/10.3354/meps11301

Newton, Christine, and Carol Thornber. (2013). Ecological impacts of macroalgal blooms on salt marsh communities. *Estuaries and Coasts* 36: 365–376. https://doi.org/10.1007/s12237-012-9565-0

Nilsson, P., Jönsson, B., Swanberg, I.L., and Sundbäck, K. (1991). Response of a marine shallow-water sediment system to an increased load of inorganic nutrients. *Marine Ecology Progress Series* 71: 275–290.

Nixon, S.W. (1995). Coastal marine eutrophication: A definition, social causes, and future concerns. *Ophelia* 41: 199–219.

Nyberg, C. and Wallentinus, I. (2005). Can species traits be used to predict marine macroalgal introduction? *Biological Invasions* 7: 265–279.

O'Connor, M.I. (2009). Warming strengthens an herbivore-plant interaction. *Ecology* 90: 388–398.

Oreska, M.P.J., Wilkinson, G.M., McGlathery, K.J. et al. (2017). Non-seagrass carbon contributions to seagrass sediment blue carbon. *Limnology and Oceanography* S3–S18. https://doi.org/10.1002/lno.10718.

Padilla, Dianna K., and Susan L. Williams. (2004). Beyond ballast water: aquarium and ornamental trades as sources of invasive species in aquatic ecosystems. *Frontiers in Ecology and the Environment* 2: 131–138. https://doi.org/10.1890/1540-9295(2004)002[0131:BBWAAO]2.0.CO;2.

Paerl, H.W. and Pinckney, J.L. (1996). A mini-review of microbial consortia: their roles in aquatic production and biogeochemical cycling. *Microbial Ecology* 31: 225–247.

Paine, R.T. (1974). Intertidal community structure: experimental studies on the relationship between a dominant competitor and its principal predator. *Oecologia* 15: 93–120.

Pandolfi, J.M., Bradbury, R.H., Sala, E. et al. (2003). Global trajectories of the long-term decline of coral reef ecosystems. *Science* 301: 955–958.

Parmesan, C. and Yohe, G. (2003). A globally coherent fingerprint of climate change impacts across natural systems. *Nature* 421: 37–42.

Peckol, P., and A. B. Putnam. (2017). Differential toxic effects of Ulva lactuca (Chlorophyta) on the herbivorous gastropods, Littorina littorea and L. obtusata (Mollusca). Journal of Phycology 53: 361–367. https://doi.org/10.1111/jpy.12507

Pedersen, M.F. (1994). Transient ammonium uptake in the macroalga *Ulva lactuca* L. (Chlorophyta): nature, regulation and consequences for choice of measuring technique. *Journal of Phycology.* 30: 980–986.

Pedersen, M.F. and Borum, J. (1996). Nutrient control of algal growth in estuarine waters. Nutrient limitation and the importance of nitrogen requirements and nitrogen storage among phytoplankton and species of macroalgae. *Marine Ecology Progress Series* 142: 261–272.

Pedersen, M., J. Borum, and F. Leck Fotel. (2010). Phosphorus dynamics and limitation of fast- and slow-growing temperate seaweeds in Oslofjord, Norway. Marine Ecology Progress Series 399: 103–115. https://doi.org/10.3354/meps08350

Pihl, L., Svenson, A., Moksnes, P.O., and Wennhage, H. (1999). Distribution of green algal mats throughout shallow soft bottoms of the Swedish Skagerrak archipelago in relation to nutrient sources and wave exposure. *Journal of Sea Research* 41: 281–294.

Pinckney, J.L., Carman, K.R., Lumsden, S.E., and Hymel, S.N. (2003). Microalgal-meiofaunal trophic relationships in muddy intertidal estuarine sediments. *Aquatic Microbial Ecology* 31: 99–108.

Platt, T., Gallegos, C.L., and Harrison, W.G. (1980). Photoinhibition of photosynthesis in natural assemblages of marine phytoplankton. *Journal of Marine Research.* 38: 687–701.

Porubsky, W.P., Velasquez, L.E., and Joye, S.B. (2008). Nutrient-replete benthic microalgae as a source of dissolved organic carbon to coastal waters. *Estuaries and Coasts* 31: 860–876. https://doi.org/10.1007/s12237-008-9077-0.

Raffaelli, D. (2000). Interactions between macro-algal mats and invertebrates in the Ythan estuary, Aberdeenshire, Scotland. *Helgoland Marine Research* 54: 71–79.

Ramus, A.P., Silliman, B.R., Thomsen, M.S., and Long, Z.T. (2017). An invasive foundation species enhances multifunctionality in a coastal ecosystem. *Proceedings of the National Academy of Sciences* 114: 8580–8585. https://doi.org/10.1073/pnas.1700353114.

Reidenbach M.A. and Timmerman R., 2019, Interactive effects of seagrass and the microphytobenthos on sediment suspension within shallow coastal bays, Estuaries and Coasts, doi:10.1007/s12237-019-00627-w

Reise, K., Herre, E., and Sturm, M. (2008). Mudflat biota since the 1930s: change beyond return? *Helgoland Marine Research* 62: 13–22.

Revsbech, N.P. (1989). An oxygen microsensor with a guard cathode. *Limnology and Oceanography* 34: 474–478.

Revsbech, N.P., Joergensen, B.B., and Brix, O. (1981). Primary production of microalgae in sediments measured by oxygen microprofile, $H^{14}CO_3^-$ fixation, and oxygen exchange methods. *Limnology and Oceanography* 26: 717–730.

Risgaard-Petersen, N. (2003). Coupled nitrification-denitrification in autotrophic and heterotrophic estuarine sediments: On the influence of benthic microalgae. *Limnology and Oceanography* 48: 93–105.

Risgaard-Petersen, N., Nicolaisen, M.H., Revsbech, N.P., and Lomstein, B.A. (2004). Competition between ammonia-oxidizing bacteria and benthic microalgae. *Applied Environmental Microbiology* 70: 5528–5537.

Roberson, L.M. and Coyer, J.A. (2004). Variation in blade morphology of the kelp *Eisenia arborea*: incipient speciation due to local water motion? *Marine Ecology Progress Series* 282: 115–128.

Round, F.E., Crawford, R.M., and Mann, D.G. (1990). *The Diatoms: Biology and Morphology of the Genera*. New York: Cambridge University Press.

Rysgaard, S., Kuhl, M., Glud, R.N., and Hansen, J.W. (2001). Biomass, production and horizontal patchiness of sea ice algae in a high-Arctic fjord (Young Sound, NE Greenland). *Marine Ecology Progress Series* 223: 15–23.

Salovius, S., Nyqvist, M., and Bonsdorff, E. (2005). Life in the fast lane: macrobenthos use temporary drifting algal habitats. *Journal of Sea Research* 53: 169–180.

Sand-Jensen, K., Binzer, T., and Middelboe, A.L. (2007). Scaling of photosynthetic production of aquatic macrophytes – a review. *Oikos* 116: 280–294.

Schaffelke, B. (1999). Particulate organic matter as an alternative nutrient source for tropical Sargassum species (Fucales, Phaeophyceae). *Journal of Phycology* 35: 1150–1157.

Scheffer, M. and van Nes, E.H. (2004). Mechanisms for marine regime shifts: can we use lakes as microcosms for oceans? *Progress in Oceanography* 60: 303–319.

Scheffer, M., Carpenter, S., Foley, J.A. et al. (2001). Catastrophic shifts in ecosystems. *Nature* 413: 591–596.

Schories, D. and Reise, K. (1993). Germination and anchorage of *Enteromorpha* spp. in the sediment of the Wadden Sea. *Helgol Meeresunters* 47: 275–285.

Schultz, J. A., R. N. Cloutier, and I. M. Côté. (2016). Evidence for a trophic cascade on rocky reefs following sea star mass mortality in British Columbia. PeerJ 4: e1980. https://doi.org/10.7717/peerj.1980

Sfriso, A., Pavoni, B., Marcomini, A., and Orio, A.A. (1992). Macroalgae, nutrient cycles, and pollutants in the lagoon of Venice. *Estuaries* 15: 517–528.

Smith, T. (2005). *The Dynamics of Coral Reef Algae in an Upwelling System*. Ph.D. dissertation, 156. Florida: University of Miami.

Smith, D.J. and Underwood, G.J.C. (2000). The production of extracellular carbohydrates by estuarine benthic diatoms: the effects of growth phase and light and dark treatments. *Journal of Phycology* 36: 321–333.

Smith, S.V., Chambers, R.M., and Hollibaugh, J.T. (1996). Dissolved and particulate nutrient transport through a coastal watershed-estuary system. *Journal of Hydrology* 176: 181–203.

Smith, J.E., Shaw, M., Edwards, R.A. et al. (2006). Indirect effects of algae on coral: algae-mediated, microbe-induced coral mortality. *Ecology Letters* 9: 835–845. https://doi.org/10.1111/j.14610248.2006.00937.x.

Smith, J.E., Kuwabara, J., Coney, J. et al. (2008). An unusual cyanobacterial bloom in Hawaii. *Coral Reefs* 27: 851.

Stal, L.J. and Caumette, P. (1992). Microbial mats. Structure, development and environmental significance. *NATO ASI Series G: Ecological Science* 35.

Steneck, R.S. (1986). The ecology of coralline algal crusts: convergent patterns and adaptive strategies. *Annual Review of Ecology and Systematics* 17: 273–303.

Steneck, R.S. and Dethier, M.N. (1994). A functional group approach to the structure of algal-dominated communities. *Oikos* 69: 476–498.

Stewart, H.L., Fram, J.P., Reed, D.C. et al. (2009). Differences in growth, morphology and tissue carbon and nitrogen of Macrocystis pyrifera within and at the outer edge of a giant kelp forest in California, USA. *Marine Ecology Progress Series* 375: 101–112.

Stimpson, J. and Larned, S.T. (2000). Nitrogen efflux from the sediments of a subtropical bay and the potential contribution to macroalgal nutrient requirements. *Journal of Experimental Marine Biology and Ecology* 252: 159–180.

Stimson, John. (2018). Recovery of coral cover in records spanning 44 yr for reefs in Kāne'ohe Bay, Oa'hu, Hawai'i. *Coral Reefs* 37: 55–69. https://doi.org/10.1007/s00338-017-1633-2

Sundbäck, K. and McGlathery, K. (2005). Interactions between benthic macroalgal and microalgal mats (review). In: *Interactions between Macro- and Microorganisms in Marine Sediments*, AGU Series: Coastal and Estuarine Studies, vol. 60 (ed. E. Kristensen, R.R. Haese and J.E. Kostka), 7–29.

Sundbäck, K., Miles, A., Hulth, S. et al. (2003). Importance of benthic nutrient regeneration during initiation of macroalgal blooms in shallow bays. *Marine Ecology Progress Series* 246: 115–126.

Sundbäck, K., Linares, F., Larson, F. et al. (2004). Benthic nitrogen fluxes along a depth gradient in a microtidal fjord: role of denitrification and microphytobenthos. *Limnology and Oceanography* 49: 1095–1107.

Sweet, M. J., J. C. Bythell, and M. M. Nugues. (2013). Algae as reservoirs for coral pathogens. PLoS ONE 8: e69717. https://doi.org/10.1371/journal.pone.0069717

Teagle, H., Hawkins, S.J., Moore, P.J., and Smale, D.A. (2017). The role of kelp species as biogenic habitat formers in coastal marine ecosystems. *Journal of Experimental Marine Biology and Ecology* 492: 81–98. https://doi.org/10.1016/j.jembe.2017.01.017.

Thompson, C.E.L. and Amos, C.L. (2002). The impact of mobile disarticulated shells of *Cerastoderma edulis* on the abrasion of a cohesive substrate. *Estuaries* 25: 204–214.

Thompson, C.E.L. and Amos, C.L. (2004). Effect of sand movement on a cohesive substrate. *Journal of Hydraulic Engineering-ASCE.* 130: 1123–1125.

Thompson, R.C., Norton, T.A., and Hawkins, S.J. (2004). Physical stress and biological control regulate the producer–consumer balance in intertidal biofilms. *Ecology* 85: 1372–1382.

Thomsen, M.S. and McGlathery, K.J. (2005). Facilitation of macroalgae by the sedimentary tube-forming polychaete *Diopatra cuprea*. *Estuarine and Coastal Shelf Science* 62: 63–73.

Thomsen, M. S., and K. McGlathery. (2006). Effects of accumulations of sediments and drift algae on recruitment of sessile organisms associated with oyster reefs. Journal of Experimental Marine Biology and Ecology 328: 22–34. https://doi.org/10.1016/j.jembe.2005.06.016

Thomsen, M.S., Gurgel, C.F.D., Fredericq, S., and McGlathery, K.J. (2005). *Gracilaria vermiculophylla* (Rhodophyta, Gracilariales) in Hog Island Bay, Virginia: a cryptic alien and invasive macroalga and taxonomic corrections. *Journal of Phycology* 42: 139–141.

Thomsen, M.S., McGlathery, K.J., and Tyler, A.C. (2006). Macroalgal distribution patterns in a shallow, soft-bottom lagoon, with emphasis on the nonnative Gracilaria vermiculophylla and Codium fragile. *Estuar Coasts* 29: 470–478.

Thomsen, M.S., McGlathery, K.J., Schwartzchild, A., and Silliman, B.R. (2009). Distribution and ecological role of the non-native macroalga *Gracilaria vermiculophylla* in Virginia salt marshes. *Biological Invasions* 11: 2303–2316.

Thomsen, M.S., Wernberg, T., Engelen, A.H. et al. (2012). A meta-analysis of seaweed impacts on seagrasses: generalities and knowledge gaps. *PLoS ONE* 7: e28595. https://doi.org/10.1371/journal.pone.0028595.

Thomsen, M.S., Wernberg, T., South, P.M., and Schiel, D.R. (2016). Non-native seaweeds drive changes in marine coastal communities around the World. In: *Seaweed Phylogeography* (ed. Z.-M. Hu and C. Fraser), 147–185. Dordrecht: Springer Netherlands https://doi.org/10.1007/978-94-017-7534-2_6.

Trimmer, M., Nedwell, D.B., Sivyer, D.B., and Malcolm, S.J. (2000). Seasonal benthic organic matter mineralisation measured by oxygen uptake and denitrification along a transect of the inner and outer River Thames estuary, UK. *Marine Ecology Progress Series* 197: 103–119.

Tyler, A.C., McGlathery, K.J., and Anderson, I.C. (2003). Benthic algae control sediment-water column fluxes of organic and inorganic nitrogen compounds in a temperate lagoon. *Limnology and Oceanography* 48: 2125–2137.

Underwood, G.J.C. and Kromkamp, J. (1999). Primary production by phytoplankton and microphytobenthos in estuaries. *Advances in Ecological Reserch* 29: 93–153.

Underwood, G.J.C., Perkins, R.G., Consalvey, M.C. et al. (2005). Patterns in microphytobenthic primary productivity: Species-specific variation in migratory rhythms and photosynthetic efficiency in mxed-species biofilms. *Limnology and Oceanography* 50: 755–767.

Valentine, J.F. and Duffy, J.E. (2006). The central role of grazing in seagrass ecology. In: *Seagrasses: Biology, Ecology, and Conservation* (ed. W.D. Larkum, R.J. Orth and C.M. Duarte), 463–501. Berlin: Springer.

Valiela, I. (1995). *Marine Ecological Processes*, 686. New York: Springer Verlag.

Valiela, I., Foreman, K., LaMontagne, M. et al. (1992). Couplings of watersheds and coastal waters: sources and consequences of nutrient enrichment in Waquoit Bay, Massachusetts. *Estuaries* 15: 443–457.

Valiela, I., McClelland, J., Hauxwell, J. et al. (1997). Macroalgal blooms in shallow estuaries: controls and ecophysiological and ecosystem consequences. *Limnology and Oceanography* 42: 1105–1118.

Vaselli, S., Bertocci, I., Maggi, and Benedetti-Cecchi, L. (2008). Effects of mean intensity and temporal variance of sediment scouring events on assemblages of rocky shores. *Marine Ecology Progress Series* 364: 57–66.

Venier, C., Figueiredo da Silva, J., McLelland, S.J. et al. (2012). Experimental investigation of the impact of macroalgal mats on flow dynamics and sediment stability in shallow tidal areas. *Estuarine, Coastal and Shelf Science* 112: 52–60. https://doi.org/10.1016/j.ecss.2011.12.035.

Verges, A., Steinberg, P.D., Hay, M.E. et al. (2014). The tropicalization of temperate marine ecosystems: climate-mediated changes in herbivory and community phase shifts. *Proceedings of the Royal Society B: Biological Sciences* 281: 20140846–20140846. https://doi.org/10.1098/rspb.2014.0846.

Veuger, B. and Middelburg, J.J. (2007). Incorporation of nitrogen from amino acids and urea by benthic microbes: role of bacteria versus algae and coupled incorporation of carbon. *Aquatic Microbial Ecology* 48: 35–46.

Viaroli, P., Bartoli, M., Bondavalli, C., and Christian, R.R. (1996). Macrophyte communities and their impact on benthic fluxes of oxygen, sulphide and nutrients in shallow eutrophic environments. *Hydrobiologia* 329: 105–119.

Volaric, M., P. Berg, M. A. Reidenbach. (2019). An invasive macroalga alters ecosystem metabolism and hydrodynamics on a tidal flat. Marine Ecology Progress Series. 628: 1-16. doi: 10.3354/meps13143

Wallentinus, I. (1984). Comparisons of nutrient uptake rates for Baltic macroalgae with different thallus morphologies. *Marine Biology* 80: 215–225.

Wernberg, T., Russell, B.D., Thomsen, M.S. et al. (2011). Seaweed communities in retreat from ocean warming. *Current Biology* 21: 1828–1832. https://doi.org/10.1016/j.cub.2011.09.028.

Whitall, D.R. and Paerl, H.W. (2001). Spatiotemporal variability of wet atmospheric nitrogen deposition to the Neuse River Estuary, North Carolina. *Journal of Environmental Quality* 30: 1508–1515.

Widdows, J., Blauw, A., Heip, C.H.R. et al. (2004). Role of physical and biological processes in sediment dynamics of a tidal flat in Westerschelde Estuary, SW Netherlands. *Marine Ecology Progress Series* 274: 41–56.

Wiencke, C., Gómez, I., and Dunton, K. (2009). Phenology and seasonal physiological performance of polar seaweeds. *Botanica Marina* 52: https://doi.org/10.1515/BOT.2009.078.

Williams, S.L. and Carpenter, R.C. (1988). Nitrogen-limited primary productivity of coral reef algal turfs: potential contribution of ammonium excreted *by Diadema antillarum*. *Marine Ecology Progress Series* 47: 145–152.

Williams, S.L. and Ruckelshaus, M.H. (1993). Effects of nitrogen availability and herbivory on eelgrass (*Zostera marina*) and epiphytes. *Ecology* 74: 904–918.

Williams, S.L. and Smith, J.E. (2007). A global review of the distribution, taxonomy, and impacts of introduced seaweeds. *Annual Review of Ecology and Systematics* 38: 327–359.

Witman, J.D. and Dayton, P.K. (2001). Rocky subtidal communities. In: *Marine Community Ecology* (ed. M.D. Bertness, S.D. Gaines and M.E. Hay), 339–366. Massachusetts: Sinauer Associates, Inc.

Worm, B. and Lotze, H.K. (2006). Effects of eutrophication, grazing, and algal blooms on rocky shores. *Limnology and Oceanography 51*: 569–576.

Worm, B., Lotze, H.K., and Sommer, U. (2000). Coastal food web structure, carbon storage, and nitrogen retention regulated by consumer pressure and nutrient loading. *Limnology and Oceanography* 45: 339–349.

Worm, B., Lotze, H.K., Hillebrand, H., and Sommer, U. (2002). Consumer versus resource control of species diversity and ecosystem functioning. *Nature 417*: 848–851.

Zedler, J.B. (1982). The ecology of southern California coastal salt marshes: a community profile. U.S. Fish and Wildlife Service, Biological Services Program, Washington, DC. FWS/OBS-81/54. pp. 110

Zedler, J.B. and West, J.M. (2008). Declining diversity in natural and restored salt marshes: a 30-year study of Tijuana Estuary. *Restoration Ecology* 16: 249–262.

CHAPTER 9

Estuarine Microbial Ecology

Byron C. Crump[1], Linda K. Blum[2], and Aaron L. Mills[2]
[1] College of Earth, Ocean, and Atmospheric Sciences, Oregon State University, Corvallis, OR, USA
[2] Laboratory of Microbial Ecology, Department of Environmental Sciences, University of Virginia, Charlottesville, VA, USA

Calcareous microbial mat comprised of cyanobacteria, diatoms, bacteria, and fungi growing attached to a submersed plant (*Bacopa* sp.) in a freshwater Everglades marsh, Florida, USA. Microbial mats provide food and hiding places for aquatic animals and control biogeochemical cycling, soil moisture, and soil formation. Photo credit: Evelyn Gaiser.

9.1 Introduction

Microbes, including Bacteria, Archaea, protists, fungi, and viruses are dominant players in estuarine ecosystems, with ecological roles that bridge the interface between ecology and chemistry. Bacteria and Archaea feature diverse metabolic pathways that control the chemical availability of essential biological elements (N, P, S, Fe, Mn, Si, etc.), and are the primary decomposers of detrital organic matter. These organisms are also prey for protists in microbial food webs that recycle indigestible detrital materials to forms more nutritious to large consumers. This chapter describes these microbes and where they are found in estuaries and discusses the ecological roles of microbes in estuarine biogeochemical cycling and food webs. It also provides an overview of how microbes metabolize organic and inorganic compounds to obtain energy and building blocks for biosynthesis, and how these processes influence nutrient cycling in estuaries.

Compared to larger organisms like zooplankton and macrobenthos, microbes in estuaries are relatively easy to sample but more difficult to analyze due to their small size. The information in this chapter was gathered using four general approaches to studying microbes in natural environments: (i) direct microscopic observation, (ii) examination of microbes grown in some type of culture medium, (iii) detection and analysis of molecules such as genes, gene products (e.g., RNA, proteins), and membrane components, and (iv) quantification of the products of metabolic activity extracted from environmental samples or experimental treatments (Kemp et al. 1993; Yates 2020). Methods relying on growth of organisms are called culture-dependent methods. Culture-independent methods rely on direct microscopy, observation or detection of molecules (e.g., phospholipid fatty acids, cell membrane proteins, cell wall components), and analysis of gene and protein sequences (e.g., metagenomics, metaproteomics). Also, the activities of microbes in natural environments are often measured as changes in the concentrations of intermediate metabolites or metabolic end products.

9.2 Diversity and Global Distribution in Estuaries

Microbes are very small, but they are also very abundant. This may seem like a trivial statement, but the reality is that the distribution of many critical ecological processes carried out

Estuarine Ecology, Third Edition. Edited by Byron C. Crump, Jeremy M. Testa, and Kenneth H. Dunton.

Companion website: www.wiley.com/go/crump/estuarine3

in estuaries is determined by the distribution of tremendous numbers of these tiny individuals of many different types. A single milliliter of estuarine water typically contains around a million individual bacterial and archaeal cells, and a square centimeter of estuarine sediment contains around a billion cells. These numbers seem remarkably high, but because the cells are so small, they occupy only a small fraction of the space in water and sediment. To visualize this, if we scale up a one-micron diameter bacteria one trillion times by volume it would be about a one-centimeter diameter sphere. If we similarly scale up one nanoliter of estuary water or sediment it would be one-square meter. At typical bacterial abundances, that square meter would contain one of the one-centimeter diameter bacteria in estuary water and about a thousand bacteria in estuary sediment.

Even though microbes are small, they exhibit a wide diversity of sizes and forms. Estuaries host all six general groups of microbes: Bacteria, Archaea, algae, protozoa, fungi, and viruses. Viruses are unique in that they are acellular and are usually considered to be nonliving entities because their reproductive capacity and energy acquisition are integrally linked to host organisms. The remaining groups fall into three Domains proposed by Woese and Fox (1977, and references therein), the Archaea, Bacteria, and Eukarya (Figure 9.1), although recently discovered novel branches of Archaea suggest that Eukarya and Archaea may belong to a single Domain (Doolittle 2020). The Archaea and Bacteria are collectively referred to as **prokaryotes** because their cells do not have a membrane-bound nucleus, and because they have a different cellular structure than eukaryotic protists, fungi, algae, plants, and animals. Prokaryotic cells have simple morphologies, but they carry out a very diverse set of processes (described below). Cyanobacteria (historically referred to as blue-green "algae") are members of the Bacteria, and are the evolutionary relatives of chloroplasts, the organelles that carry out photosynthesis in eukaryotic plants and algae. Many ecological

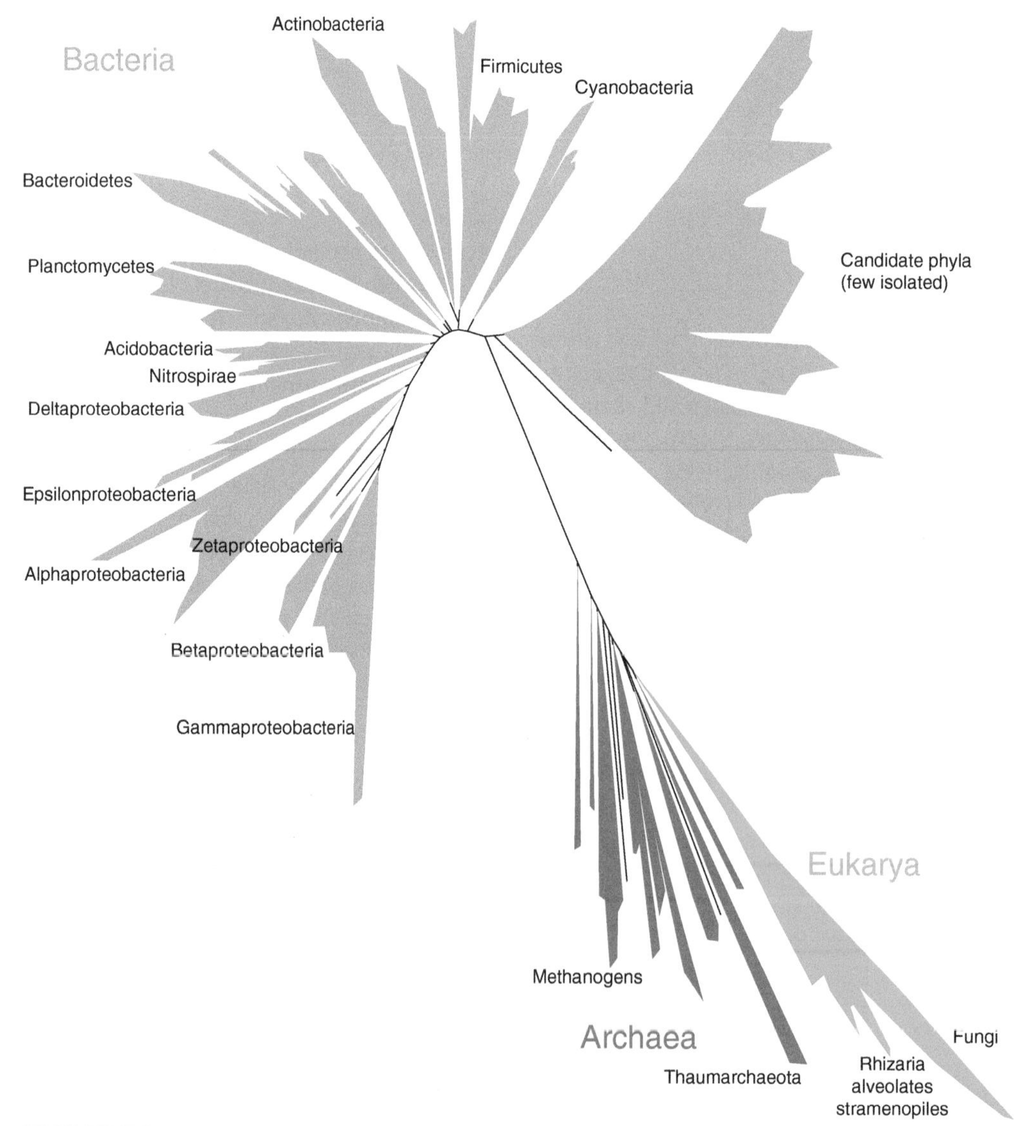

FIGURE 9.1 The three domains of life based on phylogeny as determined by similarity in the genes coding for the small subunit of the ribosome. *Source*: Modified from Hug et al. (2016).

functions are common to both Archaea and Bacteria, although several are unique to Archaea (e.g., methanogenesis). Eukaryotic microbes have diverse morphologies but carry out relatively few processes compared to the aerobic and anaerobic pathways in Bacteria and Archaea (see Section 9.4 below). The eukaryotes evolved more recently, which is consistent with their relatively low phylogenetic diversity (Figure 9.1). Together these diverse groups of microbes compose estuarine microbial communities, and each microbial group exhibits a variety of morphologies, a range of sizes, and different cell architectures.

The one common feature among this diverse collection of organisms is that they are all microscopic, meaning that a microscope is required to see individuals (with some exceptions - see below). There are a number of advantages to being microscopic in size. The acquisition of nutrients and elimination of waste is favored by high surface to volume ratios, and interactions between a cell and its environment are facilitated by a microbe's small size. Under conditions favorable for growth, the abundance of active cells can increase rapidly, altering the structure of a microbial community and the relationship between the various groups of microbes. While the environment often exerts control over rates of microbial growth (cell division) and metabolism (chemical reactions that occur within the cell), the environment also is altered by activities of the huge number of microbes, even though individual microbial cells are very small. The ability of microbes to alter their environment is a consequence of their great abundance, rapid growth, and diverse metabolisms.

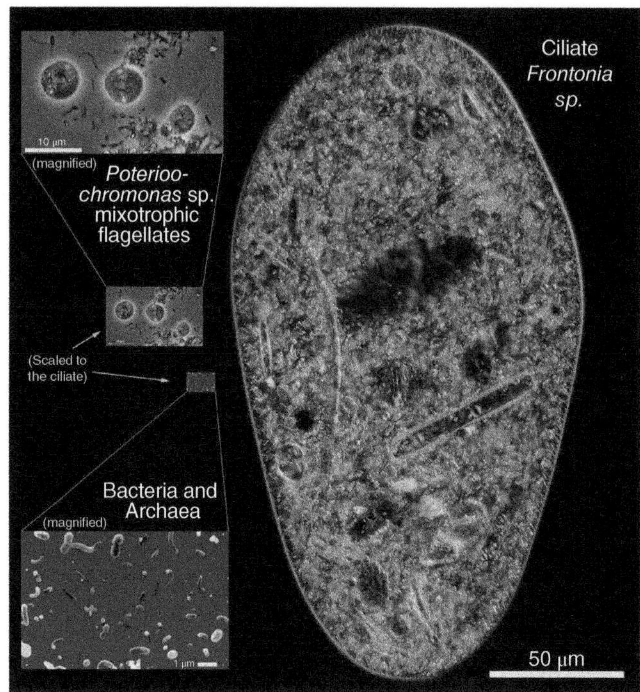

FIGURE 9.2 Comparison of representative sizes and morphologies of microbes found in estuaries shown with equal sized scale bars. Image includes a natural community of bacterioplankon (lower left) shown as comparable size and magnified (*Source*: Colorized SEM image by R. M. Morris), mixotrophic flagellates *Poterioochromonas* (upper left) which are small chrysophyte alga that photosynthesize but also consume bacteria shown as comparable size and magnified (*Source*: Image by Bob Andersen and D. J. Patterson), and a ciliate *Frontonia* sp. (right side) that recently ingested some small diatoms (*Source*: Image by Robert Berdan, **www.scienceandart.org**). Bacteria and flagellate images courtesy of **microscope.mbl.edu**.

9.2.1 Bacteria and Archaea

Bacteria and Archaea are abundant and diverse organisms that exhibit a wide range of metabolic properties, yet feature relatively simple morphologies - mainly spheres, rods, spirals, and filamentous forms (Figure 9.2). Most range in size from less than 0.2 µm to 15 µm, although some are much larger, for example the colorless sulfur oxidizing Gammproteobacteria, *Thiomargarita namibiensis*, has a diameter of 750 µm (Schulz et al. 1999). Genetically, the Archaea and Bacteria are only distantly related to each other and there is no known last common ancestor between these domains (Williams 1996, Williams et al. 2020; Figure 9.1).

Bacteria and Archaea populate all estuarine habitats, although sediments and salt marsh soils generally harbor more cells per unit volume than water (10^7–10^{10} cells cm^{-3} for typical sediments vs. 10^5–10^7 cells cm^{-3} in most estuarine waters; Table 9.1). In estuarine water, cell abundances and growth rates are generally higher than in nearby coastal marine or river waters (Figure 9.3) and tend to be highest in surface waters where productivity is elevated, and in turbid waters that support particle-attached microbes. In estuarine sediment, abundances are greatest at the sediment surface and decrease with depth (Figure 9.3) and are similar to those in freshwater sediments and coastal marine sediments.

Bacteria and Archaea are responsible for a wide range of activities from decomposition of organic matter to elemental transformations of nitrogen, sulfur, and iron. Archaea make up a small but significant fraction of prokaryotic cells in estuarine water (3–15%; Bouvier and del Giorgio 2002; Garneau et al. 2008; Mendes et al. 2014) and sediments (1–14%; Kubo et al. 2012; Webster et al. 2015). Estuarine Archaea participate in many of the same ecosystem processes as Bacteria such as organic carbon respiration, ammonia oxidation, and chemoautotrophic CO_2 fixation. Most heterotrophic Bacteria and Archaea in estuarine systems function as osmotrophs, which means they use extracellular enzymes to digest complex organic molecules into simple compounds such as sugars, amino acids, and fatty acids, and then transport these compounds through their cell walls.

Bacterial and Archaeal diversity in estuaries varies strongly with salinity in part because there is almost no overlap in composition between freshwater and marine communities (Crump et al. 1999; Doherty et al. 2017). At intermediate salinities estuarine communities form and replace communities that wash in from rivers and the coastal ocean (Herlemann et al. 2011; Campbell and Kirchman 2013). In one study of the Columbia River estuary and coastal zone, Fortunato et al. (2012) showed a

TABLE 9.1 Abundance of Bacteria and Archaea in water and sediment of estuarine and coastal marine systems as determined by direct epifluorescence microscopic counts.

Habitat	Location	Density of Bacteria + Archaea cells	References
Water	Sapelo Island, GA	$1–10 \times 10^6$ mL^{-1}	Wiebe and Pomeroy (1972)
Water	Kaneohe Bay, HI	$\geq 1 \times 10^3$ mL^{-1}	Wiebe and Pomeroy (1972)
Water	Continental Shelf, GA	$0.1–10 \times 10^4$ mL^{-1}	Wiebe and Pomeroy (1972)
Water	Kiel Fjord, Germany	$0.8–5.7 \times 10^6$ mL^{-1}	Meyer-Reil (1977)
Water	Newport R. Estuary, NC	$1.95–18.4 \times 10^6$ mL^{-1}	Palumbo and Ferguson (1978)
Water	Humber Estuary, UK	(attached) $0.46–5.6 \times 10^6$ mL^{-1} (free) $0.05–0.3 \times 10^6$ mL^{-1}	Goulder (1976)
Water	Essex, Ipswich and Parker R. estuaries, MA	$0.7–7.0 \times 10^6$ mL^{-1}	Wright and Coffin (1983)
Water	Columbia R. Estuary, OR	$1.2–5.2 \times 10^6$ mL^{-1}	Crump and Baross (1996)
Water	Columbia R. Estuary, OR	(attached) $0.0–5.1 \times 10^6$ mL^{-1} (free) $0.6–2.4 \times 10^6$ mL^{-1}	Crump et al. (1998)
Surface water	Palo Alto Salt Marsh, CA	25×10^6 mL^{-1}	Harvey and Young (1980)
Subsurface water	Palo Alto Salt Marsh, CA	5.4×10^6 mL^{-1}	Harvey and Young (1980)
Spartina raft interstitial water	Sapelo Island, GA	20×10^6 mL^{-1}	Wiebe and Pomeroy (1972)
Sediment	Sapelo Island, GA	$1–10 \times 10^6$ mL^{-1}	Wiebe and Pomeroy (1972)
Sediment	Petpaswick Inlet, Canada	$0.1–10.0 \times 10^9$ g^{-1} dry wt	Dale (1974)
Salt marsh sediment	Newport R. Estuary, NC	(surface) $8.4–10.9 \times 10^9$ cm^{-3} (20 cm) $2.2–2.6 \times 10^9$ cm^{-3}	Rublee and Dornseif (1978)
Salt marsh sediment	Ria de Aviero, Portugal	$0.4–1.7 \times 10^9$ g^{-1} dry wt	Oliveira et al. (2012)

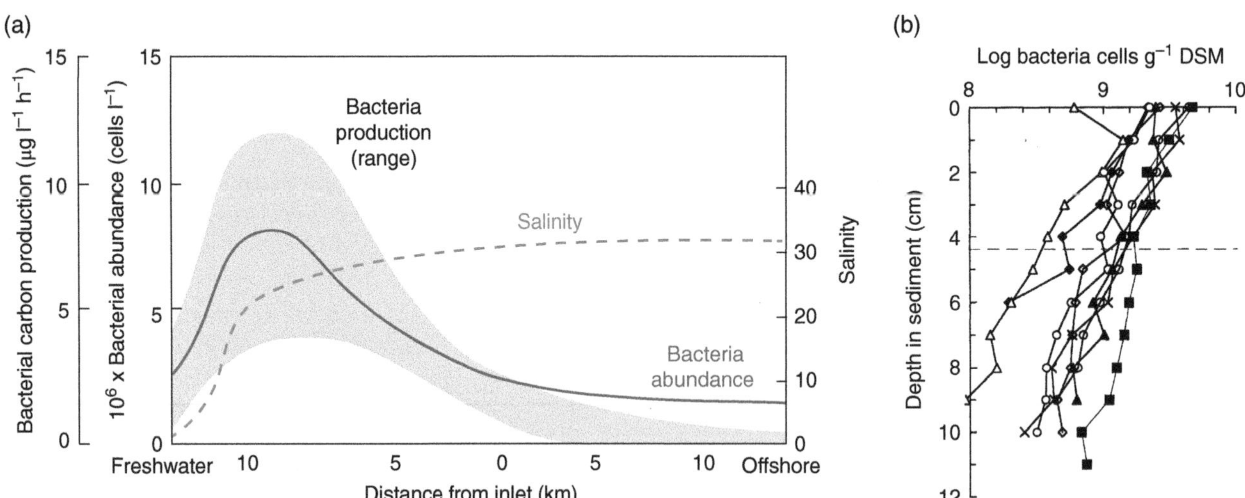

FIGURE 9.3 (a) Longitudinal distribution of bacterial abundance and production in the Essex estuary (northern Massachusetts) and connecting offshore waters. The temperature range was 15–20 °C. The yellow area represents the range of observed values for bacterial production. *Source*: Modified from Wright and Coffin (1984). (b) Bacterial cell abundance with depth in sediments expressed per gram dry sediment mass (DSM) from several different estuarine and coastal sediments. *Source*: Schmidt et al. (1998).

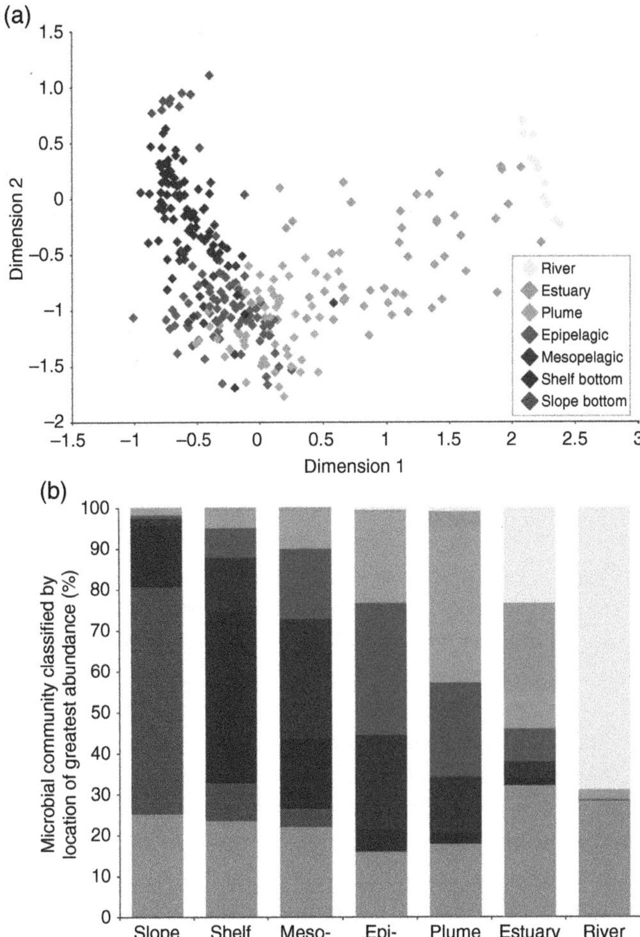

FIGURE 9.4 Prokaryotic diversity across an estuarine salinity gradient. (a) Nonmetric Multi-dimensional scaling diagram of the betadiversity of bacterial communities assessed with 16S rRNA gene amplicon sequencing of samples collected between 2007 and 2008 from the Columbia River estuary plume, and two coastal ocean transects. (b) Average percentage of bacterial 16S sequences from each location classified by most likely source environment based on maximum average relative abundance. The orange color indicates rare sequences belonging to taxa that represent <0.1% of sequences. *Source*: Modified from Fortunato et al. (2012)/Springer Nature.

strong spatial gradient in bacterial diversity from the Columbia River to the deep ocean over the continental slope, with highly variable communities in the estuary and plume (Figure 9.4a). When the bacterial taxa in these samples were classified by their most likely source environment (based on where they had highest average relative abundance), only a portion of the bacteria in the Columbia River estuary were classified as "estuarine" and the rest were allochthonous organisms that washed in from other environments (Figure 9.4b). These results contrast with those from estuaries with longer residence times, like the Baltic Sea and Chesapeake Bay, where estuarine bacterial diversity was dominated by native estuarine microbial communities that differ from river and ocean communities (Herlemann et al. 2011; Wang et al. 2020).

9.2.2 Protists

The term protist is a catch-all term that describes a diverse group of single-celled eukaryotic organisms that span three orders of magnitude in size and several trophic levels. Currently, the most common estuarine protists are classified as Stramenopiles (many flagellates, many phytoplankton and benthic microalgae), Alveolates (ciliates, dinoflagellates), and Rhizaria (some flagellates, amoeba, foraminifera, radiolaria), although some estuarine protists belong to other taxonomic groups including cryptophytes, haptophytes, and other flagellates (Burki et al. 2020). All phytoplankton and benthic microalgae are considered protists, with the exception of cyanobacteria, and are described in detail in Chapters 4 and 8.

Flagellates are the smallest protists in estuarine microbial communities. They are generally 2–20 μm in diameter or length, they move by flagellae (singular, flagellum), and many feed by phagocytosis, which involves physically engulfing prokaryotes and other small particles they encounter as they move through water and sediment. Many flagellates are purely heterotrophic while others live only by photosynthesis. The group of flagellates represented by the chrysophyte in Figure 9.2 actually carries out both processes; they are called mixotrophs. Flagellates are much less abundant than their prey. In estuaries their numbers are around 10^3 mL^{-1} of water and 10^6 cm^{-3} of sediment and are generally 1000 times less abundant than Bacteria and Archaea.

Ciliates are an order of magnitude larger than the flagellates and small algae that they feed upon. They range in size from 30 to 1000 μm, they move by cilia, and many also feed with their cilia by sweeping food into a funnel-shaped oral groove and then engulfing food via phagocytosis. The organism shown in Figure 9.2 is about 200 μm in length and has been feeding on small diatom cells. Ciliates are less abundant than flagellates: 1–10 cells mL^{-1} of water and 10^3 cells cm^{-3} of sediment. Ciliates are important consumers in estuaries and are a critical link in the food web between smaller algae and bacteria and larger organisms because they are large enough to be captured by zooplankton and filter-feeding benthic animals.

9.2.3 Fungi

Fungi are eukaryotic, heterotrophic and mostly microscopic. They have complex life cycles and sometimes have macroscopic reproductive structures called "fruiting" bodies. Mycelia of fungi inhabit plant surfaces, detrital particles, soils, sediments, and water in estuaries, but their abundance in these environments is not well understood because quantifying them is technically challenging. All fungi are heterotrophic, but they fulfill several ecological roles in estuaries. In sediments and soils they consume detrital organic matter and recycle nutrients. For example, over 100 species of filamentous fungi are associated with standing-dead plants of salt marsh grasses in the genus *Spartina* (Kohlmeyer and Volkmann-Kohlmeyer 2002), and are thought to be largely responsible for their decomposition (Newell et al. 1985). Also, many soil and sediment fungi form mutualistic mycorrhizal associations with marsh grasses (Carvalho et al. 2004), seagrasses (Vohník et al. 2019), and mangroves (Sengupta and Chaudhuri 2002). In water, estuarine fungi include flagellated, zoosporic (single-celled) organisms such as Cryptomycota and Chytridiomycota, and filamentous organisms associated with detrital particles (Blum et al. 1988). These organisms are thought to be saprotrophic (digest food extracellularly), but many are known to be parasites of cyanobacteria, diatoms, and other eukaryotes (Grossart et al. 2016). Fungi in estuarine water vary with salinity, with those in water below about 8‰ resembling freshwater communities, and those in water above 8‰ resembling marine communities (Rojas-Jimenez et al. 2019).

9.2.4 Viruses

Viruses are infectious agents with genomes of DNA or RNA that can alternate between two distinct states, intracellular and extracellular (Madigan et al. 2018). They are metabolically inert and rely on living cells (hosts) to replicate. Virus particles are usually between 20 and 200 nm long, and they vary widely in shape and size. Viruses are exceptionally diverse and probably infect all organisms in estuaries, including animals, fungi, plants, Bacteria, and Archaea.

Viruses are abundant in estuaries, ranging from 10^4 to over 10^9 mL^{-1} of water (Wommack and Colwell 2000; Suttle 2005) and from 10^7 to over 10^{10} cm^{-3} of sediment (Hewson et al. 2001; Helton et al. 2012). Also, a large but highly variable fraction (0–84%) of Bacteria in estuaries carry dormant lysogenic viruses called prophages that may be induced to become lytic, form new virus particles, and destroy the cells (Weinbauer et al. 2003). Metagenomic studies of estuarine virus communities ("viromes") suggest that most viruses in estuaries infect bacteria ("bacteriophage"), and that, like bacteria, viral diversity is strikingly different between marine and freshwater environments (Cai et al. 2016). The virus to bacteria ratio in estuarine water is usually greater than 1 and can exceed 85 in nutrient rich, highly productive waters (Hewson et al. 2001).

Ecologically, viruses that infect microbes function as grazers by contributing to the mortality of their hosts (Fuhrman 1999) and by generating detrital organic matter (Zhang et al. 2014). Since their reproduction requires host metabolic activity, they tend to cause the greatest mortality for the most rapidly growing organisms (known as the "kill the winner" hypothesis). At the same time, viruses enhance the growth of uninfected microbes by eliminating competitors and by providing a source of organic substrates from the lysed host cells. Viruses may also affect microbial community structure because viruses tend to infect specific hosts, resulting in a dynamic rearrangement of microbial communities. When the density of the natural virus assemblage was altered, Liu et al. (2017) showed dramatic changes in estuarine bacterioplankton community composition. Thus, viruses play an important role in carbon cycling and regulation of bacterial production within estuaries (Fuhrman and Noble 1995), and the structure of bacterial communities (Pradeep Ram and Sime-Ngando 2008).

9.3 Factors Controlling Productivity and Community Composition

Organisms as small as microbes have ecosystem-level impacts because they are so abundant and because the processes they mediate occur so rapidly. The primary ecosystem function of estuarine microbes is to decompose and respire non-living (detrital) organic matter. When they do this, three things happen. First, detrital carbon is converted to carbon dioxide and released to the atmosphere. Second, nutrients like nitrogen bound up in the organic matter are mineralized and released for primary producers to use again (a process called **nutrient recycling or regeneration**). And third, microbes produce new living biomass that feeds back into estuarine food webs. Decomposition of detrital organic matter is accomplished by a microbial food web of interacting Bacteria, Archaea, heterotrophic protists (flagellates and ciliates), fungi, and viruses. Bacteria and Archaea also carry out a large number of specialized metabolic processes that affect estuarine chemistry and control the cycling of nutrients including nitrogen, sulfur, and iron. This section provides an overview of microbial food webs and the ecosystem-scale processes they carry out.

9.3.1 Microbial Food Webs

Bacteria, Archaea, protists, fungi, and viruses in estuaries are organized into an intricate microbial food web that captures the carbon and energy of detrital organic matter and passes it from one trophic level to another. An idealized scheme of detrital carbon flow through a planktonic food web (Figure 9.5) emphasizes two different modes of carbon transfer: as dissolved organic carbon (DOC) that is consumed by prokaryotes and fungi, and as particles that are colonized with heterotrophic microbes and consumed by larger organisms that directly ingest particles. In estuaries, DOC is produced from many potential sources. Prokaryotes recover DOC, and these organisms, in turn, are consumed primarily by heterotrophic and mixotrophic flagellates that are then consumed by ciliates and larger organisms like copepods. Viruses infect and destroy ("lyse") organisms at all levels of this detrital food web, and when this happens, they release DOC that can subsequently be consumed by prokaryotes again.

In the laboratory, when small microcosms are set up with prokaryotes, flagellates, and ciliates the populations oscillate in classic predator-prey relationships. The same changes

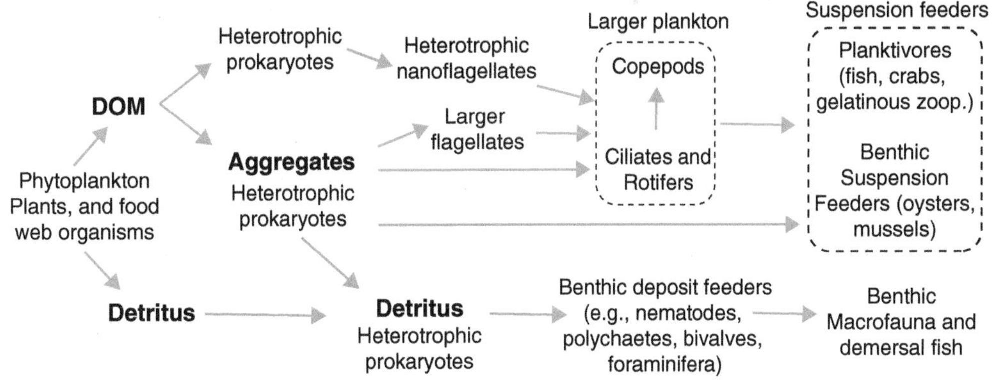

FIGURE 9.5 Idealized scheme of carbon flow from detritus and dissolved organic matter (DOM; including DOC) into estuarine microbial food webs and higher trophic levels.

likely take place in estuaries, but it is very difficult to sample because the same water mass would have to be measured regularly for weeks to identify the cycles. There is at least one site, Limfjord, a shallow marine bay in Denmark (Figure 9.6), where the cycles of prokaryotes, flagellates, and ciliates can be in synchrony for weeks. In this system, prokaryotes ranged from 2.4 to 14×10^6 mL^{-1}, flagellates from 200 to 17,000 mL^{-1}, and ciliates from 1.5 to 160 mL^{-1}. Each prokaryote peak was followed by a flagellate peak 3–7 days later, and each flagellate peak was followed by a ciliate peak within 4–6 days. These data are strong evidence for the control of prokaryote abundance by the grazing of nanoflagellates. They also indicate that nanoflagellates, too, are likely controlled by grazing by ciliates (Lynch and Hobbie 1988).

Grazing rates on prokaryotes in microbial food webs are generally similar to prokaryote growth rates, and when these rates are balanced it means that protists and viruses consume or kill all the newly produced cells and, in this way, control exert top-down control on prokaryote populations. This is a common feature of low productivity systems, shown by the points on the 1 : 1 line near the origin in Figure 9.7. In more productive estuaries where prokaryote production is higher, rates of growth and grazing can be unbalanced (Figure 9.7), which creates booms and busts in abundances of prokaryotes and their grazers (Figure 9.6). Across these cycles of abundance, however, mortality in microbial food webs due to protistan grazers and lytic viruses roughly keeps pace with production and growth, preventing huge shifts in abundance, and resulting in rapid turnover of microbial communities on the scale of days. This allows rapid reshaping of the species composition of these communities in response to any factor that influences the relative rates of growth or mortality of individual populations such as changes in resources, grazer prey selection, or grazer avoidance abilities (Strom 2008).

9.3.2 Organic Carbon Decomposition and Trophic Dynamics

One critical role of microbial food webs in estuaries is the decomposition of organic matter. Heterotrophic prokaryotes and fungi oxidize organic carbon and recycle organic nitrogen,

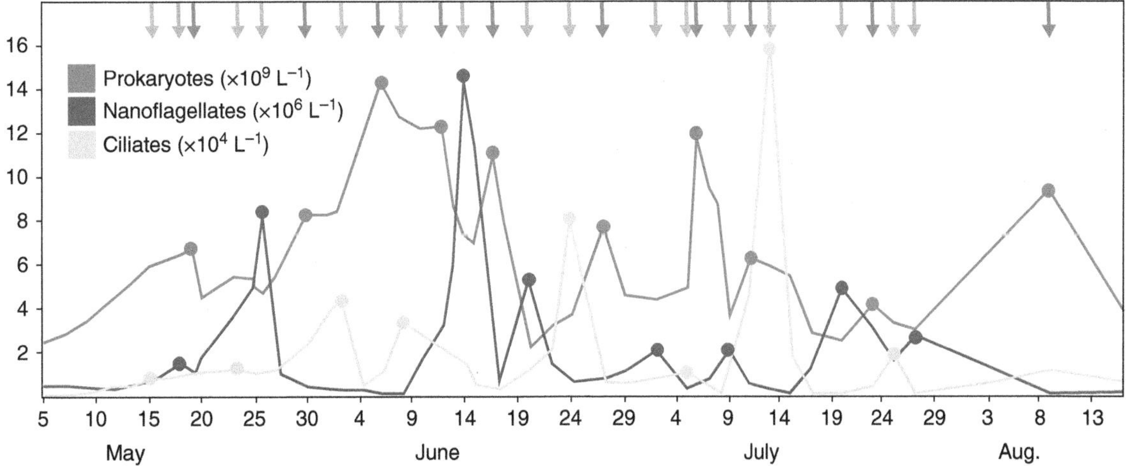

FIGURE 9.6 Prokaryotes, heterotrophic nanoflagellates, and ciliates in surface waters of Limfjord, Denmark in 1983 (mean concentrations at 1 and 2 m). Circles and arrows indicate temporal peaks in abundance. *Source:* Modified from Andersen and Sørensen (1986) and Strom (2000).

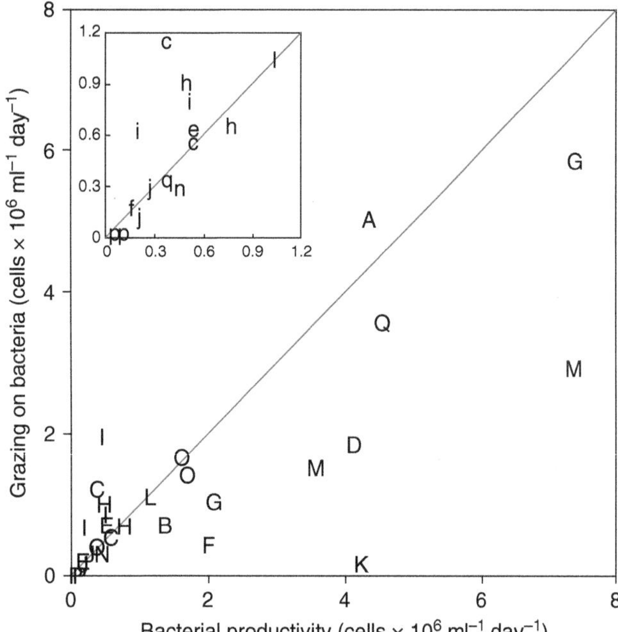

FIGURE 9.7 Bacterial productivity and grazing rates for 17 marine studies. *Source*: Summary by Strom (2000)/With Permission of John Wiley & Sons.

phosphorous and other nutrients into inorganic forms for subsequent uptake by primary producers. During decomposition, prokaryotes respire and mineralize much of their organic matter diet to acquire energy for growth, but as they grow, they incorporate a substantial portion of their diet into new biomass. Then, grazers in the microbial food web, such as heterotrophic nanoflagellates and ciliates in water and amphipods and isopods in sediment, consume this biomass, mineralizing much of the carbon for energy, and incorporating the remainder into their own biomass. Thus, all of these organisms cooperate as a community in mineralization, oxidizing most of the organic matter they consume to CO_2. In each trophic transfer, a portion of the organic carbon is mineralized. In this way, microbial food webs act as so-called "sinks" for organic matter.

Microbial food webs in estuaries are supported by dissolved organic matter (DOM) and particulate organic matter (POM) (Bauer and Bianchi 2011). **Autochthonous** (internally produced) sources of DOM to estuaries are generated by phytoplankton and benthic microalgae that release DOM as they grow, particularly under nutrient-limited conditions, and this material can account for a large fraction of the carbon fixed during photosynthesis (Azam et al. 1983). Macroalgae, seagrasses, and marsh grasses also release significant quantities of DOM. Additionally, DOM is released when primary producers and other organisms are grazed, either from the front end ("sloppy feeding") or back end (excretion) of animal grazers. Autochthonous POM (i.e., detritus) also releases DOM that can be consumed by microbes.

Besides autochthonous sources, estuaries also receive **allochthonous** DOM and POM from land and from coastal oceans. Terrestrial organic matter is often the largest allochthonous source of organic matter to estuarine microbial food webs. Estuaries typically receive runoff from watersheds that have much larger surface areas compared to the size of the estuaries. For example, Chesapeake Bay (11,600 km^2) receives input from a watershed 166,000 km^2 in area, supplying approximately 21 g of organic carbon per m^2 of the Bay per year to fuel microbial heterotrophic production and 13 g of nitrogen per m^2 of the Bay per year to support primary production (Kemp et al. 1997; Boynton et al. 1995). In some estuaries, such as the rapidly flushed Tomales Bay in California, USA, the amount of allochthonous organic matter exceeds that of the organic matter produced autochthonously (Smith and Hollibaugh 1997). Microbial food webs convert this allochthonous detritus to living biomass that can support higher trophic levels in highly productive estuarine ecosystems.

Most estuaries have fringing marshes and beds of submerged aquatic vegetation (SAV; see Chapters 5 and 6). Macroalgae, seagrasses, and marsh grasses provide large inputs of POM (i.e., detritus) during periods of senescence in the fall and winter (Bianchi and Bauer 2011). As this POM decays, the water-soluble fractions are released rapidly during a "leaching" phase (Valiela et al. 1984) and consumed by heterotrophic prokaryotes, generally within several days of death of plant material. As the water-soluble fractions decrease, the percentage of insoluble materials rises. The second stage of decay, the decomposer phase, occurs more slowly as fungi and prokaryotes colonize the tissues and consume the structural compounds hemicellulose and cellulose (Figure 9.8). This enzymatic decay weakens the tissue structure, and subsequent mechanical grinding of tissues by amphipods, grass shrimp, and other invertebrate macrofauna reduces particle size and increases surface area for further microbial colonization. During the final stage of decay, the refractory phase, very complex, difficult to decompose materials remain (lignin, waxes, resins, and pigments) and decay proceeds more slowly than in the second stage. In both of these later stages, attached microbes continue to decompose the detritus and accumulate

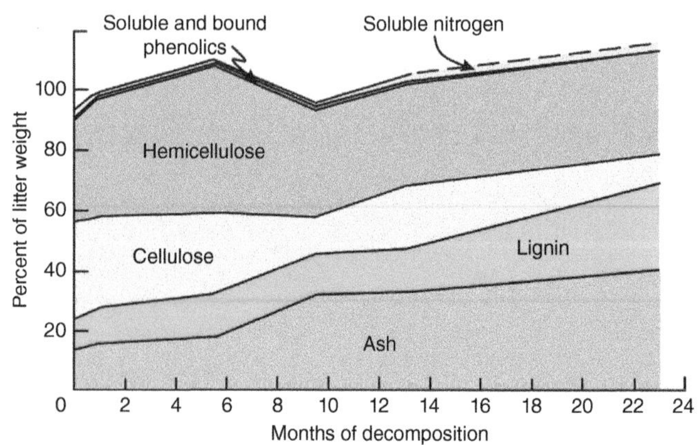

FIGURE 9.8 Cumulative percentage of the major chemical constituents of *Spartina alterniflora* litter during decay. *Source*: Modified from Valiela et al. (1984).

biomass, converting the complex organic compounds into microbial cells. Invertebrates ingest the POM, but they digest the microbes as their main source of nutrition. This process of decomposition is universal for all forms of POM, regardless of the source, although decomposition rates vary widely depending on organic matter characteristics and environmental conditions.

9.3.3 Microbes and Productivity

Microbes regulate rates of ecosystem primary production both directly and indirectly in estuaries. Eukaryotic phytoplankton and benthic microalgae (Chapters 4 and 8), compete directly with plants and macroalgae for inorganic nutrients. In addition, cyanobacteria contribute to estuarine primary production, accounting for an estimated 2–28% of the total chlorophyll *a* pool in estuarine water (Teixeira and Gaeta 1991; Iriarte 1993; Sin et al. 2000). Also, in some estuarine environments, chemoautotrophic organisms that use chemical energy to fix carbon, such as sulfur oxidizers and nitrifiers (see below), can be important sources of organic matter (Kristensen and Hansen 1995; Evrard et al. 2008).

Microbes indirectly regulate primary productivity by mineralizing organic matter and regenerating the inorganic N and P that often limits the rate of primary production in estuaries. A significant fraction of estuarine primary production is supported with regenerated nutrients, particularly in estuaries with longer residence times (Nixon et al. 1996; Christian and Thomas 2003). For example, Baird et al. (2004) calculated that molecules of N and P were recycled thousands of times in the food web of the German Wadden Sea before being exported from the system. Bacteria and other heterotrophic microbes are critical to this recycling process by degrading detrital organic matter and mineralizing nutrients for re-use by primary producers.

Microbes are also important secondary producers. In estuaries, there is a positive relationship between prokaryote abundance and protistan grazers, and between prokaryote growth and protistan grazing rates (Sanders et al. 1992). Incubation experiments using ^{14}C as a tracer of prokaryote biomass confirm that prokaryotes contribute to the diet of zooplankton such as copepods and rotifers (Havens et al. 2000; Work et al. 2005). The importance of prokaryote biomass in zooplankton diets remains a subject of debate, but any zooplankton grazing links the prokaryote-based microbial food web (supported by DOM) to higher trophic levels. This pathway of carbon and energy flow from dissolved organic matter to new microbial cells to higher trophic levels is sometimes referred to as the **microbial loop**. This loop may add as much as 30% more energy to higher trophic levels beyond that from detrital and herbivorous pathways alone (Pomeroy 1974; Azam et al. 1983). As a result of the retention of microbial community secondary production within the ecosystem, energy and organic matter may be conserved for fish and shellfish that are exploited by humans.

9.3.4 Environmental Factors Affecting Microbial Communities

Microbial abundance, diversity, and rates of ecological processes can be regulated by environmental factors including temperature, light, availability of oxygen and other electron acceptors (see Section 9.4), and organic matter quality (see above). **Temperature** regulates the rates at which metabolic processes occur, including decomposition. Increases in temperature are typically accompanied by increases in process rates over the range of temperatures supporting microbial growth in estuaries (Nedwell 1999; Cavicchioli 2002). Temperatures in estuaries typically range from −1.4 °C (the freezing point of seawater) to *ca.* 30 °C (Mongillo 2000). Seasonal increases in temperature are generally associated with elevated microbial growth and decomposition rates. Latitudinal variation in climatic temperature should also affect microbial activity and community composition, but systematic, comparative studies of estuaries with different climates are rare (e.g., Newell et al. 2000; Blum et al. 2004).

The presence or absence of **oxygen** has a major impact on rates of metabolic activity and is an important determinant of community structure (Jørgensen 1980). All aerobic members of microbial communities flourish in the presence of oxygen, using it as an electron acceptor for respiration. But, there are many estuarine environments where the rate of microbial oxygen consumption is greater than the rate of reoxygenation from the atmosphere, such as sediments and flooded soils. When microbial activity removes the oxygen from an environment (making it hypoxic or anoxic), most eukaryotic organisms can no longer survive, and the environment becomes dominated by prokaryotes that are able to use alternate electron acceptors (other than oxygen) to respire organic matter. Prokaryotes that use alternate electron acceptors for anaerobic respiration generally make use of an oxidized form of a single element to do so, such as the oxidized nitrogen in NO_3^- and the oxidized sulfur in SO_4^{2-}. There is a specific order of use for these electron acceptors based on their availability and oxidation-reduction potential (redox potential or Eh). Nitrate is the most energetically favorable alternative to oxygen, followed by Mn^{4+}, Fe^{3+}, SO_4^{2-}, CO_2, and some organic molecules (e.g., pyruvate, acetaldehyde). Carbon oxidation using each of these electron acceptors provides less energy than the previous one (Table 9.2). As electron acceptors are used up, the Eh of the environment decreases, and the species composition of prokaryotic communities becomes dominated by organisms that are able to use the most energetically-favorable and available electron acceptor. Thus, microbial community structure and the processes that communities carry out are altered dramatically by the absence of oxygen, and subsequent changes in Eh (e.g., Jørgensen 2006; Lin et al. 2007; Glud 2008).

Light (photosynthetically active radiation, including wave lengths of 400–700 nm) varies with depth and water clarity (Chapter 4), and directly influences the photosynthetic community but also indirectly affects heterotrophic microbial

TABLE 9.2 Chemical equations and energy yields of (ΔG) oxidation-reduction reactions carried out by microbes in estuaries. DNRA is dissimilatory nitrate reduction to ammonia, Anammox is anaerobic ammonia oxidation.

Nitrification (ammonia oxidation)	$NH_4^+ + 3/2\ O_2 \rightarrow NO_2^- + 2H^+ + H_2O$	$\Delta G = -275$ kJ mol^{-1} N
Nitrification (nitrite oxidation)	$NO_2^- + 1/2\ O_2 \rightarrow NO_3^-$	$\Delta G = -75$ kJ mol^{-1} N
Anammox	$NH_4^+ + NO_2^- \rightarrow N_2 + 2\ H_2O$	$\Delta G = -358$ kJ mol^{-1} NH_4^+
Sulfide Oxidation	$HS^- + H^+ + ½\ O_2 \rightarrow S^0 + H_2O$	$\Delta G = -209$ kJ mol^{-1} S
Sulfur Oxidation	$S^0 + 1.5\ O_2 + H_2O \rightarrow 2\ H^+ + SO_4^{2-}$	$\Delta G = -628$ kJ mol^{-1} S
Manganese oxidation	$Mn^{2+} + 1/2\ O_2 + H_2O \rightarrow MnO_2 + 2H^+$	$\Delta G = -68.2$ kJ mol^{-1} Mn
Iron oxidation	$Fe^{2+} + ¼\ O_2 + 5/2\ H_2O \rightarrow Fe(OH)_3 + 2H^+$	$\Delta G = -18$ kJ mol^{-1} Fe (pH 7)
Coupled S-Fe oxidation	$4\ FeS_2 + 15\ O_2 + 14\ H_2O \rightarrow 4\ Fe^{3+} + 3\ OH^- + 16\ H^+ + 8\ SO_4^{2-}$	
Photosynthetic S oxidation	$CO_2 + 2\ H_2S +$ light energy $\rightarrow CH_2O + 2S^0 + H_2O$	
Photosynthetic S oxidation	$CO_2 + S^0 + 2\ H_2O + O_2 +$ light energy $\rightarrow CH_2O + 2H^+ + SO_4^{2-}$	
Aerobic respiration	$CH_3COO^- + 2\ O_2 \rightarrow HCO_3^- + CO_2 + H_2O$	$\Delta G = -849$ kJ mol^{-1} acetate[c]
Denitrification	$5\ CH_3COO^- + 8\ NO_3^- + 8\ H^+ \rightarrow 5\ CO_2 + 4\ N_2 + 5\ HCO_3^- + 9\ H_2O$	$\Delta G = -797$ kJ mol^{-1} acetate[a]
Manganese reduction	$CH_3COO^- + 4\ MnO_2 + 2\ HCO_3 + 3\ H^+ \rightarrow 4\ MnCO_3 + 4\ H_2O$	$\Delta G = -737$ kJ mol^{-1} acetate[c]
Iron reduction	$CH_3COO^- + 8\ Fe(OH)_3 + 15\ H^+ \rightarrow 8\ Fe^{2+} + 2\ HCO_3^- + 20\ H_2O$	$\Delta G = -573$ kJ mol^{-1} acetate[b]
DNRA	$CH_3COO^- + NO_3^- + 2\ H^+ \rightarrow CO_2 + NH_4^+ + HCO_3^-$	$\Delta G = -500$ kJ mol^{-1} acetate[a]
Sulfate reduction	$CH_3COO^- + SO_4^{2-} + H^+ \rightarrow HS^- + HCO_3^- + CO_2 + H_2O$	$\Delta G = -52$ kJ mol^{-1} acetate[c]
Methanogenesis	$CH_3COO^- + H_2O \rightarrow CH_4 + HCO_3^-$	$\Delta G = -31$ kJ mol^{-1} acetate[c]
Nitrogen Fixation	$N_2 + 3H_2 \rightarrow 2NH$	$\Delta G = +160$ kJ mol^{-1}

[a] Castro-Barros et al. (2017).
[b] Wang et al. (2018).
[c] Lovely and Phillips (1988).

communities. Increases in phytoplankton and benthic microalgae generally lead to increases in heterotrophic microbes in water and sediment because these photoautotrophs provide much of the organic carbon for heterotrophic microbial growth (Dietrich and Arndt 2000; Ruiz-González et al. 2013). However, in near-surface waters high light levels, primarily in the ultraviolet region of the spectrum (UV, 280–400 nm), can have negative effects on microbial communities through direct damage to cell components, and via generation of harmful oxygen radicals during photooxidation of dissolved organic matter (Cory and Kling 2018). Small non-pigmented cells are particularly sensitive to UV damage to DNA (Jeffrey et al. 1996). Heterotrophic flagellates are inhibited by UVB (280–315 nm) and UVA (315–400 nm), and ciliate populations are reduced by UVB, which can cause shifts in the relative abundances of these grazers and their prey (Mostajir et al. 1999; Ruiz-González et al. 2013). Viruses can be inactivated by UV, but in some studies lytic virus numbers increase in response to UVB by activation of the lytic cycle (Maranger et al. 2002).

Organic matter quality also impacts microbial growth and the rate of decomposition as mentioned previously (Section 9.3.2). Freshly produced organic matter tends to have the highest quality, so the quality of organic matter in estuaries can vary with season (MacMillin et al. 1992). For example, in temperate estuaries organic matter quality peaks in spring and summer when primary production is greatest. One measure of organic matter quality is the carbon to nitrogen ratio (C:N; Bianchi and Bauer 2011). In estuaries, organic matter tends to have a high C:N (typically greater than 20:1), but microbe biomass has a C:N ratio of approximately 10:1. Heterotrophic prokaryotes can overcome this limitation by using inorganic nitrogen for growth including oxidized (NO_2^-, NO_3^-) and reduced (NH_4^+) forms of nitrogen. When heterotrophic microbes assimilate inorganic nitrogen (and other macronutrients, e.g., P, K, Ca) it can reduce the availability of these nutrients and limit plant growth. At the same time this ability gives microbial food webs greater access to low quality (i.e., low C:N) organic matter and enhances estuarine detrital food webs.

9.4 Metabolic Diversity and Element Cycling

Microbes are almost exclusively responsible for the recycling of macronutrients and micronutrients in estuaries, including those essential for primary production (Chapters 4–8). Nutrient

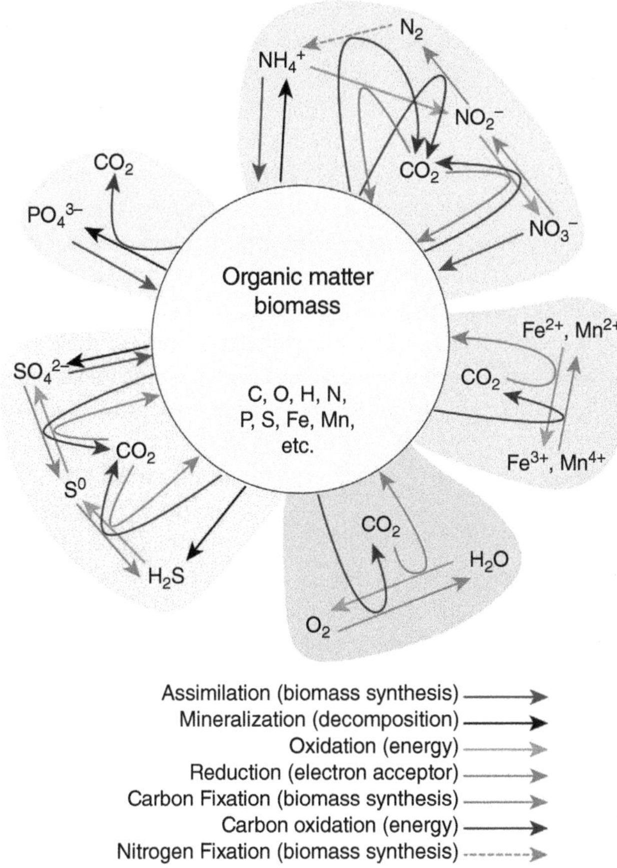

FIGURE 9.9 Material flows in an estuary emphasizing microbial reactions that dominate the processes.

cycling is closely linked with organic matter production and decomposition because all these processes require energy transfers and material transformations involving the many elements that compose biomass including N, S, P, and Fe. Moreover, the transformations of each of these elements is linked to the dominant element in biomass : carbon (Figure 9.9). Most transformations involve redox reactions in which one molecule is chemically oxidized and another reduced (Chapter 3). Inevitably, any study of microbial function in ecosystems leads to consideration of the cellular processes involved in these redox reactions.

Complete biogeochemical cycling of most elements is accomplished by the combined action of autotrophs and heterotrophs. For carbon in estuaries, this pairing is mainly accomplished by photoautotrophs (e.g., phytoplankton) that reduce the carbon in CO_2 to organic carbon, and aerobic heterotrophic prokaryotes and other heterotrophic organisms that oxidize organic carbon to CO_2. For other elements, chemoautotrophs such as nitrifying and sulfur oxidizing bacteria and archaea oxidize the reduced inorganic forms of those elements (e.g., NH_4^+, H_2S) in order to generate energy and reducing power (i.e., electrons) to fix carbon and build biomass. Subsequently, specialized heterotrophic bacteria and archaea use the oxidized forms (e.g., NO_3^-, SO_4^{2-}) produced by chemoautotrophs as electron acceptors for the oxidation of organic carbon compounds, thus recycling the elements to reduced forms. The nitrogen and sulfur cycles are excellent examples of this combination of oxidation and reduction reactions, although similar paired autotrophic and heterotrophic metabolic processes exist for the cycling of Fe, Mn, and many other elements. Phosphorus is also cycled through organic and inorganic forms, but unlike most other elements it does not change oxidation states. The following sections describe the metabolic processes that are particularly important in estuarine nutrient cycles.

9.4.1 Carbon Fixation and Mineralization

The simplest form of the carbon cycle involves two processes: carbon fixation and carbon mineralization. When carbon fixation occurs, CO_2 is converted to organic molecules (i.e., biomass) by autotrophic organisms. Autotrophs comprise two major categories: photoautotrophs and chemoautotrophs. The former obtain energy from light and convert it to chemical energy by oxidizing a reduced molecule (such as H_2O or H_2S) and using the charged electrons to generate energy stored as ATP, and reducing power stored in cofactors like NADPH. Microbial photoautotrophs include "oxygenic photosynthesizers" like phytoplankton and benthic microalgae that produce O_2 when they use light energy to oxidize the oxygen in water to obtain electrons. A second group of photoautotrophs are the prokaryotic "anoxygenic photosynthesizers" that use light energy to oxidize H_2S instead of water and produce elemental sulfur rather than O_2 (although some also use S, H_2, a few organics, or Fe^{2+} as the electron source).

Chemoautotrophs also fix CO_2 into biomass, but the source of energy for this process comes from the oxidation of a reduced chemical compound such as NH_4^+, H_2S, and Fe^{2+}. In most estuaries, the amount of carbon fixed by chemoautotrophs is far less than carbon fixed by photoautotrophs. However, chemoautotrophic processes such as ammonium oxidation, sulfur oxidation, and iron oxidation are critical processes in the biogeochemical cycling of elements other than carbon.

Organisms that use organic carbon formed during primary production are chemoheterotrophs, or more commonly, heterotrophs. Heterotrophs are responsible for the decomposition of organic matter and mineralization of associated elements to inorganic forms. In general, heterotrophs use organic compounds as sources of carbon, energy, and reducing power. When heterotrophs use oxygen to oxidize organic carbon their metabolism is called **aerobic respiration**, and when they use alternate electron acceptors it is called **anaerobic respiration**. Through mineralization, heterotrophic microbes mineralize elements such as C, N, S, and P that were incorporated in the organic matter and return them to the environment in a process called **nutrient regeneration**. Heterotrophic activity generates inorganic products such as CO_2, NH_4^+, H_2S, PO_4^{3-}, and a variety of other mineral compounds and elements.

9.4.2 Nitrogen Cycling

Nitrogen is an essential component of many different organic molecules (e.g., amino acids, chitin, peptidoglycan), and is often a limiting nutrient for primary production in estuaries (Chapter 4). Nitrogen is also involved in many different microbial processes that cycle this element through several forms and oxidation states (Figure 9.10). In estuaries, some nitrogen is "fixed" into useable forms from inert N_2 gas (see below), but most nitrogen is imported from adjacent land, ocean, and the atmosphere (Chapter 3) as detrital organic nitrogen or as dissolved inorganic nitrogen (NO_3^-, NH_4^+).

When heterotrophs mineralize detrital organic matter, nitrogen is released as NH_4^+, but in most estuaries NH_4^+ is not the most common form of inorganic nitrogen because nitrifying organisms oxidize NH_4^+ to NO_3^- using oxygen as an electron acceptor. **Nitrification** is carried out by a group of chemoautotrophic bacteria and archaea that capture the energy of this oxidation reaction to fix carbon and build biomass. Nitrification occurs in two steps, each carried out with different sets of enzymes and generally accomplished by different groups of organisms. Ammonia oxidizers, including Bacteria in the genera *Nitrosomonas* and *Nitrosococcus* and archaea in the phylum Thaumarchaeota, carry out the first step using the enzymes ammonia monooxygenase and hydroxylamine oxidoreductase to oxidize ammonium to nitrite (NO_2^-). Then nitrite oxidizers, such as Bacteria in the genus *Nitrobacter* and phylum *Nitrospirae*, use the enzyme nitrite oxidase to oxidize nitrite to nitrate. Ammonia oxidizers also generate copious amounts of the greenhouse gas N_2O (Firestone and Davidson 1989), particularly in estuaries (Maavara et al. 2019). This is thought to be due to incomplete oxidation of the intermediate product hydroxylamine by the enzyme hydroxylamine oxidoreductase, but the exact mechanism is not well understood (Prosser et al. 2020).

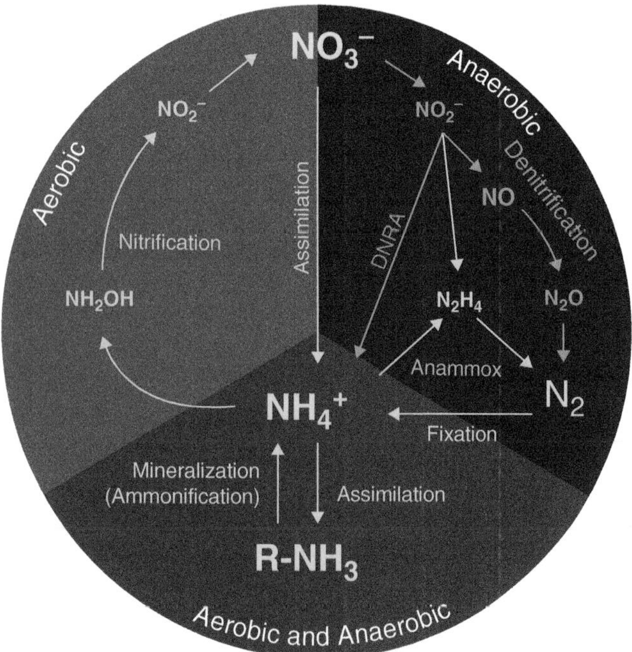

FIGURE 9.10 A generalized view of the nitrogen cycle in estuaries separated into aerobic and anaerobic processes. Autotrophic processes are in green, heterotrophic in red, and both in blue.

Nitrate produced by nitrifiers can be used by some heterotrophic bacteria and archaea to oxidize organic matter in low oxygen conditions such as in sediments. In most cases, the dominant process is **denitrification**, which generally follows the sequence $NO_3^- \rightarrow NO_2^- \rightarrow NO \rightarrow N_2O \rightarrow N_2$, with each step catalyzed by a different enzyme. Many of the organisms that respire carbon in this manner are **facultative anaerobes**. Facultative anaerobes oxidize organic matter either aerobically with oxygen or anaerobically with electron acceptors such as NO_3^-, SO_4^{2-}, and Fe^{3+}. Denitrification refers to the production of the non-reactive, gaseous, end product, dinitrogen (N_2). However, under some conditions, the intermediate gaseous compounds NO and N_2O can escape to the atmosphere before being reduced to N_2.

The contribution of denitrification to organic matter decomposition in estuaries is small compared to other forms of respiration. For example, Jørgensen (1980) attributed only 5% of total carbon mineralization in a Danish fjord to denitrification, instead attributing most mineralization to aerobic respiration and to anaerobic respiration using sulfate (see below). However, denitrification is a very important process in estuaries where it serves as an essential pathway for the removal of reactive nitrogen – particularly in highly eutrophied estuaries. Of the 48 Tg NO_3^--N year^{-1} entering estuaries from rivers globally, 10–80% is denitrified, thereby reducing greatly the amount of fixed N delivered to the ocean (Galloway et al. 2004). In estuarine sediments, nitrification and denitrification can be tightly coupled across fine-scale gradients in oxygen conditions with nitrification supplying the nitrate for denitrification (Kemp et al. 1990). Denitrification in estuaries is also an important source of the greenhouse gas N_2O (0.4 Tg N_2O-N year^{-1} in estuaries and coastal oceans). The global annual production of N_2O has at least doubled since 1860, largely as a result of N inputs from agriculture (Seitzinger et al. 2000).

Some heterotrophs can respire organic matter anaerobically by reducing NO_3^- to NH_4^+ rather than N_2 by way of **dissimilatory nitrate reduction to ammonia (DNRA)**. This process competes directly with denitrification for nitrate, and it results in recycling rather than removal of nitrogen (Gardner et al. 2006). Recent research suggests that DNRA is favored when sediments feature low nitrate to carbon ratios and high concentrations of sulfide (Kessler et al. 2018; Caffrey et al. 2019; Rahman et al. 2019). It is thought that burrowing macrobenthos that introduce oxygen to sediments (i.e., bioturbators) enhance denitrification over DNRA by stimulating nitrification and increasing the nitrate supply (Moraes et al. 2018; Magri et al. 2020). Conversely, sulfides are known to inhibit nitrification and the final two steps of denitrification (Burgin and Hamilton 2007), thus favoring DNRA in highly anoxic sediments. In addition to DNRA, a few organisms contain only the nitrate reductase enzyme, and instead of denitrifying, reduce NO_3^- only to NO_2^-.

A specialized group of autotrophic bacteria within the Planctomycetales also can remove reactive nitrogen from water and sediment through a strictly anaerobic, autotrophic process. These organisms oxidize ammonium anaerobically using nitrite as the electron acceptor in the **anammox (anaerobic ammonium oxidation)** process. They then use the energy released during the oxidation to fix carbon. The ecological function of these organisms in the nitrogen cycle is similar to that of denitrifiers because they produce nonreactive N_2 gas when they combine nitrite (N = +3) and ammonium (N = −3) to yield the elemental form (N = 0). However, these organisms are different because they are strictly anaerobic and are autotrophs rather than heterotrophs. Anammox typically accounts for less than 25% of the nitrogen removal in estuaries because conditions of high organic carbon availability select for denitrifiers in productive estuaries (Damashek and Francis 2018).

Most of the fixed nitrogen in estuaries is externally-derived material from adjacent land, ocean, and the atmosphere, but some of the nitrogen in estuaries is fixed from abundant N_2 gas by nitrogen-fixing bacteria, also called diazotrophs ("di" = two + "azo" = nitrogen). Unlike other nitrogen cycling processes described above, **nitrogen fixation** is a mechanism of nutrient acquisition rather than energy acquisition and, thus, requires energy rather than yielding energy. Organisms only fix nitrogen when their growth is nitrogen-limited. All organisms that carry out nitrogen fixation are prokaryotes, and these bacteria and archaea are a very diverse group that are capable of fixing nitrogen under many different environmental conditions. Some are autotrophs (e.g., many cyanobacteria), while others are heterotrophs. The one characteristic that all nitrogen-fixing bacteria share is the oxygen sensitive enzyme, nitrogenase. Nitrogenase is a complex enzyme consisting of several different proteins that are highly conserved (i.e., very similar) from one organism to another (Eady and Postgate 1974). These proteins are rich in Mo, Fe, Mg, and S, and so all nitrogen-fixing bacteria require these elements in greater amounts than other bacteria.

9.4.3 Sulfur Cycling

The sulfur cycle in estuaries is similar to the nitrogen cycle in several ways. Like nitrogen, sulfur is a macronutrient, and is an important component of amino acids, enzyme co-factors, and some other organic molecules. Sulfur exists in a variety of redox states, and biological processes assimilate inorganic sulfur into organic matter, mineralize organic sulfur compounds, and transform sulfur between reduced (sulfides, thiosulfate, polythionates, and elemental sulfur) and oxidized forms (sulfates and sulfites; Figure 9.11). Like nitrogen, reduced sulfur compounds can serve as energy sources for chemoautotrophs, and oxidized sulfur compounds can serve as electron acceptors for anaerobic respiration by heterotrophs. However, there are some very important differences. For example, the products of mineralization are sulfate (SO_4^{2-}) under aerobic conditions and H_2S under anaerobic conditions, whereas, regardless of

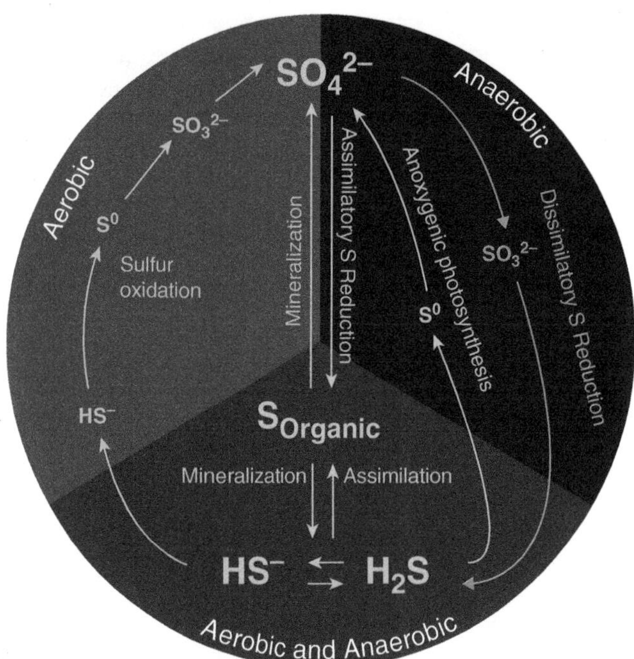

FIGURE 9.11 A generalized view of the sulfur cycle in estuaries separated into aerobic and anaerobic processes. Autotrophic processes are in green, heterotrophic in red, and both in blue.

the oxygen status of the environment, the product of organic nitrogen mineralization is always NH_4^+. Furthermore, there are no inert S compounds similar to N_2, so there is no process analogous to nitrogen fixation, nor are there S removal processes analogous to denitrification and anammox (Howarth and Teal 1980). Also, because of its high abundance in ocean water (28 mM in the ocean as SO_4^{2-}) sulfur does not limit primary production in estuarine or marine systems.

Just as reduced nitrogen serves as an energy source for some chemoautotrophic bacteria and archaea, so does reduced sulfur. The difference is that nitrifiers are primarily chemoautotrophic, but organisms that carry out **sulfur oxidation** include chemoautotrophs, facultative chemoautotrophs, and photoautotrophs. One group obligately oxidizes reduced sulfur with oxygen to provide the energy necessary to fix carbon dioxide. These organisms include members of the genera *Beggiatoa, Thiothrix*, and some species of *Thiobacillus*. Other members of *Thiobacillus* are facultative autotrophs because they derive energy from sulfur oxidation but can use organic matter as a carbon source for biosynthesis (e.g., *T. novellas, T. intermedius*). These two groups of organisms are referred to collectively as the colorless sulfur bacteria to distinguish them from the third group, the green and purple sulfur bacteria, which oxidize sulfur compounds during anoxygenic photosynthesis. This type of photosynthesis is anoxygenic because it uses reduced sulfur compounds as a source of electrons instead of water molecules and produces oxidized sulfur compounds instead of oxygen.

All sulfur oxidizers require a source of reduced sulfur compounds, which are generated by sulfate-reducing bacteria under anaerobic conditions in estuarine sediments and soils (see below). However, colorless sulfur oxidizers also require

FIGURE 9.12 Estuarine microbial mats. (a) Purple sulfur bacteria mat in Roscoff Bay, France. *Source*: Hubas et al. (2017), Frontiers Media S.A. Licensed under CC BY 4.0. (b) Cyanobacteria biofilm covering wave ripples on a Bahia Blanca Estuary tidal flat. *Source*: Perillo and Piccolo (2011), Elsevier. (c) Colorless sulfur oxidizer mat formed by Beggiotoa spp, (d) Iron oxidizer mat, *Source*: McBeth et al. (2013), John Wiley & Sons.

oxygen, and green and purple sulfur bacteria require light. Thus, these organisms typically live at the interface between dark, anaerobic conditions and sunlit, aerobic conditions. Sulfur oxidizers are commonly encountered on the surface of anaerobic sediments and salt marsh soils forming white, green, or purple mats (Figure 9.12). They are also found deeper in sediments associated with oxygenated animal burrows and associated with plant roots where they use oxygen provided by the plants to oxidize sulfides, and thus protect the plants from sulfide toxicity. In water, they are found at oxic-anoxic interfaces when the underlying water is anoxic. The significance of sulfur oxidation in estuaries is that the ultimate end product of the process is sulfate (SO_4^{2-}), a non-toxic sulfur form highly abundant in marine systems of all kinds. Indeed, Lavik et al. (2009) credit chemoautotrophic sulfur oxidation with detoxifying sulfidic waters in parts of the open ocean making the surface waters overlying sulfide-rich areas habitable for fish and other nekton.

Most primary producers take up SO_4^{2-} and reduce it to HS^- for assimilation into biomass rather than using free H_2S, which is toxic to many organisms. Some prokaryotes, however, can use SO_4^{2-} as a terminal electron acceptor for the oxidation of organic matter. This process is referred to as **dissimilatory sulfate reduction** to distinguish it from sulfate reduction for assimilation. Most of the sulfides (H_2S, HS^-, metal sulfides, etc.) that accumulate in estuaries are produced via dissimilatory sulfate reduction, although mineralization of organic sulfur compounds leads to the same products. Like the bacteria carrying out the analogous process in the nitrogen cycle (denitrifiers), the sulfate-reducing bacteria (SRB) and archaea are heterotrophs. In contrast to the diversity of denitrifiers, however, there are just a few types of SRB in estuaries, mainly in the Deltaproteobacteria and Firmicutes. Additionally, SRB are all obligate anaerobes, whereas the denitrifiers are mostly facultative anaerobes. While the presence of oxygen slows or stops denitrification, most denitrifiers can continue to respire using oxygen. The presence of O_2 also halts sulfate reduction, and prevents the SRB from mineralizing organic matter, but O_2 does not kill the SRB. When anoxic conditions reoccur sulfate reduction recommences.

Dissimilatory sulfate reduction is an important process in the largely anaerobic sediments and soils of estuaries. For example, well over half the organic matter mineralized in salt marsh sediments occurs via dissimilatory sulfate reduction (Jørgensen 1980; Howarth and Hobbie 1982; Howes et al. 1984). This may seem surprising given the low energy yield of sulfate reduction compared to denitrification (Table 9.2), but sulfate is in much higher concentration in seawater (28 mM) than nitrate or any other electron acceptor that supports anaerobic respiration. The magnitude of sulfate reduction in estuaries makes these systems important sources of hydrogen sulfide and other volatile reduced sulfur compounds. Salt marshes in particular are ideal environments for formation of volatile sulfur compounds because the soils are anaerobic just a few millimeters to a few centimeters below the surface, are rich in organic materials, and SO_4^{2-} is replenished each time the soils are flooded. Volatile sulfur compounds that enter the atmosphere are eventually oxidized to sulfur dioxide (SO_2) and sulfate (SO_4^{2-}) and contribute to the acidification of rain and snow (Andreae and Jaeschke 1992).

9.4.4 Iron and Manganese Cycling

Both iron and manganese are highly reactive elements that are cycled through various redox states by a combination of biological and abiotic processes (Madison et al. 2013; Oldham et al. 2019). Oxidation of reduced Fe^{2+} and Mn^{2+} occurs mainly in oxic environments with O_2 as the dominant electron acceptor, although there is evidence suggesting that some organisms use other oxidants in anoxic environments including NO_3^-, chlorate, and perchlorate (Weber et al. 2006). Reduction of oxidized Fe^{3+} and Mn^{4+} mainly occurs in anoxic sediments where they react abiotically with reduced sulfur compounds or are reduced by heterotrophic bacteria and archaea who use these metals to oxidize organic matter during anaerobic respiration (Table 9.2). Oxidized Fe^{3+} and Mn^{4+} form insoluble compounds and tend to settle to the bed in estuaries, making sediments the primary location for cycling of these elements.

At circumneutral pH in oxic conditions, iron and manganese exist predominantly in oxidized form (Fe^{3+} and Mn^{4+}) because at this pH, abiotic oxidation of Fe^{2+} and Mn^{2+} occurs rapidly. Since the pH of most estuarine habitats is at or above pH 7, biological **iron oxidation** and **manganese oxidation** are restricted to microaerophilic (very low oxygen) environments such as the interface of sediment oxic and anoxic zones or the rhizosphere of plant roots. At these interfaces where iron is abundant, iron oxidizing bacteria can form mats similar to those of sulfur oxidizing bacteria. These reddish-brown mats (Figure 9.12) are dominated by the Betaproteobacteria genera *Gallionella* and *Leptothrix* in freshwater environments, and by members of the Zetaproteobacteria in marine environments, although there are many other taxonomic groups of iron oxidizing bacteria (Emerson et al. 2010). Many iron oxidizers are chemoautotrophic organisms that fix carbon (similar to nitrifiers and many sulfur oxidizers), but some are mixotrophs and heterotrophs (McAllister et al. 2019). Manganese oxidizers are very diverse and belong to the Firmicutes, Proteobacteria, and Actinobacteria phyla. Most manganese oxidizers are thought to be heterotrophs (Tebo et al. 2005), although recently chemoautotrophic manganese oxidation was demonstrated in a co-culture dominated by a new organism from the Bacterial phylum Nitrospirae (Yu and Leadbetter 2020). For iron oxidation, the energy yield per mole of iron oxidized is quite low (Table 9.2), but the abundance of iron in estuarine environments is high and is thus readily recycled by microbial reduction to Fe^{2+} from Fe^{3+} (see below). By oxidizing a large amount of iron, iron oxidizing bacteria can exploit the small amount of energy released by the reaction to fuel their biosynthesis and growth. Manganese oxidation yields more energy than iron oxidation (Table 9.2), but manganese is in lower concentration than iron in estuaries, and likely fuels a smaller fraction of biological metal oxidation.

Similar to oxidation, the reduction of iron and manganese in estuaries is thought to be mainly an abiological processes. Fe^{3+} and Mn^{4+} are reduced through reactions with sulfide in anoxic estuarine sediments and soils. Biological **iron reduction** and **manganese reduction** also occur in these anoxic environments and are carried out by a diverse community of chemoheterotrophic bacteria who use Fe^{3+} and Mn^{4+} to oxidize organic matter. These organisms include Proteobacteria from the family Geobacteraceae and genera *Shewanella* and *Ralstonia*, and some Firmicutes in the orders Clostridiales and Bacillales. When these insoluble oxidized metals are reduced, they become soluble in water and thus become "mobilized" and can diffuse out of sediments. The observation of similar porewater concentrations of Fe^{2+} and Mn^{2+} suggests that rates of reduction might be similar and that the primary control on rates of reduction of both iron and manganese may be the availability of reactive oxides in the solid phase (Joye et al. 1996).

9.4.5 Phosphorous Cycling

The role of microbes in estuarine phosphorus cycles is much simpler than for any of the elements discussed so far because phosphorus does not serve as an electron acceptor or donor during microbial metabolism. As a result, the cycle can be summarized simply as one of assimilation and mineralization. Nevertheless, the phosphorus cycle is regulated both directly and indirectly by the activity of microbes. Because phosphorus is so biologically and chemically reactive, the residence time of dissolved inorganic phosphorus is short (Pomeroy 1960). Inorganic phosphorus, generally found in the form of phosphate (PO_4^{3-}), is readily assimilated into biomass by autotrophic and heterotrophic micro- and macro-organisms and is subsequently mineralized and released by heterotrophic decomposition of organic matter. Inorganic phosphorous is also subject to abiotic processes and tends to sorb to particles in oxic water, binding to oxidized Fe, Mn, Al, and organic compounds (Welsh 1980). In fact, more than 90% of the phosphorus carried by rivers is associated with suspended particles (Föllmi 1996). When this phosphorus enters estuaries, it can become concentrated in estuarine sediments.

When particle-associated phosphorus settles to estuarine sediment and encounters anoxic environments, biological and abiotic reduction of Fe^{3+}, Mn^{4+}, and other binding agents releases the phosphorus from particles, which can then diffuse back into the overlying water and become available for biological assimilation (Kemp et al. 2009). Diffusion of this released phosphorus out of sediments can be limited, however, when surface sediments are oxic, or when bioturbation increases sediment oxygen, causing oxidation of particle surfaces and re-sorption of phosphorus. Diffusion of phosphorus is enhanced if sediments are disturbed by resuspension (Pomeroy et al. 1965), or if the overlying waters are anoxic. For example, in the Baltic Sea, the concentration of dissolved inorganic phosphorus in bottom waters is strongly negatively correlated with oxygen concentration (Conley et al. 2002). When low-oxygen bottom waters enhance phosphorus release from sediments, it can cause a positive feedback on eutrophication. Phosphorus released from sediments can stimulate phytoplankton production, and subsequent aerobic microbial respiration of this phytoplankton biomass can further deplete the oxygen in bottom waters and further enhance the release of phosphorus.

9.5 Summary

Communities of microbes and the processes they carry out exhibit temporal and spatial complexity in estuarine ecosystems that is only just beginning to be understood. What is known about microbes is that they are more abundant, diverse, and active in estuaries than any other group of organisms. They are major participants in all the important ecosystem functions such as nutrient cycling, photosynthesis, and decomposition. Many of the biological transformations within the carbon, nitrogen, and sulfur cycles are uniquely microbial. These processes were occurring two billion years before present, or more, and at least one billion years before plants and animals evolved (Margulis and Dolan 2002). The early evolution of microbes, especially the Bacteria and Archaea, allowed them to become extremely diverse. As a consequence, they contribute to ecosystem biogeochemical resilience, trophic dynamics, and resistance to invasion. Although microbial communities are diverse and the numbers of individuals in the community are large, not all individuals are active any given time or place as a result of variation in the physical, chemical, and biological characteristics of estuary. This variation gives rise to differences in microbial community composition, which processes are carried out, and at what rate the processes occur.

Review Questions

Multiple choice

1. Which one of these statements about estuarine microbes is true? (Select all that apply)
   a. Bacteria and Archaea are prokaryotic organisms with similar ecological functions but are very distantly evolutionarily related.
   b. Heterotrophic flagellates are about 1000 times less abundant than their prey (bacteria and archaea).
   c. Ciliates are larger than flagellates but are not large enough to be preyed upon by zooplankton and filter-feeding benthos.
   d. The ecological roles of fungi in estuaries include detritus decomposition, nutrient recycling, and mutualisms with seagrasses, marsh grasses, and mangrove trees.
   e. Viruses are abundant in estuaries but do not influence the composition of microbial communities

2. Which one of these statements about estuarine microbial food webs is true? (Select all that apply)
   a. They mineralize detritus organic carbon via initial consumption by prokaryotes, and subsequent grazing of prokaryotes by protists.
   b. They feature predator-prey oscillations in organism abundance.
   c. In more productive estuaries microbial abundances tend to be more stable over time.
   d. Microbial species composition generally changes very slowly because grazing rates are similar to growth rates.
   e. The dynamics of viruses in microbial food webs are similar to those of grazers.

3. Sources of allochthonous organic matter to estuaries include (choose all that apply)
   a. Riverborne particles
   b. DOM released by growing benthic microalgae
   c. Mangrove leaves
   d. Diatoms from the coastal ocean
   e. Detritus produced when viruses lyse bacteria
   f. Senescing salt marsh grasses

4. Sources of autochthonous organic matter to estuaries include (choose all that apply)
   a. Riverborne particles
   b. DOM released by growing benthic microalgae
   c. Mangrove leaves
   d. Diatoms from the coastal ocean
   e. Detritus produced when viruses lyse bacteria
   f. Senescing marsh grasses

5. Heterotrophic microbes indirectly regulate primary productivity in estuaries by
   a. Grazing on phytoplankton.
   b. Releasing dissolved organic matter as they grow.
   c. Mineralizing organic N and P.
   d. Forming mutualistic relationships with primary producers.

6. The microbial loop in estuaries (choose all that apply)
   a. Produces an insignificant amount of new biomass.
   b. Causes disease in zooplankton and fish.
   c. Adds as much as 30% more energy to higher trophic levels.
   d. Cycles organic matter between the water and sediments.
   e. Captures the energy contained in dissolved organic matter.

7. When aerobic microbes use up the oxygen in an estuarine environment (e.g., bottom waters, sediments)
   a. Most heterotrophic protists continue to respire organic matter using alternate electron acceptors.
   b. Facultative anaerobic bacteria initially switch to SO_4^{2-} as an electron acceptor.
   c. Redox potential (Eh) decreases as alternate electron acceptors are used up.
   d. The species composition of microbial communities stays the same.

8. Dissolved organic matter with a low C:N ratio is considered
   a. High quality organic matter.
   b. Low quality organic matter.
   c. Easy for photosynthetic organisms to consume.
   d. Difficult for heterotrophic organisms to consume.

9. Which of these statements are correct about microbial metabolism (choose all that apply)
   a. Certain specialized protists can fix nitrogen.
   b. Some Bacteria species can create methane.
   c. All heterotrophic microbes can oxidize organic matter for energy.
   d. Nitrifiers reduce nitrate to ammonium.
   e. Some microbes can use phosphate as an electron acceptor.
   f. Chemoautotrophic microbes reduce CO_2 to organic carbon.

10. Heterotrophic microbes include (choose all that apply)
    a. Nitrifiers
    b. Denitrifiers
    c. Sulfur oxidizers
    d. Sulfate reducers
    e. Iron oxidizers
    f. DNRA organisms
    g. Anammox organisms

11. Chemoautotrophic microbes include (choose all that apply)
    a. Nitrifiers
    b. Denitrifiers
    c. Sulfur oxidizers
    d. Sulfate reducers
    e. Iron oxidizers
    f. DNRA organisms
    g. Anammox organisms

12. Denitrification is (choose all that apply)
    a. An important process by which fixed nitrogen is removed from estuarine ecosystems
    b. The second most important form of organic carbon respiration after aerobic respiration
    c. A metabolism by which microbes reduce N_2 to NH_4^+
    d. An important source of the greenhouse gas N_2O
    e. Has the same end-product as DNRA metabolism
    f. Often uses NO_3^- produced by nitrifiers

13. Denitrifiers and nitrifiers interact in estuaries (choose all that apply)
    a. In oxic water because both processes are carried out by aerobic organisms.
    b. In sediments where they can exchange NO_3^- and NH_4^+ across oxic-anoxic interfaces.
    c. On standing dead marsh plants where there is a large supply of nitrogen.
    d. In deep sediments because both processes are carried out by anaerobic organisms.

14. Sulfur oxidizers are similar to nitrifiers in that (choose all that apply)
    a. They are both chemoautotrophic organisms that fix carbon.
    b. They both respire dissolved organic matter.
    c. They both live near oxic-anoxic interfaces.
    d. They both produce extensive microbial mats on sediment surfaces.
    e. They both use oxygen as an electron acceptor.

15. Sulfate reducers are similar to denitrifiers in that (choose all that apply)
    a. They are both chemoautotrophic organisms that fix carbon
    b. They both respire dissolved organic matter
    c. They both convert fixed nitrogen to N_2 gas
    d. They both produce volatile compounds as end products of their metabolisms
    e. They both use oxygen as an electron acceptor

16. What conditions favor the accumulation of phosphorous in estuarine sediments? (choose all that apply)
    a. Oxic bottom waters
    b. Anoxic bottom waters
    c. Low turbidity
    d. High turbidity
    e. Abundant oxidized Fe and Mn
    f. Abundant reduced Fe and Mn

Short Answer

1. How do viruses influence the species composition of microbial communities?
2. What preys upon each of these organisms in estuaries: ciliates, nanoflagellates, bacteria?
3. If Bacteria and Archaea are the primary consumers of DOC and POC in estuaries, then why is the microbial food web required to respire most of this organic carbon to CO_2?
4. What nitrogen cycling processes in estuaries are accomplished by eukaryotic organisms?
5. How does the availability of light indirectly affect heterotrophic microbial communities?
6. What are the possible energy sources for microorganisms?

References

Andersen, P. and Sørensen, H.M. (1986). Population dynamics and trophic coupling in pelagic microorganisms in eutrophic coastal waters. *Marine Ecology Progress Series* 33: 99–109. https://doi.org/10.3354/meps033099.

Andreae, M.O. and Jaeschke, W.A. (1992). Exchange of sulphur between biosphere and atmosphere over temperate and tropical regions. In: *Sulphur Cycling on the Continents: Wetlands, Terrestrial Ecosystems and Associated Water Bodies* (ed. R.W. Howarth, J.W.B. Stewart and M.V. Ivanov), 27–61. Chichister, UK: Wiley.

Azam, F., Fenchel, T., Field, J. et al. (1983). The Ecological Role of Water-Column Microbes in the Sea. *Marine Ecology Progress Series* 10: 257–263. https://doi.org/10.3354/meps010257.

Baird, D., Christian, R.R., Peterson, C.H., and Johnson, G.A. (2004). Consequences of hypoxia on estuarine ecosystem function: energy diversion from consumers to microbes. *Ecological Applications* 14: 805–822. https://doi.org/10.1890/02-5094.

Bauer, J. and Bianchi, T.S. (2011). Dissolved Organic Carbon Cycling and Transformation. In: *Treatise on Estuarine and Coastal Science*, vol. 5 (ed. E. Wolanski and D.S. McLusky), 7–67. Waltham: Academic Press.

Bianchi, T.S. and Bauer, J. (2011). Particulate Organic Carbon Cycling and Transformation. In: *Treatise on Estuarine and Coastal Science*, vol. 5 (ed. E. Wolanski and D.S. McLusky), 69–117. Waltham: Academic Press.

Blum, L.K., Mills, A.L., Zieman, J.C., and Zieman, R.T. (1988). Abundance of bacteria and fungi in seagrass and mangrove detritus. *Marine Ecology Progress Series. Oldendorf* 42 (1): 73–78.

Blum, L.K., Roberts, M.S., Garland, J.L., and Mills, A.L. (2004). Microbial communities among the dominant high marsh plants and associated sediments of the United States east coast. *Microbial Ecology* 48: 375–383.

Bouvier, T.C. and del Giorgio, P.A. (2002). Compositional changes in free-living bacterial communities along a salinity gradient in two temperate estuaries. *Limnology and Oceanography* 47: 453–470.

Boynton, W.R., Garber, J.H., Summers, R., and Kemp, W.M. (1995). Inputs, transformations, and transport of nitrogen and phosphorus in Chesapeake Bay and selected tributaries. *Estuaries* 18: 285. https://doi.org/10.2307/1352640.

Burgin, A.J. and Hamilton, S.K. (2007). Have we overemphasized the role of denitrification in aquatic ecosystems? A review of nitrate removal pathways. *Frontiers in Ecology and the Environment* 5: 89–96. https://doi.org/10.1890/1540-9295(2007)5[89:HWOTRO]2.0.CO;2.

Burki, F., Roger, A.J., Brown, M.W., and Simpson, A.G.B. (2020). The new tree of Eukaryotes. *Trends in Ecology & Evolution* 35: 43–55. https://doi.org/10.1016/j.tree.2019.08.008.

Caffrey, J.M., Bonaglia, S., and Conley, D.J. (2019). Short exposure to oxygen and sulfide alter nitrification, denitrification, and DNRA activity in seasonally hypoxic estuarine sediments. *FEMS Microbiology Letters* 366: https://doi.org/10.1093/femsle/fny288.

Cai, L., Zhang, R., He, Y. et al. (2016). Metagenomic analysis of Virioplankton of the subtropical Jiulong river estuary, China. *Viruses* 8: 35. https://doi.org/10.3390/v8020035.

Campbell, B.J. and Kirchman, D.L. (2013). Bacterial diversity, community structure and potential growth rates along an estuarine salinity gradient. *ISME Journal* 7: 210–220. https://doi.org/10.1038/ismej.2012.93.

Carvalho, L.M., Correia, P.M., and Martins-Loução, M.A. (2004). Arbuscular mycorrhizal fungal propagules in a salt marsh. *Mycorrhiza* 14: 165–170. https://doi.org/10.1007/s00572-003-0247-4.

Castro-Barros, C.M., Jia, M., van Loosdrecht, M.C.M. et al. (2017). Evaluating the potential for dissimilatory nitrate reduction by anammox bacteria for municipal wastewater treatment. *Bioresource Technology* 233: 363–372. https://doi.org/10.1016/j.biortech.2017.02.063.

Cavicchioli, R. (2002). Extremophiles and the search for extraterrestrial life. *Astrobiology* 2 (3): 282–292.

Christian, R.R. and Thomas, C.R. (2003). Network analysis of nitrogen inputs and cycling in the Neuse river estuary, North Carolina, USA. *Estuaries* 26: 815–828. https://doi.org/10.1007/BF02711992.

Conley, D.J., Humborg, C., Rahm, L. et al. (2002). Hypoxia in the Baltic Sea and Basin-Scale changes in phosphorus biogeochemistry. *Environmental Science & Technology* 36: 5315–5320. https://doi.org/10.1021/es025763w.

Cory, R.M. and Kling, G.W. (2018). Interactions between sunlight and microorganisms influence dissolved organic matter degradation along the aquatic continuum. *Limnology and Oceanography Letters* 3: 102–116. https://doi.org/10.1002/lol2.10060.

Crump, B.C. and Baross, J.A. (1996). Particle-attached bacteria and heterotrophic plankton associated with the Columbia river estuarine turbidity maxima. *Marine Ecology Progress Series* 138: https://doi.org/10.3354/meps138265.

Crump, B.C., Baross, J.A., and Simenstad, C.A. (1998). Dominance of particle-attached bacteria in the Columbia river estuary, USA. *Aquatic Microbial Ecology* 14: 7–18. https://doi.org/10.3354/ame014007.

Crump, B.C., Armbrust, E.V., and Baross, J.A. (1999). Phylogenetic analysis of particle-attached and free-living bacterial communities in the Columbia river, its Estuary, and the adjacent coastal ocean. *Applied and Environmental Microbiology* 65: 3192–3204.

Dale, N.G. (1974). Bacteria in intertidal sediments: factors related to their distribution. *Limnology and Oceanography* 19: 509–518. https://doi.org/10.4319/lo.1974.19.3.0509.

Damashek, J. and Francis, C.A. (2018). Microbial nitrogen cycling in Estuaries: from genes to Ecosystem processes. *Estuaries and Coasts* 41: 626–660. https://doi.org/10.1007/s12237-017-0306-2.

Dietrich, D. and Arndt, H. (2000). Biomass partitioning of benthic microbes in a Baltic inlet: relationships between bacteria, algae, heterotrophic flagellates and ciliates. *Marine Biology.* 136: 309–322.

Doherty, M., Yager, P.L., Moran, M.A. et al. (2017). Bacterial biogeography across the Amazon River-ocean continuum. *Frontiers in Microbiology* 8: 1–17.

Doolittle, W.F. (2020). Evolution: two domains of life or three? *Current Biology* 30: R177–R179. https://doi.org/10.1016/j.cub.2020.01.010.

Eady, R.R. and Postgate, J.R. (1974). Nitrogenase. *Nature* 249: 805–810. https://doi.org/10.1038/249805a0.

Emerson, D., Fleming, E.J., and McBeth, J.M. (2010). Iron-oxidizing bacteria: an environmental and genomic perspective. *Annual Review of Microbiology* 64: 561–583. https://doi.org/10.1146/annurev.micro.112408.134208.

Evrard, V., Cook, P.L.M., Veuger, B. et al. (2008). Tracing carbon and nitrogen incorporation and pathways in the microbial community of a photic subtidal sand. *Aquatic Microbial Ecology* 53: 257–269.

Firestone, M.K. and Davidson, E.A. (1989). Microbiological basis of NO and N_2O production and consumption in soil. In: *Exchange of Trace Gases Between Terrestrial Ecosystems and the Atmosphere* (ed. M.O. Andreae and D.S. Schimel), 7–21. New York: Wiley.

Föllmi, K. (1996). The phosphorus cycle, phosphogenesis and marine phosphate-rich deposits. *Earth-Science Reviews* 40: 55–124. https://doi.org/10.1016/0012-8252(95)00049-6.

Fortunato, C.S., Herfort, L., Zuber, P. et al. (2012). Spatial variability overwhelms seasonal patterns in bacterioplankton communities across a river to ocean gradient. *ISME Journal* 6: https://doi.org/10.1038/ismej.2011.135.

Fuhrman, J.A. (1999). Marine viruses and their biogeochemical and ecological effects. *Nature* 399: 541–548.

Fuhrman, J.A. and Noble, R.T. (1995). Viruses and protists cause similar bacterial mortality in coastal seawater. *Limnology and Oceanography* 40 (7): 1236–1242.

Galloway, J.N., Dentener, F.J., Capone, D.G. et al. (2004). Nitrogen cycles: past, present, and future. *Biogeochemistry* 70: 153–226.

Gardner, W.S., McCarthy, M.J., An, S. et al. (2006). Nitrogen fixation and dissimilatory nitrate reduction to ammonium (DNRA) support nitrogen dynamics in Texas estuaries. *Limnology and Oceanography* 51: 558–568. https://doi.org/10.4319/lo.2006.51.1_part_2.0558.

Garneau, M.E., Roy, S., Lovejoy, C. et al. (2008). Seasonal dynamics of bacterial biomass and production in a coastal arctic ecosystem: Franklin Bay, western Canadian Arctic. *Journal of Geophysical Research-Oceans* 113: C07S91.

Glud, R.N. (2008). Oxygen dynamics of marine sediments. *Marine Biology Research* 4: 243–289.

Goulder, R. (1976). Relationships between suspended solids and standing crops and activities of bacteria in an estuary during a neap-spring-neap tidal cycle. *Oecologia* 24: 83–90. https://doi.org/10.1007/BF00545489.

Grossart, H.-P., Wurzbacher, C., James, T.Y., and Kagami, M. (2016). Discovery of dark matter fungi in aquatic ecosystems demands a reappraisal of the phylogeny and ecology of zoosporic fungi. *Fungal Ecology* 19: 28–38. https://doi.org/10.1016/j.funeco.2015.06.004.

Harvey, R.W. and Young, L.Y. (1980). Enrichment and association of bacteria and particulate in salt marsh surface water. *Applied and Environmental Microbiology* 39: 894–899.

Havens, K.E., Work, K., and a, and East, T.L. (2000). Relative efficiencies of carbon transfer from bacteria and algae to zooplankton in a subtropical lake. *Journal of Plankton Research* 22: 1801–1809. https://doi.org/10.1093/plankt/22.9.1801.

Helton, R.R., Wang, K., Kan, J. et al. (2012). Interannual dynamics of viriobenthos abundance and morphological diversity in Chesapeake Bay sediments. *FEMS Microbiology Ecology* 79: 474–486. https://doi.org/10.1111/j.1574-6941.2011.01238.x.

Herlemann, D.P.R., Labrenz, M., Jürgens, K. et al. (2011). Transitions in bacterial communities along the 2000 km salinity gradient of the Baltic Sea. *ISME Journal* 5: 1571–1579.

Hewson, I., O'Neil, J.M., Fuhrman, J.A., and Dennison, W.C. (2001). Virus-like particle distribution and abundance in sediments and overlying waters along eutrophication gradients in two subtropical estuaries. *Limnology and Oceanography* 46: 1734–1746.

Howarth, R.W. and Hobbie, J.E. (1982). The regulation of decomposition and heterotrophic microbial activity in salt marsh soils. In: *Estuarine Comparisons* (ed. V.S. Kennedy), 183–207. New York, NY: Academic Press.

Howarth, R.W. and Teal, J.M. (1980). Energy flow in a salt marsh ecosystem: the role of reduced inorganic sulfur compounds. *American Naturalist* 116: 862–872.

Howes, B.L., Dacey, J.W.H., and King, G.M. (1984). Carbon flow through oxygen and sulfate reduction pathways in salt marsh sediments. *Limnology and Oceanography* 29: 1037–1051.

Hubas, C., Boeuf, D., Jesus, B. et al. (2017). A nanoscale study of carbon and nitrogen fluxes in mats of purple sulfur bacteria: implications for carbon cycling at the surface of coastal sediments. *Frontiers in Microbiology* 8: 1995.

Hug, L.A., Baker, B.J., Anantharaman, K. et al. (2016). A new view of the tree of life. *Nature Microbiology* 1: https://doi.org/10.1038/nmicrobiol.2016.48.

Iriarte, A. (1993). Size-fractionated chlorophyll *a* biomass and picophytoplankton cell density along a longitudinal axis of a temperate estuary (Southampton Water). *Journal of Plankton Research* 15: 485–500.

Jeffrey, W., Pledger, R., Aas, P. et al. (1996). Diel and depth profiles of DNA photodamage in bacterioplankton exposed to ambient solar ultraviolet radiation. *Marine Ecology Progress Series* 137: 283–291. https://doi.org/10.3354/meps137283.

Jørgensen, B.B. (1980). Mineralization and the bacterial cycling of carbon, nitrogen, and sulphur in marine sediments. In: *Contemporary Microbial Ecology* (ed. D.C. Ellwood, J.N. Hedger, M.J. Latham, et al.), 239–251. London: Academic Press.

Jørgensen, B.B. (2006). Bacteria and marine biogeochemistry. In: *Marine Geochemistry* (ed. H.N. Schulz and M. Zabel), 169–205. Berlin: Springer.

Joye, S.B., Mazzotta, M.L., and Hollibaugh, J.T. (1996). Community metabolism in microbial mats: the occurrence of biologically-mediated iron and manganese reduction. *Estuarine Coastal and Shelf Science* 43: 747–766.

Kemp, W.M., Sampo, P., McCaffrey, J. et al. (1990). Ammonium recycling versus denitrification in Chesapeake Bay sediments. *Limnology and Oceanography* 35: 1545–1563.

Kemp, P.F., Sherr, B.F., Sherr, E.B., and Cole, J.J. (1993). *Handbook of Methods in Aquatic Microbial Ecology*. Boca Raton, FL: Lewis Publishers, CRC Press.

Kemp, W.M., Smith, E.M., Marvin-DiPasquale, M., and Boynton, W.R. (1997). Organic carbon balance and net ecosystem metabolism in Chesapeake Bay. *Marine Ecology-Progress Series* 150: 229–248. https://doi.org/10.3354/meps150229.

Kemp, W.M., Testa, J.M., Conley, D.J. et al. (2009). Temporal responses of coastal hypoxia to nutrient loading and physical controls. *Biogeosciences* 6: 2985–3008.

Kessler, A.J., Roberts, K.L., Bissett, A., and Cook, P.L.M. (2018). Biogeochemical controls on the relative importance of denitrification and dissimilatory nitrate reduction to ammonium in estuaries. *Global Biogeochemical Cycles* 32: 1045–1057. https://doi.org/10.1029/2018GB005908.

Kohlmeyer, J. and Volkmann-Kohlmeyer, B. (2002). Fungi on Juncus and Spartina: new marine species of Anthostomella, with a list of marine fungi known on Spartina. *Mycological Research* 106: 365–374. https://doi.org/10.1017/S0953756201005469.

Kristensen, E. and Hansen, K. (1995). Decay of plant detritus in organic-poor marine sediments: production rates and stoichiometry of dissolved C and N compounds. *Journal of Marine Research* 53: 675–702.

Kubo, K., Lloyd, K.G., Biddle, J.F. et al. (2012). Archaea of the Miscellaneous Crenarchaeotal Group are abundant, diverse and widespread in marine sediments. *ISME Journal* 6 (10): 1949–1965.

Lavik, G., Stuhrmann, T., Bruchert, V. et al. (2009). Detoxification of sulphidic African shelf waters by blooming chemolithotrophs. *Nature* 457: 581–U586.

Lin, X.J., Scranton, M.I., Varela, R. et al. (2007). Compositional responses of bacterial communities to redox gradients and grazing in the anoxic Cariaco Basin. *Aquatic Microbial Ecology* 47: 57–72.

Liu, H., Tan, S., Xu, J. et al. (2017). Interactive regulations by viruses and dissolved organic matter on the bacterial community: virus and DOM affect bacterial community. *Limnology and Oceanography* 62: S364–S380. https://doi.org/10.1002/lno.10612.

Lovely, D.R. and Phillips, E.J.P. (1988). Novel mode of microbial energy metabolism: organic carbon oxidation coupled to dissimilatory reduction of iron or manganese. *Applied and Environmental Microbiology* 54 (6): 1472–1480.

Lynch, J.M. and Hobbie, J.E. (1988). *Micro-Organisms in Action: Concepts and Applications in Microbial Ecology*. Oxford: Blackwell.

Maavara, T., Lauerwald, R., Laruelle, G.G. et al. (2019). Nitrous oxide emissions from inland waters: are IPCC estimates too high? *Global Change Biology* 25: 473–488. https://doi.org/10.1111/gcb.14504.

MacMillin, K.M., Blum, L.K., and Mills, A.L. (1992). Comparison of bacterial dynamics in tidal creeks of the lower Delmarva Peninsula, Virginia, USA. *Marine Ecology Progress Series* 11.

Madigan, M.T., Bender, K.S., Buckley, D.H. et al. (2018). *Brock Biology of Microorganisms*, 15e Global Edition. Boston, US: Benjamin Cummins.

Madison, A.S., Tebo, B.M., Mucci, A. et al. (2013). Abundant porewater Mn(III) is a major component of the sedimentary redox system. *Science* 341: 875–878. https://doi.org/10.1126/science.1241396.

Magri, M., Benelli, S., Bonaglia, S. et al. (2020). The effects of hydrological extremes on denitrification, dissimilatory nitrate reduction to ammonium (DNRA) and mineralization in a coastal lagoon. *Science of The Total Environment* 740: 140169. https://doi.org/10.1016/j.scitotenv.2020.140169.

Maranger, R., Del Giorgio, P.A., and Bird, D.F. (2002). Accumulation of damaged bacteria and viruses in lake water exposed to solar radiation. *Aquatic Microbial Ecology* 28: 213–227. https://doi.org/10.3354/ame028213.

Margulis, L. and Dolan, M.F. (2002). *Early Life: Evolution on the Precambrian Earth*. Boston: Jones and Bartlett.

McAllister, S.M., Moore, R.M., Gartman, A. et al. (2019). The Fe(II)-oxidizing *Zetaproteobacteria*: historical, ecological and genomic perspectives. *FEMS Microbiology Ecology* 95: https://doi.org/10.1093/femsec/fiz015.

McBeth, J.M., Fleming, E.J., and Emerson, D. (2013). The transition from freshwater to marine iron-oxidizing bacterial lineages along a salinity gradient on the Sheepscot river, Maine, USA: estuarine iron mat communities. *Environmental Microbiology Reports* 5: 453–463. https://doi.org/10.1111/1758-2229.12033.

Mendes, C., Santos, L., Cunha, Â. et al. (2014). Proportion of prokaryotes enumerated as viruses by epifluorescence microscopy. *Annals of Microbiology* 64 (2): 773–778.

Meyer-Reil, L.-A. (1977). Bacterial growth rates and biomass production. In: *Microbial ecology of a brackish-water environment* (ed. G. Rheinheimer), 223–235. Berlin: Springer-Verlag.

Mongillo, J.F. (2000). *Encyclopedia of Environmental Science*, 255. University Rochester Press ISBN 978-1-57356-147-1.

Moraes, P.C., Zilius, M., Benelli, S., and Bartoli, M. (2018). Nitrification and denitrification in estuarine sediments with tube-dwelling benthic animals. *Hydrobiologia* 819: 217–230. https://doi.org/10.1007/s10750-018-3639-3.

Mostajir, B., Demers, S., de Mora, S. et al. (1999). Experimental test of the effect of ultraviolet-B radiation in a planktonic community. *Limnology and Oceanography* 44: 586–596. https://doi.org/10.4319/lo.1999.44.3.0586.

Nedwell, D.B. (1999). Effect of low temperature on microbial growth: lowered affinity for substrates limits growth at low temperature. *FEMS Microbiology Ecology* 30 (2): 101–111.

Newell, S.Y., Fallon, R.D., Cal Rodriquez, R.M., and Groene, L.C. (1985). Influence of rain, tidal wetting and relative humidity on release of carbon dioxide by standing-dead salt marsh plants. *Oecologia* 68: 73–79.

Newell, S.Y., Blum, L.K., Crawford, R.E. et al. (2000). Autumnal biomass and potential productivity of salt marsh fungi from 29° to 43° north latitude along the United States Atlantic coast. *Applied and Environmental Microbiology* 66: 180–185.

Nixon, S.W., Ammerman, J.W., Atkinson, L.P. et al. (1996). The fate of nitrogen and phosphorus at the land sea margin of the North Atlantic Ocean. *Biogeochemistry* 35: 141–180. https://doi.org/10.1007/BF02179826.

Oldham, V.E., Siebecker, M.G., Jones, M.R. et al. (2019). The speciation and mobility of Mn and Fe in Estuarine sediments. *Aquatic Geochemistry* 25: 3–26. https://doi.org/10.1007/s10498-019-09351-0.

Oliveira, V., Santos, A.L., Aguiar, C. et al. (2012). Prokaryotes in salt marsh sediments of Ria de Aveiro: effects of halophyte vegetation on abundance and diversity. *Estuarine, Coastal and Shelf Science* 110: 61–68. https://doi.org/10.1016/j.ecss.2012.03.013.

Palumbo, A.V. and Ferguson, R.L. (1978). Distribution of suspended bacteria in the Newport River estuary, North Carolina. *Estuarine and*

Coastal Marine Science 7: 521–529. https://doi.org/10.1016/0302-3524(78)90062-2.

Perillo, G.M.E. and Piccolo, M.C. (2011). Global variability in estuaries and coastal settings. In: *Treatise on Estuarine and Coastal Science* (ed. E. Wolanski and D.S. McLusky), 7–36. Elsevier.

Pomeroy, L.R. (1960). Residence time of dissolved phosphorus in natural waters. *Science* 131: 1731–1732.

Pomeroy, L.R. (1974). The ocean's food web, a changing paradigm. *BioScience* 24: 499–504. https://doi.org/10.2307/1296885.

Pomeroy, L.R., Smith, E.E., and Grant, C.M. (1965). The exchange of phosphate between estuarine water and sediment. *Limnology and Oceanography* 10: 167–172.

Pradeep Ram, A.S. and Sime-Ngando, T. (2008). Functional responses of prokaryotes and viruses to grazer effects and nutrient additions in freshwater microcosms. *ISME Journal* 2: 98–509. https://doi.org/10.1038/ismej.2008.15.

Prosser, J.I., Hink, L., Gubry-Rangin, C., and Nicol, G.W. (2020). Nitrous oxide production by ammonia oxidizers: physiological diversity, niche differentiation and potential mitigation strategies. *Global Change Biology* 26: 103–118. https://doi.org/10.1111/gcb.14877.

Rahman, M.M., Roberts, K.L., Warry, F. et al. (2019). Factors controlling dissimilatory nitrate reduction processes in constructed stormwater urban wetlands. *Biogeochemistry* 142: 375–393. https://doi.org/10.1007/s10533-019-00541-0.

Rojas-Jimenez, K., Rieck, A., Wurzbacher, C. et al. (2019). A salinity threshold separating fungal communities in the Baltic Sea. *Frontiers in Microbiology* 10: 680. https://doi.org/10.3389/fmicb.2019.00680.

Rublee, P. and Dornseif, B.E. (1978). Direct counts of bacteria in the sediments of a North Carolina salt marsh. *Coastal and Estuarine Research Federation* 1: 188. https://doi.org/10.2307/1351462.

Ruiz-González, C., Simó, R., Sommaruga, R., and Gasol, J.M. (2013). Away from darkness: a review on the effects of solar radiation on heterotrophic bacterioplankton activity. *Frontiers in Microbiology* 4: 131. https://doi.org/10.3389/fmicb.2013.00131.

Sanders, R.W., Caron, D.A., and Berninger, U. (1992). Relationships between bacteria and heterotrophic nanoplankton in marine and fresh waters: an inter-ecosystem comparison. *Marine Ecology Progress Series* 86: 1–14.

Schmidt, J.L., Deming, J.W., Jumars, P.A., and Keil, R.G. (1998). Constancy of bacterial abundance in surficial marine sediments. *Limnology and Oceanography* 43: 976–982. https://doi.org/10.4319/lo.1998.43.5.0976.

Schulz, H.N., Brinkhoff, T., Ferdelman, T.G. et al. (1999). Dense populations of a giant sulfur bacterium in Namibian shelf sediments. *Science* 284: 493–495.

Seitzinger, S.P., Kroeze, C., and Styles, R.V. (2000). Global distribution of N_2O emissions from aquatic systems: natural emissions and anthropogenic effects. *Chemosphere: Global Change Science* 2: 267–279.

Sengupta, A. and Chaudhuri, S. (2002). Arbuscular mycorrhizal relations of mangrove plant community at the Ganges river estuary in India. *Mycorrhiza* 12: 169–174. https://doi.org/10.1007/s00572-002-0164-y.

Sin, Y., Wetzel, R.L., and Anderson, I.C. (2000). Seasonal variations of size-fractionated phytoplankton along the salinity gradient in the York River estuary, Virginia (USA). *Journal of Plankton Research* 22: 1945–1960.

Smith, S.V. and Hollibaugh, J.T. (1997). Annual cycle and inter-annual variability of ecosystem metabolism in a temperate climate embayment. *Ecological Monographs* 67: 509–533.

Strom, S.L. (2000). Bacterivory: interactions between bacteria and their grazers. In: *Microbial Ecology of the Oceans* (ed. D.L. Kirchman), 351–386. New York: Wiley-Liss, Inc.

Strom, S.L. (2008). Microbial ecology of ocean biogeochemistry: a community perspective. *Science* 320: 1043–1045.

Suttle, C.A. (2005). Viruses in the sea. *Nature* 437: 356–361. https://doi.org/10.1038/nature04160.

Tebo, B.M., Johnson, H.A., McCarthy, J.K., and Templeton, A.S. (2005). Geomicrobiology of manganese(II) oxidation. *Trends in Microbiology* 13: 421–428. https://doi.org/10.1016/j.tim.2005.07.009.

Teixeira, C. and Gaeta, S.A. (1991). Contribution of picoplankton to primary production in estuarine, coastal and equatorial waters of Brazil. *Hydrobiologia* 209: 117–122.

Valiela, I., Wilson, J., Buschbaum, R. et al. (1984). Importance of chemical composition of salt marsh litter on decay rates and feeding by detritivores. *Bulletin of Marine Sciences* 35: 261–269.

Vohník, M., Borovec, O., Kolaříková, Z. et al. (2019). Extensive sampling and high-throughput sequencing reveal *Posidoniomyces atricolor* gen. et sp. nov. (Aigialaceae, Pleosporales) as the dominant root mycobiont of the dominant Mediterranean seagrass Posidonia oceanica. *MycoKeys* 55: 59–86. https://doi.org/10.3897/mycokeys.55.35682.

Wang, X.-N., Sun, G.-X., Li, X.-M. et al. (2018). Electron shuttle-mediated microbial Fe(III) reduction under alkaline conditions. *Journal of Soils and Sediments* 18: 159–168. https://doi.org/10.1007/s11368-017-1736-y.

Wang, B., Zheng, X., Zhang, H. et al. (2020). Keystone taxa of water microbiome respond to environmental quality and predict water contamination. *Environmental Research* 187: 109666.

Weber, K.A., Achenbach, L.A., and Coates, J.D. (2006). Microorganisms pumping iron: anaerobic microbial iron oxidation and reduction. *Nature Reviews Microbiology* 4: 752–764.

Webster, G., O'Sullivan, L.A., Meng, Y. et al. (2015). Archaeal community diversity and abundance changes along a natural salinity gradient in estuarine sediments. *FEMS Microbiology Ecology* 91 (2): 1–18.

Weinbauer, M.G., Brettar, I., and Höfle, M.G. (2003). Lysogeny and virus-induced mortality of bacterioplankton in surface, deep, and anoxic marine waters. *Limnology and Oceanography* 48: 1457–1465. https://doi.org/10.4319/lo.2003.48.4.1457.

Welsh, B.L. (1980). Comparative nutrient dynamics of a marsh-mudflat ecosystem. *Estuarine Coastal Marine Science* 10: 143–164.

Wiebe, W.J. and Pomeroy, L.R. (1972). Microorganisms and their association with aggregates and detritus in the sea: a microscopic study. In: *Detritus and its role in aquatic ecosystems*, Mem. Ist. Ital. Idrobiol., vol. 24 (Suppl) (ed. U. Melchiorri-Santolini and J.W. Hopton), 325–352.

Williams, D. (1996). Microbial diversity: Domains and kingdoms. *Annual Review of Ecology and Systematics* 27: 569–595.

Williams, T.A., Cox, C.J., Foster, P.G. et al. (2020). Phylogenomics provides robust support for a two-domains tree of life. *Nature Ecology & Evolution* 4: 138–147. https://doi.org/10.1038/s41559-019-1040-x.

Woese, C.R. and Fox, G.E. (1977). Phylogenetic structure of the prokaryotic domain: the primary kingdoms. *Proceedings of the National Academy of Science, USA* 74: 5088–5090.

Wommack, K.E. and Colwell, R.R. (2000). Virioplankton: viruses in aquatic ecosystems. *Microbiology and Molecular Biology Reviews* 64: 69–114.

Work, K., Havens, K., Sharfstein, B., and East, T. (2005). How important is bacterial carbon to planktonic grazers in a turbid, subtropical lake? *Journal of Plankton Research* 27: 357–372. https://doi.org/10.1093/plankt/fbi013.

Wright, R. and Coffin, R. (1983). Planktonic bacteria in estuaries and coastal waters of northern Massachusetts: spatial and temporal distribution. *Marine Ecology Progress Series* 11: 205–216. https://doi.org/10.3354/meps011205.

Wright, R.T. and Coffin, R.B. (1984). Measuring microzooplankton grazing on planktonic marine bacteria by its impact on bacterial production. *Microbial Ecology* 10: 137–149. https://doi.org/10.1007/BF02011421.

Yates, M.V. (2020). *Manual of Environmental Microbiology*. Wiley.

Yu, H. and Leadbetter, J.R. (2020). Bacterial chemolithoautotrophy via manganese oxidation. *Nature* 583: 453–458. https://doi.org/10.1038/s41586-020-2468-5.

Zhang, R., Wei, W., and Cai, L. (2014). The fate and biogeochemical cycling of viral elements. *Nature Reviews Microbiology* 12: 850–851. https://doi.org/10.1038/nrmicro3384.

CHAPTER 10

Estuarine Zooplankton

James J. Pierson[1] **and Mark C. Benfield**[2]
[1]University of Maryland Center for Environmental Science, Horn Point Laboratory, Cambridge, MD, USA
[2]Louisiana State University, College of the Coast and Environment, Baton Rouge, LA, USA

Zooplankton sampling with paired vertical nets in Chesapeake Bay aboard the R/V Hugh R. Sharp. Photo credit: Diane Stoecker.

10.1 Introduction

10.1.1 Why Is this Important; What Is Unique About These Organisms?

Zooplankton are the heterotrophic planktonic organisms found in aquatic systems. As a group, zooplankton include a huge diversity of forms ranging from single-celled protists that are only a few microns in size to metazoan animals that may measure more than a meter in length. All zooplankton are considered drifters, as is the definition of the root word "plankton," and as such their observed horizontal patterns are largely correlated to the major currents and environmental conditions in an estuary. However, many zooplankton are capable swimmers and migrate vertically over hourly, daily, seasonal, or annual time scales to alter their vertical position in the water. These movements can vary by life stages for some taxa and they are often a response to a variety of cues such as tidal flow, light, food availability, and predator presence.

The key aspect of zooplankton that must be considered within the estuarine ecosystem are their roles as consumers and as prey, and the implications of those roles for energy and material cycling. Within the planktonic food web, zooplankton are often primary, secondary, and sometimes even tertiary consumers, feeding on autotrophic phytoplankton and other zooplankton. Protists have been implicated as consuming the majority of fixed carbon arising from phytoplankton photosynthesis. Nearly all of the primary production that ultimately ends up in economically important and commercially fished species has transited through zooplankton. Through their migrations and feeding activity, they are also responsible for substantial fluxes of organic material through the water column, as many taxa feed near the surface where primary production occurs and then release sinking waste products that are exported into deeper water. Thus, zooplankton are the major mediators of energy flow and organic material transfer within planktonic and pelagic food-webs.

10.1.2 Key Concepts and Definitions

Zooplankton are classified in a variety of different ways, including by life stage, called **ontogeny**. Some members of the zooplankton are **holoplankton**, meaning that they remain members of the plankton throughout their entire life history. These include a wide range of taxa that span the entire size

Estuarine Ecology, Third Edition. Edited by Byron C. Crump, Jeremy M. Testa, and Kenneth H. Dunton.

Companion website: www.wiley.com/go/crump/estuarine3

range of zooplankton, including many of the heterotrophic protists, crustaceans such as copepods, some amphipods, cumaceans, cladocerans, gelatinous zooplankton such as ctenophores and scyphozoans, and a number of other taxa from a variety of phyla. Other zooplankton are **meroplankton** and are only planktonic for a portion of their life history. Examples of meroplankton include the larvae of many crustaceans, such as crab, lobster, and barnacles, and molluscs, such as bivalves. Recall that the criterion for membership in the plankton is usually based on swimming ability relative to directed horizontal currents. This distinction does not mean that all zooplankton are poor swimmers; many have excellent swimming capabilities when moving vertically or with horizontal currents. Swimming capabilities usually improve as organisms grow in size. Consequently, small organisms are generally less competent swimmers than larger ones. For example, many zooplankton begin life as small larval forms with limited swimming abilities. As they grow, they become more capable swimmers and ultimately leave the realm of the plankton for that of the **nekton**. Many species of fish produce planktonic eggs and larvae that cease to be **ichthyoplankton** when they become juveniles and their swimming ability improves. Alternatively, many benthic or sessile organisms produce planktonic larvae. These meroplanktonic larvae cease to be plankton when they metamorphose into sessile or benthic ontogenetic stages. Examples of taxa with meroplanktonic larvae include oysters, barnacles, benthic ghost and mud shrimps, crabs, and penaeid shrimp. The patterns of vertical position of an organism are another way to categorize zooplankton. **Tychoplankton**, are benthic organisms that become transient members of the plankton when they are swept from the bottom by turbulence or currents, such as benthic hydrozoans that can subsist as plankton. Similarly, **demersal zooplankton** remain on the bottom during most of the day and actively emerge into the plankton at certain times, often at night, and include some cumaceans and ostracods (Alldredge and King 1985). **Epibenthic** plankton live just above the estuary floor, often descending to feed on benthic organisms, including some copepods and mysids, among other taxa. The differences between epibenthic and demersal zooplankton mainly concern where they are predominantly found, with demersal zooplankton found primarily in the pelagic realm but utilizing the bottom, and benthic zooplankton primarily found on the bottom but utilizing the pelagic when needed.

Zooplankton may also be categorized by their size. One widely accepted size classification system proposed by Sieburth et al. (1978) is illustrated in Figure 10.1. Most estuarine zooplankton occupy size ranges that span 20 μm to 2 cm (**microzooplankton** to **mesozooplankton**) though larger organisms such as ctenophores and cnidarians may also be abundant and may range in size up to 1 m (**macrozooplankton**

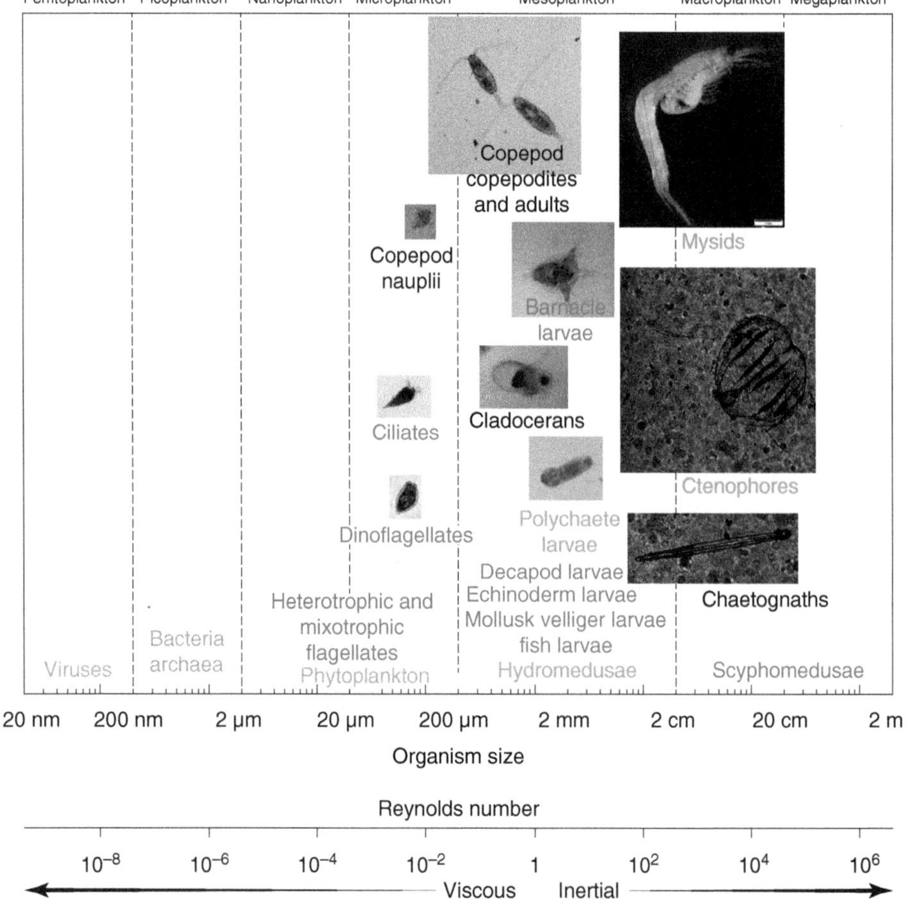

FIGURE 10.1 Size classification of plankton based on Sieburth et al. (1978) and estimates of Reynolds number for organism sizes based on Naganuma 1996. Representative zooplankton taxa are shown in their relative size classes, but images are not to scale. Estimates of Reynolds number were determined assuming a swimming speed of one body length per second. Text is color coded: grey taxa are not zooplankton, red taxa are protistan zooplankton including mixotrophs, blue taxa are meroplankton, green taxa are holoplankton or meroplankton, and black taxa are holoplankton. *Source:* Image credits are D. Stoecker (ciliate and dinoflagellate), N. Millette (copepod nauplii), J. Pierson (cladoceran, barnacle larva, copepod adults, and polychaete larva), R. Woodland (mysid), and M. Benfield (chaetognath and ctenophore). All images were collected digitally using a camera and a microscope except for the chaetognath and ctenophore which were taken using a ZOOVIS *in situ* camera system (Bi et al. 2013).

and **megazooplankton)**. The size and speed of an organism governs its interaction with the water. Smaller animals are affected more by the **viscosity** of the fluid, or the resistance of flow of the fluid. Larger animals are more affected by their own inertia, which is a function of their speed and size. The ratio of inertial forces to viscous forces is called the **Reynolds Number**. Values for Reynolds number greater than one are considered dominated by inertial forces and below one they are dominated by viscous forces. For example, consider a human and an otter in a swimming pool full of water. When they push off from the wall, the human will glide out into the pool for a longer distance than the otter because of the increased inertia of larger size. If the swimming pool was filled with honey, both the otter and the human would move much less than they did in water because of the high viscosity of the honey, but the larger human would likely be able to move further away from the wall than the otter because of the larger size. This has important ecological implications for how organisms swim, capture prey, or escape predators (Naganuma 1996). Many zooplankton have Reynolds numbers near one, meaning they exist in an environment where there is a balance between inertia and viscosity, and depending on their size and speed they can take advantage of one of the forces. For example, some copepods have been shown to use viscous forces to capture prey, creating small feeding currents with their appendages that trap small particles. At other times the same animals can use inertial forces to escape their predators, by using a coordinated movement of swimming legs and antenna to cause a rapid increase in speed that helps overcome the viscosity of the water and allows them to "jump" many body lengths in a few milliseconds. Relative estimates of the Reynolds number for different sized plankton are shown in the arrows in Figure 10.1.

10.1.3 Brief Description of Tools and Approaches (Nets, Acoustics & Optics, Models)

The study of zooplankton in estuarine systems is conducted through observations, experimentation, and modeling. In the field, observations of the abundance, distribution, and composition of zooplankton are largely done through water samplers and plankton net tows, with the specific methods chosen by the researchers depending on the target organism(s) and research question. Other means of sampling for zooplankton that have been employed to measure these quantities include active acoustics (e.g., sonar) and optical measurements with submersible equipment. The water and net sampling techniques provide a physical sample, which is necessary to gather detailed (species specific) taxonomic information, population or ecological genetic data, and physiological or biological measurements. In addition, the zooplankton collected from water sampling and net tows can provide organisms for experiments. However, water sampling and net tows may be limited in their spatial scope and temporal duration, as they are generally collected at fixed locations and require some processing time for preservation. Depending upon the abundance of zooplankton in a sample, the contents may need to be subdivided into approximately equal aliquots in a process called splitting. Mechanical plankton splitters or pipettes can perform this task and the numbers counted in each aliquot are corrected by the splitting factor to estimate the total abundance in the original sample. Acoustic and optical measurements are often conducted from moving platforms such as ships or remotely operated or autonomous vehicles, and thus can sample at higher frequency over a broader spatial domain and over a longer continuous duration. However, the increased spatial and temporal resolution comes at a cost of decreased taxonomic resolution, with measurements often constrained to broad taxonomic or size categories, or in some cases in total biomass measurements. All of these methods may also be used to collect data in the vertical dimension as well as in the horizontal dimension. In stratified estuarine systems, this may be particularly useful to discern differences in populations and communities found in the fresher and less dense surface water when compared to the saltier and denser deep water.

Experimental methods for zooplankton are often employed to measure rate processes of individuals or small numbers of copepods using incubation techniques, or the physiological or biological responses of organisms to stressors using bioassay techniques. In some cases, experimental work is conducted in the field, either on boats or near shore, and in other cases the experimental work is conducted in highly controlled laboratory settings, often with cultured organisms. Regardless of whether the research is employing observational or experimental methods, the choice of specific techniques to address the hypotheses being tested should be made based on both the target organisms and the scientific questions under scrutiny. For example, to estimate the diversity in an estuary over time, discrete samples could be taken at different locations and in different seasons. To determine if populations are controlled by birth rate, experiments can be conducted to measure the egg production rate in different locations or at different times.

Models of estuarine zooplankton range from concentration-based approaches that estimate changes in populations of different classes of organisms over time and space, to individual-based models designed to simulate how individual behaviors and variation of those behaviors within a population affect the overall population and ecosystem. All models are informed by the observations of, and experiments conducted on zooplankton in estuaries, and they are used to test hypotheses about estuarine zooplankton. Even though they are entirely virtual in practice, models provide a means to simulate and test ideas about how an ecosystem works, or our understanding of those systems, or how we expect those systems to change over time or with perturbations.

10.2 Diversity and Global Distribution in Estuaries

Taxonomically, the zooplankton assemblage in an estuary may contain a mix of marine, estuarine, and freshwater taxa, whose relative abundance depends on factors such as distance from the sea, tidal stage, and hydrological conditions in the surrounding catchment. Estuaries support high zooplankton biomasses that often exceed those found over the adjacent continental shelves. Differences in biomass and abundance of zooplankton between estuaries and their freshwater sources are less consistent, with peaks sometimes occurring in estuaries and sometimes in the rivers feeding them, depending on trophic status and flushing rate. Some overarching patterns exist with certain taxonomic groups dominating many estuaries, but the species composition of estuaries tends to vary with time (seasonally, annually), with geographical location (latitude, adjacent saltwater bodies, and freshwater sources) and with local conditions (temperature, salinity, nutrient input, dissolved oxygen, primary productivity, and other factors).

Generally, the taxonomic diversity of zooplankton in estuaries varies along the salinity gradient. Salinity zones within estuaries have typically been based on the Venice classification (Anonymous 1959). This classification scheme divides estuaries into five salinity zones: limnetic (0–0.5 psu), oligohaline (0.5–5.0 psu), mesohaline (5–18 psu), polyhaline (18–30 psu), and euhaline (>30 psu). This scheme has been criticized as lacking biologically relevant boundaries by Bulger et al. (1993) who proposed a biologically based scheme consisting of overlapping zones derived from a statistical analysis of the zonation patterns of 316 estuarine species and life stages: 0–4 psu, 2–14 psu, 11–18 psu, 16–27 psu, and 24 psu to marine waters. Whatever system is employed, estuarine zooplankton frequently manifest definite horizontal zonations that vary with the prevailing tide and freshwater discharge. The general pattern is that zooplankton species diversity is higher at the freshwater and oceanic ends and lower in the middle (mesohaline) portion of estuaries where zooplankton biomass and abundance are often dominated by a small number of species (Kimmel 2011).

Microzooplankton in estuaries are protists, and these communities often include both strictly heterotrophic taxa and **mixotrophic taxa**. Mixotrophs are organisms that can engage in both autotrophy, usually by harnessing the energy of the sun through photosynthesis, and phagotrophy, where they can also ingest organic particles that are then metabolized for energy (see Chapter 9). Mixotrophic protists include many dinoflagellate and ciliate species, and they exist along a spectrum that ranges from taxa that are primarily autotrophic to those that are primarily heterotrophic. The fraction of energy gleaned from these different processes is usually a function of the available light and nutrient conditions.

Overall, protistan microzooplankton are a highly diverse group of taxa that are responsible for much of the grazing on small primary producers and bacteria in estuaries and serve as an important food source for mesozooplankton. Global patterns of microzooplankton diversity are not well studied, but one example is given by Dolan and Gallegos (2001) who provide evidence that species richness of estuarine small microzooplanktonic ciliates, called tintinnids, is maximal near the equator and declines linearly toward both poles. While they did not provide any mechanistic explanation for the pattern, the relationship was linear suggesting that it was driven by some factor that varies in a linear manner with latitude, and similar patterns may or may not be seen in other taxa. **Rotifers** are animals that straddle the size definitions of micro- and mesozooplankton, and at times can be abundant in estuarine systems, often in fresher portions of the systems.

Holoplanktonic mesozooplankton assemblages within estuaries are frequently dominated by **copepods**, **cladocerans**, and **mysids**. These small crustaceans are important grazers on microplankton, including phytoplankton as well as mixotrophic and heterotrophic microzooplankton. In addition, all of these crustaceans have high **secondary production**, meaning that they have high growth rates driven by feeding on autotrophic primary producers. These animals serve as prey for organisms in higher trophic levels such as fish, ctenophores, medusae, chaetognaths, and various decapod crustaceans. Meroplanktonic zooplankton usually include the larval forms of barnacles and other crustaceans, and mollusks.

Copepods are a subclass of crustaceans that are the most abundant metazoans in the oceans, and that also occur in high abundances in estuarine systems. Copepods are a highly diverse group that includes parasitic, benthic, and planktonic forms from several taxonomic orders. Here we will primarily consider the planktonic forms, of which there are two dominant orders in estuarine systems: the calanoids and the cyclopoids, with a reference to the benthic harpacticoid copepods in a later section. These two groups of copepods are differentiated by their morphologic and genetic variation. Both groups are found in marine, estuarine, and freshwater environments. Calanoids generally have long antenna that extend beyond the length of the body, and are largely omnivorous or herbivorous, with a few carnivorous taxa. Cyclopoids generally have short antenna and often feed on detritus or are omnivorous. Both groups are important grazers of phytoplankton and microzooplankton and are in turn, prey for zooplanktivorous fishes and other invertebrates. Copepod development begins with eggs and depending upon the species these are either carried by the female (brooding) or released as fertilized embryos into the water (broadcast spawning). Nearly all cyclopoids carry their eggs in egg sacs on either side of their abdomen, but calanoids exhibit both broadcast spawning and brooding. Eggs hatch and pass through a series of six naupliar and six copepodite stages, with the final copepodite stage being the sexually mature adult. The rapid generation times of copepods (10–30 days in subtemperate regions; Ianora 1998) allow them to respond quickly to changing environmental conditions.

Common planktonic estuarine copepods include species of the genera *Acartia, Eurytemora*, and *Pseudodiaptomus*. In the lower reaches of estuaries, *Paracalanus* and the cyclopoid genus *Oithona* are often common. Species richness of copepod assemblages may be very high in tropical systems. For example, Revis (1988) documented 102 copepod species in a Kenyan estuary during an 11-month study, although only 12 species were dominant. In many estuarine systems observations of introduced or **invasive species** copepods are becoming more common. These introduced taxa are sometimes from the same genera as the native taxa, but not always.

Turner (1981) examined species diversity of calanoid and cyclopoid copepods over a broad latitudinal gradient in the northern hemisphere from the arctic to subtropics. There was no clear relationship between species richness of calanoid copepods over the range of latitudes examined; however, species richness peaked in temperate latitudes (30–50 °N). Cyclopoid copepods displayed a different pattern with species richness being very low in higher latitudes and increasing to a maximum in subtropical latitudes (south of 35 °N). Turner (1981) identified several groups of species with distinct latitudinal distributional patterns. The peak in calanoid species richness within temperate regions was attributed to the presence of these species groups in the estuaries at different times of the year. Calanoid and cyclopoid copepods have also been shown to vary with ecological stoichiometry in estuarine systems, with cyclopoids dominating in systems with higher N:P (less phosphorus relative to the amount of nitrogen), and calanoids dominating in systems with lower N:P (Glibert et al. 2011), possibly due to different nutritional requirements resulting from different feeding behaviors.

Recent work has suggested that there are also **cryptic species** of copepods within estuarine taxa, driven by salinity. Chen and Hare (2008, 2011) showed that in a number of estuarine systems along the US east coast, three different "lineages" of *Acartia tonsa* were identified using genetic markers, which were separated strongly by the salinity in which they were collected—an *S* lineage found in high salinity water, an *X* lineage found in mid-salinity waters and more regionally restricted, and an *F* lineage found in the lowest salinity regions. Other research has shown more diversity in populations identified as *A. tonsa* throughout the range they are found (Costa et al. 2014; Drillet et al. 2008). The *F* and *S* lineages within the Chesapeake Bay appear to be reproductively isolated, with no viable eggs produced for crosses between adults from different lineages (Plough et al. 2018). In the St. Lawrence River Estuary, cryptic species of the copepod *E. affinis* were also shown to vary in dominance seasonally and along a variety of gradients (Winkler et al. 2008; Favier and Winkler 2014). Cryptic diversity is likely to be more common in estuaries than is currently appreciated, due to the dynamic nature of estuarine systems which can support a wide variety of ecological niches that arise from the differences in salinity and other factors throughout an estuary.

Mysids are an order of small shrimp-like crustaceans that often constitute a large fraction of the zooplankton numbers and biomass in estuaries. Often called "opossum shrimp," some common estuarine genera include *Neomysis*, *Mysis*, *Mesopodopsis*, and *Rhopalophthalmus*. Mysids are among the larger estuarine zooplankton observed in estuaries. Their size means that in some cases they dominate the mesozooplankton biomass, even if they are not numerically dominant. For example, Wooldridge and Bailey (1982) reported that mysids exceeded 90% of the total mesozooplankton dry mass in the Sundays River estuary, South Africa. While mysids are generally regarded as omnivores (Mauchline 1980), individual species are able to coexist in a single estuarine system by specializing on different prey and by varying their diets as prey abundances diminish. Mysid's ability to switch prey as availability changes means that they can be important predators of a variety of different taxa including copepod nauplii and copepodites, rotifers, and other meroplanktonic larvae such as mollusc larvae (Winkler et al. 2007). Through their ability to select for different sizes of prey, mysids can play a key role in structuring estuarine food webs. Winkler et al. (2007) examined the feeding ecology of *Mysis stenolepis* and *Neomysis americana* in the St. Lawrence River estuary through a combination of controlled feeding experiments and stable isotope analyses of field collected samples. *Mysis stenolepis* was characterized as a raptorial feeder that selected for larger prey such as the copepodites of the copepod *Eurytemora affinis*. When its preferred prey was scarce, it switched to a filter-feeding mode and consumed more abundant smaller prey items such as copepod nauplii, rotifers, and gastropod veligers. In that same study, *Neomysis americana* was considered a filter-feeder that opportunistically consumed the most available, and generally smaller prey items. Thus, through their feeding activity, mysids are important in transferring carbon from the microzooplankton, mesozooplankton, and detrital pools into small zooplanktivorous fishes and other larger invertebrate predators (Vilas et al. 2007).

Other invertebrates are also important constituents of estuarine holoplankton. Freshwater cladocerans, including holoplanktonic crustaceans such as *Bosmina* and *Daphnia*, may be abundant at the fresh water end of the estuarine turbidity maximum (ETM). Toward the saltwater end of estuaries cladocerans belonging to the genera *Podon* and *Evadne* may also be common. Gelatinous zooplankton such as ctenophores and cnidarians can be important predators of mesozooplankton. Ctenophores are a phylum of gelatinous animals that are called "comb jellies" because of the long rows of sticky, specialized cells (called coloblasts) that they use for capturing prey. Most cnidarians in estuaries are in the class scyphozoa, and they have long tentacles that contain specialized stinging cells called nematocysts that are used to stun and capture prey. These gelatinous predators primarily consume copepods, fish eggs and larvae, and sometimes other gelatinous zooplankton. The ctenophore genera *Pleurobrachia* and *Beroe* are common in the middle regions of estuaries, while the ctenophore *Mnemiopsis* and cnidarian jellyfish such as those in the genus *Chrysaora* can reach high abundances in the middle and ocean ends of estuaries. *Mnemiopsis* is particularly tolerant to low salinities and the accidental introduction of *M. leidyi* into the Black Sea in the 1980s resulted in a population explosion that decimated anchovy populations until the accidental introduction of a predatory ctenophore *Beroe ovata* produced some biological control of *M. lediyi* (Purcell et al. 2001). In the Narragansett Bay, changes in the timing of ctenophore abundance have been correlated to decreased copepod abundance as the overlap in time and space between the ctenophore predator and copepod prey increased (Costello et al. 2006).

10.3 Spatial and Temporal Patterns of Biomass and Productivity

10.3.1 Spatial Patterns of Biomass

The dynamic environment of estuaries makes it difficult to assemble an accurate estimate of the distributional patterns of any estuarine zooplankton species at a given time. Estuarine circulation causes planktonic organisms to be continuously shifted toward the ocean at the surface and landwater near the bottom (see Chapter 2). Moreover, this basic pattern of dispersal varies with seasonally and annually changing freshwater runoff and ocean conditions. Thus, developing a composite picture of the distributions of organisms throughout an entire system requires sampling at the proper time and space scales. Toward the extremes of an organisms' distribution, animal densities may be too low for their abundances to be reliably estimated without extensive sampling effort, for example for freshwater species found in higher salinity water in lower estuaries. All of these factors mean that the resultant maps of estuarine species distributions are, at best, approximations of the true synoptic distributional patterns of the constituent taxa.

Given the challenges just summarized, and in spite of the importance of estuarine zooplankton, there are relatively few publications that report the distributional patterns of mesozooplankton with high spatial and taxonomic resolution in relation to hydrographic parameters. Zhang et al. (2006) used an undulating vehicle equipped with a CTD and an optical plankton counter (OPC) to map the distribution of zooplankton biovolume along the axis of the Chesapeake Bay during three seasons from 1996 to 2000. The OPC is an instrument that measures the interruption of a light beam by particles to estimates the abundance and sizes of zooplankton and other particles; however, it cannot determine the identity of each particle. This study demonstrated the dynamic and patchy nature of estuarine zooplankton distributions albeit in a taxonomically ambiguous manner. This study also demonstrates how sampling a large system can be rather time-consuming; even while using a towed, semi-automated sensing system such as the OPC, it required more than 30 hours to sample from the head of the estuary to the mouth.

The horizontal distribution patterns of zooplankton are by no means static. Individual species are concentrated within waters that contain conditions favorable for growth and reproduction. As mentioned earlier, salinity is an important determinant of the spatial distributions of most zooplankton. An example of how salinity influences the distribution of zooplankton comes from research in the Chesapeake Bay, using mesozooplankton monitoring data collected from 1984 to 2002 at stations throughout the Bay (Kimmel et al. 2006). Two copepod taxa dominate in the Chesapeake; *Eurytemora carolleeae* (recently differentiated from the European *E. affinis*) and *Acartia tonsa*. *E. carolleeae* is generally found in oligohaline water nearer to the head of the Bay and in the upper reaches of tributaries, whereas *A. tonsa* is found in mesohaline water. Years were categorized as "wet" or "dry" based on the annual discharge of freshwater in the Chesapeake Bay, and the distributions of these two dominant copepods were examined in each category of water. The findings show how in "wet" years with high freshwater input the habitat and abundance of *E. carolleeae* is increased along with the region of the bay with oligohaline water, compared to "dry" years (Figure 10.2). At the same time, in "wet" years the distribution of *A. tonsa* is reduced due to the decreased volume of mesohaline water. In dry years, the pattern is reversed, with more habitat available to support *A. tonsa* compared to *E. carolleeae*. This example highlights the importance of salinity and freshwater input in determining species distribution on an annual scale, and other studies have shown similar patterns, for example in the Elbe River Estuary, Germany (Peitsch et al. 2000), and the St. Lawrence Estuary in Canada (Bousfield et al. 1975).

During extreme precipitation and runoff events (e.g., tropical storms, floods) rapid seaward currents can flush planktonic organisms out of estuaries. Some evidence of how copepods are adapted to repopulate estuaries was provided by Ueda et al. (2004) during routine zooplankton sampling in the tidal region of the Chikugo River, Japan. This sampling fortuitously encompassed a period of heavy rain coupled with the opening of an upstream reservoir. These events create a strong, net seaward flow that flushed the majority of the population of the dominant copepods *Pseudodiaptomus inopinus* and *Sinocalanus sinensis* out of the system. Sediment sampling in the channels off the mouth of the Chikugo River revealed that many *P. inopinus* copepods were aggregated in a thin layer immediately above the bottom. Moreover, these copepods were almost entirely adults, which would be capable of stronger swimming responses than younger stages. No *S. sinensis* were detected in these samples. *S. sinensis* were subsequently detected in bottom samples during flood tides as they were displaced seaward. By aggregating in a thin layer adjacent to the sediments, *P. inopinus* were hypothesized to be exploiting a refuge with low current velocity, that was well-placed for a return to the estuary on the following flood tide. This would allow them to rebuild their population through reproduction following the high flow event. In cases where zooplankton are unable to return to the estuary from which they were flushed, they may be replaced by more rapidly growing taxa until the reestablishment of the dominant but slower growing taxa. For example, in such cases copepods could be replaced by rotifers.

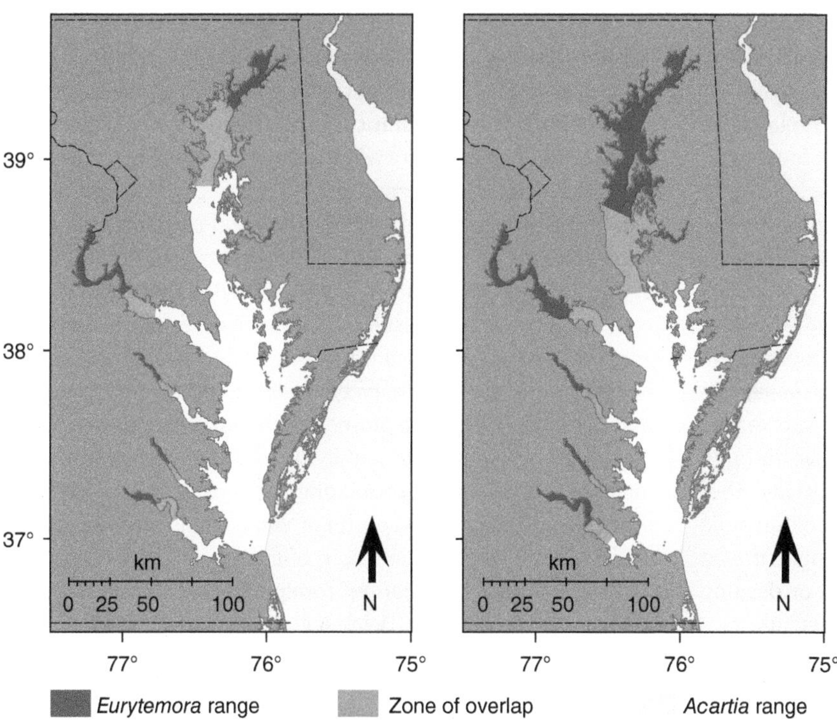

FIGURE 10.2 Conceptual diagram of the spring geographical distribution of *Acartia tonsa* (region shown in yellow), *Eurytemora carolleeae (formerly identified as E. affinis;* region shown in red), and the zone where the two species overlap (region shown in orange) in the Chesapeake Bay during a dry period (left panel) and a wet period (right panel). *Source*: Adapted from Kimmel et al. (2006). Note that these data are specific to the Chesapeake Bay, and so regions shown in light blue are not included in estimates of potential habitat.

10.3.2 Temporal Patterns in Biomass and Production

One of the most important factors affecting the rate of population growth for a species is its generation time (Cole 1954). For animals, this is the time required for an organism to develop from an egg to a mature adult and is strongly temperature dependent (Forester et al. 2011). For example, the generation time for the calanoid copepod *Eurytemora* spp. varies from approximately 68 days at 5.5 °C to 9 days at 25 °C (Heinle and Flemer 1975). Although the generation time varies by species and is influenced by a variety of factors, there are some general relationships that predict generation time as a function of body size and temperature for a variety of zooplankton taxa, including rotifers, copepods, and cladocerans among others (Gillooly 2000). Growth rates of individuals (the rate at which individuals increase in mass) are also dependent on temperature, but not in the same way as generation time. In general, growth rates increase less than generation time with each degree change in temperature, which leads to larger individuals at lower temperatures (Miller et al. 1977; Forster et al. 2011). In other words, at higher temperatures individuals spend less time in each stage, but their change in mass does not happen at the same rate, so they are not able to grow as large in each stage. By the time they reach adulthood, this leads to smaller individuals at higher temperatures (Pierson et al. 2016).

Fluctuations in the abundances of estuarine zooplankton in general, and copepods in particular, appear to be linked to freshwater discharge. In regions where there is a pronounced wet and dry season, zooplankton may increase in abundance in response to precipitation and runoff. The putative causal relationship is an increase in nutrient concentrations that stimulate phytoplankton blooms, which trigger increased production by microzooplankton and mesozooplankton. Given the preference by copepods for microzooplankton prey over phytoplankton (Gifford and Dagg 1988; Rollwagen-Bollens and Penry 2003), elevated grazing by copepods on microzooplankton may reduce grazing rates on phytoplankton and thereby enhance the standing stock of phytoplankton. Such a mechanism could account for observations of enhanced copepod abundances during the wet season in many estuaries. Pulsed discharges of water into estuaries may also enhance zooplankton production by improving the variety and nutritional value of their prey. Episodic freshwater inflows may prevent competitive exclusion among phytoplankton and lead to a more diverse phytoplankton assemblage (see summary by Miller et al. 2008). This may provide zooplankton with a more diverse and nutritious array of phytoplankton prey with consequent higher productivity.

10.4 Factors Controlling Productivity and Community Composition

10.4.1 Physiological Challenges for Zooplankton in Estuaries

The physical properties of estuaries, particularly salinity and temperature, are more variable than in the adjacent coastal ocean or freshwater systems. This is a consequence of the dynamic nature of estuaries where tides, river discharge, and winds combine to mix and shift the spatial patterns of salinity, temperature, turbidity, nutrients, dissolved oxygen, and other parameters that can affect the distributions of zooplankton. Many other chemical and biological factors such as food availability, predation, and pollutant concentrations display similar variability. The shallow depth of most estuaries compared to the coastal ocean makes estuaries subject to more frequent, higher amplitude temperature fluctuations than the coastal waters. In stratified estuaries, where a low-density freshwater layer flows down estuary over a higher density and more saline layer flowing in from the mouth, zooplankton that engage in regular vertical migration patterns will experience substantial changes in their abiotic environment as they move through strong gradients in temperature, salinity, and potentially other factors (Laprise and Dodson 1993). Estuarine zooplankton must therefore adapt to frequent physiological challenges driven by the continually changing nature of their milieu or develop strategies that reduce the magnitude of these challenges.

As noted repeatedly in this chapter and others, salinity is often the key physical factor regulating the spatial distribution and structure of estuarine zooplankton. Most estuarine zooplankton are **osmoconformers** and each taxon has a range of salinities that are within its zone of physiological tolerance. Moreover, salinity may also influence the distributions of prey items (Kimmel and Roman 2004) and predators (Kimmel et al. 2009), which will in turn, affect the distribution of those organisms that depend on them for food. Salinity fluctuates on short time scales associated with tidal cycles. In positive estuaries (those with salinity increasing toward the sea, see Chapter 2), average salinity typically increases during floods and declines over ebbs. Moreover, the spatial distribution of isohalines is displaced inland by flood tides and seaward during ebb tides. Estuarine zooplankton may engage in behavioral strategies designed to keep them within their physiologically preferred salinities. An example of such behavior that occurs in the euryhaline estuarine copepod *Eurytemora affinis* is evident in research by Devreker et al. (2008). In the Seine

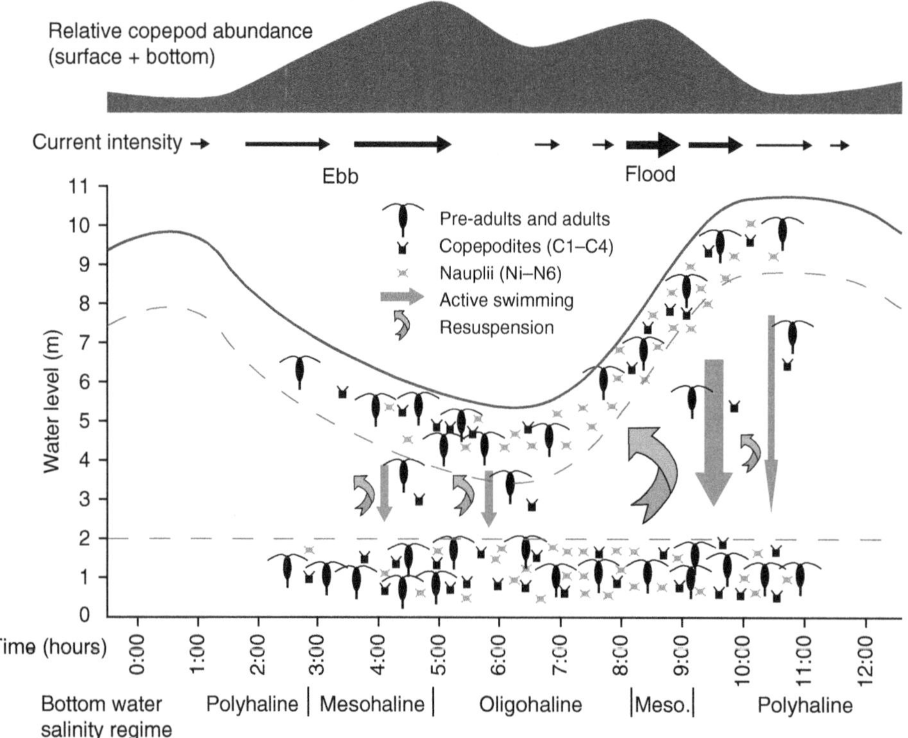

FIGURE 10.3 Schematic representation of the dispersion of developmental stages of *Eurytemora affinis* in the middle part of the Seine estuary (Normandy Bridge) as a function of the mean tidal cycle. Width of arrows at the top of the figure represents the magnitude of the water velocity during a length of time represented by their length. The bottom water masses are identified in the bottom of the figure as a function of salinity range according to McLusky (1989): oligohaline zone [0.5–5], mesohaline [5–18], and polyhaline [18–25]. The population abundance (at top of figure) increased during the ebb as copepods were advected down-estuary past the sampling station. Ebb tide featured low, constant resuspension and possible migration of adults (oval) and copepodids (square) that dominated the population from the poly- to mesohaline zone in surface and bottom water. In the oligohaline zone around low slack tide, when current velocity was low, nauplii dominated the population. At the beginning of the flood when current velocity was maximal (08:30–09:00), the population was resuspended, and adults and copepodids migrated to the bottom water while the current was decreasing. *Source*: Figure 9 from Devreker et al. (2008)/With Permission of Oxford University Press.

estuary, France, *E. affinis* is the numerically dominant copepod (90–99%) in the low salinity zone (0.5–15 psu) and it remains a dominant species in the zone for most of the year. Using very high resolution (15 minutes intervals over 50 hours) sampling at the surface and bottom, Devreker et al. (2008) demonstrated how *E. affinis* maintains high abundances in the salinity zone it favors. Densities of adults and copepodids were higher near the bottom than in surface waters during ebbs (Figure 10.3), suggesting a strategy to avoid being swept out of the estuary by ebb tides. Resuspension into surface waters occurs during the early-flood tides with consequent landward advection followed by active migration down into bottom waters during the latter part of the ebb (Figure 10.3). In this manner, *E. affinis* is able to avoid being swept out of the Seine estuary while maintaining position within its preferred salinity range. Similar behavioral and distribution patterns were shown for *E. affinis* in the Columbia River Estuary, leading to maintenance of their position in the estuary (Morgan et al. 1997). In that study, the benthic copepod *Coullana canadensis* from the order harpacticoida was also observed in the zooplankton nets and was found to maintain its position in the estuary. *C. canadensis* remained close to the bottom with distribution patterns closely aligned with sediment particles, suggesting that its position was kept through more passive means than *E. affinis*, and affirming its primarily benthic lifestyle.

Water temperature is another variable influencing estuarine zooplankton on short-term and long-term time scales. The annual succession in copepod species composition may be driven by temperature (Mauchline 1998) as well as salinity or other environmental factors. In addition to the previously mentioned temperature dependence on development and growth rates, other vital rates such as egg production rate and respiration rate are also strongly temperature dependent (Gillooly 2000). In temperate estuaries, reproduction by benthic bivalves such as oysters, resulting in release of meroplanktonic larvae, is often triggered by rising water temperatures (Ingle 1951), as is the spawning of some anadromous fish that use estuaries as nurseries (Secor and Houde 1995).

10.4.2 Trophic Interactions

Estuarine food webs can be represented by generalized marine food webs that include primary producers, primary consumers, and a series of trophic linkages culminating in top predators (Figure 10.4). Included in such food webs is a recycling step involving dissolved organic carbon (DOC), heterotrophic bacteria, and microzooplankton, termed the microbial loop (see Chapter 9). Zooplankton generally occupy the lower strata of estuarine food webs although some cnidarians and ctenophores may be important predators in some systems. Sources of food for estuarine zooplankton include phytoplankton, microzooplankton, detrital material from terrestrial plants, emergent macrophytes, submerged aquatic vegetation, other zooplankton (e.g., fecal pellets), and pollen. In addition, phytoplankton from marine sources may also be abundant in the lower regions of estuaries while freshwater phytoplankton provide an energy subsidy to the upper regions of an estuary. Global analyses suggest that microzooplankton graze on average 60% of the primary production in estuaries (Calbet and Landry 2004), and mesozooplankton consume 8–24% of primary production (Steinberg and Landry 2017), depending on the biomass and growth rate of both the mesozooplankton grazers and the phytoplankton. However, on any given day, the fraction of primary production depleted by zooplankton can be much lower or much higher depending on season and

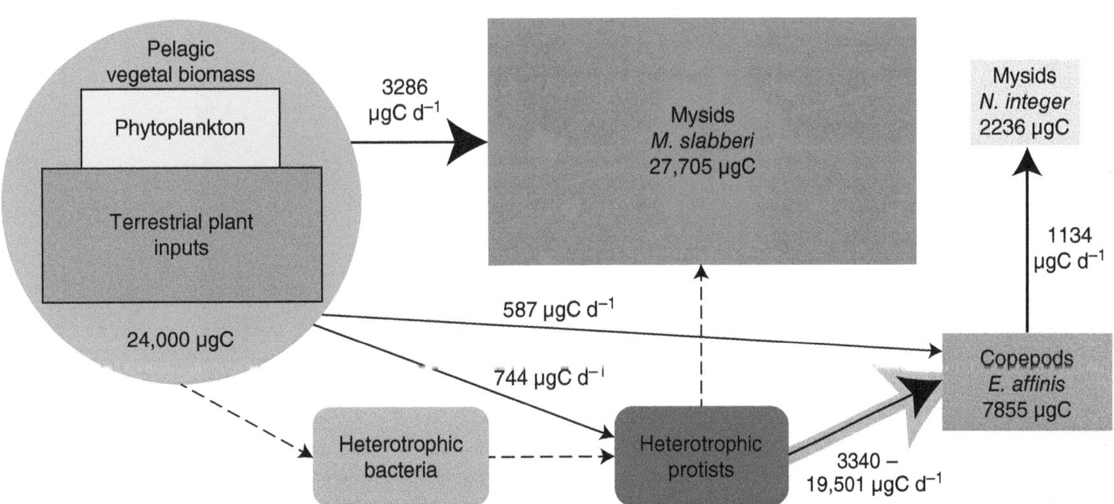

FIGURE 10.4 An example of an estuarine planktonic food web from within the turbidity maximum zone of the Gironde River estuary in July 2002 showing biomass and flux expressed as $\mu g\,C\,m^{-3}$ and $\mu g\,C\,m^{-3}\,day^{-1}$, respectively. Estimation of such a food web requires careful measurement of the standing stocks within each pool and grazing rate experiments to quantify the magnitudes of fluxes among consumers. Sizes of boxes and arrows are approximately scaled to the sizes of the pools and fluxes, respectively. Heterotrophic bacteria and protozoa biomass were not measured for this study, and dashed lines show fluxes that were not estimated in this analysis. *Source*: Adapted from David et al. (2006).

local conditions (Lonsdale et al. 1996; Millette et al. 2015). In addition, DOC released by all these organisms may be used by heterotrophic bacteria that are subsequently consumed by microzooplankton. Depending on the depth of the estuary and the degree of mixing, turbidity may be very high in estuaries, which can reduce light penetration and thus reduce the primary production in estuaries compared to adjacent marine and freshwater environments. Where primary production is reduced by turbidity, highly nutritive phytoplankton food may not be as prevalent as detrital material from other plant pools.

Certain regions of estuaries are particularly important foraging regions for zooplankton. One of these is the ETM zone. ETMs are regions of strong interaction between freshwater and saltwater, which occur in many estuaries at the landward limit of saltwater intrusion. Turbulence generated by tidal currents flowing over the bottom leads to resuspension of benthic sediments, while mixing of the overlying fresh water with the underlying saltwater causes suspended particles to flocculate and sink. These two processes combine to produce a zone that is characterized by elevated turbidity, high particulate loadings, and strong horizontal gradients in salinity. Circulation patterns produce a convergence near the bottom that leads to an accumulation of particulate material and zooplankton. In addition to the high concentration of particulates that are rich in bacteria, ETMs can also contain abundant phytoplankton derived from *in situ* production and both upstream fresh water and downstream salt water sources. Thus, ETMs can provide an enhanced foraging ground for zooplankton.

Within an ETM, copepods can attain very high abundances. For example, Winkler et al. (2003) estimated the abundance of the copepod *Eurytemora affinis* at 25,000 individuals m^{-3} in the St. Lawrence River estuary ETM during late-June. Based on measurements of body mass and carbon content and carbon-specific ingestion rates, they estimated that this population of copepods consumes approximately 50,000 kg of phytoplankton carbon per day during late-June and early-July. Other studies support the importance of copepods as grazers within the ETM but demonstrate the importance of other taxa such as mysids as grazers (Figure 10.4). Roman et al. (2001) reported abundances of *Eurytemora* from the bottom waters of the Chesapeake Bay ETM of over 200,000 individuals m^{-3}. These abundances were from acoustically derived measurements and bottle samples collected at specific depths.

Zooplankton are important prey for a broad range of predatory organisms, and their body size partly determines the predators to which they are vulnerable. Large zooplankton including adult copepods and mysids are more conspicuous and possess greater escape capabilities than smaller organisms, and are generally preyed upon by zooplanktivorous fish. Predation by these fish is usually more intense in regions where turbidity is reduced allowing more effective visual foraging. Mesocosm studies in a coastal fjord in Denmark demonstrated that zooplanktivorous fish were responsible for significant reductions in the abundances of holoplanktonic zooplankton such as the copepod *A. tonsa* and the cladoceran *Pleopis polyphemoides* (Horsted et al. 1988). In the same study, microzooplankton including tintinnids were preyed upon by suspension-feeding bivalves *Mytilus edulis* but were not impacted by fish due to their small size. In addition, larger holoplankton were unaffected by bivalves because they presumably were able to detect and avoid the incurrent siphon flows. Smaller zooplankton such as copepod nauplii and microzooplankton are often preyed upon

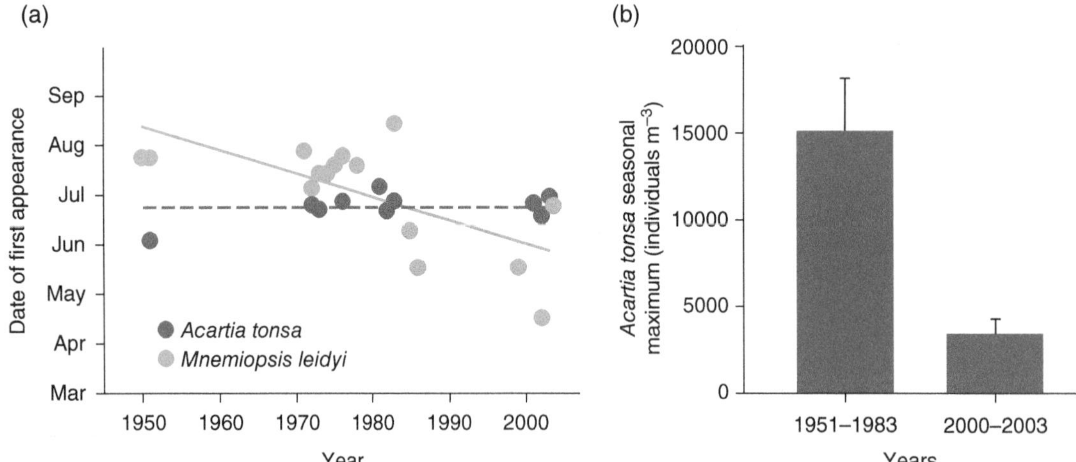

FIGURE 10.5 (a) Historical patterns in phenology of *Acartia tonsa* and *Mnemiopsis leidyi* at central estuary stations in Narragansett Bay, U.S.A. Linear regression indicates that *A. tonsa* phenology did not change ($p = 0.195$) during the period from 1950 to 2003. In contrast, the first appearance of *M. leidyi* shifted earlier in the year ($p = 0.006$) during the same period. (b) Decline in maximum seasonal concentrations of the copepod *Acartia tonsa* in Narragansett Bay during the period 2000–2003 ($n = 4$) relative to years between 1951 and 1983 ($n = 7$). All data collected from the same site; historical data (pre-2000) assembled from a variety of sources. Error bars represent standard error of the mean and, although not shown, negative bars are symmetrical with positive bars. Average values for the two time periods are significantly different (Mann–Whitney, $p = 0.008$). *Source*: From Costello et al. (2006)/With Permission of Oxford University Press.

by larger zooplankton and by filter-feeding organisms. However, foraging birds such as long-tailed ducks feed on large zooplankton, and may impact their abundance and distribution.

Ctenophores, particularly *Mnemiopsis leidyi*, are voracious consumers of copepods and other zooplankton. Condon and Steinberg (2008) found that calanoid copepod densities were inversely related to *M. leidyi* biomass in the York River estuary—a subsystem of Chesapeake Bay. These nonvisual predators can consume copepods and other prey in turbid waters that would otherwise offer some refuge from visual predators such as fishes. In the Chesapeake Bay, densities of *M. leidyi* are also under the control of other gelatinous predators such as the medusa *Chrysaora quinquecirrha* (Condon and Steinberg 2008) and potentially other medusa species, as well as the lobate ctenophore *Beroe ovata*. The absence of natural predators in systems where *M. leidyi* has been introduced likely explains why this species has undergone such dramatic population explosions.

Global climate change has the potential to alter trophic interactions in estuarine systems by altering the timing of seasonal abundance patterns. Planktonic communities may be particularly sensitive to small mean annual increases (~1 °C) in water temperature (Oviatt 2004) because individual species respond differently to altered climatic patterns. One example of this may be seen in Narragansett Bay, USA, where the lobate ctenophore *Mnemiopsis leidyi* and calanoid copepod *Acartia tonsa* are dominant members of the planktonic community (Costello et al. 2006). *Mnemiopsis leidyi* feeds on *A. tonsa* and other prey items and has the potential to drastically reduce the abundance of *A. tonsa* via predation. Both *A. tonsa* and *M. leidyi* dominate in summer with the former peaking during spring-summer and the latter during late-summer and early-fall. Both species decline to low abundances during winter with *M. leidyi* overwintering in shallow bay areas while *A. tonsa* eggs overwinter in deeper regions of the bay. As mean annual water temperatures have been warming in Narragansett Bay, the date of first appearance of *M. leidyi* at central bay stations has advanced by 59 days over the period 1951–2003 (Costello et al. 2006). Over the same period, the phenology of *A. tonsa* has not changed significantly. The differential response is presumably because the shallower waters where *M. leidyi* overwinters are more sensitive to warming and ctenophore egg production is temperature dependent. As a consequence, the peak abundance of the predator *M. leidyi* has become more closely matched to that of *A. tonsa* and extends into a period when the copepod previously enjoyed a refuge from ctenophore predation (Figure 10.5). In response to this shift, the peak abundances of *A. tonsa* have significantly declined since 2000 (Figure 10.5; Costello et al. 2006). This change has implications for the entire planktonic food web in Narragansett Bay where other organisms are dependent upon *A. tonsa*.

Other long-term changes in estuarine zooplankton have been observed. In the Chesapeake Bay, a decline in *A. tonsa* abundance has been observed over a 50-year period (Kimmel et al. 2012). During that same time period, a number of changes to the system have occurred which likely contributed to the decline in the copepod abundance, including increased temperatures, decreased abundance of the medusa *Chrysaora quinquecirrha* (now, *C. chesapeakei*) increased abundance of *M. leidyi*, and increased volume of low oxygen (**hypoxic**) bottom water (Figure 10.6; Kimmel et al. 2012). The decreases in *C. quinquecirrha* may be attributed to reduced oyster populations, as the polyp stage of *C. quinquecirrha* requires hard substrate for attachment and appears to prefer oyster shell. This decrease in a key predator of *M. leidyi* may have contributed to its increase, and the decrease in its prey, the copepod *A. tonsa*. Thus, similar patterns of long-term change may be observed in estuaries that are the ultimate result of anthropogenically induced changes in the ecosystem and in climate patterns.

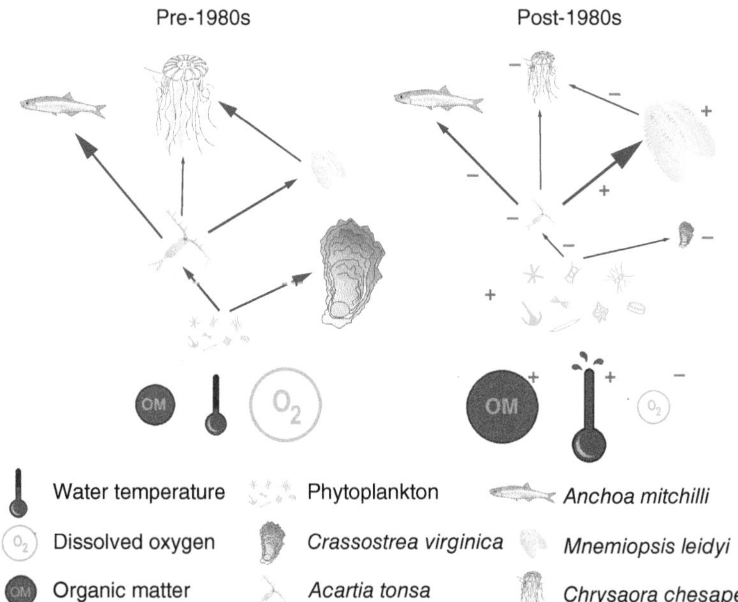

FIGURE 10.6 Conceptual diagram of environmental and trophic interactions in the middle portion of Chesapeake Bay before the 1980s (left side of diagram) and after 1980s (right side of diagram). Size of the symbols indicates the relative abundance, and thickness of the arrows indicates the relative amount of consumption. Changes in abundance and consumption from the pre-1980s to the post-1980s is also shown by the + and – symbols next to symbols and arrows in the right side of the diagram.

10.5 Special Topics Unique to Zooplankton

10.5.1 Recruitment

Many of the meroplanktonic organisms in estuaries are the larvae of species that either spawn outside of estuaries or whose larvae or juveniles leave estuaries, but some are residents of the estuaries throughout their life (Figure 10.7). Many estuarine bivalves such as oysters reside in estuaries their entire life, and their larvae are retained within estuaries. Since most estuaries have a riverine and oceanic terminus, some meroplanktonic larvae originate from freshwater species that spawn in the upper reaches of estuaries, such as the freshwater shrimp genus *Macrobrachium*, or they originate from marine species that spawn near or outside the mouth of estuaries, such as the blue crab (*Callinectes sapidus*). Marine-sourced larvae predominate in most systems. Estuarine-dependent larvae living in shelf waters face two challenges in getting back to an estuary. The first is crossing the shelf and moving toward the coast so that they are located near an estuarine mouth. The second challenge is to successfully gain entrance to an estuary. Much research has been devoted to these two topics and a consensus is beginning to develop on which factors are responsible for successfully achieving cross-shelf transport and estuarine ingress. Successful larval recruitment appears to require a combination of favorable physical and behavioral processes.

Cross-shelf transport is often facilitated by onshore circulation driven by favorable winds. On coasts where upwelling predominates, relaxation of upwelling-favorable winds can cause onshore movement of the surface layer, which brings larvae toward the coast (Farrell et al. 1991). Onshore Ekman transport induced by favorable winds can raise the slope of the waters close to the coast producing flows toward the coast and estuaries (Epifanio 1995). Internal tidal bores propagating onshore can transport larvae toward the coast (Pineda 1994). Larvae capable of stronger swimming likely undertake directed movement as has been reported for larvae of American lobsters (*Homarus americanus;* Cobb et al. 1989) and palinurid lobsters (Jeffs et al. 2005).

Gaining access to an estuary is also likely to be due to a combination of adaptive larval behaviors combined with physical transport processes and events. A good example of this is selective tidal stream transport (STST). STST describes larval behaviors which place larvae in the water column during flood tides, and near or on the bottom during ebb tides (Forward et al. 2003). This tide-hopping behavior has the net effect of progressively advecting larvae (such as zoea and megalope stages) into estuarine systems. Much recruitment occurs on nocturnal flood tides and the cues that can produce such an effect have been experimentally evaluated. As a flood tide commences, a stationary larva located on the bottom would experience an increase in hydrostatic pressure, salinity, and turbulence (Queiroga et al. 2006). These cues are thought to trigger an ascent into the flood tide water column, though curiously, it is not yet known how invertebrate larvae lacking a compressible structure, actually sense a change in hydrostatic pressure. Once in the water column, larvae need a cue or a clock that tells them when to drop out of the water column to avoid being advected seaward. A larva moving in a parcel of water would have little ability to detect changes in salinity or pressure, so a reduction in turbulence associated with the onset of slack water, has been proposed as a cue to drop out of the water column. Swimming during the ebb tide is inhibited by chemical cues associated with organic compounds of estuarine origin. Finally, inhibition of swimming

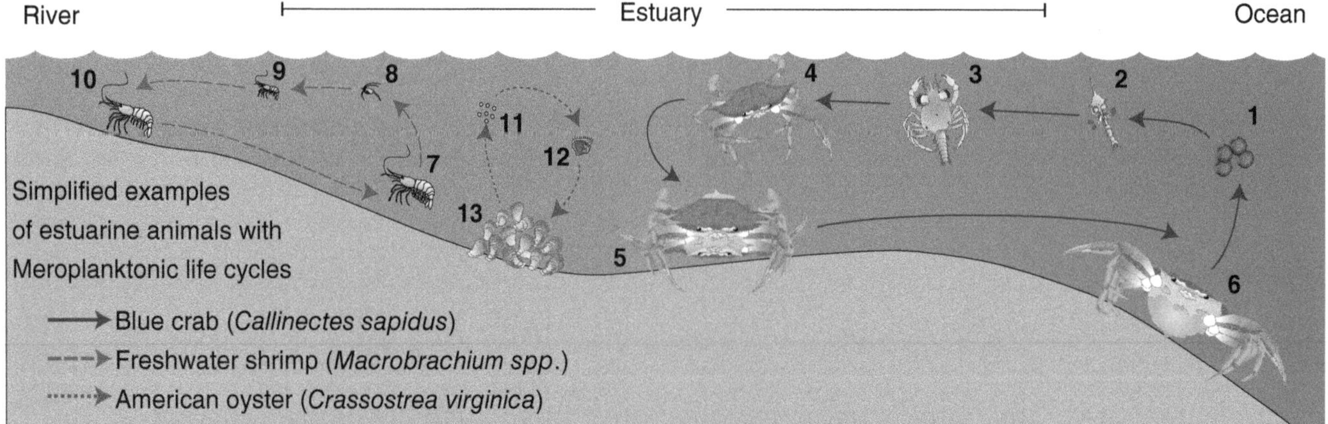

FIGURE 10.7 Examples of life history patterns for different animals that have estuarine meroplanktonic larvae. Numbers 1–6 show the life history pattern for the blue crab, *Callinectes sapidus*, which spawns outside the estuary, where it releases eggs (1) that mature into zoea (2) and megalope (3) larval stages that move back into the estuary, where they metamorphose into juveniles (4) and eventually adult stages (5) that migrate out of the estuary (6) to start the cycle again. Numbers 7–10 illustrate the life history of shrimp in the genus *Macrobrachium*, which spawns in the upper reaches of the estuary (7), and releases larvae (8) and juveniles (9) that migrate into freshwater (10) before migrating back to the estuary to spawn. Numbers 11–13 show the life history pattern of a molluscan bivalve that spends its entire life in the estuary, releasing eggs and sperm (11) into the water that will become planktonic larvae (12) that are retained within the estuary before recruiting within the estuary (13).

during daytime flood tides by higher light intensities would reduce the number of larvae present in daytime flood tides. Much of the work on these issues has been conducted on crab larvae (e.g., Forward et al. 2001).

10.5.2 Introduced Species

The introduction of zooplankton from one ecosystem into another can have a profound and adverse impact on the ecology of the affected ecosystem. In the absence of natural predators and other controlling factors, introduced species may proliferate and dramatically alter existing community structure with impacts that extend to multiple trophic levels. Introductions may be accidental or intentional; however, the transfer of very large volumes of water in the ballast of commercial ships has proved to be an effective accidental means of introducing exotic species into estuarine systems around the world. In the San Francisco estuary, Cohen and Carlton (1998) estimated that aquatic species were being introduced via ballast water transfers at the rate of one new species every 14 weeks after 1969. While not all of these introductions represent zooplankton, many were transported as meroplanktonic larvae in ballast water.

The introduction of the ctenophore *Mnemiopsis leidyi* from North American estuaries into the Black Sea and Azov Seas during the 1980s is an example of how an exotic species can rapidly alter the systems into which they have been introduced (Kideys 2002; Oguz et al. 2008). Moreover, the establishment of *M. leidyi* in the Black Sea illustrates how other anthropogenic impacts such as eutrophication and overfishing, can exacerbate the impact of an introduced species on an ecosystem. *Mnemiopsis leidyi* was first detected in the Black Sea in 1982 and subsequently underwent a rapid population increase. By the end of the decade, *M. leidyi* had replaced the anchovy *Engraulis encrasicolus* as the top predator in the Black Sea (Oguz et al. 2008). During the 1990s, *M. leidyi* populations underwent several declines and increases. Environmental factors such as a particularly cold period combined with introduction of another ctenophore (*Beroe ovata*), which is a natural predator of *M. leidyi*, were associated with declines in the abundance of the latter and the gradual recovery of anchovy stocks.

Gelatinous zooplankton are not the only examples of introduced zooplankton. Ballast water often contains abundant and diverse zooplankton assemblages (Cordell et al. 2009). In a study of the zooplankton in ballast water from ships docking in Puget Sound, Cordell et al. (2009) identified 124 different taxa. Bivalve larvae were present in almost 50% of all samples, while species of the calanoid copepod *Acartia* and cyclopoid copepod *Oithona davisae* were present in about 32% of samples and barnacle cyprid and nauplii occurred in 26 and 21% of samples, respectively. Even at low densities of a few individuals of a species per cubic meter, the large volume of water discharged by ships (mean of 9.5 million m^3 $year^{-1}$ in Washington State), means there is a high potential for the introduction of new species. Because it is so difficult to remove introduced species, the focus is on preventing their release in the first place.

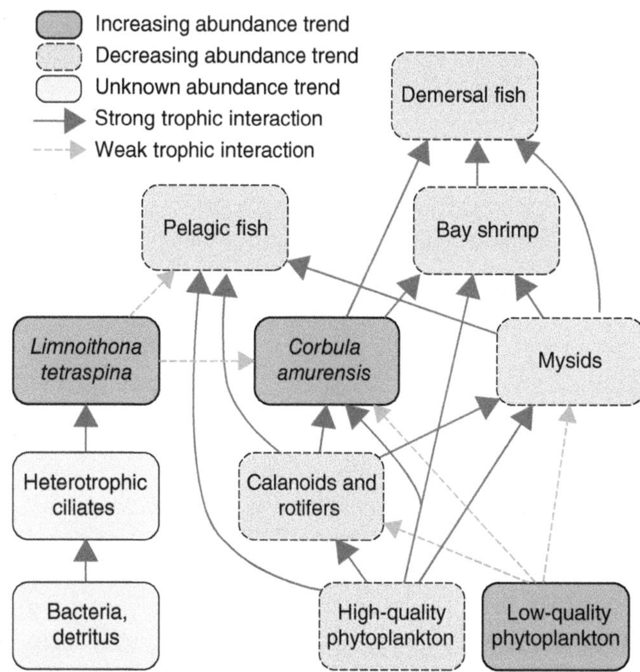

FIGURE 10.8 Example of how an estuarine food web changes over time (1972–2008) with the introduction of an introduced predator (the copepod *Limnoithona tetraspina*), which benefited the introduced clam *Corbula amurensis*, and reduced populations of high-quality phytoplankton and organisms dependent on those phytoplankton. This food web is based on a sub-region of the San Francisco Bay Delta System, and boxes represent different taxa and arrows indicate major energy flows based on gut content analysis or literature data. Solid lines for arrows represent strong trophic interactions and dashed lines indicate weak trophic interactions. Box colors reflect long-term trends for each taxa with blue boxes showing increasing abundance of a taxa, yellow boxes showing decreasing abundance of taxa, and gray boxes showing taxa with unknown trends. Phytoplankton is separated into groups of high (e.g., diatoms) and low (e.g., cyanobacteria, chlorophytes) food quality and/or availability for herbivorous organisms. *Source*: Based on Winder and Jassby (2011) and Kimmerer (2006).

How introduced species interact with endemic zooplankton ranges from competition for resources to direct predation. Kimmerer et al. (1994) provide evidence that a clam (*Potamocorbula amurensis*), which was introduced to San Francisco Bay in 1986 from eastern Siberia, was responsible for declines in the abundance of the copepod *Eurytemora affinis*. However, subsequent invasions of other non-native bivalves and mesozooplankton, including the copepod *Limnoithona tetraspina*, were coupled to anthropogenic impacts in the system such as increased freshwater usage and changes in nutrient inputs, have driven major shifts in a variety of food web components (Figure 10.8), including zooplankton abundance and composition, phytoplankton abundance and composition, and even higher trophic levels such as fish and shrimp (Winder and Jassby 2011). Such invasions often have a strong influence on the ecological communities where the invasions occur, with strongest responses on zooplankton often observed in the abundance of native species and a lesser response on the overall diversity of the system (Gallardo et al. 2016).

10.5.3 Eutrophication and Deoxygenation

The addition of nitrogen and phosphorus to estuarine systems typically produces an increase in primary production by diatoms and other phytoplankton. In addition to stimulating a bloom of phytoplankton, this increased production results in the release of additional dissolved organic matter in the form of algal exudates. This DOC stimulates bacterial production, which leads to a bloom of microplanktonic (primarily protist) grazers. At this point at least two potential outcomes are possible depending on the fate of the excess phytoplankton production. If the increase in phytoplankton food availability falls within the food limitation range of copepods and other crustacean grazers, then the stocks of these grazers should increase and graze down the excess phytoplankton production (Capriulo et al. 2002). On the other hand, if grazers cannot consume much of the excess production, then bacterial decomposition of dead phytoplankton biomass will stimulate additional microplanktonic production and the combined effect of the bacterial and microzooplankton respiration will reduce water column dissolved oxygen (**hypoxia** or **deoxygenation**), which will suppress mesozooplankton biomass and production. The latter outcome seems to be more common (Zervoudaki et al. 2009) than the former, perhaps in part due to the differences in the generation times of phytoplankton, microzooplankton, and mesozooplankton. The prevalence of these deoxygenated systems is increasing worldwide, largely tied to human influences on coastal regions (Diaz and Rosenberg 2008)

Eutrophication is frequently associated with deoxygenation that can directly impact the abundance and species composition of estuarine zooplankton. In the Bilbao estuary, Spain, improvements in wastewater treatment and a decline in industrialization resulted in higher dissolved oxygen concentrations over the period 1999–2001 compared to the early 1980s (Albaina et al. 2009). The improved water quality was accompanied by the occurrence of new species of copepods (*Calanipeda aquqedulcis* and *Eurytemora affinis*) and the establishment of several species of *Acartia*. At the same time, the composition of the zooplankton community in the Bilbao estuary began to more closely resemble that of the nearby and relatively pristine Urdaibai estuary (Albaina et al. 2009).

There is some evidence that changes in water-column dissolved oxygen can directly impact the distribution patterns of certain zooplankton taxa. For example, microzooplankton are often concentrated at the **oxycline**, where the oxygen gradient is the steepest. This location allows them to take advantage of both the oxic environment and the high bacterial production associated with lower oxygen concentrations (Coats and Revelante 1999). The copepod *E. affinis* is a euryhaline species that occurs over a wide range of salinities (Lee 1999), however, its ability to tolerate low salinities is reduced at low oxygen concentrations. In the Schelde estuary, Netherlands, Appeltans et al. (2003) documented a shift in the distribution of *E. affinis* from a brackish to freshwater region over several years that appeared to coincide with an increase in water column dissolved oxygen concentrations in the freshwater region. The copepod *Acartia tonsa* has been shown to exhibit a range of responses to deoxygenation, dependent on the severity of the deoxygenation (Elliott et al. 2013). For example, *A. tonsa* individuals have reduced egg production (Roman et al. 1993; Marcus et al. 2004), reduced escape response to predators (Decker et al. 2004), and altered vertical distribution (remaining closer to the surface) and decreased vertical migration (Elliott et al. 2013; Pierson et al. 2017) under low oxygen conditions. However, some behavioral responses appear to be adaptations based on experience. Using individuals from two populations of *A. tonsa*, one from the Chesapeake Bay that experiences extensive seasonal deoxygenation each year, and the other from an embayment in Florida where deoxygenation is not known to occur, Decker et al. (2003) conducted laboratory experiments and showed that copepods from the naïve population did not avoid the low oxygen layer. The authors concluded that these differences may also mean that different populations may have different susceptibilities to anthropogenic stressors.

10.5.4 Pollutants

Zooplankton may be particularly sensitive to inorganic and organic pollutants, including pesticides or chemicals derived from runoff and sewage outflows. In fact, the standard US Environmental Protection Agency (EPA) toxicity bioassay uses mysids (*Mysidopsis bahia*) as test organisms (USEPA 1996). While such laboratory tests are useful for estimating lethal effects of toxicants, they provide little insight into how pollutants in estuaries impact the resident zooplankton. The impact of any pollutant depends on the dose, which is dependent upon exposure time and concentration, both factors that can vary substantially in the dynamic environment of an estuary. In addition, the effect of a pollutant such as a pesticide, can be enhanced by other stresses on the animals such as low oxygen, low food concentrations, and the presence of predators (Hanazato 2001). Studies of mixed zooplankton assemblages in mesocosms of different volumes (1.5–30 m^3) treated with additions of mercury produced increased mortality of copepods (Kuiper et al. 1983). This suggests that acute exposure to high concentrations of pollutants will lead to high mortalities. A review of data from oil spills supports the existence of high mortality that is generally followed by rapid recovery, due in large part to the short generation times of zooplankton (Kennish 1997).

The impacts of pollutants may also be subtle and indirect. At sublethal concentrations, anthropogenic compounds may still influence the distribution of planktonic organisms. A laboratory study using a choice chamber demonstrated that postlarval brown shrimp (*Farfantepenaeus aztecus*) and white shrimp (*Litopenaeus setiferus*) were capable of detecting and avoiding the compound pentachlorophenol at concentrations that were well below levels that would induce mortality (Benfield and Aldrich 1994). Moreover, the presence of sublethal trace amounts of pentachlorophenol was capable of suppressing previously demonstrated attraction to estuarine water (Benfield and Aldrich 1992). Such findings suggest that the presence of sublethal concentrations of some pollutants may effectively alter distributional patterns of zooplankton in estuaries if they find such waters repelling and actively avoid them or alter their behaviors in such waters in a nonadaptive manner.

10.5.5 Freshwater Diversions and Dams

Construction of dams for hydroelectric, water conservation, or saltwater control purposes can dramatically alter the normal hydrological cycle, salinity distribution, and current velocities within an estuary. How zooplankton respond to these alterations is highly site dependent. In the Senegal River estuary, the construction of the Diama dam in 1985 appears to have altered the abundance, composition, and distribution patterns of zooplankton in the system (Champalbert et al. 2007). In comparing their results with a preconstruction study, which employed the same sampling gear, Champalbert et al. (2007) noted an increase in mesozooplankton abundance after construction of the dam by an order of magnitude. This increase was attributed to increased eutrophication in what had been an oligotrophic estuary. After construction of the dam, zooplankton abundances increased toward the dam, whereas prior to its presence, abundances increased toward the estuary mouth. In this example, the combined impacts of anthropogenic actions (water diversion, eutrophication) have a different impact on the system than what might be expected if each were considered alone. This highlights the importance of considering a broad array of factors when assessing the impacts of any single anthropogenic stressors on zooplankton communities.

Questions

Multiple Choice

1. Animals that exist for their entire lives as plankton are called
   a. Holoplankton
   b. Ichthyoplankton
   c. Zooplankton
   d. Meroplankton

2. Most zooplankton exist at a Reynolds Number less than zero.
   a. True
   b. False

3. Reynolds number is the ratio between which of the following?
   a. The number of copepods and the number of mysids in an estuary.
   b. The temperature and the salinity in a body of water.
   c. The number of holoplankton and the number of meroplankton in a coastal system.
   d. Inertial and viscous forces acting on an object.

4. Cryptic species are ones that
   a. Hide in crevices in reefs.
   b. Cannot be identified as separate species by looking at their morphology
   c. Are only found in estuaries.
   d. None of the above.

5. Which of the following lists has the zooplankton size classes ordered from smallest to largest?
   a. Megaplankton, Microplankton, Femtoplankton, Picoplankton
   b. Nanoplankton, Picoplankton, Microplankton, Mesoplankton
   c. Microplankton, Mesoplankton, Macroplankton, Megaplankton
   d. All zooplankton are mesoplankton.

6. High turbidity, which is a common feature of many estuaries, can cause (Select all that apply)
   a. Primary production to be enhanced in ETM.
   b. Detritus and non-phytoplankton material to be more important for zooplankton diets
   c. High zooplankton concentrations in ETM.
   d. Elevated predation on zooplankton by visual predators.

7. Zooplankton have been shown to avoid being washed out of estuaries during high flow events, such as floods from storms, by which of the following?
   a. Descending to the near bottom where the flow rate is low.
   b. Clinging onto vegetation.
   c. Swimming to the surface and swimming against the current.
   d. All zooplankton are washed out during such events.

8. Which of the following are responses to deoxygenation or hypoxia in copepods? (Select all that apply)
   a. Reduced vertical migration.
   b. Increased horizontal migration.
   c. Increase grazing on detritus particles.
   d. Reduced egg production.

9. The word for an organism that can generate its own nutrition and also ingest nutrition is
   a. Allelopathy
   b. Mixotrophy
   c. Osmoconformity
   d. Hungry

10. Invasive species can affect estuarine ecosystems in each of the following ways except
    a. Outcompete native species for food sources
    b. Change the salinity of the estuary.
    c. Prey on native species that are naïve to the new predator.
    d. Alter the entire biodiversity of an ecosystem.

11. Selective Tidal Stream Transport is a method for larval meroplankton to
    a. Grow fast enough to escape predation.
    b. Adapt to changing salinity conditions.
    c. Use tidal currents and vertical migration to affect their horizontal movement into or out of an estuary.
    d. None of the above.

Short Answer

1. In general, how is zooplankton species diversity related to salinity in estuaries?
2. Describe the specialized feeding ability of mysids and how it structures estuarine food webs.
3. Describe one way that feeding by zooplankton in an estuarine turbidity maximum (ETM) region might be different than by zooplankton in the open ocean?
4. Describe the food web interactions by which an increase in the population of the medusa *Chrysaora quinquecirrha* in Chesapeake Bay may decrease phytoplankton abundance.
5. How is zooplankton grazing involved in deoxygenation and the formation of hypoxia in estuaries?

Further Reading

Day, J.W. Jr. (1989). Chapter 8. Zooplankton, the Drifting Consumers. In: *Estuarine Ecology* (ed. J.W. Day, C.A.S. Hall, W.M. Kemp and A. Yáñez-Arancibia), 311–337. New York: Wiley.

Grindley, J.R. (1984). The zooplankton of mangrove estuaries. In: *Hydrobiology of the Mangal. The Ecosystem of the Mangrove Forests, Developments in Hydrobiology No. 20*, 79–87. The Hague: Dr. W. Junk Publishers.

Kimmel, D.G. (2011). Plankton Consumer Groups: Copepods. In: *Treatise on Estuarine and Coastal Science*, vol. 6 (ed. E. Wolanski and M.L. DS), 95–126. Waltham, Massachussetts: Academic Press.

Miller, C.B. (1983). The zooplankton of estuaries. In: *Estuaries and Enclosed Seas*, Ecosystems of the World. 26 (ed. B.H. Ketchum), 103–149. Amsterdam: Elsevier.

References

Albaina, A., Villate, F., and Uriarte, I. (2009). Zooplankton communities in two contrasting Basque estuaries (1999–2001): reporting changes associated with ecosystem health. *Journal of Plankton Research* 31: 739–752.

Alldredge, A.L. and King, J.M. (1985). The distance demersal zooplankton migrate above the benthos: implications for predation. *Marine Biology* 84: 253–260.

Anonymous (1959). Symposium on the classification of brackish waters. Venice 8—14th April 1958. Archivio di Oceanografia e Limnologia Volume 11. Supplement (Simposio sulla Classificazione della Acque Salmastre. Venezia 8—14 Aprile, 1958).

Appeltans, W., Hannouti, A., Van Damme, S. et al. (2003). Zooplankton in the Schelde estuary (Belgium/The Netherlands). The distribution of Eurytemora affinis: effect of oxygen? *Journal of Plankton Research* 25: 1441–1445.

Benfield, M.C. and Aldrich, D.V. (1992). Attraction of postlarval *Penaeus aztecus* Ives and *P. setiferus* (L.) (Crustacea: Decapoda: Penaeidae) to estuarine water in a laminar-flow choice chamber. *Journal of Experimental Marine Biology and Ecology* 156: 39–52.

Benfield, M.C. and Aldrich, D.V. (1994). Avoidance of pentachlorophenol by postlarval brown shrimp (Crustacea: Decapoda: Penaeidae) in a laminar-flow choice chamber. *Canadian Journal of Fisheries and Aquatic Sciences* 51: 784–791.

Bi, H., Cook, S., Yu, H. et al. (2013). Deployment of an imaging system to investigate fine-scale spatial distribution of early life stages of the ctenophore *Mnemiopsis leidyi* in Chesapeake Bay. *Journal of Plankton Research* 35 (2): 270–280.

Bousfield, E.L., Filteau, G., O'Neill, M., and Gentes, P. (1975). Population dynamics of zooplankton in the middle St. Lawrence estuary. *Estuarine Research* 1: 325–351.

Bulger, A.J., Hayden, B.P., Monaco, M.E. et al. (1993). Biologically-based estuarine salinity zones derived from a multivariate analysis. *Estuaries* 16: 311–312.

Calbet, A. and Landry, M.R. (2004). Phytoplankton growth, microzooplankton grazing, and carbon cycling in marine systems. *Limnology and Oceanography* 49 (1): 51–57.

Capriulo, G.M., Smith, G., Troy, R. et al. (2002). The planktonic food web structure of a temperate zone estuary, and its alteration due to eutrophication. *Hydrobiologia* 475/476: 263–333.

Champalbert, G., Pagano, M., Sene, P., and Corbin, D. (2007). Relationships between meso- and macro-zooplankton communities and hydrology in the Senegal River Estuary. *Estuarine, Coastal, and Shelf Science* 74: 381–394.

Chen, G. and Harc, M.P. (2008). Cryptic ecological diversification of a planktonic estuarine copepod, *Acartia tonsa*. *Molecular Ecology* 17 (6): 1451–1468.

Chen, G. and Hare, M.P. (2011). Cryptic diversity and comparative phylogeography of the estuarine copepod *Acartia tonsa* on the US Atlantic coast. *Molecular Ecology* 20 (11): 2425–2441.

Coats, D.W. and Revelante, N. (1999). Distributions and trophic implications of microzooplankton. *Ecosystems at the Land-Sea Margin: Drainage Basin to Coastal Sea* 55: 207–239.

Cobb, J.S., Wang, D., Campbell, D.B., and Rooney, P. (1989). Speed and direction of swimming by postlarvae of the American lobster. *Transactions of the American Fisheries Society* 118: 82–86.

Cohen, A.N. and Carlton, J.T. (1998). Accelerating invasion rate in a highly invaded estuary. *Science* 279: 555–557.

Cole, L.C. (1954). The population consequences of life history phenomena. *Quarterly Review of Biology* 29: 103–137.

Condon, R.H. and Steinberg, D.K. (2008). Development, biological regulation, and fate of ctenophore blooms in the York River estuary, Chesapeake Bay. *Marine Ecology Progress Series* 369: 153–168.

Cordell, J.R., Lawrence, D.J., Ferm, N.C. et al. (2009). Factors influencing densities of non-indigenous species in the ballast water of ships arriving at ports in Puget Sound, Washington, United States. *Aquatic Conservation: Marine and Freshwater Ecosystems* 19: 322–343.

Costa, K.G., Rodrigues Filho, L.F.S., Costa, R.M. et al. (2014). Genetic variability of *Acartia tonsa* (Crustacea: Copepoda) on the Brazilian coast. *Journal of Plankton Research* 36 (6): 1419–1422.

Costello, J.H., Sullivan, B.K., and Gifford, D.J. (2006). A physical-biological interaction underlying variable phenological responses to climate change by coastal zooplankton. *Journal of Plankton Research* 28: 1099–1105.

David, V., Sautour, B., Galois, R., and Chardy, P. (2006). The paradox high zooplankton biomass – low particulate organic matter in high turbidity zones: What way for energy transfer? *Journal of Experimental Marine Biology and Ecology* 333: 202–218.

Decker, M.B., Breitburg, D.L., and Marcus, N.H. (2003). Geographical differences in behavioral responses to hypoxia: local adaptation to an anthropogenic stressor? *Ecological Applications* 13 (4): 1104–1109.

Decker, M.B., Breitburg, D.L., and Purcell, J.E. (2004). Effects of low dissolved oxygen on zooplankton predation by the ctenophore *Mnemiopsis leidyi*. *Marine Ecology Progress Series* 280: 163–172.

Devreker, D., Souissi, S., Molinero, J.C., and Nkubito, F. (2008). Trade-offs of the copepod *Eurytemora affinis* in mega-tidal estuaries: insights from high frequency sampling in the Seine estuary. *Journal of Plankton Research* 30: 1329–1342.

Diaz, R.J. and Rosenberg, R. (2008). Spreading dead zones and consequences for marine ecosystems. *Science* 321 (5891): 926–929.

Dolan, J.R. and Gallegos, C.L. (2001). Estuarine diversity of tintinnids (planktonic ciliates). *Journal of Plankton Research* 23: 1009–1027.

Drillet, G., Goetze, E., Jepsen, P.M. et al. (2008). Strain-specific vital rates in four Acartia tonsa cultures, I: strain origin, genetic differentiation and egg survivorship. *Aquaculture* 280 (1-4): 109–116.

Elliott, D.T., Pierson, J.J., and Roman, M.R. (2013). Predicting the effects of coastal hypoxia on vital rates of the planktonic copepod *Acartia tonsa* Dana. *PLoS One*. 8 (5): e63987.

Epifanio, C.E. (1995). Transport of blue crab (*Callinectes sapidus*) larvae in the waters off Mid-Atlantic states. *Bulletin of Marine Science* 57: 713–725.

Farrell, T.M., Bracher, D., and Roughgarden, J. (1991). Cross-shelf transport causes recruitment to intertidal populations in central California. *Limnology and Oceanography*. 36 (2): 279–288.

Favier, J.B. and Winkler, G. (2014). Coexistence, distribution patterns and habitat utilization of the sibling species complex *Eurytemora affinis* in the St Lawrence estuarine transition zone. *Journal of Plankton Research*. 36 (5): 1247–1261.

Forster, J., Hirst, A.G., and Woodward, G. (2011). Growth and development rates have different thermal responses. *The American Naturalist* 178 (5): 668–678.

Forward, R.B. Jr., Tankersley, R.A., and Rittschof, D. (2001). Cues for metamorphosis of brachyuran crabs: an overview. *American Zoologist* 41 (5): 1108–1122.

Forward, R.B., Tankersley, R.A., and Welch, J.M. (2003). Selective tidal-stream transport of the blue crab *Callinectes sapidus*: an overview. *Bulletin of Marine Science* 72 (2): 347–365.

Gallardo, B., Clavero, M., Sánchez, M.I., and Vilà, M. (2016). Global ecological impacts of invasive species in aquatic ecosystems. *Global Change Biology* 22 (1): 151–163.

Gifford, D.J. and Dagg, M.J. (1988). Feeding of the estuarine copepod *Acartia tonsa* Dana: Carnivory vs. herbivory, in natural microplankton assemblages. *Bulletin of Marine Science* 43: 458–468.

Gillooly, J.F. (2000). Effect of body size and temperature on generation time in zooplankton. *Journal of Plankton Research* 22: 241–251.

Glibert, P.M., Fullerton, D., Burkholder, J.M. et al. (2011). Ecological stoichiometry, biogeochemical cycling, invasive species, and aquatic food webs: San Francisco Estuary and comparative systems. *Reviews in Fisheries Science* 19 (4): 358–417.

Hanazato, T. (2001). Pesticide effects on freshwater zooplankton: an ecological perspective. *Environmental Pollution* 112 (1): 1–10.

Heinle, D.R. and Flemer, D.A. (1975). Carbon requirements of a population of the estuarine copepod *Eurytemora affinis*. *Marine Biology* 31 (3): 235–247.

Horsted, S.J., Nielsen, T.G., Riemann, B. et al. (1988). Regulation of zooplankton by suspension-feeding bivalves and fish in estuarine enclosures. *Marine Ecology Progress Series* 48: 217–224.

Ianora, A. (1998). Copepod life history traits in subtemperate regions. *Journal of Marine Systems* 15: 337–349.

Ingle, R.M. (1951). Spawning and setting of oysters in relation to seasonal environmental changes. *Bulletin of Marine Science* 1: 111–135.

Jeffs, A.G., Montgomery, J.C., and Tindle, C.T. (2005). How do spiny lobster post-larvae find the coast? *New Zealand Journal of Marine and Freshwater Research* 39: 605–617.

Kennish, M.J. (1997). *Practical Handbook of Estuarine and Marine Pollution*. Inc. Boca Raton: CRC Press.

Kideys, A.E. (2002). Fall and rise of the Black Sea ecosystem. *Science* 297: 1482–1484.

Kimmel, D.G. and Roman, M.R. (2004). Long-term trends in mesozooplankton abundance in the Chesapeake Bay, USA: Influence of freshwater input. *Marine Ecology Progress Series* 267: 71–83.

Kimmel, D.G., Miller, W.D., and Roman, M.R. (2006). Regional scale climate forcing of mesozooplankton dynamics in Chesapeake Bay. *Estuaries and Coasts*. 29 (3): 375–387.

Kimmel, D.G., Miller, W.D., Harding, L.W. et al. (2009). Estuarine ecosystem response captured using a synoptic climatology. *Estuaries and Coasts*. 32 (3): 403–409.

Kimmel, D.G. (2011). Plankton consumer groups: copepods. In: *Treatise on Estuarine and Coastal Science*, vol. 6, 95–126. Waltham: Academic Press.

Kimmel, D.G., Boynton, W.R., and Roman, M.R. (2012). Long-term decline in the calanoid copepod Acartia tonsa in central Chesapeake Bay, USA: an indirect effect of eutrophication? *Estuarine, Coastal and Shelf Science* 101: 76–85.

Kimmerer, W.J., Gartside, E., and Orsi, J.J. (1994). Predation by an introduced clam as the likely cause of substantial declines in zooplankton of San Francisco Bay. *Marine Ecology Progress Series* 113: 81–93.

Kimmerer, W.J. (2006). Response of anchovies dampes effects of the invasive bivalve *Corbula amurensis* on the San Francisco Estuary foodweb. *Marine Ecology Progress Series* 324: 207–218.

Kuiper, J., Brockmann, U.H., van het Groenewoud, H. et al. (1983). Effect of mercury on enclosed plankton communities in the Rosfjord during POSER. *Marine Ecology Progress Series* 14: 93–105.

Laprise, R. and Dodson, J.J. (1993). Nature of the environmental variability experienced by benthic and pelagic animals in the St. Lawrence estuary, Canada. *Marine Ecology Progress Series* 94: 129–139.

Lee, C.E. (1999). Rapid and repeated invasions of fresh water by the copepod *Eurytemora affinis*. *Evolution* 53: 1423–1434.

Lonsdale, D.J., Cosper, E.M., and Doall, M. (1996). Effects of zooplankton grazing on phytoplankton size-structure and biomass in the Lower Hudson River Estuary. *Estuaries* 19 (4): 874–889.

Marcus, N.H., Richmond, C., Sedlacek, C. et al. (2004). Impact of hypoxia on the survival, egg production and population dynamics of *Acartia tonsa* Dana. *Journal of Experimental Marine Biology and Ecology* 301 (2): 111–128.

Mauchline, J. (1980). The biology of mysids and euphausids (Crustacea, Mysidacea). *Advances in Marine Biology* 18: 3–317.

Mauchline, J. (1998). The biology of calanoid copepods. *Advances in Marine Biology* 33: 1–710.

McLusky, D.S. (1989). *The Estuarine Ecosystem*, 2e. London: Blackie.

Miller, C.B., Johnson, J.K., and Heinle, D.R. (1977). Growth rules in the marine copepod genus. *Acartia Limnology and Oceanography* 22 (2): 326–335.

Miller, C.J., Roelke, D.L., Davis, S.E. et al. (2008). The role of inflow magnitude and frequency on plankton communities from the Guadalupe Estuary, Texas, USA: Findings from microcosm experiments. *Estuarine, Coastal and Shelf Science* 80: 67–73.

Millette, N.C., Stoecker, D.K., and Pierson, J.J. (2015). Top-down control by micro- and mesozooplankton on winter dinoflagellate blooms of *Heterocapsa rotundata*. *Aquatic Microbial Ecology* 75: 15–25.

Morgan, C.A., Cordell, J.R., and Simenstad, C.A. (1997). Sink or swim? Copepod population maintenance in the Columbia River estuarine turbidity-maxima region. *Marine Biology* 129: 309–317.

Naganuma, T. (1996). Calanoid copepods: linking lower-higher trophic levels by linking lower-higher Reynolds numbers. *Marine Ecology Progress Series.* 136 (1): 311–313.

Oguz, T., Fach, B., and Salihoglu, B. (2008). Invasion dynamics of the alien ctenophore *Mnemiopsis leidyi* and its impact on anchovy collapse in the Black Sea. *Journal of Plankton Research* 30: 1385–1397.

Oviatt, C.A. (2004). The changing ecology of temperate coastal waters during a warming trend. *Estuaries* 27 (6): 895–904.

Peitsch, A., Köpcke, B., and Bernát, N. (2000). Long-term investigation of the distribution of *Eurytemora affinis* (Calanoida; Copepoda) in the Elbe Estuary. *Limnologica* 30: 175–182.

Pierson, J.J., Kimmel, D.G., and Roman, M.R. (2016). Temperature impacts on *Eurytemora carolleeae* size and vital rates in the upper Chesapeake Bay in winter. *Estuaries and Coasts.* 39 (4): 1122–1132.

Pierson, J.J., Slater, W.C.L., Elliott, D., and Roman, M.R. (2017). Synergistic effects of seasonal deoxygenation and temperature truncate copepod vertical migration and distribution. *Marine Ecology Progress Series.* 575: 57–68.

Pineda, J. (1994). Internal tidal bores in the nearshore: Warm-water fronts, seaward gravity currents and the onshore transport of neustonic larvae. *Journal of Marine Research.* 52 (3): 427–458.

Plough, L.V., Fitzgerald, C., Plummer, A., and Pierson, J.J. (2018). Reproductive isolation and morphological divergence between cryptic lineages of the copepod *Acartia tonsa* in Chesapeake Bay. *Marine Ecology Progress Series* 597: 99–113.

Purcell, J.E., Shiganova, T.A., Decker, M.B., and Houde, E.D. (2001). The ctenophore *Mnemiopsis* in native and exotic habitats: U.S. estuaries versus the Black Sea basin. *Hydrobiologia* 451: 145–176.

Queiroga, H., Almeida, M.J., Alpuim, T. et al. (2006). Tide and wind control of megalopal supply to estuarine crab populations on the Portuguese west coast. *Marine Ecology Progress Series.* 307: 21–36.

Revis, N. (1988). Preliminary observations on the copepods of Tudor Creek, Mombasa, Kenya. *Hydrobiologia* 167 (168): 343–350.

Rollwagen-Bollens, G.C. and Penry, D.L. (2003). Feeding dynamics of *Acartia* spp. Copepods in a large, temperate estuary (San Francisco Bay). *Marine Ecology Progress Series* 257: 139–158.

Roman, M.R., Gauzens, A.L., Rhinehart, W.K., and White, J.R. (1993). Effects of low oxygen waters on Chesapeake Bay zooplankton. *Limnology and Oceanography.* 38 (8): 1603–1614.

Roman, M.R., Holliday, D.V., and Sanford, L.P. (2001). Temporal and spatial patterns of zooplankton in the Chesapeake Bay turbidity maximum. *Marine Ecology Progress Series.* 213: 215–227.

Secor, D.H. and Houde, E.D. (1995). Temperature effects on the timing of striped bass egg production, larval viability, and recruitment potential in the Patuxent River (Chesapeake Bay). *Estuaries* 18 (3): 527–544.

Sieburth, J.M.N., Smetacek, V., and Lenz, J. (1978). Pelagic ecosystem structure: Heterotrophic compartments of the plankton and their relationship to plankton size fractions. *Limnology and Oceanography* 23: 1256–1263.

Steinberg, D.K. and Landry, M.R. (2017). Zooplankton and the Ocean Carbon Cycle. *Annual Reviews in Marine Science* 9: 413–444.

Turner, J.T. (1981). Latitudinal patterns of calanoid and cyclopoid copepod diversity in estuarine waters of eastern North America. *Journal of Biogeography* 8: 369–382.

Ueda, H., Terao, A., Tanaka, M. et al. (2004). How can river-estuarine planktonic copepods survive river floods? *Ecological Research* 19: 625–632.

USEPA (1996). Ecological effects test guidelines. OPPTS 850.1035 Mysid acute toxicity test. EPA 712-C-96-136 April 1996. 8 pp.

Vilas, C., Drake, P., and Fockedey, N. (2007). Feeding preferences of estuarine mysids *Neomysis integer* and *Rhopalophthalmus tartessicus* in a temperate estuary (Guadalquivir Estuary, SW Spain). *Estuarine, Coastal, and Shelf Science* 77 (3): 345–356. In, Press.

Winder, M. and Jassby, A.D. (2011). Shifts in zooplankton community structure: implications for food web processes in the upper San Francisco Estuary. *Estuaries and Coasts* 34 (4): 675–690.

Winkler, G., Dodson, J.J., Bertrand, N. et al. (2003). Trophic coupling across the St. Lawrence River estuarine transition zone. *Marine Ecology Progress Series* 251: 59–73.

Winkler, G., Martineau, C., Dodson, J.J. et al. (2007). Trophic dynamics of two sympatric mysid species in an estuarine transition zone. *Marine Ecology Progress Series* 332: 171–187.

Winkler, G., Dodson, J.J., and Lee, C.E. (2008). Heterogeneity within the native range: population genetic analyses of sympatric invasive and noninvasive clades of the freshwater invading copepod *Eurytemora affinis*. *Molecular Ecology.* 17 (1): 415–430.

Wooldridge, T. and Bailey, C. (1982). Euryhaline zooplankton of the Sundays estuary and notes on trophic relationships. *South African Journal of Zoology* 17: 151–163.

Zervoudaki, S., Nielsen, T.G., and Carstensen, J. (2009). Seasonal succession and composition of the zooplankton community along a eutrophication and salinity gradient exemplified by Danish waters. *Journal of Plankton Research* 31 (12): 1475–1492.

Zhang, X., Roman, M., Kimmel, D. et al. (2006). Spatial variability in plankton biomass and hydrographic variables along an axial transect in Chesapeake Bay. *Journal of Geophysical Research* 111 (C05S11): 1–16.

CHAPTER 11

Estuarine Benthos

Fiddler crabs at the Elizabeth A. Morton National Wildlife Refuge, New York, USA. Photo credit: USFWS

James G. Wilson[1] **and John W. Fleeger**[2]
[1]Zoology Department, Trinity College, Dublin, Ireland
[2]Department of Biological Sciences, Louisiana State University, Baton Rouge, LA, USA

11.1 Introduction

The association of humans with estuaries goes back millennia, as indicated by the location of ancient settlements and the remains of their diet, which included shellfish such as oysters and mussels. Oysters and mussels are notable members of the estuarine benthos (collectively, the organisms that live in or on the bottom; Figure 11.1). Benthic organisms are important because they are major contributors to ecosystem functions including production, consumption, and decomposition of organic matter, nutrient regeneration, and energy transfer to higher trophic levels (Schratzberger and Ingels 2017). For example, benthos comprise by far the major fraction of consumer biomass in estuaries serving as important links in the food web and as a significant human food resource. In addition, sediment-dwelling animals increase the rate of nutrient cycling by mixing sediments, and shellfish improve water quality by filtering sediment particles, chemical pollutants, and phytoplankton from the overlying water. Oyster and mussel beds also function as hot spots of biodiversity for estuarine life, and because of their intimate association with sediments and water, benthic organisms are important indictors of environmental health. This chapter describes the characteristics and importance, as well as the factors that control the abundance and distribution of benthic animals in estuaries (see Chapters 5, 8, and 9 for discussion of the phytobenthos and microbenthos).

Most benthic animals are invertebrates that are diverse in terms of life habit and size (e.g., Snelgrove 1999, Table 11.1), and many terms are used to classify them based on living position and association with sediment. *Epibenthos* live on the surface of sediments and include mobile organisms such as crabs, snails, and starfish, and sessile organisms such as oysters and barnacles that attach to hard substrates such as rock surfaces, mangrove roots, or man-made structures (e.g., pilings). *Infauna* (sometimes called *endobenthos*) dwell in sediments as burrowers or as tube builders. *Hyperbenthos*, which are sometimes grouped with epibenthos (see epi-(2) on Table 11.1), are mobile organisms that live just above sediments and move freely between sediments and the overlying water column and include shrimps, mysids, and copepods (Table 11.1). Figure 11.1 shows how a number of characteristic estuarine benthos live in and on sediments.

Benthos are also classified by body size, based on practical (e.g., sampling) considerations (Table 11.2). *Macrobenthos*

Estuarine Ecology, Third Edition. Edited by Byron C. Crump, Jeremy M. Testa, and Kenneth H. Dunton.

Companion website: www.wiley.com/go/crump/estuarine3

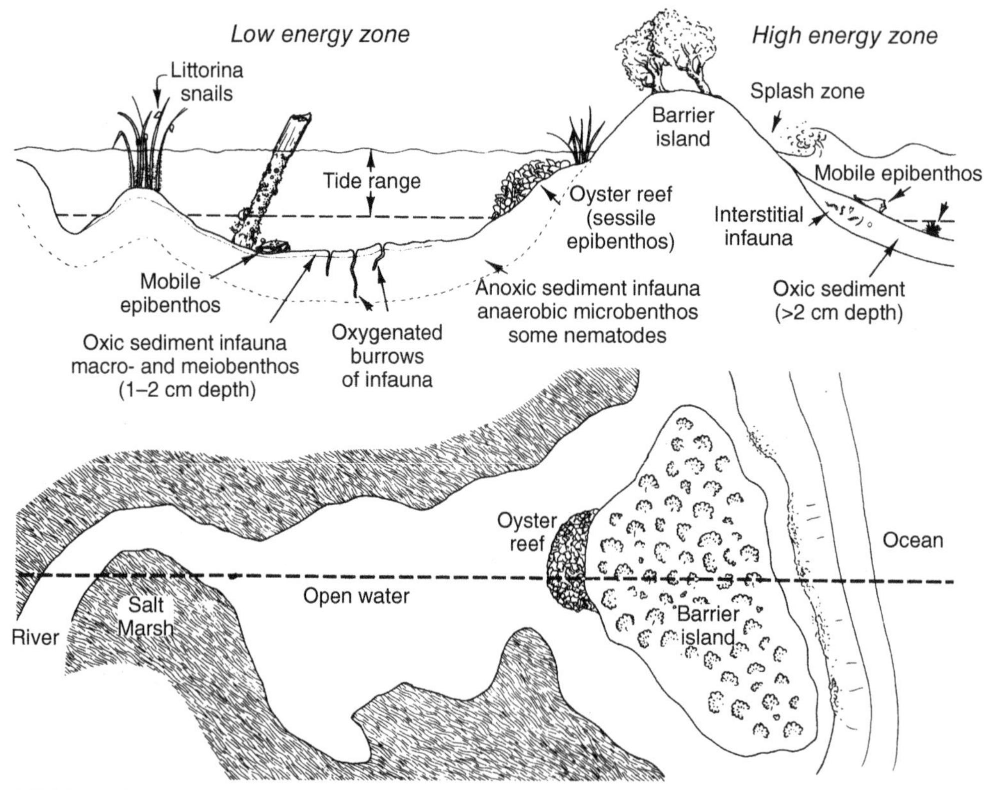

FIGURE 11.1 Diagram representing the various types of estuarine benthos. *Source*: Adapted from Day et al. (1989).

TABLE 11.1 Definition of benthos by life-style with examples.

Term	Definition	Life-Habitat	Examples
Epi-(1)	On the surface of	Epipelos—mud	Ostracods, triclads, copepods
		Epipsammon—sand	Gastropods, flatfish
		Epilithion—rock	Barnacles, sea anemones
Epi-(2)	Above the surface of	Moving between sediment and bottom water	Shrimps, mysids
Endo-	Burrowing or living in	"Infauna"	
		Endopelos—mud	Bivalves, worms
		Endopsammon—sand	Mobile worms, amphipods, nematodes
		Endolithion—rock, timber	Shipworm, gribble, paddock
Hyper-	Above the surface of	Moving between sediment and bottom water	Shrimps, mysids
Supra-	Swimming above		Fish—linked through feeding habit

TABLE 11.2 Definition of benthos by size, with references.

Term	Definition	Reference
Benthos	All organisms living on or in the sea floor	Baretta-Bekker et al. (1998)
Macrobenthos	Retained by mesh of 0.5 mm	Kramer et al. (1994) p219
Meiobenthos	Size between 0.1 and 0.5 mm or 1.0 mm	Giere (2009)
	Upper limit of size 1 mm or 0.5 mm mesh;	Higgins and Thiel (1988)
	Lower limit commonly 63 μm but sometimes 45 μm mesh	
Microbenthos	Size smaller than 0.1 mm, but see also meiobenthos lower limits	Higgins and Thiel (1988)

are defined as organisms that are retained on a 1.0 or a 0.5 mm sieve (Table 11.2). The *meiobenthos*—metazoans that pass through a 0.5 mm sieve but are retained on a 0.063 mm sieve—have representatives of most animal phyla that generally remain in the size range of meiobenthos throughout their life (e.g., nematodes), although recently settled larval stages of macrobenthos contribute to the meiobenthos. Appropriately sized single-celled organisms (e.g., foraminiferans) are also considered to be part of the meiobenthos. *Microbenthos* include single-celled eukaryotic (e.g., protozoa) and prokaryotic (e.g., Bacteria and Archaea) microorganisms.

11.1.1 Sampling

Field sampling of the estuarine benthos is often decided on pragmatic grounds, and in practice, most work is done using quantitative sampling to measure, for example, the number of individuals per m^{-2} or cm^{-2} of the sediment surface. However, this requires time-consuming and expensive field sampling, laboratory sorting, analysis, and identification. Variation among replicate samples may be high, requiring many samples and increasing sampling and processing costs. Quantitative sampling requires collecting a precise area or volume of undisturbed sediment. Some methods, including dredge and net sampling (Table 11.3) are semi-quantitative because assumptions must be made regarding how much of the calculated area or volume is passed through and retained in the sampling device.

Quantitative sampling is conducted by coring or by various remote sampling devices (Figure 11.2). Core sampling in the intertidal requires of a length of plastic tubing to enclose the required area and depth, a spade to recover the sample, and a sieve of the desired mesh to separate the biota from the enclosing sediment. Hand-held corers operated by SCUBA divers offer precise sampling in terms of sample placement and sample collection that is not attainable by other means (Figure 11.2d). However, the time that divers can remain on the bottom is limited and depends on water depth and other conditions. There are many types of remote sampling devices for use in subtidal habitats inaccessible to divers. Grab samplers (e.g., Figure 11.2b) are available in a variety of configurations and sizes, from 0.01 m^2 for hand-operated units to 0.2 m^2 (or larger) for vessels equipped with a winch. Smaller samplers (0.1 m^2) are the most common. The size of the grab sampler describes the area of its open mouth as it lands on the sediment, while the depth of the sample depends on the weight of the sampler (up to 250 kg on occasion!) and the softness of the sediment. Inconsistency among grab samples can be quality-controlled by measuring the volume taken each time, with a certain critical minimum needed to accept it as a valid sample. The operation of the grab sampler results in an uneven "bite" of the sediment, so that the vertical shape of the bite taken is not a regular rectangle but can be a semicircle. Consequently, organisms at different depths in the sediment may not be equally represented. Box corers can solve this problem by taking large volumes of sediment and can be subsampled by coring. However, grabs and box corers disrupt the sediment surface. Multicorers (Figure 11.2c) take undisturbed sediment samples necessary for fine-scale vertical profiling of infauna and chemical gradients (see Section 11.4.2).

Imaging using cameras or sonar as indirect sampling tools is becoming more popular because it makes it easier and faster to collect data on epifauna and substrate type or other features. Two examples are the REMOTS (Remote Ecological Monitoring Of The Seafloor; Rhoads and Germano 1982; Figure 11.2e), which takes a camera picture of the sediment sub-surface profile via a prism, and an ROV (Remote-controlled Underwater Vehicle; Figure 11.2f) which sends underwater video back to the surface via an undersea cable. Indirect methods (Table 11.3) are useful, but only to the extent that they can be related to data collected using direct methods (e.g., coring) to enable interpretation. In most cases, data from indirect methods should be used as a guide to further investigation and should be analyzed critically. However, the technology behind indirect methods is

TABLE 11.3 **Sampling types, sampling level and equipment with comments.**

Level	Method/Equipment	Comments
Presence/Absence	All/any	Useful for species lists; some biodiversity, habitat, or biome classifications
Qualitative	Visual, Spade, dredge	Rough estimates of abundance/cover. Can be assigned rankings with standard scales
Quantitative	Dredge	Roughly quantitative mouth*length of tow, or volume of sediment retained
	Plankton net, Epibenthic sled	Roughly quantitative: mouth*length of tow
	Grab (Van Veen, Ponar, Ekman)	Area defined *but* depth/shape of bite depends on sediment hardness
	Corers (Hand-held in the intertidal, and subtidal SCUBA and multicorers)	Can be accurately placed and area and shape are controlled. Provides fine control over sampling and undisturbed sediment. Penetration can be limited in hard sediments
	Camera, video	Useful for epibenthos: Stereo image analysis for size, biomass; non-destructive sampling
	SPI, REMOTS	Sediment Profile Imagery. Provides limited biotic data
	Multibeam, sidescan sonar	Discriminates habitat features and larger assemblages (e.g., oyster reefs)

FIGURE 11.2 Types of benthic samplers: (a) push core deployed from a boat on Elson Lagoon, Alaska, USA. *Source*: Katrin Iken, Amber Hardison, (b) Ponar grab sampler. *Source*: Aquatic BioTechnology S.L. (c) Multicorer. *Source*: Ifremer. (d) Push core deployed via SCUBA in a seagrass bed on the shore side of Buck Island, Virgin Islands, USA. *Source*: Photo by: Shaun Wolfe/NPS/OWUSS. (e) REMOTS camera and prism in compact frame (Ocean Imaging Systems, a division of EP Oceanographic, LLC) and an image of a sediment profile from West Falmouth Harbor, Massachusetts, USA. *Source*: Melanie Hayn. (f) ROV with umbilical deployed under ice by Craig Tweedie (left) and Ken Dunton (right) on Simpsons Lagoon, Alaska, USA. *Source*: Credit: Byron Crump.

improving quickly, including digital image recognition software for analyzing camera and video data and signal processing software for sonar and multibeam data.

11.2 Diversity and Global Distribution in Estuaries

One of the most ecologically meaningful ways to characterize the estuarine benthos is by feeding mode. Because estuaries are rich in organic matter, many benthic animals acquire nutrition from suspended particles or from sediments, although the more usual methods of consuming primary producers (herbivory) and primary or secondary consumers (predation) are also common. *Deposivores* consume sediment and digest the associated microorganisms and detrital particulate organic matter. Detritus is a major component of the deposivore diet and includes the remains of other biota, which can be derived from both plants and animals. Detritus includes other material such as feces, as well as living items such as fungi, microalgae, and bacteria which can either be unattached, or fixed to the sediment grains. Examples of deposivores are the lugworm *Arenicola marina*, mud-snails such as the widely distributed *Peringia* and *Potamopyrgus*, bivalves including *Limecola (Macoma) balthica* and *Mya arenaria*, and the species-rich genus of dog welks, *Nassarius*.

Suspensivores are sometimes referred to as filter feeders since many of the structures or organs for suspension-feeding have an outward appearance of a net or filter that capture particles from the water. These organisms extract their nutrition from any material suspended in the water including bacteria, phytoplankton, zooplankton, and detritus. Suspension feeding can improve water quality by removing suspended matter from the water and depositing feces and pseudofeces on the seafloor. Suspensivores may have diets similar to deposivores where high wave energy or turbulence resuspends sediments. Some benthic species function as both suspensivores and deposivores, for example, by suspension feeding when tidal currents provide a constant supply of particles and deposit feeding during slack or low tides when the supply of suspended particles is interrupted. Most benthic animals collect suspended material with the aid of a mucus covering (to which it adheres), such that particles as small as individual bacteria (<1 μm in length) can be captured. Feeding is greatly affected by the size and concentration of particles in the water. Large particles are usually rejected, or their entry barred, although recent work has shown that mussels and other suspension-feeding bivalves can consume individual zooplankton up to 3 mm in size (Davenport et al. 2000), although a size limit of 450–600 μm is more common.

Examples of suspensivores are found in almost all phyla from the simple sponges where the feeding is at the level of the individual cell, to crustaceans such as barnacles (e.g., *Austromegabalanus psittacus*) that trap suspended particles with net-like modified appendages by rapid movements through the water, to bivalves including oysters (e.g., *Crassostrea gigas*), mussels (e.g., *Mytilus edulis*), and clams (e.g., *Cerastoderma edule*) where the gills are greatly enlarged and differentiated with distinct functional roles. One functional role of the gills is to generate the feeding currents by which water is drawn into the animal and across the gills. This mechanism of food capture is enhanced by currents in the surrounding water which influences the amount of food available, with stronger currents bringing more particles into the feeding apparatus. This undoubtedly contributes to the success of suspensivores in hydrodynamically active estuaries, as evidenced by the size and extent of oyster reefs or mussel beds in many systems.

Herbivory by benthos in estuaries is mostly microherbivory, in which suspension feeders ingest phytoplankton and deposit feeders consume microphytobenthos (MPB, see Chapter 8), either by ingesting sediment with its associated MPB or by directly selecting and consuming individual cells of MPB, such as benthic diatoms, from within the sedimentary matrix. The major herbivores in estuaries, which consume macroalgae and seagrasses, are usually not benthos, but instead are vertebrates, e.g., fishes, manatees, and geese (Chapters 12 and 13). Exceptions include periwinkle snails that associate with hard substrata like rocks, pier pilings, and macrophytes, and sea-slugs such as the *Limapontia* (<2 mm long) that suck the cell sap of macroalgae. In addition, many benthic invertebrates are predators (e.g., carnivorous snails and many crab species).

Functional groupings of estuarine consumers are customarily based on trophic position (Table 11.4). However, many benthic animals have varied diets that can cause them to feed at several different trophic levels (see Chapter 15). For instance, suspensivores may consume sediments and detritus when it is resuspended and likewise deposit feeders, especially those feeding at the interface between the sediment and overlying water (the sediment–water interface), take in suspended material, including phytoplankton that settle to the bottom. The ragworm *Hediste (Nereis) diversicolor*, an omnivore in Table 11.4, is a classic estuarine opportunist feeder, using whatever food source may be present. It actively captures prey with its pair of strong chitinous jaws. It removes material from suspension and from the sediment–water interface by trapping it on the mucous lining of its burrow, which is irrigated by the animal's respiratory current. It can also feed on dissolved organic matter (e.g., amino acids), which it is able to transport across its body wall. Such differences may lead to great variation in diet from habitat to habitat for estuarine consumers (Galván et al. 2008).

Feeding type or category for most benthic consumers has typically been assigned based on structure or morphology of mouth parts, although such inferences regarding diet are not always unambiguous. For example, nematodes have been classified into feeding types depending on their mouth

TABLE 11.4 Estuarine benthic fauna and functional roles.

Group	Size range (cm)	Indiv. Life span	Population doubling rate	Functional roles/ feeding type	Example
Macro-	0.1–100	1–10 years	1 year	Deposivore	*Arenicola*
				Suspensivore	*Crassostrea*
				Predator	*Callinectes*
				Grazer	*Littoraria*
				Omnivore	*Nereis (Hediste) diversicolor*
Meio-	0.0063–0.1	1–6 months	1 month	Deposivore	*Diplosoma breviceps*
				Predator	*Onyx perfectus*
				Grazer	*Desmodora schultze*
Micro-	<0.001	1 month	1 week	Grazer/predator	*Euplotes* spp.
					Dactylamoeba spp.

parts (size of buccal cavity, buccal plates, or jaws). Recent evidence based on stable isotopes suggests that the trophic position of nematodes is broader and more variable in space and time than indicated by feeding group classifications. A case in point is that the role of marsh grass detritus vs. microalgae in the diet of benthos, and hence their respective roles in energy flow in estuaries, could not be resolved by morphological classifications. An increasing number of studies, also with stable isotopes, suggest that organic matter produced by marsh macrophytes (e.g., *Spartina* spp.), despite high abundance, is most important to the diet of only a few animals that live in the marsh proper (marsh periwinkles, the purple marsh crab (*Sesarma reticulatum*), some amphipods, some subsurface infauna, insects), while many others (fiddler crabs, and species that enter the marsh on rising tides) rely more on phytoplankton and benthic microalgae as the basal resource (Galván et al. 2011).

11.3 Spatial and Temporal Patterns of Abundance, Biomass, and Productivity

Knowledge of abundance, biomass, and secondary productivity is essential to understand the importance of benthos in energy flow pathways. Bar-On et al. (2018) found that the biomass of consumers (including predators) is more than five times greater than primary producers in marine ecosystems, which is possible only if there are high biomass turnover rates at lower trophic levels. Benthic communities exhibit several interesting allometric trends related to body size, one being that as body size decreases, abundance increases but biomass decreases (Chapter 15). The abundance of larger macrobenthos in healthy estuarine sediments (i.e., sediment that is not totally anoxic) is typically 500–10,000 ind m^{-2} while biomass ranges from ~5 to 100 gC m^{-2}. For the much smaller meiobenthos, abundances are 10^5 to 10^7 ind m^{-2}, several orders of magnitude greater, whereas biomass is 0.2–1 gC m^{-2} which is 25–100 times less than the macrobenthos (Table 11.5). Microbenthos have yet higher abundances but generally lower biomass except in anoxic sediments where larger organisms are unable to persist. Productivity by benthos is affected by the fact that turnover rate of benthic populations increases as body size decreases. Overall, because the number of individuals and the individual masses vary greatly, interactions between these factors can result in almost equal contributions to benthic secondary production from macrobenthos, meiobenthos, and microbenthos (see Table 11.5 and Schwinghamer et al. 1986; Brey 2012).

Some examples of secondary producer biomass and productivity from a range of estuaries are shown in Table 11.6. In general, estuarine benthic secondary production is highest in intertidal mudflats and vegetated sediments and low in sandy subtidal sediments. These data show a remarkable consistency among estuaries, with biomass and productivity clustered around 10–20 gC m^{-2} and 10–20 gC m^{-2} yr^{-1}, respectively. These levels of biomass and productivity are comparable to

TABLE 11.5 Biometrics for macro-, meio- and microbenthos.

Group	Individual weight (gC)	Biomass (gC m^{-2})	P:B ratio	Productivity (gC m^{-2} yr^{-1})
Macrobenthos	0.1–100	4.0–200	1 : 1	5–200
Meiobenthos	0.03×10^{-3}	0.5	10 : 1	5
Microbenthos	0.1×10^{-6}	0.05	50 : 1	2.5

1 gC ≡ 2 g AFDW ≡ 10 g wet weight (excluding shell)

TABLE 11.6 Biomass (gC m^{-2}) and secondary productivity (gC m^{-2} yr^{-1}) of estuarine consumer communities.

Location	Biomass gC m^{-2}	Productivity gC m^{-2} yr^{-1}	Main Habitat
Lynher estuary, UK			Intertidal mudflat
	5.43	5.46	Macrofauna
	2.64	20.39	Meiofauna
Baltic Sea	3.39	5.4	Macrofauna
	0.58	2.75	Meiofauna
Grevelingen, Netherlands	10.5	25 - 29	Estuary average
Forth estuary, Scotland	16.4	14.6	Intertidal upper estuary
	5.1	6.7	Intertidal middle estuary
	1.4	3.6	intertidal lower estuary
	0.7	0.6	Subtidal upper estuary
Inner Somme estuary, France	5.35	11.1	Fine muddy sand; sal 0–30.
Swartkops estuary, South Africa	34.7	38.9	Intertidal sands and muds
Berg River estuary, South Africa	9.5	44	Intertidal *Zostera* and mudflat
San Francisco Bay, USA	6.5–12	26.5–50	Intertidal
Long Island Sound, USA	27.3	10.7	Sublittoral average
Southampton Water, UK	45–95	76–113	Littoral mudflats
Kiel Bight, Germany	13.2	8.95	Sublittoral muds and sands

Source: Wilson (2002)/With permission of Elsevier.

that of estuarine zooplankton (Chapter 10) which itself may be an order of magnitude greater in estuaries than in riverine or deeper coastal water zooplankton (Heip et al. 1995).

In many shallow estuaries, benthos have more biomass and contribute more secondary productivity than plankton (Table 11.7), a key factor driving the importance of benthos in estuarine food webs. But the processes of particle deposition and resuspension can blur much of the distinction between surficial sediments and water column suspended particulate matter. For example, many benthic diatoms are commonly found in estuarine phytoplankton samples. Also, as noted above for marsh grasses, much of the autochthonous estuarine primary productivity is transferred to higher trophic levels via the decomposer (e.g., fungi and bacteria) pathway rather than directly grazed by herbivores.

Benthic food webs can be very short (e.g., primary production => benthos => vertebrate consumer) making them efficient pathways for trophic transfer to valuable fishery species. Since, as a rule of thumb, only about 10% of the energy at one trophic level is transferred to the next, the fewer the transfers,

TABLE 11.7 Biomass, gC m^{-2} and secondary productivity (gC m^{-2} yr^{-1}) of zooplankton and benthos in the estuarine environment.

Estuary	Zooplankton biomass	Zooplankton productivity	Macrobenthos biomass	Macrobenthos productivity
Chesapeake Bay, USA	0.39	0.38	5.2	5.5
Apalachee Bay, FL	0.03		1.81	
Ythan, Scotland	0.12	0.83	28.2	68.3
Dublin Bay, Ireland	0.02	0.14	14.9	15.7
Somme, France	0.04	0.24	16.0	9.5

Source: Data from Wilson et al. (2007) and references therein.

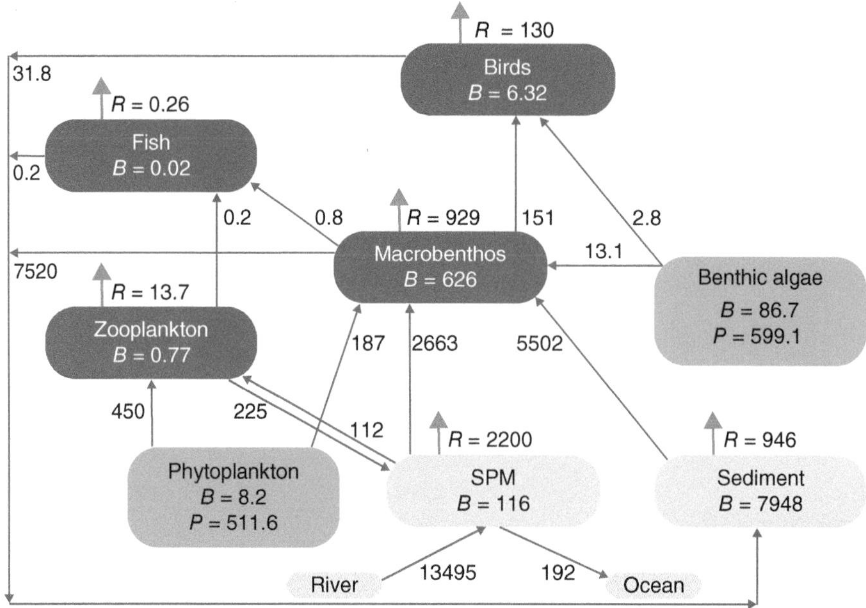

FIGURE 11.3 Simple box-model of energy in units of biomass (B; kJ m^{-2}) and productivity (P; kJ m^{-2} yr^{-1}), through the ecosystem of Dublin Bay, Ireland, showing inputs and transfers among compartments (kJ m^{-2} yr^{-1}) and losses from the system through respiration (R; kJ m^{-2} yr^{-1}). *Source*: Wilson et al. (2007)/With Permission of Springer Nature.

the greater proportion of the organic inputs that end up in the top consumers including humans (see Chapter 15). A simple box model for Dublin Bay illustrates these general principles (Figure 11.3) and highlights the central role of benthos as consumers and prey in short food chains. This estuary experiences large allochthonous inputs of organic matter from rivers that exceed primary production in the water column (Phytoplankton) and sediment (Benthic 1° producers). The food web model shows that macrobenthos are the primary consumers of this detrital organic matter and are also the most important prey for higher trophic levels (fish, birds). Also, a more complex box model of Dublin Bay (Wilson et al. 2007) showed that meiobenthos, especially harpacticoid copepods, are important prey for young-of-the-year fish and crustaceans. Some species (for example juvenile flatfish, salmon, and brown shrimp) are obligate feeders on meiobenthos, consuming thousands per day for a brief period of their life history. But these nekton species undergo feeding shifts to larger prey as they grow, and the distinction between meiobenthos and macrobenthos as prey is somewhat blurred.

11.4 Factors Controlling Productivity and Community Composition

Estuarine benthic communities are exposed to many potentially stressful physical and chemical conditions including temperature, salinity, and concentrations of dissolved oxygen, ammonium, and sulfides. In addition, benthic communities are affected by substrate composition, sediment-bound and dissolved pollutants, and biotic factors such as predation and the composition of vegetation. These factors strongly influence benthic productivity and community composition, and these controlling factors vary among biota and across space and time within estuaries.

11.4.1 Salinity

Salinity is a primary factor controlling the composition of estuarine benthic communities because benthos generally have limited physiological tolerances to salinity variation. The average as well as the range of salinity experienced by benthos varies depending on location within an estuary and with environmental factors such as tides and rainfall in the watershed, causing benthos community composition to vary within and across the estuarine gradient. When benthos are faced with rapid changes in salinity, they may either maintain constant internal salinity conditions or allow internal salinity to follow external changes. These alternative strategies are termed *regulator* and *conformer*, respectively (Figure 11.4). The balance of

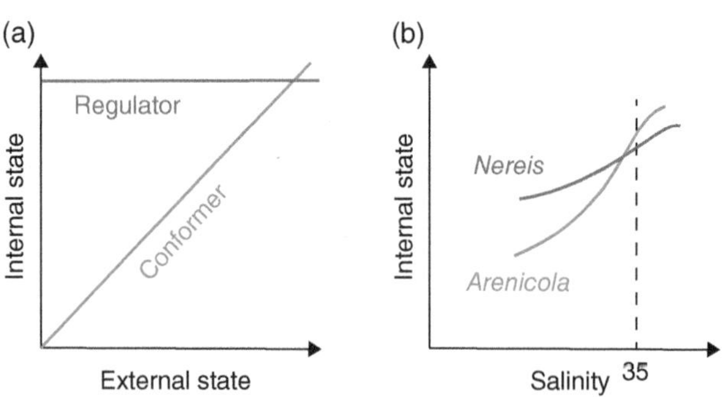

FIGURE 11.4 (a) Regulator and conformer strategies, with (b) typical responses to salinity by benthos: Hediste (*Nereis*) diversicolor osmoregulator, and *Arenicola marina* (osmoconformer). *Source*: Wilson (1988)/Springer Nature.

FIGURE 11.5 Oyster (*Crassostrea virginica*) reef. Note the habitat and refuge provided for other species (barnacles, tube-worms, anemones, mussels, and crabs) and the vertical building of the oysters themselves. *Source*: Day et al. (1989)/John Wiley & Sons.

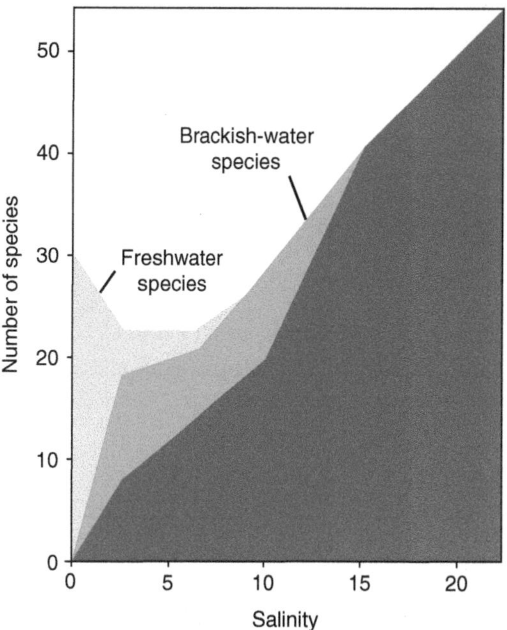

FIGURE 11.6 Remane curves, showing the number of species and the loss and replacement of marine species (blue, on right) with brackish and freshwater species as salinity declines. *Source*: Adapted from Remane and Schlieper (1958).

these strategies is determined by relative costs and benefits. The cost of the regulator strategy is both evolutionary, in the development and maintenance of necessary physical structures and metabolic processes, and bioenergetic, in terms of the energy required to maintain the internal state against external variability. The reward for implementing the regulator strategy is optimal physiological function. The conformer strategy, on the other hand, costs nothing, but comes at the expense of metabolic function. In practice, most estuarine organisms display a mix of strategies (Figure 11.4b), and the degree or range to which the regulator and conformer strategies can be viable may be rather limited. Figure 11.4b shows that the regulator *H. diversicolor* regulates over most of the estuarine salinity range but allows a small degree of conformity near full-strength seawater and cannot maintain regulation in fresh water. The osmoconforming *A. marina* on the other hand does manage to regulate near full-strength seawater, but then osmoconforms as the salinity falls further.

Many estuarine animals have evolved adaptations to salinity variation. For example, the Eastern Oyster, *Crassostrea virginica* (Figure 11.5), can survive and grow at salinities ranging from 5 to 42 psu and can withstand a rapid 15–20 psu change in salinity, but at a fitness cost. The osmoconforming lugworm, *A. marina*, lives from near full-strength seawater (>30 psu) to areas where the salinity rarely exceeds 15 psu in estuaries. However, worms in the lower salinities are smaller in size and weight, and only those in salinities ≥20 are able to reproduce, with an exponential decrease in reproduction in salinities below 30.

Remane (1935) hypothesized that because of the challenging physicochemical conditions of estuaries, the species diversity of marine benthos declines with decreasing salinity, and the species diversity of freshwater benthos is even more restricted by increasing salinity (Figure 11.6). True brackish-water species occur at salinities less than 15 psu and are particularly influential between 10 and 5 psu. This conceptual model is the foundation for much of our understanding of the distribution of benthos within estuaries. However, salinity variation also affects species diversity patterns; high levels of salinity variation are often negatively correlated with species diversity (Attrill 2002). Salinity variation also exerts an especially strong control in limiting the pool of biota available to recruit into estuarine communities, and this restraint is most forceful in mid-estuary where salinities are most changed from the end members (river, ocean) and the temporal variation in salinity is greatest.

Predators and parasites of benthic organisms are frequently less tolerant of salinity variation than their prey or hosts. The low and/or variable salinity of estuaries may reduce biological enemies, releasing species able to tolerate low salinity from biological control. For example, *C. virginica* (Figure 11.5) in the northern Gulf of Mexico is found at its highest abundance in salinities below its physiological optimum, at least partly as a result of reductions of predation (Pusack et al. 2019). Soniat (1996) concluded that this species survives well at salinities slightly lower than those tolerated by *Perkinsus marinus*, a significant oyster parasite. Thus, areas of estuaries that experience low salinities may act as refuges from harmful biological interactions for many species contributing to the high abundances and biomasses recorded for benthos in estuaries.

11.4.2 Effects of Oxygen

Like salinity, oxygen availability is another primary factor that controls the composition of estuarine benthic communities while exerting profound effects on abundance, biomass,

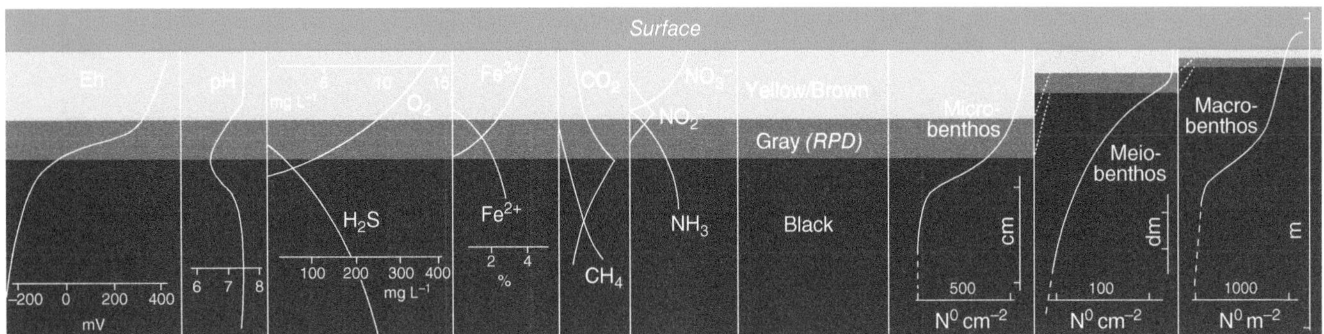

FIGURE 11.7 Chemical and biological gradients in sediment including the vertical distribution (abundance (N^0) cm^{-2} or m^{-2}) of micro-, meio-, and macro-benthos. *Source*: Adapted from Fenchel and Riedl (1970).

and productivity. Benthic animals have limited physiological tolerances to hypoxia (usually defined as oxygen levels below 2 ml L^{-1}) and anoxia which occur as episodic or chronic events in estuaries. Low bottom-water oxygen concentrations affect epibenthos as well as infauna in subtidal environments, although high concentrations of oxygen can be maintained by benthic plants and algae where light penetrates to sufficient depth, by mixing of bottom waters with oxygenated surface waters, and by tidal and other currents that bring relatively well-oxygenated water into areas with lower oxygen. However, low oxygen conditions in bottom waters occur whenever oxygen consuming processes exceed the rate of resupply of oxygen and are often seasonally related to water-column stratification. This may result from respiration in the sediments and water column or from chemical oxygen demand created by anaerobic metabolism in sediments (see Chapter 3).

In some estuaries a combination of neap tides, high temperatures, decaying phytoplankton blooms, high benthic density, and estuarine turbidity maxima can cause low oxygen conditions and benthic mortality. In many estuaries this lowered oxygen saturation, sometimes called an "oxygen sag," may be a permanent feature that extends over many kilometers. This phenomenon may be particularly pronounced when scouring is sufficient to resuspend some centimeters of the bottom sediments including those below the RPD (Figure 11.7), in which oxygen is consumed in the chemical reactions where reduced compounds are reoxidized. In estuaries prone to prolonged seasonal stratification, oxygen deprevation may result in extreme deleterious consequences for benthos. The Chesapeake Bay experiences seasonal bottom-water hypoxia, sometimes called a "dead zone," where the abundance of macrofauna is directly related to dissolved oxygen concentration, and declines to near zero levels when hypoxia is severe or prolonged (Figure 11.8). In addition, pronounced variation in sediment oxygen concentrations unrelated to bottom-water conditions (e.g., on intertidal mudflats or associated with seagrass beds) commonly occurs over daily, seasonal, or tidal time scales.

Responses of benthos to hypoxia depend on the duration, predictability, and intensity of oxygen depletion. For example, infauna living on mudflats are likely to be tolerant to hypoxia

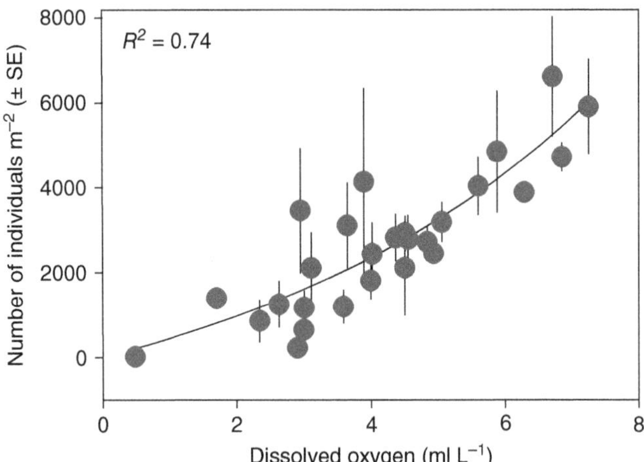

FIGURE 11.8 Relationship between the mean density of benthic organisms (total macrobenthos m^{-2}, ± SE) and dissolved oxygen concentration from random sites for the Chesapeake Bay in the summers of 1996–2004. Nonlinear regression, quadratic increase curve fit to data. Benthos were collected with a Young grab sampler and sediment sieved through a 0.5-mm screen. *Source*: Seitz et al. (2009)/ With permission of Elsevier.

on the time scale of many hours as an adaptation to living on mudflats that frequently experience nighttime hypoxia. Calcareous benthic foraminifera and nematodes are particularly tolerant of low oxygen concentrations and may attain high densities and dominance under these conditions. When oxygen is sufficient to support metazoans other than nematodes, benthic communities are often dominated by small, soft-bodied invertebrates (typically annelids), many with short generation times and elaborate branchial structures that facilitate gas exchange. Mollusks and benthic cnidarians are also tolerant of low to intermediate oxygen conditions. At higher oxygen concentrations, benthic communities can include crustaceans and echinoderms, which are typically more sensitive to hypoxia and have higher oxygen thresholds than annelids, mollusks, and cnidarians.

Natural and eutrophication-induced increases in hypoxia cause losses in benthic species diversity through attrition of intolerant taxa, and often lead to reductions in biomass and individual body size. Such shifts in benthos species composition often result in changes to the ecological function of benthic communities, including altered trophic structure, energy flow pathways, and corresponding ecosystem processes such as production, organic matter cycling, and organic C burial. For example, hypoxia reduces the diversity and abundance of bioturbators, thus limiting benthic reoxygenation at times when bottom waters are oxygenated. Hypoxia can also alter the fate of energy in sediments by reducing the energy used by benthic animals (many of which are prey for commercially important fisheries species) and leaving more energy to be used by microbes (Diaz and Rosenberg 2008).

Hypoxia also drives some motile benthos to move to other habitats, thus altering food web structure. In hypoxic conditions, benthic species such as the bivalve *L. balthica* and the polychaete *H. diversicolor* will leave their burrows in the sediment and can be seen still lying on the sediment surface when the tide is out. However, hypoxic conditions can make these organisms sluggish, often to the point of coma before actual death, which leaves them vulnerable to nektonic and avian predators. Low oxygen conditions have been shown to increase predation on benthos and to have a significant effect on the structuring and function of the benthic system (Long and Seitz 2008).

11.4.3 Substrate–Benthos Relationships

The physical substrate of the benthic environment itself can influence what kinds of benthic communities may establish, from hard substrata needed by epifauna to the soft substrata for infauna. In some cases, the fauna and the substratum are one and the same, such as when mussel and oyster beds reach such proportions as to physically dominate the habitat, turning a soft, muddy substratum into a reef of hard shells. Reefs of *C. virginica* form communities of numerous organisms (Figure 11.5) because the oyster shells act as hard substrate for epifauna (including more oysters), and the buildup of fecal deposits, including suspended material which the oysters have taken out of the water column, encourages colonization by infauna. The resulting improvement in water quality is a major ecosystem service provided by epifauna. In addition, diversity "hotspots" associated with oyster reefs and other epifaunal beds are important contributors to local patterns of marine biodiversity (Craeymeersch and Jansen 2019).

Sediment type is frequently related to infauna community composition. For example, while many species are found over a wide range of sediment types, most display clear preferences for a much narrower range. For example, the clam *L. balthica* occurs mainly in silty sediments (up to 60% silt), whereas the closely related *Tellina tenuis* is restricted to much sandier locations where silt content never exceeds 5%. The relationship between communities and sediment type may be due to differences in pore spaces and the architecture of sediment particles or to the strong indirect effect of sediment type on chemical gradients within sediments (Snelgrove and Butman 1994). In addition, infauna modify physical, chemical, and biological sediment properties that affect ecosystem function and many ecosystem services. For example, infauna displace sediment grains during burrow construction and displace organic matter and microorganisms within the sediment matrix during feeding. This process, known as *bioturbation*, affects sediment stabilization, biochemical cycling and vertical gradients, waste removal, and food web dynamics at various spatial and temporal scales (Schratzberger and Ingels 2017). Bioturbation may have positive or negative effects on ecosystem function and various benthic organisms. For example, bioturbation and burrow irrigation increase the depth of oxygen penetration (and the depth of the Redox Potential Discontinuity, see next paragraph) into the sediment contributing to higher rates of microbial decomposition while greatly increasing exchange rates of nutrients across the sediment/water interface. However, bioturbation negatively affects some benthic species by reducing sediment cohesiveness, thereby destabilizing sediments.

Within sediments, infauna is confronted with steep and varied gradients of physicochemical conditions with depth (Figure 11.7). As sediment microbes use up oxygen and produce H_2S, the sediment color changes from yellow/brown to black, redox (Eh) changes from positive to negative, denitrifying bacteria reduce NO_3^- to NO_2^- and N_2, sulfate reducers reduce SO_4^{2-} to H_2S, and methanogens reduce CO_2 to CH_4 (see Chapter 9). The depth of this Redox Profile Discontinuity (RPD) and the steepness of the gradient are controlled by the rate of diffusion of oxygen into the sediment and its rate of consumption. In coarse sediments such as sands, with large spaces between the grains, the RPD may occur many tens of centimeters below the surface, while in fine muds, it may be measured in fractions of a millimeter. The availability of oxygen is a major factor controlling the depth distribution of benthos, with most avoiding the anoxic deeper layers (Figure 11.7). The deeper macrobenthos must maintain contact with the overlying oxygenated water either through long, extensible siphons as in the larger bivalves, or through irrigated burrows. Meiofauna rely on other adaptations to live below the RPD. Some meiobenthos live at depth in the oxygenated halo around burrows of large macrobenthos. Some display body shapes that are longer and/or thinner to enhance diffusion or make their bodies long enough to extend into the oxic zone. Others tolerate anaerobic conditions for a period of time and then migrate to the oxygenated sediment surface for gas exchange.

11.4.4 Plant–Benthos Interactions

Benthos often have higher abundance and biomass in vegetated habitats, such as marshes (see Chapter 6), mangroves (see Chapter 7), and seagrass beds (see Chapter 5), than in surrounding nonvegetated areas. Plants that define these habitats provide physical structure, primary productivity, habitat cover, organic matter and detritus, food web support, nutrient cycling, and the development and stabilization of soils and sediments, and are therefore considered to be foundation species (Crain and Bertness 2006). The result is that vegetation improves the environment for consumers by providing food and shelter, and by ameliorating harsh environmental conditions (e.g., plant roots oxygenate sediment). In addition, the complex physical structure of plant communities and reefs baffle water flow, increasing the potential for larval settlement. Plants also serve as substrate for epiphytic algae and shade the sediment. The importance of foundational plant species to benthic communities is demonstrated by several studies following the 2010 oil spill in the Gulf of Mexico. For example, Fleeger et al. (2019) and Deis et al. (2020) found that the recovery of infauna after oil-induced mortality closely followed the recovery of *Spartina alterniflora*. This suggests that the recovery of meiobenthos and juvenile macrobenthos was linked to the regrowth of aboveground and belowground plant biomass.

Whereas habitat structure imposed by vascular plants as foundation species benefit benthos that live among them, benthos also provide benefits to the plants, creating a positive feedback. For example, ribbed mussels in the genus *Geukensia* have positive effects on saltmarsh function by physically binding sediment particles and limiting marsh erosion, and by excreting nitrogen-rich psuedofeces that enhance aboveground and belowground vegetation production (Bertness 1984; Bilkovic et al. 2017). Suspension-feeding benthos filter large volumes of water in estuaries, removing algae and suspended sediments, thus clarifying the water and increasing light penetration that benefits submersed plants. These positive interactions among species are increasingly becoming recognized as playing important roles in marsh ecology and in activities such as marsh restoration (Renzi et al. 2019).

11.4.5 Recruitment and Planktonic Dispersal

The productivity and diversity of estuarine benthos is tightly linked to the recruitment success of planktonic juvenile and larval life stages, and to the success of juveniles and adults following recruitment. The planktonic phase is a vital mechanism through which many benthic species are dispersed, but in the dynamic conditions in many estuaries there is a high risk that the larvae may be flushed from the system or be transported to locations that are inappropriate for settlement. Estuarine benthos have developed several strategies to favor recruitment in estuaries. Some species, such as *H. diversicolor,* evolved a modified planktonic phase in which not only is the time spent in the plankton reduced in comparison to similar fully marine species, but the larvae also stay close to the seafloor. In this way, they can take advantage of estuarine circulation by riding up-estuary movement of bottom waters to enhance estuarine retention (see Chapter 2) and by avoiding unfavorably low salinity conditions in surface waters. Other species, such as the blue crab, *Callinectes sapidus* (Bell et al. 2003), and several shrimp species migrate to the ocean to spawn and then return to estuaries as juveniles. Female blue crab migrate to the coast (near ocean salinities) to spawn, and the larvae develop through the various stages in coastal waters. Then as megalopae they drop down in the water column into the salt wedge to be transported back into estuaries where they molt into adults and continue up-estuary to their adult habitat.

Successful recruitment is also subject to stochastic processes owing to the generally narrower environmental tolerances of larvae compared to adults. For example, adults may survive low salinity conditions where larvae would not. Under these circumstances, successful recruitment and colonization depends either on mechanisms such as for the blue crab, which avoid variable estuarine conditions as larvae, or by taking the chance that spawning events coincide with favorable estuarine conditions. As a consequence, many estuarine populations show an irregular cycle of abundance, where a good recruitment year may be followed by several poor years. This results in an unbalanced population structure, in which the single successful cohort dominates and continues to dominate for a number of years (Hughes 1970).

Following recruitment, there is a broad range of biological adaptations to estuarine hydrology that influence adult benthos distribution and population densities. For many soft-sediment species, the dispersal ability of juveniles and adults contributes to the high spatial and temporal variability of benthos in estuaries. Some species are transported by tidal currents to help them find preferred habitat. For example, the pattern of tidal currents over a tidal cycle shows that the current speed required for maximum transport of the bivalve *L. balthica* juveniles (Figure 11.9) is close to the current speed that transports sediment particles that are common in their preferred habitat, causing juvenile *L. balthica* to move together with sediments to preferred habitats. The mud snail *Peringia (Hydrobia) ulvae* uses surface tension of the water and floats, upside-down, on its extended foot (Figure 11.10), allowing it to move through estuaries with relatively strong surface currents. Transport with currents and tides disperses *Peringia*

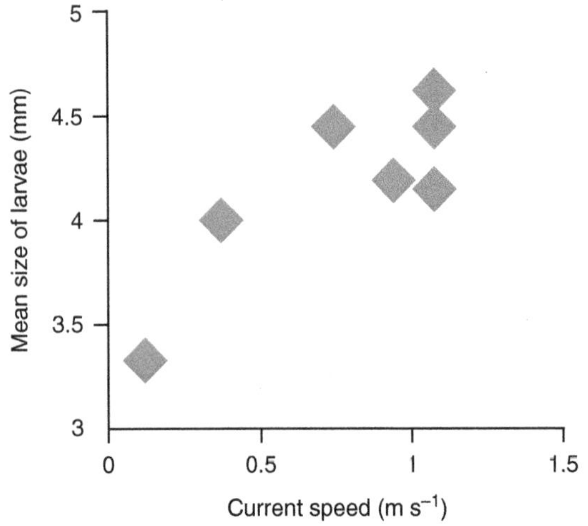

FIGURE 11.9 Mean size (mm) of *Limecola balthica* larvae transported versus tidal current (m s^{-1}). *Source*: Data from Beukema and de Vlas (1989).

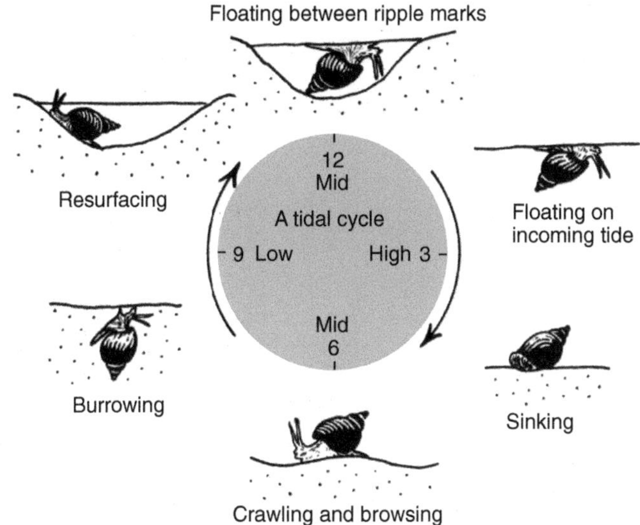

FIGURE 11.10 Flotation of *Peringia* on water surface tension and cycle of behavior. *Source*: Newell (1979)/Marine Ecological Surveys.

throughout estuaries, causing the density of these populations to vary a great deal. In addition, adults and juveniles of some benthic amphipods regularly and actively swim to seek new habitats (Drolet et al. 2012).

The distribution and abundance of meiobenthos, most of which lack dispersing larvae, are also directly influenced by the hydrodynamic regime. Instead, they are dispersed as juveniles or adults by passive mechanisms like sediment resuspension or active emergence associated with mate finding, feeding, or avoidance of harsh environmental conditions such as nighttime hypoxia or sulfide accumulation. Furthermore, the abundance of meiobenthos in sediment has been related to tidal currents wherein the highest densities are found at slack tides and the lowest densities are found during high flow after some meiobenthos emerge (Palmer and Brandt 1981). Even the location of reentry into the sediment by emerged meiobenthos is subject to flow-related constraints because meiofauna cannot swim against currents and may passively settle to the seafloor. After reentry into sediments, meiobenthos may achieve highest abundance in pits and depressions where they accumulate, along with small particles, because they are protected from small eddies and currents.

11.4.6 Density-Dependent Controls and Population Cycles

Once established, adult benthic populations in estuaries may experience mortality or great fluctuations in density for a variety of reasons. These include, but are not limited to, sudden changes in oxygen concentration or salinity, pollution events, scouring floods or storm surges, droughts, and temperature anomalies that act in a fashion independent of the density of populations (i.e., density-independent population control). However, benthic populations are also strongly affected by density-dependent controls. For example, Figure 11.11 shows a model of cyclic fluctuations in cockle (*C. edule*) populations in which both internal and external biological factors modify density following initial settlement. Successful settlement during recovery leads to rapid growth of individuals and rapid increase in numbers. Subsequently, this initial year class ("G0") spawns numerous new year classes and the population reaches a peak. Then, intraspecific competition and reduced individual growth rates restrict further growth in the population. As the population ages, mortality increases in older individuals, and there is poor recruitment due to the negative effect of high adult densities on the success of recent recruits, leading to a fall in overall population abundance and increased competition from other species.

Many benthos are able to modify the benthic environment in ways that exclude other organisms and thus compete for habitat space by taking advantage of low points in the population cycles of other taxa. For example, a cycle of abundance alternately dominated by the cockle *C. edule* and a tube-building spionid polychaete (*Pygospio elegans*) has been described, with each being able to partially exclude the other by virtue of unusually high densities at the height of the cycle shown in Figure 11.12. *P. elegans* is an opportunist that successfully colonizes the sediment when *C. edule* numbers are

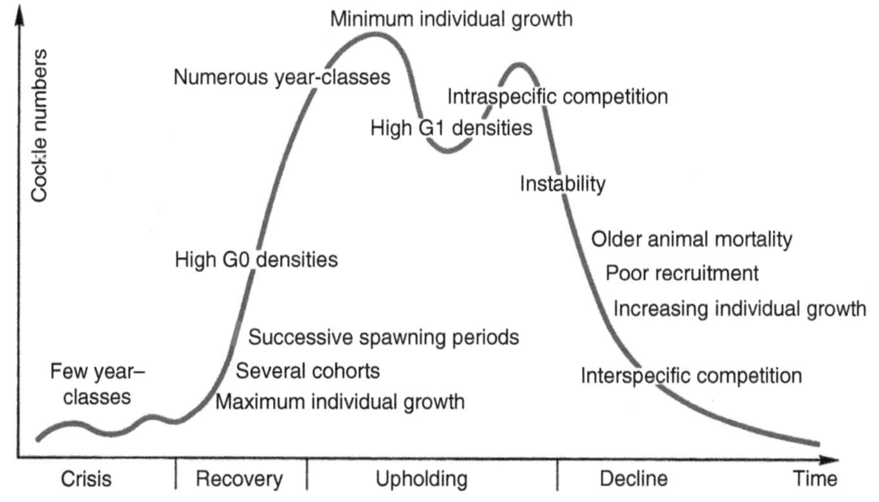

FIGURE 11.11 Population cycles in the bivalve *C. edule*, showing periods of maximum and minimum individual growth and the dominant year cohorts (G0, G1). *Source*: Adapted from Ducrotoy et al. (1991).

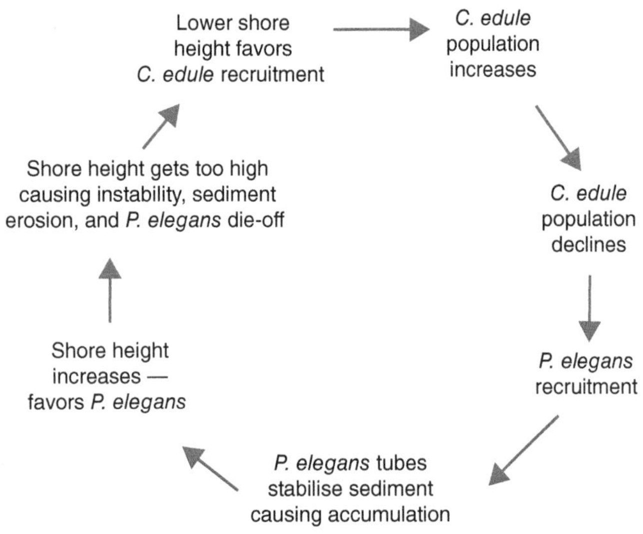

FIGURE 11.12 Model of the interactions between the cockle *C. edule* and the tube-building worm *Pygospio elegans*.

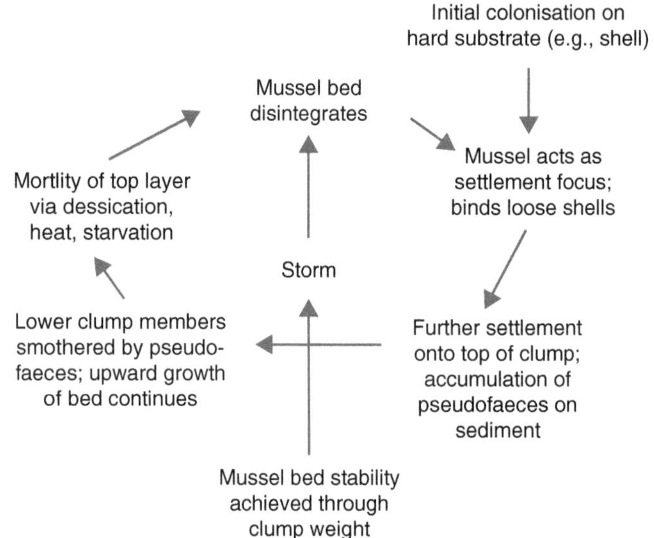

FIGURE 11.13 Diagrammatic representation of the cycle of mussel bed buildup and dieback.

low. Tubes of *P. elegans* stabilize the sediment and accelerate accretion, causing the shore level to rise and rendering the habitat less suitable for *C. edule*. *C. edule* reestablishes only when the *P. elegans* colonies have grown to such an extent that the level of the substratum has been significantly raised and the animals are subject to high-tide levels of temperature and desiccation. When the worms die off, their tubes break up and the sand bank erodes, leaving vacant space for the cockles to establish. The cockles dominate until adverse conditions (e.g., as shown in Figure 11.11) cause a decline, leaving vacant niches once again for the opportunist *P. elegans*.

A similar cycle has been shown in the growth and maturation of mussel (*M. edulis*) beds (Figure 11.13) and oyster reefs, which can overgrow soft sediment environments, radically changing the habitat for infauna. Mussels adhere to hard substrates by their byssus threads, and, as epifauna, they require something solid to first attach such as a stone or a shell. If the individual survives then its shell in turn offers hard substratum for future settlers, whose byssus threads act to bind together the clump. Once the clump reaches a certain size, then its own weight confers stability and prevents it from being displaced or otherwise washed away by storms or other events. Successive waves of settlers then increase the size and height of the mussel bed. Mussels at the top of the clump have better access to food and oxygen from the water column. Feces and pseudofeces fall to the bottom of the clump, enhancing microbial respiration and rendering the underlying sediments anoxic. Oyster reefs undergo similar dynamics and can be created by placing oyster shell on the muddy seafloor to create an "oyster cultch" or artificial oyster reef. Reef height is important for stability of mussel and oyster populations because reefs that are too short to extend above the near-bottom hypoxic zone are less productive. This is why overfishing mussel and oyster reefs can reduce reef height and threaten reef productivity. In each of these examples, the height relative to sea level is critical, and all such mechanisms will be influenced by climate-induced sea level changes.

11.4.7 Top-Down and Bottom-Up Controls

Benthic consumers are key intermediates in estuarine systems because a large fraction of the flow of energy in estuarine food webs is channeled through sediment systems and because many invertebrates, fishes, and birds consume benthic prey. Top-down trophic cascades occur when predators suppress the abundance of their prey (a direct effect), thereby releasing the next lower trophic level from predation (an indirect effect). However, the potential for trophic cascades appears to differ for infauna and epifauna. For infauna, predator exclusion studies suggest that benthic animals increase in abundance in the absence of predation, as might be expected. But this top-down control does not appear to extend to the prey of infauna because benthic microalgae rarely increase in abundance when infauna increase. This weak coupling between infauna and benthic microalgae may occur because infauna have a diverse diet that includes detritus and bacteria (Johnson and Fleeger 2009). By contrast, epifauna are often important intermediates in significant top-down trophic cascades. For example, overfishing of blue crabs may lead to increased densities of the snail *Littoraria irrorata*, and the resulting increase in snail herbivory depresses the density and biomass of the cordgrass *S. alterniflora*. Similar conclusions of the importance of trophic cascades have been reached in seagrass communities and their epiphytes in estuaries (Heck and Valentine 2007).

The shorecrab (*Carcinus maenas*) provides a classic example of predator–prey interaction in its feeding on *M. edulis*. Experiments have shown that larger shore crabs tend to prey on larger mussels (Figure 11.14b), and that this selective predation is an optimal foraging strategy. According to optimal foraging theory, the larger the mussel the greater the reward (amount of energy) to the crab, but the larger the mussel, the more time and energy it takes to open it. Thus, the optimum

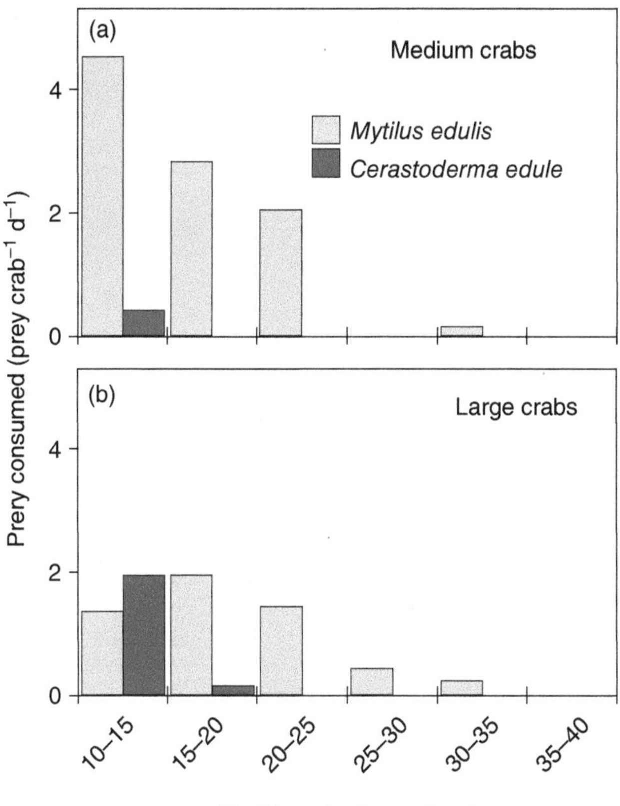

FIGURE 11.14 Optimal foraging and size selection of mussels *Mytilus edulis* (solid bars) and cockles *Cerastoderma edule* (hatched bars) by crabs *Carcinus maenas* that were (a) medium-sized (M; 40–55 mm carapace width) and (b) large-sized (L; 55–70 mm carapace width) during experiments where crabs were offered prey species in paired combinations. Vertical axes indicate the mean consumption rate in number of prey crab^{-1} d^{-1}. *Source*: Based on Mascaró and Seed (2001).

strategy is one of minimizing the costs while maximizing the return with the result that there is an optimum size of prey to be taken. Smaller mussels can be taken if the optimum size is scarce, but when the mussel reaches a certain size the shell is too thick and solid to be cracked by the crabs and they then become invulnerable to crab predation. Figure 11.14 shows that predation mortality is clearly highest on the smaller mussels, which has consequences not just in the immediate population density and abundance, but also in the longer-term balance between growth and reproductive investment which in turn feeds into some of the factors in Figures 11.12 and 11.13.

The alternative to top-down control in benthic systems is bottom-up control by factors such as food quantity/quality and nutrient (N, P) availability. Experimental evidence suggests that the density of infaunal populations can be limited by food quality of the sediments, and that adding food such as detritus increases population density especially for deposit-feeding species (Cheng et al. 1993). Questions about bottom-up control are relevant because many estuaries have experienced increases in nutrient loading and primary production. The relative importance of top-down and bottom-up controls on benthic populations has been difficult to address because large-scale manipulative experiments are difficult to conduct and are rare in estuaries (Deegan et al. 2007). Interactions between top-down and bottom-up controls appear to be rare for infaunal communities (Posey et al. 2002) but are potentially important because human activities frequently reduce top predators and simultaneously increase nutrient loading to estuaries.

11.5 Special Topics for Estuarine Benthos

11.5.1 Human Impacts on Estuarine Benthos

There is a long history of human exploitation of the estuarine benthos. The abundance of many estuarine shellfish populations has long been exploited and indeed overexploited. Table 11.8 lists some of the major estuarine species harvested by humans and the very considerable quantities taken.

The first recorded commercial lease of an oyster bed is documented from 1705 in Dublin Bay, Ireland, for which the fee for one bed alone was 10,000 oysters annually to be paid to the Mayor plus 1000 a year to each of the Sheriffs. Today, almost no trace remains of the extensive oyster beds in Dublin Bay, and while pollution may have played a part, such a high level of exploitation cannot have been sustainable. In contrast, the Dublin Bay cockles were also widely exploited throughout the nineteenth century, with some 6000 gallons in the old measures (equivalent to about 100 tonnes) taken annually, at least until their collection was forbidden in 1908 on public health grounds, yet their populations still seem to be about the same today as formerly.

C. virginica also suffered from exploitation and pollution. It was estimated that at the time of the heyday of the fishery at the end of the nineteenth century, just under a teragram of oysters (~300 million bushels) were taken in the state of Maryland alone. Prior to this era of harvesting the standing stock of oysters was by itself capable of consuming almost 80% of the phytoplankton carbon production of the system (Newell 1988). The decline of the oyster population to a small fraction of their former level, and the resulting decline in the control on the system exerted by the oyster, was suggested as a causative factor for the change in the ecosystem, and in particular the increased abundance of ctenophores and sea-nettles.

A confounding factor in the decline of *C. virginica* in Chesapeake Bay has been disease, especially the unicellular parasite MSX, and serious consideration has been given to replacing the native oyster with the MSX-resistant *Crassostrea japonicum*. However, this plan has been vigorously opposed, because the history of introductions of non-native species is littered with unintended and usually harmful consequences. The European

TABLE 11.8 **Annual harvest (1999 catch, ×10^3 Mg) and salinity tolerance of estuarine consumers with comments on status of stocks.**

Species	Comments	Salinity range	1999 Catch
Mytilus edulis	Northern Europe, cultured widely: hybridizes with *M. galloprovincialis*	3–35	122
Mytilus galloprovincialis	As *M. edulis*; southern Europe, Mediterranean and Black Sea	10–35	56
Perna viridis	As with other *Perna* species increasingly fished and cultured	18–33	22
Mussels	Global harvest		1689
Callinectes sapidus	Blue crab, No. 1 seafood product in Chesapeake Bay; Larval salinity tolerance > 20	0–35	105
Crangon crangon	Local fisheries; not cultured	5–35	37
Penaeus monodon	Widely cultivated	10–35	144
Crab, prawn, shrimp	Global harvest		4126
Crassostrea virginica	Chesapeake Bay harvests, much reduced	5–30	146
Crassostrea gigas	Most popular species for worldwide culture; spawning temperature restricted in N. Europe	8–36	12
Ostrea edulis	Beds much reduced and threatened by diseases	24–35	2
Oysters	Global harvest		3869
Cerastoderma edule	Laborious to collect: now mechanized methods (with environmental conflicts).	15–35	70
Alitta virens	Fishing bait, now small-scale culture	15–35	0.3[a]
Glycera dibranchia	Fishing bait, restrictions and quotas apply	12–47	0.3[a]

[a] Data for State of Maine only: peak catch mass (1980–1999) quoted.

flat oyster, *Ostrea edulis*, is a good case in point. As their stocks diminished throughout Europe (see example above of Dublin Bay) they were replaced first by *C. virginica*, and then by *C. gigas*, in part because these introduced species grew faster and were ready for market in less time. However, these introductions brought with them the slipper limpet, *Crepidula fornicata*, which is a competitor that also overgrows the oysters and covers them with faeces. These introductions also brought with them the predatory oyster drill *Urosalpinx cinerea*, and the oyster wasting disease *Bonamia ostreae*. Of these co-introduced species, it is the wasting disease that is having the greatest effect on the remaining *O. edulis* beds, and a great deal of effort and regulation is being implemented to control and restrict any further expansion of the condition into unaffected areas.

While many of the species listed in Table 11.8 are now grown for aquaculture, much of these harvests are still either taken from wild stock, or dependent on wild stock. For example, cultivation of the mussels *M. edulis, M. galloprovincialis,* and *P. perna* require wild stock as a recruitment source as spat or juveniles. Similarly, the oysters *O. edulis* and *C. japonicum* can be bred and raised to a small size in special facilities but need to be returned to natural estuarine systems to grow to adulthood.

Besides the direct effect of harvesting, humans also impact benthic communities indirectly by increasing nutrient inputs and causing eutrophication in estuaries. Pearson and Rosenberg (1978) developed the classic model of the effects of eutrophication-induced anoxia on benthos (Figure 11.15) showing the different zones and "typical" species associated with each. After a disturbance ("Grossly Polluted"), opportunist species, typically rapid-settling, quick-growing but small and short-lived, are the first to colonize the sediments as the situation improves ("Polluted"), followed by increasing numbers of other species ("Transitory"), until, as control shifts from physicochemical control to biological control, superior competitors with poor colonizing abilities come to dominate ("Normal"). This model of benthic succession following disturbance works well for a number of stressors of estuarine benthos, including benthic anoxia and toxins, and it can be applied both in time after a disturbance and in distance from a point source. However, since estuaries tend toward disturbance and physicochemical control, benthic communities may never actually attain the "normal" status as defined in the model. In a way, it is not surprising that species that are adapted to respond to the natural stresses

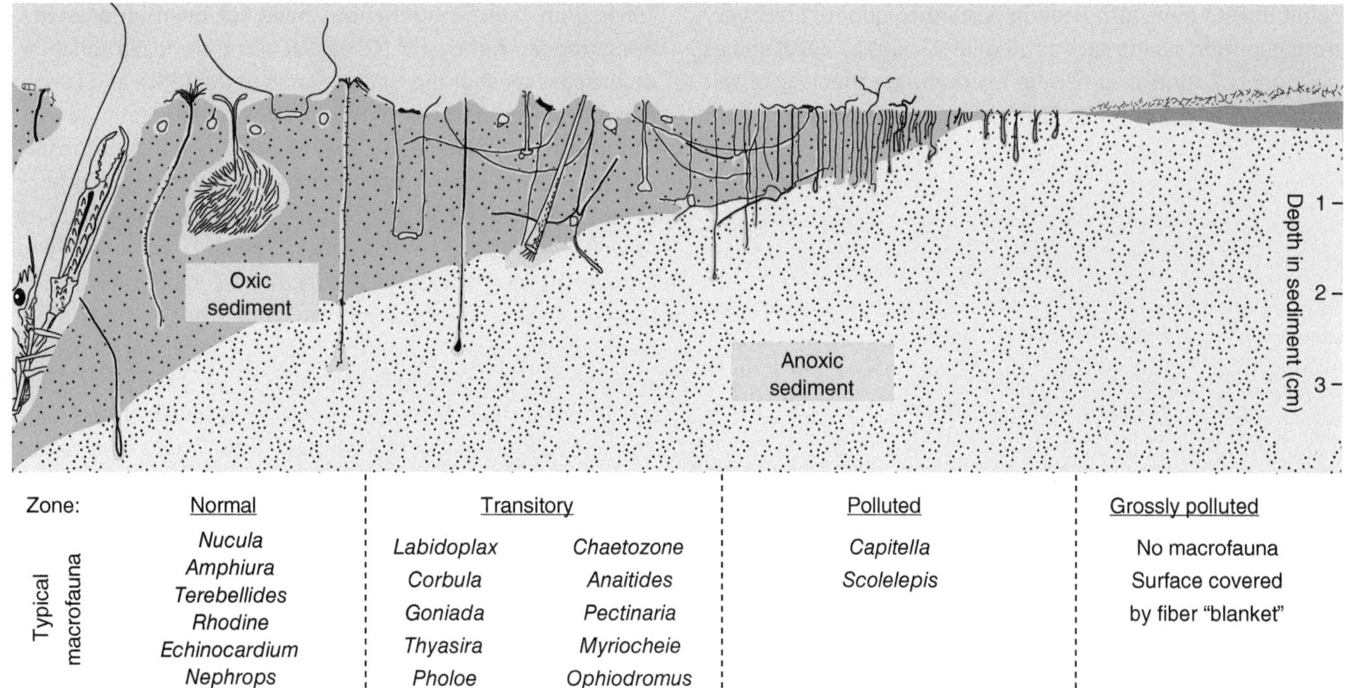

FIGURE 11.15 Benthic community response to pollution (increasing from left to right), showing decrease in diversity of species and life-forms. *Source*: Pearson and Rosenberg (1978)/FAO.

of estuarine systems have the capacity to respond to anthropogenic contamination. However, as with all stresses, the effects cannot be taken in isolation, and while estuarine benthos as a group may be relatively resistant overall, the impact of a contaminant is often much greater when salinity or other stress is added.

Many estuarine sediments serve as long-term sinks for toxic chemicals in a variety of classes, including heavy metals, hydrocarbons, herbicides, pesticides, and endocrine disruptors. Sediment quality in many estuaries is often degraded due to the presence of these toxins (Long 2000), and pollutants affect benthic communities in predictable ways. Generally, studies have shown that crustaceans and echinoderms are intolerant of most kinds of sediment contamination and that small-bodied deposit feeders such as annelids and nematodes are most tolerant. Tolerance to contaminants may differ greatly, and some species are very sensitive while others may be quite tolerant. As a result, pollution may bring about community change due to indirect effects, in which tolerant species change in abundance because competitors, predators, or prey are affected by pollutants (Fleeger et al. 2003).

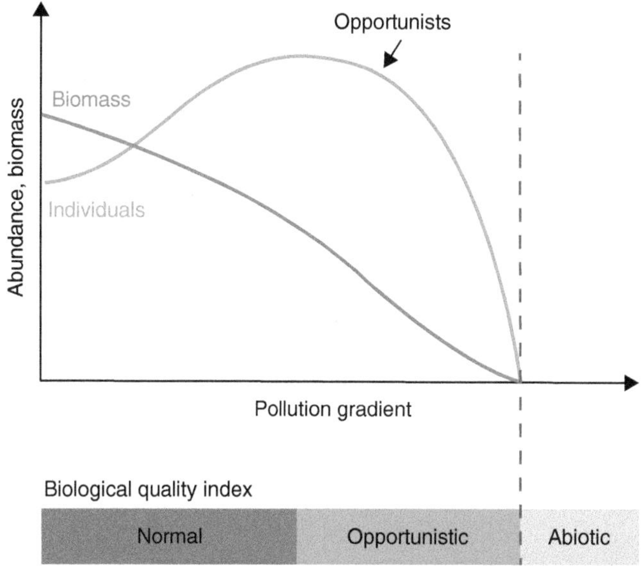

FIGURE 11.16 Biological Quality Index categories along pollution gradient with relative abundance of individuals and community biomass and showing the peak of opportunist species. Pollution increases from left to right. *Source*: Adapted from Wilson (1988).

The range of sensitivity and intimate exposure of benthos with toxins in the sediment and water column makes benthic communities useful indicators of pollution (Wilson and Elkaim 1992), and several schemes have been developed to use benthic communities in environmental monitoring programs in this way. At one end of the spectrum is the very simple Biological Quality Index (BQI) of Jeffrey et al. (1985) with just three categories (A = Abiotic, B = opportunistic, and C = Stable/Normal: see Figure 11.16), which defines diversity on a comparative scale against that expected in estuaries, with added criteria based on size and number of year classes. At the other end of the spectrum is the Benthic Index of Biological Integrity (B-IBI) of Weisberg et al. (1997), which was developed for Chesapeake Bay based on a very extensive set of databases available for the Chesapeake. This index divides an estuary into seven different habitats on the basis of salinity and substrate in which 17 different metrics are tested for deviation from reference values. Estuarine benthic communities have also been important to quantify recovery from pollution events such as oil spills (Deis et al. 2020) and as measures of success in habitat restoration, especially in salt marshes (Baumann et al. 2020).

11.5.2 Invasive Species

Estuaries have long been regarded as targets for invasive species and as locations into which non-native or alien species may be introduced before invading other environments in the region (Cohen and Carlton 1998). There are a number of reasons for this, but the central arguments are based firstly on the mechanisms for introductions and secondly on the nature of the estuarine ecosystem itself with human influences affecting both. There are many ways in which an alien species may be introduced, and these are summarized in Table 11.9.

Natural range extensions are part of the dynamic nature of ecological systems, but human influence plays its part. For example, global climate change is hastening the spread of species poleward. Fiddler crab range is extending northward in North America (Johnson 2014), and the spread of the pacific oyster (*Crassostrea gigas*) northward in Europe is no longer reliant on human transport (see also below). These northward extensions of natural ranges may be balanced at the other end of the species' ranges by the disappearance of species from estuaries. Such a change has been reported for the bivalve *L. balthica*, which disappeared from the Gironde estuary in France and the southern part of the Bay of Biscay as a whole over the last 40 years or so. In this case, exposure to high temperatures associated with climate change is suggested as the cause (Jansen et al. 2007).

Deliberate introduction of new species for food or commercial exploitation has been a human practice for centuries and estuarine benthos are no exception. Oysters are classic examples (Woolff and Reise 2002), with the North American eastern oyster (*C. virginica*) being introduced to Europe at the end of the nineteenth century and well into the twentieth, followed by the pacific oyster (*C. gigas*) from the 1960s. But, these introductions have not been all one-way: the European flat oyster (*O. edulis*) was introduced to Maine on the east coast of the United States in the 1950s and is now being cultured on the Pacific coast of the United States as well. Unfortunately, however, deliberate introductions like these—with thousands of tons being transported over the years—has meant that, inevitably, other species have been brought over as well. As mentioned earlier, when *C. virginica* was brought to Europe, it brought with it not only direct competitors like the suspensivore *C. fornicata* (slipper limpet) but the predatory oyster drill *U. cinerea* and the haplosporidian parasite *Bonamia ostreae* (Woolff and Reise 2002).

The transport itself can act as a vector for introducing other alien species, and this is a major contributor to estuaries as foci for introduction, given the siting of ports, harbors, and marinas in estuarine waters. While the most obvious pathway might seem to be with the fouling organisms on the boat hull, the practice—now discontinued in estuaries—of taking in ballast water in one estuary and releasing it in another estuary has been particularly effective in the accidental movement of organisms around the globe, resulting in both national and international protocols for ballast water (see e.g., **https://www.invasivespeciesinfo.gov**). On a smaller scale, even sport anglers may transport alien organisms with boats, with gear, with bait, or even with the catch itself.

Given the history of estuarine usage as transport hubs (Wilson 1988) it is perhaps not surprising that where there is a large volume of shipping or boat traffic, the potential to transfer organisms from one estuary to another is high. In addition, the organisms taken from an estuary may be tolerant of adverse conditions (see Study Questions) and so may be more resistant to changed or changing conditions during the transfer, and may have the ability to adapt to or tolerate the new environment.

As noted in previous sections, the low biodiversity of estuarine benthos and the abundance of certain food sources offers a great opportunity for invasive species that can tolerate the stresses. Food is abundant and biological controls such as interspecific competition, predation, parasites, and disease are often much reduced through human agency as much as the nature of the estuarine systems. Newell (1988) was one of the first to draw attention to the effect that large-scale removal of oysters had on the Chesapeake Bay. Therefore, for a potentially invasive species, ecological niches left vacant in estuaries by harvesting or other anthropogenic effects offer plentiful food sources and relatively few competitors or predators, giving strong possibilities to both establish and flourish in the new location.

The downside of these invasions is their impact on native species and on the estuaries themselves, and many of these impacts are demonstrated by the oyster examples alluded to above. In the Bay d'Oleron on the Atlantic coast of France, non-native oysters not only dominate the benthos but have shifted the system from pelagic-dominated to benthic-dominated (Leguerrier et al. 2004), while further north in the Wadden Sea the mudflats are being taken over by non-native oyster

TABLE 11.9 Invasion mechanisms for estuarine benthos.

Mechanism	Human influence	Importance
Natural range extension	Via climate change	Low
Deliberate introduction	As food or commercial species	Medium
Accidental introduction	With other species	High
Accidental introduction	Transport (boats and shipping)	High
Accidental introduction	Escapees (e.g., aquarium trade)	Low

(*C. gigas*) reefs threatening native benthos and infauna, and potentially altering estuarine hydrodynamics and navigation (Nehls and Büttger 2007). Introduced bivalves in San Francisco Bay have been highly successful and may be sufficiently abundant as to affect water column productivity (Carlton 1999). Invasive species may cause ecosystem-wide alterations that affect not just the estuarine benthos but the entire array of ecosystem services (Sousa et al. 2009).

Study Questions

Multiple Choice

1. To which group of benthic animals does the following statement best apply?

   "Abundances typically range from 10^5 to 10^7 individuals m^{-2} and biomass ranges from 0.2 to 1 gC m^{-2}"
   a. microbenthos
   b. meiobenthos
   c. macrobenthos

2. Which group of benthic animals would you expect to have the fastest biomass turnover time?
   a. microbenthos
   b. meiobenthos
   c. macrobenthos

3. Let us say you sieve a sample from an estuarine sediment and that you examine the animals that are retained on a 0.5 mm sieve. What group of animals does this size range best describe?
   a. microbenthos
   b. meiobenthos
   c. macrobenthos

4. How does bioturbation by benthic animals affect oxygen penetration into sediment?
   a. Bioturbation increases the depth to which oxygen penetrates into the sediment
   b. Bioturbation decreases the depth to which oxygen penetrates into sediment
   c. Bioturbation does not affect the depth to which oxygen penetrates into sediment

5. Which of the following is true about the redox potential discontinuity? (Select all that apply)
   a. A fine sediment should have a deeper redox potential discontinuity than a course sediment
   b. Most benthic animals live above the redox potential discontinuity
   c. Hydrogen sulfide (H_2S) is found predominantly below the redox potential discontinuity.
   d. Oxygen is found in high concentrations below the redox potential discontinuity in coarse but not fine sediments

6. Benthic animals that use a regulating strategy as the external salinity fluctuates ______.
   a. do not alter physiological functions as salinity changes.
   b. maintain constant internal physiological conditions as the external salinity changes.
   c. experience no metabolic costs as the salinity fluctuates.
   d. experience a large variation in blood or tissue salinity as the external salinity changes

7. In heavily or grossly polluted environments, sediment-dwelling animals are more likely to _____ than animals in unpolluted environments.
   a. be smaller in body size
   b. be more diverse (i.e., species rich)
   c. regulate physiological functions such as blood osmotic strength
   d. be superior competitors with poor colonizing abilities

8. Bottom-water hypoxia has many effects on benthic communities. Which of the following statements are true? (Select all that apply)
   a. Hypoxia alters the fate of energy in sediments because more of the energy entering the sediments is used by microbes and less is used by animals.
   b. Density of benthic animals decreases with increasing duration and intensity of hypoxia.
   c. Dead zones are areas under hypoxic stress that experience increased production and density of benthic animals
   d. Benthic animals may become lethargic under hypoxic conditions and predation by nektonic predators may increase

9. Which statement about benthic animals and their sensitivity to hypoxia is true? (select all that apply)
   a. Crustaceans and echinoderms are more sensitive to hypoxia than annelids and nematodes.
   b. Early life stages are less sensitive to hypoxia than older life stages.
   c. Animals that live under higher variable hypoxic conditions are more sensitive to hypoxia than animals that rarely experience hypoxia.
   d. Hypoxic bottom waters drive a positive feedback loop that enhances hypoxia in sediments by eliminating bioturbators.

10. Where would you expect to find the lowest species diversity of benthic animals in estuaries that span the range from fresh water to marine conditions?
    a. in regions of estuaries at which salinities are most similar to full-strength sea water
    b. in the middle portions of estuaries with intermediate salinities
    c. in regions of estuaries at which salinities are most similar to fresh water
    d. Species diversity usually does not vary across all regions of estuaries.

Short Answer

1. What factors contribute to the frequency of and success of invasive benthic species in estuaries?
2. Describe the contrast between top-down and bottom-up factors and how they affect benthos in estuaries.
3. What is a suspensivore? What is the food of suspensivores and how do they obtain it? In what ways do suspensivores influence estuarine ecosystems?
4. What is a deposivore and how do deposivores obtain food? In what ways do deposivores influence estuarine ecosystems?
5. Quantitative sampling of estuarine sediments and benthic animals using benthic grabs can be problematic. Discuss reasons why and how quality control can be assured.

References

Attrill, M.J. (2002). A testable linear model for diversity trends in estuaries. *J. Anim. Ecol.* 71: 262–269.

Baretta-Bekker, H.J., Duursma, E.K., and Kuipers, B.R. (ed.) (1998). *Encyclopedia of Marine Sciences*. Springer Science & Business Media.

Bar-On, Y.M., Phillips, R., and Milo, R. (2018). The biomass distribution on Earth. *Proc. Natl. Acad. Sci.* 115 (25): 6506–6511.

Baumann, M.S., Fricano, G.F., Fedeli, K. et al. (2020). Recovery of salt marsh invertebrates following habitat restoration: implications for marsh restoration in the northern Gulf of Mexico. *Estuar. Coasts* 43: 711–1721. https://doi.org/10.1007/s12237-018-0469-5.

Bell, G., Eggleston, D., and Wolcott, T. (2003). Behavioral responses of free-ranging blue crabs to episodic hypoxia. I. Movement. *Mar. Ecol. Prog. Ser.* 259: 215–225. https://doi.org/10.3354/meps259215.

Bertness, M.D. (1984). Ribbed mussels and Spartina alterniflora production in a New England salt-marsh. *Ecol.* 65: 1794–1807.

Beukema, J.J. and de Vlas, J. (1989). Tidal-current transport of thread-drifting postlarval juveniles of the bivalve Macoma balthica from the Wadden Sea to the North Sea. *Mar. Ecol. Prog. Ser.* 52: 193–200.

Bilkovic, D.M., Mitchell, M.M., Isdell, R.E. et al. (2017). Mutualism between ribbed mussels and cordgrass enhances salt marsh nitrogen removal. *Ecosphere* 8: e01795.

Brey, T. (2012). A multi-parameter artificial neural network model to estimate macrobenthic invertebrate productivity and production. *Limnol. Oceanogr. Methods* 10: 581–589.

Carlton, J.T. (1999). Molluscan invasions in marine and estuarine communities. *Malacologia* 41: 439–454.

Cheng, I.J., Levinton, J.S., McCartney, M. et al. (1993). A bioassay approach to seasonal variation in the nutritional value of sediment. *Mar. Ecol. Prog. Ser.* 94: 275–285.

Cohen, A.N. and Carlton, J.T. (1998). Accelerating invasion rate in a highly invaded estuary. *Science* 279: 555–558.

Craeymeersch, J.A. and Jansen, H.M. (2019). Bivalve assemblages as hotspots for Biodiversity. In: *Goods and Services of Marine Bivalves* (ed. A. Smaal, J. Ferreira, J. Grant, et al.). Cham: Springer.

Crain, C.M. and Bertness, M.D. (2006). Ecosystem engineering across environmental gradients: Implications for conservation and management. *Bioscience* 56: 211–218.

Davenport, J., Smith, R.J.J.W., and Packer, M. (2000). Mussels (*Mytilus edulis* L.): significant consumers and destroyers of mesozooplankton. *Mar. Ecol. Prog. Ser.* 198: 131–137.

Day, J.W. Jr., Hall, C.A.S., Kemp, W.M., and Yanez-Arancibia, A. (1989). *Estuarine Ecology*. New York: Wiley.

Deegan, L.A., Bowen, J.L., Drake, D. et al. (2007). Susceptibility of salt marshes to nutrient enrichment and predator removal. *Ecol. Appl.* 17: S42–S63.

Deis, D.R., Fleeger, J.W., Johnson, D.S. et al. (2020). Recovery of the salt marsh periwinkle (Littoraria irrorata) 9 years after the Deepwater Horizon oil spill: Size matters. *Mar. Pollut. Bull.* 160: 111581. https://doi.org/10.1016/j.marpolbul.2020.111581.

Diaz, R.J. and Rosenberg, R. (2008). Spreading dead zones and consequences for marine ecosystems. *Science* 321: 926–929.

Drolet, D., Bringloe, T.T., Coffin, M.R. et al. (2012). Potential for between-mudflat movement and metapopulation dynamics in an intertidal burrowing amphipod. *Mar. Ecol. Prog. Ser.* 449: 197–209.

Ducrotoy, J.P., Rybarczyk, H., Souprayen, J. et al. (1991). A comparison of the population dynamics of the cockle (*Cerastoderma edule* L.) in North-Western Europe. In: *Estuaries and Coasts: Spatial and Temporal Intercomparisons* (ed. M. Elliott and J.P. Ducrotoy), 173–184. Fredensborg: Olsen & Olsen.

Fenchel, T.M. and Riedl, R.J. (1970). The sulfide system: a new biotic community underneath the oxidized layer of marine sand bottoms. *Marine Biology* 7 (3): 255–268.

Fleeger, J.W., Carman, K.R., and Nisbet, R.M. (2003). Indirect effects of contaminants on aquatic ecosystems. *Sci. Total Environ.* 317: 207–233.

Fleeger, J.W., Riggio, M.R., Mendelssohn, I.A. et al. (2019). What Promotes the Recovery of Saltmarsh Infauna after Oil Spills? *Estuaries Coast.* 42: 204–217.

Galván, K.A., Fleeger, J.W., and Fry, B. (2008). Stable isotope addition reveals dietary importance of phytoplankton and benthic microalgae to saltmarsh infauna. *Mar. Ecol. Prog. Ser.* 359: 37–49.

Galván, K.A., Fleeger, J.W., Peterson, B.J. et al. (2011). Natural abundance stable isotopes and dual isotope tracer additions help to resolve resources supporting a saltmarsh food web. *J. Exp. Mar. Biol. Ecol.* 410: 1–11.

Giere, O. (2009). *Meiobenthology. The Microscopic Motile Fauna of Aquatic Sediments*, 2e, 527. Berlin: Springer-Verlag.

Heck, K.L. and Valentine, J.F. (2007). The primacy of top-down effects in shallow benthic ecosystems. *Estuaries* 30: 371–381.

Heip, C.H.R., Goosen, N.K., Herman, P.M.J. et al. (1995). Production and consumption of biological particles in temperate tidal estuaries. In: *Oceanography and Marine Biology*, 1–149. London: U C L PRESS LTD.

Higgins, R.P. and Thiel, H. (1988). *Introduction to the Study Meiofauna*, 488. Washington, DC: Smithsonian Press.

Hughes, R.G. (1970). Population dynamics of the bivalve Scrobicularia plana (da Costa) on an intertidal mudflat in North Wales. *J. Anim. Ecol.* 39: 333–356.

Jansen, J.M., Pronker, A.E., Bonga, S.W., and Hummel, H. (2007). Macoma balthica in Spain, a few decades back in climate history. *J. Exp. Mar. Biol. Ecol.* 344: 161–169. https://doi.org/10.1016/j.jembe.2006.12.014.

Jeffrey, D.W., Wilson, J.G., Harris, C.R., and Tomlinson, D.L. (1985). The Application of Two Simple Indices to Irish Estuary Pollution Status. In: *Estuarine Management and Quality Assessment* (ed. J.G. Wilson and W. Halcrow). Boston, MA: Springer https://doi.org/10.1007/978-1-4615-9418-5_14.

Johnson, D.S. (2014). Fiddler on the roof: A northern range extension for the marsh fiddler crab *Uca pugnax*. *J. Crustac. Biol.* 34 (5): 671–673.

Johnson, D.S. and Fleeger, J.W. (2009). Weak response of saltmarsh infauna to ecosystem-wide nutrient enrichment and predator reduction: a four-year study. *J. Exp. Mar. Biol. Ecol.* 373: 35–44.

Kramer, K.J.M., Brockmann, U.H., and Warwick, R.M. (1994). *Tidal Estuaries: A Manual of Sampling and Analytical Procedures*. Rotterdam: A.A. Balkema.

Leguerrier, D., Niquil, N., Petiau, A., and Bodoy, A. (2004). Modeling the impact of oyster culture on a mudflat food web in Marennes-Oléron Bay (France). *Mar. Ecol. Prog. Ser.* 273: 147–161. https://doi.org/10.3354/meps273147.

Long, E.R. (2000). Degraded sediment quality in US estuaries: a review of magnitude and ecological implications. *Ecol. Appl.* 10: 338–349.

Long, W.C. and Seitz, R.D. (2008). Trophic interactions under stress: hypoxia enhances foraging in an estuarine food web. *MEPS* 362: 59–68. https://doi.org/10.3354/meps07395.

Mascaró, M. and Seed, R. (2001). Choice of prey size and species in Carcinus maenas (L.) feeding on four bivalves of contrasting shell morphology. *Hydrobiologia* 449: 159–170. https://doi.org/10.1023/A:1017569809818.

Nehls, G. and Büttger, H., (2007). Spread of the Pacific Oyster Crassostrea gigas in the Wadden Sea: causes and consequences of a successful invasion. *HARBASINS Report, CWSS, Wilhelmshaven*, 54.

Newell, R.C. (1979). *Biology of Intertidal Animals*, 3e, 781. Faversham: Marine Ecological Surveys.

Newell, R.I.E. (1988). Ecological changes in Chesapeake Bay, are they the result of overharvesting the eastern oyster (*Crassostrea virginica*)? In: *Chesapeake Research Consortium*, Understanding the estuary, Publ 129 (ed. M.P. Lynch and E.C. Krome). Gloucester Point: VA www.vims.edu/GreyLit/crc129.pdf.

Palmer, M.A. and Brandt, R.R. (1981). Tidal variation in sediment densities of marine benthic copepods. *Mar.Ecol.Prog.Ser.* 4: 207–212.

Pearson, T.H. and Rosenberg, R. (1978). Macrobenthic succession in relation to organic enrichment and pollution of the marine environment. *Oceanogr. Mar. Biol. Annu. Rev.* 16: 229–311.

Posey, M.H., Alphin, T.D., Cahoon, L.B. et al. (2002). Top-down versus bottom-up limitation in benthic infaunal communities: direct and indirect effects. *Estuaries* 25: 999–1014.

Pusack, T.J., Kimbro, D.L., White, J.W., and Stallings, C.D. (2019). Predation on oysters is inhibited by intense or chronically mild, low salinity events: Low salinity stress reduces predation. *Limnol. Oceanogr.* 64: 81–92. https://doi.org/10.1002/lno.11020.

Remane, A. (1935). Die Brackwasserfauna. *Verhandlungen der Deutschen ZoologischenGesellschaft* 36: 34–74.

Remane, A., and Schlieper, K. (1958). Die Biologie des Brackwassers. Die Binnengewässer (The Biology of the Brackish Water. The Inland Waters), vol. 22.

Renzi, J.J., He, Q., and Silliman, B.R. (2019). Harnessing positive species interactions to enhance coastal wetland restoration. *Front. Ecol. Evol.* https://doi.org/10.3389/fevo.2019.00131.

Rhoads, D.C. and Germano, J.D. (1982). Characterization of organism-sediment relations using sediment profile imaging: an efficient method of remote ecological monitoring of the seafloor (remots™ system). *Mar. Ecol. Prog. Ser.* 8: 115–128.

Snelgrove, P. V. (1999). Getting to the bottom of marine biodiversity: sedimentary habitats: ocean bottoms are the most widespread habitat on earth and support high biodiversity and key ecosystem services. BioScience 49 (2): 129–138.

Schratzberger, M. and Ingels, J. (2017). Meiofauna matters: the roles of meiofauna in benthic ecosystems. *J. Exp. Mar. Biol. Ecol.* 502: 12–25. https://doi.org/10.1016/j.jembe.2017.01.007.

Schwinghamer, P., Hargrave, B., Peer, D., and Hawkins, C.M. (1986). Partitioning of production and respiration among size groups of organisms in an intertidal benthic community. *Mar. Ecol. Prog. Ser.* 31: 131–142.

Seitz, R.D., Dauer, D.M., Llansó, R.J., and Long, W.C. (2009). Broad-scale effects of hypoxia on benthic community structure in Chesapeake Bay, USA. *J. Exp. Mar. Biol. Ecol.* 1 (Suppl. 1): S4–S12.

Snelgrove, P.V.R. and Butman, C.A. (1994). Animal sediment relationships revisited: cause versus effect. *Oceanogr. Mar. Biol. Annu. Rev.* 32: 111–177.

Soniat, T.M. (1996). Epizootiology of *Perkinsus marinus* disease of eastern oysters in the Gulf of Mexico. *J. Shellfish Res.* 15: 35–43.

Sousa, R., Gutierrez, J.L., and Aldridge, D.C. (2009). Non-indigenous invasive bivalves as ecosystem engineers. *Biol. Invasions* 11: 2367–2385.

Weisberg, S.B., Ranasinghe, J.A., Dauer, D.M. et al. (1997). An estuarine benthic index of biotic integrity (B-IBI) for the Chesapeake Bay. *Estuaries* 20: 149–158.

Wilson, J.G. (1988). *Biology of Estuarine Management*. London: Plenum Press https://doi.org/10.1007/978-94-011-7087-1.

Wilson, J.G. (2002). Productivity, fisheries and aquaculture in temperate estuaries. *Estuar. Coast. Shelf Sci.* 55: 953–967.

Wilson, J.G. and Elkaim, B. (1992). Estuarine bioindicators - a case for caution. *Acta Oecol.* 13: 345–358.

Wilson, J.G., Rybarczyck, H., and Elkaim, B. (2007). A comparison of energy flow through the Dublin Bay and Baie de Somme intertidal ecosystems and their network analysis. *Hydrobiologia* 588: 231–243.

Woolff, W.J. and Reise, K. (2002). Oyster imports as a vector for the introduction of alien species into northern and western European coastal waters. In: *Invasive Aquatic Species of Europe. Distribution, Impacts and Management* (ed. E. Leppäkoski, S. Gollasch and S. Olenin). Dordrecht: Springer https://doi.org/10.1007/978-94-015-9956-6_21.

CHAPTER 12

Estuarine Nekton

James H. Cowan Jr[1] and Byron C. Crump[2]
[1] Department of Oceanography and Coastal Sciences, Louisiana State University, Baton Rouge, LA, USA
[2] College of Earth, Ocean, and Atmospheric Sciences, Oregon State University, Corvallis, OR, USA

Menhaden filter-feeding in Cape Cod Bay, USA. Photo credit: Ryan Collins, Myfishingcapecod.com.

12.1 Introduction

Estuarine nekton are defined as aquatic animals that are able to swim and move independently of water currents. Nekton include taxa such as squids, octopods, portunid crabs, penaeid shrimps, sharks, and rays. However, fishes constitute most of the nekton species in estuaries, both in numbers and biomass. Most estuarine nektonic species can tolerate a wide range of temperature, salinity, and tidal energy, among other estuarine conditions, and their distributions vary globally (Table 12.1, Figure 12.1).

Estuarine nekton can be obligate, facultative, or transient users of estuaries. These strategies constitute a suite of traits such as size at maturity, longevity, fecundity, and parental care, which in combination have evolved to allow a species to cope with their environment. Some estuarine nekton are **obligate** and thus complete their entire life cycle within estuaries, but there are relatively few of these. Most estuarine nekton are **facultative** estuarine organisms that occupy estuaries during specific stages of their life cycles. **Life cycle** is defined as the sequences of physiological stages that an individual must undergo as it matures from an egg to a reproductive adult.

All obligate and some facultative estuarine fishes use estuaries as spawning grounds and juvenile habitat. Estuaries often support large populations of small larvae, post-larvae, and juvenile nekton, and as such, estuaries are often referred to as **nursery areas**, which is an important function of these ecosystems for nekton. The presence of large numbers of juveniles in an estuary is thought to be due to two primary factors: (i) estuaries are regions of high primary and secondary productivity, thus providing abundant food resources for fast growing juveniles, and (ii) estuaries comprise numerous habitat types that provide refuge for small fishes and invertebrates from predators that use estuaries as feeding grounds. There are also many facultative estuarine nekton that follow the opposite pattern, such as blue crab (*Callinectes sapidus*) and the Southern Flounder (*Paralichthys lethostigma*), that spawn outside of estuaries and then move into estuaries as juveniles or adults.

Transient estuarine nekton generally use estuaries as migration routes between adult habitats and spawning grounds. These taxa are referred to as **diadromous**, and can be further divided into two groups: (i) anadromous and (ii) catadromous. **Anadromous** fishes, such as Pacific Lamprey (*Entosphenus tridentatus*), live in seawater but spawn in fresh waters, whereas **catadromous** fishes, such as European Eel

Estuarine Ecology, Third Edition. Edited by Byron C. Crump, Jeremy M. Testa, and Kenneth H. Dunton.

Companion website: www.wiley.com/go/crump/estuarine3

TABLE 12.1 Common families of estuarine fishes in four regions.

Family	Western Australia	South Africa	North America	Western Europe
Clupeidae	X	X	X	X
Atherinidae	X	X	X	X
Terapontidae	X			
Mugilidae	X	X	X	X
Gobiidae	X	X	X	X
Engraulidae	X	X	X	X
Tetraodontidae	X		X	X
Apogonidae	X			
Sillaginidae	X			
Gerreidae	X		X	
Plotosidae	X			
Sparidae	X	X	X	X
Arripidae	X			
Hemirhampidae (Exocoetidae)	X	X	X	X
Pomatomidae	X	X	X	X
Carangidae		X	X	X
Platycephalidae	X	X		
Monodactylidae		X		
Soleidae		X		X
Haemulidae		X		
Ambassidae		X		
Sciaenidae		X	X	X
Ariidae		X	X	
Clinidae		X		
Syngnathidae		X	X	X
Cichlidae		X	(X)	
Blenniidae	X	X	X	X
Galaxiidae		X		

Note: (X) for Cichlidae indicates several nonindigenous species.
Source: Based on Potter et al. (1990).

(*Anguilla anguilla*), live in fresh waters and spawn in the ocean. In both cases, fishes pass through and use estuarine resources, thus exchanging and transforming energy and biomass with neighboring ecosystems.

Nekton are usually unevenly distributed in estuaries, so quantitative sampling can be challenging. Nekton can be sampled using a variety of passive and active sampling gears. Passive sampling equipment includes fish traps and small mesh gill nets. Active equipment includes bottom and midwater trawls, beach seines, and hoop nets.

12.2 Diversity and Global Distribution in Estuaries

The diversity and distribution of estuarine nekton can be categorized by a series of physiological and behavioral adaptations that influence the distribution of fishes in estuarine environments. Many of these adaptations allow estuarine nekton to adjust to variable conditions of salinity and temperature. Others are involved in swimming, feeding, and avoiding predation. Physiological adaptations include **osmoregulation** (control of internal water and salt concentrations), placement of fins for swimming and mouthparts for feeding, and changing skin colors for camouflage. Behavioral adaptations involve active movement to optimize conditions for the organisms and their offspring. There are also several behavioral adaptations associated with feeding and predation. These adaptations of estuarine nekton help define the diversity of nekton and describe their distributions.

12.2.1 Physiological Adaptations to Environmental Conditions

Adaptations to the rapidly changing physical and chemical environment of estuaries impose high energy demands on fish, which explains why relatively few species and families of fish inhabit estuaries for their entire life cycle. There are, however, many fish species that spend at least part of their lives in estuaries. Within the relatively large environmental range in which most organisms can survive, there is a smaller range where conditions for survival are optimum. Different life history stages, especially post-larvae and juvenile stages, may be adapted to very different environmental conditions compared with adults. Thus, the concept of adaptation is complex and dynamic, especially for estuarine nekton in which young and adults live in two or more different environments. These differences may allow fishes to avoid competition, optimize growth, and reduce predation.

All nekton share many standard adaptations including the ability to extract oxygen at a sufficient rate from the surrounding water to support the energy requirements of swimming. To do this, fishes have **gills** that are well-adapted for rapid oxygen uptake from water, which permit active and sustained movements. In addition, most nekton have well-developed sensory organs that are suited to an active, often predatory, existence. For example, many have streamlined bodies and are able to control their depth through varying their specific gravity by changing the amount of gas in the **swim bladder**. Fishes also have a sensory system called the **lateral line**, which is very sensitive to sound waves and to changes in water density.

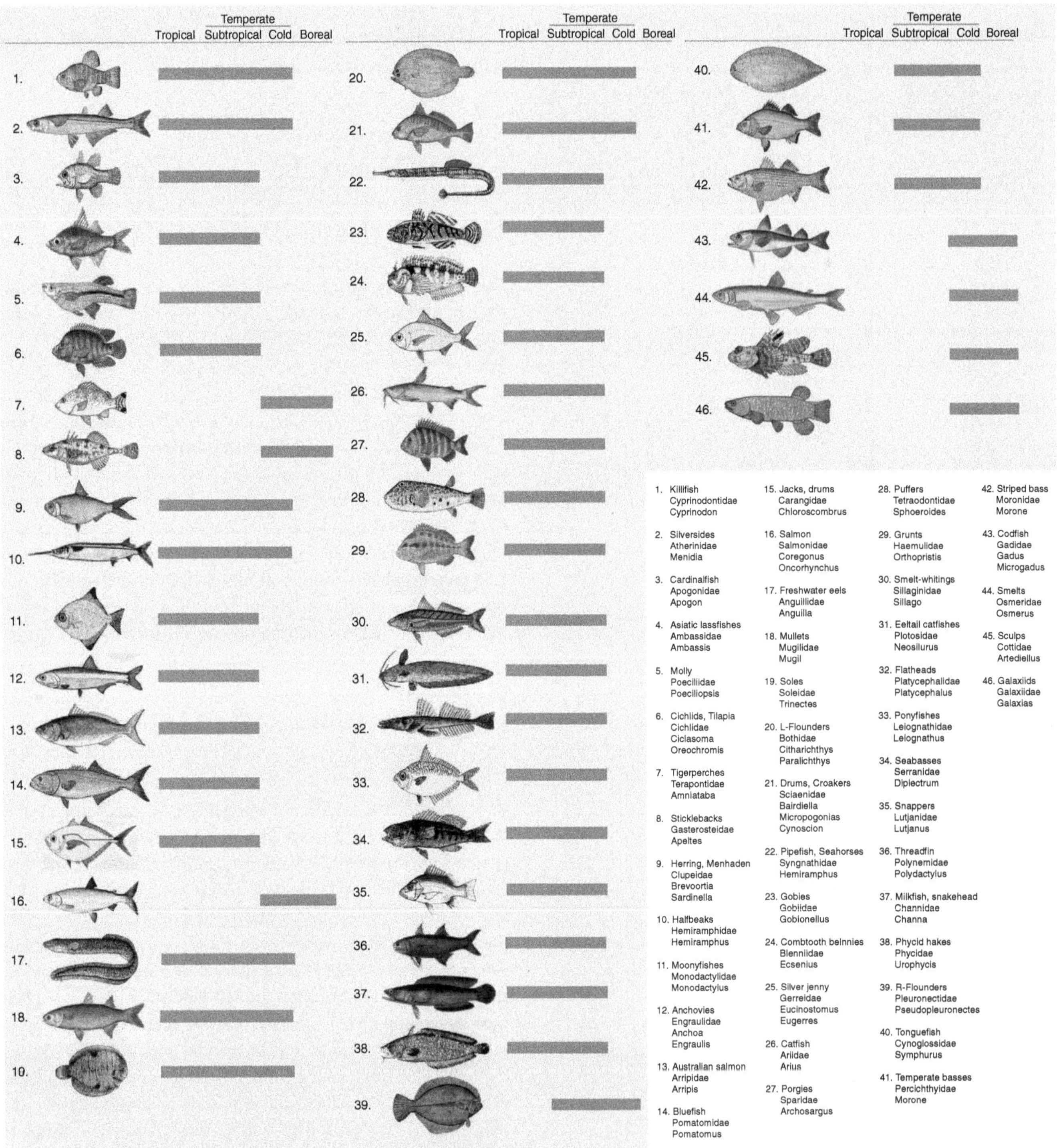

FIGURE 12.1 Examples of nektonic fishes that inhabit estuaries and approximate latitudinal distribution for each fish. Images of fishes correspond to their common and scientific names on the lower right-hand portion of the figure. *Source*: From Day et al. (2013)/John Wiley & Sons.

These adaptations are common for all fishes, but for estuarine fishes, perhaps the most difficult challenge is osmoregulation, which involves maintaining nearly constant internal osmotic pressure in a very changeable osmotic environment (Evans et al. 2005). The relatively impermeable skin, scales, and layer of mucus characteristic of most fishes help minimize ion and water exchanges associated with changes in salinity. However, the internal salt and mineral content of fishes (and most vertebrate animals) falls between that of fresh water and ocean water. This is why marine fishes are physiologically adapted to retain water they consume, release salts through their gill epithelium, and excrete excess salts concentrated in their urine. It is also why freshwater fishes retain salts from their food, actively absorb salts with their gills, and excrete excess water in their urine (Evans et al. 2005; Wurts 1998). Estuarine fishes are unusual because they are capable of using all of these physiological strategies to adapt to a range of salinities and maintain a constant internal salt content and

TABLE 12.2 **Distribution of common fishes in the Navarro River and its estuary, California, in relation to salinity. X indicates that fishes were present and common, and R indicates present but rare.**

		Salinity (ppt)						
Species	**Classification[a]**	**0[b]**	**0[c]**	**1**	**3**	**9–10**	**23–25**	**30+**
Sacramento sucker	1	X	X	X				
California roach	1	X	X	X	X			
Prickly sculpin	2	X	X	X	X	X	X	X
Rainbow trout	3	X	X	X	X	X	X	R
Threespine stickleback	2 and 3	X	X	X	X	X	X	X
Starry founder	4		X*	X*	X*	X*	X	X
Shiner perch	4			R*	R*	X*	X	X
Jacksmelt	4				R*	X*	X	X
Bay pipefish	4					X	X	X
Plainfin midshipman	4[d]					X	X	X
Penpoint gunnel	4						X	X
Pacific herring	4[d]						X	X
Surf smelt	4						X*	X
Northern anchovy	4						X	X
Lingcod	5							X*

* Predominantly young-of-year fish.
[a] The species are classified according to their salinity tolerance as follows: 1-stenohaline, freshwater; 2-euryhaline, freshwater; 3-anadromous; 4-euryhaline, marine; 5-stenohaline, marine.
[b] More than 1 km upstream from first riffle.
[c] Just above first riffle.
[d] Spawning

osmotic pressure. This dual-capability is uncommon among fishes, which explains why the lowest nekton species richness in estuaries is in low salinity regions that range from 3 to 8 psu, and the highest is near the mouth of estuaries where salinity is less variable.

Truly estuarine nekton species are usually capable of living at a wide range of salinities (including 3–8 psu), and tend to distribute themselves within estuaries by moving to the salinity ranges that are preferred by each species (Table 12.2). However, the salinity preference of some fish species changes during their life cycle. For example, young fishes tend to be able to tolerate only a narrow salinity range compared to adults due to poorly developed osmoregulatory systems.

12.2.2 Behavioral Adaptations to Environmental Conditions

Behavioral mechanisms help nekton avoid extreme situations or conditions for themselves and their offspring. Through active movements, nekton avoid large variations outside their physiological tolerances in salinity, temperature, dissolved oxygen (DO), and suspended sediments. Nekton also use active movement to spawn under conditions that are optimized for their eggs and larvae so that their planktonic offspring may survive until they grow to become nekton themselves. Some estuarine fishes such as Atlantic Menhaden (*Brevoortia tyrannus*), Bluefish (*Pomatomus saltatrix*), and Spot (*Leiostomus xanthurus*), spawn in high salinity waters offshore (Blaber 2000; Warlen and Burke 1990) presumably because their eggs and embryonic stages are more buoyant in saltwater and are intolerant to wide salinity ranges, although wave energy may also be important in keeping the eggs from settling out of the water column. After hatching, juveniles rapidly develop a greater tolerance to reduced salinity and become large enough to migrate into estuaries.

Active movements of nekton to optimize environmental conditions such as salinity and temperature help shape nekton diversity and distributions in estuaries. Nekton communities generally vary along salinity gradients as fishes seek out the salinity ranges to which they are most well adapted. These spatial distributions also tend to correlate with temperature because of temperature differences between fresh water and ocean water. In temperate estuaries, the upper reaches of estuaries are often warmer than the ocean in summer and cooler in winter, which can cause fishes to change locations within estuaries. More generally, seasonal variability in temperature often correlates with seasonal changes in nekton community composition (Tsou and Matheson 2002; Methven et al. 2001). For example, in the Rio de la Plata estuary, spatial variability of fish community composition was linked to bottom water salinity, but seasonal variability was more strongly linked to water temperature (Jaureguizar et al. 2004). Longer-term climate changes have also been linked to fish community composition, such as in the Bristol Channel estuary where the composition of dominant fish species was linked to mean annual sea-surface temperature over a 21-year period (1981–2002; Genner et al. 2004).

Variability in temperature may operate in concert with salinity to control nekton distributions. This is because reduced or fluctuating salinity in combination with varying temperature offers special problems for osmoregulation in some species. For example, the American Plaice (*Hippoglossoides platessoides*), a species of flounder, is euryhaline at low temperature and eurythermal at high salinities. These fishes succeed in estuaries provided they have either low temperature conditions or high salinity conditions. However, they do not succeed in low salinity, high temperature conditions because they cannot survive the combination of stressors (Day et al. 1989). In temperate estuaries, it may be considered a general rule that an increase in water temperature leads to reduced ability to withstand wide variations in salinity, possibly because of the effect of warm temperatures on enzymes that control osmoregulation in estuarine nekton. There are exceptions to this rule, however, such as Spotted Seatrout (*Cynoscion nebulosus*) which have been shown to withstand salinities in excess of 45 psu in the Laguna Madre on the south coast of Texas, USA, even though water temperatures exceed 27 °C (Froeschke and Froeschke 2011).

12.2.3 Physiological Feeding Adaptations

Fishes have evolved a variety of physiological feeding adaptations that influence their diversity and distributions in estuarine environments. Estuarine fishes generally have varied diets, but most possess adaptations that increase the probability of successfully handling and ingesting specific kinds of prey under different estuarine conditions. Common adaptations include differences in the size and position of the mouth, kind and position of teeth and branchial arches, and modifications of gill rakers. These adaptations allow estuarine fishes to feed at almost every trophic level in the food web, and in most estuarine habitats. They also have the potential to reduce competition among fishes for food resources in estuaries, and allow for greater estuarine nekton diversity (Larkin 1956).

Mouth size, shape, and body postion greatly influence the type of prey that fishes consume. For example, flatfishes are common inhabitants of estuaries that live in association with the sea floor (i.e., they are **demersal**) and feed on both benthic and pelagic organisms. Examples of flatfish with wide distributions in estuaries and river plumes include the hogchoker (*Trinectes maculatus*), baywhiff (*Citharichthys spilopterus*), and blackcheek tonguefish (*Symphurus plagiusa*), and the larger-bodied summer (*Paralichthys dentatus*) and winter flounders (*Pseudopleuronectus americanus*). Smaller flatfish with smaller mouths tend to prey on benthic organisms like small crustaceans, worms, and mollusks while the larger flatfish consume larger epibenthic and pelagic organisms such as shrimps, crabs, and fish (Sánchez-Gil 2009), including other small flatfish (e.g., *Scyacium gunteri*, *Cithirichthys spilopterns*, and *Etropus crossotus*).

Adaptations to mouth shape and body position also influence feeding location. Fishes with a superior mouth (located more dorsally), such as silversides, often suck in floating insects from the water surface. In contrast, fishes with an inferior mouth (located more ventrally), such as hake (*Urophyscis* sp.) feed on benthic and epibenthic organisms. Similarly, grunts like *Orthopristis crysoptera* have intermediate sized inferior mouths located below the tip of the snout and feed on small organisms on sediments and the surfaces of seagrass leaves. Top predators such as seatrout and other members of the genus *Cynoscion*, have large, terminal mouths that they use to pursue other fishes in the water column of estuaries.

After food is captured, the processing of that food begins in the throat or pharynx, where pharyngeal bones are located. The number, size, and structure of teeth on these bones differ according to food type most commonly processed. For example, the surfaces of the pharyngeal bones of phytoplankton eaters, such as shad, are covered with rows of fine recurved teeth so that when the two sets slide together, the algae are combed backward (Figure 12.2a, b). Fish-eating **Piscivores**, such as sea trouts, tend to have sharply pointed teeth that point backward to grip and force backward prey held in the mouth. Mollusk eaters, such as the freshwater drum (*Aplodinotus grunniens*) and tropical estuarine pufferfish in the Tetraodontidae (e.g., *Sphoeroides testudineus*) possess pharyngeal teeth with a number of large flat tooth plates, which crack the shells of their prey (Figure 12.2e).

Similar to teeth, adaptations of branchial arches and gill rakers aid in food selection, processing, and retention. Fish species that filter plankton from water have numerous, long, thin, and closely spaced gill rakers on the inward side of the gill arch. For example, Atlantic menhaden (*Brevortia thyrannus*) and Peruvian anchoveta (*Engraulis anchoveta*) develop gill rakers of increased number, length, and complexity during their life cycle when they change from a zooplankton diet to a phytoplankton diet (Figure 12.2d). The bay anchovy (*Anchoa mitichilli*), however, remains a planktivore throughout its life cycle, and its gill rakers remain relatively short. Predatory fishes often have tooth-shaped gill rakers that facilitate the capture and retention of large prey items (Figure 12.2a–e).

12.2.4 Behavioral Feeding Adaptations

Feeding behaviors in fishes can be categorized into two groups: (i) bottom feeding, and (ii) mobile predation. **Bottom feeders**,

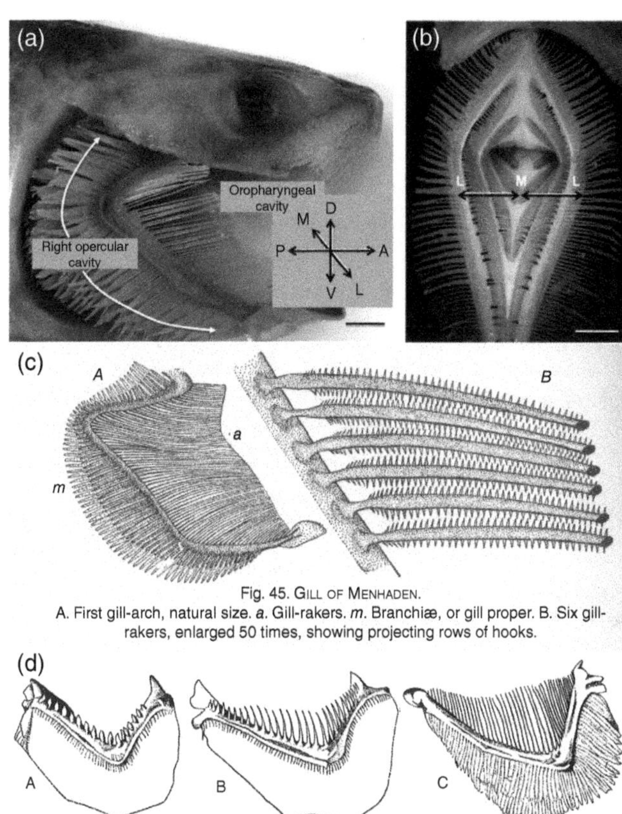

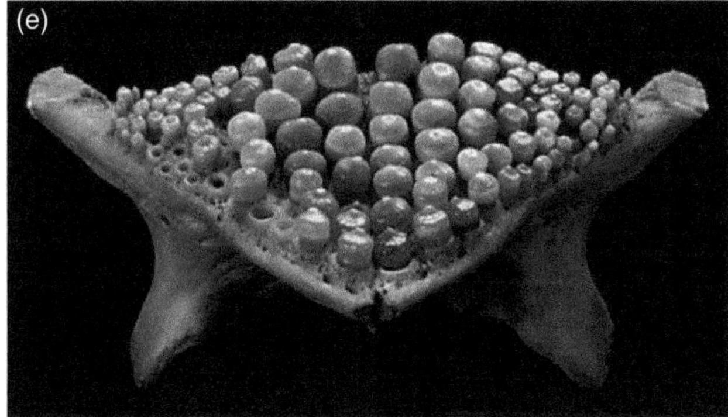

FIGURE 12.2 Examples of gill rakers and pharyngeal teeth of fishes that feed on different prey types. (a and b) Oropharyngeal morphology related to filtration mechanisms in suspension-feeding American shad (Clupeidae). *Source*: From Storm et al. (2020), John Wiley & Sons. (c) Gill structure of phytoplankton-consuming menhaden. *Source*: From Hugh McCormick Smith, Wikimedia Commons**. (d) Gill rakers of fish that feed mainly on (A and B) insect larvae, mollusks, and crustaceans, and (C) small mysid shrimp. *Source*: From Hubbs and Lagler (1964)/University of Michigan Press. (e) Pharyngeal teeth from freshwater drum (*Aplodinotus grunniens*) that prey on mollusks. *Source*: From Science Museum of Minnesota. ** **https://commons.wikimedia.org/wiki/File:FMIB_51375_Gill_of_Mehnaden.jpeg**.

such as the flatfish, freshwater drum, and pufferfish described above, eat mostly benthic invertebrates, including small crustaceans, annelids worms, and small mollusks (Sánchez-Gil 2009). Cods (*Gadus* sp.) and hake (*Urophyscis* sp.)are common members of this group in northern estuaries in North America and Europe, while mullet (*Mugil* sp.) are more typical in temperate, tropical, and subtropical estuaries. Members of this group also include Atlantic croaker (*Micropogonias undulates*), Norfolk spot (*Leiostomus xanthurus*), and silver perch (*Bairdiella* sp.) along the Atlantic coast of North America, and Yellow croaker (*Larimichthys polyactis*), Silver croaker (*Pennahia argentata*), and Bluefin gurnard (*Chelidonichthys kumu*) in China and Southeast Asian estuaries. Many of the large bottom feeders are catfish and include the Mekong catfish found in the Mekong delta, Pirarara and giant tigerfish found in the Amazon River, the Goonch found in the River Ramgangu, in India, and the Wells Catfish which is common in the Po Delta, Italy. Each of these species can grow as large as a fisherman and a few people in these regions have been eaten whole by the larger specimens.

The second feeding type of the generally smaller demersal nekton are the **mobile predators**, including top carnivores that are only loosely associated with the bottom, feeding on more motile prey such as penaeid shrimps and pelagic fishes. These fishes usually congregate in schools and forage on the bottom, or on aquatic plants (e.g., *Ulva* sp.). Some species change their bottom preference from muddy to firmer substrates as they grow older. Typically, estuarine top carnivores include the Weakfish (*Cynoscion* sp.) and Sea Basses (*Morone* sp.) in the Western Hemisphere, barramundi (*Lates calcarifer*) and dusky flathead (*Platycephalus fuscus*) in Asian and Australian estuaries, and Bluefish (*Pomatomas saltatrix*) globally. Other species include jacks and small coastal sharks that have developed a range of feeding adaptations.

12.2.5 Nekton Diets and Food Webs

Despite the many physiological and behavioral feeding adaptations that might seem to narrow the diversity of prey available to estuarine nekton, most of these animals have a wide trophic spectrum, often feeding on many different trophic groups. This is, in part, due to the diverse prey options available in many estuarine environments. For example filter feeding menhaden in estuaries consume a broad diversity of phytoplankton, zooplankton, and particulate detritus, encompassing many trophic levels. The diets of estuarine nekton are also broad because many change their preferred prey types and preferred location in estuaries as they develop from juveniles to adults. These broad feeding options aid in recruitment of juvenile stages and ensure their survival during critical stages of development (Yáñez-Arancibia 1985). Another result of having many feeding options is that the composition of nekton diets is often determined by food availability such that the most abundant prey is the most highly consumed prey. Because of these factors, trophic spectra change with (1) locality or habitat in an estuary due to different food availability, (2) the size of the fishes due to differences in food requirements, (3) the season of the year due to changes in food types available, and (4) the time of the day/night cycle due to prey behavior (Leak and Houde 1987).

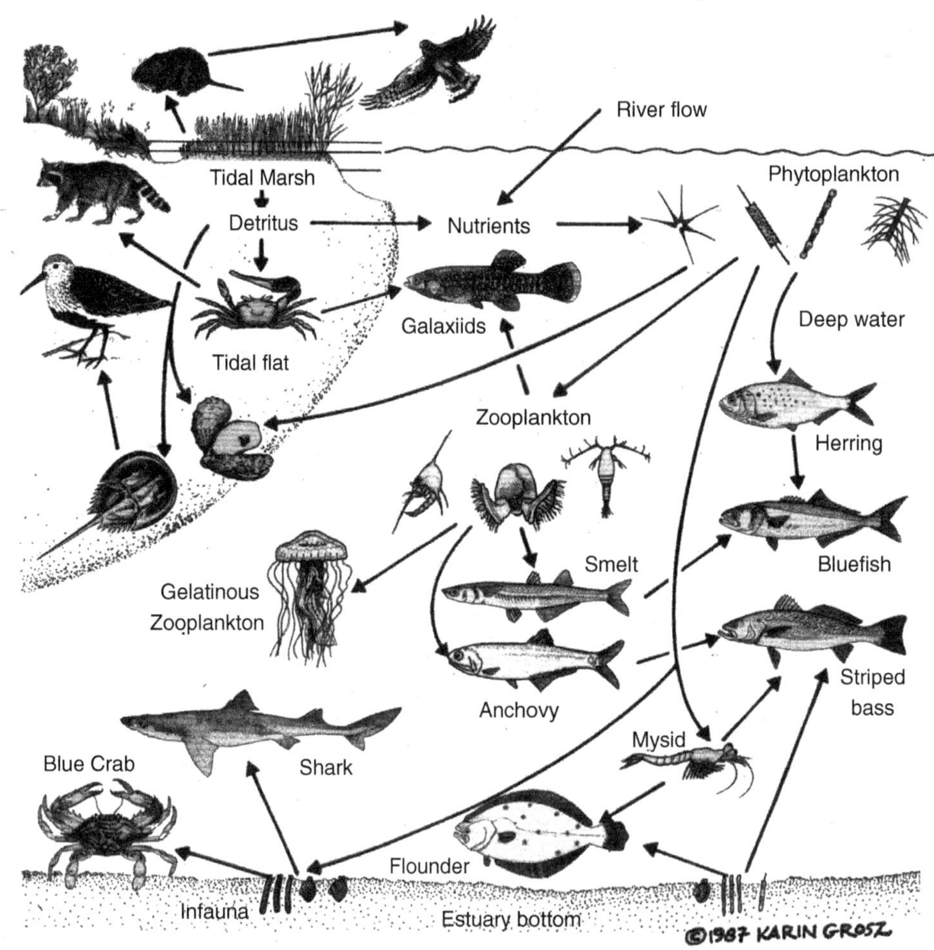

FIGURE 12.3 Nektonic food web typical of estuaries. *Source*: Adapted from Bryant and Pennock 1988.

The trophic level of nekton species varies widely, but those that function as secondary consumers (fishes that feed on primary consumers) are often the most important in determining particular characteristics of estuarine fish communities, due to their high diversity, numerical abundance, and wide trophic spectra and habitat specificity. Many nekton secondary consumers are demersal fish, including bottom feeders and mobile predators, and usually have a wide distribution inside estuarine ecosystems. For example, in tropical estuaries, as much as 50% of the fish species are secondary consumers during some stage of their life cycles (Yáñez-Arancibia 1978, 1985; Yáñez-Arancibia and Sánchez-Gil 1986).

In summary, estuaries offer both challenges and opportunities to nekton. Estuaries are dynamic environments, often with steep gradients in temperature and salinity that change rapidly in response to tides, seasonal changes in freshwater inflow, and weather. These conditions cause estuarine nekton diversity to be low and nektonic food webs to be simple (Figure 12.3), because relatively few nekton can deal with the high variability typical of estuarine environments. This does not mean, however, that estuaries are not used extensively by nekton. In fact, nekton abundance and biomass in estuaries can be among the highest in the world, and species that use estuaries for all, or a portion of their life cycle, can experience very high growth rates (>30% increase in weight each day); this is especially true for species that use estuaries as nursery areas for some portion of their early life (Yáñez-Arancibia et al. 1994; Able 2005; Cowan et al. 2008). The physical and behavioral adaptations of estuarine nekton suggest five ecological generalizations about nekton diversity and distributions: (i) distributions are influenced by environmental variability that requires physical and behavioral adaptations; (ii) the distribution of nekton in estuaries varies with ontogenetic development, growth, feeding adaptations, and prey selection; (iii) the most abundant food resources are shared among many species in a seasonally-programmed use that avoids intense interspecies competition; (iv) estuarine nekton often use both pelagic and benthic pathways taking food items from different levels of the food web; and (v) in nektonic estuarine food webs, the most diverse and abundant fishes are omnivorous and opportunistic second-order consumers of all sizes.

12.3 Spatial and Temporal Patterns of Biomass and Productivity

There are two factors that determine nekton nursery function and control biomass and productivity of juvenile nekton in estuaries: the density of prey for juvenile nekton and the availability of refuge from predation. Understanding these factors is necessary to develop restoration and preservation programs, and develop ecosystem-based management strategies. Among estuary-dependent nekton, we know that both of these factors influence the survival potential of young nekton while they are in an estuary. However, the value of these factors is variable across habitats within estuaries. For example, seagrass beds and mangroves provide far better refuge from predation for juvenile nekton than bare sediments. Thus, the importance of these factors in controlling patterns of biomass and productivity is influenced by the habitats selected by nekton, and the life history strategies they use in these habitats.

12.3.1 Habitat Usage

In their review of 26 European estuaries, Pihl et al. (2002) describe nine different estuarine habitats that are occupied by nekton species, and then identify four different functions that those habitats facilitate for nekton. The most extensive and widely distributed estuarine habitat for nekton was **subtidal soft substrate**, which accounted for about 50% of total estuarine area on average, and over 70% of the area in regions with smaller tides (Baltic/Skagerrak and Mediterranean). This unvegetated habitat, composed predominantly of sediments ranging from silts to coarse sands, hosted the highest number of fish species (Figure 12.4), presumably because of its spatial extent. **Intertidal soft substrate** was the second most extensive habitat, but only in estuaries with significant tidal ranges in the Boreal/Atlantic region where it accounted for 30% of total estuarine area. This habitat is similar to subtidal soft substrate in that it provides little protection from predation for juvenile nekton, but can provide high densities of macrobenthos prey, particularly in shallower waters where light supports growth of benthic microalgae (see Chapter 8).

Subtidal seagrass beds (see Chapter 5) accounted for more than 10% of estuarine nekton habitat in the Baltic/Skagerrak region, but very little in other studied estuaries except for the Ria de Aveiro in Portugal where it accounted for 27% of the area. Seagrass beds add considerable vertical structure (up to 1 m in height) to the soft substrate in which they develop, which enhances this habitat for juvenile nekton by providing a degree of protection from predation. In contrast to the subtidal seagrass habitat, **salt marshes** (Chapter 6) were absent from estuaries in the Baltic/Skagerrak region, but were common in several estuaries of more southern regions accounting for as much as 28% of the nekton habitat. Salt marshes and mangroves (Chapter 7) can serve as well-protected feeding grounds for nekton when tidally inundated.

Subtidal hard substrate was present in all studied estuaries in the Baltic/Skagerrak region averaging 3% of the area, and accounted for as much as 25% of the area in one Mediterranean estuary (the Messolonghi estuary in Greece). This habitat, composed of hard substrate ranging from gravel to boulders to bedrock, provides a heterogeneous mix of vegetated and unvegetated niches for nekton, and was second only to subtidal soft substrate in species richness (Figure 12.4). Other habitats (**biogenic reefs**, **reed beds, intertidal hard substrata**, and **tidal freshwater regions**) generally contributed less than 5% of the total surface area but can be locally important depending on estuary location and conditions. For example, **biogenic reefs** built by surface dwelling bivalves (e.g., oysters, mussels), corals, or sabellid and serpulid polychete worms can have high vertical relief with much complexity and provide juvenile nekton with high prey density and protection from predation.

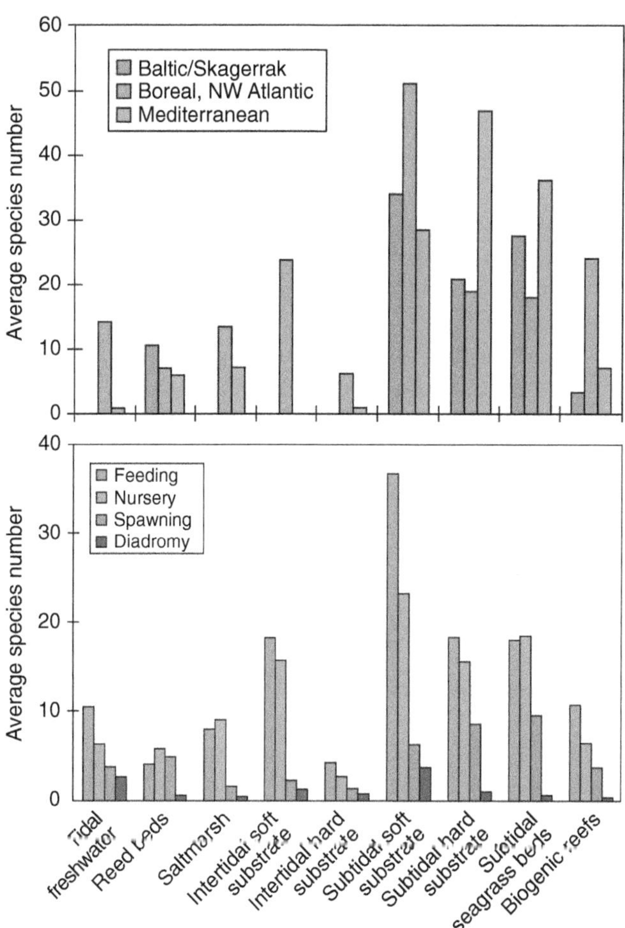

FIGURE 12.4 Average number of fish species in different habitat types for selected estuaries in three European regions (top). Habitat utilization by fish species in different habitat types in 26 selected estuaries (bottom). *Source*: From Pihl et al. (2002)/With Permission of John Wiley & Sons.

Each of these estuarine habitats can serve several functions for nekton: (i) spawning, (ii) nursery, (iii) feeding, and (iv) migration (Pihl et al. 2002). **Spawning grounds** are areas where mating occurs and eggs are deposited or released. Estuarine spawners include resident species, and other species that use estuaries only to spawn. **Nursery grounds** are areas where juveniles aggregate via active or passive transport mechanisms, but usually are spatially or temporally separated from adults. Nursery grounds generally enhance the survival of juveniles through better feeding conditions, optimal growth, and/or refuge opportunities (Nagelkerken 2009; Paterson and Whitfield 2000). As such, these sites can serve as "safe sites" from which recruitment to adult populations follows (and may depend on) nursery usage. **Feeding grounds** are used by nekton when making feeding migrations into habitats where they preferably or exclusively forage, often as adults. Many of these species are transient, migrating to feeding grounds on a tidal, diurnal, or seasonal basis. Estuarine habitats can also be used as **migration** routes for diadromous fishes as they travel between marine and freshwaters for spawning.

Most estuarine fish species used subtidal soft substrate, subtidal hard substrate, subtidal seagrass beds, and biogenic reef habitat for feeding and nursery areas, with intertidal soft substrate also being important for feeding (Figure 12.4). Fewer fishes used saltmarsh, reed beds, and tidal fresh areas for these same functions. At the species level, the percentage of the total number of fish species undertaking different habitat usage in European estuaries are in rank order as follows: feeding 76% > nursery 63% > spawning 24% > diadromous migration 9% (Pihl et al. 2020). Note that these percentages do not sum to 100% because most species use estuaries for more than one function. This ranking shows how most fishes use estuaries for feeding and nursery function, and relatively few use estuaries for spawning and diadromous migrations.

12.3.2 Habitat-Specific Use: Functional Relationships

Estuarine habitats can be assessed for their importance as nekton habitats using a habitat utilization index (HUI) calculated as the sum of life history stages (eggs, larvae, juveniles, and adults) of the fishes that use a single habitat divided by the number of sites for that habitat within an estuary. This semi-quantitative index combines form (habitat type) and function (usage) and considers whether the fishes are estuary residents or transients. In this way, the HUI evaluates a habitat on the basis of an average number of uses made by all species and all life stages. The average values for European estuaries assessed by Pihl et al. (2002) are shown in Table 12.3. These rankings were similar to patterns in species richness with highest HUI values for subtidal soft and hard substrates and subtidal seagrass beds, and lowest values for intertidal hard substrate and reed beds (Table 12.3). This suggests that the habitats hosting the greatest diversity of fishes are also the habitats that are most highly used by fishes throughout their life histories.

TABLE 12.3 The Habitat Utilization Index (HUI) calculated for European estuaries.

Habitat/Number	HUI[a]
Tidal freshwater	23.1
Reed beds	15.5
Saltmarsh	19.3
Intertidal soft substrate	37.6
Intertidal hard substrate	9.0
Subtidal soft substrate	69.7
Subtidal hard substrate	43.3
Subtidal seagrass beds	46.5
Biogenic reefs	20.7

[a] HUI index represents the sum of fish life history stages using a single habitat divided by the number of sites for that habitat in all estuaries combined (Pihl et al. 2002)

Interconnected habitats form matrices in estuaries across which nekton species travel regularly to accomplish a range of different functions. Also, individual habitats can serve multiple functions for a species across different life history stages. It is for these reasons that HUI for all habitats in an estuary does not add up to 100 because individual life history stages of a species use multiple habitats, and multiple life history stages of a species use individual habitats. Thus, the degree to which different habitats are connected and accessible greatly influences the HUI.

HUI was highest for subtidal habitats (Pihl et al. 2002), and among these habitats subtidal soft substrate was the largest by areal extent. This habitat provides excellent feeding and nursery grounds for estuarine nekton. Subtidal seagrasses and subtidal hard substrate also had high HUI even though these habitats feature far more structural complexity than subtidal soft substrate. This structural complexity also provides excellent feeding opportunities, plus it provides refuge for estuarine nekton. These patterns of multiple habitat use and habitat function are universal in estuaries and tend not to vary with latitude. This suggests that strong local gradients in temperature and salinity and high variability in food resources limits estuarine nekton diversity by favoring organisms adapted to make use of interconnected estuarine habitats.

One series of studies in Términos Lagoon in the southern Gulf of Mexico described the use of multiple habitats by several estuarine nekton (Figure 12.5). Two nekton species, Western Atlantic seabream (*Archosargus rhomboidalis*) and White Grunt (*Haemulon plumieri*), migrate into the lagoon as juveniles and occupy seagrass and mangrove habitat in the dry season (March-June) when productivity is high. There they develop into pre-spawning adults, and then return to the Gulf of Mexico during the wet season to spawn. A second pair of species, the

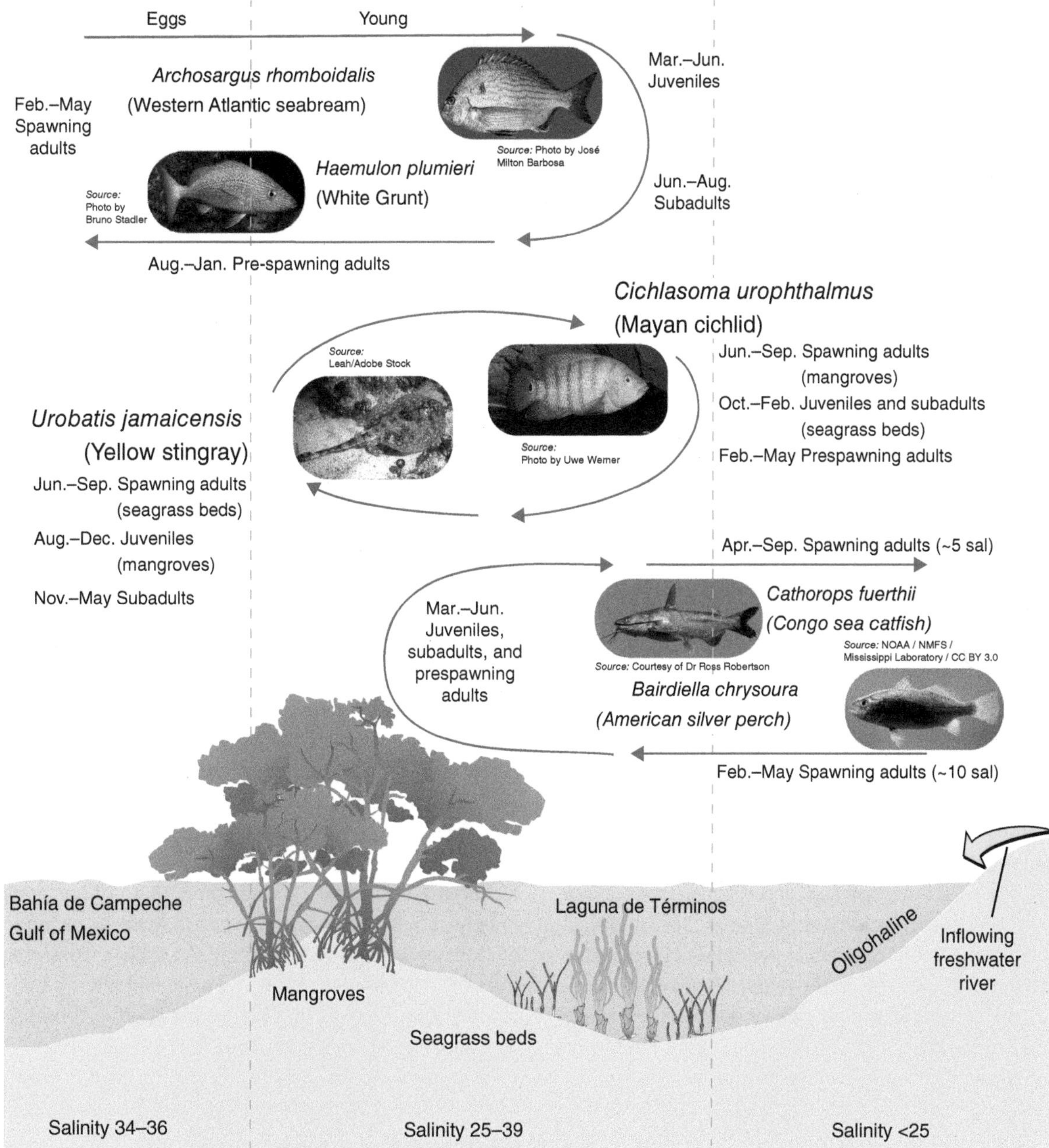

FIGURE 12.5 Life histories of six dominant fish species in Términos Lagoon, including marine-estuarine spawners *A. rhomboidalis* and *H. plumieri*, estuarine spawners *C. urophthalmus* and *U. jamaicensis*, and freshwater-estuarine spawners *C. fuerthii* and *B. chrysoura*. The fishes migrate using seagrass-mangrove habitat and fluvial lagoon system habitat during periods of highest productivity for feeding, spawning, and nursery grounds.

Congo sea catfish (*Cathorops fuerthii*) and American silver perch (*Bairdiella chrysoura*), also occupy seagrass and mangrove habitat during the dry season, but they return to oligohaline waters in the lagoon during the wet season when planktonic productivity is high. A third pair of species, the Yellow stingray (*Urobatis jamaicensis*) and Mayan cichlid (*Cichlasoma urophthalmus*), spend their whole life cycles moving between mangrove and seagrass habitat. This use of multiple habitats is typical of estuarine nekton, and it highlights the importance of maintaining mangroves and seagrass beds as part of the efforts to manage estuarine fisheries and prevent declines in sport and commercial fisheries.

12.3.3 Life History Strategies

Estuarine fishes exhibit a wide variety of life history strategies, but three endpoints with respect to parental investment and the speed at which the life cycle is completed are recognized: (i) periodic strategists, (ii) opportunistic strategists, and (iii) equilibrium strategists. These endpoints were defined by Winemiller (1989) to ordinate life history characteristics in a trilateral continuum that reflects an adaptive surface area based on fecundity, age of maturity, and juvenile survivorship (Figure 12.6). These strategies describe the many different ways nekton have evolved to cope with their environments, and they

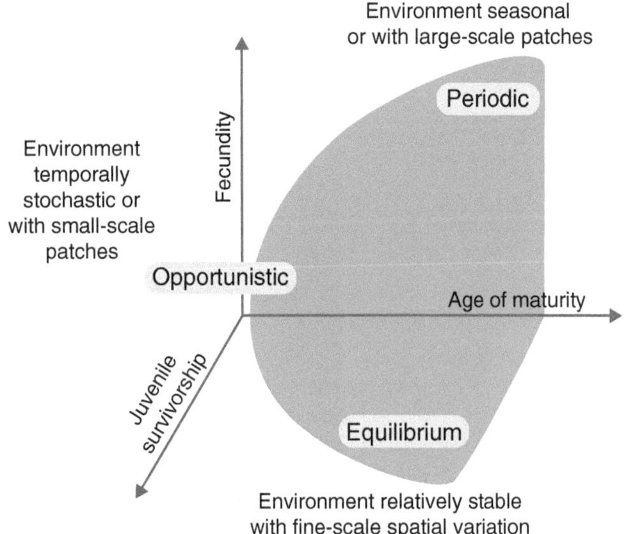

FIGURE 12.6 Trilateral life history continuum of traits that enable nekton to cope with spatially and temporally variable environments. *Source*: Adapted from Winemiller and Rose (1992).

differ primarily in the frequency of reproduction, the number of offspring, and the amount of effort adults put into caring for their young.

Periodic strategists are species that tend to be found in seasonally variable environments with large scale spatial variation. These conditions favor species with large adult body size, delayed maturation (long-lived), and large clutches of eggs spawned in one or more batches. Periodic strategists generally produce small pelagic eggs that are dispersive, and in which there is very little maternal investment. Periodic strategists are "bet-hedgers" in that they need only produce a strong **year class** (a fish cohort spawned during one year) in some years to maintain populations. Thus, in any one year it is unlikely that a given female produces any survivors. There are many periodic strategists among estuarine nekton including members of the families Cottidae (sculpins) and their relatives, Gadiidae (cods), Serranidae (basses), Sciaenidae (seatrout, drums and croakers), Sparidae (sea breams), Plueronectidae (left-eyed flounders), Chanidae (milkfish), Synodontidae (lizardfish), Belonidae (needle fishes, houndfish), Mugilidae (mullets), Polynemidae (threadfish), Cichlidae (tilapia, freshwater mojarras, and related species), Tetradontidae (puffers), Gerreidae (marine mojarras), Leiognathidae (slipmout, ponyfish), Pomadasyidae (grunts), Pomatomidae (bluefish), Carangidae (jacks), Lutjanidae (snappers), and various families of flatfishes such as Bothidae (right-eyed flounders), Soleidae (soles), and Cynoglossidae (toungefishes).

One species that represents the endpoint of the periodic strategy is the striped bass (*Morone saxatilis*). This North American species is anadromous and spawns in the tidal fresh waters of large rivers. It can live more than 25 years and begins spawning when 8–9 years old. Mature females can exceed 45 kg and have lifetime fecundity in the many millions of eggs. Striped bass need to produce a strong year class once in only 5–7 years for the population to be stable long-term. Year-class success for this species is related to seasonal and annual variability in river discharge which influences the availability of resources for juvenile striped bass. Young striped bass spend the first few years of their life in an estuary before migrating offshore to join the adult population.

Opportunistic life history strategists are often found in environments that experience stochastic variability on short temporal and small spatial scales. This strategy favors species that mature early and are small, short lived, and produce many small batches of eggs over long spawning seasons. Similar to periodic strategists, there is very little maternal investment in an individual egg, but they spawn more frequently and so can quickly recolonize an area in the event of a poor year class. There are fewer opportunistic strategists that use estuaries, but these fishes can be extremely abundant and are frequently important components of the estuarine forage base. Opportunistic strategists include members of families such as Atherinidae (silversides), Clupeidae (herringlike fish), Cyprinodontiodae (killifishes), Engraulidae (anchovies), and Polynemidae (threadfins).

One species that represents the opportunistic endpoint is the bay anchovy (*Anchoa mitchilli*). This small fish ranges from Massachusetts to Brazil in the Atlantic Ocean and Gulf of Mexico and is frequently among the most abundant fishes in estuaries over its range. Adult bay anchovy seldom reach 3 years in age and usually weigh less than 5 g. Female bay anchovy can begin to spawn during their first year of life and can spawn more than 100 times during a protracted spawning season, often spawning 50–75 days in succession during peak season. Bay anchovy are referred to as **income breeders** in that egg production is tied directly to feeding rate; it is not uncommon for a female to spawn greater than 400% of her body weight in eggs in a single spawning season, while growing in weight at rates as high as 25% per day. As such, this strategy maximizes the likelihood that some eggs will always be available to take advantage of small-scale variability in conditions that are favorable.

Equilibrium life history strategists tend to be found in resource-limited environments or in very stable environments with limited environmental variability. These conditions favor species of intermediate body size and longevity and those that produce smaller clutches of large eggs. One distinct characteristic of equilibrium strategists is a greater amount of paternal investment in each progeny. This can occur in the form of nest building and guarding, mouth brooding, and live bearing. Having large, well-developed young means that they are more likely to be competent feeders, or able swimmers, when first born. As such they avoid some of the difficulties faced by less poorly developed eggs and larvae that are associated with variable food resources and local retention within an estuarine ecosystem. Several equilibrium strategists have adapted to living in estuaries. These include members of the families Salmonidae (salmon and trout), Osmeridae (smelts and capelin), Gasterosteidae (stickleblacks), Cottidae (sculpins), and their relatives, Poecillidae (live-bearers and guppies), Ariidae (sea catfishes), Gobiidae (gobies), Syngnathidae (pipefishes and sea horses), and elasmobranchs (sharks, skates, and rays).

Salmon, sticklebacks, sculpins, and gobies build and defend nests, and both tasks are usually done by males. Smelts and capelins lay their eggs on beaches on the very highest spring tides, and the eggs develop in the sand until the next spring tide when they all hatch on the same night. Sea catfishes are mouth brooders, whereby the males scoop up recently laid eggs in their mouths where they hatch and stay until the small catfish are fully developed. Sea horse and pipe fish females lay their eggs in a pouch on the males' chest where they hatch and are released as fully capable, albeit small copies of the adults. As such, equilibrium strategists trade fecundity (lifetime egg production) for parental investment, often to the detriment of the adults.

In an examination of North American fishes for which sufficient data were available, Winemiller and Rose (1992) ordinated 68 marine species in a three-dimensional plot with axes of maturation length, investment per progeny, and clutch size, which are analogous to the traits of age of maturity, juvenile survivorship, and fecundity, respectively, in the discussions above (Figure 12.7). Most marine species that do not enter estuaries (blue circle) are more likely to fall out toward the periodic corner of the ordination, whereas estuarine species (green circle) tend to be periodic, opportunistic, and/or have traits that are intermediate between the endpoints. The red circle contains sea catfishes and pipefish, most of which occur in estuaries. Their high parental care makes them fall out toward the equilibrium side of the ordination. What should be clear is that while there are species that fit the endpoints of each strategy, there are many more that have traits that are intermediate between the endpoints. As such, it is not always clear which ecological circumstance favors which strategy, but variability in estuarine conditions and circulation patterns likely play a role. Undoubtedly, adult life history characteristics have diverged adaptively in response to environmental diversification and habitat segregation.

For estuarine-dependant organisms, there are costs and benefits involved in each of these life-history strategies, and this is why most have evolved some combination of the strategies that are most suited for the environments they occupy. For periodic strategists, spawning outside estuaries provided the stimulus for evolution of a pelagic larval period with small egg sizes that increase fecundity and prolonged larval periods during which small progeny accumulate resources from rich coastal planktonic habitat. But for these species, advective losses in the coastal ocean can be significant and many of the young may never make it to an estuary. Within estuaries, variable environmental conditions and problems related to retention of young provide the stimulus for evolution of the other life history strategies. Opportunistic strategists evolved to spawn many small batches of eggs over long spawning seasons inside estuaries, and equilibrium strategists evolved to produce big eggs with some paternal care that favors retention and to produce larvae capable of feeding immediately after birth. However, opportunistic strategists can exhibit boom or bust population dynamics given their short life spans, and the equilibrium strategy can be costly to adults, especially if survival is low.

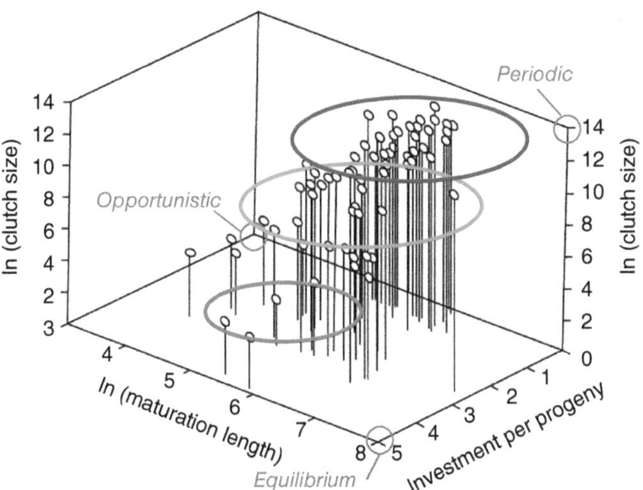

FIGURE 12.7 Three-dimensional ordination of ln(maturation length), ln(mean clutch size), and investment per progeny for 68 marine fishes. Maturation length is the modal length at maturation in mm total length, clutch size is the mean batch fecundity for a local population from a specific ecosystem or location, and investment per progeny is an index that can range from 0 = no parental care to 8 = extremely long gestation. The blue circle highlights marine fishes that do not enter estuaries, the green circle highlights estuarine fishes, and the red circle highlights sea catfishes, pipefish, and other estuarine fishes with high investment per progeny. *Source*: Adapted from Winemiller and Rose (1992).

12.4 Factors Controlling Productivity and Community Composition

12.4.1 Year Class Success

For most estuarine nekton, year class success is determined during early life through a variety of mechanisms including predation, advective loss of eggs, toxic contaminants, and lethal low temperature events (Houde 1989). This knowledge is based on the observation that there is little to no correlation between abundance estimates of successive life history stages until the late larval to juvenile stages, after which the number of survivors (often referred to as recruits) is correlated to the number of subsequent adult spawners (Figure 12.8). This lack of correlation occurs because the number of eggs and larvae decline exponentionally at very high and variable rates, especially those of nekton with periodic and opportunistic life history strategies. Eggs and newly hatched larvae have **mortality rates** (the rates at which organisms die or are grazed) exceeding 50% per day, perhaps as high as 95% per day for species

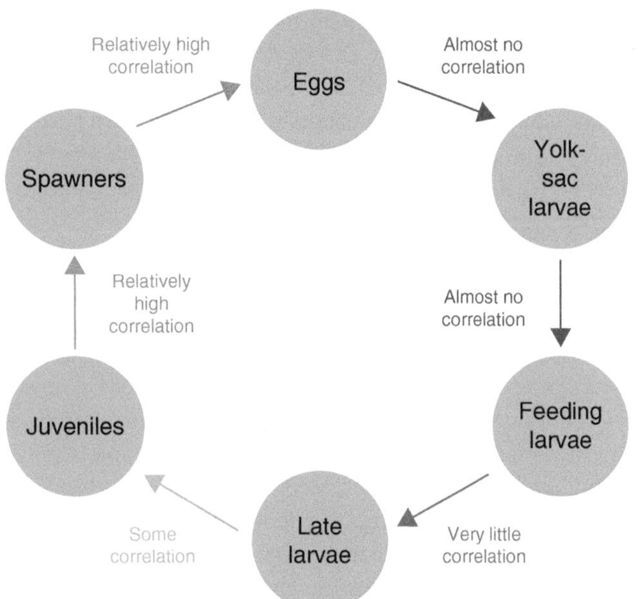

FIGURE 12.8 The degree of correlation in abundance between successive life history stages of estuarine nekton from eggs to spawning adults. *Source*: Base on Houde (1989).

such as the bay anchovy. Mortality rates of feeding-stage larvae are lower but can be as high as 30–40% per day.

For fishes, growth rates and mortality rates are related, and can differ greatly among seasons or years in response to environmental variability, including food abundance. The combined effects of growth and mortality is what David Cushing called the *single process* controlling fish productivity. Cushing (1975) inferred that the longer an individual remains vulnerable to predation, the higher the cumulative mortality rate, which means that slow growth tends to increase the likelihood of being eaten because slow growers do not outgrow their potential predators as quickly as fast growers.

Houde (1989) used scenarios presented in Table 12.4 to illustrate how the combined effects of growth and mortality can result in recruitment successes and failures without invoking episodic population losses. In these examples, the "good" year class is produced when a cohort of fishes grow rapidly. By comparison, bad year classes fail because they experience relatively high mortality (Bad-1), reach maturity at an older age (Bad-2), or both (Bad-3). These results show that relatively small changes in growth (here represented as age to metamorphosis) and mortality rates (here mortality coefficients) can generate large variability in year class success. In an estuarine setting, anything that can affect either of these rates (even subtly) can have a big effect on the number of survivors (recruits).

12.4.2 Survival and Growth

Most fish species produce cohorts of small, young fishes that have much higher mortality rates than larger, older fishes. One common index of stage specific survival (*S*) is the ratio of the instantaneous mortality coefficient (how fast they die, or *M*), to the weight specific growth rate (*G*). If one has some knowledge of the magnitudes of the ratio of *M*/*G*, it can be a useful measure of cohort dynamics that intergrates effects of mortality and growth-generating processes. As such these simple measures (*M*/*G*) may reflect size selective processes, where *S* is stage specific survivial, N_s and W_s are the number of fishes and the weight of an individual fish, respectively, and N_o and W_o are the number of fishes and the weight of an individual at the beginning of the stage, respectively.

$$S = N_s/N_o = (W_s/W_o) - (M/G) \tag{12.1}$$

Houde and colleages (1996) used a slighty different approach to show the potential for biomass proliferation in a stage specific formulation in which *B* is biomass:

$$B_s/B_o = (W_s/W_o)^{[1-(M/G)]} \tag{12.2}$$

In temperate and colder estuaries, the rate of food consumption by nekton is seasonal. Nekton that do not migrate in winter months often do not feed, or feed very little when the water is cold. This behavior is attributable to the energy budget for growth, which is called bioenergetics, and is expressed in difference form by the formula:

$$W_t = W_{t-1} + (C_{max} \times p \times A) - (R_{tot} \times \Delta_t) \tag{12.3}$$

TABLE 12.4 One good and three bad recruitment scenarios.

Condition	Initial number in a cohort[a]	Mortality Coefficient[b] (z) d^{-1}	Age at meta-Morphosis[c] (t)	Number of recruits
Good	1×10^6	0.100	45.0	11,109
Bad-1	1×10^6	0.125	45.0	3607
Bad-2	1×10^6	0.100	56.2	3625
Bad-3	1×10^6	0.125	56.2	889

[a] All cohorts initially begin with 1 million larvae.
[b] The mortality coefficient is from the formula $N_t = N_0 e^{-(zt)}$ where N_t is the number surviving to age t in days, and N_0 is the initial number of larvae (here 1 million).
[c] Age at metamorphosis is the time it takes for larvae to become juveniles (i.e., a low number = fast growth).
Source: Based on Houde (1989).

where W_t is the weight today, W_{t-1} is the weight at an earlier time, C_{max} is the maximum possible food intake, p is the proportion of the maximum food intake realized, A is the assimilation efficiency, R_{tot} is the respiration rate, and Δ_t is the change in time between t and t–1, usually 1 day. Respiration is highly dependent on temperature and weight in cold-blooded animals such as fishes and slows down in the winter. This means that in cold conditions R_{tot} is reduced and fishes do not need to eat as much (i.e., can have lower $C_{max} \times p \times A$) to maintain their weight or grow, which can be either very low or negligible in winter. Conversely, fishes must feed at high rates in summer when temperatures are high.

Resident fishes in temperate estuaries often feed actively in spring and fall in order to recover from winter conditions in the spring, and to store enough energy in fall to prepare for winter conditions. In general, however, adult fishes must consume 1–30% of their body weight per day at optimal temperatures depending on the level of normal activity. For example, a feeding rate of 3% body weight day^{-1} (calculated from Eq. 12.3 as $(C_{max} \times p \times A)/W_{t-1}$) is typically required for lurking predators, whereas higher values are required for anchovies and other constantly moving filter feeders and planktivores. Also, feeding rates are much higher for early life history stages, ranging from 3–50% body weight day^{-1} for juveniles to 50–400% body weight day^{-1} for larvae, stressing the importance of food supply on growth rates during early life for recruitment.

For clarity, we modified the bioenergetics equations as follows:

$$C = (R + A + SDA) + (F + U) + (\Delta B + G) \qquad (12.4)$$

These variables have independent functions, where C is the rate of consumption, R, A, and SDA are metabolic losses (R = respiration, A = active metabolism, and SDA = specific dynamic action such as swimming), F and U are other losses (F = egestion and U = excretion), and ΔB and G are growth (ΔB = Change in biomass as somatic growth and G = gonad growth and reproduction). If we compare these terms to those in a basic personal financial budget, consumption = income, metabolism = rent, other losses = taxes and waste, and growth = savings and investments. All of the processes represented in this equation are temperature and size dependent (Figure 12.9). In this figure, specific consumption ranges from 0.00 to 0.08 and is calculated as the ratio of grams of prey over grams of body mass per day (g_{prey}/g_{body}/d). Grams body mass stays the same so specific growth rate varies with the mass of prey consumed. Temperature is important because it ranges in this example from 5 to 30 °C.

It should be clear to the reader that growth rate is variable and depends on the amount of food consumed and the temperature of the system. When food availabity is high, so is the scope for growth as indicated by the size of the yellow area in Figure 12.9a. In nature, however, maximum consumption is almost never obtained. This is illustrated by the representation of a realistic consumption rate with $p = 0.35$ (Eq. 12.3) such that consumption equals 35% of the maximum possible rate

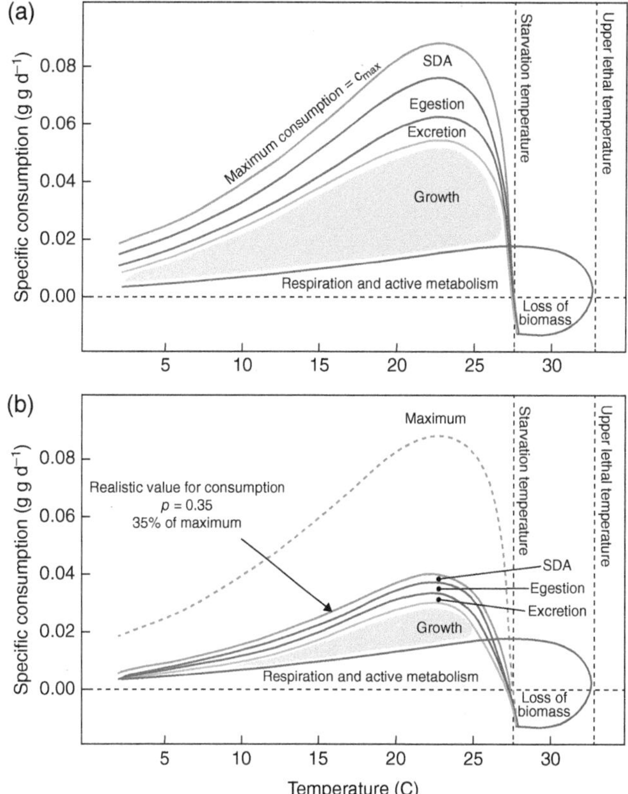

FIGURE 12.9 The theoretical specific rate of consumption of fishes at a range of temperatures under (a) maximum rate of consumption C_{max} and (b) a more realistic consumption rate (solid red line). Stacked lines indicate the portions of consumption that are used for respiration (including active metabolism), growth (proportion in yellow, including somatic growth and reproduction), excretion, egestion, and specific dynamic action (SDA). Note that in this example, above ~24 °C the animal's body gets too hot and starts to fail. *Source*: Adapted from Hanson et al. (1997) by James Kitchell, University of Wisconson, Madison.

of food consumption (Figure 12.9b). In this example, the scope for growth is reduced and the fishes represented by the small yellow area in the figure are growing very slowly.

12.4.3 Biotic and Abiotic Factors

The primary biotic factor affecting nekton productivity and community composition is the availability of prey. Although prey in estuaries is usually abundant, availability of any particular prey can show considerable short-term fluctuations. While starvation is rare, the indirect effects of food availability on recruitment via changes in prey growth rate is high. As such, most estuarine species are not specialized feeders, even though each species can show a preference for some type of food (e.g., benthic invertebrates, smaller fishes, mollusks, among many others). As such, feeding rates and habitats are related with respect to both growth and mortality (Cushing 1975). Sooner or later, almost every potential source of energy will be consumed by some type of grazer (nekton or otherwise), and at any given time, it is unlikely that any significant source of food is underutilized. Remember, nektonic food webs in estuaries tend to be simple and the number of prey species that are abundant at any one time is usually low. Also, most estuarine nekton are generalists that tend to consume different food types in the same order of abundance in which they are encountered. The bottom line is that fishes are strongly tied to their environment via food webs, and the diversity of habitats and body forms results in a diversity of feeding types.

Besides prey availability, several other factors regulate the nekton productivity and community composition including predation, competition, and environmental fluctuations. Many of the abundant transient species in estuaries become predators as adults, and can impose strong top-down control on the abundances of both invertebrates and fish nekton through predation. Moreover, large concentrations of predators can locally deplete prey populations and become food-limited. For example, demersal fishes feeding on benthic invertebrates are believed to be responsible for reducing prey abundance and biomass in subtidal soft substrate habitats. Predators can also deplete populations of young nekton that use estuaries as nursery areas. However, for some species, mortality rates of the juveniles may be compensated for by rapid growth rates. Rapid growth enables some young nekton to quickly outgrow many types of predators (Pope et al. 1994). In this way, predation on young nekton in estuaries favors species that develop rapidly into adults.

Competition can also impact community composition because most studies on feeding habits of estuarine nekton indicate a high degree of overlap in diet among the species present. This diet overlap is probably due to the low diversity of abundant prey species in estuarine soft bottoms and other estuarine habitats. However, estuarine nekton reduce interspecific competition over these resources using several strategies. One strategy is to temporally stagger the use of estuarine resources (Figure 12.10), and another is to target different resources. For example, juvenile sciaenids (e.g., Atlantic croaker, *Micropogonias undulatus*, and Norfolk spot, *Leiostomus xanthurus*) manage to segregate partially in estuaries by feeding habits owing to differences in body shape and mouth structure and partially on the basis of distribution and timing of the use (seasonality) of estuaries (Figure 12.11).

Human use of estuaries has altered nekton community composition by enhancing the abundance of invasive species, which can alter patterns of predation and competition in nekton communities. Invasive species are introduced

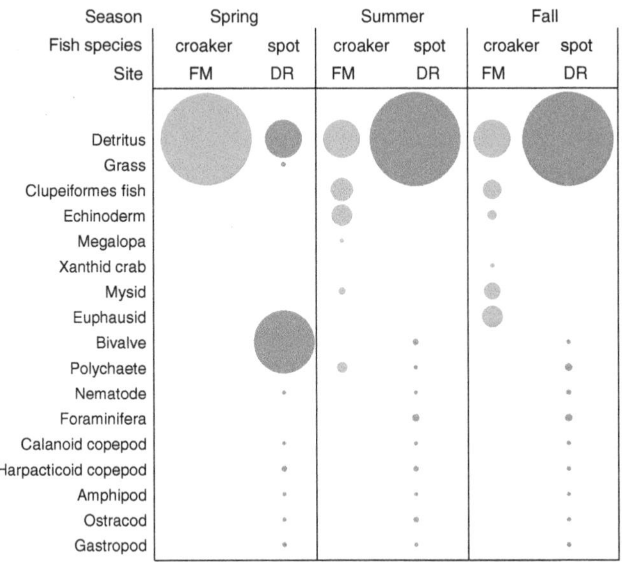

FIGURE 12.11 Diets of Atlantic croaker (*Micropogonias undulatus*) and Norfolk spot (*Leiostomus xanthurus*) in Mobile Bay, AL collected during three seasons at different locations in the Bay. The bubbles sum to 100% for each component of the overall diet. DR is a region of the mid-bay where spot were much more abundant than croaker. FM is a region in the lower bay where the inverse was true. *Source*: Data collected for Jackson and Cowan (2013).

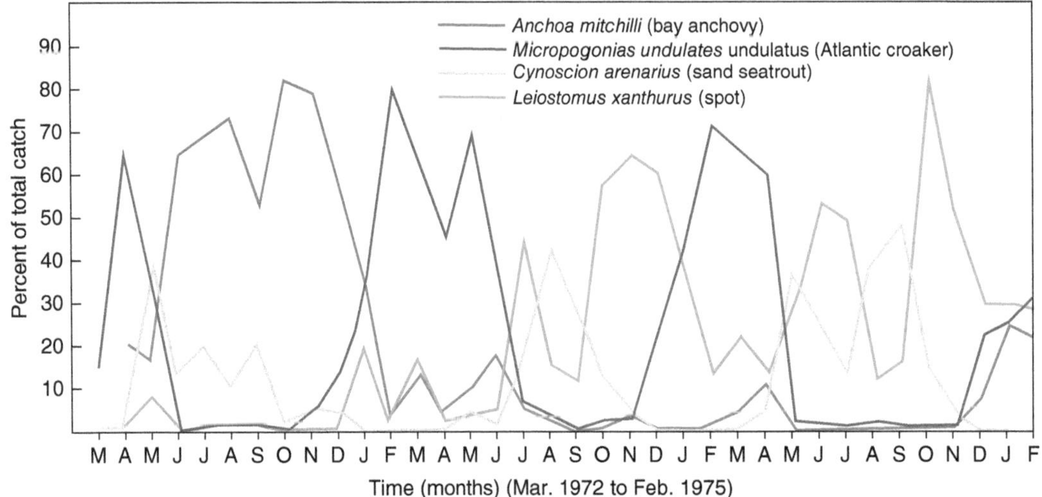

FIGURE 12.10 Relative importance (percent of total catch) of four species of estuarine fishes at different times of year in the Apalachicola Bay estuarine system. *Source*: Adapted from Livingston (1985).

to estuaries through dumping of ballast water by ships, by unintentional releases from aquaculture and aquariums, and by intentional releases of organisms. Perhaps, the most extreme example of an invaded estuary in the United States is the Sacramento-San Joaquin estuary (San Francisco Bay) of California. Several new species successfully invade this estuary each year and much of the biota is now exotic. For example, the striped bass, a native to the east coast of the United States, was intentionally introduced in the 1850s and has thrived, ultimately growing to support a substantial recreational fishery. But as often happens, the striped bass in now considered a pest because it consumes young salmon, many species of which are now endangered, as they migrate through the estuary to the ocean. Other significant introductions affect prey organisms for nekton such as the Asian clam, whose filter feeding has greatly reduced populations of zooplankton required as food by larval fishes. The result is a highly altered ecosystem with many endangered native species such as the Delta Smelt (*Hypomesus transpacificus*).

Salinity and temperature are important drivers of community responses in estuaries because of the potential for strong salinity and temperature gradients in estuaries (Deegan et al. 1986; Evans 1993; Pattillo et al. 1995). In some estuaries, there are striking differences between summer and winter fish faunas that are due to the salinity and temperature tolerances and preferences of different species. Moreover, latitudinal change in the composition of estuarine fauna is tied to differences in temperature range. Individual nekton species have salinity and temperature ranges over which they perform best. For example, the Gulf menhaden (*Brevoortia patronus*) is most abundant at temperatures ranging from 25–35 °C, and salinities ranging from near 0 to 25 ppt. In contrast, Atlantic menhaden prefer salinities ranging from 10 to 25 psu, with peak activity at 15–18 °C.

Hydrodynamics and wind stress can also regulate nekton productivity and community composition by altering recruitment patterns in estuaries. Many estuary-dependent species spawn offshore and rely on currents to transport them inshore. Hydrological conditions and coastal morphology can determine the success of this migration into estuaries. Once in an estuary, strong residual currents favor retention, but asymmetric tidal currents can be a problem. For example, many of the world's estuaries have tidal currents that are generated astronomically, but others occur in microtidal environments where currents are generated by meteorological events and are less predictable in time and amplitude. This dichotomy provides an interesting area of research because transport and retention often involves behaviorally mediated use of tidal circulation, but some estuarine-obligate nekton species are equally as successful in estuaries that are governed by astronomical tides and exhibit strong two-layered circulation and in estuaries where neither of these conditions persist.

Review Questions

Multiple Choice

1. Nekton can swim and move independently of water currents. Examples of nekton include all of these animals except:
   a. Adult Bluefish
   b. Adult Blue crab
   c. Adult copepod
   d. Penaeid shrimp

2. Osmoregulation in estuarine fish is accomplished with many different physiological adaptations. These adaptations include all these *except*:
   a. Salt release through gill epithelium
   b. Excess water excretion in urine
   c. Consumption of prey with low salt content
   d. Salt absorption through gill epithelium
   e. Salt concentration and excretion in urine

3. Juvenile fishes are generally more salt tolerant than adult fishes
   a. True
   b. False

4. Estuarine nekton generally migrate to avoid environmental conditions outside their physiological tolerances in all these factors *except*:
   a. Temperature
   b. Oxygen concentration
   c. Salinity
   d. Sunlight
   e. Turbidity

5. Nursery grounds in estuaries ------
   a. are always the same as spawning grounds.
   b. enhance survival of juveniles.
   c. are often regions where nekton migrate for feeding.
   d. are often occupied by juveniles and adults of the same species.

6. A study of habitat usage in European estuaries showed that most estuarine fish species use which habitats for feeding and nursery areas?
   a. Tidal freshwater regions and biogenic reefs
   b. Subtidal hard and soft substrates and biogenic reefs
   c. Subtidal hard and soft substrates and reed beds
   d. Intertidal and subtidal hard and soft substrates

7. Nekton generally use multiple estuarine habitats during their life cycle for all these reasons except:
   a. To avoid adult feeding in spawning grounds
   b. To vary their diets
   c. To feed in habitats that have seasonally high productivity
   d. To avoid predation

8. Winemiller's ordination of life history strategies categorizes fishes based on ____
   a. Fecundity, age at maturity, size at maturity
   b. Fecundity, size at maturity, juvenile survivorship
   c. Fecundity, age at maturity, juvenile survivorship
   d. Age at maturity, juvenile survivorship, feeding adaptations
9. Nekton with an opportunistic life history strategy often
   a. Are long-lived
   b. Protect their young
   c. Spawn often
   d. Produce large batches of eggs
10. The life history strategy of most estuarine nekton is
   a. Periodic strategy
   b. Opportunistic strategy
   c. Equilibrium strategy
   d. A combination of strategies
11. Year class success in estuarine nekton depends on the combined effects of growth and mortality (Cushing's *single process*) such that success is favored when fish
   a. Reach maturity at an older age
   b. Grow more slowly than their rate of mortality
   c. Grow more rapidly than their rate of mortality
   d. Have mortality rates below 50% per day
12. Nekton growth rate depends on ____
   a. Body size and amount of food consumed
   b. Temperature and amount of food consumed
   c. Temperature and salinity
   d. Temperature and respiration

Short Answer

1. How does temperature operate in concert with salinity to control distributions of some nekton?
2. Briefly describe how mouth body position influences the feeding location of estuarine fishes.
3. Why do estuarine nekton have a wide trophic spectrum despite species-specific phyiological and behavioral feeding adaptations? How does this impact the composition of nekton diets?
4. Describe how stage-specific survival (S) is related to the ratio of instantaneous mortality to weight specific growth (M/G) for estuarine fishes.

References

Able, K.W. (2005). A re-examination of fish estuarine dependence: evidence for connectivity between estuarine and ocean habitats. *Estuarine, Coastal and Shelf Science* 64: 5–17.

Blaber, S.J.M. (2000). *Tropical Estuarine Fishes: Ecology, Exploitation and Conservation*. Oxford: Blackwell Science.

Bryant, T.L. and Pennock, J.R. (1988). *The Delaware Estuary: Rediscovering a Forgotten Resource*. Newark, Del: University of Delaware, Sea Grant College Program.

Cowan, J.H. Jr., Grimes, C.B., and Shaw, R.F. (2008). Life history, history, hysteresis and habitat changes in Louisiana's coastal ecosystem. *Bulletin of Marine Science* 83: 197–215.

Cushing, D.H. (1975). *Marine Ecology and Fisheries*, 278. Cambridge: Cambridge University Press.

Day, J.W., Hall, C.A.S., Kemp, W.M., and Yáñez-Arancibia, A. (1989). *Estuarine Ecology*, 558. New York: Wiley.

Day, J.W.J., Yanez-Arancibia, A., Kemp, W.M., and Crump, B.C. (2013). Introduction To Estuarine. In: *Estuarine Ecology* (ed. J.W. Day Jr., B.C. Crump, W.M. Kemp and A. Yanez-Arancibia), 1–18. Wiley.

Deegan, L.A., Day, J.W., Gosselink, J.G. et al. (1986). Relationships among physical characteristics, vegetation distribution and fisheries yield in Gulf of Mexico estuaries. In: *Estuarine Variability* (ed. D.A. Wolfe), 83–100. Orlando, FL: Academic Press, Inc.

Evans, D.H. (1993). Osmotic and ionic regulation. In: *The Physiology of Fishes* (ed. D.H. Evans), 315–341. Boca Raton: CRC.

Evans, D.H., Piermarini, P.M., and Choe, K.P. (2005). The multifunctional fish gill: dominant site of gas exchange, Osmoregulation, Acid-Base Regulation, and Excretion of Nitrogenous Waste. *Physiological Reviews* 85: 97–177. https://doi.org/10.1152/physrev.00050.2003.

Froeschke, J.T. and Froeschke, B.F. (2011). Spatio-temporal predictive model based on environmental factors for juvenile spotted seatrout in Texas estuaries using boosted regression trees. *Fisheries Research* 111: 131–138. https://doi.org/10.1016/j.fishres.2011.07.008.

Genner, M.J., Sims, D.W., Wearmouth, V.J. et al. (2004). Regional climatic warming drives long-term community changes of British marine fish. *Proceedings of the Royal Society of London. Series B: Biological Sciences* 271: 655–661. https://doi.org/10.1098/rspb.2003.2651.

Hanson, P.C., Johnson, T.B., Schindler, D.E., and Kitchell, J.F. (1997). *Fish bioenergetics 3.0 software for Windows*. Madison, Wisconsin: University of Wisconsin Center for Limnology, Sea Grant Institute, Technical Report WISCUT-97-001.

Houde, E. (1989). Comparative growth, mortality, and energetics of marine fish larvae: temperature and implied latitudinal effects. *Fishery Bulletin* 87: 471–495.

Houde, E.D. (1996). Evaluating stage-specific survival during the early life of fish. In: *Survival Strategies in Early Life Stages of Marine Resources* (ed. Y. Watanabe, Y. Yamashita and Y. Oozeki), 51–66. Rotterdam: Balkema.

Hubbs, C.L. and Lagler, K.F. (1964). *Fishes of the Great Lakes Region*. Ann Arbor: The University of Michigan Press (Original work published 1947).

Jackson, J.B. and Cowan, J.H. Jr. (2013). The effects of front-associated wind events and resultant sediment resuspension on dietary habits and caloric intake of Bay Anchovy and Age-0 Atlantic Croaker in Mobile Bay, Alabama. *Marine and Coastal Fisheries* Dynamics, Manage., Ecosystem Sci. 5 (1): 103–113.

Jaureguizar, A.J., Menni, R., Guerrero, R., and Lasta, C. (2004). Environmental factors structuring fish communities of the Río de la Plata estuary. *Fisheries Research* 66: 195–211. https://doi.org/10.1016/S0165-7836(03)00200-5.

Larkin, P.A. (1956). Interspecific competition and population control in freshwater fish. *Journal of the Fisheries Board of Canada* 13 (3): 327–342.

Leak, J. and Houde, E. (1987). Cohort growth and survival of bay anchovy Anchoa mitchilli larvae in Biscayne Bay, Florida. *Marine Ecology Progress Series* 37: 109–122. https://doi.org/10.3354/meps037109.

Livingston, R.J. (1985). Organization of fishes in coastal sea-grass system: the response to stress. In: *Fish Community Ecology in Estuaries and Coastal Lagoons: Towards an Ecosystem Integration* (ed. A. Yáñez-Arancibia), 367–382. UNAM-PUAL-ICML: Editorial Universitaria, México.

Methven, D.A., Haedrich, R.L., and Rose, G.A. (2001). The fish assemblage of a newfoundland estuary: diel, monthly and annual variation. *Estuarine, Coastal and Shelf Science* 52: 669–687. https://doi.org/10.1006/ecss.2001.0768.

Nagelkerken, I. (2009). Evaluation of nursery function of mangroves and seagrass beds for tropical decapods and reef fishes: patterns and underlying mechanisms. In: *Ecological Connectivity among Tropical Coastal Ecosystems* (ed. I. Nagelkerken), 357–399. Netherlands: Springer.

Paterson, A.W. and Whitfield, A.K. (2000). Do shallow-water habitats function as refugia for juvenile fishes? *Estuarine, Coastal and Shelf Science* 51: 359–364. https://doi.org/10.1006/ecss.2000.0640.

Pattillo, M., Rozas, L.P., and Zimmerman, R.J. (1995). *A Review of Salinity Requirements for Selected Invertebrates and Fishes of U.S. Gulf of Mexico Estuaries*. Galveston: National Marine Fisheries Service, Southeast Fisheries Science Center.

Pihl, L., Cattrijsse, A., Codling, I. et al. (2002). Habitat use by fishes in estuaries and other brackish areas. In: *Fishes in Estuaries* (ed. M. Elliott and K.L. Hemingway), 10–53. Oxford: Blackwell.

Pope, J. G., J. G. Shepherd, J. Webb, A. R. D. Stebbing, M. Mangel. 1994. Philosophical Transactions Biological Sciences, Royal Society 343(1303):41–49, Generalizing across Marine and Terrestrial Ecology, The Royal Society.

Potter, I.C., Beckley, L.E., Whitfield, A.K., and Lenanton, R.C.J. (1990). Comparisons between the roles played by estuaries in the life cycles of fishes in temperate western Australia and southern Africa. *Environmental Biology of Fishes* 28: 143–178.

Sánchez-Gil, P. (2009). Ecología Demersal Tropical: Grupos Funcionales y Patrones de Utilización en Hábitats Costeros (Sur del Golfo de México). [*Tropical Demersal Ecology: Functional Groups and Pattern of Coastal Habitats Utilization*]. PhD Dissertation Universidad Autónoma Metropolitana, México D.F. pp. 109.

Storm, T.J., Nolan, K.E., Roberts, E.M., and Sanderson, S.L. (2020). Oropharyngeal morphology related to filtration mechanisms in suspension-feeding American shad (Clupeidae). *Journal of Experimental Zoology Part A: Ecological and Integrative Physiology* 333: 493–510. https://doi.org/10.1002/jez.2363.

Tsou, T.-S. and Matheson, R.E. (2002). Seasonal changes in the nekton community of the Suwannee River estuary and the potential impacts of freshwater withdrawal. *Estuaries* 25: 1372–1381. https://doi.org/10.1007/BF02692231.

Warlen, S.M. and Burke, J.S. (1990). Immigration of Larvae of fall/winter spawning marine fishes into a North Carolina estuary. *Estuaries* 13: 453. https://doi.org/10.2307/1351789.

Winemiller, K.O. (1989). Patterns of variation in life history among South American fishes in seasonal environments. *Oecologia* 81: 225–241.

Winemiller, K.O. and Rose, K.A. (1992). Patterns of life history diversification in North American fishes: implications for population regulation. *Canadian Journal of Fisheries and Aquatic Sciences* 49: 2196–2218.

Wurts, W.A. (1998). Why can some fish live in freshwater, some in salt water, and some in both. *World Aquaculture* 29: 65.

Yáñez-Arancibia, A. (1978). *Taxonomy, Ecology and Structure of Fish Communities in Coastal Lagoons with Ephemeral Inlets on the Pacific Coast of Mexico*, Inst. Cienc. del Mar y Limnol. Spec. Publ. 2, 306. México DF: UNAM Press.

Yáñez-Arancibia, A. (ed.) (1985). *Fish Community Ecology in Estuaries and Coastal Lagoons: Towards Ecosystem Integration*, Inst. Cienc. Del Mar y Limnol, 654. Mexico DF: UNAM Press.

Yáñez-Arancibia, A. and Sánchez-Gil, P. (ed.) (1986). *The Demersal Fishes of the Southern Gulf of Mexico Shelf: Environment, Ecology and Evaluation*, Inst. Cienc. del Mar y Limnol. Spec. Publ. 9, 230. México DF: UNAM Press.

Yáñez-Arancibia, A., Lara Dominguez, A.L., Rojas Galaviz, J.L. et al. (1988). Seasonal biomass and diversity of estuarine fishes coupled with tropical hábitat heterogeneity (southern Gulf of Mexico). *Journal of Fish Biology* 33 (Suppl A): 191–200.

Yáñez-Arancibia, A., Lara Dominguez, A.L., and Day, J.W. (1993). Interactions between mangrove and seagrass hábitats meadiated by estuarine nekton assemblages: Coupling of primary and secondary production. *Hydrobiologia* 264: 1–12.

Yáñez-Arancibia, A., Lara Dominguez, A.L., and Pauly, D. (1994). Coastal lagoons as fish habitats. In: *Coastal Lagoon Processes*, Elsevier Oceanography Series 60 (ed. B. Kjerfve), 363–376. Amsterdam, The Netherlands: Elsevier Science.

CHAPTER **13**

Estuarine Wildlife

Chris S. Elphick[1], W. Gregory Shriver[2], and Russel Greenberg[3]
[1]Department of Ecology and Evolutionary Biology, University of Connecticut, Storrs, CT, USA
[2]Department of Entomology and Wildlife Ecology, University of Delaware, Newark, DE, USA
[3]Smithsonian Migratory Bird Center, National Zoological Park, Washington, DC, USA

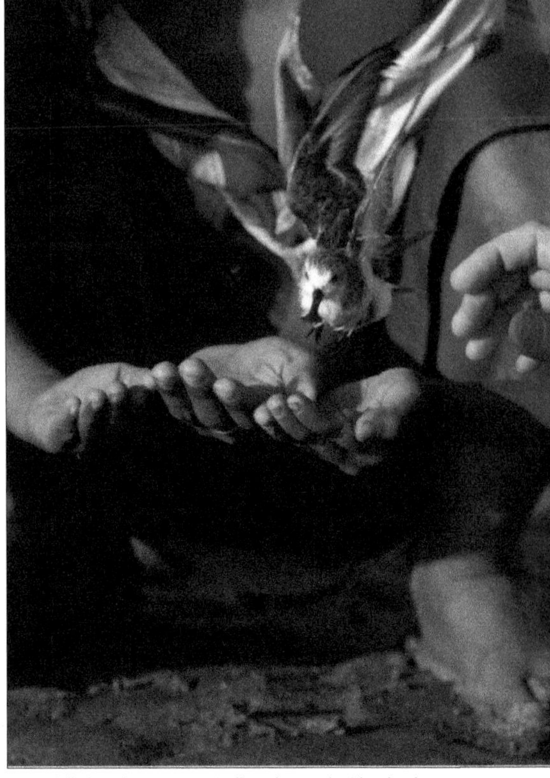

Spoon-billed sandpipers are critically endangered, with only a few hundred remaining. During the non-breeding season they migrate from their breeding grounds on the Russian tundra to estuaries throughout Asia. In Myanmar, local conservation groups have worked with local communities to explain the species plight and provide alternatives to shorebird hunting. Photo credit: R. Robinson/BTO.

13.1 Introduction

Protected from the open ocean, estuaries—the meeting places of marine, terrestrial, and riverine ecosystems—provide a home for many wildlife species, often in great abundance. From open embayments, to broad expanses of tidal mudflat, to the greenery of tidal marshes and mangrove swamps, many species of birds, mammals, reptiles, and amphibians live their lives, or visit to take advantage of an abundance of food as they move along the coast. This chapter explores the diversity of wildlife in estuaries, the patterns of wildlife abundance in time and space, the ecological roles of wildlife in estuarine ecosystems, and how the activities of wildlife and humans influence one another.

The term "wildlife" is often used to refer broadly to wild animals, and sometimes also plants. For this chapter, however, we use the term in the narrow sense used by many management agencies to mean vertebrates other than fish (Figure 13.1). These species are often the most visible animals in estuarine ecosystems, and generally receive the greatest attention from the public. For this reason, they often have disproportionate influence on the ways in which estuarine ecosystems are managed.

Many wildlife species are among the largest organisms in estuarine ecosystems, and some can have meaningful consequences for ecosystem structure and processes, via their effects as herbivores or predators. Relatively few wildlife species live in estuaries year-round and fewer breed in these ecosystems than in other types of wetland or terrestrial ecosystems. Breeding species, however, are supplemented by many migratory species that move into estuaries during non-breeding periods of their lives, often in large numbers, to take advantage of the high productivity of these ecosystems. Birds are generally the dominant wildlife group, both in terms of species variety and abundance of individual species, but a varied mixture of mammals, reptiles, and amphibians, are also found.

Estuarine wildlife communities are made up of a mix of terrestrial and oceanic species. Few wildlife species are found exclusively in estuaries; most have some overlap with other terrestrial habitats, and some with marine habitats. Common

Estuarine Ecology, Third Edition. Edited by Byron C. Crump, Jeremy M. Testa, and Kenneth H. Dunton.

Companion website: www.wiley.com/go/crump/estuarine3

FIGURE 13.1 Estuaries support a variety of wildlife species. (a) Saltmarsh sparrows are found only in coastal marshes in the eastern United States. *Source*: M. Szantyr. (b) Shorebirds gather in huge flocks during the nonbreeding period to feed on estuarine mudflats around the world, such as these plovers and sandpipers in Myanmar. *Source*: C. Elphick. (c) The endangered salt marsh harvest mouse is endemic to marshes near San Francisco, California. *Source*: K. Smith. (d) Diamondback terrapins of eastern North America are the only turtle found primarily in estuarine habitats. *Source*: M. Szantyr.

features of many estuarine wetland species are the capacity to contend with the constantly shifting conditions of tidal systems with their changing patterns of flooding, water depth, and salinity. In this chapter, we first describe the diversity of wildlife species that occur in estuaries, then describe variation in their occurrence patterns, and the ways in which they interact with each other. Finally, we discuss the major ways in which humans affect estuarine wildlife.

13.2 Patterns of Diversity

13.2.1 The Players

13.2.1.1 Birds Birds are the most diverse wildlife group associated with estuaries. The two orders that typically dominate both species richness and individual abundance are the Anseriformes (ducks, geese, and swans) and Charadriiformes (shorebirds, gulls, and terns), but waterbirds from the Gruiformes (rails, cranes), Podicipediformes (grebes), Phoenicopteriformes (flamingos), Gaviiformes (loons), Ciconiiformes (storks), Suliformes (cormorants), and Pelecaniformes (ibis, heron, and pelicans) all frequently use estuarine habitats. Although waterbirds are generally most visible, many species from groups thought of as more terrestrial are also found in estuaries. Passeriformes (songbirds), in particular, are common—not surprising since this order comprises more than half of the world's bird species—although the number of species that regularly use estuarine habitats is low compared to other habitat types. Predatory birds from the Accipitriformes (hawks, eagles, and harriers), Strigiformes (owls), and Falconiformes (falcons) also regularly occur in estuaries, often hunting ducks and shorebirds.

Estuarine bird communities throughout north temperate and Arctic regions generally comprise a similar mix of genera, and often the same species, as many northern waterfowl and shorebirds (often called wildfowl and waders, respectively, in areas outside North America) have circumpolar distributions. In North America, for example, approximately 165 bird species commonly occur in estuaries in at least a portion of their range during some time of the year. About a third of these species are coastal birds that primarily forage in deeper waters and also occur close to shore along non-estuarine coastlines. Another third are shorebirds and long-legged wading birds that feed on mudflats, beaches, and in adjacent shallow waters. About 25 species (mostly waterfowl) are associated predominantly with shallow waters of sloughs and lagoons, and about 35 species find their primary home in marshes.

Tropical and southern hemisphere estuaries often have relatively fewer bird species than their more northern counterparts. The overall richness of the estuarine avifauna in South America, for instance, is similar to that for North America, despite there being over three times the total number of bird species on the continent. Cumulatively, South American estuaries have about 120 species, plus an additional 80 or so that are associated with tropical mangrove forests (Stotz et al. 1996). The largest differences are the number of species associated with deeper water, with only about half the number of these species in South America compared to North America; there are also fewer waterfowl associated with shallow waters in South America. The major groups that show lower diversity in South America are diving piscivores (such as loons and grebes) and diving invertebrate feeders (sea and diving ducks). Moving south, estuaries have fewer species of migrants from high latitude boreal forest or tundra breeding areas, without a counterbalancing increase in the number of austral migrants. This pattern arises, in part, because there are simply fewer species from habitats equivalent to those in the far north, which is likely due to the lack of land mass at high latitudes in the southern hemisphere.

13.2.1.2 Mammals

The diversity of estuarine mammals is considerably lower than that of birds. Of the 400 species of North American mammals, for example, approximately 50 are regularly found in estuarine habitats (Greenberg and Maldonado 2006). Most estuarine mammals are associated with tidal marshes, with a small number of coastal marine mammals occurring in open water estuarine habitats. The diversity of tidal marsh mammals is lower in South America (26 species) and lower still in Europe (15 species). Once again, these numbers are augmented by a few truly aquatic species, and a total of 25–50 species of estuarine mammals can be found on each of these continents.

Terrestrial saltmarsh mammals are dominated by small species, for example members of the orders Rodentia (rodents, such as rats and mice) and Eulipotyphla (shrews) in the Northern Hemisphere. Large rodents, such as cabybaras *Hydrochoerus hydrochaeris* and beavers, also sometimes occur in estuaries. Beavers, especially, can play a role in "engineering" the structure of the habitat through the creation of pools, especially in fresher water upstream reaches of tidal systems (Hood 2012). In most parts of the world, estuarine marshes are also occupied by a few members of the order Carnivora (mustelids such as otters and mink, bears, skunks, raccoons (*Procyon lotor*), canids such as coyotes (*Canis latrans*) and foxes, etc.) and occasional species from other orders, such as the Artiodactyla (deer, pigs). Most of these species are also found in a wide variety of non-estuarine habitats. Marine species that routinely use estuaries include seals and sea lions (also in the Carnivora) and members of the Sirenia (manatees, dugongs), and Cetacea (dolphins, porpoises, and whales). Domesticated livestock such as cattle (*Bos taurus*) and sheep (*Ovis aries*) are also sometimes kept on tidal marshes, where the highly productive vegetation can provide good foraging conditions.

13.2.1.3 Reptiles

Globally, there are estimated to be approximately 10,000 reptile species, of which only about 100 are truly marine (Rasmussen et al. 2011). Various other species are also able to live in the brackish conditions of estuarine habitats and one estimate suggests that about 1.4% of reptile species have made incursions into coastal saline habitats (Heatwole 1999). Most of the reptiles found in estuaries are snakes, with at least 14 species in the family Hydrophiinae (sea snakes) associated with open estuary or mangrove habitats in southern Asia. Several species from other snake families also occur in saline or brackish habitats, such as coastal marshes, including cottonmouths (*Agkistrodon piscivorus*)—a type of pit viper—in North America, and members of the Acrochordidae and Honalopsidae families in Asia and Australasia.

The second major group of marine reptiles are the sea turtles in the Chelonioidea, which live the bulk of their lives in open ocean, but come to land to nest on beaches. Some of these species also use estuarine habitats. A few turtle species that are generally found in fresh water can also survive brackish conditions, for example members of the Geoemydidae (mangrove terrapins, painted terrapins) and Carettochelyidae (pig-nosed turtle) families from southern Asia and Australasia. Diamondback terrapins (*Malaclemys terrapin*)—found along the North American East and Gulf Coasts—are unique among turtles in that they occur only in brackish waters and are found primarily in estuarine habitats (Figure 13.1). Several crocodile species can also survive in coastal waters, in particular the saltwater crocodile (*Crocodylus porosus*), which is found in estuaries from Australia to India, and which also ventures out into open ocean.

13.2.1.4 Amphibians

Although amphibians are generally considered to be unable to tolerate saline conditions, there is growing evidence that dozens of species sometimes live in salty habitat (Hopkins and Brodie 2015)—with representatives from all three orders: Anura (frogs, toads), Caudata (salamanders), and Gymnophiona (caecilians). Many of these species can be found in tidal creeks, salt marshes, coastal pools, mangroves, and other estuarine habitats, with examples from all continents except Antarctica where no amphibians occur. Amphibian species richness at any particular location, however, is generally very low to absent, and there are few amphibian species for which estuarine habitats are especially important. Perhaps the best-known estuarine frogs are members of the Asian genus *Fejervarya*, especially the crab-eating frog (*F. cancrivora*), which occurs from eastern India through southeast Asia to China, Taiwan, and the Philippines, and can tolerate saline water by adjusting the electrolyte levels in its tissues.

13.2.2 Diversity

Although wildlife species richness is often low in estuarine ecosystems, the variety of body plans and diversity of feeding adaptations can be high. Estuarine mammals include representatives of many of the major orders and foraging guilds, and range in size from shrews (5 g) to small baleen whales (45,000 kg). Estuaries also are home to some of the world's smallest birds (hummingbirds in mangrove swamps and wrens in salt marshes), and to some of the largest flying species, including the Dalmatian pelican (*Pelicanus crispus*, 9 kg) of Eurasia. Similarly, the reptile fauna includes all of the major morphotypes, including snakes, lizards, turtles, and crocodiles, and species that range in size from small snakes and lizards up to the largest living reptile, the saltwater crocodile, the males of which can grow to 6 m in length.

Morphological variation extends beyond just size, and is often tied to behavioral differences among species (Table 13.1). In birds, for example, variation in bill morphology is considerable (Rubega 2000), and is often linked to behaviors that allow species to exploit different foraging niches. This diversity is especially pronounced within the Charadriiformes (Figure 13.2), which includes plovers and sandpipers that use their short bills to pick up invertebrates. Larger sandpipers often have much longer beaks that allow them to probe the substrate. For example, dowitchers use a rapid sewing-machine-like motion to push deep into wet mud in search of hidden prey, and curlews use their long, curved bills to probe deep into invertebrate burrows in search of crabs and other large invertebrates. Yet other shorebirds—tringine sandpipers and avocets—feed by plucking invertebrates or small fish from the shallows as they wade in the water.

The Charadriiformes also includes terns, which plunge into deeper water—often from a hovering position—to capture fish with their pointed, dagger-like bills. Skimmers have a different approach to the same problem, and fly just above the surface dragging their elongated lower jaw through the water until it hits a prey item and snaps shut on it. Also, in the Charadriiformes are the most versatile of estuarine birds, the gulls, which use their stout, medium-length beaks not only to glean

TABLE 13.1 Morphological diversity and other traits of estuarine birds are often linked to different foraging behaviors.

Foraging behavior	Prey	Advantageous traits	Examples
Grazing	Saltmarsh plants, submerged aquatic vegetation	Broad, strong bills, often with sharp edges to clip vegetation	Ducks, geese, swans
Foliage gleaning	Small invertebrates, seeds	Short, usually thin bills (stouter among seed-eaters)	Wrens, sparrows, warblers, spinetails
Surface picking	Small invertebrates	Short- to medium-length bills	Plovers, sandpipers
Tactile mud probing	Invertebrates	Long, narrow bills	Sandpipers, rails
Burrow probing	Larger invertebrates, including crabs	Very long, down-curved bills	Curlews, ibis
Wading and grabbing	Large invertebrates and fish	Long, spear-shaped bill; long legs	Herons, egrets, storks
Filter feeding	Small invertebrates	Rows of comb-like lamellae lining the bill	Flamingos, ducks
Water sweeping	Small invertebrates; small fish	Long bill with spatulate or upturned tip	Spoonbills, avocets
Swimming in circles	Small aquatic invertebrates	Needle-shaped bill, lobbed flaps along toes	Phalaropes
Surface diving	Fish, mollusks in water column or on bottom substrate	Webbed feet, set back near tail; dense bones and other features to reduce buoyancy	Cormorants, loons, grebes, some ducks
Plunge diving	Fish	Sharp, dagger-like bill or talons; high aspect ratio wings for hovering	Terns, kingfishers, osprey
Aerial piracy	Prey captured by other species	Capacity for powerful and maneuverable flight	Jaegers, skuas, gulls, frigatebirds
Scavenging	Dead animal carcasses, garbage	Good olfactory and/or visual acuity; capacity to eat rotting substances	Gulls, vultures, some eagles, crows

FIGURE 13.2 Diversity comes in various forms, as shown by this collection of shorebirds, feeding on the eggs of horseshoe crabs (pictured bottom right) on a beach in Rhode Island, USA. The four small birds are semipalmated sandpipers and have relatively short beaks that are primarily used to pick small food items off the surface (see second bird from right). The medium-sized bird in the center is a short-billed dowitcher, which have beaks more than twice the length of their heads that can be used to probe deep in the mud. Willets (top, left) are even larger, with medium-length, stouter bills that are used for a mix of pecking, probing, and manipulating varied prey such as clams, worms, small crabs, and fish. *Source*: Photo by C. Elphick.

french fries from the sidewalk, but also to rip apart scavenged fish and other detritus, harass other species into dropping their prey (a piratic behavior known as kleptoparasitism), and many other imaginative modes of foraging.

Beyond the variety of feeding behaviors within this one order, there are estuarine birds that graze on saltmarsh plants and submerged aquatic vegetation (waterfowl); that use small thin bills to glean insects from marsh plants (wrens, warblers, and spinetails) or stouter short bills to eat seeds and hard-shelled invertebrates such as snails (sparrows, finches); that use their long legs to wade in the water and sharp bills to grab fish (herons and egrets); that sweep their bills back and forth through the water to stir up prey (spoonbills); that filter-feed by using their tongue to pump water through the comb-like lamellae that line the sides of their beak (flamingos; Zweers et al. 1995); that swim in circles so tight that they generate tiny upwellings to bring small prey close to the surface for capture (phalaropes; Rubega and Obst 1993); and on and on.

13.2.3 Endemism and Specialization

Sitting at the interface between ocean and land, estuarine wildlife communities comprise species representative of both. Open water species tend to be found in a variety of coastal habitats, while species found in estuarine marshes and on mudflats often also occur in similar freshwater habitats away from the coast. Relatively few wildlife species are found only in estuaries, but the patterns that are seen among these saltmarsh endemic species are instructive about the nature of endemism and habitat specialization.

The relatively small number of endemic species is likely a product of the limited extent of estuarine habitats and the wide range of constantly changing conditions that estuaries impose on their inhabitants (Figure 13.3). Compared to many habitats, estuaries occupy little of the Earth's surface, and what habitat there is tends to occur in small, often isolated, patches associated with river mouths. Natural habitat fragmentation of this type might be expected to create many opportunities for allopatric speciation—the genetic divergence of isolated populations into new species. The small size of most estuaries relative to the typical home ranges of vertebrate species, however, means that individual patches are rarely big enough to support very large populations of a species. Since small populations are vulnerable to local extinction, estuarine systems are rarely large enough to support populations that remain viable over long periods unless there is exchange of individuals with other patches. Such immigration, however, generates gene flow among populations, removing the genetic isolation needed for species to form.

A species' ability to survive in estuaries also benefits from the ease with which it can cope with constantly changing conditions—in particular, the daily flux of the tide brings cycles of inundation and drying, and the flow of fresh water into the ocean brings a wide salinity (and sometimes temperature) gradient. Generalist species, those inherently capable of living in diverse conditions, are thus most likely to endure estuarine life. In other words, the behaviors that enable wildlife species to persist in estuaries—good dispersal capacity and the ability to tolerate a wide range of conditions—are also those that make the evolution of distinct species less likely.

Despite these opposing forces, specialization does occur (Correll et al. 2016; Figure 13.4). Population divergence also has been found in saltmarsh populations of at least 25 wildlife species (Greenberg et al. 2006) and in mangrove populations of another 69 species (Luther and Greenberg 2009). All but one of the salt marsh examples come from North America and are associated with places where theory suggests that population divergence is most likely (Figure 13.5). For example, along the eastern and southern coasts of the United States, the world's largest expanse of saltmarsh habitat connects large estuaries, such as Delaware and Chesapeake Bays, with the protected waters behind barrier islands. This region has more endemic saltmarsh wildlife species than any other region (e.g., diamondback terrapin, saltmarsh sparrow (*Ammospiza caudacuta*); Greenberg et al. 2006; Figure 13.1), presumably because it is able to better support isolated populations that are large enough to persist long enough to diverge. Similarly, much of the saltmarsh habitat in western North America is concentrated in San Francisco Bay, which is home to the endemic salt marsh harvest mouse (*Reithrodontomys raviventris*; Figure 13.1) and

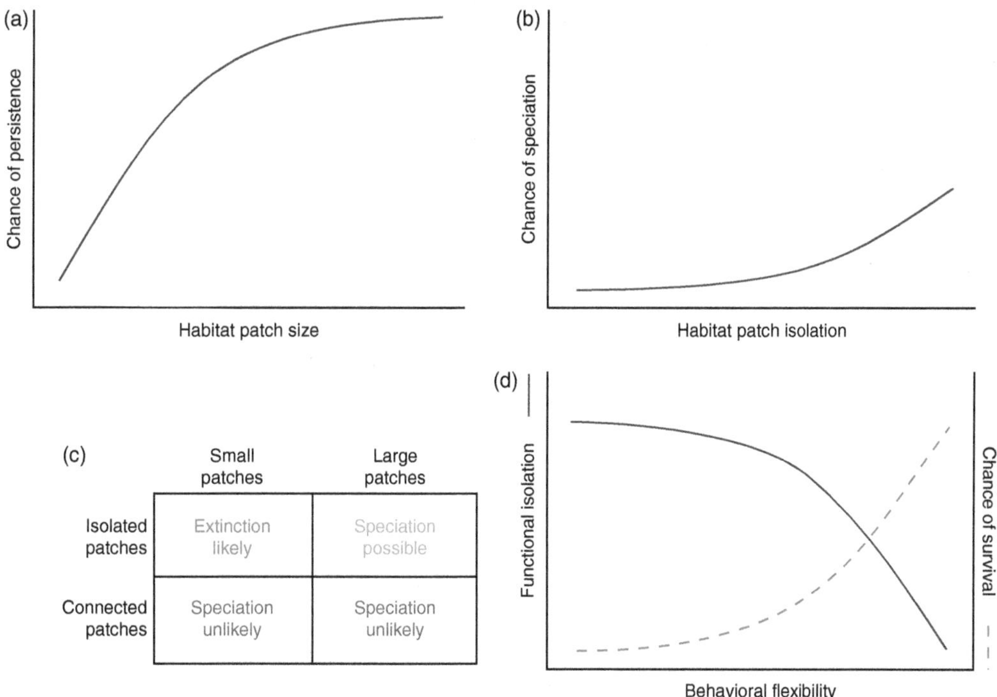

FIGURE 13.3 Features of habitats and species both affect the chance that a population will go extinct, and the chance that new species will arise. (a) Bigger habitat patches can usually support larger populations, increasing the chance of a population persisting. (b) Speciation is more likely when patches are isolated from each other because the separated populations are more likely to evolve differences that eventually accumulate to result in separate species. Consequently, (c) speciation is unlikely both in well-connected patches because populations do not become separated, and in small, isolated patches because populations often do not persist long enough for differences to accumulate. The evolution of new species is thus most likely when large areas of habitat are isolated from each other. In most places, estuaries are isolated, but also quite small, so wildlife species that are specialized to estuarine habitats do not arise. The chance of speciation among estuarine wildlife is further reduced because (d) generalist species, which are capable of adjusting to a variety of conditions, are most likely to be able both to survive in the constantly-changing environmental conditions that estuaries exhibit and to move between estuaries—because they can live in the intervening areas—thereby making estuaries functionally less isolated for those species. Exceptions involve species found in the large salt marshes of coastal North America and the extensive areas of mangrove in Asia and Australasia.

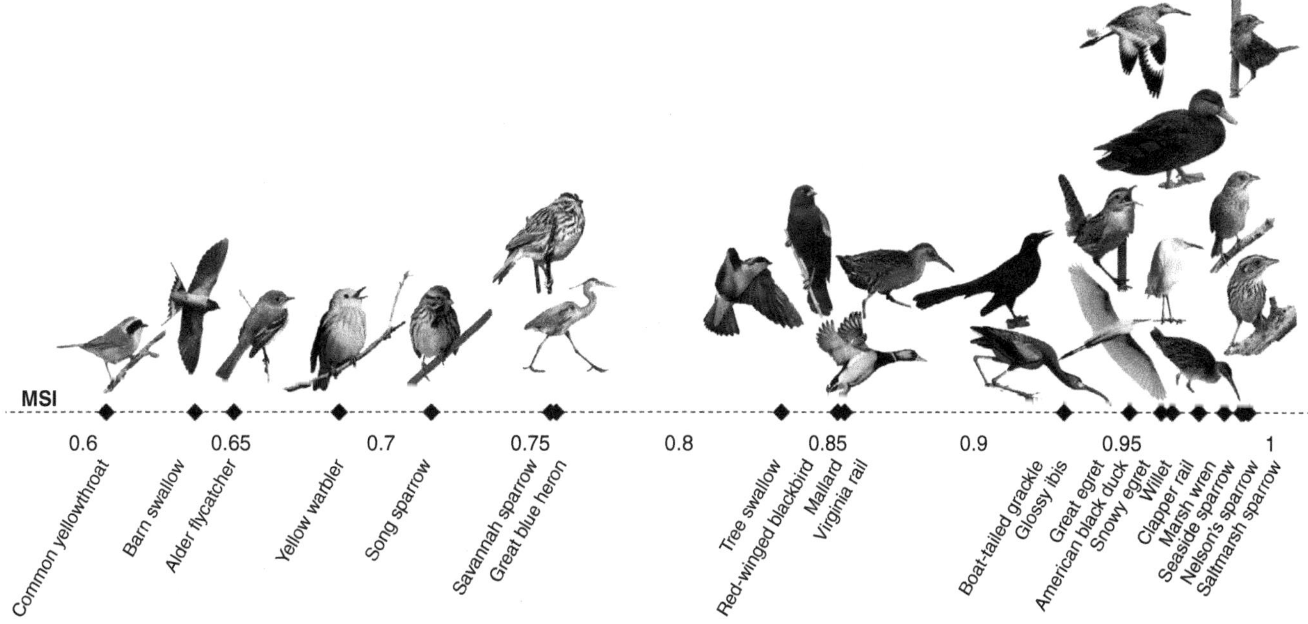

FIGURE 13.4 Many species are found in estuaries, but they vary considerably in how much they depend on them. This figure shows bird species recorded in estuary surveys from the northeastern USA, arrayed according to a measure of how much they specialize in using salt marshes (a value of 1 = entirely restricted to this habitat). *Source*: Correll et al. (2016), John Wiley & Sons/CC BY 3.0.

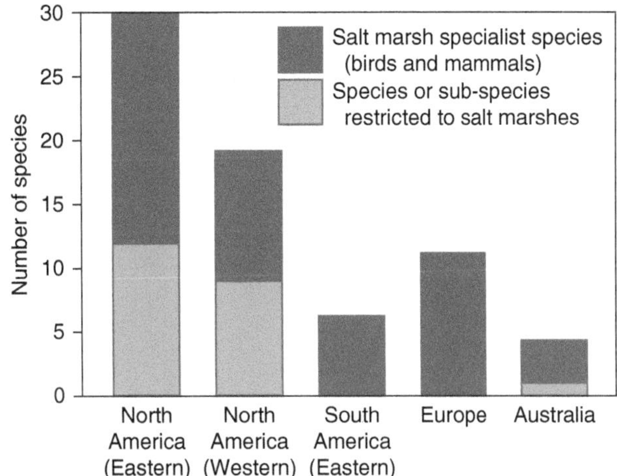

FIGURE 13.5 The diversity of specialist saltmarsh species varies among regions. Bars show the number of nesting bird species and mammals in each region; portion shaded in green represents the number of species that are restricted to salt marshes, or that have a subspecies that is. *Source*: Data from Greenberg and Maldonado (2006).

several endemic wildlife subspecies (Greenberg and Maldonado 2006). Endemic mangrove species, in contrast, are most likely to be found in Asia and Australasia, where mangrove habitat is extensive (Luther and Greenberg 2009).

Most examples of population divergence in saltmarsh species identified by Greenberg and Maldonado (2006) involve subspecies—an intermediate step in the creation of new species. The high proportion of endemic subspecies illustrates the potential for evolutionary innovation, but the small number that have become distinct species supports the idea that diverging lineages rarely survive long enough for speciation to complete. The ephemeral nature of estuarine habitats in geological time may contribute to this situation—as ice sheets encroach into temperate regions and sea levels rise and fall over millennia, the locations and sizes of marshes are constantly changing (see Chapter 6), potentially exacerbating the rate of local extinction compared to more stable habitats. Mangroves—which extend over a larger area globally and are less affected by ice ages—have more endemics, and a larger proportion that have become full species (Luther and Greenberg 2009).

When population divergence does occur, it often leads to the same suite of morphological traits, a convergence that suggests that these features may have adaptive value (Greenberg and Maldonado 2006). For example, several species of sparrows in the family Passerellidae have populations that have independently colonized salt marshes. Compared to their upland counterparts, these saltmarsh birds have greyer plumage with more black markings, larger body size, and proportionately bigger bills. Darker coloration is also found in populations of other birds, small mammals, and reptiles that are isolated in salt marshes, leading to the suggestion that it helps provide camouflage against the dark soils and turbid waters of tidal marshes (Grinnell 1913; Grenier and Greenberg 2005). Recent genomic studies have also identified differences between saltmarsh populations and their inland counterparts in genes known to be related to osmoregulatory salt tolerance and bill morphology (Walsh et al. 2019).

13.3 Temporal and Spatial Patterns in Biomass and Productivity

Biomass and productivity of estuarine wildlife varies considerably in time and space. In part, this is a response to variation in production at lower trophic levels, as is the case for many ecosystems. Places and times at which primary production is high tend to result in conditions that can support an abundance of wildlife because there is food that can sustain many individuals. Abiotic conditions also play an important role, in part mediated through their effects on primary productivity. But physical factors—especially the ebb and flow tides—also place constraints on the ability of many wildlife species to access food or reproduce successfully. Consequently, wildlife biomass and productivity may be less tightly linked to that at lower trophic levels than in some other ecosystems, especially on short time scales.

13.3.1 Time

Wildlife abundance, and consequently biomass, varies considerably over various temporal scales. Unlike many other estuarine organisms, however, this variation is driven more by the movement of individuals in and out of estuaries, than by patterns of production. Perhaps the greatest paradox of estuarine wildlife ecology is that the high productivity of estuaries can support a large biomass of vertebrates, but that the difficulty of living in estuaries means that there are times when wildlife numbers are low. This pattern plays out on an annual time scale, with relatively few wildlife species using estuaries for breeding, and on much shorter time scales as access to resources and the capacity to take advantage of them fluctuates.

13.3.1.1 Annual Cycles Wildlife abundance in estuaries can vary dramatically throughout the annual cycle. Bird richness and abundance, for example, can be extremely high during non-breeding periods, but much lower during the breeding season. Less than a third of the bird species that regularly occur in North American estuaries nest in or adjacent to estuaries, with the rest breeding in areas that are often quite different and distant. Many estuarine bird species breed on Arctic tundra or in boreal forest habitats, often hundreds of kilometers from the coast. Similarly, Lefebvre and Poulin (2000) reported that of the dozens of species that regularly inhabit the Neotropical mangrove swamps they surveyed, relatively few species bred there.

Even more striking than the changes in species richness across the year, are the fluctuations in abundance. Although most estuarine bird species breed elsewhere, the numbers of birds that use estuaries during nonbreeding periods can be spectacular, and estuaries can provide crucial habitat for many migratory species. In some cases, a single estuary can support a large portion of a species' global population during migration. For example, the Copper River Delta in southern Alaska supports up to 80% of the world's 3–4 million western sandpipers (*Calidris mauri*; Bishop et al. 2000). The two largest North American estuaries, San Francisco and Chesapeake Bays, support over two million waterfowl during winter (Perry and Deller 1995; Takekawa et al. 2006). Similar nonbreeding concentrations of migratory birds are found in estuaries worldwide, including western Europe, along the coasts of China and southern Asia, and in northern Australia. The concentrating effect of estuaries is not restricted to birds. Some estuaries, for example, the mouth of the St. Lawrence River in eastern Canada, support large numbers of foraging marine mammals at certain times of year, including many seals, beluga whales (*Delphinapterus leucas*), and blue whales (*Balaenoptera musculus)*.

Sometimes the importance of an estuary to a species is not reflected by huge numbers. For example, the spoon-billed sandpiper (*Calidris pygmaea)* is one of the world's rarest species, with no more than a few hundred individuals remaining. Although these birds breed on the tundra of far eastern Russia, they migrate south after breeding, using estuaries along the coasts of China and southern Asia, with a substantial proportion of the world population spending several months in the Gulf of Mottama, Myanmar, along with vast flocks of other species (Figure 13.1) before heading north again (Aung et al. 2020).

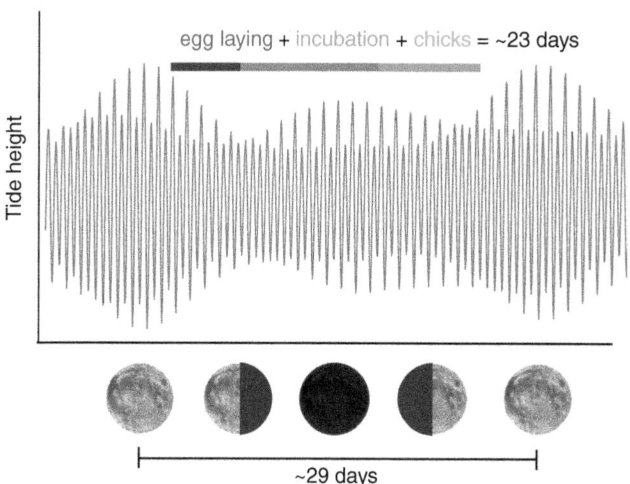

FIGURE 13.6 Some wildlife species must fit their breeding activities within the constraints of the tidal cycle. The top figure shows the daily tidal cycle and the monthly fluctuations in peak tide height, with high spring tides coincident with the new and full moon. In the saltmarsh sparrow, the highest of these spring tides typically puts nests at risk of flooding. Consequently, egg laying, incubation of eggs, and the raising of nestlings—which takes at least 23 days—must fit within this cycle. If the birds start nesting too soon, eggs are washed out of the nest; if they start too late, the chicks drown. *Source:* Top figure by C. Field; photo by J. Mielcarek.

13.3.1.2 Tidal Cycles

The migratory flow of animals in and out of estuaries over the course of the year is repeated at small scales as wildlife species respond to daily and monthly tidal cycles. The most obvious effects occur in the intertidal zone where areas are regularly exposed or inundated with water. For example, tidal mudflats provide a temporarily available habitat for foraging shorebirds. At high tide, however, birds abandon inundated areas, either to feed elsewhere or to gather at roost sites where they sleep and rest. Sanderlings (*Calidris alba)*, small sandpipers often seen chasing the waves on beaches, will move between mudflats and outer coastal beaches during high tide (Connors et al. 1981). Other shorebirds may fly to adjacent farm fields, where they probe the soil and search the grass for invertebrates, or simply roost in large flocks and wait until the tide goes down. Tidal flows also influence the foraging of birds in deeper water as well. Cormorants have been found to forage more during ebb tides, which can concentrate benthic fish and make them easier to detect (Anderson et al. 2004).

Tidal flooding also affects the productivity and behavior of breeding species. The highest spring tides, which are tied to the lunar cycle, and high tides that coincide with unusual storm surge, can wipe out entire colonies of bird nests. Tidal marsh birds and mammals regularly elevate their nests compared to their nontidal marsh relatives to reduce the risk of flooding, but there are limits to how high a species can build without making the nest visible to predators (Johnston 1957; Hunter et al. 2016). In species that regularly experience losses due to flooding, rapid renesting can improve the chance of completing nesting during the narrow window before the next set of monthly high tides (Figure 13.6). These tidal disruptions of the nesting cycle have the effect of synchronizing the behavior of some species. For example, singing rates in saltmarsh sparrows increase considerably in the days immediately after an especially high tide has caused extensive nest flooding, as males seek females that are newly available to mate (Shriver et al. 2007).

Tidal marsh mammals can also be affected by flooding as they must either walk or swim to the upland edge of the estuary, find high spots that remain free of flooding, or take refuge in floating debris. If the flooding is too severe, nests can be flooded and offspring drowned. Predators also use high tides as a cue and focus their attention on high spots in the marsh where small rodents and birds are concentrated. Large storm surges, such as those associated with hurricanes, can largely wipe out populations of small saltmarsh mammals, but the high reproductive rates of these species allow rapid recovery (Longenecker et al. 2018).

13.3.1.3 Daily Light Cycle

Wildlife activity is also affected by the day–night cycle, with some species more active at night, and others by day. For mobile species, these patterns affect the distribution of individuals within an estuary, with diurnal species often congregating at night to roost and nocturnal species often congregating by day. Shorebirds, gulls, and terns gather on elevated islands, sandbars, or sections of marsh, while herons and ibis may roost in tree-covered islands or sections of mangrove. These roosts often contain thousands of birds in a small area, and are frequently used on a daily basis, often year after year. In addition to concentrating vertebrate biomass, this behavior can also result in the movement of large quantities of nutrients such as nitrogen and phosphorus, in the form of the birds' feces, into the sites used for roosting, sometimes affecting the productivity of other organisms (Palomo et al. 1999).

Light and tidal cycles both occur on a daily time-scale, but are not synchronized, meaning that the availability of foraging habitat does not consistently occur at day or night. Species capable of finding food regardless of the light level, therefore, have an advantage when food is limited—and many species that are largely diurnal in nontidal wetlands, switch to also feeding at night when tidal conditions limit access to food during the day (Robert et al. 1998).

13.3.2 Space

Variation in wildlife abundance occurs at several spatial scales, and as the previous section shows, is closely linked to temporal variation. At large geographic scales, wildlife abundance varies considerably, with millions of birds using some estuaries at times, and very few in other estuaries or at other times. Because many wildlife species are migratory and use estuaries primarily during nonbreeding periods, spatial patterns of use are greatly influenced by proximity to the breeding grounds and time of year. The most abundant estuarine wildlife are birds that breed in boreal and tundra habitats in the northern hemisphere. Consequently, late in the northern summer, after breeding is over, biomass of estuarine birds is concentrated in high latitude estuaries in North America and Eurasia. As migration proceeds, these birds gradually move south and abundance builds throughout the autumn in the estuaries of eastern Asia, western Europe, and North America.

The onset of winter in the northern hemisphere makes the northernmost estuaries much less suitable for birds, especially as marshes and mudflats freeze over and food becomes less accessible. At this time, large numbers of birds move into the tropics and to southern hemisphere estuaries. This movement is less pronounced in regions with warmer coastal waters, as indicated by the much greater abundance of birds in western European estuaries during the northern winter, than in estuaries at similar latitudes in the western Atlantic. As conditions get colder, higher latitude estuaries also become increasingly dominated by waterfowl species that feed largely on plants, and by species that forage in open water habitats as species with other feeding strategies migrate to warmer latitudes. For example, shorebirds, which primarily eat invertebrates that become much less available during the cold, are prone to travel farther south and become more dominant in tropical and south temperate regions during the austral summer.

At smaller spatial scales, wildlife biomass also varies among habitats within estuaries. Breeding activities tend to be concentrated in salt marshes, on higher elevation sandbars and islands, and on beaches where elevation protects offspring with limited mobility from tidal flooding. Colonial waterbirds and harbor seals (*Phoca vitulina*) can gather in high densities in more open habitats, while other species of birds and small mammals usually breed in salt marshes, and turtles often nest in sandy dunes. In mangroves, the taller vegetation provides more options for breeding animals, and the smaller tidal fluctuations of many tropical areas further lessen the influence of the tides.

Offshore, the calm open waters of large estuaries are sought out by some species of cetaceans, such as the beluga and gray whales (*Eschrichtius robustus*). Marine mammal distributions within estuaries can vary seasonally in relation to the presence of food—for example, in Cook Inlet, Alaska, USA, belugas will concentrate at certain times of year in areas where anadromous fish runs provide abundant prey (Hobbs et al. 2005). Birds also move around in relation to food. Shorebirds forage on mudflats when they can, moving to saltmarsh pools and other habitats at high tide. Waterfowl often concentrate in marshes, or in areas with submerged plants such as eel grass (*Zostera spp.*,) and graze on the vegetation. Again, as water levels rise and make the vegetation less accessible, birds move to new areas to feed or settle in groups offshore to roost while waiting for the tide to fall again.

13.4 Species Interactions Affecting Productivity and Community Composition

Patterns of occurrence and abundance of wildlife species in estuaries are affected by a wide range of species interactions including herbivory, predation, and competition. In many cases, estuaries have been important locations for key studies that have informed our understanding of common ecological phenomena.

13.4.1 Herbivory

Herbivory occurs in various places within estuaries, and involves a variety of wildlife species. Saltmarsh vegetation is highly productive and often assumed to provide an abundant resource for herbivorous wildlife ranging from small rodents to ducks and geese to large mammals. Rodents are among the most abundant vertebrate herbivores in many estuarine marshes. For example, voles in the genus *Microtus* are common in marshes along both coasts of North America, and feed primarily on grass stems along well-defined runways in high-elevation areas of marsh. The creation of runways, which facilitates the movement of voles through the thick, grassy marsh vegetation requires that voles harvest the entire aboveground portion of the plant, which can begin a process of very local erosion.

Large mammalian herbivores are generally rare in salt marshes, although browsers such as deer, can be found at the upper edges of some marsh systems, and some species also occur in lower-elevation marshes. For example, sika deer *(Cervus nippon)*—native to Asia, but introduced in Europe and the United States during the 1800–1900s—can greatly reduce vegetative cover and increase open areas in *Spartina* marshes (Hannaford et al. 2006). Large grazing animals were probably much more common in estuaries during the Pleistocene, thousands of years ago, and may have had a much greater influence structuring past ecological communities (Koch et al. 1998; Levin et al. 2002). Today, much of what we know about the effects of large mammalian grazers on salt marshes comes from studies of livestock or feral species, which show a wide variety of effects on the soil, vegetation, and invertebrate communities (Davidson et al. 2017; Figure 13.7).

The most influential wildlife herbivores in many estuarine marshes today, however, are often birds, not mammals. Many waterfowl are herbivorous, and geese have received particular attention as many northern hemisphere species have undergone rapid population increases in recent decades. Although growth in goose populations is generally attributed to the increased availability of agricultural foods, especially waste grain, a consequence is that large numbers—often many thousands—congregate in salt marshes at certain times of year. New growth in late spring often results in highly nutritious saltmarsh vegetation at a time when geese are preparing to migrate to Arctic breeding grounds. Goose grazing can have complicated effects on marsh vegetation. Moderate levels of grazing can increase primary production, but where population increases lead to grazer densities that exceed the habitat's carrying capacity, overgrazing can occur, causing vegetation loss and negative repercussions for other species that use the marshes (Buij et al. 2017).

Wildlife also forage in intertidal and subtidal areas, where they eat algae and seagrass. Seagrass beds—dominated by eel grass in temperate regions and turtle grass (*Thalassia* spp.) in more tropical areas—cover the shallow substrate of estuaries throughout the world (Chapter 5). Green turtles *(Chelonia mydas)*, and manatees and dugongs, use subtidal seagrass beds in the tropics, and ducks, geese, swans, and coots use them in temperate regions (Thayer et al. 1984; Bortolus et al. 1998). The birds and turtles generally function as grazers, cropping the palatable portions of the plants. In contrast, herds of dugong (*Dugong dugon)* plough trails through seagrass beds, digging up the plants from the rhizomes and causing more extensive, longer-term disturbances.

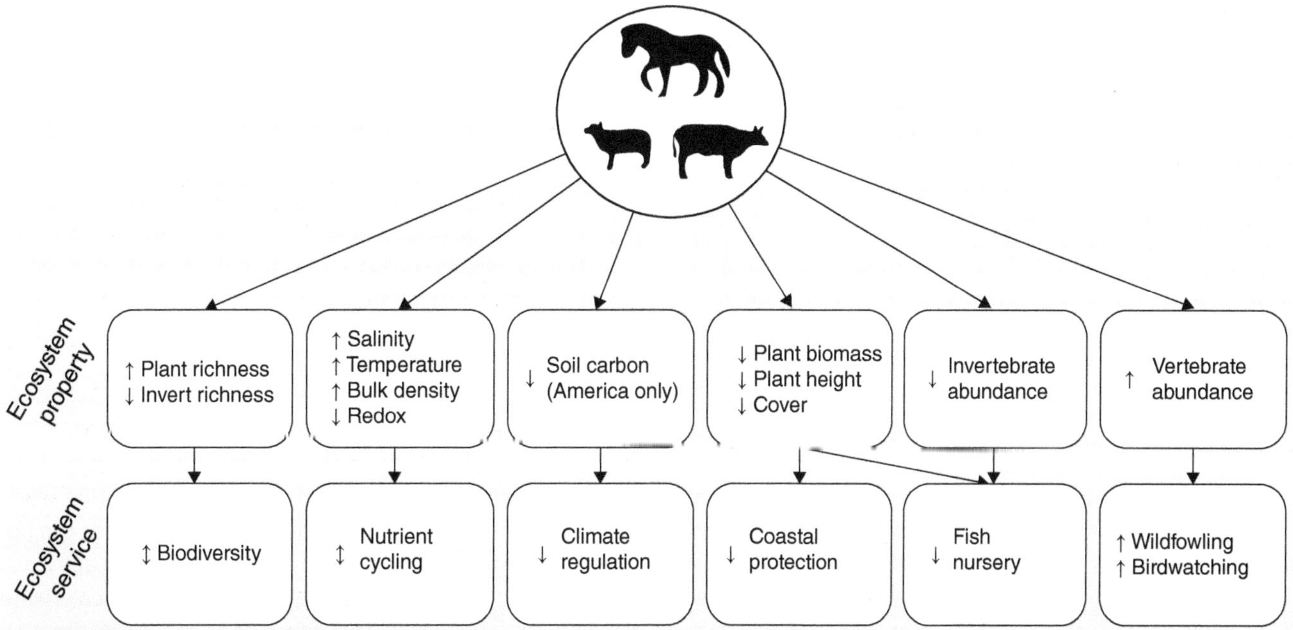

FIGURE 13.7 Grazing alters many aspects of an estuarine marsh ecosystem as shown in this summary of the documented relationships between livestock grazing and ecosystem properties, and the resulting implications for the ecosystem services that coastal marshes provide. This figure is based on a synthesis of studies from around the world, but with most coming from Europe and North America and involving cattle or sheep grazing. *Source*: Modified from Davidson et al. (2017).

Submersed aquatic vegetation (SAV) is an important food source for many waterfowl species in temperate regions. Eel grass and widgeon grass (*Ruppia maritima*) are dominant species in many waterfowl diets. Chesapeake Bay, which lies in the middle of the Atlantic American Flyway, is a globally important area for waterfowl and historically supported expansive SAV meadows estimated at 250,000 ha in area (Arnold et al. 2017). The loss of SAV along the US Atlantic Coast in the 1930s, caused by a wasting disease pathogen *Labyrinthula* sp., led to declines in many waterfowl species (Orth and Moore 1983; Perry and Deller 1996; Perry et al. 2007). Wintering populations of brant (*Bernicula branta;* also called brent goose), and ducks such as redheads (*Aythya americana*), canvasbacks (*A. valisineria*), greater scaup (*A. marila*), and lesser scaup (*A. affinis*), declined from the 1950s to the early 1990s, when populations started to recover (Perry et al. 2007). Recent nutrient management in Chesapeake Bay has been directly linked to SAV recovery (Lefcheck et al. 2018), and surveys suggest both that SAV recovery goals are on track to being met and that redhead populations have increased substantially since the 1980s.

As in salt marshes, the nature of herbivory in subtidal portions of estuaries has undoubtedly changed as populations of large vertebrate herbivores have been reduced, usually due to human exploitation. For example, green turtle populations in the Caribbean are estimated to have declined 15- to 30-fold in the past 300 years and dugong populations may be as much as 200 times smaller than a century ago (Jackson et al. 2001).

A final type of grazing that occurs in estuaries has only been recognized relatively recently. Small shorebirds, such as the western sandpiper, will feed on the biofilm comprised of microbes and organic detritus that forms on the surface of intertidal mudflats (Kuwae et al. 2008). This biofilm can provide half of the daily energy budget of these birds, which—when present at high densities—potentially deplete biofilm and compete with invertebrates that also use it as a food supply (Kuwae et al. 2012). The extent to which some shorebirds use this form of foraging also indicates the potential importance of biofilm for supporting their energy needs during migration.

In some cases, estuarine herbivores can function as **ecological engineers**—species that drastically change the basic structure of a habitat. Two examples are the muskrat (*Ondatra zibethicus*) and the snow goose (*Chen caerulescens*). Common in fresh or brackish tidal marshes, muskrats reduce the cover by dominant vegetation types through herbivory, den construction, and channel formation, which creates open water, and causes the local topography to be hummocky in nature (Errington 1961). Snow goose flocks locally denude areas of salt marsh by pulling plants out by the rhizomes, which inhibits regrowth. Long-term effects can include depletion of soil nutrients and erosion, and reduction of invertebrate diversity and abundance (Sherfy and Kirkpatrick 2003). Usually the effect is local and may serve to create some habitat heterogeneity. With rapidly increasing goose populations, though, the damage has been extensive in some areas.

13.4.2 Predation

Predation by wildlife can have ecological effects throughout estuarine food webs. When shorebirds form huge flocks during nonbreeding periods they eat a considerable biomass of mudflat invertebrates. In some cases, though not all, this predation causes depletion of prey stocks—with potential **"top-down" effects** that regulate the biomass of species at lower trophic levels (Colwell 2010; Mathot et al. 2018). Even high levels of predation, however, can have little effect on the standing numbers of prey if the reproductive rate of the prey is high enough to counter-balance the mortality caused by predation.

Predation by fish-eating wildlife has been the subject of much management concern, especially in places where commercially important fish are found. For instance, in the northwestern United States, increasing populations of Caspian terns (*Hydroprogne caspia*) and double-crested cormorants (*Phalacrocorax auritus*) have caused controversy because of their potential effects on endangered salmon and trout populations. There is little doubt that large numbers of young fish are eaten by these birds as the fish migrate between freshwater breeding grounds and the ocean. As a result, considerable effort has been put into reducing the numbers of birds nesting in the Columbia River estuary. It is less clear, however, whether this predation is **additive**, with more young fish dying when terns and cormorants are present, or **compensatory**, a scenario in which the fish that survive the predatory birds would die anyway, due to other causes. Recent studies of tagged steelhead trout (*Oncorhynchus mykiss*) have not resolved the debate, with some evidence supporting compensatory mortality (Haeseker et al. 2020), while other research suggests an additive effect (Payton et al. 2020).

Estuarine wildlife not only affect the biomass and species composition of organisms at lower trophic levels, but can also be subject to predation themselves. Large concentrations of shorebirds and waterfowl often attract predatory birds such as falcons, hawks, and owls, which can kill a substantial number of the waterbirds present at a site (Page and Whitacre 1975). These predators can show **functional responses**, in which they adjust their behavior to increase their focus on a given prey type as prey density changes. For example, studies in Scotland have shown that predation patterns vary in a way that suggests that hawks increase their focus on shorebirds as this prey type becomes more numerous (Whitfield 2003a, b).

Both the physical environment and prey behavior can influence predation levels (Dekker and Ydenberg 2004). The stage of the tidal cycle, for example, affects the success of falcons hunting shorebirds, with peak success occurring when birds are concentrated close to shore. When shorebirds can forage on wide open mudflats, it is hard for predators to approach without being seen. As tides rise, and birds get pushed toward the estuary's edge, predators can hide in the vegetation and sneak up on their prey. Shorebirds mitigate risk by seeking roost sites that afford a good view—inland farm fields, sandspits, unvegetated offshore islands, etc. If no good options are

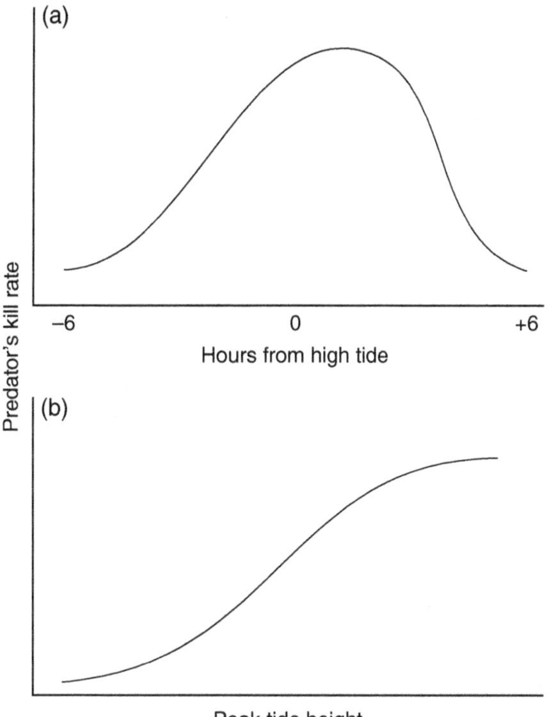

FIGURE 13.8 Falcons often kill more shorebirds (a) during the hours right after a high tide, and (b) after higher tides. These are times when their prey may be especially hungry, having not fed during the high tide period, and be more vulnerable as they spend less time watching for predators. *Source*: Figure based on data from Dekker and Ydenberg (2004).

available, they might head offshore and spend a few hours flying around. The downside to this last approach is that, when the tide recedes and birds can return to the mudflats, their need to replace lost energy can leave them so focused on feeding that they become less vigilant about watching for predators (Figure 13.8). In some cases, predators are thought to have shaped not only the daily behavior of their prey, but also larger scale behavioral patterns, such as choices about where to make migratory stops, and how long to stay at particular sites (Ydenberg et al. 2004).

13.4.3 Competition

Competition occurs when individuals negatively impact the ability of others to garner the resources necessary for survival or reproduction, either by reducing the abundance of available resources (**exploitation competition**) or by reducing the ability of others to access them (**interference competition**). For wildlife, interference competition often takes the form of territorial behavior, in which an animal invests energy into obtaining access to space and then defending the space against others. Whether to defend a territory, like many behavioral decisions, depends on the costs and benefits of doing so. If resources are concentrated and predictable, then defense may be worthwhile. In contrast, if they are spread out over a large area, or if areas of high abundance are ephemeral and hard to track, the difficulty of defending resources may simply not be worth it. Consequently, the type of competition that occurs often depends on the spatial and temporal distribution of resources. Similarly, the degree to which others are also competing for the same resources can be important—if there are many competitors, it may simply become too energetically expensive to spend time defending a territory (Figure 13.9).

Territorial relationships are common among estuarine birds, but vary considerably in their stability and the nature of the resource in the space being defended. Some animals, including many songbirds in salt marshes and mangrove forests, maintain "all-purpose" territories, where all of the needs of an individual, pair, or family group are met. Other animals maintain territories only for specific purposes. In areas where nesting habitat is limited, birds may defend tiny nesting territories and feed in larger shared feeding areas. For example, colonial breeders such as gulls and terns may concentrate on islands within estuaries, where it is hard for mammalian predators to reach the nests. In these colonies, hundreds of birds may pack in together, with each pair defending just the immediate area around their nest (Kharitonov and Siegel-Causey 1988). At the same time, because these birds feed largely on fish and invertebrates, which often have ephemeral and unpredictable spatial distributions, they generally do not defend feeding territories.

Species foraging on mudflats are more likely to be territorial than those feeding in the water, because the spatial and temporal distributions of prey are likely to be more changeable and harder to defend in the water. Even on mudflats, however, the concentration of prey can be so low, or the densities of feeding birds so high, that territoriality is not worthwhile. The assessment of whether to be territorial, however, can be complicated and even within large aggregations there may be some individuals that defend feeding territories and some that do not (Tripp and Collazo 1997).

Competition generally occurs only when resources are limited, and there are circumstances when this is not the case. For instance, the formation of stable breeding pairs in most songbirds is generally linked to the need for both parents to defend a territory, and to provide sufficient parental care to raise young. But the high productivity of coastal marshes (Chapter 6) is thought to have allowed saltmarsh sparrows to develop an unusual breeding system, in which birds neither form pairs nor defend territories, because food is sufficiently plentiful for females to raise young without a male (Post and Greenlaw 1982). In contrast, Eurasian oystercatcher (*Haematopus ostralegus)* pairs need both marsh habitat for nesting and adjacent mudflat for feeding their young, and vigorously defend territories that encompass both (Ens et al. 1992). In this system, reproductive success is so affected that birds return year after year to defend the same spot, those that lack high quality sites are constantly vying to upgrade, and a sizable portion of the population forgoes breeding altogether because they cannot obtain a territory (Heg et al. 2000).

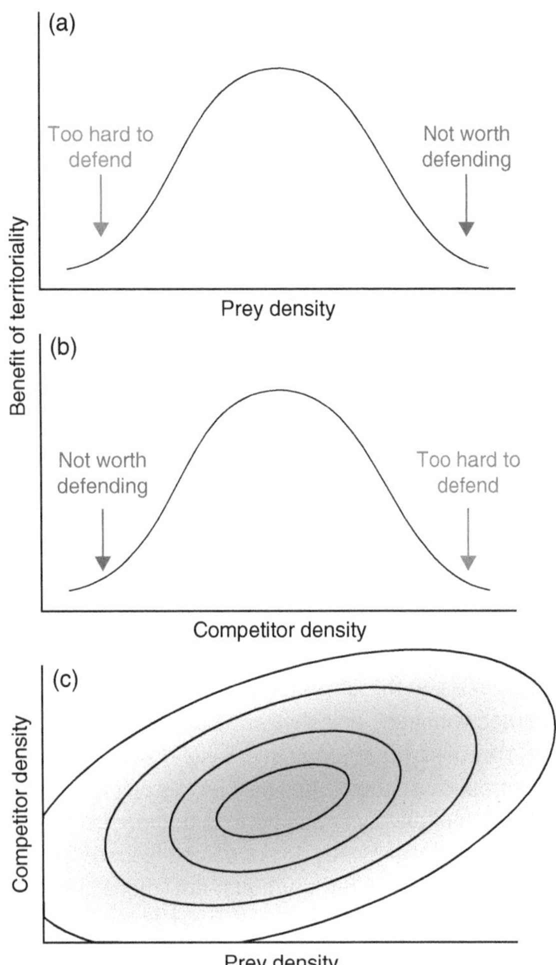

FIGURE 13.9 Whether to defend a territory can depend on both the abundance and distribution of the resource being exploited, and on the other competitors. For example, (a) if prey density is low it may be hard to defend a big enough area for territoriality to be beneficial, but if prey density is very high there may be so much food that defense is unnecessary. (b) If there are few competitors, then there's also no reason to defend. But, if competitor density gets high enough, territorial behavior may be so time consuming that individuals are unable to spend enough time on feeding and other essential activities. These two ideas can be brought together as illustrated in (c), where the green shading represents a greater benefit of territoriality and the oval lines each represent sets of points where the benefits of territoriality are equal. By determining these relationships for a given system, it is possible to determine the conditions when territoriality is more or less likely to occur.

Competitive interactions often affect different individuals within a population differently. Dominance is often associated with sex or age, which leads to segregation of habitat. For example, male bar-tailed godwits (*Limosa lapponica)* in subtropical areas of Australia feed primarily in seagrass beds, the superior habitat, whereas females are more broadly distributed across habitats (Zharikov and Skilleter 2002). In another shorebird example, common redshank (*Tringa totanus)* competition for territories causes juveniles to use riskier sites than adults, increasing their chance of being preyed upon by Eurasian sparrowhawks *(Accipiter nisus;* Whitfield 2003b). This second example illustrates how even abstract concepts like the ability to feed in a "**predator-free space**" can become the basis for competition.

Effects of competition on abundance patterns also occur at spatial scales larger than individual estuaries. For instance, as the population of black-tailed godwits (*Limosa limosa)* wintering in Great Britain has increased in recent decades, their numbers in estuaries with abundant food have remained stable. In contrast, numbers in estuaries with less food have increased, even though birds are less likely to survive the winter at these sites (Gill et al. 2001a). This pattern suggests that godwits are increasingly forced to occupy suboptimal wintering areas. The isotopic chemical signatures of feathers grown on the breeding grounds have also revealed that the birds that winter in these lower quality estuaries are often the same ones that use suboptimal habitat for nesting (Gunnarsson et al. 2005). In some cases, this displacement of individuals into poorer habitat can have broader consequences for a species. Among highly migratory species, for example, there is often a premium on migrating as short a distance as possible. In many species, this can translate into spatial segregation playing out over very large areas, with distinct latitudinal separation of males and females, and of yearling and older birds (Mathot et al. 2007).

13.4.4 Disease

The dense gatherings of waterbirds in estuaries provide high potential for the spread of disease, and waterbirds—many of which migrate over long distances—have been associated with the potential to spread and suffer from avian influenza and other pathogens (Takekawa et al. 2010; Brown et al. 2013). At the same time, Piersma (1997) has proposed that Arctic-nesting shorebirds may gather in these habitats during non-breeding periods because there may be less exposure to pathogens than in other habitats. Reconciling these ideas is hard because there is limited information on parasite occurrence in estuarine wildlife, or on their effects. Studies of blood parasites provide support for Piersma's suggestion, with a low incidence in shorebird species that predominantly use estuarine habitats (Soares et al. 2016). A similar difference between marine and freshwater habitats, however, was not found in a study of the helminth faunas of charadriiform birds (Gutiérrez et al. 2017). These parasitic worms did, however, have higher species richness in host species that use both marine and freshwater habitats, and in those that had more diverse diets—perhaps because generalist habits increase the chance of exposure to parasites.

Current evidence suggests that, although much avian influenza transmission occurs within domestic poultry populations, wild waterbirds—especially Anseriformes and some Charadriiformes—play a role in the long-distance transmission and emergence of new strains (Bahl et al. 2016; Verhagen et al. 2021). Outbreaks have occurred in wild populations, and concerns raised about their potential consequences for rarer species, but documented cases have not generally occurred in estuarine systems. Survival of the virus in the environment is lower in warmer, more saline, conditions, but much remains to be understood about its persistence in natural settings (Stallknecht and Brown 2009). Despite the many unknowns about avian influenza, our knowledge of most other wildlife diseases in estuarine systems is even more limited.

13.5 Human Effects on Estuarine Wildlife

Estuarine wildlife populations and their contributions to estuarine ecosystems are affected by human activities in a wide variety of ways. In most organisms, habitat change has likely had the greatest effects historically, but the spread of invasive species, pollution, direct human interactions through harvest and disturbance, and marsh management have all played a role. Looking to the future, however, the effects of climate change, in particular via sea level rise and coastal flooding, are likely to dominate.

13.5.1 Sea Level Rise and Climate Change

Climate change and the associated increases in sea level and storm surge (Chapter 19), affect estuarine wildlife both by changing the amount of habitat available to particular species and by altering conditions within habitats in ways that have consequences for reproduction and survival. As sea levels rise, habitats gradually change from upland into marsh, from high-elevation to low-elevation marsh, from low-elevation marsh to mudflat, and from mudflat to open water. Greater storm intensity can also increase erosion of beaches and marshes, and redistribute sediment in ways that alter the amounts of different habitats. Wildlife species that use any of these habitats are therefore potentially affected. If habitats can progressively move inland, transitioning in sequence from one type to another, species can accommodate changes. The rate of sea level rise, however, is increasingly too fast for this to happen, habitats also transition at different rates, and the upland edge of an estuary may preclude inland movement due to topography or human infrastructure. Consequently, habitat for many wildlife species is declining as certain habitats are "squeezed" into smaller and smaller areas (Field et al. 2016). At the same time, other habitats may become more common; for example, big storms may deposit large volumes of sand on top of marsh grasses, which can create nesting habitat for beach nesting species (Walker et al. 2019).

Although there is much uncertainty about the severity of habitat change due to sea level rise, projections for the northeastern United States suggest that several bird species found only in salt marshes could lose up to 85% of their nesting habitat by 2050 (Klingbeil et al. 2021). Similar analyses have suggested that sea level rise will result in substantial losses of mud flats in several estuaries that are important for migratory shorebirds in North America (Galbraith et al. 2002). In some places, however, the loss of vegetated marsh might result in an increase in mudflat, so understanding the net consequences for wildlife is often complicated.

Even when habitat is not completely altered, climate change can cause its suitability for certain species to decline. The close link between tidal flooding and breeding success in some species (Figure 13.6) make them especially vulnerable to sea level rise. The saltmarsh sparrow, for example, declined by over 70% between the 1990s and 2010s (Correll et al. 2017). Studies show that insufficient young are being produced at sites across the species range to balance the mortality rate, and population projections suggest a high likelihood of extinction at some point in the middle of the twenty-first century (Field et al. 2017). Although this species is particularly vulnerable, the same general pattern likely applies to other species that depend primarily on changing estuarine habitats.

Although rising sea level is likely to be the primary way in which climate change manifests effects on estuarine wildlife populations, it is not the only one. For instance, Maclean et al. (2008) found a northerly shift in the winter distributions of many shorebird species in western Europe, consistent with birds adjusting behavior in relation to warming temperatures and more of them spending the nonbreeding period closer to their breeding grounds. This shift has potential repercussions, not only for the affected wildlife species, but maybe also for their prey and predators. In addition to causing pole-ward range shifts in many species, warming may also affect productivity patterns within estuaries, which could alter the prey base for wildlife. Climate change may also increase the likelihood of extreme events such as heat waves and increased storm intensity. Other less obvious changes, such as shifts in prevailing wind direction which might alter flooding patterns, could have as-yet unforeseen repercussions.

13.5.2 Habitat Conversion and Fragmentation

Estuarine habitat change is caused not only by sea level rise. Historically, land development has played the dominant role, and this continues to be true in some parts of the world. In southern New England, in the United States, for example, half of all saltmarsh habitat has been lost to development since European colonization, and 90% of what remains has been modified by ditching (Bertness et al. 2009). In the countries bordering the Yellow Sea, land reclamation for development and farming caused the loss of a quarter of all inter-tidal mudflats between the 1980s and 2000s (Murray et al. 2014), and similar losses have been seen farther south along the coast of China (Sun et al. 2015). The long-term consequences of estuarine habitat loss are often not well known because of a lack of historical information on wildlife population sizes and how they have changed. The magnitude of loss, however, suggests that they must have resulted in considerable declines for many species.

Direct habitat conversion is not the sole reason for population effects. Declines in the abundance of saltmarsh breeding birds in the northeastern United States, for example, are related to the likelihood that a marsh has experienced a restriction of tidal flow due to the building of roads, railway lines, or other barriers, across marshes (Correll et al. 2017). Even when culverts allow some tidal exchange to continue through these restrictions, the upstream habitat may begin to change, becoming less saline, drier, and with less movement of sediment into the restricted areas. These changes can cause shifts in the vegetation community—and, in some cases, the incursion of non-native plant species—making the habitat less suitable for saltmarsh-dependent wildlife. Even when normal tidal flow is restored to an area, the resulting vegetation may be different from that found in downstream reference marshes, due to changes in elevation or soil properties (Burdick et al. 1997; Mossman et al. 2012). As a result, these restored areas also may lack some of the wildlife species that are most dependent on salt marshes (Elphick et al. 2015).

The subdivision of marshes into small fragments of habitat can also have negative consequences. For instance, declines of salt marsh harvest mice in San Francisco Bay have been attributed not only to habitat loss, though the filling and modification of marshes for human uses, but also to fragmentation (Smith et al. 2018). Small habitat patches contain smaller, more isolated, populations, which are affected in two ways. First, their small size makes extinction more likely, simply because small populations are prone to the loss of genetic diversity and to being wiped out by chance events. Second, the isolation that comes with habitat fragmentation reduces movement among populations, hampering gene flow, which exacerbates the loss of genetic diversity, and reduces the potential for immigrants to reinvigorate a population following losses. These are universal problems caused by fragmentation, and likely affect many estuarine species.

Loss of other estuarine habitats also has important consequences for wildlife. For instance, the extensive mudflat losses in coastal Asia, which are often coupled with pollution and the spread of introduced species, have coincided with declines of many migratory bird species that move annually between their Arctic breeding grounds and Australasia. Declines are particularly acute among those species that stop to refuel on the mudflats of the Yellow Sea coast (Studds et al. 2017).

Although habitat conversion is usually detrimental to most native species, there are cases where some species benefit. Salt marshes and mangroves, for example, have been impounded for salt production for hundreds of years in estuaries from California to southern Europe to Asia. These evaporation ponds replace natural vegetation, directly affecting some species, but can also generate considerable invertebrate production, creating foraging, nesting, and roosting habitat for large numbers of waterbirds of many species (Warnock et al. 2002). In some places, loss of natural wetlands has been so extensive that wildlife species have come to depend on these artificial habitats, and the loss of salt pans is now seen as detrimental (Green et al. 2015).

13.5.3 Invasive Species

Estuarine tidal marshes are susceptible to shifts in the dominant vegetation that come with the spread of species introduced by humans from elsewhere in the world. The effects of such changes on estuarine vertebrates are often not well documented, but research increasingly suggests a general pattern in which introduced plant species spread at the expense of wildlife that depends on tidal marshes. For example, low-elevation salt marsh and mudflats along the coasts of Tasmania, New Zealand, northern Europe, and continents bordering the Pacific Ocean are being invaded by *Spartina* species, mostly *S. alterniflora*, which often hybridizes with the local native species. In Europe, the spread into mudflats of *S. anglica*—the hybrid product of the European *S. maritima* and the North American *S. alterniflora*—reduces their suitability for foraging shorebirds (Goss-Custard and Moser 1988).

Similarly, at the upper edge of marshes along the Atlantic Coast of North America, a non-native strain of the common reed *Phragmites australis* is spreading into areas once dominated by *Juncus*, *Spartina*, and shrubs such as *Iva frutescens* and *Baccharis halimifolia*, favoring generalist bird species over tidal marsh specialists (Benoit and Askins 1999). Ironically, studies in China show that in areas where the introduced *S. alterniflora* is replacing native *P. australis*, bird diversity is also lower in the native vegetation (Gan et al. 2009).

Estuarine invertebrate assemblages, the prey base for much estuarine wildlife, are also prone to local invasions, many of which are initiated by animals transported on ships (Chapter 12). Although invasive species now comprise up to 90% of the benthic invertebrates in some estuaries (Cohen and Carlton 1995), the impact of these faunal shifts on wildlife populations are often poorly known. There is good reason to expect that there will be effects though. In San Francisco Bay, for instance, lesser scaup switched from feeding on a native clam, *Macoma balthica*, to an introduced species from Asia, *Potamocorbula amurensis*, as the latter became more dominant in the estuary. Researchers discovered that the switch likely increased the rate at which scaup are able to obtain energy while foraging because the introduced species tends to be easier for these ducks to capture and eat (Richman and Lovvorn 2004). A potential downside, however, is that the invasive species concentrates greater quantities of potentially harmful contaminants than the native species.

Wildlife species are also sometimes introduced from one region to another, and some of these species colonize estuarine habitats. Among the most widely transported species are Norway rats (*Rattus norvegicus)* and black rats (*R. rattus)*, which both prey upon and are prey for native wildlife. In general, though, these introductions are not known to have had major effects. Other introduced mammals, however, have had more serious consequences. Grazing by sika deer in southern England, for instance, alters both the plant and invertebrate composition of coastal marshes (Diaz et al. 2005). Nutria (*Myocastor coypus;* called coypu outside North America) have also been introduced from South America to coastal wetlands around the world, where they often cause overgrazing and marsh loss. Eradication efforts in some countries have been successful, and a major program in Louisiana, USA—where over 2.5 million have been killed since 2002—has been judged an important component of marsh protection efforts (Sasser et al. 2017).

13.5.4 Pollution

A wide variety of pollutants potentially affect estuarine wildlife. **Non-point source pollutants** are perhaps least obvious, because they enter estuaries diffusely from throughout a watershed rather than from a specific location, yet can have some of the most serious system-wide effects. For example, increases in nutrients due to run-off from upstream agricultural or residential areas can have indirect effects on the habitat and food resources on which wildlife species depend. Among these effects are the loss of SAV, which can reduce the food for estuarine waterfowl. Similarly, nutrient runoff can affect coastal marsh vegetation and encourage encroachment by invasive species, especially when coupled with shoreline development (Silliman and Bertness 2004). Nutrient loading, together with increased water temperature, has also been implicated in an increase in harmful algae blooms (Landsburg 2002; Chapter 4). These blooms can produce neurotoxins, which can cause illness or death in wildlife species that consume fish that have fed on the affected algae, and can cause oxygen depletion, which can reduce wildlife prey populations. Mortality linked to algal blooms has been reported for a variety of species, including various birds, cetaceans, and manatees (Landsburg 2002).

The location of estuaries at large river mouths exacerbates the risks of pollution from oil and chemical spills. These sites are often home to major ports with a lot of ship traffic and, in some places, chemical and oil refineries. Although major spills receive most media attention, the majority of the chemical and oil pollutants that enter the environment do so via small accidents and routine activities. The enclosed nature of many estuaries adds to the potential for harmful effects, as pollutants are less likely to be rapidly diluted than would be the case in the open ocean. Most importantly, because estuaries often concentrate populations of wildlife species, the consequences of spills can be severe because large numbers of individuals, often representing significant proportions of global populations, can be affected simultaneously.

Predatory wildlife are vulnerable to the effects of chemical contaminants that bioaccumulate, becoming most concentrated in species at higher trophic levels in a food web. Mercury contamination, for example, can cause acute toxicity in birds and mammals, and in some of the species they prey upon (Takekawa et al. 2006). High-level exposure damages the central nervous system, while lower-level exposure affects reproduction in vertebrates (Wolfe et al. 1998). Synthetic chemicals such as polychlorinated biphenyls (PCBs) also accumulate and harm wildlife species. PCB contamination in coastal marshes in Georgia, USA, was found to be associated with a high frequency of strand breakage in the DNA of clapper rails (*Rallus crepitans*) (Novak et al. 2006). This level of contamination has human health ramifications because the rails are game species commonly hunted and consumed by people. Broad-spectrum pesticides are also often applied to tidal marshes for mosquito control, especially in situations where the habitat is perceived to be a source of vectors for emerging diseases (Takekawa et al. 2006), potentially affecting both the prey base and the wildlife directly. Although this wide range of pollutants can affect individuals of many wildlife species, it is generally less clear whether those effects translate into population-level declines.

13.5.5 Hunting and Disturbance

Estuaries have been important locations for human settlement for thousands of years (Ames 1999; Ercolano and Carballo 2005). The concentrations of abundant waterbirds—especially during the colder months when other foods are less available—are likely to have been a major attraction, and human activity affects estuarine wildlife both directly through hunting, and indirectly when people are foraging for other resources. For millennia, these effects are likely to have been small, but over time people developed increasingly sophisticated hunting methods, culminating in nets that could capture entire flocks, and guns that could kill dozens of birds in a single shot (McPherson 1897; Hornaday 1913). These changes allowed a shift from subsistence to commercial hunting and, when combined with widespread habitat conversion, were sufficient to cause substantial population declines.

During the early twentieth century, the impacts of these depredations started to become apparent, and hunting is now closely regulated in many parts of the world. In some places, however, estuarine bird populations remain vulnerable to hunting. Increasingly, poverty is the underlying cause of detrimental hunting as people hunt to support their families. Microfinance schemes can sometimes resolve this problem, via small infusions of capital that help people develop alternative ways to obtain food and make a living. For example, nongovernmental conservation organizations in Myanmar have provided grants to local people to buy fishing equipment, with a resulting decline in the hunting of shorebirds, including the critically endangered spoon-billed sandpiper (see picture at start of this chapter).

Hunting can also have indirect effects on estuarine wildlife when human overharvest reduces prey availability. Declines of semipalmated sandpipers (*Calidris pusilla*) in the Bay of Fundy, eastern Canada, for example, have been linked to harvest of worms for fish bait from the Bay's mudflats (Shepherd and Boates 1999). Overharvest of horseshoe crabs (*Limulus polyphemus*) along the Atlantic Coast of the United States—both for bait, and for biomedical use by the pharmaceutical industry—has also resulted in substantial declines in the abundance of crab eggs, which are important food for migratory shorebirds (Mizrahi and Peters 2009).

Another way that harvest can indirectly affect estuarine wildlife arises when human activity generates disturbance that restricts a species' access to food or other resources. This can occur, for instance, when people are harvesting shellfish or other invertebrates on mudflats. How serious such disturbance is depends on the circumstances. If the disturbed wildlife can simply move elsewhere to feed, or can switch to feeding at night, or some other time when people are not present, they may be able to obtain sufficient food despite the disturbance (Gill et al. 2001b). Evaluating the consequences of human activity is thus a lot more complex than simply determining whether people displace wildlife.

13.5.6 Management of Estuaries for Wildlife

When considering human effects on nature, it is easy to dwell on the ways that people have caused harm, but many management actions are designed to benefit wildlife. Initially, much of this work was linked to hunting, though often through efforts to reverse population declines and ensure a sustainable harvest. For example, in the eastern United States, there is a long history of building impoundments in estuarine marshes. In southeastern states, many of these impoundments were created for rice farming, but are now managed to attract waterfowl. Farther north, providing waterfowl habitat was often the original goal. In some cases, management is now focused on reversing historic changes by opening up impounded areas and restoring tidal flow to benefit declining estuarine species (Brawley et al. 1998).

Fire has also long been used as a wildlife management tool. Managed burns are designed to release nutrients and promote new vegetation growth that benefits target species, such as geese, other waterfowl, and muskrats. Whether these benefits are achieved, however, is sometimes unclear, and may depend on local conditions and the way that fire is used (Flores et al. 2011). In South America, fire is also used to enhance marsh habitat for livestock. In one study in Argentina, however, marsh burning was found to favor generalist grassland species at the expense of rarer specialists (Isacch et al. 2004).

Given concerns over the rate of sea level rise, management is increasingly focused as much on ensuring that estuarine habitats persist as it is on the nature of the habitat itself. Current efforts can be divided into those focused on maintaining or restoring existing habitats and those designed to create new estuarine habitat. Intertidal habitats—mudflats and marshes—are the focus of most attention. Maintaining these habitats in place generally requires actions that increase the chance that they will be able to build elevation, which often involves ensuring an adequate supply of sediment for the vegetation to trap (Chapter 6). One way to do this is to remove upstream barriers, such as dams, that hold back sediment that would otherwise flow down rivers to estuaries. Often this is not possible, however, and managers have begun to experiment with ways to redistribute sediment within estuaries. Thin-layer sediment placement, for example, involves taking sediment—often material that has been dredged from channels or harbors nearby—and spraying it onto a marsh. The alternative to building elevation to match the rate of sea level rise, is to encourage the inland movement of marshes. Where this is impossible, managers are beginning to explore ways to speed up the process, for example, by cutting down the forest at the marsh-upland boundary.

As these examples suggest, management for estuarine wildlife often equates to estuarine management. Although estuaries support many forms of biological diversity and provide a wide variety of ecosystem services, the wildlife that is found in them frequently determines decisions about how estuaries are protected and managed. Whether those decisions are centered around providing waterfowl hunting opportunities, reducing disturbance of shorebirds feeding on mudflats, restoring saltmarsh for endangered birds and mammals, or creating ecotourism opportunities for people who want to see whales and seals, wildlife can have tremendous influence on the future of estuarine habitats.

Acknowledgments

This chapter is based on Chapter 14 from the second edition of this book written by RG. That version benefited from the assistance of A. Joyner, L. Harding, B. Peierls, and K. Rossignol and was supported by the North Carolina Department of Environment and Natural Resources, the North Carolina Sea Grant Program, the University of North Carolina Water Resources Research Institute, the US Department of Agriculture-NRI-CRGP, the US EPA-STAR-EaGLe Program, the NOAA/CSCOR Northern Gulf of Mexico Hypoxia Assessment Program, the National Science Foundation, and the NOAA Center for Sponsored Coastal Ocean Research. Remote sensing imagery was provided by NASA's SeaWiFS and MODIS Program. This new version was prepared by CSE and WGS and benefited from guidance on specific topics from N. Hill, E. Jockusch, S. Knutie, D. Prosser, K. Schwenk, K. Smith, J. Takekawa, and K. Wells. L. Aoki, B. Crump, B. Olsen, and N. Weston provided helpful comments on earlier drafts of the manuscript. We also thank M. Correll, C. Field, J. Mielcarek, R. Robinson (British Trust for Ornithology), K. Smith, and M. Szantyr for help with figures, or donating the use of their photographs.

Study Questions

Multiple Choice

1. Which of the following statements about estuarine wildlife are accurate? (Select all that apply)
   a. Few wildlife species are found only in estuaries.
   b. Many wildlife species can be extremely numerous in estuaries, but only at certain times of year.
   c. Wildlife species are most abundant in estuaries when they congregate to breed.
   d. One of the big challenges that estuarine species face is that environmental conditions are constantly changing at various temporal scales.
2. Which of the following are influential sources of herbivory in modern estuaries? (Select all that apply)
   a. Dugong feeding on tropical seagrass beds.
   b. Sandpiper foraging on mudflat biofilms.
   c. Large native mammals feeding on salt marsh grasses.
   d. Rodents grazing on salt marsh grasses.
3. Which of the following statements identify effects of climate change on estuarine wildlife? (Select all that apply)
   a. Species ranges contracting toward the equator as temperature rises.
   b. Increased nest flooding of tidal marsh breeding species.
   c. Loss of mudflats, beaches, and marsh habitat due to increased erosion.
   d. Loss of all estuarine habitats as sea level rises and storms redistribute sediment.
4. Which of the following statements about territoriality are accurate? (Select all that apply)
   a. Territoriality is more likely when resources are rare are widely dispersed.
   b. Territoriality is more likely when resources are extremely abundant everywhere.
   c. Territoriality is more likely when resources are concentrated into small areas.
   d. Territoriality is more likely when there are a limited number of places where it is possible to stay safe from predators.
5. Which of the following statements about endemic estuarine species are accurate? (Select all that apply)
   a. Estuarine endemics tend to be most common in areas with large expanses of estuarine habitat, such as the coasts of eastern North America.
   b. Any isolated patch of estuarine habitat has a high chance of containing endemic species.
   c. Many estuarine species are generalists, which makes them especially likely to become endemics.
   d. The constantly changing environmental conditions in estuaries increases the potential for endemic species to arise.
6. Estuarine birds feed using a wide variety of methods. Which of the following are real examples? (Select all that apply)
   a. Filter feeding by using the tongue to pump water in and out of the mouth.
   b. Swimming in tight circles to create upwellings.
   c. Grazing on plants underwater.
   d. Using kleptoparasitic behaviors to steal food from other birds.

Short Answer

1. How is the term "wildlife" defined, and what are some of the ways that wildlife are important in estuaries?
2. What are the dominant groups of wildlife species in estuaries, and how does wildlife diversity and abundance vary across latitudes and seasons?
3. What factors affect the development of endemic species? Why are there not more endemic estuarine wildlife species, why are they concentrated in certain parts of the world, and why are endemic subspecies so much more common than endemic species?
4. How do tides affect the use of estuarine habitats by wildlife, and what are the survival and reproduction consequences for species?
5. How have patterns of salt marsh grazing by estuarine wildlife changed over time? Consider patterns at different time scales in your answer.
6. Describe top-down effects, by which the foraging of estuarine wildlife affects other species within estuarine ecosystems? Provide at least three distinct examples. Why might predation fail to have a top-down effect?

7. What are the two main types of competition? Provide examples that involve estuarine wildlife for each.
8. Under what circumstances would one expect a species to become territorial? Why might this change over time, or differ among individuals within a population?
9. What are the main ways in which human activities are changing estuarine wildlife habitat? Provide examples of both positive and negative changes.
10. Given a specific example of a way in which human activities are increasing the risk of extinction for an estuarine wildlife species?

References

Ames, K.M. (1999). Economic prehistory of the northern British Columbia coast. *Arctic Anthropology* 35: 68–87.

Anderson, C.D., Roby, D.D., and Collis, K. (2004). Foraging patterns of male and female Double-crested Cormorants nesting in the Columbia River estuary. *Canadian Journal of Zoology-Revue Canadienne De Zoologie* 82: 541–554.

Arnold, T.M., Zimmerman, R.C., Engelhardt, K.A.M., and Stevenson, J.C. (2017). Twenty-first century climate change and submerged aquatic vegetation in a temperate estuary: the case of Chesapeake Bay. *Ecosystem Health and Sustainability* 3: 1353283.

Aung, P.-P., Moses, S., Clark, N.A. et al. (2020). Recent changes in the numbers of spoon-billed sandpipers *Calidris pygmaea* wintering on the Upper Gulf of Mottama in Myanmar. *Oryx* 54: 23–29.

Bahl, J., Pham, T.T., Hill, N.J. et al. (2016). Ecosystem interactions underlie the spread of avian influenza A viruses with pandemic potential. *PLoS Pathogens* 12: e1005620.

Benoit, L.K. and Askins, R.A. (1999). Impact of the spread of *Phragmites* on the distribution of birds in Connecticut tidal marshes. *Wetlands* 19: 194–208.

Bertness, M.D., Silliman, B.R., and Holdredge, C. (2009). Shoreline development and the future of New England salt marsh landscapes. In: *Human Impacts on Salt Marshes: A Global Perspective* (ed. B.R. Silliman, E.D. Grosholz and M.D. Bertness), 137–148. University of California Press.

Bishop, M.A., Meyers, P.M., and McNeley, P.F. (2000). A method to estimate migrant shorebird numbers on the Copper River Delta, Alaska. *Journal of Field Ornithology* 71: 627–637.

Bortolus, A., Iribarne, O.O., and Martinez, M.M. (1998). Relationship between waterfowl and the seagrass Ruppia maritima in a southwestern Atlantic coastal lagoon. *Estuaries* 21: 710–717.

Brawley, A.H., Warren, R.S., and Askins, R.A. (1998). Bird use of restoration and reference marshes within the Barn Island Wildlife Management Area, Stonington, Connecticut, USA. *Environmental Management* 22: 625–633.

Brown, V.L., Drake, J.M., Stallknecht, D.E. et al. (2013). Dissecting a wildlife disease hotspot: the impact of multiple host species, environmental transmission and seasonality in migration, breeding and mortality. *Journal of the Royal Society Interface* 10: 20120804.

Buij, R., Melman, T.C.P., Loonen, M.J.J.E., and Fox, A.D. (2017). Balancing ecosystem function, services and disservices resulting from expanding goose populations. *Ambio* 46 (Suppl. 2): S301–S318.

Burdick, D.M., Dionne, M., Boumans, R.M., and Short, F.T. (1997). Ecological responses to tidal restorations of two northern New England salt marshes. *Wetlands Ecology and Management* 4: 129–144.

Cohen, A.N. and Carlton, J.T. (1995). *Nonindigenous Aquatic Species in a United States Estuary: A Case Study of the Biological Invasions of the San Francisco Bay and Delta*. NOAA.

Colwell, M.A. (2010). *Shorebird Ecology, Conservation, and Management*. University of California Press.

Connors, P.G., Myers, J.P., Connors, C.S.W., and Pitelka, F.A. (1981). Interhabitat movements by sanderlings in relation to foraging profitability and the tidal cycle. *Auk* 98: 49–64.

Correll, M.D., Wiest, W.A., Olsen, B.J. et al. (2016). Habitat specialization explains avian persistence in tidal marshes. *Ecosphere* 7 (11): e01506.

Correll, M.D., Wiest, W.A., Hodgman, T.P. et al. (2017). Predictors of specialist avifaunal decline in coastal marshes. *Conservation Biology* 31: 172–182.

Davidson, K.E., Fowler, M.S., Skov, M.W. et al. (2017). Livestock grazing alters multiple ecosystem properties and services in salt marshes: a meta-analysis. *Journal of Applied Ecology* 54: 1395–1405.

Dekker, D. and Ydenberg, R. (2004). Raptor predation on wintering Dunlins in relation to the tidal cycle. *Condor* 106: 415–419.

Diaz, A., Pinn, E., and Hannaford, J. (2005). Ecological impacts of Sika Deer on Poole Harbour saltmarshes. *Proceedings in Marine Science* 7: 175–188.

Elphick, C.S., Meiman, S., and Rubega, M.A. (2015). Tidal-flow restoration provides little nesting habitat for a globally vulnerable saltmarsh bird. *Restoration Ecology* 23: 439–446.

Ens, B.J., Kersten, M., Brenninkmeijer, A., and Hulscher, J.B. (1992). Territory quality, parental effort and reproductive success of oystercatchers (*Haematopus ostralegus*). *Journal of Animal Ecology* 61: 703–715.

Ercolano, B. and Carballo, M.F. (2005). Hunter gatherers at the Río Gallegos estuary mouth, Santa Cruz, Argentina. *Magallania* 33: 109–126.

Errington, P.L. (1961). *Muskrats and Marsh Management*. Stackpole Company.

Field, C.R., Gjerdrum, C., and Elphick, C.S. (2016). Forest resistance to sea-level rise prevents landward migration of tidal marsh. *Biological Conservation* 201: 363–369.

Field, C.R., Bayard, T., Gjerdrum, C. et al. (2017). High-resolution tide projections reveal extinction threshold in response to sea-level rise. *Global Change Biology* 35: 2058–2070.

Flores, C., Bounds, D.L., and Ruby, D.E. (2011). Does prescribed fire benefit wetland vegetation? *Wetlands* 31: 35–44.

Galbraith, H., Jones, R., Park, R. et al. (2002). Global climate change and sea level rise: potential losses of intertidal habitat for shorebirds. *Waterbirds* 25: 173–183.

Gan, X.J., Cai, Y.T., Choi, C.Y. et al. (2009). Potential impacts of invasive *Spartina alterniflora* on spring bird communities at Chongming Dongtan, a Chinese wetland of international importance. *Estuarine Coastal and Shelf Science* 83: 211–218.

Gill, J.A., Norris, K., Potts, P.M. et al. (2001a). The buffer effect and large-scale population regulation in migratory birds. *Nature* 412: 436–438.

Gill, J.A., Norris, K., and Sutherland, W.J. (2001b). The effects of disturbance on habitat use by black-tailed godwits *Limosa limosa*. *Journal of Applied Ecology* 38: 846–856.

Goss-Custard, J.D. and Moser, M.E. (1988). Rates of change in the numbers of Dunlin, *Calidris alpina*, wintering in British estuaries in relation to the spread of *Spartina anglica*. *Journal of Applied Ecology* 25: 95–109.

Green, J.M.H., Sripanomyom, S., Giam, X., and Wilcove, D.S. (2015). The ecology and economics of shorebird conservation in a tropical human-modified landscape. *Journal of Applied Ecology* 52: 1483–1491.

Greenberg, R. and Maldonado, J.E. (2006). Diversity and endemism in tidal-marsh vertebrates. In: *Terrestrial Vertebrates of Tidal Marshes: Ecology, Evolution and Conservation*, *Studies in Avian Biology*, vol. 32 (ed. R. Greenberg, J.E. Maldonado, S. Droege and M.V. McDonald), 32–53. Cooper Ornithological Society.

Greenberg, R., Maldonado, J.E., Droege, S., and McDonald, M.V. (2006). Tidal marshes: a global perspective on the evolution and conservation of their terrestrial vertebrates. *Bioscience* 56: 675–685.

Grenier, J.L. and Greenberg, R. (2005). A biogeographic pattern in sparrow bill morphology: parallel adaptation to tidal marshes. *Evolution* 59: 1588–1595.

Grinnell, J. (1913). Notes on the palustrine fauna of west-central California. *University of California Publications in Zoology* 10: 191–194.

Gunnarsson, T.G., Gill, J.A., Newton, J. et al. (2005). Seasonal matching of habitat quality and fitness in a migratory bird. *Proceedings of the Royal Society B-Biological Sciences* 272: 2319–2323.

Gutiérrez, J.S., Rakhimberdiev, E., Piersma, T., and Thieltges, D.W. (2017). Migration and parasitism: habitat use, not migration distance, influences helminth species richness in Charadriiform birds. *Journal of Biogeography* 44: 1137–1147.

Haeseker, S.L., Scheer, G., and McCann, J. (2020). Avian predation on steelhead is consistent with compensatory mortality. *Journal of Wildlife Management* 84: 1164–1178.

Hannaford, J., Pinn, E.H., and Diaz, A. (2006). The impact of sika deer grazing on the vegetation and infauna of Arne saltmarsh. *Marine Pollution Bulletin* 53: 56–62.

Heatwole, H. (1999). *Sea Snakes*. The New South Wales University Press.

Heg, D., Ens, B.J., Van Der Jeugd, H.P., and Bruinzeel, L.W. (2000). Local dominance and territorial settlement of nonbreeding oystercatchers. *Behaviour* 137: 473–530.

Hobbs, R.C., Laidre, K.L., Vos, D.J. et al. (2005). Movements and area use of belugas, *Delphinapterus leucas*, in a subarctic Alaskan estuary. *Arctic* 58: 331–340.

Hood, W.G. (2012). Beaver in tidal marshes: dam effects on low-tide channel pools and fish use of estuarine habitat. *Wetlands* 32: 401–410.

Hopkins, G.R. and Brodie, E.D. (2015). Occurrence of amphibians in saline habitats: a review and evolutionary perspective. *Herpetological Monographs* 29: 1–27.

Hornaday, W.T. (1913). *Our Vanishing Wild Life-Its Extermination and Preservation*. Charles Scribner's Sons.

Hunter, E.A., Nibbelink, N.P., and Cooper, R.J. (2016). Threat predictability influences seaside sparrow nest site selection when facing trade-offs from predation and flooding. *Animal Behaviour* 120: 135–142.

Isacch, J.P., Holz, S., Ricci, L., and Martinez, M.M. (2004). Post-fire vegetation change and bird use of a salt marsh in coastal Argentina. *Wetlands* 24: 235–243.

Jackson, J.B.C., Kirby, M.X., Berger, W.H. et al. (2001). Historical overfishing and the recent collapse of coastal ecosystems. *Science* 293: 629–638.

Johnston, R.F. (1957). Adaptation of salt marsh mammals to high tides. *Journal of Mammalogy* 38: 529–531.

Kharitonov, S.P. and Siegel-Causey, D. (1988). Colony formation in seabirds. *Current Ornithology* 5: 223–272.

Klingbeil, B.T., Cohen, J.B., Correll, M.D. et al. (2021). High uncertainty over the future of tidal marsh birds under current sea-level rise projections. *Biodiversity and Conservation* 30: 431–443.

Koch, P.L., Hoppe, K.A., and Webb, S.D. (1998). The isotopic ecology of late Pleistocene mammals in North America. I. Florida. *Chemical Geology* 152: 119–138.

Kuwae, T., Beninger, P.G., Decottignies, P. et al. (2008). Biofilm grazing in a higher vertebrate: the western sandpiper *Calidris mauri*. *Ecology* 89: 599–606.

Kuwae, T., Miyoschi, E., Hosokawa, S. et al. (2012). Variable and complex food web structures revealed by exploring missing trophic links between birds and biofilm. *Ecology Letters* 15: 347–356.

Landsburg, J.H. (2002). The effects of harmful algal blooms on aquatic organisms. *Reviews in Fisheries Science* 10: 113–390.

Lefcheck, J.S., Orth, R.J., Dennison, W.C. et al. (2018). Long-term nutrient reductions lead to the unprecedented recovery of a temperate coastal region. *Proceedings of the National Academy of Sciences of the United States of America* 115: 3658–3662.

Lefebvre, G. and Poulin, B. (2000). Determinants of avian diversity in Neotropical mangrove forests. In: *Biodiversity in Wetlands: Assessment, Function and Conservation* (ed. B. Gopal, W.J. Junk and J.A. Davis), 161–179. Backhuys.

Levin, P.S., Ellis, J., Petrik, R., and Hay, M.E. (2002). Indirect effects of feral horses on estuarine communities. *Conservation Biology* 16: 1364–1371.

Longenecker, R.A., Bowman, J.L., Olsen, B. et al. (2018). Short-term resilience of New Jersey tidal marshes to Hurricane Sandy. *Wetlands* 38: 565–575.

Luther, D.A. and Greenberg, R. (2009). Mangroves: a global perspective on the evolution and conservation of their terrestrial vertebrates. *BioScience* 59: 602–612.

Maclean, I.M.D., Austin, G.E., Rehfisch, M.M. et al. (2008). Climate change causes rapid changes in the distribution and site abundance of birds in winter. *Global Change Biology* 14: 2489–2500.

Mathot, K.J., Smith, B.D., and Elner, R.W. (2007). Latitudinal clines in food distribution correlate with differential migration in the Western Sandpiper. *Ecology* 88: 781–791.

Mathot, K.J., Piersma, T., and Elner, R.W. (2018). Shorebirds as integrators and indicators of mudflat ecology. In: *Mudflat Ecology* (ed. P. Beninger) Aquatic Ecology Series 7, 309–338. Springer.

McPherson, H.A. (1897). *A History of Fowling*. David Douglas.

Mizrahi, D.S. and Peters, K.A. (2009). Relationships between sandpipers and horseshoe crab in Delaware Bay: a synthesis. In: *Biology and Conservation of Horseshoe Crabs* (ed. J. Tanacredi, M. Botton and D. Smith), 65–87. Springer.

Mossman, H.L., Davy, A.J., and Grant, A. (2012). Does managed coastal realignment create saltmarshes with 'equivalent biological characteristics' to natural reference sites? *Journal of Applied Ecology* 49: 1446–1456.

Murray, N.J., Clemens, R.S., Phinn, S.R. et al. (2014). Tracking the rapid loss of tidal wetlands in the Yellow Sea. *Frontiers in Ecology and the Environment* 12: 267–272.

Novak, J.M., Gaines, K.F., Cumbee, J.C. et al. (2006). The Clapper Rail as an indicator species of estuarine-marsh health. In: *Terrestrial Vertebrates of Tidal Marshes: Ecology, Evolution and Conservation*, *Studies in Avian Biology*, vol. 32 (ed. R. Greenberg, J.E. Maldonado, S. Droege and M.V. McDonald), 270–281. Cooper Ornithological Society.

Orth, R.J. and Moore, K. (1983). Chesapeake Bay: an unprecedented decline in submerged aquatic vegetation. *Science* 223: 51–52.

Page, G. and Whitacre, D.F. (1975). Raptor predation on wintering shorebirds. *Condor* 77: 73–83.

Palomo, G., Iribarne, O., and Martinez, M.M. (1999). The effect of migratory seabirds guano on the soft bottom community of a SW Atlantic coastal lagoon. *Bulletin of Marine Science* 65: 119–128.

Payton, Q., Evans, A.F., Hostetter, N.J. et al. (2020). Measuring the additive effects of predation on prey survival across spatial scales. *Ecological Applications* e02193.

Perry, M.C. and Deller, A.S. (1995). Waterfowl population trends in the Chesapeake Bay area. In: *Towards a Sustainable Coastal Watershed: the Chesapeake Bay Experiment*, 490–504. Chesapeake Research Consortium Publication 149.

Perry, M.C. and Deller, A.S. (1996). Review of factors affecting the distribution and abundance of waterfowl in shallow-water habitats of Chesapeake Bay. *Estuaries* 19: 272–278.

Perry, M.C., Wells-Berlin, A.M., Kidwell, D.M., and Osenton, P.C. (2007). Temporal changes of populations and trophic relationships of wintering diving ducks in Chesapeake Bay. *Waterbirds* 30 (sp1): 4–16.

Piersma, T. (1997). Do global patterns of habitat use and migration strategies co-evolve with relative investments in immunocompetence due to spatial variation in parasite pressure? *Oikos* 80: 623–631.

Post, W. and Greenlaw, J.S. (1982). Comparative costs of promiscuity and monogramy: a test of reproductive effort theory. *Behavioral Ecology and Sociobiology* 10: 101–107.

Rasmussen, A.R., Murphy, J.C., Ompi, M. et al. (2011). Marine reptiles. *PLoS ONE* 6 (11): e27373.

Richman, S.E. and Lovvorn, J.R. (2004). Relative foraging value to lesser scaup ducks of native and exotic clams from San Francisco Bay. *Ecological Applications* 14: 1217–1231.

Robert, M., McNeil, R., and Leduc, A. (1998). Conditions and significance of night feeding in shorebirds and other water birds in a tropical lagoon. *Auk* 106: 94–101.

Rubega, M.A. (2000). Feeding in birds: approaches and opportunities. In: *Feeding: Form, Function and Evolution in Tetrapod Vertebrates* (ed. K. Schwenk), 395–408. Academic Press.

Rubega, M.A. and Obst, B.S. (1993). Surface-tension feeding in phalaropes—discovery of a novel feeding mechanism. *Auk* 110: 169–178.

Sasser, C.E., Holm, G.O., Evers-Hebert, E., and Shaffer, G.P. (2017). The nutria in Louisiana: a current and historical perspective. In: *Mississippi Delta Restoration: Pathways to a Sustainable Future* (ed. J.W. Day and J.A. Erdman), 39–60. Springer.

Shepherd, P.C.F. and Boates, J.S. (1999). Effects of a commercial baitworm harvest on semipalmated sandpipers and their prey in the Bay of Fundy hemispheric shorebird reserve. *Conservation Biology* 13: 347–356.

Sherfy, M.H. and Kirkpatrick, R.L. (2003). Invertebrate response to snow goose herbivory on moist-soil vegetation. *Wetlands* 23: 236–249.

Shriver, W.G., Vickery, P.D., Hodgman, T.P., and Gibbs, J.P. (2007). Flood tides affect breeding ecology of two sympatric sharp-tailed sparrows. *Auk* 124: 552–560.

Silliman, B.R. and Bertness, M.D. (2004). Shoreline development drives invasion of *Phragmites australis* and the loss of plant diversity on New England salt marshes. *Conservation Biology* 18: 1424–1434.

Smith, K.R., Riley, M.K., Barthman-Thompson, L. et al. (2018). Toward salt marsh harvest mouse recovery: a review. *San Francisco Estuary and Watershed Science* 16 (2): 1–25.

Soares, L., Escudero, G., Penha, V.A.S., and Ricklefs, R.E. (2016). Low prevalence of haemosporidian parasites in shorebirds. *Ardea* 104: 129–141.

Stallknecht, D.E. and Brown, J.D. (2009). Tenacity of avian influenza viruses. *Revue scientifique et technique (International Office of Epizootics)* 28: 59–67.

Stotz, D.F., Fitzpatrick, J.W., Parker, T.A. III, and Moskovits, D.K. (1996). *Neotropical Birds: Ecology and Conservation*. University of Chicago Press.

Studds, C.E., Kendall, B.E., Murray, N.J. et al. (2017). Rapid population decline in migratory shorebirds relying on Yellow Sea tidal mudflats as stopover sites. *Nature Communications* 8: 14895.

Sun, Z., Sun, W., Tong, C. et al. (2015). China's coastal wetlands: Conservation history, implementation efforts, existing issues and strategies for future improvement. *Environment International* 79: 25–41.

Takekawa, J.Y., Woo, I., Spautz, H. et al. (2006). Environmental threats to tidal-marsh vertebrates of the San Francisco Bay Estuary. In: *Terrestrial Vertebrates of Tidal Marshes: Ecology, Evolution and Conservation, Studies in Avian Biology*, vol. 32 (ed. R. Greenberg, J.E. Maldonado, S. Droege and M.V. McDonald), 176–197. Cooper Ornithological Society.

Takekawa, J.Y., Prosser, D.J., Newman, S.H. et al. (2010). Victims and vectors: highly pathogenic avian influenza H5N1 and the ecology of wild birds. *Avian Biology Research* 3: 1–23.

Thayer, G.W., Bjorndal, K.A., Ogden, J.C. et al. (1984). Role of larger herbivores in seagrass communities. *Estuaries* 7: 351–376.

Tripp, K.J. and Collazo, J.A. (1997). Non-breeding territoriality of semipalmated sandpipers. *Wilson Bulletin* 109: 630–642.

Verhagen, J.H., Fouchier, R.A.M., and Lewis, N. (2021). Highly pathogenic avian influenza viruses at the wild-domestic bird interface in Europe: future directions for research and surveillance. *Viruses* 13: 212.

Walker, K.M., Fraser, J.D., Catlin, D.H. et al. (2019). Hurricane Sandy and engineered response created habitat for a threatened shorebird. *Ecosphere* 10 (6): e02771.

Walsh, J., Benham, P.M., Deane-Coe, P.E. et al. (2019). Genomics of rapid ecological divergence and parallel adaptation in four tidal marsh sparrows. *Evolution Letters* 3: 324–338.

Warnock, N., Page, G.W., Ruhlen, T.D. et al. (2002). Management and conservation of San Francisco Bay salt ponds: effects of pond salinity, area, tide, and season on Pacific Flyway waterbirds. *Waterbirds* 25: 79–92.

Whitfield, D.P. (2003a). Density-dependent mortality of wintering Dunlins *Calidris alpina* through predation by Eurasian Sparrowhawks *Accipiter nisus*. *Ibis* 145: 432–438.

Whitfield, D.P. (2003b). Predation by Eurasian sparrowhawks produces density-dependent mortality of wintering redshanks. *Journal of Animal Ecology* 72: 27–35.

Wolfe, M.F., Schwarzbach, S., and Sulaiman, R.A. (1998). Effects of mercury on wildlife: a comprehensive review. *Environmental Toxicology and Chemistry* 17: 146–160.

Ydenberg, R.C., Butler, R.W., Lank, D.B. et al. (2004). Western sandpipers have altered migration tactics as peregrine falcon populations have recovered. *Proceedings of the Royal Society B* 271: 1263–1269.

Zharikov, Y. and Skilleter, G.A. (2002). Sex-specific intertidal habitat use in subtropically wintering Bar-tailed Godwits. *Canadian Journal of Zoology-Revue Canadienne De Zoologie* 80: 1918–1929.

Zweers, G., Dejong, F., Berkhoudt, H., and Vandenberge, J.C. (1995). Filter feeding in flamingos (*Phoenicopterus ruber*). *Condor* 97: 297–324.

CHAPTER 14

Estuarine Ecosystem Metabolism

Ecosystem metabolism measurements at work: floating chambers in Plum Island Sound to measure air-water gas fluxes (CO_2, CH_4) in a marsh-dominated estuarine ecosystem. Photo credit: Nathaniel Weston.

Jeremy M. Testa[1], Charles S. Hopkinson[2], Peter A. Stæhr[3], and Nathaniel Weston[4]
[1]Chesapeake Biological Laboratory, University of Maryland Center for Environmental Science, Solomons, Maryland, USA
[2]Department of Marine Sciences, University of Georgia and Georgia Sea Grant, Athens, Georgia, USA
[3]Department of Ecoscience, Marine Diversity and Experimental Ecology, Roskilde, Denmark
[4]Department of Geography and the Environment, Villanova University, Villanova, Pennsylvania, USA

14.1 Introduction

In a seminal paper describing the balance between photosynthesis and respiration in aquatic environments, H.T. Odum (1956) poetically quipped "*Each day as the sun rises and retires the beautiful green bays like great creatures breathe in and out.*" While the similarities between the daily metabolic inhales and exhales of *breathing bays* and a most basic aspect of human physiology may not have been obvious to a casual observer, they allow for a convenient framework to examine ecosystem-scale dynamics in estuaries. In short, a quantification and understanding of what drives the balance between the consumption and production of oxygen and carbon dioxide—or the *Ecosystem Metabolism* of estuaries—addresses dynamic changes in organic matter availability, the rate of growth of organisms, food-web interactions, and the spatial and temporal coupling of biogeochemical cycling (Odum 1967). *Anabolic* processes use external energy to synthesize new organic matter, and *catabolic* processes derive biochemical energy from decomposition of these organic molecules. Together, these physiological processes comprise the integrated metabolism of individual organisms and when added together, the summation of anabolic and catabolic rates of all organisms in the system is equivalent to ecosystem metabolism (e.g., Kemp et al. 1997). Variations in ecosystem metabolism are constrained and driven by biological, chemical, and physical processes such as changes in the abundance of organisms, external forcing (e.g., sunlight, temperature, nutrient, and organic matter inputs), and advective and diffusive transport. Thus reflections on the overall production and respiration of an ecosystem (Hopkinson and Vallino 1995; Gazeau et al. 2005c) yield insights into several key features of estuarine ecology.

Coastal ecosystem metabolic rates provide broad measures of the ecological condition of a system including rates of biogeochemical transformations and rates of exchange with adjacent ecosystems (Heath 1995). The net production (i.e., production minus consumption) of organic matter in an ecosystem (i.e., its net metabolism) represents not only a measure of the system's balance between anabolic and catabolic processes, but reflects the import and export of biologically important substances. Organic matter production (*anabolism*) in a coastal ecosystem derives from light-driven photosynthesis of all plants, such as phytoplankton, benthic micro- and

Estuarine Ecology, Third Edition. Edited by Byron C. Crump, Jeremy M. Testa, and Kenneth H. Dunton.

Companion website: www.wiley.com/go/crump/estuarine3

macroalgae, emergent wetlands, and submersed vascular plants (seagrasses), and from chemoautotrophic production (e.g., nitrification, sulfide oxidation) driven by the chemical energy of reduced ions. Metabolic consumption organic matter (*catabolism*) in a coastal ecosystem is the sum of its respiratory processes, which come in a variety of forms; (1) aerobic respiration (characteristic of most higher organisms and many microbes) where O_2 is the terminal electron acceptor, (2) anaerobic respiration (characteristic of many bacteria and archaea) where alternative oxidants such as nitrate, sulfate, iron, manganese, and carbon dioxide are terminal electron acceptors, and (3) fermentative reactions where organic molecules serve as both oxidants and reductants of these redox reactions (Heip et al. 1995). *Net ecosystem metabolism* reflects the balance between these processes of organic matter production and consumption.

Ecologists have commonly referred to ecosystems that are net generators of organic matter as *autotrophic*, while net consumer ecosystems are referred to as *heterotrophic*. These terms reveal the analog between autotrophic organisms that generate their own energy and heterotrophic organisms that obtain energy from other organisms. The autotrophic versus heterotrophic conceptual framework is relevant to estuarine ecology because it reveals the extent to which ecosystems import organic matter from adjacent waters, watersheds, or wetlands, and identifies the system as generating an excess of organic matter associated with being a net sink of atmospheric carbon dioxide. As most aquatic systems receive more organic matter from the surrounding landscape than they produce internally, consumption tends to be higher than production leading to net heterotrophy, net carbon dioxide efflux to the atmosphere, and the potential to support food webs with externally-derived organic material. In highly productive systems receiving high loads of inorganic nutrients, primary production is stimulated leading to net autotrophic conditions that can lead to carbon dioxide influxes and organic matter excesses to support oxygen depletion in deeper waters. In this chapter, we provide several examples of how this *trophic state* concept has been used to analyze estuarine ecosystems.

Research over the last half century has demonstrated a range of applications of ecosystem metabolic rate measurements for a variety of coastal ecosystems, many of which provide insight into the controls on metabolism. In general, rates of ecosystem production of organic matter are stimulated by increases in sunlight (also water clarity) and inorganic nutrient concentrations, causing ecosystem production to be sensitive to external inputs of nutrients and light absorbing components. In contrast, ecosystem respiration tends to be enhanced by higher levels of both water temperatures and labile organic substrates (e.g., Pomeroy and Wiebe 2001); thus to some degree, organic matter production and consumption are controlled by different external forces. Net ecosystem production in shallow, clear waters tends to be higher if benthic plants that maintain large aggregations of plant biomass dominate (Stæhr et al. 2018); however, this is not true of coral reef ecosystems in which ecosystem production and respiration tend to be in balance (Smith and Gattuso 2009). Metabolic rates are generally depressed by toxic compounds from natural or human sources; however, anthropogenic contaminants (e.g., herbicides, fungicides, and insecticides) exert different effects on plants, microbes, and animals and thus shift the balance between ecosystem production and respiration (Wiegner et al. 2003). Measurements of ecosystem metabolism provide the quantitative information needed to balance food-web models (e.g., Green et al. 2006) and to partition biogeochemical processes among various aerobic and anaerobic pathways (e.g., Hopkinson et al. 1999). Ecosystem production—respiration imbalances over time and across space provide understanding of the timing and location of surplus organic matter production, which may support the development of hypoxic and anoxic zones in coastal waters (e.g., Kemp and Boynton 1992). Deviations from balance are furthermore interesting because they reveal metabolic responses to recent or nearby perturbations (e.g., Yvon-Durocher et al. 2012).

The overall aim of this chapter is to provide an introduction and overview of the concepts, methods, and ecological relevance of organic matter production and respiration processes in estuarine and coastal ecosystems, with an emphasis on temperate latitudes. We also refer the reader to several excellent reviews of ecological or biogeochemical aspects of coastal ecosystem metabolism and the future challenges for the field, including Smith and Hollibaugh (1993), Heip et al. (1995), Gattuso et al. (1998), Smith et al. (2005), Stæhr et al. (2012), and Hoellein et al. (2013). Here we summarize these reviews, but also extend them by (1) describing spatial and temporal variations in ecosystem metabolic rates and thus the important driving variables, (2) illustrating various aspects of ecosystem metabolism in four case study ecosystems, (3) presenting studies of cross-system comparisons, and (4) discussing future research challenges associated with measurements of aquatic ecosystem metabolism.

14.2 Basic Definitions and Concepts

There are many terms commonly used to define measured or computed rates of ecosystem-level primary production and respiration. Chamber incubations and open water observations of diel variations in metabolite concentrations (e.g., O_2, TCO_2 = sum of dissolved inorganic carbon in water) measure, during the daylight, rates of *net daytime production* (*NDP*, per hour), which is generally assumed to be the difference between *gross primary production* (*GPP*) and *respiration* (*R*). Changes in metabolite concentrations observed in dark incubations are often assumed to be the same as those observed *in situ* at night, and these are referred to as *night time* (or *dark*) *respiration rates* (R_n, per hour). Changes in TCO_2 must be corrected for the contribution of $CaCO_3$ precipitation and dissolution in coral reef and other calcifying systems, but inorganic carbon-based rates

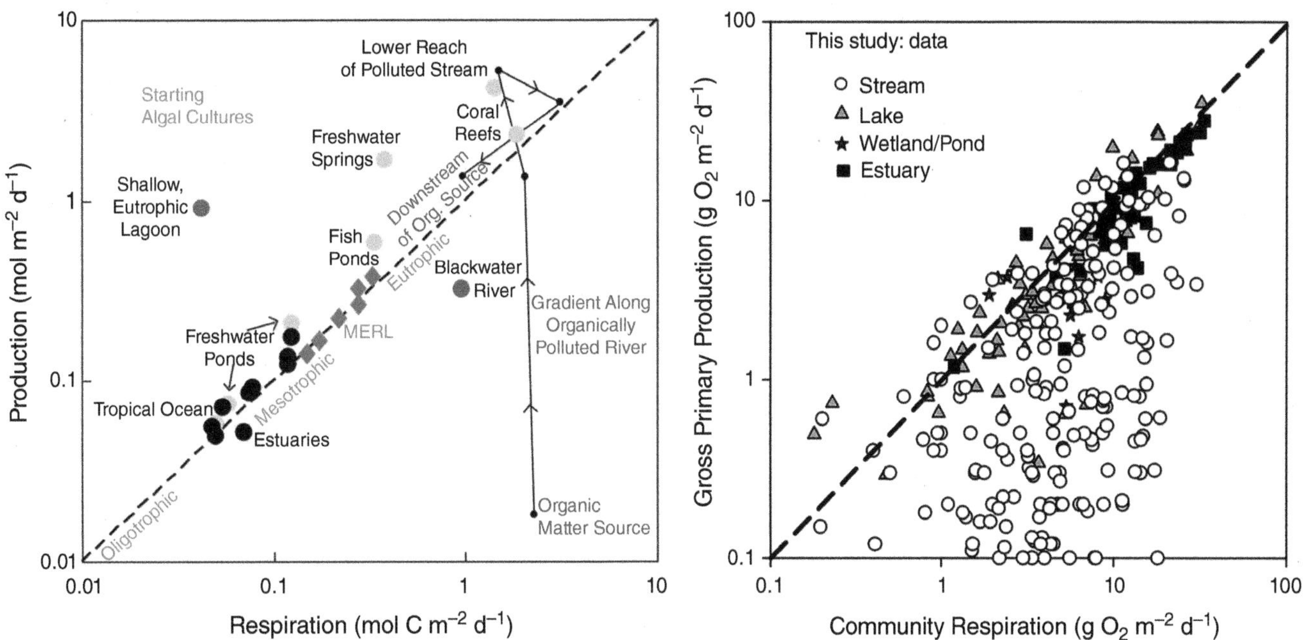

FIGURE 14.1 Regression of ecosystem respiration and ecosystem gross primary production across many types of aquatic ecosystems. The left panel is based on Figure 14.8 in Odum (1956) and the right panel is a more recent global synthesis. *Source*: Hoellein et al. (2013); reprinted with permission. Gross production and respiration tend to be balanced across many ecosystems types, including estuaries, but note that some ecosystems deviate from the 1:1 line substantially. Metabolic rates were collected from various literature sources: Non-estuarine systems (Odum 1956), estuaries (Randall and Day 1987; Day et al. 1988; Flores-Verdugo et al. 1988; Teague et al. 1988; Twilley 1988; Hoppema 1991; Heath 1995), MERL nutrient enrichment experiment (Oviatt et al. 1986), blackwater river (Caffrey 2004), shallow, eutrophic lagoon (D'Avanzo et al. 1996).

offer the advantage of measuring anaerobic metabolism. If it is assumed that night-time respiration rates (R_n) are equal to day-time respiration rates, they can be extrapolated to estimate 24-hour respiration rates (*R*, per day). The balance of both *R* and *GPP* over a full 24-hour period is then the daily *net ecosystem production* (*NEP*), which may reflect net heterotrophy (when *R* exceeds *GPP*; *NEP* < 0; net deficit of organic matter) or net autotrophy (when *GPP* exceeds *R*; *NEP* > 0; net production of organic matter). *NEP* is often used as an index of the balance between photosynthesis and respiration of organic matter for entire communities and ecosystems in coastal systems (Figures 14.1, 14.2; Box 14.1). In tidal wetland systems, rates of *R* and *GPP* are often determined by measuring the vertical exchange of O_2 or CO_2 between the wetland and atmosphere, and reflect *net ecosystem exchange* (*NEE*) rather than *NEP*, as there may be net lateral exchange of metabolites (O_2, TCO_2, organic carbon) in tidal systems that is not captured by measurements of vertical exchange. This production/respiration balance (*GPP/R*; *NEP*; *NEE*) can also be indexed by the ratio of the two measured rates, or the "P/R ratio". The concept of metabolic balance has a long history in coastal and marine science, including an emphasis on the related f-ratio, which denotes the ratio of phytoplankton production fueled by "new" external nutrient (e.g., NO_3^-) inputs to that fueled by internally "recycled" nitrogen (e.g., NH_4^+) generated from planktonic respiration (Quiñones and Platt 1991).

Gross primary production (*GPP*; often shortened to just *P*), which is defined as the total autotrophic production of organic carbon via fixation of TCO_2 (Figure 14.2), is extremely difficult to

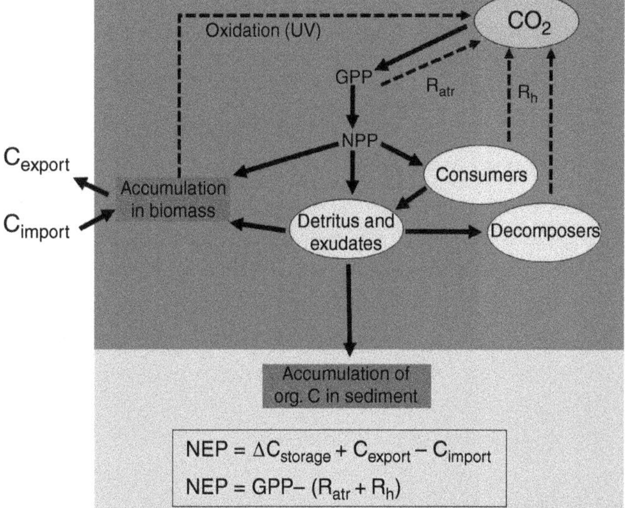

FIGURE 14.2 Fates of organic carbon (C) fixed in or imported into an aquatic ecosystem. Total ecosystem respiration (R) is the sum of autotrophic respiration (R_{atr}) and heterotrophic respiration (R_h). "Accumulation in biomass" represents all biomass (plant, animal, and microbial); the arrow is drawn from NPP in this diagram because plant biomass accumulation is generally the largest biomass term. NPP, net primary production; NEP, net ecosystem production; GPP, gross primary production; CO_2, carbon dioxide; UV, ultraviolet. Dashed lines represent mineralization of C, and solid lines represent production and/or transfer of C. Net autotrophic ecosystems (P_n > 0) have a net accumulation and/or net export of organic matter. Net heterotrophic ecosystems (NEP < 0) depend on imported organic substances. *Source*: Based on Lovett et al. (2006) and Stæhr et al. (2012).

Box 14.1 Definitions and Measures of Ecosystem Metabolism

Autotrophic organisms/communities

Autotroph	= plant or protist that produces organic matter from CO_2
Gross primary production (GPP_{atr})	= total photosynthesis in a plant or group of autotrophic organisms
Respiration (R_{atr})	= respiration of autotrophic organisms
Net primary production (*NPP*)	= $GPP_{atr} - R_{atr}$ = net organic matter production by autotrophic organisms

Heterotrophic organisms/communities

Heterotroph	= organism that consumes organic matter to produce CO_2
Respiration (R_h)	= total respiration or an organism or group of consumer organisms

Ecosystems

Net daytime production (*NDP*)	= net organic matter production by an ecosystem during sunlit hours
Nighttime respiration (R_n)	= respiration of an ecosystem during night hours
Gross primary production (*GPP*)	= total amount of photosynthesis in an ecosystem (often assumed = $NDP + R_n$)
Ecosystem respiration (*R*)	= total amount of respiration in an ecosystem
Net ecosystem production (*NEP*)	= $GPP - R = NDP + R_n$ = net organic matter production of all ecosystem components
Net ecosystem exchange (*NEE*)	= *NEP* as measured by vertical exchange of O_2 or CO_2 between system and the atmosphere

Key Ratios

P/R Ratio	= *GPP/R;* ratio of photosynthesis to respiration
Photosynthetic quotient (*PQ*)	= ratio of O_2 produced/CO_2 consumed in photosynthesis
Respiratory quotient (*RQ*)	= ratio of CO_2 produced/O_2 consumed in respiration

Definition of various terms used in the literature to describe the components of ecosystem metabolism. Many such terms are linked specifically to a particular method, while others are specific to particular components of the ecosystem (e.g., water column). Other terms include Net Community Production (NCP) = net organic matter production of an ecosystem compartment (e.g., plankton, benthos) and Gross Oxygen Production (GOP) = measurement of gross photosynthesis from the triple oxygen isotope method.

measure directly (Sarma et al. 2005). In principle, *GPP* includes both photosynthetic and chemoautotrophic production of organic matter; however, the latter rates are generally assumed to represent a small contribution for most coastal ecosystems. Nevertheless, the operational definition of gross primary production, or *gross photosynthesis*, is often assumed equal to $\alpha*NDP+(24-\alpha)*R_n$, where α is the number of hours of daylight (dawn to dusk) in the full 24-hour day and R_n is assumed to be constant during daylight hours. Box 14.1 includes a full suite of definitions for commonly used metabolic terms.

Metabolic rates (e.g., *GPP* or *R*) are often estimated by measuring changes in any reactant or metabolite involved in primary production and respiration of organic matter. For example, because O_2 is a product of photosynthesis, the quantitative increase in O_2 concentrations in a bottle over a given time represent a net photosynthetic rate for the plants within the bottle (e.g., Gaarder and Gran 1927). Measurements made using changes in concentrations of O_2 can be converted to carbon rates using an assumed or measured photosynthetic quotient (i.e., the ratio of O_2 produced/TCO_2 fixed, ranging from PQ = 1.0–1.5; Box 14.1) and respiratory quotients (RQ = 1–1.3 for aerobic conditions, which are ratios of TCO_2 released/O_2 consumed in respiratory processes (e.g., del Giorgio and Williams 2005). Similar fixed stoichiometry for inorganic nutrients and other elements (N, P, S, Fe) and metabolic gases (O_2, TCO_2) are often assumed for ecosystem metabolism methods based on mass balance calculations of the nutrients at large time/space scales (e.g., Smith 1991; Testa and Kemp 2008) or measurements of anaerobic metabolism (e.g., Kemp et al. 1997). Other methods are available for measuring metabolic rates of ecosystem components (e.g., plankton or benthos) using isotopically-labelled reactants or metabolites, including $T^{14}CO_2$, $^{18}O_2$, or $^{15}NH_4^+$ (Peterson 1980; Gazeau et al. 2007).

Ecosystem metabolism studies are implicitly concerned with carbon balance, but metabolic properties are generally measured by tracing the uptake and production of inorganic carbon or dissolved oxygen. In most cases, especially under aerobic conditions and when $CaCO_3$ precipitation/dissolution is trivial, inorganic carbon and oxygen-based measurements of primary production and respiration are equivalent (e.g., Gazeau et al. 2007) or can be converted using relatively consistent quantitative conversion factors (see Box 14.1). However, under anaerobic (no oxygen) conditions, a suite of alternative electron acceptors may be used in the decomposition of organic matter (Froelich et al. 1979), including, but not restricted to, nitrate (NO_3^-), oxidized forms of iron and manganese, and sulfate (SO_4^{2-}). These reactions generate CO_2, but do not consume oxygen directly, and as a result, O_2 measurements cannot be used to measure these respiratory processes. One particular reaction of interest is denitrification, a process where nitrate is used as the terminal electron acceptor. This anaerobic process converts NO_3^-, which is a bioavailable inorganic nitrogen ion, into N_2, a gas that escapes to the atmosphere (Seitzinger 1988), resulting in the loss of fixed nitrogen from aquatic ecosystems. This process does, however, generate phosphate during organic matter decomposition, thereby altering dissolved inorganic nitrogen to phosphorus (DIN:DIP) ratios, which may influence primary production and algal diversity. Sulfate reduction, a process where SO_4^{2-} is used as the terminal electron acceptor, is also important because bottom waters, sediments, and wetland soils in many estuaries are commonly devoid of O_2, yet rich in SO_4^{2-}, making sulfate reduction the major respiratory reaction during some seasons (e.g., Marvin-DiPasquale and Capone 1998).

Simple diagrams of ecosystem *GPP* versus *R* can be used in comparative analysis of trophic status (Odum 1956), which measures both the metabolic richness and balance of an ecosystem. Recent global syntheses of this relationship have been recently updated and generally support a tight relationship between *GPP* and *R* in estuaries (Figure 14.1; Hoellein et al. 2013). Along the *GPP* = *R* diagonal line, the system's metabolic state ranges from low rates in systems referred to as *oligotrophic* (with low nutrients and organic carbon) to high rates in systems called *eutrophic* (with high nutrients and organic carbon, Figure 14.1). Above the diagonal line, ecosystem metabolism is net autotrophic and below the line systems are net heterotrophic. Changes in trophic state along the diagonal represent changes in productivity, respiration, or general stimulation of metabolic processing, either via stimulation with additions of limiting inorganic nutrients or labile organic material. Changes in state that are perpendicular to the diagonal reflect metabolic imbalances. For example, *GPP* > *R* in eutrophic systems (Oviatt et al. 1986; Testa et al. 2008), benthic-dominated ecosystems, or beginning algal cultures, where ample light and nutrients generate short-term organic matter surpluses (Figure 14.1). Conversely, *R* > *GPP* in rivers receiving high amounts of anthropogenic organic matter (e.g., blackwater rivers), where photosynthesis is highly light-limited and respiration is fueled by high organic matter inputs (Figure 14.1). In other rivers and estuaries, upstream heterotrophy near the organic matter injection point often gives way to autotrophy in downstream waters where there is less organic matter, abundant nutrients, and often clearer water (Heath 1995; Garnier and Billen 2007), which is analogous to concepts from river and stream ecology, including the river continuum concept (RCC, Vannote et al. 1980) and nutrient spiraling (Newbold et al. 1981). At large time and space scales, ecosystem *GPP* and *R* tend to move toward a balanced condition (*GPP/R* = 1) because primary production yields organic matter that fuels respiration, while respiration releases inorganic nutrients that stimulate production (Figure 14.1). Consequently, *GPP* and *R* can fall out of balance over shorter time periods when receiving pulsed inputs of *GPP*-stimulating nutrients or *R*-stimulating labile organic material.

14.3 Approaches for Estimating Ecosystem Metabolism

Several techniques have been developed and applied to measure or calculate metabolic rates in coastal ecosystems (Table 14.1; Boxes 14.2–14.4). Methods span a range of spatial and temporal scales from litres and minutes for container incubations to cubic kilometres and multiple years for computations from whole system mass-balances. Small-scale rate measurements tend to be more controlled and precise, but more difficult to extrapolate to annual or inter-annual periods over entire ecosystems. In contrast, large-scale rates estimated from aggregated models and mass-balances, while appropriate for the scales of whole ecosystems, are generally difficult to interpret for controlling factors and contributions from constituent habitats or functional groups. Here we briefly review the range of techniques that are used to estimate primary production and respiration for estuarine and coastal ecosystems and compare and contrast the scales, advantages and disadvantages of each method.

14.3.1 Measuring Components of Ecosystem Metabolism

Perhaps the most common method to measure photosynthesis and respiration rates in the pelagic, benthic, or wetland components of coastal ecosystems is by monitoring temporal changes in concentrations of key metabolites (O_2, TCO_2, particulate organic carbon (POC), inorganic nutrients) in enclosed chambers that isolate the ecosystem component (Box 14.2). Current technologies allow for detection of low metabolic rates via precise measurements of metabolite concentrations, including O_2 concentrations (e.g., Kana et al. 1994; Glazer et al. 2004) and TCO_2 and pCO_2 (e.g., Borges et al. 2004). Changes in dissolved O_2 or TCO_2 measured in bottles sampled during typically 2–24 hour incubations are used to infer rates of apparent pelagic daytime primary production when exposed to light or to infer respiration when exposed to darkness (Box 14.2; Gaarder and Gran 1927; Gazeau et al. 2005a; del Giorgio and Williams 2005). Rates of phytoplankton photosynthesis are also commonly measured

TABLE 14.1 A comparison of aspects of the most common methods used to estimate GPP, R, NEP, and P:R in aquatic ecosystems. For isotope methods, comments specific to the triple oxygen isotope method (TI) and the ^{18}O method (^{18}O) are noted.

Method	System	Temporal scale	Advantages	Disadvantages
Diel O_2, TCO_2 Including the response surface method and the eddy correlation technique	Estuary, lake, river, ocean	Daily, seasonal, annual	• Measures all system components • Remote data collection • Straightforward computation • Precise measures • High-frequency rates • Multivariable sensors	• Air–water flux difficult to quantify • O_2:C conversion problems • Physics may obscure biology • O_2 method misses anaerobic R • Horizontal and vertical heterogeneity • Stratification causes problems • No component rates • Import of water with gas super- or sub-saturated
Oxygen isotopes	Estuary, lake, river, ocean	Daily, seasonal	• Measures all system components • Rates can be long- and short-term • Sensitive method	• Air–water flux needed (TI) • O_2:C conversion problems (TI,^{18}O) • Sampling is work intensive (TI,^{18}O) • Traces diurnal GPP and R • Known fractionations limited (TI)
Ecosystem budgets	Estuary, lake, river, ocean	Seasonal, annual	• Measures all system components • Straightforward computation • Data widely available • Formal error estimates	• Air–water flux difficult to quantify • O_2:C:DIP conversion problems • Abiotic effects on [DIP] • Large aggregation error • Net rates (NEP) only
Incubations	Estuary, lake, river, ocean	Hourly, daily	• Direct process measurement • Highly controlled • Precise measures • Can separate ecosystem components	• O_2:C conversion problems • Containment artifacts • Labor intensive • Difficult to upscale to ecosystem
Eddy covariance	Wetlands, marshes, estuary, lake	Hourly, daily, seasonal, annual	• Measures all system components • Remote data collection	• Net rates (NEP) only • Horizontal heterogeneity • Noisy data??

via uptake of ^{14}C labelled bicarbonate, and the two measures tend to be highly correlated, where rates measured by ^{14}C are generally consistent with rates of *GPP* measured with O_2 and TCO_2 incubations, as long as incubations are kept within a few hours (e.g., Peterson 1980; Williams 1993). Rates of benthic community respiration and photosynthesis, which are generally measured as dissolved TCO_2 and/or O_2 change in opaque or transparent (respectively) intact sediment cores removed for controlled incubation or *in situ* chambers inserted into ambient sediments, usually include a short (e.g., 5–25 cm) overlying water column that is sampled in a time-series over the incubation period (e.g., Hopkinson and Smith 2005; Sundbäck et al. 2006). Benthic metabolism can, however, also be assessed *in-situ* by the dissolved oxygen eddy correlation technique (Berg et al. 2003). Emergent, vegetated wetland metabolism is typically determined by measuring rates of CO_2 consumption or production in the gas headspace of clear chambers placed over the marsh, with measurements made at short timescales (~minutes) at various light levels (achieved with shade cloths) to determine wetland *NDP* and

Box 14.2 Measure Planktonic Metabolism

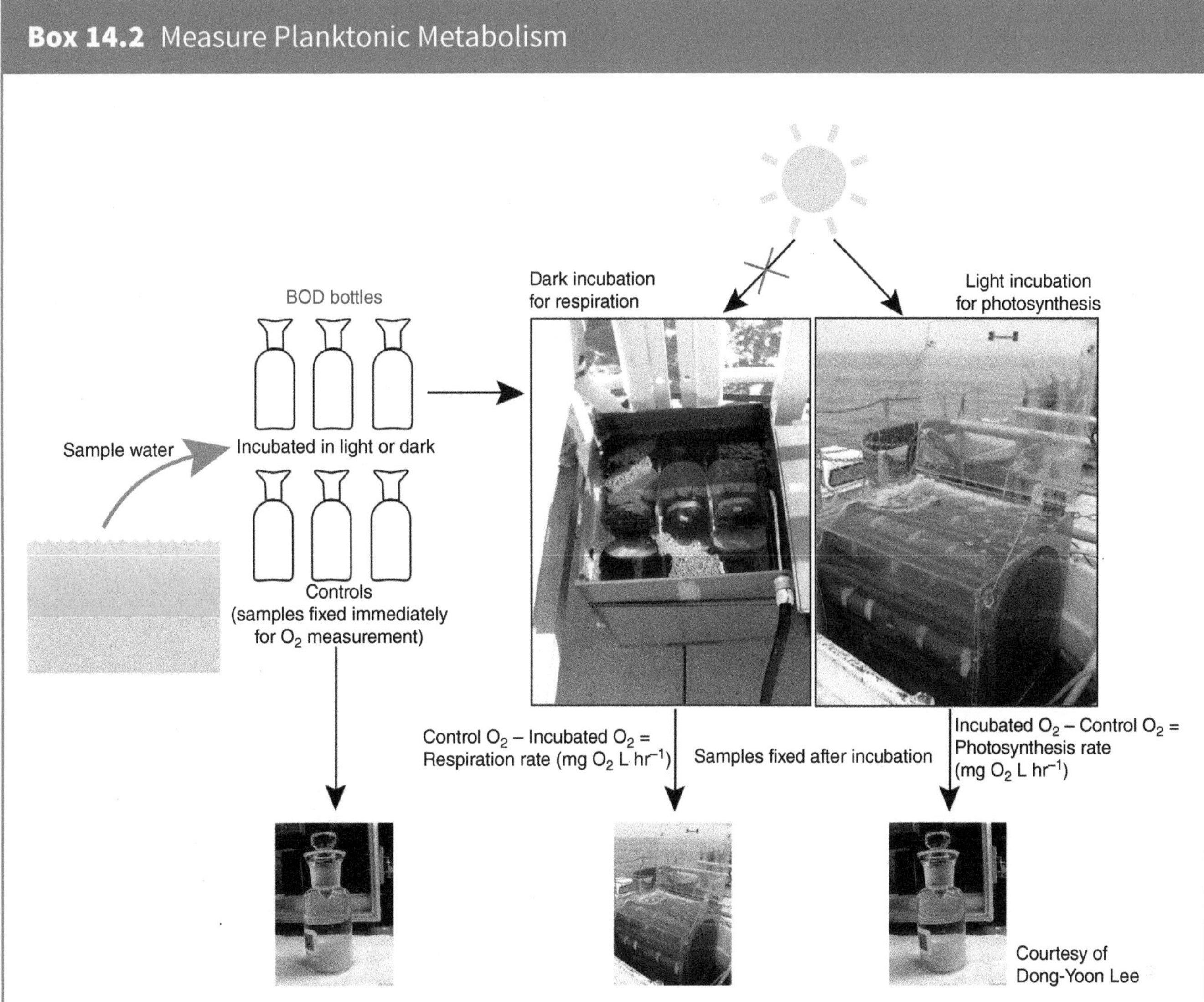

Schematic illustration of using dissolved oxygen (O_2) changes in bottle incubations to measure plankton community production and respiration. Water samples are collected and transferred to clear glass bottles (often called BOD bottles and usually 60, 125, or 300 mL in volume). For each depth and station, replicate bottles are "fixed" immediately and the dissolved O_2 in the bottle is converted into a solid compound via the Winkler titration. These bottles provide the initial O_2 concentration. Another set of bottles is incubated in the dark for 4–24 hours, after which O_2 is measured. The difference in O_2 between the mean of the dark-incubated bottles and that of the initial bottles equals the plankton community respiration rate(*R*). A third set of bottles is incubated in a rotating carousel at various light levels for 4–24 hours. Light level is usually reduced by wrapping the bottles in porous wire sheaths. The difference in O_2 between the light-incubated bottles and the initial bottles is equivalent to the net primary production rate (=*GPP* -*R*). Differences in *GPP* -*R* under different light levels are coupled to measurements of water column light availability to quantify depth-integrated *NPP*. Required durations of bottle incubations, which generally vary from 1 to 24 hours, are inversely related to both the expected ambient rates and the precision of measurement method. In principle, the same type of approach could be used with other tracers (e.g., DIC).
Source: Courtesy o.f Dong-Yoon Lee.

R (Neubauer et al. 2000; Weston et al. 2014). Note that in systems that experience lateral exchange of metabolites, such as tidal flooding of vegetated marshes that can import or export TCO_2, O_2, and organic carbon, chamber measurements typically reflect *NEE* rather than *NEP*.

The container incubation approach has several important strengths and limitations. One advantage is that it allows direct partitioning *GPP*, *R*, and *NDP* into relative contributions of different habitats. Secondly, controlled incubations allow for measurements of *NDP* and/or R_n responses to variations

Box 14.3 Measuring Metabolism

Illustration of various schemes for measuring ecosystem metabolism in coastal and estuarine waters. Physical features relevant for air-water exchange measurements are included.

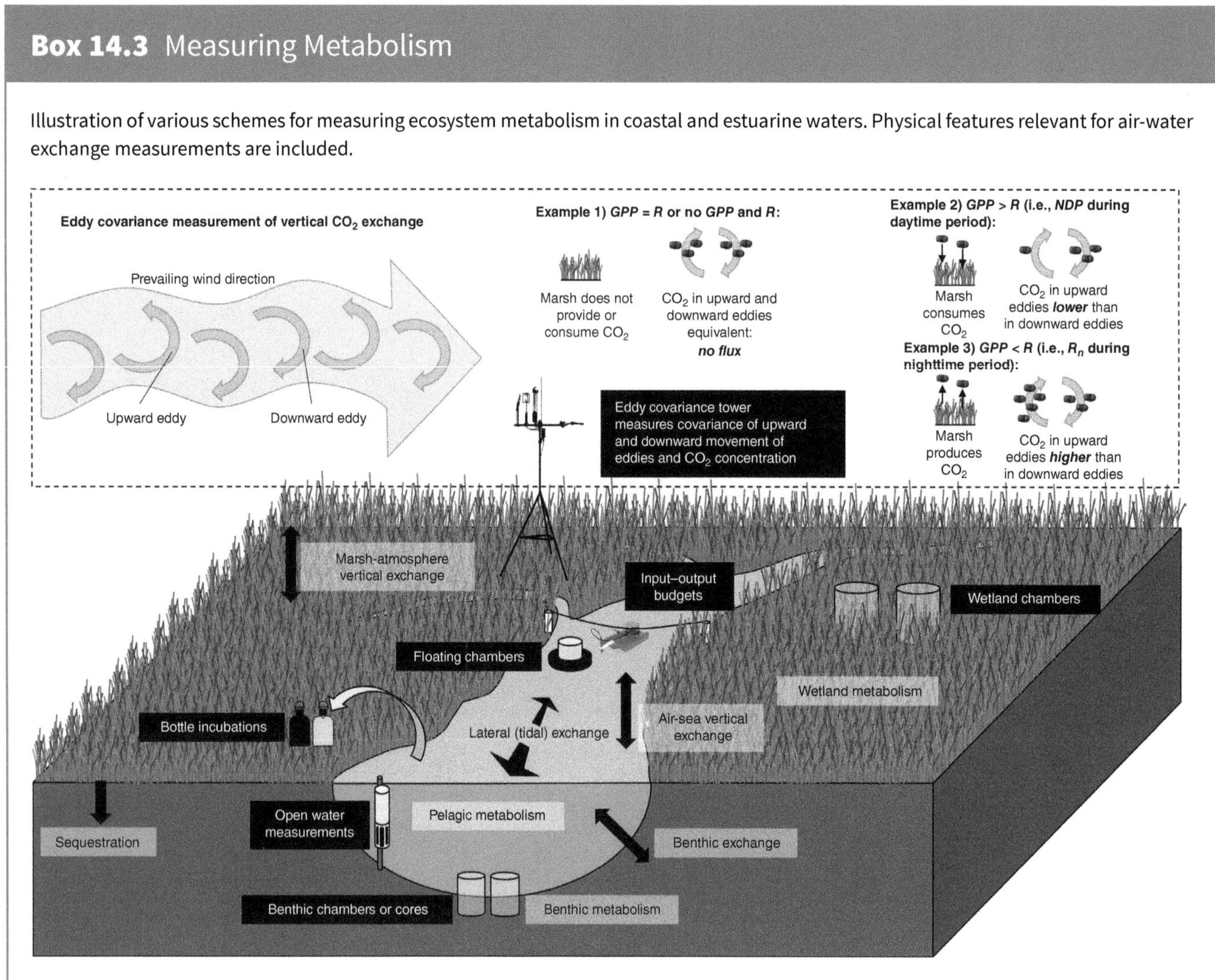

in physical (e.g., levels of temperature, light, and salinity), chemical (e.g., nutrient levels) or biological (e.g., grazers and predators) conditions. This approach has two important limitations. First, one must either make numerous individual rate measurements over large spatial and temporal scales to adequately characterize variability within the system, which can be prohibitively resource demanding. Alternatively, limited individual measurements within the system must be extrapolated over large time and space scales to yield an integrated metabolic rate, and these calculations introduce a series of individual errors that propagate to create a large and difficult-to-quantify level of uncertainty in the calculated ecosystem rate (e.g., Smith 1991). Secondly, isolation of water, benthos, or wetland within a closed space introduces what are called "container effects," in which the isolated ecosystem-unit (water, sediments, and organisms) changes over the incubation in artificial ways (e.g., predator isolation, algae growth on walls, depletion or accumulation of metabolites, etc.) that will alter the rate measurement itself. Chamber measurements in emergent wetland systems are further complicated by tidal flooding, which can substantially alter apparent rates of *GPP* and *R* as measured by changes in gaseous CO_2 production or consumption (Forbrich and Giblin 2015), necessitating alternate methodologies during periods of inundation. It may also be impractical to measure all compartments of a particular ecosystem, given constraints on the time, effort and resources involved in collecting samples within a large ecosystem.

14.3.2 Direct Air-Ecosystem Gas Exchange

The direct measurement or calculation of air-ecosystem exchange of O_2 and/or CO_2 over sufficient time and space can provide estimates of metabolism that characterize the entire ecosystem (Box 14.3) and also provide air-water exchange estimates needed for open-water approaches (Section 14.3.3). At steady state, a net efflux of O_2 and net influx of CO_2 indicate

Box 14.4 Measuring Diel Fluctuations in O2 and Computing Metabolic Parameters from O2 Time Series

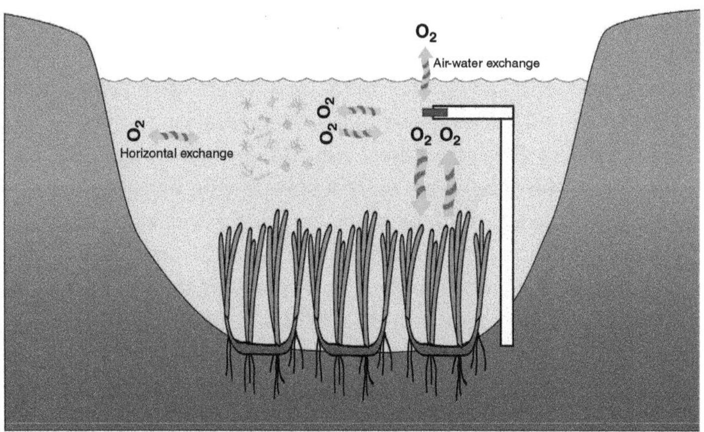

Dissolved O_2 (μM)

500
400
300
200
100
0

12 AM 4 AM 8 AM 12 PM 4 PM 8 PM 12 PM

Net primary Production

Net respiration

Metabolic computations:

O_2 Flux = $DO_t - DO_{t-1} * z - [(1 - (DOsat_t + DOsat_{t1})/ 200)) * k * dt]$

O_2 change   Air–water exchange correction

*The sum of O_2 fluxes during daytime hours = Net daytime production (*NDP*)

*The sum of O_2 fluxes during night hours = Nighttime respiration (R_n)

*24-hour respiration = R_n extrapolated to daytime hours

*Gross primary production (*GPP*) = *NDP* + extrapolated daytime respiration

*Net ecosystem production = *GPP* – 24-hour respiration

= *NDP* – R_n

*k = air-water exchange coefficient

#Note: Dirunal oxygen fluctuations can result from ecosystem metabolism, air-water exchange, and horizontal and vertical mixing. See Scully (2018) and Beck et al. (2015) for alternative methods that account for these physical transport terms.

In this box we illustrate the deployment of high-frequency dissolved O_2 sensors to measure hourly and daily rates of ecosystem production and respiration in estuarine and coastal ecosystems. Sensors can be deployed a number of ways, but one specific way is to anchor a sensor in the water column on L-shapes PVC piping (top left image). Such a deployment scheme, in a well-mixed water column, will measure the integrated change in O_2 concentration due to photosynthesis and respiration of both water column and benthic organisms (top right). As a result, short-term changes in O_2 measure net ecosystem metabolism because net primary production occurs during daylight hours causing O_2 to increase (graph, bottom left) and net respiration occurs at night, causing O_2 to decline. Hourly changes in O_2 must be corrected for exchanges of O_2 between the water and the atmosphere (equations, bottom right). Daily net ecosystem production is the difference between total O_2 increase during the daytime minus the total O_2 decline during nighttime. The advantages of this approach are that it measures the integrated metabolic activity of an entire ecosystem over a short time-period and that sensors can be deployed for weeks or more, allowing long-term measurements. The method suffers, however, from the need for an atmospheric exchange correction (which is difficult to measure accurately) and the possibility of low- or high-O_2 water being advected past the sensor, which will cause over- or under-estimates of metabolic rates. However, several methods are available to quantify advection and mixing impacts on metabolism (Scully 2018, Beck et al. 2015).
Source: Courtesy of Jessica Nagel.

positive values of *NEE* (and vice versa). Direct air-ecosystem measurements can be made for wetland/estuary systems by eddy covariance techniques (Schäfer et al. 2014), or for aquatic ecosystems by floating chambers (Frankignoulle 1988; Kremer et al. 2003), tracer additions (such as sulfur hexafluoride; SF_6) that integrate variable turbulence and weather over several days (Clark et al. 1996), or calculations based on gas transfer coefficients (e.g., Zappa et al. 2003).

Eddy covariance measurements of ecosystem metabolism are a relatively recent advance in measuring rates of ecosystem metabolism. The eddy covariance approach was adopted from use in terrestrial systems (Baldocchi 2003), and has been used to assess ecosystem metabolism in coastal marshes (Schäfer et al. 2014; Forbrich and Giblin 2015; Forbrich et al. 2018), freshwater wetlands (Schedlbauer et al. 2010), and mangroves (Barr et al. 2010). The eddy covariance technique measures the covariance of the upward and downward movements of air parcels and the CO_2 concentration of those parcels over an area of interest (Box 14.3). The horizontal movement of air (i.e., wind) over a surface (the ecosystem of interest) is composed of many rotating parcels of air, or eddies, that move in horizontal and vertical directions. By measuring the instantaneous covariance of the vertical movement and CO_2 concentration of eddies in the atmosphere above an ecosystem over time, the rate of ecosystem *NEE* can be determined. Equivalent CO_2 concentrations of the upward and downward eddies indicate that there is no net exchange of CO_2 between the ecosystem and the atmosphere (*NEE* ~ 0). When ecosystem *GPP* exceeds *R*, such as would be expected during daylight hours in the growing season over a vegetated marsh, the downward moving eddies will have higher CO_2 concentrations than the upward moving eddies. In contrast, during periods when *R* exceeds *GPP* the CO_2 concentration in the upward moving eddies will be higher than in the downward moving eddies (Box 14.3).

Instrumentation for measuring eddy covariance is typically installed on towers over the ecosystem of interest (Box 14.3). Changes in the horizontal and vertical movement of eddies during turbulent flow over an ecosystem occurs over very short timescales (~seconds), and instrumentation capable of measuring the flow direction (sonic anemometer) and CO_2 concentrations (fast gas analyzer) very rapidly (numerous times per second) is required. Eddy covariance towers, once installed, can operate continuously over long periods of time, providing direct measurements of air-ecosystem CO_2 exchange, or *NEE*. Ecosystem *R* and *GPP* are typically determined by partitioning the measured *NEE* into nighttime and daytime components, respectively, after applying a temperature correction (Reichstein et al. 2005). It is important to note that atmospheric eddy covariance is a measure of *NEE*, and does not account for the lateral flux of carbon associated with tidal exchange (Box 14.3; Forbrich and Giblin 2015). The import and/or export of carbon as DIC, DOC, or POC from coastal wetland ecosystems to adjacent estuaries and coastal oceans may be an important fate for fixed carbon that is not captured by the eddy covariance approach.

The direct field measurement of air-water gas exchange (F_{aw}) between aquatic systems and the atmosphere is often based on changes in O_2, CO_2, or inert tracer gases made in an enclosed atmosphere under floating chambers (Box 14.3; Odum 1956; Frankignoulle 1988). The product of empirically estimated gas exchange piston velocities (γ) and the difference between saturation (C_s) and measured O_2 or CO_2 concentrations (C) in surface waters ($F_{aw} = \gamma\ (C_s{-}C)$) provides an estimate of F_{aw} adequate for correcting *NDP* and R_n rates inferred from open water measurements of daily changes in gas concentration (e.g., Raymond and Cole 2001; Kremer et al. 2003; Box 14.4). Alternatively, direct estimates of *NEP* can be made from F_{aw} with precise calculations of dissolved gas partial pressure at the very surface (skin) of the water column (e.g., Najjar and Keeling 2000). Spatial interpolations of monitoring data and numerical model output generate climatologies of O_2 or CO_2 partial pressure and physical conditions to estimate seasonal patterns of *NEP* for large systems (Najjar and Keeling 2000; Cai et al. 2006). Values for the piston velocity, which generally increases with waves and turbulent mixing in surface water, are commonly computed from statistical relationships with wind speed and/or tidal currents (e.g., Wanninkhof et al. 2009; Borges and Abril 2010),

Direct measures of gas exchange are complicated by the fact that air-ecosystem gas exchange in tidal wetlands occurs over the air-water interface of both open waters and flooded marshes occupied by plants, as well as between air and unflooded marshes at low tide. Raymond and Cole (2001) and Ho et al. (2011) have argued that there is "no universally applicable parameterization" for gas exchange in estuaries, especially marsh-dominated estuaries. This argument is based on the potential for inappropriate scaling of instantaneous (in space and time) measures of γ to whole estuaries knowing that both wind speed and current induced surface water turbulence contribute to gas exchange (Zappa et al. 2003, Borges et al. 2004) *and* the fact that fetch in mesotidal estuaries varies tremendously from place to place. Consider that in a marsh there is a wide variety of habitats within the scale of meters and there are rapid changes in the surface over which the air-ecosystem flux is measured (open water to mudflat to vegetation over tidal cycles). Furthermore, there are extensive intertidal marsh platforms covered with turbulence-reducing vegetation (where air-ecosystem exchanges have yet to be measured!). While Raymond and Cole (2001) suggested that tracer additions provide optimal measures of γ and address ecosystem scales, the temporal scale of these observations (12 hours or longer) may result in wind-γ relationships that represent mean conditions. Dome measurements are straightforward to apply but do not capture spatial variability (unless many domes are deployed; e.g., Borges et al. 2004) and have been criticized for disrupting the very process they are designed to measure (wind-induced turbulence; Raymond and Cole 2001). Wind-γ relationships developed for the open ocean (e.g., Wanninkhof et al. 2009) do not explicitly account for current-induced turbulence, but have agreed with observations at low wind speeds in wetland-dominated estuaries (Carini et al. 1996, Zappa et al. 2003). More recently, an underwater eddy covariance technique has been developed (Long and

Nicholson 2018) that addresses many of the shortcomings with the floating dome and tracer techniques. Despite these efforts, air-ecosystem gas exchange is likely to remain an important source of uncertainty in metabolic studies in tidal marshes and estuaries.

14.3.3 Open Water Measurements

In systems with relatively homogeneous water masses (due to reduced mixing and/or slow water exchange), clear signals of diel changes in O_2 and TCO_2 concentration are often detectable and associated with ecosystem metabolism (Box 14.4). For example, the rate of O_2 concentration increase during the day is equivalent to *NDP* and the rate of decline during the night is equivalent to R_n. In these cases, rates of ecosystem *NDP* and R_n can be estimated by correcting water column O_2/TCO_2 changes for losses or gains via air-water exchange (e.g., Odum and Odum 1955; Odum and Hoskin 1958). The gas exchange is calculated in a similar manner as described earlier in Section 14.3.2. Tidal advection and vertical mixing may also be a significant contributor to O_2/TCO_2 time-series and obscure metabolic signals, and recent advancements have offered statistical methods for removing tidal or diel signals for improved metabolic estimates (Beck et al. 2015; Scully 2018). In coastal ecosystems with well-defined, uniform water flow during intervals of the tidal cycle, metabolic rates can, in concept, be estimated using flow velocity and differences in concentration between an upstream and a downstream station (Odum 1956). Rates of benthic production and respiration can also be measured with the eddy correlation technique, in which turbulent fluctuations in the vertical velocity and the corresponding O_2 concentration are measured simultaneously and at the same point above the sediment surface (Berg et al. 2003). Such measurements do not suffer from the shortcomings of the *in situ* chamber measurements, as O_2 uptake is determined under true *in situ* conditions, without any disturbance of the sediment and under the natural hydrodynamic conditions taking account of lateral oxygen exchanges (Hume et al. 2011). Rates of gross primary production can furthermore be inferred at broad spatial and temporal scales from the triple oxygen isotope (^{16}O, ^{17}O and ^{18}O) composition of atmospheric and dissolved O_2 using independent estimates of air-water O_2 exchange (Luz et al. 1999; Sarma et al. 2005). Measurements of $\delta^{18}O$ are also used to estimate *GPP*, *R*, and *P/R* separately in open waters over diel cycles (Tobias et al. 2007). Measurements of plankton *NDP* and R_n made in relatively large and deep coastal systems can be used to approximate total ecosystem metabolism. In shallower systems, however, vertically integrated rates of *both* plankton and benthic community metabolism (measured in chambers; Box 14.2) are sometimes combined to estimate whole ecosystem rates of *NDP* and R_n (e.g., Kemp et al. 1997; Gazeau et al. 2005a; Stæhr et al. 2018).

One advantage of the open-water technique is that long time-series of O_2 concentrations can be measured with relative ease using modern sensor systems that are generally precise, durable, and self-cleaning (Table 14.1). These sensor systems can be deployed remotely in coastal waters to provide high-frequency (e.g., 15 minutes intervals) observations for durations of weeks to months (e.g., Caffrey 2004), making this the only method to provide direct measurements of ecosystem metabolism on a wide range of temporal and spatial scales. In addition, this open-water approach provides direct measurements of ecosystem-scale production and respiration, thus avoiding artefacts associated with "container effects" or rate extrapolation. Although this method does not provide habitat-specific metabolic rates, combining it with container incubations in a single study provides both direct measurements of whole system rates and their partitioning among benthic and pelagic subsystems (e.g., Smith and Hollibaugh 1997; Kemp et al. 1997).

Among the disadvantages of this approach is the common need for separate measurements of air-water O_2 (or CO_2) exchange to correct for an inherent bias toward lower rates due to uncertainties in air-water gas exchange (Table 14.1). Direct measurements of air-water exchange for a specific system can be difficult to obtain (Raymond and Cole 2001) and gas exchange often varies substantially from system to system (Kremer et al. 2003) and with fluctuations in physical forces (Marino and Howarth 1993; Clark et al. 1996; Borges et al. 2004; Borges and Abril 2010). In dynamic tidal waters, diel changes in O_2 may be masked by physical forces (air-water gas exchange, vertical/horizontal mixing, and advection) making it difficult to detect the metabolic signal. Except for periods with very limited water movement, it seems appropriate to think of the measured open water metabolic rates as apparent, rather than absolute rates of local production and respiration (Stæhr et al. 2018). Also, metabolic estimates for large, dynamic estuarine systems with strong horizontal gradients require spatial arrays of numerous sensor systems to be adequately measured (e.g., Martz et al. 2008). Open-water time-series O_2 measurements are nevertheless assumed to provide reliable estimates of the integrated ecosystem rates in well-mixed estuarine systems (Caffrey 2004). The technique has been successfully applied to evaluate system-integrated rates across larger heterogeneous areas as compared to compartment-specific rates where incubation and eddy correlation techniques are preferable (Kemp and Testa 2011, Hume et al. 2011; Stæhr et al. 2012). *In situ* data sondes are thus an attractive option for assessing ecosystem metabolism, being generally less labor intensive and requiring less expertise than water column or benthic incubation methods. However, caveats such as stratification and lateral exchange of waters need to be carefully considered, and ancillary data (profiles of temperature and salinity, and incubations) help interpret estimates of metabolism derived from open water O_2 time series data (Murrell et al. 2018). Statistical modeling of relationships among O_2, CO_2, water temperature and/or salinity can furthermore provide a means for inferring metabolic rates in systems where the time-space variability due to physical processes is substantial (Swaney et al. 1999; Lee 2001).

14.3.4 Input-Output Budgets

At larger space and time scales, a range of mass-balance and budget approaches can be applied for computing rates of *NEP* in coastal ecosystems. The most straightforward of these approaches includes steady state nutrient, organic carbon, and/or O_2 input and output budgets that are based upon flux estimates for each material across system boundaries. The boundary fluxes can be measured and/or derived from simulation models. In this case, *NEP* is often computed as the residual of the mass balance. For estuaries and other semi-enclosed coastal systems, values of *NEP* can be inferred from measurements of differences between input and output fluxes for organic carbon, O_2, or inorganic and organic nutrients (e.g., Kemp et al. 1997; Smith et al. 2005). Metabolite inputs from the atmosphere and anthropogenic point sources (e.g., sewage effluents) are usually measured by routine monitoring programs (e.g., Carstensen et al. 2006). Inputs from diffuse watershed sources can generally be obtained from river monitoring at fall lines and/or hydrochemical modeling of runoff. Such point and diffuse inputs from the surrounding watershed are sometimes combined in mass balance calculations. For most budget calculations, direct estimates of mass exchange fluxes between the estuary and adjacent ocean must be derived using physical transport models of varying complexity (e.g., Boynton et al. 2008; Testa and Kemp 2008) because measuring such fluxes is difficult.

14.3.5 Transport-Transformation Models

A related approach for estimating values of *NEP* from mass-balances involves computing net biogeochemical transformations of non-conservative solutes within a system using hydrodynamic models or steady-state calculations of transport. These calculations are often applied to coastal systems with clear land-sea salinity gradients that permit transport computations using salt- and water-balance models (Smith et al. 1991; Gazeau et al. 2005a; Testa et al. 2008). This approach calculates the residual water fluxes between pre-defined volumes in systems where freshwater inputs, salinity distribution, and system bathymetry are known. For non-conservative solutes (e.g., PO_4^{3-}), mass balance requires an additional residual term to represent the net effect of all biogeochemical processes involving the solute. One approach for this kind of modeling focuses on calculating the net non-conservative production or consumption of dissolved inorganic phosphorus (DIP). In assuming that biological processes dominate phosphorus cycling, for example, net DIP consumption represents net biological uptake of DIP, which is converted to equivalent net carbon production assuming that stoichiometry is fixed (Smith et al. 1991). Thus, net DIP consumption is used to compute *NEP*. This approach has been supported by and widely used in the Land-Ocean-Interaction-Coastal-Zone (LOICZ) Project (Crossland et al. 2005). A related approach uses this method to balance non-conservative metabolic gases (e.g., O_2 or CO_2), also accounting for air-water gas exchange to compute *NEP* (e.g., Smith and Hollibaugh 1997; Testa and Kemp 2008). A two-layer version of this approach, which can be applied for stratified estuarine systems, provides metabolic rates that are analogous (but not equivalent) to *NDP* in the upper photic layer and *R* in the sub-photic bottom layer (e.g., Hagy et al. 2000; Testa et al. 2008).

An inherent strength for these mass-balance approaches is the fact that they are computed over large space and time scales, thus providing integrated measures of *NEP* that can be related to large time- and space-scale external forcing (e.g. Stæhr et al. 2017). These estimates of net ecosystem metabolic fluxes can also be quantitatively related to other large-scale carbon and nutrient fluxes in the ecosystem (e.g., Kemp et al. 1997). The conceptual simplicity of this approach combined with the widespread availability of monitoring data for constructing mass-balances makes it an excellent tool for comparative analysis of *NEP* across many different coastal ecosystems (e.g., Smith et al. 2005). A potential limitation of the transport-transformation modelling approach is that it requires two key assumptions, including (1) C:N:P stoichiometry is fixed and known and (2) physical-chemical reactions have no net effect on DIP dynamics (e.g., Gazeau et al. 2005a). The primary disadvantage of all mass-balance approaches is that many measurements or computations are required to estimate *NEP*, and some of these rates may be difficult to quantify and may have large associated errors. Problems of spatial and temporal extrapolation are minimized in systems with long water exchange times and relatively uniform conditions. The scales of the various flux estimates required for a complete mass-balance must be consistent, they must cover the same temporal period and spatial domain, and they must be measured at frequencies and spatial densities adequate to reflect integrated rates. Although it can be difficult to quantify the errors and uncertainties in many of these measurements, recent studies describe rigorous approaches for computing error accumulation in these mass-balance calculations for nutrients and other materials (e.g., Lehrter and Cebrian 2010).

14.4 Regulating Factors and Spatial and Temporal Patterns

An appreciation of the various internal and external factors that regulate the balance between production and consumption of organic matter is necessary for understanding ecosystem metabolism. These factors are both natural and anthropogenic, and spatial and temporal variations in these controlling factors result in distinct gradients and patterns in metabolic rates. This section highlights a subset of controlling factors that exert strong and ubiquitous controls on coastal ecosystem metabolism in varied coastal aquatic ecosystems.

14.4.1 Light and Water Clarity

Primary production in coastal waters may be limited by nutrients, temperature, and mixing, but light availability is the ultimate driver of photosynthesis. Light energy is absorbed by a suite of plant pigments, notably chlorophyll-*a*, and used to drive the light and dark reactions of photosynthesis to fix CO_2 and synthesize simple organic compounds. For example, rates of phytoplankton photosynthesis follow daily (e.g., Harding et al. 1981) and seasonal (e.g., Smith and Kemp 1995) variations in solar intensity and duration (Figure 14.3), especially in temperate climates (Smith and Hollibaugh 1997; Hashimoto et al. 2006). Photosynthetic rates generally follow a hyperbolic relationship with light up to saturation levels (e.g., Harding et al. 1981), where at higher light levels, UV-inhibition of photosynthesis is sometimes evident (e.g., Powles 1984). Such photoinhibition of phytoplankton photosynthesis is rarely observed *in situ* in most coastal ecosystems because turbid water combined with rapid vertical mixing precludes long exposures to damaging light levels (e.g., Gallegos and Platt 1985). As a result, simple empirical models are used to effectively compute phytoplankton productivity based on light availability, phytoplankton biomass, and photic zone depth across a range of diverse coastal ecosystems (Cole and Cloern 1987).

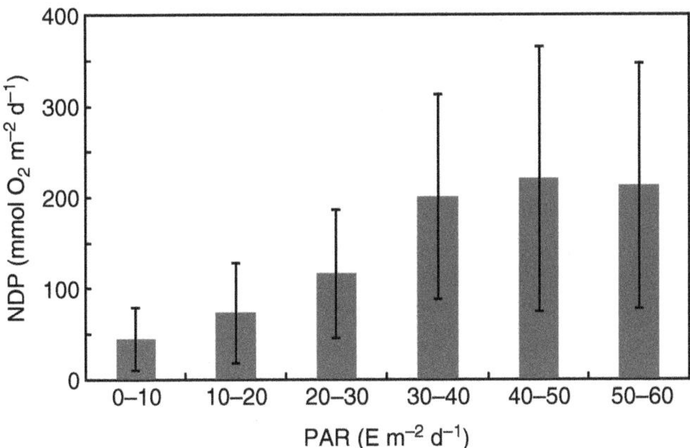

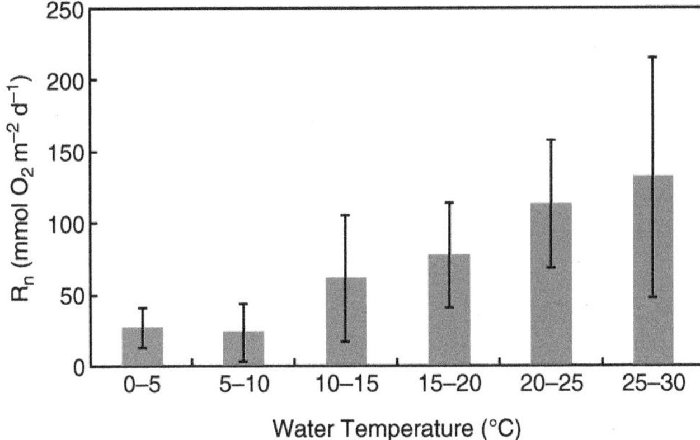

FIGURE 14.3 Plots of binned means (±SD) of (a) net daytime production(*NDP*) versus photosynthetically active radiation (PAR) and (b) nighttime respiration (R_n) versus water temperature in the Corsica River estuary, a sub-tributary of Chesapeake Bay. Metabolic rates were calculated from short-term changes in water column O_2 and represent whole-ecosystem metabolism. *NDP* correlates with PAR, but saturates at 30–40 E m^{-2} d^{-1} with high variability at high PAR, due to turbidity-induced reductions in *NDP*. R_n increases exponentially with temperature and does not appear to saturate up to 30 °C. O_2 and temperature data from Maryland Department of Natural Resources (**www.eyesonthebay.net**) and PAR data from the Horn Point Laboratory weather station in Cambridge, Maryland, U.S.A.

As a result of light control on photosynthesis, the daily rising and setting of the sun fuels one of the most clearly observed metabolic cycles in coastal ecosystems. Net increases in O_2 during sunlit hours reflect net ecosystem photosynthesis, while declines in O_2 during the night result from net respiration (Box 14.4). A common lag period (2–5 hours) between dawn and the initial switch from net respiration to net photosynthesis (Box 14.4) reflects the time required for gradually increasing photosynthetic rate to exceed baseline respiration rates. Direct bottle incubation measurements of plankton community photosynthesis and respiration indicate that mid-day maxima in *GPP* are typically followed by respiration peaks in late afternoon. Late afternoon peaks in respiration rates, however, may be driven by light-enhancement of respiration associated with high ratios of O_2:CO_2 concentrations.

Water clarity is a critical factor influencing light availability and photosynthesis that varies widely among coastal ecosystems. In natural waters, the intensity of light reaching below-surface depths is reduced rapidly due to absorption and scattering by algal cells (live and dead), dissolved organic matter (e.g., CDOM), and a variety of non-algal suspended inorganic and organic particles in the water column (Wofsy 1983; Harvey et al. 2019). Consequently, photosynthetic rates and the accumulation of phytoplankton chlorophyll-*a* are often reduced in estuaries receiving high loads of inorganic particles and/or allochthonous organic material (e.g., Wofsy 1983; Kemp et al. 1997). Photosynthesis of benthic plants is especially sensitive to decreases in water clarity (McGlathery et al. 2007) because unlike phytoplankton, benthic algae and submersed vascular plants are attached to the seabed where their lower depth limit is typically controlled by light penetration and sediment composition (e.g., Krause-Jensen et al. 2011). *GPP* is thus often lower in regions of estuaries nearest the injection point of riverine turbidity and CDOM (e.g., upper, low salinity estuarine regions; Smith and Kemp 1995; Gazeau et al. 2005c), especially during years or seasons with high river flow, such as temperate spring months and tropical monsoon periods (Ram et al. 2003).

Although respiration in coastal ecosystems is generally not directly affected by light levels, two processes can cause elevated plant respiration in the light. The first process, photorespiration, occurs in many plants when low intracellular concentrations of CO_2 allows O_2 to bind with RuBisCO forming glycolic acid. This material may be respired within the plant or excreted to surrounding waters where it used in microbial respiration (Laws et al. 2000). A second process by which O_2 consumption and CO_2 production may be stimulated in the light is the Mehler reaction (Raven and Beardall 2005). Experiments suggest that the sum of these two processes of light-stimulated respiration probably account for < 20% of *GPP* (Laws et al. 2000). Dark bottle respiration rates measured with incubations over diel cycles (e.g., Sampou and Kemp 1994) suggest that mid-day rates exceed night-time rates. However, over longer time periods (>day), high light conditions generally increase respiration indirectly by stimulating photosynthesis, which provides both organic substrate and O_2 to support elevated aerobic respiration. In benthic environments, photosynthesis by benthic microalgae and submersed vascular plants increases the depth of O_2 penetration into sediments, resulting in elevated aerobic respiration and chemoautotrophic metabolism (e.g., Epping and Jørgensen 1996).

14.4.2 Temperature

Temperature affects ecosystem metabolism through several direct and indirect interactions (Figure 14.3). Temperature has a strong effect on respiration, due to its influence on enzyme-catalyzed cellular metabolism. Metabolic rate dependencies on temperature are expressed as the ratio of rates at a given temperature and those 10 °C warmer, which is called the "Q_{10}" factor. A number of comparative analyses show that respiration of both planktonic and benthic communities follow strong exponential relationships to temperature (Hopkinson and Smith 2005), which indicates that Q_{10} should be dependent on the magnitude of temperatures in the comparison (i.e., a Q_{10} for 5–15 °C will be lower than that for 15–25 °C). Such temperature-respiration relationships were reported for a diversity of coastal ecosystems (Figure 14.3; Smith and Kemp 1995) and strong correlations between temperature and respiration also result from covariance between temperature and primary production rates. In the latter case, elevated productivity increases the availability of labile organic matter during daytime photosynthesis, which fuels increased algal respiration during daylight and may also cause respiration to be enhanced shortly after sunset compared to night-time respiration rates (which tend to be dominated by maintenance metabolism; Markager et al. 1992). In deeper, aphotic waters, respiration may peak with temperature (as in surface waters) and in many temperate ecosystems, peak benthic respiration co-occurs with peak temperature (e.g., Middelburg et al. 1996). Organic matter reactivity is thought to be more important than temperature in regulating respiration rates (Satta et al. 1996), especially in systems with large inputs of organic matter from terrestrial and anthropogenic sources, in sediments that are dependent on sinking organic material to support respiration (Graf et al. 1982), and in sediments where abundant macrobenthic fauna enhances respiration rates (Hopkinson and Smith 2005).

Seasonal and interannual temperature fluctuations can directly and indirectly affect rates of *NDP* and *GPP* in coastal ecosystems. Temperature exerts indirect effects on primary production, as planktonic and benthic nutrient recycling rates are generally enhanced in warmer water, potentially increasing the growth rate of nutrient-limited phytoplankton and benthic plants (Heip et al. 1995). Although the photochemical processes in Photosystem I and II and the initial slope of light-saturation curves (e.g., Platt and Jassby 1976) are generally unaffected by changes in temperature, temperature tends to set the maximal growth rates (μ_m) of phytoplankton under ideal light and nutrient-saturated conditions (Eppley 1972). This may be the case because the temperature sensitivity of primary production depends on the availability of nutrients (Marañón et al. 2018), where under nutrient replete conditions, internal enzymatic processes become rate limiting rather than diffusion outside the cell, but diffusion can be limiting under nutrient-poor conditions. Modeling studies similarly found that simulation performance was generally improved by using formulations for temperature-dependent μ_m rates that included modulation by light and nutrient conditions (Brush et al. 2002). Although coastal phytoplankton (and other plants) tend to adapt to local conditions, extremely high temperatures (>30 °C) can inhibit plant growth (e.g., Eppley 1972). It is generally considered that ecosystem respiration is more sensitive to temperature (as reflected in higher Q_{10} values) than ecosystem photosynthesis. Consequently, global increases in water temperature are expected to cause widespread reductions in *P*/*R* ratios and increased heterotrophy for oceanic and coastal systems (López-Urrutia et al. 2006).

Whereas both *GPP* and *R* are strongly influenced by (and correlate with) seasonal variations in temperature, temperature effects on seasonal or interannual fluctuations in *NEP* are often difficult to detect. In many temperate estuaries, *GPP* may follow the annual light cycle (peaking in June), while ecosystem respiration tends to follow the annual temperature cycle (peaking in July or August), resulting in spring annual maxima in *NEP* (Smith and Kemp 1995). Conversely, in estuaries with seasonal climate patterns driven by the monsoon cycle, *NEP* may be autotrophic during low-flow pre-monsoon periods, but reduced to heterotrophic conditions during high-flow (and high allocthonous organic matter inputs) monsoon periods (e.g., Ram et al. 2003). Thus, year-to-year variations in *NEP* may be related to a variety of factors other than temperature, as interannual variability in temperature is small relative to changes in other external forces (e.g., turbidity, nutrient inputs). Finally temperature affects the measurements of open water metabolic rates indirectly via influences on physical conditions such as water column stratification and solubility of gasses in water.

14.4.3 Inorganic Nutrient and Organic Matter Inputs and Toxic Contaminants

In many coastal ecosystems, photosynthetic production and respiratory consumption of organic matter are regulated by both inorganic nutrient and organic matter inputs. Typically, variations in land-sea input rates are driven by climatic variability or anthropogenic forcing (e.g., Smith and

Hollibaugh 1997; Testa et al. 2008; Cox et al. 2009), as changes in freshwater inputs and watershed land-use control input rates. As a result of combined anthropogenic and climatic forcing, input rates vary on seasonal to multi-decadal timescales. Primary production (both *NDP* and *GPP*) rates are often positively correlated with inorganic nutrient inputs in many types of aquatic ecosystems, as nutrients increase *GPP* and allow for increased algal and marsh biomass production (e.g., Valiela and Teal 1979; Oviatt et al. 1986; D'Avanzo et al. 1996; Caffrey 2004). For example, controlled multi-year experiments in mesocosms displayed that a large increase in inorganic nutrient inputs elicited significant increases in *NDP* and *NEP* (Oviatt et al. 1986). Such elevated organic production, in turn, induces elevated respiration rates, because much of the newly produced organic matter will likely be respired within the time-frame of a year (Oviatt et al. 1986; Smith and Hollibaugh 1997; Hashimoto et al. 2006). In tidal wetlands, there is some concern that excess nutrient availability shifts plant allocation from belowground to aboveground tissue that may ultimately result in loss in marsh elevation (Turner et al. 2009). Respiration rates vary substantially with changes in external loads of organic matter (e.g., Satta et al. 1996; Smith and Hollibaugh 1997), especially in upstream regions of estuaries adjacent to wastewater treatment plants and riverine injection points (Gazeau et al. 2005a). In some systems, ecosystem metabolism is driven by organic matter (e.g., Smith and Hollibaugh 1997) or inorganic nutrient inputs (Testa et al. 2008) from seaward sources.

The ratio of inorganic nutrient to organic matter inputs is a predictor of *NEP* and *P/R* for most coastal systems (Figure 14.4; Hopkinson and Vallino 1995; Kemp et al. 1997; Gazeau et al. 2005a; Herrmann et al. 2015) because inorganic nutrients directly stimulate *GPP* and organic nutrients fuel ecosystem respiration. For example, ecosystem metabolism in Tomales Bay is net heterotrophic, and the system is thus a sink of organic carbon and a source for CO_2. This is because the estuary's small watershed is sparsely populated, but delivers substantial terrestrial organic carbon and little inorganic nutrients to the adjacent bay (Smith et al. 1991; Smith and Hollibaugh 1997). However, the Scheldt River estuary has negative *NEP* despite high inorganic nutrient loads, due to light limitation and excessive inputs of allochthonous organic material (Gazeau et al. 2005a). Narragansett Bay, in contrast, receives higher loads of inorganic nutrients relative to organic carbon from its watershed and ecosystem metabolism is slightly autotrophic in this system (Figure 14.4; Nixon et al. 1995). The Patuxent River estuary is highly enriched with inorganic nutrients and is autotrophic, but interannual variability indicates shifts between autotrophy and heterotrophy caused by variations in other external forces (Testa et al. 2008).

The effect of organic matter inputs on the metabolic balance of coastal waters depends in large part on the bioavailability and elemental composition of this material. In general, organic matter produced by phytoplankton and benthic microalgae tends to be very labile (high bioavailability) with relatively low molar ratios of C:N (~7) and C:P (~106), so that respiration of 106 moles of C will result in release of 16 moles of inorganic N and 1 mole of inorganic P. In contrast, organic material produced by seagrass and salt-marsh grass tends to be more refractory with higher C:N ratios (~20 and ~40, respectively) because a larger portion of the plant biomass is associated with structural tissues (e.g., Twilley et al. 1986). Without the buoyancy effects of water, terrestrial plants have even more biomass devoted to structural carbon, and C:N ratios can exceed 150 (e.g., Cloern et al. 2002). Although cellulose, lignin, and other structural compounds tend to be relatively resistant to degradation, they do support respiration of bacteria and fungi, but with very low rates of inorganic nutrient release. Inputs of allochthonous terrestrial organic matter thus tends to support lower respiratory rates that release far less inorganic N and P, compared with a similar input of algal-derived organic matter produced by phytoplankton. Therefore, increased inputs of terrestrial organic matter generally increases estuarine respiration with relatively little effect on photosynthesis from associated nutrient recycling, whereas enhanced inputs of organic matter from algal-derived material will tend to increase both *R* and *GPP*, while driving a net increase in heterotrophy.

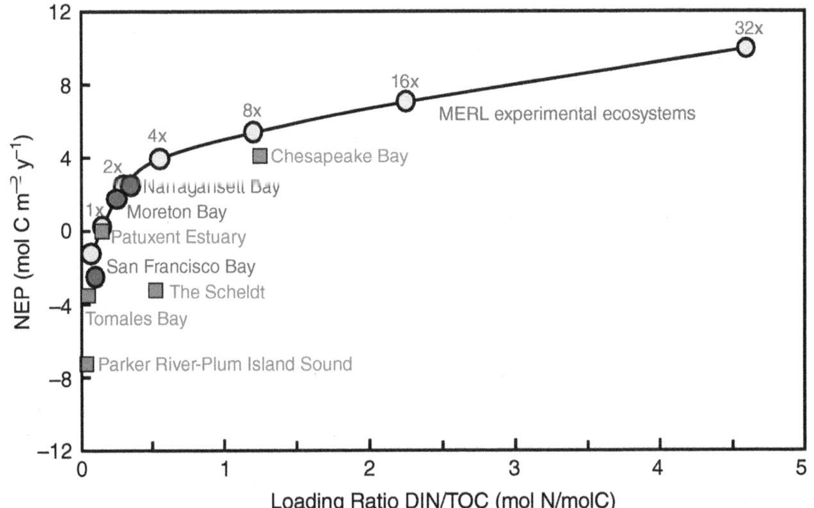

FIGURE 14.4 Comparative analysis of net ecosystem metabolism for estuarine systems in relation to the loading ratio of DIN:TOC (dissolved inorganic nitrogen: total organic carbon). Data from experimental marine ecosystems (yellow, filled circles, MERL, Oviatt et al. 1986, C=control and 1×, 2×, etc. are scales of nutrient enrichment relative to control), selected estuaries for which sufficient information is available (blue circles, Kemp et al. 1997; Moreton Bay from Eyre and McKee 2002), and the case study systems from this chapter (red squares), including the Patuxent River estuary (Testa et al. 2008), Chesapeake Bay (Kemp et al. 1997), the Scheldt (Hofmann et al. 2008), Tomales Bay (as in Kemp et al. 1997), and the Parker River-Plum Island Sound ecosystem (Vallino et al. 2005).

Inputs of toxins to coastal ecosystems may also affect *NEP*, *GPP*, and *R*. The toxicological literature is filled with data reporting experimental ecosystem metabolic responses to diverse organic and inorganic contaminants (e.g., Graney et al. 1994). Response patterns vary for different toxic chemicals, but there is a general concern that the homeostatic nature of ecosystem metabolism makes these rates poor indicators of low-level ecosystem stress. A study contrasting estuarine ecosystem metabolic responses to inputs of nutrients (N and P) and heavy metals (arsenic, copper, cadmium) reported that ecosystem production was significantly stimulated by nutrients and inhibited by metal toxins, while input of both sets of chemicals resulted in slight enhancement of *GPP* and *NEP* (Wiegner et al. 2003). In contrast, herbicide additions to seagrass-dominated mesocosms resulted in sharp declines in *NDP* that abated within weeks as herbicides degraded (e.g., Cunningham et al. 1984). It is interesting that, despite these large impacts on *GPP*, community respiration showed little response to these phytotoxic compounds over the full course of 6–10 week treatments (e.g., Kemp et al. 1985). In rare cases, extreme levels of NH_4^+ may be toxic to phytoplankton and reduce *NDP* (Cox et al. 2009).

14.4.4 Exchanges of Organic Matter within Ecosystems

Analyses of metabolic balance in ecosystems open to exchange (especially streams and macrotidal estuaries) must consider how organic material is exchanged between adjacent aquatic and wetland ecosystems (Box 14.3). Measurements of *NEP* can be used to estimate the direction and magnitude of organic transport across system boundaries, as organic matter is generally imported to fuel heterotrophic ecosystems and exported from autotrophic ecosystems at steady-state. Of course, organic carbon export may be driven by a combination of both autotrophic *NEP* and the throughput of unused imported materials, but it is nonetheless useful to consider potential relationships between net biological production and net physical transport.

Several studies linked net ecosystem production in upper euphotic layers of the water column to net *vertical transport* (i.e., sinking) of particulate organic carbon (POC) and to subsequent bottom respiration supported by these sinking organic materials. For instance, in stratified estuarine ecosystems there can be a strong inter-annual correlation between surface-layer net O_2 production and net O_2 consumption in the underlying layer (Testa and Kemp 2008) and an analogous relationship between metabolism and POC deposition (e.g., Bozec et al. 2006). Direct measurements of *NEP* and sinking of POC were also strongly correlated in mesocosm experiments (Oviatt et al. 1993), field studies in estuaries (Kemp and Boynton 1992), and on the North American continental shelf (Kemp et al. 1994). Perhaps the most important example of downward vertical sinking of surface water POC on a global scale is the potential export of newly-fixed carbon to the deep oceans as part of the "biological pump" (e.g., Lutz et al. 2007).

Changes in *NEP* are also related to variations in net *horizontal transport* between adjacent coastal regions. For example, despite the high net autotrophy of the seagrass plants themselves, net heterotrophy is often observed for the integrated ecosystems associated with seagrass beds (Barron et al. 2004) because of the tendency for allochthonous POC to be trapped within the bed and respired by benthic microbes and metazoan communities. The box-modeling analyses described above (Bozec et al. 2006; Testa and Kemp 2008) showed that regions of net autotrophy were often adjacent to regions of net heterotrophy, and that seaward horizontal transport of excess production in the net autotrophic areas fueled heterotrophy in downstream regions. Horizontal transport of carbon between vegetated tidal wetlands and adjacent estuarine waters is likely an important but understudied linkage between typically autotrophic (wetland) and heterotrophic (estuary) components of wetland ecosystems. A significant portion of the carbon fixed in vegetated wetlands may be exported to estuarine waters (Najjar et al. 2018).

Horizontal transport of organic matter also determines regional metabolic balance. Landward transport of organic matter from autotrophic coastal upwelling areas can also drive heterotrophy in nearby estuaries (Smith and Hollibaugh 1997). On a similar note, high production in nearshore net autotrophic environments by seagrasses may provide organic matter to deeper more central parts of an estuary where net heterotrophy dominates (Stæhr et al. 2018). Previous conceptual models of "river to sea" interactions of organic matter and nutrient cycles explained observed horizontal patterns of *NEP* along estuarine land-sea gradients in terms of linkages between *NEP* and net organic matter transport (Heath 1995; Hopkinson and Vallino 1995). Such estuarine-centric models can be related to the river continuum concept (Vannote et al. 1980), which describes a pattern of heterotrophic low-order streams giving way to autotrophic high-order river systems. It is suggested that forest cover in low-order streams has the dual function of restricting light to stream beds and providing ample amounts of organic material to fuel respiration, while higher-order streams receive nutrients from upstream respiration and are wide enough to provide ample light for photosynthesis, permitting net autotrophy. This pattern is intimately linked with the concept of "nutrient spiraling" in streams, which has recently been documented along a salinity gradient in an estuary (Asmala et al. 2018). Autotrophic uptake and heterotrophic release of nutrients co-occurs with unidirectional, downstream flow, thus decoupling regions of organic matter degradation/nutrient release from those of organic matter production/nutrient uptake (Ensign and Doyle 2006).

14.4.5 River Flow, Flushing, and Wind

Because many estuaries, by their nature, receive freshwater, organic matter, and nutrient inputs from one or many sources, variations in freshwater flow exert strong, if not dominant, controls over metabolic rates (Paerl et al. 1998; Howarth et al. 2000;

Russell et al. 2006). Depending on the estuary and its watershed, inter-annual changes in river flow can increase or decrease metabolic rates, or lead to spatial and temporal shifts in peak metabolism. In estuaries receiving large amounts of inorganic nutrients with river flow, elevated flow leads to increases in phytoplankton production (e.g., Paerl et al. 1998). Conversely, elevated river flow can lead to reduced *GPP* by importing suspended material that reduces light availability (e.g., Howarth et al. 2000). If the estuary is large enough, flow may reduce *GPP* via turbidity in landward estuarine regions, while elevating *GPP* in seaward waters where dissolved nutrients are delivered, but solids had settled to sediments in landward reaches. Large seasonal pulses of freshwater and organic material (e.g., monsoon) can also reduce *GPP* and *NEP* via light reductions and elevated respiration due to organic matter imports (Ram et al. 2003). More complex interactions occur in naturally turbid estuaries, where freshwater discharge can affect both water column stability and turbidity (Cloern et al. 1983; Howarth et al. 2000), and large stream-flow pulses can radically alter *NEP* with differential effects on *GPP* and *R* resulting from mixed inputs of DOC and/or turbidity (Russell et al. 2006).

Day-to-day fluctuations in *GPP*, *R*, and *NEP* are often large, due to daily variance in other external forcing variables, such as wind and sunlight. In the few instances when continuous daily measurements of *NEP* were reported over a season (e.g., Figure 14.3; Stæhr and Sand-Jensen 2007), variations in *NEP* were greater than 10-fold. Large daily variability in *GPP* is due to variations in cloud-cover and resulting changes in sunlight (Figure 14.3). Wind-induced changes in vertical mixing (e.g., Stæhr and Sand-Jensen 2007) and lateral exchange (Hume et al. 2011; Stæhr et al. 2018) can also cause fluctuations in ecosystem production. In shallow ecosystems, short-term increases in wind stress lead to higher rates of sediment resuspension, which reduces light penetration and thus *GPP*. Wind-induced resuspension may also reintroduce relatively labile organic matter into the water column (Demers et al. 1987), stimulating plankton community respiration and reducing rates of *NEP* (Dokulil 1994). However, wind-induced resuspension can transport benthic diatoms into the water column where they can access light and achieve higher rates of photosynthesis (e.g., Demers et al. 1987). Wind-induced resuspension, however, typically reduces light availability to elevate the light requirements for benthic vegetation (Duarte et al. 2007) and decrease productivity (Stæhr et al. 2018).

The exchange time of water parcels in estuarine ecosystems (i.e., the average time a given water parcel remains in the estuary) can also regulate the balance between photosynthesis and respiration by controlling the time that organic material may be processed within the system (Hopkinson and Vallino 1995). Exchange time can vary within an estuary over seasonal and interannual time scales, particularly depending on changes in freshwater inputs (Hagy et al. 2000), tidal cycles, and offshore climatic variability. Modeling studies suggest that short exchange times would promote net autotrophy as dissolved inorganic nutrients are rapidly assimilated but particulate organic material is flushed from the system

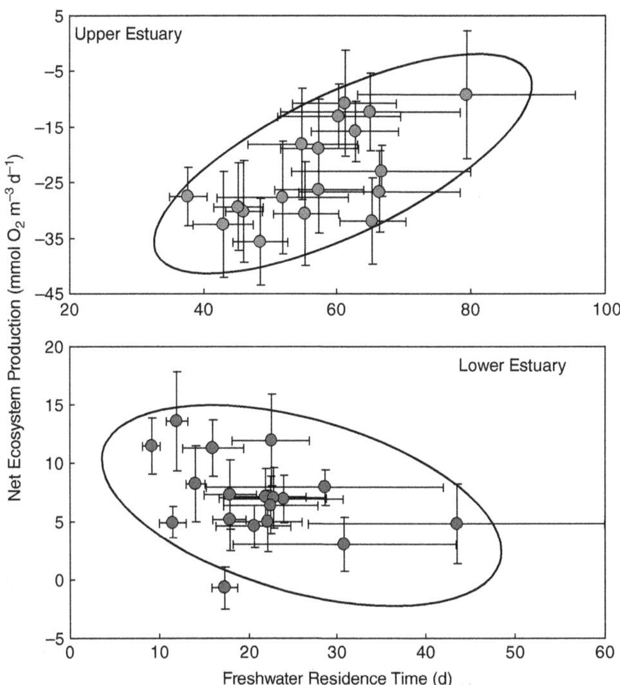

FIGURE 14.5 Relationships of freshwater residence time (FRT) with net ecosystem production (*NEP* = production-respiration) in two regions of the Patuxent River estuary (±SE; FRT from Hagy et al. 2000 and *NEP* data from Testa and Kemp 2008). The data indicate that *NEP* is positively related to FRT in the upper estuary, but in contrast, *NEP* and FRT are negatively correlated in the lower estuary.

before being respired (Hopkinson and Vallino 1995). In the LOICZ program, *NEP* rates and estuarine exchange times were computed for roughly 200 estuarine ecosystems (Crossland et al. 2005). These data indicate that *NEP* was negatively correlated with exchange time (Smith et al. 2005; Borges and Abril 2010), with more variability in *NEP* with reduced exchange time. This suggests that longer exchange times increase net heterotrophy (or decrease net autotrophy) by increasing the amount of time that organic material can be respired within the ecosystem (Smith et al. 2005). Similarly, in the lower Patuxent River estuary, low exchange times are associated with elevated autotrophy (Figure 14.5) due to high flushing of organic material out of the ecosystem (Vallino et al. 2005) and stimulation of *GPP* via associated elevated nutrient input rates. In contrast, long exchange times lead to more autotrophic conditions in the turbid and shallow upper Patuxent River estuary, where reduced sediment input and enhanced stability of the water column allow elevated rates of *GPP* (Figure 14.5). Recent syntheses have documented consistent scaling between ecosystem metabolism and estuarine volume, concluding that the residence time of relatively new and nutrient-rich water (which scales to size) is a key factor for metabolism (Nidzieko 2018).

14.4.6 Water Depth

The general depth-dependence of metabolic balance (Odum 1967; Alderman et al. 1995) regulates the relative dominance

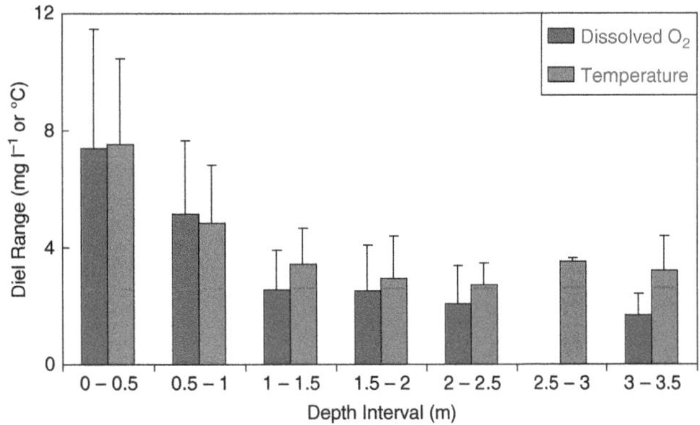

FIGURE 14.6 Average diel ranges (daily maximum - daily minimum) in water temperature and O_2 in 0.5-m depth categories in several Texas coastal bays during summer. *Source*: Redrawn from Odum (1967).

of primary producers and the relative contributions of benthic and pelagic rates to total ecosystem respiration. Because light levels reaching the sediment surface in shallow estuaries are often sufficient to support submerged vascular plants, *NEP* rates in these systems will likely be autotrophic because (1) such rooted plants are able to fix more CO_2 per unit nutrient input because vascular plant tissues have much higher ratios of C:N or C:P than microalgae, and (2) water column respiration rates tend to be low in clear-water environments (Ziegler and Benner 1998; Stæhr et al. 2018). Thus, *NEP* shifts from net autotrophy to net heterotrophy with increases in depth because water-column-integrated plankton respiration increases while *GPP* does not change (Ryther 1961; Wofsy 1983) and benthic photosynthesis is inhibited by light limitation (Van de Bogert et al. 2007). In studies of the shallow, turbid bays of the Texas coast, Odum (1967) noted an exponential decline in the daily excursions of O_2 (a measure of *GPP*) and temperature with depth (Figure 14.6). This decline in variability indicated that large diel shifts in O_2 were representative of high metabolic rates found in shallow waters, where high O_2 excursions may be related to the presence of rooted autotrophs or to temperature stimulation of *GPP* and *R* (Figure 14.6).

Depth interacts with nutrient status and habitat types to determine the dominant primary producers in an ecosystem and their effects on metabolic rates. Measuring individual components of an ecosystem using container incubations (e.g., Gazeau et al. 2005a) allows partitioning the contribution of various biotic units to *NEP*. Such an approach has been applied to studies focusing on the ecological role and metabolic contribution of seagrass and related submersed plants in shallow coastal environments (Stæhr et al. 2018). For example, seagrass growing in oligotrophic waters generally dominate shallow-water *NEP* over an annual cycle both in comparison to adjacent areas without seagrass and to other biotic components within the community. Seagrass beds also produce a clear metabolic signature in nutrient-poor, clear-water environments like the Mediterranean Sea margins (Gazeau et al. 2005b). In nutrient-enriched environments, such as the coastal Texas lagoons, seagrass contributions to *NDP* are often less important. In these lagoons, seagrass beds accounted for only ~30–40% of ecosystem *GPP*, with macroalgae (~30–45%), phytoplankton (~15–20%) and benthic microalgae (~10–35%) accounting for the remaining primary production (Kaldy et al. 2002). Additionally, mesocosm experiments suggested that the relative contribution of seagrass to *NEP* tends to decline with high inputs of nutrients, because seagrass are either forced to compete with phytoplankton and epiphytic algae for nutrients or because water column nutrients cause algal and epiphyte shading that reduces light availability for seagrass growth.

Flooding depth also exerts a strong influence on rates of metabolism and the fate of metabolites in vegetated tidal wetlands. Tidal flooding on marsh surfaces sufficient to inundate the plant community strongly reduces *apparent* rates of *GPP* and *R* as measured by vertical exchange; i.e., *NEE* measured with chambers (Zawatski 2018) or eddy covariance towers (Forbrich and Giblin 2015; Box 14.3). The diffusion of CO_2 for photosynthesis and O_2 for plant respiratory processes between plants and their environment is highly constrained during submergence, which reduces plant production and respiration (Mommer and Visser 2005). Light availability is also attenuated, especially when flood waters are turbid, further limiting photosynthesis (Vervuren et al. 2003). Some plants are capable of performing underwater photosynthesis (Winkel et al. 2011), though it is unclear how important this process is during tidal flooding in coastal wetland systems. Flooding further limits oxygen diffusion into soils thereby reducing rates of aerobic respiration (Froelich et al. 1979). However, anaerobic respiratory processes in marsh soils, which are typically the dominant pathway of organic matter decomposition in these systems (Howarth and Teal 1979), do not cease upon flooding. CO_2 produced during respiration may not escape to the atmosphere when the wetland is flooded, but rather diffuse into the overlying flood water as DIC which can be carried into the estuary as the tide ebbs. The export of DIC from wetlands to estuarine waters has been shown to be an important component of coastal ecosystem carbon budgets (Neubauer and Anderson 2003; Wang et al. 2016), though it remains poorly constrained and is an area of active research.

Although wetland flooding may reduce rates of *GPP* and *R* briefly during periods of inundation, it is important to note that the overall impact of tidal flooding on marsh productivity is positive. Tidal flooding is important in flushing wetland soils, which would otherwise become hypersaline (as water evaporates leaving behind salts) and sulfidic (from the buildup of sulfides from sulfate reduction), conditions that are toxic to plants (King et al. 1982). Tidal flooding also delivers nutrients that can fuel primary production in marshes (Gribsholt et al. 2005). Moderate increases in tidal inundation have been shown to increase salt marsh productivity, an important mechanism that may allow saltmarshes to keep pace with modest increases in sea level (Morris et al. 2002). A key feature of most tidal wetlands is that the tallest, most productive vegetation can be found along the lower elevation, more frequently flooded creek-banks (King et al. 1982).

14.5 Ecosystem Metabolism Applications and Case Studies

Over the last several decades, certain coastal ecosystems have been sites of extensive integrated research focused on ecosystem metabolism and its relationship to key ecological, biogeochemical, and management factors. A subset of these sites represents an interesting combination of habitat types and physical settings, including a wide range of circulation patterns, depths, habitat types, dominant primary producers, and degrees of human perturbation. The following *case studies* of ecosystem metabolism provide an opportunity to explore a variety of metabolic patterns, while underscoring some important principles of coastal ecosystem ecology.

14.5.1 Tomales Bay

Tomales Bay is located at 38°N, 123°W in northern California and is a hydrographically simple estuary, with a linear basin of tectonic origin, modest rate of water exchange with the adjacent Pacific Ocean, and relatively shallow and uniform depths (Smith and Hollibaugh 1997; Box 14.5). The watershed has a Mediterranean climate (cool and wet winters, dry and hot summers) and a low population density, with grassland, forest, and chaparral ecosystems dominant (Smith and Hollibaugh 1997). Land clearing for agriculture (primarily potato farming) and logging in the latter half of the 19th century led to dramatically increased erosion and sediment transport to the bay through the first half of the twentieth century (Rooney and Smith 1999). Otherwise, human impact on the system is low. Watershed carbon inputs are strongly related to inter-annual variations in winter rainfall, and are primarily in the form of particles (Smith et al. 1996).

Primary production in Tomales Bay is dominated by phytoplankton, but the seagrass *Zostera marina* contributes measurably to ecosystem metabolism (Fourqurean et al. 1997). Throughout the bay, sediments without seagrass are colonized by a robust infaunal community and contribute substantially to ecosystem respiration (Figure 14.7; Smith and Hollibaugh 1997). Additional site characteristics are listed in Table 14.2. Tomales Bay was the site of extensive studies concerning metabolic rates of various community types over the course of a decade as part of the United States National Science Foundation's Land Margin Ecosystem Research (LMER) program. Among other accomplishments, the diverse studies at Tomales Bay provided insights into the external drivers of the bay's metabolism, and accomplished perhaps the first analysis of inter-annual variations in metabolism by applying a systematic methodology using biogeochemical budgets (See Section 14.3.5). This study also quantified relative contributions of particular ecosystem components to whole-ecosystem metabolism (Dollar et al. 1991; Smith and Hollibaugh 1997).

The partitioning of the metabolism of distinct habitats in Tomales Bay allowed an understanding of the contribution of the various habitats to whole-ecosystem production. Seasonal metabolic cycles followed the seasonal temperature cycle, with net heterotrophy dominating during the July to September period (Smith and Hollibaugh 1997). The plankton and heterotrophic benthic communities dominated Tomales Bay metabolism (Figure 14.7). Although plankton dominated *GPP* in Tomales Bay, benthic communities dominated respiration, and net system heterotrophy indicated that a carbon source aside from internal phytoplankton production must exist (Figure 14.7). A subsequent analysis of organic carbon imports indicated that watershed inputs (via stream flow) and tidal influxes of labile organic material from the Pacific Ocean, associated with seasonal variations in upwelling, contributed about equally to this additional carbon (Smith and Hollibaugh 1997). This study was one of the first to emphasize and quantitatively support the idea that ecosystem metabolism for small estuaries adjacent to regions of coastal upwelling are often strongly influenced by import of marine detrital organic matter, and upwelling events appear to be better correlated to *NEP* than other carbon sources, at least over part of the time series (Figure 14.8; Smith and Hollibaugh 1997).

14.5.2 Parker River-Plum Island Sound

Plum Island Sound is located at 43°N, −71°W on the northeast coast of Massachusetts and is fed by the Parker, Ipswich, and Rowley Rivers (Box 14.5). The suburban watershed is within the Boston Metropolitan Area with about 40% of the area classified as "urban," and the watershed retains large areas of forest, wetlands, and ponds. A bar-built estuary, Plum Island Sound exchanges with the Gulf of Maine through a single inlet near its southern terminus. A small connection between Plum Island Sound and the Merrimack River to the north creates a net residual circulation from the ocean to the Merrimack (Zhao et al. 2010). The sound and river system are shallow (mean depth ~2 m) and vertically well-mixed (Hopkinson et al. 1999). The estuary is ~25 km long, with a tidal river in the upper half (wetland:water area ≈ 9:1) and a broad shallow bay in the lower half (wetland:water area ≈2:1). Although the population density in the watershed is moderate, a mostly intact riverine riparian zone with extensive wetlands helps minimize inorganic nutrient levels and eutrophication in the system. In fact, dissolved organic nitrogen concentrations exceed inorganic N levels (Williams et al. 2004). The temperate climate in this region results in cold winters and mild summers, with seasonal peaks in discharge occurring in spring during snow melt, high precipitation, and low evapotranspiration. Plum Island Sound and its associated estuaries continue to be extensively studied via the United States National Science Foundation's long term ecological research (PIE-LTER; **http://pie-lter.ecosystems.mbl.edu**) program. This program investigates the long-term response of watershed and estuarine ecosystems at the land-sea interface to changes in climate, land use, and sea level.

Research in Plum Island Sound has addressed metabolism within each of the major ecosystem components (vegetated

Box 14.5 Maps of Case Study Estuaries

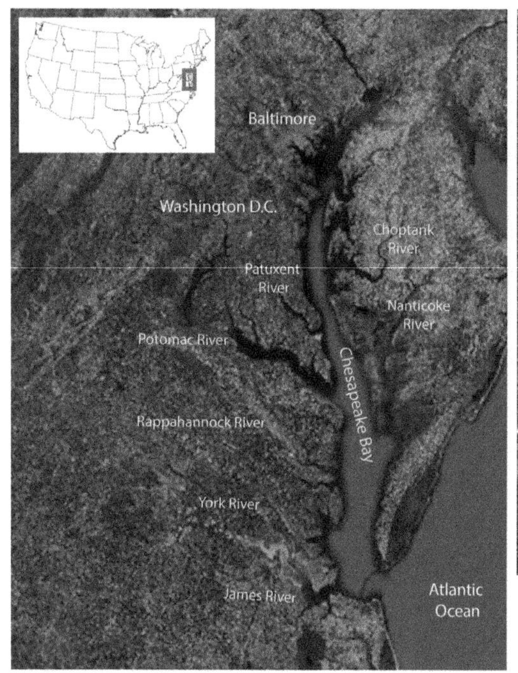

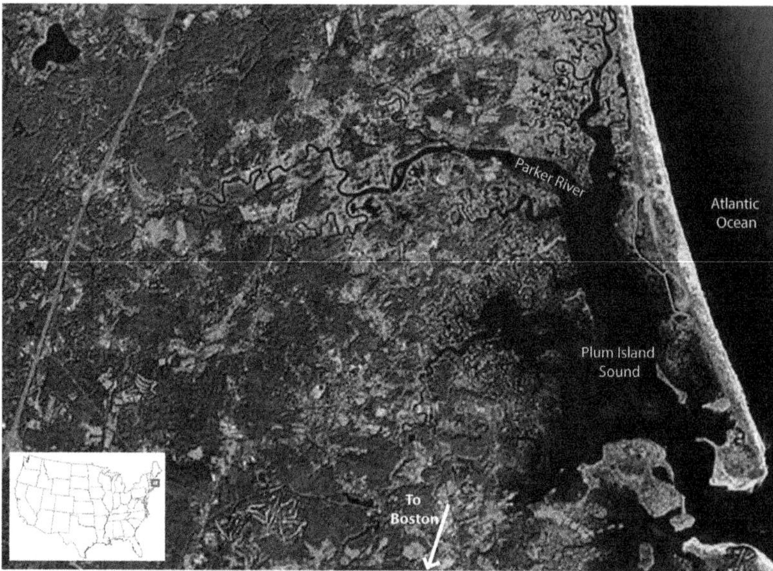

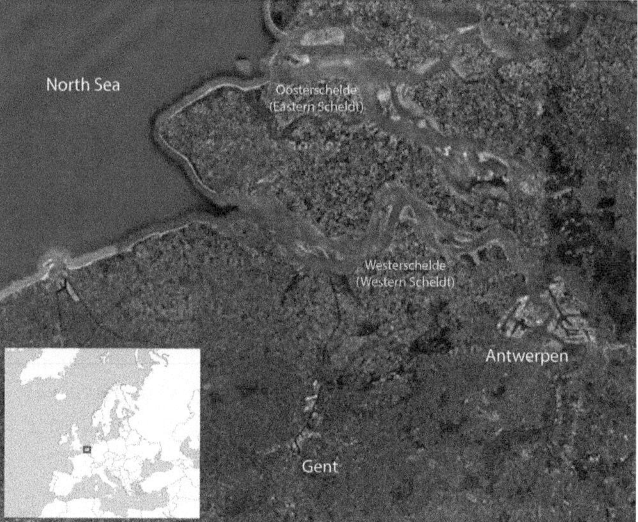

(Top left) Arial map of Chesapeake Bay and its tributaries, which are located on the Mid-Atlantic Coast of the United States (inset). Chesapeake Bay is a large, partially-stratified estuary that receives freshwater from several large rivers, most notably the Susquehanna. The metropolitan areas of Baltimore and Washington, D.C. are adjacent to Chesapeake Bay, which exchanges water with the Atlantic Ocean at its southern end. (Top right) Arial map of the Plum Island Sound/Parker River estuarine system, which is a bar-built estuary on the northeastern coast of Massachusetts, USA (inset). Plum Island Sound receives freshwater from several rivers and exchanges with the Atlantic Ocean through an inlet at its southern end. Its watershed is relatively forested, and the sound is fringed by extensive tidal salt marshes (light brown areas). (Bottom left). Arial map of Tomales Bay, which is a relatively shallow, lagoon-like estuary in northern California, USA (inset). Tomales Bay is a linearly-oriented system with an open exchange with the Pacific Ocean at its seaward terminus. Its watershed is sparsely populated with mixed land uses. (Bottom right) Arial map of the Scheldt estuary, which resides on the Belgian-Dutch border of northern Europe (inset). The Scheldt and its delta include eastern and western sections (Oosterschelde and Westerschelde) and the Scheldt's watershed includes parts of France, Belgium, and The Netherlands. The estuary is highly turbid and receives high loads of organic matter and inorganic nutrients from its densely populated watershed. The estuary includes several intertidal mudflats and exchanges with the southern North Sea. Arial maps from Google®.

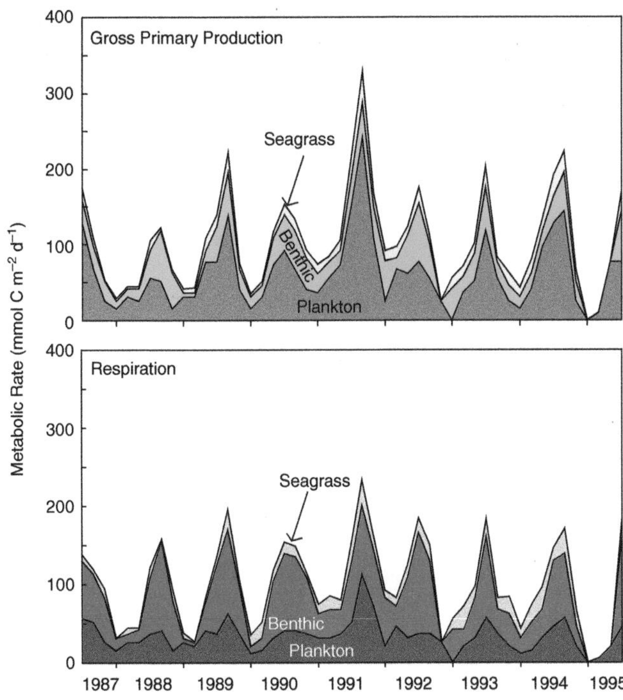

FIGURE 14.7 Time series of measurements of *GPP* (top panel) and *R* (bottom panel) for planktonic communities and both unvegetated and seagrass-dominated benthic communities in Tomales Bay, California, U.S.A. *Source*: Data from Smith and Hollibaugh (1997).

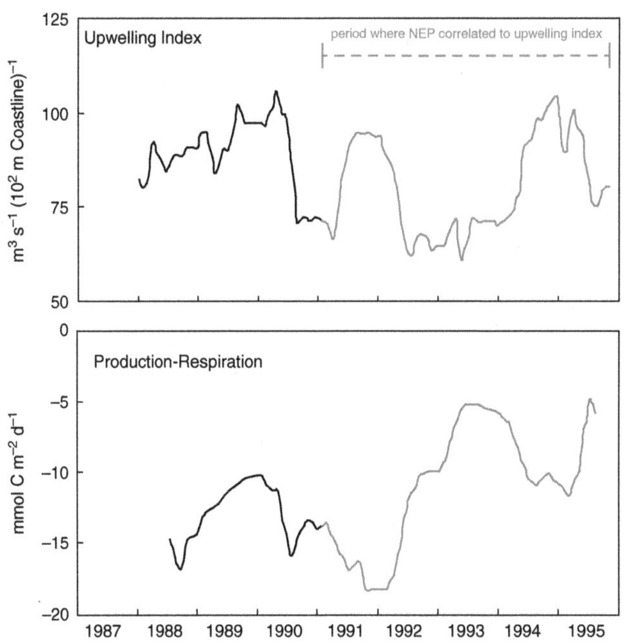

FIGURE 14.8 Time series of (top panel) 1-year moving averages for an upwelling index on the Pacific coast and (bottom panel) net ecosystem production (*P-R*) in Tomales Bay, California, U.S.A. *Source*: Data from Smith and Hollibaugh (1997). The negative correlation between the upwelling index and *P-R* from 1991–1996 (and lack of correlation between watershed POC and *P-R* over the same time period) suggests that upwelling-derived organic inputs fuel net heterotrophy in this ecosystem.

TABLE 14.2 Coastal systems that are highlighted as case studies in Section 14.5.

System name	Estuary type	Salinity	Mean depth (m)	Watershed population	Nutrient and organic matter load source	Population density (km^{-2})
Chesapeake Bay	Coastal Plain	0–30	6.5	~16,000,000	Agricultural runoff wastewater atmospheric	100
Plum Island Sound	Coastal Plain, Bar-Built	0-32	1.4 (head) 4.7 (mouth)	~150,000	Watershed runoff some wastewater	250
The Scheldt	Well-mixed	0.5–28	7 (Gent) 14 (mouth)	~10,000,000	Wastewater	450
Tomales Bay	Lagoon	20–35	3	~11,000	Coastal upwelling watershed runoff	20

tidal marsh, marsh ponds, estuarine water-column, and benthos), and the exchanges of metabolic substrates and products between these habitat types. The salt marsh cordgrass *Spartina alterniflora* dominates portions of the vegetated marsh that are relatively low in elevation, whereas *S. patens* is found at higher elevations. *S. alterniflora* and *S. patens* peak live aboveground biomass is approximately 540 and 1200 g dry mass m^{-2}yr^{-1}, respectively, though there is high spatial and annual variability in biomass production (Morris et al. 2013). Measurements of marsh-atmosphere CO_2 exchange made using chambers indicate the extensive tidal wetlands in the Plum Island system are strongly autotrophic and exhibit a seasonal pattern of carbon uptake from the atmosphere during the growing season (June through October) and moderate carbon loss to the atmosphere in other months (Figure 14.9; Forbrich et al. 2018). The production of organic matter in tidal marshes is the key process driving carbon sequestration in coastal marsh ecosystems. While much of the carbon fixed into organic matter by *GPP* is respired by plants and heterotrophs, *NEE* on an annual basis reflects the net uptake balance of CO_2 from the atmosphere (Figure 14.9). Eddy covariance towers, which measure air-water gas exchange from the aquatic habitat within the footprint of the instrument installation along with marsh-atmosphere exchange, have provided high-resolution, long-term measurements of *NEE* in

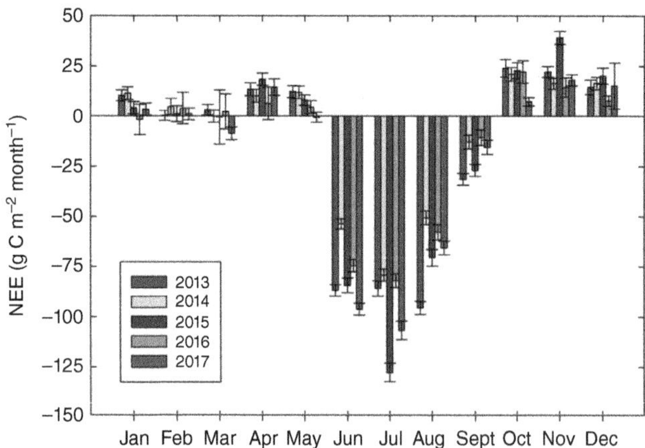

FIGURE 14.9 Rates of tidal marsh ecosystem *NEE* measured by eddy covariance tower at a tidal marsh site in Plum Island from 2013 through 2017. The marsh ecosystem is strongly autotrophic from June through September, with some modest return of carbon to the atmosphere from October through May. *Source*: Reprinted with permission from Forbrich et al. (2018).

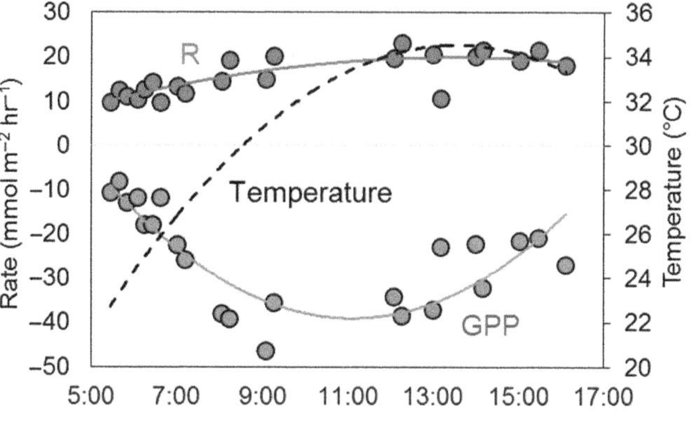

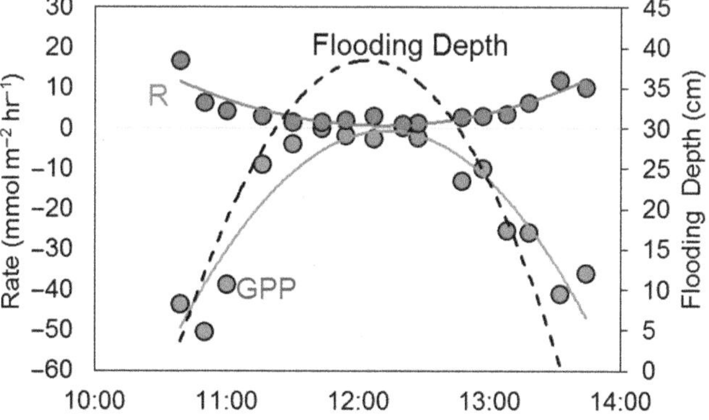

FIGURE 14.10 Hourly, time-course rates of vegetated marsh *apparent R* and *GPP* measured using chambers that assess marsh-atmosphere exchange during daytime periods when the marsh does not flood (top) and when the marsh is flooded (bottom). Rates of vertical exchange change during the unflooded daytime period as light and temperature change throughout the day (top). The vertical exchange of CO_2 is attenuated during marsh flooding, suggesting that rates of *GPP* and respiration decline during flooding and/or that the exchange of CO_2 occurs between the marsh and flood-water rather than with the atmosphere (bottom). *Source*: Data from Zawatski (2018).

the Plum Island marshes (Forbrich and Giblin 2015). The tidal marsh ecosystem exhibits a net uptake of about 180 g C m^{-2} from the atmosphere on an annual basis. Considerable inter-annual variation in carbon uptake rates may be linked to spring rainfall, where relatively low spring rainfall can lead to elevated marsh soil salinities during a critical period of plant growth that reduces the annual rate of productivity, resulting in lower annual *NEE* (Forbrich et al. 2018).

NEE measured by eddy covariance towers (Box 14.3) reflects the net exchange of carbon with the atmosphere, and does not account for lateral exchange of carbon or other metabolites. Depending on elevation relative to sea-level, marshes can flood on nearly every high tide (twice daily) or only on several spring tides per month. During marsh flooding, the vertical marsh-atmosphere exchange of carbon is attenuated (Figure 14.10; Zawatski 2018). Changes in vertical exchange may reflect actual changes in rates of metabolism; for instance, *GPP* becomes limited by availability of CO_2 when plants are submerged, and turbid waters reduce the light available to fuel photosynthesis (Vervuren et al. 2003). However, metabolites may be exchanged directly between the marsh and the flood water, instead of the atmosphere, resulting in an apparent decline in metabolic rate as measured by chambers (Figure 14.10) and eddy covariance towers (Figure 14.9). These reductions in marsh-atmosphere exchange do not necessarily reflect actual declines in metabolic activity. The products of soil organic matter oxidation in a flooded tidal marsh, for instance, may be exported as dissolved inorganic carbon on the ebb tide rather than fluxing vertically off the marsh into the atmosphere. Measurements over tidal cycles in small creeks in Plum Island indicate higher dissolved CO_2 at low tide relative to high tide, with generally consistent supersaturation of CO_2 relative to atmospheric equilibrium, indicating that marsh drainage supplies inorganic carbon to estuarine waters or that organic carbon is exported from tidal marshes and respired in the tidal creek waters. Therefore, understanding the lateral exchange of metabolites is required for a complete understanding of coastal ecosystem metabolism. Forbrich et al. (2018) estimated rates of carbon burial in Plum Island marsh soils to be 110 g C $m^{-2} y^{-1}$ from radiometric dating of soil cores. The difference between net uptake of atmospheric carbon (180 g C $m^{-2} y^{-1}$) and the long-term burial of carbon may be due, in part, to the lateral, tidal exchange of metabolites between the marsh and estuarine waters that is not captured by vertical measurements of marsh-atmosphere exchange (Forbrich et al. 2018).

In addition to the lateral flux of inorganic products of metabolism (CO_2 and O_2) between the marsh and estuary that reflect *GPP* and *R* in the marsh, marshes in Plum Island also deliver organic matter (both particulate and dissolved) to estuarine waters and benthic environments that tends to fuel heterotrophy in these habitats. High rates of productivity in tidal marshes, and the relatively large wetland to water ratio in Plum Island Sound, result in an estuarine system typically dominated by heterotrophy. A number of studies in Plum Island Sound have focused on spatial variations in water-column and benthic metabolism along the salinity gradient of the system (Figure 14.11; Alderman et al. 1995; Hopkinson et al. 1999). Free-water diurnal measures of metabolism (dissolved O_2 and/or TCO_2) along the entire length of the estuary within a 45 minute period at several consecutive dawn and dusk periods have enabled estimates of *NDP* and R_n once corrected for air-water gas exchange and tidal advection/dispersion (Figure 14.11; Vallino et al. 2005). *GPP* and *NEP* can then be calculated

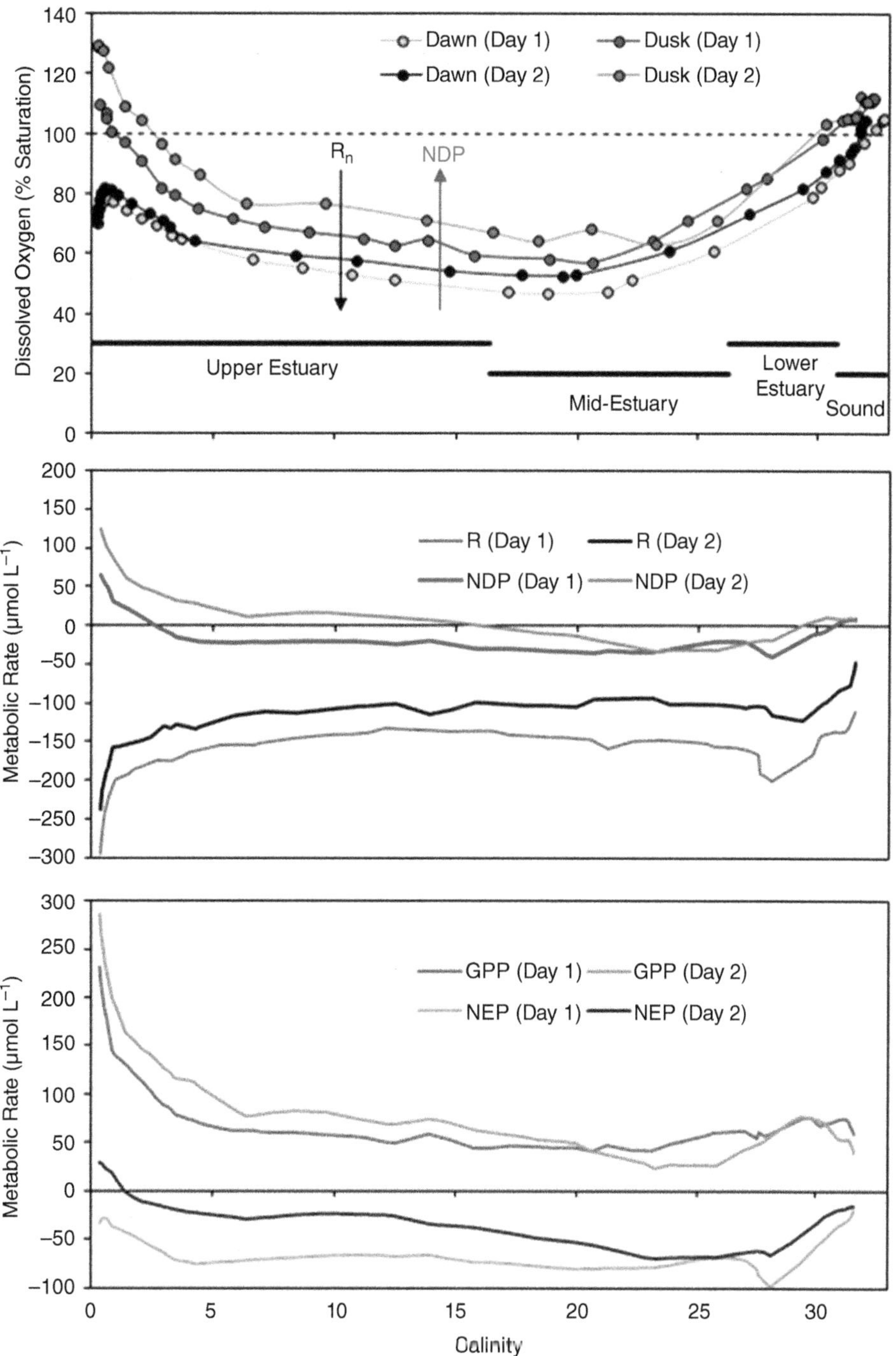

FIGURE 14.11 Transects of dissolved oxygen along the salinity gradient in Plum Island Sound on two consecutive days at dawn and dusk demonstrate the decline in dissolved oxygen during nighttime periods reflecting R_n and the increase in dissolved oxygen during daytime periods reflecting *NDP*, but also physical processes such as air-water gas exchange. These data are from July 2000, and were retrieved from the Plum Island Ecosystem Long Term Ecological Research project website (**http://pie-lter.ecosystems.mbl.edu**). The measured changes in oxygen, once corrected for air-water gas exchange, are then used to calculate (middle panel) biologically-mediated *R* and *NDP*, which are subsequently used to calculate (bottom panel) *GPP* and *NEP* along the estuary. Note the strong heterotrophy along most of the estuary, with the exception of the lowest salinity regions in the upper estuary where high *GPP* is fueled by watershed nutrient inputs and salinity-mediated release of benthic nutrients.

from R_n and *NDP* (Box 14.4), allowing for estimates of whole-ecosystem metabolism (Figure 14.11). Despite high turbidity in the upper reaches of the Parker River, *GPP* rates in this region exceed those of more seaward waters (Figure 14.11; Alderman et al. 1995; Vallino et al. 2005). Such high *GPP*, despite high turbidity, is mostly due to phytoplankton blooms during periods of low river runoff and, consequently, long water exchange times. Elevated per-cell chlorophyll-*a* levels in the region (phytoplankton increase chlorophyll-*a* to harvest more light in turbid conditions) combined with salinity-enhanced rates of benthic nutrient flux and shallow depths promote these high rates of production (Alderman et al. 1995; Hopkinson et al. 1999; Weston et al. 2010). Net autotrophy in these upper, low salinity regions gives way to net heterotrophy in most of the rest of the estuarine system (Figure 14.11; Alderman et al. 1995; Vallino et al. 2005), which is due to down-estuary transport of carbon into this region (from both the watershed and the low-salinity region) and import of labile organic matter from the extensive tidal marshes flanking the estuary. These inputs of organic matter fuel water column and benthic respiration that exceed *GPP* (Figure 14.11). Benthic respiration, in particular, is higher in the mid-Parker River than sites upstream and downstream and accounts for nearly all of the ecosystem respiration at the site (Hopkinson et al. 1999). Although the mid-river site may receive exceptional amounts of organic carbon from the adjacent marshes, this site also had an abundant population of the soft shelled clam *Mya arenaria*, which likely contributed significantly to the benthic respiration rates (Hopkinson et al. 1999).

Previous (Hopkinson and Vallino 1995; Forbrich et al. 2018; Hopkinson et al. 2018) and ongoing work in both the marsh and estuary system in Plum Island has provided us with estimates of important carbon cycling linkages between the wetland and aquatic portions of the ecosystem that accounts for the estuarine heterotrophy. The metabolism of both tidal wetland and estuarine water bodies is linked bys the relatively small but important movement of carbon between the two habitats via the deposition of estuarine-derived organic matter onto the marsh surface and the erosion of marsh edges, and the larger export of "excess" (i.e., not sequestered in marsh soils) OC from the marsh to the estuary during tidal flooding. We find that OC loading from the watershed (labile DOC) meets only a small fraction (<2%) of the overall estuarine heterotrophy, and water-column GPP provides about 60% of the OC required to support estuarine respiration. Inputs of OC from the marsh must then fuel a large portion (~40%) of estuarine respiration. Estimates of export of 'excess' OC from the marsh and erosion of OC from the marsh edge are more than capable of supporting estuarine heterotrophy. This suggests, however, that estuarine metabolism may have responded to recent changes in rates of sea-level rise, as the rate of lateral marsh erosive loss is likely higher now than over much of the past several thousand years. Will the eroded sediment and OC be redeposited onto the marsh surface helping to maintain marsh elevation with increasing rates of sea level rise (SLR), will it rapidly decompose in the presence of a new microbial community and oxygenated water, or will it be exported to the coastal ocean? Answers to these questions are required to examine the impact of tidal wetlands on atmospheric CO_2 levels over the next century or two. The fate of eroded OC from the marsh is an active area of marsh research. Continuing studies at Plum Island Sound are focused on modeling the long-term effects of land-use change, climate change, and SLR on metabolic processes in tidal marshes and estuarine waters.

14.5.3 Chesapeake Bay System

Chesapeake Bay is located at 38°N, 76°W in the mid-Atlantic coast of the United States and is fed by several large rivers, most notably the Susquehanna River, which supplies more than half of the freshwater flow (Box 14.5; Kemp et al. 2005). The temperate climate in this region results in cold winters and warm summers, with seasonal peaks in discharge occurring during spring, as snow melts and precipitation is high. A partially-stratified estuary, Chesapeake Bay is characterized by two-layer circulation, including a seaward-flowing surface current and a landward-flowing bottom current. The bottom current generally moves through a deep, central channel, which is flanked by extensive shallow areas such that ~75% of the bay surface area is shallower than 10 m (Kemp et al. 2005). Relatively large freshwater inputs drive a strong longitudinal salinity gradient and induce stratification in the deep, central parts of the bay, which prevents mixing of deeper waters with surface water and the atmosphere. Chesapeake Bay exchanges with the Atlantic Ocean at its southern terminus (Box 14.5). Although the population density in the watershed is moderate, the dominance of agriculture throughout much of the estuary and large population centers results in large nutrient loads to the estuary, leading to elevated nutrient levels and extensive eutrophication in the system (Kemp et al. 2005). Eutrophication in this estuary has been reinforced by the loss of once extensive sediment-stabilizing seagrass beds and water-filtering oyster reefs throughout the bay's shallow flats, as well as a large watershed to water volume ratio (Table 14.2). Chesapeake Bay and its tributary estuaries were extensively studied for over seven decades, revealing many key processes and external drivers in the bay's ecosystem.

One of the more pressing ecological concerns over recent decades in Chesapeake Bay and its tributaries is the seasonal development of low-O_2 waters in deeper sections of the central channel (Hagy et al. 2004). *Hypoxia* (O_2 < 20-30% saturation, 2 mg L^{-1}, or 62.5 µM) and *anoxia* (0% saturation or < 0.2 mg L^{-1}, 6.25 µM) are important consequences of human activities and eutrophication of estuaries and bays worldwide. Eutrophication-induced hypoxia occurs in stratified coastal systems through nutrient stimulation of surface layer phytoplankton production and sinking, which in turn leads to stimulation of bottom-layer respiration (e.g., Testa et al. 2008, Rabalais and Gilbert 2009). These metabolic processes provide a link between nutrient inputs and seasonal depletion of bottom water O_2 (Kemp et al. 2009). Hence, coastal hypoxia is often associated with increases in ecosystem production and respiration, with elevated net autotrophy in surface layers (and associated CO_2 uptake) and higher net heterotrophy in underlying waters (Kemp et al. 2009). Considering that the organic matter produced in surface waters supports much of the benthic and plankton respiration, the coupling of surface production and bottom respiration is a key driver of bottom water O_2 depletion in Chesapeake Bay and many similar systems (Rowe 2001; Conley et al. 2009) and associated DIC accumulation (Cai et al. 2017).

In Chesapeake Bay, oxygen and carbon budgets, as well as seasonal cycles of ecosystem metabolism, were used to infer sources of organic material and sinks of oxygen that drive hypoxia (Figure 14.12; Kemp et al. 1997). Such budgets support two primary conclusions: (1) planktonic production and respiration dominate bay C budgets and (2) surface waters (and the whole bay) are slightly net autotrophic (Kemp et al. 1997). More recent modeling analysis and investigations of DIC budgets have generally confirmed this (Feng et al. 2015; Shen et al. 2019). Further investigations of the spatial and seasonal variations in planktonic metabolic balance indicate several key features of planktonic contributions to particle sinking and bottom layer respiration. Rates of planktonic *GPP* and *R* are lowest in the upper bay, where high flushing and elevated rates of allochthonous organic matter inputs from the Susquehanna River cause light limitation of *GPP* (Figure 14.12). This region of the estuary is also net heterotrophic in all months, as imported organic matter fuels respiration over production (Figure 14.12; Smith and Kemp 1995). In the middle and lower bay, elevated light availability allows for seasonal peaks in *GPP* and *R* coincident with summer peaks in photosynthetically active radiation (PAR) and water temperature (Figure 14.12). However, *GPP* increases more rapidly than *R* during the March to May period, resulting in spring peaks in *GPP* that correspond to the spring diatom bloom in Chesapeake Bay (Figure 14.12, green areas). A second peak in *NEP* occurs in late summer, after a ~2 month period of net heterotrophy during the crash of the spring diatom bloom and transition into blooms of smaller dinoflagellates during July and August (Figure 14.12). The net result of these spatial/temporal dynamics is that spring and summer periods of net autotrophy in the middle and lower estuary generate the excess carbon that can fuel bottom water O_2 consumption (Smith and Kemp 1995). Similar patterns exist in the adjacent Patuxent River estuary, where biogeochemical budget computations suggest a spring peak in surface water autotrophy and an extended period of net heterotrophy in bottom waters (Testa and Kemp 2008).

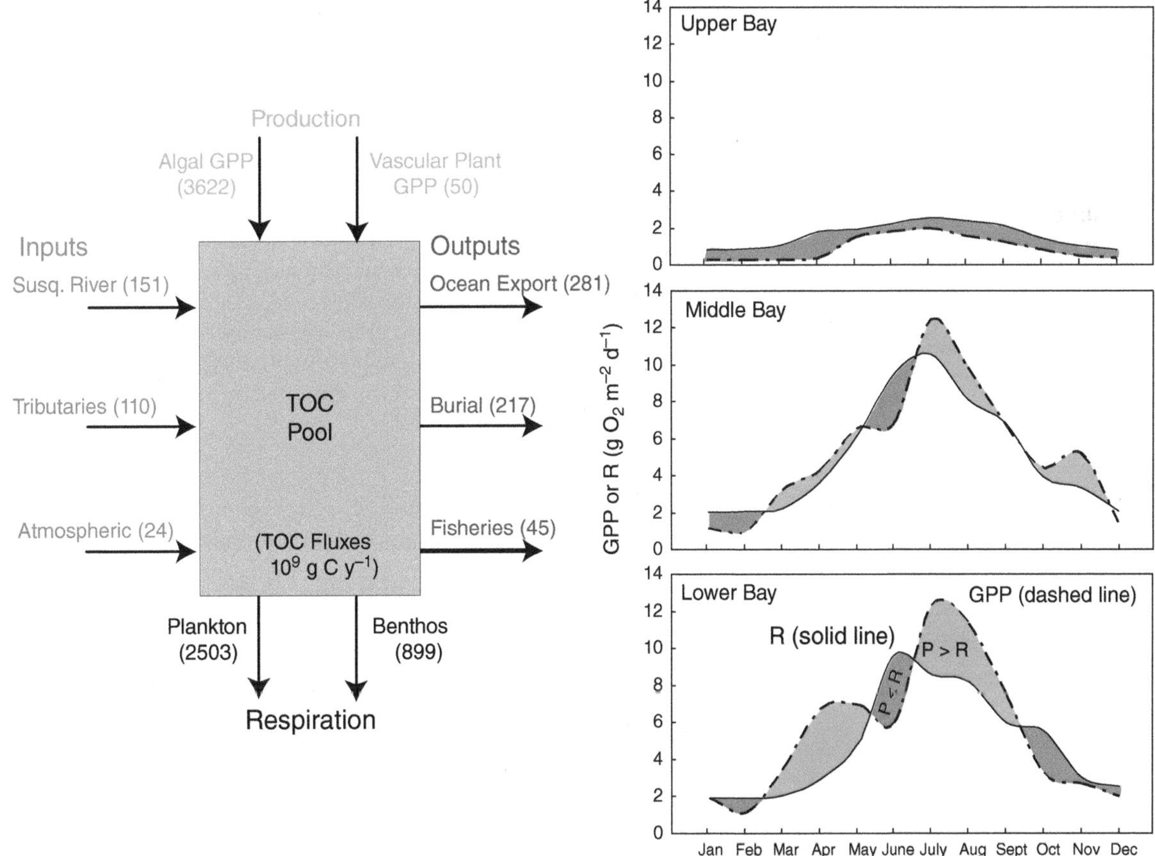

FIGURE 14.12 (left panel) Steady state organic carbon budget for the main-stem of Chesapeake Bay, including biogeochemical rates and physical transports. *Source*: Data from Kemp et al. (1997). (right panel) Seasonal distribution of *GPP* and *R* for the main-stem of Chesapeake Bay in the upper, middle, and lower regions of the Bay. The upper estuary is net heterotrophic, while the middle and lower Bay are progressively net autotrophic, with March to May (the spring diatom bloom) and July-September (picoplankton dominated) periods dominating the net production.

In stratified estuaries like Chesapeake Bay and the Patuxent estuary, a "metabolic separation" exists where watershed nutrient inputs in the surface layer stimulate excess *NEP* (i.e., organic matter production), some of which sinks to bottom layers where net respiration (e.g., organic matter consumption) prevails. Consequently, surface layer *NEP* in these estuaries correlates strongly with the sinking of organic material and subsequent bottom layer respiration. In the Patuxent River, surface layer *NEP* peaks in spring (Testa and Kemp 2008) when nutrient inputs are high and temperature (and thus respiration) is relatively low; this is the period of the conventional "spring bloom" in temperate estuaries. Bottom layer respiration (water column and sediments) lags behind the period of peak surface layer *NEP*, as respiration is delayed until June-August when temperature peaks. Although surface layer *NEP* declines over this summer period as respiration increases, depth-integrated surface (above-pycnocline) and bottom (below-pycnocline) layer *NEP* are proportional to each other over annual time scales (Testa and Kemp 2008).

In main-stem Chesapeake Bay, *NEP* in the surface layer correlates with measurements of sinking particulate organic carbon (POC) captured in sediment traps (Figure 14.13). Thus, both the amount of POC sinking to bottom waters and the resulting respiration rates appear to be proportional to excess C production in surface waters (Figure 14.13). Similarly, *NEP* measured in mesocosms of Narragansett Bay correlated strongly with the production of flocculent organic material. Both *NEP* and flocculent production were stimulated by inorganic nitrogen additions (Figure 14.13). Additional inputs of dissolved silica (DSi) further enhanced the flocculent material production per unit *NEP*. This occurred because diatoms tend to rapidly take up nutrients and their large cells lead to rapid organic matter accumulation, which is much like their growth in natural estuaries during the spring bloom (Figure 14.13). These examples illustrate how elevated nutrient inputs to coastal systems stimulate the production and subsequent consumption of organic matter, yielding O_2 deficiencies in the bottom waters of Chesapeake Bay and many other ecosystems (Kemp et al. 2009).

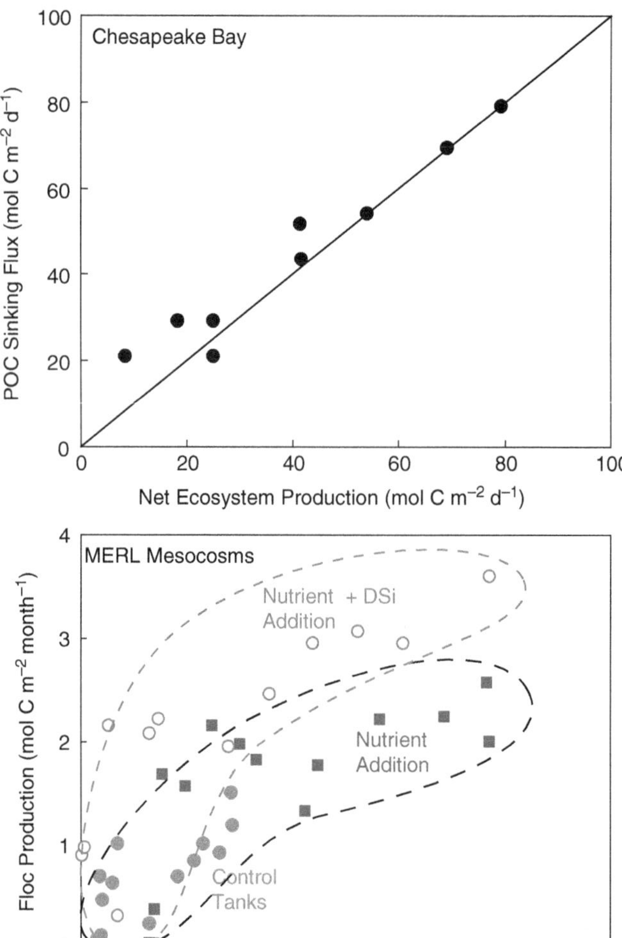

FIGURE 14.13 Correlations between net ecosystem production and indexes of organic particle formation in (top panel) Chesapeake Bay, where net ecosystem production (primarily plankton) is highly correlated to sediment trap collection rates (Boynton and Kemp, unpublished data) and (bottom panel) MERL mesocosms, where net ecosystem production is highly correlated to the accumulation of organic flocs. *Source*: Based on Oviatt et al. (1993). DSi = dissolved silica.

14.5.4 Scheldt Estuary

The Scheldt estuary is located at 51°N, 4°W in northern Europe and its 19,500 km^2 watershed covers parts of France, Belgium and The Netherlands (Box 14.5, Gazeau et al. 2005c). The river Scheldt contributes the majority of freshwater to the estuary, which empties into the southern part of the North Sea (Box 14.5). The temperate climate in this region results in cold winters and mild summers, with seasonal peaks in discharge occurring during spring. Because tidal currents are strong, the estuary is well-mixed, with axial salinity ranging from 0.5 to 28. The mean depth of the estuary varies along its axis, from 7 m near Gent and 14 m near the mouth. The watershed includes the Belgian cities of Antwerp and Gent, and the total watershed population approaches nearly 10,000,000 people. As a result of high human population densities (450 people km^{-2}), the Scheldt estuary is one of the most polluted rivers in Europe, receiving large quantities of nutrients and organic matter from agriculture and wastewater sources (Gazeau et al. 2005c; Soetaert et al. 2006). Due to the high level of anthropogenic disturbance in its watershed, the Scheldt estuary provides a useful example of the effects of pollution on metabolism.

Although tidal flats flank the main estuary channel in the lower reaches of the estuary, the Scheldt's metabolism is dominated by phytoplankton and bacteria along most of its axis. Despite the fact that the Scheldt is highly turbid, net phytoplankton production occurs along the length of the estuarine axis, and peaks in low salinity regions of the estuary and near the seaward terminus (Figure 14.14; Gazeau et al. 2005c). Interestingly, net phytoplankton production has been found to be possible in other well-mixed, highly turbid estuaries, where nutrient limitation is absent (i.e., exceptionally high

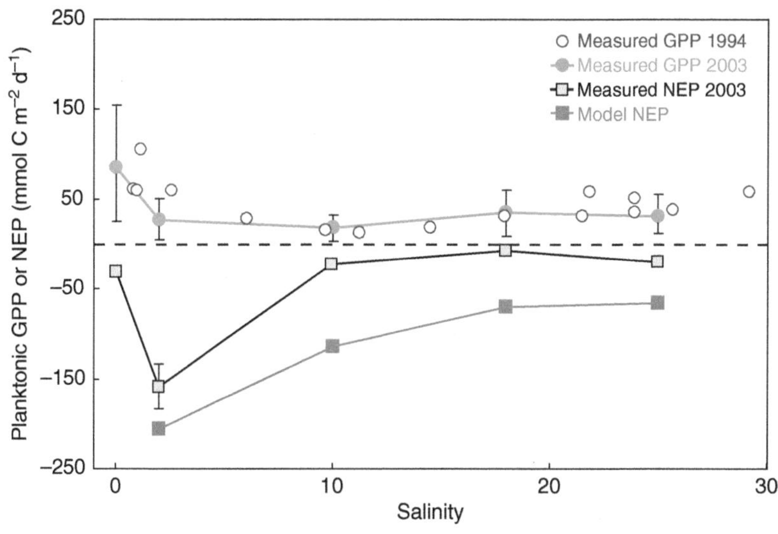

FIGURE 14.14 Distribution of *GPP* and *NEP* (of plankton community) along the salinity gradient of the Scheldt estuary. Yellow squares are measured *NEP* in 2003 (Gazeau et al. 2005c) and are compared to modeled *NEP* rates (filled, red squares) from Soetaert and Herman (1995). Open blue circles are measured *GPP* rates from Kromkamp and Peene (1995) and teal circles are measured *GPP* rates from 2003 (Gazeau et al. 2005c). *Source*: Figure adapted from Gazeau et al. (2005c).

nutrient levels: DIN > 200 μM, DIP > 5 μM) and water depth is shallow enough to allow phytoplankton to access light for sufficient amounts of time.

However, due to high inputs of relatively labile organic matter from wastewater treatments plants (Soetaert et al. 2006) and watershed runoff, bacterial respiration is high, leading to highly heterotrophic net ecosystem metabolism (Figure 14.14; Gazeau et al. 2005c). Net respiration peaks in low salinity water, where organic materials accumulate due to close proximity to wastewater discharges (Figure 14.14). Consequently, despite extremely high levels of inorganic nutrient inputs to the Scheldt, the system is net heterotrophic, due to high levels of respiration of imported organic matter combined with light limitation of algal growth (Kromkamp and Van Engeland 2010). Thus, this estuary does not fit the relationship between *NEP* and the ratio of DIN:TOC that characterizes many temperate estuaries that are primarily nutrient-limited (Figure 14.4).

In addition to high organic matter inputs, extensive nutrient inputs to the Scheldt may have reduced primary production and promoted heterotrophy. Extremely high wastewater NH_4^+ inputs caused extremely high NH_4^+ concentrations in the freshwater region of the estuary (>900 μM, Figure 14.15). At

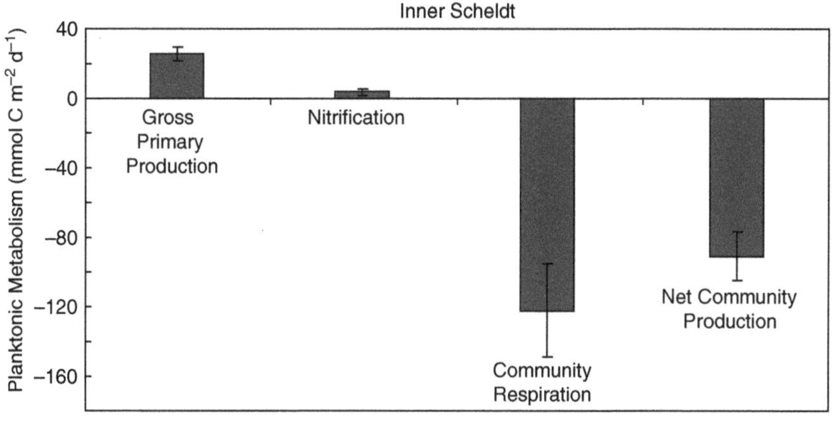

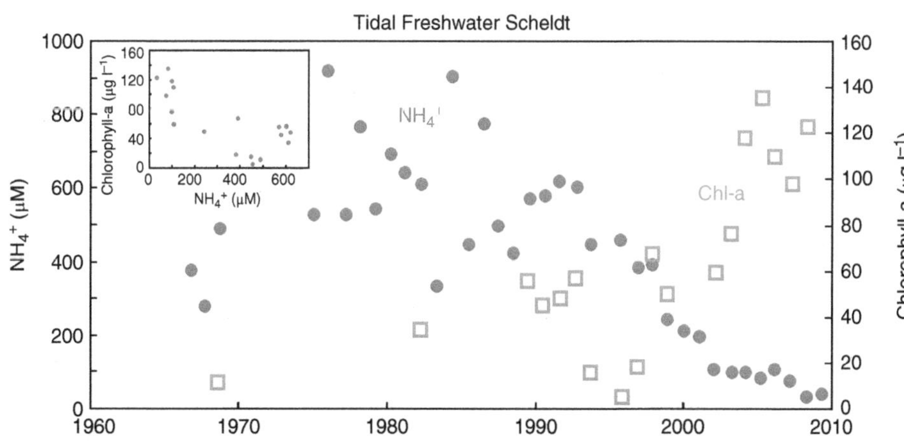

FIGURE 14.15 (top panel) Annual mean components of metabolism in the inner Scheldt estuary. *Source*: Data from Gazeau et al. (2005c). The data indicate a high level of community respiration, due to high organic matter inputs rates to this system, and also a non-negligible contribution of nitrification (a chemoautotrophic process) to organic carbon production in the estuary. (bottom panel) Time series of NH_4^+ and chlorophyll-a concentrations averaged over the tidal freshwater region of the Scheldt estuary from 1967 to 2008. Inset is correlation between the annual data. *Source*: Based on Cox et al. (2009). High NH_4^+ concentrations prevailed through the mid-1990s, until management actions dramatically reduced NH_4^+ concentrations after 2000. Both figures illustrate the importance of nitrification in highly polluted estuaries, both as a sink of O_2 and a source of carbon.

these levels, NH_4^+ is toxic to algae, leading to inhibition of photosynthesis and resultant low levels of chlorophyll-a in the estuary (Figure 14.15; Cox et al. 2009). As management actions reduced NH_4^+ concentrations in the Scheldt over the past two decades, chlorophyll-a concentrations have tripled (Soetaert et al. 2006; Cox et al. 2009). Another consequence of high NH_4^+ is the stimulation of nitrification under oxic conditions. Nitrification-based O_2 demand caused hypoxia throughout much of the inner Scheldt estuary, which may have further inhibited photosynthesis in the 1960–1990 period (Soetaert et al. 2006). Nitrification, however, is a chemoautotrophic process that generates organic material; thus unlike many estuaries, high NH_4^+ concentrations support nitrification at sufficient rates to contribute to carbon production (just as *GPP*) in the Scheldt (Figure 14.15; Gazeau et al. 2005c). It is unclear how the balance between NH_4^+-driven reductions in photosynthesis and NH_4^+-driven nitrification-based carbon production has changed *NEP* over the past 50 years. Considering the tendency for net heterotrophy in this system in all seasons, potential shifts toward elevated autotrophy over time would surely have been overwhelmed by high respiration rates (Figure 14.15).

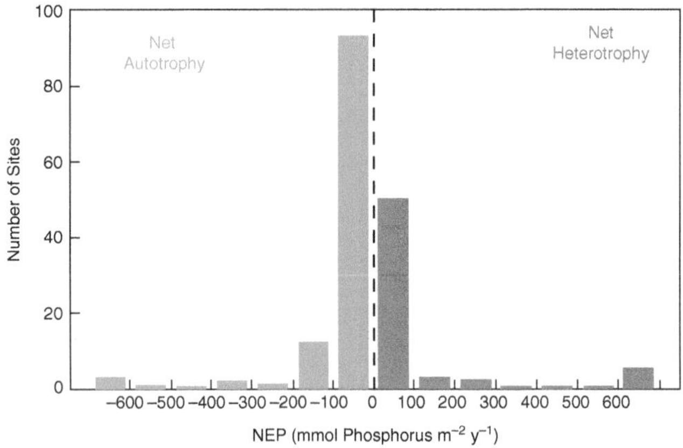

FIGURE 14.16 Frequency distribution of *NEP* derived from computations of the net non-conservative production rates of dissolved inorganic phosphorus from budget estimates for coastal systems around the world. *Source*: Rates are from Smith et al. (2005) based on the LOICZ biogeochemical budget model analysis.

14.6 Cross-Ecosystem Comparisons

Over the past two decades, a few efforts have examined the state, dynamics, and drivers of ecosystem metabolism across many different ecosystems. We review here a few examples of cross-system comparisons of coastal ecosystem metabolism and the lessons learned from the efforts.

14.6.1 Trophic State and Dominant Controls

Under the auspices of the Land-Ocean Interactions in the Coastal Zone program (LOICZ, **http://www.loicz.org**, a project of the International Geosphere-Biosphere Program, IGBP), a major scientific effort was to compile carbon, nitrogen, and phosphorus data for coastal systems around the world and apply the data to construct carbon and nutrient budget models using a consistent and relatively simple methodology (**http://nest.su.se/mnode/**). Among other system properties, these analyses produced estimates of net ecosystem production, based on the net non-conservative production of inorganic phosphorus (Smith et al. 2005; see Section 14.3.5). As a result, comparable estimates of *NEP* were made for approximately 200 ecosystems across the globe, and analyses of these estimates provided key insights into the global trophic tendency of coastal systems, as well as insight into patterns and controls of ecosystem metabolism.

The system-wide synthesis of LOICZ *NEP* rates for ~200 coastal ecosystems across the globe (Smith et al. 2005) revealed that net metabolism of these estuaries and bays tends to be close to zero, and with slightly more systems being autotrophic than heterotrophic (Figure 14.16). This tendency toward net autotrophic ecosystems may reflect the rapid global expansion of coastal eutrophication, driven primarily by inorganic nutrients from runoff of agricultural fertilizers and from secondary-treated sewage, which would favor primary production over respiration. Despite the impressive number of systems analyzed in LOICZ, the sample size is still too small to infer generalizations about the tendency for estuaries to be net autotrophic or heterotrophic on a global scale. The fact that other studies have found estuaries to be generally heterotrophic (e.g., Borges and Abril 2010) suggests that the metabolic state of any particular estuary is not only specific to its watershed, climate, and bathymetry, but also to the particular season and year in which measurements were made, with their characteristic human impact and year-specific climatic regime. Although many coastal systems across the world are receiving large inputs of inorganic nutrients, the large number of systems characterized by heterotrophic metabolism suggests that (1) organic matter inputs from surrounding watersheds and wetlands tends to fuel high respiration rates and that (2) characteristically high turbidity in estuaries often restricts photosynthesis. For example, a comparative analysis of coastal ecosystem metabolism (Caffrey 2004) suggested a predominance of heterotrophic systems, most of which were flanked by mangrove or marsh ecosystems, which tend to export large amounts of relatively labile organic matter.

14.6.2 Eutrophication Effects on Metabolism

As discussed in Section 14.4, ecosystem metabolism has been shown to respond strongly to a wide range of anthropogenic disturbances to coastal systems including increases in inputs of toxic contaminants, suspended sediments, inorganic nutrients, and labile organic matter. In this chapter we emphasized how measurements and calculations of coastal ecosystem net production (*NEP*) are useful for distinguishing the relative importance of external inputs of inorganic nutrients versus labile organic matter. The former tends to increase *NEP*, while the latter drives *NEP* negative. In several important papers in the 1990s, budgets (Smith and Hollibaugh 1993) and data syntheses (Heip et al. 1995) were used to understand the trophic state of the coastal zone. It was suggested that the coastal zone was net heterotrophic, due to high inputs of organic matter from adjacent watersheds, marshes, and human population centers. These analyses emphasized that estuaries, despite being net heterotrophic, were also highly productive because primary production tends to be stimulated by nutrients regenerated from respiration of imported organic matter (Smith and Hollibaugh 1993; Heath 1995; Heip et al. 1995). This initial perception was likely influenced by the types of systems included in this analysis, which included few deep, eutrophic estuaries (Figure 14.17). Because forests tend to conserve inorganic nutrients, but often export DOC and POC from plant debris and soil materials, drainage from forested or disturbed watersheds would tend to support moderately heterotrophic estuaries (Figure 14.17). Estuarine waters that are surrounded by tidal marshes and mangroves tend to be even more heterotrophic despite high *GPP*, given high nutrient regeneration from substantial labile DOC and POC exports from those wetlands (Figure 14.17). Conversely, in plankton-dominated estuaries receiving high levels of inorganic nutrients but moderate levels of organic matter, nutrient-driven increases in *GPP* would tend to cause elevated *NEP* (Figure 14.17). As human populations and impacts of land-use continue to grow and development, the future trajectory of metabolic balance in coastal ecosystems is unclear.

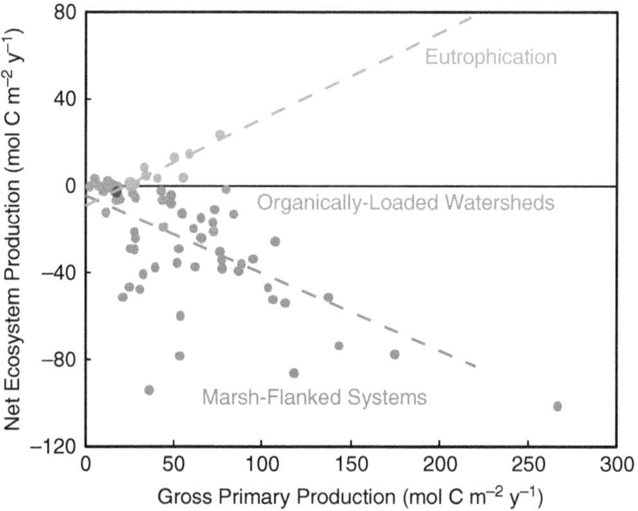

FIGURE 14.17 Relationships between net ecosystem production (*NEP*) versus gross primary production (*GPP*) for ecosystems under various land-use and anthropogenic impacts. *Source*: Based on Nixon and Pilson (1984), Oviatt et al. (1986), Smith and Hollibaugh (1993), Ver et al. (1994), Kemp et al. (1997), and Caffrey (2004); and Testa, unpublished. Organically-loaded watersheds (brown circles) are those impacted by human perturbation and some naturally impacted by high organic matter loads and display a negative relationship between *GPP* and *NEP*. Systems with increasingly negative *NEP* (higher respiration) are associated with higher nutrient remineralization rates to fuel elevated *GPP* (Smith and Hollibaugh 1993). This is also true for the shallow, turbid systems analyzed by Caffrey (2004), many of which are flanked by tidal marshes ("Marsh-Flanked Systems") or other ecosystems that tend to export organic material to open water regions of estuaries. For eutrophic systems (teal circles) with elevated nutrient inputs and relatively high light availability, *NEP* is positively related to *GPP* because nutrient loads stimulate *GPP*. A 1000-year simulation of organic inputs and metabolism for the coastal zone (dark blue circle) suggests low, heterotrophic NEP (see Ver et al. 1994).

As just mentioned, a number of publications have linked elevated *NEP* to increases in inorganic nutrient inputs (Oviatt et al. 1986; Testa et al. 2008). Such increases are expected for relatively clear water systems that are generally nutrient limited. The consequences of elevated *NEP* in the face of global increases in anthropogenic nutrient inputs include elevated carbon fluxes from coastal surface waters to adjacent ecosystems. A primary sink for excess carbon production in surface waters is the underlying bottom waters and eutrophication has been linked to increasing bottom water and sediment respiration rates, and associated depletion of O_2 (Kemp et al. 2009). Thus, elevated nutrient input not only increases surface water *NEP*, but also enhances metabolic processing within the whole ecosystem, as fresh organic material is respired at higher rates.

14.6.3 Coastal Ecosystems and Global Carbon Balance

Syntheses of site-level metabolism studies have provided interesting perspectives on coastal contributions to global CO_2 dynamics (Crossland et al. 1991; Smith and Hollibaugh 1993; Duarte et al. 2005; Borges and Abril 2010). Early budgets of the global CO_2 balance suggested that the net metabolism of the world's oceans represent a major sink for atmospheric CO_2 (e.g., Woodwell and Pecan 1973), and subsequent measurements suggested that vertical sinking of surface water net production of organic matter could by an important mechanism driving this sink. Later analyses using bottle incubations (Duarte and Augusti 1998; Williams 1998) and ocean-scale distributions of surface CO_2 and O_2 partial pressures (Najjar and Keeling 2000; Takahashi et al. 2002; Cai et al. 2006) have revealed strong seasonal and regional variations in oceanic *NEP* suggesting a variable trophic state for the open ocean. Regardless of the magnitude and sign of global ocean *NEP*, computations of integrated net metabolic rates for particular coastal biomes provide perspective on their potential contributions to CO_2 balance in the global ocean or in specific ocean basins. For example, estimates for coral reef metabolism indicated that their global gross production amounts to ~2% of oceanic net production; however, *NEP* for coral reefs comprises ~0.05% of net biological CO_2 influx in the global ocean (Crossland et al. 1991). More importantly, as calcification in coral reefs generates CO_2, the biological *and* geochemical processes on coral reefs generate a net efflux of CO_2 to the atmosphere on the order of 0.02–0.08 Gt C y^{-1} (Ware et al. 1991). As another example, the combined *NEP* for all shallow vegetated habitats in the coastal ocean (e.g., mangroves, salt marshes, seagrass, macroalgae) is estimated to exceed that of the pelagic ocean, and computed carbon burial in these vegetated biomes approximates that estimated for the entire open ocean (Duarte et al. 2005). These analyses suggest that coastal biomes play a critical role in the global CO_2 balance.

The most recent global C budgets illustrate the relative importance of all aquatic systems to the atmospheric CO_2 balance as well as spatial patterns of CO_2 flux for tidal wetlands, estuaries, shelves and the open ocean (Figure 14.18; e.g., Borges and Abril 2010; Regnier et al. 2013; Bauer et al. 2013). Globally, the open ocean is a large C sink (~ −1.5 Pg yr^{-1}), lakes and rivers are a large source (~1.0 Pg C yr^{-1}) and the coastal ocean (estuaries, tidal wetlands and continental shelves) is a net sink (~ −0.45 Pg C yr^{-1}). Estuaries are net sources of CO_2 to the atmosphere (0.25 Pg C yr^{-1}). Large spatial gradients are found in continental shelves with upwelling areas and the open shelf at low latitudes being net sources of CO_2 to the atmosphere and enclosed seas, polar shelves and shelves between 30 and 60° being net sinks, especially polar shelves (Bauer et al. 2013). The average shelf pCO_2 level is currently about 350 ppm under an average atmospheric CO_2 level nearly 400 ppm. Clearly shelves will continue to act as CO_2 sinks as the atmospheric CO_2 gradient increases further (Bauer et al. 2013).

14.6.4 Estuaries and Blue Carbon Sequestration

A most striking aspect of the overall carbon balance of estuarine ecosystems is the dominance of water flow-mediated fluxes of organic and inorganic carbon through estuaries, linking adjacent terrestrial and continental shelf systems (Figure 14.19). River runoff contributes 0.85 Pg C yr^{-1}, and estuaries export 0.95 Pg C yr^{-1} to the ocean, with OC representing just over 50% of the total in both directions. This is not surprising considering the location of estuaries at the land-sea interface and that virtually all lateral connections between land and ocean are hydrologically mediated. Regnier et al. (2013) estimated that anthropogenic perturbation has increased the flux of carbon to inland waters and the coast by as much as 1.0 Pg C yr^{-1}, but that export to the ocean has increased only about 0.1 Pg C yr^{-1}. The processing, transformation, and sequestration of terrestrial-derived C in the physical-biogeochemical estuarine reactors is an important characteristic of estuarine ecosystems, and the fates of various sources and their complex interactions remain poorly understood (Bauer et al. 2013).

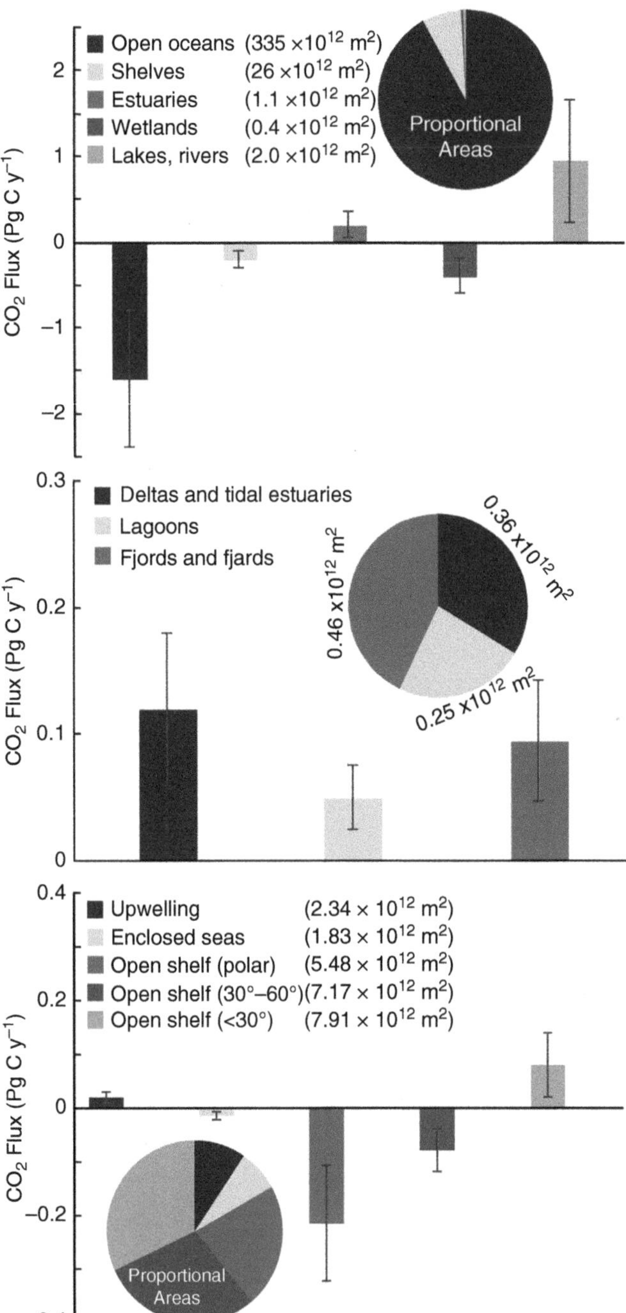

FIGURE 14.18 Air-water CO_2 exchange fluxes (bars) and surface areas (pie chart insets) for different aquatic systems, including (top) major aquatic systems, (middle) deltas and tidal estuaries, lagoons, and fjords and fjards, and (bottom) enclosed seas and coastal shelves (including upwelling zones). Note that large river plumes extending onto and across continental shelves are not included, although it is known that they are CO_2 sinks for the atmosphere. An uncertainty of <50% was assigned to each flux term except inland waters, fjords and fjards, enclosed seas, and low latitude open shelves, which have an assigned uncertainty of 50–100% because of very sparse data coverage. *Source*: Redrawn from Bauer et al. (2013).

Estuaries are more than a black boxes mediating the exchange of terrestrial C with the ocean. Globally, estuaries are heterotrophic and net sources of CO_2 to the atmosphere and DIC to the ocean (Figure 14.19). The two sources of organic

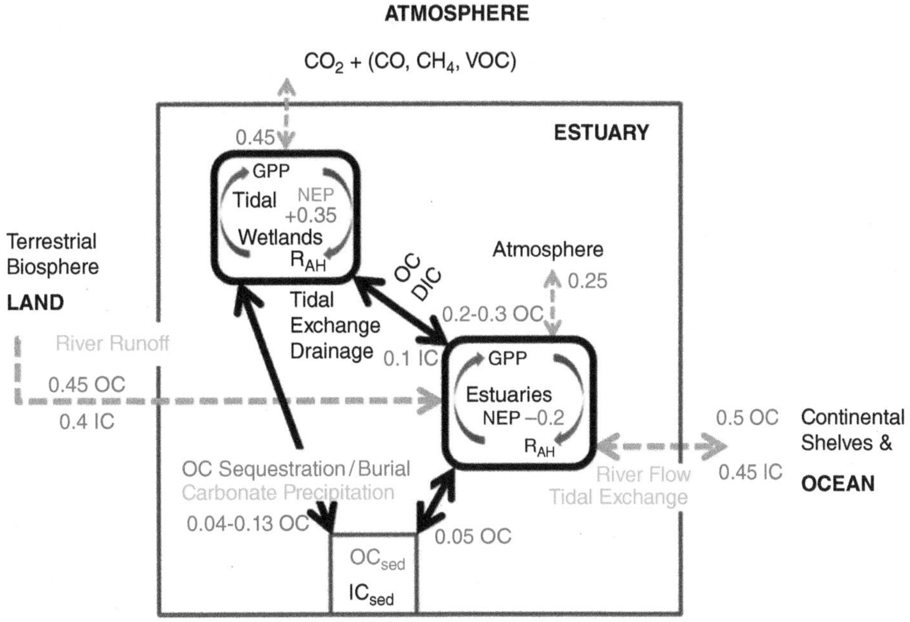

FIGURE 14.19 Net ecosystem carbon balance for inorganic and organic carbon for estuaries of the world (all values in Pg C yr^{-1}). This model illustrates the key importance that tidal wetlands (and seagrasses) play in controlling patterns of autotrophy and heterotrophy in estuaries. Sequestration of organic carbon in sediments associated with tidal wetlands and seagrasses is known as blue carbon sequestration. Globally, tidal wetlands help to remove CO_2 from the atmosphere. OC—organic carbon, IC—inorganic carbon, sed—sediments, *GPP*—gross primary production, *NEP*—net ecosystem production, R_{AH}—respiration of autotrophs and heterotrophs, VOC—volatile organic carbon, a gaseous form of C exchange with the atmosphere. *Source*: Modified from Bauer et al. (2013), Hopkinson et al. (2012), and Hopkinson (2018).

carbon fueling estuarine heterotrophy are terrestrial systems and adjacent tidal wetlands and seagrasses. Terrestrial loading of OC is high—equivalent to about 400 g OC $m^{-2} yr^{-1}$ (Bauer et al., 2013). Tidal wetlands themselves are highly autotrophic ecosystems taking up a net 0.45 Pg CO_2-C yr^{-1} from the atmosphere. About 0.1 Pg C yr^{-1} is exported to estuaries as DIC leaving a net tidal wetland NEP of 0.35 Pg C yr^{-1}. A portion of this NEP is is buried or sequestered as OC in wetland soils and the remainder is exported as OC to estuarine waters. This is the OC that largely fuels estuarine heterotrophy. The other OC source fueling heterotrophy is that brought in from land.

It has become increasingly evident in the past 10 years that the combined tidal wetland/estuarine ecosystems are a globally significant net sink for atmospheric carbon dioxide, sequestering between 0.1 and 0.2 Pg C yr^{-1} in sediments of open water (including deltas), seagrass and tidal wetland habitats as organic carbon (Figure 14.19). The sink associated with tidal wetlands and seagrasses is called blue carbon sequestration (Nellemann et al. 2009). The magnitude of this sink is surprising considering that the areal extent of these ecosystems (~1×10^6 km^2) is but a sliver of the ocean or land areas (335×10^6 and 790×10^6 km^2). The basis of blue carbon sequestration in tidal wetlands and seagrasses is high *NEP* and burial of undecomposed root, rhizome, and aboveground plant tissues in sediments of these systems (Figure 14.20). The aboveground plant structure efficiently traps suspended (organic and inorganic) particles from the water column (during tidal inundation in the case of tidal wetlands). Continual accretion and capping of the sediment surface results in a deepening of previously buried materials, which contributes to less porewater flushing, reduced oxygen and other terminal electron acceptors enabling organic matter decomposition, the accumulation of toxic byproducts of anaerobic metabolism, and a general slowing of decomposition. Rates of burial are therefore high in these environments, averaging on the order of 210, 226, and 138 g OC $m^{-2} yr^{-1}$, in mangrove, salt marsh, and seagrass systems, respectively (Hopkinson et al. 2012, Nellemann et al. 2009). While there is wide variability in these rates between systems, there is also great uncertainty in the areal extents of these systems globally. Blue carbon sequestration is calculated as the product of burial rate and areal extent. Thus there is large uncertainty in the globally-scaled estimates of blue carbon sequestration. For instance, most recent estimates for the areal extent of mangroves ranges from about 83,495 to 137,760 km^2 (Giri et al. 2011; Hamilton and Casey 2016). For salt marshes, global estimates are between 200,000 and 400,000 km^2, while Mcowen et al. (2017) revised this to only 54,950 km^2, a decrease of 75–87%! The uncertainty in areal extent is highest for seagrasses, ranging from 165,000 to over 700,000 km^2 (Green and Short 2003; Waycott et al. 2009; Nellemann et al. 2009; Hopkinson et al. 2012).

The basis for tidal wetland blue carbon sequestration is important to understand (Figure 14.20) because the major controls of sequestration, including anthropogenic activities,

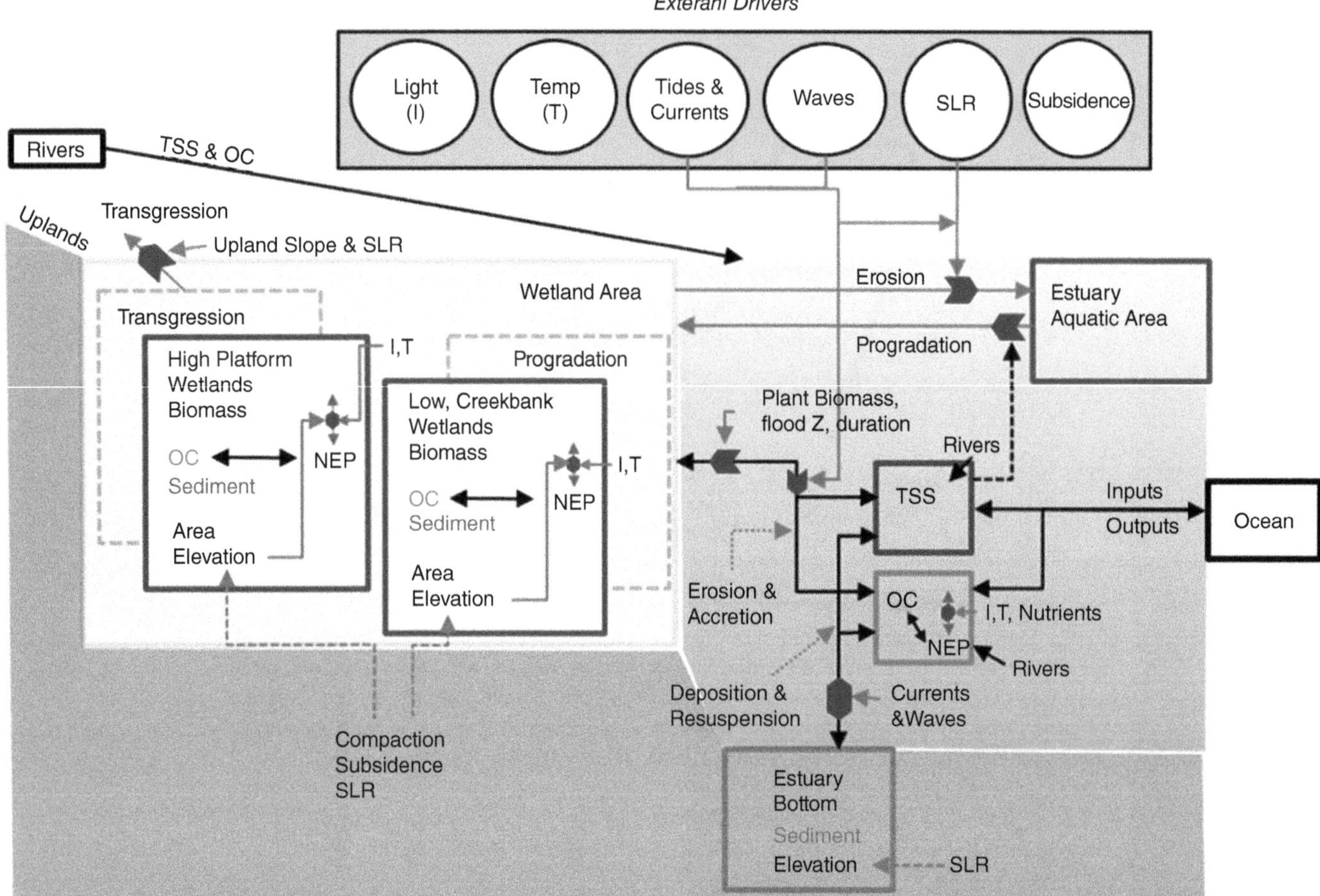

FIGURE 14.20 Conceptual model of blue carbon sequestration for tidal wetlands, illustrating the factors contributing to marsh expansion vertically and laterally and the key role played by sediment solids availability, marsh biomass, and the linkages between watersheds and estuaries. Key drivers that control the structure and function of tidal wetland-dominated estuaries are shown as circles in the External Drivers box above the overall ecosystem. Many of these drivers will likely change substantially over the remainder of the twenty-first century, thereby affecting the long-term survival of tidal wetlands and the blue carbon sequestration. TSS = total suspended solids, Z = water depth, OC = organic carbon, SLR = sea level rise. *Source*: Modified from Hopkinson (2018).

climate change, and sea level rise, are rapidly changing (Weston 2013; Hopkinson 2018; Hopkinson et al. 2018; Fagherazzi et al. 2013; Kirwan et al. 2016; Morris 2016; Ganju et al. 2015). Tidal wetlands have been developing ever since the rate of SLR following the last glacial maximum slowed to less than about 1 mm yr^{-1}, roughly 3000–5000 years ago (Hopkinson et al. 2018). They now contain a vast store of organic carbon that is attributable to three processes—transgression, vertical accretion, and progradation (Redfield 1967). As sea level rises, tidal waters flood previously upland areas, enabling the lateral expansion of tidal wetlands, called transgression. The upland slope relative to SLR dictates the areal expansion rate. As estuarine bay bottoms trap ocean- and land-derived sediments, bottom elevations rise. If the rise brings them above mean sea level in elevation, tidal wetlands may prograde into formerly open water areas, thereby decreasing the area of water and increasing the area of tidal wetland (Figure 14.20). The primary factors controlling progradation are sediment availability, wave and current regimes, SLR, and the overall balance between erosion, deposition, and resuspension (Mariotti and Fagherazzi 2010). The rate of vertical accretion and wetland elevation gain is controlled by the accumulation of undecomposed plant tissues, sediment availability, flooding depth and duration, and wetland biomass which controls sediment trapping efficiency (Kirwan et al. 2016; Morris 2016; Ganju et al. 2015; Fagherazzi et al. 2013). There is an optimum flooding duration and depth for wetland plant biomass and production (Morris 2016). If the rate of SLR (combined with rates of sediment compaction and land subsidence) results in flooding depths greater than the optimum for plant growth, the ability of a wetland to maintain elevation relative to SLR will decline and wetlands will drown. For seagrasses, an additional factor has become increasingly important—light transparency. Nutrient enrichment, eutrophication, and increased suspended solids concentrations all decrease light penetration into the water column and therefore decrease the area potentially occupied by seagrasses.

An important question for predicting future rates of blue carbon sequestration and thus the ability of these systems to contribute to the reduction of CO_2 levels in the atmosphere is how anthropogenic change (land use, wetland reclamation, river damming), climate change, and SLR will influence *NEP*. What happens to blue carbon sequestration if *NEP* decreases?

How will the absolute and relative magnitudes of OC burial and OC export to tidal waters change? With increased tidal flooding, will *GPP* and hence *NEE* decline (Nahrawi et al. 2017)? Will DIC and OC export increase, leaving less for burial? These are currently active areas of research.

There is strong evidence that tidal wetlands are being degraded by anthropogenic activities, SLR, and climate change: marshes are drowning, edges are eroding, and estuaries are exporting rather than importing and burying sediments (Ganju et al. 2015; Fagherazzi et al. 2013; Kirwan et al. 2016; Hopkinson 2018; Weston 2013). How will these changes influence metabolism of entire estuarine systems?

14.7 Metabolic Responses to Climate Change and Variability

Climatic change and variability are expected to alter both respiration and primary production of aquatic ecosystems through a variety of mechanisms and pathways. Although it is widely expected that the climate of the earth system will change over the next century, the anticipated magnitude and nature of these changes varies widely depending on geographic location and modeling approach. Most model predictions of global climate change, however, predict regional increases in ambient temperature, shifts in precipitation and storm frequency (e.g., Mulholland et al. 1998), and increased acidity of the ocean (e.g., Hoegh-Guldberg et al. 2007).

These changes in climatic and biogeochemical conditions are all likely to alter regional and global rates and balances between *P* and *R*. Elevated precipitation tends to increase ecosystem production in temperate, deeper estuaries due to increased stream-flow and associated nutrient delivery (e.g., Justić et al. 2003) as well as reduced estuarine exchange time (Hagy et al. 2000; Smith et al. 2005). Elevated freshwater runoff associated with climatic events may, however, lead to more net heterotrophic conditions in shallow lagoons (Russell et al. 2006) and estuaries receiving high carbon loads from adjacent marshes or rivers (Jiang et al. 2008). Elevated temperature is expected to increase rates of evapotranspiration, which may balance or exceed increases in precipitation, causing reduced runoff. Conversely, changes in atmospheric pressure fields and associated wind patterns could decrease the intensity of upwelling-induced ecosystem production along many coastal shelf areas (e.g., Barth et al. 2007); changing winds could also contribute to changes in stratification strength, as well as bottom O_2 pools and respiration rates. For example, historical analysis of *NEP* estimated for European coastal shelf communities revealed a positive relationship with climate-induced vertical mixing (Heath and Beare 2008), while climate-induced changes in vertical mixing were associated with a decadal scale increase in subsurface R_n in the North Pacific (Emerson et al. 2004). Loss of coral reefs associated with ocean acidification could greatly diminish metabolic rates in these important ecosystems, but knowledge about coral reef resilience and climate-induced species shifts is lacking and it is uncertain what alternative primary producer communities might replace corals. Because photosynthesis is generally less sensitive to temperature than respiration, global temperature increases will likely lead to lower *NEP* in the coastal ocean via elevated respiration (e.g., López-Urrutia et al. 2006). Other estuarine studies have linked elevated temperatures to reduced spring phytoplankton biomass, as temperature increases stimulated zooplankton grazing (van Beusekom et al. 2009). In the open ocean, elevated temperature will likely decrease phytoplankton production, in part by enhancing stratification and reducing the mixing of nutrients into the photic layer (Behrenfeld et al. 2006).

Although it is clear that climate change will have significant impacts on organic matter production and consumption, the magnitudes, trajectories, and geographic distributions of these changes are poorly understood. The lack of understanding of these metabolic consequences of climate change is distressing, particularly because of the potential for positive and negative feedbacks that could reinforce or stabilize climatic changes. Future empirical and modeling studies should focus on improving knowledge about how rates and balance in ecosystem metabolism will respond to changing environmental conditions and alter the nature of estuaries and other aquatic environments.

14.8 Summary and Conclusions

Ecosystem metabolism represents an important measure of trophic status of estuarine and coastal regions, reflecting the combined catabolic and anabolic processes of all biological components of an ecosystem. A wide variety of methods are available to measure different processes that contribute to ecosystem metabolism, ranging from continuous incubations, to open water changes in O_2 or TCO_2, to mass-balance and model calculations. Each method has its own advantages and disadvantages. Although there are many drivers of metabolic processes in estuarine ecosystems, photosynthetic components of ecosystem metabolism are primarily driven by sunlight, water clarity, nutrient availability, temperature, the presence of toxic contaminants, and water exchange times; respiratory components of ecosystem metabolism are primarily driven by temperature and organic matter quantity and quality, but are also influenced by water exchange time, especially when considering an entire system. The magnitude and variability of ecosystem metabolism thus differs among coastal ecosystems, given the variety of water depth, species composition, and external organic matter sources across estuaries, as well as the relative importance of climatic controls and human perturbations. Despite this variety, cross-system comparisons indicate a strong tendency for *NEP* = 0 across ecosystem types, but coastal ecosystem metabolism varies widely between autotrophic and heterotrophic conditions. Shallow, clear-water systems (especially those with macrophyte communities) and larger, open-water systems receiving high loads of inorganic nutrients relative to organic matter tend to be autotrophic, while enclosed, poorly-flushed, and turbid ecosystems adjacent to wetlands or receiving high loads of anthropogenic organic material tend to be heterotrophic. Most systems show a predictable seasonality in anabolic and catabolic processes, dominated by net autotrophy during warm sunny periods with limited wind, and net heterotrophy during periods of light limitation. Future increases in coastal eutrophication will likely stimulate organic carbon production and likely push coastal ecosystems toward net autotrophy. The effects of future climatic changes on ecosystem metabolism are uncertain, however, given alterations in precipitation, evapotranspiration, river flow, temperature, and atmospheric circulation that could interact to cause large changes in coastal metabolism, either toward more autotrophic or heterotrophic conditions. Future research should continue methodological development, while pushing for further investigations into how global environmental changes will affect metabolic balance and carbon sequestration in coastal ecosystems.

Acknowledgements

We would like to acknowledge the substantial contributions of Steven Smith and W. Michael Kemp to the version of this chapter presented in the second edition of *Estuarine Ecology*.

This chapter is University of Maryland Center for Environmental Science Contribution #6121 and Ref. No. [UMCES] CBL 2022-044.

Study Questions

Multiple choice

1. Which of the following elements can be used as a tracer of ecosystem metabolism?
   - **a.** Carbon
   - **b.** Iron
   - **c.** Oxygen
   - **d.** a & c
2. What word can be used to describe an estuary where respiration exceeds primary production?
   - **a.** Autotrophy
   - **b.** P/R Ratio < 1
   - **c.** Heterotrophy
   - **d.** b and c
3. Under which conditions would you expect to measure net heterotrophy?
   - **a.** A highly illuminated water body enriched with nutrients
   - **b.** An estuary receiving high imports of organic matter from nearby rivers
   - **c.** A shallow, well illuminated estuary with dense submerged vegetation
   - **d.** All of the Above
4. An excess of organic matter via net autotrophy could have which ecosystem impacts?
   - **a.** The excess organic carbon is exported to adjacent estuaries to support heterotrophy there
   - **b.** The excess organic carbon is deposited to underlying deeper waters to support hypoxia
   - **c.** The excess organic carbon serves as an organic matter source for heterotrophs, contributing to secondary production
   - **d.** All of the Above
5. Which of the following statements are false?
   - **a.** Estuaries contribution to blue carbon sequestration is small relative to their area
   - **b.** Globally, estuarine open waters release carbon dioxide to the atmosphere.
   - **c.** Estuaries are a location where substantial blue carbon sequestration occurs.
   - **d.** Estuarine ecosystem metabolism is highly variable of over space and time

Short answer

1. Please name four physical variables that influence ecosystem metabolism and state how you would expect these variables to change (i.e., increase or decrease) net ecosystem production.
2. Imagine that you would like to estimate gross primary production, ecosystem respiration, and net ecosystem production in a small inlet with low tidal advection. Suggest a method that you would

choose to generate these estimates, describe the timescales, and explain why.

3. Describe the features of an ecosystem that you would expect to be net autotrophic.
4. Describe how the relative importance of metabolism in benthic and pelagic habitats is expected to change when moving from shallow towards the deeper waters and how this will influence the balance between gross primary production (PP) and ecosystem respiration (ER).
5. How are measurement approaches for net ecosystem metabolism different and/or similar in wetlands versus tidal estuarine waters?
6. What is blue carbon? Identify two sources of blue carbon and how these sources might change under conditions associated with an altered future climate?

References

Alderman, D.W.M., Balsis, B.R., Buffam, I.D. et al. (1995). Pelagic metabolism in the Parker River/Plum Island Sound estuarine system. *Biological Bulletin* 189: 250–251.

Asmala, E., Haraguchi, L., Markager, S. et al. (2018). Eutrophication leads to accumulation of recalcitrant autochthonous organic matter in coastal environment. *Global Biogeochemical Cycles* https://doi.org/10.1029/2017GB005848.

Baldocchi, D.D. (2003). Assessing the eddy covariance technique for evaluating carbon dioxide exchange rates of ecosystems: past, present and future. *Global Change Biology* 9: 479–492.

Barr, J.G., Engel, V., Fuentes, J.D. et al. (2010). Controls on mangrove forest-atmosphere carbon dioxide exchanges in western Everglades National Park. *Journal of Geophysical Research* 115: G02020.

Barron, C., Marba, N., Terrados, J. et al. (2004). Community metabolism and carbon budget along a gradient of seagrass (*Cymodocea nodosa*) colonization. *Limnology and Oceanography* 49: 1642–1651.

Barth, J.A., Menge, B.A., Lubchenco, J. et al. (2007). Delayed upwelling alters nearshore coastal ocean ecosystems in the northern California current. *Proceeding of the National Academy of Sciences USA* 104: 3719–3724.

Bauer, J., Raymond, P., Cai, W.-J. et al. (2013). The changing C cycle of the coastal ocean. *Nature* 504: 61–70.

Beck, M.W., Hagy, J.D., and Murrell, M.C. (2015). Improving estimates of ecosystem metabolism by reducing effects of tidal advection on dissolved oxygen time series. *Limnology and Oceanography: Methods* 13: 731–745.

Behrenfeld, M.J., O'Malley, R.T., Siegel, D.A. et al. (2006). Climate-driven trends in contemporary ocean productivity. *Nature* 444: 752–755.

Berg, P., Røy, H., Janssen, F. et al. (2003). Oxygen uptake by aquatic sediments measured with a novel non-invasive eddy-correlation technique. *Marine Ecology Progress Series* 261: 75–83.

van Beusekom, J.E.E., Loebl, M., and Martens, P. (2009). Distant riverine nutrient supply and local temperature drive the long-term phytoplankton development in a temperate coastal basin. *Journal of Sea Research* 61: 26–33.

Borges, A.V. and Abril, G. (2010). Carbon dioxide and methane dynamics in estuaries. In: *Treatise on Estuarine and Coastal Science*, vol. 5 (ed. E.W. Wolansky and D. McLusky), 119–161. Waltham: Academic Press.

Borges, A.V., Vanderbrought, J.-P., Schiettecatte, L.-S. et al. (2004). Variability of the gas transfer velocity of CO_2 in a macrotidal estuary (the Scheldt). *Estuaries* 27: 593–603.

Boynton, W.R., Hagy, J.D., Cornwell, J.C. et al. (2008). Nutrient budgets and management actions in the Patuxent River estuary, Maryland. *Estuaries and Coasts* 31: 623–651.

Bozec, Y., Thomas, H., Schiettecatte, L.-S. et al. (2006). Assessment of the processes controlling the seasonal variations of dissolved inorganic carbon in the North Sea. *Limnology and Oceanography* 51: 2746–2762.

Brush, M.J., Brawley, J.W., Nixon, S.W., and Kremer, J.N. (2002). Modeling phytoplankton production: problems with the Eppley curve and an empirical alternative. *Marine Ecology Progress Series* 238: 31–45.

Caffrey, J.M. (2004). Factors controlling net ecosystem metabolism in U.S. estuaries. *Estuaries* 27: 90–101.

Cai, W.-J., Dai, M., and Wang, Y. (2006). Air-sea exchange of carbon dioxide in ocean margins: a province-based synthesis. *Geophysical Research Letters* 33: L12603.

Cai, W.-J., Huang, W.-J., Luther, G.W. et al. (2017). Redox reactions and weak buffering capacity lead to acidification in the Chesapeake Bay. *Nature Communications* 8: 369.

Carini, S., Weston, N., Hopkinson, C. et al. (1996). Gas exchange rates in the Parker River estuary, Massachusetts. *Biological Bulletin* 191: 333–334.

Carstensen, J., Conley, D.J., Andersen, J.H., and Ærtebjerg, G. (2006). Coastal eutrophication and trend reversal: a Danish case study. *Limnology and Oceanography* 51: 398–408.

Clark, J.F., Schlosser, P., Stute, M., and Simpson, H. (1996). SF_6-^3He tracer release experiment: a new method of determining longitudinal dispersion coefficients in large rivers. *Environmental Science and Technology* 30: 1527–1532.

Cloern, J.E., Alpine, A.E., Cole, B.E. et al. (1983). River discharge controls phytoplankton dynamics in the Northern San Francisco Bay estuary. *Estuarine, Coastal and Shelf Science* 16: 415–429.

Cloern, J.E., Canuel, E.A., and Harris, D. (2002). Stable carbon and nitrogen isotope composition of aquatic and terrestrial plant of the San Francisco Bay Estuarine System. *Limnology and Oceanography* 47: 713–729.

Cole, B.E. and Cloern, J.E. (1987). An empirical model for estimating phytoplankton productivity in estuaries. *Marine Ecology Progress Series* 36: 299–305.

Conley, D.J., Björck, S., Bonsdorff, E. et al. (2009). Hypoxia-related processes in the Baltic Sea. *Environmental Science and Technology* 43: 3412–3420.

Cox, T.J.S., Maris, T., Soetaert, K. et al. (2009). A macro-tidal freshwater ecosystem recovering from hypereutrophication: the Schelde case study. *Biogeosciences* 6: 2935–2948.

Crossland, C.J., Hatcher, B.G., and Smith, S.V. (1991). Role of coral reefs in global ocean production. *Coral Reefs* 10: 55–64.

Crossland, C.J., Kremer, H.H., Lindeboom, H.J. et al. (2005). *Coastal fluxes in the anthropocene—The Land-Ocean Interactions in the Coastal Zone Project of the International Geosphere-Biosphere Programme*. Berlin, Germany: Springer.

Cunningham, J., Kemp, W.M., Lewis, M., and Stevenson, J.C. (1984). Temporal responses of the macrophyte, *Potamogeton perfoliatus* L., and its associated autotrophic community to Atrazine exposure in estuarine microcosms. *Estuaries* 7: 519–530.

D'Avanzo, C., Kremer, J.N., and Wainright, S.C. (1996). Ecosystem production and respiration in response to eutrophication in shallow temperate estuaries. *Marine Ecology Progress Series* 141: 263–274.

Day, J.W., Madden, C.J., Ley-Lou, F. et al. (1988). Aquatic primary productivity in Terminos Lagoon. In: *Ecology of Coastal Ecosystems in the Southern Gulf of Mexico: The Terminos Lagoon Region* (ed. A. Yañez-Arancibia and J.W. Day), 221–236. Autónoma de Mexico, México City, Mexico: Universidad Nacional.

Demers, S., Therriault, J.-C., Bourget, E., and Bah, A. (1987). Resuspension in the shallow sublittoral zone of a macrotidal estuarine environment: wind influence. *Limnology and Oceanography* 32: 327–339.

Dokulil, M.T. (1994). Environmental control of phytoplankton productivity in turbulent turbid systems. *Hydrobiologia* 289: 65–72.

Dollar, S.J., Smith, S.V., Vink, S.M. et al. (1991). Annual cycle of benthic nutrient fluxes in Tomales Bay, California, and contribution of the benthos to total ecosystem metabolism. *Marine Ecology Progress Series* 79: 115–125.

Duarte, C.M. and Augusti, S. (1998). The CO_2 balance of unproductive aquatic ecosystems. *Science* 281: 234–236.

Duarte, C.M., Middleburg, J.J., and Caraco, N. (2005). Major role of marine vegetation in the oceanic carbon cycle. *Biogeosciences* 2: 1–8.

Duarte, C.M., Marbà, N., Krause-Jensen, D., and Sánchez-Camacho, M. (2007). Testing the predictive power of seagrass depth limit models. *Estuaries and Coasts* 30: 652.

Emerson, S., Watanabe, Y.W., Ono, T., and Mecking, S. (2004). Temporal trends in apparent oxygen utilization in the upper pycnocline of the North Pacific: 1980–2000. *Journal of Oceanography* 60: 139–147.

Ensign, S.H. and Doyle, M.W. (2006). Nutrient spiraling in streams and river networks. *Journal of Geophysical Research* 111: G04009.

Epping, E.H.G. and Jørgensen, B.B. (1996). Light-enhanced oxygen respiration in benthic phototrophic communities. *Marine Ecology Progress Series* 139: 193–203.

Eppley, R.W. (1972). Temperature and phytoplankton growth in the sea. *Fishery Bulletin* 70: 1063–1085.

Eyre, B.D. and McKee, L. (2002). Carbon, nitrogen and phosphorus budgets for a shallow subtropical coastal embayment (Moreton Bay, Australia). *Limnology and Oceanography* 47: 1043–1055.

Fagherazzi, S., Mariotti, G., Wiberg, P., and McGlathery, K. (2013). Marsh collapse does not require sea level rise. *Oceanography* 26: 70–77.

Feng, Y., Friedrichs, M.A.M., Wilkin, J. et al. (2015). Chesapeake Bay nitrogen fluxes derived from a land-estuarine ocean biogeochemical modeling system: Model description, evaluation, and nitrogen budgets. *Journal of Geophysical Research: Biogeosciences* 120: 1666–1695.

Flores-Verdugo, F.J., Day, J.W., Mee, L., and Briseno-Dueñas, R. (1988). Phytoplankton production and seasonal biomass variation of seagrass, *Ruppia maritima* L., in a tropical Mexican lagoon with an ephemeral inlet. *Estuaries* 11: 51–56.

Forbrich, I. and Giblin, A.E. (2015). Marsh-atmosphere CO_2 exchange in a New England salt marsh. *Journal of Geophysical Research Biogeosciences* 120: 1825–1838.

Forbrich, I., Giblin, A., and Hopkinson, C. (2018). Constraining marsh carbon budgets using long-term C burial and contemporary atmospheric CO_2 fluxes. *Journal of Geophysical Research Biogeosciences* 123: 867–878.

Fourqurean, J.W., Moore, T.O., Fry, B., and Hollibaugh, J.T. (1997). Spatial and temporal variation in C:N:P ratios, $\delta^{15}N$, and $\delta^{13}C$ of eelgrass *Zostera marina* as indicators of ecosystem processes, Tomales Bay, California, USA. *Marine Ecology Progress Series* 157: 145–157.

Frankignoulle, M. (1988). Field measurements of air-sea CO_2 exchange. *Limnology and Oceanography* 33: 313–322.

Froelich, P.N., Klinkhammer, G.P., Bender, M.L. et al. (1979). Early oxidation of organic matter in pelagic sediments of the eastern equatorial Atlantic: suboxic diagenesis. *Geochimica et Cosmochimica acta* 43: 1075–1090.

Gaarder, T. and Gran, H. (1927). *Investigations of the Production of Plankton in the Oslo Fjord*, Rapports Et Procès-Verbaux des Rèunions Vol=XLII, vol. 42, 1–48. Conseil Permanent International Pour L'Exploration de la Mer.

Gallegos, C.L. and Platt, T. (1985). Vertical advection of phytoplankton and productivity estimates: a dimensional analysis. *Marine Ecology Progress Series* 26: 125–134.

Ganju, N., Kirwan, M., Dickhudt, P. et al. (2015). Sediment transport-based metrics of wetland stability. *Geophysical Research Letters* 42: 7992–8000.

Garnier, J. and Billen, G. (2007). Production vs. respiration in river systems: an indicator of an "ecological status". *Science of the Total Environment* 375: 110–124.

Gattuso, J.-P., Frankignoulle, M., and Wollast, R. (1998). Carbon and carbonate metabolism in coastal aquatic ecosystems. *Annual Review of Ecology and Systematics* 29: 405–434.

Gazeau, F., Borges, A.V., Barron, C. et al. (2005a). Net ecosystem metabolism in a micro-tidal estuary (Randers Fjord, Denmark): evaluation of methods. *Marine Ecology Progress Series* 301: 23–41.

Gazeau, F., Duarte, C.M., Gattuso, J.-P. et al. (2005b). Whole-system metabolism and CO_2 fluxes in a Mediterranean Bay dominated by seagrass beds (Palma Bay, NW Mediterranean). *Biogeosciences* 2: 43–60.

Gazeau, F., Gattuso, J.-P., Middelburg, J.J. et al. (2005c). Planktonic and whole system metabolism in a nutrient-rich estuary (the Scheldt estuary). *Estuaries and Coasts* 28: 868–883.

Gazeau, F., Middelburg, J.J., Loijens, M. et al. (2007). Planktonic primary production in estuaries: comparison of ^{14}C, O_2, and ^{18}O methods. *Aquatic Microbial Ecology* 46: 95–106.

del Giorgio, P.A. and Williams, P.J.l. (2005). *Respiration in Aquatic Ecosystems*. Oxford, UK: Oxford University Press.

Giri, C., Ochieng, E., Tieszen, L. et al. (2011). Status and distribution of mangrove forests of the world using earth observation satellite data (version 1.3, updated by UNEP-WCMC). *Global Ecology and Biogeography* 20: 154–159. Data URL: http://data.unep-wcmc.org/datasets/4.

Glazer, B., Marsh, A., Stierhoff, K., and Luther, G.W. (2004). The dynamic response of optical oxygen sensors and voltammetric electrodes to temporal changes in dissolved oxygen concentration. *Analytica Chimica Acta* 518: 93–100.

Graf, G., Bengtsson, W., Diesner, U. et al. (1982). Benthic response to sedimentation of a spring phytoplankton bloom: process and budget. *Marine Biology* 67: 201–208.

Graney, R., Kennedy, J., and Rodgers, J. (ed.) (1994). *Aquatic Mesocosm Studies in Ecological Risk Assessment*. Boca Raton, Florida, USA: CRC Press.

Green, E.P. and Short, F.T. (2003). *World Atlas of Seagrasses*. Berkeley, USA: University of California Press.

Green, R.E., Bianchi, T.S., Dagg, M. et al. (2006). An organic carbon budget for the Mississippi River turbidity plume and plume contributions to air-sea CO_2 fluxes and bottom water hypoxia. *Marine Ecology Progress Series* 278: 35–51.

Gribsholt, B., Boschker, H.T.S., Struyf, E. et al. (2005). Nitrogen processing in a tidal freshwater marsh: a whole-ecosystem ^{15}N labeling study. *Limnology and Oceanography* 50: 1945–1959.

Hagy, J.D., Sanford, L., and Boynton, W.R. (2000). Estimation of net physical transport and hydraulic residence times for a coastal plain estuary using box models. *Estuaries* 23: 328–340.

Hagy, J.D., Boynton, W.R., Keefe, C.W., and Wood, K.V. (2004). Hypoxia in Chesapeake Bay, 1950–2001: long-term change in relation to nutrient loading and river flow. *Estuaries* 27: 634–658.

Hamilton, S. and Casey, D. (2016). Creation of a high spatio-temporal resolution global database of continuous mangrove forest cover for the 21st century (CGMFC-21). *Global Ecology and Biogeography* 25: 729–738.

Harding, L.W. Jr., Meeson, B.W., Prézelin, B.B., and Sweeney, B.M. (1981). Diel periodicity of photosynthesis in marine phytoplankton. *Marine Biology* 61: 95–105.

Harvey, E.T., Walve, J., Andersson, A. et al. (2019). The effect of optical properties on Secchi Depth and implications for eutrophication management. *Frontiers in Marine Science* 10: https://doi.org/10.3389/fmars.2018.00496.

Hashimoto, S., Horimoto, N., Ishimaru, T., and Saino, T. (2006). Metabolic balance of gross primary production and community respiration in Sagami Bay, Japan. *Marine Ecology Progress Series* 321: 31–40.

Heath, M. (1995). An holistic analysis of the coupling between physical and biological processes in the coastal zone. *Ophelia* 42: 95–125.

Heath, M. and Beare, D. (2008). New primary production in northwest European shelf seas, 1960–2003. *Marine Ecology Progress Series* 363: 183–203.

Heip, C.H.R., Goosen, N.K., Herman, P.M.J. et al. (1995). Production and consumption of biological particles in temperate tidal estuaries. *Oceanography and Marine Biology: An Annual Review* 33: 1–149.

Herrmann, M., Najjar, R.G., Kemp, W.M. et al. (2015). Net ecosystem production and organic carbon balance of U.S. East Coast estuaries: a synthesis approach. *Global Biogeochemical Cycles* 29: 96–111.

Ho, D.T., Wanninkhof, R., Schlosser, P. et al. (2011). Toward a universal relationship between wind speed and gas exchange: gas transfer velocities measured with $^3He/SF_6$ during the Southern Ocean Gas Exchange Experiment. *Journal of Geophysical Research: Oceans* 116: C00F04. https://doi.org/10.1029/2010JC006854.

Hoegh-Guldberg, O., Mumby, P.J., Hooten, A.J. et al. (2007). Coral reefs under rapid climate change and ocean acidification. *Science* 318: 1737–1742.

Hoellein, T.J., Bruesewitz, D.A., and Richardson, D.C. (2013). Revisiting Odum (1956): a synthesis of aquatic ecosystem metabolism. *Limnology and Oceanography* 58: 2089–2100.

Hofmann, A.F., Soetaert, K., and Middelburg, J.J. (2008). Present nitrogen and carbon dynamics in the Scheldt estuary using a novel 1-D model. *Biogeosciences* 5: 981–1006.

Hopkinson, C.S. (2018). Net Ecosystem Carbon balance of coastal wetland-dominated estuaries: where's the blue carbon? In: *A Blue Carbon Primer: The State of Coastal Wetland Carbon Science, Policy, and Practice* (ed. L. Windham-Myers, S. Crooks and T. Troxler), 51–66. Boca Raton, FL: CRC Press.

Hopkinson, C.S. and Smith, E.M. (2005). Estuarine respiration: an overview of benthic, pelagic, and whole system respiration. In: *Respiration in Aquatic Ecosystems* (ed. P. del Giorgio and P.J.l. Williams), 122–146. Oxford, UK: Oxford University Press.

Hopkinson, C. and Vallino, J. (1995). The relationships among man's activities in watersheds and estuaries: a model of runoff effects on patterns of estuarine community metabolism. *Estuaries* 18: 598–621.

Hopkinson, C.S., Giblin, A.E., Tucker, J., and Garritt, R.H. (1999). Benthic metabolism and nutrient cycling along an estuarine salinity gradient. *Estuaries* 22: 863–881.

Hopkinson, C.S., Cai, W.-J., and Hu, X. (2012). Carbon sequestration in wetland dominated coastal systems—a global sink of rapidly diminishing magnitude. *Current Opinion On Environmental Sustainability* 4: 1–9.

Hopkinson, C., Morris, J., Fagherazzi, S. et al. (2018). Lateral marsh edge erosion as a source of sediments for vertical marsh accretion. *Journal of Geophysical Research Biogeosciences* 123: 2444–2465.

Hoppema, J.M.J. (1991). The oxygen budget of the western Wadden Sea, the Netherlands. *Estuarine, Coastal and Shelf Science* 32: 483–502.

Howarth, R.W. and Teal, J.M. (1979). Sulfate reduction in a New England salt marsh. *Limnology and Oceanography* 24: 999–1013.

Howarth, R.W., Swaney, D.P., Butler, T.J., and Marino, R. (2000). Climatic control on eutrophication of the Hudson River estuary. *Ecosystems* 3: 210–215.

Hume, A.C., Berg, P., and McGlathery, K.J. (2011). Dissolved oxygen fluxes and ecosystem metabolism in an eelgrass (*Zostera marina*) meadow measured with the eddy correlation technique. *Limnology and Oceanography* 56: 86–96.

Jiang, L.-Q., Cai, W.-J., and Wang, Y. (2008). A comparative study of carbon dioxide degassing in river- and marine-dominated estuaries. *Limnology and Oceanography* 53: 2603–2615.

Justić, D., Turner, R.E., and Rabalais, N.N. (2003). Climatic influences on riverine nitrate flux: implications for coastal marine eutrophication and hypoxia. *Estuaries* 26: 1–11.

Kaldy, J., Onuf, C., Eldridge, P., and Cifuentes, L. (2002). Carbon budget for a subtropical seagrass dominated coastal lagoon: how important are seagrasses to total ecosystem net primary production? *Estuaries* 25: 528–539.

Kana, T.M., Darkangelo, C., Hunt, C. et al. (1994). A membrane inlet mass spectrometer for rapid high precision determination of N_2, O_2, and Ar in environmental water samples. *Analytical Chemistry* 66: 4166–4170.

Kemp, W.M. and Boynton, W.R. (1992). Benthic-pelagic interactions: nutrient and oxygen dynamics. In: *Oxygen Dynamics in Chesapeake Bay: A Synthesis of Research* (ed. D. Smith, M. Leffler and G. Mackiernan), 149–221. College Park, Maryland, USA: Maryland Sea Grant Publications.

Kemp, W.M. and Testa, J.M. (2011). Metabolic balance between ecosystem production and consumption. In: *Treatise on Estuarine and Coastal Science* (ed. E. Wolansky and D. McLusky), 83–118. Waltham, Massachusetts: Academic Press.

Kemp, W.M., Boynton, W.R., Cunningham, J. et al. (1985). Effects of Atrazine and Linuron on photosynthesis and growth of the macrophytes, *Potamogeton perfoliatus* L. and *Myriophyllum spicatum* L. in an estuarine environment. *Marine Environmental Research* 16: 255–280.

Kemp, P.F., Falkowski, P., Flagg, C. et al. (1994). Modeling vertical oxygen and carbon flux during stratified spring and summer conditions on the continental shelf. *Deep-Sea Research Part II* 41: 620–655.

Kemp, W.M., Smith, E.M., Marvin-DiPasquale, M., and Boynton, W.R. (1997). Organic carbon balance and net ecosystem metabolism in Chesapeake Bay. *Marine Ecology Progress Series* 150: 229–248.

Kemp, W.M., Boynton, W.R., Adolf, J. et al. (2005). Eutrophication of Chesapeake Bay: historical trends and ecological interactions. *Marine Ecology Progress Series* 303: 1–29.

Kemp, W.M., Testa, J.M., Conley, D.J. et al. (2009). Temporal responses of coastal hypoxia to nutrient loading and physical controls. *Biogeosciences* 6: 2985–3008.

King, G.M., Klug, M.J., Wiegert, R.G., and Chalmers, A.G. (1982). Relation of soil water movement and sulfate concentration to *Spartina alterniflora* production in a Georgia salt marsh. *Science* 218: 61–63.

Kirwan, M., Walters, D., Reay, W., and Carr, J. (2016). Sea level driven marsh expansion in a coupled model of marsh erosion and migration. *Geophysical Research Letters* 43: 4366–4373.

Krause-Jensen, D., Carstensen, J., Nielsen, S.L. et al. (2011). Sea bottom characteristics affect depth limits of eelgrass *Zostera marina*. *Marine Ecology Progress Series* 425: 91–102.

Kremer, J.N., Reischauer, A., and D'Avanzo, C. (2003). Estuary-specific variation in the air-water gas exchange coefficient for oxygen. *Estuaries* 26: 829–836.

Kromkamp, J. and Peene, J. (1995). Possibility of net phytoplankton primary production in the turbid Schelde estuary (SW Netherlands). *Marine Ecology Progress Series* 121: 249–259.

Kromkamp, J.C. and Van Engeland, T. (2010). Changes in phytoplankton biomass in the western Scheldt estuary during the period 1978–2006. *Estuaries and Coasts* 33: 270–285.

Laws, E., Landry, M., Barber, R. et al. (2000). Carbon cycling in primary production bottle incubations: inferences from grazing experiments and photosynthetic studies using ^{14}C and ^{18}O in the Arabian Sea. *Deep-Sea Research II* 47: 1339–1352.

Lee, K. (2001). Global net community production estimated from the annual cycle of surface water total dissolved inorganic carbon. *Limnology and Oceanography* 46: 1287–1297.

Lehrter, J.C. and Cebrian, J. (2010). Uncertainty propagation in ecosystem nutrient budgets. *Ecological Applications* 20: 508–524.

Long, M.H. and Nicholson, D.P. (2018). Surface gas exchange determined from an aquatic eddy covariance floating platform. *Limnology and Oceanography: Methods* 16: 145–159.

López-Urrutia, Á., San Martin, E., Harris, R., and Irigoien, X. (2006). Scaling the metabolic balance of the ocean. *Proceedings of the National Academy of Science USA* 103: 8739–8744.

Lovett, G.M., Cole, J.J., and Pace, M.L. (2006). Is net ecosystem production equal to ecosystem carbon accumulation? *Ecosystems* 9: 1–14.

Lutz, M.J., Caldeira, K., Dunbar, R.B., and Behrenfeld, M.J. (2007). Seasonal rhythms of net primary production and particulate organic carbon flux to depth describe the efficiency of biological pump in the global ocean. *Journal of Geophysical Research* 112: C10011.

Luz, B., Barkan, E., Bender, M.L. et al. (1999). Triple-isotope composition of atmospheric oxygen as a tracer of biosphere productivity. *Nature* 400: 547–550.

Marañón, E., Lorenzo, M.P., Cermeño, P., and Mouriño-Carballido, B. (2018). Nutrient limitation suppresses the temperature dependence of phytoplankton metabolic rates. *The ISME Journal* 12: 1836–1845.

Marino, R. and Howarth, R. (1993). Atmospheric oxygen exchange in the Hudson River: dome measurements and comparison with other natural waters. *Estuaries* 16: 433–445.

Mariotti, G. and Fagherazzi, S. (2010). A numerical model for the coupled long-term evolution of salt marshes and tidal flats. *Journal of Geophysical Research Earth Surface* 115: F01004.

Markager, S., Jespersen, A.-M., Madsen, T.V. et al. (1992). Diel changes in dark respiration in a plankton community. *Hydrobiologia* 238: 119–130.

Martz, T.R., Johnson, K.S., and Riser, S.C. (2008). Ocean metabolism observed with oxygen sensors on profiling floats in the South Pacific. *Limnology and Oceanography* 53: 2094–2111.

Marvin-DiPasquale, M.C. and Capone, D.G. (1998). Benthic sulfate reduction along the Chesapeake Bay central channel. I. Spatial trends and controls. *Marine Ecology Progress Series* 168: 213–228.

McGlathery, K.J., Sundbäck, K., and Anderson, I.C. (2007). Eutrophication in shallow coastal bays and lagoons: the role of plants in the coastal filter. *Marine Ecology Progress Series* 348: 1–18.

Mcowen, C., Weatherdon, L., Bochove, J. et al. (2017). A global map of saltmarshes. *Biodiversity Data Journal* 5: e11764. https://doi.org/10.3897/BDJ.5.e11764.

Middelburg, J.J., Klaver, G., Nieuwenhuize, J. et al. (1996). Organic matter mineralization in intertidal sediments along an estuarine gradient. *Marine Ecology Progress Series* 132: 157–168.

Mommer, L. and Visser, E.J. (2005). Underwater photosynthesis in flooded terrestrial plants: a matter of leaf plasticity. *Annals of Botany* 96: 581–589.

Morris, J.T. (2016). Marsh equilibrium theory. In: *4th International Conference on Invasive Spartina, ICI-Spartina 2014* (ed. M. Ainouche), 67–71. Rennes, France: University of Rennes Press.

Morris, J.T., Sundareshwar, P.V., Nietch, C.T. et al. (2002). Responses of coastal wetlands to rising sea level. *Ecology* 83: 2869–2877.

Morris, J.T., Sundberg, K., and Hopkinson, C.S. (2013). Salt marsh primary production and its responses to relative sea level and nutrients in estuaries at Plum Island, Massachusetts and North Inlet, South Carolina, USA. *Oceanography* 26: 78–84.

Mulholland, P.J., Best, G.R., Coutant, C.C. et al. (1998). Effects of climate change on freshwater ecosystems of the south-eastern United States and the Gulf Coast of Mexico. *Hydrological Processes* 11: 949–970.

Murrell, M.C., Caffrey, J.M., Marcovich, D.T. et al. (2018). Seasonal oxygen dynamics in a warm temperate estuary: effects of hydrologic variability on measurements of primary production, respiration, and net metabolism. *Estuaries and Coasts* 41: 690–707.

Nahrawi, H., Leclerc, M., Zhang, G., and Pahari, R. (2017). Influence of tidal inundation on CO_2 exchange between salt marshes and the atmosphere. *Biogeosciences Discussions* https://doi.org/10.5194/bg-2017-356.

Najjar, R.G. and Keeling, R.F. (2000). Mean annual cycle of the air-sea oxygen flux: a global view. *Global Biogeochemical Cycles* 14: 573–584.

Najjar, R.G., Herrmann, M., Alexander, R. et al. (2018). Carbon budget of tidal wetlands, estuaries, and shelf waters of Eastern North America. *Global Biogeochemical Cycles* 32: 389–416.

Nellemann, C., Corcoran, E., Duarte, C. et al. (2009). *Blue Carbon A Rapid Response Assessment*. Arendal: United Nations Environment Programme, GRID https://cld.bz/bookdata/WK8FNPt/basic-html/index.html#1.

Neubauer, S.C. and Anderson, I.C. (2003). Transport of dissolved inorganic carbon from a tidal freshwater marsh to the York River estuary. *Limnology and Oceanography* 48: 299–307.

Neubauer, S.C., Miller, W.D., and Anderson, I.C. (2000). Carbon cycling in a tidal freshwater marsh ecosystem: a carbon gas flux study. *Marine Ecology Progress Series* 199: 13–30.

Newbold, D., Elwood, J.W., O'Neill, V., and Winkle, W.V. (1981). Measuring nutrient spiralling in streams. *Canadian Journal of Fisheries and Aquatic Sciences* 38: 860–863.

Nidzieko, N.J. (2018). Allometric scaling of estuarine ecosystem metabolism. *Proceedings of the National Academy of Sciences* 115: 6733–6738.

Nixon, S.W. and Pilson, M. (1984). Estuarine total system metabolism and organic exchange calculated from nutrient ratios an example from Narragansett Bay. In: *The Estuary as a Filter* (ed. V.S. Kennedy), 261–290. New York, New York, USA: Academic Press.

Nixon, S.W., Granger, S.L., and Nowicki, B.L. (1995). An assessment of the annual mass balance of carbon, nitrogen, and phosphorus in Narragansett Bay. *Biogeochemistry* 31: 15–61.

Odum, H.T. (1956). Primary production in flowing waters. *Limnology and Oceanography* 1: 102–117.

Odum, H.T. (1967). Biological circuits and the marine systems of Texas. In: *Pollution and Marine Ecology* (ed. T.A. Olson and F.J. Burgess), 99–157. New York, New York, USA: Wiley.

Odum, H.T. and Hoskin, C.M. (1958). Comparative studies of the metabolism of marine waters. *Publications of the Institute of Marine Science-University of Texas* 5: 16–46.

Odum, H.T. and Odum, E.P. (1955). Trophic structure and productivity of a windward coral reef community on Eniwetok Atoll. *Ecological Monographs* 25: 291–320.

Oviatt, C.A., Keller, A.A., Sampou, P., and Beatty, L. (1986). Patterns of productivity during eutrophication: a mesocosm experiment. *Marine Ecology Progress Series* 28: 69–80.

Oviatt, C., Doering, P., Nowicki, B., and Zoppini, A. (1993). Net system production in coastal waters as a function of eutrophication, seasonality and benthic macrofaunal abundance. *Estuaries* 16: 247–254.

Paerl, H.W., Pinckney, J.L., Fear, J.M., and Peierls, B.L. (1998). Ecosystem responses to internal and watershed organic matter loading: consequences for hypoxia in the eutrophying Neuse River estuary, North Carolina, USA. *Marine Ecology Progress Series* 166: 17–25.

Peterson, B.J. (1980). Aquatic primary productivity and the $^{14}C\text{-}CO_2$ method—a history of the productivity problem. *Annual Review of Ecology and Systematics* 11: 359–385.

Platt, T. and Jassby, A.D. (1976). The relationship between photosynthesis and light for natural assemblages of coastal marine phytoplankton. *Journal of Phycology* 12: 421–430.

Pomeroy, L.R. and Wiebe, W.J. (2001). Temperature and substrates as interactive limiting factors for marine heterotrophic bacteria. *Aquatic Microbial Ecology* 23: 187–204.

Powles, S.B. (1984). Photoinhibition of photosynthesis induced by visible light. *Annual Review of Plant Physiology* 35: 15–44.

Quiñones, R.A. and Platt, T. (1991). The relationship between the *f*-ratio and the *P:R* ratio in the pelagic ecosystem. *Limnology and Oceanography* 36: 211–213.

Rabalais, N.N. and Gilbert, D. (2009). Distribution and consequences of hypoxia. In: *Watersheds, Bays and Bounded Seas* (ed. E. Urban, B. Sundby, P. Malanotte-Rizzoli and J.M. Melillo), 209–226. Washington, DC, USA: Island Press.

Ram, A.S.P., Nair, S., and Chandramohan, D. (2003). Seasonal shift in net ecosystem production in a tropical estuary. *Limnology and Oceanography* 48: 1601–1607.

Randall, J.M. and Day, J.W. (1987). Effects of river discharge and vertical circulation on aquatic primary production in a turbid Louisiana (USA) estuary. *Netherlands Journal of Sea Research* 21: 231–242.

Raven, J.A. and Beardall, J. (2005). Respiration in aquatic photolithotrophs. In: *Respiration in Aquatic Ecosystems* (ed. P. del Giorgio and P.J.l. Williams), 36–46. Oxford, UK: Oxford University Press.

Raymond, P.A. and Cole, J.J. (2001). Gas exchange in rivers and estuaries: choosing a gas transfer velocity. *Estuaries* 24: 312–317.

Redfield, A. (1967). The ontogeny of a salt marsh estuary. In: *Estuaries* (ed. G. Lauff), 108–114. AAAS Publication 83.

Regnier, P., Friedlingstein, P., Ciais, P. et al. (2013). Anthropogenic perturbation of the carbon fluxes from land to ocean. *Nature Geoscience* 6: 597.

Reichstein, M., Falge, E., Baldocchi, D. et al. (2005). On the separation of net ecosystem exchange into assimilation and ecosystem respiration: review and improved algorithm. *Global Change Biology* 11: 1424–1439.

Rooney, J.J. and Smith, S.V. (1999). Watershed landuse and bay sedimentation. *Journal of Coastal Research* 15: 478–485.

Rowe, G. (2001). Seasonal hypoxia in the bottom water off the Mississippi River delta. *Journal of Environmental Quality* 30: 281–290.

Russell, M.J., Montagna, P.A., and Kalke, R.D. (2006). The effect of freshwater flow on net ecosystem metabolism in Lavaca Bay, Texas. *Estuarine, Coastal and Shelf Science* 68: 231–244.

Ryther, J.H. (1961). Organic production by plankton algae, and its environmental control. In: *The Ecology of Algae* (ed. C.A. Tryon and R.T. Hartman), 72–83. Pittsburgh, PA, USA: Special Publication No. 2, Pymatuning Laboratory of Field Biology, University of Pittsburgh.

Sampou, P. and Kemp, W.M. (1994). Factors regulating plankton community respiration in Chesapeake Bay. *Marine Ecology Progress Series* 110: 249–258.

Sarma, V.S.S., Abe, O., Hashimoto, S. et al. (2005). Seasonal variations in triple oxygen isotopes and gross oxygen production in the Sagami Bay, central Japan. *Limnology and Oceanography* 50: 544–552.

Satta, M.P., Agustí, S., Mura, M.P. et al. (1996). Microplankton respiration and net community metabolism in a bay on the NW Mediterranean coast. *Aquatic Microbial Ecology* 10: 165–172.

Schäfer, K.V.R., Tripathee, R., Artigas, F. et al. (2014). Carbon dioxide fluxes of an urban tidalmarsh in the Hudson-Raritan estuary. *Journal of Geophysical Research Biogeosciences* 119: 2065–2081.

Schedlbauer, J.L., Oberbauer, S.F., Starr, G., and Jimenez, K.L. (2010). Seasonal differences in the CO_2 exchange of a short-hydroperiod Florida Everglades marsh. *Agricultural and Forest Meteorology* 150: 994–1006.

Scully, M.E. (2018). A diel method of estimating gross primary production: 1. Validation with a realistic numerical model of Chesapeake Bay. *Journal of Geophysical Research: Oceans* 123: 8411–8429.

Seitzinger, S.P. (1988). Denitrification in freshwater and coastal marine ecosystems: ecological and geochemical significance. *Limnology and Oceanography* 33: 702–724.

Shen, C., Testa, J.M., Li, M. et al. (2019). Controls on carbonate system dynamics in a coastal plain estuary: a modeling study. Journal of Geophysical Research. *Biogeosciences* https://doi.org/10.1029/2018JG004802.

Smith, S.V. (1991). Stoichiometry of C:N:P fluxes in shallow-water marine ecosystems. In: *Comparative Analysis of Ecosystems: Patterns, Mechanisms and Theories* (ed. J. Cole, G. Lovett and S. Findlay), 259–286. New York, New York, USA: Springer-Verlag.

Smith, S.V. and Gattuso, J.-P. (2009). Coral reefs. In: *The Management of Natural Coastal Carbon Sinks* (ed. D.d.'A. Laffoley and G. Grimsditch), 39–45. Gland, Switzerland: IUCN.

Smith, S.V. and Hollibaugh, J.T. (1993). Coastal metabolism and the oceanic organic carbon balance. *Reviews of Geophysics* 31: 75–89.

Smith, S.V. and Hollibaugh, J.T. (1997). Annual cycle and interannual variability of ecosystem metabolism in a temperate climate embayment. *Ecological Monographs* 67: 509–533.

Smith, E.M. and Kemp, W.M. (1995). Seasonal and regional variations in plankton community production and respiration for Chesapeake Bay. *Marine Ecology Progress Series* 116: 217–231.

Smith, S.V., Hollibaugh, J.T., Dollar, S.J., and Vink, S. (1991). Tomales Bay metabolism—C-N-P stoichiometry and ecosystem heterotrophy at the land sea interface. *Estuarine, Coastal and Shelf Science* 33: 223–257.

Smith, S.V., Chambers, R.M., and Hollibaugh, J.T. (1996). Dissolved and particulate nutrient transport through a coastal watershed-estuary system. *Journal of Hydrology* 176: 181–203.

Smith, S.V., Buddemeier, R.W., Wulff, F., and Swaney, D.P. (2005). C, N, P fluxes in the coastal zone. In: *Coastal Fluxes in the Anthropocene* (ed. C.J. Crossland, H.H. Kremer, H.J. Lindeboom, et al.), 95–143. Berlin, Germany: Springer.

Soetaert, K. and Herman, P.M.J. (1995). Carbon flows in the Westerschelde estuary (The Netherlands) evaluated by means of an ecosystem model (MOSES). *Hydrobiologia* 311: 247–266.

Soetaert, K., Middelburg, J.J., Heip, C. et al. (2006). Long-term change in dissolved inorganic nutrients in the heterotrophic Scheldt estuary (Belgium, The Netherlands). *Limnology and Oceanography* 51: 409–423.

Stæhr, P.A. and Sand-Jensen, K. (2007). Temporal dynamics and regulation of lake metabolism. *Limnology and Oceanography* 52: 108–120.

Stæhr, P.A., Testa, J.M., Kemp, W.M. et al. (2012). The metabolism of aquatic ecosystems: history, methods, and applications. *Aquatic Sciences* 74: 15–29.

Stæhr, P.A., Testa, J.M., and Carstensen, J. (2017). Decadal changes in water quality and net productivity of a shallow Danish estuary following significant nutrient reductions. *Estuaries and Coasts* 40: 63–79.

Stæhr, P.A., Asmala, E., Carstensen, J. et al. (2018). Ecosystem metabolism of benthic and pelagic zones of a shallow productive estuary: spatio-temporal variability. *Marine Ecology Progress Series* 601: 15–32.

Sundbäck, K., Miles, A., and Linares, F. (2006). Nitrogen dynamics in nontidal littoral sediments: role of microphytobenthos and denitrification. *Estuaries and Coasts* 29: 1196–1211.

Swaney, D.P., Howarth, R.W., and Butler, T.J. (1999). A novel approach for estimating ecosystem production and respiration in estuaries: application to the oligohaline and mesohaline Hudson River. *Limnology and Oceanography* 44: 1509–1521.

Takahashi, T., Sutherland, S., Sweeney, C. et al. (2002). Global sea-air CO_2 flux based on climatological surface ocean pCO_2, and seasonal biological and temperature effects. *Deep Sea Research Part II* 49: 1601–1622.

Teague, K.G., Madden, C.J., and Day, J.W. (1988). Sediment-water oxygen and nutrient fluxes in a river-dominated estuary. *Estuaries* 11: 1–9.

Testa, J.M. and Kemp, W.M. (2008). Variability of biogeochemical processes and physical transport in a partially stratified estuary: a box-modeling analysis. *Marine Ecology Progress Series* 356: 63–79.

Testa, J.M., Kemp, W.M., Boynton, W.R., and Hagy, J.D. (2008). Long-term changes in water quality and productivity in the Patuxent River estuary: 1985–2003. *Estuaries and Coasts* 31: 1021–1037.

Tobias, C.R., Bölke, J.K., and Harvey, W. (2007). The oxygen-18 isotope approach for measuring aquatic metabolism in high-productivity waters. *Limnology and Oceanography* 52: 1439–1453.

Turner, R.E., Howes, B.L., Teal, J.M. et al. (2009). Salt marshes and eutrophication: an unsustainable outcome. *Limnology and Oceanography* 54: 1634–1642.

Twilley, R.R. (1988). Coupling of mangroves to the productivity of estuarine and coastal waters. In: *Coastal-Offshore Ecosystem Interactions* (ed. B.O. Jansson), 155–180. Berlin, Germany: Springer-Verlag.

Twilley, R.R., Ejdung, G., Romare, P., and Kemp, W.M. (1986). A comparative study of decomposition, oxygen consumption and nutrient release for selected aquatic plants occurring in an estuarine environment. *Oikos* 47: 190–198.

Valiela, I. and Teal, J.M. (1979). The nitrogen budget of a salt marsh ecosystem. *Nature* 280: 652–656.

Vallino, J.J., Hopkinson, C.S., and Garritt, R.H. (2005). Estimating estuarine gross production, community respiration and net ecosystem production: a nonlinear inverse technique. *Ecological Modelling* 187: 281–296.

Van de Bogert, M.C., Carpenter, S.R., Cole, J.J., and Pace, M.L. (2007). Assessing pelagic benthic metabolism using free water measurements. *Limnology and Oceanography: Methods* 5: 145–155.

Vannote, R.L., Minshall, G.W., Cummins, K.W. et al. (1980). The river continuum concept. *Canadian Journal of Fisheries and Aquatic Science* 37: 130–137.

Ver, L.M.B., Mackenzie, F.T., and Lerman, A. (1994). Modeling pre-industrial C-N-P-S biogeochemiocal cycling in the land-coastal margin system. *Chemosphere* 29: 855–887.

Vervuren, P.J.A., Blom, C.W.P.M., and De Kroon, H. (2003). Extreme flooding events on the Rhine and the survival and distribution of riparian plant species. *Journal of Ecology* 91: 135–146.

Wang, Z.A., Kroeger, K.D., Ganju, N.K. et al. (2016). Intertidal salt marshes as an important source of inorganic carbon to the coastal ocean. *Limnology and Oceanography* 61: 1916–1931.

Wanninkhof, R., Asher, W.E., Ho, D.T. et al. (2009). Advances in quantifying air-sea gas exchange and environmental forcing. *Annual Review of Marine Science* 1: 213–244.

Ware, J.R., Smith, S.V., and Reaka-Kudla, M.L. (1991). Coral reefs: sources or sinks of atmospheric CO_2? *Coral Reefs* 11: 127–130.

Waycott, M., Duarte, C.M., Carruthers, T.J.B. et al. (2009). Accelerating loss of seagrasses across the globe threatens coastal ecosystems. *Proceedings of the National Academy of Sciences* 106: 12377–12381.

Weston, N. (2013). Declining sediments and rising seas: an unfortunate convergence for tidal wetlands. *Estuaries and Coasts* 37: 1–23.

Weston, N.B., Giblin, A.E., Banta, G. et al. (2010). The effects of varying salinity on ammonium exchange in estuarine sediments of the Parker River, Massachusetts. *Estuaries and Coasts* 33: 985–1003.

Weston, N.B., Neubauer, S.C., Velinsky, D.J., and Vile, M.A. (2014). Net ecosystem carbon exchange and the greenhouse gas balance of tidal marshes along an estuarine salinity gradient. *Biogeochemistry* 120: 163–189.

Wiegner, T.N., Seitzinger, S.P., Breitburg, D.L., and Sanders, J.G. (2003). The effects of multiple stressors on the balance between autotrophic and heterotrophic processes in an estuarine system. *Estuaries* 26: 352–364.

Williams, P.J.l. (1993). Chemical and tracer methods of measuring plankton production: what do they in fact mean? The ^{14}C technique reconsidered. *ICES Marine Science Journal* 197: 20–36.

Williams, P.J.l. (1998). The balance of plankton respiration and photosynthesis in the open oceans. *Nature* 394: 55–57.

Williams, M., Hopkinson, C., Rastetter, E., and Vallino, J. (2004). N budgets and aquatic uptake in the Ipswich River Basin, northeastern Massachusetts. *Water Resources Research* 40: 1–12.

Winkel, A., Colmer, T.D., and Pedersen, O. (2011). Leaf gas films of *Spartina anglica* enhance rhizome and root oxygen during tidal submergence. *Plant, Cell and Environment* 34: 2083–2092.

Wofsy, S.C. (1983). A simple model to predict extinction coefficients and phytoplankton biomass in eutrophic waters. *Limnology and Oceanography* 28: 1144–1155.

Woodwell, G. and Pecan, E. (ed.) (1973). *Carbon in the Biosphere*. Springfield, Virginia, USA: National Technical Information Service.

Yvon-Durocher, G., Caffrey, J.M., Cescatti, A. et al. (2012). Reconciling the temperature dependence of respiration across timescales and ecosystem types. *Nature* 487: 472–476.

Zappa, C.J., Raymond, P.A., Terray, E.A., and McGillis, W.R. (2003). Variation in surface turbulence and the gas transfer velocity over a tidal cycle in a macro-tidal estuary. *Estuaries* 26: 1401–1415.

Zawatski, M. H. 2018. Carbon exchange and sediment deposition in a heterogeneous New England salt marsh. MS Thesis. Villanova University, Villanova, PA, USA.

Zhao, L., Chen, C., Vallino, J. et al. (2010). Wetland-estuarine-shelf interactions in the Plum Island Sound and Merrimack River in the Massachusetts coast. *Journal of Geophysical Research* 115: C10039.

Ziegler, S. and Benner, R. (1998). Ecosystem metabolism in a subtropical, seagrass-dominated lagoon. *Marine Ecology Progress Series* 173: 1–12.

CHAPTER 15

Estuarine Food Webs

Ryan J. Woodland[1] and James D. Hagy III[2]
[1] Chesapeake Biological Laboratory, University of Maryland Center for Environmental Science, Solomons, MD, USA
[2] US Environmental Protection Agency, Center for Environmental Measurement and Modeling, Atlantic Coastal Environmental Sciences Division, Narragansett, RI, USA

Many estuarine predators feed in and around structural habitats, such as this yellow stingray (*Urobatis jamaicensis*) foraging in a mixed seagrass bed in a Belizean lagoon. Photo credit: Theresa Murphy.

15.1 Introduction

Estuaries provide habitat for abundant plants, animals, and microorganisms, ranging from microscopic plankton (bacteria, yeasts, algae, and protozoa) to larger benthic and pelagic organisms (seagrass, clams, crabs, sea trout, pelicans, and dolphins). Estuarine biota can be characterized in a variety of ways, including by taxonomy (e.g., algae, crustaceans, and flounder), ecological function (e.g., primary producers, decomposers), and habitat and/or "niche" (e.g., benthic suspension feeders). Whereas these groupings are often the subject of distinct disciplines within estuarine ecology, *food web ecology* examines their arrangements and interactions as a whole. Feeding habits are an integrative way to organize interactions among a diverse spectrum of organisms and functional groups, creating food webs. Like many areas of estuarine ecology, food web ecology developed first elsewhere and was later applied to estuaries. Important early studies in the field, such as Lindeman's (1942), focused on lakes. Early studies also examined planktonic food webs in the open ocean and interactions involving ocean fisheries (Ryther 1969; Pomeroy 1974).

Studying estuarine food web dynamics requires a multi-faceted approach that accounts for structural elements (e.g., plants and animals), processes such as predation and competition, and interactions between the organisms and their environment. Because food webs are typically very complex, a *holistic* approach (i.e., accounting for all of the component parts of a system) is often taken to analyze food webs after an understanding of individual structures and processes has been acquired. Analyzing food webs holistically is valuable from several perspectives, ranging from theoretical to real-world applications. Estuarine ecosystems function as an interactive whole and cannot be fully understood by examining only a subset of the whole. Analysis of estuarine food webs can help identify errors or gaps in our ecological understanding. This is important because quantifying estuarine food webs usually involves assembling many kinds of information. Uncertainty is reduced by reconciling independent observations via the mass balance

Estuarine Ecology, Third Edition. Edited by Byron C. Crump, Jeremy M. Testa, and Kenneth H. Dunton.

Companion website: www.wiley.com/go/crump/estuarine3

Box 15.1 An Introduction to Chesapeake Bay

Chesapeake Bay is located in the mid-Atlantic region of the eastern United States of America. Chesapeake Bay's 164,200 km^2 watershed covers land in six different states, including New York, Pennsylvania, Maryland, Delaware, Virginia, West Virginia, and the District of Columbia. The estuary has a surface area of 11,500 km^2, extending up-estuary from the most saline region at the mouth of Chesapeake Bay to the tidal freshwater region near the confluence of the Susquehanna River. The dynamic mixing of nutrient-laden freshwater and coastal marine waters make Chesapeake Bay a highly productive ecosystem across a wide range of living (e.g., oyster reefs, seagrass beds, saltmarshes) and non-living habitats (e.g., rocks, unconsolidated sediments). Chesapeake Bay serves as an important nursery and foraging area for many species of fish, invertebrates, and birds. The estuary has a long history of human use, starting long before Europeans arrived in North America and continuing through the present day. Human alterations to Chesapeake Bay are considerable, ranging from eutrophication, overfishing, habitat loss, and shoreline development.

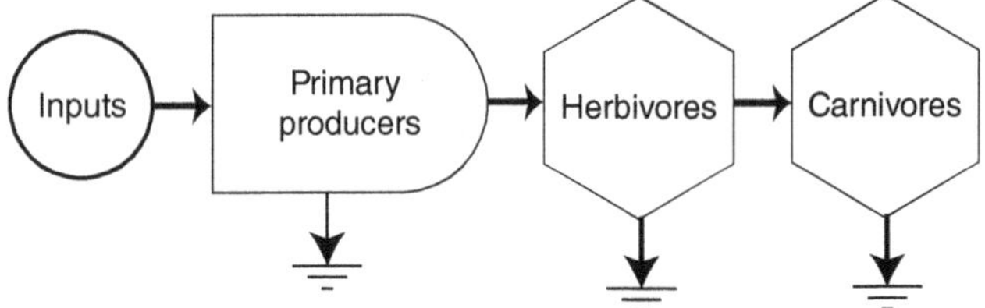

FIGURE 15.1 A simple food chain with three trophic levels illustrated using energy systems language (Odum 1967; Odum 1971). Arrows indicate transfer of matter or energy. Circles indicate inputs, which could include matter, such as nutrients, or energy. A "bullet" symbol indicates a primary producer, whereas hexagons indicate consumers. Horizontal lines (i.e., electrical "ground" symbols) indicate energy dissipation or recycling.

and other constraints associated with quantifying food webs (Section 15.6.1; Christensen and Walters 2004; Ulanowicz 2004). Quantifying the whole may also reveal if one or more components has been overlooked or mischaracterized (Guénette et al. 2008). In this sense, analysis of whole food webs is useful in the same way as construction of estuarine nutrient budgets, which relate independent measurements of nutrient inputs, fluxes, and transformations via a mass balance.

In this chapter, we examine some of the terminology, concepts, theory, and models that are used in the study of food webs in general and in food webs of estuaries in particular. After reading this chapter, students should be able to knowledgeably discuss key concepts of food web ecology as they relate to estuarine ecosystems. Students will also be exposed to basic yet important calculations that are commonly used to analyze different elements of food webs. Throughout the chapter, we will provide examples by walking through calculations or providing results of analyses using actual and artificial food web data to highlight key topics. Chesapeake Bay was selected as a case study to serve as the focus for several of these examples (Box 15.1).

15.1.1 Food Chains and Food Webs

The transfer of food from its ultimate source (e.g., photosynthesis, import from outside ecosystem) through a series of organisms or groups of organisms eating each other is referred to as a *food chain* (Elton 1927). Figure 15.1 shows a simple food chain using energy systems language (Odum 1967). At the base of the chain are the primary producers, depicted with a "bullet" symbol. In an estuary, the primary producers could be phytoplankton or other plants such as seagrasses and benthic algae. Primary producers require inputs of matter and energy from external sources (depicted with a circle symbol). Arrows represent flows of either mass or energy, first to the herbivores and then to the carnivores, both of which are represented by hexagon symbols. Only a portion of the mass or energy produced in one compartment is transferred to the next, the remainder being transferred out of the food chain as export, emigration, or dissipated as respiratory end-products. In ecology, the term *trophic* (from the Greek word *trophikós*) refers to feeding or food. Thus, a group of organisms that are considered together because they are similar in their feeding habits (e.g., consuming plants) and relationships may be called a *trophic group*. Transfers from one trophic group to another are called *trophic transfers*. *Trophic level* indicates the average number of trophic transfers required for food (measured as energy, nutrients, or organic matter) to reach the group starting from the primary producers, which are considered trophic level 1. In Figure 15.1, the trophic groups are at trophic levels 1 (primary producers), 2 (herbivores), and 3 (carnivores).

Rarely are trophic interactions (i.e., feeding interactions) as simple as depicted by a short food-chain (Figure 15.1). In most ecosystems, organisms feed on many different foods, often including both plants and animals. As our understanding of trophic relationships has increased, it has become necessary to think of trophic interactions as a *food web* rather than as a branching collection of food chains. One major reason is *omnivory*, which refers to the habit of consuming more than one type of food. Trophic level omnivory occurs when these different foods also occupy different trophic levels (Pimm 1980). Species exhibiting omnivory are referred to as *omnivores* and are said to be *omnivorous*. A simple diagram of a food web (Figure 15.2) illustrates omnivory and how it links food chains together into a web. An increased appreciation of the importance of recycling of non-living organic matter (food energy), or *detritus*, is another reason for thinking of trophic relationships in estuaries in terms of a food web. Organic matter from either inside or outside an ecosystem may cycle several times between living organisms and detritus before being exported, buried, or respired. Early ecologists used the term *food cycle* to refer to these repeating or cycling flows of matter and energy.

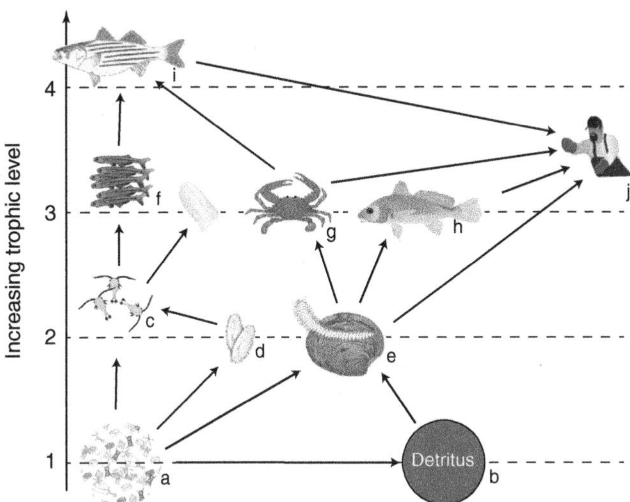

FIGURE 15.2 Energy flow in an aggregated Chesapeake Bay food web, illustrating the complexity of a more realistic, but still simple food web aggregated by ecological function rather than taxonomy. Functional groups include: a—pelagic primary producers, b—detritus, c—mesozooplankton, d—microzooplankton, e—benthic infauna, f—pelagic zooplanktivores, g—demersal invertebrate predators, h—demersal fish predators, i—piscivorous fish, j—humans (fisheries). Some predators (e.g., piscivorous fish) consume prey from several prey groups. Similarly, some groups (e.g., pelagic producers, benthic infauna) are consumed by several different consumers. Symbols from Integration and Application Network (ian.umces.edu/media-library).

Detritus may be particularly important in estuaries because these ecosystems often receive significant inputs of organic matter from upstream terrestrial and freshwater ecosystems as well as from the coastal ocean.

15.1.2 Functional Groups and Trophic Guilds

Various terms are commonly used to describe the functional roles of organisms within food webs. *Primary producers*, also known as *autotrophs*, capture energy from an external source, usually the sun, to synthesize organic matter from carbon dioxide and water. Producers that use photosynthesis to generate organic matter are known as *photoautotrophs*. Because light is usually abundant in estuaries, photosynthesis tends to be by far the dominant organic matter source to estuarine food webs. Some bacteria are *chemoautotrophs* that utilize energy available from reduced inorganic chemical substrates, such as reduced sulfur compounds, to synthesize organic matter. Although there are entire food webs (e.g., in the deep sea) based on chemoautotrophy, it is much less important than photoautotrophy in estuaries. *Consumers*, also known as *heterotrophs*, obtain nutrition by consuming organic matter. Neglecting omnivory for the moment, consumers can be categorized according to their predominant trophic level relative to producers or detritus. Organisms functioning as *primary consumers* (i.e., 1° consumers) are herbivores that consume primary producers directly. *Secondary consumers* (i.e., 2° consumers) are carnivores that consume the primary consumers. Trophic levels can also be defined for higher-order consumers (e.g., 3°, 4°). *Mixotrophs* meet their nutritional and energetic requirements through a combination of autotrophy and heterotrophy, for example, grazing to obtain nutrients and utilizing primary production to support energetic requirements (Anderson et al. 2008). In estuaries, mixotrophs are most common within planktonic food webs (e.g., dinoflagellates).

Idealized trophic levels become somewhat blurred in real food webs because of omnivory. For example, suppose a group of organisms consumes phytoplankton (a primary producer, trophic level = 1) as well as herbivores that graze on phytoplankton cells. This group would function as both a primary consumer (trophic level = 2) and a secondary consumer (trophic level = 3). Ecologists can determine a *fractional* or *average* trophic position by weighting, or apportioning, the feeding of the consumer to trophic level 2 and 3 (Box 15.2). Supposing that this consumer obtained 75% of its intake from phytoplankton directly and the rest from the other primary consumer species, its average trophic position would be 2.25 [= $(0.75 \times 2) + (0.25 \times 3)$]. The concept of fractional trophic positions has proven to be very useful for analyzing trophic relationships in estuarine food webs. As one example, the mean trophic position can be used to estimate how much primary production is needed to support a given production of consumers, or to consider the impact of human activities like fishing on trophic structure (Pauly and Christensen 1995; Pauly et al. 1998; Knight and Jiang 2009).

Trophic guilds provide another useful way to describe the role of a group of organisms within a food web. These terms describe how the organisms feed, or on what type of food, with varying levels of specificity. For example, *scavengers* are *detritivores* that consume nonliving organic matter obtained from carcasses of dead animals. Important trophic guilds in estuaries include *planktivores* (plankton eaters), *benthivores* (consumers of bottom dwelling animals), and *piscivores* (fish eaters). *Suspension feeders* comprise a guild that obtains food by filtering out plankton and other small particles suspended in the water. Estuarine food webs may include both benthic suspension feeders, such as oysters or mussels, and pelagic suspension feeders, including specialized fish (e.g., Atlantic menhaden (*Brevoortia tryannus*), striped mullet (*Mugil cephalus*) or plankton such as the appendicularian (*Oikopleura dioica*; Thomson 1966; Alldredge 1981). *Deposit feeders* ingest sediments, extracting the available nutrition and energy in the form of algae, bacteria, protozoa, or detritus. Feeding habits of some fish and invertebrates can change either with size and age (i.e., ontogenetically), or according to environmental conditions (Kanou et al. 2004). Others, such as the clam *Limecola balthica* (previously *Macoma balthica*), can switch feeding guild (e.g., from suspension feeding to deposit feeding) depending on the availability of different foods (Hummel 1985).

Box 15.2 Calculating Trophic Position Using Stomach Contents

Trophic position refers to the location of a consumer along a food chain. Simply speaking, the number of trophic transfers that occur between primary producers and the consumer is equal to the consumer's trophic position. By convention, primary producers are considered to have a trophic position of 1, so if a consumer feeds only on plants, then that consumer has a trophic position of 2. Trophic position differs from trophic level in that trophic positions can take fractional values; whereas, a trophic level is restricted to whole number values. Fractional values are very useful and, typically, more realistic. This is because consumers often feed on many types of prey. The resulting trophic position of the consumer is an average of the trophic positions of its prey, plus 1 to account for the trophic transfer between the consumer and its prey. These average trophic position calculations are often weighted by the biomass of each prey that a consumer ingests. The resulting calculation provides a nuanced estimate of trophic position that reflects not only the types of prey that a consumer ingests, but also the relative amounts of each prey ingested.

Estimating trophic position requires information on the diet of a consumer. This often takes the form of direct measurements, such as the identification of stomach contents or field-observations of feeding. Below, we walk through the equation used to calculate trophic position using stomach contents data.

Assume we conducted a study on the stomach contents data of juvenile bluefish (*Pomatomus saltatrix*) from Chesapeake Bay. Bluefish are a piscivore and top predator within the finfish assemblage. Averaging across all of the individual bluefish, we summarize the prey types and amounts of each prey we've observed. We then consult the published literature to obtain an estimate of the trophic position of each of these prey types:

Prey type	% biomass	Prey trophic position
bay anchovy (*Anchoa mitchilli*)	50	3.2
sand shrimp (*Crangon septemspinosa*)	25	3.0
Atlantic menhaden (*Brevoortia tyrannus*)	25	2.5

Using the information in the table above, we can use the following equation to calculate trophic position:

$$T_i = 1.0 + \sum_{j=1}^{n} T_j \left(p_{ij} \right)$$

where T_i is the trophic position of consumer i, T_j is the trophic position of prey type j and p_{ij} is the proportion that prey j contributed to the diet of consumer i (Akin and Winemiller 2008). Filling in data from the table, we get:

$$T_i = 1.0 + (3.2 \times 0.50) + (3.0 \times 0.25) + (2.5 \times 0.25).$$

Solving for T_i, we calculate a trophic position for bluefish of 3.975, rounded to 4.0.

15.1.3 Trophic Efficiencies

The efficiencies by which organisms assimilate food, grow, and transfer energy to organisms at higher trophic levels, are important measures of food web structure and function. These are easily defined and understood as ratios of inputs/outputs for the bioenergetic budget. A simple equation can be written to describe this budget for an organism or group of organisms:

$$C = P + R + U$$

Consumption (C) is the amount of food ingested. A fraction of C is usually not fully utilized by the animal and is expelled (e.g., as feces) or excreted (e.g., as urine). Since excretion is often not estimated separately from egestion, the sum is represented as *unassimilation* (U). A portion of the assimilated matter and energy is used to produce new tissue (growth) or new organisms (i.e., reproduction) and is called *production* (P) of new biomass. *Respiration (R)* is the fraction devoted to supporting metabolic needs. Each of these quantities must be expressed in terms of a common measure, or currency, which for food webs in estuaries is usually energy, carbon, or nitrogen. Early studies of trophic transfers were focused on energy flow (Lindeman 1942; Odum 1957), whereas more recent studies have emphasized carbon (Baird and Ulanowicz 1989), nitrogen (Baird et al. 2011), or other currency units (e.g., toxic contaminants; Braune and Gaskin 1987). Three measures of efficiency are related to these terms. G*ross growth efficiency* (*GGE*) is the ratio of production (i.e., growth and reproduction) to consumption (P/C). *Assimilation efficiency* (*AE*) is the fraction of food consumed that crosses the gut of the consumer and contributes to its metabolism. This may be thought of equivalently as $(C-U)/C$ or $(P+R)/C$. The efficiency by which assimilated food is converted to production is net growth efficiency (*NGE*). *NGE* is equal to *GGE/AE* but is most commonly thought of as $P/(P+R)$. For example, *bacterial growth efficiency* (see Chapter 9) is used in reference to bacteria and is defined in exactly this way (Valiela 1995).

Whereas the terms defined above apply to the bioenergetics of single populations or trophic groups in isolation, other related terms describe efficiency in the context of a food web. For example, *ecotrophic efficiency* (*EE*) refers to the fraction of production by a population or trophic group that is utilized within the food web at the next higher trophic level, harvested, or exported to another food web (Christensen and Walters 2004). Low *EE* indicates that a large fraction of production is not utilized in the food web and is likely passed to detritus. *Trophic transfer efficiency (TTE)* is the fraction of production at one trophic level that results in production at the next higher trophic level. It may be helpful to think of *TTE* as the product of *EE* (expressing production not used in the food web), *AE* (expressing food consumed but not retained), and *NGE* (expressing retained food not converted to production). Other common terms related to food webs include *consumption efficiency* which is nearly identical to *EE*, and *production efficiency*, which is the same as *NGE*. Slobodkin (1960) estimated several efficiencies using the freshwater zooplankter *Daphnia pulex* feeding on the flagellated green alga *Chamydomonas reinhardtii* as a test case, concluding that "ecological efficiency is effectively constant" at approximately 10%. Although Slobodkin's ecological efficiency was defined in slightly different terms than the efficiencies defined here, and the estimate was based on relatively thin evidence, the 10% "rule-of-thumb" has proven to be remarkably enduring. In a meta-analysis, Slobodkin's 10% estimate was close to the mean trophic transfer efficiency for 140 aquatic food webs (Pauly and Christensen 1995).

15.1.4 Food Chain Length

An interesting question about food webs is how many trophic levels are present and what limits the maximum number possible. Early ecologists suggested that there are usually at most five trophic levels (Elton 1927). However, later studies found trophic pathways with as many as eight steps (Baird and Ulanowicz 1989; Hagy 2002; Hussey et al. 2014). Some ecologists have noted that explicitly accounting for microbial food webs and biomass can increase estimated trophic positions among detritivores (Steffan et al. 2017). A common convention to maintain detritus at a trophic level of one limits this possibility. Accordingly, the longest food chain within the simplified food web shown in Figure 15.3 has a length of 8. However, no trophic group has an average trophic level greater than 5 because feeding via relatively short pathways is always important. One obvious reason for this limitation on food chain length is that substantial energy is lost from flow to higher trophic levels with each trophic transfer. Even if the average trophic transfer efficiency is a relatively high 20%, 1.6×10^4 units of primary production are required to support 1 unit of production at trophic level 6. Practically speaking, a species feeding exclusively at such a high trophic level is likely to be rare and energetically unimportant. From the standpoint of energetic limitation, one might expect more trophic levels in food webs with high primary productivity. However, other factors could also affect the number of trophic levels. One hypothesis is that high trophic level pathways are likely to be unstable, limiting their prevalence (Pimm and Lawton 1977), but this view remains a matter of debate (Sterner et al. 1997). Another consideration is the size

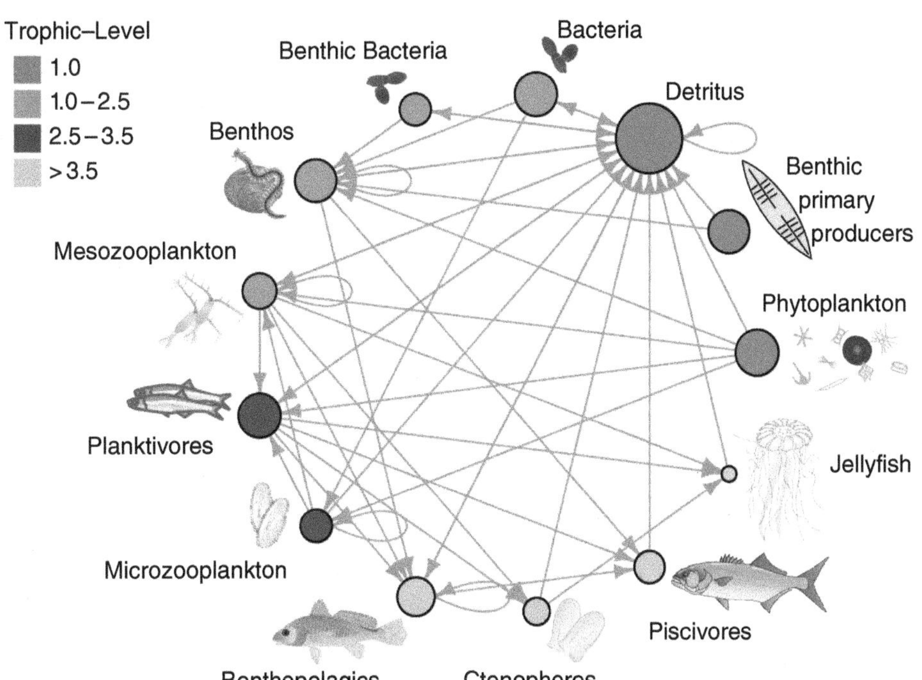

FIGURE 15.3 A simplified network model of the Chesapeake Bay food web. Network nodes (trophic groups) are scaled by relative biomass and color-coded by average trophic position (see inset color scale). Flows between nodes are depicted by arrows which point toward the consumer (or the receiving detritus pool), and internal cycles (a.k.a., intraguild predation: feeding on taxa within the same trophic group) are shown with looping arrows. Note the higher trophic level fish (Benthopelagics, Piscivores) consume prey associated with benthic (bottom) and pelagic (open-water) food webs. This allows higher trophic levels to facultatively switch among different prey types and serves as a stabilizing mechanism for food webs. *Source*: Based on Hagy (2002) and Rooney et al. (2006). Symbols from Integration and Application Network (ian.umces.edu/media-library).

of food. Particularly in the plankton, organisms usually capture and consume food only if it falls within a certain size range (Elton 1927; Brooks and Dodson 1965). The number of trophic levels and stability of the resulting food webs could therefore depend upon the size and feeding preferences of the important species (Post 2002a; Otto et al. 2007). Because the species composition of phytoplankton in high productivity waters tends to be dominated by larger cells (e.g. diatoms), this may actually result in fewer trophic levels needed to produce carnivorous fish (Ryther 1969; Landry 1977). Many productive estuarine and coastal waters support stable fisheries that harvest large consumers at low trophic levels. Examples include filter-feeding fish such as the Atlantic menhaden (*Brevoortia tyrannus*) and Pacific sardines (*Sardinops sagax*) and suspension-feeding bivalves such as Pacific oysters (*Crassostrea gigas*) and New Zealand green-lipped mussels (*Perna canaliculus*). These species feed principally as herbivores (trophic level = 2) or primary consumers (trophic level = 3) and therefore have an average trophic level between 2 and 3 (Pauly et al. 1998). These filter-feeding animals have a "telescoping effect" (Elton 1927), wherein relatively large organisms (e.g., blue whales and whale sharks) feed on very small ones, leading to increased trophic transfer efficiencies and potentially shorter food chains.

Quantifying aspects of food web structure such as food chain length or the biomass of different functional groups allows contrasts across food webs and assessments of change over time. These changes may then be related to factors causing change. Food web changes may be reflected in total systems properties, such as total systems throughput or average path length, or alternatively, properties that pertain to one or more nodes, such as average trophic level of a species or of the total fisheries harvest (Christian et al. 2005; Shannon et al. 2009). Such changes can be caused by a variety of factors, including eutrophication (Baird et al. 2004) and fishing pressure (Breitburg et al. 2009), both potentially important in estuaries. Major storms and river flow events may also impact food webs. Whereas studies have characterized effects of such perturbations on components of estuarine food webs (Paerl et al. 2006), more holistic analyses appear to be lacking. In general, understanding and predicting how many of the factors that impact estuaries (e.g., river flow, nutrient loading, storms, habitat modifications, and climate change) impact the size, structure, and function of whole food webs remains fertile ground for research.

15.2 Portraying Food Webs

15.2.1 Ecological Pyramids

Different approaches have been used to examine the structure of food webs. One approach is to examine graphically the relative magnitude of abundance, biomass, or productivity across trophic levels using *ecological pyramids* (Odum 1971).

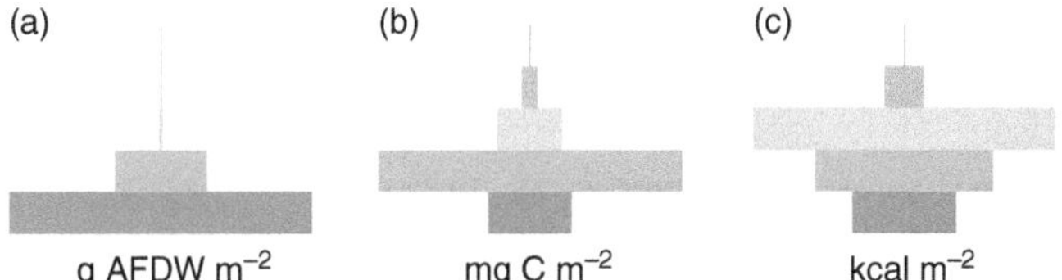

FIGURE 15.4 Pyramids of (a) biomass [AFDW: ash free dry weight], (b) carbon, and (c) energy storage for food webs in (a) a seagrass dominated area of Mondego Estuary, (b) the Neuse River estuary in late summer, and (c) the English Channel. Each stacked block represents a different trophic level. Biomass or energy storage at integer trophic levels was calculated from food web models included in the enaR package in R. *Source*: (a) Patrício and Marques (2006); (b) Based on Baird et al. (2004).

The shape of a pyramid reflects differences in the food web and characteristics of the dominant taxa. The "pyramid of abundance (numbers)" is particularly affected by the dominant taxa and provides less insight into community structure than other types of pyramids. On the other hand, the pyramid of biomass is more interesting because it can potentially reveal substantial differences in structure. For example, a biomass pyramid for a seagrass dominated area of Mondego Estuary has a significant biomass of primary producers, and much lower biomass of consumers (Figure 15.4). In contrast, the biomass of primary consumers exceeds the biomass of producers in the Neuse River. Similarly, the energy stored in secondary consumers exceeds both producers *and* primary consumers in the English Channel. These "inverted pyramids" suggest that biomass or energy storage at the lowest trophic levels is rapidly replaced by growth (i.e., higher productivity, lower biomass, and less energy storage), while biomass or energy accumulates slowly in longer-lived consumer organisms. Unlike biomass or energy storage pyramids, production or energy flow pyramids are always wider at the bottom trophic level because a fraction of energy and biomass is lost from the food web with each transfer to a higher trophic level.

15.2.2 Projecting Food Webs to Food Chains

One problem with ecological pyramids is that they are predicated upon the existence of discrete trophic levels (i.e., 1, 2, 3), when we have already seen that many organisms in fact feed at more than one trophic level and therefore occupy noninteger or fractional trophic positions. Consequently, ecological pyramids have not been used extensively in recent work on estuarine food webs. One way to approach the problem of representing feeding at fractional trophic positions is to apportion feeding to one of several *canonical trophic levels* (Ulanowicz and Kemp 1979). For example, similar to the example in Section 15.1.2, a species that feeds 50% at trophic level 2 and 50% at trophic level 3 would have a fractional trophic level of 2.5 [=(0. 5 × 2) + (0.5 × 3)]. In this case, 50% of the biomass and trophic demand of this species would be allocated to each of

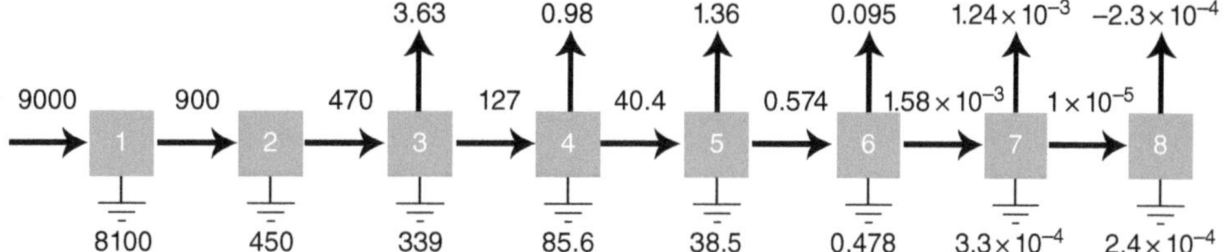

FIGURE 15.5 The effective food chain for the North Sea food web network (Steele 1974). Flows, which in this case are energy flows in kcal m^{-2} year^{-1}, are apportioned to integer trophic levels following the method of Ulanowicz and Kemp (1979). Flows from each integer trophic level to the next indicate upward transfer of energy within the food web. Flows to "ground" are defined as in Figure 15.1, where respiration is combined with flows to detritus, and lost from the food web. Additional flows can be included in the effective food chain to explicitly represent the return of detritus to the food web. Upward arrows represent the yield to humans at each effective trophic level. Not surprisingly, the largest yield to humans occurs via the shortest and most efficient trophic pathway (Boxes 15.1–15.3), which in this example is Primary Producers → Pelagic Herbivores → Pelagic Fish. Note that the negative flow associated with the eighth box results from the mathematical transformations used to linearize this food web and does not indicate a reversed flow. *Source*: Ulanowicz and Kemp (1979)/University of Chicago Press.

the canonical trophic levels 2 and 3. Mathematical procedures for this analysis, which include both apportioning feeding to canonical trophic levels and calculating canonical flows for the food web (Ulanowicz and Kemp 1979), are now part of the methods known collectively as Ecological Network Analysis (see Section 15.6). The effective food chain obtained by projecting the North Sea food web (Steele 1974) onto a series of canonical trophic levels (Figure 15.5) illustrates several key features of the food web, such as the length of the longest trophic pathway (i.e., eight trophic steps) and the fact that most flow, in this instance measured as energy flows (kcal m^{-2} year^{-1}), is confined to the first five trophic levels. Additionally, Figure 15.5 shows that 10% of primary production (900 of 9000 kcal m^{-2} year^{-1}) is transferred to the second trophic level, whereas more than 50% of the input to trophic level 2 (470 of 900 kcal m^{-2} year^{-1}) is transferred to trophic level 3. We also see that much of the yield to humans (3.63 kcal m^{-2} year^{-1}) occurs at trophic level 3, the lowest exploited trophic level (i.e., the lowest trophic level subject to fishing), even though fishing extracts a relatively small fraction of total production at the lower trophic levels. Projecting food webs into food chains provides a useful approach for characterizing overall food web structure.

15.2.3 Trophic Spectra

Darnell (1961) introduced the *trophic spectrum* as a tool to illustrate graphically his detailed data on the food habits of organisms in Lake Pontchartrain, an estuarine lake (Figure 15.6). Among his key observations was that food habits of species often change as they develop through different life stages and as they increase in size (i.e., diet changes *ontogenetically*). Darnell's trophic spectrum also showed the great diversity of prey types in most diets (i.e., the prevalence of omnivory), with few species being highly specialized in their food habits. Many species included significant quantities of organic detritus in their diets. Omnivory and detritivory are usually important in estuarine food webs. This appears to result from several factors: (1) the unpredictable availability of specific food items, (2) the fact that estuaries often receive significant inputs of detritus from terrestrial sources, and (3) the tendency for physical processes in estuaries to retain and concentrate particulate detritus. An additional methodological factor is that unidentifiable food items may erroneously be considered "detritus."

A more quantitative application of the trophic spectrum concept involves overlaying the biomass of individual species across the range of trophic positions occupied by that species. By summing overlapping biomass in each trophic position increment (e.g., 0.1 trophic level), a biomass trophic spectrum is created (Gascuel 2005). This same approach can be used to create catch trophic spectra or abundance trophic spectra, using fisheries catch or species abundance data, respectively. Biomass trophic spectra provide information on the structure and function of a food web. Theoretical biomass trophic spectra can be simulated based on expected trophic transfer efficiencies, and any deviations of empirical spectra from theoretical spectra can provide insight into how local processes are shaping the food web. For example, losses in key primary consumer groups (e.g., parrot fishes) at a heavily fished coral reef would result in a biomass trophic spectrum with a very low peak at trophic levels 2–3 (due to omnivory). Biomass trophic spectra have been used to assess the effects of human activities at local (Laurans et al. 2004; Valls et al. 2012) and global scales (Maureaud et al. 2017).

15.2.4 Biomass Body-Size Spectra

Yet another type of trophic spectra,biomass body-size spectra (hereafter *size spectra*), show the distribution of biomass as a function of logarithmic body-size classes for all organisms in an ecosystem or habitat. Ecological analysis based on size spectra has developed from early observational work into a well-developed body of theory and predictive science

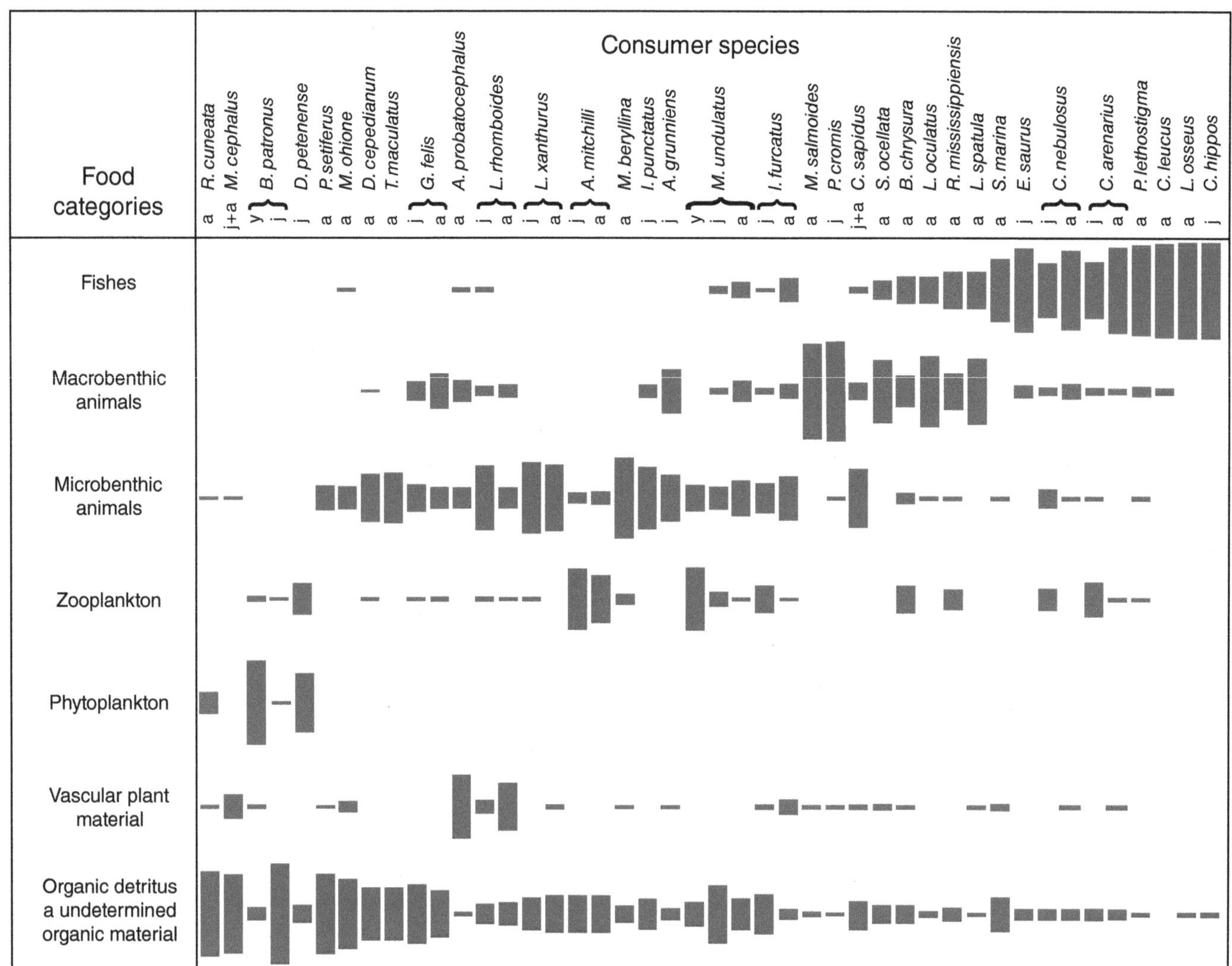

FIGURE 15.6 Trophic spectra for the most important consumer species of Lake Pontchartrain, indicating the feeding habits of each of the consumer species, listed across the top (y = young, j = juvenile, a = adult). As an example, the bivalve *Rangia cuneata*, in the first column, largely consumes organic detritus, plus some phytoplankton. In contrast, *Caranx hippos*, a marine fish in the jack family, consumes fish almost exclusively. *Micropogonias undulatus*, or Atlantic croaker, has more varied food habits than either of the other species, consuming a mixture of fish, large and small benthic animals, zooplankton, and organic detritus. *Source*: Darnell (1961)/With Permission of John Wiley & Sons.

(Kerr and Dickie 2001). Interpretation of size spectra is based on the assumption that these analyses reflect transmission of matter and energy through trophic interactions in which large organisms generally consume smaller organisms as prey (Elton 1927). Early size spectra focused on microscopic plankton in the world's oceans (Sheldon et al. 1972) which showed surprising uniformity in biomass across size classes (i.e., flat spectra), prompting further empirical and theoretical investigations. Systematic differences in size spectra across a gradient of latitude in the South Atlantic Ocean suggested that size spectra might reveal important characteristics of ecosystems, as well as differences among and within them (Sheldon et al. 1972). Platt and Denman (1977) introduced the normalized biomass spectrum, in which the biomass in a size-class is divided by the width of the size class. The slopes of normalized spectra tend to have a characteristic value of approximately −1.0, rather than the relatively low slope (i.e., flat) for size spectra that are not normalized. An advantage of normalized spectra is that their shape is independent of the width of the intervals, making these spectra well-suited to comparing different ecological systems.

Research using size spectra has shown that they often have one or more size ranges in which there is very little biomass, creating the appearance of distinct "domes" or biomass modes within the spectrum (i.e., specific size classes that occur more frequently). The investigators that first identified these structures within size spectra for plankton communities suggested that they represent physical limits on the possible sizes of plankton species. Subsequent investigators noted that domes of high biomass in benthic spectra complemented size ranges with low biomass in the plankton, suggesting coupling of planktonic and benthic food webs (Kerr and Dickie 2001). Jung and Houde (2005) observed this pattern in size spectra for fish communities in Chesapeake Bay (Figure 15.7), finding that the "dome" of abundance (and related biomass) representing piscivorous fish could not be explained solely on the biomass of

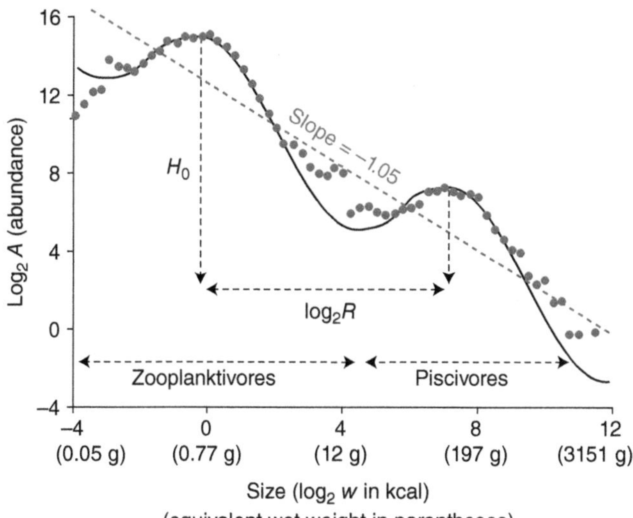

FIGURE 15.7 A normalized biomass size spectrum for fish captured in Chesapeake Bay. In this example, the graph depicts the normalized abundance of fish (instead of biomass) within energy intervals (or equivalently, body weight intervals as shown in parentheses). The normalized spectrum is characterized as having two distinct trophic groups, which appear as "domes." The first dome corresponds to bay anchovy (*Anchoa mitchilli*), the dominant zooplanktivore, and the second dome corresponds to the larger piscivores such as striped bass (*Morone saxatilis*). Bay anchovy account for a significant fraction of the diet of the piscivores. *Source*: Jung and Houde (2005)/With Permission of Springer Nature.

smaller zooplanktivorous fish. Rather, they suggested that the diet of the larger species was supplemented by consumption of benthic invertebrates (Jung and Houde 2005).

15.3 Trophic Theory

The populations and communities that compose estuarine food webs interact in numerous ways, including both direct and indirect feeding relationships. Here, we focus on trophic interactions as they appear in estuarine food webs. Although interactions are described for convenience as occurring between "species," the terminology applies equally to groups of species such as trophic guilds.

15.3.1 Direct Trophic Interactions

There are five types of direct trophic interactions: *competition*, *predation* or *parasitism*, *mutualism*, *commensalism*, and *amensalism*. These interactions are defined based on the effect of the interaction on each of the species involved (Table 15.1). *Competition* is an important interaction that has a negative effect on each species. In the context of food webs, direct trophic competitors are two or more species that derive a substantial fraction of their diet from the same food resource. Intraspecific competition is also an important factor structuring population dynamics (e.g., density-dependence in fish populations).

TABLE 15.1 Ecological relationships, defined in terms of the effect of each species involved in the relationship on the other species. + = beneficial effect, − = negative effect, 0 = no effect.

	Effect on species b	Effect on species a
Competition	−	−
Predation/Herbivory	−	+
Parasitism	+	−
Mutualism	+	+
Commensalism	+	0
Amensalism	−	0

Predation and *parasitism* each describe interactions in which one species benefits from the interaction, to the detriment of the other. In the case of predation, the predator kills and consumes some or all of the prey. Predation is ubiquitous in nature and is a key factor structuring most estuarine food webs. Parasites negatively affect the species they live in, on, or in association with (i.e., their *host*), but they generally do not immediately kill the host. Instead, parasites obtain food by consuming a portion of the host or a portion of the diet obtained by the host. While parasitism is very common and potentially very important (Thompson et al. 2005), the effect of parasite–host interactions on food webs (and the best way to represent parasites in food web models), are not well understood (Lafferty et al. 2008). For example, a parasite might simply impair its host and contribute little to the food web, or it could support additional trophic pathways within the food web as a prey species.

Mutualism is a direct interaction in which both species benefit from the interaction. *Commensalism* is a related interaction, but one species benefits from the interaction and the other is unaffected. *Amensalism* is an interaction that is detrimental to one species and has no effect on the other. These interactions are not as ubiquitous within estuarine food webs as competition and predation, but may still be very important. Often they appear as non-trophic relationships. For example, the eastern oyster *Crassostrea virginica*, creates an entire habitat structure, providing ample opportunity for *commensals*, species that utilize the structure and excess particulate organic matter concentrated by the oysters, but have little direct effect on the oysters. Similarly, fiddler crabs and marsh cordgrass (i.e., *Spartina*) each benefit by the other's presence, an example of mutualism related to habitat structure rather than trophic interactions (Bertness 1985).

15.3.2 Indirect Trophic Interactions

Indirect trophic interactions involve interactions between two species that are mediated by interactions involving other species. Analysis of estuarine food webs often reveals many indirect trophic interactions. In an example of "top-down control" in a food web, feeding activity of predators on planktonic herbivores can result in positive effects on phytoplankton by reducing grazing pressure. Indirect trophic interactions can sometimes overwhelm direct interactions. For example, a predator can have an indirect net positive impact on one or more of its own prey, making it a "beneficial predator" (Ulanowicz and Puccia 1990). Consider the sea nettle *Chrysaora quinquecirrha*, a predator that consumes zooplankton, fish eggs, and larvae (Purcell 1992), but also preys very efficiently on the ctenophore *Mnemiopsis leidyii* (Purcell and Cowan 1995). Since *Mnemiopsis* can become very abundant and also is a voracious predator on zooplankton (Purcell et al. 2001), the net effect of *Chrysaora* on zooplankton can be positive due to its effective control of *Mnemiopsis* (Breitburg and Fulford 2006; Testa et al. 2008). Selective predation on diseased prey can also benefit a prey population by preventing disease epidemics, a well-recognized dynamic among terrestrial wildlife, but also identified in lake food webs (Duffy et al. 2005). Other examples of beneficial predation have been identified, illustrating the importance of predators for maintaining healthy ecosystems (Bondavalli and Ulanowicz 1999).

The concept of *keystone species* also relates principally to indirect trophic interactions. As originally suggested by Paine (1966, 1969), a keystone species is a high trophic level predator whose presence or abundance has a large effect on community structure and its persistence through time. A keystone predator is not necessarily abundant, but it may be. Libralato et al. (2006) define keystone species as "a relatively low biomass species with a structuring role in their food web." A keystone species generally alters the outcome of competitive interactions among species at lower trophic levels, possibly preventing exclusion of one or more competitors by the otherwise dominant competitor. Although ecologists have further developed this concept (see Mills et al. 1993), applications in the context of estuarine food webs have been limited. Libralato et al. (2006) developed a quantitative measure of "keystone-ness," derived from analyses of trophic networks, which they argue could improve application of the concept to conservation biology. Tsagarakis et al. (2010) applied this approach to coastal systems in the northern Mediterranean and concluded that mesozooplankton were quantitatively most like a keystone species.

15.3.3 Trophic Cascades

The idea of cascading trophic controls traces back at least as far as Hairston et al. (1960), who explained, among other observations, how trophic controls may be responsible for terrestrial plant communities generally being intact and green rather than defoliated by grazers (i.e., the green world hypothesis). Carpenter et al. (1985) used the term "trophic cascade" referring to how trophic interactions impact algal biomass in lakes. An example three-trophic-level pelagic food chain might include phytoplankton as primary producers, zooplankton as herbivores grazing on phytoplankton, and small carnivorous fish consuming the zooplankton (e.g., Figure 15.1). In this example, fish biomass and production are limited only by production of its prey and is said to have *bottom-up control*. In contrast, the zooplankton grazer is also subject to *top-down control*, since predation by fish can limit its biomass. Control of grazers via predation is likely to leave phytoplankton abundance controlled by resource limitation (i.e., availability of nutrients). This theory has been generalized as the "exploitation ecosystem hypothesis," which predicts that even numbers of trophic levels (e.g., 2 or 4) will lead to lower standing biomass of plants compared to odd numbers of trophic levels (1 or 3; Polis and Strong 1996). Trophic cascade theory also suggests that increased predation by organisms at the top of a 5-trophic-level food chain will cause decreases and increases, respectively, in alternating levels down to primary producers (levels 4 and 2 decrease and levels 3 and 1 increase); these responses will also be increasingly attenuated down the food chain (Carpenter et al. 1985).

Although the concepts of trophic cascades and top-down vs. bottom-up control have their place in the literature describing trophic dynamics of estuaries, two key questions are how important these are in the dynamics of food webs as compared to food chains, and to what extent food webs in estuaries should be expected to exhibit trophic cascades. Several attributes of estuarine food webs may in fact tend to reduce the importance of trophic cascades in estuaries (Borer et al. 2004). Omnivory, which we have suggested often characterizes the food habits of estuarine animals, may reduce cascades, which are strongest in food chains involving trophic specialists (i.e., species with narrow and unchanging diet composition) and weaker when species have omnivorous and dynamic (i.e., opportunistic) food habits (Polis and Strong 1996). Strong coupling of benthic and pelagic food webs may also limit trophic cascades. This is because energy often travels at relatively different rates through benthic (slow) and pelagic (fast) food webs and this asymmetry provides a natural buffer against trophic changes in the face of perturbations. Other factors limiting trophic cascades in estuaries are their openness and habitat complexity. Predators and prey in estuaries can migrate between estuaries and open coastal waters (Sackett et al. 2007) and are subject to external controls during their period of non-residence. The complex, three-dimensional habitats that often occur in estuaries can mediate strong trophic interactions by altering predator–prey behavior (e.g., predator avoidance by an intermediate consumer can reduce predation pressure on its prey; Grabowski 2004). Strong physical controls such as changes in the inflow of freshwater and nutrients into estuaries can also impose short-term variability on lower trophic levels, overwhelming internal biological controls and preventing clear expression of top-down effects on phytoplankton via trophic cascades.

These considerations notwithstanding, there are at least a few examples of trophic cascades in coastal marine and estuarine ecosystems. A well-cited example is the decimation of kelp forests due to sea urchin grazing, which followed a decline in sea otters (Estes et al. 1999). In another example, an increase in phytoplankton abundance in Patuxent River estuary occurred despite decreased nutrient inputs (Testa et al. 2008). The authors' analysis suggested that a trophic cascade was involved, wherein decreased abundance of a top gelatinous predator led to increased abundance a zooplanktivorous intermediate predator, reduced abundance of herbivorous zooplankton, and finally increased phytoplankton abundance. In general, estuarine food webs are affected by both bottom-up and top-down controls (Heck and Valentine 2007), but examples of strong cascades are relatively rare (Micheli 1999).

15.4 Attributes of Estuarine Food Webs

15.4.1 Spatial Mosaic of Coastal Habitats

In many instances, the structure of estuarine food webs is regulated by the complex spatial mosaic of different habitat types that characterize the ecosystem (Heck et al. 2008). The "coastal ecosystem mosaic" (Sheaves 2009) includes distinct habitat types such as open-water pelagic, tidal creeks and channels, seagrass meadows, oyster or mussel beds, coral reefs, intertidal sand or mud flats, rocky shores and deeper platforms, and emergent tidal wetlands including salt marshes and mangroves. Distinct habitats occur across the estuarine salinity gradient that are defined by salinity and inhabited by organisms possessing the physiological tolerances needed to survive the local salinity regime. Although each of these habitats supports distinct biological communities, they are also connected within the overall food web by fluxes of matter and energy across the spatial mosaic. Mechanisms of spatial coupling frequently include physical exchange of water and associated plankton and detritus. Other mechanisms include directed movements or migration for the purpose of feeding, spawning, seeking refuge, or ontogenetic change. The latter has been referred to as the "trophic relay" hypothesis, which has been used to describe mechanisms coupling intertidal salt marshes and adjacent subtidal habitats (Kneib 1997).

One of the most widely relevant examples of spatial linkage within estuarine food webs is coupling of benthic and planktonic (or pelagic) food webs. Whereas extreme water depth prevents such coupling in the deep sea, the relatively shallow depth of most estuaries favors benthic–pelagic coupling. The estuarine benthos usually consumes a significant fraction of total organic inputs to estuaries (~25%; Nixon 1982). Benthic filter feeders, which benefit from the constant motion of water associated with tides, can exert grazing control over phytoplankton production and biomass (e.g., Cloern 1982; Cohen et al. 1984; Nixon 1988). Suspension and deposit feeding benthos contribute significantly to the diets of demersal predatory fishes and invertebrates. Some of these benthivorous (i.e., benthos-eating) species are prey for top pelagic predators, some of which also consume macrobenthos directly (Hartman and Brandt 1995; Tsagarakis et al. 2010). Finally, direct herbivory by fishes, turtles, marine mammals, and birds (Jackson 2001); Heck and Valentine 2006) on benthic primary producers, including microalgae, macroalgae, and seagrasses links shallow benthic communities to overall estuarine food webs (Choy et al. 2008).

Heck et al. (2008) illustrated how highly productive shallow-water seagrass habitats are often relatively important to the food webs of surrounding habitats within estuaries (Figure 15.8). For example, some fish live in mangroves but feed in seagrass meadows, resulting in trophic flows from seagrass to mangrove habitats. Similarly, some reef resident fishes feed in seagrass meadows, generating net fluxes of organic matter and nutrients to reef areas and their associated food webs. In temperate estuaries, various fishes feed in both salt marshes and seagrass beds, while juvenile blue crabs utilize seagrasses for feeding and shelter before migrating elsewhere as adults (Heck et al. 2008). Tidal marshes also provide physical habitat and food for omnivorous consumers. For example, in Louisiana coastal environments, regional distributions of shrimp abundance and production are proportional to the length of marsh/open-water interface (Haas et al. 2004). In some cases the value of different habitats for a particular group in a food web can be defined in terms of competing needs for refuge from predation and access to food, an important way that the spatial habitat mosaic structures trophic interactions in estuarine food webs (Walters et al. 2000; Grabowski 2004)

15.4.2 The Importance of Detritus

Early work investigating productivity of marine food webs was strongly influenced by work describing a grazing food chain in which phytoplankton production is directly grazed by herbivorous zooplankton, linking microplankton production to pelagic food webs. Research along the southeast coast of North America, however, showed that this model was inadequate to describe trophic flows in some estuaries. Large tides in these coastal systems, which are fringed by extensive salt marshes, promoted vigorous material exchanges between marshes and open water habitats. Significant quantities of production by marsh grasses was not grazed as living material, but instead was passed to a "detritus food chain" where the decaying material is consumed and directed into productive trophic pathways (Odum 1971; Choy et al. 2008; Galván et al. 2008). Similarly, Darnell (1961), whose work we have already mentioned (Figure 15.6), observed that detritus accounted for a fraction of the diet in nearly all the species analyzed, with the exception of the top piscivorous (i.e., fish-eating) predators. By the early 1960s, detritus was recognized as an important energy

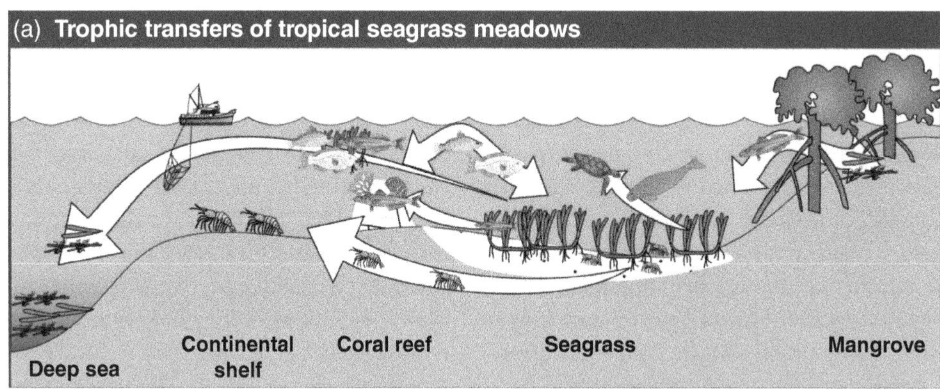

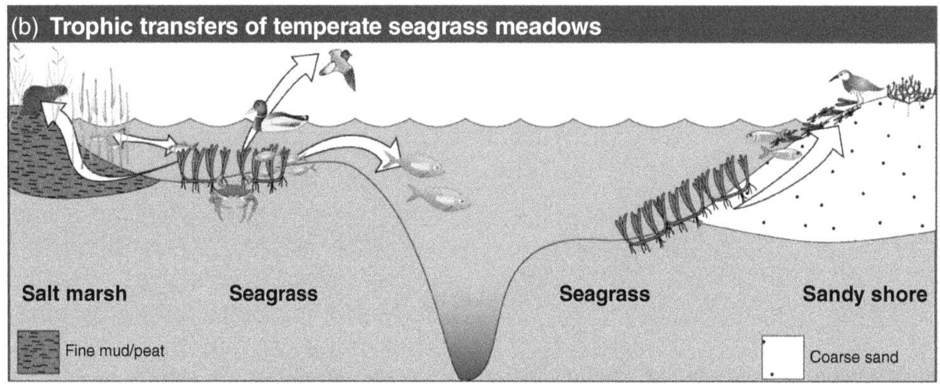

FIGURE 15.8 Mechanisms of trophic transfers from seagrass meadows to adjacent coastal habitats in (a) tropical regions and (b) temperate regions. *Source*: Heck et al. (2008)/With Permission of Springer Nature.

source in nearly every type of aquatic ecosystem. Detritus may be especially important in estuaries because they receive significant external inputs of detritus (i.e., allochthonous carbon) from adjacent terrestrial and oceanic ecosystems. Estuarine ecosystems may also generate significant quantities of vascular plant detritus from seagrass meadows, marshes, and mangroves within the littoral and intertidal zones (Teal 1962; Odum and de la Cruz 1963). Food webs in shallow tropical estuaries fringed by mangroves tend to be supported by decomposing plant material and associated microbes, with the types and sizes of materials utilized varying by consumer (Odum and Heald 1975; Figure 15.8). The slow processing and consistent replenishment of detrital material provides much of the temporal stability associated with energy flows through benthic food webs as described previously. Ultimately, lability and nutritional value are as important as quantity available for determining the importance of contribution of detritus to activity in an estuarine food web (Cebrian 1999).

While simple in concept, quantifying trophic flows involving detritus can become surprisingly complicated. For example, particulate detritus is often colonized by algae, bacteria, and meiofauna, forming a complex of biota and detritus whose nutritional and elemental composition (e.g., C:N ratio) differs from that of the detritus alone. The complex and numerous functions of microbes in estuarine ecosystems are addressed in Chapter 9. Here we note that microbes mediate a very slow biogeochemical "aging" of detritus, breaking down the most recalcitrant compounds (e.g., lignin) and eventually improving its nutritional value to consumers (Tenore et al. 1984). Given adequate time for processing, detritus can be processed repeatedly until the material is ultimately buried or otherwise exported, thereby increasing the potential importance of detritus in estuarine food webs.

15.5 Constructing Food Web Models

Estuarine food webs are complex, dynamic, and open. To understand these complex systems better, it is often useful to construct models of the system that can be analyzed quantitatively. A good first step in modeling estuarine food webs is to select the temporal and spatial scales of interest. This section emphasizes models that address at least a seasonal time scale and some sort of spatial average. This leaves for the future a discussion of explicit modeling of patchy and variable food web dynamics. A second step is defining research questions to be addressed with the food web model. Next, the modeler must choose the currency for the food web (e.g., energy, carbon, and nitrogen), the *nodes* (feeding groups) to include, and the important trophic interactions or *flows*. Finally, the investigator must compile relevant data and ecological information to quantify the food web. These steps may all be taken iteratively and interactively. For example, ecological information being collected to describe the food web as initially planned may reveal that an additional node is needed, or that a node may reasonably be omitted.

There are two basic approaches to calculating trophic flows for a food web model. The first, which is called the *a priori* approach, involves estimating all the flows independently from empirical data. For example, the total consumption by a group of fish might be estimated based on a bioenergetics approach, while the diet composition is defined by analysis of gut contents or by biochemical markers. The same analysis could quantify production by the same group of fish, which provides an upper limit on the combined consumption of those fish by all that prey on it. This approach provides the investigator with the greatest opportunity to apply expert ecological knowledge in an informal, intuitive way. The second approach, called *inverse modeling,* utilizes some of the same information to define a series of equations or inequalities, for which the solution is the needed estimates of flows between food web compartments. Inverse modeling provides a more formalized way to utilize empirical data to estimate the unknown trophic flows. In some applications of inverse modeling, the known quantities are primarily biomasses plus estimates of a few flows (e.g., primary production and sedimentation) and a set of physiological constraints (Vézina and Platt 1988). Other applications have addressed more complex systems with more constraints, capitalizing on the availability of new analytical and computational resources (Leguerrier et al. 2004; van Oevelen et al. 2009; Middelburg 2014). Inverse methods have also been used to solve for biomasses, rather than flows (Polovina 1984), and to estimate solutions when several different types of information were unknown (Christensen and Pauly 1992). In some cases, statistical methods can be used to find a best fit solution given the data and constraints (Christensen and Pauly 1992). Other analyses have employed elements of both approaches, for example by developing initial *a priori* estimates of flows, and then adjusting flows using a "balancing" algorithm to achieve mass balance (Allesina and Bondavalli 2003; Christensen and Walters 2004). The next several sections present details of constructing estuarine food webs.

15.5.1 Choosing the Appropriate Scale and Currency

Key decisions in food web model development involve defining the temporal and spatial extent of the model and the resolution of its trophic groups. The appropriate choices depend on the research questions of interest. The spatial extent is the area (or water volume) under consideration for the analysis and may be an entire estuary (Baird and Ulanowicz 1989; Rybarczyk et al. 2003) or a region of an estuary (Carrer and Opitz 1999; Tecchio et al. 2015). Temporal extent defines a period of time over which data are being considered, usually ranging from seasonal to decadal time frames (Baird and Ulanowicz 1989; Heymans et al. 2016). The spatial and temporal scales should be large enough to encompass the ecological features of interest and the associated food webs. In general, the "currency" of the food web can be flows of energy (e.g., Dame and Patten 1981) or flows of mass such as carbon (Baird and Ulanowicz 1989), nitrogen (Christian et al. 1996), or wet-weight biomass (Manickchand-Heileman et al. 1998). Regardless of the choice, models developed in one currency may reflect implicitly the importance of flows in other currencies. For example, if growth is limited by low nitrogen content of food, rather than carbon, this may be reflected in a carbon-based food web analysis as low carbon growth efficiency. The discussion below is most applicable for food webs using carbon or energy as currency. Generally, the availability of information for constructing food webs in terms of nitrogen or phosphorus is more limited than for carbon or energy.

15.5.2 Composition of the Food Web (Nodes)

Another important step in developing a food web model is deciding what to include in the food web and how to represent it. An almost universal attribute of food web models is their uneven level of aggregation of species across communities and trophic levels. Whereas lower trophic levels are usually represented as broad groups of similar species (e.g., zooplankton) or feeding guilds (e.g., suspension-feeding benthos), upper trophic levels are often resolved to the species level. This tendency is driven primarily by availability of data but may also reflect the questions of interest. To represent a trophic node (group of organisms with similar diets) in the food web model, it is necessary to quantify the magnitude of production, amount and composition of diet, and fate of the biomass that is produced (e.g., who consumes it). This information is usually not available for individual planktonic or benthic species, but may be estimated for groups of species or feeding guilds. Species-level definition of organisms at higher trophic levels can provide key insights into food web dynamics. For example, by combining two distinct groups of sharks (pelagic- and benthic-feeding) into a single aggregated node, the resulting "super shark" group had an unrealistically broad impact on the food web, even though neither type of shark has as large trophic effect (Friere et al. 2007). Interactions among subsets of nodes often display recurring patterns, called motifs, which can provide information on common trophic relationships or pathways (food web substructure). The presence, prevalence, and identity of motifs will depend on how species or taxonomic groups are aggregated. Examples of some motifs include linear pathways among nodes (food chains) or shared single- and double-step pathways linking a shared resource to a consumer and a higher predator (omnivory; McCann 2012). Food web models are sometimes focused on particular groups of consumers, ranging for example, from microplankton (Baird and Ulanowicz 1989), to gelatinous zooplankton (Tsagarakis et al. 2010), to fisheries (Manickchand-Heileman et al. 1998). In other cases, food web models reflect broad interest in ecosystem level structure (Rybarczyk et al. 2003). Although a variety of food web structures and approaches may be legitimate, the level of aggregation chosen for the food web model can have a substantial impact on the results (Martinez and Lawton 1995).

15.5.3 Organism Abundance and Biomass

Even though trophic networks are defined by flows between nodes, estimation of the abundance or biomass of organisms is a key step in constructing food webs. The reason is that, for most trophic groups, flows are often computed as a function of biomass. Exceptions include phytoplankton and bacterial production rates, for which direct measurements are often available for community production rates (Parsons et al. 1984; Kemp et al. 1993). Biomass of zooplankton and benthos may be based on plankton net tows or sediment grab samples. Net-based sampling (e.g., trawls, seines) is commonly used to quantify "relative abundance" and biomass of fish, but other survey methods such as underwater video, multi-frequency sonar, or passive gear deployment (e.g., traps, gill nets, and fyke nets) can be used to supplement active net sampling. Direct measurement of fish abundance is difficult because all these methods require careful experimental design and an understanding of the inherent bias and assumptions in each method. Sometimes it is possible to infer fish abundances indirectly, such as from production of fish eggs (Rilling and Houde 1999). For exploited fish populations, virtual population analysis (Gulland 1965) can be used to infer population size given the size and age structure of the catch and other parameters (e.g., natural mortality rate).

15.5.4 Bioenergetic Rates

The flows of food between the nodes (feeding groups) effectively define the food web. As discussed in Section 15.1.3, critical rates are related to each other by the bioenergetic equation for an organism, population, or feeding group, where consumption (C) is equal to the sum of production (P), respiration (R), and excretion plus egestion (U). All these rates are needed for each node. A good approach (called *allometry*) is to use physiological equations and ratios that vary consistently with organism size. For example, annual production (P) of individual populations or whole communities can be computed from average population biomass (B) and an estimate of average individual organism biomass (M; Figure 15.9). The allometric equation $P/B = 0.65M_s^{-0.37}$ applies across a 10^5-fold range in body size (M_s, units are body mass converted to energy equivalents in kilocalories) for a diversity of taxa. Similar relationships between organism physiology and size have been reported for all bioenergetic rates (Harris et al. 2006). Thus, respiration can be estimated using allometric equations or

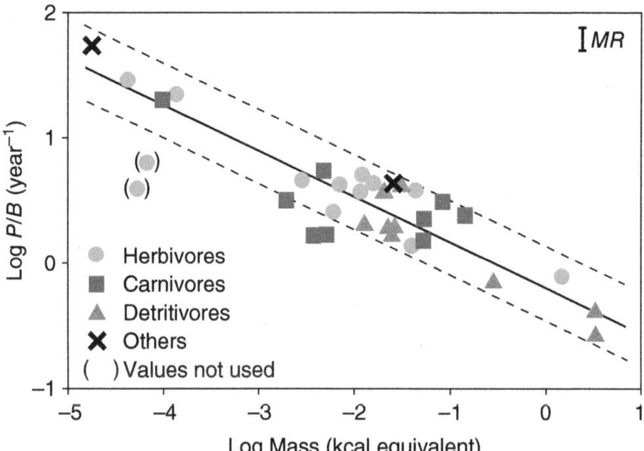

FIGURE 15.9 Relationship between log body size at maturity and log P/B (annual production/biomass) for a variety of functional groups of species. The equation for the solid regression line is $\log(P/B) = -0.19 - 0.27 \log M$. Dotted lines indicate 50 and 200% of the predicted P/B. Body size is converted to energy content (i.e., caloric) equivalents. The relationship spans five orders of magnitude in body size. *MR* refers to the mean range in P/B for 16 species with multiple observations. *Source*: Banse and Mosher (1980)/With Permission of John Wiley & Sons.

empirical relationships between respiration and consumption. Net growth efficiency ($NGE = P/(P+R)$) can also be used to compute R given an estimate of P. Detailed review of the literature on NGE for aquatic organisms suggests a typical value of ~60% (Schroeder 1981).

The final term needed to describe a bioenergetic balance is the combined egestion and excretion rate, sometimes referred to as unassimilation, or U. Like net growth efficiency, generalizations about U are possible for different classes of organisms (Schroeder 1981), depending on diet and particularly food quality. For aquatic organisms, reported values for assimilation efficiency ($=((C-U)/C)A$) are 60–70%, with lower values for herbivores and higher values for carnivores, especially fish (Schroeder 1981). Diets that include large fractions of indigestible molecules (e.g., cellulose) or animal parts (e.g., bones or feathers) leads to lower assimilation efficiency. Differences in elemental composition of a consumer and its food may limit assimilation, where, for example, an herbivore with a lower carbon to nitrogen ratio than its diet (e.g., herbivorous fish) will tend to have lower carbon assimilation efficiency (Elser and Urabe 1999). Note that depending on the application, this combined treatment (as U) of egested food, which has not been fully processed, and excretion of metabolic wastes may be unsatisfactory.

15.5.5 Food Habits

Constructing a food web requires quantitative data on the diets of each feeding group that can be used to partition total consumption for each node into contributions obtained from each of the other nodes. Three broad categories of empirical methods for obtaining diet information include (1) analysis of stomach contents, (2) feeding experiments, and (3) chemical biomarkers. Chemical biomarkers such as stable isotopes, fatty acids, and genetic material are particularly useful forms of chemical markers for analysis of estuarine food webs, and these are examined separately in the next section.

Quantitative analysis of the contents of stomachs of individual organisms, referred to as "gut contents analysis," involves visual identification and measurement of all food materials in stomachs of sample organisms. This is one of the earliest and most widely applied approaches for examining food habits. For example, the study by Darnell (1961) on food habits of consumers in Lake Pontchartrain, Louisiana (largely fish) was based on extensive analysis of gut contents. Similarly, Purcell (1992) used gut contents analysis to describe the diet of a gelatinous predator (*C. quinquecirrha*). Analysis of gut contents is widely applied in examining the food habits of larger organisms such as fish (Hyslop 1980). Such studies can provide very specific information regarding food habits and may address the sizes and types of prey consumed and how prey choice varies with an individual's size and age or the season or location of capture (Hartman and Brandt 1995). Drawbacks to relying solely on traditional stomach contents analysis include the relatively short interval between ingestion and excretion of prey residence in the gut (often hours to several days), difficulties in identifying highly digested prey, and bias that can arise from differences in the rate of digestion of dissimilar types of prey (e.g., crustaceans with hard carapaces versus polychaetes dominated by soft-tissues).

The application of genetic or genomic approaches (e.g., DNA barcoding) can help alleviate issues with prey identification during stomach contents analysis. DNA barcoding involves amplifying and sequencing short tracts of genetic material, often the mitochondrial cytochrome c oxidase subunit I (COI) region, and then searching for the unique COI sequences in gene databases to match sequences to specific species (or close genetic relatives; Bucklin et al. 2011). Applied to diet analysis, this technique provides a powerful method of keying out partially digested or difficult to identify prey. For example, Aguilar et al. (2017) combined DNA barcoding and traditional visual prey identification of stomach contents of several catfish species (*Ictaluridae*) in Chesapeake Bay. Using DNA barcoding methods, they were able to increase their identification of prey collected from stomachs of catfish to the species level from 9.4% using visual identification alone to 91.6% using a combination of visual and genetic methods (Aguilar et al. 2017). Genetic-based approaches for studying diet are improving and are being applied to a wide range of estuarine and marine taxa (e.g., crustaceans, worms, fish; Carreon-Martinez and Heath 2010).

Measures of the relative preference of a consumer for particular diet components are useful for interpreting data from studies of food habits. One such measure is the index of electivity, $E_i = (p_i - P_i)/(p_i + P_i)$, where for prey item *i*, p_i is frequency in the diet and P_i is frequency in the environment (Ivlev 1961). E_i varies from −1 for an abundant food that is not eaten to +1 if the item is much more abundant in the diet than in the environment. The *standardized forage ratio* is another, possibly better index (Chesson 1978, 1983). Food preference indices can be useful for evaluating diet composition in food webs. When the diet preference is unknown, one may estimate it by assuming neutral electivity (i.e., equal preference) for potential prey items. Given independent diet data, electivity can be used to evaluate the data and possibly improve the food web model.

In feeding experiments, another source of diet information for construction of food webs, consumers are presented with potential prey items from the natural environment in a tank or enclosure. Selectivity is estimated by examining which items are consumed. Feeding experiments can quantify prey-specific clearance rates for plankton, which are useful in food web models. For example, a large medusa (*C. quinquecirrha)*, can consume all the ctenophores from 2000 L of water each day (Purcell and Cowan 1995), but can only clear copepods from 48 L day^{-1} (Purcell 1992) indicating a preference for ctenophores. Experimental approaches have also been utilized to examine prey preferences for small zooplankton. For example, plankton dilution experiments, used to infer grazing rates, can utilize photopigments to quantify which phytoplankton are grazed (McManus and Ederington-Cantrell 1992). Selective grazing has also been measured using flow cytometers, which are instruments that determine numbers and sizes of particles (Jochem 2003).

A variety of chemical markers, including fatty acids and sterols (Bianchi and Canuel 2011) and heavy metals (Chen et al. 2009) have been used to study trophic relationships in estuarine food webs. Chemical markers may be passed through one or more trophic transfers, relating presence in a consumer back to a source in the diet. Ideal biomarkers are relatively unique to a particular organic matter source and are transferred with little modification. St. John and Lund (1996) illustrated how the ratio of two fatty acids could indicate the importance of a diatom and a flagellate as the base of the food chain supporting North Sea cod (*Gadus morhua*) larvae (Dalsgaard et al. 2003; Figure 15.10). Copepods were fed one phytoplankton species, and then cod larvae were fed nauplii from each treatment group. The observed fatty acid ratio in the cod larvae resolved the phytoplankton source, even after two trophic transfers. Fatty acid and sterol biomarkers have also resolved transfers of organic matter from bacteria to ciliates to copepods, quantifying the significance of the bacterial food source (Ederington et al. 1995). This type of information is particularly useful in combination with other biochemical data, such as stable isotopes.

15.5.6 Stable Isotopes in Estuarine Food Webs

Analysis of stable isotopes of major elements in organic matter can provide information on average food habits and direct, integrative information on trophic position and relative dependence on different types of primary producers. Stable

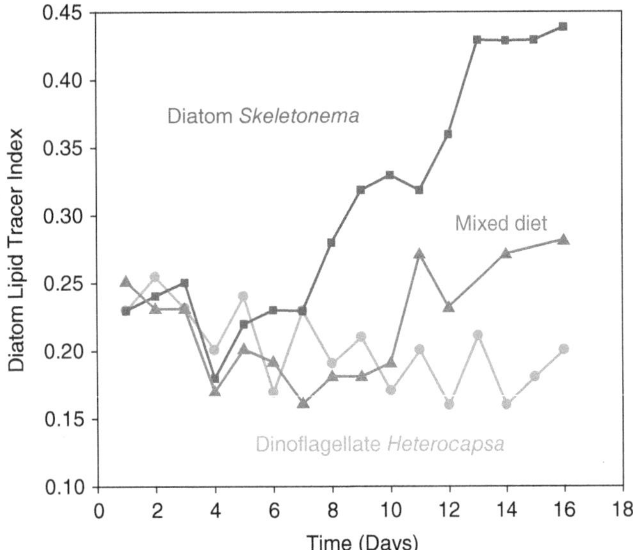

FIGURE 15.10 Results of an experiment following incorporation of lipid biomarkers in a food chain. In each of three treatments, North Sea cod larvae were fed copepod nauplii hatched from adult copepods raised on a (1) the diatom *Skeletonema costatum*, (2) the dinoflagellate *Heterocapsa triquetra* or (3) a 50:50 mixture of each. The lipid tracer index is the ratio of abundance of two specific lipid markers in the cod larvae. *Source*: St. John and Lund (1996)/With Permission of Inter-Research.

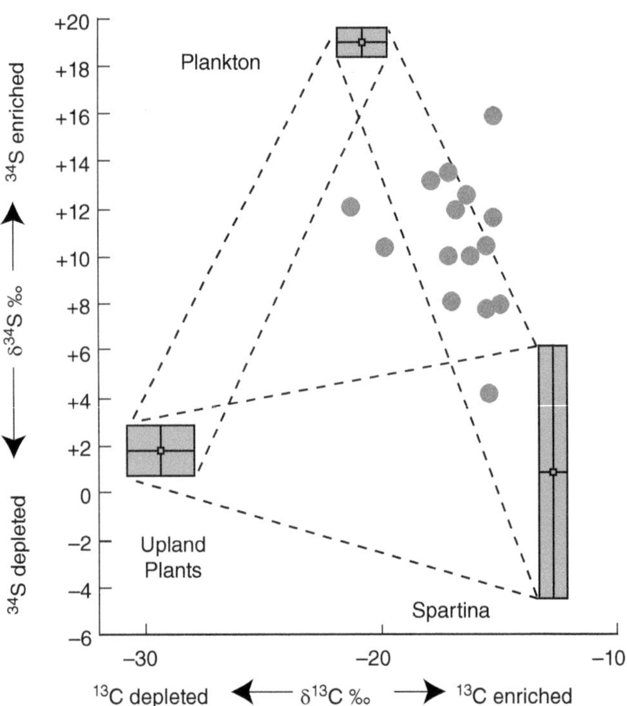

FIGURE 15.11 Graph illustrating the use of stable carbon and sulfur data from biota collected near Sapelo Island, Georgia, USA to evaluate the organic matter sources for the food web. In this case, stable isotope data exclude upland plants as an important source for the food web. *Source*: Peterson and Howarth (1987)/With Permission of John Wiley & Sons.

isotopes of carbon (^{12}C, ^{13}C), nitrogen (^{14}N, ^{15}N), and sulfur (^{32}S, ^{34}S) have been widely used in studies of estuarine ecology. Fry (2006) provides an excellent introduction, illustrating both the power of analyzing stable isotopes and some of the complications inherent in their application to food webs. Very basically, the ratio (R) of the heavy isotope to the light isotope for a consumer organism reflects that of the organic matter it eats. This relationship is modified by the process of *isotope fractionation* (a.k.a., *trophic enrichment* or *trophic discrimination* in trophic applications), which occurs because reactions involving the lighter isotope usually proceed slightly faster than those involving the heavier isotope. Isotope ratios are most commonly expressed as a deviation from that of a standard reference material using the double ratio $\delta = [(R_{sample}/R_{standard}) - 1] \times 1000$, where the ratio R is as defined above (Fry 2006). For stable isotopes of C, N, and S, the elements most commonly used in studies of estuarine food webs, the δ values are $\delta^{13}C$, $\delta^{15}N$, and $\delta^{34}S$, respectively, and have units of "per mil" (‰).

Stable isotopes vary across the biosphere in relatively predictable ways (Peterson and Fry 1987). In the best-known examples from estuaries, $\delta^{13}C$ and $\delta^{34}S$ are the most reliable indicators of organic matter sources at the base of the food web. Terrestrially derived organic matter and organic matter from freshwater sources is more depleted in ^{13}C ($\delta^{13}C \sim -28$‰) than marine phytoplankton ($\delta^{13}C \sim -21$‰), whereas seagrasses, microphytobenthos, and marsh plants tend to be less depleted with $\delta^{13}C \sim -18$ to −10‰. These values approximate the range of values often reported in the literature but it is important to note that local producer $\delta^{13}C$ values can differ widely in mean and variance through both space and time (Michener and Lajtha 2007). This variability can propagate to consumers (Rolff 2000; Woodland et al. 2012) and it is important to be aware of the specific conditions present in a study system. Another complication is that mixtures of carbon from multiple sources (e.g., the marsh plant *Spartina* and terrestrial plants) could have the same $\delta^{13}C$ as a different source (e.g., marine phytoplankton). This problem can be addressed by simultaneously examining additional isotopes, such as $\delta^{34}S$, which helps differentiate among multiple potential sources (Figure 15.11). In the case of the salt marsh at Sapelo Island, the organic matter source for most organisms includes a mixture of plankton and *Spartina*, but no significant contribution from upland plants (Peterson and Howarth 1987).

Whereas $\delta^{13}C$ and $\delta^{34}S$ are the most useful stable isotopes for examining organic matter sources, $\delta^{15}N$ is often useful for characterizing average trophic position because of the relatively large N isotope fractionation associated with each trophic transfer (Box 15.3). Fry (1988) observed that a relatively consistent 3.6‰ increase in $\delta^{15}N$ was associated with an increase of one trophic level (Figure 15.12). In an example from Fry (1988), planktivorous fish were found to have an average $\delta^{15}N$ of 10–11‰, and based on known feeding habits were known to have an average trophic level of ~3.2. Samples from their plankton diet had an average $\delta^{15}N$ of 6–7‰, 4‰ less than the planktivores and reflecting

Box 15.3 Calculating Trophic Position Using Stable Isotopes

Another option for calculating trophic position is to use a natural biomarker, such as stable isotopes, as proxies for diet. Stable isotope approaches can be less time consuming than stomach contents analysis but require additional sampling and an *a priori* understanding of the major primary producers in an ecosystem. Below, we walk through the equations used to calculate trophic position using stable isotope data. Please refer to Section 16.5.6 for a refresher on how stable isotopes are used in trophic ecology.

Assume we collected samples of white muscle tissue from several individual bluefish during our sampling project in Chesapeake Bay. We elected to analyze the muscle tissue of individual bluefish for nitrogen ($\delta^{15}N$) stable isotope composition. At the time and location of sampling, we also collected several filter-feeding bivalves to serve as our isotopic baseline. Based on an initial reading of the literature, we decide a single baseline is sufficient for this study because our initial impression is that bluefish feed primarily within pelagic food webs. After preparing and analyzing our samples for $\delta^{15}N$ on an isotope-ratio mass spectrometer coupled to an elemental analyzer, we summarize our findings in a table.

Taxon	Identity	$\delta^{15}N$
bluefish	Consumer	17.2
Filter-feeding bivalve	Baseline	10.0

Using this stable isotope information, we can use the following equation to calculate trophic position:

$$T_i = \lambda + \left(\delta^{15}N_i - \delta^{15}N_{baseline}\right) / \Delta\delta^{15}N_{TEF}$$

where T_i is the trophic position of consumer i, λ is the trophic position of the baseline organism, $\delta^{15}N_{baseline}$ is the average $\delta^{15}N$ value of the baseline organism, and $\Delta\delta^{15}N_{TEF}$ is the step-wise enrichment of $\delta^{15}N$ across each trophic transfer, termed the trophic enrichment factor (TEF). In this case, our λ is 2 because we are using a primary consumer (bivalve) as our trophic baseline, and we select a $TEF = 3.3$ as an appropriate estimate based on a review of the literature (e.g., Vander Zanden and Rassmussen 2001; McCutchan et al. 2003). Substituting our values into this equation, we get:

$$T_i = 2 + \left(17.2 - 10.0\right) / 3.3$$

Solving for T_i, we calculate a trophic position for bluefish of 4.182, rounded to 4.2.

This model can be extended to include two or more trophic baselines. For example, if we took a closer look at the diet literature on bluefish, we would have realized that bluefish will readily feed on bottom-associated organisms such as polychaetes or benthic crustaceans. Given this better resolved information, we should conclude that a two baseline model is needed to account for the fact that our bluefish were likely deriving some of their tissue from the benthic food web in addition to the pelagic food web. A two-baseline model that includes a separate isotopic baseline for both pelagic and benthic trophic pathways would look like this:

$$T_i = \lambda + \left(\delta^{15}N_i - \left[\alpha \times \delta^{15}N_{baseline,1} + (1-\alpha) \times \delta^{15}N_{baseline,2}\right]\right) / \Delta\delta^{15}N_{TEF}$$

where T_i is the trophic position of consumer i, λ is the trophic position of the baseline organisms, α is the proportion that each baseline is contributing to the diet of the consumer, $\delta^{15}N_{baseline,n}$ is the average $\delta^{15}N$ value of baseline n, and $\Delta\delta^{15}N_{TEF}$ is the selected TEF value (Post 2002b). Values for alpha can be estimated using $\delta^{13}C$ isotope data, stomach contents data, or previous literature on feeding habitats.

an average difference of just over one trophic level. Fry (1988) also observed variable $\delta^{15}N$ for fish characterized as "opportunistic generalists," likely reflecting their varied diet (Figure 15.12). In another study, Wells et al. (2008) used stable N and C isotopes to explore ontogenetic changes in the feeding ecology, including organic matter sources and average trophic level for red snapper (*Lutjanus campechanus*) in the northern Gulf of Mexico. The study found an abrupt increase in $\delta^{15}N$ between larval fish and juveniles, followed by a slower increase in $\delta^{15}N$ with age. These results were interpreted to indicate an increase in average trophic level, which was supported by results from analysis of gut contents. Although most applications of $\delta^{15}N$ focus on analyzing trophic position, $\delta^{15}N$ can provide source information in some instances. This is particularly true in estuaries with primary producer groups that use nitrogen fixation of gaseous N_2 to supply their biological nitrogen requirements. Nitrogen fixation is the process by which certain bacteria or archea, called diazotrophs, transform N_2 gas to biologically available forms of nitrogen such as ammonium. As this fixed nitrogen from microbes becomes available to other plants, it imparts very low $\delta^{15}N$ values (~0‰). This unique signature of fixed $\delta^{15}N$ can be used as a tracer of fixed N through space and time in estuarine ecosystems (Woodland and Cook 2014).

Methods are now available to analyze stable isotope composition of specific biochemical compounds such as individual amino acids or fatty acid methyl esters in addition to the 'bulk' composites of tissue traditionally used in trophic applications of stable isotopes. While the methods for 'compound-specific' stable isotope analysis (CSIA) have been available for decades, it is only within the last 15–20 years that CSIA has become a well-recognized tool for trophic ecologists (Ruess

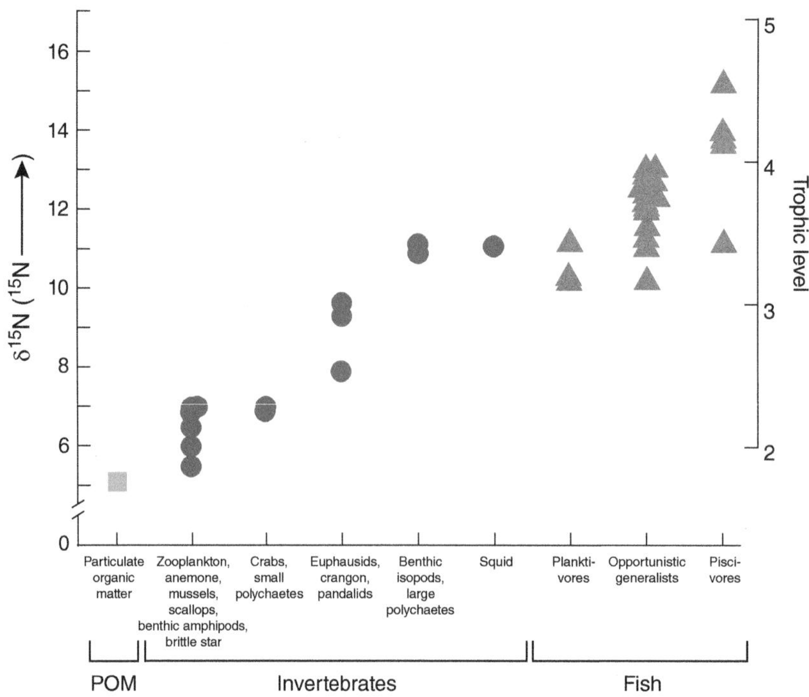

FIGURE 15.12 The relationship between $\delta^{15}N$ and trophic level in the food web of Georges Bank. *Source*: Fry (1988)/With Permission of John Wiley & Sons. Later research revealed that the "piscivore" with particularly low $\delta^{15}N$ consumes lower trophic level prey, consistent with the stable isotope results.

and Chamberlain 2010; McMahon and Newsome 2019; Whiteman et al. 2019). A key advantage of CSIA over traditional bulk stable isotope analysis is the ability to differentiate between compounds that are assimilated directly from the diet, called source or essential compounds, versus compounds synthesized de novo by the consumer, called trophic or nonessential compounds. The terms *source* and *trophic* or *essential* and *nonessential* are used to refer to amino acids or fatty acids, respectively (Whiteman et al. 2019). Distinguishing between these two classes of compounds can provide benefits such as increased resolution of trophic relationships or relaxing the need to have simultaneous measurements of isotope composition at the base of the food web.

15.6 Quantitative Analysis of Food Web Network Models

Development of computational methods for analyzing food webs began in earnest in the early 1970s (Hannon 1973) with adaptation of Leontief's (1951) economic input–output analysis to analyze ecological networks, rather than economic goods and services. Additional methods developed during the 1970s and 1980s coalesced as *ecological network analysis*, or ENA (Kay et al. 1989). ENA contributed to an evolving quantitative theory of ecosystem development (Ulanowicz 1986) that was implemented in a computer program called NETWRK and later programs such as WAND which added new procedures (Allesina and Bondavalli 2003, 2004; Ulanowicz 2004). Software applications for analyzing food web networks continue to be developed (e.g., enaR, an R software package; Borrett et al. 2014).

Another approach called "Ecopath" (Polovina 1984) emerged independently in the mid-1980s from research on fish ecology of French Frigate Shoals. A main objective was to estimate the biomasses of the coral reef food web components. Ecopath was later expanded to include more sophisticated inverse modeling capabilities and other tools to assist with model construction, and to implement the ENA procedures from NETWRK (Christensen and Pauly 1992), ultimately becoming "Ecopath with Ecosim" (**www.ecopath.org**; Christensen and Walters 2004), a comprehensive program for construction, analysis and dynamic simulation (i.e., simulating changes over time, see Section 15.6.5) of food webs. Ecopath with Ecosim can implement spatially explicit simulations, which may be useful for examining the spatial mosaic of habitats in estuaries (i.e., as in Section 15.4). Although the software continues to emphasize applications in fish ecology and fisheries management, it can also be applied to other kinds of food webs.

To illustrate quantitative analysis of a trophic network, consider a hypothetical food web with six compartments representing distinct trophic groups (Figure 15.13). The flows include

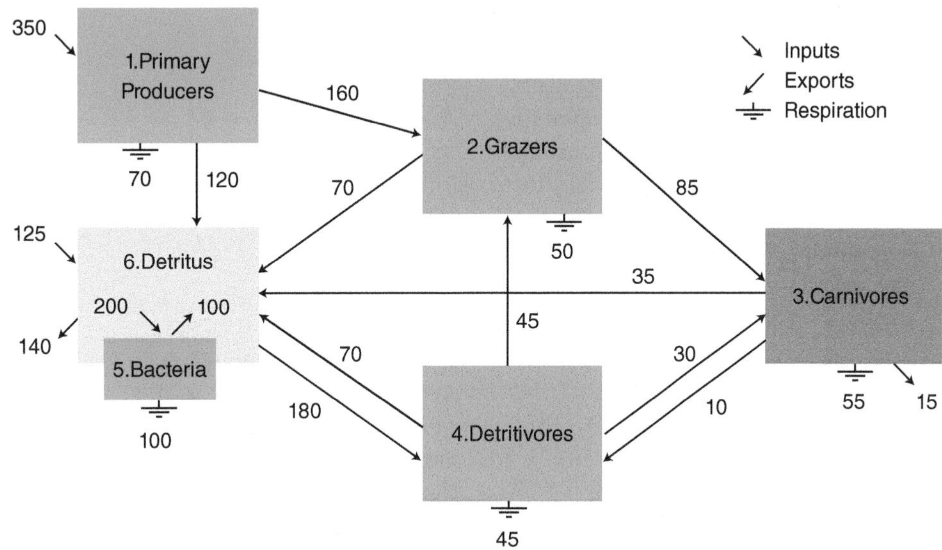

FIGURE 15.13 A simple hypothetical carbon food web for an estuary. Inputs, outputs, respiration, and internal flows have units of $gC\,m^{-2}\,year^{-1}$. Major components are: primary producers (e.g., phytoplankton), grazers (e.g., copepods) feeding on a mixture of phytoplankton and heterotrophic microplankton, carnivores (organisms feeding on zooplankton or small benthic species), and detritivores. Bacteria are represented as a sub-component of the detritus pool, returning biomass produced back to detritus.

major trophic pathways that occur in estuaries, including a grazing chain (primary producers, grazers, and carnivores) and a detritus-based food web (particulate and dissolved organic matter, detritivores, and carnivores). The particulate and dissolved organic matter group (i.e., detritus) includes bacteria as a sub-component that consumes detritus and generates respiratory losses, but returns all the produced biomass back to the detritus pool. Each compartment contributes to detritus as a result of natural mortality and unassimilated food (e.g., egestion). Organic matter inputs include primary production and external inputs of particulate and dissolved organic matter. Exports include particulate and dissolved organic matter and carnivore biomass associated with fish harvest and emigration.

15.6.1 Mass Balance in Food Web Networks

A fully quantified food web model satisfies several mass balance or "master" (Christensen and Walters 2004) equations. One such equation is the bioenergetic mass balance, $C = P + R + U$, introduced in section 15.1.3, which applies to the aggregate flows to and from a node. A second mass balance accounts for the fate of production (P). The second Ecopath master equation (Christensen and Walters 2004), balances production for a trophic node with outputs, including catch (i.e., for fisheries), predation losses (i.e., to other nodes), net migration (i.e., immigration minus emigration), and natural mortality. The mass balance also includes a term for the net change in biomass, which applies when food web models consider significant long-term or seasonal changes in biomass. A more generic mass balance equation simply states that, for each node, the sum of external inputs and inflows from other nodes equals the sum of exports, respiration, and flows to other nodes (Ulanowicz 2004). The Ecopath equations can be related to generic mass balance. For example, the flow from detritivores to carnivores in Figure 15.13 can also be expressed as the product of total predation by carnivores multiplied by the fraction of the carnivore diet obtained from detritivores. Mathematical expressions for the analyses described below are provided by Ulanowicz (Ulanowicz 1986, 1997, 2004) and others (Christensen and Walters 2004), and are partially presented in an Appendix.

15.6.2 Total System Properties

Once a food web has been quantified, ecological network analysis can be used to examine *total system properties*, which pertain to the food web as a whole. One question is "how big is the food web?" Although it is intuitive to think of "size" in terms of biomass, it is often more informative to consider the magnitude of trophic flows. The term *total system throughput* (TST) refers to the sum all the internal and external flows, including inputs, exports, respirations, and flows among the compartments in the food web. Summing the flows in Figure 15.13, the TST is $2055\,gC\,m^{-2}\,year^{-1}$ (or $5630\,mg\,C\,m^{-2}\,day^{-1}$) making this food web fall between the smallest (St. Marks,

Florida, USA in January—1900 mg C m^{-2} day^{-1}) and largest (Chesapeake Bay in summer—1,700,000 mg C m^{-2} day^{-1}) food webs examined in a comparative study (Christian et al. 2005). Total system throughput is usually larger than the sum of the flows into the food web because each unit of input is usually involved in several trophic transfers before being finally lost to the system. The average number of trophic transfers that occur after a unit of material or energy enters the food web is called the *average path length*. Average path length can be computed by subtracting the sum of all inputs from TST, then dividing by the sum of inputs, the result of which is 3.3 for the food web in Figure 15.13.

Another question about total system properties is "how is the food web organized." Drawing from early efforts to quantify species diversity (MacArthur 1955), ecological network analysis makes it possible to compute *flow diversity* and to decompose flow diversity into two components, *residual diversity/freedom* and *average mutual information* (AMI). AMI quantifies the degree of organization of the flows and is often scaled by TST to compute *network ascendency* or simply *ascendency* (Ulanowicz 2004). The upper limit for ascendency is called *development capacity* or *capacity,* while the difference between ascendency and capacity is called *overhead*. Ascendency is 30,225 g C m^{-2} year^{-1} for the example food web in Figure 15.13, which is 37% of development capacity (82,370 g C m^{-2} year^{-1}), similar to the ratio ascendancy/capacity for many estuarine and other food webs (Christian et al. 2005; Ulanowicz 2009). Given sufficient data, these quantities can be used to evaluate changes in food webs (Heymans et al. 2007) and to evaluate theories related to ecosystem development (Ulanowicz 1997, 2009).

15.6.3 Trophic Structure

Quantitative analysis of flows in a food web provides a way to understand the structure of trophic flows, or simply "trophic structure." Flow *cycles*, present in most food webs, involve trophic pathways that return to the trophic node from which they originated. Cycles often involve a non-living compartment (i.e., detritus, Figure 15.13). The food web in Figure 15.13 includes seven cycles, five of which involve recycling via detritus. One longer cycle involves four compartments: carnivores → detritus → detritivores → grazers → carnivores. Cycles among living components are relatively rare, perhaps due to size-based limitations on trophic interactions. However, to illustrate cycles, a flow has been introduced from carnivores to detritivores in Figure 15.13. This could represent a suspension feeding detritivore feeding on an early life stage of a carnivore. This additional flow creates two cycles, one of which is grazers → carnivores → detritivores → grazers. Complex food webs often have many cycles. The importance of cycling can be expressed via the ratio of flows associated with cycling relative to the magnitude of all flows (i.e., TST; Finn 1976). In the example food web, the sum of flows due to cycling is 535 g C m^{-2} year^{-1}, which amounts to 28% of TST.

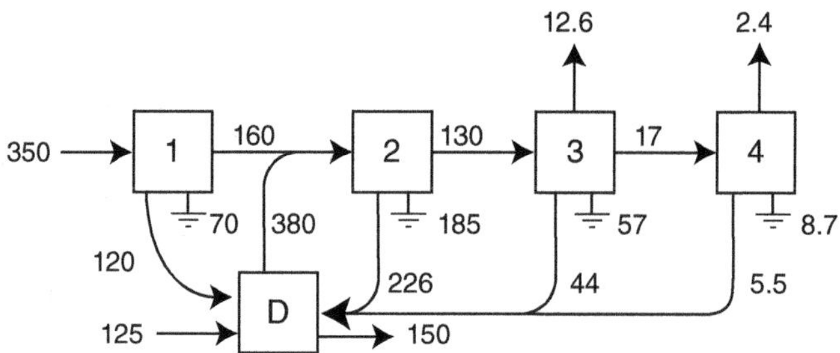

FIGURE 15.14 The effective food chain obtained by projecting the food web in Figure 15.13 into a chain of "canonical trophic levels." Flows to detritus are collected in a detritus compartment (labeled "D") and assigned a trophic level of 1. Exports from trophic level 3 and 4 total to 15, the export from compartment 3 (carnivores), revealing the "apportionment" of activity between trophic level 3 and 4. The effective trophic level for carnivores is 3.13 because most of the export is at trophic level 3.

As indicated earlier (Section 15.2.2) complex food webs can be represented as an "effective food chain" composed of virtual or "canonical" trophic levels (Figure 15.5) by apportioning the network of flows among a series of integer trophic levels (Ulanowicz and Kemp 1979). This representation, sometimes called a "Lindeman trophic aggregation" or "Lindeman spine," can provide insight into trophic structure. A key step involves mathematically removing flows associated with cycles among living compartments, as these create, in theory, an infinite number of trophic levels. Cycles involving detritus may remain since by convention the trophic level of detritus is equal to 1 (i.e., primary producer). The effective food chain generated from the food web in Figure 15.13 has four virtual trophic levels (Figure 15.14). The longest path begins with either primary producers or detritus and involves flow from detritus, to detritivores, to grazers, and then to carnivores (i.e., 6→4→2→3; Figure 15.13). Only a small fraction (13%) of consumption by carnivores occurs via this pathway. As a result, the yield to humans of carnivores is at an average trophic level of only 3.16, or 16% at trophic level 4 and 84% at trophic level 3. This is typical of the world's fisheries harvests (Pauly et al. 1998).

The effective food chain may also describe average efficiency of the food web at each virtual trophic level. Of the 750 gC m^{-2} year^{-1} input to trophic level 1 (phytoplankton [350] plus detritus [70+35+70+100+125]), 540 gC m^{-2} year^{-1}, or 72% flows to the second trophic level (Figure 15.14). Upper trophic levels tend to be less efficient. In Figure 15.14, only 24% of the flow to trophic level 2 is passed to the third trophic level, and only 13% is subsequently passed to the fourth. The highest trophic level has, by definition, a trophic efficiency of zero.

15.6.4 Indirect Interactions

Analysis of food web models enables quantification of how flows depend on any other flows, whether directly or indirectly (Ulanowicz 2004). For example, an important question is how much the flows within a food web ultimately depend on a specific input, such as primary production. Inputs in the example food web (Figure 15.13) include primary production and detritus. Network analysis reveals that 84% of the export from compartment 3 (carnivores) depends on primary production, even though there are no flows from primary producers to compartment 3. The *total dependency matrix* (Szyrmer and Ulanowicz 1987) generalizes this kind of analysis, quantifying the fraction of each flow that previously flowed through each of the other nodes via all pathways. A related concept is the *total contribution matrix*, which quantifies the fraction of flow from each node that eventually reaches each of the other nodes, again via all pathways. As an example, the total contribution matrix for the example food web (Figure 15.13) indicates that 50% of phytoplankton production reaches grazers via all pathways, even though only 46% of primary production flows directly from primary producers to grazers. The contribution matrix is related to the concept of *primary production required* (PPR; Christensen et al. 2004), which illustrates the magnitude of PPR to sustain a particular fisheries yield. In the example, 26% of the primary production flows to carnivores via all pathways. Thus, $90\,g\,C\,m^{-2}\,year^{-1}$ of primary production is required to sustain the harvest of carnivores.

A useful way to understand both direct and indirect trophic impacts in a food web is to utilize a procedure called *mixed trophic impacts* (Ulanowicz and Puccia 1990), which expresses the net effect of a node on another node via a coefficient between −1 (strong negative effect) and +1 (strong positive effect). For example, the direct effect of primary producers on grazers (Figure 15.13) is strongly positive (+0.78) and the net effect of grazers on phytoplankton is, not surprisingly, strongly negative (−0.57). Carnivores have no direct effect on phytoplankton; their indirect trophic impact is positive (+0.17) because they exert top-down control on the grazers, implying a subtle trophic cascade (Section 15.3.3).

15.6.5 Dynamic Simulation of Food Webs

Thus far, all the food web models that have been discussed are *static* models. That is, they describe the magnitude of trophic flows in a food web as either observed or calculated (e.g., via inverse analysis). Although much can be inferred from static food web models, particularly by comparing models for different times and places, they cannot address "what if" questions, such as "how would the food web change if" an important predator or grazer were eliminated, or if an important habitat area was made unavailable. These questions can be addressed using *dynamic* food web models, which simulate changes in biomass and flows through time. Although dynamic simulation models are not new, successful applications to complex food webs are relatively new. One problem that has been encountered is that dynamic food web models tended to "self-simplify," unrealistically losing trophic groups (i.e., their biomass goes to zero), whereas real food webs tend to maintain diversity (Walters et al. 1997). To address this, Walters et al. (1997) observed that predator–prey encounters can be mediated by behavioral or physical mechanisms that can limit the availability of prey. A model structure representing these processes performed much better and incorporated as "Ecosim" into the previously mentioned Ecopath with Ecosim software. The software has since been generalized by adding "Ecospace" to support spatially explicit simulations, broadening the range of "what if" questions that may be addressed (Pauly et al. 2000). Dynamic food web models require all the information needed to quantify static food web models, plus additional information to characterize trophic dynamics (e.g., how flows change in relation to changes in biomasses of each trophic node). The additional data requirements for parameterizing dynamic food web models poses a challenge for ensuring that these models are well constrained by data.

15.7 Estuarine Food Webs in a Changing Climate

Climate change is expected to bring about a number of changes in coastal marine ecosystems. Among these, the most obvious are changes in the seasonal and spatial patterns of water temperature and salinity. These are already causing observable shifts in the seasonal and spatial distributions of coastal and marine species, although natural climate variability can make these patterns difficult to identify (Perry et al. 2005; Booth et al. 2011; Morley et al. 2018). Further, shifts in the timing of seasonal temperature or precipitation cycles due to climate change can have different effects on individual trophic levels or trophic pathways (e.g., benthic versus pelagic pathways) within an ecosystem (Edwards and Richardson 2004). Beyond temperature and salinity, climate change will affect coastal habitats through sea level rise, altered storm surges regimes, and by changing the acidity of coastal waters through a process known as ocean acidification that arises from increased atmospheric CO_2 concentrations.

In the context of food webs, a key question is whether the effects of introducing or removing species from an ecosystem propagate through food webs and especially whether these changes are attenuated or amplified via trophic interactions. Most likely, a climate change-driven effect on one species will also affect other species, due to the presence of important predator–prey or producer–grazer relationships. Alternatively, changes in local conditions can shift the outcome of competitive interactions from favoring one species to another (Helland et al. 2011). Some of these trophic relationships may be quantitatively very important, which can be examined with trophic network models or by coupling climate habitat models with trophic models (Fernandes et al. 2013; Niiranen et al. 2013; Ihde and Townsend 2017; Chevillot et al. 2019). We introduced the concept of the keystone species, which has been defined as a species that has an outsized effect on community structure and its persistence through time given the species' abundance (or biomass). Given that such trophic roles exist, one can predict that climate-induced changes affecting keystone species can have especially wide-ranging top-down effects on estuarine food webs (Harley et al. 2006; Citadin et al. 2018). Disruption of key trophic pathways can lead to compensatory changes to other trophic pathways, further propagating climate effects through food webs.

In the face of mounting climate change pressures, food webs can change to accommodate some of these disturbances. For example, many coastal and marine food webs have changed in response to sustained fishing pressure (Pauly et al. 1998; Lotze et al. 2011; Altieri et al. 2012), yet many of these regions are still able to support mid- and top-trophic level predators (albeit sometimes at greatly reduced abundances). However, if the history of biological invasions caused by artificial species introductions is a guide, simple replacement of displaced species or accommodation of new arrivals without structural change is unlikely. In the extreme, climate change can cause dramatic shifts in the structure and function of important marine food webs (Kortsch et al. 2015, Ullah et al. 2018).

15.8 Summary

Estuarine food web ecology focuses on the flow of biomass or energy throughout the living (and some non-living) components of estuary ecosystems. These flows represent interactions among a diverse spectrum of organisms and their environment, creating a food web. Food webs are very complex and *holistic* analytical approaches such as network analysis are often used to analyze their structures and processes. These approaches allow us to study food webs as interactive whole systems and thereby help us find errors or gaps in our ecological understanding, reduce uncertainty in our estimates, and identify overlooked or mischaracterized nodes or flows.

Many different techniques can be used to portray food webs. Some of these techniques include the simplification of trophic relationships into discrete levels and subsequent representation of biomass, abundance, or energy as ecological pyramids. Other techniques maintain the trophic complexity of fractional trophic positions (rather than levels) and explicitly depict the distribution of diets, biomass, or organisms across the spectrum of trophic positions or a close proxy (e.g., body size). These approaches seek to integrate the vast complexity of information present in a given food web into an approachable and understandable representation that captures a critical, often structural, aspect of the food web.

A critical element of food web structure is the network of flows that link different nodes. These flows represent direct interactions between nodes (i.e., *competition*, *predation* or *parasitism*, *mutualism*, *commensalism*, and *amensalism*). These direct interactions differ from indirect interactions which are consequences realized by nodes that are not connected by a direct flow path. Trophic cascades, while seemingly rare in estuarine systems, are an example of a circumstance in which indirect interactions have a strong effect on the architecture of a food web and dynamics of an ecosystem.

The spatial heterogeneity of estuarine habitats that arise from variable physical and environmental conditions also leads to spatially heterogeneous food webs. These different food webs within the boundaries of an estuary are intimately linked by hydrodynamic mixing and the movement of organisms. These linkages between habitats provide important pathways for energy or biomass exchange between habitats and can extend beyond the boundaries of an estuary, for example through the reproductive migrations of fish or the advection of detrital vegetation to the adjacent coastal ocean.

An understanding of what different organisms eat is necessary to understand how food webs function. Historical methods of describing feeding habits of estuarine organisms that emphasized direct observation of diet composition are now routinely supplemented with biomarker analysis. The most robust methods involve a combination of direct observation, such as stomach contents analysis, and natural biomarker analysis, such as stable isotope analysis or genetic barcoding. The combination of these methods is powerful because it allows researchers to examine diet at different taxonomic resolutions and over different temporal resolutions. The power of biomarkers such as stable isotopes, fatty acids, and genetic or genomic material to strengthen our inferences regarding trophic relationships or to uncover previously unknown relationships is difficult to overstate.

Quantifying and subsequently analyzing estuary food webs requires a modeling approach capable of capturing the complexity of these systems while helping to balance the many knowledge gaps that often exist in our understanding of these systems. Network models have become the tool of choice for analyzing food web properties. Nodes form the components (often species or functional groups) of the food web and flows (a.k.a. edges) form the connections between nodes. Designating the currency for a food web is a critical first step in the process of constructing a food web and nodes represent "pools" of that currency present within a specific component. Flow values can be populated using empirical data, termed *a priori* approaches, or estimated using algorithms that constrain the possible range of flow values based on the need for the model to balance flows into, within, and out of the system (*inverse modeling*). Once constructed, food web attributes can be calculated based on web characteristics such as flow patterns, material or energy cycling, food chain length, and inputs and outputs. Dynamic modeling, in which the structure of the food web is allowed to change according to a set of stipulated rules, has become increasingly important as a tool for evaluating the effects of temporal or spatial changes to food webs.

Review Questions

Multiple Choice

1. What value is the most reasonable "rule of thumb" for estimating trophic transfer efficiency in estuarine food webs?
   a. 1%
   b. 10%
   c. 50%
   d. 100%
2. Which of the following functional trophic groups is quantitatively least important to energy flows in estuarine food webs?
   a. Mixotrophs
   b. Primary consumers
   c. Autotrophs
   d. Chemoautotrophs
3. What can cause estuarine biomass pyramids to become inverted?
   a. Water circulation patterns
   b. Longer average lifespan of higher trophic level organisms
   c. Loss of habitat complexity
   d. Increased trophic connectivity
4. Which of the following is considered an indirect trophic interaction?
   a. Predator consumes a grazer and its producer food increases
   b. Predator consumes a grazer and the grazer declines
   c. Omnivore competes with a predator for food and the predator increases
   d. Parasite consumes part of a consumer but does not kill it
5. Top-down control in food webs refers to:
   a. Atmospheric deposition
   b. Benthic–pelagic coupling
   c. Channel–marsh edge exchange
   d. Change in abundance of lower trophic level organisms caused by a predator

Short Answer

1. What is a food web and how does it differ from a food chain?
2. What is a trophic guild? Describe several key trophic guilds that occur in estuarine food webs.
3. What are the major components of a bioenergetics budget? How can these components be used to explore trophic efficiencies?
4. How do direct trophic interactions differ from indirect trophic interactions? Give an example of each that you might find in an estuarine food web.
5. Estuaries often contain a wide variety of habitat types. Discuss some of the consequences of this spatial variability on estuarine food webs.
6. What are two common methods used to study diet of estuarine organisms? List several advantages and disadvantages to each approach.
7. Compare ecological pyramids and food web networks—how do they differ and how are they similar?
8. What does mass balance mean from a quantitative food web model perspective?
9. How do flow cycles differ from flow pathways in a food web?
10. What is meant by a "static" food web model? What is an alternative to analyzing a static model and how does it differ?

References

Aguilar, R., Ogburn, M.B., Driskell, A.C. et al. (2017). Gutsy genetics: identification of digested piscine prey items in the stomach contents of sympatric native and introduced warmwater catfishes via DNA barcoding. *Environmental Biology of Fishes* 100: 325–336.

Akin, S. and Winemiller, K.O. (2008). Body size and trophic position in a temperate estuarine food web. *Acta Oecologica* 33: 144–153.

Alldredge, A. (1981). The impact of appendicularian grazing on natural food concentrations in situ. *Limnology and Oceanography* 26: 247–257.

Allesina, S. and Bondavalli, C. (2003). Steady state of ecosystem flow networks: a comparison between balancing procedures. *Ecological Modelling* 165: 221–229.

Allesina, S. and Bondavalli, C. (2004). WAND: an ecological network analysis user-friendly tool. *Environmental Modelling and Software* 19: 337–340.

Altieri, A.H., Bertness, M.D., Coverdale, T.C. et al. (2012). A trophic cascade triggers collapse of a salt-marsh ecosystem with intensive recreational fishing. *Ecology* 93: 1402–1410.

Anderson, D., Burkholder, J., Cochlan, W. et al. (2008). Harmful algal blooms and eutrophication: examining linkages from selected coastal regions of the United States. *Harmful Algae* 8: 39–53.

Baird, D. and Ulanowicz, R. (1989). The seasonal dynamics of the Chesapeake Bay ecosystem. *Ecological Monographs* 59: 329–364.

Baird, D., Christian, R., Peterson, C., and Johnson, G. (2004). Consequences of hypoxia on estuarine ecosystem function: energy diversion from consumers to microbes. *Ecological Applications* 14: 805–822.

Baird, D., Asmus, H., and Asmus, R. (2011). Carbon, nitrogen and phosphorus dynamics in nine sub-systems of the Sylt-Rømø Bight ecosystem, German Wadden Sea. *Estuarine, Coastal and Shelf Science* 91: 51–68.

Banse, K. and Mosher, S. (1980). Adult body mass and annual production/biomass relationships of field populations. *Ecological Monographs* 50: 355–379.

Bertness, M. (1985). Fiddler crab regulation of *Spartina alterniflora* production on a New England salt marsh. *Ecology* 66: 1042–1055.

Bianchi, T. and Canuel, E. (2011). *Chemical Biomarkers in Aquatic Ecosystems*. Princeton: Princeton University Press.

Bondavalli, C. and Ulanowicz, R. (1999). Unexpected effects of predators upon their prey: the case of the American Alligator. *Ecosystems* 2: 49–63.

Booth, D.J., Bond, N., and Macreadie, P. (2011). Detecting range shifts among Australian fishes in response to climate change. *Marine and Freshwater Research* 62: 1027–1042.

Borer, E., Seabloom, E., Shurin, J. et al. (2004). What determines the strength of a trophic cascade. *Ecology* 86: 528–537.

Borrett, S.R., Lau, M.K., and Dray, S. (2014). enaR: an r package for ecosystem network analysis. *Methods in Ecology and Evolution* 5: 1206–1213.

Braune, B.M. and Gaskin, D.E. (1987). A mercury budget for the Bonaparte's gull during autumn moult. *Ornis Scandinavica* 18: 244–250.

Breitburg, D. and Fulford, R. (2006). Oyster-sea nettle interdependence and altered control within the Chesapeake Bay ecosystem. *Estuaries and Coasts* 29: 776–784.

Breitburg, D., Craig, J., Fulford, R. et al. (2009). Nutrient enrichment and fisheries exploitation: interactive effects on estuarine living resources and their management. *Hydrobiologia* 629: 31–47.

Brooks, J. and Dodson, S. (1965). Predation, body size, and composition of plankton. *Science* 150: 28–35.

Bucklin, A., Steinke, D., and Blanco-Bercial, L. (2011). DNA barcoding of marine metazoa. *Annual Review of Marine Science* 3: 471–508.

Carpenter, S., Kitchell, J., and Hodgson, J. (1985). Cascading trophic interactions and lake productivity. *BioScience* 35: 634–639.

Carreon-Martinez, L. and Heath, D.D. (2010). Revolution in food web analysis and trophic ecology: diet analysis by DNA and stable isotope analysis. *Molecular Ecology* 19: 25–27.

Carrer, S. and Opitz, S. (1999). Trophic network model of a shallow water area in the northern part of the Lagoon of Venice. *Ecological Modelling* 124: 193–219.

Cebrian, J. (1999). Patterns in the fate of production in plant communities. *The American Naturalist* 154: 449–468.

Chen, C.Y., Dionne, M., Mayes, B.M. et al. (2009). Mercury bioavailability and bioaccumulation in estuarine food webs in the Gulf of Maine. *Environmental Science and Technology* 43: 1804–1810.

Chesson, J. (1978). Measuring preference in selective predation. *Ecology* 59: 211–215.

Chesson, J. (1983). The estimation and analysis of preference and its relationship to foraging models. *Ecology* 64: 1297–1304.

Chevillot, X., Tecchio, S., Chaalali, A. et al. (2019). Global changes jeopardize the trophic carrying capacity and functioning of estuarine ecosystems. *Ecosystems* 22: 473–495.

Choy, E., An, S., and Kang, C.-K. (2008). Pathways of organic matter through food webs of diverse habitats in the regulated Nakdong River estuary (Korea). *Estuarine, Coastal and Shelf Science* 78: 215–226.

Christensen, V. and Pauly, D. (1992). ECOPATH II—a software for balancing steady-state ecosystem models and calculating network characteristics. *Ecological Modelling* 61: 169–185.

Christensen, V. and Walters, C. (2004). Ecopath with Ecosim: methods, capabilities and limitations. *Ecological Modelling* 172: 109–139.

Christensen, V., Walters, C., and Pauly, D. (2004). *Ecopath with EcoSim: a User's Guide. Fisheries Centre Research Reports*. Vancouver, Canada: University of British Columbia.

Christian, R., Forés, E., Comin, F. et al. (1996). Nitrogen cycling networks of coastal ecosystems: influence of trophic status and primary producer form. *Ecological Modelling* 87: 111–129.

Christian, R., Baird, D., Luczkovich, J. et al. (2005). Role of network analysis in comparative ecosystem ecology of estuaries. In: *Aquatic Food Webs* (ed. A. Belgrano, U. Scharler, J. Dunne and R. Ulanowicz). Oxford: Oxford University Press.

Citadin, M., Costa, T.M., and Netto, S.A. (2018). Response of estuarine meiofauna communities to shifts in spatial distribution of keystone species: an experimental approach. *Estuarine, Coastal and Shelf Science* 212: 365–371.

Cloern, J. (1982). Does the benthos control phytoplankton biomass in South San Francisco Bay? *Marine Ecology Progress Series* 9: 191–202.

Cohen, R.R.H., Dresler, P.V., Phillips, E.J.P., and Cory, R.L. (1984). The effect of the Asiatic clam, Corbicula-fluminea, on phytoplankton of the Potomac River, Maryland. *Limnology and Oceanography* 29: 170–180.

Dalsgaard, J., St. John, M., Kattner, G. et al. (2003). Fatty acid trophic markers in the pelagic marine environment. *Advances in Marine Biology* 46: 225–340.

Dame, R. and Patten, B. (1981). Analysis of energy flows in an intertidal oyster reef. *Marine Ecology Progress Series* 5: 115–124.

Darnell, R. (1961). Trophic spectrum of an estuarine community, based on studies of Lake Pontchartrain, Louisiana. *Ecology* 42: 553–568.

Duffy, M., Hall, S., Tessier, A., and Huebner, M. (2005). Selective predators and their parasitized prey: are spidemics in zooplankton under top-down control. *Limnology and Oceanography* 50: 412–420.

Ederington, M., McManus, G., and Harvey, H. (1995). Trophic transfer of fatty acids, sterols, and a triterpenoid alcohol between bacteria, a ciliate, and the copepod *Acartia tonsa*. *Limnology and Oceanography* 40: 860–867.

Edwards, M. and Richardson, A. (2004). Impact of climate change on marine pelagic phenology and trophic mismatch. *Nature* 430: 881–884.

Elser, J. and Urabe, J. (1999). The stoichiometry of consumer-driven nutrient recycling: theory, Observations and Consequences. *Ecology* 80: 735–751.

Elton, C. (1927). *Animal Ecology*. New York: Macmillan.

Estes, J., Tinker, M., Williams, T., and Doak, D. (1999). Killer whale predation on sea otters linking oceanic and nearshore ecosystems. *Science* 282: 473–476.

Fernandes, J.A., Cheung, W.W.L., Jennings, S. et al. (2013). Modelling the effects of climate change on the distribution and production of marine fishes: accounting for trophic interactions in a dynamic bioclimate envelope model. *Global Change Biology* 19: 2596–2607.

Finn, J. (1976). Measure of ecosystem structure and function derived from the analysis of flows. *Journal of Theoretical Biology* 56: 363–380.

Friere, K., Christensen, V., and Pauly, D. (2007). Assessing fishing policies for northeastern Brazil. *Pan-American Journal of Aquatic Sciences* 2: 113–130.

Fry, B. (1988). Food web structure on Georges Bank from stable C, N, and S isotopic compositions. *Limnology and Oceanography* 33: 1182–1190.

Fry, B. (2006). *Stable Isotope Ecology*. New York: Springer.

Galván, K., Fleeger, J., and Fry, B. (2008). Stable isotope addition reveals dietary importance of phytoplankton and microphytobenthos to saltmarsh infauna. *Marine Ecology Progress Series* 359: 37–49.

Gascuel, D. (2005). The trophic-level based model: a theoretical approach of fishing effects on marine ecosystems. *Ecological Modelling* 189: 315–332.

Grabowski, J.H. (2004). Habitat complexity disrupts predator–prey interactions but not the trophic cascade on oyster reefs. *Ecology* 85: 995–1004.

Guénette, S., Christensen, V., and Pauly, D. (2008). Trophic modelling of the Peruvian upwelling ecosystem: towards reconciliation of multiple datasets. *Progress in Oceanography* 79: 326–335.

Gulland, J. 1965. Estimation of mortality rates. Annex to Arctic Fisheries Working Group Report, ICES CM 15 Doc No. 3: 9 pp.

Haas, H., Rose, K., Fry, B. et al. (2004). Brown shrimp on the edge: linking habitat to survival using an individual-based simulation model. *Ecological Applications* 14: 1232–1247.

Hagy, J. III (2002). *Eutrophication, Hypoxia and Trophic Transfer Efficiency in Chesapeake Bay*. University of Maryland-College Park.

Hairston, N., Smith, F., and Slobodkin, L. (1960). Community structure, population control and competition. *The American Naturalist* 94: 421–425.

Hannon, B. (1973). The structure of ecosystems. *Journal of Theoretical Biology* 41: 535–546.

Harley, C.D.G., Randall Hughes, A., Hultgren, K.M. et al. (2006). The impacts of climate change in coastal marine systems. *Ecology Letters* 9: 228–241.

Harris, L., Duarte, C., and Nixon, S. (2006). Allometric laws and prediction in estuarine and coastal ecology. *Estuaries and Coasts* 29: 340–344.

Hartman, K. and Brandt, S. (1995). Trophic resource partitioning, diets, and growth of sympatric estuarine predators. *Transaction of the American Fisheries Society* 124: 520–537.

Heck, K. Jr. and Valentine, J. (2006). Plant-Herbivore interactions in seagrass meadows. *Journal of Experimental Marine Biology and Ecology* 330: 420–436.

Heck, K. Jr. and Valentine, J. (2007). The Primacy of top-down effects in shallow benthic ecosystems. *Estuaries and Coasts* 30: 371–381.

Heck, K. Jr., Carruthers, T., Duarte, C. et al. (2008). Trophic transfers from seagrass meadows subsidize diverse marine and terrestrial consumers. *Ecosystems* 11: 1198–1210.

Helland, I.P., Finstad, A.G., Forseth, T. et al. (2011). Ice-cover effects on competitive interactions between two fish species. *Journal of Animal Ecology* 80: 539–547.

Heymans, J., Guénette, S., and Christensen, V. (2007). Evaluating network analysis indicators of ecosystem status in the Gulf of Alaska. *Ecosystems* 10: 488–502.

Heymans, J.J., Coll, M., Link, J.S. et al. (2016). Best practice in Ecopath with Ecosim food-web models for ecosystem-based management. *Ecological Modelling* 331: 173–184.

Hummel, H. (1985). Food intake of *Macoma balthica* (Mullusca) in relation to seasonal changes in its potential food on a tidal flat in the Dutch Wadden Sea. *Netherlands Journal of Sea Research* 19: 52–76.

Hussey, N.E., Macneil, M.A., McMeans, B.C. et al. (2014). Rescaling the trophic structure of marine food webs. *Ecology Letters* 17: 239–250.

Hyslop, E.J. (1980). Stomach contents analysis-a review of methods and their application. *Journal of Fish Biology* 17: 411–429.

Ihde, T.F. and Townsend, H.M. (2017). Accounting for multiple stressors influencing living marine resources in a complex estuarine ecosystem using an Atlantis model. *Ecological Modelling* 365: 1–9.

Ivlev, V. (1961). *Experimental Ecology of the Feeding of Fishes*. New Haven, CT: Yale University Press.

Jackson, J. (2001). What was natural in the coastal oceans. *Proceedings of the National Academy of Sciences* 98: 5411–5418.

Jochem, F. (2003). Photo- and heterotrophic pico- and nanoplankton in the Mississippi River plume: distribution and grazing activity. *Journal of Plankton Research* 25: 1201–1214.

Jung, S. and Houde, E. (2005). Fish biomass size spectra in Chesapeake Bay. *Estuaries* 28: 226–240.

Kanou, K., Sano, M., and Kohno, H. (2004). Food habits of fishes on unvegetated tindal mudflats in Tokyo Bay, central Japan. *Fisheries Science* 70: 978–987.

Kay, J., Graham, L., and Ulanowicz, R. (1989). A detailed guide to network analysis. In: *Network Analysis in Marine Ecology* (ed. F. Wulff, J. Field and K. Mann), 15–61. New York: Springer-Verlag.

Kemp, P., Sherr, B., Sherr, E., and Cole, J. (1993). *Handbook of Methods in Aquatic Microbial Ecology*. Boca Raton, FL: Lewis Publishers.

Kerr, S. and Dickie, L. (2001). *The Biomass Spectrum: A Predator-Prey Theory of Aquatic Production*. New York: Columbia University Press.

Kneib, R. (1997). The role of tidal marshes in the ecology of estuarine nekton. *Oceanography and Marine Biology: An Annual Review* 35: 163–220.

Knight, B. and Jiang, W. (2009). Assessing primary production constraints in New Zealand Fisheries. *Fisheries Research* 100: 15–25.

Kortsch, S., Primicerio, R., Fossheim, M. et al. (2015). Climate change alters the structure of arctic marine food webs due to poleward shifts of boreal generalists. *Proceedings of the Royal Society B* 282: 20151546.

Lafferty, K., Allesina, S., Arim, M. et al. (2008). Parasites in food webs: the ultimate missing links. *Ecology Letters* 11: 533–546.

Landry, M. (1977). A review of important concepts in the trophic organization of pelagic ecosystems. *Helgoländer wiss. Meeresunters* 30: 8–17.

Laurans, M., Gascuel, D., Chassot, E., and Thiam, D. (2004). Changes in the trophic structure of fish demersal communities in West Africa in the three last decades. *Aquatic Living Resources* 17: 163–173.

Leguerrier, D., Niquil, N., Petiau, A., and Bodoy, A. (2004). Modeling the impact of oyster culture on a mudflat food web in Marennes-Oléron Bay (France). *Marine Ecology Progress Series* 273: 147–162.

Leontief, W. (1951). *The Structure of the American Economy, 1919–1939*, 2e. New York: Oxford University Press.

Libralato, S., Christensen, V., and Pauly, D. (2006). A method for identifying keystone species in food web models. *Ecological Modelling* 195: 153–171.

Lindeman, R. (1942). The trophic-dynamic aspect of ecology. *Ecology* 23: 399–417.

Lotze, H.K., Coll, M., and Dunne, J.A. (2011). Historical changes in marine resources, food-web structure and ecosystem functioning in the Adriatic Sea, Mediterranean. *Ecosystems* 14: 198–222.

MacArthur, R. (1955). Fluctuations of animal populations and a measure of community stability. *Ecology* 54: 533–536.

Manickchand-Heileman, S., Soto, L., and Escobar, E. (1998). A preliminary trophic model of the continental shelf, south-western Gulf of Mexico. *Estuarine, Coastal and Shelf Science* 46: 885–899.

Martinez, N. and Lawton, J. (1995). Scale and food-web structure—from local to global. *Oikos* 73: 148–154.

Maureaud, A., Gascuel, D., Colleter, M. et al. (2017). Global change in the trophic functioning of marine food webs. *PLoS One* 12: e0182826.

McCann, K. (2012). *Food Webs*. Princeton, NJ: Princeton University Press.

McCutchan, J.H. Jr., Lewis, W.M., Kendall, C., and McGrath, C.C. (2003). Variation in trophic shift for stable isotope ratios of carbon, nitrogen, and sulfur. *Oikos* 102: 378–390.

McMahon, K.W. and Newsome, S.D. (2019). Amino acid isotope analysis: a new frontier in studies of animal migration and foraging ecology. In: *Tracking Animal Migration with Stable Isotopes*, 2e (ed. K.A. Hobson and L.I. Wassenaar), 173–190. Acadmic Press.

McManus, G. and Ederington-Cantrell, M. (1992). Phytoplankton pigments and growth rates, and microzooplankton grazing in a large temperate estuary. *Marine Ecology Progress Series* 87: 77–85.

Micheli, F. (1999). Consumer-resource dynamics in marine pelagic ecosystems. *Science* 285: 1396–1398.

Michener, R. and Lajtha, K. (2007). *Stable Isotopes in Ecology and Environmental Science*, 2e. Malden, MA: Blackwell Publishers.

Middelburg, J.J. (2014). Stable isotopes dissect aquatic food webs from the top to the bottom. *Biogeosciences* 11: 2357–2371.

Mills, L., Soulé, M., and Doak, D. (1993). The keystone-species concept in ecology and conservation. *BioScience* 43: 219–224.

Morley, J.W., Selden, R.L., Latour, R.J. et al. (2018). Projecting shifts in thermal habitat for 686 species on the North American continental shelf. *PLOS One* 13: e0196127.

Niiranen, S., Yletyinen, J., Tomczak, M.T. et al. (2013). Combined effects of global climate change and regional ecosystem drivers on an exploited marine food web. *Global Change Biology* 19: 3327–3342.

Nixon, S. (1982). Nutrient dynamics, primary production and fisheries yields of lagoons. *Oceanologica Acta* Proceedings of the International Symposium on coastal lagoons SCOR/IABO/UNESCO: 357–371.

Nixon, S. (1988). Physical energy inputs and the comparative ecology of lake and marine ecosystems. *Limnology and Oceanography* 33: 1005–1025.

Odum, H. (1957). Trophic structure and productivity of Silver Springs, Florida. *Ecological Monographs* 27: 55–112.

Odum, H. (1967). Biological circuits and the marine systems of Texas. In: *Pollution and Marine Ecology* (ed. T. Olson and F. Burgess), 99–157. New York: Wiley Interscience.

Odum, E. (1971). *Fundamentals of Ecology*. Philadelphia: W.B. Saunders.

Odum, E. and de la Cruz, A. (1963). Detritus as a major component of ecosystems. *AIBS Bulletin* 13: 39–40.

Odum, W. and Heald, E. (1975). The detritus-based food web of an estuarine mangrove community. In: *Estuarine Research, Volume I. Chemistry, Biology and the Estuarine System* (ed. L. Cronin), 265–286. New York: Academic Press.

van Oevelen, D., Van den Meersche, K., Meysman, F.J.R. et al. (2009). Quantifying food web flows using linear inverse models. *Ecosystems* 13: 32–45.

Otto, S.B., Rall, B.C., and Brose, U. (2007). Allometric degree distributions facilitate food-web stability. *Nature* 450: 1226–1229.

Paerl, H., Valdes, L., Joyner, A. et al. (2006). Ecological response to hurricane events in the Pamlico Sound system, North Carolina, and implications for assessment and management in a regime of increased frequency. *Estuaries and Coasts* 29: 1033–1045.

Paine, R. (1966). Food web complexity and species diversity. *The American Naturalist* 100: 65–75.

Paine, R. (1969). A note on trophic complexity and community stability. *The American Naturalist* 103: 91–93.

Parsons, T., Maita, Y., and Lalli, C. (1984). *A manual of Chemical and Biological Methods for Seawater Analysis*. New York: Pergamon Press.

Patrício, J. and Marques, J.C. (2006). Mass balanced models of the food web in three areas along a gradient of eutrophication symptoms in the south arm of the Mondego Estuary (Portugal). *Ecological Modelling* 197 (1): 21–34.

Pauly, D. and Christensen, V. (1995). Primary production required to sustain global fisheries. *Nature* 374: 255–257.

Pauly, D., Christensen, V., Dalsgaard, J. et al. (1998). Fishing down marine food webs. *Science* 279: 860–863.

Pauly, D., Christensen, V., and Walters, C. (2000). Ecopath, Ecosim, and Ecospace as tools for evaluating ecosystem impact of fisheries. *ICES Journal of Marine Science* 57: 697–706.

Perry, A.L., Low, P.J., Ellis, J.R., and Reynolds, J.D. (2005). Climate change and distribution shifts in marine fishes. *Science* 308: 1912–1915.

Peterson, B. and Fry, B. (1987). Stable isotopes in ecosystem studies. *Annual Review of Ecology and Systematics* 18: 293–320.

Peterson, B. and Howarth, R. (1987). Sulfur, carbon, and nitrogen isotopes used to trace organic matter flow in the salt-marsh estuaries of Sapelo Island, Georgia. *Limnology and Oceanography* 32: 1195–1213.

Pimm, S.L. (1980). Properties of food webs. *Ecology* 61: 219–225.

Pimm, S. and Lawton, J. (1977). Number of trophic levels in ecological communities. *Nature* 268: 329–331.

Platt, T. and Denman, K. (1977). Organisation in the pelagic ecosystem. *Helgoländer wiss. Meeresunters (Helgoland Marine Research)* 30: 575–581.

Polis, G. and Strong, D. (1996). Food web complexity and community dynamics. *The American Naturalist* 147: 813–846.

Polovina, J. (1984). Model of a coral reef ecosystem. *Coral Reefs* 3: 1–11.

Pomeroy, L. (1974). The ocean's food web, a changing paradigm. *BioScience* 24: 499–504.

Post, D.M. (2002a). The long and short of food-chain length. *Trends in Ecology & Evolution* 17: 269–277.

Post, D.M. (2002b). Using stable isotopes to estimate trophic position: models, methods, and assumptions. *Ecology* 83: 703–718.

Purcell, J. (1992). Effects of predation by the scyphomedusan Chrysaora quinquecirrha on zooplankton populations in Chesapeake Bay, USA. *Marine Ecology Progress Series* 87: 65–76.

Purcell, J. and Cowan, J. Jr. (1995). Predation by the scyphomedusan *Chrysaora quinquecirrha* on *Mnemiopsis leidyi* ctenophores. *Marine Ecology Progress Series* 129: 63–70.

Purcell, J., Shiganova, T., Decker, M., and Houde, E. (2001). The ctenophone Mnemiopsis in native and exotic habitats: U.S. estuaries versus the Black Sea basin. *Hydrobiologia* 451: 145–176.

Rilling, G. and Houde, E. (1999). Regional and temporal variability in distribution and abundance of bay anchovy (Anchoa mitchilli) eggs, larvae, and adult biomass in the Chesapeake Bay. *Estuaries* 22: 1096–1109.

Rolff, C. (2000). Seasonal variation in $\delta^{13}C$ and $\delta^{15}N$ of size-fractionated plankton at a coastal station in the northern Baltic proper. *Marine Ecology Progress Series* 203: 47–65.

Rooney, N., McCann, K., Gellner, G., and Moore, J.C. (2006). Structural asymmetry and the stability of diverse food webs. *Nature* 442: 265–269.

Ruess, L. and Chamberlain, P.M. (2010). The fat that matters: Soil food web analysis using fatty acids and their carbon stable isotope signature. *Soil Biology & Biochemistry* 42: 1898–1910.

Rybarczyk, H., Elkaim, B., Ochs, L., and Loquet, N. (2003). Analysis of the trophic network of a macrotidal ecosystem: the Bay of Somme (Eastern Channel). *Estuarine, Coastal and Shelf Science* 58: 405–421.

Ryther, J. (1969). Photosynthesis and fish production in the sea. *Science* 166: 72–76.

Sackett, D., Able, K., and Grothues, T. (2007). Dynamics of summer flounder, *Paralichthys dentatus*, seasonal migrations based on ultrasonic telemetry. *Estuarine, Coastal and Shelf Science* 74: 119–130.

Schroeder, L. (1981). Consumer growth efficiencies: their limits and relationships to ecological energetics. *Journal of Theoretical Biology* 93: 805–828.

Shannon, L., Coll, M., and Neira, S. (2009). Exploring the dynamics of ecological indicators using food web models fitted to time series of abundance and catch data. *Ecological Indicators* 9: 1078–1095.

Sheaves, M. (2009). Consequences of ecological connectivity: the coastal ecosystem mosaic. *Marine Ecology Progress Series* 391: 107–115.

Sheldon, R., Prakash, A., and Sutcliffe, W. Jr. (1972). The size distribution of particles in the ocean. *Limnology and Oceanography* 17: 327–340.

Slobodkin, L. (1960). Ecological energy relationships at the population level. *The American Naturalist* 94: 213–236.

St. John, M. and Lund, T. (1996). Lipid biomarkers: linking the utilization of fontal plankton biomass to enhanced condition of juvenile North Sea cod. *Marine Ecology Progress Series* 131: 75–85.

Steele, J. (1974). *The Structure of Marine Ecosystems*. Cambridge, MA: Harvard University Press.

Steffan, S.A., Chikaraishi, Y., Dharampal, P.S. et al. (2017). Unpacking brown food-webs: animal trophic identity reflects rampant microbivory. *Ecology and Evolution* 7: 3532–3541.

Sterner, R., Bajpai, A., and Adams, T. (1997). The enigma of food chain length: absence of theoretical evidence for dynamic constraints. *Ecology* 78: 2258–2262.

Szyrmer, J. and Ulanowicz, R. (1987). Total flows in ecosystems. *Ecological Modelling* 35: 123–136.

Teal, J. (1962). Energy flow in the salt marsh ecosystem of Georgia. *Ecology* 43: 614–624.

Tecchio, S., Rius, A.T., Dauvin, J.C. et al. (2015). The mosaic of habitats of the Seine estuary: insights from food-web modelling and network analysis. *Ecological Modelling* 312: 91–101.

Tenore, K., Hanseon, R., McClain, J. et al. (1984). Changes in composition and nutritional value to a benthic deposit feeder of decomposing detritus pools. *Bulletin of Marine Science* 35: 299–311.

Testa, J., Kemp, W., Boynton, W., and Hagy, J. III (2008). Long-term changes in water quality and productivity in the Patuxent River estuary: 1985–2003. *Estuaries and Coasts* 31: 1021–1037.

Thompson, R., Mouritsen, K., and Poulin, R. (2005). Importance of parasites and their life cycle characteristics in determining the structure of a large marine food web. *Journal of Animal Ecology* 74: 77–85.

Thomson, J. (1966). The grey mullets. *Oceanography and Marine Biology Annual Review* 4: 301–335.

Tsagarakis, K., Coll, M., Giannoulaki, M. et al. (2010). Food web traits of the north Aegean sea ecosystem (Eastern Mediterranean) and comparison with other Mediterranean ecosystems. *Estuarine, Coastal and Shelf Science* 88: 233–248.

Ulanowicz, R. (1986). *Growth and Development, Ecosystems Phenomenology*. Lincoln, NE: ToExcell Press.

Ulanowicz, R. (1997). *Ecology, the Ascendent Perspective*. New York: Columbia University Press.

Ulanowicz, R. (2004). Quantitative methods for ecological network analysis. *Computation Biology and Chemistry* 28: 321–339.

Ulanowicz, R. (2009). The dual nature of ecosystem dynamics. *Ecological Modelling* 220: 1886–1892.

Ulanowicz, R. and Kemp, W. (1979). Toward canonical trophic aggregations. *The American Naturalist* 114: 871–883.

Ulanowicz, R. and Puccia, C. (1990). Mixed trophic impacts in ecosystems. *Coenoses* 5: 7–16.

Ullah, H., Nagelkerken, I., Goldenberg, S.U., and Fordham, D.A. (2018). Climate change could drive marine food web collapse through altered trophic flows and cyanobacterial proliferation. *PLOS Biology* 16: e2003446.

Valiela, I. (1995). *Marine Ecological Processes*, 2e. New York: Springer.

Valls, A., Gascuel, D., Guénette, S., and Francour, P. (2012). Modeling trophic interactions to assess the effects of a marine protected area: case study in the NW Mediterranean Sea. *Marine Ecology Progress Series* 456: 201–214.

Vander Zanden, M.J. and Rasmussen, J.B. (2001). Variation in δ^{15}N and δ^{13}C trophic fractionation: implications for aquatic food web studies. *Limnology and Oceanography* 46: 2061–2066.

Vézina, A. and Platt, T. (1988). Food web dynamics in the ocean. I. best-estimates of flow networks using inverse methods. *Marine Ecology Progress Series* 42: 269–287.

Walters, C., Christensen, V., and Pauly, D. (1997). Structuring dynamic models of exploited ecosystems from trophic mass-balance assessments. *Reviews in Fish Biology and Fisheries* 7: 139–172.

Walters, C., Pauly, D., and Christensen, C. (2000). Ecospace: prediction of mesoscale spatial patternsin trophic relationships of exploited ecosystems, with emphasis on the impacts of marine protected areas. *Ecosystems* 2: 539–554.

Wells, R., Cowan, J., and Fry, B. (2008). Feeding ecology of red snapper *Lutjanus campechanus* in the northern Gulf of Mexico. *Marine Ecology Progress Series* 361: 213–225.

Whiteman, J.P., Smith, E.A.E., Besser, A.C., and Newsome, S.D. (2019). A guide to using compound-specific stable isotope analysis to study the fates of molecules in organisms and ecosystems. *Diversity-Basel* 11 (1): 8.

Woodland, R.J. and Cook, P.L.M. (2014). Using stable isotope ratios to estimate atmospheric nitrogen fixed by cyanobacteria at the ecosystem scale. *Ecological Applications* 24: 539–547.

Woodland, R.J., Magnan, P., Glemet, H. et al. (2012). Variability and directionality of temporal changes in δ^{13}C and δ^{15}N of aquatic invertebrate primary consumers. *Oecologia* 169: 199–209.

CHAPTER 16

Estuarine Ecological Modeling

NOAA Ship Okeanos Explorer photographed in the Gulf of Mexico in 2012. Photo credit: NOAA / Flickr / CC BY-SA 2.0.

Kenneth A. Rose[1], Enrique Reyes[2], and Dubravko Justić[3]
[1]University of Maryland Center for Environmental Science, Horn Point Laboratory, Cambridge, MD, USA
[2]Department of Biology, East Carolina University, Greenville, NC, USA
[3]Department of Oceanography & Coastal Sciences, School of the Coast & Environment, Louisiana State University, Baton Rouge, LA, USA

16.1 Introduction

Estuarine ecosystems are composed of many biotic and abiotic components (e.g., nutrients, organisms) that interact and affect each other in complex ways. The movement of water, cycling of nutrients and sediments, and food web dynamics of microbes, phytoplankton, zooplankton, benthos, and fish vary in time and space within estuaries. All of these components (expressed as densities, concentrations, or abundances) show variability from scales of seconds and millimeters to decades and kilometers. The processes that affect these components (often expressed as rates per unit time) provide linkages that dictate how one component affects another component. Examples include water velocities affecting the transport of phytoplankton, freshwater inflows from rivers causing stratification of the water column (reduced mixing between surface and bottom layers) that determines oxygen concentrations near the bottom, and zooplankton grazing on phytoplankton affecting zooplankton growth and phytoplankton mortality. As these components and their process linkages interact with each other, they create a hierarchy of dynamics in which space and time play critical parts. A major challenge in the study of ecology is to determine which of the components and processes are important to examine for a given question, as the possibilities are practically infinite.

One common approach for studying this complexity is to develop mathematical (simulation) models of these ecosystems. Simulation or mechanistic models are by definition simplifications of reality and, as such, describe only the most important features and interactions (process linkages) in a system. Models provide a methodology that allows for data synthesis and hypothesis testing under the strictly controlled conditions of the virtual world, which is not possible with field measurements. Simulations can be performed by varying single factors (e.g., warming), as well as combinations of factors (e.g., warming and flooding), in ways that have never been observed in nature. Mechanistic models are also very useful in exploring future scenarios (e.g., climate change) and can also be used via hindcasting to better understand what caused the past behavior of the ecosystem.

In addition to generating predictions, mechanistic modeling provides tests of our knowledge and understanding about a system. Modeling forces one to propose a mathematical representation of our understanding of an ecosystem for all components and processes that have been deemed to be important

Estuarine Ecology, Third Edition. Edited by Byron C. Crump, Jeremy M. Testa, and Kenneth H. Dunton.

Companion website: www.wiley.com/go/crump/estuarine3

enough to be included in the model. This representation is created by first defining the components to be tracked through time and space, and then defining the processes that link the components to the environment and to each other. Such model development eliminates vagueness in what one thinks is important because it must be written down (often in mathematical terms that are very precise). This allows the modeler to communicate the assumptions to others and ensures there are no missing gaps. Calculations to solve a model do not allow for missing values or for some process calculations to be skipped. Modeling thus highlights what we know about ecosystems while also clearly identifying those processes and components that we do not know enough about.

Research and development in the mechanistic modeling of estuaries has paralleled developments in estuarine physics, chemistry, and biology. Holistic perspectives of estuarine ecosystems in the 1970s required the construction of conceptual and simulation models that tested current knowledge and directed future research. As this holistic approach gained popularity (Odum 1960), estuarine systems became proving grounds for modeling efforts. Among the first mechanistic modeling studies were those of Teal (1962) working in salt (or saline) marshes, Pomeroy (1972) examining phosphorus flux in estuaries, and Nixon and Oviatt (1973) modeling *Spartina.* Wiegert et al. (1975) and Ulanowicz et al. (1975) combined an empirical microcosm phytoplankton experiment with a simulation model. Dame (1979) is an excellent source of information on these early estuarine modeling efforts.

Due to developments in computer technology and our ability to measure components and processes at multiple time and space scales, the use of simulation models has grown explosively over the last four decades. For example, in their review of estuarine biogeochemistry models over 20 years ago, Jorgensen and Bendoricchio (2001) listed more than 50 different models. Most of these models were concerned with specific problems, such as predicting the effects of eutrophication on water quality or simulating the cycling and fate of toxic compounds in an ecosystem. Now there are hundreds to thousands of ecological models of estuarine systems, including some developed from earlier versions and others newly developed, and depositories for models (e.g., GitHub) to encourage sharing among scientists.

16.2 Classes of Models

The discussion above was focused on mechanistic models, but there are a variety of modeling approaches that are commonly used in estuarine ecosystems. Models can be generally divided into two categories: ***conceptual*** (qualitative, diagrammatic, and textual) and ***mathematical*** (quantitative, statistical, and mechanistic). Conceptual models often use flow diagrams to illustrate the relationships among model components (shown as boxes) and processes that show energy or mass transfers among components (arrows from one box to another box). Conceptual models have also been used to summarize or communicate scientific research in estuaries, although they are not always formally described. For example, when researchers are deciding where and when to make measurements of nutrients or fish in the field, they often have a conceptual model in mind that guides their decisions about locating sampling stations and how often to sample, but rarely is the conceptual model documented in writing. An example of a conceptual model is shown in Figure 16.5 as part of example 1. Formally describing conceptual models with diagrams and descriptive text about why the components and processes were included and which ones were not included, and the sources of information used, can greatly help all studies of estuaries, not just modeling. Formally described conceptual models have also become a communication tool to explain the basis for projects that involve large public investment of money, such as ecosystem restoration efforts (Ogden et al. 2005; Twilley 2004; DiGennaro et al. 2012). Various conceptual models are presented in this book, for example Box 14.3 and Figures 3.8, 3.12, 12.3, 12.5, 14.2, 15.2, 15.3, and 15.8. The idea of requiring modelers and collaborators to carefully document and agree on the structure and details of a formal conceptual model helps clarify ideas, assumptions, and certainties and uncertainties (what we know well and what we do not know).

The second type of models are mathematical models that typically include equations to represent the dynamic interactions within ecosystems. Mathematical models can be further subdivided into statistical and mechanistic models. Statistical models are often single equations that are built primarily from empirical data like those obtained from field monitoring. Statistical modeling focuses on deriving relationships (equations) among the components (boxes) identified in the conceptual diagram, such as a correlation between the abundance of prey (one box) and the abundance of a predator (another box) over time or a regression equation relating the two (i.e., $\text{Predator}_t = \beta_o + \beta_1 \cdot \text{Prey}_t$). Mechanistic models are process-based simulation models that are built upon how we believe a population or ecosystem functions. Mechanistic models focus on the process linkages (the arrows connecting the boxes) and represent the rates of flow of material or energy among the boxes as mathematical equations and then solve those equations to obtain the values of the variables shown as boxes. These two approaches are not completely distinct from each other, as they really form two clusters of approaches that overlap with each other in various ways. For example, some statistical equations include simplified representations of processes (i.e., functions beyond the linear form), and mechanistic models often use fitted statistical equations to represent specific processes as part of their equations of rates of change.

Mechanistic models can be viewed as going to the next step of assigning numerical values to the components and process flows (boxes and arrows) in a conceptual model. The step of going from conceptual to mechanistic model is not as simple as it sounds and involves synthesis of many studies, decisions about the structure of the equations, and knowledge about the mathematics to be used. The benefit is that when successfully formulated, mechanistic models allow for quantitative predictions of how components will respond to stressors and other changes in the environment, and enable the evaluation of the consequences and effectiveness of management actions. Such quantitative predictions are impossible with conceptual models and possible but limited with statistical models. Statistical equations are excellent ways to describe the patterns and correlations among the components; however, the relationships among variables represented in statistical equations are not necessarily causal and thus predicting system responses to new conditions can be limited. Mechanistic models try to go beyond statistical models and include cause-and-effect relationships in the equations.

While there are recommended steps for developing mechanistic models (described below), the selection of a specific mechanistic model is much more dependent on judgments by the model developer than for statistical modeling, which has standard protocols. Further, demonstrating how well a mechanistic model performs (calibration and validation) is not as clean and straightforward as reporting the goodness of fit of a statistical equation. Judging mechanistic models often involves comparing components of the model to data that come from multiple sources thereby hindering simple model versus data comparisons. The general role of modeling in advancing ecology is well established (Caswell 1988; Breckling et al. 2011); however, the debate between statistical versus mechanistic approaches (i.e., how complex to make models) continues to this day (Weisberg 2006; Getz et al. 2018; Larsen et al. 2016).

Mechanistic models are the focus in this chapter because they are less familiar than statistical modeling to scientists and because they serve a valuable role in estuarine ecology. The general approach for developing and using mechanistic models is described below, followed by several examples. Real-world examples are used because the general guidelines for developing and using mechanistic models are vague. These real-world examples show very clearly how mechanistic models are developed and used, and they illustrate several key concepts about mechanistic modeling.

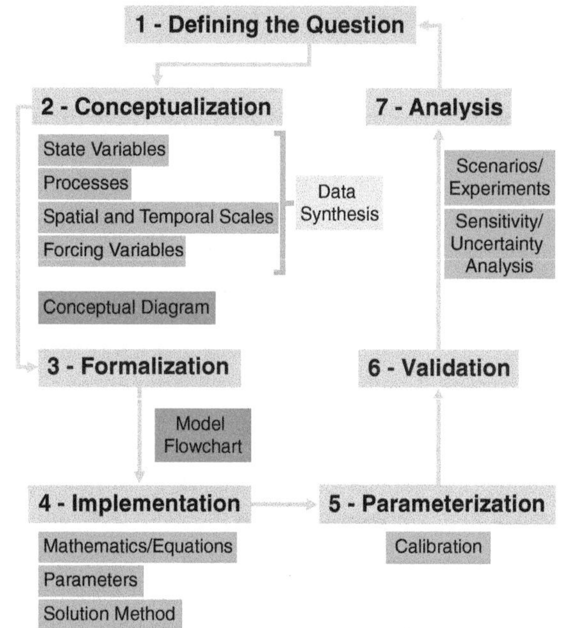

FIGURE 16.1 General steps in the development and use of mechanistic ecological models. The blue boxes are the major steps, with sub-steps under each shown in teal. The red boxes show some specific products that should be formally produced and documented.

16.3 Developing and Using Mechanistic models

Estuarine models are typically constructed using a series of highly interrelated steps (Figure 16.1), and common terms used with mechanistic modeling are defined in Box 16.1. These steps are often repeated and, in practice, models are developed by going through these steps many times and sometimes not in the same logical order as shown. Developing a mechanistic model is an iterative procedure.

There have been multiple attempts to formalize how mechanistic models should be developed, calibrated, validated, and used to make predictions. If there were a recipe that could be used by everyone, it would help standardize mechanistic models, reduce the role of judgment by model developers, and ease the communication of the results. All modeling uses judgment, including statistical modeling, but it is the high degree of judgment involved that can hinder the confidence in mechanistic models. Many of the decisions by model developers also go undocumented so it can be a bit mysterious to others, especially non-modelers, how and why a specific mechanistic model was developed (i.e., they just see the final version). To many audiences, one can think of each mechanistic model as being a special situation. However, to model developers, there is recognition of some general principles, repeatedly used formulations, and mathematics that are universally known.

Examples of proposed ways to develop mechanistic models, often referred to as "best practices," are numerous (Acevedo 2013; Jakeman et al. 2006). These all involve some version of following a series of general steps, sometimes tailored to the specific goal of the modeling. For example, the proposed scheme by Rose et al. (2015) was specific to models of fish and shellfish population and food web dynamics being used to quantify the benefits of habitat restoration; other schemes are for more general situations. Regardless, they all attempt to provide a way to document and make transparent the development of a mechanistic model through these general steps: (1) define the question, (2) conceptualize the model, (3) formalize the model often by adding many details to the conceptual model on the way to a set of equations (like a flowchart of calculations), (4) implement the model by finishing the formalization by specifying exact mathematical equations and then converting those into computer code, (5) parameterize the model via calibration, (6) validate the model, and, finally, and (7) analyze the model to answer the question.

Box 16.1 Key terms used with describing mechanistic models.

Mechanistic models are often described using several types of variables that define the role played by those variables within the model. These variables are:

State variables—These are the quantities being simulated and will typically appear on the left hand side of equations that define their rates of change.

Forcing or driving variables—These are often environmental variables (e.g., temperature, light), but can be other variables such as the flux of a state variable into a modeled area. The key distinction that defines forcing variables is that they affect state variables and processes, but are not themselves affected by the dynamics of the model.

Processes—These are the mechanisms that appear on the right side of equations that use parameters, state variables, and forcing variables to determine the rate of change of the state variable that is on the left-hand side of the equation. For example, feeding rate and metabolic rate are processes that update the state variable of organism body weight.

Parameters—These are the constants that appear in the mathematical representations of the processes. Values of parameters are fixed and generally do not vary in time or space.

Formalization—The step in model development when more and more details are added to the conceptual vision of the model, including explicitly defining the state variables, forcing functions, processes, and spatial and temporal dimensions.

Implementation—The step in model development often done simultaneously with Formalization when the details are converted to a system of coupled mathematical equations and further details about consistency of units, temporal and spatial scales, and feedbacks are confirmed and solution techniques are tested.

Difference and differential equations—Two types of mathematical systems of equations that are commonly used in many mechanistic ecology models. Difference means the values of state variables are solved in discrete time steps, while differential involves solutions that generate values of state variables in continuous time (and space).

Calibration (Parameterization)—The step in model development where values for all parameters in the model are determined. Models consist of a mix of parameters with known values (e.g., physical constants) and those obtained from studies that range from well-known to highly uncertain. The model is run many times (either by user or automatically with a search algorithm) until model predictions are judged sufficiently similar to data and other empirically based information (e.g., known habitat preferences).

Validation—The step in model development when a calibrated model is run under new conditions (which are different from those used for calibration) and its performance is evaluated by how well model predictions agree with the data.

Prediction—A general term used when the output of a model is examined and includes more specific terms of hindcasting, projection, and forecasting. A key distinguishing aspect is how time is displayed with the model predictions. A projection is typically labeled as years of the simulation (1, 2, 3, . . .), while forecasting indicates results for actual years (2010, 2011, . . .). Hindcasting is more often treated like forecasting, but simulations are for an historical time period.

Factors—A general term that refers to inputs or processes of a model when they are being manipulated in simulation experiments. Factors can be forcing functions (e.g., warming, river discharge) and processes (e.g., faster growth, impose higher mortality from hypoxia).

Conceptual model—A list or graphic image that shows the state variables, forcing functions, and processes, and their connections, of the system under study.

Feedbacks—Feedback dynamics emerge when the response of a state variable is amplified (positive) or dampened (negative) by a change in the system. Positive feedbacks are unstable because they encourage ever increasing changes in the system, while negative feedbacks are stabilizing because they work to keep the state variables near the original (before the changes) values.

Interactions—These occur between two variables when the response of one of the variables depends on the value of the other variable. For example, a mortality rate increases by only 20% when temperature at 10°C is increased by 2°C versus increasing by a much greater percentage (50%) when the same 2°C increase occurs at 15°C.

Sensitivity and Uncertainty analysis—A procedure by which parameters and other inputs are varied in their values to assess how state variables respond. Sensitivity analysis typically uses a set of simulations with user-defined incremental (e.g., ±10%) changes in a parameters or inputs, while uncertainty analysis often uses Monte Carlo to randomly vary many parameters and express state variable outputs as probability distributions.

Model currency—Units of the state variables, such as density (g/m^3) and abundance (total number of individuals on the grid).

16.3.1 Step 1: Defining the Question

The first step in the development of a mechanistic model is to define the question (or questions) to be answered and how those results (answers) will be used by others. The modeler has so much flexibility in developing a mechanistic model that the question is critical in order to know what components and processes are important to include. The more specific the question, the more guidance is given to narrow down the options that could be considered by the modeler and thereby helps make decisions about which components and processes to include and how to represent them. Real-world questions and how mechanistic models attempted to answer them are illustrated in the three examples described later in the chapter. If one does not start with the question, then it is likely that a model will not be the best for answering the questions defined later. While collaboration with experts about an ecosystem of interest are vital throughout the model development, getting good input from people (e.g., collaborators, managers, and stakeholders) on defining the question is equally critical.

Also important in defining the question is how the answers provided by the models will be used and by whom. At one extreme, the results may be used to advance theory, which can be addressed with a wide range of models and formulations, some well-grounded in data and others highly speculative. At the other extreme, the results are used directly by resource managers and regulators to make decisions that affect an ecosystem and people for years to decades. Such models must adhere more closely to the available data so that people (stakeholders and general public) have sufficient confidence in the model results. Many models fall near the middle of these two extremes in that they inform management decisions, but are not the only or even the major source of information used. Thus, model development must find the correct balance between high flexibility when testing new ideas versus ensuring the model is carefully checked with data for informing management.

16.3.2 Step 2: Conceptualization

The second step, model conceptualization, consists of many smaller steps: defining the important components (state variables), specifying the spatial and temporal scales, and deciding on model currency, key processes, and forcing variables (see Box 16.1). The main questions we use to conceptualize the model are: How do we think the system of interest works? What are the components and processes that greatly influence the dynamics of the system? An extremely important part of the conceptualization step is to continually refer back to the question being asked. Specific aspects of the conceptualization step are: (a) decide on the state variables of interest that one wants to predict over time or in space, (b) identify the major processes that need to be included to sufficiently predict how the state variables will change, (c) decide on the spatial and temporal scales of the model, and (d) identify environmental variables (forcing variables) that regulate these processes. One begins to develop the model in this step and often, as this is being done, developers look at the available data and empirical information. Conceptualization often involves developing a conceptual diagram of what should be represented in the model.

16.3.3 Step 3: Formalization

While free-thinking is encouraged in conceptualization, the formalization step is more constrained by available information and knowledge since it is when many details get added to the conceptual diagram. This step involves extensive review of field and laboratory studies, consideration of other similarly developed models, and often many discussions with experts about the system and the key organisms. First, we ask: What exactly are the state variables, processes, and forcing variables needed to sufficiently describe the conceptualized model? What information is known about these, both in general and for the specific system of interest? What aspects of the ecosystem can be ignored and not included in the model and why can they be ignored? Remember a model is, by definition, a simplification of the real ecosystem; thus, things known to operate in the real ecosystem must be left out or greatly simplified in a model in order to avoid unnecessarily complicated models. An important aspect of the formalization step is to list the assumptions that need to be made as a result of the simplifications that underlie the model.

Once the list of state variables is selected, the next step is to ask: How are the state variables affected by processes and do the desired spatial and temporal scales match what is known about their dynamics and the processes that affect and link them? Specification of the spatial scale involves deciding whether one, a few, or many spatial boxes need to be represented, the dimensions of each of these boxes, the total area to be covered by the spatial boxes (e.g., entire estuary or just the near-shore zone), and how to deal with the boundaries in terms of how the modeled area is linked to adjacent areas. For example, do materials or organisms enter and exit the modeled area, and if so, what is their rate of exchange? Specification of the temporal scales involves deciding on the time step of the model (how often the values of the state variables are updated) and how long (days, months, and years) should model simulations be.

The result of formalization is often a detailed flowchart of how calculations (with descriptions of equations) can be made that match the conceptual diagram. The flowchart is critical to determine (a) which components are important to include in the model, (b) how these components relate to each other, and (c) will these components contribute to a model that will be able to answer the question of interest. Sometimes the question of interest evolves during the development of the model. One must always ensure the conceptualized and formalized model remains appropriate for the current question of interest.

16.3.4 Step 4: Implementation

When a conceptualized and formalized model is ready to be expressed in mathematical terms, we begin the implementation step. The implementation step involves translating words, diagrams, and flowcharts into model equations. The model is a set of equations, often one per state variable, that have explicit mathematical representations of the processes and their links to the forcing variables and one another. Several types of mathematics can be used to express these equations, which are often grounded in calculus. Commonly used mathematics are differential equations and difference equations. Differential equations treat time as continuous so that predictions can be made at any time, while difference equations update or predict the dynamics through time in discrete time steps (e.g., hourly, daily).

Usually, the implementation step results in an equation (differential or difference) that describes the rate of change with time (and also space) of each state variable. With multiple state variables, one therefore obtains a system of equations in which the equation for a state variable may include terms that involve the values of other state variables. The specific types of mathematics used then determine the numerical techniques needed to solve these equations, and to thereby obtain values of the state variables over time and, if appropriate, over space. While the details of the mathematics and solution methods are often not described, it is important to be sure that one is obtaining the correct solution to the model equations. Also, as part of the implementation step, model parameters or coefficients that determine the rates of the processes, and thus the rates of change of the state variables, are carefully defined. Keeping track of the units throughout the calculations of the model is critical.

The appearance of state variables on the right-hand-side of the equations is how state variables affect each other and enables the model to represent the whole system. If state variables did not appear on the right-hand-side, then the model would be a set of independent predictor equations for each state variable. Not only do forcing variables affect state variables, but other state variables can affect state variables. The combination of complicated dependencies on forcing variables and on other state variables leads to interaction and feedback effects that cause the complex behavior of ecological systems and their associated mechanistic models (Anand et al. 2010).

Interaction effects occur when the response of a state variable to a change depends on the specific values of a forcing variable or the values of other state variables, often in a non-additive or nonlinear way. For example, decreases in dissolved oxygen in estuarine water (hypoxia) can become so severe that they cause a sudden change in benthic communities (a state variable) that then cascade to changes in other state variables in the ecosystem, such as water-column nutrient concentrations (Conley et al. 2009).

Feedback dynamics emerge when the response of a state variable is amplified (positive) or dampened (negative) by a change in the system. Positive feedbacks are unstable because they encourage ever increasing changes in the system, while negative feedbacks are stabilizing because they work to keep the state variables near the original (before the changes) values. An example of a positive feedback is how seagrasses trap suspended sediments, which in turn improves water clarity and increases seagrass growth, which leads to even more trapping of sediments and further increases growth (van der Heide et al. 2011). A common negative feedback is described by the relationship between a predator and its prey; increased prey abundance leads to more predators which then reduces prey abundance and eventually reduces predator abundance. A major consideration about how to include state variables in the equations is whether a feedback may emerge and, if so, whether the feedback should be positive or negative.

Computer-related questions are also dealt with in the implementation step, such as whether off-the-shelf computer software can be used to solve the equations or which programming language should be used to develop a customized code that solves the model equations. Some very simple models can be solved using mathematical techniques in which equations for rates of change in state variables are expressed as the values of state variables over time (a closed form solution). However, most mechanistic models are solved numerically by repeated calculations. A simple illustration of how models can be solved is shown in Box 16.2. While there are a limited number of ways to develop the model equations, it is good to work with or consult with quantitatively trained people to ensure the implementation is well grounded and solutions are numerically correct. The focus should be on getting the physics and biology as realistic as possible; the mathematical aspects should not get in the way.

16.3.5 Step 5: Parameterization

While values for some parameters can be estimated with great accuracy and precision, often some or many parameters are only known approximately. Some processes, and the associated parameters, can be specified from first principles, such as the number of calories in a gram of respired oxygen. Many processes use measurable parameters, but often these parameters either have not been measured, or have reported values for only a few conditions (e.g., one year in one place in the estuary), in the laboratory, or only in other ecosystems or for other species. Rarely are mechanistic models developed for which all parameters are known with great confidence. An important aspect of the parameterization step is therefore model calibration, in which values of imperfectly known model parameters are determined.

Calibration can involve simply making educated guesses at parameter values, or repeated adjustment of parameter values until the model behaves reasonably. Another approach is to apply automated calibration methods that use algorithms to search for a set of parameter values that work best for predicting state variables that closely match observed state variables (Janssen and Heuberger 1995; Rose et al. 2007). A key consideration in how to calibrate a model is the time required to run automated calibration methods, as they often involve thousands of model runs. For example, the model in Box 16.2 takes less than seconds to run, while other models can require hours to days, even on fast computers. The decision about how to calibrate a model depends on the model's mathematics, the complexity of its equations, the purpose of the model, and the quality and quantity of the available data. Often, parameter values are simply tried until model results sufficiently "agree" with the data. This step, whether done manually or with optimization methods, often leads to debate as to whether a model is adequately calibrated for generating predictions, especially those that affect management decisions.

16.3.6 Step 6: Validation

Confidence in models often depends on how well a model predicts values of state variables that are determined with data that were not used to construct or calibrate the model (i.e., measurements of state variables not used in the previous steps). This is commonly known as an independent test

Box 16.2 Lotka–Volterra Predator–Prey Model.

A common model that is used for teaching about ecological modeling is the Lotka–Volterra predator–prey model. The model has a long history in ecology (Wangersky 1978), including biographies and information about how the two scientists (Lotka and Volterra) separately developed the model (Anisiu 2014). The model consists of a pair of differential equations for the two state variables of number of prey (N) and number of predators (P) in an area.

$$\frac{dN}{dt} = rN - \alpha PN \tag{16.1}$$

$$\frac{dP}{dt} = c\alpha PN - qP \tag{16.2}$$

The processes in the model are the growth of the prey population in the absence of predators (r) and the decline of the predators in the absence of prey (q). They are coupled by how much the predator eats the prey. The assumption is that the predator encounters the prey in proportion to both the number of prey and predators present ($\alpha \cdot P \cdot N$), and then a fraction of that amount (c) become new predators.

Assigning values to the parameters and starting numbers of prey and predator enables illustration of how numerical solutions operate and also some simple behavior of a negative feedback. The solution to the equations can be approximated by using small increments of time (Δt):

$$N(t+\Delta t) = N(t) + \left[rN(t) - \alpha P(t)N(t)\right] \cdot \Delta t \tag{16.3}$$

$$P(t+\Delta t) = P(t) + \left[c\alpha P(t)N(t) - qP(t)\right] \cdot \Delta t \tag{16.4}$$

Equations (16.3) and (16.4) can be evaluated and updated repeatedly to obtain the prey and predator population abundances over time.

Assume that in absence of the other, the prey grows at rate of $r = 0.3$/day and the predators die at a rate $q = 0.1$/day. Further assume that $\alpha = 0.04$ and $c = 0.25$. Start the prey population at 50 individuals and the predator population at 15 individuals. Using a time step of 0.1 to step-through the equations, then the first two time steps are computed as follows:

Calculate the rates of change of prey and predator at time 1 (the brackets) using the parameter values and 50 prey and 15 predators.

The rate of change of the prey is: $0.3 \times 50 - 0.04 \times 50 \times 15 = 15.0 - 30.0 = -15.0$

The rate of change of the predator is: $0.25 \times 0.04 \times 50 \times 15 - 0.1 \times 15 = 7.5 - 1.5 = 6.0$

Using equations (16.3) and (16.4), update the populations at time 1.1 to be: $50 - 15.0 = 35.0$ and $15 + 6.0 = 21.0$

Now repeat and calculate the rates of changes at time 1.1 (−18.9 for prey and 5.25 for predators), and update again to get prey and predators for the next time 1.2: $35 - 18.9 = 16.1$ for prey and $21 + 5.25 = 26.25$ for predators. This evaluation of rates of change and updating of state variables continues until the solution is obtained.

Notice how the components (processes) in each equation are computed, then are combined to get the rates of changes of the state variables, and finally the last time step's values are updated to get values for the next time step. This type of solution method is called Euler and although it is helpful to demonstrate numerical approximation methods, is not recommend for use because the solution obtained is inaccurate. There are many more accurate methods than Euler.

Continuing with the example above, and using a more accurate numerical method, a simulation of N (prey) and P (predator) for 200 days is shown in Figure 16.2. Another way to look at the model dynamics, especially cycling like this example, is to plot the predator versus the prey for each day (Figure 16.3).

The negative feedback emerges when predators increase, their consumption causes the prey to decrease, and when prey get low enough, this causes the predators to decrease, which allows the prey to increase.

The Lotka–Volterra model has been modified to reflect more realism in the prey and predator growth, death, and interaction terms (Oaten and Murdoch 1975; Křivan 2007), has been extended to three interacting species (Korobeinikov and Wake 1999), and, in addition to a teaching tool, has been used to address management questions (May et al. 1979).

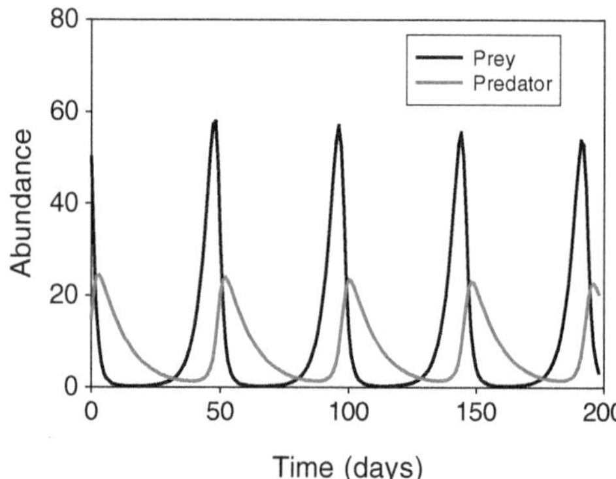

FIGURE 16.2 Solution of the Lotka-Volterra model for 200 days using an accurate numerical method.

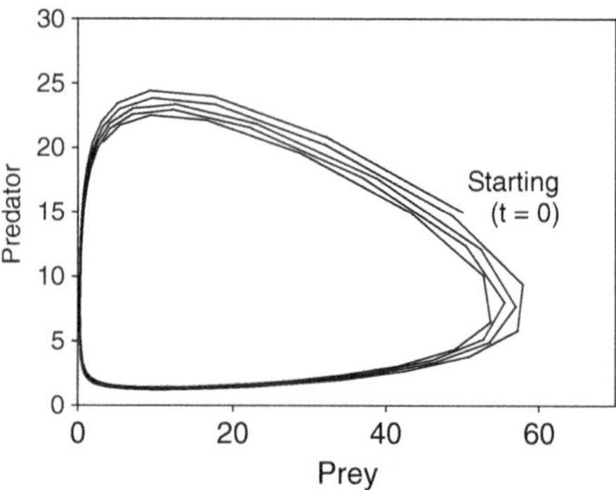

FIGURE 16.3 Phase plane plot (predator versus prey) of the 200-day solution of the Lotka-Volterra model shown in Figure 16.2.

of a model (Rykiel 1996). It is not always possible to rigorously test a model against independent data because there is not enough data to both calibrate and validate the model. But a good validation demonstrates the degree to which model predictions can be generalized and applicable beyond the conditions underlying the calibration (parameterization) step. While it is reassuring to have a model that accurately predict the values of state variables under new conditions, it is also useful when a model does poorly for some comparisons and thereby helps define the limits of the model and identifies aspects and assumptions that need further investigation.

Model validation almost always raises questions about the quality and quantity of the data. For example, when one models long-lived animals such as fish, the data are rarely sufficient for rigorous model validation and sometimes validation results in the question: Is the model wrong or are the data not representative of general conditions?

Despite the apparent simplicity of the term "validation," this is a major step that can often determine whether model predictions will be accepted by others. A plan for calibration and validation, sometimes even agreed to beforehand by the interested parties, is an effective way to ensure that the results will be accepted and properly interpreted.

16.3.7 Step 7: Analysis

A calibrated and validated model is then ready to be used to address the question of interest. Model analysis can take many forms ranging from simple simulations under a few sets of conditions (e.g., warming effects under a low and high freshwater input years) to a coordinated set of simulations using many variations in factors as if a laboratory experiment was being performed. Factors can include changes in forcing functions that generate the forcing variables (e.g., temperature and warming), parameters (e.g., increased mortality rate of key species), and management actions that can involve changes in multiple forcing functions and many parameters (e.g., increased habitat due to restoration, reduced harvest). Concepts from experimental design used with laboratory and field experiments can be applied to the analysis of a mechanistic model (Lawton 1995; Kleijen 1998). Use of formal design enables determination of the effects of individual factors on ecosystem response and the effects of changing combinations of factors simultaneously. Of course, these experiments are done in the virtual world, which has the advantage of being able to control everything in the simulated ecosystem. Predicting responses to combinations of factors not previously observed in the historical data are possible. The results of the analysis are only as good as the model is realistic, which goes back to how well all of the previous steps were done.

The analysis step often also includes sensitivity analysis (Cariboni et al. 2007; Saltelli et al. 2008), which involves varying model process parameters (and sometimes forcing functions) in a systematic way to determine which of the inputs varied have a large influence on model predictions, and to identify the range of parameter values under which the model generates realistic outcomes. Sensitivity analysis is typically done by varying each parameter in the model by a small, fixed amount. The change in model predictions of state variables between the original simulation and with a parameter increased by, for example 10%, is an index of the sensitivity of the model to that parameter. Ten percent changes in some parameters will cause large changes in model predictions, while 10% changes in other parameters will have little effect. Complex models generate predictions for multiple variables, and the importance of parameters can vary among the prediction variables. Uncertainty analysis is an extension of sensitivity analysis that uses more sophisticated ways to change parameter values, both singly and in combination, and changes parameter values by realistic amounts that reflect how the parameters vary in nature. Uncertainty analysis attempts to generate realistic variability in model predictions. The results of model experiments, combined with sensitivity and uncertainty analyses, are used to answer the question of interest.

16.3.8 Going Through the Steps

In practice, the development of a model is not a simple path from step 1 to step 7 as shown in Figure 16.1. Rather, model development is highly iterative; the results (successes and failures) at each step lead to revisiting earlier steps. For example, the parameterization and validation steps very often show differences between the model predictions and data that cause a reevaluation of the conceptualization and formalization steps.

The next section consists of three modeling examples that illustrate how models have been used to address important questions about estuaries and their responses to change. In all three examples, models were used because other methods (such as data collection) to answer the questions would be impossible, difficult, or expensive. Each example involved multiple passes through the model development steps. These models are still undergoing modifications, as more data and information become available and as questions of interest evolve and new questions are posed.

16.4 Example 1: Gulf of Mexico Hypoxia

16.4.1 Defining the Question

The large-scale hypoxic zone (<2 mg O_2 L^{-1}; Figure 16.4) in the northern Gulf of Mexico (NGOM), recently exceeding an area of 22,000 km^2 (**www.gulfhypoxia.net**), overlaps with the habitat and fishing grounds of commercially important fish and shrimp species. Hypoxia develops as the consequence of the high stability (strong stratification) of the water column combined with high primary productivity; high stability prevents mixing of oxygen-rich surface waters with the isolated bottom layer and high productivity in surface waters results in high carbon flux to the bottom layer where decomposition results in the depletion of oxygen (Rabalais et al. 1996). Hypoxia in the NGOM typically occurs from March through October in the bottom waters below the pycnocline, and extends between 5 and 60 km offshore (Rabalais et al. 2007). Retrospective analyses (Sen Gupta et al. 1996; Turner and Rabalais 1994) and model simulations (Justić et al. 2002) suggest that the Gulf's hypoxia intensified during the past five decades due to increased nutrient inputs from the Mississippi River and more balanced ratios of nutrients in the river water (Justić et al. 2005; Rabalais et al. 2007).

Because the NGOM supports important U.S. fisheries, NGOM hypoxia has received considerable attention from scientists and policy makers (Rabalais et al. 2002; Langseth et al. 2014; Smith et al. 2017). In 2001, the Mississippi River Watershed/Gulf of Mexico Hypoxia Task Force set a goal to reduce the 5-year running average of the NGOM's hypoxic zone to less than 5000 km² by the year 2015 (Task Force 2001, 2008, 2015). The proposed action plan suggested that a 30% decrease in the Mississippi River nitrogen loading to the NGOM would be required to reach this goal.

This plan addressed the linkage between river-borne anthropogenic nutrients and hypoxia, but it also recognized that hypoxia in the NGOM is sensitive to inter-annual variability in Mississippi River discharge (Figure 16.4). For example, during the drought of 1988 (a 52-year low discharge record for the Mississippi River), the areal extent of midsummer hypoxia was minimal. In contrast, during the flood of 1993 (a 62-year maximum discharge for August and September) the hypoxic zone was twice the 1985 to 1990 average area. Thus, the question to be addressed by this modeling example is, given the wide interannual variation in the size of the hypoxic zone due to climate, weather, river discharges, and other environmental factors, how will reducing nutrient loadings from the Mississippi River affect the size and duration of the hypoxic zone?

Several different hypoxia models have been developed for the NGOM over the past 20 years, ranging from simple statistical and box models to complex three-dimensional (3D) coupled hydrodynamic–biogeochemical models (Justić et al. 2007; Scavia et al. 2013; Greene et al. 2009; Forrest et al. 2011; Fennel et al. 2016). The examples presented here use two of these models, one simple and one complex, to illustrate modeling of hypoxia dynamics in the NGOM. Simple mechanistic models (Justić et al. 2002) can provide long-range (i.e., decadal) forecasts at the expense of understanding the importance of key parameters and forcing variables, while complex mechanistic models (Justić and Wang 2014) can help describe spatial patterns and dissect controls and consequences of hypoxia over hourly to annual time-scales.

16.4.2 Simple Two-layer Hypoxia Model

16.4.2.1 Conceptualization To address the question of how reducing nutrient loadings from the Mississippi River will affect hypoxia, a simple two-layer model for a specific location in the NGOM was developed (Figure 16.5; Justić et al. 1996, 2002).

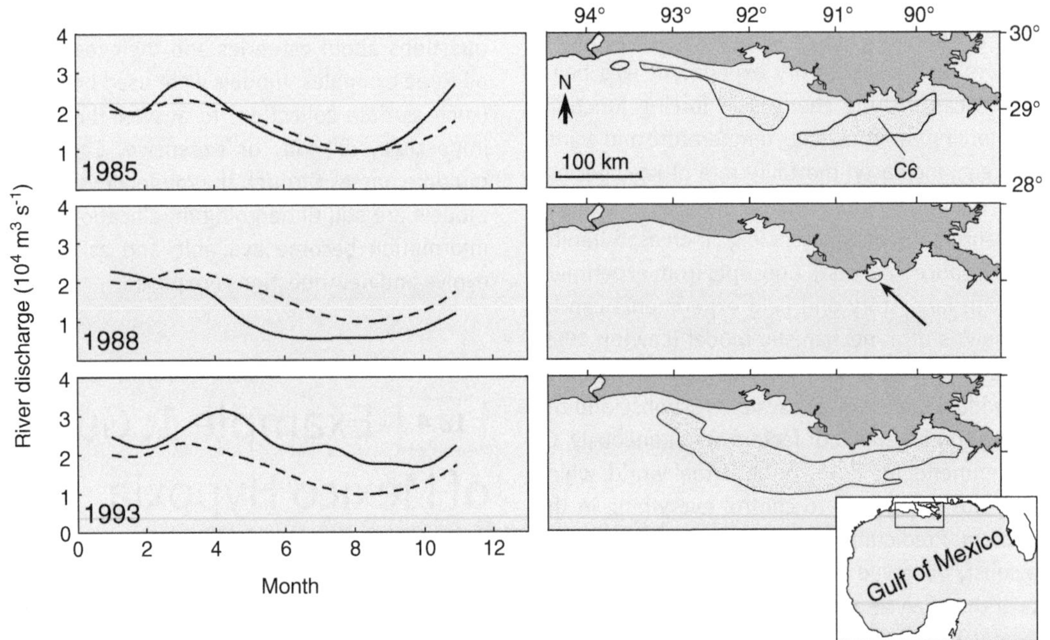

FIGURE 16.4 The Mississippi River discharge (Q, left panels) and corresponding areas of summertime bottom hypoxia (right panels) in the northern Gulf of Mexico during 1985, 1988, and 1993. The solid line represents the mean monthly discharge for a given year and the dashed line is the mean monthly discharge for the 1985–1993 period. The yellow shaded areas represent the distributions of bottom waters with dissolved oxygen concentration below 2 mg O_2 L^{-1}. Note that during 1988 hypoxia was observed only at one location off the Louisiana coast. The reference station C6 is indicated in the upper right panel. *Source*: From Justić et al. 2005, reprinted with permission from Elsevier.

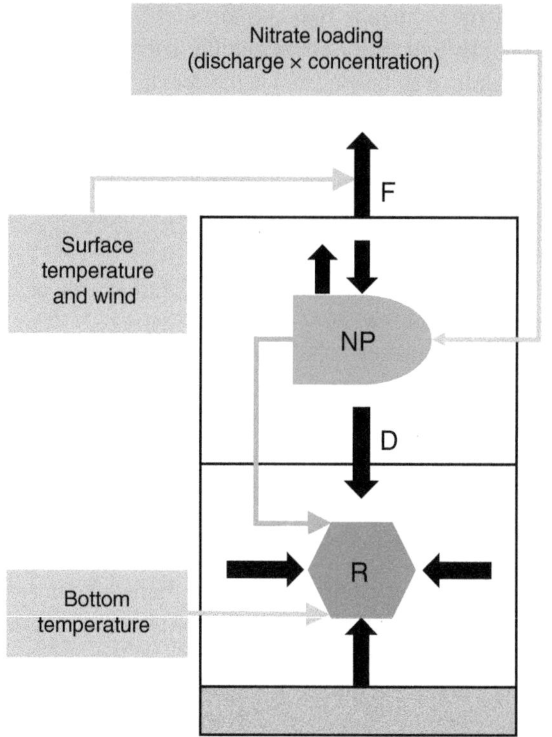

FIGURE 16.5 A conceptual model of oxygen cycling in a shallow, stratified, coastal ecosystem. F denotes the total air-sea oxygen flux, NP is the net rate of oxygen production in the upper layer (production—respiration), D is the diffusive oxygen flux through the pycnocline, and R is the rate of oxygen consumption by respiration in the lower water column. *Source*: Modified from Justić et al. 1996, with permission from the Association for the Sciences of Limnology and Oceanography and John Wiley & Sons.

The approach used a very simplified set of spatial regions (only two layers at a point in space) and a simple biological model (relatively few processes) on a monthly time step that was tightly linked to the available field data. As a consequence, its predictions, while limited in generality and scope, would have a high degree of confidence. There were a total of 45 years of field data collected between 1955 and 2000 available to validate and calibrate the model. A portion was used for calibration and the entire time series was used to explore scenarios.

Once calibrated, it was envisioned to use this model to simulate the effects of six hypothetical future scenarios on the frequency of hypoxia and the areal extent of the summertime hypoxic zone. Future scenarios were based on observed and projected changes in Mississippi River discharge and nitrate concentrations, and in ambient water temperatures in the NGOM (Table 16.1). Total loading of nitrogen from the river ($kg\,day^{-1}$) is the product of the flow rate of the water (discharge, $m^3\,second^{-1}$) and the concentration assumed for nitrate ($mg\,L^{-1}$) in that water. Management actions in the watershed, such as reduced use of fertilizer in agriculture, can be considered to affect the nitrate concentration; climate change will affect the discharge through changes in temperature and precipitation. One scenario corresponded to a proposed management measure and examined what would happen to hypoxia with a 30% decrease in nitrogen loading due to a 30% lower nitrate concentration in river water.

16.4.2.2 Formalization and Implementation

The two-layer simple model assumes uniform properties within the surface and bottom layers (Figure 16.5). The surface

TABLE 16.1 **Simulated changes in the frequency of hypoxia in the bottom layer in the core of the Gulf of Mexico hypoxic zone over a 45-year period from 1955 to 2000 for a number of climatic and nitrogen loading scenarios. The baseline model simulation used as its forcing functions the time series of observed monthly values of the Mississippi River discharge and nitrate concentration (N-NO_3). The investigated model scenarios are based on the available projections of climate, Mississippi River flows, water temperatures in the northern Gulf of Mexico, and proposed nutrient management goals. Hypoxia is defined as water with DO less than 2 mg O_2 L^{-1}. The percent change is calculated as 100 × ($S-B$)/B where S is the number of years of hypoxia in the scenario simulation and B is the number of years of hypoxia in baseline.**

Scenario		Number of years with hypoxia	Percent change relative to baseline
Baseline	1955–2000 conditions	19	–
1	Nitrate concentrations characteristic of 1955–1967	0	−100
2	30% reduction in nitrate concentration	12	−37
3	30% reduction in river discharge	8	−58
4	20% increase in river discharge	26	+37
5	4 °C warmer Gulf waters	25	+32
6	20% increase in river discharge and 4 °C warmer Gulf waters	31	+63

Source: Adapted from Justić et al., (2003).

layer corresponds to the waters above the average depth of the pycnocline (zero to ~10 m) and the bottom layer corresponds to the waters below the average depth of the pycnocline (10 m to ~20 m). The model has two main state variables, one equation for the oxygen concentration in the surface, and one equation for the bottom and the terms in the equations for each process are relatively simple. Surface layer oxygen (O_S, g m^{-3}) is the result of air-water gas exchange (flux) with the atmosphere, loss of oxygen due to diffusion across the pycnocline into the bottom layer, and biological oxygen production minus respiration (i.e., Net Production; NP): $\partial O_S/\partial t = -F_t - D_t + NP_t$. Oxygen in the bottom layer (O_B, g m^{-3}) depended on the balance between oxygen consumption by respiration (benthic and water column) and the increase from the oxygen diffusion in from the surface layer: $\partial O_B/\partial t = -R_t + D_t$. The detailed equations for each of these terms, which depend on the monthly values of the forcing variables of river discharge, nitrate concentration, surface and bottom layer temperatures, and surface wind data, are found in Justić et al. (2003). In addition to oxygen concentrations appearing in many of these terms, other interesting dependencies include how NP in the surface layer affects R in the bottom layer and that a lagged value of nitrate loading one month earlier determines NP: $NP_t = -0.34 + 3.93 \times 10^{-7} \times [Nitrate\text{-}Loading]_{t-1}$. Many software programs can be used to solve this simple model (differential equations), including the statistical package R.

16.4.2.3 Parameterization and Validation

The model was calibrated using 1985–1993 data from the 1955 to 2000 record. The 1985–1993 period included three average hydrologic years (1985, 1986, and 1989), a record flood year (1993, a 62-year record-high discharge), two years with above average discharge (1990 and 1991), three years with below average discharge (1987, 1988, and 1992), and a record drought year (1988, a 52-year record-low discharge). Given the timespan of the data, the 1985–1993 data subset was well suited for model calibration because the data contained a wide variety of river discharge conditions. Biweekly to monthly values of temperature, salinity, and dissolved oxygen (by 1–2 m depth intervals) were used from a station located in the core of the hypoxic zone. Process parameters were adjusted singly and in combinations until time series plots of predicted oxygen concentration values matched the time series of observations for the top and bottom layers as well as for vertical profiles at selected times.

16.4.2.4 Analysis

Baseline and future scenarios investigated with the simple two-layer model (Figure 16.5, Table 16.1) were based on observed and projected changes in the Mississippi River discharge, the assumed concentration of nitrate in Mississippi River water, and ambient water temperatures in the NGOM. Model predictions were compared to baseline simulation results that used as forcing functions the observed time-series of temperature, river freshwater discharge, and nitrate flux for the 45-year period from 1955 to 2000.

Predictions of the calibrated simple model under baseline conditions identified the mid-1970s as the start of the recurring hypoxia in the bottom layer and predicted a total of 19 years with hypoxia between 1955 and 2000 (Figure 16.6; Table 16.1). These results are in good agreement with the timing of first reports documenting hypoxia in the NGOM (Rabalais and Turner 2001) and are additionally supported by the retrospective analyses of sedimentary records (Turner and Rabalais 1994).

Simulated changes in the loading of nitrogen from the Mississippi River greatly affected the frequency of hypoxia in the Gulf of Mexico. Under a scenario in which nitrate concentrations in the Mississippi River were kept at 1955–1967 levels (Scenario 1), the model predicted no years of hypoxia, and a 30% decrease in nitrate concentration (Scenario 2) resulted in a 37% decrease in the frequency of hypoxia. Only varying Mississippi River discharge showed a similar magnitude of effects on hypoxia: a 30% reduction (Scenario 3) resulted in 8 years with hypoxia (a 58% decrease) and a 20% increase in river discharge (Scenario 4)

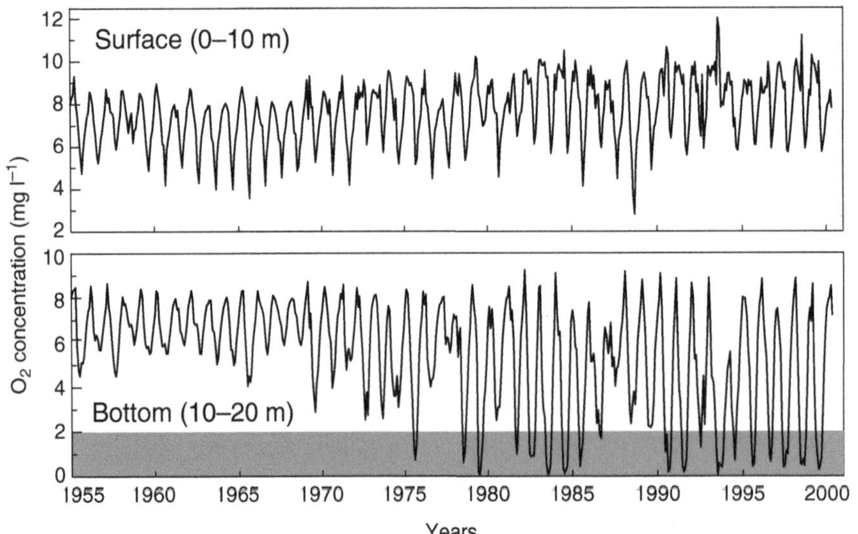

FIGURE 16.6 Simulated changes in the average surface (0–10 m) and bottom (10–20 m) oxygen concentrations at station C6 for the period January 1955 to May 2000. The red shaded area in the lower panel denotes hypoxic conditions (<2 mg O_2 L^{-1}) in bottom waters. *Source:* From Justić et al. 2002, reprinted with permission from Elsevier.

led to a 37% increase in hypoxia frequency. When the temperature of the NGOM was increased by 4°C (Scenario 5) the model predicted a 32% increase in hypoxia frequency. Finally, the combination of increased discharge with warmer temperature (Scenario 6) resulted in a 63% increase in the frequency.

These simulation results suggest that the frequency of hypoxia in the Gulf of Mexico is very sensitive to the loading of nitrogen from the Mississippi River and to water temperatures in the Gulf, and that increased discharge and warming can offset some of the beneficial reductions in hypoxia from reduced nitrate concentrations in the river water.

16.4.3 Complex 3D-Coupled Hydrodynamic–Biogeochemical Model

16.4.3.1 Conceptualization To address the same question of how nutrient loadings affect hypoxia in the Gulf of Mexico, a very complex model was envisioned. A three-dimensional (3D), coupled hydrodynamic–biogeochemical model (FVCOM-WASP; Figure 16.7) was developed to more precisely

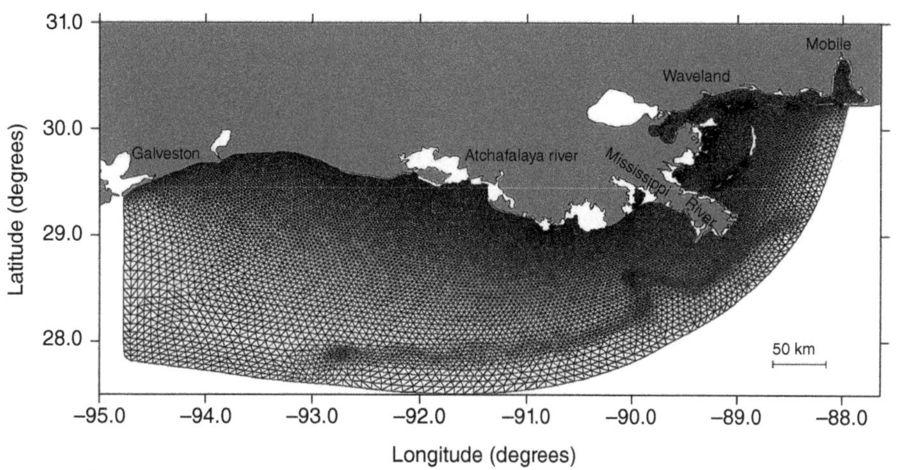

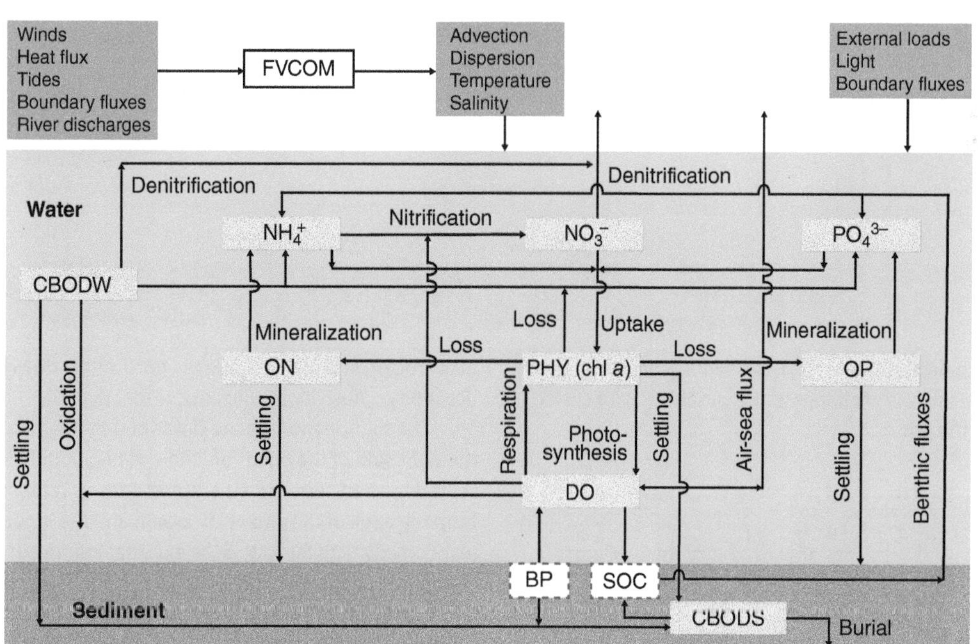

FIGURE 16.7 Upper panel—Map of the northern Gulf of Mexico showing the FVCOM computational grid. Lower panel—Conceptual framework of the FVCOM model showing forcing variables (in red), state variables (in yellow), processes and biochemical sources and sinks explicitly included in the model (arrows); DO—dissolved oxygen, PHY (chl *a*)—phytoplankton biomass expressed as chlorophyll *a*, CBODW—carbonaceous biochemical oxygen demand in the water column, NH_4^+—ammonium, NO_3^-—nitrate, PO_4^{3-}—soluble reactive phosphorus, ON—organic nitrogen, OP—organic phosphorus, BP—benthic photosynthesis, SOC—sediment oxygen consumption, and CBODS—carbonaceous biochemical oxygen demand in the sediment. *Source:* From Justić and Wang 2014, reprinted with permission from Elsevier. The terms BP (Benthic Photosynthesis) and SOC (Sediment Oxygen Consumption) are processes like Photosynthesis (arrows); they are shown in dotted boxes because they occur in the sediments and not within the water column.

describe how physical and biological processes control dissolved oxygen dynamics in the NGOM and to predict spatial changes in the hypoxic area and volume in response to various nutrient reduction scenarios (Wang and Justić 2009; Justić and Wang 2014). Many others have developed models to address very similar questions using coupled hydrodynamic–water quality models (Ji 2017). Therefore, the conceptualization was based on the investigators (Justić and Wang 2014) examining available models to determine which, if any, represent the state variables and processes appropriate for this specific system and question (i.e., match their conceptual model). They determined that the hydrodynamic properties of the system (water depths, velocities) would be well simulated using the high-resolution, 3D, unstructured grid Finite Volume Coastal Ocean Model (FVCOM; Chen et al. 2003, 2006). They also examined available models for simulating biogeochemistry, and especially dissolved oxygen, and selected (and then revised) a version of the off-the-shelf model called the Water Analysis Simulation Program (WASP).

Because both the FVCOM and WASP models were developed by others, an important aspect is to check and confirm that the models represent the system in agreement with the conceptual model (as well as the details added with the formalization step) and would therefore be well suited to address the question being asked. Like many circulation-type models, FVCOM represents hydrodynamics using laws of physics; what is important is that FVCOM divides the grid into triangles (rather than rectangles) and thus can capture fine-scale coastlines and other features. A conceptual model of WASP is shown in Figure 16.7. The model includes nine water column state variables, including dissolved oxygen (DO; mg O_2 L^{-1}), phytoplankton biomass (PHY) expressed as chlorophyll (Chl *a*; μg Chl *a* L^{-1}), water column carbonaceous biochemical oxygen demand (CBODW; mg O_2 L^{-1}), ammonium nitrogen (NH_4^+; mg N L^{-1}), nitrate and nitrite nitrogen (herein referred to as nitrate, NO_3^-; mg N L^{-1}), soluble reactive phosphorus (PO_4^{3-}; mg P L^{-1}), and organic nitrogen (ON; mg N L^{-1}), organic phosphorus (OP; mg P L^{-1}), and one sediment state variable of carbonaceous biochemical oxygen demand (CBODS; g O_2 m^{-2}).

The 3D model expands on the simple two-layer model by putting the effects of forcing variables in the context of a simulated circulation model that recreates the physical oceanographic conditions of the NGOM over space. This model provides an accurate description of the offshore circulation generated by the westerly winds during summer months, as well as the prevalent westward flow along the coast caused by the easterly winds during the rest of the study period (Wang and Justić 2009). The seasonal stratification cycle is also well represented, and simulations can predict the total area of hypoxia (like the simple model) and also the spatial–temporal dynamics of the hypoxic region in three dimensions.

16.4.3.2 Formalization and Implementation

For each water column state variable, *j*, the same general equation is used to calculate the rate of change in that variable (C_j) for each of the spatial cells of the water column (i.e., for each triangular cell at each depth). This equation is based on the conservation of mass and is shown in Box 16.3. The S terms in the general state variable equation that denote biological or chemical source and sink terms can be quite complicated. The expanded equation for the "S terms" for DO in Box 16.4 shows the many parameters that are needed, how a state variable can depend on environmental conditions (i.e., forcing variables

Box 16.3 The Terms in a Mass Balance Equation in the WASP-FVCOM Model.

For each water column state variable, j, the same general equation is used to calculate the rate of change of the variable (C_j) for each of the spatial cells in the grid.

$$\underbrace{\frac{\partial C_j}{\partial t}}_{\text{Rate of change of solute}}+\underbrace{\frac{\partial (uC_j)}{\partial x}+\frac{\partial (vC_j)}{\partial y}+\frac{\partial (wC_j)}{\partial z}}_{\text{Advection (3 directions)}}=\underbrace{\frac{\partial}{\partial x}\left(A_h\frac{\partial C_j}{\partial x}\right)+\frac{\partial}{\partial y}\left(A_h\frac{\partial C_j}{\partial y}\right)}_{\text{Horizontal Mixing in 2 directions}}+\underbrace{\frac{\partial}{\partial z}\left(K_h\frac{\partial C_j}{\partial z}\right)}_{\text{Vertical Mixing (e.g. turbulence)}}+\underbrace{S_j}_{\text{Bio- or chemical source or sink term}}+\underbrace{W_j+B_j}_{\text{External inputs (river, ocean)}}$$

where *u*, *v*, and *w* are the water velocity components corresponding to the x, y, and z coordinates in that cell (x and y are horizontal while z is vertical), and A_h and K_h are the horizontal eddy viscosity and vertical eddy diffusivity coefficients, respectively.

The hydrodynamics are described by the first three terms on the right side of the equation, the chemistry and biology are represented by the *S* term, and the *W* and *B* terms represent sources of C_j from the edge of the grid at the ocean and the input from the rivers, respectively. The term W_j is the external loading of C_j from the Mississippi and Atchafalaya Rivers and B_j is the input of C_j at the ocean boundary. The equation states that the change in C_j at a location is the net effects of all sources and sinks for C_j.

Notice that the symbols used for the rate of change (left side of the equation) denote a partial differential equation, which can be thought of as a rate of change (like in Box 16.2); the symbols denote that the rate of change depends on more than one dimension (time is shown by the t term, and location in space by the x, y, and z terms).

Box 16.4 Full Equation for the S term for the DO State Variable in the Example Model (Box 16.3).

The dynamics of dissolved oxygen (DO) in a water column cell is described by the equation S_{DO} = Re-aeration – CBODW oxidation – Phytoplankton respiration – Nitrification + Phytoplankton photosynthesis + Benthic photosynthesis – Sediment oxygen consumption. Each of these terms is a process. When the processes are put into mathematical form, the equation is as expressed here. Below the equation is an illustration of the specific mathematical expressions used in the computation of oxygen consumption associated with nitrification.

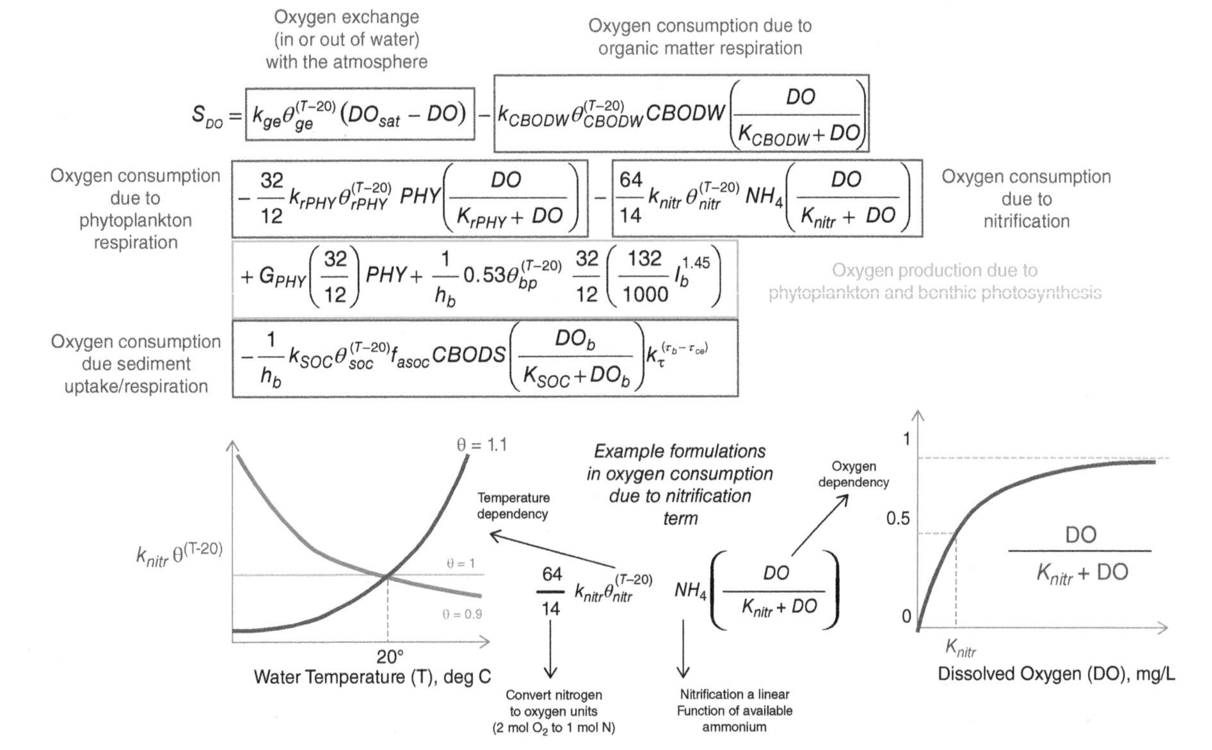

such as temperature), and also on other state variables (e.g., CBODW). Now picture similar equations as for DO for each of the other eight S terms that represent the remaining state variables. The coupled FVCOM-WASP model is driven by the forcing variables of surface wind forcing, tidal forcing, offshore remote forcing, heat fluxes, air-water gas exchanges, solar radiation, and freshwater and nutrient (nitrogen and phosphorus) loadings from the Mississippi and Atchafalaya Rivers. In the general equation (Box 16.3), the external loadings of all nine state variables from the two rivers is denoted W_j [j=1,. . .,9] and the external loadings of state variables at the open boundary to the ocean is denoted B_j [j=1,. . .,9].

The year 2002 was chosen as the reference year for model calibration. The measured extent of the hypoxic zone during July 2002 was the second highest on record (22,000 km^2) and provides a good reference value for model-data comparisons. Mississippi and Atchafalaya River discharge peaked three times during 2002, in February, April, and early June. The study area also experienced frequent high-wind events during winter and spring, and increased tropical storm activity during summer and fall. Model simulations for 2002 were carried out from 1 January until 4 October. Thus, the model simulations captured the entire stratification/hypoxia cycle that included the initial development of hypoxia, persistent summertime hypoxic conditions, and the dissipation of hypoxia.

The FVCOM model is publically available as source code and comes with extensive installation and user manuals (**http://fvcom.smast.umassd.edu/fvcom**). The user then provides the inputs, such as grid information and forcing functions according to the format required by the code. The investigators in our example also obtained the WASP model (similar to FVCOM) as source code (**https://www.epa.gov/ceam/water-quality-analysis-simulation-program-wasp**) and coupled them for the application. Both were coded in FORTRAN and are in a generalized form so others can configure them for their system. In the example here, the FVCOM grid was configured with over 28,320 triangular cells, which vary from 1 to 10 km in one horizontal dimension (e.g., width) and 30 vertical layers (Figure 16.7), which is a total of over 800,000 cells. Setting-up FVCOM requires expertise in data analysis to prepare the inputs and forcing functions, deftness with large spatial files to correctly set-up the grid using bathymetry data for your location, knowledge of computer programming to compile and make any changes, knowledge of software (R, Matlab) to post-process the output as graphs, and access to a high performance computer to make model runs. Many aspects of the modeling are options already built-in that the user selects from. The complex 3-D model in this example involved all of these activities over time as the model was tested and compared to data, and then periodically updated as our knowledge improved about the processes in the model. Indeed, based on analyses with the FVCOM-WASP coupled models, the investigators, at various times, revised the conceptualization, formalization, and implementation steps and modified the WASP equations.

16.4.3.3 Parameterization and Validation Calibration involved the modeler adjusting values of parameters known to be uncertain until model predictions reasonably matched field data. The hydrodynamics (FVCOM) were first calibrated to get realistic advection (i.e., transport associated with fluid flow) and mixing/diffusion rates that affect nutrients and dissolved oxygen, and then the water quality model (WASP) was calibrated. There was extensive data available on time series values of water heights and water velocities at key locations, and for many state variables along several transects (within and outside of the core hypoxia zone) that start from near-shore and go out onto the shelf away from the coast. Measurements were made in vertical layers at locations along these transects. Data on DO included time series from a transect station within the core of the NGOM hypoxic zone, measurements collected along many transects during the mid-summer cruises, and vertical DO profiles throughout the year. The predicted hypoxic area for 2002 after calibration (Figure 16.8) was within 15% of the observed extent (Justić and Wang 2014). In this case, true independent validation of DO predictions was not included.

16.4.3.4 Analysis Nine month (January to September) simulations were performed under present-day conditions for 2002, as well as the same 2002 conditions but with a 20% reduced nitrate concentration in Mississippi River water, and a 50% reduced nitrate concentration in river water. Summarizing model results by the area and volume of the mid-summer hypoxic zone (Figure 16.9) showed that decreasing nitrate concentrations would reduce both metrics of hypoxia. The hypoxic area for baseline (Present day) conditions shown in Figure 16.9 was 18,550 km^2, which was reduced to 12,493 (33% lower) and 10,017 (35% lower) for the 25 and 50% decreases in nitrate simulations. A similar reduction was predicted for volume of the hypoxic zone for the 25% reduction in nitrate concentration (40–26 km^3, 35%), while a much larger reduction in volume was predicted for the 50% reduction (40–14 km^3, or 65%).

Detailed examination of the 3D model results indicate that hypoxia originates in bottom waters on the mid-continental shelf, where isolated pockets of hypoxic water develop during early spring and later coalesce into a larger continuous hypoxic zone (Figure 16.8). The set-up and dynamics of the hypoxic zone are influenced by several features and by weather events. Bathymetric features of the shelf, namely the presence of shallow and deep water shoals in certain regions determine the location of the hypoxic zone. Also, the subsequent hypoxia dynamics are strongly modulated by the frequency and intensity of cold fronts and tropical storms. High winds associated with these events disturb stratification, causing partial or complete breakdown of hypoxia. Cold fronts and tropical storms also cause significant sediment resuspension that fuels respiration in the lower water column, and in this manner promotes redevelopment of hypoxia. The temporal and spatial variation in the extent of hypoxia (Fig. 16.8) is a product of these random and episodic physical events that disrupt vertical stratification of the water column.

16.4.4 Conclusions from Both Models

Both simple two-layer and complex 3D hypoxia model simulations suggest that NGOM hypoxia is highly sensitive to variations in freshwater discharge, riverine nitrate flux, and the ambient water temperatures. Results of the two-layer model indicate that both major increases and major decreases in the frequency of hypoxia are possible in the future depending on the effects of climate change on these forcing variables (Table 16.1). Further, model simulations reveal that, if potential climatic variations are taken into account, a 30% decrease in the nitrogen flux of the Mississippi River by reducing nitrate concentration may not be sufficient to accomplish the proposed hypoxia management goal. For instance, a 20% increase in the Mississippi River discharge, projected under some climate change scenarios due to changes in precipitation, would completely offset a decrease in the frequency of hypoxia resulting from a 30% decrease in the nitrate concentration. However, one key limitation of the simple two-layer model is that it describes the dynamics of hypoxia only at a single location within the core of the hypoxic zone (station C6; Figure 16.8). In this respect, this model could not answer the question whether a 30% decrease in nitrate flux would reduce the average areal extent of hypoxia below 5000 km^2, as suggested by the Mississippi River Watershed/Gulf of Mexico Hypoxia Task Force (Task Force 2001, 2008, 2015).

Results from the 3D model show that it would be possible to meaningfully reduce the areal extent of hypoxia

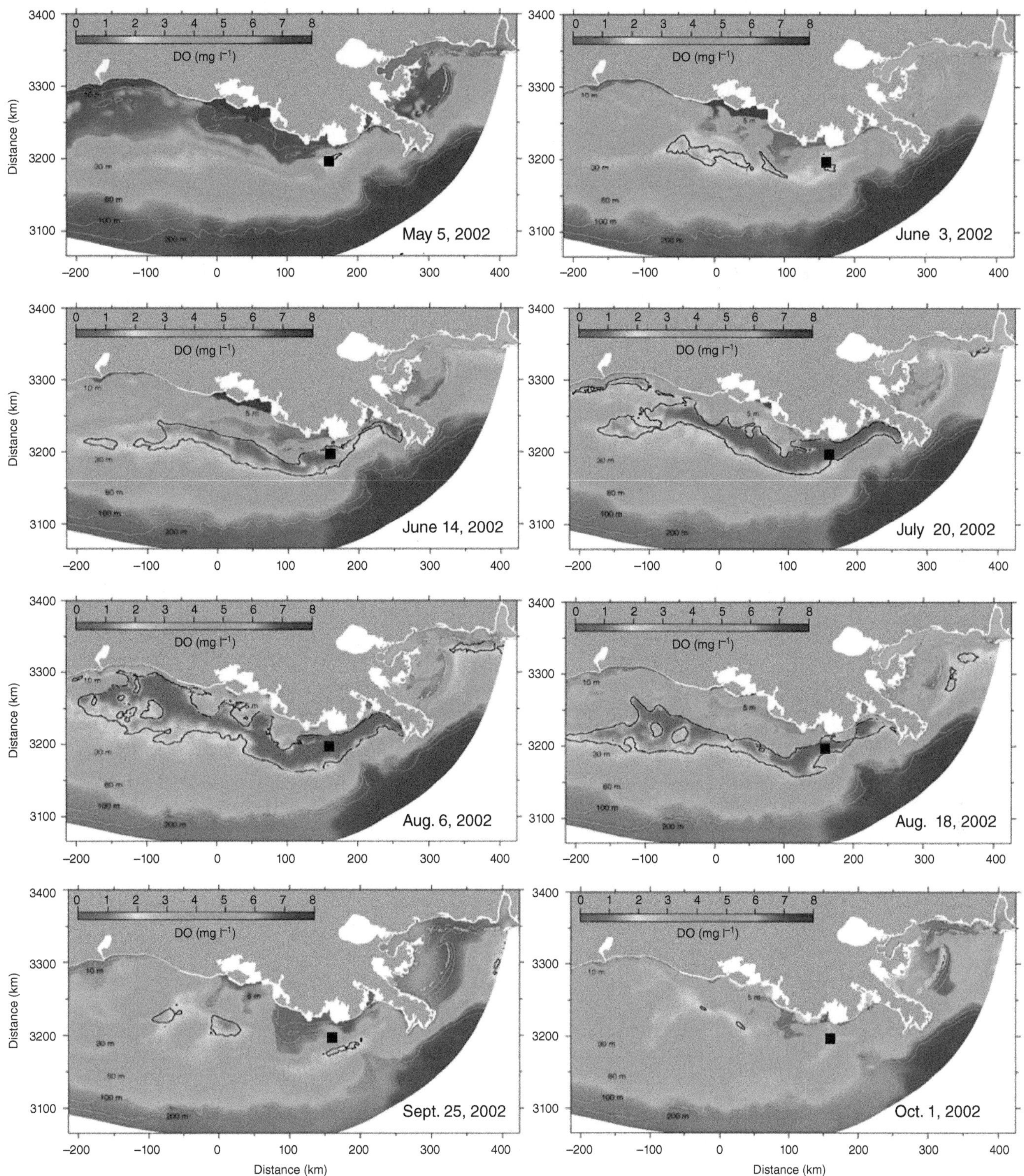

FIGURE 16.8 Simulated changes in the extent of the Gulf of Mexico hypoxic zone for May 11—October 1, 2002. Black square denotes the location of station C6. *Source*: From Justić and Wang 2014, reprinted with permission from Elsevier.

if nitrate concentration in the Mississippi River is reduced by at least 50% (Figure 16.9). Further, the model results also indicate that hypoxic volume appears to be more sensitive to nutrient load reduction than hypoxic area. For example, even under a modest 25% nitrogen load reduction, the thickness of the hypoxic layer in the northern Gulf of Mexico decreases markedly, and hypoxia remains localized to a relatively thin layer near the bottom that most fish and other mobile organisms can more effectively avoid (Figure 16.9).

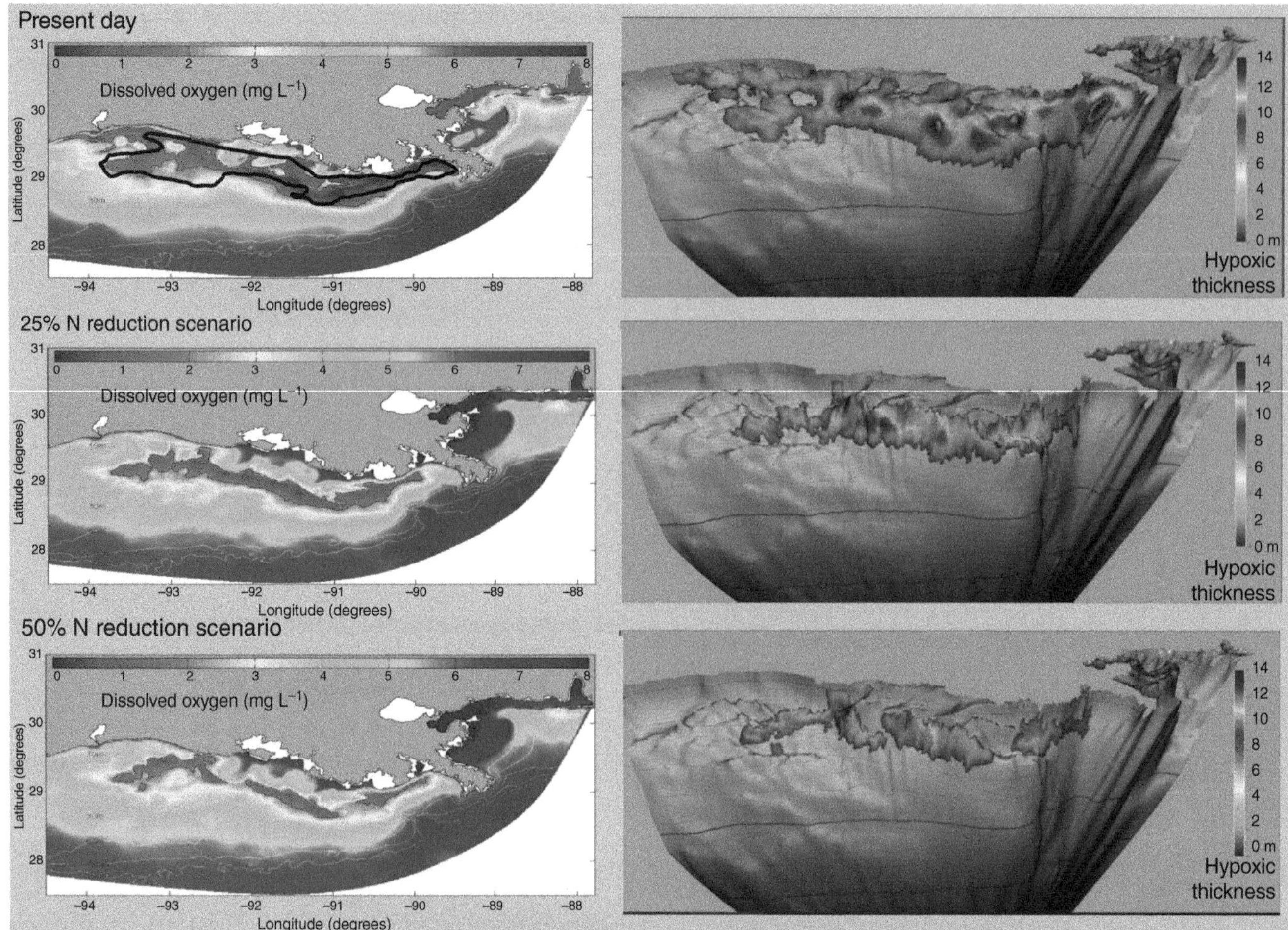

FIGURE 16.9 Midsummer (July 21–26) hypoxic area (left) and hypoxic thickness (right) in the northern Gulf of Mexico hypoxic zone simulated by the Justić and Wang model for present day conditions (2002, upper panels), for a 25% N reduction scenario (middle panels), and for a 50% N reduction scenario (lower panels). Color bars on the left indicate bottom water dissolved oxygen concentrations; color bars on the right indicate the thickness of the hypoxic layer (in meters). Because the footprint of the hypoxic area shown is the same in both panels, the right panel as a 3-D image shows the volume of the hypoxic zone. The solid black line in the upper left panel outlines the areal extent of hypoxic (dissolved oxygen < 2 mg L^{-1}) bottom oxygen waters observed on July 21–26, 2002. *Source*: From Scavia et al. 2019, reprinted with permission from IOP Publishing.

16.5 Example 2: Marsh Habitat and Brown Shrimp Production

16.5.1 Defining the Question

Many fish and shellfish use marshes in estuaries as nursery areas (Minello 1999). For example, brown shrimp (*Farfantepenaeus aztecus*) spawn offshore in the ocean during winter, and in the spring very young shrimp move shoreward until they reach estuaries. When they are about 15 mm in length (February to May) they settle to the bottom in the bays and marshes of estuaries, where they grow rapidly. Sampling with nets has shown that young shrimp congregate in edge habitats where marshes meet the water (Rozas et al. 2000). Upon reaching about 70 mm in length in the fall, shrimp leave estuaries and return to the ocean. Harvesting of adult shrimp in the offshore waters is a very important fishery in the Gulf of Mexico (Bourgeois et al. 2016).

In many coastal areas, including the Gulf of Mexico, marshes are disappearing because of rising sea levels caused by climate change, reductions in sediment supply associated with flood prevention measures, and several natural processes including subsidence (compaction of the soils under a marsh; Boesch et al. 1984). It is difficult to investigate this issue with field sampling because comparison of shrimp growth and mortality among marshes could be affected by other differences between the marshes then just their degree of degradation. Thus, Haas et al. (2004) developed a simulation model of shrimp during their time in marshes with the purpose of predicting how degradation of marshes, with all else equal, affects the growth and mortality of young brown shrimp. Their results can also be used to describe how shrimp growth and mortality may respond if degraded marshes are restored.

16.5.2 Conceptualization

Haas et al. (2004) constructed an agent-based model of brown shrimp in which each individual was moved around on a square grid of spatial cells. Following individuals or "agents" in ecological models has gained popularity in the past several decades (DeAngelis and Grimm 2014). The idea is that, for some questions, rather than following the change in mass or concentration (Boxes 16.2 and 16.3) as state variables, it is better to follow individuals as they move in space and experience changing environmental and habitat conditions and interact with other individuals (e.g., prey, predators, and mates). This general approach is called Lagrangian (following particles or individuals through space and time) as compared to Eulerian (net changes in mass or concentrations). It was decided that following individuals was a better way to simulate how shrimp move around within a fine-scale marsh and thus would better answer the question about how fine-scale arrangement of vegetation within a marsh (e.g., healthy versus degraded) would affect shrimp productivity.

The Haas et al. model needed to cover January through September when shrimp are found in marshes. A time step of 6 hours was selected to enable representation of important phenomena, such as water level changes within cells and shrimp moving in and out of vegetation at marsh edges. The state variables were the size (length in mm), status (dead or alive), and location (cell) of each individual shrimp in the spatial grid. The idea is to then simulate thousands of individuals and obtain predictions of total shrimp by summing over all alive shrimp, and keeping track of the numbers of shrimp surviving to 70-mm when they leave the estuary and return to the ocean.

The processes thought important for each individual shrimp were growth rate that determined the body size, mortality that determined the status of each shrimp (alive or dead), and movement that determined their location on the grid each time step. The environmental forcing variables needed for these growth, mortality, and movement processes were water temperature (for growth), and flooding which affected whether cells in the grid had water on them and were therefore accessible to the shrimp (affected movement and mortality). Based on available data and patterns of shrimp arrivals in estuaries, very young shrimp (about 15 mm in length) could be added to the simulated population every week between January and May and started by randomly placing them in cells that had water. The investigators decided that an important output would include the number of shrimp that survived to 70 mm, because this is the size where most shrimp leave marshes to enter deeper waters and the shelf habitat.

16.5.3 Formalization and Implementation

Spatial grids (maps) representing marshes at different stages of degradation were developed from aerial photographs of actual marshes. Within these 100 cell × 100 cell grids, 1-meter square cells were assigned as either vegetated or open water. The

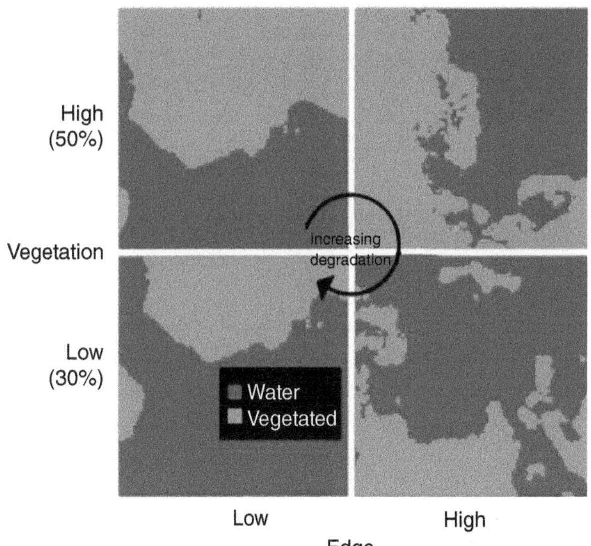

FIGURE 16.10 The four spatial maps of vegetated and water cells in the 100 × 100 grids used for simulating individual brown shrimp growth, mortality, and movement. The two maps on the top row had similar total amount of vegetated cells (50%) but the left-side map had this arranged in large clumps and thus low edge while the right-side map has smaller clumps and thus higher edge. The bottom row had fewer vegetated cells (about 30% of the map) and were arranged so the left-side map had low edge and the right-side map had high edge. The four maps can be thought of as showing a progression of degradation as one goes clockwise starting at the top left map.

boundary between vegetated cells and water cells is considered "edge." Vegetated cells were therefore of two types: vegetated cells adjacent to other vegetated cells or edge-vegetated cells that were adjacent to open water cells. Four maps of marsh distributions were selected to provide a nice contrast in the health of the marshes (Figure 16.10). Two habitat maps were about 50% water cells but one map had a large cluster of vegetated cells with relatively few edges, while the other map had the same number of vegetated cells but was more fragmented and therefore had more edges. The other two habitat maps had fewer vegetated cells (30%) but also differed by vegetated cells being highly clustered or fragmented. One could view the degradation of a marsh as starting with the high-vegetation, low edge grid, progressing through fragmented vegetation conditions, and ending with the low-vegetation, low edge grid.

A flowchart of the model calculations is shown in Figure 16.11. Initial conditions were the habitat map to be used, and number of young shrimp (post-larvae) to be added during the simulation, and the temperature and flooding conditions. The outer loop is over time and proceeds in 6-hour time steps to capture the within-day flooding and drying of marsh cells. Simulations were from January through September. Every 6 hours, the age of each shrimp was incremented. Water temperature was updated each day from a function fitted to field data, and all cells were assumed to have the same water temperature.

Flooding was represented in a more complicated way. Each cell could be in one of four stages (low, rising, high tide, and falling). A function was fit to the field data on the probability of

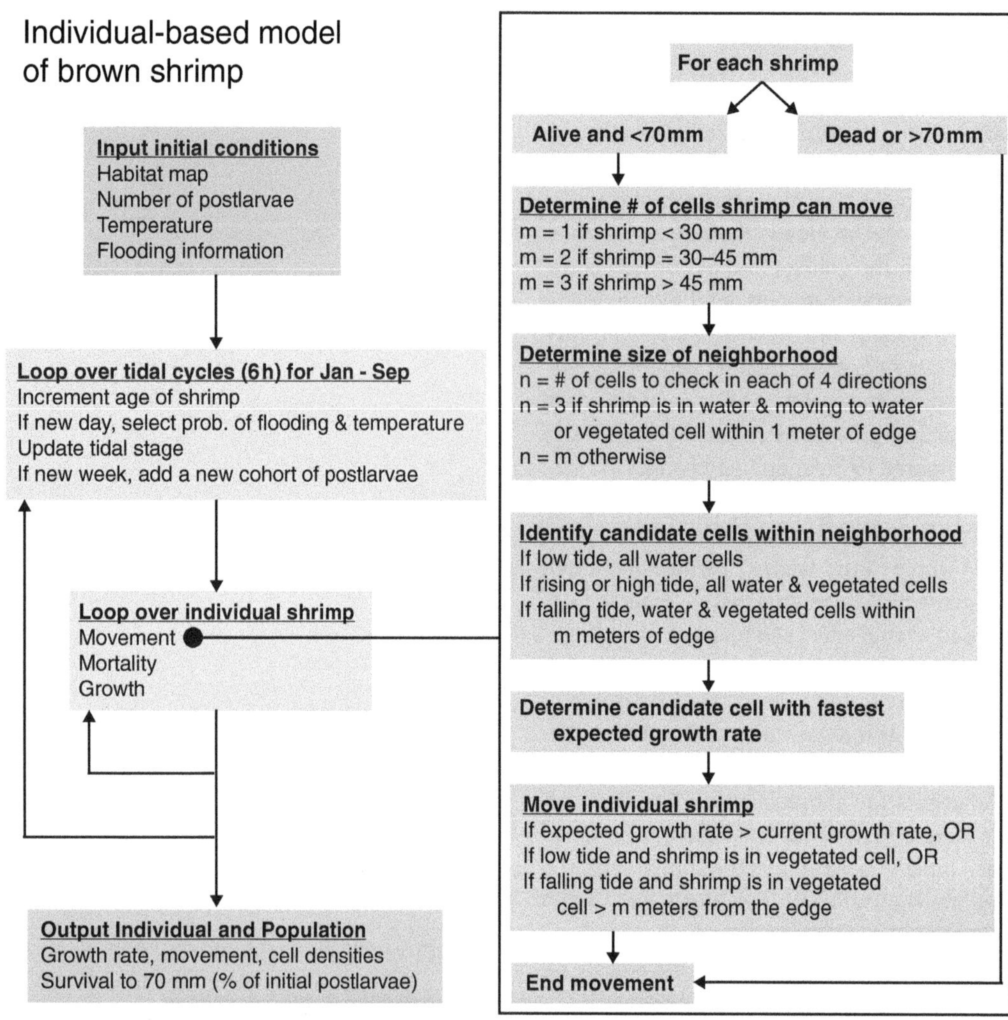

FIGURE 16.11 A flowchart showing the major calculation steps (left side) in the individual-based model of brown shrimp. The insert (right side) is how the movement of shrimp on the model grid was done and is a blow-up (detailed diagram) of the "movement" in the "Loop over individual shrimp" box on the left. Yellow boxes show loops, blue boxes are movement, and green boxes are inputs and outputs.

flooding occurring on the entire marsh grid on a day. For each day in the model, the function was used to determine the probability of flooding, and the same probability was used throughout that day. During each time step, a random number was generated to determine whether or not flooding actually occurred in each cell. Thus, a single probability of flooding was used each day and there were four chances (every 6 hours) for flooding to occur. If flooding occurred on a time step, then the stages of all cells were changed for the next time step to mimic rising water associated with marsh-wide flooding (low went to rising, rising to high, falling to high, and high stayed high). If flooding did not occur on a time step, then the stages of all cells were changed to mimic falling water levels (high to falling, falling to low, rising to low, and low remained low). Thus, all spatial cells on the marsh grid were always in the same flooding stage, and the stages changed every time step in unison. When enough time had passed to start a new week, the next batch of post-larval shrimp were added to the simulation.

The next loop was an iteration over each individual shrimp (one-at-a-time until all were evaluated) on each time step (Figure 16.11). Haas et al. (2004) applied a series of rules to move individual shrimp around the grid of cells every 6 hours, and then compared the spatial patterns of their distributions to field data (insert to flowchart in Figure 16.11). The process of movement was modeled by allowing each shrimp to look at neighboring cells to its present location and try to go to the cell that would give it the fastest growth rate. Larger shrimp examined larger neighborhoods, and shrimp in water cells examined larger neighborhoods than those in vegetated cells to mimic shrimp wanting to be in vegetated cells. Shrimp could only move to water or vegetated cells with sufficient water (rising, falling, high). Water cells were always available for shrimp to move to. However, the availability of vegetated cells varied with stage: at low stage on the marsh, all vegetated cells were unavailable to shrimp, at rising and high stages, vegetated cells within 3 m of the edge were available, and at falling stage, vegetated cells within 1–3 m of the edge (depending on shrimp size) were available. The movement rules were programmed in the computer code as complicated series of IF-THEN conditional statements resulting in each shrimp either staying in its current cell or moving to a neighboring cell.

The growth rate and mortality rate of each shrimp was evaluated every 6 hours (Box 16.5). Growth rate was assigned as a baseline growth rate that was then adjusted based on

Box 16.5 Computing Growth and Mortality of Individual Shrimp.

Figure 16.11 shows the algorithm for determining what cell an individual shrimp moves to every 6 hours. Once a shrimp is located in its new cell, the model computes the realized growth rate (RGROW) and realized mortality rate (RMORT) at that location and then updates the state variables of shrimp length (LEN, mm) and status (STATUS, alive or dead). As part of initial conditions, each shrimp is assigned a baseline growth rate (BGROW, mm/6 hours) and a baseline mortality rate (BZMOR, fraction dying/6-hours). Each shrimp gets a unique value of BGROW from a normal probability distribution (mean = 0.25 mm/6 hours, standard deviation = 0.25) and keeps it throughout the simulation; BZMOR was set to 0.005 mm/6 hours for every individual shrimp and it also did not vary in the simulation. These baseline values then were adjusted every 6 hours (Δt) for each shrimp based on the forcing function of temperature, whether the cell is vegetated or open water, shrimp length, how crowded the cell is with other shrimp, and how many cells the shrimp moved to get from its location during the last time step to its location this time step. The shrimp has moved from t-Δt to time t and now its growth and mortality are evaluated to update its length and status for time t in its new cell location.

$RGROW = BGROW \times G_T \times G_H \times G_D$ (mm/6-hours)

$G_T = v^z e^{z(1-v)}$

where:

$w = \ln(\theta) \cdot (T_m - T_o)$

$y = \ln(\theta) \cdot (T_m - T_o + 2)$

$$z = \frac{w^2\left(1+\sqrt{1+40/y}\right)^2}{400}$$

$$v = \frac{T_m - T(t)}{T_m T_o}$$

$\theta = 2.5;\ T_o = 32;\ T_m = 45$

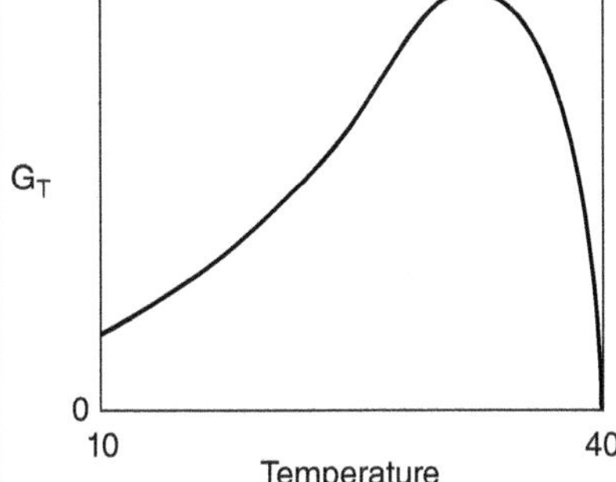

Temperature effect. This function is commonly used because it is a dome-shape that mimics the results of temperature effects on growth in many laboratory experiments on fish and invertebrates. The function uses three commonly reported features of temperature effects: optimal temperature for growth (T_o), temperature at which growth becomes zero (T_m), and a parameter (θ) that determines how quickly growth increases with temperature before it reaches the optimal temperature. The terms z, y, w, and v are expressions that shape the temperature function.

$$G_H = \begin{cases} 1 \text{ in open water} \\ 2 \text{ in vegetated} \end{cases}$$

Cell habitat effect. Extensive data and experiments showed that the increased food in vegetated conditions was roughly twice the rate as in open water.

$$G_D = 1.041 - \frac{0.0815}{1+e^{\frac{-(B(t)-7.9)}{1.98}}} S$$

where:

$B(t) = \sum_j W_j$

for all j shrimp in the same cell

$W_j = 0.000006 \times LEN^{3.071}$

(mm to grams)

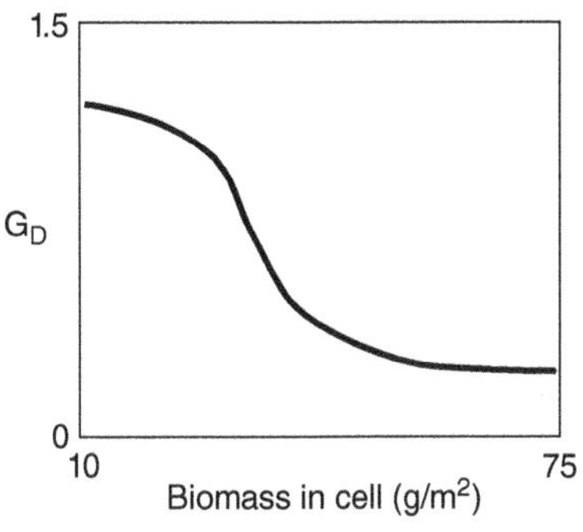

Density-dependence (crowding) effect. As the biomass (number × weight) of shrimp increases in a small area (1 m × 1 m cell), the food intake per shrimp decreases. This acts as a negative feedback by decreasing growth when cells are crowded. *B(t)* is the density of shrimp in each cell in total grams.

$RMORT = BMORT \times M_S \times M_H \times M_M$ (per 6 hours)

$M_S = 53.1 \times LEN(t-\Delta t)^{-1.1163}$

3

M_S

0

15

75

Shrimp Length (mm)

Size effect. Extensive field sampling and plots of shrimp densities versus average length showed that mortality rates declined with increasing size. Large shrimp are better at escaping predators and can also become too big for some predators to pursue.

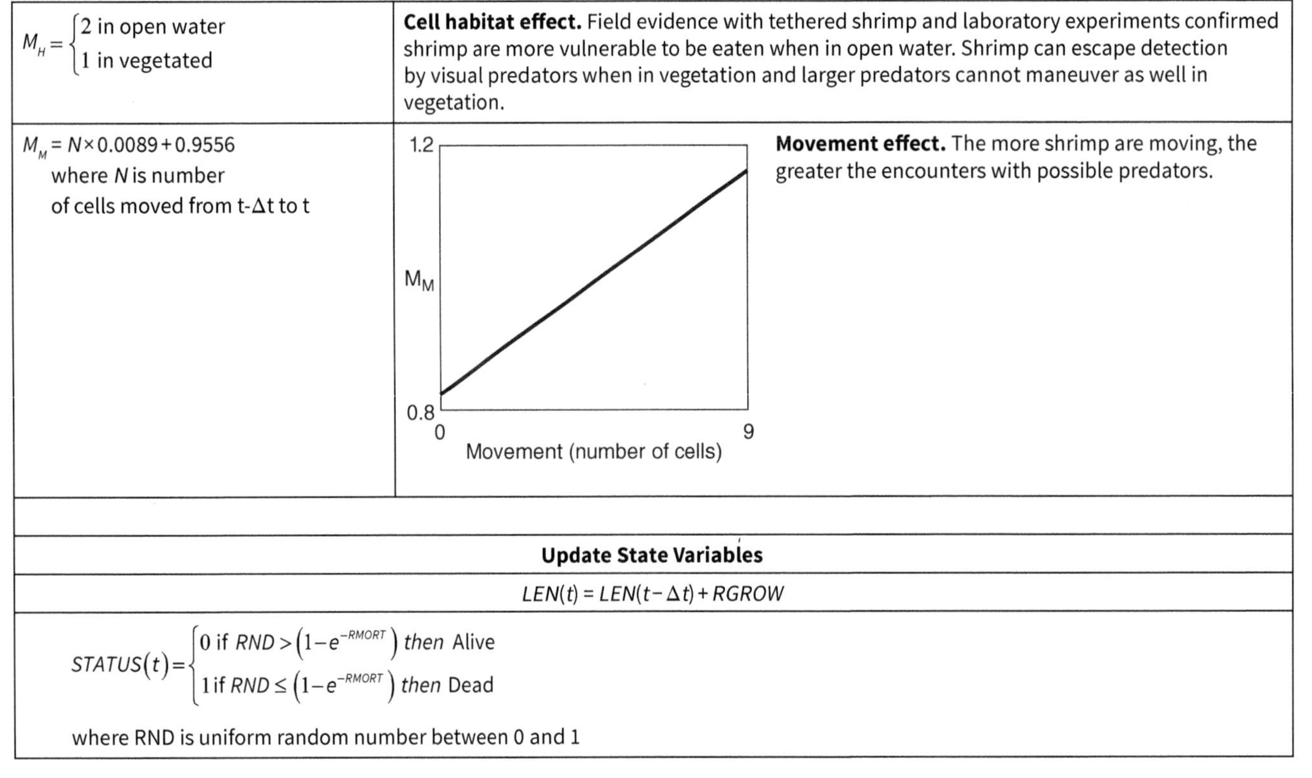

$M_H = \begin{cases} 2 \text{ in open water} \\ 1 \text{ in vegetated} \end{cases}$	**Cell habitat effect.** Field evidence with tethered shrimp and laboratory experiments confirmed shrimp are more vulnerable to be eaten when in open water. Shrimp can escape detection by visual predators when in vegetation and larger predators cannot maneuver as well in vegetation.
$M_M = N \times 0.0089 + 0.9556$ where N is number of cells moved from t-Δt to t	1.2; M_M; 0.8; 0; 9; Movement (number of cells) — **Movement effect.** The more shrimp are moving, the greater the encounters with possible predators.
Update State Variables	
$LEN(t) = LEN(t-\Delta t) + RGROW$	
$STATUS(t) = \begin{cases} 0 \text{ if } RND > \left(1-e^{-RMORT}\right) \textit{ then } \text{Alive} \\ 1 \text{ if } RND \leq \left(1-e^{-RMORT}\right) \textit{ then } \text{Dead} \end{cases}$ where RND is uniform random number between 0 and 1	

temperature in the cell, whether the cell was vegetated or open water, and depending on how many other shrimp were also in the same cell with them (more shrimp lowered the growth rate due to competition over food). The assigned baseline mortality rate of an individual was adjusted by the length of the individual, whether the cell was vegetated or open water, and depended on the number of cells the individual moved from its previous location to its present location. Every 6 hours, the shrimp were moved and the adjusted growth and mortality rates were evaluated to update shrimp length and status for the next time step. Length was calculated as the previous length plus growth rate. Because an individual is either alive or dead, a random number between zero and one was generated and compared to the probability of dying that was based on the adjusted mortality rate. Dead shrimp were removed from the simulation.

The mathematics of the model is a series of difference equations for each shrimp, with an equation for the length of a shrimp, an equation for the status (dead or alive) of a shrimp, and two equations for the current cell location (row number, and column number) of a shrimp. With difference equations, the solution technique is simply to go through each shrimp each 6 hours and update the equations based on the values from the previous time step. This sounds easy, but can be tricky, because the order in which the state variables are updated can affect the predictions. So, despite the mathematics being simple (compared to the FVCOM-WASP example), the computer code to set-up the model, read in the maps, solve the four equations for each shrimp, and process the results into output files was about 2000 lines. These types of relatively simple models can be coded in R, Excel, and most other software languages that allow for numerical calculations and looping.

16.5.4 Parametrization and Validation

The model was calibrated mainly by tweaking the movement rules until realistic model behavior about shrimp distributions in water and on marsh surfaces was obtained. The calibration used extensive field data of shrimp densities in open water and on marsh surfaces, and measurements of shrimp distances from the marsh edge. Haas et al. (2004) also adjusted the sizes of the neighborhoods searched and other aspects of the movement, as well as the mortality rates, until reasonable densities (numbers of shrimp m^{-2}) were obtained over time, and shrimp seemed congregated in edge habitat in agreement with the field data. They used the map with high vegetation and high edge for calibration because most of the field data was from field sampling on this type of marsh.

Starting with 100,000 shrimp, the simulated growth, mortality, and movement, resulted in peak densities in April of about 20 shrimp m^{-2} at high tide in edge cells, and average densities in vegetated cells were about eight times greater than in open water cells. Also, densities were about three times higher in edge cells compared to interior marsh cells. Haas et al. (2004) examined the model output looking for patterns in shrimp densities by tide stage and season in the different habitats (on the marsh, in the edge, and in open water). These values and patterns matched field data quite well. While such agreement is encouraging, there may be several sets of rules that result in similar spatial patterns. Some caution is needed, however, because agreement between predicted and observed patterns does not guarantee that the movement rules are correct.

Haas et al. (2004) used two available datasets to validate the model. The first was long-term (1970–1997) field monitoring of shrimp in Louisiana estuaries that was statistically analyzed to obtain annual index values (Haas et al. 2004). These were then grouped into warm versus cold years, with the average annual abundance index estimated to be 1.5 times higher in warm years. Using the same map as used for calibration (high vegetation and high edge), temperature was increased and decreased by 1 °C, and the fraction surviving to 70 mm compared between simulations. The model generated survival that was 1.25 times higher in warm versus cold years; similar to the 1.5 times observed in the data.

The second dataset used for validation included stable isotopes measured in the muscle tissue of dozens of individual shrimp in different habitats within a pond (Fry et al. 2003). Stable isotopes show the mix of food eaten from the three habitats (vegetation, pond perimeter, and pond interior). Haas et al. added stable isotopes as two new state variables ($\delta^{13}C$ and $\delta^{15}N$) to each individual shrimp to enable the model to predict the same stable isotopes values as was measured. The model was roughly set-up to simulate the same pond as was sampled and predicted isotope values were compared to the data. While the data were within the wider range of model predictions, more rigorous comparisons were not possible because there were too many unknowns relative to the layout of the habitats within the pond, the limited number of shrimp measured, and how shrimp had moved around in the pond prior to their measurement.

16.5.5 Analysis

Once calibrated and validated (to the extent possible), Haas et al. (2004) used the model to simulate the number of shrimp reaching 70 mm on the four maps. All other factors, such as temperature, pattern of weekly introduction of newly arrived shrimp, and flooding patterns, were maintained the same; the only difference was the maps.

Predicted percent of the 100,000 initial shrimp that survived to reach 70 mm was higher in maps with high edge (Table 16.2). The model results suggest that the higher survival was due to shrimp moving less (less vulnerable to being eaten) and spending more time in vegetated cells where they had lower mortality and faster growth. In the two maps with high edge, the average number of cells moved by a shrimp during each tidal stage was 3.14 (high vegetation) and 3.27 (low vegetation) versus 5.08 and 5.60 on the maps with low edge. Shrimp moved less because they more easily found the preferred edge habitat when it was more available, and this was reflected in the elevated time spent in vegetation (edge cells count as vegetated cells). However, the signal for spending more time in vegetated cells was strong only for when extremes of the high edge with high vegetation map (76%) were compared to the low edge with low vegetation map (66%).

Haas et al. (2004) went on to perform many more simulations using the model and the four habitat maps. They ran simulations introducing 300,000 and 600,000 individuals to determine the effect of crowding on growth, survival, and movement. They also ran simulations with alternative rules about movement, such as making the neighborhood searched smaller and larger.

TABLE 16.2 **Model predictions of shrimp on the four different marsh habitat maps (see Figure 16.10). Results are for simulations initiated with 100,000 shrimp. Survival is the percentage of these shrimp alive at the end of the simulation. Growth rate, number of cells moved per tidal stage, and time spent in vegetated cells was recorded each time step (tidal cycle) for each alive shrimp and then averaged over all tidal cycles in the simulation.**

Habitat map		Survival (%) to 70 mm	Growth rate (mm/d)	Number of cells moved per tidal stage	Time spent (%) in vegetation
Edge	Vegetation				
Low	High	23	1.33	5.08	69
High	High	31	1.37	3.14	76
High	Low	28	1.34	3.27	71
Low	Low	24	1.33	5.60	66

16.5.6 Conclusions

The model of Haas et al. (2004) provided valuable information on the role of edge habitat and marsh loss on shrimp survival and growth. The similarity of the predicted survival to 70-mm among the four maps was surprising to people knowledgeable about shrimp and wetlands. If the model is sufficiently realistic, the shrimp, at least in the virtual world of this model, are especially good at finding the edge habitat, even in degraded marshes with relatively little edge because of a limited vegetation. However, Haas et al. go on to illustrate that even this seemingly small reduction in survival to 70-mm can be ecologically important. They extrapolated the model results to the entire coast of Louisiana and showed that the difference between the lowest and highest survival, if applied to the entire coast, would be a difference of about 36 million pounds, or about the annual catch of shrimp reported for Louisiana (Bourgeois et al. 2016).

The modeling of Haas et al. (2004) also showed that measurements of the movement of shrimp on small scales (meters and hours) would be a worthwhile area to pursue. Putting tags and other devices on individual shrimp to record their movement is not quite possible yet, but likely will be in the next few years as the technology gets smaller and smaller. The ability of shrimp to find edge habitat is essential to the modeling results, so this behavior must be quantified with field data. Based these modeling results, a second generation version of the model was developed to improve the movement aspects of the model. Roth et al. (2008) revised the flooding algorithms by specifying an elevation for each cell to allow continuous water depths and refined the movement by switching to a one-hour time step. Finally, the analysis of Haas et al. (2004) is a good example of the power of the individual- or agent-based approach in estuarine modeling. Following thousands of individuals, while computationally complicated, made it relatively easy to include shrimp movement behavior and the effect of fine-scale spatial arrangement of marsh cells on shrimp growth and mortality.

16.6 Example 3: Coastal Habitat Under Sea Level Rise

16.6.1 Defining the Question

The need for a regional scale model of coastal habitats is based on the fact that the habitat changes and wetland flooding due to increased sea level rise into the coastal zone are driven by long-term and complex physical and biological processes. As sea level rise continues to accelerate, substantial hydrological modifications will occur across diverse scales. The question of interest for this modeling example is how will rising sea levels affect the amount and spatial arrangement of vegetated habitats in a coastal area? Also, how can the model results be easily used by local and state resource managers and planners? The model would need to be capable of simulating the responses in terms of habitat switching to both realistic and extreme rates of sea level rise because rates are highly variable and bounding possible responses using high rates is way to deal with some of the uncertainties.

North Carolina's estuaries are broad and shallow with mean depths of about 3 m. The upper ends of these systems often consist of a region of complex and connected marshes intersected by creeks, tidal channels, and streams. The regional topographic gradient is very small (i.e., small changes in elevation as one goes inland), so the marsh is highly susceptible to flooding and erosion. These estuaries are generally well mixed, and circulation is driven by tides and wind (Hart and Murray 1978). While tidal range in the region is small (about 60 cm), the shallowness of those bays makes tidal currents an important aspect of bay circulation (Hacker 1973). Wind forcing is potentially important to the hydrodynamics and shows a pattern of steady northeasterly winds in winter and fall, winds associated with the recurrence of cold air outbreaks in winter, southwesterly winds during the summer, and strong diurnal sea breeze systems along the coast. Local rainfall as well as evapotranspiration could also be important.

The following example uses an analysis of the Neuse River area that includes multiple small and large estuaries and bays of coastal North Carolina. Using habitat types estimated from satellite imagery and other sources, the change in habitats between 1991 and 2001 in this region shows indications of the transition from vegetated habitats to open water (Figure 16.12). The bottom panel is designed to highlight how sea level rise would affect switching of habitat in individual cells: "saline or brackish (marsh) to open water" cells reflects flooding and water depth, saline marsh to brackish marsh "indicates freshening of the water in the cell, and brackish marsh to saline marsh" shows saltwater intrusion. The changes appear small (green and purple in a few selected cells and orange on the barrier islands) because vegetation in these coastal habitats respond on the scales of years to decades to environmental conditions. The analysis of the imagery data was limited to certain years and suggests that even within 10 years, there are changes occurring in coastal habitats on the land–water edges, with vegetated cells becoming saltier marsh and some places switching from vegetated cells to open water.

16.6.2 Conceptualization

A long-term, spatially explicit model was designed to simulate processes that determine habitat types for a broad coastal area on the scale of decades. Vegetated habitats are often classified into types based on the amount of plant biomass

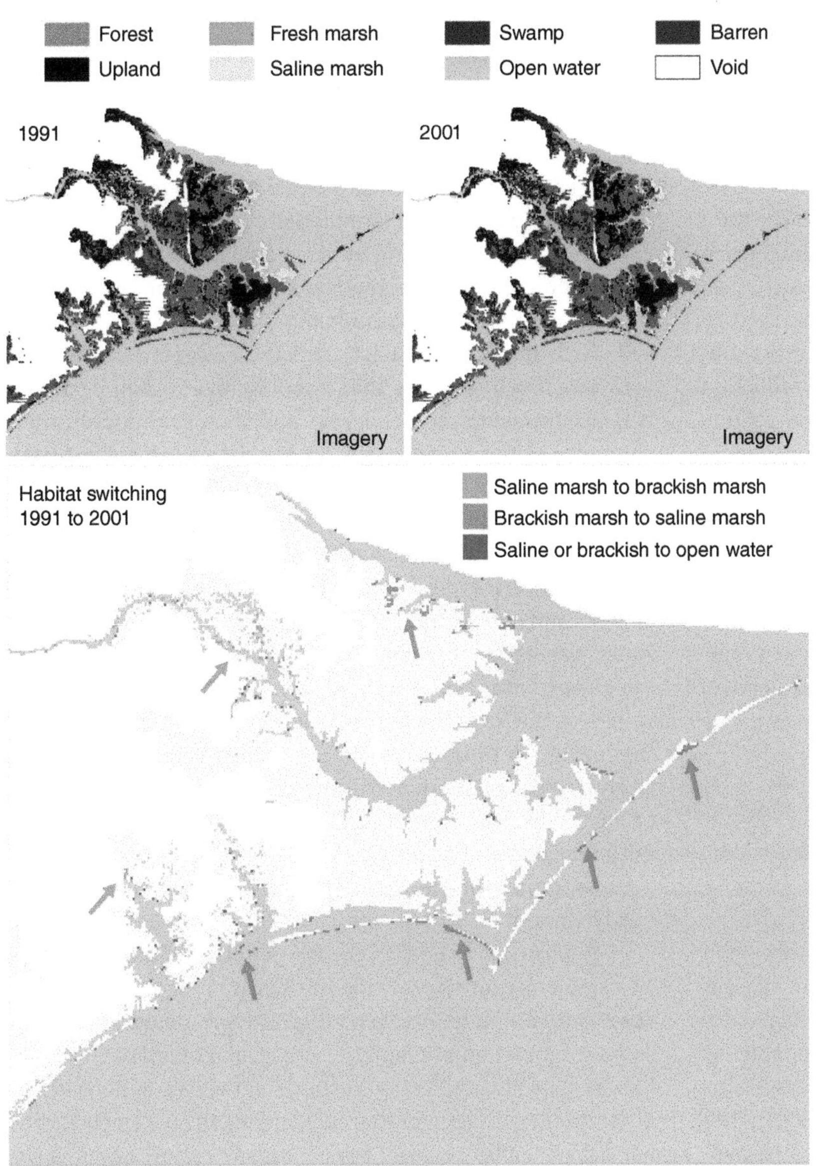

FIGURE 16.12 The upper panel shows the habitat map for the study area determined from imagery at 25 m^2 scale for 1991 (left) and 2001 (right). The Intermediate and Brackish Marsh habitats are relatively rare and cannot be seen in the figure and so they are not shown with a color in the legend. The lower panel is the difference between these habitat maps. The colors on the habitat-switching map (1991–2001) indicate the type of habitat changes that occurred related to sea level rise: saline or brackish marsh to open water (purple), saline marsh to brackish marsh (green), and brackish to saline marsh (orange).

present, salinity tolerances, and degree of flooding (e.g., Forest, Swamp, and Fresh, Intermediate, Brackish, and Saline Marsh). The marsh types show a progression in terms of salinity: Fresh (0 ppt), Intermediate (0–5 ppt), Brackish (5–15 ppt), and Saline (greater than 15 ppt). Because switching among these habitat types is based on local conditions, a spatially explicit approach is needed. Thus, most of these types of modeling studies use an approach in which the landscape is divided into a horizontal grid of spatial cells and each cell is modeled individually with exchanges of water and materials among its neighbors (Sklar and Costanza 1991). Thus, hydrodynamic information is needed to determine water depths in cells and transport material among cells. Within each cell, how much sediment is added (land building) is important to determine water depth, and biological productivity of the vegetation is needed to know the biomass of vegetation. This information can then be used to determine if the habitat type in a cell (say each day) is favorable to a different habitat type. Because the transport operates on faster time steps than the habitat changes, a coupled modeling approach is appropriate. The modules that are coupled are treated somewhat independently and exchange information so that outputs of one are used as inputs to another. Using modules allows for flexibility; if desired, each module can use its own spatial grid and time step. In this case, hydrodynamics would need a faster time step then the other modules.

The investigators had used such a coupled modeling system in other applications and so adapted that model to the Neuse River and called this version the Neuse Landscape Model (NLM). The NLM was then examined for its ability to answer the question of how will sea level rise affect habitat in the Neuse River estuary. The NLM, like its predecessor versions, has four modules: hydrodynamics, land building, biological productivity, and habitat switching. The hydrodynamic model provides the flooding of cells and the transport and exchange of nutrients, salinity, and sediments among cells. Land building takes the results of the hydrodynamic model and uses the contribution of vegetation (belowground biomass), with the net effects of sedimentation and erosion, to determine the elevation of each cell. The balance of sediment inputs and outputs is critical for predicting how wetland succession and productivity is affected by natural and human activities (Day et al. 1997; Pezeshki and DeLaune 1996; Kirby-Smith et al. 2003). The third module of biological production uses salinity and flooding to compute changes in vegetation biomass. Similar mathematical structures for daily production of vegetation were used in other models, but in this case the authors modeled the limiting effects of flooding and salinity on vegetation production rate (maximum production rate × flooding effect × salinity effect). Finally, the fourth module represents habitat switching and uses functions and rules based on the daily conditions (water quality, flooding, salinity, depth) in each cell to determine the annual habitat type of each cell. With this conceptual view of the modeling, changes in sea level rise would result in the switching of habitat types from those habitats that prefer lower water levels and less flooding to habitats that prefer more water or, if extreme enough so no vegetation can survive, to open water.

16.6.3 Formalization and Implementation

In the NLM example used here, the investigators leveraged a long history of model development designed to address similar questions by the modeling team. An initial version of the model (Costanza et al. 1988, 1990) has been periodically updated and modified as it was applied to other ecosystems (e.g., coastal Louisiana; Reyes et al. 2000, 2003, 2004). The development of the NLM described here continued the evolution of the modeling approach with its application to a new system.

The NLM was therefore constructed by reviewing the specific equations and parameter values for the four modules from the earlier applications. The developers of the NLM version then went through the modules and made modifications and decisions (e.g., cell size, grid, forcing functions, and time steps) so the model (coupled modules) was appropriate for the Neuse River system and for answering questions about the effects of sea level rise on habitat. For example, the developers considered the appropriateness of the hydrodynamic module being two dimensional (horizontal) and vertically averaged, the realism of parameters determining vegetation biomass, and the rules for habitat switching.

The domain of the grid was defined and 100 m × 100 m cells were determined to be appropriate given the rates of historical changes in habitats and the availability of data. Forcing functions were obtained from monitoring data and estimated as needed to be used with the modules. For example, the hydrodynamic module needed the forcing variables of wind, sea level rise, precipitation, river discharge, and sediment and salt concentrations in the river waters. It was determined that the hydrodynamic module could use the same grid but a shorter time step (as short as 12 minutes if rates are fast), while the land building and productivity modules operated on daily changes. Model results were accumulated for the year to then invoke the annual habitat switching formulation.

The habitat switching module was adapted to the vegetation and conditions of the Neuse River region. A key state variable predicted by the modeling is the habitat type of each cell. Each day, the biomass density of vegetation, salinity, and duration of the day that flooding (over 0.3 m) occurred was accumulated over the year and then the modal (most common) values of each was used to define the habitat type of the cell for the following year. For example, Brackish Marsh was assigned if the modal daily value of salinity was 4.5–12.5 and biomass was 0.4–2.2 kg organic matter m^{-2}, whereas Saline Marsh was assigned for modal values of salinity of 12.5–40 and of biomass of 1.2–6.0 kg organic matter m^{-2}. Open water was assigned when a cell was flooded for 24 hours of the day and modal biomass was less than all of the minimum values of the vegetated types. Flooding duration influenced the maximum production rate of the vegetation by progressively decreasing production once flooding exceeded 12 hours or more. This decrease of coefficient values was the same for both Brackish and Saline Marsh types given the similar flooding adaptation for the vegetation (*Sporobolus* sp. mostly) in both habitats.

Of particular interest with the coupled modules and the NLM implementation were the feedbacks among the modules. For example, the habitat type of a cell was determined from some combination of salinity, flooding, and biomass, and then habitat type affected various aspects of the calculations in the other three modules. For example, among the parameters that habitat type affected were the hydrodynamic calculations via habitat types having different Manning coefficients (which affect the drag of water along the bottom due to roughness) and the calculations of land building by having different values for sedimentation rate coefficients.

The model was coded in FORTRAN, partly because earlier versions were coded that way. If the model was new and developed today, NLM could be coded in similar programming languages (e.g., C, Python) and in R. Computing speed would be an important consideration and this is determined by the programming language and power of the computer being used to solve the equations.

A major effort in the formalization and implementation of the NLM was gathering field data to use for environmental variables and determining what habitat classification maps looked like through time. Types of data brought together and analyzed for the model included: (a) meteorological and environmental data from NOAA weather service archives, (b) USGS stream flow data (waterdata.usgs.gov/nc/nwis/current/?type=flow), (c) various data from The North Carolina Coastal Ocean Observing System (UNCH; ncoos.org), (d) local data from other studies to derive rating curves and sediment loads for the Neuse River, and (e) historical changes of the shoreline and shore zone over time from satellite photographs to determine the recent effects of sea level rise.

Multiple data sources had to be merged in a consistent way to avoid introducing artifacts or losing important information. For example, different data sources used different numbers of habitat classifications and different spatial resolutions. Once the habitat maps were developed, they were overlaid on an elevation map created for the same grid. Boundary conditions were enlarged and this required the use of new offshore tidal data. After evaluation, the source of boundary data that was finally determined to be best was a tidal model rather than observed data.

Given the many sources used and variety of data available and how they can be obtained in different formats and units, model development included the writing of several software utilities to aid in data preprocessing. These included the development of a utility to reformat tide information for input to the NLM, several PERL and JAVA scripts to format the NML files, and several tools to split and merge large maps so that they could be manipulated with spreadsheet programs.

TABLE 16.3 **Number of cells of each habitat type for 2001 from imagery data and from the model calibration simulation of 1991 to 2001. The habitat types of Upland and Barren included in Figure 16.13 are not shown because the model does not calculate these as part of the simulation and so there are no changes to report.**

Habitat Type	Simulated	Imagery
Intermediate Marsh	225	99
Fresh Marsh	6	8
Swamp	2292	2181
Brackish Marsh	65	0
Water (open)	54,037	54,254
Forest	9120	9950
Saline Marsh	2292	1411

16.6.4 Parameterization and Validation

A three-step calibration was performed using several types of simulations to calibrate land/water ratios, habitat type proportions, and habitat distributions. First, the model was run repeatedly year-after-year until the land/water ratio matched the ratio observed on an early imagery-based habitat map. Second, the habitat type proportions across the entire map were calibrated (1991–2001), and finally the spatial distribution of the habitat types was determined.

For land/water ratios (first step), the NLM was run using 1988 values of the forcing functions and the coefficients that determine the variation in flooding. The parameters that determine the flooding tolerance of plants were adjusted by the modeler within reasonable ranges until long-term predictions of land/water ratio matched the values determined from the imagery-based habitat maps for 1991. Predicted values of land/water ratio were nearly identical to the observed values (0.32 vs. 0.33). For habitat type proportions (second step), the model was run for 1991–2001 and the coefficients that determine salinity in model cells and the plant growth parameters were adjusted until the predicted number of cells by habitat type for 2001 was similar to the summed values from the imagery-based map for 2001 (Table 16.3). Using a method to compare spatial models to data (Costanza 1989), calibration yielded a fit of 92.4 for the NLM. This is interpreted as 92.4% agreement between the number of cells (not necessarily their spatial arrangement) of different habitat types between the predicted and imagery-based maps. The difference between the simulated and imagery maps in Figure 16.13 should be small and the goodness-of-fit high because that indicates the calibrated model results agree with the data.

Finally, to calibrate the spatial distribution of habitat types on the map (step 3), the calibration used the same adjustment strategy as in steps 1 and 2 (adjusting forcing functions that generate flooding and salinity) but now focused on adjustment until the simulated spatial arrangement of habitat types were similar to imagery-based habitat maps for 2001 (Figure 16.13). Once step 3 was accomplished, the calibration results obtained previously for step 1 (land/water ratio) and step 2 (summed cells of each habitat type) were confirmed as reasonable with the step 3 coefficients and parameters. Thus, a single set of coefficients for the forcing functions and values of parameters were obtained that satisfied the comparisons to data in all three steps.

16.6.5 Analysis

After calibration, two scenarios were developed to investigate potential consequences of sea level rise. The first scenario was implemented using the present North Carolina rate of sea level rise (3.8 mm year^{-1}). The simulation run was initiated in 2001 and ran for 40 years. Due to limitations of having only 10 years worth of data, this was done by repeating the 10-year time-series of forcing functions 4 times, but alternating the sequence of data (i.e., 2001 to 2010, then 2010 to 2001, then 2001 to 2010, and then 2010 to 2001. The simulated habitat map for 2041 (year 40 in the simulations) was compared to the simulated map for 2001 (year 1) to see how today's estimate of sea level rise would affect the distribution of habitat

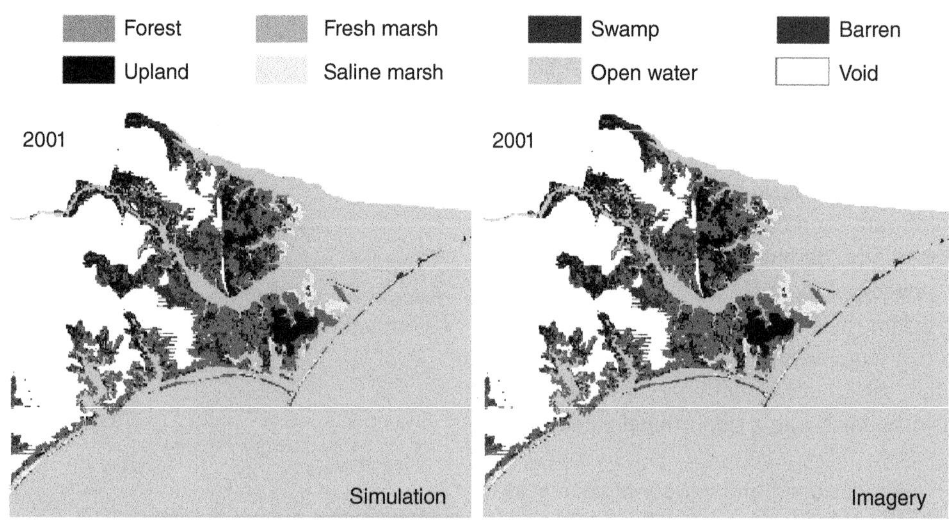

FIGURE 16.13 Calibration results of the Neuse Landscape Model comparing model simulation results for 2001 (left panel) to a 2001 habitat map based on imagery (right panel). The Intermediate and Brackish Marsh habitats are relatively rare and cannot be seen in the figure and so they are not shown with a color in the legend.

types into the future (Figure 16.14). Table 16.4 summarizes the number of cells of each habitat type in both the simulated 2001 map and the simulated 2041 map. These results indicate that under the present rate of sea level rise the changes in habitat types, when viewed over the entire map, is relatively small (Table 16.4). There is some switching of saline or brackish marsh to open water and saline marsh to brackish marsh along the smaller estuaries and in interior low-lying areas (Figure 16.14). Salt water intrusion causes switching of brackish marsh to saline marsh along the barrier islands.

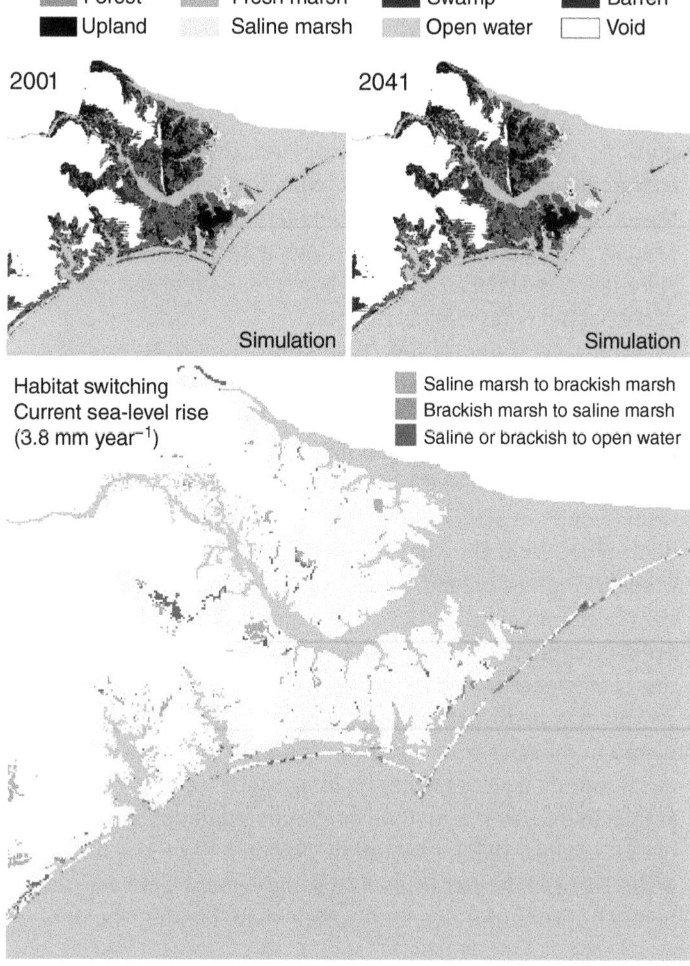

FIGURE 16.14 Simulation results of the Neuse Landscape Model under the present rate of sea level rise (3.8 mm year^{-1}). The simulated map for 2001 (left) is compared to the simulated map for 2041 (right). The Intermediate and Brackish Marsh habitats are relatively rare and cannot be seen in the figure and so they are not shown with a color in the legend. The colors on the habitat-switching map (2001–2041) indicate the type of habitat changes that occurred related to sea level rise: brackish or saline or brackish (marsh) to open water (purple), saline marsh to brackish marsh (green), and brackish marsh to saline marsh (orange). The small differences between the two maps and scattered habitat switching show the effects of a continuation of present-day sea level rise into the future will only slightly change the types and spatial distribution of habitats.

TABLE 16.4 Number of cells of each habitat type on habitat maps for 2001 and 2041 from 50-year model simulations under the present day rate of sea level rise (3.8 mm yr^{-1}) and a catastrophic rate of sea level rise (50 mm yr^{-1}). The model simulation started in year 1991 so the year 2001 is 10 years into the simulation and 2041 is the final year of the simulation. The habitat types of Upland and Barren included in Figures 16.14 and 16.15 are not shown because the model does not calculate these as part of the simulation and so there are no changes to report.

Habitat type	Rate of sea level rise			
	Baseline (3.8 mm yr^{-1})		Catastrophic (50 mm yr^{-1})	
	2001	2041	2001	2041
Intermediate Marsh	225	393	365	250
Fresh Marsh	6	5	5	3
Swamp	2292	2302	2200	1942
Brackish Marsh	65	114	515	266
Water (open)	54,037	54,106	55,463	58,405
Forest	9120	8961	8056	6331
Saline Marsh	1477	1463	668	132

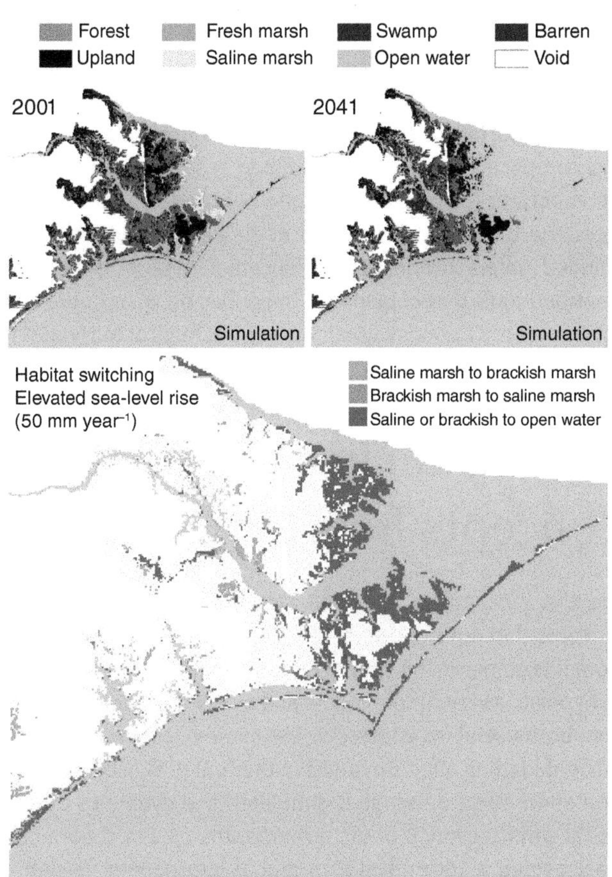

FIGURE 16.15 Simulation results of the Neuse Landscape Model under an assumed catastrophic rate of sea level rise (50 mm $year^{-1}$). The Intermediate and Brackish Marsh habitats are relatively rare and cannot be seen in the figure and so they are not shown with a color in the legend. The colors on the habitat-switching map (2001–2041) indicate the type of habitat changes that occurred related to sea level rise: or saline or brackish marsh to open water (purple), saline marsh to brackish marsh (green), and brackish marsh to saline marsh (orange). The major differences between the two maps, and the many cells switching, show that a catastrophic rate of sea level rise will cause extensive changes in the vegetation habitats, with a major effect of their conversion to open water.

To further explore model behavior under sea level rise, a second scenario of an extreme (catastrophic) rate of sea level rise of 50 mm $year^{-1}$ was used. This is presently unrealistic but puts an upper bound on model results and helps the investigators understand model dynamics. Kopp et al. (2015) reported a range of sea level rise values, with 15 cm $year^{-1}$ for 2000–2100 as an upper value; however, estimates continue to increase with each update (Grinsted and Christensen 2021). Much larger changes in habitats were predicted for this catastrophic scenario (Table 16.4; Figure 16.15). This caused much greater frequency and extent of flooding in the simulation compared to today's sea level rise rate of 3.8 mm $year^{-1}$. By 2041, there were predicted to be decreases of all vegetated habitats that were mostly replaced by open water cells (almost 3000 more water cells) and these losses were widely distributed on the land–water interface along the coastline and barrier islands. In addition to the loss of all marsh habitat types, there was also a net loss of Forest habitat cells that were located along the shores of the bays and embayments.

16.6.6 Conclusions

Predicted responses to baseline and catastrophic sea level rise to 2041 showed that changes in habitat types under the present rate of sea level rise will be small, but that the catastrophic sea level rise would result in reduction of all vegetated habitats via their conversion to open water. This catastrophic rate of sea level rise exceeds current estimates (up to 100 cm by 2100; NCRCSP 2010), but estimates of sea level rise are recalculated periodically and the rates increase with each update (Kopp et al. 2015). Knowing model responses between today's baseline value and the catastrophic rate is helpful by illustrating how the NLM would react to very high rates of sea level rise. Such clear results help interpret the smaller model-predicted changes in vegetation habitats obtained with the slower, present-day rate of 3.8 mm $year^{-1}$. Realistic expectations lie somewhere between these two assumptions about sea level rise.

Faced with some degree of potential land loss, as a result of increased sea-level rise and subsidence, the state of North Carolina requires tools to analyze coastal ecosystem response. The NLM aims to integrate and develop a landscape model to assess the quality and quantity of spatial and temporal habitat change for the eastern North Carolina coast under diverse sea level rise scenarios. NLM results could be used to help management, such as informing Coastal Habitat Protection Plans (CHPPs; **http://portal.ncdenr.org/web/mf/habitat/CHPP**). The complicated nature of predicted responses emphasizes the important role of communication of model results to managers and the public (Cartwright et al. 2016). The NLM project included outreach and public education through easy-to-use maps and tools that help inform the public on what future habitat landscapes may look like under various rates of sea level rise.

16.7 Some Important Concepts from the Examples

The three examples illustrate several important modeling concepts that apply to almost all situations. First, mechanistic modeling comes in many forms and sizes. Indeed, in Example 1 on hypoxia, a very simple model and a very complicated model were both useful for addressing the same question about how nutrient loading affects hypoxia in the Gulf of Mexico. Example 2 on shrimp use of marshes simulated the dynamics of individual shrimp every 6 hours for 9 months on 1 m square cells that covered a 100 m by 100 m grid. A subsequent version of the same shrimp model switched to an hourly time step (Roth et al. 2008). Compare this to Example 3 on habitat types that used 0.25 km^2 (500 m by 500 m) cells that covered the 5950 km^2 area with simulations that updated conditions every day for 40 years.

Second, while the models in the examples differed greatly, the mathematics used were all various types of differential or difference equations. Although it may not be obvious, the mathematics used in the three examples were standard types of equations that are not specific to their models. All of the models used numerical methods of stepping through the equations in small time steps to solve their system of equations.

Third, the models differed in their intentions to inform management. Example 2 on shrimp was developed to answer ecological questions about shrimp use of habitat on fine scales, whereas Examples 1 and 3 (hypoxia, vegetation types) had direct implications for management. Model results for Example 3 were specifically designed to enable clear communication to managers and expressed in a way to directly inform coastal management plans. Model results were presented for incorporation into Coastal Habitat Protection Plans through easy-to-use maps and visualization tools accessible to the public. Other models go further and are used as a basis for making management decisions, such as in ecosystem restoration, biological opinions for endangered species, environmental impact assessments, and setting of annual harvest levels for fisheries and wildlife populations (Jorgensen and Fath 2011).

All examples went through the steps shown in Figure 16.1, although every model may not be described or presented exactly as that specific series of steps. Figure 16.7 (Example 1) illustrates the detailed conceptual model as part of the Formalism step; Figure 16.11 (Example 2) is a flowchart as part of the Implementation step. Models are described in their final version in reports and publications which do not document the many alternatives tried and the many adjustments and modifications that were done to models to get to the final versions.

One common issue in model development (Figure 16.1) is how calibration and validation approaches can vary greatly, depending on the question to be answered, who will use the results, and the availability of data. Although not shown, the hypoxia analysis (Example 1) tested the model on independent data and used many graphs that compared model results to field data, while the shrimp model (Example 2) relied on calibration until model results, in the judgment of the modelers, roughly resembled the known general spatial patterns of shrimp affinity for edge habitat on marshes. Example 3 on vegetation habitat used a calibration approach that was intermediate between Examples 1 and 2 in terms of how rigorously field data were used. Example 3 compared model results to field data that was part of a three-step approach to calibration and used visual comparison of observed and simulated spatial habitat maps (Figure 16.13) and tabular summary of the total number of cells of each habitat type (Table 16.3). Such comparisons can be considered less rigorous than the many predicted versus observed comparisons of Example 1 but more rigorous than the simple qualitative comparison of model results to published patterns of shrimp distributions relative to edge habitat used in Example 2. Noteworthy is the way available data were used for validation of the shrimp model: field data comparing shrimp abundance between warm and cold years and adding stable isotopes to the model and applying it to the same pond as fields measurements were made.

All the models used different types of data for their development, parameterization (especially calibration), and validation (Data Synthesis box in Figure 16.1). These included field data collected at the location of the study, published laboratory and field papers on processes rates and controlling factors in many different systems, and how other models dealt with the same questions or represented the same or similar systems. Most models, other than highly theoretical analyses, involve extensive analysis of data to understand what is important in the system (including interactions and feedbacks) and to formulate the representations of processes and forcing variables (Steps 2 and 3), and to calibrate (Step 5) and validate (Step 6) the models. All steps use information and data, often from unrelated studies, that is then synthesized and creatively used throughout all of the steps.

Describing the details of a complicated model is challenging as illustrated with all three of the examples. More details for the examples can be found in their original publications but model descriptions in general do not use a universal format and organization for their description. This was a major problem with agent-based models and has initially hindered their acceptance, leading to a standard reporting format (ODD, Grimm et al. 2010) that has gained popularity. Effective communication of a model and its results to other modelers and scientists, and especially to nontechnical audiences, is critical.

A strength of mechanistic models is to not only predict state variables (which statistical models also do quite well) but also to provide an explanation of WHY the results occurred. The shrimp analysis (Example 2) illustrated this by not only reporting survival for the four maps but by using other model variables (e.g., growth rate, movement) to explain why the differences in survival occurred. The other examples also provided "mechanistic" explanations, but they were not included here in these brief summaries.

Finally, the examples show the usefulness of mechanistic ecological modeling of estuarine systems. All were used to simulate conditions not possible to observe with field data. Example 1 predicted how climate conditions expected in the future would act to increase hypoxia and maybe offset reductions in hypoxia associated with management efforts to lower nutrient loadings. Example 2 on shrimp was able to isolate the effects of different spatial arrangements of marsh by keeping all other factors, such as temperature and flooding patterns, the same. Example 3 simulated the spatial arrangement of vegetation habitats under two different scenarios of sea level rise.

16.8 The Future

The use of mechanistic modeling is accelerating as our knowledge of estuarine ecosystems expands, as computing power continues to increase, and as our questions become more complicated and involve previously unobserved conditions (e.g., the future under climate change). Mechanistic modeling was briefly described in this chapter, organized by a general set of steps for model development and use and illustrated with three examples. The examples showed how models are constructed, the commonality of the mathematics that are used to express models, and the power of modeling in addressing important environmental issues. We barely discussed the management applications of these and other mechanistic models (see Schuwirth et al. 2019).

Advances in ecological modeling will continue to accelerate in both the technical aspects and management arenas. Computing power will continue to increase and software for coding models will increasingly be more accessible. Many of the stressors that motivated ecological modeling of estuaries in the past (e.g., eutrophication, contaminants, land-use) continue to put pressure on coastal ecosystems. Overlain on these local and regional stressors is global climate change. The situation requires the use of ecological models to separate stressor effects (and thus design effective management) and to predict the state of estuaries into the future.

Continued efforts in development and use of modeling will lead to improved understanding of estuarine ecosystems, more informative field and laboratory data, and better informed management decisions (Schmolke et al. 2010). A trend is to couple models of natural systems (like our case studies) with models of people (e.g., economics, fishers, and local communities) and include dynamic two-way interactions between the models (Liu et al. 2007). This challenge must be met as environmental problems become more complex and we recognize more and more that not only are the physics, chemistry, and biology of estuaries highly coupled but so are the people to their environment.

16.9 Further Reading: Getting Started

How does one go from the general steps shown in Figure 16.1 to actually developing and using a model in their research? There are hundreds to thousands of estuarine ecological models. This suggests that there is no single standard way to develop models and shows the importance of modeling in estuarine ecology. Undoubtedly, there is a model somewhere likely similar to what you need to get you started answering your question. There are also likely many technical questions about formulating the processes and the mathematics and solution methods.

There are several ways to get started on ecological modeling. For those just curious and who want to explore ecological modeling further, there are user-friendly model packages designed for education and teaching the principles of modeling. Going further with modeling can involve taking a formal or short course in ecological modeling, just like being trained in statistics. There are also textbooks dedicated to environmental and ecological modeling (Acevedo 2013; Hannon and Ruth 2014; Jopp et al. 2011). After getting some background from a course or textbook, there is "learning from practitioners" by collaborating with other scientists that have skills in ecological modeling. In addition, the software now available for solving many ecological models is quite user-friendly and likely familiar. For example, solving differential or difference equations can be done with already written packages in R, MATLAB, Stella, and other software, and some have books on ecological modeling (e.g., R; Soetaert and Herman 2009, Stella; Ford 2010). Special software is also available for solving specific types of models; Ecopath with Ecosim (EwE) is commonly used for food web modeling of estuarine systems (Heymans et al. 2012) and there are specific packages for agent-based models (NetLogo, SWARM, and Julia).

A cautionary note: off-the-shelf models allow for relatively easy use but it is important to carefully evaluate the model in its details to ensure the model formulations are well suited for your question. You should still go through the basic steps in Figure 16.1 and never use an off-the-shelf model blindly. Such models do not always allow for easy customization, but they try to overcome this by offering different pre-programmed options for the user to select from. This caution also applies to models obtained from others and from depositories. Always be sure to go through a systematic evaluation of the model relative to the question being asked using a version of Figure 16.1.

Review Questions

Multiple Choice

1. What are reasons to use a mechanistic simulation model?
   a. To predict future conditions of ecosystem states
   b. Testing hypotheses under "controlled" model conditions
   c. To test our knowledge of ecosystems
   d. To discover new species in coastal ecosystems
2. Which of the following are steps in the modeling process?
   a. Formalization
   b. Parameterization
   c. Conceptualization
   d. All of the above
3. Which statement about model parameterization is false?
   a. Few model parameters are known with great confidence.
   b. Model parameters can be estimated from measurements
   c. Models often have a complete set of accurate and high-confidence parameter values.
   d. One approach to estimating the value of an uncertain parameter is through model calibration.
4. Which of the terms in the dissolved oxygen mass balance (Box 16.3) include parameters?
   a. The S terms (chemistry and biology)
   b. The W terms (riverine inflows)
   c. The B terms (boundary conditions)
   d. None of the above
5. Which statements about Model Example #3 (Coastal Habitat Under Sea Level Rise) are true? (Select all that apply)
   a. The model coupled hydrodynamics, land building biological productivity, and habitat switching modules.
   b. No model calibration was performed.
   c. The model was used to make future projections
   d. No future improvements to the model were suggested.
   e. Fish growth was simulated as part of the model

Short Answer Questions

1. How do mechanistic methods differ in their development from statistical models?
2. What are some advantages and disadvantages to mechanistic models over statistical models?
3. Go through the three examples and state the question and answer addressed by the modeling.
4. What step(s) in the development of a mechanistic model are usually at the center of discussions about the confidence we have in model predictions?
5. Using the Lotka-Volterra model (Box 16.2 and the 3-D Model of Oxygen Dynamics (Boxes 16.3 and 16.4), answer the following:

Lotka-Volterra Model

a. Differential or difference equations?
b. Eulerian or Lagrangian?
c. Name the state variables
d. List 3 parameters
e. What are the forcing functions?

3-D Model of Oxygen Dynamics

f. Differential or difference equations?
g. Eulerian or Lagrangian?
h. Name five of the state variables
i. List 3 parameters
j. What are the forcing functions for the S term of the DO state variable equation (Box 16.4)?

References

Acevedo, M.F. (2013). *Simulation of Ecological and Environmental Models*. Boca Raton: CRC Press.

Anand, M., Gonzalez, A., Guichard, F. et al. (2010). Ecological systems as complex systems: challenges for an emerging science. *Diversity 2* (3): 395–410.

Anisiu, M.C. (2014). Lotka, Volterra and their model. *Didáctica mathematica* 32: 9–17.

Boesch, D.F., Turner, R.E., and Day, J.W. (1984). Deterioration of coastal environments in the Mississippi Deltaic Plain: options for riverine and wetland management. In: *The Estuary as a Filter* (ed. V. Kennedy), 447–466. New York, NY: Academic Press.

Bourgeois, M., Chapiesky, K., Landry, L. et al. (2016). *Louisiana Shrimp Fishery Management Plan*. Baton Rouge: Louisiana Department of Wildlife and Fisheries Office of Fisheries.

Breckling, B., F. Jopp, and H. Reuter. 2011. Historical background of ecological modelling and its importance for modern ecology. Pages 29–40, Jopp, F., Reuter, H., Breckling, B. (editors). 2011. Modelling Complex Ecological Dynamics. Springer-Verlag, Berlin.

Cariboni, J., Gatelli, D., Liska, R., and Saltelli, A. (2007). The role of sensitivity analysis in ecological modelling. *Ecological Modelling 203* (1–2, 182): 167.

Cartwright, S.J., Bowgen, K.M., Collop, C. et al. (2016). Communicating complex ecological models to non-scientist end users. *Ecological Modelling 338*: 51–59.

Caswell, H. (1988). Theory and models in ecology: a different perspective. *Ecological Modelling 43* (1–2): 33–44.

Chen, C., Liu, H., and Beardsley, R.C. (2003). An unstructured grid, finite-volume, three-dimensional, primitive equations ocean model: application to coastal ocean and estuaries. *Journal of Atmospheric and Oceanic Technology* 20 (1): 159–186.

Chen, C., Beardsley, R.C., and Cowles, G. (2006). An unstructured grid, finite-volume coastal ocean model (FVCOM) system. *Oceanography* 19 (1): 78–89.

Conley, D.J., Carstensen, J., Vaquer-Sunyer, R., and Duarte, C.M. (2009). Ecosystem thresholds with hypoxia. In: *Eutrophication in Coastal Ecosystems*, 21–29. Dordrecht: Springer.

Costanza, R. (1989). Model goodness of fit: a multiple resolution procedure. *Ecological Modelling* 47: 199–215.

Costanza, R., Sklar, F.H., White, M.L., and Day, J.W. Jr. (1988). A dynamic spatial simulation model of land loss and marsh succession in coastal Louisiana. In: *Developments in Environmental Modelling*, vol. 12 (ed. W.J. Mitsch, M. Straškraba and S.E. Jørgensen), 99–114. Elsevier.

Costanza, R., Sklar, F.H., and White, M.L. (1990). Modeling coastal landscape dynamics. *BioScience* 40: 91–107.

Dame, R.F.E. (1979). *Marsh-Estuarine System Simulation*. Columbia, SC: University of South Carolina Press.

Day, J.W., Martin, J.F., Cardoch, L.C., and Templet, P.H. (1997). System functioning as a basis for sustainable management of deltaic ecosystems. *Coastal Management* 25: 115–153.

DeAngelis, D.L. and Grimm, V. (2014). Individual-based models in ecology after four decades. *F1000Prime Reports* 6: 39.

DiGennaro, B., Reed, D., Swanson, C. et al. (2012). Using conceptual models in ecosystem restoration decision making: an example from the Sacramento-San Joaquin River Delta, California. *San Francisco Estuary and Watershed Science 10* (3): 1–15.

Fennel, K., Laurent, A., Hetland, R., Justić, D., Ko, D. S., Lehrter, J., Murrell, M., Wang, L., Yu, L., Zhang, W. 2016. Effects of model physics on hypoxia simulations for the northern Gulf of Mexico: a model inter-comparison. Journal of Geophysical Research: Oceans 121, doi:https://doi.org/10.1002/2015JC011577.

Ford, A. (2010). *Modeling the Environment.* Washington, DC: Island Press.

Forrest, D.R., Hetland, R.D., and DiMarco, S.F. (2011). Multivariable statistical regression models of the areal extent of hypoxia over the Texas-Louisiana continental shelf. *Environmental Research Letters* 6 (4): 045002.

Fry, B., Baltz, D.M., Benfield, M.C. et al. (2003). Stable isotope indicators of movement and residency for brown shrimp (Farfantepenaeus aztecus) in coastal Louisiana marshscapes. *Estuaries 26* (1): 82–97.

Getz, W.M., Marshall, C.R., Carlson, C.J. et al. (2018). Making ecological models adequate. *Ecology Letters* 21 (2): 153–166.

Greene, R.M., Lehrter, J.C., and Hagy, J.D. (2009). Multiple regression models for hindcasting and forecasting midsummer hypoxia in the Gulf of Mexico. *Ecological Applications* 19: 1161–1175.

Grimm, V., Berger, U., DeAngelis, D.L. et al. (2010). The ODD protocol: a review and first update. *Ecological Modelling.* 221 (23): 2760–2768.

Grinsted, A. and Christensen, J.H. (2021). The transient sensitivity of sea level rise. *Ocean Science 17* (1): 181–186.

Haas, H.L., Rose, K.A., Fry, B. et al. (2004). Brown shrimp on the edge: linking habitat to survival using an individual-based model. *Ecological Applications* 14: 1232–1247.

Hacker, S. (1973). Transport phenomena in Estuaries. PhD dissertation. Louisiana State University, Baton Rouge, La.

Hannon, B. and Ruth, M. (2014). *Modeling Dynamic Biological Systems*, 2e. New York: Springer Cham.

Hart, W.E. and Murray, S.P. (1978). Energy balance and wind effects in a shallow sound. *Journal of Geophysical Research: Oceans 83* (C8): 4097–4106.

van der Heide, T., van Nes, E.H., van Katwijk, M.M. et al. (2011). Positive feedbacks in seagrass ecosystems–evidence from large-scale empirical data. *PLoS One 6* (1): e16504.

Heymans, S., Coll, M., Libralato, S., and Christensen, V. (2012). Ecopath theory, modelling and application to coastal ecosystems. In: *Treatise on Estuarine and Coastal Science*, 1e, vol. 9 (ed. D. McLusky and E. Wolanski), 93–111. Elsevier.

Jakeman, A.J., Letcher, R.A., and Norton, J.P. (2006). Ten iterative steps in development and evaluation of environmental models. *Environmental Modelling & Software 21* (5): 602–614.

Janssen, P.H.M. and Heuberger, P.S.C. (1995). Calibration of process-oriented models. *Ecological Modelling 83* (1-2): 55–66.

Ji, Z.-G. (2017). *Hydrodynamics and Water Quality: Modeling Rivers, Lakes, and Estuaries.* Hoboken, NJ: Wiley.

Jopp, F., Reuter, H., and Breckling, B. (ed.) (2011). *Modelling Complex Ecological Dynamics.* Berlin: Springer-Verlag.

Jorgensen, S.E. and Bendoricchio, G. (2001). *Fundamentals of Ecological Modelling.* New York: Elsevier Science & Technology Books.

Jorgensen, S.E. and Fath, B.D. (2011). *Fundamentals of Ecological Modelling: Applications in Environmental Management and Research.* Amsterdam: Elsevier. ISBN 9780444535672.

Justić, D. and Wang, L. (2014). Assessing temporal and spatial variability of hypoxia over the inner Louisiana–upper Texas shelf: Application of an unstructured-grid three-dimensional coupled hydrodynamic-water quality model. *Continental Shelf Research* 72: 163–179.

Justić, D., Rabalais, N.N., and Turner, R.E. (1996). Effects of climate change on hypoxia in coastal waters: a doubled CO_2 scenario for the northern Gulf of Mexico. *Limnology and Oceanography* 41: 992–1003.

Justić, D., Rabalais, N.N., and Turner, R.E. (2002). Modeling the impacts of decadal changes in riverine nutrient flux on coastal eutrophication near the Mississippi River Delta. *Ecological Modelling* 153: 33–46.

Justić, D., Rabalais, N.N., and Turner, R.E. (2003). Simulated responses of the Gulf of Mexico hypoxia to variations in climate and anthropogenic nutrient loading. *Journal of Marine Systems 42* (3-4): 115–126.

Justić, D., Rabalais, N.N., and Turner, R.E. (2005). Coupling between climate variability and marine coastal eutrophication: historical evidence and future outlook. *Journal of Sea Research* 54: 25–35.

Justić, D., Bierman, J.V., Scavia, D., and Hetland, D.R. (2007). Forecasting Gulf's hypoxia: the next 50 years? *Estuaries and Coasts* 30 (5): 791–801.

Kirby-Smith, W.W., Lebo, M.E., and Herrmann, R.B. (2003). Importance of water quality to nekton habitat use in a North Carolina branch estuary. *Estuaries 26* (6): 1480–1485.

Kleijen, J.P.C. (1998). Experimental design for sensitivity analysis, optimization, and validation of simulation models. In: *Handbook of Simulation: Principles, Methodology, Advances, Applications, and Practice* (ed. J. Banks), 173–223. New York: Wiley.

Kopp, R.E., Horton, B.P., Kemp, A.C., and Tebaldi, C. (2015). Past and future sea-level rise along the coast of North Carolina, USA. *Climatic Change* 132: 693–707.

Korobeinikov, A. and Wake, G.C. (1999). Global properties of the three-dimensional predator-prey Lotka-Volterra systems. *Journal of Applied Mathematics and Decision Sciences 3* (2): 155–162.

Křivan, V. (2007). The Lotka-Volterra predator-prey model with foraging-predation risk trade-offs. *The American Naturalist 170* (5): 771–782.

Langseth, B.J., Purcell, K.M., Craig, J.K. et al. (2014). Effect of changes in dissolved oxygen concentrations on the spatial dynamics of the Gulf Menhaden fishery in the northern Gulf of Mexico. *Mar Coast Fish* 6: 223–234.

Larsen, L.G., Eppinga, M.B., Passalacqua, P. et al. (2016). Appropriate complexity landscape modeling. *Earth-Science Reviews 160*: 111–130.

Lawton, J.H. (1995). Ecological experiments with model systems. *Science 269* (5222): 328–331.

Liu, J., Dietz, T., Carpenter, S.R. et al. (2007). Complexity of coupled human and natural systems. *Science 317* (5844): 1513–1516.

May, R.M., Beddington, J.R., Clark, C.W. et al. (1979). Management of multispecies fisheries. *Science 205* (4403): 267–277.

Minello, T.J. (1999). *Nekton Densities in Shallow Estuarine Habitats of Texas and Louisiana and the Identification of Essential Fish Habitat*, 43–75. American Fisheries Symposium.

NCRCSP (North Carolina Coastal Resources Commission's Science Panel) (2010). *North Carolina Sea-Level Rise Assessment Report.* Division of Coastal Management. 16: North Carolina Department of Environment and Natural Resources.

Nixon, S.W. and Oviatt, C. (1973). Ecology of a New England salt marsh. *Ecological Monographs* 43: 463–498.

Oaten, A. and Murdoch, W.W. (1975). Functional response and stability in predator-prey systems. *The American Naturalist 109* (967): 289–298.

Odum, H.T. (1960). Ecological potential and analogue circuits for the ecosystems. *American Journal of Science* 48: 1–8.

Ogden, J.C., Davis, S.M., Jacobs, K.J. et al. (2005). The use of conceptual ecological models to guide ecosystem restoration in South Florida. *Wetlands 25* (4): 795–809.

Pezeshki, S.R. and DeLaune, R.D. (1996). Factors controlling coastal wetland formation and losses in the northern Gulf of Mexico, USA. *Recent Research and Developments in Coastal Resources* 1: 13–27.

Pomeroy, L.R., Shenton, L.R., Jones, R.D.H., and Reimold, R.J. (1972). Nutrient flux in estuaries. *Limnology and* Oceanography:274-291.

Rabalais, N.N. and Turner, R.E. (2001). Hypoxia in the northern Gulf of Mexico: description, causes and change. In: *Coastal Hypoxia: Consequences for Living Resources and Ecosystems* (ed. N.N. Rabalais and R.E. Turner), 1–36. Washington, D.C: American Geophysical Union.

Rabalais, N.N., Turner, R.E., Justić, D. et al. (1996). Nutrient changes in the Mississippi River and system responses on the adjacent continental shelf. *Estuaries* 19: 386–407.

Rabalais, N.N., Turner, R.E., and Scavia, D. (2002). Beyond science into policy: Gulf of Mexico hypoxia and the Mississippi River. *BioScience* 52: 129–142.

Rabalais, N.N., Turner, R.E., Sen Gupta, B.K. et al. (2007). Hypoxia in the northern Gulf of Mexico: does the science support the plan to reduce, mitigate, and control hypoxia? *Estuaries and Coasts* 30 (5): 753–772.

Reyes, E., White, M.L., Martin, J.F. et al. (2000). Landscape modeling of coastal habitat change in the Mississippi delta. *Ecology* 81: 2331–2349.

Reyes, E., Martin, J.F., White, M.L. et al. (2003). Habitat changes in the Mississippi Delta: future scenarios and alternatives. In: *Explicit Landscape Modeling* (ed. A. Voinov and R. Costanza), 119–142. New York: Springer-Verlag.

Reyes, E., Day, J.W., Lara-Domínguez, A.L. et al. (2004). Assessing coastal management plans using watershed spatial models for the Mississippi delta, USA, and the Ususmacinta–Grijalva delta, Mexico. *Ocean & Coastal Management 47* (11–12): 693–708.

Rose, K.A., Megrey, B.A., Werner, F.E., and Ware, D.M. (2007). Calibration of the NEMURO nutrient–phytoplankton–zooplankton food web model to a coastal ecosystem: Evaluation of an automated calibration approach. *Ecological Modelling 202* (1-2): 38–51.

Rose, K.A., Sable, S., DeAngelis, D.L. et al. (2015). Proposed best modeling practices for assessing the effects of ecosystem restoration on fish. *Ecological Modelling 300*: 12–29.

Roth, B.M., Rose, K.A., Rozas, L.P., and Minello, T.J. (2008). Relative influence of habitat fragmentation and inundation on brown shrimp Farfantepenaeus aztecus production in northern Gulf of Mexico salt marshes. *Marine Ecology Progress Series 359*: 185–202.

Rozas, L.P., Minello, T.J., and Henry, C.B. (2000). An assessment of potential oil spill damage to salt marsh habitats and fishery resources in Galveston Bay, Texas. *Marine Pollution Bulletin* 40: 1148–1160.

Rykiel, E.J. Jr. (1996). Testing ecological models: the meaning of validation. *Ecological Modelling* 90 (3): 229–244.

Saltelli, A., Ratto, M., Andres, T. et al. (2008). *Global Sensitivity Analysis: The Primer.* Wiley.

Scavia, D., Evans, M.A., and Obenour, D. (2013). A scenario and forecast model for Gulf of Mexico hypoxic area and volume. *Environmental Science & Technology* 47: 10423–10428.

Scavia, D., Justić, D., Obenour, D.R. et al. (2019). Hypoxic volume is more responsive than hypoxic area to nutrient load reductions in the northern Gulf of Mexico—and it matters to fish and fisheries. *Environmental Research Letters 14* (2): 024012.

Schmolke, A., Thorbek, P., DeAngelis, D.L., and Grimm, V. (2010). Ecological models supporting environmental decision making: a strategy for the future. *Trends in Ecology & Evolution 25* (8): 479–486.

Schuwirth, N., Borgwardt, F., Domisch, S. et al. (2019). How to make ecological models useful for environmental management. *Ecological Modelling 411*: 108784.

Sen Gupta, B.K., Turner, R.E., and Rabalais, N.N. (1996). Seasonal oxygen depletion in continental-shelf waters of Louisiana: historical record of benthic foraminifers. *Geology* 24: 227–230.

Sklar, F.H. and Costanza, R. (1991). *Dynamic Spatial Models Quantitative Methods in Landsccape Ecology* (ed. M.G. Turner and R.H. Gardner). Springer.

Smith, M.D., Oglend, A., Kirkpatrick, J. et al. (2017). Seafood prices reveal impacts of a major ecological disturbance. *Proceedings of the National Academy of Science* 114 (7): 1512–1517.

Soetaert, K. and Herman, P.M. (2009). *A Practical Guide to Ecological Modelling: Using R as a Simulation Platform.* Springer Science & Business Media.

Task Force (2001). *Mississippi River/Gulf of Mexico Watershed Nutrient Task Force Action Plan for Reducing, Mitigating, and Controlling Hypoxia in the Northern Gulf of Mexico; Office of Wetlands, Oceans, and Watersheds.* Washington, DC: U.S. Environmental Protection Agency 2001; http://water.epa.gov/type/watersheds/named/msbasin/history.cfm.

Task Force (2008). *Mississippi River/Gulf of Mexico Watershed Nutrient Task Force Gulf Hypoxia Action Plan 2008 for Reducing Mitigating, and Controlling Hypoxia in the Northern Gulf of Mexico and Improving Water Quality in the Mississippi River Basin.* Washington, DC: U.S. Environmental Protection Agency, Office of Wetlands, Oceans, and Watersheds 2008; http://water.epa.gov/type/watersheds/named/msbasin/actionplan.cfm.

Task Force 2015. Mississippi River/Gulf of Mexico Watershed Nutrient Task Force. Report to Congress. (Mississippi River/Gulf of Mexico

Watershed Nutrient Task Force, Washington, DC). https://www.epa.gov/sites/production/files/2015-10/documents/htf_report_to_congress_final_-_10.1.15.pdf. (accessed 1 March 2017).

Teal, T.M. (1962). Energy flow in the salt marsh ecosystem of Georgia. *Ecology* 43: 614–624.

Turner, R.E. and Rabalais, N.N. (1994). Coastal eutrophication near the Mississippi river delta. *Nature* 368: 619–621.

Twilley, R.R. (ed.) (2004). *Appendix C—Hydrodynamic and Ecological Modeling*. New Orleans: US Army COE and LA Department of Natural Resources.

Ulanowicz, R.E., Flemer, D.A., Heinle, D.R., and Mobley, C.D. (1975). The a posteriori aspects of estuarine modelling. In: *Estuarine Research* (ed. L.E. Cronin), 602–616. New York: Academic Press, Inc.

Wang, L. and Justić, D. (2009). A modeling study of the physical processes affecting the development of seasonal hypoxia over the inner Louisiana-Texas shelf: circulation and stratification. *Continental Shelf Research* 9: 1464–1476.

Wangersky, P.J. (1978). Lotka-Volterra population models. *Annual Review of Ecology and Systematics* 9 (1): 189–218.

Weisberg, M. (2006). Forty years of "the strategy": Levins on model building and idealization. *Biology and Philosophy 21* (5): 623–645.

Wiegert, R.G., Christian, R.R., Gallagher, J.L. et al. (1975). A preliminary ecosystem model of coastal Georgia Spartina marsh. In: *Estuarine Research* (ed. L.E. Cronin). New York: Academic Press.

CHAPTER 17

Estuarine Fisheries and Aquaculture

Floats for aquaculture of scallop, mussel, oyster, and sea squirt in Geoje-Hansan Bay, South Korea. Photo credit: Geneviève Nesslage.

Geneviève Nesslage[1] and Daniel Pauly[2]
[1]Chesapeake Biological Laboratory, University of Maryland Center for Environmental Science, Solomons, MD, USA
[2]Institute for the Oceans and Fisheries & Department of Zoology, The University of British Columbia, Vancouver, BC, Canada

17.1 Introduction

Estuaries are highly productive ecological systems that support extensive fisheries and aquaculture activities around the world (Kapetsky 1984). The term "**fisheries**" refers to the capture of wild aquatic organisms which include not only finfish, but also shellfish and other invertebrates (Box 17.1). In contrast, **aquaculture** involves the farming of animals or plants in aquatic environments.

Estuarine fisheries and aquaculture play an essential role in global seafood production (FAO 2018d). Over half of humanity lives in coastal areas (Vitousek et al. 1997) and many coastal populations rely on fisheries and aquaculture products as a major source of animal protein and micronutrients (Kawarazuka and Béné 2011; Hall et al. 2013; Tacon and Metian 2018). Estuaries are the source of an estimated 16 percent of non-oceanic, global fisheries **yield** (Houde and Rutherford 1993), and aquaculture is one of the fastest growing food animal production industries in the world, accounting for almost half of global seafood supply (Campbell and Pauly 2013; FAO 2018d). Current and anticipated impacts of human population growth and climate change on estuarine-based fisheries and aquaculture pose a threat to global food security and nutrition, particularly in developing countries (FAO 2018c, d; Teh and Pauly 2018).

Estuaries are critical to the success of many fisheries because they serve as habitat during all or certain stages in the life cycle of many wild populations (Pihl et al. 2002; Able 2005). Some species found in estuaries are lifelong residents of estuaries, whereas other species use estuaries as nursery or juvenile habitat while spending the majority of their lives in the ocean or freshwater (see Chapter 12). Also, high secondary productivity makes estuaries important foraging grounds for many piscivorous stocks (see Chapter 15). Some of the largest and economically important fisheries in the world target estuarine-dependent species such as menhadens, herrings, breams, mullets, temperate basses, salmons, crabs, flatfish (e.g., sole, plaice, and flounder), eels, shads, sturgeons, shrimp, and clams (Costa et al. 2002; Wilson 2002; Pérez-Ruzafa and Marcos 2012). Pihl et al. (2002) estimated that nearly 40% of commercially exploited fish species in Europe use estuaries at some stage in their life cycle. In the United States, estuaries serve as habitat for over 80% of commercially caught fish,

Estuarine Ecology, Third Edition. Edited by Byron C. Crump, Jeremy M. Testa, and Kenneth H. Dunton.

Companion website: www.wiley.com/go/crump/estuarine3

Box 17.1 Fisheries and Aquaculture Terms

FISHERIES—the capture of wild aquatic organisms which include not only finfish, but also shellfish and other invertebrates for use as human food, fishmeal (animal feed, often for farmed fish), oil, medicine, and other products. Fisheries may be subdivided into commercial (or "industrial"), recreational, artisanal (small-scale commercial), and subsistence sectors.

AQUACULTURE—the breeding, rearing, and harvesting of aquatic organisms in water, primarily for use as human food. MARICULTURE is a subset of all aquaculture that is restricted to brackish and marine waters.

STOCK—a wild fished population of aquatic organisms that is typically self-sustaining, geographically isolated from others of the same species, and managed as a distinct unit.

YIELD—the amount of products generated from fisheries or aquaculture activities, typically measured by weight.

RECRUITMENT—fish in a given stock that have grown large enough or migrated to new areas such that they have become vulnerable to capture fisheries. This term may also be used more generally to describe production of juvenile fish.

CATCH-PER-UNIT-EFFORT—the amount (number or biomass) of fish caught as a function of a standard unit of fishing effort.

STOCK ASSESSMENT—the process of collecting and analyzing biological and fishery information to estimate stock abundance and the stock's response to fishing.

COHORT—a group of fish born during the same spawning season or other time frame meaningful to that species.

REFERENCE POINTS—A benchmark used in fisheries management against which stock biomass and fishing mortality rate can be compared. Reference points often focus on either the desired state of the fishery called the "target" (e.g., fishing mortality that will produce sustainable yield), or a state of the fishery that should be avoided called a "threshold" or "limit."

NATURAL MORTALITY—Natural mortality includes any cause of death that cannot be attributed to the effects of fishing, especially such processes as predation and disease.

FISHING MORTALITY—The rate at which fish are being removed from the population due to fishing activities.

STOCK STATUS—a formal determination made describing the condition of the stock based on the results of a stock assessment or other analyses of available data. Although multiple stock status determinations exist, the most common are: (i) the stock is/is not experiencing "overfishing" (i.e., the rate of fishing mortality is higher than our fishing mortality threshold or limit **reference point**), and (ii) the stock is/is not "overfished" (the stock size is lower than our biomass threshold or limit **reference point**).

BYCATCH—Incidental catch of nontarget species during fishing activities that is either retained or discarded at sea.

LANDINGS—Portion of the commercial fishery catch that is brought back to port and sold (or given away in case of subsistence or recreational fisheries).

and over 50% of recreational catches come from estuaries (National Research Council 1997; National Marine Fisheries Service 2017).

Given the migratory nature of many fish populations, estuaries have broad, regional impacts that extend far beyond their physical borders. For migratory species, conditions in estuaries can affect the entire population. For example, if growth and survival conditions for juvenile grey mullet (*Mugil cephalus*) in Chinese estuaries is poor, mullet **recruitment** in the purse seine and trail net fisheries along the coasts of both China and Taiwan will be negatively impacted (Lan et al. 2017). Thus, the management of fisheries that rely on estuarine-dependent species often require complex regional and interjurisdictional systems (see Section 17.4).

Estuaries are also home to a large portion of global finfish and shellfish aquaculture activities (FAO 2018d). The most commonly farmed groups in estuaries include penaeid shrimps and bivalves such as mussels, clams, and oysters that can take advantage of their highly productive ecosystem by filtering phytoplankton for food. Estuarine aquaculture typically involves the use of artificial substrates (e.g., shell piles, racks, ropes, and lines) or fish cages and pens (Macintosh 1994; Bostock et al. 2010). Culture in estuaries is advantageous because they are typically shallow, protected, and located in close proximity to land (Macintosh 1994). However, nearness of estuarine aquaculture operations to shore generates management challenges, including conflicts with fisheries, negative impacts on biodiversity and coastal health, and competition with shoreline development interests (see Section 17.4). As aquaculture continues to outpace fisheries in seafood production (FAO 2018d), research has highlighted the complex and situational nature of the impacts aquaculture can have on estuaries (Cranford et al. 2006; Dumbauld et al. 2009; Forrest et al. 2009; Kellogg et al. 2013; Kellogg et al. 2014; Testa et al. 2015; Murphy et al. 2016).

In this chapter, we begin by providing an overview of global patterns and potential drivers of estuarine fisheries and aquaculture yields. We then transition into a description of the types of information used to monitor and assess wild fish dynamics in estuaries. We conclude with an overview of fisheries and aquaculture management in estuaries with a focus on current and future challenges.

17.2 Estuarine Yield

Although fisheries and aquaculture yields may vary, estuaries are typically more productive than other types of ecosystems (Kapetsky 1984; Pérez-Ruzafa and Marcos 2012). High yields are driven by a number of physical, biological, and anthropogenic factors, including high primary productivity and proximity to land (Houde and Rutherford 1993; Macintosh 1994). In this section, we compare fisheries and aquaculture yield across estuaries and between estuaries and other ecosystems. We then discuss a suite of potential biotic and abiotic drivers of estuarine yield.

17.2.1 Global Patterns

Estuaries are thought to be highly productive in terms of fisheries and aquaculture yields, yet quantifying that yield is a challenge because many countries do not distinguish estuarine from coastal ocean or freshwater production statistics (Pérez-Ruzafa and Marcos 2012; FAO 2018d). Houde and Rutherford (1993) estimated that approximately 16% of non-oceanic global fisheries yield is derived from estuaries. Approximately 36% of global aquaculture yield can be attributed to coastal and marine areas, which include estuaries; however, extensive farming of species that are estuarine-dependent in the wild also occurs in inland ponds in which salinity is controlled (Macintosh 1994; FAO 2018d).

Highly productive estuaries are found in many tropical, subtropical and temperate regions around the globe (Kapetsky 1984; Pérez-Ruzafa and Marcos 2012). Some of the highest reported estuarine yields come from the large estuaries of the Northwest Atlantic (Pérez-Ruzafa and Marcos 2012). For example, the 250,000 km^2 of the Gulf of St. Lawrence, Canada, yielded fisheries catches of 400,000 tonnes (hereafter t), plus 90,000 t from aquaculture in 2016 (Fisheries and Oceans Canada). In that same year, approximately 200,000 t of fisheries yield was reported from the 165,000 km^2 area of the Chesapeake Bay in the USA (NOAA Fisheries). Among coastal lagoons, the highest **catch-per-unit-effort** from capture fisheries has been reported in the Northwest Atlantic, East Indian Ocean, Pacific Southwest (Australia), and Mediterranean Sea regions (Pérez-Ruzafa and Marcos 2012).

Although aquaculture is expanding worldwide, Asia far exceeds other regions in terms of marine and brackish seafood production (Campbell and Pauly 2013; FAO 2018d). China's coastal provinces of Liaoning, Shandong, Fujian, and Guangdong alone produced more farmed seafood than any other maritime country between 2000 and 2010 (Campbell and Pauly 2013). Even when China is excluded, Asia remains the continent producing the most farmed seafood due to high yields in countries such as India, Indonesia, and Viet Nam (Campbell and Pauly 2013).

17.2.2 Physical, Biological, and Evolutionary Drivers

High fisheries and aquaculture yields from estuaries are generally explained by their high primary production (Nixon 1982, 1988; see also Chapters 1 and 4). High primary production is driven by physical factors such as: (i) the availability of organic matter inputs via rivers (Yáñez-Arancibia et al. 1985a, b; Day et al. 1997; Bianchi et al. 1999); (ii) shallowness, which is conducive to rapid remobilization of nutrients (Jones 1982; Nixon 1982; Deegan 2002); and (iii) velocity and volume of water exchanges between the sea and the estuarine system, which also affects fish populations directly via recruitment (Deegan et al. 1986; Yáñez-Arancibia et al. 2007). In addition, high yields are driven by proximity of estuaries to ports and human settlements which makes them convenient for aquaculturists in particular, but also for fishers (Houde and Rutherford 1993; Macintosh 1994).

TABLE 17.1 Fisheries yields of coastal lagoons (all groups included) as compared with the yields of other aquatic ecosystems.

Systems	Yield (t km^{-2} year^{-1}) Median	Mean	n
Coastal lagoons	5.1	11.3	107
Continental shelves	4.8	5.9	20
African/Asian reservoirs	4.2	7.5	41
Coral reefs	4.1	4.9	15
River floodplains	3.2	4.0	33
Reservoirs (United States)	1.3	2.4	148
Natural lakes	0.5	2.8	43

Source: Kapetsky (1984)/FAO.

The impact of primary production and other environmental influences on fisheries and aquaculture yields vary across ecosystems. Estuaries have been found to be twice as productive for aquaculture as coastal seas and freshwater systems (Macintosh 1994). Fisheries yields from estuaries also are, overall, higher than the yields from other exploited marine and fresh water ecosystems, whether one considers the mean or the median as a measure of central tendency (Table 17.1; Kapetsky 1984). However, estuaries do not have uniformly high fisheries yields, and, indeed, the frequency distribution is strongly skewed, indicating numerous instances of unproductive estuaries (Figure 17.1). The extremely productive estuaries in Figure 17.1 may benefit from a number of positive factors, including:

1. Nearby coastal habitat supplying high recruitment (Pauly 1986a);
2. Fertilization from rivers, agricultural runoff, or human sewage, and through water exchanges with the ocean (Kapetsky 1984);
3. A management regime that makes the best of incoming recruitment.

Some factors that likely explain the occurrence of the unproductive estuaries in the long tail of this nearly log-normal distribution are:

1. Extreme salinity and temperature fluctuations, turbidity, anoxic conditions, or toxic discharges;

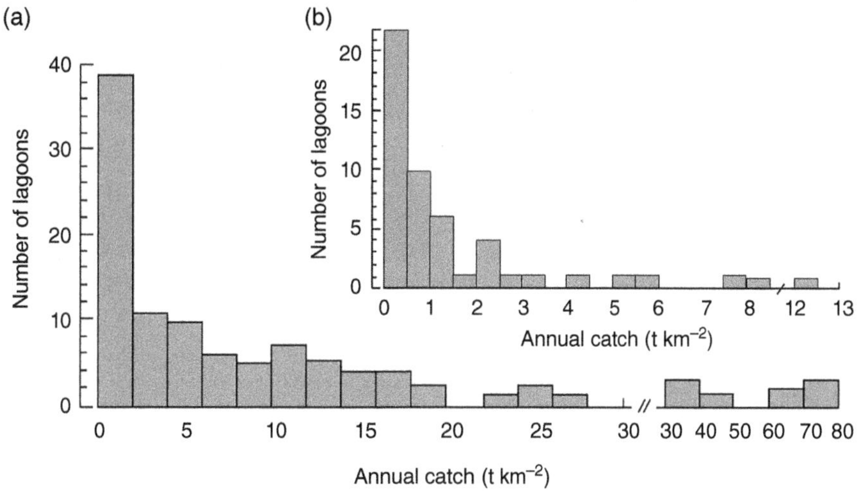

FIGURE 17.1 Frequency distribution of fishery catches from coastal lagoons.
Source: Kapetsky (1984)/FAO. (a) Finfish, $n = 106$. (b) Penaeid shrimp, $n = 51$.

2. Very shallow sills, preventing high recruitment;
3. Excessive illumination or turbidity, both of which can lead to reduced primary productivity (Qasim 1970, 1973);
4. Insufficient fishing effort (Bayley 1988), a condition that is increasingly rare.

A major component of high fisheries production in many estuaries is high natural recruitment (Day et al. 1989; Yáñez-Arancibia et al. 1993, 1994; Mann 2009). High recruitment in estuaries generally leads to high fishery catches once these animals are large enough to interact with fishing gear. One reason for this high recruitment is that many of the fish species of importance in estuarine environments are either (i) r-selected sensu (Pianka 2011), that is, relatively small, fast growing, with a high production/biomass (P/B) ratio or (ii) the juveniles of *K*-selected species, that is, the fast growing, high production stage fishes whose adult form, however, typically occurs outside estuaries. The high P/B ratios of the constituent species are not the only explanation for the generally high production of estuarine fish communities. The high biomasses themselves, that is, the high carrying capacities of these ecosystems in terms of seasonal food availability to fishes due to high primary and secondary production, also play a crucial role.

In areas where estuarine systems have maintained themselves over long periods (i.e., the Gulf of Mexico), evolutionary mechanisms have emerged that have stabilized and refined such seasonal programming, making the fish population in question gradually more dependent on the estuarine system for the maintenance of high biomass. In other areas where the estuarine system does not persist or is not regularly open to juvenile migration (i.e., along the coast of Northwestern Africa, the Pacific coast of Mexico, and the eastern coast of East Australia), the use of lagoons seems to be more a matter of random movements along the coast and of inshore movements. This implies (i) a lower conversion of primary and secondary production into fish flesh and hence (ii) lower biomasses of coastal fishes. However, the practical difficulties in separating random along-shore/inshore movements from evolutionary, fine-tuned, aimed movements toward and within estuaries, and the difficulties involved in precise field estimation of biomass and conversion efficiencies, make rigorous testing of the hypothesis difficult. Indeed, this may be the main reason for the continuing debate on the degree of dependence of coastal fishes on estuaries (Able 2005; Litvin et al. 2018).

17.3 Fish Population Dynamics and Its Four Factors

The yield from both fisheries and aquaculture practices are impacted by estuarine conditions and the life history characteristics of the fished species. In the case of aquaculture, yields are a function of the quality of seed animals, culture practices, and environmental conditions, such as the quality of the water in the production area. However, determining the yield that can be sustainably removed from a wild population is much more difficult because the animals are not constrained to one location and the initial conditions of the stock are often unknown.

To estimate fishing mortality rates and set sustainable quotas for a given fishery, scientists conduct a **stock assessment** in which fish populations are monitored and the resulting data either examined or, when possible, used to build a statistical model that estimates fish stock size, fishing mortality rates, and management **reference points**. The goal of this section is to provide an overview of the basic concepts of fish population dynamics used in fisheries stock assessment. Although we refer to the dynamics of fish populations throughout this section, many of these same concepts and the same or similar techniques can be applied to shellfish and other exploited animal populations as well.

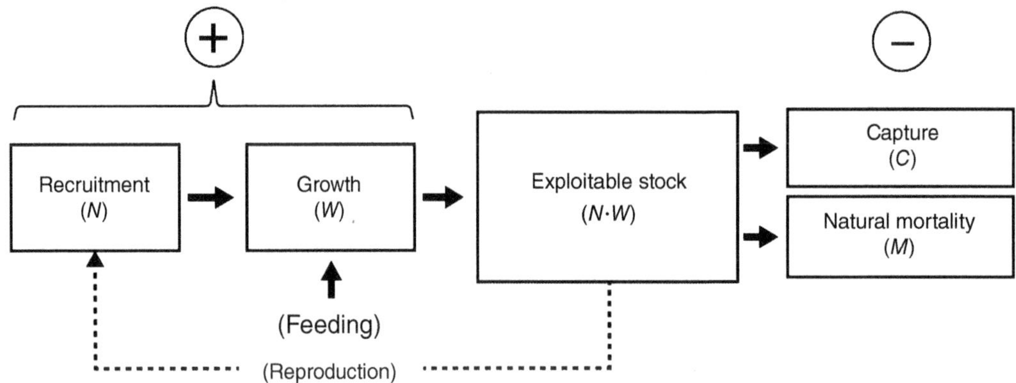

FIGURE 17.2 Schematic representation of four key factors determining the future biomass of and catches from a stock: Recruitment where N = number of fish added to the stock, individual growth in weight (W) of the fish recruited into the exploitable stock, M = natural mortality, and C = capture via fishing mortality. *Source*: Adapted from Russel (1931) and Ricker (1975).

17.3.1 Russell's Axiom

Fish population dynamics are often represented by means of Russell's axiom, that is:

$$B_2 = B_1(R+G)-(M+F), \tag{17.1}$$

which states that a well-defined stock of biomass (B_1) will have, after an arbitrary period Δt, the biomass B_2 as a result of positive processes (R = recruitment; G = growth) that have added to the stock, while negative processes (M = **natural mortality**; F = **fishing mortality**) have reduced it (Russell 1931; Ricker 1975). The four processes included in Eq. (17.1), plus some peripheral processes also considered by fishery biologists (Figure 17.2), have been studied extensively and described in a mathematically tractable form by numerous authors (Ricker 1954; Beverton and Holt 1957; Schaefer 1957; Ricker 1975; Gulland 1983; Pauly 1984; Hilborn and Walters 1992; Pauly 1998; Quinn and Deriso 1999). In this section, we present a few of the models that have resulted from these efforts, specifically those that have been used to characterize fish dynamics in estuarine systems. The four factors shall be examined in the sequence: growth, natural mortality, fishing mortality, and related factors (mainly catch/effort), with recruitment being last, because it is the most complex factor to investigate, model, and predict.

17.3.2 Growth of Fishes

Information on how fish grow is an important component of fish population dynamics for several reasons. First, the relationship between length and age tells us how early in life an animal might become susceptible to fishing mortality and, if associated maturity data are available, how many chances the fish might have to spawn before being caught. Differences in growth between sexes and across a species' range are important for managers to consider when setting regulations such as minimum size limits. Second, quantitative relationships between the length and age of fish are commonly used to convert length data to age data for use in stock assessment models. The most common data collected on animals that have been caught is length because it is easy and inexpensive to obtain. However, most stock assessment models used to estimate changes in stock size and fishing mortality over time are constructed such that they track fish by age because it is easy to mark the transition from one age to another at the turn of the year.

The relationship between length and age can be quantified using the standard von Bertalanffy curve (von Bertalanffy 1938) which has the form:

$$L_t = L_\infty\left(1-e^{-K(t-t_0)}\right), \tag{17.2}$$

where L_t is length at age t; L_∞ is the asymptotic length, that is, the average length the fish would attain if they lived indefinitely; K, is a parameter expressing how fast L_∞ is approached and has the dimension of time^{-1} (e.g., year^{-1}); t is age; and t_0 is the theoretical age at length zero if the fish always grew according to the equation. An example is presented in Box 17.2.

Although the von Bertalanffy growth equation works well in many circumstances, it does not account for seasonality in fish growth patterns. Estuarine environments are highly seasonal, more so than the open marine environment to which they are connected. Thus, the food types (Chavance et al. 1984; Aguirre-Leon and Yañez-Arancibia 1986), food consumption (Figure 17.3), and hence, the growth of estuarine fishes are bound to oscillate seasonally, whether the fish in question undertake seasonal migrations in and out of estuaries or not. The typical dominant source of seasonal oscillations, however, is water temperature, which exerts a profound influence on the metabolic processes of poikilotherms and hence on their growth (Pauly 2010). Even in tropical and subtropical waters, animals such as the checkered puffer fish (*Sphoeroides testudineus*) display seasonal growth oscillations (Pauly and Ingles 1981; Longhurst and Pauly 1987).

Various authors have modified the von Bertalanffy growth equation to accommodate seasonal growth (Hoenig and Chaudhury Hanumara 1982; Longhurst and Pauly 1987; Soriano and

Box 17.2 Case study: changes in growth of southern black bream in response to environmental degradation.

Southern black bream (*Acanthopagrus butcheri*) is a fish endemic to the southern coast of Australia that spends its entire life in its natal estuary and that has highly plastic biological characteristics, including growth. Since the early 1990s, the estuaries in which southern black bream reside have experienced extensive habitat degradation. Fisheries scientists have observed concurrent, significant changes in southern black bream growth. For example, estimated values for L_∞ and K parameters of the von Bertalanffy growth equation (Eq. 17.2) for male bream have both declined over time.

Parameter	1993–1995	2003–2004	2007–2011
L_∞ (TL, mm)	424	351	288
K (year^{-1})	0.29	0.28	0.25
t_0 (year)	−0.19	−0.28	−0.76

Resulting growth curves illustrate how southern black bream growth rates have slowed over time and the average asymptotic length is smaller.

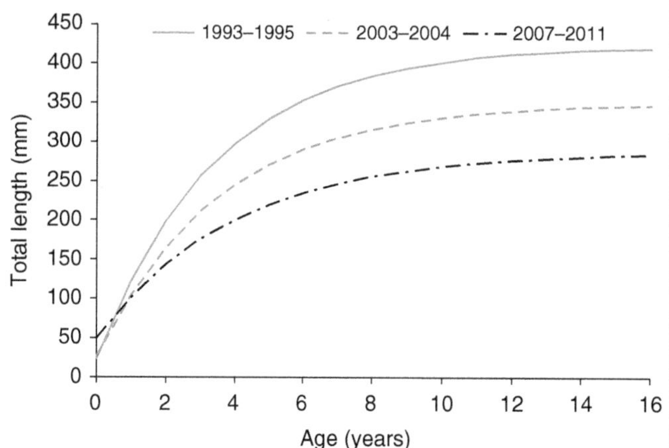

One potential driver of this change in growth is thought to be hypoxia, which typically results in reduced metabolism (Pauly 2010). Also, density-dependent effects may be at play given that the fish in question have increasingly tended to aggregate inshore, where oxygen levels remain high.

Source: Adapted from Cottingham (2014).

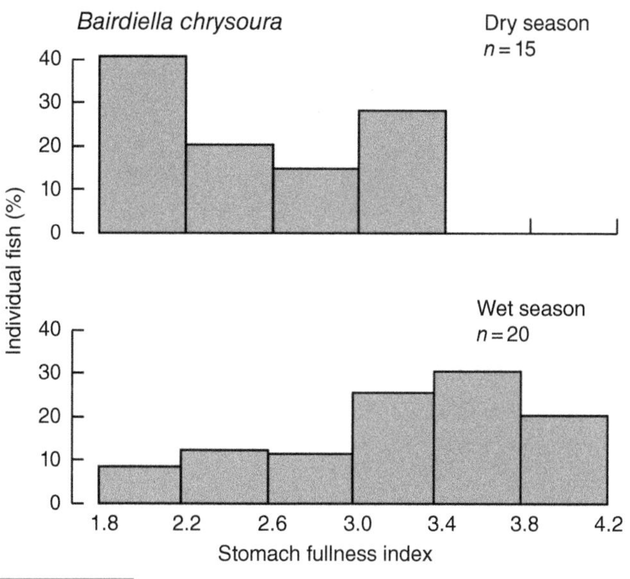

FIGURE 17.3 An example of seasonal changes in the stomach contents and hence, presumably the food consumption of a lagoon fish, the silver perch *Biardiella chrysoura*, in Terminos Lagoon, Mexico. *Source*: Adapted from data in Chavance et al. (1984).

Jarre 1988; Pauly 2010). These growth models can accommodate stronger growth oscillations in "winter," all the way up to a very short period of zero growth, for example, when temperatures are lowest (Pauly 2010). However, these models cannot accommodate longer periods of zero growth. An alternative model that can accommodate a period of growth stagnation, i.e., *no-growth time* (NGT; Pauly et al. 1992) is proposed below. To fit this model, the time axis is divided into one growth and one NGT over each period of one year. Then, during growth time, we have

$$L_t = L_\infty\left[1-exp(-\omega)\right], \tag{17.3}$$

in which L_t is the length at age t, and where,

$$\omega = K'(t'-t_0) + \frac{\frac{K'}{2\pi}}{(1-NGT)}\left[\sin\frac{2\pi}{1-NGT}(t'-t_s) - \sin\frac{2\pi}{1-NGT}\right](t_0-t_s), \tag{17.4}$$

where, t' is obtained by subtracting the total amount of NGT that the fish experienced from the age t since $t = 0$, and t_s is the parameter adjusting a seasonal cycle to start at $t = 0$. Note that the seasonal growth itself (outside of NGT) is described by a sine wave curve with period 1–NGT, and that the unit of K' is (year-NGT)$^{-1}$ instead of year^{-1}. A program in R to fit Eq. (17.4) to length-at-age data was published by Ogle (2017). An application example for this model is given in Figure 17.4. The model predicts a no-growth time of about three months (January to March) for the European seabass (*Dicentrarchus labrax*) in l'Etang d'Or, France (Quignard 1984; Beauchot 1987), a feature that other seasonal growth models could not have detected.

The growth of fishes within estuaries relative to that of conspecifics (members of the same species) growing in other habitats appears to be a function of (i) the type of estuary and/or of habitats being compared, (ii) the species of fish, and (iii) the life stage of the fish species. Shallow, eutrophic estuaries lead to improved growth compared with that of deep

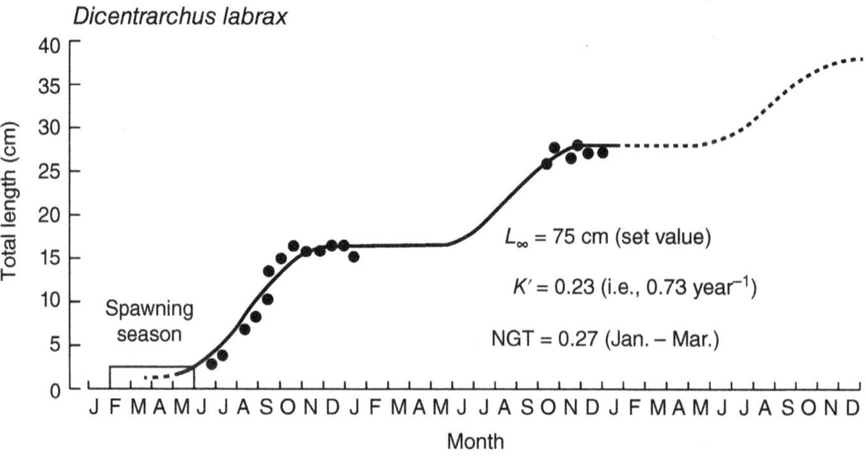

FIGURE 17.4 Seasonally oscillating growth of European sea bass, important in Mediterranean lagoon catches. *Source*: Monthly size data from Quignard (1984). Maximum size ($\approx L_\infty$) and spawning season from Beauchot (1987).

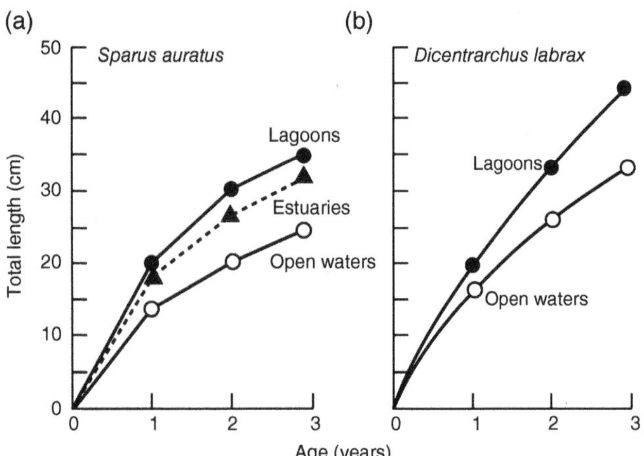

FIGURE 17.5 Growth differences among juveniles of two species of fishes important to Mediterranean lagoon fisheries. (a) Gilthead sea bream. (b) European sea bass. Note improved growth in the lagoon habitat. *Source*: Adapted from Chauvet (1984) and Chauvet (1988).

estuaries which are strongly influenced by the marine regime (Chauvet 1984; Chauvet 1988). With regard to Mediterranean species such as gilthead sea bream (*Sparus aurata*) and European sea bass, estuarine habitats appear to lead to higher growth rates among juveniles and young adults (Figure 17.5), while among the larger, older specimens of these two species, growth within estuaries is at best equivalent to, and generally less than, that in the marine environment (Chauvet 1988). Fast growth of juveniles and smaller maximum sizes of adults are not incompatible. Indeed, rapid juvenile growth due to availability of abundant food and high habitat temperatures generally goes along with smaller maximum adult sizes (Longhurst and Pauly 1987; Pauly 2010).

The West African coast offers further and clearer indications of at least small coastal lagoons being habitats that lead to reduced maximum lengths among resident fishes that are independent of fishing effects. For example, the maximum observed length of the blackchin tilapia *Sarotherodon melanotheron* (Cichlidae) in the small poikilohaline Sakumo Lagoon was 19.5 cm (Pauly 1976) versus 25 cm in the much larger estuarine Lagos Lagoon (Fagade 1974). Also, two forms of the "bonga" *Ethmalosa fimbriata* (Clupeidae) occur in West Africa. One form reaches lengths of up to 30 cm and is found along the coast, large estuaries, and open coastal lagoons; the other form occurs only within closed lagoons and is limited to a length of about 15 cm (Longhurst and Pauly 1987). Note that these effects are independent of, and added to, the size-based artificial selection that is imposed by a long-term fishery, and which ultimately results in small adults (Conover and Munch 2002; Pauly 2002).

Habitat quality may also affect fish growth rates. In some French estuaries, fish size was found to be greater in systems with high eutrophication and ecotoxicity levels than in less impacted systems, suggesting that low habitat quality may increase fish growth rate at the expense of larval survival (Brehmer et al. 2013). In other systems, recent growth was found to be slower for fish in waters with high sediment metal concentrations (Gilliers et al. 2006; Amara et al. 2007; Amara et al. 2009). However, these results are not universal and may depend on the species of interest and types of contaminants to which fish are exposed (Gilliers et al. 2004). Other poor habitat conditions such as low dissolved oxygen levels have been shown to reduce growth rates (Campbell and Rice 2014; Cottingham et al. 2014). However, water temperature still has an overwhelming impact on growth, as mentioned above, and must be accounted for first before using fish growth as a metric of habitat quality (Searcy et al. 2007).

17.3.3 Natural Mortality

Natural mortality includes any cause of death that cannot be attributed to the effects of fishing, including such processes as disease and predation. The effects of natural mortality on a **cohort** of fish can be modeled using:

$$N_2 = N_1 e^{-M\Delta t}, \tag{17.5}$$

where N_1 and N_2 are the numbers of fish at the beginning and end, respectively, of a period Δt, and M is the instantaneous rate of natural mortality during that period. In fisheries science, mortality rates (both natural and fishing mortality) are typically reported as either annual or instantaneous rates; annual mortality is the percentage that dies in one year, whereas, instantaneous mortality refers to the fraction dying over a very short time interval.

Estimates of M are typically derived from either empirical data collection programs such as tagging or other mark-recapture studies (Pine et al. 2003), or M can be estimated using information about the species' life history (Pauly 1980; Then et al. 2014). Across fish species, natural mortality of fishes is strongly correlated with other aspects of their life history such as growth, which is demonstrated in the model

$$M \approx \left(K^{0.65}\right)\frac{T^{0.46}}{L_{\infty}^{0.28}}, \quad (17.6)$$

where L_{∞} is total length in centimeters, K is in year^{-1} and T, the environmental temperature, is in degree Celsius (Pauly 1980). Hence, estuarine fishes which tend to have higher K and lower L_{∞} values than conspecifics in open waters, can be expected to have generally higher natural mortalities than their open-water counterparts (Chauvet 1988). It is important, however, to distinguish between the relatively low natural mortality rates affecting late juveniles and adults, which are fairly constant and largely predictable, from those affecting larval and early juveniles, which are extremely high and essentially unpredictable.

Moreover, an important distinction with regard to estuaries is that between natural mortality rates, as discussed above, and catastrophic mortalities, as caused, for example, by sudden cold snaps or hypoxic events. The former may be seen, at least as far as populations rather than individuals are concerned, as a gradual process. Thus, natural mortality, as influenced or determined by predation, will be more or less continuous over a certain period and can be compensated for by population growth, leading to the observed narrow range of M/K values (Pauly 1980; Thorson et al. 2017; Froese et al. 2018). Otherwise, populations could not maintain themselves over evolutionary timescales.

Catastrophic mortalities, on the other hand, which are quite frequent in estuarine systems, are episodic events, usually connected with sudden changes in water characteristics such as salinity, dissolved O_2, H_2S content, and temperature, which can induce large-scale death among resident communities (i.e., "fish kills") and sometimes their total annihilation. Many natural or anthropogenic factors may cause catastrophic mortalities in estuaries, including the following:

1. Eutrophication, leading to nighttime depletion of oxygen and/or benthic production of H_2S, which can be released into the water column by storms (Luther et al. 2004; Pollock et al. 2007);
2. Harmful algal blooms (Landsberg 2002);
3. Hypersalinity (Gunter 1952; Whitfield 2005; Whitfield et al. 2006);
4. Cold or hot spells, particularly effective in shallow estuaries (Gunter 1952; Thronson and Quigg 2008);
5. Pollution from land-based sources such as agricultural run-off and pesticides (Davis et al. 2017; O'Mara et al. 2017).

Catastrophic mortalities are difficult to incorporate into standard population dynamics models and have indeed not generally been considered explicitly in fisheries management. Their probability of occurrence and prevention are complicating aspects of estuarine fisheries and aquaculture management.

17.3.4 Fishing Mortality and Catch-per-unit-effort

Fishing mortality is the rate at which fish are being removed from the population due to fishing whether commercial, subsistence or recreational. The mortalities considered here include any death caused by fishing activities whether the animal is caught and retained, dies after being discarded (Zeller et al. 2018), or dies after interacting with the fishing gear in some other way (Avila et al. 2018). Catch statistics and biological data such as size and age of fish caught are collected, whenever possible, to monitor fishing activities and help quantify the effect that fishing is having on the population.

In the previous section, we used Eq. (17.5) to describe the effects of natural mortality on a population; however, that model assumed fishing mortality was not an influential driver of population dynamics. In situations where fishing is occurring, the rates of natural mortality (M) and fishing mortality (F) can be added to yield the total mortality rate (Z) and used to model the exponential decline in population abundance over time such that:

$$N_2 = N_1 e^{-Z\Delta t}, \quad (17.7)$$

where

$$Z = M + F. \quad (17.8)$$

We can convert this from an instantaneous rate to an annual catch rate (A), using:

$$A = 1 - e^{-Z}. \quad (17.9)$$

Equations (17.7–17.9) form the backbone of many stock assessment models used to estimate fishing mortality and stock size (Quinn and Deriso 1999). Changes in the age composition of catch data are often used to estimate mortality with Eqs. (17.7–17.9), a simple example of which is shown in Box 17.3. In many situations, though, age composition of the catch is unknown. Thus, fisheries scientists often examine trends in catch to monitor changes in the underlying population (Froese et al. 2012).

Fishery catch is a function of both the size of the population available to be caught and the amount of effort expended. Catch can be described mathematically as follows:

$$C = q \cdot f \cdot B, \quad (17.10)$$

Box 17.3 Estimation of fishing mortality using catch-at-age data: a case study of yellow perch (*Perca flavescens*) in the upper Chesapeake Bay.

Typically, fish produce many juveniles, but few survive to old age given mortality factors such as predation, disease, and fishing. The model presented in Eq. (17.7) describes the exponential decline in abundance of fish within a given cohort over time. To obtain a rough estimate of the total mortality experienced by a stock, we apply a method called a "catch curve" which assumes the rate of decline in catch-at-age either for a cohort of fish or within a given year is indicative of the total mortality (Quinn and Deriso 1999).

For example, the distribution of yellow perch catch-at-age from the 2015 commercial harvest in the upper Chesapeake Bay shows that catch of fish is low for ages 2–3, increases for age 4, then largely declines across the remaining ages 5–10+. Yellow perch younger than age 5 are not yet fully exposed to the fishery, meaning they are typically too small to be reliably caught in the nets used by fishermen. Therefore, we will focus our analysis on catch of ages 5+ fish that are fully selected to the gear.

Taking the natural logarithm of catches for ages 5+ linearizes the data such that the slope of a line drawn through these points, with the sign changed, is an estimate of the total instantaneous mortality rate on the population; in this example, total mortality (Z) = 0.55 year^{-1}.

Based on observed catch-at-age in other areas of the Chesapeake Bay that are closed to fishing, the natural mortality rate for yellow perch in this area was estimated to be 0.25 year^{-1}. Therefore, according to Eqs. (17.8 and 17.9), the instantaneous fishing mortality rate is equal to 0.30 year^{-1}, and the annual harvest rate is 0.42 or 42% of the population.

Source: Paul Piavis, Maryland Department of Natural Resources.

where C is the fisheries catch in weight during a given period, B is the mean biomass during that same period, and q is a scaling coefficient called catchability that represents the fraction of a fish stock caught by a defined unit of the fishing effort, f. This definition implies that C/f (the catch per unit of effort, or CPUE) is proportional to the biomass, an assumption that generally holds, but which does not apply in some cases, especially in schooling fish (Hilborn and Walters 1992). If q remains constant, CPUE can be used to monitor stock trends when absolute biomass is unknown (Figure 17.6). Effort is often gear-specific and can be based on a variety of metrics such as number of fishers, days at sea, number of hooks set, hours trawled, or hours of trap soak time. In estuaries, fishery CPUE tends to decline rapidly as the density of fishers increases, implying that, past a certain level, increased fishing effort results in smaller catches (Figure 17.7a).

CPUE should be interpreted with some caution, though, because q often changes over time due to the tendency of fishers to move to areas of high density, environmental changes that drive movement of fish outside the surveyed range, etc. (Wilberg et al. 2010). In such circumstances, the relationship between catch and biomass is not proportional across the time

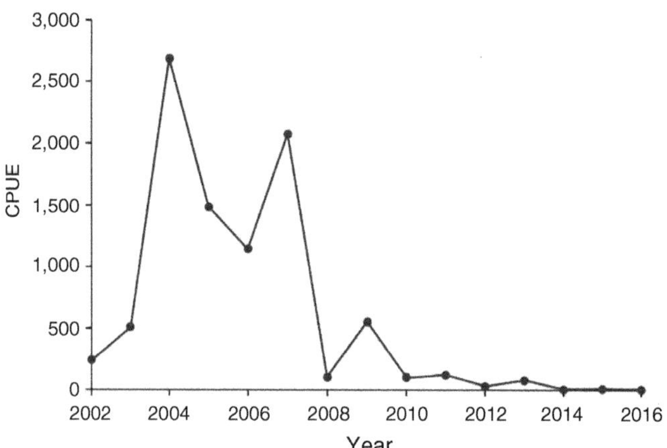

FIGURE 17.6 Trend in catch-per-unit-effort (CPUE) of Atlantic croaker (*Micropogonias undulatus*) in a trawl survey of the Chesapeake Bay mainstem, 2002–2016. Effort is measured as the area swept by the trawl net. If catchability (q) has been constant over the time series, this CPUE trend should reflect trends in total abundance in the survey area, indicating that croaker have largely been in decline since 2004. *Source*: Based on Virginia Institute of Marine Science (2016).

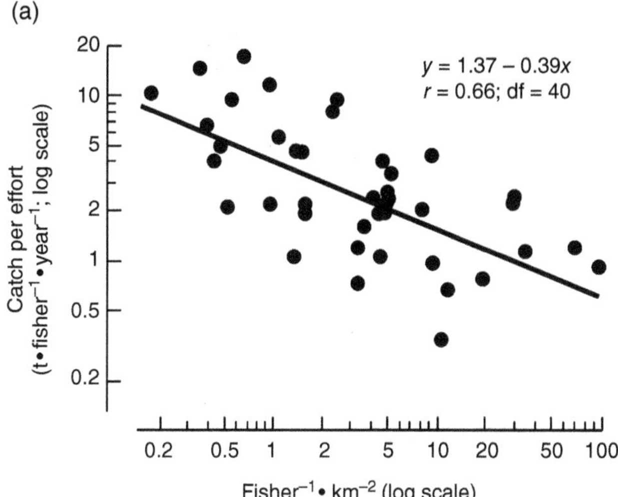

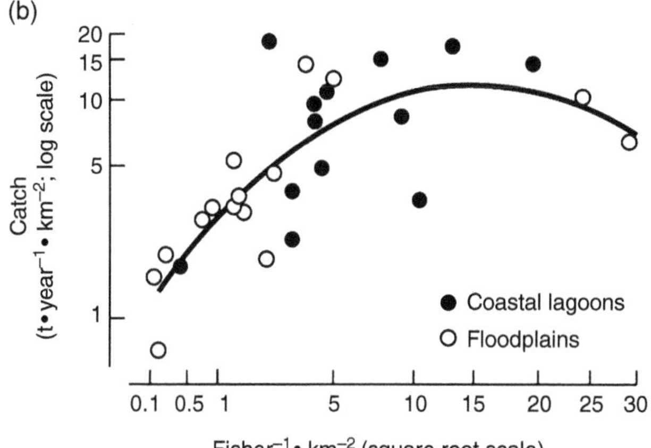

FIGURE 17.7 Examples of the relationship between fishing effort per area, that is, fishing intensity, and dependent variables. (a) Relationship between C/f and fishing intensity in 42 coastal lagoons. *Source*: Kapetsky (1984)/FAO. (b) Relationship between yield, C, and fishing intensity in 13 coastal lagoons and 15 river floodplains, which appear to have similar yield. *Source*: Based on Bayley (1988).

series and CPUE may provide misleading information about true stock trends (Erisman et al. 2011). In general, though, CPUE information is welcome when interpreting fishery statistics; indeed, the inclusion of fishing effort has been shown to add considerably to model precision when predicting yields in estuarine systems (Figure 17.7b).

Despite the simplicity of Eqs. (17.7) and (17.9), estimating the fishing mortality that occurs in estuaries is a significant challenge for several reasons, including:

1. Estuaries do not encompass the complete life cycle of many species that inhabit them. Species that are not permanent residents of estuaries may be subject to different natural and fishing mortalities in the riverine systems or coastal oceans that they inhabit during different times in their lives (Bayley 1988). Thus, to estimate the amount of fishing mortality that occurs in a given estuary, we must know both the proportion of the total fishery that occurs in an estuary and the rates and times at which fish move in and out of estuaries throughout their lives. However, estuarine migration rates require extensive, expensive tagging studies and are thus unknown for many species.
2. Most fisheries statistics collection programs do not distinguish between fish of the same species caught in estuaries versus the nearby coastal ocean, as mentioned in Section 17.2.1 (Bayley 1988). Thus, the application of classical methods of fish population dynamics and fishing management to estuarine fisheries without consideration of the entire stock is often inappropriate, even if isolated elements of these resource systems can be described by these classical methods (Lam Hoai and Lasserre 1984).
3. Similarly, the scattered and often small-scale nature of fisheries operations in estuaries generally makes the routine collection of CPUE data too costly. Hence, such data are lacking for most estuaries (Kapetsky 1984) or are largely unreliable (Bayley 1988).
4. The methods that may be most appropriate for assessing estuarine stocks, which have the advantage of not requiring estimates of fishing effort, do require information on the age of fish caught, which can be obtained in a cost-effective manner only for the most important species in major estuarine fisheries.

In summary, the methodologies to monitor and assess the dynamics of estuarine fish stocks have largely been developed; however, the data required to conduct these analyses are often lacking due to funding restrictions and the complex nature of many estuarine fish stocks and small-scale estuarine fisheries.

17.3.5 Recruitment

Fisheries scientists use information on recruitment to determine how many new fish will become available to the fishery and how sustainable catch advice provided to managers might change as a result of high or low juvenile fish production. This often proves to be a difficult task because fish recruitment is highly variable and can fluctuate 10-fold among years in an apparently random or even chaotic fashion (Hjort 1914; Houde 2009).

Recruitment is thought to be a function of the size of the adult spawning portion of the stock. In general, recruitment increases with increasing spawning stock size until it either levels off (Figure 17.8a; Beverton and Holt 1957) or declines again at large stock sizes due to processes such as cannibalism or nest site competition (Ricker 1954). A suite of different mathematical models linking production of recruits to size of the spawning stock have been described, all of which assume recruitment declines at low stock size (Sharp and Csirke 1983; Quinn and Deriso 1999).

In practice, though, fisheries scientists have discovered that recruitment is highly unpredictable and is often poorly correlated with spawning stock size (Houde 2009). In addition, extended periods of unexplained high or low recruitment produced by a range of spawning stock sizes often baffle fishermen,

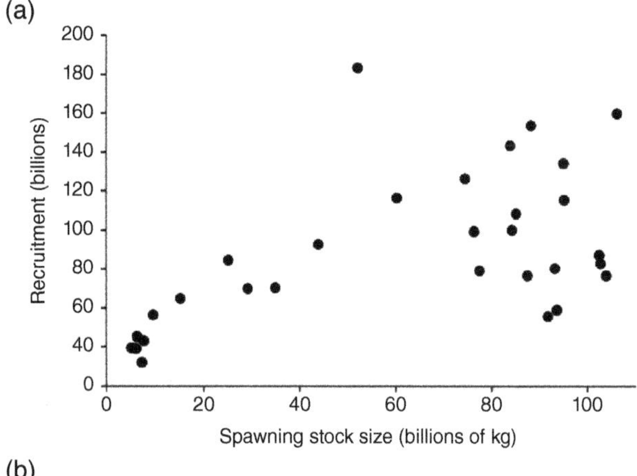

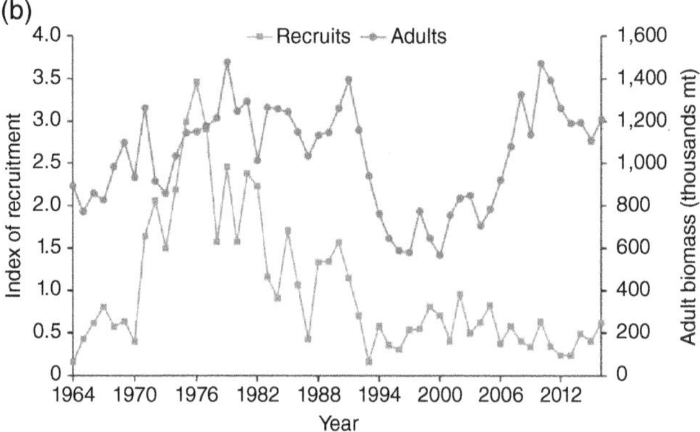

FIGURE 17.8 (a) Relationship between female spawning stock size and recruitment of Atlantic striped bass (*Morone saxatilis*). Note decreased recruitment at low spawning stock size and high variability in recruitment across larger stock sizes. *Source*: Based on NEFSC (2013). (b) Trends in Atlantic menhaden recruitment (*Brevoortia tyrannus*) and adult biomass. Note sustained periods of both high and low recruitment at high adult stock sizes. *Source*: Based on ASMFC (2017).

scientists, and managers (Figure 17.8b; ASMFC 2017). Advances have been made in understanding and modeling drivers and patterns of past recruitment events (Houde 2008; Subbey et al. 2014). For example, consideration of parental size, growth history, and environmental conditions such as temperature and oceanographic patterns have improved our understanding of the stock-recruit relationship in some cases (Hare et al. 2010; Ottersen et al. 2013; Subbey et al. 2014). However, providing reliable predictions of fish recruitment for use in fishery management is something that has largely eluded fishery biologists, despite massive efforts being put in since the very beginning of fishery science as a discipline of its own near the end of the last century (Pauly 1986b; Houde 2008). Most attempts to predict future recruitment have utilized simple correlations between environmental variables and recruitment success but ultimately fail because drivers controlling recruitment are not simple and often change over time (Houde 2008). If forecasting of recruitment is to be successful, several daunting challenges must be faced, including, but not limited to, collection of long-term observational data and development of complex, linked biophysical models (Houde 2008; Subbey et al. 2014).

Understanding fish recruitment in estuaries is further complicated by the diversity of ways in which fish species use estuaries in the recruitment portion of their life cycles (see also Section 17.1 and Chapter 12). Three main groups of fishes occur in estuaries: (i) resident species: those which spend their entire life cycles within an estuary; (ii) seasonal migrants: those which enter estuaries during a more or less well-defined season from either the marine or the fresh water side and leave it during another season; and (iii) occasional visitors: those which enter and leave an estuary without a clear pattern within and among years. To these three basic groups, two other groups may be added: (iv) marine, estuarine-related species, which spend their entire life cycle in the upper shelf under the influence of an estuarine plume; and (v) fresh water, estuarine-related species, which spend their entire life cycle in the fluvial-deltaic zone in the upper reaches of estuarine systems.

The degree to which fish use estuaries can affect our ability to monitor and predict recruitment dynamics. Estuarine residents pose the least challenge because both the spawning stock and recruits can be directly monitored within estuaries if resources are sufficient. However, most estuarine fishes are seasonal migrants that spawn outside estuaries; young fish are either flushed into estuaries while still in the planktonic stage, or swim as early juveniles into estuaries against the outgoing current (Figure 17.9), either due to their effort to stay close inshore or due to coastal wanderings in search of food (Pauly 1982; Quignard 1984; Chauvet 1988). The relative level of recruitment into estuaries by seasonal migrants is thus determined by a combination of the overall number of potential recruits along the coast and oceanographic and estuarine conditions that allow fish to enter estuaries. For other migratory species, recruits may not rely on estuaries as nursery sites, but estuaries serve as important foraging grounds during a later stage in their life cycle; by associating with productive estuarine systems at some life stage, a high standing stock can be maintained (Yáñez-Arancibia and Sanchez-Gil 1988).

Outside estuaries, numerous marine species spend their entire life cycle in the upper shelf deriving benefits from estuarine plumes. For example, a significant fraction of the secondary production in the western Gulf of Mexico's "fertile crescent" (Mississippi River mouth to the northern tip of the Yucatan Peninsula, and including some areas the Atlantic coast of Central America) is derived from estuarine ecosystems, including areas on the shallow shelf influenced by estuarine plumes (Darnell 1990; Sánchez-Gil and Yáñez-Arancibia 1997; Chesney and Baltz 2001; Yáñez-Arancibia 2005; Sánchez-Gil et al. 2008). Characteristics of these estuaries are high riverine discharge rates, large fresh water surpluses, and low water residence times. Much of the production and subsequent trophic transfer may therefore occur outside the physical boundaries of the estuaries, that is, in association with plumes of fresh water over the inner continental shelves. These contrasting sources—estuary and shelf—of trophic delivery to the fishery forage base, and ultimately to larger consumers, is one cause of uncertainty on how we view the functions of estuaries and the shelf ecosystem they influence.

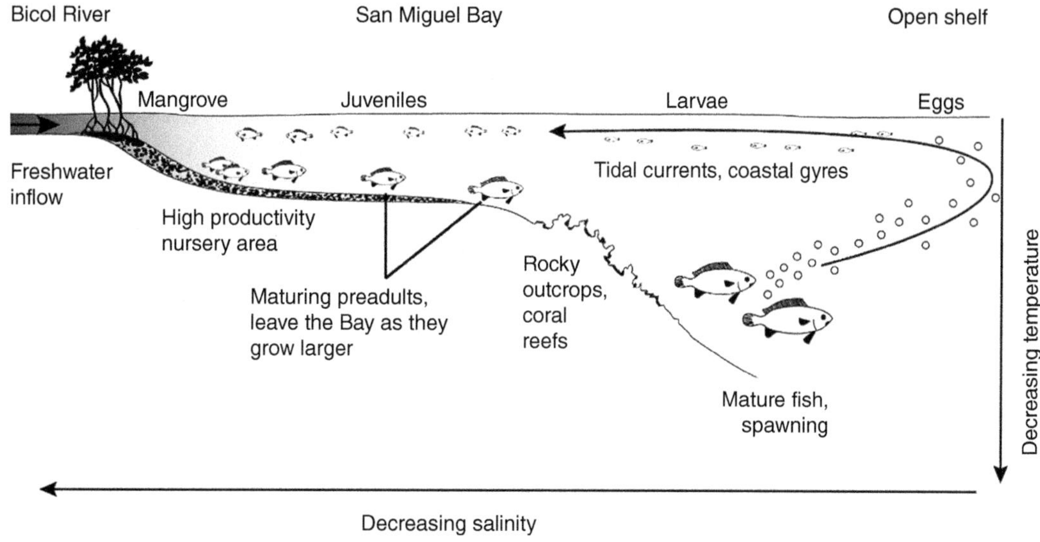

FIGURE 17.9 Schematic representation of the role of San Miguel Bay, Philippines as a highly productive nursery ground for marine fishes hatched further offshore. *Source*: Pauly (1982)/With Permission of WorldFish.

Thus, high fisheries production may be attributable in part to the estuary-like conditions prevailing in large parts of the inner continental shelf during high river discharge periods, as relatively a few fish species are wholly adapted to life cycles within estuarine systems.

17.4 Management of Estuarine Fisheries and Aquaculture

Both fishing and aquaculture activities in estuaries involve the use of natural resources. This implies the need for management because common-property, open-access natural resources systems, given competing users, cannot produce high, sustained yields if left to themselves (Hardin 1968). Since the advent of modern industrial fishing fleets, fish stocks have tended to quickly become overfished unless they are managed, and in some cases even when they are (Pauly et al. 2002). When unregulated, aquaculture practices can also result in low yields, polluted waters, and toxic products (Ottinger et al. 2016). This section outlines the general principles of fisheries and aquaculture management in estuaries and concludes with a summary of future challenges.

17.4.1 Fisheries Management

Fisheries management is a process by which the current and future behavior of participants in the fishery is controlled in an attempt to ensure sustainable use of the resource. The process, which often starts too late when stocks are already depleted (Ludwig et al. 1993), typically begins with a recognition or statement of a problem and the gathering and analysis of available data, sometimes in the form of a formal stock assessment. Estimated trends in quantities such as fishery catch and effort, stock biomass, fishing mortality, and recruitment are used by fisheries scientists to establish the **stock status** and suggest sustainable catch levels. Managers then consider that information when making decisions about setting regional quotas, or catch limits, and how to allocate the resource among participants.

Fisheries management action is implemented through the establishment and enforcement of regulations. Fisheries agencies that are granted management authority to set fishery regulations vary by country, but can include local, provincial/state, and national governments as well as international organizations with multiple countries as members. Fisheries management actions typically involve putting restrictions on: (i) access to the resource either through licensing or closure of specific areas to fishing, (ii) number of gears, (iii) type of gear deployed, (iv) timing of effort deployment, or (v) some combination of these (Pauly et al. 2002). Without such restrictions, stocks often collapse, leading to negative ecological and socioeconomic consequences, including food web alterations (Pauly et al. 1998) and food insecurity (Pauly et al. 2005).

Estuarine fisheries possess several characteristics that complicate management compared with many freshwater or marine fisheries. First, the seasonality of estuaries poses unique management challenges. A number of investigations have demonstrated the existence of complex, seasonally changing relationships between fisheries yields and high nutrient loads, fresh water inputs, shallow depths, large areas of tidal mixing, coastal vegetated area, surface area of estuarine systems, and the resulting high productivities that are typical of estuaries and estuarine plume ecosystems (Deegan et al. 1986; Nixon 1988; Sánchez-Gil and Yáñez-Arancibia 1997; Yáñez-Arancibia et al. 2007). Thus, management decisions must account for this seasonal pulsing habitat, and the protection of its different components, including the aquatic vegetation, in the context of comprehensive environmental planning. Coastal fisheries resources are an expression of ecosystem functioning and to assure the persistence of such resources, the protection and conservation of essential habitats is the key.

However, fisheries management agencies often have limited or no jurisdiction over many activities that affect fish habitat in estuaries such as channel dredging, agricultural runoff, shoreline development, or hydropower plant operations. In some cases, controlling fishing practices alone may not be sufficient to maintain robust stocks. Thus, fishery management plans often include habitat management goals that other agencies can use to guide their actions. In some circumstances, fisheries scientists or managers are consulted before certain actions are taken that might affect fish habitat. However, in many circumstances, fisheries professionals are not consulted on estuary management issues or given authority to maintain or enhance fish habitat, making management of fish habitat in estuaries particularly challenging given they often contain or are in close proximity to major ports, shipping lanes, hydropower plants, agriculture production, and human population centers.

Also, the predominance of migratory species in estuaries that inhabit these systems for only part of their life cycle means that estuarine fisheries often require complex inter-jurisdictional management (i.e., the cooperative or complementary management of the same fishery across multiple agencies). For example, many part-time residents, particularly anadromous and catadromous species, in the Chesapeake Bay require the coordination of up to six different agencies to regulate all the different fisheries that interact with the species across its entire range and life history, both in and outside estuaries (Table 17.2; Chesapeake Bay Fisheries Ecosystem Advisory Panel (NCBO) 2006). If any one of these agencies, including those with jurisdictions limited to areas outside estuaries, does not maintain sustainable fishing practices, the entire stock and the estuarine fisheries that rely on it may be negatively affected.

Finally, it is also important to recognize that many fisheries around the globe, particularly in estuaries, are not industrial but small in scale. Approximately 25% of current global fisheries catch are from small scale fisheries sectors, either artisanal, subsistence, or recreational (Figure 17.10; Pauly and Zeller 2016). One of the major trends in global fisheries is increased competition between small-scale and large-scale fisheries due to overfishing (Pauly 2006; FAO 2015). Yet small-scale fisheries tend to directly benefit more people and their livelihoods, and generally have fewer negative environmental

TABLE 17.2 **A summary of fishery management plans (FMPs) adopted, and regulations established by the 8 fisheries management agencies and organizations with authority to regulate the harvest of 10 example estuarine fishes/crustaceans that inhabit the Chesapeake Bay, USA. FMP indicates which agency/organization has adopted an FMP in their region. C and R indicate an agency has set commercial or recreational fisheries regulations, respectively, in their waters. Note that several different agencies may have complementary (or competing) FMPs for the same species and that multiple agencies can set their own independent regulations for the same species within their jurisdictions. Blue crab and white perch are the only lifelong resident species in the Chesapeake Bay on this list with all others being diadromous or seasonal migrant species.**

Species	ASMFC	MAFMC	SAFMC	DCF	MDDNR	PFBC	PRFC	VMRC
American eel	FMP				C/R	R	C/R	C/R
Atlantic croaker	FMP				C/R		C	
Atlantic sturgeon	FMP			C/R	C/R	C/R	C/R	C/R
Black sea bass	FMP	FMP			C/R		C/R	C/R
Blue crab[a]					C/R			C/R
Bluefish	FMP	FMP			C/R		C/R	C/R
Spanish mackerel	FMP		FMP		C/R		C/R	C/R
Striped bass	FMP			R	C/R		C/R	C/R
Summer flounder	FMP	FMP			C/R		C/R	C/R
White perch[b]					C/R	R	C/R	

[a] The Chesapeake Bay Program regional partnership maintains a Blue Crab Fishery Management Plan, but does not have management authority to set regulations.
[b] An FMP for white perch was drafted in 1990, but never formally adopted by Maryland Department of Natural Resources.
ASMFC Atlantic States Marine Fisheries Commission; MAFMC Mid-Atlantic Fishery Management Council; SAFMC South Atlantic Fishery Management Council; DCF District of Columbia Fisheries Management Branch; MD DNR Maryland Department of Natural Resources; PFBC Pennsylvania Fish and Boat Commission; PRFC Potomac River Fisheries Commission; VMRC Virginia Marine Resources Commission.
Source: Adapted from Chesapeake Bay Fisheries Ecosystem Advisory Panel (2006).

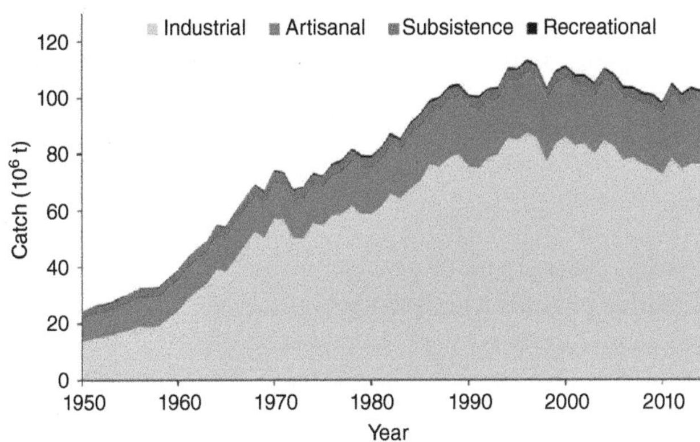

FIGURE 17.10 Reconstructed global marine catch broken down into large-scale (industrial) and small-scale (artisanal, subsistence, recreational) sectors. *Source*: Data from Sea Around Us.

and socioeconomic side effects (Zeller et al. 2016; Zeller et al. 2017). Given the difficulty of monitoring small-scale fisheries, their contribution to national fish supply tends to be greatly underestimated compared to that of industrial fisheries, leading to undervaluation of their importance to communities (Zeller et al. 2007). For example, reconstructed marine fish catch statistics from Cambodia, Malaysia, Thailand, and Viet Nam showed that small-scale sector catches were underestimated by an average of two times (Teh and Pauly 2018). Until small-scale fisheries catches are consistently monitored independent of industrial catches, small-scale fisheries will continue to be undervalued (Pauly and Charles 2015) and overall fishing pressure underestimated (Teh and Sumaila 2011) in estuaries and other aquatic systems.

Estuarine fisheries management involves not just the regulation of targeted fishing on individual stocks, but also monitoring and control of the potential negative impacts fisheries can have on the broader ecosystem, including nontargeted catch, estuarine habitats, and the estuarine food web.

Fishers often catch species that were not targeted and this is called **bycatch**. Bycatch may be retained, or may be discarded because they are undesirable, usually for economic reasons (Zeller et al. 2018). Note that not all bycatch is discarded, and that targeted species may be discarded as well. When the amount of catch allowed for a targeted species is limited by a quota, some fishers may discard all or a portion of their catch to make space for bigger and hence more valuable fish of the same species; this process is called high-grading, and it is illegal.

Discarded bycatch can include not just fish, but many species of shellfish, marine mammal, coral, seabird, or sea turtle. The degree to which discarded animals can survive capture and handling varies by species and life stage and is a significant source of uncertainty in estimation of fishing mortality (Davis 2002). Some animals are quite resilient, while others such as Mexico's endemic porpoise, the vaquita (*Phocoena sinus*), are at the brink of extinction due to them being bycatch of a fishery for a giant croaker, the totoaba (Taylor et al. 2017). Also, the bycatch of juvenile fish in estuaries varies seasonally but can have serious negative impacts on the recruitment of both target and nontarget stocks (Blaber et al. 2000). The monitoring and regulation of discard mortality is a major challenge in fisheries management because fishers are typically not required to report fish they do not land and sell. In particular, many estuarine-related fisheries employ gears or techniques that are nonspecific for their target species; for example, shrimp trawling results in high discard to **landings** ratios that range from 2 to 10+ kg discarded per kg shrimp kept (Alverson et al. 1994; Pauly and Zeller 2016). Also, derelict fishing gear that has been lost or abandoned can continue to "ghost fish" or cause bycatch mortality due to accidental ingestion of plastic debris or entanglement (Coe and Rogers 2012).

Another way in which fishing affects estuarine ecosystems is through direct habitat disturbance or destruction. Although the degree of impact may vary across gear types, dredges and trawls tend to have the largest effect on benthic communities, turbidity, and disturbance or destruction of sensitive habitats such as seagrass beds and oyster reefs that serve as nursery and foraging habitat (Blaber et al. 2000). The rate of recovery of estuarine habitats from fishery disturbance depends largely on the substrate type, frequency of disturbance, and degree of tidal flow (DeAlteris 1988; DeAlteris et al. 1999; Collie et al. 2000).

Without careful management, fisheries can induce unintended trophic alterations that impact community structure and functioning of estuarine food webs. Aquatic systems have experienced a gradual transition in the catch of largely long-lived, high trophic level, piscivorous fish toward short-lived, low trophic level invertebrates and planktivorous pelagic fish, a process that has been termed "fishing down marine food webs" (Pauly et al. 1998; Pauly et al. 2000; Liang and Pauly 2017). Northern temperate regions of the globe such as the North Atlantic where fisheries are most developed have experienced the most severe sequential collapse and replacement of upper trophic level fisheries (Pauly et al. 1998; Essington et al. 2006), although some countries in lower latitudes such as India and Brazil have also experienced striking rates of trophic level declines in estuarine and marine fisheries (Bhathal and Pauly 2008; Freire and Pauly 2010).

The overall functioning of estuarine ecosystems can be altered by fishing activities as well. For example, the Chesapeake Bay estuary was once home to extensive reefs of the eastern oyster (*Crassostrea virginica*) which helped to filter vast quantities of water (Newell 1988). With the advent of oyster dredging in the 1870s, oyster reefs and subsequent fishery catches rapidly declined to low levels within 60 years (Rothschild et al. 1994; Jackson et al. 2001). Given continued overfishing, habitat destruction, and disease, the population has not recovered and is estimated to be at approximately 0.3% of its historic, unfished state (Wilberg et al. 2011), leaving the ability of oysters to help mitigate the effects of anthropogenic eutrophication greatly diminished. Ending overfishing and rebuilding oyster populations in the

Bay could help restore habitat for numerous finfish species (e.g., reef-obligate gobies, blennies, and toadfish), improve submerged aquatic vegetation abundance, increase denitrification, and reduce the quantity of suspended solids and phytoplankton (Cerco and Noel 2007).

17.4.2 Aquaculture Management

Management of aquaculture in estuaries is in many ways similar to land-based agriculture in that it involves activities such as regular stocking, intervention in the rearing process to enhance production, feeding, and predation control (NOAA 2006). Unlike most capture fisheries, aquaculture can involve the permitting or leasing of some static area of the aquatic ecosystem to culturists, and the ownership of the farmed stock by individuals or corporations.

Aquaculture regulations focus on a few key issues: (i) species to be cultivated; (ii) water quality, (iii) water use, (iv) land use for associated hatchery/processing activities, (v) hatchery management; (vi) feed types; (vii) types and placement of artificial substrates or cages; and (viii) processing (DeVoe and Hodges 2002). In some countries, aquaculture management involves control of hatchery and culture practices on land for seeding or stocking, aquatic zoning and location siting procedures, leasing and permitting of the water column or bottom to private companies by government agencies, monitoring of water quality impacts, and regular testing of seafood products to protect human health. Like fisheries, the degree of regulatory control over aquaculture practices varies widely among nations (Boyd and Schmittou 1999). Although aquaculture management standards are increasing around the world, enforcement of regulations is often lax (Wood and Mayer 2007).

Aquaculture management has several issues that often pose greater challenges in estuaries than fresh or marine waters, including use conflicts and coordination across multiple jurisdictions. Estuarine aquaculture activities are typically stationary farms that involve the use of artificial structures such as racks, stakes and lines, or cages and pens (Macintosh 1994; Bostock et al. 2010). These semipermanent structures placed in the water create use conflicts with other traditional estuary activities and interests such as commercial fisheries, navigation, shoreline development, and recreation (DeVoe and Hodges 2002; Harvey and McKinney 2002; Klinger and Naylor 2012). Balancing the competing interests of aquaculture and other human activities in estuaries requires comprehensive coastal zone management and planning (DeVoe and Hodges 2002; Cataudella et al. 2015).

As with estuarine fisheries management, aquaculture activities, when regulated, are often subject to oversight by multiple authorities that regulate everything from navigational hazards posed by aquaculture gear to treatment of farmed stock with pharmaceuticals. For example, aquaculture practices in US estuaries must comply not only with local and state government regulations, but also with that of seven different federal agencies, including the US Army Corps of Engineers, US Environmental Protection Agency, US Fish and Wildlife Service, NOAA Fisheries, US Food and Drug Administration, US Department of Agriculture, and the US Coast Guard (DeVoe and Hodges 2002).

The management of aquaculture in estuaries involves not only the regulation of culturing activities, but also the monitoring and control of potential impacts on the broader ecosystem. Primary concerns include changes in water quality and estuarine habitats, structure and functioning of the estuarine community, and human health. See Box 17.4 for a case study example of the aquaculture of milkfish (*Chanos chanos*) and its ecosystem impacts.

Estuarine habitats can be altered by aquaculture activities particularly when gear displaces or destroys local sea grass beds, mangroves, or wetlands (Páez-Osuna 2001; DeVoe and Hodges 2002). Aquaculture also has the potential to disturb the local sediment, induce changes in benthic community structure (Findlay et al. 1995; Simenstad and Fresh 1995), and cause localized eutrophication and sediment deposition through the concentrated release of feces by sedentary or penned farmed animals, particularly in areas with low flow (Páez-Osuna 2001; Testa et al. 2015). When localized eutrophication occurs, the potential for lowered dissolved oxygen and increased turbidity and frequency of algal blooms also increases (Frankic and Hershner 2003; McLusky et al. 2004). In some circumstances, shellfish aquaculture can ameliorate poor water quality by improving light penetration for seagrasses and enhancing removal of nitrogen (Newell 2004; Dumbauld et al. 2009; Kellogg et al. 2014). However, the extent to which aquaculture combats eutrophication varies based on a number of factors, including tidal exchange conditions, proximity to phytoplankton sources, residence time, and shellfish density (Dumbauld et al. 2009; Kellogg et al. 2014; Murphy et al. 2016). Sediment deposition by bivalves may actually be higher in farmed areas if currents are not strong enough to redistribute fecal deposits (Cranford et al. 2006; Testa et al. 2015).

In addition to the effects on benthic organisms mentioned above, aquaculture can directly impact many estuarine populations. Aquaculture involves the introduction of new gear and hatchery-raised animals into the estuarine ecosystem which creates an opportunity for the introduction of potentially harmful diseases to wild conspecifics as well as the introduction of invasive species that could alter community structure and functioning (Naylor et al. 2001; Thorstad et al. 2008). Farmed species such as salmon (*Salmo salar*) have been known to escape cages and become established as invasive species or interbreed with local wild stocks, resulting in lowered individual fitness, reduced lifetime success, and decreased production across generations (McKinnell and Thomson 1997; Thorstad et al. 2008). Non-native sturgeon (*Acipenser sturio*) in Western and Central Europe introduced as escapees from fish farms and the pet trade have thrived and may interfere with endemic sturgeon restoration efforts

Box 17.4 Aquaculture of an estuarine-dependent species: a case study of milkfish (*Chanos chanos*) in Southeast Asia.

Milkfish, known as *bangus* in the Philippines and bandeng in Indonesia, is a euryhaline fish native to tropical and subtropical waters of the Pacific and Indian Oceans that relies on estuaries as a nursery ground (Froese and Pauly 2018). Adult milkfish school in marine waters above 20 °C near coasts and around islands.

Spawning occurs in marine waters on sand or coral reefs. Larvae remain at sea for 2–3 weeks before migrating *en masse* into estuaries, mangroves, and, occasionally, freshwater lakes. Subadults return to the ocean to mature and complete the life cycle.

Milkfish are an important food fish in Southeast Asia, primarily in the Philippines, Indonesia, and Taiwan. Milkfish culture began primarily in brackish water ponds and has spread to freshwater and marine sites as well given the species' high tolerance for a wide range of salinities (Macintosh 1994).

Culture of milkfish in estuaries occurs in shallow ponds often excavated from nipa palm beds located in mangroves. To prepare ponds for culture, chicken manure and other fertilizers are commonly added (FAO 2018a). Mature breeders spawn in hatchery ponds without the use of hormones.

Traditionally, young seed fish called fry are harvested wild in estuaries and sold to culturists to be raised to harvestable size. Since the 1970s, milkfish have also been bred in hatcheries (FAO 2018a). However, the supply of wild fry, as with most fish recruitment, is often unpredictable, and the wild capture of fry cannot satisfy the demand from aquaculture. Catches of wild milkfish fry in recent years have diminished to the point that the Philippines has begun importing hatchery-raised fry from Taiwan and Indonesia (FAO 2018b).

Milkfish typically consume natural food (benthic diatoms, epiphytic algae, and detritus) in culture, but are often supplemented with artificial feed when natural productivity of the water cannot sustain optimum fish growth, more so in marine than brackish sites. Artificial feed options used include *lab-lab* (a mixture of cyanobacteria, diatoms, filamentous algae, and invertebrates) or rice bran mixed with fish meal or other animal byproducts (FAO 2018a). Fish meal and animal byproducts may be produced locally or imported and include a variety of sources including anchoveta-based fish meal, tuna or shrimp offals, snail meat, mussel, and poultry (FAO 2018b). The use of fish meal in milkfish aquaculture is small relative to other farmed species and is gradually being replaced with vegetable-based protein sources (Tacon and Metian 2008). However, the production of farmed milkfish is rising, resulting in a total increase in artificial feed use.

Source: Pauly et al. (2020).

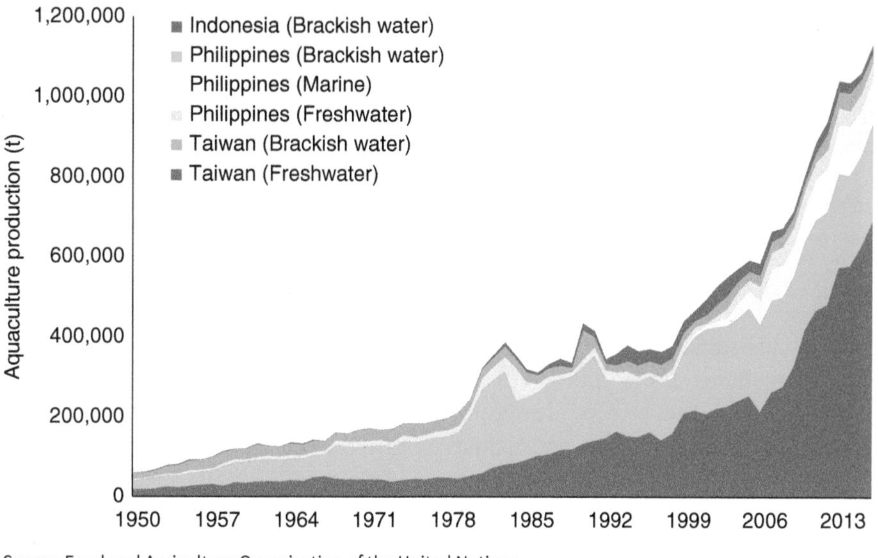

Source: Food and Agriculture Organization of the United Nations

by competing for habitat, introducing disease, and potentially hybridizing (Arndt et al. 2002).

Aquaculture sites are seen as irresistible concentrations of prey to many aquatic predators that not only cause economic losses for the culturist, but also increase opportunities for bycatch mortality. For example, Australian fur seals (*Arctocephalus pusillus*) are often drawn to forage on farmed salmonids where they harass and eat fish and damage nets in which they sometimes become entangled, injured, and drown (Kemper et al. 2003). Also, cownose rays (*Rhinoptera bonasus*) are known to ransack bivalve aquaculture sites in US estuaries; in this situation, fisheries managers must balance the interests of culturists who advocate for "harvest control" of rays with the concern that this species has extremely low reproductive rates (one pup per year), making high catches unsustainable (Fisher et al. 2011).

Aquaculture can also have widespread effects on ecosystem and human health when pharmaceuticals and other chemicals are used. Many aquaculture operations employ the use of antibiotics, hormones, and biocides to treat disease and enhance production (Subasinghe et al. 2000; Hoga et al. 2018). In addition, pollutants such as pesticide run off from land-based agriculture can be taken up by the farmed animals from their aquatic environment and bioaccumulate (Wood and Mayer 2007). Chemicals and pollutants are later consumed by predators and humans or dispersed into sediments and the water column, having localized or broad effects depending on the size and location of the aquaculture operation (Frankic and Hershner 2003; Guardiola et al. 2012).

Finally, some forms of estuarine aquaculture can impact wild fish stock dynamics due to their reliance on wild-caught broodstock and forage fish as a source of protein in aquaculture feed. For example, many aquaculture operations that raise tropical shrimp are stocked using wild-caught juveniles and are sustained using fish-derived feeds (Klinger et al. 2013). Even low trophic level species farmed inland or in estuaries and which require very little protein in their diet such as tilapia, are being produced at such high and increasing rates that demand for fishmeal is placing growing pressures on fisheries exploiting fish to turn into fishmeal and fish oil (Chiu et al. 2013). Although the use of fishmeal in aquaculture has declined in recent years, aquaculture production and demand for omega-3 oils has grown (Naylor et al. 2009). Advances in the production and use of alternative plant- and animal-based feeds may alleviate this problem in the near future (Frankic and Hershner 2003; Hasan and Halwart 2009; Naylor et al. 2009).

17.4.3 Future Challenges

Looking ahead, the major challenges facing managers of estuarine fisheries and aquaculture in the future will be rebuilding wild fish populations in the face of increasing human population size and climate change. With anticipated population growth, global demand for seafood may exceed our capacity to fish sustainably (Pauly et al. 2002). Although some regions have made improvements, overall global stock status trends are worsening, and approximately 40% of stocks are estimated to be overexploited or collapsed (Froese and Kesner-Reyes 2002; Froese et al. 2012; Kleisner et al. 2013; Pauly and Zeller 2017). To reverse this trend, significant reductions in fishing effort will be required (Pauly et al. 2002). Countries identified as being the most dependent on fisheries and vulnerable to marine production losses are also some of the same countries expected to have the highest population growth (Bradshaw and Brook 2014; Fulton et al. 2018). If truly precautionary fisheries management practices are adopted worldwide, fisheries alone will not be able to meet the current and growing global demand for seafood.

It has been suggested that aquaculture could, in some circumstances, complement, or supplant fisheries catches to help address growing demand (Kobayashi et al. 2015). However, most aquaculture production, particularly in estuaries, is currently aimed at producing specific products for export to wealthy markets in developed countries (Little et al. 2016; FAO 2018d). Many developing countries do not currently have the infrastructure to support aquaculture production at levels sufficient to meet nutritional demands, particularly in sub-Saharan Africa and the Pacific Islands where seafood production consists predominantly of subsistence and artisanal fishing (Golden et al. 2016; Blanchard et al. 2017; Tacon and Metian 2018).

Further complicating the issues of overfishing and population growth is climate change (Mohanty et al. 2010). Future plans for maintaining or enhancing the production of seafood from estuaries must take into account anticipated effects of climate change on these fragile ecosystems. Shifts in species distributions and abundances are expected which could result in lower catches for some and higher for others (Kennedy 1990; Wood et al. 2002; Roessig et al. 2004; Barange and Perry 2009; De Silva and Soto 2009). The highly seasonal nature of estuarine biological processes such as migration may be altered with unpredictable consequences for fish yields (Cochrane et al. 2009). Also, both fisheries and aquaculture will likely be challenged by higher incidence of disease caused by even a small increase in temperature (Roessig et al. 2004; Mohanty et al. 2010). The erosion of coastal margins as a result of sea level rise will likely cause increased turbidity and decreased extent of vegetative habitats upon which estuarine fishes rely (Wood et al. 2002).

Changes in freshwater flow will affect nutrient and dissolved oxygen levels, salinity, and estuarine circulation patterns that affect seasonal migrations as well as the quality and quantity of fish nursery and foraging habitats. Climate change-induced ocean acidification already threatens shellfish aquaculture in temperate regions because acidic waters can corrode shell. For example, hatcheries in the Pacific Northwest, the United States have begun buffering seawater and relocating operations to more optimal sites (Clements and Chopin 2017). Changes in precipitation patterns will also significantly affect fish habitat conditions in estuaries and require adaptive measures to minimize impacts on fisheries and aquaculture systems (Cochrane et al. 2009). If management of fisheries and aquaculture in estuaries is going to effectively rise to these challenges, novel solutions generated through interdisciplinary cooperation will be required that involves stakeholders, fisheries and aquaculture scientists, ecosystem managers, public health specialists, environmental engineers, and policymakers (Golden et al. 2016; Fulton et al. 2018).

17.5 Summary

Estuarine fisheries and aquaculture activities are an important component of global food security and nutrition. Fisheries yields are typically higher in estuaries than other types of ecosystems, and aquaculture yields are typically twice as high in estuaries as in coastal seas or freshwater systems.

Estuaries are highly productive systems that are critical to the success of many fisheries because they serve as habitat during all or certain stages in the life cycle of many fish and invertebrate populations. Also, given the migratory nature of many estuarine fish populations, estuaries have broad, regional impacts that extend far beyond their physical borders.

Assessing wild fish population dynamics primarily involves quantifying recruitment, growth, natural mortality, and fishing mortality. Fish growth in estuaries is influenced by several factors including temperature/seasonality, species, sex, estuary size, and habitat quality. Estuarine fishes display relatively high natural recruitment and natural mortality compared with their open-water counterparts. Natural mortality is strongly correlated with other aspects of fish life history such as growth, and thus can be estimated from life history information. Changes in the age composition of fishery catch and tagging data are often used to estimate fishing mortality. When proportional to biomass, CPUE can be used to monitor stock trends when absolute biomass is unknown.

Fisheries management involves controlling the current and future behavior of participants in a fishery in an attempt to ensure sustainable use of the resource. Establishment and enforcement of fishery regulations is complicated by numerous factors including seasonality, migration, jurisdictional limitations of oversight agencies, and the small-scale nature of many estuarine fisheries.

Management of aquaculture in estuaries is more similar to land-based agriculture, and involves activities such as regular stocking, broodstock maintenance, feeding, and predation control. Aquaculture management challenges in estuaries include use conflicts with other activities such as fishing and navigations, and coordination across multiple natural resources agency jurisdictions.

Ecosystem impacts of fisheries and aquaculture activities in estuaries include bycatch, habitat disturbance or destruction, and alterations in community structure and functioning of food webs. Ecosystem impacts of aquaculture may also include alteration of local water quality, introduction of harmful diseases and invasive species, the use of pharmaceuticals and other chemicals, and the reliance on wild-caught broodstock and forage fish.

The future of estuarine fisheries and aquaculture will involve tackling global problems such as overfishing, a growing demand for seafood worldwide, and the negative impacts of climate change.

Review Questions

Multiple Choice

1. Which of the following factors promote high fisheries yields in estuaries? (Select all that apply)
   a. High primary productivity
   b. High turbidity
   c. High natural recruitment
   d. Proximity to human population centers and ports
   e. Extreme temperature and salinity fluctuations

2. Which of the following factors affect the growth of fish in estuaries? (Select all that apply)
   a. Dissolved oxygen levels
   b. Size and type of estuary
   c. Water salinity
   d. Water temperature
   e. Moon phase

3. When comparing estuarine fishes with conspecifics in the coastal ocean, which of the following is generally true?
   a. Estuarine fishes have higher natural mortality rates
   b. Estuarine fishes migrate longer distances
   c. Estuarine fishes have lower incidence of disease
   d. Estuarine fishes have higher catch per unit effort
   e. None of the above

4. What are the two components of total mortality (Z) in fish population dynamics modeling?
   a. Fishing mortality and bycatch mortality
   b. Natural mortality and fishing mortality
   c. Natural mortality and migration mortality
   d. Natural mortality and discard mortality
   e. None of the above

5. Which of the following are aspects of estuarine aquaculture that are often regulated by managers?
   a. Feed type
   b. Location of aquaculture site
   c. Use of pharmaceuticals
   d. Water quality impacts
   e. All of the above

6. Which of the following statements is false? (Select all that apply)
   a. Most estuarine fisheries are industrial in scale.
   b. Most fish species in estuaries are seasonal migrants.
   c. Estuarine fisheries can be affected by shoreline development and agricultural run-off.
   d. Fish recruitment in estuaries is stable from year to year
   e. Estuarine fisheries often compete for space and market share with aquaculture.
7. Which of the following statements is true? (Select all that apply)
   a. Spawning stock size is a good predictor of recruitment.
   b. Recruitment is highly variable in estuarine fishes.
   c. Most fish species that occupy estuaries are year-round residents.
   d. Recruitment is most easily predicted for year-round estuarine resident fishes.
   e. Temperature has little effect on fish growth.
8. Fishery catch is a function of (Select all that apply)
   a. The age of the stock
   b. The size of the stock
   c. The amount of effort expended
   d. Catchability
   e. Water temperature
9. Which of the following statements is true?
   a. Globally, aquaculture production is declining.
   b. Most estuarine-dependent species spawn in freshwater.
   c. The impact of estuarine fisheries is limited in spatial scope.
   d. Fisheries yields are typically higher in estuaries than other types of ecosystems.
10. Which continent has the largest aquaculture industry?
    a. North America
    b. Europe
    c. Asia
    d. Australia

Short Answer Questions

1. What are the primary causes of catastrophic mortality in estuarine fishes?
2. Describe the four main challenges to estimating fishing mortality in estuaries.
3. In what way are the management of estuarine fisheries and aquaculture similar?
4. Describe three ecological impacts of fisheries activities on estuarine systems.
5. Describe three ecological impacts of aquaculture activities on estuarine systems.
6. Given the following catch-at-age data from a 2009 survey of white perch (*Morone americana*) in Chesapeake Bay, estimate the instantaneous fishing mortality rate assuming natural mortality is equal to 0.2 $year^{-1}$. Assume all fish ages provided are fully selected by the gear.

Age (years)	4	5	6	7	8	9	10
Numbers caught	2916	710	1614	884	896	50	153

Source: Paul Piavis, Maryland Department of Natural Resources.

7. Define fisheries bycatch and describe why it is important to characterize.
8. List three ways in which climate change will likely impact estuarine aquaculture.

Spreadsheet Exercises

1. In the Northern Gulf of Mexico, a growth study of red snapper (*Lutjanus campechanus*) yielded the parameters $L_\infty = 90.4$ cm (total length, or TL), $K = 0.19\ year^{-1}$ and $t_0 = -0.48$ year.

   Using information available in FishBase (fishbase.de), present a von Bertalanffy growth function (VBGF) that incorporates these parameters, but with length expressed as fork length (FL).

   Convert your VBGF for length into weight.

   Plot the 10 first years of the VBGF for length (cm FL) and for weight (kg).

   Hints: Consult the FishBase glossary for an explanation of fork vs. total length measurements and the length-length and length-weight pages for additional parameters. Above each table, click on the "More Info" link for detailed explanations of each equation.
2. Using data available from SeaLifeBase (sealifebase.ca), view growth parameters for all available species. Hint: search "Information by topic".

   Create a spreadsheet containing K and L_∞ growth parameters for any 5 marine mammals, seabirds, crustaceans, and cephalopods, being sure to choose a combination of both large and small species in each group. Visit FishBase (fishbase.de) and select 5 large and small fish species and add their growth parameters to your spreadsheet. In total, you should have 25 sets of parameters. Be sure to include a column containing the species group name (e.g., "fish", "seabirds") as well.

   Plot log(K) vs log(L_∞). Label your points using the species group name (e.g., seabirds). Hint: To add custom data labels, right click on any single point on your graph, select "Add Data Labels", right click on any data label, select "Value From Cells", and select the column containing your species group names. Unclick "Y value".

   What is the main pattern that emerges? How do fish, crustacean, and cephalopod growth rates and max compare with seabirds and marine mammals?

3. Using information obtained from the Sea Around Us (seaaroundus.org) tool and the links provided in the tool to FishBase (fishbase.de) data,

   Examine data queries for the Beaufort Sea and Gulf of Mexico Large Marine Ecosystems (LMEs). Scroll down to the "More Info" section and create a table of the number of fish species exploited (i.e., caught by fisheries) in each LME, both in total and on a per-area (km^2) basis.

   Compare statistics gathered between the two LMEs and provide an ecological explanation for these differences based on what you know about fisheries in estuaries.

References

Able, K.W. (2005). A re-examination of fish estuarine dependence: evidence for connectivity between estuarine and ocean habitats. *Estuarine, Coastal and Shelf Science* 64 (1): 5–17.

Aguirre-Leon, A. and Yañez-Arancibia, A. (1986). Las mojarras de la laguna de Terminos: taxonomía, biología, ecología y dinámica trófica (Pisces: Gerreidae). *Anales del Instituto Ciencias del Mar y Limnologia, Universidad Nacional Autonoma de Mexico* 13: 369–444.

Alverson, D.L., Freeberg, M., Pope, J., and Murawski, S. (1994). *A global assessment of fisheries bycatch and discards*, FAO Fisheries Technical Paper. No. 339. Rome: Food and Agriculture Organization of the United Nations.

Amara, R., Méziane, T., Gilliers, C. et al. (2007). Growth and condition indices in juvenile sole *Solea solea* measured to assess the quality of essential fish habitat. *Marine Ecology Progress Series* 351: 201–208.

Amara, R., Selleslagh, J., Billon, G., and Minier, C. (2009). Growth and condition of 0-group European flounder, *Platichthys flesus* as indicator of estuarine habitat quality. *Hydrobiologia* 627 (1): 87.

Arndt, G.M., Gessner, J., and Raymakers, C. (2002). Trends in farming, trade and occurrence of native and exotic sturgeons in natural habitats in Central and Western Europe. *Journal of Applied Ichthyology* 18 (4–6): 444–448.

ASMFC (2017). Atlantic menhaden stock assessment update, Arlington, VA.

Avila, I.C., Kaschner, K., and Dormann, C.F. (2018). Current global risks to marine mammals: taking stock of the threats. *Biological Conservation* 221: 44–58.

Barange, M. and Perry, R.I. (2009). Physical and ecological impacts of climate change relevant to marine and inland capture fisheries and aquaculture. In: *Climate Change Implications for Fisheries and Aquaculture: Overview of Current Scientific Knowledge*, FAO Fisheries and aquaculture technical paper No. 530 (ed. K. Cochrane, C. De Young, D. Soto and T. Bahri), 212. Rome: FAO.

Bayley, P.B. (1988). Accounting for effort when comparing tropical fisheries in lakes, river-floodplains, and lagoons. *Limnology and Oceanography* 33 (33): 963–972.

Beauchot, M. (1987). Poissons osseux. In: Volume II. In: *Fiches FAO d'Identification des Espèces pour les Besoins de la Pêche. Méditerranée et Mer Noire Zone de Pêche 37.* Rev.1 Vertébrés (ed. W. Fischer, M. Schneider and M. Beauchot). Rome: FAO.

von Bertalanffy, L. (1938). A quantitative theory of organic growth (inquiries on growth laws II). *Human Biology* 10 (2): 181–213.

Beverton, R.J. and Holt, S.J. (1957). *On the Dynamics of Exploited Fish Populations*, Fish and Fisheries Series 11. Springer Science & Business Media.

Bhathal, B. and Pauly, D. (2008). 'Fishing down marine food webs' and spatial expansion of coastal fisheries in India, 1950–2000. *Fisheries Research* 91 (1): 26–34.

Bianchi, T.S., Pennock, J.R., and Twilley, R.R. (1999). *Biogeochemistry of Gulf of Mexico Estuaries*. Wiley.

Blaber, S.J., Cyrus, D., Albaret, J.-J. et al. (2000). Effects of fishing on the structure and functioning of estuarine and nearshore ecosystems. *ICES Journal of Marine Science* 57 (3): 590–602.

Blanchard, J.L., Watson, R.A., Fulton, E.A. et al. (2017). Linked sustainability challenges and trade-offs among fisheries, aquaculture and agriculture. *Nature Ecology & Evolution* 1 (9): 1240.

Bostock, J., McAndrew, B., Richards, R. et al. (2010). Aquaculture: global status and trends. *Philosophical Transactions of the Royal Society of London B: Biological Sciences* 365 (1554): 2897–2912.

Boyd, C.E. and Schmittou, H. (1999). Achievement of sustainable aquaculture through environmental management. *Aquaculture Economics & Management* 3 (1): 59–69.

Bradshaw, C.J. and Brook, B.W. (2014). Human population reduction is not a quick fix for environmental problems. *Proceedings of the National Academy of Sciences* 111 (46): 16610–16615.

Brehmer, P., Laugier, T., Kantoussan, J. et al. (2013). Does coastal lagoon habitat quality affect fish growth rate and their recruitment? Insights from fishing and acoustic surveys. *Estuarine, Coastal and Shelf Science* 126: 1–6.

Campbell, B. and Pauly, D. (2013). Mariculture: a global analysis of production trends since 1950. *Marine Policy* 39: 94–100.

Campbell, L.A. and Rice, J.A. (2014). Effects of hypoxia-induced habitat compression on growth of juvenile fish in the Neuse River Estuary, North Carolina, USA. *Marine Ecology Progress Series* 497: 199–213.

Cataudella, S., Crosetti, D., and Massa, F. (2015). Mediterranean coastal lagoons: sustainable management and interactions among aquaculture, capture fisheries and the environment. Studies and Reviews. General Fisheries Commission for the Mediterranean. No. 95, FAO. 288 p.

Cerco, C.F. and Noel, M.R. (2007). Can oyster restoration reverse cultural eutrophication in Chesapeake Bay? *Estuaries and Coasts* 30 (2): 331–343.

Chauvet, C. (1984). Fisheries in the Lake of Tunis: fishery biology and increasing production by means other than regulation. In: *Management of Coastal Lagoon Fisheries*, vol. 2 (ed. J. Kapetsky and G. Lasserre), 615–694. Studies and Reviews-General Fisheries Council for the Mediterranean (FAO) no. 61.

Chauvet, C. (1988). *Manuel sur l'aménagement des pêches dans les lagunes côtières: la bordigue méditerranéenne*, Fish Technical Document: 290. Food & Agriculture Organization.

Chavance, P., Flores, D., Yañez-Arancibia, A., and Amezcua, L.F. (1984). Ecologia, biología y dinámica de las poblaciones de *Bairdiella chrysoura* en la Laguna de Terminos, Sur del Golfo de Mexico (Pisces: Sciaenidae). *Anales del Instituto de Ciencias del Mar y Limnología* 11: 123–162.

Chesapeake Bay Fisheries Ecosystem Advisory Panel (NCBO) (2006). *Fisheries ecosystem planning for Chesapeake Bay*, Trends in Fisheries Science and Management 3. Bethesda, MD: American Fisheries Society.

Chesney, E.J. and Baltz, D.M. (2001). The effects of hypoxia on the northern Gulf of Mexico coastal ecosystem: a fisheries perspective. In: *Coastal hypoxia—Consequences for Living Resources and Ecosystems*, Coastal and Estuarine Studies 58 (ed. N. Rabalais and R. Turner), 321–354. Washington DC: American Geophysical Union.

Chiu, A., Li, L., Guo, S. et al. (2013). Feed and fishmeal use in the production of carp and tilapia in China. *Aquaculture* 414: 127–134.

Clements, J.C. and Chopin, T. (2017). Ocean acidification and marine aquaculture in North America: potential impacts and mitigation strategies. *Reviews in Aquaculture* 9 (4): 326–341.

Cochrane, K., De Young, C., Soto, D., and Bahri, T. (2009). *Climate Change Implications for Fisheries and Aquaculture: Overview of Current Scientific Knowledge*, FAO Fisheries and aquaculture technical paper No. 530, 212. Rome: FAO.

Coe, J.M. and Rogers, D. (2012). *Marine Debris: Sources, Impacts, and Solutions*. New York: Springer-Verlag.

Collie, J.S., Hall, S.J., Kaiser, M.J., and Poiner, I.R. (2000). A quantitative analysis of fishing impacts on shelf-sea benthos. *Journal of Animal Ecology* 69 (5): 785–798.

Conover, D.O. and Munch, S.B. (2002). Sustaining fisheries yields over evolutionary time scales. *Science* 297 (5578): 94–96.

Costa, M., Cabral, H., Drake, P. et al. (2002). Recruitment and production of commercial species in estuaries. In: *Fishes in Estuaries* (ed. M. Elliott and K. Hemingwa), 54–123. Wiley.

Cottingham, A., Hesp, S.A., Hall, N.G. et al. (2014). Marked deleterious changes in the condition, growth and maturity schedules of *Acanthopagrus butcheri* (Sparidae) in an estuary reflect environmental degradation. *Estuarine, Coastal and Shelf Science* 149: 109–119.

Cranford, P.J., Anderson, R., Archambault, P., et al. (2006). Indicators and thresholds for use in assessing shellfish aquaculture impacts on fish habitat. DFO Canadian Scientific Advisory Secretariat Research Document 34.

Darnell, R.M. (1990). Mapping of the biological resources of the continental shelf. *American Zoologist* 30 (1): 15–21.

Davis, M.W. (2002). Key principles for understanding fish bycatch discard mortality. *Canadian Journal of Fisheries and Aquatic Sciences* 59 (11): 1834–1843.

Davis, A.M., Pearson, R.G., Brodie, J.E., and Butler, B. (2017). Review and conceptual models of agricultural impacts and water quality in waterways of the Great Barrier Reef catchment area. *Marine and Freshwater Research* 68 (1): 1–19.

Day, J., Hall, C., Kemp, W., and Yañez-Arancibia, A. (1989). *Estuarine Ecology*, 558. New York: Wiley.

Day, J.W., Martin, J.F., Cardoch, L., and Templet, P.H. (1997). System functioning as a basis for sustainable management of deltaic ecosystems. *Coastal Management* 25 (2): 115–153.

De Silva, S.S. and Soto, D. (2009). Climate change and aquaculture: potential impacts, adaptation and mitigation. In: *Climate Change Implications for Fisheries and Aquaculture: Overview of Current Scientific Knowledge*, FAO Fisheries and aquaculture technical paper No. 530 (ed. K. Cochrane, C. De Young, D. Soto and T. Bahri), 212. Rome: FAO.

DeAlteris, J.T. (1988). The geomorphic development of Wreck Shoal, a subtidal oyster reef of the James River, Virginia. *Estuaries and Coasts* 11 (4): 240–249.

DeAlteris, J., Skrobe, L., and Lipsky, C. (1999). The significance of seabed disturbance by mobile fishing gear relative to natural processes: a case study in Narragansett Bay, Rhode Island, pp. 224–237.

Deegan, L.A. (2002). Lessons learned: the effects of nutrient enrichment on the support of nekton by seagrass and salt marsh ecosystems. *Estuaries* 25 (4): 727–742.

Deegan, L.A., Day, J.W., Gosselink, J.G. et al. (1986). Relationships among physical characteristics, vegetation distribution and fisheries yield in Gulf of Mexico estuaries. In: *Estuarine Variability* (ed. W. DA), 83–100. Elsevier.

DeVoe, M.R. and Hodges, C.E. (2002). Management of marine aquaculture: the sustainability challenge. In: *Responsible Aquaculture Development* (ed. R. Stickney and J. McVey), 21–24. CABI Publishing, New York, USA: World Aquaculture Society.

Dumbauld, B.R., Ruesink, J.L., and Rumrill, S.S. (2009). The ecological role of bivalve shellfish aquaculture in the estuarine environment: a review with application to oyster and clam culture in West Coast (USA) estuaries. *Aquaculture* 290 (3): 196–223.

Erisman, B.E., Allen, L.G., Claisse, J.T. et al. (2011). The illusion of plenty: hyperstability masks collapses in two recreational fisheries that target fish spawning aggregations. *Canadian Journal of Fisheries and Aquatic Sciences* 68 (10): 1705–1716.

Essington, T.E., Beaudreau, A.H., and Wiedenmann, J. (2006). Fishing through marine food webs. *Proceedings of the National Academy of Sciences* 103 (9): 3171–3175.

Fagade, S. (1974). Age determination in Tilapia melanotheron (Ruppell) in the Lagos Lagoon, Lagos, Nigeria with a discussion of the environmental and physiological basis of growth markings in the tropics. In: *Ageing of Fish* (ed. T. Bagenal), 71–77. London.

FAO (2015). *Voluntary Guidelines for Securing Sustainable Small-scale Fisheries in the Context of Food Security and Poverty Eradication*. Rome: Food and Agriculture Organization of the United Nations.

FAO. (2018a). Aquaculture Feed and Fertilizer Resources Information System. http://www.fao.org/fishery/affris/affris-home/en/.

FAO. (2018b). Cultured Aquatic Species Information Programme. http://www.fao.org/fishery/culturedspecies/Chanos_chanos/en.

FAO (2018c). *Impacts of climate change on fisheries and aquaculture: Synthesis of current knowledge, adaptation and mitigation options*, Fisheries and Aquaculture Technical Paper No. 627. Rome, Italy: Food and Agriculture Organization of the United Nations.

FAO (2018d). The State of World Fisheries and Aquaculture 2018—Meeting the sustainable development goals. Rome. Licence: CC BY-NC-SA 3.0 IGO.

Findlay, R.H., Watling, L., and Mayer, L.M. (1995). Environmental impact of salmon net-pen culture on marine benthic communities in Maine: a case study. *Estuaries* 18 (1): 145.

Fisher, R.A., Call, G.C., and Grubbs, R.D. (2011). Cownose ray (Rhinoptera bonasus) predation relative to bivalve ontogeny. *Journal of Shellfish Research* 30 (1): 187–196.

Forrest, B.M., Keeley, N.B., Hopkins, G.A. et al. (2009). Bivalve aquaculture in estuaries: review and synthesis of oyster cultivation effects. *Aquaculture* 298 (1): 1–15.

Frankic, A. and Hershner, C. (2003). Sustainable aquaculture: developing the promise of aquaculture. *Aquaculture International* 11 (6): 517–530.

Freire, K.M. and Pauly, D. (2010). Fishing down Brazilian marine food webs, with emphasis on the east Brazil large marine ecosystem. *Fisheries Research* 105 (1): 57–62.

Froese, R., and Kesner-Reyes, K. (2002). Impact of fishing on the abundance of marine species, ICES Council Meeting Report ICES CM 2002/L2, 12 p.

Froese, R., and Pauly, D. (2018). FishBase. World Wide Web electronic publication. www.fishbase.org, version (accessed 06/2018).

Froese, R., Zeller, D., Kleisner, K., and Pauly, D. (2012). What catch data can tell us about the status of global fisheries. *Marine Biology* 159 (6): 1283–1292.

Froese, R., Winker, H., Coro, G. et al. (2018). A new approach for estimating stock status from length frequency data. *ICES Journal of Marine Science* 75 (6): 2004–2015.

Fulton, E.A., Plagányi, É., Cheung, W. et al. (2018). Marine systems, food security, and future Earth. In: *Global Change and Future Earth: The Geoscience Perspective 3* (ed. T. Beer, J. Li and K.D. Alverson), 296. Cambridge University Press.

Gilliers, C., Amara, R., and Bergeron, J.-P. (2004). Comparison of growth and condition indices of juvenile flatfish in differentcoastal nursery grounds. *Environmental Biology of Fishes* 71 (2): 189–198.

Gilliers, C., Le Pape, O., Desaunay, Y. et al. (2006). Are growth and density quantitative indicators of essential fish habitat quality? An application to the common sole Solea solea nursery grounds. *Estuarine, Coastal and Shelf Science* 69 (1-2): 96–106.

Golden, C., Allison, E.H., Cheung, W.W. et al. (2016). Fall in fish catch threatens human health. *Nature* 534 (7607): 317–320.

Guardiola, F.A., Cuesta, A., Meseguer, J., and Esteban, M.A. (2012). Risks of using antifouling biocides in aquaculture. *International Journal of Molecular Sciences* 13 (2): 1541–1560.

Gulland, J.A. (1983). *Fish stock assessment: a manual of basic methods*. New York: Wiley.

Gunter, G. (1952). The import of catastrophic mortalities for marine fisheries along the Texas coast. *The Journal of Wildlife Management* 16 (1): 63–69.

Hall, S.J., Hilborn, R., Andrew, N.L., and Allison, E.H. (2013). Innovations in capture fisheries are an imperative for nutrition security in the developing world. *Proceedings of the National Academy of Sciences* 110 (21): 8393–8398.

Hardin, G. (1968). The Tragedy of the Commons. *Science* 162 (3859): 1243–1248.

Hare, J.A., Alexander, M.A., Fogarty, M.J. et al. (2010). Forecasting the dynamics of a coastal fishery species using a coupled climate–population model. *Ecological Applications* 20 (2): 452–464.

Harvey, W.D. and McKinney, L.D. (2002). Recreational fishing and aquaculture: throwing a line into the pond. In: *Responsible Marine Aquaculture* (ed. R. Stickney and J. McVey), 61. New York, USA: World Aquaculture Society, CABI Publishing.

Hasan, M.R. and Halwart, M. (2009). *Fish and Feed Inputs for Aquaculture. Practices, Sustainability and Implications*, FAO Fisheries and aquaculture technical paper(518). Food and Agriculture Organization of the United Nations.

Hilborn, R. and Walters, C.J. (1992). *Quantitative Fisheries Stock Assessment: Choice*. Dynamics and Uncertainty/Book and Disk: Springer Science & Business Media.

Hjort, J. (1914). *Fluctuations in the Great Fisheries of Northern Europe Viewed in the Light of Biological Research*, vol. XX. ICES: Rapports et Procès-Verbaux.

Hoenig, N. and Chaudhury Hanumara, R. (1982). A statistical study of seasonal growth model for fishes. Technical Report of the Department of Computer Science and Statistics, Kingston, RI, University of Rhode Island.

Hoga, C.A., Almeida, F.L., and Reyes, F.G. (2018). A review on the use of hormones in fish farming: Analytical methods to determine their residues. *CyTA-Journal of Food* 16 (1): 679–691.

Houde, E.D. (2008). Emerging from Hjort's shadow. *Journal of Northwest Atlantic Fishery Science* 41: 53–70.

Houde, E.D. (2009). Recruitment variability. In: *Fish Reproductive Biology: Implications for Assessment and Management* (ed. T. Jakobsen, M.J. Fogarty, B.A. Megrey and E. Moksness), 91–171. Chichester, UK: Wiley-Blackwell Scientific Publications.

Houde, E.D. and Rutherford, E.S. (1993). Recent trends in estuarine fisheries: predictions of fish production and yield. *Estuaries* 16 (2): 161–176.

Jackson, J.B., Kirby, M.X., Berger, W.H. et al. (2001). Historical overfishing and the recent collapse of coastal ecosystems. *Science* 293 (5530): 629–637.

Jones, R. (1982). Ecosystems, food chains and fish yields. In: *Theory and Management of Tropical Fisheries.*, ICLARM Conference Proceedings (ed. D. Pauly and G. Murphy), 195–239. Manila: The WorldFish Center.

Kapetsky, J.M. (1984). Coastal lagoon fisheries around the world: some perspectives on fishery yields, and other comparative fishery characteristics. In: *Management of Coastal Lagoon Fisheries*, FAO Stud. Rev. GFCM No. 61, vol. 1 (ed. J. Kapetsky and G. Lasserre), 97–139. Rome: FAO.

Kawarazuka, N. and Béné, C. (2011). The potential role of small fish species in improving micronutrient deficiencies in developing countries: building evidence. *Public Health Nutrition* 14 (11): 1927–1938.

Kellogg, M.L., Cornwell, J.C., Owens, M.S., and Paynter, K.T. (2013). Denitrification and nutrient assimilation on a restored oyster reef. *Marine Ecology Progress Series* 480: 1–19.

Kellogg, M.L., Smyth, A.R., Luckenbach, M.W. et al. (2014). Use of oysters to mitigate eutrophication in coastal waters. *Estuarine, Coastal and Shelf Science* 151: 156–168.

Kemper, C.M., Pemberton, D., Cawthorn, M. et al. (2003). Aquaculture and marine mammals: co-existence or conflict. In: *Marine Mammals: Fisheries, Tourism and Management Issues* (ed. N. Gales, M. Hindell and R. Kirkwood), 209–225. Melbourne, Australia: CSIRO.

Kennedy, V.S. (1990). Anticipated effects of climate change on estuarine and coastal fisheries. *Fisheries* 15 (6): 16–24.

Kleisner, K., Zeller, D., Froese, R., and Pauly, D. (2013). Using global catch data for inferences on the world's marine fisheries. *Fish and Fisheries* 14 (3): 293–311.

Klinger, D. and Naylor, R. (2012). Searching for solutions in aquaculture: charting a sustainable course. *Annual Review of Environment and Resources* 37: 247–276.

Klinger, D.H., Turnipseed, M., Anderson, J.L. et al. (2013). Moving beyond the fished or farmed dichotomy. *Marine Policy* 38: 369–374.

Kobayashi, M., Msangi, S., Batka, M. et al. (2015). Fish to 2030: the role and opportunity for aquaculture. *Aquaculture Economics & Management* 19 (3): 282–300.

Lam Hoai, T. and Lasserre, G. (1984). Stock assessment methods in coastal lagoon fisheries. In: *Management of Coastal Lagoon Fisheries*, Studies and Reviews-General Fisheries Council for the Mediterranean, No. 61 (ed. J. Kapetsky and G. Lasserre). Rome: FAO.

Lan, K.-W., Zhang, C.I., Kang, H.J. et al. (2017). Impact of fishing exploitation and climate change on the grey mullet *mugil cephalus* stock in the Taiwan Strait. *Marine and Coastal Fisheries* 9 (1): 271–280.

Landsberg, J.H. (2002). The effects of harmful algal blooms on aquatic organisms. *Reviews in Fisheries Science* 10 (2): 113–390.

Liang, C. and Pauly, D. (2017). Fisheries impacts on China's coastal ecosystems: unmasking a pervasive 'fishing down'effect. *PLoS One* 12 (3): e0173296. https://doi.org/10.1371/journal.pone.0173296.

Little, D.C., Newton, R., and Beveridge, M. (2016). Aquaculture: a rapidly growing and significant source of sustainable food? Status, transitions and potential. *Proceedings of the Nutrition Society* 75 (3): 274–286.

Litvin, S.Y., Weinstein, M.P., Sheaves, M., and Nagelkerken, I. (2018). What makes nearshore habitats nurseries for nekton? An emerging view of the nursery role Hypothesis. *Estuaries and Coasts* 41: 1–12.

Longhurst, A. and Pauly, D. (1987). *Ecology of Tropical Oceans*. San Diego: Academic Press.

Ludwig, D., Hilborn, R., and Walters, C. (1993). Uncertainty, resource exploitation, and conservation: lessons from history. *Science* 260 (17): 36.

Luther, G.W., Ma, S., Trouwborst, R. et al. (2004). The roles of anoxia, H2S, and storm events in fish kills of dead-end canals of Delaware inland bays. *Estuaries* 27 (3): 551–560.

Macintosh, D.J. (1994). Aquaculture in coastal lagoons. Chapter 14. In: *Coastal Lagoon Processes* (ed. B. Kjerfve), 401–442. Amsterdam, The Netherlands: Elsevier.

Mann, K.H. (2009). *Ecology of Coastal Waters: With Implications for Management*. Wiley.

McKinnell, S. and Thomson, A. (1997). Recent events concerning Atlantic salmon escapees in the Pacific. *ICES Journal of Marine Science* 54 (6): 1221–1225.

McLusky, D.S., Elliott, M., and Elliott, M. (2004). *The Estuarine Ecosystem: Ecology, Threats and Management*. Oxford University Press on Demand.

Mohanty, B., Mohanty, S., Sahoo, J., and Sharma, A. (2010). Climate change: impacts on fisheries and aquaculture. In: *Climate Change and Variability* (ed. S. Simard), 119–138. Croatia: InTech Open.

Murphy, A.E., Emery, K.A., Anderson, I.C. et al. (2016). Quantifying the effects of commercial clam aquaculture on C and N cycling: an integrated ecosystem approach. *Estuaries and Coasts* 39 (6): 1746–1761.

National Marine Fisheries Service (2017). Fisheries of the United States, 2016. U.S. Department of Commerce, NOAA Current Fishery Statistics No. 2016. https://www.st.nmfs.noaa.gov/commercial-fisheries/fus/fus16/index.

National Research Council (1997). *Striking a Balance: Improving Stewardship of Marine Areas*. National Academies Press.

Naylor, R.L., Williams, S.L., and Strong, D.R. (2001). Aquaculture--A gateway for exotic species. *American Association for the Advancement of Science* 294 (5547): 1655–1656.

Naylor, R.L., Hardy, R.W., Bureau, D.P. et al. (2009). Feeding aquaculture in an era of finite resources. *Proceedings of the National Academy of Sciences* 106 (36): 15103–15110.

NEFSC (2013). 57th Northeast Regional Stock Assessment Workshop (57th SAW) Assessment Report. US Dept Commer, Northeast Fish Sci Cent Ref Doc. 13-16; 967 p. National Marine Fisheries Service, 166 Water Street, Woods Hole, MA 02543-1026, or online at http://nefsc.noaa.gov/publications.

Newell, R.I. (1988). Ecological changes in Chesapeake Bay: are they the result of overharvesting the American oyster, *Crassostrea virginica. Understanding the Estuary: Advances in Chesapeake Bay research* 129: 536–546.

Newell, R.I. (2004). Ecosystem influences of natural and cultivated populations of suspension-feeding bivalve molluscs: a review. *J. Shellfish Res.* 23 (1): 51–62.

Nixon, S.W. (1982). Nutrient dynamics, primary production and fisheries yields of lagoons. *In* Actes du Symposium International sur les Lagunes Côtières, SCOR/IABO/UNESCO, Bordeaux, 8–14 Septembre 1981. Oceanologica Acta, Special Issue 5. pp. 357–371.

Nixon, S.W. (1988). Physical energy inputs and the comparative ecology of lake and marine ecosystems. *Limnology and Oceanography* 33: 1005–1025.

NOAA (2006). NOAA Fisheries Glossary. NOAA Technical Memorandum NMFS-F/SPO-69.

O'Mara, K., Miskiewicz, A., and Wong, M.Y. (2017). Estuarine characteristics, water quality and heavy metal contamination as determinants of fish species composition in intermittently open estuaries. *Marine and Freshwater Research* 68 (5): 941–953.

Ogle, D.H. (2017). An algorithm for the von Bertalanffy seasonal cessation in growth function of Pauly et al. (1992). *Fisheries Research* 185: 1–5.

Ottersen, G., Stige, L.C., Durant, J.M. et al. (2013). Temporal shifts in recruitment dynamics of North Atlantic fish stocks: effects of spawning stock and temperature. *Marine Ecology Progress Series* 480: 205–225.

Ottinger, M., Clauss, K., and Kuenzer, C. (2016). Aquaculture: relevance, distribution, impacts and spatial assessments–a review. *Ocean & Coastal Management* 119: 244–266.

Páez-Osuna, F. (2001). The environmental impact of shrimp aquaculture: causes, effects, and mitigating alternatives. *Environmental Management* 28 (1): 131–140.

Pauly, D. (1976). The biology, fishery and potential for aquaculture of Tilapia melanotheron in a small West African lagoon. *Aquaculture* 7 (1): 33–49.

Pauly, D. (1980). On the interrelationships between natural mortality, growth parameters, and mean environmental temperature in 175 fish stocks. *ICES Journal of Marine Science* 39 (2): 175–192.

Pauly, D. (1982). The fishes and their ecology. In: *Small-scale fisheries of San Miguel Bay, Philippines: Biology and Stock Assessment*, ICLARM Technical Report, vol. 7 (ed. D. Pauly and A.N. Mines), 15–33. Manila, Philippines: International Center for Living Aquatic.

Pauly, D. (1984). *Fish population dynamics in tropical waters: a manual for use with programmable calculators*. Manila, Philippines: ICLARM Studies and Reviews.

Pauly, D. (1986a). Problems of tropical inshore fisheries: Fishery research on tropical soft-bottom communities and the evolution of its conceptual base. In: *Ocean Yearbook 1986* (ed. E. Borgese and N. Ginsburg), 29–37. Chicago: University of Chicago Press.

Pauly, D. (1986b). Towards appropriate concepts and methodologies for the study of recruitment in tropical demersal communities. Proceedings of the IREP, OSLR Workshop on the Recruitment of Tropical Coastal Demersal Communities, IOC (UNESCO) Workshop Report No. 44. Campeche Mexico, April 21–25, Paris, France.

Pauly, D. (1998). Beyond our original horizons: the tropicalization of Beverton and Holt. *Reviews in Fish Biology and Fisheries* 8 (3): 307–334.

Pauly, D. (2002). Spatial modelling of trophic interactions and fisheries impacts in coastal ecosystems: a case study of Sakumo Lagoon, Ghana. In: *The Gulf of Guinea Large Marine Ecosystem: Environmental Forcing and Sustainable Development of Marine Resources* (ed. J. McGlade, P. Cury, K. Koranteng and N. Hardman-Mountford), 289–295. Amsterdam: Elsevier Science.

Pauly, D. (2006). Major trends in small-scale marine fisheries, with emphasis on developing. Countries, and some implications for the social sciences. *Maritime Studies* 4: 7–22.

Pauly, D. (2010). *Gasping Fish and Panting Squids: Oxygen, Temperature and the Growth of Water-Breathing Animals. Oldendorf/Luhe, Germany: Excellence in Ecology (22)*, xxviii. International Ecology Institute.

Pauly, D. and Charles, A. (2015). Counting on small-scale fisheries. *Science* 347 (6219): 242–243.

Pauly, D. and Ingles, J. (1981). Aspects of the growth and natural mortality of exploited coral reef fishes. In: *Proceedings of the 4th International Coral Reef Symposium*, vol. 1 (ed. E. Gomez, C. Birkeland, R. Buddemeier, et al.), 89–98. Manila: Philippines.

Pauly, D. and Zeller, D. (2016). Catch reconstructions reveal that global marine fisheries catches are higher than reported and declining. *Nature Communications* 7: ncomms10244.

Pauly, D. and Zeller, D. (2017). Comments on FAOs state of world fisheries and aquaculture (SOFIA 2016). *Marine Policy* 77: 176–181.

Pauly, S., Soriano-Bartz, M., Moreau, J., and Jarre-Teichmann, A. (1992). A new model accounting for seasonal cessation of growth in fishes. *Marine and Freshwater Research* 43 (5): 1151–1156.

Pauly, D., Christensen, V., Dalsgaard, J. et al. (1998). Fishing down marine food webs. *Science* 279 (5352): 860–863.

Pauly, D., Froese, R., and Palomares, M.L. (2000). Fishing down aquatic food webs: Industrial fishing over the past half-century has noticeably depleted the topmost links in aquatic food chains. *American Scientist* 88 (1): 46–51.

Pauly, D., Christensen, V., Guénette, S. et al. (2002). Towards sustainability in world fisheries. *Nature* 418 (6898): 689.

Pauly, D., Watson, R., and Alder, J. (2005). Global trends in world fisheries: impacts on marine ecosystems and food security. *Philosophical Transactions of the Royal Society of London B: Biological Sciences* 360 (1453): 5–12.

Pauly, D., Zeller, D., and Palomares, M.L.D. (eds.) (2020). Sea around us concepts, design and data. seaaroundus.org.

Pérez-Ruzafa, A. and Marcos, C. (2012). Fisheries in coastal lagoons: an assumed but poorly researched aspect of the ecology and functioning of coastal lagoons. *Estuarine, Coastal and Shelf Science* 110: 15–31.

Pianka, E.R. (2011). *Evolutionary ecology*, 6th edition New York: HarperCollins College Div.

Pihl, L., Cattrijsse, A., Codling, I. et al. (2002). Habitat use by fishes in estuaries and other brackish areas. *Fishes in Estuaries* 10–53.

Pine, W.E., Pollock, K.H., Hightower, J.E. et al. (2003). A review of tagging methods for estimating fish population size and components of mortality. *Fisheries* 28 (10): 10–23.

Pollock, M., Clarke, L., and Dubé, M. (2007). The effects of hypoxia on fishes: from ecological relevance to physiological effects. *Environmental Reviews* 15 (NA): 1–14.

Qasim, S. (1970). Some problems related to the food chain in a tropical estuary. In: *Marine food chains* (ed. J. Steele), 45–51. Edinburgh: Oliverand Boyd.

Qasim, S. (1973). Productivity of backwaters and estuaries. In: *The biology of the Indian Ocean*, Ecological Studies 3, 143–154. Berlin: Springer Verlag.

Quignard, J. (1984). The biological and environmental characteristics of lagoons as the biological basis of fisheries management. In: *Management of Coastal Lagoon Fisheries*, FAO Stud. Rev. GFCM No. 61 (ed. J. Kapetsky and G. Lasserre), 3–38. Rome: FAO.

Quinn, T.J. and Deriso, R.B. (1999). *Quantitative Fish Dynamics*. Oxford University Press.

Ricker, W.E. (1954). Stock and recruitment. *Journal of the Fisheries Board of Canada* 11 (5): 559–623.

Ricker, W. (1975). *Computation and Interpretation of Biological Statistics of Fish Populations*, Bulletin of the Fisheries Research Board of Canada. Bulletin 191, 382. Blackburn Press.

Roessig, J.M., Woodley, C.M., Cech, J.J., and Hansen, L.J. (2004). Effects of global climate change on marine and estuarine fishes and fisheries. *Reviews in Fish Biology and Fisheries* 14 (2): 251–275.

Rothschild, B.J., Ault, J., Goulletquer, P., and Heral, M. (1994). Decline of the Chesapeake Bay oyster population: a century of habitat destruction and overfishing. *Marine Ecology Progress Series* Ser: 29–39.

Russell, E.S. (1931). Some theoretical Considerations on the "Overfishing" Problem. *ICES Journal of Marine Science* 6 (1): 3–20.

Sánchez-Gil, P. and Yáñez-Arancibia, A. (1997). Grupos ecológicos funcionales y recursos pesqueros tropicales. In: *UAC, Análisis y Diagnóstico de los Recursos Pesqueros Críticos del Golfo de México*, EPOMEX Serie Científica 7 (ed. D. Flores, P. Sanchez-Gil, J. Seijo and F. Arreguin), 357–389.

Sánchez-Gil, P., Yáñez-Arancibia, A., Tapia-García, M. et al. (2008). Ecological and biological strategies of *Etropus crossotus* and *Citharichthys spilopterus* (Pleuronectiformes: Paralichthyidae) related to the estuarine plume, Southern Gulf of Mexico. *Journal of Sea Research* 59 (3): 173–185.

Schaefer, M.B. (1957). A study of the dynamics of the fishery for yellowfin tuna in the eastern tropical Pacific Ocean. *Inter-American Tropical Tuna Commission Bulletin* 2 (6): 243–285.

Searcy, S.P., Eggleston, D.B., and Hare, J.A. (2007). Is growth a reliable indicator of habitat quality and essential fish habitat for a juvenile estuarine fish? *Canadian Journal of Fisheries and Aquatic Sciences* 64 (4): 681–691.

Sharp, G., and Csirke, J. (1983). Proceedings of the expert consultation to examine changes in abundance and species composition of neritic fish resources, San José, Costa Rica, 1983, 18–29 April, FAO Fisheries Report No. 291, Rome, FAO.

Simenstad, C.A. and Fresh, K.L. (1995). Influence of intertidal aquaculture on benthic communities in Pacific Northwest estuaries: scales of disturbance. *Estuaries* 18 (1): 43–70.

Soriano, M. and Jarre, A. (1988). On fitting Somer's equation for seasonally oscillating growth, with emphasis on T-subzero. *Fishbyte (ICLARM)* 6 (2): 13–14.

Subasinghe, R., Barg, U., and Tacon, A. (2000). Chemicals in Asian aquaculture: need, usage, issues and challenges. SEAFDEC. https://repository.seafdec.org.ph/handle/10862/612

Subbey, S., Devine, J.A., Schaarschmidt, U., and Nash, R.D.M. (2014). Modelling and forecasting stock–recruitment: current and future perspectives. *ICES Journal of Marine Science* 71 (8): 2307–2322.

Tacon, A.G. and Metian, M. (2008). Global overview on the use of fish meal and fish oil in industrially compounded aquafeeds: trends and future prospects. *Aquaculture* 285 (1-4): 146–158.

Tacon, A.G. and Metian, M. (2018). Food matters: fish, income, and food supply—a comparative analysis. *Reviews in Fisheries Science & Aquaculture* 26 (1): 15–28.

Taylor, B.L., Rojas-Bracho, L., Moore, J. et al. (2017). Extinction is imminent for Mexico's endemic porpoise unless fishery bycatch is eliminated. *Conservation Letters* 10 (5): 588–595.

Teh, L.C. and Pauly, D. (2018). Who Brings in the Fish? The Relative Contribution of Small-Scale and Industrial Fisheries to Food Security in Southeast Asia. *Frontiers in Marine Science* 5: 44.

Teh, L.C. and Sumaila, U.R. (2011). Contribution of marine fisheries to worldwide employment. *Fish and Fisheries* 14 (1): 77–88.

Testa, J.M., Brady, D.C., Cornwell, J.C. et al. (2015). Modeling the impact of floating oyster (*Crassostrea virginica*) aquaculture on sediment-water nutrient and oxygen fluxes. *Aquaculture Environment Interactions* 7 (3): 205–222.

Then, A.Y., Hoenig, J.M., Hall, N.G. et al. (2014). Evaluating the predictive performance of empirical estimators of natural mortality rate using information on over 200 fish species. *ICES Journal of Marine Science* 72 (1): 82–92.

Thorson, J.T., Munch, S.B., Cope, J.M., and Gao, J. (2017). Predicting life history parameters for all fishes worldwide. *Ecological Applications* 27 (8): 2262–2276.

Thorstad, E.B., Fleming, I.A., McGinnity, P. et al. (2008). Incidence and impacts of escaped farmed Atlantic salmon Salmo salar in nature. *NINA Special Report* 36 (6).

Thronson, A. and Quigg, A. (2008). Fifty-five years of fish kills in coastal Texas. *Estuaries and Coasts* 31 (4): 802–813.

VIMS Multispecies Research Group (2016). *Fishery Analyst Online Catch Data Maps*. Gloucester Point, VA: Virginia Institute of Marine Science http://www.vims.edu/fisheries/mrg/gis.

Vitousek, P.M., Mooney, H.A., Lubchenco, J., and Melillo, J.M. (1997). Human domination of Earth's ecosystems. *Science* 277 (5325): 494–499.

Whitfield, A.K. (2005). Fishes and freshwater in southern African estuaries–a review. *Aquatic Living Resources* 18 (3): 275–289.

Whitfield, A.K., Taylor, R.H., Fox, C., and Cyrus, D.P. (2006). Fishes and salinities in the St Lucia estuarine system—a review. *Reviews in Fish Biology and Fisheries* 16 (1): 1.

Wilberg, M.J., Thorson, J.T., Linton, B.C., and Berkson, J. (2010). Incorporating time-varying catchability into population dynamic stock assessment models. *Reviews in Fisheries Science* 18 (1): 7–24.

Wilberg, M.J., Livings, M.E., Barkman, J.S. et al. (2011). Overfishing, disease, habitat loss, and potential extirpation of oysters in upper Chesapeake Bay. *Marine Ecology Progress Series* 436: 131–144.

Wilson, J. (2002). Productivity, fisheries and aquaculture in temperate estuaries. *Estuarine, Coastal and Shelf Science* 55 (6): 953–967.

Wood, A. and Mayer, J. (2007). *Changing the face of the waters: the promise and challenge of sustainable aquaculture*. Washington, DC: World Bank.

Wood, R.J., Boesch, D.F., and Kennedy, V.S. (2002). Future consequences of climate change for the Chesapeake Bay ecosystem and its fisheries. *American Fisheries Society Symposium* 32: 171–184.

Yáñez-Arancibia, A. (2005). Middle America, coastal ecology and geomorphology. In: *The Encyclopedia of Coastal Sciences* (ed. M. Schwartz), 639–645. Dordrecht, The Netherlands: Kluwer/Springer Scientific Publ.

Yáñez-Arancibia, A. and Sanchez-Gil, P. (1988). *Ecologia de los Recursos Pesqueros Demersales Tropicales*. Mexico DF: AGT Editorial.

Yáñez-Arancibia, A., Lara-Domínguez, A., Aguirre-León, A. et al. (1985a). Ecology of dominant fish populations in tropical estuaries: environmental factors regulating biological strategies and production. In: *Fish Community Ecology in Estuaries and Coastal Lagoons: Towards an Ecosystem Integration* (ed. A. Yáñez-Arancibia), 311–365. Mexico: Universidad Nacional Autónoma de México Press.

Yáñez-Arancibia, A., Soberon-Chavez, G., and Sanchez-Gil, P. (1985b). Ecology of control mechanisms of natural fish production in the coastal zone. In: *Fish Community Ecology in Estuaries and Coastal Lagoons: Towards an Ecosystem Integration* (ed. A. Yáñez-Arancibia), 571–594. Mexico DF: Universidad Nacional Autónoma de México Press.

Yáñez-Arancibia, A., Lara-Domínguez, A.L., and Day, J.W. (1993). Interactions between mangrove and seagrass habitats mediated by estuarine nekton assemblages: coupling of primary and secondary production. *Hydrobiologia* 264 (1): 1–12.

Yáñez-Arancibia, A., Lara-Domínguez, A.L., and Pauly, D. (1994). Coastal lagoons as fish habitat. In: *Coastal Lagoon Processes* (ed. B. Kjerfve), 339–351. Elsevier Oceanography Series.

Yáñez-Arancibia, A., Lara-Domínguez, A.L., Sánchez-Gil, P., and Day, J.W. (2007). Estuary-sea ecological interactions: a theoretical framework for the management of coastal environment. In: *Environmental analysis of the Gulf of Mexico*, 271–301. The Harte Research Institute for Gulf of Mexico Studies. Special Publication (1).

Zeller, D., Booth, S., Davis, G., and Pauly, D. (2007). Re-estimation of small-scale fishery catches for US flag-associated island areas in the western Pacific: the last 50 years. Fish. Bull, 105: 266–277.

Zeller, D., Palomares, M.D., Tavakolie, A. et al. (2016). Still catching attention: sea around Us reconstructed global catch data, their spatial expression and public accessibility. *Marine Policy* 70: 145–152.

Zeller, D., Cashion, T., Palomares, M., and Pauly, D. (2017). Global marine fisheries discards: a synthesis of reconstructed data. *Fish and Fisheries* 19 (1): 30–39.

Zeller, D., Cashion, T., Palomares, M., and Pauly, D. (2018). Global marine fisheries discards: a synthesis of reconstructed data. *Fish and Fisheries* 19 (1): 30–39.

CHAPTER **18**

Global Climate Change and Estuarine Systems

Flooding in Venice's Piazza de San Marco, with its iconic basilica, has increased from just a few times a year to 100 times a year over the past century due to rising sea levels in the Venice Lagoon. Photo credit: ChiccoDodiFC / Shutterstock.

John M. Rybczyk[1], John W. Day[2], and Alejandro Yáñez-Arancibia[3,†]
[1] Department of Environmental Science, Western Washington University, Bellingham, WA, USA
[2] Department of Oceanography and Coastal Sciences, College of the Coast and Environment, Louisiana State University, Baton Rouge, LA, USA
[3] Instituto de Ecologia, A.C. (CPI-CONACYT), Xalapa, Veracruz, México

> *"It is interesting to note that practically all of the estuaries discussed in this book did not exist 10,000–15,000 years ago and that they will cease to exist in the near geologic future. The world's present estuaries were formed when sea level rose after the last glaciation."*
> *From the first edition of Estuarine Ecology (Day et al. 1989)*

18.1 Introduction

It is ironic that the very process that led to the formation of modern estuaries, sea level rise, may now lead, geologically, to their rapid demise. Indeed, of all the manifestations of climate change and global warming, none have captured the imagination as much as the specter of increasing rates of sea level rise. Perhaps this is because we have already witnessed the effects that high rates of relative sea level rise (RSLR; sea level rise + land subsidence) have had on coastal cities such as Venice and New Orleans, as well as on major deltas such as the Mississippi and Mekong. Or perhaps it is just that it is easier for us to imagine the dramatic effects of coastal flooding on human infrastructure as opposed to the more subtle, but no less troubling effects of increased atmospheric carbon dioxide on ocean acidification or reductions in freshwater input on estuarine productivity.

Sea level rise, however, is only one of many processes influenced by climate change that can affect estuaries. Pritchard (1967) defined an estuary as "a semi-enclosed body of water which has a free connection with the open sea and within which sea water is measurably diluted with fresh water derived from land drainage." Embodied within that definition is the fact that an estuary is the manifestation of interactions among land, river, and ocean processes. Therefore, any changes in either oceans or watersheds caused by climate change has the potential to affect estuaries. Indeed, climate change is currently raising sea levels and increasing temperatures in the oceans and estuaries and changing freshwater runoff in watersheds (Wuebbles et al. 2017).

As a way of introduction, and an interesting way to illustrate the complexities of how climate change affects estuaries and estuarine dependent organisms (including us!), we can examine the potential effects of climate change on estuary dependent salmon on the Pacific coast of North America. Because they are anadromous (they return to freshwater streams and rivers to spawn after spending their adult life in

† Deceased.

Estuarine Ecology, Third Edition. Edited by Byron C. Crump, Jeremy M. Testa, and Kenneth H. Dunton.

Companion website: www.wiley.com/go/crump/estuarine3

the ocean) any change in freshwater stream flow, estuarine habitat, or ocean conditions, due to climate change, has the potential to negatively affect population levels.

Salmon are certainly a cultural and economic icon of the Pacific Northwest (PNW) region of the United States. Historic spawning runs in the Columbia River, for example, once numbered in the millions (Figure 18.1). Unfortunately, overfishing, pollution, habitat modification and loss, and dam construction over the past 150 years have greatly reduced salmon populations in the PNW (Meengs and Lackey 2005). The US Environmental Protection Agency (EPA) currently lists nine genetically distinct stocks of Chinook Salmon on the West Coast as either threatened or endangered.

There are six recognized salmon species in the PNW; Chinook (*Oncorhynchus tshawytscha*), Chum (*Oncorhynchus keta*), Coho (*Oncorhynchus kisutch)*, Pink (*Oncorhynchus gorbuscha*), Sockeye (*Oncorhynchus nerka*), and Steelhead Trout (*Oncorhynchus mykiss*). All salmon species use estuaries to some extent for migration, adult residence, or juvenile rearing (Williams and Thom 2001), none more so than the Chinook, which may spend up to six months in estuaries as juveniles. Past studies have linked the survival of Chinook salmon to pristine estuarine habitat in the PNW (Greene et al. 2005). Any loss of estuarine habitat due to sea level rise then, has the potential to reduce population levels. For example, Hood (2005) estimated that, for a sea level rise of 45 centimeters, juvenile Chinook salmon in the delta of the Skagit River, the largest river flowing into the Puget Sound, would decline by 211,000 fish due to habitat loss. Rising sea levels could also inundate estuarine and beach spawning habitat for the forage fish that salmon depend on (Glick et al. 2007).

FIGURE 18.1 Salmon piled high at a Seattle, WA cannery around 1900. Some studies suggest that climate change will further reduce already decimated populations. *Source*: University of Washington Libraries.

Finally, many salmon streams in the PNW depend on a winter snowpack in the mountains to provide adequate flow in the summer (Figure 18.2). Regional climate change models predict that a larger proportion of winter precipitation will fall as rain, and less as snow for the PNW, leading to reduced winter snowpack, lower summer stream flows due to reduced snowpack, higher summer temperature under a low summer flow regime, and increased fall and winter flooding (Rybczyk

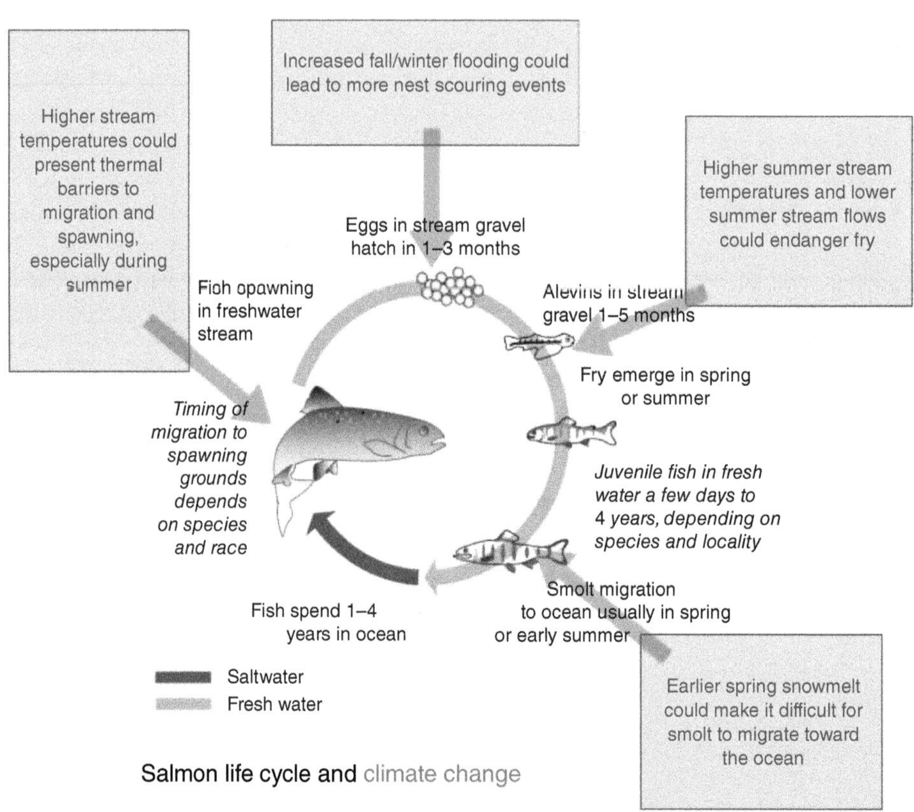

FIGURE 18.2 Sea level rise due to climate change can directly affect salmon populations by reducing critical estuarine habitat. Additionally, the red boxes explain how changes in stream temperature and flow may also affect salmon during various phases of their life cycle. *Source*: Based on Casola et al. (2005).

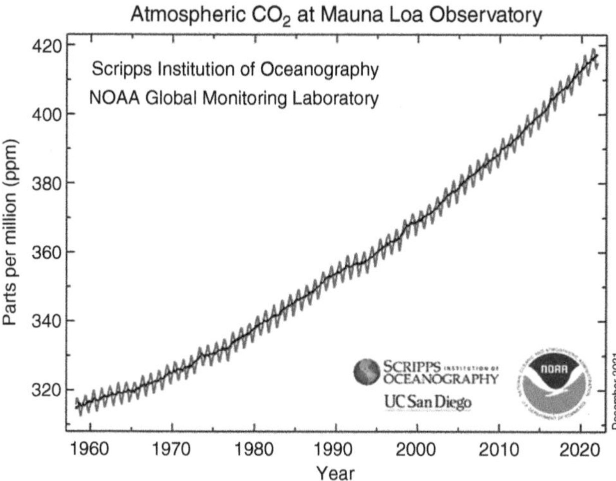

FIGURE 18.3 Atmospheric carbon dioxide concentrations from Mauna Loa, Hawaii as first measured by David Keeling staring in 1958. *Source*: Figure courtesy of NOAA **Climate.gov**. The smaller, seasonal oscillations against the overall upward trend are due to summertime CO_2 uptake due to photosynthesis in the northern hemisphere (where most of the earth's land mass is).

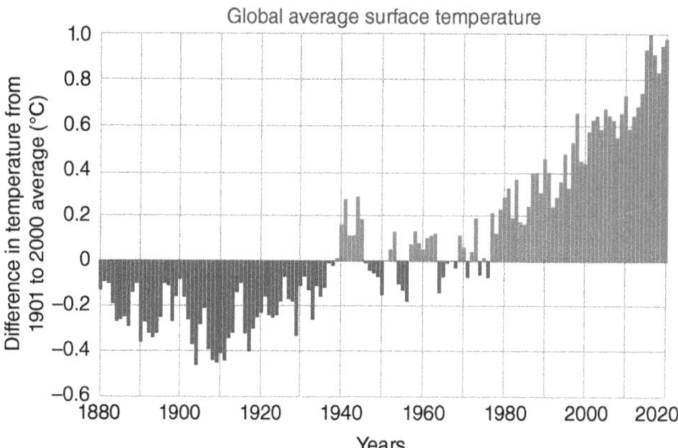

FIGURE 18.4 Yearly surface temperature compared to the twentieth century average from 1880 to 2020. Blue bars indicate cooler-than-average years; red bars show warmer-than-average years. *Source*: Figure courtesy of NOAA **Climate.gov**.

et al. 2016). One recent paper suggests the possibility of a "no snow future" condition for the western United States (Siirila-Woodburn et al. 2021). All of these scenarios have the potential to negatively affect salmon populations along with the overall impacts of changing ocean conditions.

To conclude this introductory section, in the first edition of Estuarine Ecology (Day et al. 1989), climate change was a minor topic and the atmospheric concentrations of carbon dioxide, the most abundant greenhouse gas and the one responsible for most global warming, stood at 353 ppm (Figure 18.3). But in the ensuing decades, climate change impacts on coastal ecosystems have become a major topic of research and policy. By the time the second edition of Estuarine Ecology was published (Day et al. 2012a), carbon dioxide concentration had risen to 394 ppm and, for the first time, an entire chapter was devoted to climate change and the focus was largely on climate impacts to estuarine systems. Unfortunately, as this third edition comes to press, carbon dioxide concentrations stand at 412 ppm and show no sign of abating.

18.2 Climate Change: Historic Patterns and Projections

18.2.1 Historic Patterns and Projections

It has been noted for decades that CO_2 levels (Figure 18.3) in the atmosphere and global mean temperatures have been rising (Figure 18.4) at rates that exceed anything measured directly or by proxy in the past 800,000 years. Because of these concerns, the Intergovernmental Panel on Climate Change (IPCC) was established in 1988 by the World Meteorological Organization and the United Nations Environment Programme. The goal of the IPCC is to objectively synthesize the ever-growing body of technical climate change literature into a format that is accessible to policy makers and scientists across disciplines. The panel consists of three "working groups" that assess, respectively, I) the scientific aspects of climate change, II) the effects of climate change on socio-economic and natural systems and, III) options for reducing greenhouse gases and mitigating the effects of climate change. The IPCC produces a variety of special reports and technical papers, but the most anticipated and widely read publications are the Comprehensive Assessment Reports published in 1990, 1996, 2001, 2007, and 2014. The sixth assessment report will be published in 2022 and draft reports are already available. These peer-reviewed reports, the product of hundreds of experts from all three working groups, assess and synthesize the current scientific and socio-economic literature concerning climate change. In a synthesis of the most recent 2014 report (IPCC 2014a), the IPCC concluded:

> *"Human influence on the climate system is clear, and recent anthropogenic emissions of greenhouse gases are the highest in history. Recent climate changes have had widespread impacts on human and natural systems."*
>
> *"Warming of the climate system is unequivocal, and since the 1950's many of the observed changes are unprecedented over decades to millennia. The atmosphere and ocean have warmed, the amounts of land-based snow and ice have diminished, and sea levels have risen."*

More recent climate reports support these conclusions (Oppenheimer et al. 2019, IPCC 2019).

TABLE 18.1 IPCC Representative Concentration Pathways (RCP) scenarios.

RCP Scenario	CO_2 concentrations by 2100 (ppm)	Predicted temperature increase over the next century[a]
RCP 2.6	450 or lower	0.5–1.7 °C
RCP 4.5	550 or lower	1.1–2.6
RCP 6.0	620	1.4–2.1
RCP 8.5	950	2.6–4.8

[a] Global mean, likely range.

The IPCC makes projections based upon representative concentration pathways (RCPs), which are essentially potential greenhouse gas concentration trajectories based on projections of population size, economic activity, energy use, lifestyle, land use patterns, technology, and climate policy. These RCPs range from RCP2.6, a stringent mitigation scenario that aims to keep global warming below 2°C above preindustrial temperature levels, to RCP8.5, an unlikely "worst case" scenario where CO_2 emissions continue to rise through the twenty-first century from current-day levels of 415–950 ppm (Table 18.1). The numbers after the "RCP" indicate the change in total radiative forcing (watts per square meter) in the troposphere by 2100, relative to preindustrial levels. Intermediate scenarios include RCP4.5 and RCP6.0. RCP4.5 represents a scenario where greenhouse emissions increase through 2040 and then decline to 550 ppm by the end of the century. In the RCP6.0 scenario, emissions peak in 2080. In the next section, we summarize climate predictions by the IPCC and other studies, based on these RCP projections.

18.2.2 Temperature

The IPCC predicts that, as atmospheric greenhouse gas concentrations increase, global temperatures will rise between 2.6 and 4.8 °C over the next century under an RCP8.5 worst case scenario and between 0.5 and 1.7 °C under the much more optimistic RCP2.6 scenario (Hayhoe et al. 2017). This translates to a rise of about one degree in the twentieth century. By comparison, the mean global temperature increased by about 6 °C from the height of the ice age 15,000 years ago to about 5000 years ago when the oceans approached their present level. Thus, human activity may lead to a temperature increase of a similar magnitude in one century. Temperature directly affects many vital life processes, and a change in the thermal regime (extreme temperatures, duration of extreme temperature events, and seasonal rates of temperature change) can directly regulate rates of growth, reproduction, and migratory patterns of many species. Increasing temperature will also lead to changes in precipitation patterns and an acceleration of sea level rise and will likely affect tropical storm activity (IPCC 2014a).

18.2.3 Sea Level Rise (Eustatic and Relative)

Sea level, as it appears to an ecologist standing in an estuarine wetland, or to an individual *Spartina* plant growing in that wetland, is a function of both the volume of water in the ocean (the eustatic sea level) and the vertical displacement of the land surface due to processes such as shallow and deep subsidence or accretion (Rybczyk and Callaway 2009; Cahoon et al. 2019). The apparent sea level rise in an estuary that results from both eustatic sea level rise (ESLR) and vertical land displacement is referred to as RSLR.

Eustatic sea levels rise and fall as a function of long-term climate variation. During cold glacial periods, water is locked up in land-based glaciers and ice caps, and sea levels are relatively low. For example, during the last glacial maximum, which ended approximately 18,000 years ago, sea levels were more than 100 m below current levels (Valiela 2006) and have been rising ever since. Additionally, under a regime of global warming, the volume of water in the ocean expands as the ocean itself warms. Currently, about a third of ESLR is due to this thermal expansion (Table 18.2). The relative contribution of thermal expansion will decrease with continued warming because there is enough water locked in land-based ice to raise sea levels by 80 m if it all melted (Sweet et al. 2017; Emery and Aubrey 1991).

Recent observations have revealed an ESLR rate of 1.8 mm $year^{-1}$ for the period 1961–2003, 3.1 mm $year^{-1}$ for the period 1993–2003, and 3.4 mm $year^{-1}$ currently (2021), compared to the background rate of 1 to 2 mm $year^{-1}$ for the nineteenth and early twentieth century (Meehl et al. 2007). Although it is generally agreed that the rate of ESLR will increase in the future, predicting the exact future rate of sea level rise is uncertain because it is a function of numerous complex processes, both physical (e.g., the thermal expansion of water and the melting of glaciers and ice caps) and political (e.g., future carbon emissions).

The IPCC (2014b) predicts that sea level rise will range from 42 to 98 cm by the end of the twenty-first century (Figure 18.5). However, many feel that this estimate is conservative. For example, the observed rate of ESLR from 1993 to 2003 has already exceeded the lower limit predicted by the fourth IPCC assessment. Recent studies suggest that ESLR will likely be a meter or more (Horton et al. 2020). These conclusions are

TABLE 18.2 Sea level rise (SLR) and estimated contributions from different sources. Uncertainties are 5 and 95%.

	Rate of ESLR (mm year^{-1})	
	1901–1990	**1993–2010**
Actual observed rate of SLR	1.5 (1.3–1.7)	3.2 (2.8–3.6)
Estimated contribution from sources		
• Thermal expansion	–	1.1 (0.8–1.4)
• Glaciers except in Greenland and Antarctica	0.54 (0.47–0.61)	0.76 (0.39–1.13)
• Greenland glaciers	0.15 (0.10–0.19)	0.10 (0.07–0.13)
• Greenland ice sheet	–	0.33 (0.25–0.41)
• Antarctic ice sheet	–	0.27 (0.16–0.38)
• Land water storage	−0.11 (−0.16–0.06)	0.38 (0.26–0.49)
• Sum of estimated contribtions[a]	–	2.8 (2.3–3.4)

[a] The sum of estimated contributions, derived from various sources and methodologies do not necessarily add up to the actual observed rate of SLR.
Source: Based on data from Section B.4, Sea Level, in IPCC (2014b).

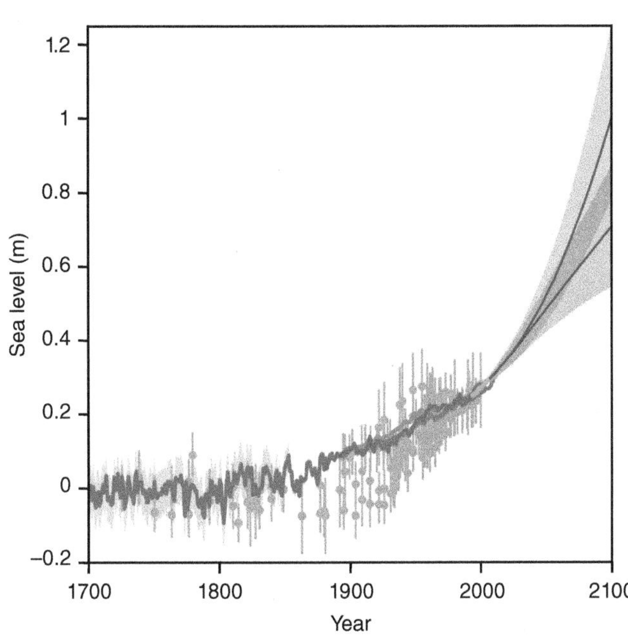

FIGURE 18.5 Past, current, and likely predicted ranges for global mean sea level rise for RCP2.6 (blue) and RCP8.5 (red) scenarios. *Source*: From IPCC (2014b).

based on temperature–sea level rise relationships in the twentieth century and better understanding of ice sheet dynamics in a warming world.

RSLR is the combination of ESLR and subsidence. Where there are high rates of subsidence, RSLR can be much greater than ESLR. High rates of geological subsidence commonly occur in deltas due to compaction, consolidation, and dewatering of sediments. Compared to a global ESLR of 1–2 mm year^{-1} for the twentieth century, RSLR for the Mississippi delta was in excess of 10 mm year^{-1} (Day et al. 2007). RSLR in the Nile delta region is as high as 5 mm year^{-1} (Stanley and Warne 1998) and is between 2 and 6 mm year^{-1} for the Rhone and Ebro deltas (Sánchez-Arcilla et al. 1996; Pont et al. 2002). Humans have accelerated RSLR by drainage and withdrawal of water, oil, and gas (Sestini 1992; Ko and Day 2004). An understanding of vegetation response in areas with high RSLR can provide insights into the effects of accelerated ESLR in the future (Stagg et al. 2017; Schile et al. 2014). What are the limits to accretion of coastal wetlands and how may human activities affect this accretion? We will address this question in a later section.

As an aside, in any discussion of wetland elevation relative to sea level there is often ambiguity regarding the terminology itself. The term "sea level rise" alone is ambiguous as it could refer to the absolute rise in sea level (i.e., ESLR), or the apparent rise caused by both ESLR and land subsidence. Similarly, land subsidence could include deep geologic subsidence, shallow subsidence caused by compaction and organic matter decomposition, or both. We define the terms, as used in this chapter, in Table 18.3.

18.2.4 Changes in Storm Frequency and Intensity

Tropical cyclones impact the western basins of the North Atlantic and North Pacific, the northern part of the Indian Ocean, and parts of east Africa and Australia. Recent research has suggested that a warming ocean surface has led to an increase in the frequency of tropical cyclones and will continue to do so (IPCC 2014a). Mei et al. (2015), for example, predicted climate change will increase the already high average typhoon intensity in the Pacific by 14% by 2100. Tropical cyclones are also intensifying more rapidly (Bhatia et al. 2019), as are large storm surge events, and cyclones are retaining their strength

TABLE 18.3 Definition of terms.

Term	Definition
Accretion	Surface vertical accumulation of mineral and organic sediment, usually over some marker horizon. May also integrate processes occurring on and within the upper part of marsh and mangrove substrates (e.g., root growth, decomposition).
Shallow Subsidence	Primary compaction, decomposition, and dewatering that occurs in upper sediments (up to10 meters).
Deep Subsidence	Deep primary compaction, secondary compaction, and other processes such as geosynclinic downwarping and tectonic activity.
Eustatic sea level rise (ESLR)	Global sea level rise caused by changes in the volumes of glaciers and ice caps and by water density/temperature dependent relationships.
Relative sea level rise (RSLR)	Long term, absolute vertical relationship between the land and water surface. On the marsh surface, RSLR should be calculated as ESLR + deep subsidence+shallow subsidence. However, RSLR, measured using tidal gauge records, represents only ESLR+deep subsidence.
Net accretion balance	= Accretion—shallow subsidence—deep subsidence—ESLR or = Accretion—RLSR

longer after landfall in inverse proportion to the rise in sea surface temperature (Li and Charkraborty 2020). These changes will cause greater flooding concerns in coastal and inland cities by increasing both precipitation and storm surges. Changes in the intensity and frequency of storms can have a variety of impacts on estuarine wetlands that lie in the path of these storms, which will be covered later in the chapter in our section on *Effects of Climate Change on Estuarine Ecosystems*.

18.2.5 Freshwater Input, Sediment Transport, and Nutrient Delivery

Changes in precipitation patterns have the potential to greatly alter patterns in freshwater runoff, river discharge, and associated riverine material fluxes to estuaries. In general, IPCC climate models predict increased precipitation at high latitudes and in the moist tropics close to the equator, but decreased precipitation in subtropics and mid-latitudes. In other regions, it is less that the quantity of precipitation will change, but rather that the kind of precipitation that falls in estuarine watersheds will change under a regime of global warming. For example, projected decreases in interannual snowpack in the coastal Pacific Northwest, USA as more winter precipitation falls as rain rather than snow will increase winter river flows and decrease summer flows (Figure 18.1), affecting downstream estuarine systems (Rybczyk et al. 2016).

These changes in precipitation are likely to be strongly site-specific and highly variable, but some models suggest that the Mediterranean and Mid East, southern Africa, large parts of Australia, and much of the U.S. southwest and northwest Mexico will experience much lower freshwater runoff. Though there is uncertainty in these predictions, there is no doubt reduced freshwater runoff will impact estuarine ecosystems in these regions by decreasing the supply of sediments and decreasing the flux of nutrients. The degree to which these alterations in supply are affected depends on both the degree to which human's control the flow of water to the coast and our management of land and waters.

An interesting example of the complexity of this issue is the Mississippi River and delta. IPCC models suggest that there may be lower local freshwater runoff along the northern coast of the Gulf of Mexico, including the region around the Mississippi River delta. But over 90% of Mississippi discharge is derived from the Ohio and upper Mississippi River basins, which are much farther north where precipitation is forecast to increase. NOAA (2021) indicates that there has already been a 10–15% increase in precipitation over the Ohio and upper Mississippi basins compared to the twentieth century average. Some models predict that these changes in precipitation will increase Mississippi River discharge by up to 40% by 2100 (i.e., Day et al. 2005).

18.3 Effects of Climate Change on Estuarine Ecosystems

What then are the impacts to estuaries of these current and predicted changes in climate? We address these below.

18.3.1 Temperature

Of course, temperature is the forcing function that accelerates rates of sea level rise, alters freshwater flow, and influences patterns of cyclone activity and extreme weather events. However, in this section we focus on the effects of increasing temperature within estuarine ecosystems.

Strong geographic gradients in temperature exist from the tropics to the poles. Many species, such as mangrove trees, have thermal niches that allow them to inhabit only portions of these temperature gradients. Correspondingly, climate-induced changes in regional temperature regimes will almost certainly induce a wide range of ecological responses, ranging from local extinction of individual species and changes in biodiversity to changes in the rates of ecosystem processes, such as primary production and bacterially mediated decomposition. For example, net primary production of some estuarine species, such as the seagrass *Zostera marina*, may decrease as leaf respiration rates increase more rapidly than photosynthesis rates with increasing temperature (Short and Neckles 1999).

Changing temperatures will also cause many estuarine species to shift their geographic ranges to the north.

One of the most interesting areas to study the effects of temperature is the tropical-temperate interface. In the Gulf of Mexico, for example, much of the northern Gulf is dominated by salt marshes with mangroves occurring only in south Florida, sporadically in the Mississippi delta, and from south Texas into the southern Gulf. In the twenty-first century, it is highly likely that the entire coastal zone of the Gulf of Mexico will become tropical in the twenty-first century (Osland et al. 2017), allowing mangroves to migrate and replace salt marshes. Saintilan et al. (2014) have documented mangrove poleward expansion in Mexico, the Gulf of Mexico, Peru, China, and Australia.

Temperature also has a strong positive effect on rates of organic matter degradation (i.e., respiration), especially in the temperate climates, and warming is projected to enhance oxygen depletion rates in estuaries. When combined with the effect of warming on reductions in oxygen solubility, future warming is expected to expand hypoxic zones in many estuaries worldwide (Breitburg et al. 2018).

18.3.2 Accelerated Sea Level Rise

Of all climate change-related threats to estuaries, perturbations caused by rising sea levels are certainly the most obvious and the most studied. Undoubtedly, this is because we have already observed the dramatic effects of high rates of RSLR in regions such as the Mississippi River delta (2019) where high rates of land subsidence, combined with ESLR, led to wetland loss rates of 60 km^2 year^{-1} in the 1980s and 1990s (Boesch et al. 1994). In fact, high rates of RSRL have already led to significant geomorphological changes of coastal systems, salinity intrusion in estuaries, and loss of associated wetlands around the world including Chesapeake Bay (Stevenson et al., 1985), the Grijalva/Usumacinta delta in the southern Gulf of Mexico (Ortiz-Perez et al. 2013), Rhone (Pont et al. 2002), Nile and Ganges (Stanley and Warne 1998), Indus (Snedaker 1984) and Ebro (Ibàñez et al. 1997) deltas, and Venice Lagoon (Day et al. 1999). Observed rates of RSLR and predicted increases in the rate of ESLR have led to concerns regarding the resiliency of coastal wetlands worldwide, with the essential questions being: how will estuarine habitats change, and can wetland elevation keep pace with rising sea levels?

Estuarine wetlands exist in a dynamic equilibrium between the forces that lead to their establishment and maintenance, such as sediment accretion, and forces that lead to their deterioration, such as increasing rates of ESLR and deep and shallow subsidence (Day et al. 1999; Keogh and Törnqvist 2019). Rising sea levels and concomitant increasing marsh inundation, for example, can lead to increased sediment delivery and deposition on the marsh surface, and to decreased rates of organic matter decomposition because of more rapid burial and expanded anoxic conditions in sediments. Both increased deposition and decreased decomposition contribute to gains in surface elevation, which would counter sea level rise. However, there are limits to how fast marshes can accrete. Rapid sea level rise could lead to changes in the inundation regime and shifts in habitat types (Morris et al. 2021).

As sea levels rise, estuarine wetlands can respond in one of three ways. *First*, they can accrete sediments at a rate that equals the rate of RSLR, resulting in no net change in habitat area or type. Estuarine wetland elevation, relative to sea level, is a function of numerous processes including mineral and organic matter accretion, sediment compaction, deep subsidence, and ESLR, all operating at different time scales (Törnqvist et al. 2021; Figure 18.6; Box 18.1).

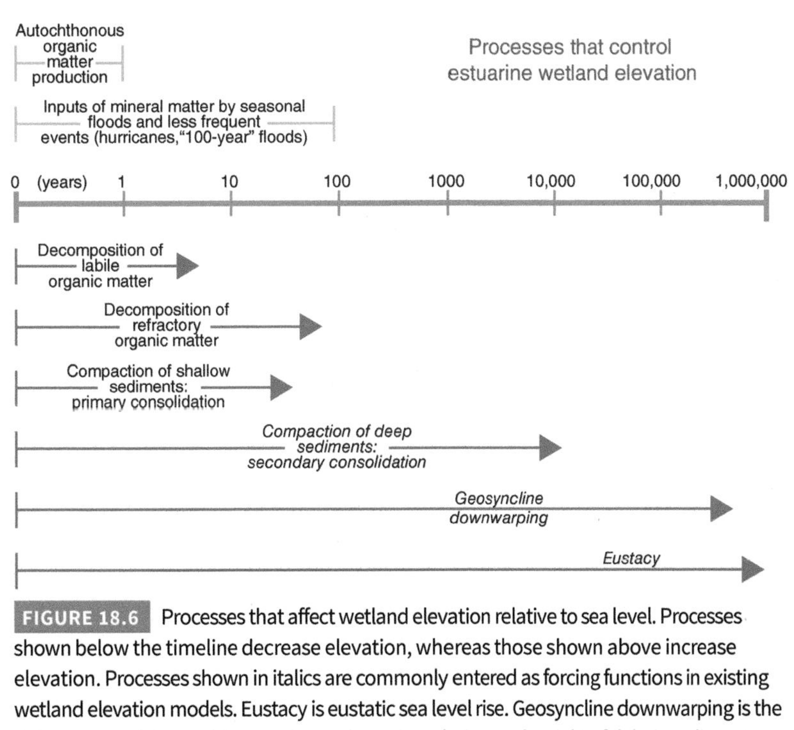

FIGURE 18.6 Processes that affect wetland elevation relative to sea level. Processes shown below the timeline decrease elevation, whereas those shown above increase elevation. Processes shown in italics are commonly entered as forcing functions in existing wetland elevation models. Eustacy is eustatic sea level rise. Geosyncline downwarping is the deformation of the earth's crust due to the accumulation and weight of deltaic sediments.

A number of studies have shown that estuarine wetlands can persist for long periods of time (thousands of years) in the face of rising sea levels when sediment accretion equals or exceeds the rate of land subsidence plus ELSR, as is the case for most wetlands worldwide under current rates of ESLR (Morris et al. 2002; Kirwan et al. 2016).

The rate of vertical accretion is a function of the inputs of both inorganic and organic material to the soil (Morris et al. 2016). Organic material is mostly derived from the growth of plant roots, whereas inorganic material is mostly supplied as sediments from either ocean or freshwater sources. Many rivers flowing into estuaries now carry only a fraction of the inorganic sediment that they did historically because of reservoir construction (e.g., Po, Ebro, and Nile Rivers). For example, sediment discharge to the Mississippi delta has decreased by at least 50% since 1860 (Kesel 1989). Giosan et al. (2014) reported that all of the world's deltas larger than 10,000 square kilometres, and most deltas larger than 1000 square kilometers, will not grow/accrete fast enough to keep pace with sea level rise of one meter in the next 100 years, due to lack of available sediment (Figure 18.7).

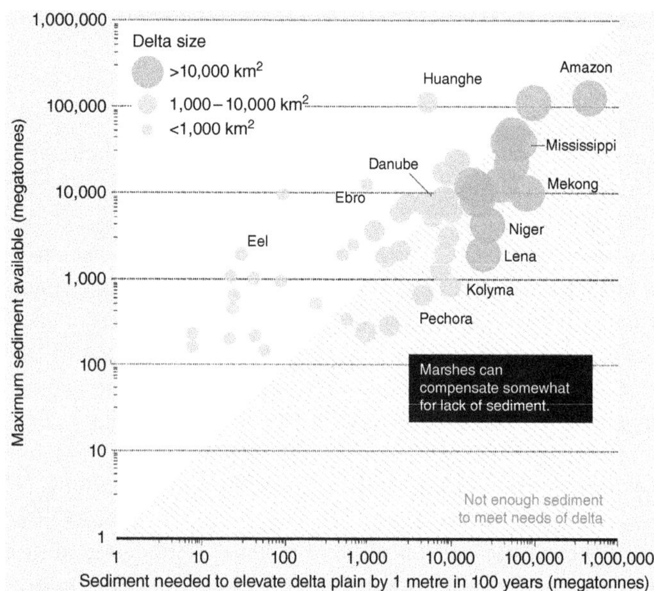

FIGURE 18.7 Many of the world's largest deltas do not have the sediment supply to keep pace with predicted rates of eustatic sea level rise (shown here as deltas that fall into the red zone on this graph. *Source*: Giosan et al. (2014)/Springer Nature.

Box 18.1 Measuring the processes that affect wetland elevation.

While many coastal wetlands accumulate sediment at a rate that keeps pace with current rates of ESLR (Kirwan and Megonigal 2013), the focus of many managers is the loss of relative elevation and the submergence of tidal wetlands. The potential for coastal wetland submergence has traditionally been determined by calculating a net accretion balance (Table 18.3). This is accomplished by comparing rates of vertical accretion to rates of RSLR (ESLR plus deep subsidence). Rates of accretion are typically estimated by measuring the accumulation of sediments, both organic and mineral, over some known and dated marker horizon such as feldspar clay, ^{137}Cs, or ^{210}Pb (Figure 18.8). Estimates of deep subsidence are usually based on long-term records from tide gauges that are mounted on stable piers, bridges, or pilings that extend through the shallow subsidence zone (and thus do not include shallow subsidence). A tidal gauge record spanning at least 18.6 years is required to factor out variations due to the moon's nodal cycle (Hiatt et al. 2019). Typically, mean annual or monthly water levels are regressed against time to yield a rate of RSLR. To estimate the deep subsidence component of RSLR, current ESLR is subtracted from the water-level rise recorded from pier-mounted tide gauges. ESLR is derived from the analysis of tide gauge data from coasts worldwide that are assumed to be experiencing no deep subsidence or, more recently, from satellite data.

However, shallow subsidence caused by compaction in the upper five to ten meters of sediment is often overlooked when calculating RSLR (Cahoon et al. 1995). This results in an underestimation of net accretion balance. Shallow subsidence is the result of primary sediment consolidation and the decomposition of organic matter. It is an especially important process in coastal systems under stress (e.g., from flooding or salt) where belowground plant structures, such as roots and rhizomes, die and decompose, leading to rapid sub-surface collapse (Chambers et al. 2019). It is also an active

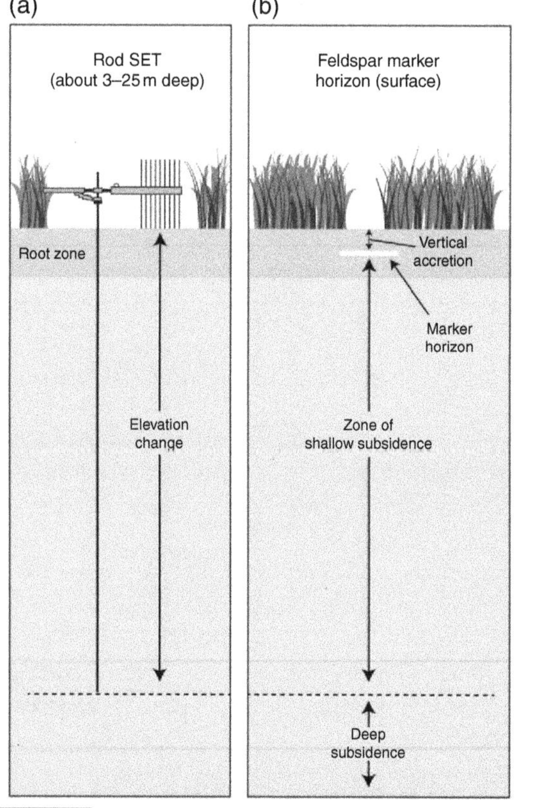

FIGURE 18.8 The original (a) Surface Elevation Table shown here being used in conjunction with (b) a feldspar marker horizon. When used simultaneously, shallow subsidence can be calculated as: Change in Elevation (as measured by the SET)—accretion (as measured by feldspar marker horizon). *Source*: From Lynch et al. (2015)/public domain.

process in coastal systems that are rapidly accreting, such as deltas, as surface sediments consolidate (Day et al. 1999). In the Mississippi River delta, for example, high rates of RSLR in coastal marshes have led to increasingly long periods of flooding, plant mortality, and rates of shallow subsidence that sometimes exceed 2.5 cm year^{-1}.

To this end, surface elevation tables (SETs) have become widely used for measuring both changes in elevation and shallow subsidence in coastal estuaries (Lynch et al. 2015).

The SET, introduced from the Netherlands into the United States in the early 1990s by Boumans and Day (1993) is a portable leveling device that is attached to a permanent benchmark pipe or rod driven to the "point of refusal" (e.g., bedrock) in marsh sediments (Figure 18.8) and provides a fixed reference position from which precise changes in elevation (within mm) can be measured. At any one point in time, the distance from the fixed SET to the marsh surface is measured by lowering a set of pins (usually nine) from the SET to the marsh surface (Figure 18.8). Furthermore, the SET can be rotated around several fixed positions (usually four or eight) on the benchmark pipe or rod, allowing for up to 9 × 8 precise measurements. Repeated measurements, from exactly the same positions, yield change in elevation over time.

SETs are especially useful when used simultaneously with feldspar marker horizons, or other methods that measure recent sediment accretion, because it allows for the estimation of shallow subsidence.

A *second* possible response of estuarine wetlands to sea level rise is for estuarine habitat to migrate upslope. At the wetland–upland transition, there could be conversion of upland areas to wetlands as frequent flooding reaches farther upslope and wetlands "migrate" inland (Brinson et al. 1995). However, in many locations, coastal wetlands are bordered by development or steep habitats, such that migration would not be possible (Titus 1991), in which case remaining wetland habitat would be slowly "squeezed" out of existence.

Finally, if wetlands cannot keep pace with rates of sea level rise, an increase in inundation frequency and duration could lead to a shift in the distribution of vegetated habitats across a wetland (Warren and Niering 1993; Kirwan and Murray 2007). For example, sea level rise is thought to be responsible for species shifts in New England salt marshes including replacement of the high marsh grass *Spartina patens* with low marsh *Spartina alterniflora* (Watson et al. 2017). Over longer time periods, these types of changes would lead to conversion of more and more wetland area to unvegetated mudflats or even subtidal open water (Schile et al. 2014) as happened in the Mississippi delta. During the process of wetland loss, a key transition occurs when wetland plants can no longer continue to grow. During this transition, wetland plants experience flooding stress due to increased flooding duration and salinity stress due to higher water elevation (Mendelssohn and Morris 2000). Morris et al. (2002) developed, and later generalized, a theoretical feedback framework (Morris et al. 2021) and model, the marsh equilibrium model (MEM), that predicts change in salt marsh plant production and elevation in response to sea level rise. Their simulations and field work revealed that the optimal rate of sea level rise maximizes sediment accretion and productivity but also represents the upper limit of flooding tolerance for a given plant species. The model predicted that their study site, a *S. alterniflora* marsh in South Carolina, could tolerate a rate of RSLR of up to 1.2 cm year^{-1}.

18.3.3 Modeling the Effects of Sea Level Rise on Estuarine Marshes

Given the complexity of estuarine systems, the time scales involved, and non-linear feedbacks between the many processes that contribute to wetland elevation sustainability in the face of sea level rise, researchers have turned to computer modeling to explore and predict long-term wetland sustainability in the face of climate change. Over the past 40 years, a variety of coastal wetland/estuarine models have been developed to address relationships between climate change, sea level rise, and coastal wetland sustainability (see Chapter 16 for an example). These models differ in spatial scales considered, processes simulated, and model forcing functions. A forcing function is an independent input variable that drives a model but is not modified by the model itself. For example, the rate of ESLR would typically be a forcing function in a model that simulates the effects of climate change on estuarine vegetation. We can divide these models into "landscape models" that simulate processes over large regions (i.e., entire estuaries or coastlines) and "geomorphic and ecogeomorphic models" that simulate physical and ecological processes at a single point in a marsh or across an individual marsh platform or transect.

In general, landscape models excel at simulating general trends at large spatial scales, but often do not mechanistically simulate processes, instead inputting processes that are measured or predicted separately as forcing functions. One of the most well-known landscape-scale models used to address climate change in coastal marshes is the sea level affecting marsh model (SLAMM). The model is initialized with digital elevation maps, local tidal data, potential rates of accretion, and vegetation distributions, and is forced by predicted rates of sea level rise. Model output includes maps of coastal habitat change in response to sea level rise. Glick et al. (2007), for example, used SLAMM to predict the effects of several sea level rise scenarios on habitat change and loss in the estuaries of Washington and Oregon, USA. They found that, for an ESLR of 0.69 m, 52% of estuarine brackish marsh in the region would be converted to either tidal flats or salt marshes.

At the other end of the spatial scale are models that simulate wetland elevation dynamics for one specific location in a coastal wetland (point models) or across a marsh transect or platform. These models are said to be "ecogeomorphic" (or "biogeomorphic") if they include feedbacks between marsh vegetation and physical processes such as sedimentation, decomposition, and erosion. One example of these feedbacks occurs when standing vegetation limits mixing of water during flooding, which reduces erosion and increases rates of sediment trapping. In an early example of a point model, Kirwan and Murray (2007) developed an exploratory, ecogeomorphic model for Westham Island, a salt marsh/tidal flat system in the Fraser River delta, British Columbia. Output was generated as two-dimensional elevation and vegetation zone maps of the island under various sea-level rise scenarios. They found that an RSLR of 1.036 meters in the next 100 years resulted in a loss of approximately 38% of total vegetated area.

Another widely used ecogeomorphic model is the MEM, which is a mechanistic point model (Morris et al. 2002) that is widely used for assessing the effects of RSLR on marsh systems. The model simulates marsh elevation as a function of autochthonous primary productivity and mineral deposition. There is also a family of other ecogeomorphic models that simulate surface elevation change dynamics by simulating how sediment accretes and compact. These are called "Sediment Cohort Models" because they track layers (or cohorts) of sediment over time. These models were derived from the early cohort models of Morris and Bowden (1986) and later modified and re-purposed, for climate change simulations, first by Callaway et al. (1996). Sediment cohort models are widely used to predict and assess coastal wetland responses to sea level rise (Table 18.4). These models excel at simulating belowground dynamics that affect wetland elevation, including compaction, root production, and sediment organic matter decomposition, and as output, they simulate changes in wetland surface elevation.

It is also important to consider some of the weakness and limitations of both landscape and ecogeomorphic models. First, both emphasize dynamics in the vertical dimension, i.e., simulating change in wetland elevation. Neither type emphasizes the processes that control lateral migration in the face of sea level rise (Kirwan et al. 2016). Second, these models, for the most part, simulate the effects of rising sea levels and ignore other climate change related phenomena that might also affect resiliency, including, for example, the effects of increasing temperature and CO_2 concentrations on primary productivity and organic matter decomposition. For an in-depth review of numerical, ecogeomorphic models of salt marsh evolution, with attention paid to climate change, see Fagherazzi et al. (2012).

TABLE 18.4 **The family of sediment cohort models used to predict the effect of rising sea levels on coastal wetland elevation.**

Model	Applications
WARMER: Wetland Accretion Rate Model for Ecosystem Resilience	Norfolk, England, Biloxi Bay, Mississippi, (Callaway et al. 1996), California Coastal Marshes (Swanson et al. 2015)
REM: Relative Elevation Model	Louisiana cypress/tupelo swamps (Rybczyk et al. 1998), Venice Lagoon, Italy (Day et al., 1999), Honduran mangrove systems (Cahoon et al. 2003), Eelgrass meadow, Washington State (Kairis and Rybczyk, 2010).
NUMAN: Mangrove Nutrient Model	South Florida mangroves (Chen and Twilley 1999; Twilley and Rivera- 2005)

18.3.4 Impacts of Changes in Freshwater Input on Coastal Ecosystems

For different areas of the world, climate projections indicate both increases and decreases in freshwater input. Increased freshwater input can have both beneficial and detrimental impacts on coastal systems. The benefit of freshwater input via enhancement of vertical soil accretion in coastal wetlands was highlighted above. Another benefit is increased fisheries production in coastal systems through stimulation of primary production from the inputs of freshwater nutrients and associated boosts to estuarine food webs that support fish (Nixon 1988).

One negative impact to coastal ecosystems associated with an increase in freshwater runoff is the potential for excessive increases in nutrients and the eutrophication of coastal waters. There is already considerable evidence that agricultural runoff and sewage wastewater in tributary watersheds has degraded many coastal ecosystems. For example, nuisance algal blooms and low oxygen in bottom waters caused by eutrophication have been well documented in Chesapeake Bay (Harding 1994), the Baltic Sea (Carstensen et al. 2014), the northern Gulf of Mexico adjacent to the Mississippi River delta (Rabalais et al. 2002), and world-wide (Fennel and Testa 2019). The negative effects of low oxygen and harmful algae on fish and macroinvertebrates are well known (Rabalais and Turner, 2019), and fish kill events are often reported

A number of management suggestions have been made to reduce high nutrient loading to streams and coastal waters. For the Mississippi River basin, for example, these recommendations include changes in farming practices, buffer strips along streams, use of wetlands to improve water quality, and reduction of nitrate in river water by diversions into riparian ecosystems and the Mississippi delta (Mitsch et al. 2001). The increased runoff may also lead to problems with toxic pollutants (e.g., heavy metals, organic chemicals) if there are high levels of these chemicals in the runoff.

There is also concern that diversions carried out to restore coastal wetlands and enhance their resiliency to sea level rise will lead to algal blooms because of the added nutrients. Thus, diversions will have to be studied and managed carefully to avoid problems. On the other hand, if freshwater input decreases, it will likely lead to less accretion, lowered productivity, and saltwater intrusion. Day et al. (2016) recommended that diversions be opened infrequently, every few years, to maximize land building and minimize the detrimental impacts, such as excessive nutrient inputs, that a continuous or semi-continuous diversion might bring (Xu et al. 2019).

Finally, Day et al. (2021b) reviewed the impact of climate change on arid estuaries, which are particularly sensitive to reductions in freshwater input. These ecologically important estuaries, which include the Colorado River and Indus and Nile deltas, are already threatened by increased water storage and diversion in their watersheds, political conflicts over water rights, eutrophication, increased salinization, and wetland loss. Further reductions in precipitation will lower freshwater input, especially with overuse in arid river basins.

18.3.5 Storms and Extreme Weather Events

Increasing warming of the ocean surface has contributed to an increase in the frequency and intensity of tropical cyclones, and the IPCC projects a continued increase in storms and associated precipitation and freshwater runoff. Changes in the intensity and frequency of storms can have a variety of impacts, especially on coastal wetlands. In general, long-term changes in the frequency and intensity of strong storms will most likely alter the species composition and biodiversity of coastal wetlands and may impact important ecosystem processes such as nutrient cycling and primary and secondary productivity, leading to both positive and negative effects. For example, tropical cyclones can introduce pulses of sediment to coastal marshes, helping to offset accelerated SLR (Yeates et al. 2020; Rybczyk and Cahoon 2002).

Runoff generated by tropical cyclones introduces freshwater and nutrients which can enhance coastal wetland productivity (Conner et al. 1989). On the other hand, high runoff from tropical cyclones can also lead to excessive nutrient loading and eutrophication problems. For example, record runoff from Hurricane Floyd into the Pamlico Sound estuary, North Carolina, USA, led to water quality problems (Paerl et al. 2006).

Tropical cyclones can reduce the structural complexity of forested wetlands such as mangroves and tidal freshwater forested wetlands as well as coastal marshes (Rybczyk et al. 1995; Cahoon et al. 2003). In 1998, category 4 Hurricane Mitch caused the destruction of 97% of the mangroves on the island of Guanaja, Honduras (Cahoon et al. 2003). During Hurricane Andrew in 1992, an extensive swath of mangrove trees was downed and defoliated across southern Florida (Doyle et al. 1995) and nearly 10% of trees in a freshwater forested wetland in Louisiana were blown down in this single storm (Rybczyk et al. 1995).

Other types of extreme weather events due to global warming may also affect estuarine ecosystems, and evidence does indicate that the intensification of extreme weather events will continue in a warming climate. Most intuitively, a shift toward warmer weather conditions will increase evaporation and surface drying, therefore increasing drought events (Dai 2011), and this will further stress estuaries in arid regions (Day et al. 2021a). Several recent examples are available for South Louisiana and Texas. In 2000–2001, an extreme drought raised salinities in western Lake Pontchartrain and Lake Maurepas from an average of 2–3 to 10–12 and led to mortality of cypress over a wide area (Day et al. 2012b; Shaffer et al. 2016). Such droughts, in combination with increasing sea level and strong tropical cyclones, will lead to salinity stress on broad areas of freshwater wetlands.

Alternatively, warmer air also can hold more moisture if moisture is available. Heavy precipitation events, and consequently flooding, are expected to increase in intensity with climate change (Schiermeier 2011; Prein et al. 2016). In August 2016, nearly a meter of rain fell in 3 days during a stalled front that was not associated with a tropical cyclone, leading to extensive flooding east of the Mississippi River. Hurricane Harvey dumped over a meter of rain on the Houston area in 2017 over a few days when its forward movement was stalled by a frontal system (Day et al. 2021b).

18.3.6 Ocean Acidification

When carbon dioxide dissolves in sea water, carbonic acids are formed, which then dissociate to hydrogen and bicarbonate ions, leading to ocean acidification (see Chapter 3). It is expected that pH in the oceans will decrease from 8.1 to 7.8 by 2100 if atmospheric CO_2 increases to 750 ppm under the more dire RCP scenarios. Ocean acidification reduces the ability of many marine organisms, including corals, to secrete their calcium carbonate skeletons (Doney et al. 2009). Recent studies suggest that reef sediments globally will probably transition from net precipitation to net dissolution by 2050 under higher emission scenarios (Eyre et al., 2018). Similarly, lobsters, mussels, shrimp, and oysters, especially in their larval forms, will not be able to form strong enough shells in acidified waters. The ecological effects of ocean acidification in estuaries and coastal waters may thus be profound. However, recent research has described the complexity of acidification patterns in estuaries, known as "Estuarine Acidification" (Cai et al. 2021), where pH can either increase or decrease over time as a result of altered riverine chemistry, eutrophication (Laurent et al. 2018), and internal biogeochemical cycling (Duarte et al. 2013). Thus, predicting future changes in pH in estuaries is complex and uncertain.

18.4 Human Activity and Coastal Management Implications

Anthropogenic impacts interact with physical perturbations related to climate change, leading to events that are often more severe than either impact acting alone. For example, in deltas such as the Mississippi, Rhone, Po, Ebro, and Nile, high rates of coastal wetland loss have resulted from isolation of the delta from the river (via levees and dikes), pervasive hydrologic alteration, and conversion to agriculture (Day et al. 2007, 2016, 2019; Ibàñez et al. 1997; Pont et al. 2002). These changes have also made the deltas more vulnerable to accelerated sea level rise and reduced freshwater input due to climate change. In the tropical wetlands of the Everglades in southern Florida, USA, hydrological alterations and diversion of freshwater flows for agriculture, human consumption, and flood control have reduced freshwater input, resulting in saltwater intrusion, wetland loss, eutrophication, and habitat changes (Sklar et al. 2005). In addition, agricultural runoff has caused nutrient enrichment in freshwater marshes of the Everglades, resulting in replacement of native vegetation with *Typha* spp. In arid and semi-arid watersheds, freshwater withdrawals have resulted in lowered freshwater inputs and salinization of soils, as for example in the Indus River delta (Day et al. 2021a). These types of changes will make the effects of climate change worse, and management is needed to counter these changes

Management for climate change will necessitate working with natural systems to enhance their ability to survive altered conditions. A fundamental concept in doing this is that sustainable management should be based on system functioning (Day et al. 1997, 2000). A basic management approach is that of ecological engineering, which is defined as "the design of sustainable ecosystems that integrate human society with its natural environment for the benefit of both" (Mitsch and Jorgensen 2003). This approach combines basic and applied science for the restoration, design, and construction of ecosystems. Ecological engineering relies primarily on the energies of nature, with human energy used in design and control of key processes.

For example, one management strategy to offset sea level rise and promote continued coastal wetland productivity is to make use of rivers as resources rather than letting most freshwater, sediments, and nutrients flow directly to the sea or be trapped and used upstream. An example of this is the Mississippi delta where levees cause most river water to discharge directly to the Gulf. This has led to widespread wetland loss (Day et al. 2007). In an effort to solve this problem, river diversions, such as the Caernarvon Freshwater Diversion, are being used where structures allow river water and sediments to flow back into coastal wetlands (Lane et al. 2006).

Another example is Puget Sound in Washington State, USA, where only 25% of the original estuarine habitat of its 16 major river deltas remain (Figure 18.9). Most of this loss is due to diking and draining to convert former estuaries into agricultural and urban land and most of that conversion happened in the late 1800s and early 1900s (Figure 18.10). Over time, the land behind the dikes has subsided due to oxidation of organic matter and compaction until, in some situations, the land can no longer drain, salt water intrudes, and farming becomes untenable. Rising sea levels outside the levees further exacerbate the problem. Restoration efforts for these estuaries are focused on either removing dikes entirely or breaching dikes to return tidal water to former estuarine wetlands (Ramirez 2019). In the Stillaguamish River delta, for example, a 60-hectare marsh was reintroduced to the tidal regime in 2012 from its previous use as diked and drained farmland, by lowering and

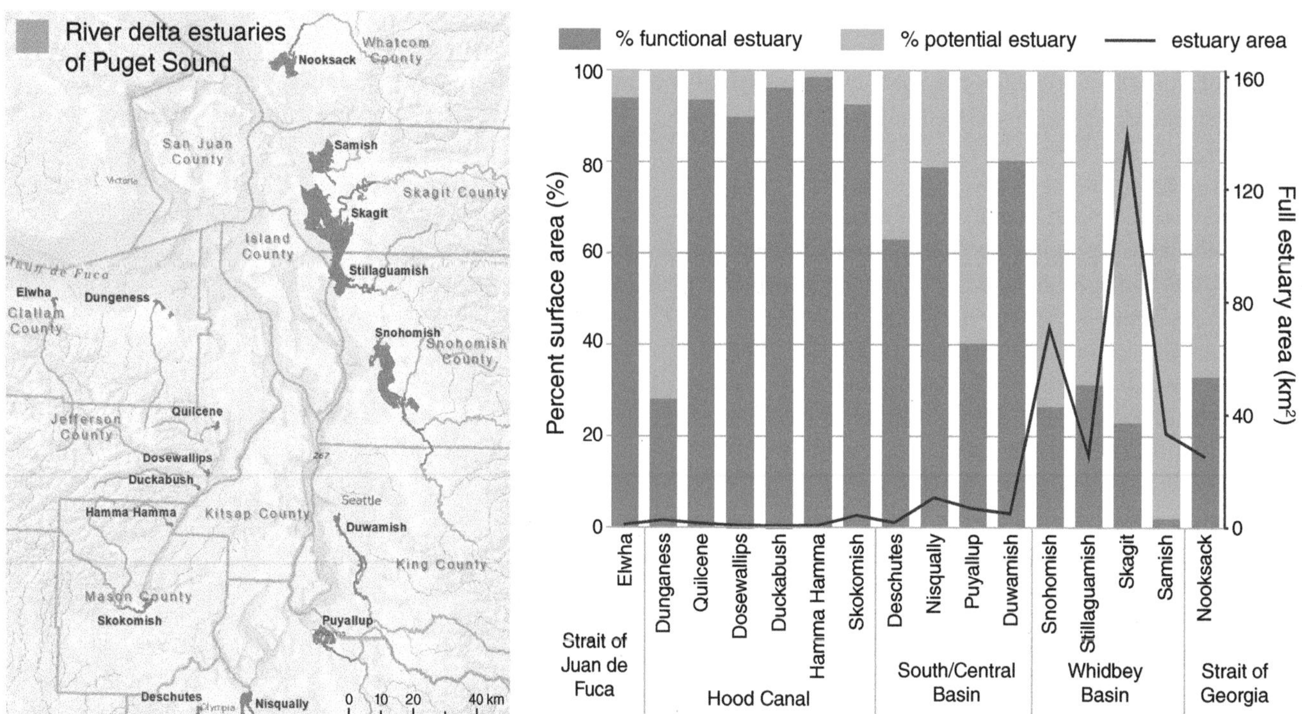

FIGURE 18.9 Percent composition of river deltas in the Puget Sound in Washington State, USA that is functional and fully connected to tidal inundation (blue bar) or disconnected (green bar), and the surface area of full estuary extent (black line). *Source*: Data from Ramirez (2019).

FIGURE 18.10 The construction of seaward dikes for the conversion of former estuaries into agricultural land has resulted in the loss of extensive estuarine habitat in the Puget Sound, WA. Pictured here is the diked Stillaguamish Estuary, Washington State, USA.

breaching a dike constructed in the early 1950. With the return of nutrients and sediment, the restored marsh is gaining elevation at a rate of 2.74 cm $year^{-1}$, a rate that greatly exceeds both the current rate of ESLR and the rate of accretion outside of the restoration area (1.03 cm $year^{-1}$; Poppe and Rybczyk 2021).

18.5 Coastal Wetlands for Climate Change Mitigation: Blue Carbon

It is probably safe to say that most of the discussion and research related to estuaries and climate change centers around some aspect of climate change *adaptation* (understanding, quantifying, and reducing the vulnerability of estuaries to the negative impacts of climate change) rather than *mitigation* (reducing the flow of greenhouse gasses to the atmosphere). For example, many recent estuarine restoration efforts, such as dike and levee removal, are designed with enhanced resiliency to sea level rise as a project outcome. Reconnecting estuaries to riverine and tidal inputs via dike removal can enhance rates of accretion in restored marshes and allow those wetlands to keep pace with rising seas (Poppe and Rybczyk, 2021).

However, estuarine wetlands also play a role in climate change mitigation. In particular, estuarine wetlands are able to rapidly accumulate and store "blue carbon." Here blue carbon is defined as "the atmospheric carbon captured during photosynthesis and accumulating in vegetated, tidally influenced coastal ecosystems such as tidal forests, tidal marshes, and intertidal seagrass meadows (Windham-Myers et al. 2018)." This includes carbon in plants, live or dead, and, most significantly, carbon stored in sediments (Figure 18.11).

FIGURE 18.11 A sediment core taken from an estuarine salt marsh reveals approximately 20 cm of organic matter and associated carbon sequestered over an underlaying sand layer.

There are four reasons why estuarine wetlands are especially good at sequestering blue carbon. First, of all plant communities in the world, coastal marsh habitats rank among the highest in net primary production (Mitsch and Gosselink 2015), thus capturing large amounts of atmospheric carbon dioxide through photosynthesis relative to other plant communities on a square meter basis. Second, characteristically flooded and anaerobic sediments inhibit the decomposition and oxidation of soil organic matter and associated carbon, thus preserving it in the sediments. Third, estuarine wetlands, existing in a dynamic equilibrium with sea level rise (Day et al. 2012a), are commonly long-term depositional environments, sequestering both organic and mineral matter over time. Finally, in saltwater systems, abundant sulfates suppress methane production, a powerful greenhouse gas, from the decomposition of organic matter, which could otherwise negate the climate benefit of carbon dioxide removal (Kroeger et al. 2017).

With regards to blue carbon, climate change mitigation can be enhanced in two ways. First, we can work toward keeping carbon stocks in place by protecting existing estuaries and specifically coastal wetlands (Spivak et al. 2019). Pendleton et al. (2012) estimated that degradation of coastal ecosystems (mangroves, seagrasses, and tidal marshes) worldwide releases 0.45 billion tons of stored carbon as CO_2 to the atmosphere each year. Furthermore, this estimate does not include the loss of sequestration potential, i.e., future annual uptake of carbon lost when estuarine habitat is lost. Second, we can work to restore estuaries and their embodied ecosystem services, including carbon sequestration. Recent studies have shown that estuary restoration can enhance sedimentation and concomitant carbon sequestration (Drexler et al. 2020). For example, Poppe and Rybczyk (2021) found that four years after restoration via levee removal, mean carbon accumulation rates in the restored marsh were nearly twice as high (230.49 g C m^{-2} $year^{-1}$) as the rates in adjacent natural marshes (123.00 g C m^{-2} $year^{-1}$) due to high rates of accretion in the recently restored marsh (1.57 cm $year^{-1}$).

18.6 Summary

Perhaps the biggest challenge for estuarine ecosystem management in the face of climate change is to wisely allocate our limited resources toward estuarine restoration and maintenance. Day et al. (2016, 2019) discussed approaches to deltaic sustainability, both methodological and philosophical, in the face of climate change and other anthropogenic perturbations and provided a framework for considering future climate adaptation efforts in estuaries. They considered a variety of factors to classify the potential sustainability of deltas worldwide including decreased freshwater and sediments, subsidence rates on the delta plain, increased rates of sea level rise, resource scarcity, population pressure, elevation capital (current elevation above sea level), predicted cyclone impacts, and more. This resulted in the classification of the world's major deltas into four categories, ranging from those with a high potential for sustainability to systems in an advanced state of deterioration with little hope for sustainability, even with substantial inputs of energy and money (Table 18.5). They concluded that with accelerating climate change impacts, growing resource scarcity, growing populations, and increasing costs of energy, efforts toward sustainability should be focused on those systems that have at least some potential for sustainability in the face of rising seas and climate change.

In concept, this is similar to the National Park Service's climate change management paradigm that centers around the resist-accept-direct (RAD) decision framework (Table 18.6; Lynch et al. 2022). In effect, the Park Service recognizes, in the face of accelerated warming and changing disturbance regimes associated with climate change, that "holding the line" would become difficult, if not impossible for many park ecosystems in an increasingly non-stationary world. This is a fundamental and dramatic change in the Service's historic core mission of preserving a baseline "natural condition" in U.S. national parks. In other words, the Park Service recognizes that not all park ecosystems can be managed to maintain historical conditions, given climate change.

Circling back to the Day et al. (2016) classification (Table 18.5), the sustainability of coastal systems requires management plans that incorporate climate resiliency at the lowest operating energy cost. This includes, most importantly, the incorporation and return of natural, ordering energies. Specifically, we can promote estuarine resiliency to climate change by enhancing and restoring connectivity with river and tidal energies, as discussed earlier in this chapter. Connectivity to rivers provides for much needed sediments that can compensate for rising sea levels. Where possible, the removal of shoreline dikes and armoring can allow for upslope migration of estuaries as sea levels rise. Conversely, energy intensive management, such as diking and pumping, will become less and less practical, affordable, or advisable.

TABLE 18.5 Sustainability of major world estuarine deltas, classified according to potential for sustainability in the face of climate change.

Potential for Sustainability	Criteria	Examples
1. High	Estuarine deltas currently able to deal with forces challenging sustainability. Adequate sediment and water supply. Open to tidal and riverine pulses.	Amazon, Congo, Orinioco, Volga
2. Moderate	Potential for low-energy management to lessen forces threatening sustainability. Typically, systems with low sediment deficits and subsidence rates.	Yukon, Rhone, Danube
3. Low	Strongly impacted by human activity. Some of the estuary behind dikes and levees and below sea level. High energy management required for sustainability.	Mississippi, Ebro, Rhine, Po, Mekong, Ganges
4. Not Sustainable	Systems in advanced states of deterioration with greatly reduced water and sediment inputs. Little if any functioning estuary remaining. Arid regions	Nile, Colorado, Tigres-Euprhrates, Indus

Source: Adapted from Day et al. (2016).

TABLE 18.6 The resist–accept–direct management options.

1. **Resist** the trajectory of change, by working to maintain or restore ecosystem processes, function, structure, or composition based upon historical or acceptable current conditions.
2. **Accept** the trajectory of change, by allowing ecosystem processes, function, structure, or composition to change, without intervening to alter their trajectory.
3. **Direct** the trajectory of change, by actively shaping ecosystem processes, function, structure, or composition toward desired new conditions.

Source: From Lynch et al. (2022).

Study Questions

Multiple Choice

1. Which of the following process must be considered to calculate a rate of Relative Sea Level Rise (RSLR) for any given location? (Select all that apply)
   a. Eustatic sea level rise
   b. Shallow subsidence
   c. Deep Subsidence
   d. The decomposition of organic matter in the sediments
   e. Tectonic activity

2. Why do estuarine wetlands excel at sequestering sediment carbon compared to terrestrial ecosystems? (Select all that apply)
   a. Estuarine wetlands are among the most productive ecosystems on earth.
   b. Estuarine wetlands are commonly depositional environments.
   c. The typically anaerobic environment in sediments inhibits organic matter decomposition.
   d. The carbon fixed by wetland plants does not decompose as fast as carbon fixed by terrestrial plants.
   e. There is abundant sulfate in estuarine wetlands.

3. For the first 10 years of the twenty-first century, the largest fraction of the observed increase in ESLR was due to
   a. the melting of Greenland icesheets and glaciers.
   b. the thermal expansion of sea water as the earth and ocean warms.
   c. the melting of the Antarctic ice.
   d. changes in land water storage.
   e. the melting of sea ice.

4. For parts of the world, including the Pacific Coast of the United States, salmon are an iconic, estuarine dependent species. How is their future survival threatened by global climate change? (Select all that apply)
   a. A warming climate may result in higher stream temperatures.
   b. A warming climate may increase winter flooding.
   c. Rates of ESLR will increase under a regime of global warming
   d. A warming climate would likely increase summer stream flow in the Pacific Northwest.
   e. A warming climate would reduce spring snowpack.

5. Current atmospheric carbon dioxide concentrations (Select all that apply)
   a. exceeds any concentration measured directly or by proxy in the past 800,000 years.
   b. exceeds 400 ppm
   c. are lower than they were during the end of the first Industrial Revolution (1830).
   d. are causing the earth to warm faster than it has in the past 800,000 years.
   e. are as low as they ever will be in the next 100 years, under the IPCC RCP 8.5 scenario.

6. With regards to climate change related observations and predictions, which of the following are we the *least* certain about right now?
   a. Atmospheric carbon dioxide concentrations are increasing and are the highest they have been the past 800,000.
   b. That the human influence on the climate system is clear.
   c. Global average surface temperatures have increased over the past 100 years.
   d. Anthropogenic climate change has and will increase the size and intensity of tropical cyclones

7. Estuarine wetlands can respond to sea level rise in which of the following ways. (Select all that apply)
   a. They can migrate upland as seas rise.
   b. They can accrete at a rate that matches the rate of sea level rise, thus maintaining their position.
   c. Habitat switching
   d. They can accrete at a rate that greatly exceeds sea level rise and convert to terrestrial systems.

8. For different areas of the world, climate projections indicate both increases and decreases in freshwater inputs to coastal systems. Increased freshwater inputs can lead to both beneficial and detrimental impacts. Which of the following would generally be considered a detrimental impact? (Select all that apply)
   a. Increased sediment delivery to coastal estuaries
   b. Excessive nutrient delivery to the coast, leading to eutrophication
   c. Excessive nutrient delivery to the coast could increase fisheries production.
   d. The increased runoff may also lead to problems with toxic pollutants.

9. The IPCC's Representative Concentration Pathways (Select all that apply)
   a. represent projected warming trajectories based on predicted future population size and energy use.
   b. indicate the change in total radiative forcing (watts per square meter) in the troposphere by 2100, relative to pre-industrial levels.
   c. vary from overly optimistic predictions to, perhaps, overly pessimistic predictions regarding future emission scenarios.
   d. are unlikely to ever actually happen in the future.
   e. assume a steady state scenario in which greenhouse gas emissions do not change in the future.

Short answer

1. How does Eustatic Sea Level Rise differ from Relative Sea Level Rise?
2. What is the difference between climate change mitigation and climate change adaptation?
3. Give an example of how estuaries can serve to mitigate climate change.
4. Surface elevation tables (SET's) are devices that can precisely measure change in marsh surface elevation. How does this differ from simply measuring short term rates of sediment accretion over a marsh surface to calculate rates of elevation change.
5. Estuarine wetlands have some ability to maintain their elevation relative to sea level as sea levels rise or, in other words, to keep pace with sea level rise. By what processes can they do so?
6. What kind of climate future does the RCP2.6 project?
7. Describe a sustainable coastal management approach that would maximize estuarine resiliency to climate change and sea level rise.

References

Bhatia, K.T., Vecchi, G.A., Knutson, T.R. et al. (2019). Recent increases in tropical cyclone intensification rates. *Nature Communications 10* (1): 1–9.

Boesch, D.F., Josselyn, M.N., Mehta, A.J. et al. (1994). Scientific assessment of coastal wetland loss, restoration and management in Louisiana. *Journal of Coastal Research*, Special Issue 20 i–103.

Boumans, R.M. and Day, J.W. (1993). High precision measurements of sediment elevation in shallow coastal areas using a sedimentation-erosion table. *Estuaries 16* (2): 375–380.

Breitburg, D., Levin, L.A., Oschlies, A. et al. (2018). Declining oxygen in the global ocean and coastal waters. *Science 359* (6371).

Brinson, M.M., Christian, R.R., and Blum, L.K. (1995). Multiple states in the sea-level induced transition from terrestrial forest to estuary. *Estuaries 18* (4): 648–659.

Cahoon, D.R., Reed, D.J., and Day, J.W. Jr. (1995). Estimating shallow subsidence in microtidal salt marshes of the southeastern United States: Kaye and Barghoorn revisited. *Marine geology 128* (1–2): 1–9.

Cahoon, D.R., Hensel, P., Rybczyk, J.M. et al. (2003). Mangrove peat collapse following mass tree mortality: implications for forest recovery from Hurricane Mitch. *Journal of Ecology* 91: 1093–1105.

Cahoon, D.R., Lynch, J.C., Roman, C.T. et al. (2019). Evaluating the relationship among wetland vertical development, elevation capital, sea-level rise, and tidal marsh sustainability. *Estuaries and Coasts 42* (1): 1–15.

Cai, W.J., Feely, R.A., Testa, J.M. et al. (2021). Natural and anthropogenic drivers of acidification in large estuaries. *Annual Review of Marine Science 13*: 23–55.

Callaway, J.C., Nyman, J.A., and DeLaune, R.D. (1996). Sediment accretion in coastal wetlands: a review and a simulation model of processes. *Current Topics in Wetland Biogeochemistry 2* (2): 23.

Carstensen, J., Andersen, J.H., Gustafsson, B.G., and Conley, D.J. (2014). Deoxygenation of the Baltic Sea during the last century. *Proceedings of the National Academy of Sciences 111* (15): 5628–5633.

Casola, J.H., Kay, J.E., Snover, A.K., Norheim, R.A. and Whitely Binder, L.C., (2005). Climate Impacts on Washington's Hydropower, Water Supply, Forests, Fish, and Agriculture. A report prepared for King County (Washington) by the Climate Impacts Group; 44 pp.

Chambers, L.G., Steinmuller, H.E., and Breithaupt, J.L. (2019). Toward a mechanistic understanding of "peat collapse" and its potential contribution to coastal wetland loss. *Ecology 100* (7): e02720.

Chen, R. and Twilley, R.R. (1999). A simulation model of organic matter and nutrient accumulation in mangrove wetland soils. *Biogeochemistry 44* (1): 93–118.

Conner, W.H., Day, J.W., Baumann, R.H., and Randall, J.M. (1989). Influence of hurricanes on coastal ecosystems along the northern Gulf of Mexico. *Wetlands Ecology and Management 1* (1): 45–56.

Dai, A. (2011). Drought under global warming: a review. *Wiley Interdisciplinary Reviews*: Climate Change 2 (1): 45–65.

Day, J.W. Jr., Rybczyk, J., Scarton, F. et al. (1999). Soil accretionary dynamics, sea-level rise and the survival of wetlands in Venice Lagoon: a field and modelling approach. *Estuarine, Coastal and Shelf Science 49* (5): 607–628.

Day, J.W. Jr., Barras, J., Clairain, E. et al. (2005). Implications of global climatic change and energy cost and availability for the restoration of the Mississippi delta. *Ecological Engineering 24* (4): 253–265.

Day, J.W. Jr., Kemp, W.M., Yáñez-Arancibia, A., and Crump, B.C. (ed.) (2012a). *Estuarine Ecology*. Wiley.

Day, J.W. Jr., Hall, C.A.S., Kemp, W.M., and Yáñez-Arancibia, A. (1989). *Estuarine Ecology*, 1e. Hoboken, N.J.: Wiley-Blackwell.

Day, J., Martin, J., Cardoch, L., and Templet, P. (1997). System functioning as a basis for sustainable management of deltaic ecosystems. *Coastal Management*. 25: 115–154.

Day, J., Psuty, N., and Perez, B. (2000). The role of pulsing events in the functioning of coastal barriers and wetlands: Implications for human impact, management and the response to sea level rise. In: *Concepts and Controversies in Salt Marsh Ecology* (ed. M. Weinstein and D. Kreeger), 633–660. Dordrecht, The Netherlands: Kluwer Academic Publishers.

Day, J.W., Boesch, D.F., Clairain, E.J. et al. (2007). Restoration of the Mississippi Delta: lessons from hurricanes Katrina and Rita. *Science 315* (5819): 1679–1684.

Day, J., Hunter, R., Keim, R.F. et al. (2012b). Ecological response of forested wetlands with and without large-scale Mississippi River input: implications for management. *Ecological Engineering 46*: 57–67.

Day, J.W., Agboola, J., Chen, Z. et al. (2016). Approaches to defining deltaic sustainability in the 21st century. *Estuarine, Coastal and Shelf Science 183*: 275–291.

Day, J., Ramachandran, R., Giosan, L. et al. (2019). Delta winners and losers in the anthropcene. In: *Coasts and Estuaries—the Future* (ed. E. Wolanski, J. Day, M. Elliott and R. Ramachandran), 153–172. Amsterdam: Elsevier.

Day, J., Goodman, R., Chen, Z. et al. (2021a). Deltas in arid environments. *Water 13* (12): 1677.

Day, J.W., Hunter, R., Kemp, G.P. et al. (2021b). The "Problem" of New Orleans and diminishing sustainability of Mississippi River management—future options. *Water 13* (6): 813.

Doney, S.C., Fabry, V.J., Feely, R.A., and Kleypas, J.A. (2009). Ocean acidification: the other CO_2 problem. *Annual Review of Marine Science* 1: 169–192.

Doyle, T.W., Smith, T.J. III, and Robblee, M.B. (1995). Wind damage effects of Hurricane Andrew on mangrove communities along the southwest coast of Florida, USA. *Journal of Coastal Research* 159–168.

Drexler, J.Z., Davis, M.J., Woo, I., and De La Cruz, S. (2020). Carbon sources in the sediments of a restoring vs. historically unaltered salt marsh. *Estuaries and Coasts* 43 (6): 1345–1360.

Duarte, C.M., Hendriks, I.E., Moore, T.S. et al. (2013). Is ocean acidification an open-ocean syndrome? Understanding anthropogenic impacts on seawater pH. *Estuaries and Coasts* 36 (2): 221–236.

Emery, K.O. and Aubrey, D.G. (1991). *Sea Levels, Land Levels, and Tide Gauges*. New York: Springer-Verlag.

Eyre, B.D., Cyronak, T., Drupp, P. et al. (2018). Coral reefs will transition to net dissolving before end of century. *Science* 359 (6378): 908–911.

Fagherazzi, S., Kirwan, M., Mudd, S. et al. (2012). Numerical models of salt marsh evolution: ecological, geomorphic and climatic factors. *Reviews of Geophysics* https://doi.org/10.1029/2011RG000359.

Fennel, K. and Testa, J.M. (2019). Biogeochemical controls on coastal hypoxia. *Annual Review of Marine Science 11*: 105–130.

Giosan, L., Syvitski, J., Constantinescu, S., and Day, J. (2014). Climate change: protect the world's deltas. *Nature News 516* (7529): 31.

Glick, P., Clough, J., and Nunley, B. (2007). *Sea-level Rise and Coastal Habitats in the Pacific Northwest: An Analysis for Puget Sound, Southwestern Washington, and Northwestern Oregon*. Reston, VA: National Wildlife Federation.

Greene, C.M., Jensen, D.W., Pess, G.R. et al. (2005). Effects of environmental conditions during stream, estuary, and ocean residency on Chinook salmon return rates in the Skagit River, Washington. *Transactions of the American Fisheries Society 134* (6): 1562–1581.

Harding, L.W. (1994). Long-term trends in the distribution of phytoplankton in Chesapeake Bay: roles of light, nutrients and streamflow. *Marine Ecology-Progress Series 104*: 267–267.

Hayhoe, K., Edmonds, J., Kopp, R., LeGrande, A., Sanderson, B., Wehner, M. and Wuebbles, D., 2017. Climate models, scenarios, and projections. 2017. In: Climate Science Special Report: Fourth National Climate Assessment, Volume I [Wuebbles, D.J., D.W. Fahey, K.A. Hibbard, D.J. Dokken, B.C. Stewart, and T.K. Maycock(eds.)]. U.S. Global Change Research Program, Washington, DC, USA, pp. 133-160, https://doi.org/10.7930/J0WH2N54.

Hiatt, M., Snedden, G., Day, J.W. et al. (2019). Drivers and impacts of water level fluctuations in the Mississippi River delta: Implications for delta restoration. *Estuarine, Coastal and Shelf Science 224*: 117–137.

Hood, W.G. (2005). *Sea Level Rise in the Skagit Delta*. Mount Vernon (WA): Skagit River Tidings, Skagit Watershed Council.

Horton, B.P., Khan, N.S., Cahill, N. et al. (2020). Estimating global mean sea-level rise and its uncertainties by 2100 and 2300 from an expert survey. *npj Climate and Atmospheric Science 3* (1): 1–8.

Ibàñez, C., Canicio, A., Day, J.W., and Curcó, A. (1997). Morphologic development, relative sea level rise and sustainable management of water and sediment in the Ebre Delta, Spain. *Journal of Coastal Conservation 3* (1): 191–202.

IPCC (2014a). Summary for policymakers. In: *Climate Change 2014: Mitigation of Climate Change. Contribution of Working Group III to the Fifth Assessment Report of the Intergovernmental Panel on Climate Change* (ed. O. Edenhofer, R. Pichs-Madruga, Y. Sokona, et al.). Cambridge: Cambridge University Press United Kingdom and New York, NY, USA.

Intergovernmental Panel on Climate Change. (2014b). Sea level change. In Climate Change 2013 – The Physical Science Basis: Working Group I Contribution to the Fifth Assessment Report of the Intergovernmental Panel on Climate Change (pp. 1137–1216). Cambridge: Cambridge University Press. doi:https://doi.org/10.1017/CBO9781107415324.026

IPCC (2019). Summary for policymakers. In: *IPCC Special Report on the Ocean and Cryosphere in a Changing Climate* (ed. H.-O. Pörtner, D.C. Roberts, V. Masson-Delmotte, et al.). Geneva, Switzerland: IPCC.

Kairis, P.A. and Rybczyk, J.M. (2010). Sea level rise and eelgrass (Zostera marina) production: a spatially explicit relative elevation model for Padilla Bay, WA. *Ecological Modelling 221* (7): 1005–1016.

Keogh, M.E. and Törnqvist, T.E. (2019). Measuring rates of present-day relative sea-level rise in low-elevation coastal zones: a critical evaluation. *Ocean Science 15* (1): 61–73.

Kesel, R.H. (1989). The role of the Mississippi River in wetland loss in southeastern Louisiana, USA. *Environmental Geology and Water Sciences 13* (3): 183–193.

Kirwan, M.L. and Megonigal, J.P. (2013). Tidal wetland stability in the face of human impacts and sea-level rise. *Nature 504* (7478): 53–60.

Kirwan, M.L. and Murray, A.B. (2007). A coupled geomorphic and ecological model of tidal marsh evolution. *Proceedings of the National Academy of Sciences 104* (15): 6118–6122.

Kirwan, M.L., Temmerman, S., Skeehan, E.E. et al. (2016). Overestimation of marsh vulnerability to sea level rise. *Nature Climate Change 6* (3): 253–260.

Ko, J.Y. and Day, J.W. (2004). A review of ecological impacts of oil and gas development on coastal ecosystems in the Mississippi Delta. *Ocean & Coastal Management 47* (11–12): 597–623.

Kroeger, K.D., Crooks, S., Moseman-Valtierra, S., and Tang, J. (2017). Restoring tides to reduce methane emissions in impounded wetlands: a new and potent blue carbon climate change intervention. *Scientific Reports* 7 (1): 1–12.

Lane, R.R., Day, J.W., and Day, J.N. (2006). Wetland surface elevation, vertical accretion, and subsidence at three Louisiana estuaries receiving diverted Mississippi River water. *Wetlands* 26 (4): 1130–1142.

Laurent, A., Fennel, K., Ko, D.S., and Lehrter, J. (2018). Climate change projected to exacerbate impacts of coastal eutrophication in the northern Gulf of Mexico. *Journal of Geophysical Research: Oceans* 123 (5): 3408–3426.

Li, L. and Chakraborty, P. (2020). Slower decay of landfalling hurricanes in a warming world. *Nature 587* (7833): 230–234.

Lynch, J.C., Hensel, P., and Cahoon, D.R. (2015). *The surface elevation table and marker horizon technique: A protocol for monitoring wetland elevation dynamics* (No. NPS/NCBN/NRR—2015/1078). National Park Service.

Lynch, A.J., Thompson, L.M., Morton, J.M. et al. (2022). RAD adaptive management for transforming ecosystems. *BioScience* 72 (1): 45–56.

Meehl, G.A., Stocker, T.F., Collins, W.D., Friedlingstein, P., Gaye, A.T., Gregory, J.M., Kitoh, A., Knutti, R., Murphy, J.M., Noda, A. and Raper, S.C., 2007. *Global Climate Projections*. Chapter 10.

Meengs, C.C. and Lackey, R.T. (2005). Estimating the size of historical Oregon salmon runs. *Reviews in Fisheries Science 13* (1): 51–66.

Mei, W., Xie, S.P., Primeau, F. et al. (2015). Northwestern Pacific typhoon intensity controlled by changes in ocean temperatures. *Science Advances 1* (4): e1500014.

Mendelsshon, I. and Morris, J. (2000). Eco-physiological controls on the productivity of Spartina alterniflora Loisel. In: *Concepts and Controversies in Tidal Marsh Ecology* (ed. M. Weinstein and D. Kreeger), 59–80. Dordrecht, The Netherlands: Kluwer Academic Publishers.

Mitsch, W.J. and Gosselink, J.G. (2015). *Wetlands*. New York, NY: Wiley.

Mitsch, W.J. and Jørgensen, S.E. (2003). Ecological engineering: a field whose time has come. *Ecological Engineering 20* (5): 363–377.

Mitsch, W., Day, J., Gilliam, J. et al. (2001). Reducing nitrogen loading to the Gulf of Mexico from the Mississippi River basin: Strategies to counter a persistent problem. *BioScience.* 51 (5): 373–388.

Morris, J.T. and Bowden, W.B. (1986). A mechanistic, numerical model of sedimentation, mineralization, and decomposition for marsh sediments. *Soil Science Society of America Journal 50* (1): 96–105.

Morris, J.T., Sundareshwar, P.V., Nietch, C.T. et al. (2002). Response of coastal wetlands to rising sea-levels. *Ecology* 83 (10): 2869–2877.

Morris, J.T., Barber, D.C., Callaway, J.C. et al. (2016). Contributions of organic and inorganic matter to sediment volume and accretion in tidal wetlands at steady state. *Earth's Future 4* (4): 110–121.

Morris, J., Cahoon, D., Callaway, J. et al. (2021). Marsh equilibrium theory: implications for responses to rising sea level. In: *Salt Marshes: Function, Dynamics, and Stresses* (ed. D. FitzGerald and Z. Hughes), 157–177. Cambridge: Cambridge University Press https://doi.org/10.1017/9781316888933.009.

Nixon, S.W. (1988). Physical energy inputs and the comparative ecology of lake and marine ecosystems. *Limnology and Oceanography 33* (4part2): 1005–1025.

NOAA (2021). The New Climate Normals are Here. What do they tell us about climate change? NOAA News and Features. (accessed 4 May 2021).

Oppenheimer, M., Glavovic, B.C., Hinkel, J. et al. (2019). Sea level rise and implications for low-lying islands, coasts and communities. In: *IPCC Special Report on the Ocean and Cryosphere in a Changing Climate* (ed. H.O. Pörtner, D.C. Roberts, V. Masson-Delmotte, et al.). Geneva, Switzerland: IPCC.

Ortiz-Perez, M., Mendez, A., and Hernandez, J. (2013). Sea-level rise and vulnerability of coastal lowlands in the Mexican areas of the Gulf of Mexico and the Caribbean Sea. In: *Gulf of Mexcio Origin, Waters, and Biota*, vol. 4 (ed. J. Day and A. Yanez), 273–290. College Station: Ecosystem Based Managemen. Texas A&M University Press.

Osland, M.J., Day, R.H., Hall, C.T. et al. (2017). Mangrove expansion and contraction at a poleward range limit: climate extremes and land-ocean temperature gradients. *Ecology 98* (1): 125–137.

Paerl, H.W., Valdes, L.M., Joyner, A.R. et al. (2006). Ecological response to hurricane events in the Pamlico Sound system, North Carolina, and implications for assessment and management in a regime of increased frequency. *Estuaries and Coasts* 29 (6): 1033–1045.

Pendleton, L., Donato, D.C., Murray, B.C. et al. (2012). Estimating global "blue carbon" emissions from conversion and degradation of vegetated coastal ecosystems. *PLoS One* 7 (9): e43542.

Pont, D., Day, J.W., Hensel, P. et al. (2002). Response scenarios for the deltaic plain of the Rhône in the face of an acceleration in the rate of sea-level rise with special attention to Salicornia-type environments. *Estuaries* 25 (3): 337–358. http://www.jstor.org/stable/1352959.

Poppe, K.L. and Rybczyk, J.M. (2021). Tidal marsh restoration enhances sediment accretion and carbon accumulation in the Stillaguamish River Estuary, Washington. *PLoS One 16* (9): e0257244.

Prein, A.F., Holland, G.J., Rasmussen, R.M. et al. (2016). Running dry: the US Southwest's drift into a drier climate state. *Geophysical Research Letters* 43 (3): 1272–1279.

Pritchard, D.W. (1967). What is an Estuary: physical viewpoint. American Association for the Advancement of Science.

Rabalais, N.N. and Turner, R.E. (2019). Gulf of Mexico hypoxia: past, present, and future. *Limnology and Oceanography Bulletin 28* (4): 117–124.

Rabalais, N.N., Turner, R.E., and Wiseman, W.J. Jr. (2002). Gulf of Mexico hypoxia, aka "The dead zone". *Annual Review of Ecology and Systematics 33* (1): 235–263.

Ramirez, M. (2019). Tracking Estuarine Wetland Restoration in Puget Sound; Reporting on the Puget Sound Estuaries Vital Sign Indicator. Report prepared for the Puget Sound Partnership at the University of Washington, Seattle, WA. pdf.

Rybczyk, J.M. and Cahoon, D.R. (2002). Estimating the potential for submergence for two wetlands in the Mississippi River Delta. *Estuaries 25* (5): 985–998.

Rybczyk, J.M. and Callaway, J.C. (2009). Surface elevation models. In: *Coastal Wetlands: An Ecosystem Integrated Approach* (ed. G.M.E. Perillo, E. Wolanski, D.R. Cahoon and C.S. Hopkinson), 834–853. Amsterdam: Elsevier.

Rybczyk, J.M., Zhang, X.W., Day, J.W. Jr. et al. (1995). The impact of Hurricane Andrew on tree mortality, litterfall, nutrient flux, and water quality in a Louisiana coastal swamp forest. *Journal of Coastal Research* (21): 340–353.

Rybczyk, J.M., Callaway, J.C., and Day, J.W. Jr. (1998). A relative elevation model for a subsiding coastal forested wetland receiving wastewater effluent. *Ecological Modelling 112* (1): 23–44.

Rybczyk, J.M., Hamlet, A.F., MacIlroy, C., and Wasserman, L. (2016). Introduction to the Skagit issue—from glaciers to Estuary: assessing climate change impacts on the Skagit River Basin. *Northwest Science 90* (1): 1–4.

Saintilan, N., Wilson, N.C., Rogers, K. et al. (2014). Mangrove expansion and salt marsh decline at mangrove poleward limits. *Global Change Biology 20* (1): 147–157.

Sánchez-Arcilla, A., Jiménez, J.A., Stive, M.J.F. et al. (1996). Impacts of sea-level rise on the Ebro Delta: a first approach. *Ocean & Coastal Management 30* (2–3): 197–216.

Schiermeier, Q. (2011). Increased flood risk linked to global warming. *Nature* 470: 316. https://doi.org/10.1038/470316a.

Schile, L.M., Callaway, J.C., Morris, J.T. et al. (2014). Modeling tidal marsh distribution with sea-level rise: evaluating the role of vegetation, sediment, and upland habitat in marsh resiliency. *PLoS One 9* (2): e88760.

Sestini, G. (1992). Implications of climatic changes for the Po delta and Venice lagoon. In: *Climatic Change and the Mediterranean* (ed. L. Jeftic, J. Milliman and G. Sestini). London: Edward Arnold.

Shaffer, G., Day, J., Kandalepas, D. et al. (2016). Decline of the maurepas swamp, pontchartrain basin, louisiana and approaches to restoration. *Water* 7: https://doi.org/10.3390/w70x000x.

Short, F.T. and Neckles, H.A. (1999). The effects of global climate change on seagrasses. *Aquatic Botany 63* (3–4): 169–196.

Siirila-Woodburn, E.R., Rhoades, A.M., Hatchett, B.J. et al. (2021). A low-to-no snow future and its impacts on water resources in the western United States. *Nature Reviews Earth & Environment 2* (11): 800–819.

Sklar, F.H., Chimney, M.J., Newman, S. et al. (2005). The ecological–societal underpinnings of Everglades restoration. *Frontiers in Ecology and the Environment 3* (3): 161–169.

Snedaker, S. (1984). Mangroves: a summary of knowledge with emphasis on Pakistan. In: *Marine Geolgoy and Oceanography of Arabian Sea and Coastal Pakistan* (ed. B.H. Haq and J.D. Milliman), 255–262. New York: Van Nostrand Reinhold Co.

Spivak, A.C., Sanderman, J., Bowen, J.L. et al. (2019). Global-change controls on soil-carbon accumulation and loss in coastal vegetated ecosystems. *Nature Geoscience 12* (9): 685–692.

Stagg, C.L., Schoolmaster, D.R., Piazza, S.C. et al. (2017). A landscape-scale assessment of above-and belowground primary production in coastal wetlands: implications for climate change-induced community shifts. *Estuaries and Coasts 40* (3): 856–879.

Stanley, D.J. and Warne, A.G. (1998). Nile Delta in its destruction phase. *Journal of Coastal Research 14* (3): 795–825.

Stevenson, J.C., Kearney, M.S., and Pendleton, E.C. (1985). Sedimentation and erosion in a Chesapeake Bay brackish marsh system. *Marine Geology 67* (3–4): 213–235.

Swanson, K.M., Drexler, J.Z., Fuller, C.C., and Schoellhamer, D.H. (2015). Modeling tidal freshwater marsh sustainability in the Sacramento—san Joaquin Delta under a broad suite of potential future scenarios. *San Francisco Estuary and Watershed Science 13* (1).

Sweet, W.V., R.E. Kopp, C.P. Weaver, J. Obeysekera, R.M. Horton, E.R. Thieler, and C. Zervas (2017). Global and Regional Sea Level Rise Scenarios for the United States. NOAA Technical Report NOS CO-OPS 083. NOAA/NOS Center for Operational Oceanographic Products and Services.

Titus, J.G. (1991). Greenhouse effect and coastal wetland policy: how Americans could abandon an area the size of Massachusetts at minimum cost. *Environmental Management 15* (1): 39–58.

Törnqvist, T.E., Cahoon, D.R., Morris, J.T., and Day, J.W. (2021). Coastal wetland resilience, accelerated sea-level rise, and the importance of timescale. *AGU Advances 2* (1): e2020AV000334.

Twilley, R.R. and Rivera-Monroy, V.H. (2005). Developing performance measures of mangrove wetlands using simulation models of hydrology, nutrient biogeochemistry, and community dynamics. *Journal of Coastal Research* 40: 79–93.

Valiela, I. (2006). *Global Coastal Change*, 368. Malden (MA): Blackwell Publishing.

Warren, R.S. and Niering, W.A. (1993). Vegetation change on a northeast tidal marsh: interaction of sea-level rise and marsh accretion. *Ecology 74* (1): 96–103.

Watson, E.B., Wigand, C., Davey, E.W. et al. (2017). Wetland loss patterns and inundation-productivity relationships prognosticate widespread salt marsh loss for southern New England. *Estuaries and Coasts 40* (3): 662–681.

Williams, G.D. and Thom, R.M. (2001).Marine and Estuarine Shoreline Modification Issues. Battelle Marine Sciences Laboratory White Paper. WA State Department of Ecology Report; 102 pp.

Windham-Myers, L., Crooks, S., and Troxler, T.G. (ed.) (2018). *A Blue Carbon Primer: The State of Coastal Wetland Carbon Science, Practice and Policy*. Boca Raton, FL: CRC Press.

Wuebbles, D.J., Fahey, D.W., Hibbard, K.A. et al. (2017). *Climate Science Special Report: Fourth National Climate Assessment (NCA4), Volume I*. Washington, DC: U.S. Global Change Research Program.

Xu, K., Bentley, S.J., Day, J.W., and Freeman, A.M. (2019). A review of sediment diversion in the Mississippi River Deltaic Plain. *Estuarine, Coastal and Shelf Science 225*: 106241.

Yeates, A.G., Grace, J.B., Olker, J.H. et al. (2020). Hurricane Sandy effects on coastal marsh elevation change. *Estuaries and Coasts* 43 (7): 1640–1657.

Appendix 1

Multiple Choice Question Answers

Chapter 2 Estuarine Geomorphology, Circulation, and Mixing

1c, 2c, 3c, 4a, 5d, 6d, 7b, 8b, 9b, 10b

Chapter 3 Estuarine Chemistry

1c, 2b, 3abc, 4e, 5c, 6abe, 7c, 8c, 9ad, 10abc

Chapter 4 Estuarine Phytoplankton

1c, 2b, 3d, 4c, 5b, 6b, 7bcdf, 8ace, 9ce, 10ace, 11ad, 12ade

Chapter 5 Estuarine Seagrasses

1c, 2c, 3c, 4d, 5c, 6c, 7c, 8b, 9c, 10c

Chapter 6 Coastal Marshes

1c, 2d, 3ab, 4d, 5abc, 6a, 7b, 8ae, 9d, 10b

Chapter 7 Mangrove Wetlands

1b, 2c, 3a, 4d, 5a, 6b, 7c, 8b, 9d, 10b

Chapter 8 Estuarine Benthic Algae

1a, 2c, 3ace, 4a, 5ac, 6bc, 7ae, 8ac, 9b

Chapter 9 Estuarine Microbial Ecology

1abd, 2abe, 3ad, 4bcef, 5c, 6ce, 7c, 8a, 9cf, 10bdf, 11aceg, 12adf, 13b, 14ace, 15bd, 16ade

Chapter 10 Estuarine Zooplankton

1a, 2b, 3d, 4b, 5c, 6bc, 7a, 8abd, 9b, 10b, 11c

Chapter 11 Estuarine Benthos

1b, 2a, 3c, 4a, 5bc, 6b, 7a, 8abd, 9ad, 10b

Chapter 12 Estuarine Nekton

1c, 2c, 3b, 4d, 5b, 6b, 7b, 8c, 9c, 10d, 11c, 12b

Chapter 13 Estuarine Wildlife

1abd, 2abd, 3bc, 4cd, 5a, 6abcd

Chapter 14 Estuarine Ecosystem Metabolism

1d, 2d, 3b, 4d, 5a

Chapter 15 Estuarine Food Webs

1b, 2d, 3b, 4a, 5d

Chapter 16 Estuarine Ecological Modeling

1abc, 2d, 3c, 4a, 5ac

Chapter 17 Estuarine Fisheries and Aquaculture

1acd, 2abd, 3a, 4b, 5e, 6ad, 7bd, 8bcd, 9d, 10c

Chapter 18 Global Climate Change and Estuarine Systems

1abcde, 2abce, 3b, 4abce, 5abde, 6d, 7abc, 8bd, 9abc

Estuarine Ecology, Third Edition. Edited by Byron C. Crump, Jeremy M. Testa, and Kenneth H. Dunton.

Companion website: www.wiley.com/go/crump/estuarine3

Index

Figure references are in italics. Table references in bold. Locations of common terms are listed by chapter.

Estuarine Ecology, Third Edition. Edited by Byron C. Crump, Jeremy M. Testa, and Kenneth H. Dunton.

Companion website: www.wiley.com/go/crump/estuarine3

R

S

T